Rules

Divisibility Rules

If the last digit (ones digit) of a whole number is $0, 2, 4, 6,$ or 8 (an even digit), then the number is divisible by **2**. (The number is even.)

If the sum of the digits of a whole number is divisible by 3, then the number is divisible by **3**.

If the last two digits of a whole number form a number that is divisible by 4, then the number is divisible by **4**. (00 is considered to be divisible by 4.)

If the last digit (ones digit) of a whole number is 0 or 5, then the number is divisible by **5**.

If a whole number is divisible by both 2 and 3, then the number is divisible by **6**.

If the sum of the digits of a whole number is divisible by 9, then the number is divisible by **9**.

If the last digit (ones digit) of a whole number is 0, then the number is divisible by **10**.

Rules for Exponents

For nonzero real numbers a and b and integers m and n:

The exponent 1: $a = a^1$

The exponent 0: $a^0 = 1$

The product rule: $a^m \cdot a^n = a^{m+n}$

The quotient rule: $\dfrac{a^m}{a^n} = a^{m-n}$

Negative exponents: $a^{-n} = \dfrac{1}{a^n}$

Power rule: $\left(a^m\right)^n = a^{mn}$

Power of a product: $(ab)^n = a^n b^n$

Power of a quotient: $\left(\dfrac{a}{b}\right)^n = \dfrac{a^n}{b^n}$

Linear Equations

Summary of Formulas and Properties of Lines

Standard Form:

$$Ax + By = C \quad \text{where } A \text{ and } B \text{ do not both equal } 0$$

Slope of a line:

$$m = \frac{y_2 - y_1}{x_2 - x_1} \quad \text{where } x_1 \neq x_2$$

Slope-intercept form:

$$y = mx + b \quad \text{with slope } m \text{ and } y\text{-intercept } (0, b)$$

Point-slope form:

$$y - y_1 = m(x - x_1) \quad \text{with slope } m \text{ and point } (x_1, y_1) \text{ on the line}$$

Horizontal line, slope 0: $y = b$

Vertical line, undefined slope: $x = a$

1. **Parallel lines** have the same slope.
2. **Perpendicular lines** have slopes that are negative reciprocals of each other.

Cartesian Coordinate System

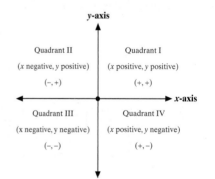

Factoring Polynomials

Special Factoring Techniques

1. $x^2 - a^2 = (x + a)(x - a)$: difference of two squares

2. $x^2 + 2ax + a^2 = (x + a)^2$: square of a binomial sum

3. $x^2 - 2ax + a^2 = (x - a)^2$: square of a binomial difference

4. $x^3 + a^3 = (x + a)(x^2 - ax + a^2)$: sum of two cubes

5. $x^3 - a^3 = (x - a)(x^2 + ax + a^2)$: difference of two cubes

Inequalities

Linear Inequalities

Linear Inequalities have the following forms where a, b, and c are real numbers and $a \neq 0$:

$$ax + b < c \quad \text{and} \quad ax + b \leq c$$
$$ax + b > c \quad \text{and} \quad ax + b \geq c$$

Compound Inequalities

The inequalities $c < ax + b < d$ and $c \leq ax + b \leq d$ are called **compound linear inequalities**.
(This includes $c < ax + b \leq d$ and $c \leq ax + b < d$ as well.)

Interval Notation

Type of Interval	Algebraic Notation	Interval Notation	Graph
Open Interval	$a < x < b$	(a, b)	
Closed Interval	$a \leq x \leq b$	$[a, b]$	
Half-open Interval	$\begin{cases} a \leq x < b \\ a < x \leq b \end{cases}$	$\begin{matrix} [a, b) \\ (a, b] \end{matrix}$	
Open Interval	$\begin{cases} x > a \\ x < b \end{cases}$	$\begin{matrix} (a, \infty) \\ (-\infty, b) \end{matrix}$	
Half-open Interval	$\begin{cases} x \geq a \\ x \leq b \end{cases}$	$\begin{matrix} [a, \infty) \\ (-\infty, b] \end{matrix}$	

Systems of Linear Equations

Systems of Linear Equations (Two Variables)

Consistent
(One solution)

Inconsistent
(No solution)

Dependent
(Infinite number of solutions)

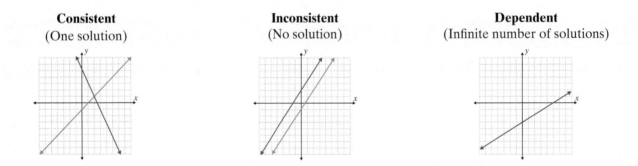

Quadratic Equations

Quadratic Formula

The solutions of the general quadratic equation $ax^2 + bx + c = 0$, where $a \neq 0$, are $x = \dfrac{-b \pm \sqrt{b^2 - 4ac}}{2a}$.

General Information on Quadratic Functions

For the quadratic function $y = ax^2 + bx + c$

1. If $a > 0$, the parabola "opens upward."

2. If $a < 0$, the parabola "opens downward."

3. $x = -\dfrac{b}{2a}$ is the line of symmetry.

4. The vertex (turning point) occurs where $x = -\dfrac{b}{2a}$.

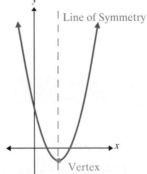

Substitute this value for x in the function and find the y-value of the vertex. The vertex is the lowest point on the curve if the parabola opens upward or it is the highest point on the curve if the parabola opens downward.

HAWKES
LEARNING
SYSTEMS

DEVELOPMENTAL
MATHEMATICS

D. FRANKLIN WRIGHT

Editor: Nina Waldron
Co-Editor: Susan Niese
Vice President, Development: Marcel Prevuznak
Production Editor: Kim Cumbie
Editorial Assistant: Joseph Miller
Layout Design: Tracy Carr, Nancy Derby, Rachel A. I. Link, Jennifer Moran, Tee Jay Zajac
Layout Production: E. Jeevan Kumar, D. Kanthi, U. Nagesh, B. Syamprasad
Copy Editors: Jessica Ballance, Danielle C. Bess, Margaret Gibbs, Mandy Glover, Taylor Hamrick, Debbie Rogina, Tristan Vogler, Colin Williams
Answer Key Editors: Jessica Ballance, Caroline Bauknecht, Joshua Falter, Rebecca Hughes, Akshi Kakar, Jaime Miller, Bill Radjewski , James von der Lieth
Art: Jennifer Guerette, Rachel A. I. Link, Jennifer Moran, Ayvin Samonte
Cover Design: Jessica Cokins, Tee Jay Zajac

Photograph Credits:
BigStockPhoto.com, iStockPhoto.com, and Digital Vision

HAWKES
LEARNING
SYSTEMS

A division of Quant Systems, Inc.
546 Long Point Road, Mt. Pleasant, SC 29464

Library of Congress Control Number: 2010931078

Printed in the United States of America

ISBN:
Student Textbook: 978-1-932628-83-8
Student Textbook and Software Bundle: 978-1-932628-84-5

Table of Contents

Chapter 4 Ratios and Proportions, Percent, and Applications

Chapter 5 Geometry

Chapter 6 Statistics, Graphs, and Probability

Chapter 10 Systems of Linear Equations

Chapter 11 Exponents and Polynomials

Chapter 12 Factoring Polynomials and Solving Quadratic Equations

Chapter 13 Rational Expressions

Chapter 14 Radicals

Chapter 15 Quadratic Equations

Appendices Further Topics in Algebra

Preface

Purpose and Style

The purpose of Developmental Mathematics is to provide students with a review of basic arithmetic, an introduction to algebra, and a learning tool that will help them:

1. review basic arithmetic skills,
2. develop reasoning and problem-solving skills,
3. become familiar with algebraic notation,
4. understand the connections between arithmetic and algebra,
5. develop basic algebra skills,
6. provide a smooth transition from arithmetic through prealgebra to algebra, and
7. achieve satisfaction in learning so that they will be encouraged to continue their education in mathematics.

The writing style gives carefully worded, thorough explanations that are direct, easy to understand, and mathematically accurate. The use of color, boldface, subheadings, and shaded boxes helps students understand and reference important topics.

Each topic is developed in a straightforward step-by-step manner. Each section contains many detailed examples to lead students successfully through the exercises and help them develop an understanding of the related concepts. Practice Problems with answers are provided in almost every section to allow students to "warm up" and to provide instructors with immediate classroom feedback.

Reading graphs and topics from geometry are integrated within the discussions and problems. In particular, Chapter 5 provides an in-depth study of geometrical concepts (lines, triangles, rectangles, circles, and so on). Chapter 6 introduces statistical concepts (mean, median, mode, and range), goes deeper into graphing, and provides an interesting introduction to probability. (What is the probability of getting two heads if a coin is tossed twice?) From Chapter 7 on, the text concentrates on developing useful algebraic skills and concepts.

Students are encouraged to use calculators when appropriate and explicit directions and diagrams are provided as they relate to a simple four-function calculator, as well as to a TI-84 Plus graphing calculator.

The NCTM and AMATYC curriculum standards have been taken into consideration in the development of the topics throughout the text.

Special Features

In each chapter:

- Mathematics at Work! presents a brief discussion related to a concept developed in the coming chapter. Sometimes these sections are challenging and may be better understood after the student completes a portion, or all, of the chapter.
- Learning Objectives are listed at the beginning of each section and are used to highlight headers in the section to emphasize the topic being discussed.
- Chapter Review Exercises appear in each chapter and are keyed to each section.
- Cumulative Review exercises appear in each chapter beyond Chapter 1 to provide continuous, cumulative review.

In the exercise sets:

- Writing and Thinking About Mathematics exercises encourage students to express their ideas, interpretations, and understanding in writing.
- Collaborative Learning Exercises are designed to be done in interactive groups.

Features

Chapter Openers:

Each chapter begins with a chapter table of contents and an engaging preview of an application of the chapter's material.

Statistics, Graphs, and Probability

Mathematics at Work!

Assessing the likelihood of certain situations is a useful tool in everyday life. For example, say you were playing cards. How would you wager differently if you knew you had only a 3% chance of winning a hand versus if you knew you had a 53% chance of winning? In both cases, you need to be able to count the number of possible outcomes and then calculate the likelihood of the one situation you have.

A standard deck of cards is 52 cards with four suits (hearts, diamonds, spades, and clubs) and 13 cards in each suit. The cards are ace, king, queen, jack, 10, 9, 8, 7, 6, 5, 4, 3, and 2. If one card is drawn from a deck of cards, find the probability of each of the following events. (See Exercises 29 – 36 in Section 6.4.)

1. The card is an ace.
2. The card is the 3 of diamonds.
3. The card is a club.
4. The card is a queen or a jack.
5. The card is red.
6. The card is a 10, 9, or 8.
7. The card is the king of hearts.
8. The card is a 1.

Math at Work:

A practical application of the chapter's material designed to peak student interest and give an idea of why the topics to be studied are useful.

Objectives:

The objectives provide students with a clear and concise list of the main concepts and methods taught in each section, enabling students to focus their time and effort on the most important topics. Objectives have corresponding buttons located in the section text where the topic is introduced for ease of reference.

Objectives

A Understand the basic concepts of fractions.

B Multiply fractions.

C Find equivalent fractions.

D Reduce fractions to lowest terms.

E Change mixed numbers to improper fractions.

F Change improper fractions to mixed numbers.

2.1 Introduction to Fractions and Mixed Numbers

Objective A   Fractions

Numbers such as $\frac{2}{3}$ are said to be in **fraction form**. The top number, 2, is called the **numerator** and the bottom number, 3, is called the **denominator**.

$$\frac{a}{b} \quad \longleftarrow \quad \text{numerator}$$
$$\quad \longleftarrow \quad \text{denominator}$$

Fractions can be used to indicate parts of a whole. For example, if a whole candy bar has 7 equal parts, then the fraction $\frac{3}{7}$ indicates that we are considering 3 of those parts.

7 parts

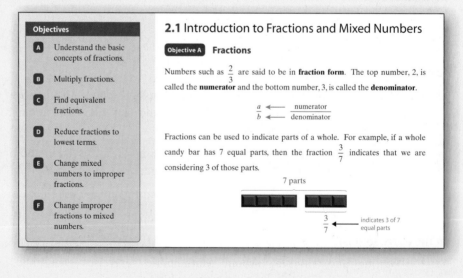

$\frac{3}{7}$ ← indicates 3 of 7 equal parts

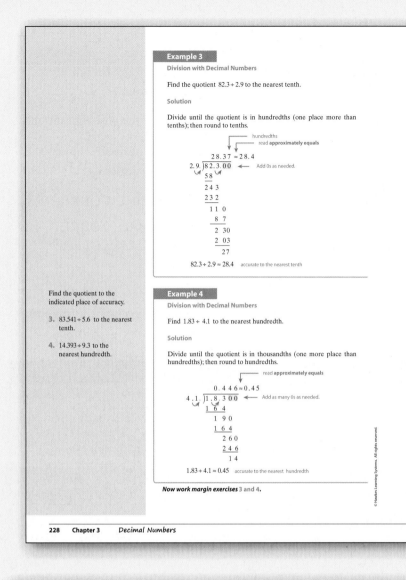

Examples:

Examples are denoted with titled headers indicating the problem-solving skill being presented. Each section contains many carefully explained examples with appropriate tables, diagrams, and graphs. Examples are presented in an easy to understand, step-by-step fashion and annotated with notes for additional clarification.

Example 3

Division with Decimal Numbers

Find the quotient $82.3 \div 2.9$ to the nearest tenth.

Solution

Divide until the quotient is in hundredths (one place more than tenths); then round to tenths.

hundredths
read **approximately equals**

$$
\begin{array}{r}
28.37 \approx 28.4 \\
2.9\,)\overline{82.3.00} \quad \leftarrow \text{Add 0s as needed.} \\
58 \\
\overline{243} \\
232 \\
\overline{110} \\
87 \\
\overline{230} \\
203 \\
\overline{27}
\end{array}
$$

$82.3 \div 2.9 \approx 28.4$ accurate to the nearest tenth

Example 4

Division with Decimal Numbers

Find $1.83 \div 4.1$ to the nearest hundredth.

Solution

Divide until the quotient is in thousandths (one more place than hundredths); then round to hundredths.

read **approximately equals**

$$
\begin{array}{r}
0.446 \approx 0.45 \\
4.1.\,)\overline{1.8.300} \quad \leftarrow \text{Add as many 0s as needed.} \\
164 \\
\overline{190} \\
164 \\
\overline{260} \\
246 \\
\overline{14}
\end{array}
$$

$1.83 \div 4.1 \approx 0.45$ accurate to the nearest hundredth

Now work margin exercises 3 and 4.

Find the quotient to the indicated place of accuracy.

3. $83.541 \div 5.6$ to the nearest tenth.

4. $14.393 \div 9.3$ to the nearest hundredth.

---

notes

Notes boxes throughout the text point out important information that will help deepen student understanding of the topics. Often these will be helpful hints about subtle details in the definitions that many students do not notice upon first glance.

---

notes

Note that in Rule 3 neither multiplication nor division has priority over the other. Whichever of these operations occurs first, moving **left to right**, is done first. In Rule 4, addition and subtraction are handled in the same way. Unless they occur within grouping symbols, **addition and subtraction are the last operations to be performed**.

Multiplicative Identity Property

For any whole number a, $a \cdot 1 = a$.

For example, $6 \cdot 1 = 6$.

(The product of any number and 1 is that same number.)

The number 1 is called the **multiplicative identity**.

Definition Boxes:

Straightforward definitions are presented in highly-visible boxes for easy reference.

Common Error:

These hard-to-miss boxes highlight common mistakes and how to avoid them.

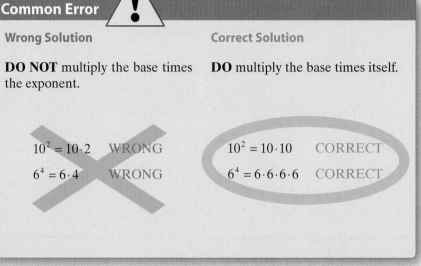

Common Error

Wrong Solution

DO NOT multiply the base times the exponent.

$$10^2 = 10 \cdot 2 \qquad \text{WRONG}$$

$$6^4 = 6 \cdot 4 \qquad \text{WRONG}$$

Correct Solution

DO multiply the base times itself.

$$10^2 = 10 \cdot 10 \qquad \text{CORRECT}$$

$$6^4 = 6 \cdot 6 \cdot 6 \cdot 6 \qquad \text{CORRECT}$$

Practice Problems:

Practice Problems are presented at the end of almost every section giving the students an opportunity to practice their newly acquired skills. The answers to the problems are provided at the bottom of the page so students can immediately assess their understanding of the topic at hand.

Practice Problems

Find the value of each expression.

1. 5^2 **2.** 3^4 **3.** 4^0 **4.** 1^5

Use the rules for order of operations to evaluate each expression.

5. $14 \div 7 + 3 \cdot 2^3$ **6.** $9 \div 3 \cdot 2 + 3\left(6 - 2^2\right)$ **7.** $6\left[(6-1)^2 - \left(4^2 - 9\right)\right]$

Practice Problem Answers

1. 25   **2.** 81   **3.** 1   **4.** 1

5. 26   **6.** 12   **7.** 108

Calculator:

As many students are visual learners, we provide students with key strokes and screen shots when appropriate for visual reference. We also provide step-by-step instructions for using a simple four-function calculator for more basic operations, as well as a TI-84 Plus for graphing skills.

Example 4

Adding Whole Numbers

Stan bought a television set for $859, a stereo for $697, and a computer (with printer) for $1285. What total amount did he spend?

Solution

The total amount spent is the sum:

```
  ¹ ² ²
    8 5 9
    6 9 7
+ 1 2 8 5
─────────
  2 8 4 1
```

Therefore Stan spent $2841.

Now work margin exercise 4.

Using a Calculator to Add Whole Numbers

At times you may want to use a simple scientific calculator to add numbers. To add numbers on a calculator, you will need to find the addition key, ⊞ , and the equal sign key, ⊟ .
To add the values given above in Example 4, press the keys
8 5 9 ⊞ 6 9 7 ⊞ 1 2 8 5 .
Then press ⊟ . The display will read 2841.

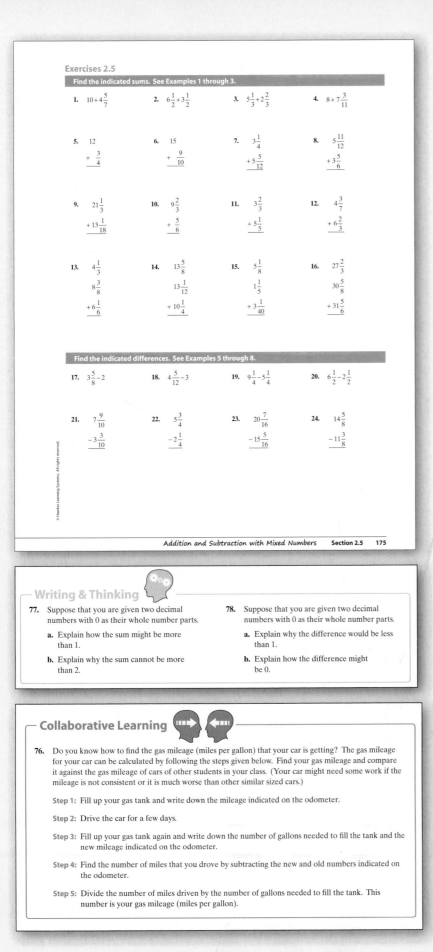

Exercises 2.5

Find the indicated sums. See Examples 1 through 3.

1. $10 + 4\frac{5}{7}$
2. $6\frac{1}{2} + 3\frac{1}{2}$
3. $5\frac{1}{3} + 2\frac{2}{3}$
4. $8 + 7\frac{3}{11}$

5. $\begin{array}{r} 12 \\ + \ \frac{3}{4} \\ \hline \end{array}$
6. $\begin{array}{r} 15 \\ + \ \frac{9}{10} \\ \hline \end{array}$
7. $\begin{array}{r} 3\frac{1}{4} \\ + 5\frac{5}{12} \\ \hline \end{array}$
8. $\begin{array}{r} 5\frac{11}{12} \\ + 3\frac{5}{6} \\ \hline \end{array}$

9. $\begin{array}{r} 21\frac{1}{3} \\ + 13\frac{1}{18} \\ \hline \end{array}$
10. $\begin{array}{r} 9\frac{2}{3} \\ + \ \frac{5}{6} \\ \hline \end{array}$
11. $\begin{array}{r} 3\frac{2}{3} \\ + 5\frac{1}{5} \\ \hline \end{array}$
12. $\begin{array}{r} 4\frac{3}{7} \\ + 6\frac{2}{3} \\ \hline \end{array}$

13. $\begin{array}{r} 4\frac{1}{3} \\ 8\frac{3}{8} \\ + 6\frac{1}{6} \\ \hline \end{array}$
14. $\begin{array}{r} 13\frac{5}{8} \\ 13\frac{1}{12} \\ + 10\frac{1}{4} \\ \hline \end{array}$
15. $\begin{array}{r} 5\frac{1}{8} \\ 1\frac{1}{5} \\ + 3\frac{1}{40} \\ \hline \end{array}$
16. $\begin{array}{r} 27\frac{2}{3} \\ 30\frac{5}{8} \\ + 31\frac{5}{6} \\ \hline \end{array}$

Find the indicated differences. See Examples 5 through 8.

17. $3\frac{5}{8} - 2$
18. $4\frac{5}{12} - 3$
19. $9\frac{1}{4} - 5\frac{1}{4}$
20. $6\frac{1}{2} - 2\frac{1}{2}$

21. $\begin{array}{r} 7\frac{9}{10} \\ - 3\frac{3}{10} \\ \hline \end{array}$
22. $\begin{array}{r} 5\frac{3}{4} \\ - 2\frac{1}{4} \\ \hline \end{array}$
23. $\begin{array}{r} 20\frac{7}{16} \\ - 15\frac{5}{16} \\ \hline \end{array}$
24. $\begin{array}{r} 14\frac{5}{8} \\ - 11\frac{3}{8} \\ \hline \end{array}$

Writing & Thinking

77. Suppose that you are given two decimal numbers with 0 as their whole number parts.
    a. Explain how the sum might be more than 1.
    b. Explain why the sum cannot be more than 2.

78. Suppose that you are given two decimal numbers with 0 as their whole number parts.
    a. Explain why the difference would be less than 1.
    b. Explain how the difference might be 0.

Collaborative Learning

76. Do you know how to find the gas mileage (miles per gallon) that your car is getting? The gas mileage for your car can be calculated by following the steps given below. Find your gas mileage and compare it against the gas mileage of cars of other students in your class. (Your car might need some work if the mileage is not consistent or it is much worse than other similar sized cars.)

    Step 1: Fill up your gas tank and write down the mileage indicated on the odometer.

    Step 2: Drive the car for a few days.

    Step 3: Fill up your gas tank again and write down the number of gallons needed to fill the tank and the new mileage indicated on the odometer.

    Step 4: Find the number of miles that you drove by subtracting the new and old numbers indicated on the odometer.

    Step 5: Divide the number of miles driven by the number of gallons needed to fill the tank. This number is your gas mileage (miles per gallon).

Exercises:

Each section includes a variety of exercises to give the students much-needed practice applying and reinforcing the skills they learned in the section. The exercises progress from relatively easy problems to more difficult problems.

Writing and Thinking:

In this feature, students are given an opportunity to independently explore and expand on concepts presented in the chapter. These questions will foster a better understanding of the concepts learned within each section.

Collaborative Learning:

In this feature, students are encouraged to work with others to further explore and apply concepts learned in the chapter. These questions will help students realize that they see many mathematical concepts in the world around them every day.

Index of Key Terms and Ideas:

Each chapter contains an index highlighting the main concepts within the chapter. This summary gives complete definitions and concise steps to solve particular types of problems. Page numbers are also given for easy reference.

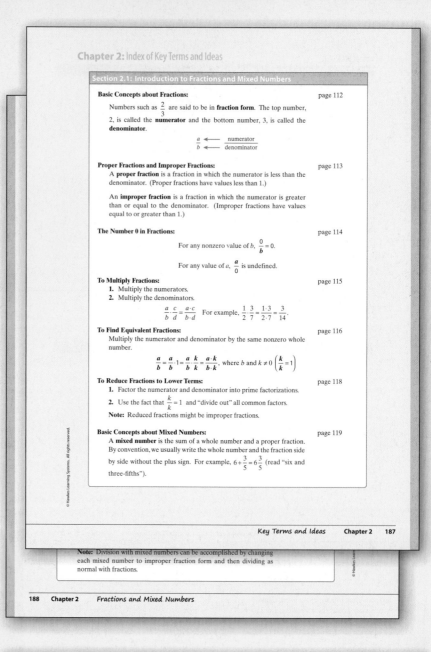

Section 2.1: Introduction to Fractions and Mixed Numbers

Basic Concepts about Fractions: page 112

Numbers such as $\frac{2}{3}$ are said to be in **fraction form**. The top number, 2, is called the **numerator** and the bottom number, 3, is called the **denominator**.

$$\frac{a}{b} \longleftarrow \begin{array}{l}\text{numerator}\\\text{denominator}\end{array}$$

Proper Fractions and Improper Fractions: page 113

A **proper fraction** is a fraction in which the numerator is less than the denominator. (Proper fractions have values less than 1.)

An **improper fraction** is a fraction in which the numerator is greater than or equal to the denominator. (Improper fractions have values equal to or greater than 1.)

The Number 0 in Fractions: page 114

For any nonzero value of b, $\frac{0}{b} = 0$.

For any value of a, $\frac{a}{0}$ is undefined.

To Multiply Fractions: page 115
1. Multiply the numerators.
2. Multiply the denominators.

$$\frac{a}{b} \cdot \frac{c}{d} = \frac{a \cdot c}{b \cdot d} \quad \text{For example, } \frac{1}{2} \cdot \frac{3}{7} = \frac{1 \cdot 3}{2 \cdot 7} = \frac{3}{14}.$$

To Find Equivalent Fractions: page 116

Multiply the numerator and denominator by the same nonzero whole number.

$$\frac{a}{b} = \frac{a}{b} \cdot 1 = \frac{a}{b} \cdot \frac{k}{k} = \frac{a \cdot k}{b \cdot k}, \text{ where } b \text{ and } k \neq 0 \left(\frac{k}{k} = 1\right)$$

To Reduce Fractions to Lower Terms: page 118
1. Factor the numerator and denominator into prime factorizations.
2. Use the fact that $\frac{k}{k} = 1$ and "divide out" all common factors.

Note: Reduced fractions might be improper fractions.

Basic Concepts about Mixed Numbers: page 119

A **mixed number** is the sum of a whole number and a proper fraction. By convention, we usually write the whole number and the fraction side by side without the plus sign. For example, $6 + \frac{3}{5} = 6\frac{3}{5}$ (read "six and three-fifths").

Note: Division with mixed numbers can be accomplished by changing each mixed number to improper fraction form and then dividing as normal with fractions.

Answer Key:

Located in the back of the book, the answer key provides all answers to the margin exercises, odd answers to all section exercises, and all answers to exercises in the Chapter Reviews, Chapter Tests, and Cumulative Reviews. This allows students to check their work to ensure that they are accurately applying the methods and skills they have learned.

Chapter 5: Geometry

Section 5.1: Angles

Margin Exercises   **1.** $m\angle 1 = 120°$; $m\angle 2 = 60°$ **2. a.** right   **b.** obtuse   **c.** straight   **3.** neither; because $m\angle 2 + m\angle 3 = 110°$ **4. a.** $110°$ **b.** no   **5.** $\angle ROS \cong \angle TOU$ and $\angle ROU \cong \angle SOT$ **6.** $40°$ **b.** $90°$ **c.** $50°$ **d.** $50°$ **7. a.** $\angle VQZ$ or $\angle WQX$ **b.** $m\angle WQV = 75°$ **8.** $m\angle 4 = 80°$; $m\angle 6 = 80°$; $m\angle 5 = 100°$

Exercises   **1.** an angle measuring $90°$ **3.** two lines that intersect to form right angles   **5.** $35°$ **7.** $120°$ **9.** acute   **11.** obtuse   **13.** right   **15. a.** obtuse   **b.** acute   **c.** right   **17. a.** $180°$ **b.** $90°$ **c.** $30°$ **d.** $150°$ **19.** $30°$ **21.** $30°$ **23.** $120°$ **25. a.** $135°$ **b.** $90°$ **c.** $70°$ **d.** $45°$ **27. a.** $\angle AOF$ and $\angle FOE$; $\angle AOF$ and $\angle AOB$; $\angle FOC$ and $\angle COB$; $\angle DOF$ and $\angle DOB$; $\angle EOF$ and $\angle EOB$; $\angle EOB$ and $\angle BOA$; $\angle EOC$ and $\angle COA$; $\angle EOD$ and $\angle DOA$ **b.** $\angle AOB$ and $\angle BOC$; $\angle COD$ and $\angle DOE$; $\angle BOC$ and $\angle EOF$ **29. a.** $\angle CPB$ and $\angle BPD$; $\angle CPA$ and $\angle APD$; $\angle CPA$ and $\angle CPB$; $\angle BPD$ and $\angle DPA$ **b.** There are none.   **31. a.** $m\angle 2 = 70°$; $m\angle 3 = 90°$; $m\angle 4 = 20°$; $m\angle 5 = 70°$

Chapter 5: Review

Section 5.1: Angles

Use a protractor to measure all the angles in each figure. Each line segment may be extended as a ray to form the side of an angle.

1.

2.

Classify each of the following angles as acute, right, obtuse, or straight.

3.

4.

Chapter 11: Test

Use the rules for exponents to simplify each expression. Each answer should have only positive exponents. Assume that all variables represent nonzero numbers.

1. $\left(5a^2b^5\right)\left(-2a^3b^{-5}\right)$

2. $\left(-7x^4y^{-3}\right)^0$

3. $\dfrac{\left(-8x^2y^{-3}\right)^2}{16xy}$

4. $\dfrac{3x^{-2}}{9x^{-3}y^2}$

5. $\left(\dfrac{4xy^2}{x^3}\right)^{-1}$

6. $\left(\dfrac{2x^0y^3}{x^{-1}y}\right)^2$

Solve the following problems.

Cumulative Review: Chapters 1 - 3

Write the following numbers in their English word equivalents.

1. a. 5,612,009   b. 5612.009

Name the properties illustrated.

2. a. $15 + 6 = 6 + 15$ b. $2(6 \cdot 7) = (2 \cdot 6)7$ c. $51 + 0 = 51$

Round as indicated.

	Nearest Hundredth	Nearest Tenth	Nearest Integer	Nearest Hundred
3. 4591.057	a. _____	b. _____	c. _____	d. _____

Solve the following problems.

4. Change the following mixed number to a fraction: $12\dfrac{4}{5}$

5. Change the following fraction to a mixed number: $\dfrac{35}{8}$

6. Change each fraction to decimal form. If the decimal number is nonterminating, write it using the bar notation over the repeating pattern of digits.

   a. $\dfrac{7}{8}$ b. $\dfrac{3}{11}$

7. Find the prime factorization of 420.

8. Find the LCM for each set of numbers.

   a. 28, 70   b. 8, 16, 32, 64

Perform the indicated operations.

9. $3 + 15 + 9 + 7$

10. $\begin{array}{r} 2023 \\ -\ 579 \end{array}$

11. $612 \div 24$

12. $\begin{array}{r} 45 \\ \times\ 27 \end{array}$

Chapter Review:

At the end of each chapter, a chapter review provides extra problems organized by section. These problems give students an opportunity to review concepts presented throughout the chapter and to identify strengths and potential weaknesses before taking an exam.

Chapter Test:

Each chapter also includes a chapter test that provides an opportunity for the students to practice the skills presented in the chapter in a test format.

Cumulative Review:

As new concepts build on previous knowledge, the cumulative review at the end of each chapter provides students with an opportunity to continually reinforce existing skills while also practicing more recent material.

Content

Chapter 1, Whole Numbers, reviews the fundamental operations of addition, subtraction, multiplication, and division with whole numbers. Estimating results is used to develop better understanding of whole number concepts, and applications help to reinforce the need for these ideas and skills in problems such as finding averages and making purchases. Exponents are introduced and the rules for order of operations are used to evaluate expressions with more than one operation. The last two sections of the chapter deal with short tests for divisibility and prime numbers. Students will find prime numbers particularly useful in working with fractions in Chapter 2.

Chapter 2, Fractions and Mixed Numbers, discusses the operations of multiplication, division, addition, and subtraction with fractions and mixed numbers. Emphasis is placed on using prime numbers to reduce fractions and to find the least common denominator (LCD) when adding or subtracting fractions. The rules for order of operations are applied to evaluate expressions with fractions and mixed numbers.

Chapter 3, Decimal Numbers, reviews reading and writing decimal numbers as well as operating with decimal numbers. Rounding decimals is used to help in estimating results in addition, subtraction, multiplication, and division. The last section shows how to change decimals to fraction form and to change fractions to decimal form.

Chapter 4, Ratios and Proportions, Percent, and Applications, develops an understanding of ratios and presents techniques for solving equations through finding the unknown term in a proportion. Percent is approached in two ways: solving proportions of the form $\dfrac{P}{100} = \dfrac{A}{B}$ and solving equations of the form $R \cdot B = A$. Applications with percent are related to discount, sales tax, commission, profit, simple interest, compound interest, percent increase, and percent decrease.

Chapter 5, Geometry, introduces geometric concepts through angles and angle measurement. Topics include perimeter, area, volume, and surface area. Triangles are shown to be classified in terms of angle measure and lengths of sides. Similar triangles and congruent triangles are discussed. Square roots are presented and shown to be related to right triangles through the Pythagorean Theorem.

Chapter 6, Statistics, Graphs, and Probability, begins with discussions of the statistical concepts of mean, median, mode, and range. There are two sections on graphs: reading graphs (bar graphs, circle graphs, line graphs, and histograms) and constructing graphs from given data. The last section, on probability, introduces basic probability terminology and shows how experiments can be illustrated with tree diagrams.

Chapter 7, Introduction to Algebra, develops the algebraic concepts of integers and real numbers and discusses operations with real numbers. Absolute value is defined and expressions are evaluated by using the rules for order of operations. The properties of addition and multiplication with real numbers are listed and their importance is emphasized. The chapter closes with simplifying algebraic expressions (combining like terms) and translating English phrases into algebraic expressions.

Chapter 8, Solving Linear Equations and Inequalities, is designed to develop the skills needed to solve equations in a step-by-step manner. Equations are presented in the following forms: $x + b = c$, $ax = c$, $ax + b = c$, and $ax + b = cx + d$. Applications involving number problems, consecutive integers, distance-rate-time, interest, and average are presented and solved by using the related equation solving skills. Intervals of real numbers and methods of solving linear inequalities are discussed in the last section.

Chapter 9, Linear Equations and Inequalities in Two Variables, allows for the introduction of graphing in two dimensions and the ideas and notation related to functions. Included are complete discussions on the three basic forms for equations of lines in a plane: the standard form, the slope-intercept form, and the point-slope form. Slope is discussed for parallel and perpendicular lines and treated as a rate of change. Functions are introduced and the vertical line test is used to tell whether or not a graph represents a function. Use of a TI-84 graphing calculator is an integral part of this introduction to functions as well as part of graphing linear inequalities in the last section.

Chapter 10, Systems of Linear Equations, covers solving systems of two equations in two variables. The basic methods of graphing, substitution, and addition are presented. Applications involve mixture, interest, work, algebra, and geometry.

Chapter 11, Exponents and Polynomials, studies the properties of exponents in depth and shows how to read and write scientific notation. The remainder of the chapter is concerned with definitions and operations related to polynomials. Included are the FOIL method of multiplication with two binomials, special products of binomials, and the division algorithm.

Chapter 12, Factoring Polynomials and Solving Quadratic Equations, discusses methods of factoring polynomials, including finding common monomial factors, factoring by grouping, factoring trinomials by grouping and by trial-and-error, and factoring special products. A special section provides students with tips on determining which type of factoring to use and provides extra exercises to allow them to practice related skills. The topic of solving quadratic equations is introduced, and quadratic equations are solved by factoring only. Applications with quadratic equations are included and involve topics such as the use of function notation to represent area, the Pythagorean Theorem, and consecutive integers.

Chapter 13, Rational Expressions, provides still more practice with factoring and shows how to use factoring to operate with rational expressions. Included are the topics of multiplication, division, addition, and subtraction with rational expressions, simplifying complex fractions, and solving equations containing rational expressions. Applications are related to work, distance-rate-time, and variation.

Chapter 14, Radicals, discusses simplifying radicals along with the use of calculators to find approximate values of expressions with radicals. Arithmetic with radicals includes addition, subtraction, multiplication, and rationalizing denominators. Methods for solving equations with radicals are developed. The rules for using rational exponents (fractional exponents) and simplifying expressions with rational exponents are discussed. A section on functions with radicals shows how to analyze the domain and range of radical functions and how to graph these functions by using a graphing calculator.

Chapter 15, Quadratic Equations, reviews solving quadratic equations by factoring and introduces the methods of using the square root property and completing the square. The quadratic formula is developed by completing the square and students are encouraged to use the most efficient method for solving any particular quadratic equation. Applications are related to the Pythagorean Theorem, work, and distance-rate-time. The last section presents quadratic functions and the graphs of parabolas. The graphing calculator provides a valuable aid to understanding the nature of these graphs.

Appendices, Further Topics in Algebra

- U. S. Measurements
- The Metric System
- U. S. to Metric Conversions
- Absolute Value Equations and Inequalities
- Synthetic Division and the Remainder Theorem
- Graphing Systems of Linear Inequalities
- Systems of Linear Equations Three Variables
- Introduction to Complex Numbers
- Multiplication and Division with Complex Numbers

Practice and Review

There are more than 9500 margin exercises, completion examples, practice problems, section exercises, and review exercises overall. Section exercises are carefully chosen and graded, proceeding from easy exercises to more difficult ones. Each chapter includes Review Exercises grouped by section title, a Chapter Test, and a Cumulative Review (beginning with Chapter 2). Also, each chapter contains an Index of Key Terms and Ideas.

Many sections have special exercises entitled "Writing and Thinking about Mathematics" and "Collaborative Learning Exercises." These exercises are an important part of the text and provide a chance for each student to improve communication skills, to develop a deeper understanding of general concepts, and to communicate his or her ideas to the instructor. Written responses can be a great help to the instructor in identifying just what students do and do not understand. Many of these questions are designed for the student to investigate ideas other than those presented in the text with responses that are to be based on each student's own experiences and perceptions.

Answers to the odd-numbered exercises, all margin exercises, all Review Exercises, all Chapter Test questions, and all Cumulative Review questions are provided in the back of the book.

Acknowledgements

I would especially like to acknowledge and thank my Editor Nina Waldron for her hard work and attention to detail that only she knows how to do. Also, thanks to Editor Susan Niese, Production Editor Kim Cumbie, and Vice President of Development Marcel Prevuznak for their invaluable assistance in the development and production of this text.

Many thanks to the following manuscript reviewers who offered their constructive and critical comments: Elizabeth Chu at Suffolk County Community College, Chris Copple at Northwest State Community College, Richard Getso at South Texas College, Jeffrey Grell at Baltimore City Community College, Lutchmiparsad Hazareesingh at Mesabi Range Community and Technical College, Heather Huntington at Nassau Community College, L. John Jerome at Suffolk County Community College, David Lung at South Texas College, Jo McCormick at Northwest State Community College, Nathan Mercer at Holyoke Community College, Michelle Toni Parsons at San Diego Mesa College, Jim Sheff at Spoon River College, and Merrie Van Loy at South Texas College.

Furthermore, I would like to thank Henry Miller for his hard work and efforts in finding interesting application problems.

Finally, special thanks go to Dr. James Hawkes for his support in this first edition and his willingness to commit so many resources to guarantee a top-quality product for students and teachers.

To the Student

The goal of this text and of your instructor is for you to succeed in your study of mathematics. Certainly, you should make this your goal as well. What follows is a brief discussion about developing good work habits and using the features of this text to your best advantage. For you to achieve the greatest return on your investment of time and energy you should practice the following three rules of learning:

1. Reserve a block of time for study every day.
2. Study what you don't know.
3. Don't be afraid to make mistakes.

How to Use This Book

The following eight-step guide will not only make using this book a more worthwhile and efficient task, but it will help you benefit more from classroom lectures or the assistance that you receive in a math lab.

1. Try to look over the assigned section(s) before attending class or lab.

2. Read examples carefully.

3. Work margin exercises when asked to do so throughout the lessons.

4. Work the section exercises faithfully as they are assigned.

5. Study and work the Review Exercises as you feel the need for extra practice and to help you prepare for an exam.

6. Use the Writing and Thinking About Mathematics questions as an opportunity to explore the way you think about mathematics.

7. Use the Chapter Tests to practice for the tests that are actually given in class or lab.

8. Study the Cumulative Reviews to help you retain the skills that you have acquired in studying earlier chapters.

How to Prepare for an Exam

Gaining Skill and Confidence

The stress that many students feel while trying to succeed in mathematics is what you have probably heard called "math anxiety." It is a real-life phenomenon, and many students experience such a high level of anxiety during mathematics exams in particular that they simply cannot perform to the best of their abilities. It is possible to overcome this stress simply by building your confidence in your ability to do mathematics and by minimizing your fears of making mistakes.

No matter how much it may seem that in mathematics you must either be right or wrong, with no middle ground, you should realize that you can be learning just as much from the times that you make mistakes as you can from the times that your work is correct. Success will come. Don't think that making mistakes at first means that you'll never be any good at mathematics. Learning mathematics requires lots of practice. Most importantly, it requires a true confidence in yourself and in the fact that with practice and persistence the mistakes will become fewer, the successes will become greater, and you will be able to say, "I can do this."

Showing What You Know

If you have attended class or lab regularly, taken good notes, read your textbook, kept up with homework exercises, and asked for help when it was needed, then you have already made significant progress in preparing for an exam and conquering any anxiety. Here are a few other suggestions to maximize your preparedness and minimize your stress.

1. Give yourself enough time to review. You will generally have several days advance notice before an exam. Set aside a block of time each day with the goal of reviewing a manageable portion of the material that the test will cover. Don't cram!

2. Work many problems to refresh your memory and sharpen you skills. Go back and redo selected exercises from all of your homework assignments.

3. Reread the text and your notes, and use the Chapter Index of Key Terms and Ideas to recap major ideas and test yourself by going back over problems.

4. Be sure that you are well-rested so that you can be alert and focused during the exam.

5. Don't study up to the last minute. Give yourself some time to wind down before the exam. This will help you to organize your thoughts and feel more calm as the test begins.

6. As you take the test, realize that its purpose is not to trick you, but to give you and your instructor an accurate idea of what you have learned. Good study habits, a positive attitude, and confidence in your own ability will be reflected in your performance on any exam.

7. Finally, you should realize that your responsibility does not end with taking the exam. When your instructor returns your corrected exam, you should review your instructor's comments and any mistakes that you might have made. Take the opportunity to learn from this important feedback about what you have accomplished, where you could work harder, and how you can best prepare for future exams.

HAWKES LEARNING SYSTEMS:
Developmental Mathematics

Hawkes Learning Systems specializes in interactive courseware with a unique mastery-based approach to student learning. The courseware is designed to help you develop a solid foundation of skills and has been proven to increase your overall success. Within each homework lesson you will find three learning modes: Learn, Practice, and Certify.

Learn: Learn is a multimedia presentation that includes the information you need to successfully answer each question in your assignment. Each lesson includes definitions, rules, properties, and examples, along with instructional videos.

Practice: Practice gives you unlimited opportunities to practice the types of problems you will receive in Certify. In Practice, you have access to learning aids through the Interactive Tutor. Step-By-Step breaks a problem down into smaller steps; Solution offers guided solutions to every problem; and Explain Error gives targeted feedback specific to your mistake.

Certify: This is the credit component of your homework! You will answer your problem set by using your knowledge and the foundation you built in Learn and Practice. You will have the opportunity to try again with no penalty if you do not demonstrate Mastery in your initial attempt(s). Pay close attention to any due dates or benchmarks assigned by your instructor.

Video

View instructional videos anytime, anywhere at **HawkesTV.com**.

Minimum Requirements

To view the minimum requirements for *HLS: Developmental Mathematics*, please visit:

hawkeslearning.com/requirements

Getting Started

Before you can run *HLS: Developmental Mathematics*, you will need an access code. This 30 character code is *your* personal access code. To obtain an access code, go to **hawkeslearning.com** and follow the links to the access code request page (unless directed otherwise by your instructor).

Support

Feel free to contact support for questions or technical help.

Support Center: support.hawkeslearning.com

Chat: chat.hawkeslearning.com

E-mail: support@hawkeslearning.com

Phone: 843.571.2825

Whole Numbers

Mathematics at Work!

We see whole numbers everywhere. People in various businesses, such as carpenters, teachers, auto mechanics, and airline pilots, all deal with whole numbers on a daily basis. As a student, you may be interested in determining your average score on your tests in your math class.

For example, say you received the following scores on four tests:

78	85	94	83

What is your test average for the class?

(See Section 1.7).

A   Understand the place value system.

B   Write a whole number in standard and expanded notation.

C   Write a whole number as its English word equivalent.

1.1 Reading and Writing Whole Numbers

Objective A   **The Place Value System**

The symbols used in our number system are called **digits**, and the value of a digit depends on its position to the left of a beginning point, called a **decimal point**. Figure 1 shows the value of the first ten places in the place value system we use. Every three places constitutes a **period** and periods are separated with commas. A number with 4 or fewer digits need not have any commas.

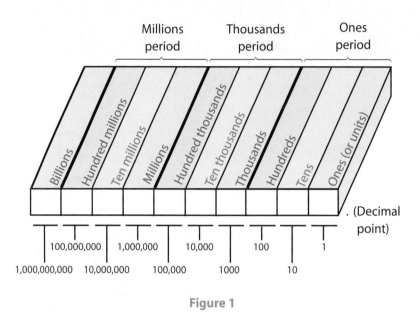

Figure 1

The **whole numbers** are the number 0 and the **natural numbers** (also called the counting numbers). We use $\mathbb{N}$ to represent the set of natural numbers and $\mathbb{W}$ to represent the set of whole numbers.

Whole Numbers

The **whole numbers** are the **natural numbers** (or counting numbers) and the number 0.

Natural numbers = $\mathbb{N}$ = { 1, 2, 3, 4, 5, 6, 7, 8, 9, 10, 11, . . . }
Whole numbers = $\mathbb{W}$ = { 0, 1, 2, 3, 4, 5, 6, 7, 8, 9, 10, 11, . . . }

Note that 0 is a whole number but not a natural number.

The three dots (called an *ellipsis*) in the definition indicate that the pattern continues without end.

To write a whole number in **standard notation** (or **standard form**) we use a **place value system**.

Place Value System

The decimal system (or base ten system) is a **place value system** that depends on three things:

1. the **ten digits**: 0, 1, 2, 3, 4, 5, 6, 7, 8, 9;
2. the **placement** of each digit; and
3. the **value** of each place.

Objective B **Standard and Expanded Notation**

Standard notation is the notation you are familiar with.

$$865 \quad \text{standard notation}$$

In **expanded notation**, the values represented by each digit are written as a sum.

$$865 = 800 + 60 + 5 \quad \text{expanded notation}$$

Example 1

Expanded Notation

Write 954 in expanded notation.

954   (standard notation)

9	5	4	← digits (standard notation)
100	10	1	← place values

$954 = 900 + 50 + 4$ (expanded notation)

Example 2

Expanded Notation

Write 6507 in expanded notation.

6507   (standard notation)

6	5	0	7	← digits (standard notation)
1000	100	10	1	← place values

$6507 = 6000 + 500 + 0 + 7$ (expanded notation)

***Now work margin exercises* 1 and 2.**

1. Write 463 in expanded notation.

2. Write 7302 in expanded notation.

3. Write 29,524 in expanded notation.

Completion Example 3

Expanded Notation

Complete the expanded notation form of 32,081.

Solution

$$32,081 = 30,000 + \underline{\hspace{1cm}} + \underline{\hspace{1cm}} + \underline{\hspace{1cm}} + 1$$

Note: Completion Examples are answered at the bottom of the page on which they appear.

Now work margin exercise 3.

Objective C **English Word Equivalents**

To read (or write) a whole number, consider one period at a time, moving from left to right. State the name of the number in each period (this number will be three digits or less), followed by the period name. Disregard any period containing three zeros.

Expanded notation can help you to translate a whole number into its English word equivalent. For example,

$$1397 = 1000 + 300 + 90 + 7 \text{ is read as}$$
one thousand three hundred ninety-seven.

As another example,

9,076,532 is read as
nine million, seventy-six thousand, five hundred thirty-two.

---

notes

■ A **hyphen** is used when writing two-digit numbers larger than 20 that do not end in a zero.

■

■ Note that the word "**and**" does not appear as part of reading (or writing) any whole number. The word "**and**" indicates the decimal point. You may put a decimal point to the right of the digits in a whole number if you choose, but it is not necessary unless digits are written to the right of the decimal point.

■

Completion Example Answers
3. $32,081 = 30,000 + 2000 + 0 + 80 + 1$

Example 4

Reading and Writing Whole Numbers

Given the number 350,472, which digit indicates the number of

a. thousands? **b.** tens? **c.** hundreds?

Solutions

a. 0 **b.** 7 **c.** 4

Now work margin exercise 4.

Example 5

Reading and Writing Whole Numbers

The following numbers are written in standard notation. Write them in words.

a. 25,380

Solution

twenty-five thousand, three hundred eighty

b. 3,000,562

Solution

three million, five hundred sixty-two

Now work margin exercise 5.

Example 6

Reading and Writing Whole Numbers

The following numbers are written in words. Rewrite them in standard notation.

a. twenty-seven thousand, three hundred thirty-six

Solution

27,336

b. three hundred forty million, sixty-two thousand, forty-eight

Solution

340,062,048

Note: 0's must be used to fill out a three-digit period.

Now work margin exercise 6.

4. Given the number 498,651 which digit indicates the number of

a. ten thousands?

b. hundreds?

c. ones?

5. Write each of the following numbers in words.

a. 32,450,090

b. 5784

6. Write each of the following numbers in standard notation.

a. six thousand forty-one

b. one million, four hundred eighty-three thousand, seven

Practice Problems

1. Write the following numbers in expanded notation and in their English word equivalents.

   a. 512   **b.** 6394   **c.** 50,690

2. Write one hundred eighty thousand, five hundred forty-three in standard notation.

3. Write two million, five hundred forty thousand, three hundred ten in standard notation.

Practice Problem Answers

   1. a. 500 + 10 + 2; five hundred twelve

     b. 6000 + 300 + 90 + 4; six thousand three hundred ninety-four

     c. 50,000 + 0 + 600 + 90 + 0; fifty thousand, six hundred ninety

   2. 180,543

   3. 2,540,310

Exercises 1.1

1. For whole numbers, the value of a digit depends on its position to the _____ of the beginning point, called a _____.

2. The word _____ does not appear in English word equivalents for whole numbers.

3. Use a comma to separate groups of three digits if a number has more than _____ digits.

4. Hyphens are used to write English words for numbers from _____ to _____ that do not end in _____.

5. List the ten digits used to write whole numbers.

6. Given the number 284,065, which digit indicates the number of
   a. tens?
   b. ten thousands?
   c. hundreds?

7. Given the number 13,476,582, which digit indicates the number of
   a. thousands?
   b. millions?
   c. ten millions?

8. Name the position of each nonzero digit in the following number: 2,403,189,500.

Write the following whole numbers in expanded notation. See Examples 1 through 3.

9. 37
10. 84
11. 56
12. 821

13. 1892
14. 2059
15. 25,658
16. 32,341

Write the following whole numbers in their English word equivalents. See Example 5.

17. 83
18. 122

19. 10,500
20. 683,100

21. 592,300

22. 16,302,590

23. 71,500,000

Write the following numbers in standard notation. See Example 6.

24. seventy-six

25. five hundred eighty

26. seven hundred fifty-seven

27. two thousand five

28. three thousand eight hundred thirty-four

29. ten thousand, eleven

30. seventy-eight thousand, nine hundred two

31. four hundred thousand, seven hundred thirty-six

32. five hundred thirty-seven thousand, eighty-two

33. eighty-two million, seven hundred thousand

34. sixty-three million, two hundred fifty-one thousand, sixty-five

35. two hundred eighty-one million, three hundred thousand, five hundred one

Write the English word equivalent for the number(s) in each sentence. See Example 5.

36. **Population:** The population of Los Angeles is 3,849,378.

37. **Astronomy:** The average distance from the earth to the sun is about 93,000,000 miles, or 149,730,000 kilometers.

38. **Geography:** The country of Chile averages about 110 miles in width and is about 2650 miles long.

39. **Geography:** The Republic of Venezuela covers an area of 352,143 square miles, or about 912,050 square kilometers.

40. **Architecture:** One of the world's tallest buildings is the Taipei Financial Center in Taipei, Taiwan. The building has 101 stories and reaches 1761 feet tall.

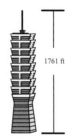

1761 ft

41. **Oceans:** The Pacific Ocean has an area of approximately 63,800,000 square miles and has an average depth of 14,040 feet.

1.2 Addition and Subtraction with Whole Numbers

Objective A **Addition with Whole Numbers**

An electronics store sold 2 cell phones one day and 3 the next day. The total number of cell phones sold in those two days is 5. The process of finding the total is called **addition**.

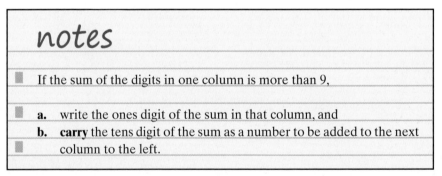

Addition is indicated either by writing the numbers horizontally, separated by (+) signs, or by writing the numbers vertically in columns. The numbers being added are called **addends**, and the result of the addition is called the **sum**.

$$
\begin{array}{ccccccc}
2 & + & 3 & = & 5 & \text{or} & 2 \quad \text{addend} \\
\text{addend} & & \text{addend} & & \text{sum} & & \underline{+\,3} \quad \text{addend} \\
& & & & & & 5 \quad \text{sum}
\end{array}
$$

1. Add.

   a. $162 + 35$

   b. $\quad\ \ 41$
   $\underline{+\,116}$

Example 1

Adding Whole Numbers

Add $14 + 132$.

Solution

$$
\begin{array}{rl}
14 & \text{addend} \\
\underline{+\,132} & \text{addend} \\
146 & \text{sum}
\end{array}
$$

Now work margin exercise 1.

notes

▪ If the sum of the digits in one column is more than 9,

▪ **a.** write the ones digit of the sum in that column, and
  b. **carry** the tens digit of the sum as a number to be added to the next column to the left.

Example 2 illustrates these ideas.

Example 2

Adding Whole Numbers

Add 475 + 59.

Solution

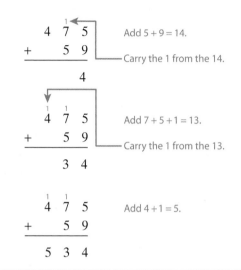

$$\begin{array}{ccc} & & {}^{1} \\ 4 & 7 & 5 \\ + & 5 & 9 \\ \hline & & 4 \end{array}$$

Add 5 + 9 = 14.

Carry the 1 from the 14.

$$\begin{array}{ccc} {}^{1} & {}^{1} & \\ 4 & 7 & 5 \\ + & 5 & 9 \\ \hline & 3 & 4 \end{array}$$

Add 7 + 5 + 1 = 13.

Carry the 1 from the 13.

$$\begin{array}{ccc} {}^{1} & {}^{1} & \\ 4 & 7 & 5 \\ + & 5 & 9 \\ \hline 5 & 3 & 4 \end{array}$$

Add 4 + 1 = 5.

Now work margin exercise 2.

Completion Example 3

Adding Whole Numbers

Add 328 + 604 + 517 + 192.

Solution

$$\begin{array}{ccc} 3 & 2 & 8 \\ 6 & 0 & 4 \\ 5 & 1 & 7 \\ + 1 & 9 & 2 \\ \hline & - & \end{array} \qquad \begin{array}{ccc} 3 & 2 & 8 \\ 6 & 0 & 4 \\ 5 & 1 & 7 \\ + 1 & 9 & 2 \\ \hline & & \end{array} \qquad \begin{array}{ccc} 3 & 2 & 8 \\ 6 & 0 & 4 \\ 5 & 1 & 7 \\ + 1 & 9 & 2 \\ \hline & & \end{array}$$

Now work margin exercise 3.

2. Find the sum.

$$\begin{array}{r} 463 \\ + 38 \\ \hline \end{array}$$

3. Add.

$$\begin{array}{r} 348 \\ 851 \\ 279 \\ + 136 \\ \hline \end{array}$$

Completion Example Answers

3 1; 41; 1641

4. Diedra bought a couch for $315, a coffee table for $199, and a washer and dryer for $863. What total amount did she spend?

Example 4

Adding Whole Numbers

Stan bought a television set for $859, a stereo for $697, and a computer (with printer) for $1285. What total amount did he spend?

Solution

The total amount spent is the sum:

$$
\begin{array}{r}
\overset{1}{} \ \overset{2}{8} \ \overset{2}{5} \ 9 \\
6 \ 9 \ 7 \\
+ \ 1 \ 2 \ 8 \ 5 \\
\hline
2 \ 8 \ 4 \ 1
\end{array}
$$

Therefore Stan spent $2841.

Now work margin exercise 4.

 Using a Calculator to Add Whole Numbers

At times you may want to use a simple scientific calculator to add numbers. To add numbers on a calculator, you will need to find the addition key, [+], and the equal sign key, [=].
To add the values given above in Example 4, press the keys
[8] [5] [9] [+] [6] [9] [7] [+] [1] [2] [8] [5].
Then press [=]. The display will read 2841.

Objective B **The Properties of Addition**

There are several properties of the operation of addition. To state the general form of each property, we introduce the notation of a **variable**. As we will see, variables not only allow us to state general properties, they also enable us to set up equations to help solve many types of applications.

Variable

A **variable** is a symbol (generally a letter of the alphabet) that is used to represent an unknown number or any one of several numbers.

In the following statements of the properties of addition, the set of whole numbers is the **replacement set** for each variable. (The replacement set for a variable is the set of all possible values for that variable.)

Commutative Property of Addition

For any whole numbers a and b, $a + b = b + a$.

For example, $3 + 14 = 14 + 3$.

(The **order** of the numbers in addition can be reversed.)

Associative Property of Addition

For any whole numbers a, b and c, $(a+b)+c = a+(b+c)$.

For example, $(6+12)+5 = 6+(12+5)$.

(The **grouping** of the numbers in addition can be changed.)

Additive Identity Property

For any whole number a, $a + 0 = a$.

For example, $8 + 0 = 8$.

(The sum of a number and 0 is that same number.)

The number 0 is called the **additive identity**.

Example 5

The Properties of Addition

Each of the properties of addition is illustrated.

a. $40 + 3 = 3 + 40$ commutative property of addition
As a check, we see that $40 + 3 = 43$ and $3 + 40 = 43$.

b. $2 + (5+9) = (2+5) + 9$ associative property of addition
As a check, we see that $2 + (14) = (7) + 9 = 16$.

c. $86 + 0 = 86$ additive identity property

Now work margin exercise 5.

5. Find each sum and state which property of addition is illustrated.

a. $(9+1)+7 = 9+(1+7)$

b. $25 + 4 = 4 + 25$

c. $39 + 0$

Objective C **Calculating the Perimeter of Geometric Figures**

The **perimeter** of a geometric figure is the distance around the figure. The perimeter is found by adding the lengths of the sides of the figure and is measured in linear units such as inches, feet, yards, miles, centimeters, and meters. The perimeter of the triangle (a geometric figure with three sides) shown in Example 6 is found by adding the measures of the three sides. Be sure to label the answers for the perimeter of any geometric figure with the correct units of measurement.

6. Find the perimeter of the rectangle.

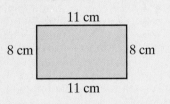

Example 6

Finding the Perimeter of a Triangle

To find the perimeter, find the sum of the lengths of the sides:

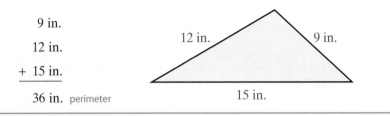

$$\begin{array}{r} 9 \text{ in.} \\ 12 \text{ in.} \\ + \ 15 \text{ in.} \\ \hline 36 \text{ in.} \end{array}$$ perimeter

Now work margin exercise 6.

7. Find the perimeter of the figure shown below.

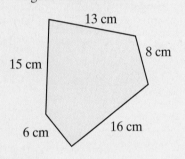

Example 7

Finding the Perimeter of a Geometric Figure

Find the perimeter of the figure shown.
(**Note:** A 5-sided geometric figure is called a **pentagon**.)

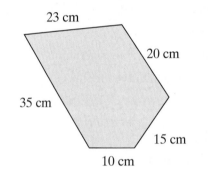

Solution

The perimeter of any geometric figure is found by adding the lengths of the sides. Thus, for this pentagon we have

$$10 + 15 + 20 + 23 + 35 = 103 \text{ cm.}$$

Now work margin exercise 7.

Objective D **Subtraction with Whole Numbers**

If you have $2, and you need $6, the question is "How many more dollars do you need?" That is, what do you add to 2 to get 6? We can represent the problem in the following format.

$$2 \quad + \quad \underline{\hspace{1cm}} \quad = \quad 6$$

addend   missing   sum

                   addend

Finding the missing addend is called **subtraction** and can be indicated with a **minus sign** (−) as follows.

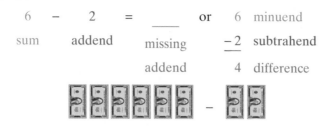

$$6 \quad - \quad 2 \quad = \quad \underline{} \qquad \text{or} \qquad \begin{array}{r} 6 \\ -\,2 \\ \hline 4 \end{array} \begin{array}{l} \text{minuend} \\ \text{subtrahend} \\ \text{difference} \end{array}$$

sum   addend   missing addend

Subtraction is reverse addition. To subtract, we must know how to add. The missing addend is called the **difference**. The sum is now called the **minuend**, and the given addend is called the **subtrahend**.

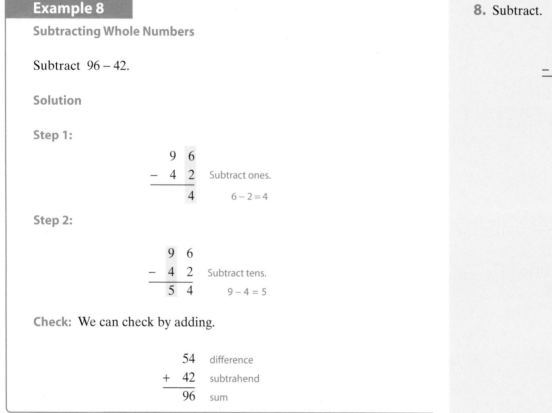

Example 8

Subtracting Whole Numbers

Subtract 96 − 42.

Solution

Step 1:

$$\begin{array}{r} 9\;6 \\ -\;4\;2 \\ \hline 4 \end{array} \quad \text{Subtract ones.}$$

            $6 - 2 = 4$

Step 2:

$$\begin{array}{r} 9\;6 \\ -\;4\;2 \\ \hline 5\;4 \end{array} \quad \text{Subtract tens.}$$

            $9 - 4 = 5$

Check: We can check by adding.

$$\begin{array}{r} 54 \\ +\;42 \\ \hline 96 \end{array} \begin{array}{l} \text{difference} \\ \text{subtrahend} \\ \text{sum} \end{array}$$

Now work margin exercise 8.

As illustrated in Example 9, finding the difference 742 − 259 involves "borrowing" because the 9 in the ones place of 259 is larger than the 2 in 742.

8. Subtract.

$$\begin{array}{r} 654 \\ -\;421 \end{array}$$

Borrowing

1. **Borrowing** is necessary when a digit is smaller than the digit being subtracted.
2. The process starts from the rightmost digit. **Borrow** from the digit to the left.

9. Find the difference.

$$
\begin{array}{r}
342 \\
-\ 187 \\
\end{array}
$$

Example 9

Subtracting Whole Numbers

Find the difference.

$$
\begin{array}{r}
742 \\
-\ 259 \\
\end{array}
$$

Solution

Step 1: Since 2 is smaller than 9, borrow 10 from 40 and add this 10 to 2 to get 12. This leaves 30 in the tens place; cross out 4 and write 3.

$$
\begin{array}{r}
7 \quad \overset{3}{\cancel{4}} \quad {}^{1}2 \quad \text{10 borrowed from 40}\\
-\ 2 \quad 5 \quad 9 \\
\end{array}
$$

Step 2: Since 3 is smaller than 5, borrow 100 from 700. This leaves 600, so cross out 7 and write 6.

$$
\begin{array}{r}
\overset{6}{\cancel{7}} \quad \overset{{}^{1}3}{\cancel{4}} \quad {}^{1}2 \\
-\ 2 \quad 5 \quad 9 \\
\end{array}
$$

Step 3: Now subtract.

$$
\begin{array}{r}
\overset{6}{\cancel{7}} \quad \overset{{}^{1}3}{\cancel{4}} \quad {}^{1}2 \\
-\ 2 \quad 5 \quad 9 \\
\hline
4 \quad 8 \quad 3 \\
\end{array}
$$

Check: Add the difference to the subtrahend. The sum should be the minuend.

$$
\begin{array}{r}
483 \\
+\ 259 \\
\hline
742 \\
\end{array}
$$

Now work margin exercise 9.

Example 10

Subtracting Whole Numbers

Subtract:

$$
\begin{array}{r}
8000 \\
-\ 657 \\
\end{array}
$$

Solution

Step 1: Trying to borrow from 0 each time, we end up borrowing 1000 from 8000. Cross out 8 and write 7.

$$
\begin{array}{r}
\overset{7}{\cancel{8}} \quad {}^{1}0 \quad 0 \quad 0 \\
-\ \quad\quad 6 \quad 5 \quad 7 \\
\end{array}
$$

Step 2: Borrow 100 from 1000. Cross out 10 and write 9.

$$
\begin{array}{r}
\overset{7}{\cancel{8}}\ \overset{9}{\cancel{10}}\ {}^{1}0\ \ 0 \\
-\ \ \ \ \ 6\ \ 5\ \ 7 \\
\hline
\end{array}
$$

Step 3: Borrow 10 from 100. Cross out 10 and write 9.

$$
\begin{array}{r}
\overset{7}{\cancel{8}}\ \overset{9}{\cancel{10}}\ \overset{9}{\cancel{10}}\ {}^{1}0 \\
-\ \ \ \ \ 6\ \ 5\ \ 7 \\
\hline
\end{array}
$$

Step 4: Now subtract.

$$
\begin{array}{r}
\overset{7}{\cancel{8}}\ \overset{9}{\cancel{10}}\ \overset{9}{\cancel{10}}\ {}^{1}0 \\
-\ \ \ \ \ 6\ \ 5\ \ 7 \\
\hline
7\ \ 3\ \ 4\ \ 3
\end{array}
$$

Check:

$$
\begin{array}{r}
7343 \\
+\ \ 657 \\
\hline
8000
\end{array}
$$

Now work margin exercise 10.

Example 11

Subtracting Whole Numbers

What number should be added to 546 to get a sum of 732?

Solution

We know the sum and one addend. To find the missing addend, subtract 546 from 732.

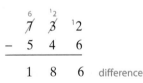

$$
\begin{array}{r}
\overset{6}{\cancel{7}}\ \overset{{}^{1}2}{\cancel{3}}\ {}^{1}2 \\
-\ 5\ \ 4\ \ 6 \\
\hline
1\ \ 8\ \ 6 \quad \text{difference}
\end{array}
$$

The number to be added is 186.

Now work margin exercise 11.

10. Subtract:

$$
\begin{array}{r}
7000 \\
-\ \ 423 \\
\hline
\end{array}
$$

11. What number should be added to 268 to get a sum of 654?

12. The cost of repairing Robert's used DVD player will be $165. To buy a new DVD player, Robert will have to pay $129. How much more would it cost to fix the old DVD player than to buy a new one?

Example 12

Subtracting Whole Numbers

The cost of repairing Ed's used TV set is $395. To buy a new set, he will have to pay $447. How much more would Ed have to pay for a new set than to have his old set repaired?

Solution

$$
\begin{array}{r}
\overset{3}{\cancel{4}}\ {}^{1}4\ 7 \\
-\ 3\ 9\ 5 \\
\hline
5\ 2
\end{array}
$$

Ed would pay $52 more for a new TV set than to have his old set repaired.

Now work margin exercise **12.**

Using a Calculator to Subtract Whole Numbers

To subtract numbers on a calculator, you will need to find the subtraction key, ⊟ , and the equal sign key, ⊒ .
To subtract the values given above in Example 12, press the keys
4 4 7 ⊟ 3 9 5 . Then press ⊒ .
The display will read 52.

13. In pricing a new motorcycle, Alex found that he would have to pay a base price of $10,470 plus $630 in taxes and $70 for the license fee. If the bank loaned him $7530, how much cash would Alex need to buy the motorcycle?

Example 13

Subtracting Whole Numbers

In pricing a new car, Jason found that he would have to pay a base price of $15,200 plus $1025 in taxes and $575 for license fees. If the bank loaned him $10,640, how much cash would Jason need to buy the car?

Solution

The solution involves both addition and subtraction. First, we add Jason's expenses and then we subtract the amount of the bank loan.

Expenses		Cash Needed	
15,200 base price		1 6, $\overset{7}{\cancel{8}}$ ${}^{1}0$ 0 total expenses	
1 025 taxes		− 1 0, 6 4 0 bank loan	
+ 575 license fees		6 1 6 0 cash	
16,800 total expenses			

Jason would need $6160 in cash to buy the car.

Now work margin exercise **13.**

Completion Example 14

Subtracting Whole Numbers

After selling their house for $132,000, the owners paid the realtor $7920, back taxes of $450, and $350 in other fees. If they also paid off the bank loan of $57,000, how much cash did the owners receive from the sale?

Solution

In this problem, experience tells us that we must add and subtract even though there are no specific directions to do so. We add the expenses and then subtract this sum from the selling price to find the cash that the owners received.

$$\begin{array}{r} \$\ 7920 \\ 450 \\ 350 \\ +57,000 \\ \hline \underline{\qquad} \quad \text{total expenses} \end{array}$$

$$\begin{array}{r} \$132,000 \quad \text{selling price} \\ - \underline{\qquad} \quad \text{total expenses} \\ \underline{\qquad} \quad \text{cash to owners} \end{array}$$

Now work margin exercise 14.

14. You have $40,000 invested in stocks, $38,500 invested in a CD, and $16,480 in a money market account. If you decide to sell your stock and withdraw your money from the other two accounts to buy a sailboat for $79,750, how much cash will you have after the purchase?

Practice Problems

Add.

1. $\begin{array}{r} 36 \\ +\ 92 \\ \hline \end{array}$
2. $\begin{array}{r} 5791 \\ +\ 6342 \\ \hline \end{array}$
3. $\begin{array}{r} 124,564 \\ 345,025 \\ +\ 1,671,000 \\ \hline \end{array}$

Subtract.

4. $\begin{array}{r} 375 \\ -\ 52 \\ \hline \end{array}$
5. $\begin{array}{r} 7642 \\ -\ 6181 \\ \hline \end{array}$
6. $\begin{array}{r} 7,000,000 \\ -\ 352,911 \\ \hline \end{array}$

Identify the property being illustrated.
7. $8+5=5+8$
8. $3+(4+5)=(3+4)+5$

Practice Problem Answers

1. 128 **2.** 12,133 **3.** 2,140,589
4. 323 **5.** 1461 **6.** 6,647,089
7. commutative property of addition
8. associative property of addtition

Completion Example Answers

14. total expenses = $65,720
cash to owners = $66,280

Exercises 1.2

1.   24
     + 54

2.   15
     + 143

3.   5341
     + 2154

4.   9745
     +   22

5. $8 + 6$

6. $72 + 63$

7. $184 + 6498$

8. $1304 + 8766$

9.   268
     + 93

10.   981
      + 146

11.   7527
      + 2050

12.   7518
      + 2077

13.   9315
      + 1185

14.   8324
      + 1958

15.   6530
      + 9542

16.   1731
      + 5934

17. $213,116$
       $+ 116,018$

18. $21,442$
       $+ 32,462$

19. $438,966$
         $572,486$
       $+ 227,462$

20. $123,456$
           $456,123$
       $+ 1,879,282$

21.   6530
         $93,800$
         $12,465$
         $19,324$
       $+ 45,090$

22. $1,458,853$
           $13,680,099$
       $+ 46,700,894$

23. $2,892,003$
             $370,022$
           $1,763,987$
       $+ 37,095,634$

24. $275,683$
             $413,645$
               1446
             $73,866$
       $+ 135,070$

Identify the property being illustrated. See Example 5.

25. $9 + 3 = 3 + 9$

26. $8 + (5 + 2) = 8 + (2 + 5)$

27. $4 + (5 + 3) = (4 + 5) + 3$

28. $4 + 8 = 8 + 4$

29. $2 + (1 + 6) = (2 + 1) + 6$

30. $(8 + 7) + 3 = 8 + (7 + 3)$

31. $9 + 0 = 9$

32. $7 + (6 + 0) = 7 + 6$

33. $(2 + 3) + 4 = (3 + 2) + 4$

34. $8 + 20 + 1 = 8 + 21$

35. Travel: Mr. Juarez kept the mileage records indicated in the table shown here. How many miles did he drive during the six months?

Month	Mileage
January	546
February	378
March	496
April	357
May	503
June	482

36. Profits: The Modern Products Corp. showed profits as indicated in the table for the years 2006–2009. What were the company's total profits for the years 2006–2009?

Year	Profits
2006	$ 1,078,416
2007	$ 1,270,842
2008	$ 2,000,593
2009	$ 1,963,472

37. Budget: The Magley family has the following monthly budget: $815 mortgage, $69 electric, $47 water, and $122 phone bills (including cell phones). What is the family's budget for each month for these expenses?

38. Enrollment: The following numbers of students at South Junior College are enrolled in mathematics courses: 303 in arithmetic; 476 in algebra; 293 in trigonometry; 257 in college algebra; and 189 in calculus. Find the total number of students taking mathematics.

39. Manufacturing: In one year, Stream Line Appliance Co. made 4217 gas stoves; 3947 electric stoves; and 9576 toasters. What was the total number of appliances Stream Line Appliance Co. produced that year?

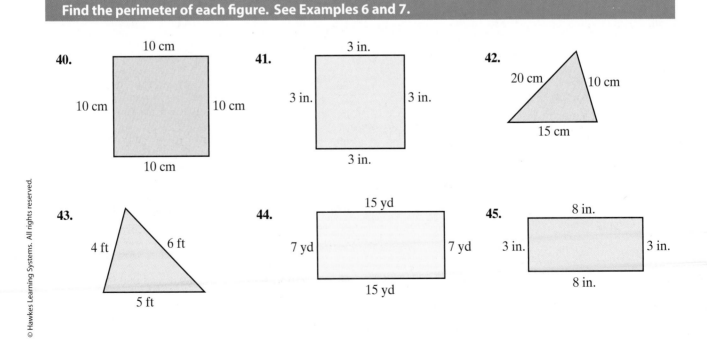

40. 10 cm / 10 cm / 10 cm / 10 cm

41. 3 in. / 3 in. / 3 in. / 3 in.

42. 20 cm / 10 cm / 15 cm

43. 4 ft / 6 ft / 5 ft

44. 15 yd / 7 yd / 7 yd / 15 yd

45. 8 in. / 3 in. / 3 in. / 8 in.

46.

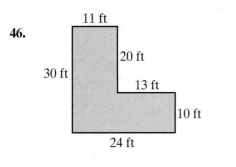

47.

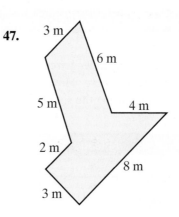

48. Perimeter: Find the perimeter of a parking lot that is in the shape of a rectangle 50 yards wide and 75 yards long.

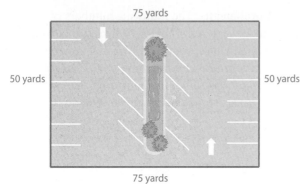

49. Perimeter: A window is in the shape of a triangle placed on top of a square. The length of each of two equal sides of the triangle is 24 inches and the third side is 36 inches long. The length of each side of the square is 36 inches long. Find the perimeter of the window.

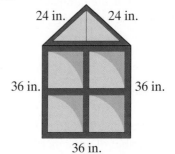

Subtract. See Examples 8 through 10.

50. 89 − 76	**51.** 53 − 33	**52.** 96 − 27	**53.** 126 − 32

54. 19 − 6 **55.** 572 − 41 **56.** 739 − 562 **57.** 8736 − 3527

58. 692 − 217	**59.** 3275 − 1744	**60.** 6793 − 5827	**61.** 7843 − 6274

62. 4900 − 3476	**63.** 5070 − 4376	**64.** 7602 − 2985	**65.** 7,085,076 − 4,278,432

66. 830,900 − 74,985	**67.** 6,543,222 − 2,742,663	**68.** 716,924 − 574,185	**69.** 8,000,000 − 647,561

70. What number should be added to 978 to get a sum of 1200?

71. What number should be added to 860 to get a sum of 1000?

72. Business: An ice cream shop began the day with 32 gallons of mint chocolate chip ice cream. By closing time they had sold 14 gallons. How many gallons of mint chocolate chip ice cream were left?

73. Travel: James decided to take a road trip from Wilmington, NC to Los Angeles, CA. The total distance of the trip is 2590 miles. He stops for a break after he has traveled 438 miles. How many more miles must James travel before he reaches Los Angeles?

74. Farming: Farmer Pat raises chickens and guinea fowl. If she has a total of 6372 birds and 4685 of them are chickens, how many guinea fowl does she have?

75. Construction: The Kingston Construction Co. made a bid of $7,043,272 to build a stretch of freeway, but the Beach City Construction Co. made a lower bid of $6,792,868. How much lower was the Beach City bid?

76. Real estate: A couple sold their house for $135,000. They paid the realtor $8100, and other expenses of the sale came to $800. If they owed the bank $87,000 for the mortgage, what were their net proceeds from the sale?

77. Schooling: A man is 36 years old, and his wife is 34 years old. Together, they have attended 28 years of school, including college. If the man attended 12 years of school, how many years did his wife attend?

78. Checking: In June, Ms. White opened a checking account and deposited $1342, $238, $57, and $486. She also wrote checks for $132, $76, $42, $480, $90, and $327. What was her balance at the end of June?

A Understand the parts of a multiplication problem.

B Multiply whole numbers.

C Understand the properties of multiplication.

D Understand the distributive property.

E Calculate the area of a rectangle.

1.3 Multiplication with Whole Numbers

Objective A **The Parts of a Multiplication Problem**

If you buy five 6-packs of soda, to determine the total number of cans of soda purchased, you can add five 6's:

$$6 + 6 + 6 + 6 + 6 = 30.$$

The process of addition with the same number is also known as **repeated addition**. Or you can **multiply** 5 times 6 and get the same result:

$$5 \cdot 6 = 30.$$

The raised dot indicates multiplication.

In either case, you know that you have 30 cans of soda.

The result of multiplication is called a **product**. The two numbers being multiplied are called **factors** of the product. In the following example, the repeated addend (175) and the number of times it is used (4) are both **factors** of the **product** 700.

$$175 + 175 + 175 + 175 = 4 \cdot 175 = 700$$

factor factor product

Several notations can be used to indicate multiplication. In this text, to avoid possible confusion between the letter x (used as a variable) and the $\times$ sign, we will use the raised dot and parentheses most of the time.

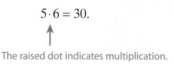

Symbols for Multiplication	
Symbol	**Example**
· raised dot	$4 \cdot 175$
( ) numbers inside or next to parentheses	$4(175)$ or $(4)175$ or $(4)(175)$
× cross sign	4×175 or 175 $\times 4$

Objective B **Multiplication with Whole Numbers**

Multiplication is explained step-by-step in Examples 1 and 2, and then shown in the standard form in Example 3.

Example 1

Multiplying Whole Numbers

The steps of multiplication are shown in finding the product $73 \cdot 4$.

Step 1:

$$
\begin{array}{r}
\overset{1}{7}3 \\
\times \quad 4 \\
\hline
2
\end{array}
$$

multiply: $4 \cdot 3 = 12$
1 carried from 12

Step 2:

$$
\begin{array}{r}
\overset{1}{7}3 \\
\times \quad 4 \\
\hline
292
\end{array}
$$

now multiply: $4 \cdot 7 = 28$
and add the 1: $28 + 1 = 29$

Example 2

Multiplying Whole Numbers

The steps of multiplication are shown in finding the product $37 \cdot 27$.

Step 1:

$$
\begin{array}{r}
\overset{4}{3}7 \\
\times \ 27 \\
\hline
9
\end{array}
$$

multiply: $7 \cdot 7 = 49$
4 carried from 49

Step 2:

$$
\begin{array}{r}
\overset{4}{3}7 \\
\times \ 27 \\
\hline
259
\end{array}
$$

now multiply: $7 \cdot 3 = 21$
and add the 4: $21 + 4 = 25$

Step 3:

$$
\begin{array}{r}
\overset{1}{3}7 \\
\times \ 27 \\
\hline
259 \\
4
\end{array}
$$

multiply: $2 \cdot 7 = 14$
1 carried from 14

Write the 4 in the tens column because you are actually multiplying $20 \cdot 7 = 140$. Generally, the 0 is not written.

Use the step-by-step method shown in Examples 1 and 2 to find the products.

1.   18
       × 5

2.   26
       × 34

3. Use the standard form of multiplication, as shown in Example 3, to find the product 15 · 32.

Step 4:

$$\overset{1}{37}$$

$$\times\ 27$$

$$259$$

$$74$$

now multiply: 2 · 3 = 6

and add the 1: 6 + 1 = 7

Step 5:

$$37$$

$$\times\ 27$$

$$259$$

$$74$$

$$999$$

Add to find the final product.

Now work margin exercises 1 and 2.

Now work margin exercises 1 and 2.

Example 3

Multiplying Whole Numbers

The standard form of multiplication is used here to find the product 93 · 46.

$$\overset{11}{93}$$

$$\times\ 46$$

$$558$$ ← 6 · 3 = 18; write 8, carry 1; 6 · 9 = 54; add 1: 54 + 1 = 55

$$372$$ ← 4 · 3 = 12; write 2, **carry 1**; 4 · 9 = 36; add 1: 36 + 1 = 37

$$4278$$ ← product

Now work margin exercise 3.

Using a Calculator to Multiply Whole Numbers
To multiply numbers on a calculator, you will need to find the multiplication key, ✕ , and the equal sign key, ═ .
To multiply the values given above in example 3, press the keys
9 3 ✕ 4 6 . Then press ═ .
The display will read 4278.

Objective C The Properties of Multiplication

As with addition, the operation of multiplication has several properties. Multiplication is **commutative**, **associative**, and has an identity. The number 1 is the **multiplicative identity**. Multiplication by 0 always gives a product of 0, and this fact is called the **multiplication property of 0**.

Commutative Property of Multiplication

For any whole numbers a and b, $a \cdot b = b \cdot a$.

For example, $3 \cdot 4 = 4 \cdot 3$.

(The **order** of the numbers in multiplication can be reversed.)

Associative Property of Multiplication

For any whole numbers a, b, and c, $(a \cdot b) \cdot c = a \cdot (b \cdot c)$.

For example, $(7 \cdot 2) \cdot 5 = 7 \cdot (2 \cdot 5)$.

(The **grouping** of the numbers in multiplication can be changed.)

Multiplicative Identity Property

For any whole number a, $a \cdot 1 = a$.

For example, $6 \cdot 1 = 6$.

(The product of any number and 1 is that same number.)

The number 1 is called the **multiplicative identity**.

Multiplication Property of 0 (or Zero-Factor Law)

For any whole number a, $a \cdot 0 = 0$.

For example, $63 \cdot 0 = 0$.

(The product of a number and 0 is always 0.)

4. Find the product and identify which property of multiplication is being illustrated.

a. $8 \times 12 = 12 \times 8$

b. $0 \cdot 57$

c. $35 \cdot 1$

d. $7 \cdot (3 \cdot 2) = (7 \cdot 3) \cdot 2$

Example 4

The Properties of Multiplication

Each of the properties of multiplication is illustrated.

a. $5 \times 6 = 6 \times 5$ commutative property of multiplication
As a check, we see that $5 \times 6 = 30$ and $6 \times 5 = 30$.

b. $2 \cdot (5 \cdot 9) = (2 \cdot 5) \cdot 9$ associative property of multiplication
As a check, we see that $2 \cdot (45) = 90$ and $(10) \cdot 9 = 90$.

c. $8 \cdot 1 = 8$ multiplicative identity property

d. $196 \cdot 0 = 0$ multiplication property of 0

Now work margin exercise 4.

Objective D   **Understand the Distributive Property**

To understand the technique for multiplying two whole numbers, we use expanded notation and the following property, called the **distributive property of multiplication over addition** (or simply the **distributive property**).

Distributive Property of Multiplication over Addition

For any whole numbers a, b, and c, $a(b+c) = a \cdot b + a \cdot c$.

For example, to multiply $3(50+2)$, we can add first and then we can multiply.

$$3(50+2) = 3(52) = 156$$

But we can also multiply first and then add, in the following manner.

$$3(50+2) = 3 \cdot 50 + 3 \cdot 2 \quad \text{This step is called the distributive property.}$$
$$= 150 + 6 \quad \text{150 and 6 are called partial products.}$$
$$= 156$$

Example 5

The Distributive Property

Evaluate each of the following expressions by using the distributive property.

a. $6(3+5)$

Solution

$$6(3+5) = 6 \cdot 3 + 6 \cdot 5 = 18 + 30 = 48$$

b. $12(9+2)$

Solution

$$12(9+2) = 12 \cdot 9 + 12 \cdot 2 = 108 + 24 = 132$$

c. $4(7+8)$

Solution

$$4(7+8) = 4 \cdot 7 + 4 \cdot 8 = 28 + 32 = 60$$

Now work margin exercise 5.

5. Rewrite the following expression using the distributive property and then evaluate.

$$3(10+6)$$

Objective E **Calculate the Area of a Rectangle**

Area is the measure of the interior, or enclosed region, of a plane surface and is measured in **square units**.

Finding the Area of a Rectangle

The **area** of a rectangle (measured in square units) is found by multiplying its length by its width.

The concept of area is illustrated in Figure 1 in terms of square inches.

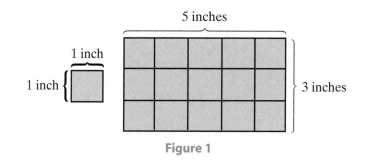

Figure 1

The area is $5 \cdot 3 = 15$ square inches.

Some of the units of area in the metric system are square meters, square decimeters, square centimeters, and square millimeters. In the U.S. customary system, some of the units of area are square yards, square feet, and square inches.

6. Find the area of an elementary school playground that is in the shape of a rectangle 52 yards wide and 73 yards long.

Example 6

Finding the Area of a Rectangle

Find the area of a rectangular plot of land with dimensions as shown here.

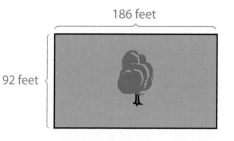

Solution

To find the area, we multiply the length and the width: $186 \cdot 92$.

$$
\begin{array}{r}
186 \\
\times \quad 92 \\
\hline
372 \\
16\,74 \quad \\
\hline
17{,}112 \text{ square feet}
\end{array}
$$

Now work margin exercise 6.

Practice Problems

Multiply.

1. 37
   × 8

2. 781
   × 9

3. 116
   × 25

4. 207
   × 83

5. 1486
   × 7004

Identify the property being illustrated.

6. $17 \cdot 0 = 0$

7. $3(1 \cdot 7) = (3 \cdot 1) \cdot 7$

8. $19 \cdot 1 = 19$

Practice Problem Answers

1. 296

2. 7029

3. 2900

4. 17,181

5. 10,407,944

6. multiplication property of 0

7. associative property of multiplication

8. multiplicative identity

Exercises 1.3

1. 27
   × 6

2. 48
   × 9

3. 84
   × 3

4. 72 × 1

5. 37 · 0

6. (0)(19)

7. 1 · 56

8. 3 · 21

9. 84 × 4

10. 427
    × 2

11. 691
    × 6

12. 702
    × 4

13. 108
    × 3

14. 300
    × 7

15. 8 × 372

16. 437 · 7

17. 3007
    × 5

18. 2342
    × 8

19. 5031
    × 9

20. 6795
    × 3

21. 6(2345)

22. 4 × 8075

23. 27,831
    × 6

24. 37,205
    × 3

25. 42
    × 56

26. 93
    × 30

27. 16
    × 26

28. 20
    × 44

29. 51 × 83

30. 35 · 84

31. (70)(30)

32. 126
    × 41

33. 232
    × 76

34. 114
    × 25

35. 72
    × 106

36. 900
    × 20

37. 37 · 574

38. (430)(50)

39. 30 × 810

40. 348 · 29

41. 2706
    × 30

42. 8411
    × 40

43. (12,030)20

44. 207
    × 143

45. 420
    × 104

46. 673
    × 186

47. 192
    × 467

48. 500
    × 600

49.
$$\begin{array}{r} 210 \\ \times\ 301 \\ \hline \end{array}$$

50.
$$\begin{array}{r} 650 \\ \times\ 400 \\ \hline \end{array}$$

51.
$$\begin{array}{r} 4321 \\ \times\ \ 765 \\ \hline \end{array}$$

52.
$$\begin{array}{r} 7913 \\ \times\ \ 104 \\ \hline \end{array}$$

53.
$$\begin{array}{r} 1892 \\ \times\ \ 200 \\ \hline \end{array}$$

54.
$$\begin{array}{r} 8527 \\ \times\ \ 111 \\ \hline \end{array}$$

55. 500×3000

56. $300 \cdot 8600$

57.
$$\begin{array}{r} 3463 \\ \times\ 1743 \\ \hline \end{array}$$

58.
$$\begin{array}{r} 1512 \\ \times\ 2065 \\ \hline \end{array}$$

59.
$$\begin{array}{r} 1376 \\ \times\ 6005 \\ \hline \end{array}$$

60.
$$\begin{array}{r} 2592 \\ \times\ 1106 \\ \hline \end{array}$$

61.
$$\begin{array}{r} 27{,}000 \\ \times\ \ \ 4000 \\ \hline \end{array}$$

62.
$$\begin{array}{r} 53{,}000 \\ \times\ \ \ 7000 \\ \hline \end{array}$$

63. $13 \cdot 7 \cdot 0$

64. $(96)(0)(4)$

65. $832 \times 17 \times 0$

66. $(260)(70)(1)$

67. $830 \cdot 10 \cdot 1$

68. $694 \cdot 0 \cdot 72$

69. $(537)(38)(0)$

Identify the property of multiplication being illustrated. See Example 4.

70. $4 \cdot 7 = 7 \cdot 4$

71. $2(1 \cdot 6) = (2 \cdot 1)6$

72. $5 \cdot 1 = 5$

73. $8 \cdot 0 = 0$

Evaluate each expression by using the distributive property. See Example 5.

74. $6(3 + 11)$

75. $3(9 + 7)$

76. $7(8 + 4)$

77. $9(2 + 9)$

Find the area of the following rectangles. See Example 6.

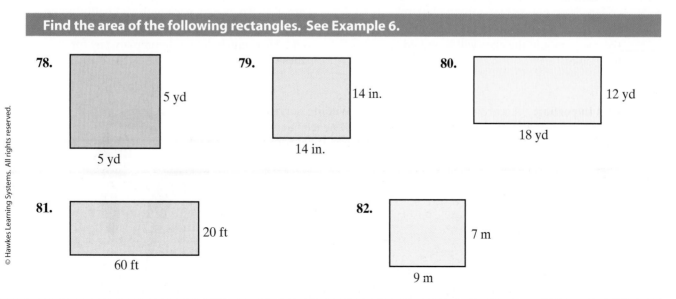

78. 5 yd, 5 yd

79. 14 in., 14 in.

80. 12 yd, 18 yd

81. 20 ft, 60 ft

82. 7 m, 9 m

83. Area: A rectangular pool measures 36 feet long by 18 feet wide. Find the area of the pool in square feet.

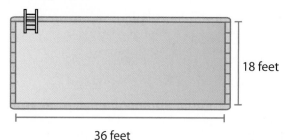

18 feet

36 feet

84. Area: A rectangular lot for a house measures 210 feet long by 175 feet wide. Find the area of the lot in square feet.

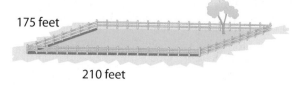

175 feet

210 feet

85. Area: A painting is mounted in a rectangular frame (16 inches by 24 inches) and hung on a wall. How many square inches of wall space will the framed painting cover?

24 in.

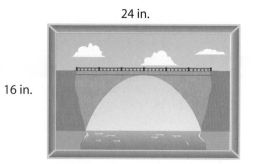

16 in.

86. Fundraising: The Math Club members decided to attend the national meeting of the NCTM (National Council of Teachers of Mathematics) and had a book sale to raise money for the event. Registration fees were $85 per member and the club had 35 members. How much money did the club need to raise for registration fees?

87. Diet: If one regular pack of candy contains 250 calories, how many calories are there in 37 packs of the same candy?

88. Business: A sandwich shop buys 372 loaves of bread for the week. If each loaf of bread has 24 slices, how many slices of bread were purchased?

89. Education: Students at the local community college must pay $83 for a math textbook. If there are 43 students in the class, find the total amount the class will spend on textbooks.

90. Automobile Purchase: Your company bought 18 new cars, each with air conditioning and anti-lock brakes, at a price of $15,800 per car. How much did your company pay for these cars?

91. Bird Importing: According to the U.S. Fish and Wildlife Service, migratory birds are imported at a value of about $19 each. Suppose that about 800,000 live birds are imported each year. What is the total value of these imported birds?

1.4 Division with Whole Numbers

Objective A **Division with Whole Numbers**

George is having a meeting with three other people. He has decided to bring a dozen donuts as part of the refreshments. If each person is to get the same number of donuts, then the 12 donuts are **divided** into 4 parts so that each person will get 3 donuts.

The process of separating the donuts into equally-sized groups is called **division**. We can use the division sign $(\div)$ to indicate the division procedure as follows.

$$12 \ \div \ 4 \ = \ 3 \qquad \text{(Read "12 divided by 4 equals 3.")}$$

$$\underset{\text{dividend}}{\uparrow} \quad \underset{\text{divisor}}{\uparrow} \quad \underset{\text{quotient}}{\uparrow}$$

Additionally, the following two notations can be used to indicate division.

We know that multiplication can be related to addition in the following way.

$$6 + 6 + 6 + 6 + 6 = 5 \cdot 6 = 30$$

So there are five 6's in 30.

In a similar way, division can be related to repeated subtraction. For example, to find how many 8's are in 32, we can repeatedly subtract 8 as follows.

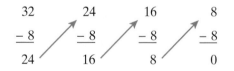

8 is subtracted four times. So there are four 8's in 32.

The number left after division is called the **remainder**. In the above illustration, the remainder was 0 because there are exactly four 8's in 32. Example 1 illustrates repeated subtraction with a remainder that is not 0. It is important to note that **the remainder must be less than the divisor**.

Terms Used in Division

The **dividend** is the number being divided.
The **divisor** is the number doing the dividing.
The **quotient** is the result of division.
The **remainder** is the number left after division.

Objectives

A Understand the process of division with whole numbers.

B Know how division is related to multiplication.

C Understand division involving 0.

D Learn the long division process.

1. Divide $23 \div 4$ using
repeated subtraction.

Example 1

Division using Repeated Subtraction

Divide using repeated subtraction: $27 \div 6$.

Solution

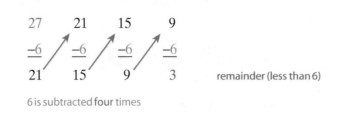

remainder (less than 6)

6 is subtracted **four** times

Thus $27 \div 6 = 4$ with R3. (**Note:** R is a notation for the remainder.)

Check: The division can be checked as follows. Multiply the divisor times the quotient and then add the remainder. The result must be the dividend.

$$6 \quad \cdot \quad 4 \quad + \quad 3 \quad = \quad 27$$

divisor $\cdot$ quotient $+$ remainder $=$ dividend

Now work margin exercise **1.**

Objective B **Division Related to Multiplication**

We know that $5 \cdot 7 = 35$ and that 5 and 7 are **factors** of 35. Factors are also called **divisors** because each factor will divide evenly into the product. That is, 5 and 7 will both divide evenly into 35. As another example:

We know that $4 \cdot 9 = 36$.
So $36 \div 4 = 9$ and $36 \div 9 = 4$.

Thus, 4 and 9 are divisors of 36. This means that if 36 is divided by either 4 or 9, the remainder will be 0.

The relationship between multiplication and division can be seen from the following table.

	Division				Multiplication		
Dividend		**Divisor**		**Quotient**	**Factors**		**Product**
28	÷	7	=	4	4 · 7	=	28
30	÷	5	=	6	6 · 5	=	30
90	÷	10	=	9	9 · 10	=	90
37	÷	37	=	1	1 · 37	=	37
64	÷	1	=	64	64 · 1	=	64

Division by 1

For any number a, $\dfrac{a}{1} = a$.

Division of a Number by Itself

For any nonzero number a, $\dfrac{a}{a} = 1$.

Example 2

Division with Whole Numbers

a. $24 \div 6 = 4$ because $4 \cdot 6 = 24$

b. $26 \div 2 = 13$ because $2 \cdot 13 = 26$

c. $\dfrac{72}{8} = 9$ because $8 \cdot 9 = 72$

d. $\dfrac{35}{1} = 35$ because $1 \cdot 35 = 35$

e. $\dfrac{17}{17} = 1$ because $17 \cdot 1 = 17$

Now work margin exercise 2.

2. Find the following quotients.

a. $28 \div 7$

b. $\dfrac{42}{6}$

c. $\dfrac{57}{57}$

d. $86 \div 1$

Objective C   **Division Involving 0**

There are two cases to consider with division involving 0. One is when 0 is divided by a number. The other is dividing by 0. Both can be explained in terms of multiplication. For example,

$$\dfrac{0}{7} = 0 \text{ because } 0 = 0 \cdot 7. \quad \text{(by the zero factor law)}$$

However, if we divide by 0, we have

$$\dfrac{7}{0} = \boxed{?} \text{ which would indicate } 7 = 0 \cdot \boxed{?}.$$

But this is not possible because $0 \cdot \boxed{?} = 0$ for any value of $\boxed{?}$. What this means is that we cannot divide by 0. We say that

$$\dfrac{7}{0} \text{ is } \textbf{undefined}.$$

These ideas lead to the following two rules about division involving 0.

Division Involving 0

Case 1: If a is any nonzero whole number, then $\dfrac{0}{a} = 0$.

Case 2: If a is any whole number, then $\dfrac{a}{0}$ is **undefined**.

3. Find the following quotients.

a. $29 \div 0$

b. $\dfrac{0}{8}$

Example 3

Division Involving 0

a. $0 \div 23 = 0$ or $\dfrac{0}{23} = 0$ or $23\overline{)0}^{\,0}$

(0 can be the quotient.)

b. $14 \div 0$ is undefined or $\dfrac{14}{0}$ is undefined or $0\overline{)14}^{\,\text{undefined}}$

(Division by 0 is not possible.)

Now work margin exercise 3.

Objective D Long Division

Repeated subtraction helps in understanding the division process. This works well when the numbers are small. However, with larger numbers, the familiar **long division** process (called the **division algorithm**[1]) is generally used. This process can be written in the following format.

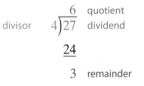

The following examples illustrate long division in a step-by-step way. Study these examples carefully. Remember that the remainder must be less than the divisor.

Example 4

Using Long Division

Find the quotient and remainder for $33 \div 8$

Solution

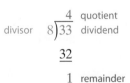

---

1   An algorithm is a process or pattern of steps to be followed in working with numbers.

Check: If the following equation is not true, then there is an error.

divisor	·	quotient	+	remainder	=	dividend
8	·	4	+	1	=	33
		32	+	1	=	33
				33	=	33

So $33 \div 8 = 4$ R1.

Example 5

Using Long Division

Find the quotient and remainder for $683 \div 7$.

Solution

Step 1:

$$\begin{array}{r} 9 \\ 7\overline{)683} \\ \underline{63} \\ 5 \end{array}$$

Trial divide 7 into 68.

Subtract 63.

Step 2:

$$\begin{array}{r} 97 \\ 7\overline{)683} \\ \underline{63} \\ 53 \\ \underline{49} \\ 4 \end{array}$$

quotient

Bring down the 3 from the dividend, then trial divide 7 into 53.

Subtract 49.

remainder

Check:

divisor	·	quotient	+	remainder	=	dividend
7	·	97	+	4	=	683
		683			=	683

So $683 \div 7 = 97$ R4.

Example 6

Using Long Division

Find the quotient and remainder for $696 \div 30$.

Solution

Step 1:

$$\begin{array}{r} 2 \\ 30\overline{)696} \\ \underline{60} \\ 9 \end{array}$$

Trial divide 30 into 69, or 3 into 6. This gives 2 in the tens position.

Subtract 60.

Step 2:

$$\begin{array}{r} 23 \\ 30\overline{)696} \\ \underline{60} \\ 96 \\ \underline{90} \\ 6 \end{array}$$

quotient

96   Bring down the 6 from the dividend then trial divide 30 into 96, or 3 into 9. This gives 3 in the ones position. Subtract 90.

6   remainder

Check:

divisor	·	quotient	+	remainder	=	dividend
30	·	23	+	6	=	696
			696		=	696

So $696 \div 30 = 23 \text{ R}6$.

Example 7

Using Long Division

Find $9325 \div 45$.

Solution

Step 1:

$$\begin{array}{r} 2 \\ 45\overline{)9325} \\ \underline{90} \\ 3 \end{array}$$

Trial divide 45 into 93 or 4 into 9, this gives 2 in the hundreds position.

Step 2:

$$\begin{array}{r} 20 \\ 45\overline{)9325} \\ \underline{90} \\ 32 \\ \underline{0} \\ 32 \end{array}$$

45 does not divide into 32. So write 0 in the tens column and multiply $0 \cdot 45 = 0$.

Step 3:

$$\begin{array}{r} 208 \\ 45\overline{)9325} \\ \underline{90} \\ 32 \\ \underline{0} \\ 325 \\ \underline{360} \end{array}$$

Trial divide 45 into 325 or 4 into 32. But the trial quotient of 8 is too large since $8 \cdot 45 = 360$ is larger than 325.

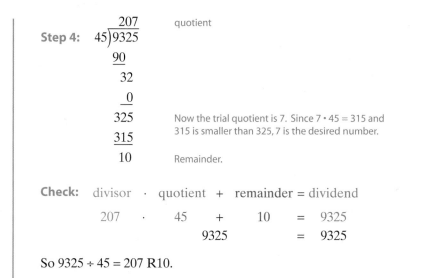

Step 4:

$$45\overline{)9325}$$ quotient = 207

90

32

0

325   Now the trial quotient is 7. Since 7 · 45 = 315 and
         315 is smaller than 325, 7 is the desired number.

315

10   Remainder.

Check: divisor · quotient + remainder = dividend

207   ·   45   +   10   =   9325

9325   =   9325

So 9325 ÷ 45 = 207 R10.

Now work margin exercises 4 through 7.

Find the quotients and remainders

4. $9\overline{)23}$

5. $415 \div 6$

6. $340 \div 16$

7. $31\overline{)9571}$

⚠ Common Error

In step 2 of Example 7, we wrote 0 in the quotient because 45 did not divide into 32. Many students fail to write the 0. This error obviously changes the value of the quotient.

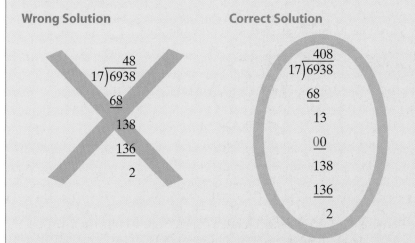

Wrong Solution

$$17\overline{)6938}$$ = 48

68

138

136

2

Correct Solution

$$17\overline{)6938}$$ = 408

68

13

00

138

136

2

Writing the 0 in the quotient gives the correct quotient of 408, which is considerably different from 48.

Completion Example 8

Using Long Division

Finish the long division process to find the quotient and remainder.

$$4\overline{)334}$$ = 8

32

14

Find the quotients and remainders.

8. $12\overline{)1869}$

9. $39\overline{)8370}$

10. Show that 21 and 14 are divisors of 294.

Completion Example 9

Using Long Division

Finish the long division process to find the quotient and remainder.

$$
\begin{array}{r}
2 \\
12\overline{)2451} \\
\underline{24} \\
05
\end{array}
$$

Now work margin exercises 8 and 9.

Remainder of 0

If the remainder is 0, then the following statements are true:

1. Both the divisor and quotient are **factors** of the dividend.
2. We say that both factors **divide exactly** into the dividend.
3. Both factors are called **divisors** of the dividend.

Example 10

Using Long Division

Show that 17 and 36 are factors (or divisors) of 612.

Solution

$$
\begin{array}{r}
36 \\
17\overline{)612} \\
\underline{51} \\
102 \\
\underline{102} \\
0
\end{array}
$$

Because 0 is the remainder, both 17 and 36 are factors (or divisors) of 612. Just to double-check, we also find $612 \div 36$.

$$
\begin{array}{r}
17 \\
36\overline{)612} \\
\underline{36} \\
252 \\
\underline{252} \\
0
\end{array}
$$

Yes, both 17 and 36 divide exactly into 612.

Now work margin exercise 10.

Completion Example Answers
  8. 83 R2
  9. 204 R3

Example 11

Using Long Division

A plumber purchased 18 pipe fittings. What was the price of one fitting if the bill was $630 for all of the fittings?

Solution

We need to know how many times 18 goes into 630.

$$
\begin{array}{r}
35 \\
18\overline{)630} \\
\underline{54} \\
90 \\
\underline{90} \\
0
\end{array}
$$

The price for one fitting was $35.

Now work margin exercise 11.

Example 12

Using Long Division

In a typical day, Quality Call Center receives approximately 3860 calls. If 20 employees are working, and each employee answers the same number of calls, how many calls does each employee receive?

Solution

The number of calls for each employee is found by dividing the total number by 20.

$$
\begin{array}{r}
193 \\
20\overline{)3860} \\
\underline{20} \\
186 \\
\underline{180} \\
60 \\
\underline{60} \\
0
\end{array}
$$

Each employee receives 193 calls.

Now work margin exercise 12.

11. If 8 identical bicycles cost $856, what was the cost of each bike?

12. It costs $1428 for a group of 6 friends to rent a beach house for a week. How much will each person pay?

13. A box of cookies contains a total of 1872 calories. If there are 8 servings in the box of cookies, how many calories are in each serving of cookies?

Example 13

Using Long Division

Tasty Treat Bakery just received an order for 1296 cupcakes. When the order is shipped, 6 cupcakes will be put into 1 box. How many boxes will the bakery need to fill the order?

Solution

The number of boxes can be found by dividing the total number of cupcakes by the number of cupcakes in each box.

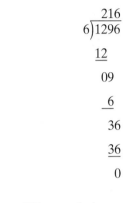

$$\begin{array}{r} 216 \\ 6\overline{)1296} \\ \underline{12} \\ 09 \\ \underline{6} \\ 36 \\ \underline{36} \\ 0 \end{array}$$

Thus 216 boxes will be needed.

Now work margin exercise 13.

 Using a Calculator to Divide Whole Numbers

To divide numbers on a calculator, you will need to find the division key, ÷ , and the equal sign key, = .
To divide the values given above in Example 13, press the keys
 1 2 9 6 ÷ 6 . Then press = .
The display will read 216.

Practice Problems

Perform each division. If the division is not possible, write "undefined."

1. $35 \div 7$ **2.** $\dfrac{0}{16}$ **3.** $3\overline{)54}$ **4.** $13\overline{)305}$

5. $26{,}580 \div 20$ **6.** $107{,}120 \div 301$

Practice Problem Answers
1. 5   **2.** 0   **3.** 18
4. 23 R6   **5.** 1329   **6.** 355 R265

Exercises 1.4

Find each quotient. If the division is not possible, write "undefined." See Examples 2 and 3.

1. $21 \div 7$

2. $\dfrac{24}{4}$

3. $54 \div 6$

4. $\dfrac{25}{5}$

5. $0\overline{)17}$

6. $\dfrac{23}{1}$

7. $\dfrac{14}{14}$

8. $\dfrac{0}{3}$

9. $48 \div 8$

10. $\dfrac{18}{0}$

11. $35 \div 7$

12. $16\overline{)0}$

13. $\dfrac{32}{1}$

14. $5 \div 5$

15. $8\overline{)56}$

16. $22 \div 0$

Perform each division. Check your answer using multiplication. See Examples 4 through 9.

17. $70 \div 5$

18. $3\overline{)51}$

19. $\dfrac{120}{4}$

20. $\dfrac{210}{7}$

21. $8\overline{)52}$

22. $44 \div 6$

23. $25\overline{)600}$

24. $413 \div 20$

25. $\dfrac{161}{15}$

26. $12\overline{)108}$

27. $305 \div 20$

28. $206 \div 18$

29. $19\overline{)7619}$

30. $\dfrac{8338}{22}$

31. $\dfrac{1472}{64}$

32. $9963 \div 27$

33. $16\overline{)4813}$

34. $4406 \div 11$

35. $3917 \div 13$

36. $50\overline{)3065}$

37. $\dfrac{8844}{33}$

38. $26,964 \div 28$

39. $\dfrac{25,506}{39}$

40. $60,696 \div 72$

41. $\dfrac{77,742}{63}$

42. $14\overline{)28,923}$

43. $47\overline{)56,821}$

44. $87,583 \div 29$

45. $84\overline{)10,477}$

46. $\dfrac{18,583}{19}$

47. $\dfrac{33,591}{76}$

48. $97\overline{)47,031}$

49. $98,762 \div 502$

50. $317\overline{)70,365}$

51. $\dfrac{169,719}{417}$

52. $105,123 \div 201$

53. Find the quotient of 98 and 7.

54. Find the quotient of 292 and 4.

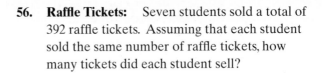

55. Cookies: A mother has 12 cookies which she will equally divide among 4 children. How many cookies will each child get?

56. Raffle Tickets: Seven students sold a total of 392 raffle tickets. Assuming that each student sold the same number of raffle tickets, how many tickets did each student sell?

57. Camp: The Cedarville Baseball Camp has 198 youths. How many 9-member teams can this club have?

58. Tutoring: Jane Scott tutors students in reading and makes $25 per student. If she makes $475 in a week, how many students did she tutor?

59. Ice Cream: One pint of Ben and Jerry's Crème Brûlée Ice Cream has 68 grams of fat. If there are 4 servings per pint, how many grams of fat are in each serving? **Source:** Ben and Jerry's

60. Painting: If one person can paint a small house in 48 hours, how long will it take a crew of 8 people to paint the house, assuming that all 8 work at the same speed, and do not interfere with each other?

61. Education: For 2009–2010, the average tuition cost for four years at a public, 4-year institution is $28,080. If tuition did not increase each year, how much would you pay per year for the four years you were in college? **Source:** www.collegeboard.com

62. Pianos: Smithfield High School paid $29,022 for six grand pianos. How much did each piano cost?

63. Paper: An instructor has 100 pieces of scratch paper to pass out to her students. If there are 30 students in the class, how many pieces of paper will each student receive and how many pieces of paper will the instructor have left over?

64. Landscaping: The area of every NFL football field is 57,600 square feet. If a bag of grass seed covers 50 square feet, how many bags of grass seed will be needed to cover one grass football field? **Source:** National Football League

65. Space: U.S. Astronaut Peggy Whitson orbited the Earth 6032 times during her space flights on the International Space Station. If the International Space Station orbits the Earth 16 times a day, how many days was Peggy Whitson in space? **Source:** National Aeronautics and Space Administration

66. Gardening: A community has 5978 square feet available for individual gardens which will be evenly distributed among 14 people. How much space will each person get?

67. Boating: Thirteen men purchase a boat together. If the cost of the boat is $33,462, how much will each man contribute if each contributes an equal amount?

68. Farming: A large chicken farm produces 65,076 eggs during a typical week. How many dozen eggs does this represent?

69. Pasta: 262,800 pounds of pasta is served every year at Mama Melrose's Ristorante Italiano at Disney MGM Studios. How many pounds of pasta are served every day, assuming there are 365 days in each year?
Source: www.diningindisney.com

70. Pizza: Katelyn is having a birthday party and has invited 11 friends. For lunch, Katelyn's parents bought 3 large pizzas that have 8 slices per pizza. How many slices of pizza will each child get?

1.5 Rounding and Estimating with Whole Numbers

Objective A **Rounding Whole Numbers**

You owe a friend $38 for a concert ticket. You don't have cash and stop at an ATM. However, the ATM can only give you money in increments of $10. You need to decide how much money to withdraw. In order to have enough money to pay your friend, you must round $38 up to $40.

Rounded numbers are simply approximations of given numbers, and are used in many everyday applications, like the one just discussed.

Rounding Numbers

To **round** a given number means to find another number close to the given number. The desired place of accuracy must be stated.

For example, if you were asked to round 762, you might say 760 or 800. Either answer could be correct, depending on whether the accuracy was to be the nearest ten or the nearest hundred.

In Example 1, we use **number lines** as visual aids in understanding the rounding process. On number lines, the whole numbers are used to label equally-spaced points to the right of the point labeled 0.

Example 1

Rounding Whole Numbers

Round the following numbers to the indicated place of accuracy.

a. 43, tens

Solution

To round 43 to the nearest ten:

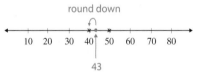

We see that 43 is closer to 40 than to 50. Thus 43 rounds to 40 (to the **nearest ten**).

b. 762, hundreds

Solution

To round 762 to the nearest hundred:

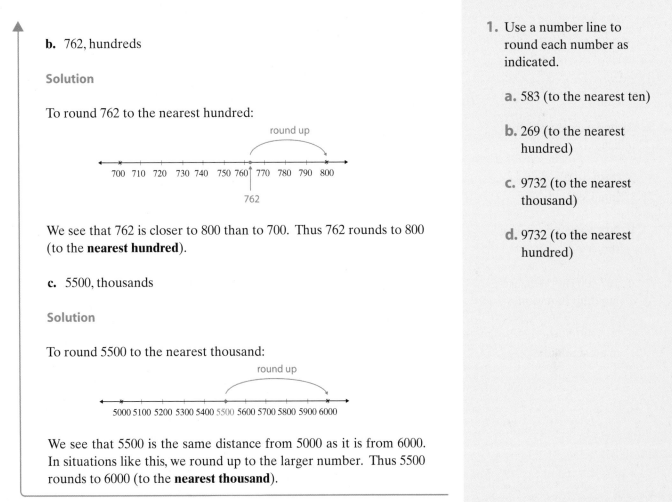

round up

700 710 720 730 740 750 760 770 780 790 800

762

We see that 762 is closer to 800 than to 700. Thus 762 rounds to 800 (to the **nearest hundred**).

c. 5500, thousands

Solution

To round 5500 to the nearest thousand:

round up

5000 5100 5200 5300 5400 5500 5600 5700 5800 5900 6000

We see that 5500 is the same distance from 5000 as it is from 6000. In situations like this, we round up to the larger number. Thus 5500 rounds to 6000 (to the **nearest thousand**).

Now work margin exercise 1.

Using number lines as aids for understanding is fine, but for practical purposes, the following rule is more useful.

Rounding Rule for Whole Numbers

1. Look at the single digit just to the right of the digit that is in the place of desired accuracy.
2. **If this digit is 5 or greater**, make the digit in the desired place of accuracy one larger and replace all digits to the right with zeros. All digits to the left remain unchanged unless a 9 is made one larger; then the next digit to the left is increased by 1.
3. **If this digit is less than 5**, leave the digit that is in the place of desired accuracy as it is, and replace all digits to the right with zeros. All digits to the left remain unchanged.

1. Use a number line to round each number as indicated.

   a. 583 (to the nearest ten)

   b. 269 (to the nearest hundred)

   c. 9732 (to the nearest thousand)

   d. 9732 (to the nearest hundred)

2. Round each number as indicated.

a. 7350 (to the nearest hundred)

b. 29,736 (to the nearest thousand)

c. 137,800 (to the nearest hundred thousand)

d. 28,379,200 (to the nearest million)

Example 2

Rounding Whole Numbers

Round the following numbers to the indicated place of accuracy.

a. 6849, hundreds

Solution

6849
↑
Place of desired accuracy.

6849
↑
Look at one digit to the right; 4 is less than 5.

6800
↑
Leave 8 and fill in zeros.

So 6849 rounds to 6800 (to the nearest hundred).

b. 3500, thousands

Solution

3500
↑
Place of desired accuracy.

3500
↑
Look at 5; 5 is 5 or greater.

4000
↑
Increase 3 to 4 (one larger) and fill in zeros.

So 3500 rounds to 4000 (to the nearest thousand).

c. 597, tens

Solution

597
↑
Place of desired accuracy.

597
↑
Look at 7; 7 is 5 or greater.

600
↑↑
Increase 9 to 10 (this changes the 5 to a 6).

So 597 rounds to 600 (to the nearest ten).

d. 20,560, ten thousands

Solution

20,560
↑
Place of desired accuracy.

20,560
↑
Look at 0; 0 is less than 5.

20,000
↑
Leave 2 and fill in zeros.

So 20,560 rounds to 20,000 (to the nearest ten thousand).

Now work margin exercise 2.

Example 3

Rounding Whole Numbers

A jar of jelly beans at a local candy store is filled with 2709 jelly beans and put on display in the store window. Round this figure to the nearest thousand.

Solution

To round 2709 to the nearest thousand:

2709
↑
Place of desired accuracy.

2709
↑
Look at 7; 7 is 5 or greater.

3000
↑
Increase 2 to 3 (one larger) and fill in zeros.

So 2709 rounds to 3000 (to the nearest thousand).

Now work margin exercise 3.

3. A redwood tree measures 285 feet in height. Round the height of the tree to the nearest ten.

Objective B **Estimating Sums and Differences**

One use for rounded numbers is to **estimate** an answer (or to find an **approximate** answer) before any calculations are made with the given numbers.

To estimate an answer means to use rounded numbers in a calculation to form an idea of what the size of the actual answer should be. In some situations an estimated answer may be sufficient. For example, a shopper may simply estimate the total cost of purchases to be sure that he or she has enough cash to cover the cost.

To Estimate a Sum or Difference

1. Round each number to the place of the **leftmost** digit.
2. Perform the addition or subtraction with these rounded numbers.

Example 4

Estimating a Sum of Whole Numbers

Estimate the sum; then find the sum.

$$
\begin{array}{r}
68 \\
925 \\
+\ 487 \\
\end{array}
$$

Solution

Note that in this example, numbers are rounded to different places because they are of different sizes. That is, the leftmost digit is not in the same place for all numbers.

a. Estimate the sum by first rounding each number to the place of the leftmost digit and then adding.

68	$\longrightarrow$	70	rounded value of 68
925	$\longrightarrow$	900	rounded value of 925
+ 487	$\longrightarrow$	+ 500	rounded value of 487
		1470	estimated sum

b. Now find the sum, knowing that the answer should be close to 1470.

$$
\begin{array}{r}
\overset{1\ 2}{68} \\
925 \\
+\ 487 \\
\hline
1480 \\
\end{array}
$$
This sum is very close to 1470.

Note that the sum in Example 4 is very close to our estimate. If we had arrived at a sum that was far away from our estimate, like 14,000 in the example above, we would know that we made an error and that we should check our work.

Completion Example 5

Estimating a Sum of Whole Numbers

Estimate the sum; then find the sum.

$$
\begin{array}{r}
5483 \\
232 \\
+\ 657 \\
\end{array}
$$

Solution

a. First, estimate the sum by rounding each number to the place of the leftmost digit and adding these rounded numbers.

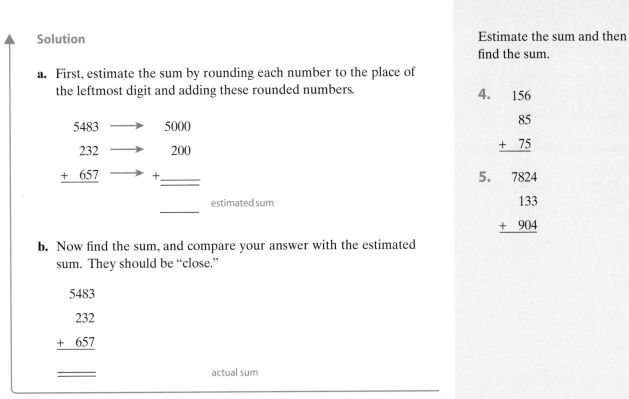

5483 $\longrightarrow$ 5000

232 $\longrightarrow$ 200

+ 657 $\longrightarrow$ +_____

_____

estimated sum

b. Now find the sum, and compare your answer with the estimated sum. They should be "close."

5483

232

+ 657

_____

actual sum

Now work margin exercises 4 and 5.

Estimate the sum and then find the sum.

4. 156
   85
+ 75

5. 7824
   133
+ 904

Example 6

Estimating a Difference of Whole Numbers

Estimate the difference; then find the difference.

2783
− 975

Solution

a. Round each number to the place of the leftmost digit and subtract using these rounded numbers.

2783 $\longrightarrow$ 3000   rounded value of 2783

− 975 $\longrightarrow$ − 1000   rounded value of 975

2000   estimated difference

b. Now we find the difference, knowing that the difference should be close to 2000.

$$\begin{array}{cccc} \overset{1}{\cancel{2}} & {}^{1}7 & \overset{7}{\cancel{8}} & {}^{1}3 \\ - & 9 & 7 & 5 \\ \hline 1 & 8 & 0 & 8 \end{array}$$ The difference is close to 2000.

Now work margin exercise 6.

6. Estimate the difference and then find the difference.

2685
− 847

Completion Example Answers

5. a. 700, 5900

   b. 6372

7. The World of Pets pet store orders 1598 bags of dog food in September. During the month, the store sells 597 bags. Estimate the remaining number of unsold bags of dog food at the end of the month. Then find the actual number of bags left.

Example 7

Estimating a Difference of Whole Numbers

At the beginning of July, your bank account balance is $3859. Over the course of the month, you spend $823. Estimate the remaining balance in your account. Then find the actual balance.

Solution

In order to find the balance, we must subtract the amount of money you spent throughout the month from the starting balance. First, estimate this difference.

$$
\begin{array}{rcll}
3859 & \longrightarrow & 4000 & \text{rounded value of 3859} \\
-\ \ \ 823 & \longrightarrow & -\ \ 800 & \text{rounded value of 823} \\
\hline
& & 3200 & \text{estimated difference}
\end{array}
$$

Now, find the actual difference. This should be close to 3200.

$$
\begin{array}{r}
3859 \\
-\ \ \ 823 \\
\hline
3036
\end{array}
$$

Thus, the remaining balance in your account is $3036.

Now work margin exercise 7.

Objective C **Estimating Products Using Rounded Numbers**

Estimating the product of two numbers before performing the multiplication can be of help in detecting errors. Just as with addition and subtraction, estimations can be done with rounded numbers.

To Estimate a Product

1. Round each number to the place of the **leftmost** digit.
2. Multiply the rounded numbers.

Example 8

Estimating Products of Whole Numbers

Find the product 62·38, but first estimate the product by multiplying the numbers in rounded form.

Solution

First, estimate the product.

$$62 \longrightarrow 60 \quad \text{rounded value of 62}$$
$$\times 38 \longrightarrow \times 40 \quad \text{rounded value of 38}$$
$$2400 \quad \text{estimated product}$$

Now, we find the product, keeping in mind that our answer should be close to 2400.

$$
\begin{array}{r}
\overset{1}{6}2 \\
\times \quad 38 \\
\hline
496 \\
+186 \\
\hline
2356 \quad \text{The actual product is close to 2400.}
\end{array}
$$

Now work margin exercise 8.

Example 9

Estimating Products of Whole Numbers

An apartment complex is planning to furnish all of its 93 apartments with new dishwashers. The owner can purchase the dishwashers for $267 each. Estimate the total cost of buying dishwashers for every apartment. Then find the actual total cost to the apartment complex.

Solution

In order to find the total cost to the apartment complex, we must multiply the number of dishwashers purchased by the cost of each dishwasher.

a. We begin by estimating the product 267·93.

$$267 \longrightarrow 300 \quad \text{rounded value of 267}$$
$$\times 93 \longrightarrow \times 90 \quad \text{rounded value of 93}$$
$$27{,}000 \quad \text{estimated product}$$

b. The product should be near 27,000.

$$
\begin{array}{r}
267 \\
\times \quad 93 \\
\hline
801 \\
+24\,03 \\
\hline
24{,}831
\end{array}
$$

Thus, the actual total cost to the apartment complex is $24,831.

Now work margin exercise 9.

8. First estimate, and then find the following product.

$$18 \cdot 74$$

9. The principal of North Valley High School wants to buy 185 computers for the computer lab. Each new computer costs $472. Estimate the total cost of the new computers. Then find the actual cost.

Estimating Quotients Using Rounded Numbers

By rounding both the divisor and the dividend, we can estimate the quotient. Estimation can help identify unreasonable answers when the actual value is calculated.

To Estimate a Quotient

1. Round both the divisor and dividend to the place of the **leftmost** digit.
2. Divide with the rounded numbers. (When calculating an estimate, ignore the remainder if there is one.)

Example 10

Estimating Quotients of Whole Numbers

Estimate the quotient $8875 \div 25$ by using rounded values; then find the quotient.

Solution

a. Estimation: $8875 \div 25 \longrightarrow 9000 \div 30$

$$
\begin{array}{r}
300 \quad \text{estimated quotient} \\
30\overline{)9000} \\
\underline{90} \\
00 \\
\underline{0} \\
00 \\
\underline{0} \\
0
\end{array}
$$

b. The quotient should be near 300.

$$
\begin{array}{r}
355 \quad \text{quotient} \\
25\overline{)8875} \\
\underline{75} \\
137 \\
\underline{125} \\
125 \\
\underline{125} \\
0
\end{array}
$$

Example 11

Estimating Quotients of Whole Numbers

Estimate the quotient $325 \div 42$ by using rounded values; then find the quotient.

Solution

a. Estimation: $325 \div 42 \longrightarrow 300 \div 40$

$$
\begin{array}{r}
7 \quad \text{estimated quotient}\\
40\overline{)300}\\
\underline{280}\\
20
\end{array}
$$

Thus the estimated quotient is 7.

b. The quotient should be near 7.

$$
\begin{array}{r}
7 \quad \text{quotient}\\
42\overline{)325}\\
\underline{294}\\
31 \quad \text{remainder}
\end{array}
$$

The quotient is 7 and the remainder is 31. In this case, the quotient is the same as the estimated value. The true remainder is different.

Completion Example 12

Estimating Quotients of Whole Numbers

Estimate the quotient $6461 \div 21$. Then finish the division to find the quotient and remainder.

Solution

$$
\begin{array}{r}
\text{estimate}\\
20\overline{)6000}
\end{array}
\qquad
\begin{array}{r}
3\\
21\overline{)6461}\\
\underline{63}\\
16
\end{array}
$$

Is your estimate close to the actual quotient?_____

What is the difference between your estimate and the actual quotient?_____

Now work margin exercises 10 through 12.

Estimate the following quotients by using rounded values; then find the quotient.

10. $378 \div 18$

11. $882 \div 76$

12. $36\overline{)1340}$

Completion Example Answers

12. estimate: 300; quotient: 307; remainder: 14; yes; difference: 7

13. The Sunnyside Elementary School is taking a field trip to the zoo. If one school bus can seat 22 people, and there are 462 students and teachers going on the field trip, estimate the number of buses needed. Then find the actual number of buses.

Example 13

Estimating Quotients of Whole Numbers

A group of 11 friends bought tickets to a local jazz concert. The total price of all 11 tickets was $341.

a. Estimate the cost of each ticket.

Solution

In order to find the cost of each person's ticket, we must divide the total cost of the tickets by the number of tickets that were purchased.

We begin by estimating the quotient.

$$341 \div 11 \longrightarrow 300 \div 10$$

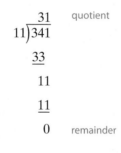

$$\begin{array}{r} 30 \\ 10\overline{)300} \\ \underline{300} \\ 0 \end{array}$$ estimated quotient

Thus, the estimated cost of each ticket is $30.

b. Then, calculate the actual cost of each ticket.

Solution

The quotient should be near 30.

$$\begin{array}{r} 31 \\ 11\overline{)341} \\ \underline{33} \\ 11 \\ \underline{11} \\ 0 \end{array}$$ quotient

remainder

So the actual cost of each ticket is $31.

Now work margin exercise 13.

Practice Problems

Round each number as indicated.

1. 1832 (nearest ten)

2. 14,751 (nearest hundred)

3. 289,300 (nearest ten thousand)

4. Estimate the sum; then find the sum.

$$
\begin{array}{r}
246 \\
359 \\
+\ 486 \\
\hline
\end{array}
$$

5. Estimate the difference; then find the difference.

$$
\begin{array}{r}
9652 \\
-\ 3357 \\
\hline
\end{array}
$$

6. Estimate the product; then find the product.

$$
\begin{array}{r}
739 \\
\times\ 206 \\
\hline
\end{array}
$$

7. Estimate the quotient; then find the quotient and remainder.

$$18\overline{)773}$$

Practice Problem Answers

1. 1830 **2.** 14,800 **3.** 290,000

4. estimate: 1100; sum: 1091 **5.** estimate: 7000; difference: 6295

6. estimate: 140,000; product: 152,234

7. estimate: 40; quotient 42 R17

Exercises 1.5

Round each number to the nearest ten. See Examples 1 and 2.

1. 763

2. 31

3. 85

4. 296

5. 5347

6. 1722

7. 3503

8. 995

Round each number to the nearest hundred. See Examples 1 and 2.

9. 4475

10. 795

11. 12,637

12. 43,789

13. 7007

14. 76,523

15. 805

16. 5958

Round each number to the nearest thousand. See Examples 1 and 2.

17. 4912

18. 6200

19. 7499

20. 3495

21. 10,397

22. 27,501

23. 99,920

24. 39,497

Round each number to the nearest ten thousand. See Example 2.

25. 78,419

26. 184,900

27. 295,321

28. 325,396

Solve the following word problems. See Example 3.

29. **Population:** According to U.S. News and World Report, the U.S. population estimate for January 1, 2009 was 305,529,237. Round this number to the nearest million. **Source:** U.S. News and World Report

30. **Religion:** In 1990 there were 1,582,580 Southern Baptists in Georgia. Round this number to the nearest ten thousand. **Source:** adherents.com

31. **Budget:** The 2010 General Fund for the Virginia state budget was: $16,222,462,810. Round this number to the nearest billion dollars. **Source:** virginia.com

32. **Debt:** The average U.S. household debt was $668,621. Round this number to the nearest thousand dollars. **Source:** virginia.com

33. **Automobiles:** A Honda Accord EX sedan with automatic transmission has a sticker price of $25,380. Round this figure to the nearest thousand dollars. **Source:** automobiles.honda.com

34. **Productivity:** A certain manufacturer pays his workers for every ten widgets that a person produces, rounded to the nearest ten. If Jim produced 4564 widgets, how many widgets will he be paid for?

35. Debt: As of August 2009, average credit card debt in Tennessee was $7054. Round this number to the nearest hundred dollars. **(Source:** cardtrack.com)

36. Traffic: On a typical day, 46,931 vehicles are driven through the northbound toll gates at a busy bridge. Round this figure to the nearest hundred vehicles.

37. Mileage: A salesperson drives 96,469 miles a year. Round this figure to the nearest ten thousand miles.

38. Income: XYZ company had a total income of $3,475,849 for 2009. Round this figure to the nearest hundred thousand dollars.

Find the estimated sum using rounded numbers. Then find the actual sum. See Examples 4 and 5.

39.
$$\begin{array}{r} 146 \\ 259 \\ + 384 \\ \hline \end{array}$$

40.
$$\begin{array}{r} 475 \\ 953 \\ + 705 \\ \hline \end{array}$$

41.
$$\begin{array}{r} 22,506 \\ 38,700 \\ + 10,465 \\ \hline \end{array}$$

42.
$$\begin{array}{r} 10,531 \\ 3789 \\ + 9733 \\ \hline \end{array}$$

Find the estimated difference using rounded numbers. Then find the actual difference. See Examples 6 and 7.

43.
$$\begin{array}{r} 8742 \\ - 3275 \\ \hline \end{array}$$

44.
$$\begin{array}{r} 6421 \\ - 1652 \\ \hline \end{array}$$

45.
$$\begin{array}{r} 10,531 \\ - 4600 \\ \hline \end{array}$$

46.
$$\begin{array}{r} 63,504 \\ - 42,700 \\ \hline \end{array}$$

Find the estimated product using rounded numbers. Then find the actual product. See Examples 8 and 9.

47.
$$\begin{array}{r} 849 \\ \times 205 \\ \hline \end{array}$$

48.
$$\begin{array}{r} 72 \\ \times 163 \\ \hline \end{array}$$

49.
$$\begin{array}{r} 3592 \\ \times 95 \\ \hline \end{array}$$

50.
$$\begin{array}{r} 8672 \\ \times 53 \\ \hline \end{array}$$

Find the estimated quotient using rounded numbers. Then find the actual quotient. See Examples 10 through 12.

51. $19\overline{)6783}$

52. $22,506 \div 33$

53. $72\overline{)5328}$

54. $\dfrac{57,888}{67}$

55. Education: College costs for a private four-year college in the 2008-09 academic year are as follows.

Tuition & Fees	$25,243
Room & Board	$8996
Books & Supplies	$1077

Estimate the total cost to attend for a year using rounded numbers. Then calculate the actual cost. **Source:** umuc.com

56. Driving: Carol is driving from Daytona Beach, FL to Minneapolis, MN using the following route: from Daytona Beach to Jacksonville (93 miles), from Jacksonville to Atlanta (345 miles), and from Atlanta to Minneapolis (1176 miles). Estimate the total mileage of her trip using rounded numbers. Then calculate the actual mileage. **Source:** 2010 American Automobile Association Road Atlas, Chart 3

57. Photography: The Smithville School purchased a Canon XF300 professional camcorder for $6799. In addition, they purchased a Barber STP1 tripod for $195 and a Canon HC-4200 case for $310. Estimate the total purchase price using rounded numbers. Then calculate the actual purchase price.

58. Office Supplies: Jerry purchased the following items for his home office: one Toshiba laptop computer for $483, one Microsoft Office Home and Student software for $99, one Kodak 5020 inkjet printer for $138, and a 100-pack of blank DVDs for $18. Estimate the total purchase price using rounded numbers. Then calculate the actual purchase price.

59. Landscaping: Geoffrey went on a buying spree at Sears Lawn and Garden department, and purchased the following items to maintain his lawn: a Craftsman YT4000 46" Lawn Tractor for $1699, a Craftsman 675 Series 22" Mower for $247, a Craftsman gas weed trimmer for $121, and a lawn tractor cover for $49. In addition, he purchased a cover for his zero-turn mower for $58. Estimate the total purchase price using rounded numbers. Then calculate the actual purchase price.

60. Shopping: John purchased a wrinkle-resistant sports shirt and a chino blazer through a mail order catalogue. The sports shirt cost $44, and the total cost of the order was $160. Estimate the purchase price of the chino blazer using rounded numbers. Then calculate the actual purchase price.

61. Automobile: Elizabeth paid $2191 to add on special paint, 18" alloy wheels, and a cargo net to the Ford Fusion automobile she just purchased. If she paid a total of $23,766 for the car with the add-ons, estimate the base base price of the car using rounded numbers. Then find the actual base price.

Source: fordvehicles.com

62. Fundraising: The ABC charity received a total income of $874,927 for the year 2008. In 2009, the charity received a total income of $997,354. Estimate the increase in income from 2008 to 2009 using rounded numbers. Then calculate the actual increase.

63. Microphones: A total of 22 microphones are to be purchased. Each microphone costs $347. Estimate the total purchase price by using rounded numbers. Then calculate the actual purchase price.

64. Jewelry: A jewelry store purchased 11 fine diamond-studded watches at a wholesale cost of $1,586 each. Estimate the total cost for all of the watches by using rounded numbers. Then calculate the actual cost for all of the watches.

65. Camp: A summer camp has 396 campers who will be going on an outing. Estimate the number of vans needed to accommodate the entire camp if each van can comfortably hold 11 passengers. Then calculate the actual number of vans that will be needed.

66. Beach House: Four people decide to purchase equal shares of a beach house which costs $312,760. Estimate the contribution each person makes by using rounded numbers. Then calculate the actual contribution.

67. Equipment: A technical school plans on purchasing four pieces of test equipment which cost $648 each plus two work benches at $284 each. Estimate the total cost of the purchase using rounded numbers. Then calculate the actual cost.

1.6 Exponents and Order of Operations

Objectives

A Understand the terms base and exponent.

B Know how to evaluate expressions containing exponents.

C Understand how to evaluate expressions with 1 and 0 as exponents.

D Know the rules for order of operations.

1. In each exponential expression, identify the base and the exponent.

   a. 8^3

   b. 14^6

2. How would you read each of the following equations?

   a. $7^2 = 49$

   b. $4^3 = 64$

   c. $3^5 = 243$

Objective A Understand the Terms Base and Exponent

Repeated multiplication by the same number can be shortened by using **exponents**. For example, if 5 is used as a factor three times, we can write

$$\underbrace{5 \cdot 5 \cdot 5}_{\text{factors}} = \overset{\text{exponent}}{\underset{\text{base}}{5^3}} = \underset{\text{product}}{125} \qquad \text{The base 5 is used as a factor three times.}$$

When looking at $5^3 = 125$, 5 is the **base**, 3 is the **exponent**, and 125 is the **product**. Exponents are written slightly to the right and above the base. The expression 5^3 is an **exponential expression**.

Example 1

Understanding the Terms Base and Exponent

In each exponential expression, identify the base and the exponent.

a. 6^2: 6 is the base, and 2 is the exponent.

b. 10^4: 10 is the base, and 4 is the exponent.

Now work margin exercise 1.

In expressions with exponent 2, the base is said to be **squared**. In expressions with exponent 3, the base is said to be **cubed**. With other exponents, the base is said to be "**to the _____ power.**"

Example 2

Reading Equations Containing Exponential Expressions

a. $8^2 = 64$ is read "eight squared is equal to sixty-four"

b. $6^3 = 216$ is read "six cubed is equal to two hundred sixteen"

c. $5^4 = 625$ is read "five to the fourth power is equal to six hundred twenty-five."

Now work margin exercise 2.

Evaluate Expressions Containing Exponents

Example 3

Writing and Evaluating Exponential Expressions

Rewrite each expression in exponential form and then evaluate the expression.

a. $7 \cdot 7$

Solution

$7 \cdot 7 = 7^2 = 49$

b. $2 \cdot 2 \cdot 2$

Solution

$2 \cdot 2 \cdot 2 = 2^3 = 8$

c. $2 \cdot 2 \cdot 2 \cdot 2 \cdot 2$

Solution

$2 \cdot 2 \cdot 2 \cdot 2 \cdot 2 = 2^5 = 32$

Now work margin exercise 3.

3. Rewrite each expression in exponential form, then evaluate the expression.

a. $8 \cdot 8 \cdot 8$

b. $9 \cdot 9$

c. $4 \cdot 4 \cdot 4 \cdot 4$

d. $2 \cdot 2 \cdot 2 \cdot 2 \cdot 2 \cdot 2 \cdot 2$

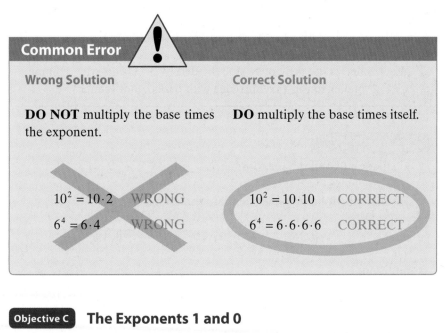

Common Error

Wrong Solution	**Correct Solution**
DO NOT multiply the base times the exponent.	**DO** multiply the base times itself.

$10^2 = 10 \cdot 2$ WRONG $10^2 = 10 \cdot 10$ CORRECT

$6^4 = 6 \cdot 4$ WRONG $6^4 = 6 \cdot 6 \cdot 6 \cdot 6$ CORRECT

Objective C **The Exponents 1 and 0**

If there is no exponent written with a number, then the exponent is understood to be 1. That is, any number is equal to itself raised to the first power.

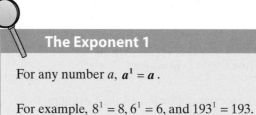

The Exponent 1

For any number a, $a^1 = a$.

For example, $8^1 = 8$, $6^1 = 6$, and $193^1 = 193$.

When the exponent 0 is used for any base except 0, the value of the expression is defined to be 1.

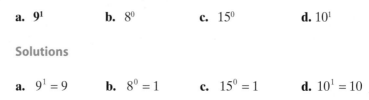

The Exponent 0

For any nonzero number a, $a^0 = 1$.

For example, $2^0 = 1$, $5^0 = 1$, and $46^0 = 1$.

Note: The expression 0^0 is undefined.

---

notes

To help in understanding the exponent 0, note the following pattern of powers of the base 3:

$$3^4 = 81, \quad 3^3 = 27, \quad 3^2 = 9, \quad 3^1 = 3, \quad \text{and} \quad 3^0 = 1.$$

Each value is the previous value divided by 3. In this way defining $3^0 = 1$ makes sense. (Try this idea with bases other than 3.)

---

4. Evaluate each of the following exponential expressions.

   a. 6^0

   b. 1^1

   c. 813^1

   d. 54^0

Example 4

Evaluating Exponential Expressions

Evaluate the following exponential expressions.

a. 9^1 **b.** 8^0 **c.** 15^0 **d.** 10^1

Solutions

a. $9^1 = 9$ **b.** $8^0 = 1$ **c.** $15^0 = 1$ **d.** $10^1 = 10$

Now work margin exercise 4.

Know the Rules for Order of Operations

How would you evaluate the expression $2+5\cdot6$? Would you add 2 and 5 and then multiply by 6? This would give $2+5\cdot6=7\cdot6=42$. Or would you multiply 5 times 6 and then add 2? This would give $2+5\cdot6=2+30=32$. This second answer is the correct one. To ensure that everyone gets the same correct answer, mathematicians have agreed on a set of **rules for order of operations**. These rules are used in all branches of mathematics and computer science.

Rules for Order of Operations

1. First, simplify within grouping symbols, such as parentheses ( ), brackets [ ], or braces { }. Start with the innermost group.
2. Second, evaluate any numbers or expressions indicated by exponents.
3. Third, moving from **left to right**, perform any multiplication or division in the order in which it appears.
4. Fourth, moving from **left to right**, perform any addition or subtraction in the order in which it appears.

notes

Note that in Rule 3 neither multiplication nor division has priority over the other. Whichever of these operations occurs first, moving **left to right**, is done first. In Rule 4, addition and subtraction are handled in the same way. Unless they occur within grouping symbols, **addition and subtraction are the last operations to be performed**.

The following examples show how to apply the rules for order of operations.

Example 5

Order of Operations

Evaluate $14\div7+3\cdot2-5$.

Solution

In this example there are no grouping symbols or exponents, so we begin with multiplication and division (left to right).

	$14\div7+3\cdot2-5$	Divide before multiplying in this case.
$=$	$2\ +3\cdot2-5$	Multiply before adding or subtracting.
$=$	$2\ +\ 6\ -5$	Add before subtracting in this case.
$=$	$8\ \ \ -5$	Subtract.
$=$	3	

Example 6

Order of Operations

Evaluate $(6+2)+(8+1)\div 9$.

Solution

$$\underbrace{(6+2)}+\underbrace{(8+1)}\div 9 \qquad \text{Operate within parentheses.}$$
$$= \quad 8 \quad + \quad \underbrace{9 \div 9} \qquad \text{Divide.}$$
$$= \quad \underbrace{8 \quad + \quad 1} \qquad \text{Add.}$$
$$= \qquad\qquad 9$$

Example 7

Order of Operations

Evaluate $2\cdot 3^2 + 18\div 3^2$.

Solution

$$2\cdot\underbrace{3^2} + 18\div\underbrace{3^2} \qquad \text{Evaluate the exponents.}$$
$$= \underbrace{2\cdot 9} + \underbrace{18\div 9} \qquad \text{Multiply and divide.}$$
$$= \underbrace{18 \quad + \quad 2} \qquad \text{Add.}$$
$$= \qquad 20$$

Example 8

Order of Operations

Evaluate $30\div 10\cdot 2^3 + 3(6-2)$.

Solution

$$30\div 10\cdot 2^3 + 3\underbrace{(6-2)} \qquad \text{Operate within parentheses.}$$
$$= 30\div 10\cdot\underbrace{2^3} + 3\ (4) \qquad \text{Evaluate the exponent.}$$
$$= \underbrace{30\div 10}\cdot 8 + 3\ (4) \qquad \text{Divide.}$$
$$= \underbrace{3\ \cdot 8} + \underbrace{3\ (4)} \qquad \text{Multiply in each part.}$$
$$= \underbrace{24 \quad + \quad 12} \qquad \text{Add.}$$
$$= \qquad 36$$

Example 9

Order of Operations

Evaluate $2\left[5^2 + \left(2 \cdot 3^2 - 10\right)\right]$.

Solution

$2\left[5^2 + \left(2 \cdot 3^2 - 10\right)\right]$ Evaluate the exponents.

$= 2\left[25 + \left(2 \cdot 9 - 10\right)\right]$ Multiply inside the parentheses.

$= 2\left[25 + \left(18 - 10\right)\right]$ Subtract inside the parentheses.

$= 2\left[25 + 8\right]$ Add inside the brackets.

$= 2 \quad \left[33\right]$ Multiply.

$= \quad 66$

Completion Example 10

Order of Operations

Evaluate $3\left(2 + 2^2\right) - 6 - 3 \cdot 2^2$.

Solution

$3\left(2 + 2^2\right) - 6 - 3 \cdot 2^2$

$= 3\left(2 + \underline{\quad}\right) - 6 - 3 \cdot \underline{\quad}$

$= 3\left(\underline{\quad}\right) - 6 - \underline{\quad}$

$= \underline{\quad} - 6 - \underline{\quad}$

$= \underline{\quad} - \underline{\quad}$

$= \underline{\quad}$

Now work margin exercises 5 through 10.

Evaluate the following.

5. $6 \cdot 2 - 4 \cdot 3 + 8$

6. $\left(9 + 3\right) \div 4 + \left(9 - 1\right)$

7. $6^2 \div 9 + 3 - 14 \div 7$

8. $2\left(5^2 - 1\right) - 4 + 3 \cdot 2^3$

9. $4 \cdot 3 - \left[\left(2 + 3^2\right) - 10\right]$

10. $6^2 - 4\left(3^2 - 2\right) - 5 + 4 \cdot 2^2$

Practice Problems

Find the value of each expression.

1. 5^2 **2.** 3^4 **3.** 4^0 **4.** 1^5

Use the rules for order of operations to evaluate each expression.

5. $14 \div 7 + 3 \cdot 2^3$ **6.** $9 \div 3 \cdot 2 + 3\left(6 - 2^2\right)$ **7.** $6\left[\left(6 - 1\right)^2 - \left(4^2 - 9\right)\right]$

Completion Example Answers

10. $4, 4$; $6, 12$; $18, 12$; $12, 12$; 0

Practice Problem Answers

1. 25 **2.** 81 **3.** 1 **4.** 1

5. 26 **6.** 12 **7.** 108

Exercises 1.6

Identify the base and exponent in each exponential expression. See Example 1.

1. 12^7

base _____

exponent _____

2. 2^6

base _____

exponent _____

3. 10^4

base _____

exponent _____

4. 5^2

base _____

exponent _____

5. 19^0

base _____

exponent _____

6. 1^{50}

base _____

exponent _____

7. 4^{72}

base _____

exponent _____

8. 33^3

base _____

exponent _____

Write each of the following expressions in exponential form. See Example 3.

9. $2 \cdot 2$

10. $5 \cdot 5 \cdot 5 \cdot 5$

11. $9 \cdot 9 \cdot 9$

12. $6 \cdot 6 \cdot 6 \cdot 6 \cdot 6$

13. $7 \cdot 7 \cdot 7 \cdot 7$

14. $3 \cdot 3 \cdot 3 \cdot 3 \cdot 3 \cdot 3 \cdot 3 \cdot 3$

15. $8 \cdot 8 \cdot 8 \cdot 8 \cdot 8 \cdot 8$

16. $13 \cdot 13 \cdot 13 \cdot 13 \cdot 13$

17. $2 \cdot 2 \cdot 7 \cdot 7$

18. $2 \cdot 2 \cdot 3 \cdot 3 \cdot 3$

19. $5 \cdot 5 \cdot 5 \cdot 11 \cdot 11$

20. $2 \cdot 3 \cdot 3 \cdot 11 \cdot 11$

Evaluate each of the following exponential expressions. See Examples 3 and 4.

21. 2^3

22. 4^2

23. 9^2

24. 3^3

25. 20^1

26. 2^6

27. 0^0

28. 6^2

29. 37^0

30. 5^3

31. 8^1

32. 4^3

33. 1^{17}

34. 7^0

35. 6^3

36. 10^3

37. 30^2

38. 11^3

39. $5 \cdot 3^2$

40. $4 \cdot 2^4$

Evaluate each expression by using the rules for order of operations. See Examples 5 through 10.

41. $6 + 5 \cdot 3$

42. $6 - 15 \div 3$

43. $32 - 14 + 10$

44. $4^2 \div 8 \cdot 2$

45. $2^2 \div 2 + 7 - 3 \cdot 2$

46. $6 + 3 \cdot 2 - 10 \div 2$

47. $2 + 3 \cdot 7 - 10 \div 2$

48. $5 \cdot 1 \cdot 3 - 4 \div 2 + 6 \cdot 3$

49. $(2 + 3 \cdot 4) \div 7 + 3$

50. $(2 + 3) \cdot 4 \div 5 + 3 \cdot 2$

51. $35 \div (6 - 1) - 5 + 6 \div 2$

52. $(33 - 2 \cdot 6) \div 7 + 3 - 6$

53. $14(2+3)-65+5$

54. $16(2+4)-90-3\cdot 2$

55. $30-4(8-3)\div 10+7$

56. $4+(9-3)\div(6-3)+10\div 2$

57. $3+(4+2)\cdot(3+1)\div 12-2$

58. $\left[(2+3)(5-1)\div 2\right](10+1)$

59. $3\left[4+2(6\div 3\cdot 2)\right]$

60. $5+2\left[9+(9+1\cdot 9)\right]\div 6$

61. $3-\left[8-(3\cdot 5-9)\right]\div 2$

62. $2\cdot 5^2-8\div 2$

63. $3\cdot 2^3-8\div 2+3\cdot 4$

64. $3+3\cdot 2\div 6+4^2$

65. $9\div 3+2^2\cdot 5$

66. $20\div 2\cdot 3+1\cdot 3^2\cdot 2$

67. $16\div 2^4-9\div 3^2$

68. $3^2-5+2^3\cdot 2$

69. $4^2-2^4+5^2\cdot 2-6^2$

70. $\left(2^3+2\right)\div 5+7^2\div 7$

71. $18-9\left(3^2-2^3\right)\div 3$

72. $(2+1)^2+(4+1)^2$

73. $16+3\left(17+2^3\div 2^2-4\right)$

74. $\left(2^5+1\right)\div 11-3+7\left(3^3-7\right)$

75. $\left(4^2-7\right)\cdot 2^3-8\cdot 5-10$

76. $\left(6+8^2-10\div 2\right)\div 5+3^2\cdot 5$

77. $\left(3^3+3\right)\div 5+\left(4^2\div 4\right)$

78. $5\left[3^2+\left(8+2^3\right)\right]-15$

79. $(4+3)^2-(2+3)^2$

80. $2(15-6+4)\div 13\cdot 2+1^0$

81. $\left(20\div 2^2\cdot 5\right)+\left(51\div 17\right)^2$

82. $\left(3\cdot 2^2-5\cdot 2+2\right)-\left(2-1^2+5\cdot 2-10\right)$

83. $\left(2^4-16\right)+\left[13+5^2-20\right]$

84. $3^2\cdot 2+5\cdot 3^2+15^2-\left(3^2\cdot 21+6\right)$

85. $5^2-1\left[\left(3^2-5\right)^2\right]$

86. $100+2\left[\left(4^2-9\right)(2+1)^2\right]$

87. $(3+2)^2\left[(2+1)^2-14\div 2\right]$

88. $2+5\left[10\div 5\cdot 2+3^2-(6+4)\right]$

89. $(2+5)\left[10\div 5\cdot 2+3^2-(6+4)\right]$

90. $(6+2)\left[10^2\cdot 2+\left(3\cdot 5^2-4^2\right)\right]$

91. $6+2\left[10^2\cdot 2+\left(3\cdot 5^2-4^2\right)\right]$

Writing & Thinking

92. Perfect Squares: Make a list of the squares of the whole numbers from 1 to 20. Such numbers are called **perfect squares**.

93. Perfect Cubes: Make a list of the cubes of the whole numbers from 1 to 10. These numbers are called **perfect cubes**.

Objectives

A Learn the basic strategy for solving word problems.

B Analyze and solve word problems involving numbers.

C Analyze and solve word problems involving consumer items.

D Analyze and solve word problems involving checking accounts.

E Analyze and solve word problems involving average.

1.7 Problem Solving with Whole Numbers

Objective A **Basic Strategy for Solving Word Problems**

The problems discussed in this section fall under the following headings: number problems, consumer items, checking accounts, and average. The steps in the basic strategy listed here will help give an organized approach regardless of the type of problem.

Basic Strategy for Solving Word Problems

1. Read the problem carefully.
2. Draw any type of figure or diagram that might be helpful and decide what operations are needed.
3. Perform the operations to solve the problem.
4. Check your work.

Objective B **Problem Solving: Number Problems**

Number problems usually contain key words or phrases that tell what operations are to be performed with given numbers. Look for these key words.

Key Words that Indicate Operations

Addition	Subtraction	Multiplication	Division
add	subtract	multiply	divide
sum	difference	product	quotient
plus	minus	times	ratio
more than	less than	twice	
increased by	decreased by	double	
total			

Example 1

Number Problem

Find the **sum** of 78 and 93. Then **double** the sum. What is the result?

Solution

The key word **sum** indicates addition.

$$\begin{array}{r} 78 \\ + 93 \\ \hline 171 \end{array} \quad \text{sum}$$

Double means to multiply by 2.

$$171$$
$$\times\ 2$$
$$342 \quad \text{product}$$

Example 2

Number Problem

If the **quotient** of 265 and 5 is **decreased by** 36, what is the **difference**?

Solution

The key word **quotient** indicates division.

$$
\begin{array}{r}
53 \quad \text{quotient} \\
5\overline{)265} \\
\underline{25} \\
15 \\
\underline{15} \\
0
\end{array}
$$

Decreased by indicates subtraction.

$$53$$
$$-\ 36$$
$$17 \quad \text{difference}$$

Example 3

Number Problem

If the **product** of 15 and 32 is **added to** 1500, what is the **total**?

Solution

The key word **product** indicates multiplication.

$$32$$
$$\times 15$$
$$160$$
$$\underline{32}$$
$$480 \quad \text{product}$$

Added to indicates addition.

$$1500$$
$$+\ 480$$
$$1980 \quad \text{total}$$

Now work margin exercises 1 through 3.

1. Find the **quotient** of 176 and 4. Then **increase** the **quotient** by 3.

2. If the **difference** between 279 and 150 is **doubled**, what is the result?

3. If the **sum** of 89 and 169 is **divided by** 6, find the **quotient**.

4. Mrs. Spencer bought a new car for her daughter for $27,550. The salesperson added $1600 for taxes and $475 for license fees. If she made a down payment of $4500 and financed the rest through her bank, how much did she finance?

Example 4

Consumer Items

Mr. Lukin bought a used car for $8000. Taxes of $640 and license fees of $320 were then added to the purchase price. He made a down payment of $2000 and financed the rest through his credit union. What was the amount of his loan from the credit union?

Solution

First, find his total cost by adding the expenses. Then subtract his down payment. This difference will be the amount of his loan.

Add expenses		**Subtract down payment**	
$8000	price	$8960	total expenses
640	taxes	− 2000	down payment
+ 320	license fees	$6960	amount of loan
$8960	total expenses		

After a down payment of $2000, his loan will be $6960.

Now work margin exercise 4.

Objective D   **Problem Solving: Checking Accounts**

Example 5

Checking Accounts

In January, Kathleen opened a checking account and deposited $2500. During the month, she made another deposit of $800 and wrote checks for $132, $425, $196, and $350. What was the balance in her account at the end of the month?

Solution

First, find the sum of her deposits.

$2500

+   800

$3300   total deposits

Then find the sum of her checks.

$132

425

196

+   350

$1103   total of checks

Finally, find the balance by subtracting.

$$\begin{array}{r} \$3300 \\ -\ \ 1103 \\ \hline \$2197 \end{array}\quad\text{balance}$$

Now work margin exercise 5.

Objective E **Problem Solving: Average**

A topic related to addition and division is **average**. Your grade in this course may be based on the average of your exam scores. The average of a set of numbers is a sort of a "middle number" of the set.

To Find the Average of a Set of Numbers

Step 1: Find the sum of the given set of numbers.
Step 2: Divide this sum by the number of numbers in the set. This quotient is called the **average** of the given set of numbers.

Example 6

Average

Find the average of the following set of numbers: $15, 8, 90, 35, 27$.

Solution

Step 1: First, find the sum of the numbers.

$$\begin{array}{r} 15 \\ 8 \\ 90 \\ 35 \\ +\ 27 \\ \hline 175 \end{array}$$

Step 2: Now, divide the sum by 5, since we have a list of five numbers.

$$\begin{array}{r} 35 \quad\text{average} \\ 5\overline{)175} \\ \underline{15} \\ 25 \\ \underline{25} \\ 0 \end{array}$$

The average of the set of numbers is 35.

Now work margin exercise 6.

5. In December, Kurt opened a checking account and deposited $2000 into it. He also made another deposit of $200. During that month, he wrote checks for $57, $120, and $525, and used his debit card for purchases totaling $630. What was his balance at the end of the month?

6. Find the average of the following set of numbers: $18, 29, 6, 33, 14, 26$.

A useful way to look at numbers in a set is a diagram called a **bar graph**. In Example 7, a bar graph is used to display a company's profits for each month over a 6-month period.

7. A baseball player hits 30, 48, 31, and 35 home runs in four consecutive seasons. What is his average number of home runs per season?

Example 7

Average

Top-Notch Sporting Goods recorded its profits for tennis rackets for six months. The profits were: January, $5380; February, $7590; March, $6410; April, $4530; May, $5840; June, $6250. Below is a bar graph with this information.

a. In what month were the profits the most?
b. In what month were the profits the least?
c. What is the average monthly profit over the six months?

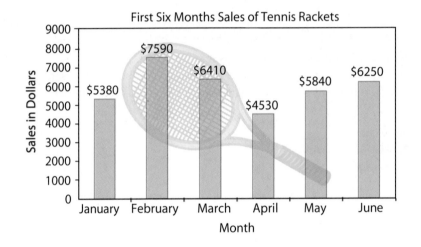

Solution

a. From the bar graph, we can see that the month with the most profits was February with $7590.
b. April was the month with the least profits with $4530.
c. The average monthly profit can be found by finding the sum of the profits for each month and dividing by 6.

Add profits	Divide by 6
$5380	6 000 average
7590	6)36,000
6410	36 000
4530	0
5840	
+ 6250	
$36,000	

The average profit each month for the six months was $6000.
(In this case, we see that the average can be very useful. The store manager can use the monthly profits for planning and budgeting.)

Now work margin exercise 7.

Example 8

Average

Five people in a survey reported the following incomes for one year: $18,000; $28,000; $20,000; $20,000; $214,000. What was the average annual income for these five people?

Solution

Add incomes	Divide by 5
$18,000	$\underset{\text{average}}{60\ 000}$
28,000	$5\overline{)300,000}$
20,000	$\underline{300,000}$
20,000	0
+ 214,000	
$300,000	

The average income was $60,000.
(Because of one large income, the average income was much higher then the income of the other four people. Judging the importance of an average, particularly in a case like this, is up to the reader of the information.)

Now work margin exercise 8.

Example 9

Average

On an English exam, two students scored 95, five scored 86, one scored 82, one scored 78, and six scored 75. What was the average score for the class?

Solution

There were fifteen students in the class. We can multiply as follows rather than add all fifteen scores.

95	86	82	78	75
$\times\ 2$	$\times\ 5$	$\times\ 1$	$\times\ 1$	$\times\ 6$
190	430	82	78	450

Next, we add the five products to find the sum of all of the scores.

190
430
82
78
+ 450
1230

8. Five cities have populations of 5000, 10,000, 15,000, 20,000 and 950,000. What is the average population of the cities?

9. In a recent high school basketball game, two of the starters scored 18 points each, two scored 10 points, and one scored 4 points. What was the average number of points scored among the starters?

Finally, divide by 15 because the total represents 15 scores.

$$\begin{array}{r} 82 \quad \text{average} \\ 15\overline{)1230} \\ \underline{120} \\ 30 \\ \underline{30} \\ 0 \end{array}$$

The average score was 82.

Now work margin exercise 9.

Practice Problems

1. Find the **sum** of the three numbers 915, 862, and 453. Then **subtract** 580. What is the **quotient** if the **difference** is **divided by** 3?

2. David bought an iPhone® for $250. He had a coupon that gave him $10 off. He also bought 3 video games for $48 each. What did he pay overall for the iPhone® and the video games?

3. Mr. Morton opened a checking account and deposited $4000. He wrote two checks for $175 each and one for $300. What was his balance after writing these checks?

4. The Lee family spent the following amounts for groceries: $338 in June; $307 in July; $318 in August. What was the average amount they spent for groceries for these three months?

Practice Problem Answers
 1. 550
 2. $384
 3. $3350
 4. $321

Exercises 1.7

Solve the following number problems. See Examples 1 through 3.

1. Find the **sum** of the three numbers 745, 860, and 355. Then **subtract** 390. What is the **difference**?

2. The **difference** between 8000 and 1766 is **added to** 545. What is the **sum**?

3. The **product** of 24 and 45 is **increased by** the **product** of 240 and 3. What is the **total**?

4. If the **quotient** of 660 and 5 is **decreased by** 80, what is the **difference**?

5. What is the result if the **quotient** of 1050 and 15 is **doubled**?

6. Find the **product** of 32 and 92 and the **sum** of 433 and 1037. What is the **difference** between the **product** and the **sum**?

Solve the following problems involving consumer items. See Example 4.

7. **Kitchen Appliances:** To purchase a new refrigerator for $1200 including tax, Mr. Kline paid $240 down and the remainder in six equal monthly payments. What were his monthly payments?

8. **Shopping:** Miguel decided to go shopping for school clothes before college started in the fall. How much did he spend if he bought four pairs of pants for $21 each, five shirts for $18 each, three pairs of socks for $4 a pair, and two pairs of shoes for $38 a pair?

9. **Furniture:** To purchase a new dining room set for $1200, Mrs. Steel had to pay an additional $72 in sales tax. If she made a deposit of $486, how much did she still owe?

10. **Automobile:** Alan wants to buy a new car. He could buy a red one for $8500 plus $510 in sales tax and $135 in fees, or he could buy a blue one for $8700 plus $522 in sales tax and $140 in fees. If the manufacturer is giving a $250 rebate on the blue model, which car would be cheaper for Alan? How much cheaper would it be?

11. **Surfing:** Lynn decided to take up surfing. She bought a new surfboard for $675, a wet suit for $130, a beach towel for $12, and a new swimsuit for $57. How much money did she spend? (Sales tax was included in the prices.)

12. **Education:** Pat needed art supplies for a new course at the local community college. She bought a portfolio for $32, a zinc plate for $44, etching ink for $12, and three sheets of rag paper for a total of $6. She received a student discount of $9. How much did she spend on art supplies?

13. Budgeting: The circle graph shown here indicates the budgeted expenses for one year in the Young family. Use the numbers in the graph to answer the following questions.

a. What amount of money did the Young family budget for rent?

b. How much more did they budget for food than for entertainment?

c. What was the total amount the Youngs budgeted for this year?

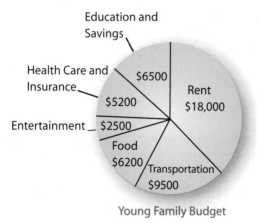

Young Family Budget

14. Medicine: The circle graph shown here shows the number of different types of operations performed at Southern Hospital last year. Use the numbers in the graph to answer the following questions.

a. How many abdominal operations were performed last year?

b. How many more general operations than urologic operations were performed?

c. What was the total number of operations performed last year at Southern?

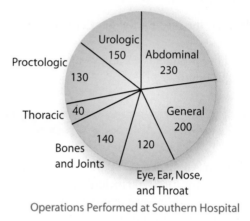

Operations Performed at Southern Hospital

15. Population: The bar graph shown here indicates the populations of four American colonies in 1630. Use this graph to answer the following questions.

a. Which of these colonies had the highest population?

b. Which of these colonies had the lowest population?

c. What was the total population of these four colonies in 1630?

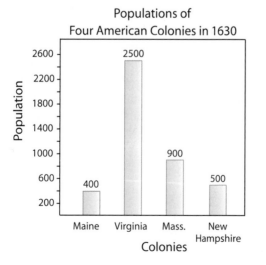

16. Entertainment: The bar graph shown below indicates the number of tickets sold for five movies in 2009. Use this graph to answer the following questions. **Source:** Boxofficemojo.com. 2009.

a. Which of these movies had the highest number of tickets sold?

b. How many tickets were sold for Star Trek in 2009?

c. What was the total number of tickets sold for these five movies?

Number of Tickets Sold for Five Movies in 2009

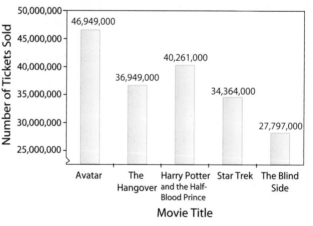

17. Checking Accounts: If you opened a checking account with $875, then wrote checks for $20, $35, $115, $8, and $212, what would be your balance?

18. Checking Accounts: Your friend had a checking account balance of $1250 and wrote checks for $375, $52, $83, and $246. What was her new balance?

19. Checking Accounts: On August 1, Matt had a balance of $250 in his checking account. During August, he made deposits of $200, $350, and $236. He wrote checks for $487, $25, $33, and $175. What was his balance on September 1?

20. Checking Accounts: Melissa deposited $500, $2470, $800, $3562, and $2875 in her checking account over a five-month period. She wrote checks totaling $6742. If her beginning balance was $1400, what was her balance at the end of the five months?

21. 102, 113, 97, 100

22. 56, 64, 38, 58

23. 6, 8, 7, 4, 4, 5, 6, 8

24. 5, 4, 5, 6, 5, 8, 9, 6

25. 512, 618, 332, 478

26. 436, 520, 630, 422

27. Cell Phones: If Rina's cell phone bills for the past five months have been $56, $63, $52, $85, and $49, what was her average cell phone bill for the past 5 months?

28. Insurance: Bill wanted to compare car insurance rates to find out the average amount he should pay for car insurance. If he looked at four different companies and the monthly rates were $164, $107, $131, and $98, what is the average monthy rate of car insurance?

29. Trivia Scores: During a sports trivia game, one team scored 35 points, three teams scored 23 points, two teams scored 18 points, and two teams scored 14 points. What was the average score of the teams?

30. Below is a bar graph showing the number of hybrid cars sold in the U.S. from the year 2005 to 2009.
 a. In what year was the fewest number of hybrid cars sold?
 b. What was the average number of hybrid cars sold over the five years?

Hybrid Car Sales Over Five Years

352,183
313,486
292,528
254,545
199,148

Number of Cars Sold

400,000
340,000
280,000
220,000
160,000
100,000

2005 2006 2007 2008 2009
Year

31. Test Scores: On a history exam, two students scored 95, six scored 90, three scored 80, and one scored 50. What was the class average?

32. Sales: A salesman sold items from his sales list for $972, $834, $1005, $1050, and $799. What was the average price per item?

33. Stocks: Ms. Lee bought 150 shares of stock in Microsoft at $24 per share. Two months later, she bought another 100 shares at $29 per share. What average price per share did she pay? If she sold all 250 shares at $28 per share, what was her profit?

34. Income: Three families, each with two children, had incomes of $56,000. Two families, each with four children, had incomes of $62,000. Four families, each with two children, had incomes of $45,000. One family had no children and an income of $37,000. What was the average income per family?

35. Checking Accounts: During July, Mr. Rodriguez made deposits in his checking account of $400 and $750 and wrote checks totaling $625. During August, his deposits were $632, $322, and $798, and his checks totaled $978. In September, his deposits were $520, $436, $200, and $376, and his checks totaled $836.

   a. What was the average monthly difference between his deposits and his withdrawals?

   b. What was his bank balance at the end of September if he had a balance of $500 on July 1?

36. Flight Time: In one month (30 days), an airline pilot spent the following number of hours in preparation for and flying each of 12 flights: 6, 8, 9, 6, 7, 7, 7, 5, 6, 6, 6, and 11 hours. What was the average amount of time the pilot spent per flight?

37. Population: The 10 largest cities in South Carolina have the following approximate populations. What is the average population of these cities?

City	Population
Columbia	129,333
Charleston	115,638
North Charleston	97,601
Rock Hill	69,210
Mount Pleasant	66,420
Greenville	61,782
Sumter	59,180
Summerville	45,240
Spartanburg	40,387
Hilton Head	34,249

38. Geography: The five longest rivers in the world are given in the following table. What is the average length of these rivers?

River	Length (miles)
Nile	4132
Amazon	3980
Yangtze	3917
Mississippi/Missouri	3902
Yenisei	3434

39. College Tuition: The costs to attend ten well-known colleges and universities are given in the following table.

Institution	Enrollment	Room/Board	Tuition Resident	Tuition Nonresident
Univ. of Arizona	38,060	$8610	$6860	$22,260
Boston College	13,900	$11,040	$39,130	$39,130
Univ. of Colorado	32,190	$10,380	$7930	$28,190
Florida A & M	11,590	$6980	$4120	$15,510
Purdue Univ.	40,090	$8710	$8640	$25,120
Johns Hopkins	6440	$12,040	$39,650	$39,650
Ohio State Univ.	53,720	$8410	$8710	$22,280
Princeton Univ.	7500	$11,680	$35,340	$35,340
Univ. of Tennessee	30,410	$7250	$6860	$20,650
Univ. of Wisconsin	42,030	$8040	$8310	$23,060

For these universities and colleges, find:

a. The average enrollment,

b. The average tuition for residents,

c. The average tuition for nonresidents, and

d. The average cost of room and board.

1.8 Tests for Divisibility (2, 3, 4, 5, 6, 9, and 10)

Objectives

A Know the tests for checking divisibility by 2, 3, 4, 5, 6, 9, and 10.

B Apply the concept of divisibility to products of whole numbers.

Objective A **Tests for Divisibility by 2, 3, 4, 5, 6, 9, and 10**

It can be very helpful to be able to divide quickly and easily by small numbers. We will want to know if a number is **exactly divisible** (remainder 0) by some number **before** actually dividing. For example, we will see that, by knowing the test for divisibility by 3, we can tell that 297 is exactly divisible by 3. In fact, $297 \div 3 = 99$.

Divisibility

If a number can be divided by another number so that the remainder is 0, then we say:

The dividend is **exactly divisible by** (or is **divisible by**) the divisor. Or, the divisor **divides** the dividend.

For example, $46 \div 2 = 23$, so 46 is divisible by 2. Or, 2 divides 46.

Even and Odd Whole Numbers

Even whole numbers are divisible by 2.
(If a whole number is divided by 2 and the remainder is 0, then the whole number is even.)

Odd whole numbers are not divisible by 2.
(If a whole number is divided by 2 and the remainder is 1, then the whole number is odd.)

(**Note:** Every whole number is either even or odd.)

The even whole numbers are
0, 2, 4, 6, 8, 10, 12, ...

The odd whole numbers are
1, 3, 5, 7, 9, 11, 13, ...

There are simple tests that can be performed mentally to determine whether a number is divisible by 2, 3, 4, 5, 6, 9, or 10 **without actually dividing**.

Divisibility by 2

If the last digit (ones digit) of a whole number is 0, 2, 4, 6, or 8 (an even digit), then the number is divisible by 2 (the number is even).

Example 1

Divisibility by 2

Determine which of the following numbers are divisible by 2.

a. 674

Solution

674 is divisible by 2 because the ones digit is 4 (an even digit).

b. 357

Solution

357 is not divisible by 2 because the ones digit is not 0, 2, 4, 6, or 8.

Now work margin exercise 1.

Divisibility by 3

If the sum of the digits of a whole number is divisible by 3, then the number is divisible by 3.

Example 2

Divisibility by 3

Determine which of the following numbers are divisible by 3.

a. 6801

Solution

6801 is divisible by 3 because $6 + 8 + 0 + 1 = 15$, and 15 is divisible by 3.

b. 356

Solution

356 is not divisible by 3 because $3 + 5 + 6 = 14$, and 14 is not divisible by 3.

Now work margin exercise 2.

Divisibility by 4

If the last two digits of a whole number form a number that is divisible by 4, then the number is divisible by 4. (00 is considered to be divisible by 4.)

1. Is 548 divisible by 2? Explain why or why not.

2. Is 7912 divisible by 3? Explain why or why not.

3. Is 2476 divisible by 4? Explain why or why not.

Example 3

Divisibility by 4

Determine which of the following numbers are divisible by 4.

a. 9036

Solution

9036 is divisible by 4 because 36 (the number formed by the last two digits) is divisible by 4.

b. 6700

Solution

6700 is divisible by 4 because 00 is considered to be divisible by 4.

c. 15,031

Solution

15,031 is not divisible by 4 because 31 is not divisible by 4.

Now work margin exercise 3.

Divisibility by 5

If the last digit (ones digit) of a whole number is 0 or 5, then the number is divisible by 5.

4. Is 6827 divisible by 5? Explain why or why not.

Example 4

Divisibility by 5

Determine which of the following numbers are divisible by 5.

a. 1365 **b.** 970 **c.** 1863

Solutions

a. 1365 is divisible by 5 because the ones digit is 5.

b. 970 is divisible by 5 because the ones digit is 0.

c. 1863 is not divisible by 5 because the ones digit is not 0 or 5.

Now work margin exercise 4.

Divisibility by 6

If a whole number is divisible by both 2 and 3, then the number is divisible by 6.

Example 5

Divisibility by 6

Determine which of the following numbers are divisible by 6.

a. 9054

Solution

9054 is divisible by 2 because the ones digit is 4. 9054 is divisible by 3 because $9 + 0 + 5 + 4 = 18$, and 18 is divisible by 3. Therefore, 9054 is divisible by 6.

b. 17,000

17,000 is divisible by 2 because the ones digit is 0. 17,000 is not divisible by 3 because $1 + 7 + 0 + 0 + 0 = 8$, and 8 is not divisible by 3. Therefore, 17,000 is not divisible by 6.

Now work margin exercise 5.

5. Is 1576 divisible by 6? Explain why or why not.

Divisibility by 9

If the sum of the digits of a whole number is divisible by 9, then the number is divisible by 9.

Example 6

Divisibility by 9

Determine which of the following numbers are divisible by 9.

a. 2530

Solution

2530 is not divisible by 9 because $2 + 5 + 3 + 0 = 10$, and 10 is not divisible by 9.

b. 873

Solution

873 is divisible by 9 because $8 + 7 + 3 = 18$, and 18 is divisible by 9.

Now work margin exercise 6.

6. Is 4653 divisible by 9? Explain why or why not.

If the last digit (ones digit) of a whole number is 0, then the number is divisible by 10.

7. Is 8510 divisible by 10? Explain why or why not.

Example 7

Divisibility by 10

Determine which of the following numbers are divisible by 10.

a. 12,530

Solution

12,530 is divisible by 10 because the ones digit is 0.

b. 841

Solution

841 is not divisible by 10 because the ones digit is not 0.

Now work margin exercise **7.**

8. Is 612 divisible by 9? Explain why or why not.

Completion Example 8

Divisibility Rules

a. 250 is divisible by 10 because _____

_____

b. 5712 is divisible by 4 because _____

_____

c. 5402 is not divisible by 3 because _____

_____

d. 6036 is divisible by 6 because _____

_____

Now work margin exercise **8.**

Completion Example Answers

8. a. the last digit is 0.

   b. the number formed by the last 2 digits (12) is divisible by 4.

   c. $5 + 4 + 0 + 2 = 11$, and 11 is not divisible by 3.

   d. 6036 is divisible by both 2 and 3.

Objective B **Divisibility of Products**

Consider that $300 = 3 \cdot 4 \cdot 5 \cdot 5$. Because 3 is one of the factors, 3 will divide into 300. By grouping factors, we see that:

$$300 = 3 \cdot 4 \cdot 5 \cdot 5 = 3 \cdot (4 \cdot 5 \cdot 5) = 3 \cdot 100.$$

Similarly, we find that 12 divides 300, 25 times:

$$300 = 3 \cdot 4 \cdot 5 \cdot 5 = (3 \cdot 4) \cdot (5 \cdot 5) = 12 \cdot 25.$$

Also, 20 divides 300, 15 times:

$$300 = 3 \cdot 4 \cdot 5 \cdot 5 = (4 \cdot 5) \cdot (3 \cdot 5) = 20 \cdot 15.$$

Example 9

Divisibility of Products

Does 36 divide the product $3 \cdot 4 \cdot 5 \cdot 7 \cdot 9$? If so, how many times?

Solution

Because $36 = 4 \cdot 9$, we have

$$3 \cdot 4 \cdot 5 \cdot 7 \cdot 9 = (4 \cdot 9)(3 \cdot 5 \cdot 7)$$
$$= 36 \cdot 105$$

Thus 36 divides the product 105 times.

Example 10

Divisibility of Products

Does 15 divide the product $5 \cdot 7 \cdot 2 \cdot 3 \cdot 2$? If so, how many times?

Solution

Because $15 = 3 \cdot 5$, we have

$$5 \cdot 7 \cdot 2 \cdot 3 \cdot 2 = (3 \cdot 5)(7 \cdot 2 \cdot 2)$$
$$= 15 \cdot 28$$

Thus 15 divides the product 28 times.

9. Does 45 divide the product $2 \cdot 5 \cdot 3 \cdot 6 \cdot 9$? If so, how many times?

10. Does 55 divide the product $3 \cdot 5 \cdot 4 \cdot 11 \cdot 8$? If so, how many times?

11. Does 30 divide the product $3 \cdot 7 \cdot 5 \cdot 2 \cdot 3$? If so, how many times?

12. Does 21 divide the product $2 \cdot 3 \cdot 5 \cdot 6 \cdot 2$? If so, how many times?

Example 11

Divisibility of Products

Does 35 divide the product $3 \cdot 4 \cdot 5 \cdot 11$? If so, how many times?

Solution

We know that $35 = 5 \cdot 7$ and even though 5 is a factor of the product, 7 is not. Therefore, 35 does not divide the product $3 \cdot 4 \cdot 5 \cdot 11$. In other words, $3 \cdot 4 \cdot 5 \cdot 11 = 660$; 660 is not divisible by 35.

Completion Example 12

Divisibility of Products

Does 77 divide the product $3 \cdot 11 \cdot 6 \cdot 7 \cdot 2$? If so, how many times?

Solution

Because $77 = \underline{\quad} \cdot \underline{\quad}$, we have

$$3 \cdot 11 \cdot 6 \cdot 7 \cdot 2 = \left(\underline{\quad} \cdot \underline{\quad}\right)\left(\underline{\quad} \cdot \underline{\quad} \cdot \underline{\quad}\right)$$
$$= \left(\underline{\quad}\right)\left(\underline{\quad}\right)$$

Thus 77 divides the product $\underline{\quad}$ times.

Now work margin exercises 9 through 12.

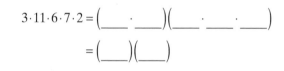

Practice Problems

1. Using the techniques of this section, determine which of the numbers 2, 3, 4, 5, 6, 9, and 10 (if any) will divide into each of the following numbers.
   a. 842   **b.** 675   **c.** 9030   **d.** 4031

2. Does 16 divide the product $3 \cdot 5 \cdot 4 \cdot 7 \cdot 4$? If so, how many times?

Completion Example Answers
   12. $7 \cdot 11; (7 \cdot 11)(2 \cdot 3 \cdot 6); (77)(36); 36$ times

Practice Problem Answers
   1. a. 2
      b. 3, 5, 9
      c. 2, 3, 5, 6, 10
      d. none
   2. yes, 105 times

1.9 Prime Numbers and Prime Factorizations

Objectives

A Understand the terms prime number and composite number.

B Understand the Sieve of Eratosthenes. *(Optional)*

C Determine whether a number is prime.

D Find the prime factorization of a composite number.

E Find all of the factors of a composite number.

Objective A   **Prime Numbers and Composite Numbers**

Every counting number, except 1, has two or more factors (or divisors). The following list shows examples of the factors of some counting numbers.

Counting Numbers	Factors
3 ⟶	1, 3
14 ⟶	1, 2, 7, 14
17 ⟶	1, 17
18 ⟶	1, 2, 3, 6, 9, 18
21 ⟶	1, 3, 7, 21
23 ⟶	1, 23
36 ⟶	1, 2, 3, 4, 6, 9, 12, 18, 36

Note that in this list, 3, 17, and 23 have exactly two different factors (or divisors). Such numbers are called **prime numbers**. The other numbers in the list, 14, 18, 21, and 36, have more than two different factors and are called **composite numbers**.

Prime Number

A **prime number** is a counting number greater than 1 that has exactly two different factors (or divisors) — itself and 1.

Composite Number

A **composite number** is a counting number with more than two different factors (or divisors).

Note: The number 1 is neither prime nor composite.

Example 1

Prime and Composite Numbers

The following numbers are prime.

- 2:   2 has exactly two different factors, 1 and 2.
- 7:   7 has exactly two different factors, 1 and 7.
- 11:   11 has exactly two different factors, 1 and 11.
- 29:   29 has exactly two different factors, 1 and 29.

The following numbers are composite.

- 12: $1 \cdot 12 = 12$, $2 \cdot 6 = 12$, and $3 \cdot 4 = 12$. 1, 2, 3, 4, 6, and 12 are all factors of 12. Thus 12 has more than two different factors.

- 33: $1 \cdot 33 = 33$ and $3 \cdot 11 = 33$. So 1, 3, 11, and 33 are all factors of 33, and 33 has more than two different factors.

Now work margin exercise 1.

1. Determine whether each of the following numbers is prime or composite. Explain your reasoning.

   a. 13

   b. 25

   c. 32

Objective B The Sieve of Eratosthenes (*Optional*)

One method used to find prime numbers involves the concept of **multiples**. To find the multiples of a counting number, multiply each of the counting numbers by that number. The multiples of 2, 3, 5, and 7 are listed here.

Counting Numbers:	**1,**	**2,**	**3,**	**4,**	**5,**	**6,**	**7,**	**8, …**
Multiples of 2:	2,	4,	6,	8,	10,	12,	14,	16, …
Multiples of 3:	3,	6,	9,	12,	15,	18,	21,	24, …
Multiples of 5:	5,	10,	15,	20,	25,	30,	35,	40, …
Multiples of 7:	7,	14,	21,	28,	35,	42,	49,	56, …

All of the multiples of a number have that number as a factor. Therefore, none of the multiples of a number, except possibly that number itself, can be prime. A process called the **Sieve of Eratosthenes** involves eliminating multiples to find prime numbers as the following steps describe. The description here illustrates finding the prime numbers less than 50.

Step 1: List the counting numbers from 1 to 50 as shown. 1 is neither prime nor composite, so cross out 1. Then, circle the next listed number that is not crossed out, and cross out all of its listed multiples. In this case, circle 2 and cross out all the multiples of 2.

~~1~~	②	3	~~4~~	5	~~6~~	7	~~8~~	9	~~10~~
11	~~12~~	13	~~14~~	15	~~16~~	17	~~18~~	19	~~20~~
21	~~22~~	23	~~24~~	25	~~26~~	27	~~28~~	29	~~30~~
31	~~32~~	33	~~34~~	35	~~36~~	37	~~38~~	39	~~40~~
41	~~42~~	43	~~44~~	45	~~46~~	47	~~48~~	49	~~50~~

Step 2: Circle 3 and cross out all the multiples of 3. (Some of these, such as 6 and 12, will already have been crossed out.)

~~1~~	②	③	~~4~~	5	~~6~~	7	~~8~~	~~9~~	~~10~~
11	~~12~~	13	~~14~~	~~15~~	~~16~~	17	~~18~~	19	~~20~~
~~21~~	~~22~~	23	~~24~~	25	~~26~~	~~27~~	~~28~~	29	~~30~~
31	~~32~~	~~33~~	~~34~~	35	~~36~~	37	~~38~~	~~39~~	~~40~~
41	~~42~~	43	~~44~~	~~45~~	~~46~~	47	~~48~~	49	~~50~~

Step 3: Circle 5 and cross out the multiples of 5. Then circle 7 and cross out the multiples of 7. Continue in this manner and the prime numbers will be circled and composite numbers crossed out (up to 50).

~~1~~	②	③	~~4~~	⑤	~~6~~	⑦	~~8~~	~~9~~	~~10~~
⑪	~~12~~	⑬	~~14~~	~~15~~	~~16~~	⑰	~~18~~	⑲	~~20~~
~~21~~	~~22~~	㉓	~~24~~	~~25~~	~~26~~	~~27~~	~~28~~	㉙	~~30~~
㉛	~~32~~	~~33~~	~~34~~	~~35~~	~~36~~	㊲	~~38~~	~~39~~	~~40~~
㊶	~~42~~	㊸	~~44~~	~~45~~	~~46~~	㊼	~~48~~	~~49~~	~~50~~

This last table shows that the prime numbers less than 50 are

2, **3**, **5**, **7**, **11**, **13**, **17**, **19**, **23**, **29**, **31**, **37**, **41**, **43**, and **47**.

Two important facts about prime numbers are

1. **Even Numbers:** 2 is the only even prime number.
2. **Odd Numbers:** All other prime numbers are odd numbers. But, not all odd numbers are prime.

Determining Whether a Number is Prime

The following procedure of dividing by prime numbers can be used to determine whether or not a number is prime. (Computers are used to determine whether or not very large numbers are prime.)

To Determine Whether a Number is Prime

Divide the number by progressively larger prime numbers $(2, 3, 5, 7, 11,$ and so forth) until:

1. **The remainder is 0.** This means that the prime number is a factor and **the given number is composite**; or
2. **You find a quotient smaller than the prime divisor.** This means that **the given number is prime** because it has no smaller prime factors.

Example 2

Determining Whether a Number is Prime

Is 605 a prime number?

Solution

The ones digit is 5. Therefore, 605 is divisible by 5 and is not prime. The number 605 is a composite number and $605 = 5 \cdot 121 = 5 \cdot 11 \cdot 11$.

Example 3

Determining Whether a Number is Prime

Is 103 a prime number?

Solution

Tests for 2, 3, and 5 fail. (The number 103 is not even; $1 + 0 + 3 = 4$ and 4 is not divisible by 3; and the last digit is not 0 or 5.)

Divide by 7:

$$
\begin{array}{r}
14 \\
7{\overline{\smash{\big)}\,103}} \\
\underline{7} \\
33 \\
\underline{28} \\
5
\end{array}
$$
The quotient is greater than the divisor.

The remainder is not 0.

Divide by 11:

$$
\begin{array}{r}
9 \\
11{\overline{\smash{\big)}\,103}} \\
\underline{99} \\
4
\end{array}
$$
The quotient is less than the divisor, so we are done.

The remainder is not 0.

The number 103 is prime.

2. Is 404 prime or composite? Explain your reasoning.

3. Is 113 prime or composite? Explain your reasoning.

4. Is 247 prime or composite? Explain your reasoning.

5. Find two factors of 84 such that their product is 84 and their sum is 25.

Example 4

Determining Whether a Number is Prime

Is 221 prime or composite?

Solution

Tests for 2, 3, and 5 fail.

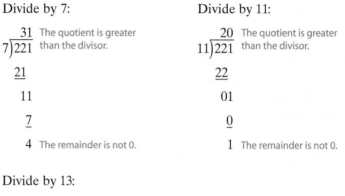

Divide by 7:

$$\begin{array}{r} 31 \\ 7\overline{)221} \\ \underline{21} \\ 11 \\ \underline{7} \\ 4 \end{array}$$ The quotient is greater than the divisor.

The remainder is not 0.

Divide by 11:

$$\begin{array}{r} 20 \\ 11\overline{)221} \\ \underline{22} \\ 01 \\ \underline{0} \\ 1 \end{array}$$ The quotient is greater than the divisor.

The remainder is not 0.

Divide by 13:

$$\begin{array}{r} 17 \\ 13\overline{)221} \\ \underline{13} \\ 91 \\ \underline{91} \\ 0 \end{array}$$ The remainder is 0.

The number 221 is composite.

Note: $221 = 13 \times 17$; that is, 13 and 17 are factors of 221.

Now work margin exercises 2 through 4.

Example 5

Application of Factors of Counting Numbers

One interesting application of factors of counting numbers (very useful in beginning algebra) involves finding two factors whose sum is some specified number. For example, find two factors of 70 such that their product is 70 and their sum is 19.

Solution

$$1 \cdot 70 = 70, \quad 2 \cdot 35 = 70, \quad 5 \cdot 14 = 70, \quad 7 \cdot 10 = 70.$$

Thus the numbers we are looking for are 5 and 14 because

$$5 \cdot 14 = 70 \quad \text{and} \quad 5 + 14 = 19.$$

Now work margin exercise 5.

Objective D **The Prime Factorization of a Composite Number**

When we operate with fractions in Chapter 2, we will want to factor whole numbers so that all of the factors are prime numbers. This is called finding the prime factorization of a whole number. For example, to find the **prime factorization** of 70, we start with any two factors of 70.

$$70 = 7 \cdot 10 \qquad \text{7 is prime, but 10 is not prime.}$$

So, by factoring 10, we find the following.

$$70 = 7 \cdot 2 \cdot 5 \qquad \text{Now all of the factors are prime.}$$

This last product $(7 \cdot 2 \cdot 5)$ is the prime factorization of 70.

Also, we could have started with different factors as follows.

$$70 = 2 \cdot 35 = 2 \cdot 5 \cdot 7 \qquad \text{The prime factorization is the same.}$$

Because multiplication is commutative, the factors may be written in any order. For consistency, we will generally write the factors in ascending order, from smallest to largest.

Regardless of the factors used in the beginning, **there is only one prime factorization for any composite number**. This important fact is called the **fundamental theorem of arithmetic**.

The Fundamental Theorem of Arithmetic

Every composite number has exactly one prime factorization.

To Find the Prime Factorization of a Composite Number

1. Factor the composite number into any two factors.
2. Factor each factor that is not prime.
3. Continue this process until all factors are prime.

The prime factorization is the product of all of the prime factors.

Example 6

Finding the Prime Factorization of a Composite Number

Find the prime factorization of 60.

Solution

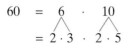

$$60 \;=\; 6 \;\cdot\; 10 \qquad \text{Since the last digit is 0, we know 10 is a factor.}$$
$$\;=\; 2 \cdot 3 \;\cdot\; 2 \cdot 5 \qquad \text{6 and 10 can both be factored so that each factor is a prime number. This is the prime factorization of 60.}$$

We can also start with a different set of factors.

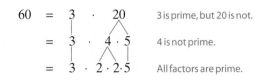

$60 = 3 \cdot 20$ 3 is prime, but 20 is not.

$= 3 \cdot 4 \cdot 5$ 4 is not prime.

$= 3 \cdot 2 \cdot 2 \cdot 5$ All factors are prime.

Writing the factors in order, we see the prime factorization of 60 is $2 \cdot 2 \cdot 3 \cdot 5$ or, using exponents, $2^2 \cdot 3 \cdot 5$.

Example 7

Finding the Prime Factorization of a Composite Number

Find the prime factorization of 72.

Solution

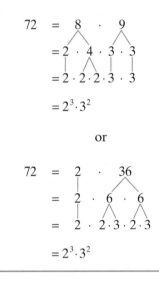

$72 = 8 \cdot 9$

$= 2 \cdot 4 \cdot 3 \cdot 3$

$= 2 \cdot 2 \cdot 2 \cdot 3 \cdot 3$

$= 2^3 \cdot 3^2$

or

$72 = 2 \cdot 36$

$= 2 \cdot 6 \cdot 6$

$= 2 \cdot 2 \cdot 3 \cdot 2 \cdot 3$

$= 2^3 \cdot 3^2$

6. Find the prime factorization of 84.

7. Find the prime factorization of 80.

8. Find the prime factorization of 165.

Completion Example 8

Finding the Prime Factorization of a Composite Number

Find the prime factorization of 196.

Solution

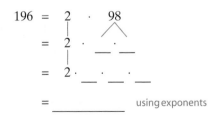

$196 = 2 \cdot 98$

$= 2 \cdot \underline{\quad} \cdot \underline{\quad}$

$= 2 \cdot \underline{\quad} \cdot \underline{\quad} \cdot \underline{\quad}$

$= \underline{\quad\quad\quad}$ using exponents

Now work margin exercises 6 through 8.

Completion Example Answers

8. $2 \cdot 7 \cdot 14$ or $2 \cdot 2 \cdot 49$; $2 \cdot 2 \cdot 7 \cdot 7$; $2^2 \cdot 7^2$

Once the prime factorization of a composite number is known, all of the factors (or divisors) of that number can be found. For a number to be a factor of a composite number, it must be either 1, the number itself, one of the prime factors, or the product of two or more of the prime factors.

Factors of a Composite Number

The only factors (or divisors) of a composite number are:

1. 1 and the number itself;
2. each prime factor; and
3. products formed by all combinations of the prime factors (including repeated factors).

Example 9

Finding the Factors of a Composite Number

Find all the factors of 30.

Solution

Because $30 = 2 \cdot 3 \cdot 5$, the factors are

a. 1 and the number itself, 30.

b. Each prime factor: 2, 3, 5.

c. Products of all combinations of the prime factors:
$$2 \cdot 3 = 6, \qquad 2 \cdot 5 = 10, \qquad 3 \cdot 5 = 15.$$

The only factors are 1, 2, 3, 5, 6, 10, 15, and 30.

Now work margin exercise 9.

9. Find all the factors of 42.

Example 10

Finding the Factors of a Composite Number

Find all the factors of 140.

Solution

The prime factorization of 140 is as follows

$$140 = 14 \cdot 10$$
$$= 2 \cdot 7 \cdot 2 \cdot 5$$
$$= 2 \cdot 2 \cdot 5 \cdot 7.$$

10. Find all the factors of 160.

The factors of 140 are

a. 1 and the number itself: 1 and 140.

b. Each prime factor: 2, 5, 7.

c. Products of all combinations of the prime factors:
$2 \cdot 2 = 4$, $2 \cdot 5 = 10$, $2 \cdot 7 = 14$, $5 \cdot 7 = 35$,
$2 \cdot 2 \cdot 5 = 20$, $2 \cdot 2 \cdot 7 = 28$, $2 \cdot 5 \cdot 7 = 70$.

The factors are 1, 2, 4, 5, 7, 10, 14, 20, 28, 35, 70, and 140.

There are no other factors (or divisors) of 140.

Now work margin exercise 10.

Practice Problems

Determine whether each number is prime or composite.

1. 39   **2.** 79   **3.** 151   **4.** 143

5. Find two factors of 80 such that their product is 80 and their sum is 21.

Find the prime factorization of each number.

6. 42   **7.** 56   **8.** 230

9. Using the prime factorization of 63, find all of the factors of 63.

Practice Problem Answers

1. composite   **2.** prime   **3.** prime
4. composite   **5.** 5 and 16   **6.** $2 \cdot 3 \cdot 7$
7. $2^3 \cdot 7$ **8.** $2 \cdot 5 \cdot 23$ **9.** 1, 3, 7, 9, 21, 63

Exercises 1.9

In your own words, write the definition of the following terms.

1. prime number

2. composite number

List the first five multiples of each of the following numbers.

3. 5

4. 8

5. 13

6. 20

7. 25

8. 30

9. 31

10. 50

Use the process described in the text to construct the following Sieve of Eratosthenes.

11. Construct a Sieve of Eratosthenes for the numbers from 1 to 100. List the prime numbers from 1 to 100.

1	2	3	4	5	6	7	8	9	10
11	12	13	14	15	16	17	18	19	20
21	22	23	24	25	26	27	28	29	30
31	32	33	34	35	36	37	38	39	40
41	42	43	44	45	46	47	48	49	50
51	52	53	54	55	56	57	58	59	60
61	62	63	64	65	66	67	68	69	70
71	72	73	74	75	76	77	78	79	80
81	82	83	84	85	86	87	88	89	90
91	92	93	94	95	96	97	98	99	100

Determine whether each of the following numbers is prime or composite. If the number is composite, find at least three factors of the number. See Examples 1 through 4.

12. 28

13. 32

14. 47

15. 59

16. 63

17. 51

18. 67

19. 89

20. 73

21. 61

22. 52

23. 57

24. 98

25. 205

26. 103

27. 117

Two numbers are given. Find two factors of the first number such that their product is the first number and their sum is the second number. See Example 5.

28. 24, 10

29. 12, 7

30. 16, 10

31. 12, 13

32. 14, 9

33. 50, 27

34. 20, 9

35. 24, 11

36. 48, 19

37. 36, 15

38. 7, 8

39. 63, 24

Find the prime factorization of each number. Use the tests for divisibility for 2, 3, 4, 5, 6, 9, and 10 whenever they help to find beginning factors. See Examples 6 through 8.

40. 24   **41.** 28   **42.** 27   **43.** 16

44. 36   **45** 60   **46.** 72   **47.** 78

48. 81   **49.** 105   **50.** 125   **51.** 160

52. 75   **53.** 150   **54.** 210   **55.** 40

56. 250   **57.** 93   **58.** 168   **59.** 360

Using the prime factorization of each number, find all of the factors (or divisors) of the number. See Examples 9 and 10.

60. 12   **61.** 18   **62.** 28   **63.** 98   **64.** 121

65. 45   **66.** 105   **67.** 54   **68.** 97   **69.** 275

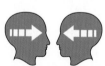

Writing & Thinking

70. Find all prime numbers less than 1000 that are not odd

71. **a.** Explain why the number 1 is not prime and not composite.

   b. Explain why the number 0 is not prime and not composite.

72. Numbers of the form $2^N - 1$, where N is a prime number, are sometimes prime. These prime numbers are called "Mersenne primes" (after Marin Mersenne, 1588 – 1648). Show that for $N = 2, 3, 5,$ and 7 the numbers $2^N - 1$ are prime.

Collaborative Learning

73. Mathematicians have been interested since ancient times in a search for **perfect numbers**. A **perfect number** is a counting number that is equal to the sum of its proper divisors (divisors not including itself). For example, the first perfect number is 6. The proper divisors of 6 are 1, 2, and 3, and $1 + 2 + 3 = 6$. With the class separated into groups of 2 to 4 students, each team is to try to find the second and third perfect numbers. (**Hint:** The second perfect number is between 20 and 30, and the third perfect number is between 450 and 500.)

Section 1.1: Reading and Writing Whole Numbers

Whole Numbers   page 2

The **whole numbers** are the **natural numbers** (or counting numbers) and the number 0.

Natural numbers = ℕ = { **1, 2, 3, 4, 5, 6, 7, 8, 9, 10, 11,** . . . }

Whole numbers = 𝕎 = { **0, 1, 2, 3, 4, 5, 6, 7, 8, 9, 10, 11,** . . . }

Note that 0 is a whole number but not a natural number.

The Place Value System   page 3

The decimal system (or base ten system) is a **place value system** that depends on three things:
1. the **ten digits**: 0, 1, 2, 3, 4, 5, 6, 7, 8, 9;
2. the **placement** of each digit; and
3. the **value** of each place.

Standard and Expanded Notation   page 3

Standard notation is the notation you are familiar with:

865   standard notation.

In **expanded notation**, the values represented by each digit are written as a sum:

$865 = 800 + 60 + 5$ expanded notation.

Comments on Notation   page 4

1. A **hyphen** is used when writing two-digit numbers larger than 20 that do not end in a zero.
2. Note that the word "**and**" does not appear as part of reading (or writing) any whole number. The word "**and**" indicates the decimal point. You may put a decimal point to the right of the digits in a whole number if you choose, but it is not necessary unless digits are written to the right of the decimal point.

Section 1.2: Addition and Subtraction with Whole Numbers

Addition with Whole Numbers   page 10

The numbers being added are called **addends**, and the result of the addition is called the **sum**.

To Add Whole Numbers   page 10

If the sum of the digits in one column is more than 9,
1. write the ones digit in that column, and
2. **carry** the other digit as a number to be added to the next column to the left.

Variable   page 12

A **variable** is a symbol (generally a letter of the alphabet) that is used to represent an unknown number or any one of several numbers.

Properties of Addition   page 12

Commutative Property of Addition: For any whole numbers a and b, $a + b = b + a$.

Associative Property of Addition: For any whole numbers a, b and c, $(a + b) + c = a + (b + c)$.

Additive Identity Property: For any whole number a, $a + 0 = a$.

Perimeter   page 13

The **perimeter** of a geometric figure is the distance around the figure. The perimeter is found by adding the lengths of the sides of the figure and is measured in linear units such as inches, feet, yards, miles, centimeters, and meters.

Subtraction with Whole Numbers   page 15

Subtraction is reverse addition. To subtract, we must know how to add. The missing addend is called the **difference**. The sum is now called the **minuend**, and the given addend is called the **subtrahend**.

Borrowing   page 15

1. **Borrowing** is necessary when a digit is smaller than the digit being subtracted.
2. The process starts from the rightmost digit. **Borrow** from the digit to the left.

Section 1.3: Multiplication with Whole Numbers

Multiplication with Whole Numbers   page 24

The result of multiplication is called a **product**. The two numbers being multiplied are called **factors** of the product.

Properties of Multiplication   page 27

Commutative Property of Multiplication: For any whole numbers a and b, $a \cdot b = b \cdot a$.

Associative Property of Multiplication: For any whole numbers a, b and c, $(a \cdot b) \cdot c = a \cdot (b \cdot c)$.

Multiplicative Identity Property: For any whole number a, $a \cdot 1 = a$.

Multiplication Property of 0:   For any whole number a, $a \cdot 0 = 0$.

Distributive Property of Multiplication over Addition: For any whole numbers a, b and c, $a(b + c) = a \cdot b + a \cdot c$.

Area   page 29

The **area** of a rectangle (measured in square units) is found by multiplying its length by its width.

Division with Whole Numbers   page 35

The **dividend** is the number being divided.
The **divisor** is the number doing the dividing.
The **quotient** is the result of division.
The **remainder** is the number left after division.

Division by 1   page 37

For any number $a, \dfrac{a}{1} = a$.

Division of a Number by Itself   page 37

For any nonzero number $a, \dfrac{a}{a} = 1$.

Division Involving 0   page 38

Case 1: If a is any nonzero whole number, then $\dfrac{0}{a} = 0$.

Case 2: If a is any whole number, then $\dfrac{a}{0}$ is **undefined**.

Remainder of 0   page 42

If the remainder is 0, then the following statements are true:
1. Both the divisor and quotient are **factors** of the dividend.
2. We say that both factors **divide exactly** into the dividend.
3. Both factors are called **divisors** of the dividend.

Rounding Numbers   page 48

To **round** a given number means to find another number close to the given number. The desired place of accuracy must be stated.

Rounding Rule for Whole Numbers   page 49
1. Look at the single digit just to the right of the digit that is in the place of desired accuracy.
2. **If this digit is 5 or greater**, make the digit in the desired place of accuracy one larger and replace all digits to the right with zeros. All digits to the left remain unchanged unless a 9 is made one larger; then the next digit to the left is increased by 1.
3. **If this digit is less than 5**, leave the digit that is in the place of desired accuracy as it is, and replace all digits to the right with zeros. All digits to the left remain unchanged.

To Estimate a Sum or Difference   page 51
1. Round each number to the place of the **leftmost** digit.
2. Perform the addition or subtraction with these rounded numbers.

To Estimate a Product   page 54
1. Round each number to the place of the **leftmost** digit.
2. Multiply the rounded numbers.

To Estimate a Quotient   page 56
   1. Round each number to the place of the leftmost digit.
   2. Divide with the rounded numbers. (When calculating an estimate, ignore the remainder if there is one.)

Section 1.6: Exponents and Order of Operations

Exponents   page 64
   When looking at $5^3 = 125$, 5 is the **base**, 3 is the **exponent**, and 125 is the **product**. Exponents are written slightly to the right and above the base. The expression 5^3 is an **exponential expression**.

The Exponent 1   page 66
   For any number a, $a^1 = a$.

The Exponent 0   page 66
   For any nonzero number a, $a^0 = 1$.
   Note: The expression 0^0 is **undefined**.

Rules for Order of Operations   page 67
   1. First, simplify within grouping symbols, such as parentheses ( ), brackets [ ], or braces { }. Start with the innermost group.
   2. Second, evaluate any numbers or expressions indicated by exponents.
   3. Third, moving from **left to right**, perform any multiplication or division in the order in which it appears.
   4. Fourth, moving from **left to right**, perform any addition or subtraction in the order in which it appears.

Section 1.7: Problem Solving with Whole Numbers

Basic Strategy for Solving Word Problems   page 72
   1. Read the problem carefully.
   2. Draw any type of figure or diagram that might be helpful and decide what operations are needed.
   3. Perform the operations to solve the problem.
   4. Check your work.

Key Words that Indicate Operation   page 72
   Addition: add, sum, plus, more than, increased by, total
   Subtraction: subtract, difference, minus, less than, decreased by
   Multiplication: multiply, product, times, twice, double
   Division: divide, quotient, ratio

To Find the Average of a Set of Numbers   page 75
   1. Find the sum of the given set of numbers.
   2. Divide this sum by the number of numbers in the set. This quotient is called the **average** of the given set of numbers.

Divisibility   pages 84

If a number can be divided by another number so that the remainder is 0, then we say:

The dividend is **exactly divisible by** (or is **divisible by**) the divisor.

Or, the divisor **divides** the dividend.

For example, $46 \div 2 = 23$, so 46 is divisible by 2. Or, 2 divides 46.

Even and Odd Whole Numbers   page 84

Even whole numbers are divisible by 2.

(If a whole number is divided by 2 and the remainder is 0, then the whole number is even.)

Odd whole numbers are not divisible by 2.

(If a whole number is divided by 2 and the remainder is 1, then the whole number is odd.)

(**Note:** Every whole number is either even or odd.)

Divisibility by 2   page 84

If the last digit (ones digit) of a whole number is 0, 2, 4, 6, or 8 (an even digit), then the number is divisible by 2 (the number is even).

Divisibility by 3   page 85

If the sum of the digits of a whole number is divisible by 3, then the number is divisible by 3.

Divisibility by 4   page 85

If the last two digits of a whole number form a number that is divisible by 4, then the number is divisible by 4. (00 is considered to be divisible by 4.)

Divisibility by 5   page 86

If the last digit (ones digit) of a whole number is 0 or 5, then the number is divisible by 5.

Divisibility by 6   page 87

If a whole number is divisible by both 2 and 3, then the number is divisible by 6.

Divisibility by 9   page 87

If the sum of the digits of a whole number is divisible by 9, then the number is divisible by 9.

Divisibility by 10   page 88

If the last digit (ones digit) of a whole number is 0, then the number is divisible by 10.

Prime Numbers   page 93

A **prime number** is a counting number greater than 1 that has exactly two different factors (or divisors) — itself and 1.

Composite Numbers   page 93

A **composite number** is a counting number with more than two different factors (or divisors).

Multiples of a Counting Number   page 94

To find the **multiples** of a counting number, multiply each of the counting numbers by that number.

To Determine Whether a Number is Prime   page 95

Divide the number by progressively larger prime numbers (2, 3, 5, 7, 11, and so forth) until:

1. **The remainder is 0.** This means that the prime number is a factor and **the given number is composite**; or
2. **You find a quotient smaller than the prime divisor.** This means that **the given number is prime** because it has no smaller prime factors.

The Fundamental Theorem of Arithmetic   page 97

Every composite number has exactly one prime factorization.

To Find the Prime Factorization of a Composite Number   page 97

1. Factor the composite number into any two factors.
2. Factor each factor that is not prime.
3. Continue this process until all factors are prime.

The prime factorization is the product of all of the prime factors.

Factors of a Composite Number   page 99

The only factors (or divisors) of a composite number are:

1. 1 and the number itself;
2. Each prime factor; and
3. Products formed by all combinations of the prime factors (including repeated factors).

Chapter 1: Chapter Review

Section 1.1: Reading and Writing Whole Numbers

Write the following whole numbers in expanded notation and in their English word equivalents.

1. 495

2. 1975

3. 60,308

4. 2,460,000

Write the following numbers in standard notation.

5. eight hundred seven

6. four thousand six hundred fifty-six

7. seventeen thousand, two

8. seventy-two million, three hundred forty thousand, eighty-three

Section 1.2: Addition and Subtraction with Whole Numbers

State which property of addition is illustrated.

9. $17 + 32 = 32 + 17$

10. $28 + (2 + 13) = (28 + 2) + 13$

11. $63 + 0 = 63$

Add.

12.
$$\begin{array}{r} 27 \\ 9 \\ +\ 36 \\ \hline \end{array}$$

13.
$$\begin{array}{r} 8445 \\ 267 \\ 1351 \\ +\ \ 478 \\ \hline \end{array}$$

14.
$$\begin{array}{r} 39 \\ 487 \\ 966 \\ +\ 182 \\ \hline \end{array}$$

Subtract.

15.
$$\begin{array}{r} 647 \\ -\ 139 \\ \hline \end{array}$$

16.
$$\begin{array}{r} 7036 \\ -\ 4652 \\ \hline \end{array}$$

17.
$$\begin{array}{r} 5000 \\ -\ 2898 \\ \hline \end{array}$$

Find the perimeter of the rectangle with the given dimensions.

18. width 20 inches and length 43 inches

Section 1.3: Multiplication with Whole Numbers

Multiply (mentally).

19. $0 \cdot 36$

20. $70 \cdot 80$

21. $90 \cdot 4000$

22. $17,394 \cdot 1$

Find each product.

23. $\begin{array}{r} 38 \\ \times\ 41 \\ \hline \end{array}$

24. $\begin{array}{r} 98 \\ \times\ 52 \\ \hline \end{array}$

25 $\begin{array}{r} 8975 \\ \times\ 436 \\ \hline \end{array}$

26. $\begin{array}{r} 4837 \\ \times\ 5000 \\ \hline \end{array}$

Section 1.4: Division with Whole Numbers

Divide (mentally).

27. $0 \div 16$

28. $19 \div 0$

29. $75 \div 5$

30. $48 \div 6$

Find each quotient.

31. $7 \overline{)2065}$

32. $38 \overline{)23,028}$

33. $25 \overline{)14,082}$

34. $529 \overline{)71,496}$

Section 1.5: Rounding and Estimating with Whole Numbers

Round as indicated.

35. 625 (nearest ten)

36. 749 (nearest hundred)

37. 2570 (nearest hundred)

38. 14,620 (nearest thousand)

First estimate the answers by using rounded numbers; then find the actual sum, difference, product, or quotient.

39. $\begin{array}{r} 600 \\ 542 \\ +\ 483 \\ \hline \end{array}$

40. $\begin{array}{r} 10,531 \\ -\ 4\ 600 \\ \hline \end{array}$

41. $\begin{array}{r} 168 \\ \times\ 29 \\ \hline \end{array}$

42. $53 \overline{)3551}$

Section 1.6: Exponents and Order of Operations

43. Identify the base and exponent for the following exponential expression.

3^5

base _____

exponent _____

44. Write the following expression in exponential form.

$2 \cdot 3 \cdot 3 \cdot 3 \cdot 5 \cdot 5$

Evaluate each of the following exponential expressions.

45. 2^4

46. 13^0

Evaluate each of the following expressions.

47. $7 + 3 \cdot 2 - 1 + 9 \div 3$

48. $3 \cdot 2^5 - 2 \cdot 5^2$

49. $14 \div 2 + 2 \cdot 8 + 30 \div 5 \cdot 2$

50. $\left(16 \div 2^2 + 6\right) \div 2 + 8$

51. $\left(75 - 3 \cdot 5\right) \div 10 - 4$

52. $\left(7^2 \cdot 2 + 2\right) \div 10 \div \left(2 + 3\right)$

Section 1.7: Problem Solving with Whole Numbers

Solve the following word problems.

53. If the product of 17 and 51 is added to the product of 16 and 12, what is the sum?

54. Find the average of 33, 42, 25, and 40.

55. If the quotient of 546 and 6 is subtracted from 100, what is the difference?

56. **Stock:** Two years ago, Ms. Miller bought five shares of stock at $353 per share. One year ago, she bought another ten shares at $290 per share. Yesterday, she sold all her shares at $410 per share. What was her total profit? What was her average profit per share?

57. **Grades:** On a history exam, two students scored 98 points, five students scored 87 points, one student scored 81 points, and six students scored 75 points. What was the average score in the class?

58. **Missing Addend:** What number should be added to seven hundred forty-three to get a sum of eight hundred thirteen?

59. **Buying a Car:** If you buy a car for $10,000 plus taxes and license fees totaling $1200 and make a down payment of $3500, how much will you finance?

60. **Checking Account:** You opened a checking account with a deposit of $3000. What is your balance if you wrote checks for $630, $45, $80, and $630 and you made two more deposits of $500 and $235?

Section 1.8: Tests for Divisibility (2, 3, 4, 5, 6, 9, and 10)

Determine which of the numbers 2, 3, 4, 5, 6, 9, and 10 (if any) will divide exactly into the following numbers.

61. 45

62. 72

63. 479

64. 5040

65. 8836

66. 575,493

Determine the following: a. Whether each of the given numbers divides the given product and b. How many times the number divides the product, if it does.

67. 6; $2 \cdot 3 \cdot 3 \cdot 7$

68. 28; $2 \cdot 2 \cdot 3 \cdot 5 \cdot 7$

Section 1.9: Prime Numbers and Prime Factorizations

Solve the following problems.

69. List the first 10 multiples of 3. Are any of these numbers prime?

70. Is 223 a prime number? Explain.

71. List the prime numbers less than 60.

72. Find two factors of 24 whose product is 24 and whose sum is 10.

Find the prime factorization of each number.

73. 150

74. 65

75. 384

76. 990

Chapter 1: Test

Solve the following problems.

1. Write 8952 in expanded notation and in its English word equivalent.

2. The number 1 is called the multiplicative _____.

3. Give an example that illustrates the commutative property of multiplication.

Round each number as indicated.

4. 997 to the nearest hundred

5. 135,721 to the nearest ten-thousand

Add.

6.
$$\begin{array}{r} 9586 \\ 345 \\ + \ 2078 \\ \hline \end{array}$$

7.
$$\begin{array}{r} 37 \\ 486 \\ 493 \\ 162 \\ + \ 557 \\ \hline \end{array}$$

8.
$$\begin{array}{r} 1,480,900 \\ + \ 2,576,850 \\ \hline \end{array}$$

Subtract.

9.
$$\begin{array}{r} 850 \\ - \ 362 \\ \hline \end{array}$$

10.
$$\begin{array}{r} 5097 \\ - \ 3868 \\ \hline \end{array}$$

11.
$$\begin{array}{r} 6000 \\ - \ 293 \\ \hline \end{array}$$

Multiply.

12.
$$\begin{array}{r} 34 \\ \times \ 76 \\ \hline \end{array}$$

13.
$$\begin{array}{r} 2593 \\ \times \ 85 \\ \hline \end{array}$$

14.
$$\begin{array}{r} 793 \\ \times \ 266 \\ \hline \end{array}$$

Divide.

15. $25\overline{)10{,}075}$

16. $462\overline{)79{,}852}$

17. $603\overline{)1{,}209{,}015}$

Solve the following word problems.

18. Find the average of 82, 96, 49, and 69.

19. If the quotient of 51 and 17 is subtracted from the product of 19 and 3, what is the difference?

20. TV: Robert and his brother were saving money to buy a new TV for their parents.

    a. If Robert saved $23 a week and his brother saved $28 a week, how much did they save in six weeks?

    b. What was their average weekly savings?

    c. If the TV they wanted to buy cost $530 including tax, how much did they still need after the six weeks?

21. Checking Accounts: You open a checking account with a deposit of $2500. What is your balance if you write checks for $520, $35, $70, and $230 and make further deposits of $200 and $180?

22. Perimeter: Find the perimeter of the figure shown below.

23. Rectangles: Refer to the figure below to find:

    a. the perimeter,

    b. the area.

Solve the following problems.

24. In the expression 7^3, 7 is called the _____ and 3 is called the _____. The value of 7^3 is _____.

25. a. List the prime numbers between 6 and 20.

    b. List the squares of these prime numbers.

Find the value of each expression using the rules for order of operations.

26. $12 + 9 \div 3 - 2$

27. $60 \div 4(4-1) + 3$

28. $2 \cdot 3^2 - (2^2 \cdot 3 \div 2) - 3(5-1)$

Use the tests for divisibility to tell which of the numbers 2, 3, 4, 5, 6, 9, and 10 (if any) will divide the following numbers.

29. 90

30. 324

31. 1700

Solve the following problem.

32. Does 42 divide the product $2 \cdot 2 \cdot 3 \cdot 3 \cdot 5 \cdot 7$? If so, how many times?

Find the prime factorizations of each number.

33. 124

34. 165

35. 148

Fractions and Mixed Numbers

Mathematics at Work!

Politics and the political process affect everyone in some way. In local, state or national elections, registered voters make decisions about who will represent them and make choices about various ballot measures.

In major issues at the state and national levels, pollsters use mathematics (in particular, statistics and statistical methods) to indicate attitudes and to predict, within certain percentages, how the voters will vote. When there is an important election in your area, read the papers and magazines and listen to the television reports for mathematically related statements predicting the outcome.

Consider the following situation: There are 8000 registered voters in Brownsville, and $\frac{3}{8}$ of these voters live in neighborhoods on the north-side of town. A survey indicates that $\frac{4}{5}$ of these north-side voters are in favor of a bond measure for constructing a new recreation facility that would largely benefit their neighborhoods. Also, $\frac{7}{10}$ of the registered voters from all other parts of town are in favor of the measure. We might then want to know: (See Exercise 99, Section 2.2).

    a.   How many north-side voters favor the bond measure?

    b.   How many voters in the town favor the bond measure?

Objectives

A   Understand the basic concepts of fractions.

B   Multiply fractions.

C   Find equivalent fractions.

D   Reduce fractions to lowest terms.

E   Change mixed numbers to improper fractions.

F   Change improper fractions to mixed numbers.

1. Write a fraction that indicates the shaded parts in each of the following diagrams.

a.

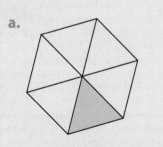

b.

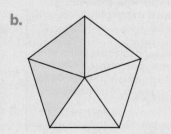

2.1 Introduction to Fractions and Mixed Numbers

Objective A   Fractions

Numbers such as $\dfrac{2}{3}$ are said to be in **fraction form**. The top number, 2, is called the **numerator** and the bottom number, 3, is called the **denominator**.

$$\dfrac{a}{b} \longleftarrow \quad \begin{array}{l} \text{numerator} \\ \hline \text{denominator} \end{array}$$

Fractions can be used to indicate parts of a whole. For example, if a whole candy bar has 7 equal parts, then the fraction $\dfrac{3}{7}$ indicates that we are considering 3 of those parts.

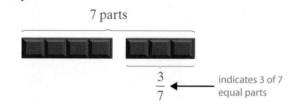

7 parts

$\dfrac{3}{7}$ indicates 3 of 7 equal parts

Example 1 shows several fractions indicating parts of a whole.

Example 1

Understanding Fractions

a.

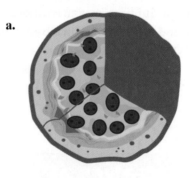

If a whole pizza is cut into 3 equal pieces, then 2 of these pieces represent $\dfrac{2}{3}$ of the pizza (see pizza portion of the figure). The remaining piece (missing portion of the pizza) represents $\dfrac{1}{3}$ of the pizza.

b.

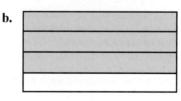

In the rectangle, 3 of the 4 equal parts are shaded. Thus $\dfrac{3}{4}$ of the rectangle is shaded and $\dfrac{1}{4}$ is not shaded.

Now work margin exercise 1.

Figure 1 below shows how a whole can be separated into equal parts in several ways. The shaded parts represent the same "fraction" of the whole.

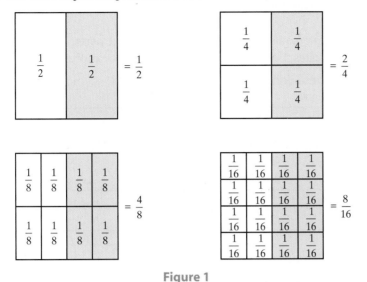

Figure 1

From the shading in Figure 1 we see that $\dfrac{1}{2} = \dfrac{2}{4} = \dfrac{4}{8} = \dfrac{8}{16}$. These fractions are **equal fractions**. Equal fractions are said to be **equivalent**.

Proper Fractions and Improper Fractions

A **proper fraction** is a fraction in which the numerator is less than the denominator. (Proper fractions have values less than 1.)

Examples of proper fractions: $\dfrac{2}{3}$, $\dfrac{7}{8}$, and $\dfrac{32}{60}$

An **improper fraction** is a fraction in which the numerator is greater than or equal to the denominator. (Improper fractions have values greater than or equal to 1.)

Examples of improper fractions: $\dfrac{15}{8}$, $\dfrac{14}{14}$, and $\dfrac{250}{100}$

Example 2

Proper Fractions

$\dfrac{5}{6}$ indicates 5 of 6 equal parts.

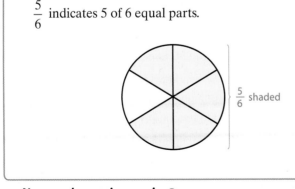

Now work margin exercise 2.

2. Write a fraction that indicates the shaded parts in the diagram.

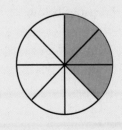

3. Write a fraction that indicates the shaded parts of the diagram.

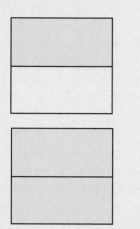

4. Find the values of the expressions below.

a. $\dfrac{0}{45}$

b. $\dfrac{10}{0}$

Example 3

Improper Fractions

Each whole square is separated into 3 equal parts. The shading here indicates 5 of these equal parts and can be represented by the improper fraction $\dfrac{5}{3}$.

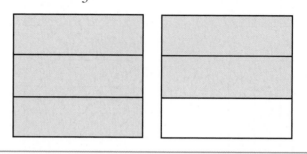

Now work margin exercise 3.

Whole numbers can be thought of as fractions with denominator 1. Thus, in fraction form:

$$0 = \frac{0}{1}, \qquad 1 = \frac{1}{1}, \qquad 2 = \frac{2}{1}, \qquad 3 = \frac{3}{1}, \qquad \text{and so on.}$$

Fraction notation indicates division. For example, $24 \div 8$ can be written in the fraction form $\dfrac{24}{8}$ which indicates that the numerator is to be divided by the denominator. Thus

$$\frac{24}{8} = 3, \qquad \frac{45}{5} = 9, \qquad \text{and} \qquad \frac{0}{5} = 0.$$

Because we know that division by 0 is **undefined**, no denominator can be 0. Thus, in the fraction form $\dfrac{a}{b}$, we write $b \neq 0$ (read, "b is not equal to 0").

The Number 0 in Fractions

For any nonzero value of b, $\dfrac{0}{b} = 0$.

For any value of a, $\dfrac{a}{0}$ is undefined.

Example 4

The Number 0 in Fractions

a. $\dfrac{0}{36} = 0$

b. $\dfrac{0}{124} = 0$

c. $\dfrac{17}{0}$ is undefined

d. $\dfrac{1}{0}$ is undefined

Now work margin exercise 4.

Objective B **Multiplication with Fractions**

Now we state the rule for multiplying fractions and discuss the use of the word "of" to indicate multiplication by fractions. Remember that any whole number can be written in fraction form with a denominator of 1 and no denominator can be 0.

> ### To Multiply Fractions
>
> **1.** Multiply the numerators.
> **2.** Multiply the denominators.
>
> $$\frac{a}{b} \cdot \frac{c}{d} = \frac{a \cdot c}{b \cdot d}$$ For example, $\frac{1}{2} \cdot \frac{3}{7} = \frac{1 \cdot 3}{2 \cdot 7} = \frac{3}{14}$.

Finding the product of two fractions can be thought of as finding one fractional part **of** another fraction. For example, when we multiply $\frac{1}{2}$ by $\frac{1}{4}$ we are finding $\frac{1}{2}$ of $\frac{1}{4}$.

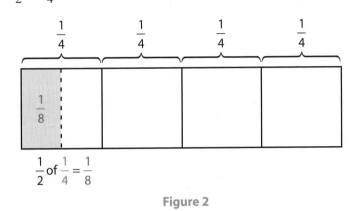

Figure 2

Thus we see from the diagram that $\frac{1}{2}$ of $\frac{1}{4}$ is $\frac{1}{8}$, and from the definition, $\frac{1}{2} \cdot \frac{1}{4} = \frac{1 \cdot 1}{2 \cdot 4} = \frac{1}{8}$.

Example 5

Multiplication with Fractions

Find $\frac{2}{5}$ of $\frac{7}{3}$.

Solution

$$\frac{2}{5} \cdot \frac{7}{3} = \frac{2 \cdot 7}{5 \cdot 3} = \frac{14}{15}$$

Now work margin exercise 5.

5. Find $\frac{3}{4}$ of $\frac{5}{8}$.

6. Find each product.

a. $\dfrac{4}{5} \cdot 7$

b. $\dfrac{0}{4} \cdot \dfrac{11}{16}$

c. $\dfrac{1}{7} \cdot \dfrac{3}{5} \cdot 3$

Example 6

Multiplication with Fractions

a. $\dfrac{4}{13} \cdot 3 = \dfrac{4}{13} \cdot \dfrac{3}{1} = \dfrac{4 \cdot 3}{13 \cdot 1} = \dfrac{12}{13}$ **b.** $\dfrac{9}{8} \cdot 0 = \dfrac{9}{8} \cdot \dfrac{0}{1} = \dfrac{9 \cdot 0}{8 \cdot 1} = \dfrac{0}{8} = 0$

c. $4 \cdot \dfrac{5}{3} \cdot \dfrac{2}{7} = \dfrac{4}{1} \cdot \dfrac{5}{3} \cdot \dfrac{2}{7} = \dfrac{4 \cdot 5 \cdot 2}{1 \cdot 3 \cdot 7} = \dfrac{40}{21}$

Now work margin exercise 6.

Objective C   **Finding Equivalent Fractions**

We know that any nonzero whole number divided by itself is 1. In fraction notation, this can be written $1 = \dfrac{1}{1} = \dfrac{2}{2} = \dfrac{3}{3} = \dfrac{4}{4} = \dfrac{19}{19} = \dfrac{63}{63} = \dfrac{k}{k}$. We know that 1 is the multiplicative identity for whole numbers; that is, for any whole number k, $k \cdot 1 = k$. The number 1 is the multiplicative identity for fractions as well. For example,

$$\dfrac{3}{4} = \dfrac{3}{4} \cdot 1 = \dfrac{3}{4} \cdot \dfrac{5}{5} = \dfrac{15}{20}.$$

The fractions $\dfrac{3}{4}$ and $\dfrac{15}{20}$ are **equivalent** and we say that we have **found an equivalent fraction** for $\dfrac{3}{4}$.

To Find an Equivalent Fraction

Multiply the numerator and denominator by the same nonzero whole number.

$$\dfrac{a}{b} = \dfrac{a}{b} \cdot 1 = \dfrac{a}{b} \cdot \dfrac{k}{k} = \dfrac{a \cdot k}{b \cdot k}, \text{ where } b \text{ and } k \neq 0 \ \left(\dfrac{k}{k} = 1 \right)$$

The following examples illustrate the use of this technique of finding equivalent fractions. Note carefully the importance of the choice of the form $\dfrac{k}{k}$.

Example 7

Finding Equivalent Fractions

Find the missing numerator that will make the fractions equal.

$$\dfrac{3}{4} = \dfrac{?}{28}$$

Solution

Because $4 \cdot 7 = 28$, multiply by $\dfrac{7}{7}$.

$$\dfrac{3}{4} = \dfrac{3}{4} \cdot 1 = \dfrac{3}{4} \cdot \dfrac{7}{7} = \dfrac{3 \cdot 7}{4 \cdot 7} = \dfrac{21}{28}$$

Example 8

Finding Equivalent Fractions

Find the missing numerator that will make the fractions equal.

$$\frac{9}{10} = \frac{?}{30}$$

Solution

Because $10 \cdot 3 = 30$, multiply by $\frac{3}{3}$.

$$\frac{9}{10} = \frac{9}{10} \cdot 1 = \frac{9}{10} \cdot \frac{3}{3} = \frac{9 \cdot 3}{10 \cdot 3} = \frac{27}{30}$$

Completion Example 9

Finding Equivalent Fractions

Find the missing numerator that will make the fractions equal.

$$\frac{3}{5} = \frac{?}{55}$$

Solution

Because $5 \cdot \underline{\hspace{1cm}} = 55$, we have $\frac{3}{5} = \frac{3}{5} \cdot 1 = \frac{3}{5} \cdot \underline{\hspace{1cm}} = \underline{\hspace{1cm}}$.

***Now work margin exercises* 7 through 9.**

Example 10

Application of Multiplication with Fractions

In a certain voting district, $\frac{3}{5}$ of the eligible voters are actually registered to vote. Of those registered voters, $\frac{2}{7}$ are independents (have no party affiliation). What fraction of the eligible voters are registered independents?

Solution

Since the independents are a fraction **of** the eligible voters, we multiply.

$$\frac{2}{7} \cdot \frac{3}{5} = \frac{2 \cdot 3}{7 \cdot 5} = \frac{6}{35}$$

Thus $\frac{6}{35}$ of the eligible voters are registered as independents.

***Now work margin exercise* 10.**

Completion Example Answers

9. $11;\ \dfrac{11}{11} = \dfrac{33}{55}$

---

7. Find the missing numerator that will make the fractions equal.

$$\frac{2}{3} = \frac{?}{12}$$

8. Find the missing numerator that will make the fractions equal.

$$\frac{5}{8} = \frac{?}{24}$$

9. Find the missing numerator that will make the fractions equal.

$$\frac{4}{7} = \frac{?}{63}$$

10. For a certain piece of furniture, $\frac{1}{8}$ of the included hardware is wingnuts. Of these wingnuts, $\frac{1}{20}$ are defective. What fraction of the included hardware is defective?

Objective D Reducing Fractions to Lowest Terms

A fraction is **reduced to lowest terms** if the **numerator and denominator have no common factors other than 1**. To reduce a fraction we reverse the process used when finding an equivalent fraction. For example,

$$\frac{21}{35} = \frac{3 \cdot 7}{5 \cdot 7} = \frac{3}{5} \cdot \frac{7}{7} = \frac{3}{5} \cdot 1 = \frac{3}{5}$$

The use of prime factors is very helpful in reducing to lowest terms.

11. Reduce the fractions to lowest terms.

a. $\dfrac{8}{36}$

b. $\dfrac{18}{27}$

c. $\dfrac{49}{28}$

To Reduce a Fraction to Lowest Terms

1. Factor the numerator and denominator into prime factors.

2. Use the fact that $\dfrac{k}{k} = 1$ and **"divide out"** all common factors.

Note: Reduced fractions might be improper fractions.

Example 11
Reducing Fractions to Lowest Terms

a. $\dfrac{15}{20} = \dfrac{3 \cdot 5}{2 \cdot 2 \cdot 5} = \dfrac{3}{4} \cdot \dfrac{5}{5} = \dfrac{3}{4} \cdot 1 = \dfrac{3}{4}$

b. $\dfrac{4}{36} = \dfrac{2 \cdot 2 \cdot 1}{2 \cdot 2 \cdot 3 \cdot 3} = \dfrac{2}{2} \cdot \dfrac{2}{2} \cdot \dfrac{1}{3 \cdot 3} = 1 \cdot 1 \cdot \dfrac{1}{9} = \dfrac{1}{9}$

c. $\dfrac{35}{21} = \dfrac{5 \cdot 7}{3 \cdot 7} = \dfrac{5}{3} \cdot \dfrac{7}{7} = \dfrac{5}{3} \cdot 1 = \dfrac{5}{3}$

Now work margin exercise 11.

As illustrated in Example 11b, 1 is a factor of any whole number. So, **if all of the factors in the numerator or denominator are divided out, 1 must be used as a factor**.

12. Reduce the fraction to lowest terms.

$\dfrac{81}{45}$

Example 12
Reducing Fractions to Lowest Terms

Reduce $\dfrac{44}{20}$ to lowest terms.

Solution

We can divide out common factors (prime or not) with the understanding that a number divided by itself equals 1.

$$\frac{44}{20} = \frac{\cancel{4} \cdot 11}{\cancel{4} \cdot 5} = \frac{11}{5}$$

Now work margin exercise 12.

Example 12 illustrates the fact that when reducing, common factors in the numerator and denominator need not be prime. However, the advantage of using prime factors is that you can be certain that the fraction is reduced to lowest terms.

Completion Example 13

Reducing Fractions to Lowest Terms

Reduce $\dfrac{52}{65}$ to lowest terms.

Finding a common factor could be difficult here. Prime factoring helps.

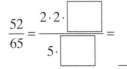

$$\frac{52}{65} = \frac{2 \cdot 2 \cdot \boxed{}}{5 \cdot \boxed{}} = \underline{\qquad}$$

Now work margin exercise 13.

13. Reduce the fraction to lowest terms.

$$\frac{48}{66}$$

Objective E **Changing Mixed Numbers to Improper Fractions**

A **mixed number** is the sum of a whole number and a proper fraction. By convention, we usually write the whole number and the fraction side by side without the plus sign. For example, $6 + \dfrac{3}{5} = 6\dfrac{3}{5}$ (read "six and three-fifths"). Typically, people are familiar with mixed numbers and use them frequently. For example, a carpenter might measure a board to be $2\dfrac{1}{4}$ feet long.

Multiplication and division with mixed numbers are generally easier if the mixed numbers are first changed to improper fractions.

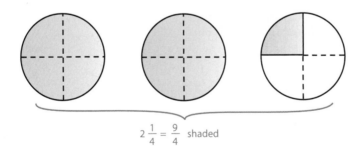

$2\dfrac{1}{4} = \dfrac{9}{4}$ shaded

Figure 3

Completion Example Answers

13. $\dfrac{2 \cdot 2 \cdot \cancel{13}}{5 \cdot \cancel{13}} = \dfrac{4}{5}$

To Change a Mixed Number to an Improper Fraction

1. Multiply the whole number by the denominator of the proper fraction.
2. Add the numerator of the proper fraction to this product.
3. Write this sum over the denominator of the fraction.

For example: $2\dfrac{7}{8} = 2 + \dfrac{7}{8} = \dfrac{16}{8} + \dfrac{7}{8} = \dfrac{16+7}{8} = \dfrac{23}{8}$

Thus, $2\dfrac{7}{8} = \dfrac{16+7}{8} = \dfrac{23}{8}$.

Example 14

Changing a Mixed Number to an Improper Fraction

Change $8\dfrac{9}{10}$ to an improper fraction.

Solution

Step 1: Multiply the whole number by the denominator: $8 \cdot 10 = 80$.

Step 2: Add the numerator: $80 + 9 = 89$.

Step 3: Write this sum over the denominator: $8\dfrac{9}{10} = \dfrac{89}{10}$.

Change each of the following mixed numbers to improper fractions.

14. $10\dfrac{4}{9}$

15. $3\dfrac{5}{32}$

Completion Example 15

Changing a Mixed Number to an Improper Fraction

Change $11\dfrac{2}{3}$ to an improper fraction.

Solution

Step 1: Multiply $11 \cdot 3 = $ _____.

Step 2: Add the numerator: _____ + _____ = _____.

Step 3: Write this sum, _____, over the denominator _____.

Therefore, $11\dfrac{2}{3} = $ _____

Now work margin exercises 14 and 15.

Completion Example Answers

15. Step 1: $11 \cdot 3 = 33$; **Step 2:** $33 + 2 = 35$; **Step 3:** 35; 3; $11\dfrac{2}{3} = \dfrac{35}{3}$

Objective F **Changing Improper Fractions to Mixed Numbers**

To reverse the process (that is, to change an improper fraction to a mixed number), we use the fact that a fraction can indicate division.

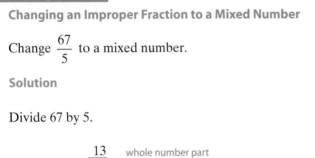

To Change an Improper Fraction to a Mixed Number

1. Divide the numerator by the denominator. The quotient is the whole number part of the mixed number.
2. Write the remainder over the denominator as the fraction part of the mixed number.

Example 16

Changing an Improper Fraction to a Mixed Number

Change $\frac{67}{5}$ to a mixed number.

Solution

Divide 67 by 5.

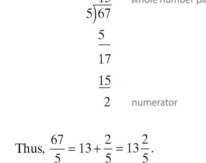

Thus, $\frac{67}{5} = 13 + \frac{2}{5} = 13\frac{2}{5}$.

Now work margin exercise 16.

16. Change the improper fraction to a mixed number.

$$\frac{20}{11}$$

Practice Problems

1. Find the products.

   a. $\dfrac{1}{4} \cdot \dfrac{1}{4}$ b. $\dfrac{3}{5} \cdot \dfrac{4}{7}$ c. $\dfrac{1}{2} \cdot \dfrac{3}{7} \cdot \dfrac{9}{2}$ d. Find $\dfrac{1}{2}$ of $2\dfrac{1}{3}$.

2. Reduce to lowest terms.

   a. $\dfrac{25}{55}$ b. $\dfrac{42}{63}$ c. $\dfrac{390}{260}$

3. Change $7\dfrac{2}{3}$ to an improper fraction.

4. Change $\dfrac{19}{4}$ to a mixed number.

5. Write $\dfrac{77}{20}$ as a mixed number.

Practice Problem Answers

1. a. $\dfrac{1}{16}$ b. $\dfrac{12}{35}$ c. $\dfrac{27}{28}$ d. $\dfrac{7}{6}$ or $1\dfrac{1}{6}$

2. a. $\dfrac{5}{11}$ b. $\dfrac{2}{3}$ c. $\dfrac{3}{2}$ or $1\dfrac{1}{2}$

3. $\dfrac{23}{3}$ 4. $4\dfrac{3}{4}$ 5. $3\dfrac{17}{20}$

Exercises 2.1

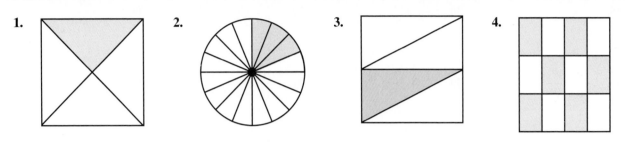

1. 2. 3. 4.

What is the value, if any, of each of the following numbers? See Example 4.

5. $\dfrac{0}{6}$

6. $\dfrac{0}{35}$

7. $\dfrac{15}{0}$

8. $\dfrac{2}{0}$

Find the indicated products. See Examples 5 and 6.

9. $\dfrac{3}{16} \cdot \dfrac{1}{2}$

10. $\dfrac{2}{5} \cdot \dfrac{2}{5}$

11. $\dfrac{3}{7} \cdot \dfrac{3}{7}$

12. $\dfrac{1}{2} \cdot \dfrac{3}{4}$

13. $\dfrac{5}{8} \cdot \dfrac{3}{4}$

14. $\dfrac{1}{9} \cdot \dfrac{4}{9}$

15. $\dfrac{0}{3} \cdot \dfrac{5}{7}$

16. $\dfrac{0}{4} \cdot \dfrac{7}{6}$

17. $\dfrac{7}{6} \cdot \dfrac{5}{2}$

18. $\dfrac{4}{1} \cdot \dfrac{3}{1}$

19. $\dfrac{2}{1} \cdot \dfrac{5}{1}$

20. $\dfrac{14}{1} \cdot \dfrac{0}{2}$

21. $\dfrac{15}{1} \cdot \dfrac{3}{2}$

22. $\dfrac{6}{5} \cdot \dfrac{7}{1}$

23. $\dfrac{8}{5} \cdot \dfrac{4}{3}$

24. $\dfrac{5}{6} \cdot \dfrac{11}{3}$

25. $\dfrac{9}{4} \cdot \dfrac{11}{5}$

26. $\dfrac{1}{5} \cdot \dfrac{2}{7} \cdot \dfrac{3}{11}$

27. $\dfrac{4}{13} \cdot \dfrac{2}{5} \cdot \dfrac{6}{7}$

28. $\dfrac{7}{8} \cdot \dfrac{7}{9} \cdot \dfrac{7}{3}$

29. Find $\dfrac{1}{5}$ of $\dfrac{1}{2}$.

30. Find $\dfrac{1}{3}$ of $\dfrac{1}{2}$.

31. Find $\dfrac{1}{4}$ of $\dfrac{1}{2}$.

32. Find $\dfrac{2}{3}$ of $\dfrac{2}{15}$.

33. Find $\dfrac{4}{7}$ of $\dfrac{3}{5}$.

34. Find $\dfrac{1}{3}$ of $\dfrac{2}{3}$.

In each equation, find the missing numerator that will make the fractions equal. See Examples 7 through 9.

35. $\dfrac{5}{8} = \dfrac{?}{24}$

36. $\dfrac{1}{16} = \dfrac{?}{64}$

37. $\dfrac{2}{5} = \dfrac{?}{25}$

38. $\dfrac{6}{7} = \dfrac{?}{49}$

39. $\dfrac{9}{16} = \dfrac{?}{96}$

40. $\dfrac{7}{2} = \dfrac{?}{20}$

41. $\dfrac{10}{11} = \dfrac{?}{44}$

42. $\dfrac{3}{16} = \dfrac{?}{80}$

43. $\dfrac{11}{12} = \dfrac{?}{48}$

44. $\dfrac{5}{21} = \dfrac{?}{42}$

45. $\dfrac{2}{3} = \dfrac{?}{48}$

46. $\dfrac{1}{13} = \dfrac{?}{39}$

47. $\dfrac{9}{10} = \dfrac{?}{100}$

48. $\dfrac{3}{10} = \dfrac{?}{100}$

49. $\dfrac{7}{10} = \dfrac{?}{70}$

50. $\dfrac{9}{10} = \dfrac{?}{90}$

Reduce each fraction to lowest terms. If it is already in lowest terms, simply rewrite the fraction. See Examples 11 through 13.

51. $\dfrac{3}{9}$

52. $\dfrac{16}{24}$

53. $\dfrac{9}{12}$

54. $\dfrac{6}{20}$

55. $\dfrac{16}{40}$

56. $\dfrac{24}{30}$

57. $\dfrac{14}{36}$

58. $\dfrac{5}{11}$

59. $\dfrac{0}{25}$

60. $\dfrac{75}{100}$

61. $\dfrac{22}{55}$

62. $\dfrac{60}{75}$

63. $\dfrac{30}{36}$

64. $\dfrac{7}{28}$

65. $\dfrac{26}{39}$

66. $\dfrac{27}{56}$

67. $\dfrac{34}{51}$

68. $\dfrac{36}{48}$

69. $\dfrac{24}{100}$

70. $\dfrac{16}{32}$

71. $\dfrac{30}{45}$

72. $\dfrac{28}{42}$

73. $\dfrac{12}{35}$

74. $\dfrac{66}{84}$

75. $\dfrac{48}{12}$ **76.** $\dfrac{150}{135}$ **77.** $\dfrac{140}{112}$ **78.** $\dfrac{84}{42}$

79. $\dfrac{108}{96}$ **80.** $\dfrac{85}{51}$ **81.** $\dfrac{165}{121}$

Change each of the following mixed numbers to improper fractions. See Examples 14 and 15.

82. $4\dfrac{3}{4}$ **83.** $3\dfrac{5}{8}$ **84.** $1\dfrac{2}{15}$ **85.** $5\dfrac{3}{5}$

86. $2\dfrac{1}{4}$ **87.** $15\dfrac{1}{3}$ **88.** $10\dfrac{2}{3}$ **89.** $12\dfrac{1}{2}$

Change each of the following improper fractions to mixed numbers. See Example 16.

90. $\dfrac{4}{3}$ **91.** $\dfrac{5}{2}$ **92.** $\dfrac{13}{2}$ **93.** $\dfrac{17}{8}$

94. $\dfrac{43}{7}$ **95.** $\dfrac{31}{15}$ **96.** $\dfrac{35}{17}$ **97.** $\dfrac{31}{9}$

Solve the following word problems. Write all fractions in lowest terms. See Example 10.

98. Class Grades: In a class of 30 students, 6 received a grade of A. What fraction of the class did not receive an A?

99. Pizza: A pizza pie is to be cut into fourths. Each of these fourths is to be cut into thirds. What fraction of the pie is each of the final pieces?

100. Food Budget: If you had $20 and you spent $9 for a hamburger, fries, and a soft drink, what fraction of your money did you spend? What fraction would you still have?

101. Class Grades: In a class of 35 students, 6 students received As on a mathematics exam. What fraction of the class did not receive an A?

102. **Cooking:** A recipe calls for $\frac{3}{4}$ cup of flour. How much flour should be used if only half the recipe is to be made?

103. **Candy:** One of Maria's birthday presents was a box of candy. Half of the candy was chocolate covered and one-fourth of the chocolate covered candy had cherries inside. What fraction of the candy was chocolate covered cherries?

104. **Office Supplies:** In a box of mixed colored ink cartridges for a desk printer, $\frac{5}{6}$ of the cartridges are not black ink. Of these non-black ink cartridges, $\frac{1}{2}$ are magenta. What fraction of the box is magenta?

105. **Books:** Of the books in a personal library, $\frac{3}{4}$ are fiction. Of these books, $\frac{3}{5}$ are paperback. What fraction of the books in the library are fiction paperbacks?

Writing & Thinking

106. Explain, in your own words, why no denominator can be 0.

2.2 Multiplication and Division with Fractions and Mixed Numbers

Objective A Multiplication with Mixed Numbers

The simplest way to multiply mixed numbers is to change each mixed number to an improper fraction and then multiply. The result can be changed back to mixed number form.

Example 1

Multiplication with Mixed Numbers

Find the product: $1\dfrac{1}{2} \cdot 2\dfrac{1}{5}$

Solution

Change the mixed numbers to improper fractions, then multiply the fractions.

$$1\frac{1}{2} \cdot 2\frac{1}{5} = \frac{3}{2} \cdot \frac{11}{5} = \frac{3 \cdot 11}{2 \cdot 5} = \frac{33}{10} \text{ or } 3\frac{3}{10}$$

Example 2

Multiplication with Fractions and Mixed Numbers

Find $\dfrac{3}{5}$ of $5\dfrac{3}{4}$.

Solution

$$\frac{3}{5} \cdot 5\frac{3}{4} = \frac{3}{5} \cdot \frac{23}{4} = \frac{69}{20} \text{ or } 3\frac{9}{20}$$

Now work margin exercises 1 and 2.

1. Find the product.

$$3\frac{1}{2} \cdot 2\frac{1}{3}$$

2. Find $\dfrac{1}{2}$ of $3\dfrac{2}{5}$.

Objective B Multiplication and Reducing with Fractions

Now we will see how to multiply fractions and mixed numbers and reduce at the same time. If you have any difficulty understanding how to multiply and reduce, use prime factors. By using prime factors, you can be sure that you have not missed a common factor and that your answer is reduced to lowest terms. **Remember that if all the factors in the numerator or denominator divide out, then 1 must be used as a factor.**

Example 3

Multiplying and Reducing with Fractions

Multiply and reduce to lowest terms: $\dfrac{15}{28} \cdot \dfrac{7}{9}$

Solution

Using prime factors, we have

$$\frac{15}{28} \cdot \frac{7}{9} = \frac{15 \cdot 7}{28 \cdot 9} = \frac{\cancel{3} \cdot 5 \cdot \cancel{7}}{2 \cdot 2 \cdot \cancel{7} \cdot \cancel{3} \cdot 3} = \frac{5}{2 \cdot 2 \cdot 3} = \frac{5}{12}.$$

Example 4

Multiplying and Reducing with Fractions

Multiply and reduce to lowest terms: $\dfrac{9}{10} \cdot \dfrac{25}{32} \cdot \dfrac{44}{33}$

Solution

Using prime factors, we have

$$\frac{9}{10} \cdot \frac{25}{32} \cdot \frac{44}{33} = \frac{9 \cdot 25 \cdot 44}{10 \cdot 32 \cdot 33} = \frac{\cancel{3} \cdot 3 \cdot \cancel{5} \cdot 5 \cdot \cancel{2} \cdot \cancel{2} \cdot \cancel{11}}{2 \cdot \cancel{5} \cdot 2 \cdot 2 \cdot 2 \cdot \cancel{2} \cdot \cancel{2} \cdot \cancel{3} \cdot \cancel{11}} = \frac{3 \cdot 5}{2 \cdot 2 \cdot 2 \cdot 2} = \frac{15}{16}.$$

Completion Example 5

Multiplying and Reducing with Fractions

Multiply and reduce to lowest terms: $\dfrac{55}{26} \cdot \dfrac{8}{44} \cdot \dfrac{91}{35}$

Solution

$$\frac{55}{26} \cdot \frac{8}{44} \cdot \frac{91}{35} = \frac{55 \cdot 8 \cdot 91}{26 \cdot 44 \cdot 35} = \underline{\hspace{2cm}} = \underline{\hspace{1.5cm}} = \underline{\hspace{1.5cm}}$$

Now work margin exercises 3 through 5.

By changing mixed numbers to improper fraction form, we can multiply and reduce mixed numbers just as we have with fractions. The answer can be left as an improper fraction or changed to mixed number form.

Multiply and reduce to lowest terms.

3. $\dfrac{14}{27} \cdot \dfrac{6}{8}$

4. $\dfrac{5}{6} \cdot \dfrac{8}{7} \cdot \dfrac{14}{10}$

5. $\dfrac{15}{6} \cdot \dfrac{4}{19} \cdot \dfrac{38}{10}$

Completion Example Answers

5. $\dfrac{\cancel{5} \cdot \cancel{11} \cdot \cancel{2} \cdot \cancel{2} \cdot \cancel{2} \cdot \cancel{7} \cdot \cancel{13}}{\cancel{2} \cdot \cancel{13} \cdot \cancel{2} \cdot \cancel{2} \cdot \cancel{11} \cdot \cancel{5} \cdot \cancel{7}} = \dfrac{1}{1} = 1$

Example 6

Multiplying and Reducing with Mixed Numbers

Multiply and reduce to lowest terms: $4\frac{2}{3} \cdot 1\frac{1}{7} \cdot 2\frac{1}{16}$.

Solution

Using prime factors, we have

$$4\frac{2}{3} \cdot 1\frac{1}{7} \cdot 2\frac{1}{16} = \frac{14}{3} \cdot \frac{8}{7} \cdot \frac{33}{16} = \frac{\cancel{2} \cdot \cancel{7} \cdot \cancel{2} \cdot \cancel{2} \cdot \cancel{2} \cdot \cancel{3} \cdot 11}{\cancel{3} \cdot \cancel{7} \cdot \cancel{2} \cdot \cancel{2} \cdot \cancel{2} \cdot \cancel{2}} = \frac{11}{1} = 11.$$

Now work margin exercise 6.

Another method frequently used to multiply and reduce at the same time is to divide numerators and denominators by common factors whether they are prime or not. If these factors are easily determined, then this method is probably faster. But common factors are sometimes missed with this method whereas they are not missed with the prime factorization method.

Example 7

Multiplying and Reducing with Fractions

Multiply and reduce to lowest terms: $\frac{9}{10} \cdot \frac{25}{64} \cdot \frac{8}{3}$.

Solution

$$\frac{\overset{3}{\cancel{9}}}{\underset{2}{\cancel{10}}} \cdot \frac{\overset{5}{\cancel{25}}}{\underset{8}{\cancel{64}}} \cdot \frac{\overset{1}{\cancel{8}}}{\underset{1}{\cancel{3}}} = \frac{15}{16}$$

3 divides both 3 and 9.
5 divides both 25 and 10.
8 divides both 8 and 64.

Example 8

Multiplying and Reducing with Fractions

Multiply and reduce to lowest terms: $\frac{36}{49} \cdot \frac{14}{75} \cdot \frac{15}{18}$.

Solution

$$\frac{\overset{2}{\cancel{36}}}{\underset{7}{\cancel{49}}} \cdot \frac{\overset{2}{\cancel{14}}}{\underset{5}{\cancel{75}}} \cdot \frac{\overset{1}{\cancel{15}}}{\underset{1}{\cancel{18}}} = \frac{4}{35}$$

18 divides both 36 and 18.
7 divides both 14 and 49.
15 divides both 15 and 75.

Another approach is to use factors that are not all prime:

$$\frac{36}{49} \cdot \frac{14}{75} \cdot \frac{15}{18} = \frac{4 \cdot \cancel{9} \cdot \cancel{2} \cdot \cancel{7} \cdot \cancel{15}}{\cancel{7} \cdot 7 \cdot 5 \cdot \cancel{15} \cdot \cancel{2} \cdot \cancel{9}} = \frac{4}{35}$$

Now work margin exercises 7 and 8.

6. Multiply and reduce to lowest terms.

$$3\frac{3}{4} \cdot 1\frac{1}{6} \cdot 4\frac{2}{5}$$

Find the product by dividing the numerators and denominators by common factors.

7. $\frac{14}{15} \cdot \frac{25}{21} \cdot \frac{3}{10}$

8. $\frac{16}{27} \cdot \frac{45}{6} \cdot \frac{2}{5}$

9. A delivery service found that $\frac{3}{4}$ of its drivers wore their seatbelts. If this company employed 500 drivers, how many drivers wore their seatbelts?

Example 9

Application of Multiplying and Reducing with Fractions

A study showed that $\frac{5}{8}$ of the members of a public service organization were in favor of a new set of bylaws. If the organization had a membership of 200 people, how many were in favor of the changes in the bylaws?

Solution

We want to find $\frac{5}{8}$ of 200, so we multiply:

$$\frac{5}{8} \cdot 200 = \frac{5}{8} \cdot \frac{200}{1} = \frac{5 \cdot \cancel{2} \cdot \cancel{2} \cdot 5 \cdot \cancel{2} \cdot 5}{\cancel{2} \cdot \cancel{2} \cdot \cancel{2} \cdot 1} = \frac{5 \cdot 5 \cdot 5}{1} = 125.$$

Thus there are 125 members in favor of the bylaw changes.

Now work margin exercise 9.

Objective C   Reciprocals

If the product of two fractions is 1, then the fractions are called **reciprocals** of each other. (Remember that whole numbers can also be written in fraction form.) For example,

$$\frac{5}{8} \text{ and } \frac{8}{5} \text{ are reciprocals because } \frac{5}{8} \cdot \frac{8}{5} = \frac{40}{40} = 1.$$

Reciprocals

The reciprocal of $\frac{a}{b}$ is $\frac{b}{a}$ ($a \neq 0$ and $b \neq 0$). The product of a nonzero number and its reciprocal is always 1.

$$\frac{a}{b} \cdot \frac{b}{a} = 1$$

Note: $0 = \frac{0}{1}$, but $\frac{1}{0}$ is undefined. That is, **the number 0 has no reciprocal**.

Example 10

Reciprocals

The reciprocal of $\frac{2}{3}$ is $\frac{3}{2}$.

$$\frac{2}{3} \cdot \frac{3}{2} = \frac{\cancel{2} \cdot \cancel{3}}{\cancel{3} \cdot \cancel{2}} = 1$$

Example 11

Reciprocals

The reciprocal of 10 is $\dfrac{1}{10}$.

$$10 \cdot \dfrac{1}{10} = \dfrac{10}{1} \cdot \dfrac{1}{10} = \dfrac{10 \cdot 1}{1 \cdot 10} = 1$$

Now work margin exercises **10 and 11.**

Find the reciprocal.

10. $\dfrac{7}{8}$

11. 16

Objective D **Division with Fractions and Mixed Numbers**

In the division problem $14 \div 2$ we are asking how many 2s are in 14. Similarly the division problem $\dfrac{2}{3} \div \dfrac{1}{6}$ is asking how many $\dfrac{1}{6}$s there are in $\dfrac{2}{3}$. The figure below shows that there are four $\dfrac{1}{6}$s in $\dfrac{2}{3}$.

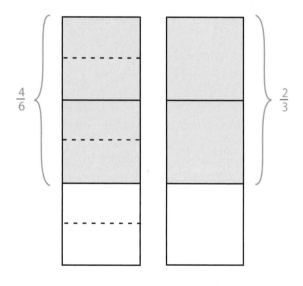

Figure 1

The division can be accomplished in the following way:

$$\dfrac{2}{3} \div \dfrac{1}{6} = \dfrac{2}{3} \cdot \dfrac{6}{1} = \dfrac{12}{3} = 4 \qquad \text{Multiply by the reciprocal of } \dfrac{1}{6}, \text{ namely } \dfrac{6}{1}.$$

To Divide Fractions

To divide by any nonzero number, multiply by its reciprocal. In general,

$$\dfrac{a}{b} \div \dfrac{c}{d} = \dfrac{a}{b} \cdot \dfrac{d}{c} \text{ where } b, c, d \neq 0. \text{ For example, } \dfrac{1}{2} \div \dfrac{4}{3} = \dfrac{1}{2} \cdot \dfrac{3}{4} = \dfrac{3}{8}.$$

12. Divide.

$$\frac{4}{9} \div \frac{1}{2}$$

Example 12

Dividing with Fractions

Divide: $\dfrac{3}{4} \div \dfrac{2}{3}$

Solution

The reciprocal of $\dfrac{2}{3}$ is $\dfrac{3}{2}$, so we multiply by $\dfrac{3}{2}$.

$$\frac{3}{4} \div \frac{2}{3} = \frac{3}{4} \cdot \frac{3}{2} = \frac{9}{8}$$

Now work margin exercise **12.**

Division with mixed numbers can be accomplished by changing each mixed number to improper fraction form and then dividing as normal with fractions.

Example 13

Dividing and Reducing with Fractions and Mixed Numbers

Divide and reduce to lowest terms: $\dfrac{16}{27} \div 2\dfrac{8}{9}$

Solution

First, change the mixed number $2\dfrac{8}{9}$ to the improper fraction $\dfrac{26}{9}$.

The reciprocal of $\dfrac{26}{9}$ is $\dfrac{9}{26}$, so multiply by $\dfrac{9}{26}$ and reduce by factoring as follows:

$$\frac{16}{27} \div 2\frac{8}{9} = \frac{16}{27} \div \frac{26}{9} = \frac{16}{27} \cdot \frac{9}{26} = \frac{8 \cdot \cancel{2} \cdot \cancel{9}}{3 \cdot \cancel{9} \cdot \cancel{2} \cdot 13} = \frac{8}{39}$$

Divide and reduce to lowest terms.

13. $\dfrac{7}{5} \div 2\dfrac{2}{3}$

14. $3\dfrac{3}{4} \div 2\dfrac{2}{5}$

Completion Example 14

Dividing and Reducing with Mixed Numbers

Divide and reduce to lowest terms: $3\dfrac{1}{4} \div 19\dfrac{1}{2}$

Solution

$$3\frac{1}{4} \div 19\frac{1}{2} = \frac{13}{4} \div \frac{39}{2} = \frac{13}{4} \cdot \underline{\hspace{2cm}} = \underline{\hspace{2cm}} = \underline{\hspace{2cm}}$$

Now work margin exercises **13** and **14.**

Suppose the product of two whole numbers is 45 and one number is 5. What is the other number? To find this number we can divide $45 \div 5 = 9$. The other number is 9. This procedure of dividing the product to find a second number is used if the numbers are fractions or mixed numbers as well as whole numbers.

Completion Example Answers

14. $3\dfrac{1}{4} \div 19\dfrac{1}{2} = \dfrac{13}{4} \div \dfrac{39}{2} = \dfrac{13}{4} \cdot \dfrac{2}{39} = \dfrac{\cancel{13} \cdot \cancel{2} \cdot 1}{\cancel{2} \cdot 2 \cdot 3 \cdot \cancel{13}} = \dfrac{1}{6}$

Example 15

Multiplying and Dividing with Fractions and Mixed Numbers

If the product of $1\frac{1}{2}$ and another number is $\frac{5}{18}$, what is the other number?

Solution

Divide the product by the given number to find the other number.

$$\frac{5}{18} \div 1\frac{1}{2} = \frac{5}{18} \div \frac{3}{2} = \frac{5}{18} \cdot \frac{2}{3} = \frac{5 \cdot \cancel{2}}{\cancel{2} \cdot 9 \cdot 3} = \frac{5}{27}$$

$\frac{5}{27}$ is the other number. Check: $1\frac{1}{2} \cdot \frac{5}{27} = \frac{3}{2} \cdot \frac{5}{27} = \frac{\cancel{3} \cdot 5}{2 \cdot \cancel{3} \cdot 9} = \frac{5}{18}$.

Now work margin exercise 15.

Example 16

Multiplying and Dividing with Fractions and Mixed Numbers

A box contains 30 pieces of candy. This is $\frac{3}{5}$ of the maximum amount of candy the box can hold.

a. Is the maximum amount of candy the box can hold more or less than 30 pieces?

Solution

The maximum number of pieces of candy is more than 30.

b. If you want to multiply $\frac{3}{5}$ times 30, would the product be more or less than 30?

Solution

The product would be less than 30.

c. What is the maximum number of pieces of candy the box can hold?

Solution

To find the maximum number of pieces, divide:

$$30 \div \frac{3}{5} = \frac{30}{1} \cdot \frac{5}{3} = \frac{2 \cdot \cancel{3} \cdot 5 \cdot 5}{1 \cdot \cancel{3}} = \frac{50}{1} = 50$$

The maximum number of pieces the box will hold is 50.

Now work margin exercise 16.

15. If the product of $\frac{7}{5}$ and another number is $\frac{4}{9}$, what is the other number?

16. A truck is towing 3000 pounds. This is $\frac{2}{3}$ of the truck's towing capacity.

   a. Can the truck tow 4000 pounds?

   b. If you were to multiply $\frac{2}{3}$ times 3000, would the product be more or less than 3000?

   c. What is the maximum towing capacity of the truck?

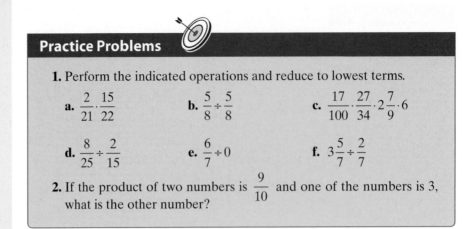

Practice Problems

1. Perform the indicated operations and reduce to lowest terms.

a. $\dfrac{2}{21} \cdot \dfrac{15}{22}$

b. $\dfrac{5}{8} \div \dfrac{5}{8}$

c. $\dfrac{17}{100} \cdot \dfrac{27}{34} \cdot 2\dfrac{7}{9} \cdot 6$

d. $\dfrac{8}{25} \div \dfrac{2}{15}$

e. $\dfrac{6}{7} \div 0$

f. $3\dfrac{5}{7} \div \dfrac{2}{7}$

2. If the product of two numbers is $\dfrac{9}{10}$ and one of the numbers is 3, what is the other number?

Practice Problem Answers

1. a. $\dfrac{5}{77}$

b. 1

c. $\dfrac{9}{4}$ or $2\dfrac{1}{4}$

d. $\dfrac{12}{5}$ or $2\dfrac{2}{5}$

e. undefined

f. 13

2. $\dfrac{3}{10}$

Exercises 2.2

Find each product and write the answers as mixed numbers. See Examples 1 and 2.

1. $2\dfrac{3}{5} \cdot 1\dfrac{1}{7}$

2. $2\dfrac{1}{3} \cdot 3\dfrac{1}{3}$

3. $5\dfrac{1}{3} \cdot 2\dfrac{1}{3}$

4. $2\dfrac{1}{4} \cdot 3\dfrac{1}{8}$

5. $9\dfrac{1}{3} \cdot 3\dfrac{3}{4}$

6. $6\dfrac{3}{8} \cdot 2\dfrac{1}{3}$

Find each product and reduce to lowest terms.   See Examples 3 through 8.

7. $\dfrac{2}{3} \cdot \dfrac{4}{3}$

8. $\dfrac{1}{5} \cdot \dfrac{4}{7}$

9. $\dfrac{3}{7} \cdot \dfrac{5}{3}$

10. $\dfrac{2}{11} \cdot \dfrac{3}{2}$

11. $\dfrac{5}{16} \cdot \dfrac{16}{15}$

12. $\dfrac{7}{8} \cdot \dfrac{9}{14}$

13. $\dfrac{10}{18} \cdot \dfrac{9}{5}$

14. $\dfrac{11}{22} \cdot \dfrac{6}{8}$

15. $\dfrac{15}{27} \cdot \dfrac{9}{30}$

16. $\dfrac{35}{20} \cdot \dfrac{36}{14}$

17. $\dfrac{25}{9} \cdot \dfrac{3}{100}$

18. $\dfrac{30}{42} \cdot \dfrac{7}{100}$

19. $\dfrac{18}{42} \cdot \dfrac{14}{75}$

20. $\dfrac{42}{70} \cdot \dfrac{20}{12}$

21. $8 \cdot \dfrac{5}{12}$

22. $9 \cdot \dfrac{7}{24}$

23. $\dfrac{6}{85} \cdot \dfrac{34}{9}$

24. $\dfrac{13}{91} \cdot \dfrac{34}{65}$

25. $1\dfrac{13}{23} \cdot \dfrac{20}{48}$

26. $5\dfrac{1}{4} \cdot 1\dfrac{1}{7}$

27. $1\dfrac{6}{9} \cdot 1\dfrac{6}{18}$

28. $\dfrac{32}{20} \cdot \dfrac{13}{9} \cdot \dfrac{7}{26}$

29. $\dfrac{70}{15} \cdot \dfrac{30}{8} \cdot \dfrac{48}{32}$

30. $\dfrac{42}{52} \cdot 1\dfrac{5}{22} \cdot 3\dfrac{6}{9}$

31. $1\dfrac{1}{3} \cdot 18 \cdot 3\dfrac{1}{2} \cdot 1\dfrac{18}{36}$

32. $1\dfrac{3}{7} \cdot 1\dfrac{5}{35} \cdot 4\dfrac{10}{15}$

33. $1\dfrac{18}{66} \cdot 2\dfrac{2}{5} \cdot 1\dfrac{5}{28}$

34. $4\dfrac{4}{24} \cdot 1\dfrac{12}{36} \cdot \dfrac{9}{15}$

35. $\dfrac{10}{17} \cdot 8\dfrac{2}{5} \cdot 2\dfrac{15}{18} \cdot \dfrac{1}{4}$

36. $9\dfrac{3}{8} \cdot \dfrac{16}{36} \cdot 9 \cdot \dfrac{7}{25}$

37. $\dfrac{2}{3} \div \dfrac{3}{4}$

38. $\dfrac{1}{5} \div \dfrac{3}{4}$

39. $\dfrac{3}{7} \div \dfrac{3}{5}$

40. $\dfrac{2}{11} \div \dfrac{2}{3}$

41. $\dfrac{3}{5} \div \dfrac{3}{7}$

42. $\dfrac{2}{3} \div \dfrac{2}{11}$

43. $\dfrac{5}{16} \div \dfrac{15}{16}$

44. $\dfrac{7}{18} \div \dfrac{3}{9}$

45. $\dfrac{3}{14} \div \dfrac{2}{7}$

46. $\dfrac{13}{40} \div \dfrac{26}{35}$

47. $\dfrac{5}{12} \div \dfrac{15}{16}$

48. $\dfrac{12}{27} \div \dfrac{10}{18}$

49. $\dfrac{17}{48} \div \dfrac{51}{90}$

50. $\dfrac{3}{5} \div \dfrac{7}{8}$

51. $\dfrac{13}{16} \div \dfrac{2}{3}$

52. $\dfrac{3}{4} \div \dfrac{5}{6}$

53. $\dfrac{14}{15} \div \dfrac{21}{25}$

54. $\dfrac{6}{13} \div \dfrac{6}{13}$

55. $\dfrac{16}{27} \div \dfrac{7}{18}$

56. $\dfrac{20}{21} \div \dfrac{15}{42}$

57. $\dfrac{25}{36} \div \dfrac{5}{24}$

58. $\dfrac{17}{20} \div \dfrac{3}{14}$

59. $\dfrac{26}{35} \div \dfrac{39}{40}$

60. $\dfrac{5}{6} \div \dfrac{13}{4}$

61. $1\dfrac{1}{7} \div 7\dfrac{1}{2}$

62. $1\dfrac{11}{14} \div 2\dfrac{2}{7}$

63. $4\dfrac{1}{5} \div 3\dfrac{1}{3}$

64. $2\dfrac{1}{17} \div 1\dfrac{1}{4}$

65. $4\dfrac{1}{5} \div 3$

66. $6\dfrac{5}{6} \div 2$

67. $3 \div 4\dfrac{1}{5}$

68. $2 \div 6\dfrac{5}{6}$

69. $5 \div 1\dfrac{7}{8}$

70. $14 \div \dfrac{1}{7}$

71. $1\dfrac{7}{8} \div 5$

72. $\dfrac{1}{7} \div 14$

73. $56 \div \dfrac{1}{8}$

74. $24 \div \dfrac{1}{4}$

75. $1\dfrac{1}{32} \div 2\dfrac{3}{4}$

76. $13\dfrac{1}{7} \div 4\dfrac{2}{11}$

77. **a.** $\frac{3}{4} \div 2$ **b.** $\frac{3}{4} \div 3$ **78.** **a.** $\frac{4}{7} \div 4$ **b.** $\frac{4}{7} \div \frac{1}{4}$

79. **a.** $\frac{5}{8} \div 2$ **b.** $\frac{5}{8} \div \frac{1}{2}$ **80.** **a.** $\frac{3}{10} \div 10$ **b.** $\frac{3}{10} \div \frac{1}{10}$

Solve the following word problems. See Examples 9, 15, and 16.

81. **Missing Number:** The product of $\frac{9}{10}$ and another number is $\frac{5}{3}$. What is the other number? (**Hint:** Think $\frac{9}{10} \cdot \boxed{} = \frac{5}{3}$.)

82. **Missing Number:** The result of multiplying two numbers is 150. If one of the numbers is $\frac{5}{7}$, what is the other number? (**Hint:** Think $\frac{5}{7} \cdot \boxed{} = 150$.)

83. **Product:** The product of $\frac{5}{6}$ and another number is $\frac{2}{5}$.
   a. Which number is the product?
   b. What is the other number?

84. **Product:** The result of multiplying two numbers is 150.
   a. Is 150 a product or a quotient?
   b. If one of the numbers is $\frac{15}{7}$, what is the other one?

85. **Quotient:** What is the quotient if $\frac{1}{4}$ of 36 is divided by $\frac{2}{3}$ of $\frac{5}{8}$?

86. **Quotient:** What is the quotient if $\frac{5}{8}$ of 64 is divided by $\frac{3}{5}$ of $\frac{15}{16}$?

87. **Baseball:** Major league baseball teams play 162 games each season. If a team has played $\frac{5}{9}$ of its games by the All-Star break (around mid-season), how many games has it played by that time?

88. **Planetary Orbits:** Venus orbits the sun in $\frac{45}{73}$ the time that the earth takes to orbit the sun. Assuming that one "earth-year" is 365 days long, how long is a "Venus-year" in terms of "earth-days"?

89. **Budgeting:** If you have $35 and you spend $7 for a deli sandwich and an iced tea, what fraction of your money have you spent on food? What fraction do you still have?

90. **Water Glass:** A glass is 8 inches tall. If the glass is $\frac{3}{4}$ full of water, what is the height of the water in the glass?

$\frac{3}{4}$ full   8 inches

Multiplication and Division with Fractions and Mixed Numbers   **Section 2.2**

91. Falling Object: Suppose that a ball is dropped from a height of 40 feet and that each time the ball bounces it bounces back to $\frac{1}{2}$ the height it dropped. How high will the ball bounce on its third bounce?

40 ft

1/2 of previous bounce

92. Bicycling: If you go on a bicycle trip of 75 miles in the mountains and $\frac{1}{5}$ of the trip is downhill, what fraction of the trip is not downhill? How many miles are not downhill?

93. Enrollment: A small private college has determined that about $\frac{11}{25}$ of the students that it accepts will actually enroll. If the college wants 550 freshmen to enroll, how many should it accept?

94. Geology: The floor of the Atlantic Ocean is spreading apart at an average rate of $\frac{3}{50}$ of a meter per year. About how long will it take for the sea floor to spread 12 meters?

95. Airplane Capacity: An airplane is carrying 180 passengers. This is $\frac{9}{10}$ of the capacity of the airplane.

   a. Is the capacity of the airplane more or less than 180?

   b. If you were to multiply 180 times $\frac{9}{10}$, would the product be more or less than 180?

   c. What is the capacity of the airplane?

96. Land Area: The continent of Africa covers approximately 11,707,000 square miles. This is $\frac{1}{5}$ of the land area of the world. What is the approximate total land area of the world?

97. Mileage: If you drive your car to work $6\frac{3}{10}$ miles one-way each weekday (5 days a week), how many miles do you drive each week going to and from work?

98. Reading: A man can read $\frac{1}{5}$ of a book in 3 hours.

   a. What fraction of the book can he read in 6 hours?

   b. If the book contains 450 pages, how many pages can he read in 6 hours?

   c. How long will he take to read the entire book?

99. Tennis Club: The tennis club has 250 members, and they are considering putting in a new tennis court. The cost of the new court is going to involve a charge of $200 for each member. Of the seven-tenths of the members who live close to the club, $\frac{3}{5}$ of them are in favor of the assessment. However, only $\frac{1}{3}$ of the members who do not live close to the club are in favor of the assessment.

a. If a vote were taken today, would more than one-half of the members vote for or against the new court?

b. By how many votes would the question pass or fail if more than one-half of the members must vote in favor for the question to pass?

100. Politics: The student senate has 75 members, and $\frac{7}{15}$ of them are women. A change in the senate constitution is being considered, and at the present time (before debating has begun), a survey shows that $\frac{3}{5}$ of the women and $\frac{4}{5}$ of the men are in favor of this change.

a. How many women are on the student senate?

b. How many women on the senate are in favor of the change?

c. If the change requires a $\frac{2}{3}$ majority vote in favor to pass, would the change pass if the vote were taken today?

d. By how many votes would the change pass or fail?

101. Manufacturing: A manufacturing plant is currently producing 6000 steel rods per week. Because of difficulties getting materials, this number is only $\frac{3}{4}$ of the plant's potential production.

a. Is the potential production number more or less than 6000 rods?

b. If you were to multiply $\frac{3}{4}$ times 6000, would the product be more or less than 6000?

c. What is the plant's potential production?

102. Agriculture: A grove of orange trees was struck by an off-season frost and the result was a relatively poor harvest. This year's crop was 10,000 tons of oranges which is about $\frac{4}{5}$ of the usual crop.

a. Is the usual crop more or less than 10,000 tons of oranges?

b. If you were to multiply 10,000 times $\frac{4}{5}$, would the product be more or less than 10,000?

c. About how many tons of oranges are usually harvested?

103. Debate Team: There are 3000 students at Mountain High School, and $\frac{1}{4}$ of these students are seniors. If $\frac{3}{5}$ of the seniors are in favor of the school forming a debating team and $\frac{7}{10}$ of the remaining students (not seniors) are also in favor of forming a debating team, how many students do not favor this idea?

104. Voting: There are 4000 registered voters in Roseville, and $\frac{3}{8}$ of these voters are registered Democrats. A survey indicates that $\frac{2}{3}$ of the registered Democrats are in favor of Bond Measure A and $\frac{3}{5}$ of the other registered voters are in favor of this measure.

a. How many of the voters are registered Democrats?

b. How many of the voters are not registered Democrats?

c. How many of the registered Democrats favor Measure A?

d. How many of the registered voters favor Measure A?

105. Show that the phrases "15 divided by three" and "15 divided by one-third" have different meaning.

106. Show that the phrases "12 divided by three" and "12 times one-third" have the same meaning.

Collaborative Learning

107. With the class divided into teams of 2 to 4 students, each team is to analyze the following problem and discuss its solutions in class.

One-third of the hexagon is shaded. Copy the hexagon and shade one-third as many different ways as you can. How many ways do you think that this can be done?

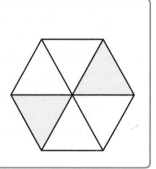

2.3 Least Common Multiple (LCM)

Objective A **Least Common Multiple**

In this section we discuss how to find the least common multiple (LCM) of a set of whole numbers, and then we will apply these ideas in our work with adding and subtracting fractions.

The **multiples** of a number are the products of that number with the counting numbers. The multiples of 8 and 12 are shown here.

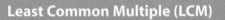

Counting Numbers: **1, 2, 3, 4, 5, 6, 7, 8, 9, 10, 11, ...**

Multiples of 8:   8, 16, (24) 32, 40, (48) 56, 64, (72) 80, 88, ...

Multiples of 12:   12, (24) 36, (48) 60, (72) 84, (96) 108, (120) 132, ...

The common multiples of 8 and 12 are 24, 48, 72, 96, and so on. The smallest of these, 24, is the **least common multiple (LCM)**.

Least Common Multiple (LCM)

The **least common multiple (LCM)** of two (or more) whole numbers is the smallest number that is a multiple of each of these numbers.

Objective B **Finding the Least Common Multiple (LCM)**

The following method, involving prime factorizations, is an efficient way to find the least common multiple of a set of counting numbers.

To Find the LCM of a Set of Counting Numbers

1. Find the prime factorization of each number.
2. Identify the prime factors that appear in any one of the prime factorizations.
3. Form the product of these primes using each prime the most number of times it appears in any one of the prime factorizations.

Note: There are other methods for finding the LCM, maybe even easier to use at first. However, the method here allows for a solid understanding that carries over to use with fractions.

Objectives

A Understand the meaning of the term least common multiple.

B Use prime factorizations to find the LCM of a set of numbers.

C Recognize the application of the LCM concept in a word problem.

Example 1

Finding the Least Common Multiple (LCM)

Find the LCM of 20 and 45.

Solution

Step 1: Prime factorizations:

$$20 = 4 \cdot 5 = 2 \cdot 2 \cdot 5 \quad \text{two 2s, one 5}$$
$$45 = 9 \cdot 5 = 3 \cdot 3 \cdot 5 \quad \text{two 3s, one 5}$$

Step 2: 2, 3, and 5 are the only prime factors.

Step 3: Most of each factor in any one factorization:

two 2s	(in 20)
two 3s	(in 45)
one 5	(one in 20 and one in 45)

So, LCM $= 2 \cdot 2 \cdot 3 \cdot 3 \cdot 5 = 2^2 \cdot 3^2 \cdot 5 = 180$.

180 is the smallest number divisible by both 20 and 45. (Note also that the LCM, 180, contains all the factors of the numbers 20 and 45.)

1. Find the LCM of 12 and 25.

2. Find the LCM of 18, 21, and 42.

Example 2

Finding the Least Common Multiple (LCM)

Find the LCM of 12, 18, and 48.

Solution

Step 1: Prime factorizations:

$$12 = 4 \cdot 3 = 2^2 \cdot 3 \quad \text{two 2s, one 3}$$
$$18 = 2 \cdot 9 = 2 \cdot 3^2 \quad \text{one 2, two 3s}$$
$$48 = 16 \cdot 3 = 2^4 \cdot 3 \quad \text{four 2s, one 3}$$

Step 2: 2 and 3 are the only prime factors.

Step 3: Most of each factor in any one factorization:

four 2s	(in 48)
two 3s	(in 18)

So, LCM $= 2 \cdot 2 \cdot 2 \cdot 2 \cdot 3 \cdot 3 = 2^4 \cdot 3^2 = 144$.

Now work margin exercises 1 and 2.

Completion Example 3

Finding the Least Common Multiple (LCM)

Find the LCM of 36, 24, and 48.

Solution

Step 1: Prime factorizations:

$$36 = \underline{\hspace{1.5cm}} = \underline{\hspace{1.5cm}}$$

$$24 = \underline{\hspace{1.5cm}} = \underline{\hspace{1.5cm}}$$

$$48 = \underline{\hspace{1.5cm}} = \underline{\hspace{1.5cm}}$$

Step 2: _____ and _____ are the only prime factors.

Step 3: Most of each factor in any one factorization:

_____   (in 48)

_____   (in 36)

So, LCM = _____ = _____.

_____ is the smallest number divisible by the numbers 36, 24, and 48.

Example 4

Finding the Least Common Multiple (LCM)

Find the LCM of 27, 30, and 42.

Solution

$$\left. \begin{array}{l} 27 = 3 \cdot 9 = 3 \cdot 3 \cdot 3 = 3^3 \\ 30 = 6 \cdot 5 = 2 \cdot 3 \cdot 5 \\ 42 = 6 \cdot 7 = 2 \cdot 3 \cdot 7 \end{array} \right\} \text{LCM} = 2 \cdot 3^3 \cdot 5 \cdot 7 = 1890$$

Example 5

Finding the Least Common Multiple (LCM)

Find the LCM of 8 and 25.

Solution

$$\left. \begin{array}{l} 8 = 2 \cdot 2 \cdot 2 = 2^3 \\ 25 = 5 \cdot 5 = 5^2 \end{array} \right\} \text{LCM} = 2^3 \cdot 5^2 = 200$$

In this case, where the two numbers have no common prime factors, the LCM is the product of the two numbers.

***Now work margin exercises* 3 through 5.**

3. Find the LCM of 12, 30, and 40.

4. Find the LCM of 36, 45, and 60.

5. Find the LCM of 9 and 49.

Completion Example Answers

3. Step 1: $36 = 4 \cdot 9 = 2^2 \cdot 3^2$; $24 = 8 \cdot 3 = 2^3 \cdot 3$; $48 = 16 \cdot 3 = 2^4 \cdot 3$

Step 2: 2 and 3   **Step 3:** four 2s; two 3s; LCM $= 2^4 \cdot 3^2 = 144$; 144

In Example 4, we found that 1890 is the LCM for the three numbers 27, 30, and 42. This tells us that 1890 is the smallest number that is divisible by 27, 30, and 42. The next question is how many times each of these numbers divides into 1890. In particular, how many times does 27 divide into 1890? We could simply divide by using long division; however, this is not necessary because we know the prime factorizations of 27 and 1890.

$$1890 = 2 \cdot 3 \cdot 3 \cdot 3 \cdot 5 \cdot 7 = \underbrace{(3 \cdot 3 \cdot 3)} \cdot \underbrace{(2 \cdot 5 \cdot 7)}$$
$$= \quad 27 \quad \cdot \quad 70$$

Thus, if we were to divide 1890 by 27, we would get a quotient of 70.

The following examples illustrate how to use prime factorizations to tell how many times each number in a set divides into the LCM for that set of numbers. **This method will be very useful when adding and subtracting fractions.**

6. Find the LCM of 15, 35, and 42; and then state how many times each number divides into the LCM.

Example 6

Prime Factorizations and the Least Common Multiple (LCM)

Find the LCM for 27, 30, and 42, and then state how many times each number divides into the LCM.

Solution

Recall from Example 4, LCM $= 2 \cdot 3^3 \cdot 5 \cdot 7 = 1890$.

$$1890 = 2 \cdot 3 \cdot 3 \cdot 3 \cdot 5 \cdot 7 = \underbrace{(3 \cdot 3 \cdot 3)} \cdot \underbrace{(2 \cdot 5 \cdot 7)}$$
$$= \quad 27 \quad \cdot \quad 70$$

$$1890 = 2 \cdot 3 \cdot 3 \cdot 3 \cdot 5 \cdot 7 = \underbrace{(2 \cdot 3 \cdot 5)} \cdot \underbrace{(3 \cdot 3 \cdot 7)}$$
$$= \quad 30 \quad \cdot \quad 63$$

$$1890 = 2 \cdot 3 \cdot 3 \cdot 3 \cdot 5 \cdot 7 = \underbrace{(2 \cdot 3 \cdot 7)} \cdot \underbrace{(3 \cdot 3 \cdot 5)}$$
$$= \quad 42 \quad \cdot \quad 45$$

So, 27 divides into 1890 70 times;
30 divides into 1890 63 times;
42 divides into 1890 45 times.

Now work margin exercise 6.

Example 7

Prime Factorizations and the Least Common Multiple (LCM)

Find the LCM for 12, 18, and 66, and then state how many times each number divides into the LCM.

Solution

$$12 = 2^2 \cdot 3$$
$$18 = 2 \cdot 3^2 \left.\begin{array}{l}\\\\\\\end{array}\right\} \text{LCM} = 2^2 \cdot 3^2 \cdot 11 = 396$$
$$66 = 2 \cdot 3 \cdot 11$$

$$396 = 2 \cdot 2 \cdot 3 \cdot 3 \cdot 11 = \underbrace{(2 \cdot 2 \cdot 3)} \cdot \underbrace{(3 \cdot 11)}$$
$$= \qquad 12 \quad \cdot \quad 33$$

$$396 = 2 \cdot 2 \cdot 3 \cdot 3 \cdot 11 = \underbrace{(2 \cdot 3 \cdot 3)} \cdot \underbrace{(2 \cdot 11)}$$
$$= \qquad 18 \quad \cdot \quad 22$$

$$396 = 2 \cdot 2 \cdot 3 \cdot 3 \cdot 11 = \underbrace{(2 \cdot 3 \cdot 11)} \cdot \underbrace{(2 \cdot 3)}$$
$$= \qquad 66 \quad \cdot \quad 6$$

So, 12 divides into 396 33 times;
18 divides into 396 22 times;
66 divides into 396   6 times.

Now work margin exercise 7.

7. Find the LCM for 10, 18, and 75; and then state how many times each number divides into the LCM.

Objective C **Applications of LCM in Word Problems**

Many events occur at regular intervals of time. Weather satellites may orbit the Earth once every 10 hours or once every 12 hours. Delivery trucks arrive once a day or once a week at department stores. Traffic lights change once every 3 minutes or once every 2 minutes. The periodic frequency with which such events occur can be explained in terms of the least common multiple, as illustrated in Example 8.

8. A walker and two joggers begin using the same track at the same time. Their lap times are 6, 3, and 5 minutes, respectively.

a. In how many minutes will they be together at the starting place?

b. How many laps will each person have completed at this time?

Example 8

Application with Least Common Multiple (LCM)

Suppose three weather satellites – A, B, and C – are orbiting the Earth at different times. Satellite A takes 24 hours, B takes 18 hours, and C takes 12 hours. If they are directly above each other now, as shown in part **a.** of the figure below, in how many hours will they again be directly above each other in the position shown in part **a.**? How many orbits will each satellite have made in that time?

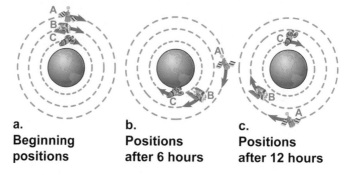

a.
Beginning positions

b.
Positions after 6 hours

c.
Positions after 12 hours

Solution

Study the diagram shown above. When the three satellites are again in the position shown in part **a.**, each will have made some number of complete orbits. Since A takes 24 hours to make one complete orbit, the solution must be a multiple of 24. Similarly, the solution must be a multiple of 18 and a multiple of 12 to allow for complete orbits of satellites B and C.

The solution is the LCM of 24, 18, and 12.

$$\left.\begin{array}{l} 24 = 2^3 \cdot 3 \\ 18 = 2 \cdot 3^2 \\ 12 = 2^2 \cdot 3 \end{array}\right\} \qquad \text{LCM} = 2^3 \cdot 3^2 = 72$$

Thus, the satellites will align again at the position shown in 72 hours (or 3 days).

Note that: Satellite A will have made 3 orbits: $24 \cdot 3 = 72$;
Satellite B will have made 4 orbits: $18 \cdot 4 = 72$;
Satellite C will have made 6 orbits: $12 \cdot 6 = 72$.

Now work margin exercise 8.

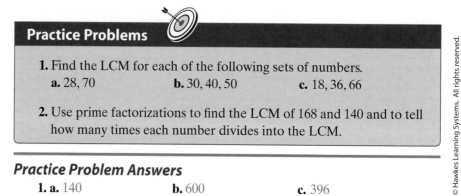

Practice Problems

1. Find the LCM for each of the following sets of numbers.
   a. 28, 70   **b.** 30, 40, 50   **c.** 18, 36, 66

2. Use prime factorizations to find the LCM of 168 and 140 and to tell how many times each number divides into the LCM.

Practice Problem Answers

1. a. 140   **b.** 600   **c.** 396
2. 840; 168 divides into 840 5 times, 140 divides into 840 6 times.

Exercises 2.3

List the first six multiples of each number below.

1. 5   **2.** 6   **3.** 10

4. 15   **5.** 25   **6.** 30

Find the LCM for each of the following sets of counting numbers. See Examples 1 through 5.

7. 8, 12   **8.** 3, 5, 7   **9.** 4, 6, 9   **10.** 3, 5, 9

11. 2, 5, 11   **12.** 4, 14, 18   **13.** 6, 10, 12   **14.** 6, 8, 27

15. 25, 40   **16.** 40, 75   **17.** 28, 98   **18.** 30, 75

19. 30, 80   **20.** 16, 28   **21.** 25, 100   **22.** 20, 50

23. 35, 100   **24.** 144, 216   **25.** 36, 42   **26.** 40, 100

27. 2, 4, 8   **28.** 10, 15, 35   **29.** 6, 12, 15   **30.** 8, 10, 120

31. 6, 15, 80   **32.** 13, 26, 169   **33.** 45, 125, 150   **34.** 34, 51, 54

35. 33, 66, 121   **36.** 36, 54, 72   **37.** 45, 145, 290   **38.** 54, 81, 108

39. 45, 75, 135   **40.** 35, 40, 72   **41.** 8, 13, 15   **42.** 25, 35, 49

43. 10, 20, 30, 40   **44.** 15, 25, 30, 40

45. 8, 10, 15   **46.** 6, 15, 30   **47.** 10, 15, 24   **48.** 8, 10, 120

49. 6, 18, 27, 45   **50.** 12, 95, 228   **51.** 45, 63, 98   **52.** 40, 56, 196

53. 99, 143, 363   **54.** 125, 135, 225

55. Security Guards: Three security guards meet at the front gate for coffee before they walk around inspecting buildings at a manufacturing plant. The guards take 15, 20, and 30 minutes, respectively, for the inspection trip.

  a. If they start at the same time, how many minutes will it take for them to meet again at the front gate?

  b. How many trips will each guard have made?

56. Astronauts: Two astronauts miss their first connection in space.

  a. If one astronaut circles Earth every 15 hours and the other every 18 hours, in how many hours will they rendezvous again at the same time.

  b. After meeting, how many orbits will each astronaut make before they meet again?

57. Truck Drivers: Three truck drivers have dinner together whenever all three are at the routing station at the same time. The first driver's route takes 6 days, the second driver's route takes 8 days, and the third driver's route takes 12 days.

  a. How frequently do the three drivers have dinner together?

  b. How frequently do the first two drivers meet?

58. Lawn Mowing: Three neighbors mow their lawns at different intervals during the summer months. The first one mows every 5 days, the second every 7 days, and the third every 10 days.

  a. How frequently do they mow their lawns on the same day?

  b. How many times does each neighbor mow his lawn in between the times when they all mow together?

59. Ships: Four ships leave port on the same day. They take 12, 15, 18, and 30 days respectively to sail their routes and reload cargo. How frequently do the four ships leave port on the same day?

60. Saleswomen: Four women travel continuously selling text books for the same publishing company. It takes them 8 days, 12 days, 18 days, and 15 days respectively to visit schools in their sales regions.

  a. If they all leave on the same day, how many days will it be until they meet again at the home office?

  b. How many sales trips will each have made in this time?

61. Swimming: Three swimmers decide to swim laps together, and they will quit when they reach the starting end of the pool together. The first swimmer can swim a lap in 35 seconds, the second will take 40 seconds, and the third takes 42 seconds.

    a. How many seconds will it take before they quit?

    b. How many laps will each swimmer swim in that interval?

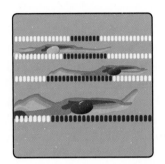

62. Fruit: A fruit production company has three packaging facilities, each of which uses different-sized boxes as follows: 24 pieces/box, 36 pieces/box, and 45 pieces/box.

    a. Assuming that the truck provides the same quantity of uniformly-sized pieces of fruit to all three packaging facilities, what is the minimum number of pieces of fruit that will be delivered so that no fruit will be left over?

    b. How many boxes will each facility package?

63. Bus Routes: Three bus routes originate from the same station at the same time at the beginning of the day. The round trip times for each of the routes is 1 hour, 2 hours, and 3 hours respectively.

    a. How many hours will it take for all three buses to return to the terminal at the same time?

    b. If they all leave at 1 PM, what time will they be back together?

64. Satellites: Two satellites are in polar orbit around the earth. One takes 8 hours and the other 15 hours to complete its orbit.

    a. Assuming that they start over the North Pole, how long will it take before they are both over the North Pole again?

    b. How many orbits does each satellite make during this time?

65. Cars: Three cars are being driven around the same road circuit. Car 1 can drive it in 10 min, car 2 takes 20 min, and car 3 takes 40 min.

    a. Assuming that all the cars start at the same time, how long will it take for them to reach the starting point together?

    b. How long will it take for the first two cars to meet?

    c. How long will it take for cars 1 and 3 to meet?

66. Clocks: Two analog clocks are sitting next to each other. The first clock keeps perfect time, where the minute hand takes exactly 60 min to travel completely around the dial. The second clock runs fast, and the minute hand will make one complete revolution in 55 min.

    a. Assuming that both clocks are started so that the minute hands are at 12, how many minutes will it take until both minute hands return to 12 at the same time?

    b. How many hours does this represent?

67. Planets: A certain star has four planets, all moving in the same plane. The first one takes 4 years to go around the star, the second takes 6 years, the third 10 years, and the fourth takes 15 years. Assuming that they are initially aligned, how many years will it take until they are aligned again?

68. College Courses: A small college offers some courses at intervals greater than one year. In one department of this college, four courses are each offered at different intervals as follows. One course is offered every 2 years, a second course is held every 3 years, a third course is given every 4 years, and a special elective is offered only once every 6 years. How long will it take for these courses to be offered on the same year after the previous time this happened?

69. Food Manufacturer: A certain food manufacturer packages 6 frozen toaster waffles in a box. If a family eats 8 waffles at a given meal, what is the minimum number of boxes to purchase if the family desires to completely empty all of the boxes before making the next purchase? (**Note:** Solving this problem takes two steps. First compute the total number of waffles consumed, and then divide this number by the number of waffles in a box.)

70. Tour Bus: Four tour buses depart the terminal on the same day. The itinerary of the first bus takes 6 days, for the second bus it takes 12 days, for the third bus it takes 14 days, and the fourth bus takes 21 days. Assuming that another tour begins the same day that the bus returns to the terminal, how many days will it take until all four buses return to the terminal on the same day?

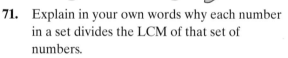

Writing & Thinking

71. Explain in your own words why each number in a set divides the LCM of that set of numbers.

72. Explain in your own words why the LCM of a set of numbers is greater than or equal to each number in the set.

Collaborative Learning

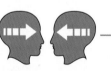

73. With the class separated into teams of two to four students, each team is to write one or two paragraphs explaining what topic they have found most interesting so far in this chapter and why. Each team leader is to read the paragraph(s), with classroom discussion to follow.

2.4 Addition and Subtraction with Fractions

Objective A Adding Fractions with the Same Denominator

Objectives

A Add fractions with the same denominator.

B Add fractions with different denominators.

C Subtract fractions with the same denominator.

D Subtract fractions with different denominators.

To add two or more fractions with the same denominator, we can think of this denominator as the "name" of each fraction. The sum has this common name. Just as 3 *oranges* plus 2 *oranges* gives a total of 5 *oranges*, 3 *eighths* plus 2 *eighths* gives a total of 5 *eighths*. This denominator is called a **common denominator**. The following figure illustrates how the sum of the two fractions $\frac{3}{8}$ and $\frac{2}{8}$ might be diagrammed.

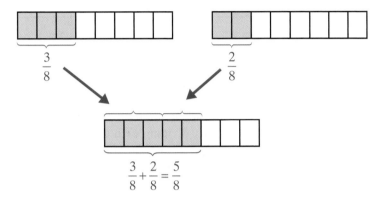

Figure 1

To Add Fractions with the Same Denominator

1. Add the numerators.
2. Keep the common denominator.
3. Reduce, if possible.

$$\frac{a}{b} + \frac{c}{b} = \frac{a+c}{b}, b \neq 0$$

Example 1

Adding Fractions with the Same Denominator

$$\frac{1}{5} + \frac{2}{5} = \frac{1+2}{5} = \frac{3}{5}$$

Example 2

Adding Fractions with the Same Denominator

$$\frac{1}{11} + \frac{5}{11} + \frac{3}{11} = \frac{1+5+3}{11} = \frac{9}{11}$$

Now work margin exercises 1 and 2.

Find the sum.

1. $\frac{1}{7} + \frac{3}{7}$

2. $\frac{1}{9} + \frac{4}{9} + \frac{3}{9}$

Objective B **Adding Fractions with Different Denominators**

We know that fractions to be added will not always have the same denominator. Figure 2 illustrates how the two fractions $\frac{1}{3}$ and $\frac{2}{5}$ might be diagrammed.

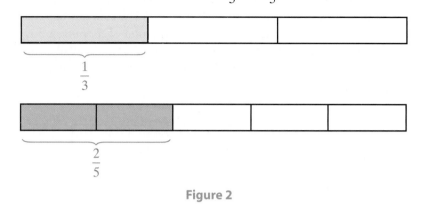

Figure 2

To find the result of adding these two fractions, a common unit (that is, a common denominator) is needed. By dividing each third into five parts and each fifth into three parts, we find that a common unit is fifteenths. Now the two fractions can be added as shown in Figure 3.

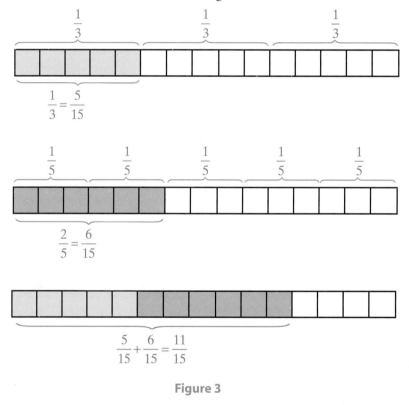

Figure 3

Of course, drawing diagrams every time two or more fractions are to be added would be difficult and time-consuming. A better way is to find the least common denominator of all the fractions, change each fraction to an equivalent fraction with this common denominator, and then add. In Figure 3, the least common denominator was the product. In general, the **least common denominator (LCD)** is the least common multiple (LCM) of the denominators (see Section 2.3).

To Add Fractions with Different Denominators

1. Find the least common denominator (LCD).
2. Change each fraction into an equivalent fraction with that denominator.
3. Add the new fractions.
4. Reduce, if possible.

Example 3

Adding Fractions with Different Denominators

Find the sum: $\dfrac{3}{8} + \dfrac{13}{12}$

Solution

Step 1: Find the LCD. Remember that the least common denominator (LCD) is the least common multiple (LCM) of the denominators.

$$\left. \begin{array}{l} 8 = 2 \cdot 2 \cdot 2 \\ 12 = 2 \cdot 2 \cdot 3 \end{array} \right\} \quad \text{LCD} = 2 \cdot 2 \cdot 2 \cdot 3 = 24$$

Note: You might not need to use prime factorizations to find the LCD. If the denominators are numbers that are familiar to you, then you might be able to find the LCD simply by inspection.

Step 2: Find fractions equal to $\dfrac{3}{8}$ and $\dfrac{13}{12}$ with denominator 24.

$$\frac{3}{8} = \frac{3}{8} \cdot 1 = \frac{3}{8} \cdot \frac{3}{3} = \frac{9}{24} \qquad \text{Multiply by } \frac{3}{3} \text{ because } 8 \cdot 3 = 24.$$

$$\frac{13}{12} = \frac{13}{12} \cdot 1 = \frac{13}{12} \cdot \frac{2}{2} = \frac{26}{24} \qquad \text{Multiply by } \frac{2}{2} \text{ because } 12 \cdot 2 = 24.$$

Step 3: Add.

$$\frac{3}{8} + \frac{13}{12} = \frac{3}{8} \cdot \frac{3}{3} + \frac{13}{12} \cdot \frac{2}{2} = \frac{9}{24} + \frac{26}{24} = \frac{9 + 26}{24} = \frac{35}{24}$$

Step 4: The fraction $\dfrac{35}{24}$ is in lowest terms because 35 and 24 have only 1 as a common factor.

Example 4

Adding Fractions with Different Denominators

Find the sum: $\dfrac{7}{45} + \dfrac{7}{36}$

Add and reduce to lowest
terms.

3. $\dfrac{1}{6} + \dfrac{3}{10}$

4. $\dfrac{2}{21} + \dfrac{9}{28}$

Solution

Step 1: Find the LCD.

$$\left.\begin{array}{l} 45 = 3 \cdot 3 \cdot 5 \\ 36 = 2 \cdot 2 \cdot 3 \cdot 3 \end{array}\right\} \begin{array}{l} \text{LCD} = 2 \cdot 2 \cdot 3 \cdot 3 \cdot 5 = 180 \\ \phantom{\text{LCD}} = (3 \cdot 3 \cdot 5)(2 \cdot 2) = 45 \cdot 4 \\ \phantom{\text{LCD}} = (2 \cdot 2 \cdot 3 \cdot 3)(5) = 36 \cdot 5 \end{array}$$

Step 2: Steps 2, 3, and 4 from Example 3 can be written in one step.

$$\frac{7}{45} + \frac{7}{36} = \frac{7}{45} \cdot \frac{4}{4} + \frac{7}{36} \cdot \frac{5}{5}$$
$$= \frac{28}{180} + \frac{35}{180} = \frac{63}{180}$$
$$= \frac{3 \cdot 3 \cdot 7}{2 \cdot 2 \cdot 3 \cdot 3 \cdot 5} = \frac{7}{20}$$

Note that, in adding fractions, we also may choose to write them vertically. The process is the same.

$$\begin{array}{l} \dfrac{7}{45} = \dfrac{7}{45} \cdot \dfrac{4}{4} = \dfrac{28}{180} \\[2mm] + \dfrac{7}{36} = \dfrac{7}{36} \cdot \dfrac{5}{5} = \dfrac{35}{180} \\ \hline \end{array}$$
$$\frac{63}{180} = \frac{3 \cdot 3 \cdot 7}{2 \cdot 2 \cdot 3 \cdot 3 \cdot 5} = \frac{7}{20}$$

Now work margin exercises 3 and 4.

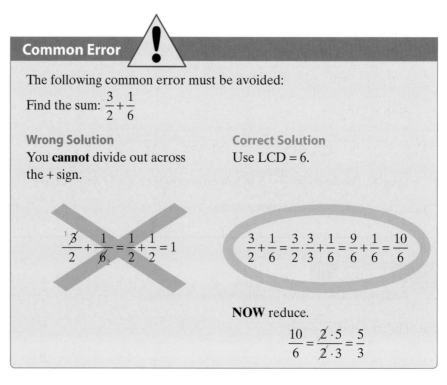

Common Error

The following common error must be avoided:

Find the sum: $\dfrac{3}{2} + \dfrac{1}{6}$

Wrong Solution
You **cannot** divide out across
the + sign.

$$\frac{{}^{1}\cancel{3}}{2} + \frac{1}{\cancel{6}_{2}} = \frac{1}{2} + \frac{1}{2} = 1$$

Correct Solution
Use LCD = 6.

$$\frac{3}{2} + \frac{1}{6} = \frac{3}{2} \cdot \frac{3}{3} + \frac{1}{6} = \frac{9}{6} + \frac{1}{6} = \frac{10}{6}$$

NOW reduce.

$$\frac{10}{6} = \frac{2 \cdot 5}{2 \cdot 3} = \frac{5}{3}$$

Example 5

Application of Adding Fractions with Different Denominators

Rachel is mailing two letters at the post office. One letter weighs $\frac{1}{3}$ ounce and the other weighs $\frac{3}{7}$ ounce. What is the combined weight of the letters?

Solution

The LCD of 3 and 7 is 21. Using this information we can add the two fractions.

$$\frac{1}{3} + \frac{3}{7} = \frac{1}{3} \cdot \frac{7}{7} + \frac{3}{7} \cdot \frac{3}{3}$$

$$= \frac{7}{21} + \frac{9}{21}$$

$$= \frac{16}{21}$$

Thus the total weight of the two letters is $\frac{16}{21}$ ounce.

Now work margin exercise 5.

Example 6

Application of Adding Fractions with Different Denominators

If Keith's total income for the year was $36,000, and he spent $\frac{1}{4}$ of his income on rent and $\frac{1}{12}$ of his income on his car, what total amount did he spend on these two items?

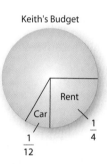

Keith's Budget

Rent $\frac{1}{4}$

Car $\frac{1}{12}$

Solution

We can add the two fractions, and then multiply the sum by $36,000. (Or, we can multiply each fraction by $36,000, and then add the results. We will get the same answer either way.) The LCD is 12.

$$\frac{1}{4} + \frac{1}{12} = \frac{1}{4} \cdot \frac{3}{3} + \frac{1}{12} = \frac{3}{12} + \frac{1}{12} = \frac{4}{12} = \frac{\cancel{4} \cdot 1}{\cancel{4} \cdot 3} = \frac{1}{3}$$

Now multiply $\frac{1}{3}$ times $36,000.

$$\frac{1}{\cancel{3}} \cdot \overset{12,000}{\cancel{36,000}} = 12,000$$

Keith spent a total of $12,000 on rent and his car.

Now work margin exercise 6.

5. Catherine is at the grocery store buying fruit. She buys one grapefruit that weighs $\frac{3}{5}$ lb, one pomegranate that weighs $\frac{1}{2}$ lb, and a pear that weighs $\frac{3}{8}$ lb. What is the total weight of fruit that she will be purchasing?

6. Fred's total income for the year was $30,000, and he spent $\frac{1}{3}$ of his income on traveling and $\frac{1}{10}$ of his income on entertainment. What total amount did he spend on these two items?

Objective C **Subtracting Fractions with the Same Denominator**

Finding the difference between two fractions with a common denominator is similar to finding the sum. The numerators are simply subtracted instead of added.

Figure 4 shows how the difference between $\dfrac{4}{5}$ and $\dfrac{1}{5}$ might be diagrammed.

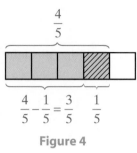

Figure 4

Find the difference.

7. $\dfrac{11}{15} - \dfrac{8}{15}$

8. $\dfrac{7}{12} - \dfrac{5}{12}$

To Subtract Fractions with the Same Denominator

1. Subtract the numerators.
2. Keep the common denominator. $\dfrac{a}{b} - \dfrac{c}{b} = \dfrac{a-c}{b}, \ b \neq 0$
3. Reduce, if possible.

Example 7

Subtracting Fractions with the Same Denominator

Find the difference: $\dfrac{9}{10} - \dfrac{7}{10}$

Solution

$$\frac{9}{10} - \frac{7}{10} = \frac{9-7}{10} = \frac{2}{10} = \frac{\cancel{2} \cdot 1}{\cancel{2} \cdot 5} = \frac{1}{5}$$

The difference is reduced just as any fraction is reduced.

Example 8

Subtracting Fractions with the Same Denominator

Find the difference: $\dfrac{7}{8} - \dfrac{3}{8}$

Solution

$$\frac{7}{8} - \frac{3}{8} = \frac{7-3}{8} = \frac{4}{8} = \frac{\cancel{4} \cdot 1}{\cancel{4} \cdot 2} = \frac{1}{2}$$

The difference is reduced just as any fraction is reduced.

Now work margin exercises 7 and 8.

Subtracting Fractions with Different Denominators

To Subtract Fractions with Different Denominators

1. Find the least common denominator (LCD).
2. Change each fraction into an equivalent fraction with that denominator.
3. Subtract the new fractions.
4. Reduce, if possible.

Example 9

Subtracting Fractions with Different Denominators

Find the difference: $\dfrac{9}{10} - \dfrac{2}{15}$

Solution

Step 1: Find the LCD.

$$\left.\begin{array}{l} 10 = 2 \cdot 5 \\ 15 = 3 \cdot 5 \end{array}\right\} \text{LCD} = 2 \cdot 3 \cdot 5 = 30$$

Step 2: Find equivalent fractions with denominator 30.

$$\frac{9}{10} = \frac{9}{10} \cdot \frac{3}{3} = \frac{27}{30}$$

$$\frac{2}{15} = \frac{2}{15} \cdot \frac{2}{2} = \frac{4}{30}$$

Step 3: Subtract.

$$\frac{9}{10} - \frac{2}{15} = \frac{27}{30} - \frac{4}{30} = \frac{23}{30}$$

Example 10

Subtracting Fractions with Different Denominators

Find the difference: $\dfrac{7}{20} - \dfrac{3}{28}$

Solution

Step 1: Find the LCD.

$$\left.\begin{array}{l} 20 = 2 \cdot 2 \cdot 5 \\ 28 = 2 \cdot 2 \cdot 7 \end{array}\right\} \text{LCD} = 2 \cdot 2 \cdot 5 \cdot 7 = 140$$

Find the difference.

9. $\dfrac{7}{11} - \dfrac{3}{5}$

10. $\dfrac{7}{12} - \dfrac{2}{15}$

11. Find the difference.

$$\dfrac{4}{15} - \dfrac{3}{25}$$

Step 2: Steps 2 and 3 from Example 8 can be written in one step.

$$\frac{7}{20} - \frac{3}{28} = \frac{7}{20} \cdot \frac{7}{7} - \frac{3}{28} \cdot \frac{5}{5} = \frac{49}{140} - \frac{15}{140}$$

$$= \frac{49 - 15}{140} = \frac{34}{140} = \frac{\cancel{2} \cdot 17}{\cancel{2} \cdot 70} = \frac{17}{70}$$

Or, writing the fractions vertically,

$$\frac{7}{20} = \frac{7}{20} \cdot \frac{7}{7} = \frac{49}{140}$$

$$-\frac{3}{28} = \frac{3}{28} \cdot \frac{5}{5} = \frac{15}{140}$$

$$\frac{34}{140} = \frac{\cancel{2} \cdot 17}{\cancel{2} \cdot 70} = \frac{17}{70}.$$

Now work margin exercises 9 *and* 10.

Completion Example 11

Subtracting Fractions with Different Denominators

Find the difference: $\dfrac{11}{12} - \dfrac{13}{20}$.

Solution

Step 1: Find the LCD.

$$\left.\begin{array}{l} 12 = 2 \cdot 2 \cdot 3 \\ 20 = 2 \cdot 2 \cdot 5 \end{array}\right\} \quad \text{LCD} = \underline{\hspace{2cm}}$$

Step 2:

$$\frac{11}{12} - \frac{13}{20} = \frac{11}{12} \cdot \underline{\hspace{0.5cm}} - \frac{13}{20} \cdot \underline{\hspace{0.5cm}} = \underline{\hspace{0.5cm}} - \underline{\hspace{0.5cm}}$$

$$= \underline{\hspace{0.5cm}} = \underline{\hspace{0.5cm}} = \underline{\hspace{0.5cm}} = \underline{\hspace{0.5cm}}$$

Now work margin exercise 11.

Completion Example Answers

11. Step 1: LCD $= 2 \cdot 2 \cdot 3 \cdot 5 = 60$

Step 2: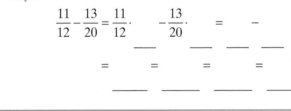

$$= \frac{11}{12} \cdot \frac{5}{5} - \frac{13}{20} \cdot \frac{3}{3} = \frac{55}{60} - \frac{39}{60}$$

$$= \frac{55 - 39}{60} = \frac{16}{60} = \frac{\cancel{4} \cdot 4}{\cancel{4} \cdot 15} = \frac{4}{15}$$

Example 12

Application of Subtracting Fractions with Different Denominators

The Narragansett Grays baseball team lost 90 games in one season. If $\frac{1}{5}$ of their losses were by 1 or 2 runs and $\frac{4}{9}$ of their losses were by 3 or fewer runs, what fraction of their losses were by exactly 3 runs?

Solution

The losses that were by 3 or fewer runs include those that were by 1 or 2 runs. So to find the fraction that were by exactly 3 runs, we need to subtract $\frac{1}{5}$ from $\frac{4}{9}$. (The LCD is 45.)

$$\frac{4}{9} - \frac{1}{5} = \frac{4}{9} \cdot \frac{5}{5} - \frac{1}{5} \cdot \frac{9}{9} = \frac{20}{45} - \frac{9}{45} = \frac{11}{45}$$

Thus the Grays lost $\frac{11}{45}$ of their games by exactly 3 runs.

Now work margin exercise **12.**

12. The Cavalier baseball team lost 30 games in one season. If $\frac{1}{3}$ of their losses were by more than 5 runs, and $\frac{2}{5}$ of their losses were by more than 4 runs, what fraction of their losses were by exactly 5 runs?

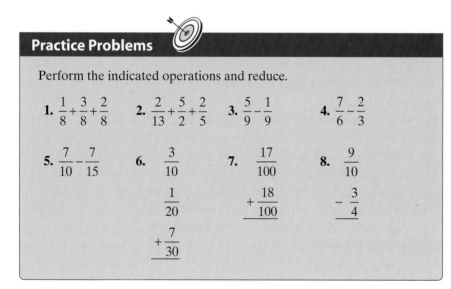

Practice Problems

Perform the indicated operations and reduce.

1. $\frac{1}{8} + \frac{3}{8} + \frac{2}{8}$ **2.** $\frac{2}{13} + \frac{5}{2} + \frac{2}{5}$ **3.** $\frac{5}{9} - \frac{1}{9}$ **4.** $\frac{7}{6} - \frac{2}{3}$

5. $\frac{7}{10} - \frac{7}{15}$ **6.** $\begin{array}{r} \frac{3}{10} \\ \frac{1}{20} \\ + \frac{7}{30} \end{array}$ **7.** $\begin{array}{r} \frac{17}{100} \\ + \frac{18}{100} \end{array}$ **8.** $\begin{array}{r} \frac{9}{10} \\ - \frac{3}{4} \end{array}$

Practice Problem Answers

1. $\frac{3}{4}$ **2.** $\frac{397}{130}$ **3.** $\frac{4}{9}$ **4.** $\frac{1}{2}$

5. $\frac{7}{30}$ **6.** $\frac{7}{12}$ **7.** $\frac{7}{20}$ **8.** $\frac{3}{20}$

Exercises 2.4

Find the indicated sums and reduce if possible. See Examples 1 through 4.

1. $\dfrac{3}{14} + \dfrac{2}{14}$

2. $\dfrac{3}{4} + \dfrac{3}{4}$

3. $\dfrac{7}{5} + \dfrac{3}{5}$

4. $\dfrac{7}{9} + \dfrac{8}{9}$

5. $\dfrac{7}{90} + \dfrac{37}{90} + \dfrac{21}{90}$

6. $\dfrac{14}{32} + \dfrac{7}{32} + \dfrac{1}{32}$

7. $\dfrac{5}{8} + \dfrac{3}{4}$

8. $\dfrac{1}{4} + \dfrac{5}{6}$

9. $\dfrac{2}{7} + \dfrac{1}{8}$

10. $\dfrac{7}{9} + \dfrac{3}{5}$

11. $\dfrac{5}{6} + \dfrac{2}{3}$

12. $\dfrac{2}{5} + \dfrac{7}{20}$

13. $\dfrac{1}{12} + \dfrac{2}{3} + \dfrac{1}{4}$

14. $\dfrac{2}{5} + \dfrac{3}{10} + \dfrac{3}{20}$

15. $\dfrac{2}{7} + \dfrac{4}{21} + \dfrac{1}{3}$

16. $\dfrac{2}{39} + \dfrac{1}{3} + \dfrac{4}{13}$

17. $\dfrac{1}{27} + \dfrac{4}{18} + \dfrac{1}{6}$

18. $\dfrac{1}{8} + \dfrac{1}{12} + \dfrac{1}{9}$

19. $\dfrac{2}{3} + \dfrac{3}{4} + \dfrac{9}{14}$

20. $\dfrac{1}{5} + \dfrac{2}{15} + \dfrac{1}{6}$

21. $\dfrac{1}{5} + \dfrac{7}{40} + \dfrac{1}{4}$

22. $\dfrac{1}{4} + \dfrac{1}{20} + \dfrac{8}{15}$

23. $\dfrac{5}{8} + \dfrac{4}{27} + \dfrac{1}{24}$

24. $\dfrac{72}{105} + \dfrac{2}{45} + \dfrac{15}{21}$

25. $\dfrac{3}{10} + \dfrac{1}{100} + \dfrac{7}{1000}$

26. $\dfrac{0}{27} + \dfrac{0}{16} + \dfrac{1}{5}$

27. $\dfrac{17}{1000} + \dfrac{1}{100} + \dfrac{1}{10,000}$

28. $8 + \dfrac{1}{10} + \dfrac{9}{100} + \dfrac{1}{1000}$

29. $\dfrac{7}{10} + \dfrac{5}{100} + \dfrac{3}{1000}$

30. $\dfrac{1}{4} + \dfrac{1}{8} + \dfrac{7}{100}$

31. $\dfrac{11}{100} + \dfrac{1}{2} + \dfrac{3}{1000}$

32. $5 + \dfrac{1}{10} + \dfrac{3}{100} + \dfrac{4}{1000}$

33.
$\dfrac{3}{4}$
$\dfrac{1}{2}$
$+\dfrac{5}{12}$

34.
$\dfrac{7}{8}$
$\dfrac{2}{3}$
$+\dfrac{1}{9}$

35.
$\dfrac{3}{20}$
$\dfrac{1}{100}$
$+\dfrac{3}{100}$

36.
$\dfrac{7}{12}$
$\dfrac{1}{9}$
$+\dfrac{2}{3}$

37.
$\dfrac{9}{16}$
$\dfrac{5}{48}$
$+\dfrac{3}{32}$

38. $\dfrac{4}{7} - \dfrac{1}{7}$

39. $\dfrac{9}{10} - \dfrac{3}{10}$

40. $\dfrac{5}{8} - \dfrac{1}{8}$

41. $\dfrac{11}{12} - \dfrac{7}{12}$

42. $\dfrac{13}{15} - \dfrac{4}{15}$

43. $\dfrac{5}{6} - \dfrac{1}{3}$

44. $\dfrac{11}{15} - \dfrac{3}{10}$

45. $\dfrac{3}{4} - \dfrac{2}{3}$

46. $\dfrac{15}{16} - \dfrac{21}{32}$

47. $\dfrac{5}{4} - \dfrac{3}{5}$

48. $\dfrac{14}{27} - \dfrac{7}{18}$

49. $\dfrac{8}{45} - \dfrac{11}{72}$

50. $\dfrac{5}{36} - \dfrac{1}{45}$

51. $\dfrac{4}{1} - \dfrac{5}{8}$

52. $\dfrac{9}{10} - \dfrac{3}{100}$

53. $\dfrac{76}{100} - \dfrac{7}{10}$

54. $\begin{array}{r} \dfrac{54}{100} \\[2mm] -\dfrac{5}{10} \\ \hline \end{array}$

55. $\begin{array}{r} \dfrac{31}{40} \\[2mm] -\dfrac{5}{8} \\ \hline \end{array}$

56. $\begin{array}{r} \dfrac{20}{35} \\[2mm] -\dfrac{24}{42} \\ \hline \end{array}$

57. $\begin{array}{r} \dfrac{1}{10} \\[2mm] -\dfrac{8}{100} \\ \hline \end{array}$

58. Postage: Three pieces of mail weigh $\dfrac{1}{2}$ ounce, $\dfrac{1}{5}$ ounce, and $\dfrac{3}{10}$ ounce. What is the total weight of the letters?

$\dfrac{1}{2}$ ounce

$\dfrac{1}{5}$ ounce

$\dfrac{3}{10}$ ounce

59. Machine Shop: A machinist drills four holes in a straight line. Each hole has a diameter of $\dfrac{1}{10}$ inch and there is $\dfrac{1}{4}$ inch between the holes. What is the distance between the outer edges of the first and last holes?

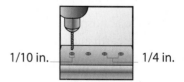

1/10 in.   1/4 in.

60. Measurement: Using a microscope, a scientist measures the diameters of three hairs to be $\dfrac{1}{1000}$ inch, $\dfrac{3}{1000}$ inch, and $\dfrac{1}{100}$ inch. What is the total of these three diameters?

61. Stationery: A notebook contains a piece of cardboard as a back cover that is $\dfrac{1}{16}$ inch thick. It has a front cover that is $\dfrac{1}{4}$ inch thick. All together, the sheets of paper between the front and back are $\dfrac{3}{10}$ inch thick. What is the total thickness of the notebook?

62. Carpentry: A carpenter is installing baseboard and toe molding. If the baseboard is $\frac{3}{8}$ inch thick and the toe molding (to be put in front of the baseboard) is $\frac{1}{4}$ inch thick, what is the total thickness of the two trim pieces?

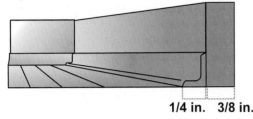

1/4 in.   3/8 in.

63. Investment: Beth's investment strategy is to put $\frac{1}{6}$ of her paycheck into a savings account and another $\frac{1}{9}$ into a retirement account.

a. What fraction of her salary does Beth invest each month?

b. If she maintains this strategy for 24 paychecks and receives $900 per paycheck, how much money will she have saved?

64. Ancient Number Systems: In ancient Egypt, fractions were described as sums with a numerator of 1. What fraction would be described by the sum: $\frac{1}{20} + \frac{1}{124} + \frac{1}{155}$?

65. Pipe: A $\frac{7}{8}$ inch pipe is to be shortened to $\frac{7}{12}$ inch. How much must be removed?

66. Road Maintenance: Near the end of the snow season, the road salt supply for a small college had dwindled down to $\frac{7}{10}$ ton. When the next snow storm came, $\frac{1}{2}$ ton of salt was used for the roads. How much road salt was left?

67. Gas Stations: Mark has driven to a national park with no gas stations, and he wants to drive around some before leaving the park. He knows he can safely make it the nearest gas station on $\frac{5}{12}$ of a tank of gas. If the tank is currently $\frac{13}{20}$ full, what fraction of a tank of gasoline does he have to use for touring the park?

68. Astronomy: About $\frac{1}{2}$ of all incoming solar radiation is absorbed by the earth, $\frac{1}{5}$ is absorbed by the atmosphere, and $\frac{1}{20}$ is scattered by the atmosphere. The rest is reflected by the earth and clouds.

a. What fraction of solar radiation is absorbed or scattered?

b. What fraction of solar radiation is reflected by the earth and clouds?

69. Cooking: A recipe calls for the following spices: $\frac{1}{2}$ teaspoon of turmeric, $\frac{1}{4}$ teaspoon of ginger, and $\frac{1}{8}$ teaspoon of cumin. What is the total quantity of these three spices?

70. Budgeting: Sam's income is \$3300 a month and he plans to budget $\frac{1}{3}$ of his income for rent and $\frac{1}{10}$ of his income for food.

a. What fraction of his income does he plan to spend on these two items?

b. What amount of money does he plan to spend each month on these two items?

Sam's Budget

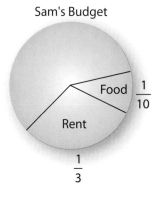

71. Computers: Of the personal computers (PCs) in use worldwide, the U.S. has $\frac{272}{1000}$, Japan $\frac{84}{1000}$, China $\frac{65}{1000}$, and Germany $\frac{56}{1000}$. **Source:** *2006 World Almanac*

a. What total fraction of the world's PCs are used in these four countries?

b. What fraction is used in the rest of the world?

PCs Worldwide

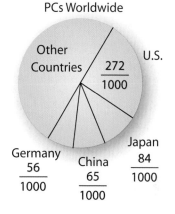

72. Pie: Jenny has $\frac{3}{4}$ of an apple pie left over from a party last night. Her roommates found it and cut themselves three unequal sized pieces in the following amounts: $\frac{1}{3}$ of a pie, $\frac{1}{4}$ of a pie, and $\frac{1}{6}$ of a pie.

a. What fraction of a full pie did Jenny's roommates take?

b. What fraction of the pie is left over?

73. Rod: A $\frac{5}{8}$ inch rod has a $\frac{1}{20}$ inch coating around it. What is the total diameter of the coated rod? Remember that the coating is on all sides of the rod.

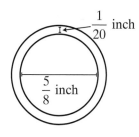

74. Demographics: According to a 2009 demographic study of San Francisco, CA, the fractions of the three largest racial minorities are as follows: $\frac{1}{15}$ African American, $\frac{3}{10}$ Asian, and $\frac{3}{20}$ Hispanic. **Source:** U.S. Census Bureau

a. What fraction of San Francisco's population belong to these three racial groups?

b. What fraction of this city's population belong to all other racial groups?

75. Savings: John has a monthly income of \$3,000. $\frac{1}{10}$ of it goes to college savings, $\frac{2}{15}$ to general savings, and $\frac{1}{12}$ to retirement.

a. What fraction of John's income is being saved?

b. How much money is being saved each month?

c. What is the total amount saved in a 12-month year?

Collaborative Learning

76. With the class separated into teams of two to four students, each team is to write one or two paragraphs explaining what topic in this chapter they found to be most difficult and what techniques they used to learn the related material. Each team leader is to read the paragraph(s) with classroom discussion to follow.

2.5 Addition and Subtraction with Mixed Numbers

Objectives

A Add mixed numbers.

B Subtract mixed numbers.

C Subtract mixed numbers with borrowing.

Objective A **Addition with Mixed Numbers**

Because a mixed number itself represents the sum of a whole number and a proper fraction, two or more mixed numbers can be added by adding the whole number parts and fraction parts separately.

To Add Mixed Numbers

1. Add the fraction parts.
2. Add the whole numbers.
3. Write the answer as a mixed number with the fraction part less than 1.

Example 1

Adding Mixed Numbers

Find the sum: $5\frac{2}{9} + 8\frac{5}{9}$

Solution

We can write each number as a sum and then use the commutative and associative properties of addition to treat the whole numbers and fraction parts separately.

$$5\frac{2}{9} + 8\frac{5}{9} = 5 + \frac{2}{9} + 8 + \frac{5}{9}$$

$$= (5+8) + \left(\frac{2}{9} + \frac{5}{9}\right)$$

$$= 13 + \frac{7}{9} = 13\frac{7}{9}$$

Or, vertically,

$$5\frac{2}{9}$$
$$+ 8\frac{5}{9}$$
$$\overline{\quad 13\frac{7}{9}}$$

Now work margin exercise 1.

1. Find the sum: $8\frac{1}{5} + 3\frac{3}{5}$

2. Find the sum: $5\dfrac{1}{2}+3\dfrac{1}{5}$.

Example 2

Adding Mixed Numbers

Find the sum: $35\dfrac{1}{6}+22\dfrac{7}{18}$

Solution

In this case, the fractions do not have the same denominator.

$$35\dfrac{1}{6}+22\dfrac{7}{18}=35+\dfrac{1}{6}+22+\dfrac{7}{18}$$

$$=\left(35+22\right)+\left(\dfrac{1}{6}\cdot\dfrac{3}{3}+\dfrac{7}{18}\right)\qquad \text{LCD}=18$$

$$=57+\left(\dfrac{3}{18}+\dfrac{7}{18}\right)$$

$$=57\dfrac{10}{18}=57\dfrac{5}{9}\qquad\qquad \text{Reduce the fraction part.}$$

Now work margin exercise **2.**

If the sum of the proper fractions is more than 1, rewrite this sum as a mixed number and add it to the sum of the whole numbers. Example 3 illustrates this situation.

3. Find the sum:
$$3\dfrac{4}{5}$$
$$8$$
$$+7\dfrac{2}{7}$$

Example 3

Adding Mixed Numbers

Find the sum: $5\dfrac{3}{4}+9\dfrac{3}{10}+2$

Solution

Since the whole number 2 has no fraction part, we must be careful to align the whole numbers if we write the numbers vertically.

$$5\dfrac{3}{4}\quad=5\dfrac{3}{4}\cdot\dfrac{5}{5}\quad=5\dfrac{15}{20}$$
$$9\dfrac{3}{10}\quad=9\dfrac{3}{10}\cdot\dfrac{2}{2}\quad=9\dfrac{6}{20}\qquad \text{LCD}=20$$
$$+2\quad\quad\ =2\qquad\qquad\ =2$$
$$16\dfrac{21}{20}=16+1\dfrac{1}{20}=17\dfrac{1}{20}$$

Fraction is greater than 1.   Change to a mixed number.

Now work margin exercise **3.**

Example 4

Application of Adding with Mixed Numbers

A triangle has sides measuring $3\frac{1}{3}$ meters, $6\frac{4}{5}$ meters, and $6\frac{14}{15}$ meters. Find the perimeter (total distance around) of the triangle.

Solution

We find the perimeter by adding the lengths of the three sides.

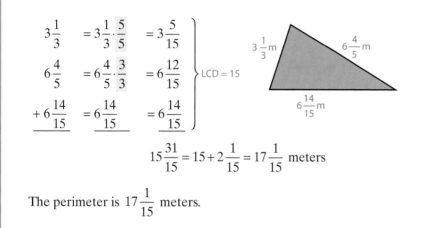

$$\left.\begin{array}{l} 3\frac{1}{3} \quad = 3\frac{1}{3}\cdot\frac{5}{5} \quad = 3\frac{5}{15} \\[2ex] 6\frac{4}{5} \quad = 6\frac{4}{5}\cdot\frac{3}{3} \quad = 6\frac{12}{15} \\[2ex] +6\frac{14}{15} \quad = 6\frac{14}{15} \quad = 6\frac{14}{15} \end{array}\right\} \text{LCD} = 15$$

$$15\frac{31}{15} = 15 + 2\frac{1}{15} = 17\frac{1}{15} \text{ meters}$$

The perimeter is $17\frac{1}{15}$ meters.

Now work margin exercise 4.

Objective B **Subtraction with Mixed Numbers**

Subtraction with mixed numbers is similar to addition in that we subtract the fraction parts and the whole numbers separately.

To Subtract Mixed Numbers

1. Subtract the fraction parts.
2. Subtract the whole numbers.

Example 5

Subtracting Mixed Numbers

Find the difference: $10\frac{3}{7} - 6\frac{2}{7}$

Solution

$$10\frac{3}{7} - 6\frac{2}{7} = (10 - 6) + \left(\frac{3}{7} - \frac{2}{7}\right)$$

$$= 4 + \frac{1}{7} = 4\frac{1}{7}$$

4. A triangle has sides measuring $4\frac{1}{4}$ meters, $6\frac{1}{4}$ meters, and $6\frac{1}{2}$ meters. Find the perimeter of (total distance around) the triangle.

Or, we can subtract vertically.

$$10\dfrac{3}{7}$$
$$-\ 6\dfrac{2}{7}$$
$$\overline{\ 4\dfrac{1}{7}}$$

Find the difference:

5. $6\dfrac{3}{5} - 2\dfrac{2}{5}$

6. $11\dfrac{1}{2} - 4\dfrac{2}{7}$

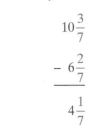

Example 6

Subtracting Mixed Numbers

Find the difference: $13\dfrac{4}{5} - 7\dfrac{1}{3}$

Solution

$$\left. \begin{array}{rcl} 13\dfrac{4}{5} &= 13\dfrac{4}{5}\cdot\dfrac{3}{3} &= 13\dfrac{12}{15}\\[2mm] -\ 7\dfrac{1}{3} &= -7\dfrac{1}{3}\cdot\dfrac{5}{5} &= -7\dfrac{5}{15} \end{array} \right\} \text{LCD} = 15$$
$$6\dfrac{7}{15}$$

Now work margin exercises 5 and 6.

Objective C **Subtracting Mixed Numbers with Borrowing**

If the fraction part being subtracted is larger than the fraction part of the first number, then we must rewrite the first number by "borrowing" 1 from the whole number as illustrated in the following examples.

Example 7

Subtracting Mixed Numbers with Borrowing

Find the difference: $6 - 2\dfrac{1}{3}$

Solution

Borrow 1 from 6 in the form of $\dfrac{3}{3}$ as follows:

$$6 \quad = \quad 5\dfrac{3}{3}$$
$$-2\dfrac{1}{3} \quad = \quad -2\dfrac{1}{3}$$
$$\overline{3\dfrac{2}{3}}$$

Example 8

Subtracting Mixed Numbers with Borrowing

Find the difference: $76\dfrac{5}{12} - 29\dfrac{13}{20}$

Solution

First, find the LCD and then borrow 1 if necessary.

$$\left.\begin{array}{l} 12 = 2\cdot2\cdot3 \\ 20 = 2\cdot2\cdot5 \end{array}\right\} \text{LCD} = 2\cdot2\cdot3\cdot5 = 60$$

$$76\dfrac{5}{12} \;=\; 76\dfrac{5}{12}\cdot\dfrac{5}{5} \;=\; 76\dfrac{25}{60} \;=\; 75\dfrac{85}{60} \qquad \text{Borrow } 1 = \dfrac{60}{60}.$$

$$-29\dfrac{13}{20} \;=\; -29\dfrac{13}{20}\cdot\dfrac{3}{3} \;=\; -29\dfrac{39}{60} \;=\; -29\dfrac{39}{60}$$

$$46\dfrac{46}{60} = 46\dfrac{23}{30}$$

Example 9

Subtracting Mixed Numbers with Borrowing

Find the difference: $4\dfrac{2}{9} - 1\dfrac{5}{9}$

Solution

$$4\dfrac{2}{9}$$
$$-1\dfrac{5}{9} \qquad \dfrac{5}{9} \text{ is larger than } \dfrac{2}{9}, \text{ so "borrow" 1 from 4.}$$

Rewrite $4\dfrac{2}{9}$ as $3 + 1 + \dfrac{2}{9} = 3 + 1\dfrac{2}{9} = 3\dfrac{11}{9}$.

$$4\dfrac{2}{9} = 3 + 1\dfrac{2}{9} = 3\dfrac{11}{9}$$
$$-1\dfrac{5}{9} = -1\dfrac{5}{9} = -1\dfrac{5}{9}$$
$$2\dfrac{6}{9} = 2\dfrac{2}{3}$$

Now work margin exercises 7 through 9.

Find the following differences:

7. $10\dfrac{1}{8} - 3\dfrac{5}{8}$

8. $21\dfrac{3}{4}$
   $-17\dfrac{7}{8}$

9. 16
   $-8\dfrac{2}{5}$

Practice Problems

Find the following sums.

1. $11\dfrac{1}{9}$

$+9\dfrac{2}{9}$

2. $\dfrac{7}{8}$

$1\dfrac{3}{5}$

$+7\dfrac{1}{3}$

3. $1\dfrac{2}{7}$

$8\dfrac{1}{2}$

$+2\dfrac{3}{5}$

Find the following differences.

4. $3\dfrac{3}{4}$

-2

5. $6\dfrac{1}{3}$

$-4\dfrac{1}{2}$

6. $5\dfrac{2}{5}$

$-3\dfrac{7}{10}$

Practice Problem Answers

1. $20\dfrac{1}{3}$

2. $9\dfrac{97}{120}$

3. $12\dfrac{27}{70}$

4. $1\dfrac{3}{4}$

5. $1\dfrac{5}{6}$

6. $1\dfrac{7}{10}$

Exercises 2.5

1. $10 + 4\dfrac{5}{7}$

2. $6\dfrac{1}{2} + 3\dfrac{1}{2}$

3. $5\dfrac{1}{3} + 2\dfrac{2}{3}$

4. $8 + 7\dfrac{3}{11}$

5. $\begin{aligned} &12 \\ + &\dfrac{3}{4} \\ \hline \end{aligned}$

6. $\begin{aligned} &15 \\ + &\dfrac{9}{10} \\ \hline \end{aligned}$

7. $\begin{aligned} &3\dfrac{1}{4} \\ + &5\dfrac{5}{12} \\ \hline \end{aligned}$

8. $\begin{aligned} &5\dfrac{11}{12} \\ + &3\dfrac{5}{6} \\ \hline \end{aligned}$

9. $\begin{aligned} &21\dfrac{1}{3} \\ + &13\dfrac{1}{18} \\ \hline \end{aligned}$

10. $\begin{aligned} &9\dfrac{2}{3} \\ + &\dfrac{5}{6} \\ \hline \end{aligned}$

11. $\begin{aligned} &3\dfrac{2}{3} \\ + &5\dfrac{1}{5} \\ \hline \end{aligned}$

12. $\begin{aligned} &4\dfrac{3}{7} \\ + &6\dfrac{2}{3} \\ \hline \end{aligned}$

13. $\begin{aligned} &4\dfrac{1}{3} \\ &8\dfrac{3}{8} \\ + &6\dfrac{1}{6} \\ \hline \end{aligned}$

14. $\begin{aligned} &13\dfrac{5}{8} \\ &13\dfrac{1}{12} \\ + &10\dfrac{1}{4} \\ \hline \end{aligned}$

15. $\begin{aligned} &5\dfrac{1}{8} \\ &1\dfrac{1}{5} \\ + &3\dfrac{1}{40} \\ \hline \end{aligned}$

16. $\begin{aligned} &27\dfrac{2}{3} \\ &30\dfrac{5}{8} \\ + &31\dfrac{5}{6} \\ \hline \end{aligned}$

17. $3\dfrac{5}{8} - 2$

18. $4\dfrac{5}{12} - 3$

19. $9\dfrac{1}{4} - 5\dfrac{1}{4}$

20. $6\dfrac{1}{2} - 2\dfrac{1}{2}$

21. $\begin{aligned} &7\dfrac{9}{10} \\ - &3\dfrac{3}{10} \\ \hline \end{aligned}$

22. $\begin{aligned} &5\dfrac{3}{4} \\ - &2\dfrac{1}{4} \\ \hline \end{aligned}$

23. $\begin{aligned} &20\dfrac{7}{16} \\ - &15\dfrac{5}{16} \\ \hline \end{aligned}$

24. $\begin{aligned} &14\dfrac{5}{8} \\ - &11\dfrac{3}{8} \\ \hline \end{aligned}$

25. $9\dfrac{5}{16}$

$-2\dfrac{1}{4}$

26. $4\dfrac{7}{8}$

$-1\dfrac{1}{4}$

27. $10\dfrac{5}{6}$

$-4\dfrac{2}{3}$

28. $5\dfrac{11}{12}$

$-1\dfrac{1}{4}$

29. $15\dfrac{5}{8}$

$-11\dfrac{3}{4}$

30. $8\dfrac{5}{6}$

$-2\dfrac{1}{4}$

31. 7

$-2\dfrac{1}{2}$

32. 12

$-5\dfrac{2}{3}$

33. 14

$-3\dfrac{5}{8}$

34. 30

$-10\dfrac{1}{6}$

Solve the following word problems. See Example 4.

35. **Travel:** A bus trip is separated into three parts. The first part takes $2\dfrac{1}{3}$ hours, the second part takes $2\dfrac{1}{2}$ hours, and the third part takes $3\dfrac{3}{4}$ hours. How long does the entire trip take?

36. **Construction:** A construction company was contracted to build three sections of highway. One section was $20\dfrac{7}{10}$ kilometers, the second section was $3\dfrac{4}{10}$ kilometers, and the third section was $11\dfrac{6}{10}$ kilometers. What was the total length of highway built?

37. **Geometry:** A triangle has sides that measure $42\dfrac{3}{4}$ feet, $23\dfrac{1}{2}$ feet, and $22\dfrac{7}{8}$ feet. Find the perimeter of the triangle.

38. **Geometry:** A quadrilateral (four-sided figure) has sides that measure $3\dfrac{1}{2}$ inches, $2\dfrac{1}{4}$ inches, $3\dfrac{5}{8}$ and $2\dfrac{3}{4}$ inches. What is the perimeter of the quadrilateral?

39. **Construction:** A carpenter buys five boards at the lumber yard. The widths of the boards are $3\dfrac{1}{2}$, $9\dfrac{1}{4}$, $11\dfrac{1}{4}$, $7\dfrac{1}{4}$, and $5\dfrac{1}{2}$ inches, respectively. What is the total width of these boards?

40. **Painting:** Sara can paint a room in $3\dfrac{3}{5}$ hours, and Emily can paint a room of the same size in $4\dfrac{1}{5}$ hours.

   a. How many hours are saved by having Sara paint a room of this size?

   b. How many minutes are saved?

41. Grading: A teacher graded two sets of test papers. The first set took $3\frac{3}{4}$ hours to grade, and the second set took $2\frac{3}{5}$ hours. How much faster did he grade the second set?

42. Cleaning: Mike takes $1\frac{1}{2}$ hours to clean a pool, and Tom takes $2\frac{1}{3}$ hours to clean the same pool. How much longer does Tom take?

43. Running: When she first started training, a long-distance runner could run 10 miles in $70\frac{3}{10}$ minutes. Three months later she ran the same 10 miles in $63\frac{7}{10}$ minutes. By how much did her time improve?

44. Stock Market: A certain stock was selling for $43\frac{7}{8}$ dollars per share. One month later it was selling for $48\frac{1}{2}$ dollars per share. By how much did the stock increase in price?

45. Weight Loss: Sally needs to lose 10 pounds. If she weighs 180 pounds now and she loses $3\frac{1}{4}$ pounds during the first week and $3\frac{1}{2}$ pounds during the second week, how much more weight does she need to lose?

46. Travel: A salesman drove $5\frac{3}{4}$ hours one day and $6\frac{1}{2}$ hours the next day. How much more time did he spend driving on the second day?

47. Physiology: On average, the air that we inhale includes $1\frac{1}{4}$ parts water, and the air we exhale includes $5\frac{9}{10}$ parts water. How many more parts water are in exhaled air?

48. Running: A person who is running will burn about $14\frac{7}{10}$ calories each minute, and a person who is walking will burn about $5\frac{1}{2}$ calories each minute. How many more calories does a runner burn in a minute than a walker?

Objectives

A Compare fractions by finding a common denominator and comparing the numerators.

B Evaluate expressions with fractions.

C Simplify complex fractions.

D Evaluate expressions with mixed numbers.

E Find the average of a group of fractions or mixed numbers

2.6 Order of Operations with Fractions and Mixed Numbers

Objective A **Comparing Two or More Fractions**

Many times we want to compare two (or more) fractions to see which is smaller or larger. Then we can subtract the smaller from the larger, or possibly make a decision based on the relative sizes of the fractions. Related word problems will be discussed in detail in later chapters.

To Compare Two Fractions (to Find which is Larger or Smaller):

1. Find the least common denominator (LCD).
2. Change each fraction to an equivalent fraction with that denominator.
3. Compare the numerators.

Example 1

Comparing Fractions

Which is larger: $\frac{5}{6}$ or $\frac{7}{8}$? How much larger?

Solution

Step 1: Find the LCD.

$$\left.\begin{array}{l} 6 = 2 \cdot 3 \\ 8 = 2 \cdot 2 \cdot 2 \end{array}\right\} \text{LCD} = 2 \cdot 2 \cdot 2 \cdot 3 = 24$$

Step 2: Find equal fractions with denominator 24.

$$\frac{5}{6} = \frac{5}{6} \cdot \frac{4}{4} = \frac{20}{24} \text{ and } \frac{7}{8} = \frac{7}{8} \cdot \frac{3}{3} = \frac{21}{24}$$

Step 3: $\frac{7}{8}$ is larger than $\frac{5}{6}$, since 21 is larger than 20.

$$\frac{7}{8} - \frac{5}{6} = \frac{21}{24} - \frac{20}{24} = \frac{1}{24}$$

$\frac{7}{8}$ is larger by $\frac{1}{24}$.

Example 2

Comparing Fractions

Which is larger: $\dfrac{8}{9}$ or $\dfrac{11}{12}$? How much larger?

Solution

Step 1: LCD $= 2 \cdot 2 \cdot 3 \cdot 3 = 36$

Step 2: $\dfrac{8}{9} = \dfrac{8}{9} \cdot \dfrac{4}{4} = \dfrac{32}{36}$ and $\dfrac{11}{12} = \dfrac{11}{12} \cdot \dfrac{3}{3} = \dfrac{33}{36}$

Step 3: $\dfrac{11}{12}$ is larger than $\dfrac{8}{9}$, since 33 is larger than 32.

$$\dfrac{11}{12} - \dfrac{8}{9} = \dfrac{33}{36} - \dfrac{32}{36} = \dfrac{1}{36}$$

$\dfrac{11}{12}$ is larger by $\dfrac{1}{36}$.

Now work margin exercises 1 and 2.

Example 3

Comparing Fractions

Arrange $\dfrac{2}{3}$, $\dfrac{7}{10}$, and $\dfrac{9}{15}$ in order, from smallest to largest. Then find the difference between the smallest and the largest.

Solution

Step 1: LCD $= 30$

Step 2: $\dfrac{2}{3} = \dfrac{2}{3} \cdot \dfrac{10}{10} = \dfrac{20}{30}$; $\quad \dfrac{7}{10} = \dfrac{7}{10} \cdot \dfrac{3}{3} = \dfrac{21}{30}$; $\quad \dfrac{9}{15} = \dfrac{9}{15} \cdot \dfrac{2}{2} = \dfrac{18}{30}$

Step 3: Smallest to largest: $\dfrac{9}{15}$, $\dfrac{2}{3}$, $\dfrac{7}{10}$.

$$\dfrac{7}{10} - \dfrac{9}{15} = \dfrac{21}{30} - \dfrac{18}{30} = \dfrac{3}{30} = \dfrac{1}{10}$$

Now work margin exercise 3.

---

1. Which is larger: $\dfrac{9}{22}$ or $\dfrac{10}{33}$? How much larger?

2. Which is larger: $\dfrac{7}{9}$ or $\dfrac{19}{24}$? How much larger?

3. Arrange $\dfrac{7}{12}$, $\dfrac{5}{9}$, and $\dfrac{2}{3}$ in order, from smallest to largest. Then find the difference between the smallest and the largest.

Evaluating Expressions with Fractions

An expression with fractions may involve more than one operation. To evaluate such an expression, use the **rules for order of operations** just as they were discussed for whole numbers in Section 1.6. The rules are restated here for easy reference. Remember that to add or subtract fractions you need a common denominator; to divide, multiply by the reciprocal of the divisor.

Rules for Order of Operations

1. Simplify within grouping symbols, such as parentheses ( ), brackets [ ], and braces { }. Start with the innermost grouping.
2. Evaluate any numbers or expressions indicated by exponents.
3. Moving from **left to right**, perform any multiplications or divisions in the order in which they appear.
4. Moving from **left to right**, perform any additions or subtractions in the order in which they appear.

Example 4

Order of Operations with Fractions

Evaluate the expression: $\dfrac{1}{2} \div \dfrac{3}{4} + \dfrac{5}{6} \cdot \dfrac{1}{5}$

Solution

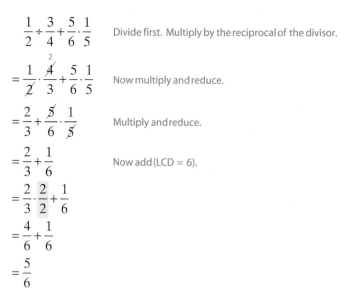

$$\frac{1}{2} \div \frac{3}{4} + \frac{5}{6} \cdot \frac{1}{5}$$ Divide first. Multiply by the reciprocal of the divisor.

$$= \frac{1}{2} \cdot \frac{\overset{2}{\cancel{4}}}{3} + \frac{5}{6} \cdot \frac{1}{5}$$ Now multiply and reduce.

$$= \frac{2}{3} + \frac{\cancel{5}}{6} \cdot \frac{1}{\cancel{5}}$$ Multiply and reduce.

$$= \frac{2}{3} + \frac{1}{6}$$ Now add (LCD = 6).

$$= \frac{2}{3} \cdot \frac{2}{2} + \frac{1}{6}$$

$$= \frac{4}{6} + \frac{1}{6}$$

$$= \frac{5}{6}$$

Example 5

Order of Operations with Fractions

Evaluate the expression: $\dfrac{9}{10}-\left(\dfrac{1}{4}\right)^2+\dfrac{1}{2}$

Solution

$\dfrac{9}{10}-\left(\dfrac{1}{4}\right)^2+\dfrac{1}{2}$ Use the exponent first. Remember, $\left(\dfrac{1}{4}\right)^2=\left(\dfrac{1}{4}\right)\cdot\left(\dfrac{1}{4}\right)$.

$=\dfrac{9}{10}-\dfrac{1}{16}+\dfrac{1}{2}$ Now add and subtract from left to right (LCD = 80).

$=\dfrac{9}{10}\cdot\dfrac{8}{8}-\dfrac{1}{16}\cdot\dfrac{5}{5}+\dfrac{1}{2}\cdot\dfrac{40}{40}$

$=\dfrac{72}{80}-\dfrac{5}{80}+\dfrac{40}{80}$

$=\dfrac{107}{80}$ or $1\dfrac{27}{80}$

Example 6

Order of Operations with Fractions

Evaluate the expression: $\dfrac{3}{5}\cdot\dfrac{1}{2}\div\dfrac{1}{4}+\left(\dfrac{1}{3}\right)^2$

Solution

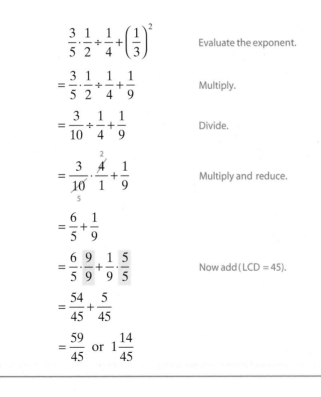

$\dfrac{3}{5}\cdot\dfrac{1}{2}\div\dfrac{1}{4}+\left(\dfrac{1}{3}\right)^2$ Evaluate the exponent.

$=\dfrac{3}{5}\cdot\dfrac{1}{2}\div\dfrac{1}{4}+\dfrac{1}{9}$ Multiply.

$=\dfrac{3}{10}\div\dfrac{1}{4}+\dfrac{1}{9}$ Divide.

$=\dfrac{3}{\overset{}{\underset{5}{10}}}\cdot\dfrac{\overset{2}{\cancel{4}}}{1}+\dfrac{1}{9}$ Multiply and reduce.

$=\dfrac{6}{5}+\dfrac{1}{9}$

$=\dfrac{6}{5}\cdot\dfrac{9}{9}+\dfrac{1}{9}\cdot\dfrac{5}{5}$ Now add (LCD = 45).

$=\dfrac{54}{45}+\dfrac{5}{45}$

$=\dfrac{59}{45}$ or $1\dfrac{14}{45}$

Evaluate the expressions.

4. $\dfrac{2}{5}\cdot\dfrac{1}{4}+\dfrac{1}{9}\div\dfrac{2}{5}$

5. $\dfrac{1}{4}+\left(\dfrac{1}{3}\right)^{2}\div 3$

6. $\left(\dfrac{3}{2}\right)^{2}-\dfrac{3}{4}\cdot\dfrac{5}{6}\div\dfrac{3}{8}$

7. $\dfrac{2}{3}\div\left(\dfrac{4}{5}-\dfrac{4}{7}\right)$

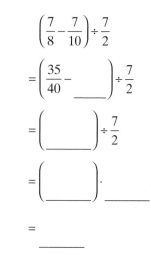

Completion Example 7

Order of Operations with Fractions

Evaluate the expression: $\left(\dfrac{7}{8}-\dfrac{7}{10}\right)\div\dfrac{7}{2}$

Solution

$$\left(\dfrac{7}{8}-\dfrac{7}{10}\right)\div\dfrac{7}{2}$$

$$=\left(\dfrac{35}{40}-\underline{\quad}\right)\div\dfrac{7}{2}$$

$$=\left(\underline{\quad}\right)\div\dfrac{7}{2}$$

$$=\left(\underline{\quad}\right)\cdot\underline{\quad}$$

$$=\underline{\quad}$$

***Now work margin exercises* 4 through 7.**

Objective C   Simplifying Complex Fractions

A **complex fraction** is a fraction in which the numerator or denominator or both contain one or more fractions or mixed numbers. To simplify a complex fraction, we treat the fraction bar as a symbol of inclusion for both the numerator and denominator. The procedure is outlined as follows:

To Simplify Complex Fractions

1. Simplify the numerator so that it is a single fraction, possibly an improper fraction.
2. Simplify the denominator so that it also is a single fraction, possibly an improper fraction.
3. Divide the numerator by the denominator, and reduce if possible.

Completion Example Answers

7. $\dfrac{28}{40}$; $\dfrac{7}{40}$; $\left(\dfrac{7}{40}\right)\cdot\dfrac{2}{7}$; $\dfrac{1}{20}$

Example 8

Simplifying Complex Fractions

Simplify the complex fraction: $\dfrac{\dfrac{2}{3}+\dfrac{1}{6}}{1-\dfrac{1}{3}}$

Solution

$$\frac{2}{3}+\frac{1}{6}=\frac{4}{6}+\frac{1}{6}=\frac{5}{6} \quad \text{numerator}$$

$$1-\frac{1}{3}=\frac{3}{3}-\frac{1}{3}=\frac{2}{3} \quad \text{denominator}$$

So,

$$\frac{\dfrac{2}{3}+\dfrac{1}{6}}{1-\dfrac{1}{3}}=\frac{\dfrac{5}{6}}{\dfrac{2}{3}}=\frac{5}{6}\div\frac{2}{3}=\frac{5}{\overset{}{\underset{2}{6}}}\cdot\frac{\overset{1}{3}}{2}=\frac{5}{4} \text{ or } 1\frac{1}{4}.$$

Now work margin exercise 8.

notes

The complex fraction in Example 8 could have been written as

$$\frac{\dfrac{2}{3}+\dfrac{1}{6}}{1-\dfrac{1}{3}}=\left(\frac{2}{3}+\frac{1}{6}\right)\div\left(1-\frac{1}{3}\right).$$

Thus the fraction bar in a complex fraction serves the same purpose as two sets of parentheses, one surrounding the numerator and the other surrounding the denominator.

Objective D **Evaluating Expressions with Mixed Numbers**

The rules for order of operations can be used with mixed numbers just as with whole numbers and fractions. In some cases a good strategy in evaluating expressions with several operations is to change each mixed number to an improper fraction and then perform the operations. Study the following examples carefully.

Example 9

Order of Operations with Mixed Numbers

Evaluate the expression: $3\dfrac{2}{5} \div \left(\dfrac{1}{4} + \dfrac{3}{5}\right)$

Solution

$3\dfrac{2}{5} = \dfrac{17}{5}$ 　　　　　Change the mixed number to an improper fraction.

$\dfrac{1}{4} + \dfrac{3}{5} = \dfrac{1}{4} \cdot \dfrac{5}{5} + \dfrac{3}{5} \cdot \dfrac{4}{4} = \dfrac{5}{20} + \dfrac{12}{20} = \dfrac{17}{20}$ 　　Add the numbers inside parentheses.

Now divide.

$$3\dfrac{2}{5} \div \left(\dfrac{1}{4} + \dfrac{3}{5}\right) = \dfrac{17}{5} \div \dfrac{17}{20} = \dfrac{17}{5} \cdot \dfrac{20}{17} = \dfrac{\cancel{17} \cdot 4 \cdot \cancel{5}}{\cancel{5} \cdot \cancel{17}} = \dfrac{4}{1} = 4$$

Example 10

Order of Operations with Mixed Numbers

Evaluate the expression: $\left(5\dfrac{2}{3} - 2\dfrac{1}{3}\right) \div \left(1\dfrac{1}{2} + 1\right)$

Solution

Work inside each set of parentheses:

$$5\dfrac{2}{3} - 2\dfrac{1}{3} = 3\dfrac{1}{3} = \dfrac{10}{3} \quad \text{and} \quad 1\dfrac{1}{2} + 1 = 2\dfrac{1}{2} = \dfrac{5}{2}$$

Now divide.

$$\left(5\dfrac{2}{3} - 2\dfrac{1}{3}\right) \div \left(1\dfrac{1}{2} + 1\right) = \dfrac{10}{3} \div \dfrac{5}{2} = \dfrac{10}{3} \cdot \dfrac{2}{5} = \dfrac{2 \cdot \cancel{5} \cdot 2}{3 \cdot \cancel{5}} = \dfrac{4}{3} \text{ or } 1\dfrac{1}{3}$$

Example 11

Order of Operations with Mixed Numbers

Evaluate the expression: $2\frac{1}{2} \cdot 1\frac{1}{6} + 7 \div \frac{3}{4}$

Solution

$$2\frac{1}{2} \cdot 1\frac{1}{6} + 7 \div \frac{3}{4} = \frac{5}{2} \cdot \frac{7}{6} + \frac{7}{1} \div \frac{3}{4}$$

Change the mixed numbers to improper fractions.

$$= \frac{5}{2} \cdot \frac{7}{6} + \frac{7}{1} \cdot \frac{4}{3}$$

Multiply and divide from left to right.

$$= \frac{35}{12} + \frac{28}{3}$$

Find the LCD.

$$= \frac{35}{12} + \frac{28}{3} \cdot \frac{4}{4}$$

Now add.

$$= \frac{35}{12} + \frac{112}{12}$$

$$= \frac{147}{12} = \frac{\cancel{3} \cdot 49}{\cancel{3} \cdot 4} = \frac{49}{4} \text{ or } 12\frac{1}{4}$$

Solution

Or, working with separate parts, we can write:

$$2\frac{1}{2} \cdot 1\frac{1}{6} = \frac{5}{2} \cdot \frac{7}{6} = \frac{35}{12} = 2\frac{11}{12}$$

Multiply.

$$7 \div \frac{3}{4} = \frac{7}{1} \cdot \frac{4}{3} = \frac{28}{3} = 9\frac{1}{3}$$

Divide.

$$2\frac{11}{12} = 2\frac{11}{12}$$

$$+9\frac{1}{3} = 9\frac{4}{12}$$

Add the results.

$$11\frac{15}{12} = 12\frac{3}{12} = 12\frac{1}{4} \text{ or } \frac{49}{4}$$

Now work margin exercises 9 through 11.

Evaluate the expressions.

9. $3\frac{1}{4} \div \left(\frac{3}{4} + \frac{7}{8} \right)$

10. $\left(4\frac{1}{3} + 2\frac{2}{3} \right) \div \left(3\frac{3}{5} - \frac{2}{5} \right)$

11. $3\frac{1}{5} \div \frac{2}{3} + 1\frac{1}{2} \cdot 2\frac{1}{3}$

Objective E **Applications: Average**

Another topic related to order of operations is that of average. Remember that the **average** of a set of numbers can be found by adding the numbers and then dividing this sum by the quantity of numbers in the set.

12. Find the average of

$$3\frac{1}{4}, \ 4\frac{1}{12}, \text{ and } 2\frac{5}{6}.$$

Example 12

Average of Mixed Numbers

Find the average of $1\frac{1}{2}$, $2\frac{3}{4}$, and $3\frac{5}{8}$.

Solution

Finding the average is the same as evaluating the expression

$$\left(1\frac{1}{2}+2\frac{3}{4}+3\frac{5}{8}\right)\div 3.$$

Find the sum first and then divide by 3.

$$1\frac{1}{2}=1\frac{4}{8}$$

$$2\frac{3}{4}=2\frac{6}{8}$$

$$+\ 3\frac{5}{8}=3\frac{5}{8}$$

$$6\frac{15}{8}=7\frac{7}{8}$$

$$7\frac{7}{8}\div 3=\frac{63}{8}\cdot\frac{1}{3}=\frac{\cancel{3}\cdot 3\cdot 7}{8\cdot\cancel{3}}=\frac{21}{8}\text{ or }2\frac{5}{8}$$

The average is $2\frac{5}{8}$.

Now work margin exercise **12.**

Practice Problems

Evaluate each of the following expressions by using the rules for order of operations.

1. $2\div\left(\dfrac{1}{2}+\dfrac{3}{7}\right)$ **2.** $\dfrac{2}{3}\cdot\dfrac{9}{10}-\dfrac{1}{8}\div\dfrac{1}{2}$ **3.** $5\dfrac{1}{2}\div\left(2-\dfrac{1}{3}\right)^{2}$

4. Simplify the complex fraction: $\dfrac{\dfrac{3}{4}+\dfrac{5}{8}}{2}$

5. Find the average of $3\dfrac{1}{2}$, $5\dfrac{1}{6}$, and $7\dfrac{1}{4}$.

Practice Problem Answers

1. $\dfrac{28}{13}$ or $2\dfrac{2}{13}$ **2.** $\dfrac{7}{20}$ **3.** $\dfrac{99}{50}$ or $1\dfrac{49}{50}$

4. $\dfrac{11}{16}$ **5.** $\dfrac{191}{36}$ or $5\dfrac{11}{36}$

Exercises 2.6

Find the larger number of each pair and state how much larger it is. See Examples 1 and 2.

1. $\frac{2}{3}, \frac{3}{4}$

2. $\frac{7}{10}, \frac{8}{15}$

3. $\frac{4}{5}, \frac{17}{20}$

4. $\frac{4}{10}, \frac{3}{8}$

5. $\frac{13}{20}, \frac{5}{8}$

6. $\frac{13}{16}, \frac{21}{25}$

7. $\frac{14}{35}, \frac{12}{30}$

8. $\frac{10}{36}, \frac{7}{24}$

9. $\frac{17}{80}, \frac{11}{48}$

10. $\frac{37}{100}, \frac{24}{75}$

Arrange the numbers in order from smallest to largest. Then find the difference between the largest and smallest numbers. See Example 3.

11. $\frac{1}{3}, \frac{2}{7}, \frac{3}{8}$

12. $\frac{8}{9}, \frac{9}{10}, \frac{11}{12}$

13. $\frac{7}{6}, \frac{11}{12}, \frac{19}{20}$

14. $\frac{1}{3}, \frac{5}{42}, \frac{3}{7}$

15. $\frac{1}{2}, \frac{1}{3}, \frac{1}{4}$

16. $\frac{2}{3}, \frac{3}{4}, \frac{5}{8}$

17. $\frac{7}{9}, \frac{31}{36}, \frac{13}{18}$

18. $\frac{17}{12}, \frac{40}{36}, \frac{31}{24}$

19. $\frac{1}{100}, \frac{3}{1000}, \frac{20}{10,000}$

20. $\frac{32}{100}, \frac{298}{1000}, \frac{3}{10}$

Evaluate each expression by using the rules for order of operations. See Examples 4 through 6.

21. $\frac{1}{2} \div \frac{7}{8} + \frac{1}{7} \cdot \frac{2}{3}$

22. $\frac{1}{2} \div \frac{1}{2} + \frac{2}{3} \cdot \frac{2}{3}$

23. $6 - \frac{5}{8} \div 4$

24. $\frac{5}{8} \cdot \frac{1}{10} \div \frac{3}{4} + \frac{1}{6}$

25. $\frac{2}{15} \cdot \frac{1}{4} \div \frac{3}{5} + \frac{1}{27}$

26. $\left(\frac{1}{2} - \frac{1}{3}\right) \div \left(\frac{5}{8} + \frac{3}{16}\right)$

27. $\left(\frac{1}{2}\right)^2 - \left(\frac{1}{4}\right)^3$

28. $\frac{2}{3} + \frac{3}{4} + \left(\frac{1}{2}\right)^2$

29. $\left(\frac{1}{3}\right)^2 + \left(\frac{1}{6}\right)^2 + \frac{2}{3}$

30. $\left(\frac{3}{4} - \frac{1}{2}\right) \div \left(1 + \frac{1}{3}\right)$

31. $\left(\frac{1}{5} + \frac{1}{6}\right) \div \left(2 + \frac{1}{3}\right)$

32. $\left(\frac{2}{3} - \frac{1}{4}\right) \div \left(\frac{3}{5} - \frac{1}{4}\right)$

33. $\left(\dfrac{7}{8} - \dfrac{3}{16}\right) \div \left(\dfrac{1}{3} - \dfrac{1}{4}\right)$ 34. $\left(\dfrac{5}{8} - \dfrac{1}{8}\right) \div \left(\dfrac{1}{2} - \dfrac{1}{4}\right)$ 35. $16 + \left(\dfrac{1}{3} \div \dfrac{2}{3}\right)$ 36. $\left(\dfrac{2}{3}\right)^2 \div \left(\dfrac{1}{3}\right)^2 - \dfrac{5}{9}$

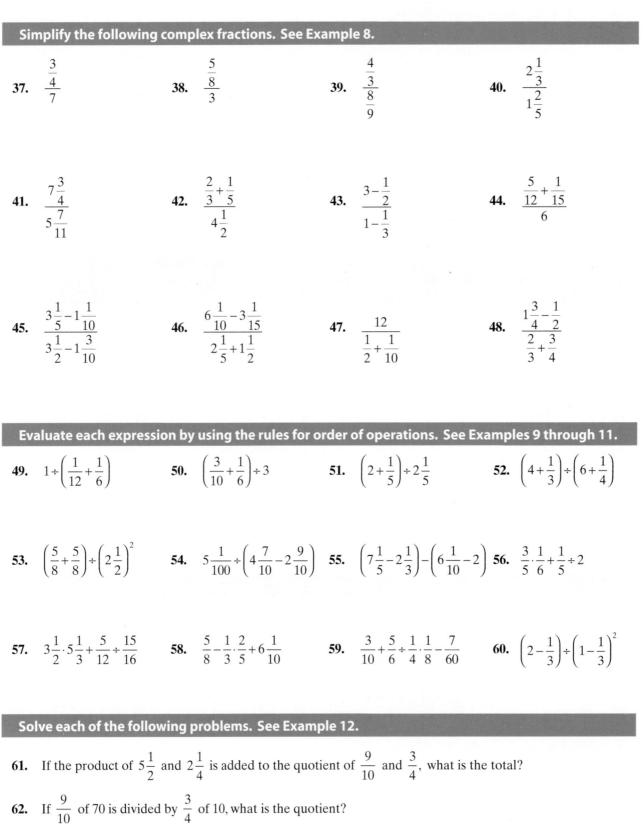

Simplify the following complex fractions. See Example 8.

37. $\dfrac{\frac{3}{4}}{7}$ 38. $\dfrac{\frac{5}{8}}{3}$ 39. $\dfrac{\frac{4}{3}}{\frac{8}{9}}$ 40. $\dfrac{2\frac{1}{3}}{1\frac{2}{5}}$

41. $\dfrac{7\frac{3}{4}}{5\frac{7}{11}}$ 42. $\dfrac{\frac{2}{3} + \frac{1}{5}}{4\frac{1}{2}}$ 43. $\dfrac{3 - \frac{1}{2}}{1 - \frac{1}{3}}$ 44. $\dfrac{\frac{5}{12} + \frac{1}{15}}{6}$

45. $\dfrac{3\frac{1}{5} - 1\frac{1}{10}}{3\frac{1}{2} - 1\frac{3}{10}}$ 46. $\dfrac{6\frac{1}{10} - 3\frac{1}{15}}{2\frac{1}{5} + 1\frac{1}{2}}$ 47. $\dfrac{12}{\frac{1}{2} + \frac{1}{10}}$ 48. $\dfrac{1\frac{3}{4} - \frac{1}{2}}{\frac{2}{3} + \frac{3}{4}}$

Evaluate each expression by using the rules for order of operations. See Examples 9 through 11.

49. $1 \div \left(\dfrac{1}{12} + \dfrac{1}{6}\right)$ 50. $\left(\dfrac{3}{10} + \dfrac{1}{6}\right) \div 3$ 51. $\left(2 + \dfrac{1}{5}\right) \div 2\dfrac{1}{5}$ 52. $\left(4 + \dfrac{1}{3}\right) \div \left(6 + \dfrac{1}{4}\right)$

53. $\left(\dfrac{5}{8} + \dfrac{5}{8}\right) \div \left(2\dfrac{1}{2}\right)^2$ 54. $5\dfrac{1}{100} \div \left(4\dfrac{7}{10} - 2\dfrac{9}{10}\right)$ 55. $\left(7\dfrac{1}{5} - 2\dfrac{1}{3}\right) - \left(6\dfrac{1}{10} - 2\right)$ 56. $\dfrac{3}{5} \cdot \dfrac{1}{6} + \dfrac{1}{5} \div 2$

57. $3\dfrac{1}{2} \cdot 5\dfrac{1}{3} + \dfrac{5}{12} \div \dfrac{15}{16}$ 58. $\dfrac{5}{8} - \dfrac{1}{3} \cdot \dfrac{2}{5} + 6\dfrac{1}{10}$ 59. $\dfrac{3}{10} + \dfrac{5}{6} \div \dfrac{1}{4} \cdot \dfrac{1}{8} - \dfrac{7}{60}$ 60. $\left(2 - \dfrac{1}{3}\right) \div \left(1 - \dfrac{1}{3}\right)^2$

Solve each of the following problems. See Example 12.

61. If the product of $5\dfrac{1}{2}$ and $2\dfrac{1}{4}$ is added to the quotient of $\dfrac{9}{10}$ and $\dfrac{3}{4}$, what is the total?

62. If $\dfrac{9}{10}$ of 70 is divided by $\dfrac{3}{4}$ of 10, what is the quotient?

63. Find the average of the numbers $\dfrac{7}{8}, \dfrac{9}{10}$, and $1\dfrac{3}{4}$.

64. Find the average of the numbers $\dfrac{5}{6}$, $1\dfrac{2}{3}$, and $\dfrac{3}{4}$.

65. Find the average of the numbers $5\dfrac{1}{8}$, $7\dfrac{1}{2}$, $4\dfrac{3}{4}$, and $10\dfrac{1}{2}$.

66. Find the average of the numbers $8\dfrac{1}{4}$, $6\dfrac{7}{8}$, $9\dfrac{4}{5}$, and $8\dfrac{7}{10}$.

Solve the following word problems.

67. Picture Frame: An $8\dfrac{1}{2}$ inch by $10\dfrac{3}{4}$ inch picture is placed in a frame which is $\dfrac{9}{16}$ inch wide. Find the perimeter of the outside edges of the frame. (**Reminder:** The perimeter of a rectangle is computed by adding twice the length to twice its width.)

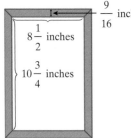

68. Recipe: A recipe calls for $2\dfrac{1}{5}$ cup of stewed tomatoes, and $1\dfrac{2}{3}$ cup of fully cooked beans. If the recipe is multiplied by $2\dfrac{1}{2}$, how much tomato/bean mixture will there be?

69. Snakes: A person has a collection of four snakes with the following lengths: $1\dfrac{1}{4}$ feet, $3\dfrac{5}{8}$ feet, $\dfrac{7}{8}$ feet, and $2\dfrac{1}{3}$ feet. What is the average length of the four snakes?

70. Snowfall: A town which normally does not receive significant snow experienced measurable snow for three weeks in a row. The first week it snowed $2\dfrac{1}{4}$ inches, the second week it snowed $10\dfrac{2}{3}$ inches, and the third week it snowed $1\dfrac{5}{6}$ inches. What was the average weekly snow fall for this three-week period?

71. Force: The force between two charged particles is a product of terms divided by the square of the distance between the two particles. Assume that the product of the terms is $\dfrac{3}{5}$ and the distance between the two particles is $\dfrac{1}{25}$. What is the force?

72. Painting: Two painters paint $76\dfrac{1}{2}$ feet of fencing in one day. The first painter contributes $2\dfrac{1}{2}$ hours, and the second works for $5\dfrac{3}{5}$ hours. How many feet of fencing are painted in each hour? (**Hint:** Add the number of hours, then divide into the length of fencing.)

73. Mail Order: A customer orders some items from a mail order store. Included in this order are five boxes of candy which weigh $1\frac{1}{3}$ lb each plus two each of the following three items: $\frac{2}{3}$ lb box of chewing gum, $2\frac{1}{2}$ lb can of peanuts, and $1\frac{1}{4}$ lb box of gourmet popping corn. If the shipping materials add $1\frac{1}{2}$ lb, what is the total shipping weight?

74. Books: An art book has 40 two-sided pages of pictures, each of which is $\frac{1}{32}$ inch thick, where each two-sided page is protected by a $\frac{1}{80}$ inch thick piece of paper, and each side of the book is bound by a $\frac{1}{6}$ inch cover. What is the total thickness of the book?

75. Buffets: Emma goes to a buffet where the cost of the meal is determined by the weight of the food. Among the items currently on the buffet are $\frac{8}{9}$ kg of potato salad and $\frac{3}{4}$ kg of chicken salad. Emma puts $\frac{3}{16}$ of the potato salad and $\frac{2}{15}$ of the chicken salad onto her plate. In addition, she puts 2 slices of bread which weigh $\frac{1}{30}$ kg each.

  a. What is the total weight of the food taken (the weight of the plate is not included)?

  b. Assuming that the restaurant charges \$18 for each kg, how much did Emma spend (before taxes)?

76. Building Demolition: A building was partially destroyed by a bad storm, and it had to be torn down because it was condemned. At the beginning of the demolition project, $\frac{8}{9}$ of the building was standing, and after the first day of work, $\frac{1}{3}$ of the building remained standing. Assuming that the crew spent $6\frac{2}{3}$ hours work on the project this first day, what fraction of the building was torn down each hour? (**Hint:** To solve the problem, first subtract the ending fraction from the starting fraction to find how much was torn down that day, and then divide by the total number of hours spent tearing it down.)

Writing & Thinking

77. a. If two fractions are between 0 and 1, can their sum be more than 1? Explain.

  b. If two fractions are between 0 and 1, can their product be more than 1? Explain.

78. If a fraction is between 0 and 1 and the fraction is squared, will the result be larger or smaller than the original fraction? Explain.

Collaborative Learning

79. With the class separated into teams of two to four students, each team is to write one to two paragraphs on the following two topics. Then the team leader is to read the paragraphs with classroom discussion to follow.

  a. What topic have you found to be the most interesting in the text so far? Why?

  b. What topic have you found to be the most useful in the text so far? Why?

Section 2.1: Introduction to Fractions and Mixed Numbers

Basic Concepts about Fractions   page 116

Numbers such as $\dfrac{2}{3}$ are said to be in **fraction form**. The top number, 2, is called the **numerator** and the bottom number, 3, is called the **denominator**.

$$\dfrac{a}{b} \quad \longleftarrow \quad \text{numerator} \\ \quad \longleftarrow \quad \text{denominator}$$

Proper Fractions and Improper Fractions   page 117

A **proper fraction** is a fraction in which the numerator is less than the denominator. (Proper fractions have values less than 1.)

An **improper fraction** is a fraction in which the numerator is greater than or equal to the denominator. (Improper fractions have values equal to or greater than 1.)

The Number 0 in Fractions   page 118

For any nonzero value of b, $\dfrac{0}{b} = 0$.

For any value of a, $\dfrac{a}{0}$ is undefined.

To Multiply Fractions   page 119

1. Multiply the numerators.
2. Multiply the denominators.

$$\dfrac{a}{b} \cdot \dfrac{c}{d} = \dfrac{a \cdot c}{b \cdot d} \quad \text{For example, } \dfrac{1}{2} \cdot \dfrac{3}{7} = \dfrac{1 \cdot 3}{2 \cdot 7} = \dfrac{3}{14}.$$

To Find Equivalent Fractions   page 120

Multiply the numerator and denominator by the same nonzero whole number.

$$\dfrac{a}{b} = \dfrac{a}{b} \cdot 1 = \dfrac{a}{b} \cdot \dfrac{k}{k} = \dfrac{a \cdot k}{b \cdot k}, \text{ where } b \text{ and } k \neq 0 \left(\dfrac{k}{k} = 1 \right)$$

To Reduce Fractions to Lowest Terms   page 122

1. Factor the numerator and denominator into prime factorizations.
2. Use the fact that $\dfrac{k}{k} = 1$ and "**divide out**" all common factors.

Note: Reduced fractions might be improper fractions.

Basic Concepts about Mixed Numbers   page 123

A **mixed number** is the sum of a whole number and a proper fraction. By convention, we usually write the whole number and the fraction side by side without the plus sign. For example, $6 + \dfrac{3}{5} = 6\dfrac{3}{5}$ (read "six and three-fifths").

To Change a Mixed Number to an Improper Fraction   page 124
1. Multiply the whole number by the denominator of the proper fraction.
2. Add the numerator of the proper fraction to this product.
3. Write this sum over the denominator of the fraction.

For example: $2\dfrac{7}{8} = \dfrac{16+7}{8} = \dfrac{23}{8}$

To Change an Improper Fraction to a Mixed Number   page 125
1. Divide the numerator by the denominator. The quotient is the whole number part of the mixed number.
2. Write the remainder over the denominator as the fraction part of the mixed number.

Section 2.2: Multiplication and Division with Fractions and Mixed Numbers

Multiplication with Mixed Numbers   page 131

The simplest way to multiply mixed numbers is to change each mixed number to an improper fraction and then multiply. The result can be changed back to mixed number form.

Strategies for Multiplying and Reducing Fractions and Mixed Numbers   page 131

Use prime factors. With this method you can be sure that you have not missed a common factor and that your answer is reduced to lowest terms. **Remember that if all the factors in the numerator or denominator divide out, then 1 must be used as a factor.** Another method frequently used is to divide numerators and denominators by common factors whether they are prime or not. If these factors are easily determined, then this method is probably faster. But common factors are sometimes missed with this method.

Reciprocals   page 134

The reciprocal of $\dfrac{a}{b}$ is $\dfrac{b}{a}$ ($a \neq 0$ and $b \neq 0$). The product of a nonzero number and its reciprocal is always 1.

$$\frac{a}{b} \cdot \frac{b}{a} = 1$$

Note: $0 = \dfrac{0}{1}$, but $\dfrac{1}{0}$ is undefined. That is, **the number 0 has no reciprocal.**

To Divide Fractions   page 135

To divide by any nonzero number, multiply by its reciprocal. In general,

$$\frac{a}{b} \div \frac{c}{d} = \frac{a}{b} \cdot \frac{d}{c}$$ where $b, c, d \neq 0$. For example, $\dfrac{1}{2} \div \dfrac{4}{3} = \dfrac{1}{2} \cdot \dfrac{3}{4} = \dfrac{3}{8}$.

Note: Division with mixed numbers can be accomplished by changing each mixed number to improper fraction form and then dividing as normal with fractions.

Multiples   page 145

The **multiples** of a number are the products of that number with the counting numbers. The common multiples of 8 and 12 are 24, 48, 72, 96, and so on.

Least Common Multiple   page 145

The **least common multiple (LCM)** of two (or more) whole numbers is the smallest number that is a multiple of each of these numbers.

To Find the Least Common Multiple   page 145

1. Find the prime factorization of each number.
2. Identify the prime factors that appear in any one of the prime factorizations.
3. Form the product of these primes using each prime the most number of times it appears in any one of the prime factorizations.

Section 2.4: Addition and Subtraction with Fractions

Common Denominator   page 155

To add two or more fractions with the same denominator, we can think of this denominator as the "name" of each fraction. The sum has this common name. Just as 3 *oranges* plus 2 *oranges* gives a total of 5 *oranges*, 3 *eighths* plus 2 *eighths* gives a total of 5 *eighths*. This denominator is called a **common denominator**.

To Add Fractions with the Same Denominator   page 155

1. Add the numerators.
2. Keep the common denominator. $\dfrac{a}{b}+\dfrac{c}{b}=\dfrac{a+c}{b}, b \neq 0$
3. Reduce, if possible.

Least Common Denominator   page 156

The **least common denominator (LCD)** is the least common multiple (LCM) of the denominators (see Section 2.3).

To Add Fractions with Different Denominators   page 157

1. Find the least common denominator (LCD).
2. Change each fraction into an equivalent fraction with that denominator.
3. Add the new fractions.
4. Reduce, if possible.

To Subtract Fractions with the Same Denominator   page 160

1. Subtract the numerators.
2. Keep the common denominator. $\dfrac{a}{b}-\dfrac{c}{b}=\dfrac{a-c}{b}, b \neq 0$
3. Reduce, if possible.

To Subtract Fractions with Different Denominators　　　　　　page 161
  1. Find the least common denominator (LCD).
  2. Change each fraction into an equivalent fraction with that denominator.
  3. Subtract the new fractions.
  4. Reduce, if possible.

Section 2.5: Addition and Subtraction with Mixed Numbers

To Add Mixed Numbers　　　　　　page 169
  1. Add the fraction parts.
  2. Add the whole numbers.
  3. Write the answer as a mixed number with the fraction part less than 1.

To Subtract Mixed Numbers　　　　　　page 171
  1. Subtract the fraction parts.
  2. Subtract the whole numbers.

Section 2.6: Order of Operations with Fractions and Mixed Numbers

To Compare Two Fractions (to Find Which is Larger or Smaller)　　　　　　page 178
  1. Find the least common denominator (LCD).
  2. Change each fraction to an equivalent fraction with that denominator.
  3. Compare the numerators.

Order of Operations with Fractions and Mixed Numbers　　　　　　page 180

Complex Fractions　　　　　　page 182
  A **complex fraction** is a fraction in which the numerator or denominator or both contain one or more fractions or mixed numbers.

To Simplify Complex Fractions　　　　　　page 182
  1. Simplify the numerator so that it is a single fraction, possibly an improper fraction.
  2. Simplify the denominator so that it also is a single fraction, possibly an improper fraction.
  3. Divide the numerator by the denominator and reduce, if possible.

Chapter 2: Review

Section 2.1: Introduction to Fractions and Mixed Numbers

Complete the statements below.

1. The denominator of a fraction cannot be _____ .

2. $\dfrac{0}{7} = 0$, but $\dfrac{7}{0}$ is _____ .

In each equation, find the missing numerator that will make the fractions equal.

3. $\dfrac{1}{6} = \dfrac{?}{12}$

4. $\dfrac{9}{10} = \dfrac{?}{60}$

5. $\dfrac{15}{13} = \dfrac{?}{65}$

Reduce each fraction to lowest terms.

6. $\dfrac{15}{30}$

7. $\dfrac{99}{88}$

8. $\dfrac{150}{120}$

Change each mixed number to improper fraction form.

9. $5\dfrac{1}{10}$

10. $2\dfrac{11}{12}$

11. $13\dfrac{2}{5}$

Change each improper fraction to mixed number form.

12. $\dfrac{53}{8}$

13. $\dfrac{91}{13}$

14. $\dfrac{171}{50}$

Section 2.2: Multiplication and Division with Fractions and Mixed Numbers

Find each product.

15. $\dfrac{1}{7} \cdot \dfrac{3}{7}$

16. $4\dfrac{1}{3} \cdot 2\dfrac{1}{3}$

17. $1\dfrac{1}{5} \cdot 1\dfrac{1}{5}$

18. Find $\dfrac{3}{4}$ of $3\dfrac{2}{5}$.

Find each product and reduce to lowest terms.

19. $\dfrac{1}{3} \cdot \dfrac{1}{2} \cdot \dfrac{1}{5}$

20. $\dfrac{35}{56} \cdot \dfrac{4}{15} \cdot \dfrac{5}{10}$

21. $\dfrac{5}{12} \cdot 6\dfrac{3}{10} \cdot 7\dfrac{1}{9}$

22. $7\dfrac{1}{11} \cdot 2\dfrac{3}{4} \cdot 5\dfrac{1}{3}$

Find each quotient and reduce to lowest terms.

23. $\dfrac{7}{12} \div \dfrac{7}{12}$

24. $2\dfrac{2}{5} \div 4$

25. $5\dfrac{1}{4} \div 2\dfrac{1}{3}$

26. $\dfrac{15}{16} \div \dfrac{3}{4}$

Section 2.3: Least Common Multiple (LCM)

List the multiples as indicated below.

27. List the first six multiples of 11.

Find the LCM of each set of numbers and tell how many times each number divides into the LCM.

28. $4, 14, 21$

29. $6, 15, 60$

30. $8, 10, 15, 28$

31. $3, 5, 7, 11$

32. $3, 9, 27, 81$

Solve the following word problem.

33. **Sales Travel:** Three continuously traveling salespeople have lunch together each time they are in the home office on the same day.

    a. If it takes Andrew 6 days to cover his territory, Lisa 9 days, and Abby 12 days, how often do they have lunch together?

    b. If Andrew's route is changed so that he takes only 5 days to cover his territory, how often do they have lunch together?

Section 2.4: Addition and Subtraction with Fractions

Find the indicated sums and differences. Reduce each answer to lowest terms.

34. $\dfrac{3}{7} + \dfrac{2}{7}$

35. $\dfrac{5}{6} - \dfrac{1}{6}$

36. $\dfrac{5}{8} - \dfrac{3}{8}$

37. $\dfrac{1}{12} + \dfrac{5}{36} + \dfrac{11}{24}$

38. $\dfrac{13}{22} - \dfrac{9}{33}$

39. $\dfrac{5}{27} + \dfrac{5}{18}$

40. $1 - \dfrac{13}{20}$

41. $\dfrac{3}{4} - \dfrac{5}{12}$

42. $\begin{array}{r} \dfrac{2}{3} \\[4pt] \dfrac{1}{8} \\[4pt] + \dfrac{1}{12} \\ \hline \end{array}$

43. **Candy:** Sue bought 3 bags of candy weighing $\frac{1}{4}$ pound each. She gave one bag to Tom. How many pounds of candy did she have left?

44. **Career Plans:** Ms. Clarke teaches statistics. This semester, $\frac{2}{3}$ of her statistics class of 42 students are planning to be elementary school teachers. How many of her students do not plan to be elementary school teachers?

Section 2.5   Addition and Subtraction with Mixed Numbers

Find the indicated sums and differences. Reduce each answer to lowest terms.

45.
$$13\frac{5}{9}$$
$$+\,14\frac{1}{9}$$

46.
$$6\frac{5}{8}$$
$$+\,2\frac{3}{8}$$

47.
$$9\frac{1}{2}$$
$$-\,7$$

48.
$$20\frac{3}{4}$$
$$-\,7\frac{7}{8}$$

49.
$$9$$
$$-\,5\frac{1}{4}$$

50.
$$12\frac{5}{6}$$
$$-\,6\frac{1}{4}$$

51.
$$6\frac{1}{4}$$
$$4\frac{1}{3}$$
$$+\,8\frac{1}{2}$$

52.
$$42\frac{7}{9}$$
$$53\frac{4}{15}$$
$$+\,24\frac{9}{10}$$

53. If the sum of $5\frac{1}{2}$ and $2\frac{1}{3}$ is subtracted from $15\frac{2}{3}$, what is the difference?

Section 2.6: Order of Operations with Fractions and Mixed Numbers

Answer the questions below.

54. Which is larger, $\frac{2}{3}$ or $\frac{4}{5}$? How much larger?

55. Arrange the numbers in order from smallest to largest. Then find the difference between the largest and smallest numbers. See Example 3.

$$\frac{7}{12},\ \frac{5}{9},\ \frac{11}{20}$$

Evaluate each expression by using the rules for order of operations.

56. $\dfrac{5}{8}\cdot\dfrac{3}{10}+\dfrac{1}{14}\div 2$

57. $\left(\dfrac{3}{5}-\dfrac{1}{3}\right)\div\left(\dfrac{1}{6}\div\dfrac{7}{8}\right)$

58. $\dfrac{7}{15}+\dfrac{5}{9}\div\dfrac{2}{3}-\dfrac{2}{3}$

59. $\left(\dfrac{2}{3}\right)^2 - \left(\dfrac{1}{3}\right)^2 + \dfrac{1}{18}$

60. $\left(\dfrac{3}{8} + \dfrac{1}{2}\right) \div \left(\dfrac{1}{2} - \dfrac{1}{10}\right)$

Simplify the complex fractions.

61. $\dfrac{\left(1\dfrac{1}{9} + \dfrac{5}{18}\right)}{1 + \dfrac{1}{3}}$

62. $\dfrac{\left(2\dfrac{1}{2} - 2\dfrac{1}{3}\right)}{1\dfrac{1}{2} + \dfrac{1}{6}}$

Evaluate each expression by using the rules for order of operations.

63. $4\dfrac{2}{7} \div 3\dfrac{3}{5} + 4\dfrac{1}{6} \cdot 2\dfrac{4}{5}$

64. $2\dfrac{3}{10} - 5\dfrac{3}{5} \div 4\dfrac{2}{3} + 2\dfrac{1}{6}$

65. $\left(\dfrac{7}{8} - \dfrac{3}{16}\right) \div \left(\dfrac{1}{3} - \dfrac{1}{4}\right)$

66. $\left(\dfrac{1}{5} + \dfrac{1}{6}\right) \div 2\dfrac{1}{3}$

Solve the following problem.

67. Find the average of $1\dfrac{3}{10}$, $2\dfrac{1}{5}$, and $1\dfrac{3}{4}$.

Chapter 2: Test

Reduce to lowest terms.

1. $\dfrac{90}{108}$

2. $\dfrac{117}{156}$

Change the numbers as indicated.

3. Change $3\dfrac{1}{10}$ to improper fraction form.

4. Change $\dfrac{35}{3}$ to mixed number form.

Perform the indicated operations and reduce to lowest terms.

5. $\dfrac{3}{7} \cdot \dfrac{14}{27}$

6. $10 \div \dfrac{2}{5}$

7. $\dfrac{3}{5} \cdot \dfrac{1}{2} \cdot \dfrac{3}{8}$

8. $6\dfrac{2}{5} \cdot 3\dfrac{1}{8}$

9. $2\dfrac{2}{3} \cdot 2\dfrac{1}{4} \cdot \dfrac{5}{8}$

10. $\dfrac{5}{6} \div 2\dfrac{1}{2}$

11. $4\dfrac{7}{8} \div 3\dfrac{1}{4}$

Solve the following problem.

12. Find the LCM for the following set of numbers and tell how many times each number divides into the LCM: $20, 25, 35$.

Perform the indicated operations and reduce to lowest terms.

13. $\dfrac{4}{35} + \dfrac{2}{7} + \dfrac{1}{10}$

14. $\dfrac{17}{19} + \dfrac{6}{19} + \dfrac{15}{19}$

15. $\begin{aligned} 9\dfrac{3}{8} \\ + \dfrac{5}{8} \\ \hline \end{aligned}$

16. $\begin{aligned} 7 \\ - 5\dfrac{3}{5} \\ \hline \end{aligned}$

17. $4\dfrac{3}{10} + 2\dfrac{3}{4}$

Simplify the complex fraction.

18. $\dfrac{6\dfrac{2}{3}}{\dfrac{1}{2} + 1\dfrac{3}{4}}$

Evaluate each expression by using the rules for order of operations.

19. $\left(\dfrac{1}{2}\right)^3 - \left(\dfrac{1}{4}\right)^2$ **20.** $\left(\dfrac{51}{16} - 3\right) \div \dfrac{3}{8}$ **21.** $\left(6\dfrac{1}{5} + 3\dfrac{7}{10}\right) \div \left(5\dfrac{3}{10} - 4\dfrac{1}{2}\right)^2$ **22.** $\dfrac{5}{16} + \left(\dfrac{1}{4}\right)^2$

Arrange the numbers in order from smallest to largest. Then find the difference between the largest and smallest numbers

23. $\dfrac{7}{8}$, $\dfrac{3}{4}$, and $\dfrac{2}{3}$

Solve the following word problems.

24. Running: During three days of practice, a high school cross country team member ran $8\dfrac{3}{10}$ miles, $14\dfrac{1}{5}$ miles, and $6\dfrac{1}{4}$ miles.

    a. How far did she run during those three days?

    b. What was her average distance per day?

25. Geometry: Find the perimeter of a rectangle that is 8 inches wide and $11\dfrac{1}{2}$ inches long.

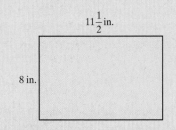

Cumulative Review: Chapters 1 - 2

Solve the problems below.

1. Write 53,460 in expanded notation and in its English word equivalent.

2. Round 265,400 to the nearest ten thousand.

3. Name the property of addition illustrated: $16 + 35 = 35 + 16$.

4. Multiply mentally: $800(7000)$.

5. Use the rules for order of operations to evaluate the following expression: $7^2 + 2(12 \cdot 4 \div 2 \cdot 3) - 7 \cdot 5$.

6. List all the prime numbers less than 30.

7. Find the prime factorization of each number.
   a. 170   **b.** 305

8. The value of $0 \div 36$ is _____, while $36 \div 0$ is _____.

9. Find the average of $44, 35, 53,$ and 40.

Perform the indicated operations and reduce all fractions to lowest terms.

10. $\begin{array}{r} 8597 \\ + 4653 \\ \hline \end{array}$

11. $\begin{array}{r} 3782 \\ - 1255 \\ \hline \end{array}$

12. $\begin{array}{r} 732 \\ \times \ 68 \\ \hline \end{array}$

13. $13\overline{)2639}$

14. $\begin{array}{r} 125 \\ + 57 \\ \hline \end{array}$

15. $\dfrac{7}{8} \div 6$

16. $\dfrac{7}{18} \cdot \dfrac{3}{10} \cdot \dfrac{12}{28}$

17. $\dfrac{7}{8} - \dfrac{7}{12}$

18. $\dfrac{9}{10} + \dfrac{4}{5} + \dfrac{1}{6}$

19. $\begin{array}{r} 13\frac{4}{5} \\ - 8\frac{4}{7} \\ \hline \end{array}$

20. $4\dfrac{3}{8} \div 2\dfrac{1}{2}$

21. $5 \cdot 5\dfrac{2}{5} \cdot 5\dfrac{5}{6}$

22. $\left(1\dfrac{3}{10} + 2\dfrac{3}{5}\right) \div \left(2\dfrac{4}{5} - 1\dfrac{1}{2}\right)$

23. $\left(\dfrac{3}{4}\right)^2 - \dfrac{1}{8} + \dfrac{5}{2} \cdot \dfrac{1}{3}$

24. $\dfrac{5+3\dfrac{1}{5}}{2-1\dfrac{1}{3}}$

25. Without actually dividing, determine which of the numbers 2, 3, 4, 5, 6, 9, and 10 will divide 8190. Give a brief reason for each.

26. Determine whether or not the number 431 is prime.

27. Find the LCM for each set of numbers and tell how many times each number divides into the LCM.

    a. 18, 42, 90   **b.** 30, 60, 84, 96

28. **Banking:** Samantha opened her checking account with a deposit of \$5380. She wrote checks for \$95, \$265, \$107, and \$1573 and made another deposit of \$340. What was the new balance in her account?

29. **Number Problem:** If the quotient of 119 and 17 is subtracted from the product of 23 and 34, what is the difference?

30. **Perimeter:** A triangle has dimensions as shown in the figure. Find the perimeter of the triangle.

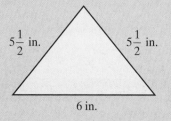

$5\dfrac{1}{2}$ in. $5\dfrac{1}{2}$ in.

6 in.

31. **Perimeter:** A soccer field is in the shape of a rectangle 55 yards by 100 yards. Find the perimeter of the soccer field.

32. **Manufacturing:** A machinist starts with an $\dfrac{11}{12}$ meter long rod. How long will the remaining rod be after a $\dfrac{4}{9}$ meter piece is cut off?

33. **Cooking:** A person prepares $4\dfrac{1}{8}$ quarts of soup. How many $\dfrac{3}{8}$ quart servings will this provide?

34. **Grocery Shopping:** A person has just obtained a piece of beef weighing $8\dfrac{3}{4}$ pounds. If she gives $\dfrac{2}{5}$ of this piece of beef to a friend, how many pounds of beef does she give?

35. **Maple Syrup:** After boiling sap from some maple trees, Aaron ends up with $5\dfrac{1}{2}$ gallons of maple syrup. One neighbor takes $2\dfrac{1}{3}$ gallon of it, while another neighbor takes $\dfrac{3}{4}$ gallon.

    a. How many gallons of maple syrup is taken from the original batch?

    b. How many gallons of maple syrup remains?

Decimal Numbers

Mathematics at Work!

In 1996, Lance Armstrong was diagnosed with testicular cancer that had spread to his brain and lungs. Against all odds, in 1999 he won his first Tour de France bicycling race by 7 minutes 37 seconds. In 2005 he won the Tour for a record seventh consecutive time. The race covered 2242 miles in all kinds of weather conditions and over mountain passes, switchback roads and winding paths. His time was 86 hours 15 minutes 2 seconds. This time was 4 minutes 40 seconds better than the second place finisher, Ivan Basso of Italy, and 6 minutes 21 seconds better than the third place finisher, Jan Ullrich of Germany. How would you find the average speed of these three cyclists? (See Exercises 74 and 75 in Section 3.4).

3.1 Introduction to Decimal Numbers

Objective A **Reading and Writing Decimal Numbers**

The common **decimal notation** uses a **place value system** and a **decimal point**, with whole numbers written to the left and fractions written to the right of the decimal point. Numbers written in decimal notation are said to be **decimal numbers** (or simply **decimals**). The values of several places in this decimal system are shown in Figure 1.

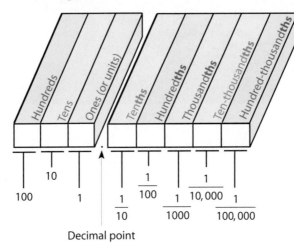

Decimal point

Figure1

Familiar decimal numbers involve dollars and cents. For example, you might pay $45.25 for a new shirt or $3.50 for a taco. These numbers are mixed numbers and can be written as:

$$\$45\frac{25}{100} = \$45.25 \quad \text{Read "forty-five dollars and 25 cents."}$$

$$\$3\frac{50}{100} = \$3.50 \quad \text{Read "three dollars and fifty cents."}$$

In general, decimal numbers are read (and written) according to the following convention.

To Read or Write a Decimal Number

1. Read (or write) the whole number.
2. Read (or write) the word "**and**" in place of the decimal point.
3. Read (or write) the fraction part as a whole number. Then name the fraction with the name of the place of the last digit on the right.

notes

■ If there is no whole number part, then 0 can be written to the left of the decimal point. For example, 0.6 can be written for six tenths.

■

Example 1

Reading and Writing Decimal Numbers

Write the mixed number $48\frac{6}{10}$ in decimal notation and in words.

Solution

$$\underset{\text{forty-eight}}{\underline{4\ 8}}\ .\ \underset{\text{six tenths}}{\underline{6}}$$

     forty-eight and six tenths   in words

And indicates the decimal point; the digit 6 is in the tenths position.

Example 2

Reading and Writing Decimal Numbers

Write the mixed number $12\frac{75}{10,000}$ in decimal notation and in words.

Solution

Two 0s must be inserted as placeholders.

1 2 . 0 0 7 5   in decimal notation

twelve   and   seventy-five ten-thousandths   in words

And indicates the decimal point; the digit 5 is in the ten-thousandths position.

Now work margin exercises 1 and 2.

> ## notes
>
> The letters **th** at the end of a word indicate a fraction part (a part to the right of the decimal point).
>
> six hundred = 600   two hundred thousand = 200,000
> six hundred**ths** = 0.06   two hundred-thousand**ths** = 0.00002

Example 3

Reading and Writing Decimal Numbers

Write **four hundred and two thousandths** in decimal notation. (Note how **and** indicates the decimal point.)

Solution

Two 0s are inserted as placeholders.

400.002   The digit 2 is in the thousandths position.

Write each mixed number in decimal notation and in words.

1. $19\frac{3}{10}$

2. $39\frac{184}{10,000}$

Write each of the following in decimal notation.

3. one thousand two hundred and thirty-five ten-thousandths

4. one thousand two hundred thirty-five ten-thousandths

Example 4

Reading and Writing Decimal Numbers

Write **four hundred two thousandths** in decimal notation.

Solution

0.402

Note carefully how the use of **and** in the phrase in Example 3 gives it a completely different meaning from the phrase in this example.

Now work margin exercises 3 and 4.

Objective B ## Comparing Decimal Numbers

Determining the relative size of numbers, as we did with fractions in Chapter 2, can be a very useful skill. The following technique can be used to compare two decimal numbers to see which is smaller or larger.

To Compare Two Decimal Numbers

1. Moving **left to right**, compare digits with the same place value. (Insert 0s to the right to continue comparison if necessary.)
2. When one compared digit is larger, then the corresponding number is larger.

Example 5

Comparing Decimal Numbers

Which number is larger: 3.126 or 3.14?

Solution

Comparing digits (from left to right) gives:

$$3.1\overset{\updownarrow}{2}6$$
$$3.140$$

Because $4 > 2$, the number 3.14 is greater than 3.126. That is, $3.14 > 3.126$.

Example 6

Comparing Decimal Numbers

Which number is larger: 0.08 or 0.085?

Solution

Comparing digits (from left to right) gives:

0.080

$\updownarrow$

0.085

Because $5 > 0$, $0.085 > 0.08$.

Now work margin exercises 5 and 6.

Example 7

Comparing Decimal Numbers

Write the following three numbers in order, smallest to largest:
6.37, 5.14, 6.28

Solution

Comparing digits from left to right:

In the ones place $5 < 6$ so 5.14 is the smallest number.
Then, in the tenths place $2 < 3$ so 6.28 is smaller than 6.37.

Thus, from smallest to largest, the numbers are 5.14, 6.28, 6.37.

Now work margin exercise 7.

Objective C **Rounding Decimal Numbers**

Measuring devices such as rulers, meter sticks, speedometers, micrometers, and surveying transits give only approximate measurements (See Figure 2). Whether the units are large (such as miles and kilometers) or small (such as inches and centimeters), there are always smaller, more accurate units (such as eighths of an inch and millimeters) that could be used to indicate a measurement. We are constantly dealing with approximate (or rounded) numbers in our daily lives. If a recipe calls for 1.5 cups of flour and the cook puts in 1.53 cups (or 1.47 cups), the result will still be reasonably tasty. In fact, the measures of all ingredients will have been approximations.

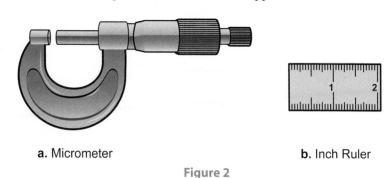

a. Micrometer   **b.** Inch Ruler

Figure 2

5. Which number is larger:
6.438 or 6.44?

6. Which number is larger:
9.25 or 9.251?

7. Write the following three numbers in order, smallest to largest.

2.1, 2.01, 2.11

When rounding a decimal number, the desired place of accuracy must be stated. For example, looking at the number line below, we can see that 5.77 is closer to 5.8 than it is to 5.7. So (to the nearest tenth), 5.77 rounds to 5.8.

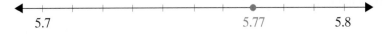

5.7   5.77   5.8

There are several rules for rounding decimal numbers. The IRS, for example, allows rounding to the nearest dollar on income tax forms. The rule chosen in a particular situation depends on the use of the numbers and whether there might be some sort of penalty, such as a spaceship not docking properly.

In this text, we will use the following rule for rounding decimal numbers.

Rules for Rounding Decimal Numbers

1. Look at the digit to the right of the place of desired accuracy.
2. If this digit is 5 or greater, make the digit in the desired place of accuracy one larger and replace all digits to the right with zeros. All digits to the left remain unchanged unless a 9 is made one larger. This effectively changes the 9 to 10 which means the next digit to the left must be increased by 1.
3. If this digit is less than 5, leave the digit in the desired place of accuracy as it is and replace all digits to the right with zeros. All digits to the left remain unchanged.
4. Zeros **to the right of the place of accuracy** and to the right of the decimal point must be dropped. In this way the place of accuracy is clearly understood. If a rounded number has a 0 in the desired place of accuracy, then that 0 remains.

Example 8

Rounding Decimal Numbers

Round 18.649 to the nearest tenth.

Solution

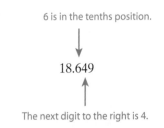

6 is in the tenths position.

18.649

The next digit to the right is 4.

Since 4 is less than 5, leave the 6 and replace 4 and 9 with 0s.

Note that the 0s in 18.600 are dropped to indicate the position of accuracy. Thus 18.649 rounds to 18.6 to the nearest tenth.

Example 9

Rounding Decimal Numbers

Round 5.83971 to the nearest thousandth.

Solution

9 is in the thousandths position.

5.83971

The next digit to the right is 7.

Since 7 is greater than 5, make 9 one larger and replace 7 and 1 with 0s. (Making 9 one larger gives 10, which affects the digit 3 too.)

5.83971 rounds to 5.840 to the nearest thousandth, and only two 0s are dropped.

Completion Example 10

Rounding Decimal Numbers

Round 2.00643 to the nearest ten-thousandth.

Solution

a. The digit in the ten-thousandths position is _____.

b. The next digit to the right is _____.

c. Since ____ is less than 5, leave ____ as it is and replace ____ with a 0.

d. 2.00643 rounds to _____ to the nearest _____.

Completion Example Answers

10. a. The digit in the ten-thousandths position is 4.
  b. The next digit to the right is 3.
  c. Since 3 is less than 5, leave 4 as it is and replace 3 with a 0.
  d. 2.00643 rounds to 2.0064 to the nearest ten-thousandth.

Round each number to the place indicated.

8. 8.637 (nearest tenth)

9. 5.042 (nearest hundredth)

10. 0.01792 (nearest thousandth)

11. 239.53 (nearest ten)

Completion Example 11

Rounding Decimal Numbers

Round 9653 to the nearest hundred.

Solution

a. The decimal point is understood to be to the right of _____.

b. The digit in the hundreds position is _____.

c. The next digit to the right is _____.

d. Since _____ is equal to 5, change the _____ to _____ and replace _____ and _____ with 0s.

e. So, 9653 rounds to _____ (to the nearest hundred).

Now work margin exercises **8 through 11.**

Practice Problems

1. Write 20.7 in words.

2. Write 18.051 in words.

3. Write $4\dfrac{6}{100}$ in decimal notation.

4. Write eight hundred and three tenths in decimal notation.

Round as indicated.

5. 572.3 (to the nearest ten)

6. 6.749 (to the nearest tenth)

7. A penny dated from 1959 through 1982 had an original weight of 3.11 grams. A penny dated 1983 or later had an original weight of 2.5 grams. Write the numbers representing weights in words.

Completion Example Answers

11. a. The decimal point is understood to be to the right of 3.
b. The digit in the hundreds position is 6.
c. The next digit to the right is 5.
d. Since 5 is equal to 5, change the 6 to 7 and replace 5 and 3 with 0s.
e. So 9653 rounds to 9700 (to the nearest hundred).

Practice Problem Answers

1. twenty and seven tenths
2. eighteen and fifty-one thousandths
3. 4.06   **4.** 800.3   **5.** 570   **6.** 6.7
7. three and eleven hundredths; two and five tenths

Exercises 3.1

Write the following mixed numbers in decimal notation. See Examples 1 and 2.

1. $6\dfrac{5}{10}$ 2. $82\dfrac{3}{100}$ 3. $19\dfrac{75}{1000}$ 4. $100\dfrac{25}{100}$ 5. $62\dfrac{547}{1000}$

6. $2\dfrac{57}{100}$ 7. $13\dfrac{2}{100}$ 8. $38\dfrac{4}{1000}$ 9. $200\dfrac{6}{10}$ 10. $50\dfrac{1}{1000}$

Write the following decimal numbers in words. See Examples 1 and 2.

11. 0.9   12. 0.53   13. 6.05   14. 6.004

15. 50.007   16. 19.102   17. 800.009   18. 0.809

19. 5000.005   20. 25.4538

Write the following numbers in decimal notation. See Examples 3 and 4.

21. four tenths

22. fifteen thousandths

23. twenty-three hundredths

24. five and twenty-eight hundredths

25. five and twenty-eight thousandths

26. seventy-three and three hundred forty-one thousandths

27. six hundred and sixty-six hundredths

28. six hundred and sixty-six thousandths

29. three thousand four hundred ninety-five and three hundred forty-two thousandths

30. seven thousand five hundred and eighty-three ten-thousandths

Find the larger number in each pair. See Examples 5 and 6.

31. 0.26, 0.27   32. 0.153, 0.163   33. 0.01, 0.009   34. 7.006, 7.015

35. 0.0499, 0.0488   36. 0.00576, 0.00476   37. 23.521, 24.645   38. 14.158, 14.358

Arrange the numbers in order, from smallest to largest.   See Example 7.

39. 0.03, 0.003, 0.33

40. 0.25, 0.26, 0.19

41. 1.762, 2.51, 2.3644

42. 0.157, 0.2611, 0.192, 0.26

43. 10.113, 11.4, 9.52, 9.523

44. 0.031, 0.0521, 0.006, 0.0078

Fill in the blanks to correctly complete each statement. See Examples 10 and 11.

45. Round 34.78 to the nearest tenth.
   a. The digit in the tenths position is ____.
   b. The next digit to the right is ____.
   c. Since ____ is greater than 5, change ____ to ____ and replace ____ with 0.
   d. So 34.78 rounds to _____ to the nearest tenth.

46. Round 3.00652 to the nearest ten-thousandth.
   a. The digit in the ten-thousandths position is ____.
   b. The next digit to the right is ____.
   c. Since ____ is less than 5, leave ____ as it is and replace ____ with 0.
   d. So 3.00652 rounds to _____ to the nearest _____.

Round each of the following decimal numbers to the nearest tenth See Examples 8 through 11.

47. 89.016

48. 8.555

49. 18.123

50. 0.076

51. 19.961

52. 46.444

Round each of the following decimal numbers to the nearest hundredth.

53. 0.385

54. 0.296

55. 7.997

56. 13.1345

57. 0.0764

58. 6.0035

Round each of the following decimal numbers to the nearest thousandth.

59. 0.0572

60. 0.6338

61. 0.00191

62. 20.76953

63. 32.4578

64. 1.66666

65. 479.32

66. 7.8

67. 163.5

68. 701.41

69. 300.5

70. 29.999

71. 15,963

72. 6475

73. 76,523.2

74. 435.7

75. 453.7

76. 1572.36

77. 62,375

78. 75,544

79. 103,947

80. 4,500,766

81. 7,305,438

82. 573,333.15

83. **Unicycles:** The tallest unicycle ever ridden, by Steve McPeak for 376 feet in 1980 in Las Vegas, was 101.75 feet tall.

84. **Measuring:** One foot is equal to 12 inches. One foot is also equal to 30.48 centimeters. One square foot is approximately 0.093 square meter.

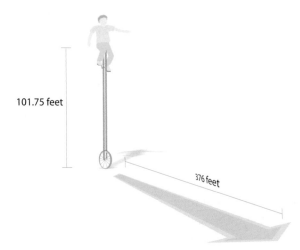

101.75 feet

376 feet

85. **Math Symbols:** The number π is approximately equal to 3.14159.

86. **Math Symbols:** The number e is approximately equal to 2.71828.

87. Weighing: One quart of water weighs approximately 2.0825 pounds.

2.0825 lbs.

88. Banking: Suppose you went to the bank and deposited three checks. The first check was for $656.43. The second check was for $268.68. Finally, the third check was for $65.59.

89. Life Expectancy: An interesting fact is that your life expectancy increases the older you become. A white male of age 40 can expect to live 35.8 more years; of age 50, can expect to live 26.9 more years; of age 60 can expect to live 18.9 more years; of age 70 can expect to live 12.3 more years; and of age 80 can expect to live 7.2 more years. (This same phenomenon is true of men and women of all races.)

90. Astronomy: One period of revolution of the Earth around the Sun takes 365.2 days, and one period of revolution of Venus around the Sun takes 224.7 days.

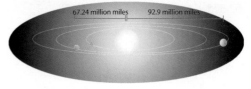

67.24 million miles   92.9 million miles

91. Measuring: One yard is equal to 36 inches. One yard is also approximately equal to 0.914 meter. One meter is approximately equal to 1.09 yards. One meter is also approximately equal to 39.37 inches. (Thus a meter is longer than a yard by about 3.37 inches.)

92. World records: 9.58 seconds for 100 meter dash (by Usain Bolt, Jamaica, 2009); 19.19 seconds for 200 meters dash (by Usain Bolt, Jamaica, 2009); 43.18 seconds for 400 meters dash (by Michael Johnson, USA, 1999).

Writing & Thinking

93. In your own words, state why the word "and" is so commonly misused when numbers are spoken and/or written. Bring an example of this from a newspaper, magazine, or television show to class for discussion.

3.2 Addition and Subtraction with Decimal Numbers

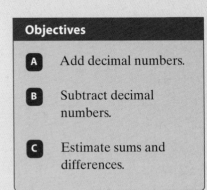
Objective A **Addition with Decimal Numbers**

Addition with decimal numbers can be accomplished by writing the decimal numbers in a vertical format with the decimal points aligned. In this way, whole numbers are added to whole numbers, tenths to tenths, hundredths to hundredths, and so on. The decimal point in the sum is in line with the decimal points in the addends. For example,

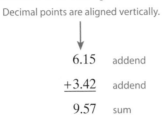

Decimal points are aligned vertically.

$$
\begin{array}{r}
6.15 \quad \text{addend} \\
+3.42 \quad \text{addend} \\
\hline
9.57 \quad \text{sum}
\end{array}
$$

Any number of 0s may be written to the right of the last digit in the fraction part of a number to help keep the digits in the correct alignment. This will not change the value of any number or the value of the sum.

To Add Decimal Numbers

1. Write the addends vertically.
2. Keep the decimal points aligned vertically.
3. Keep digits with the same position value aligned. (Zeros may be written in to help keep the digits aligned properly.)
4. Add, just as with whole numbers, keeping the decimal point in the sum aligned with the other decimal points.

Example 1

Adding Decimal Numbers

Find the sum: $6.3 + 5.42 + 14.07$

Solution

Decimal points are aligned vertically.

$$
\begin{array}{r}
6.3\,0 \quad \longleftarrow \quad \text{0 may be written in to help keep the digits aligned.} \\
5.4\,2 \\
+\ 14.07 \\
\hline
25.79
\end{array}
$$

Example 2

Adding Decimal Numbers

Find the sum: $9 + 4.86 + 37.479 + 0.6$

Solution

$$
\begin{array}{r}
9.000 \\
4.860 \\
37.479 \\
+\ 0.600 \\
\hline
51.939
\end{array}
$$

The decimal point is understood to be to the right of 9, as in 9.0, 9.00, or 9.000.

0s are written in to help keep the digits aligned properly.

Find each sum.

1. $14.86 + 5 + 9.1$

2. $23.8 + 4.2567 + 11 + 3.01$

3.
$$
\begin{array}{r}
1.093 \\
37 \\
+\ 24.58 \\
\hline
\end{array}
$$

Example 3

Adding Decimal Numbers

Find the sum:
$$
\begin{array}{r}
56.2 \\
85.75 \\
+\ 29.001 \\
\hline
\end{array}
$$

Solution

You can write
$$
\begin{array}{r}
56.200 \\
85.750 \\
+\ 29.001 \\
\hline
170.951
\end{array}
$$

0s are written in to keep the digits aligned.

Now work margin exercises 1 through 3.

4. Mr. Riley went to the furniture store and bought a rug for $50.79, a lamp for $33.68 and a nightstand for $72.11. How much did he spend? (Tax was included in the prices.)

Example 4

Adding Decimal Numbers

Mrs. Finn went to the local store and bought a pair of shoes for $42.50, a blouse for $25.60, and a skirt for $37.55. How much did she spend? (Tax was included in the prices.)

Solution

$$
\begin{array}{r}
\$\ 42.50 \\
25.60 \\
+\quad 37.55 \\
\hline
\$105.65
\end{array}
$$

She spent $105.65.

Now work margin exercise 4.

Objective B   **Subtraction with Decimal Numbers**

> ### To Subtract Decimal Numbers
>
> 1. Write the numbers vertically.
> 2. Keep the decimal points aligned vertically.
> 3. Keep digits with the same position value aligned. (Zeros may be written in to help keep the digits aligned properly.)
> 4. Subtract, just as with whole numbers, keeping the decimal point in the difference aligned with the other decimal points.

Example 5

Subtracting Decimal Numbers

Find the difference: $16.715 - 4.823$

Solution

$$
\begin{array}{r}
16.\overset{5}{\cancel{7}}\overset{1}{1}5 \\[-2pt]
-\ \ 4.823 \\[-2pt]
\hline
11.892
\end{array}
$$

Example 6

Subtracting Decimal Numbers

Find the difference: $21.2 - 13.716$

Solution

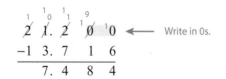   ← Write in 0s.

$$
\begin{array}{r}
21.2\,0\,0 \\[-2pt]
-13.716 \\[-2pt]
\hline
7.484
\end{array}
$$

***Now work margin exercises* 5 and 6.**

Objective C   **Estimating Sums and Differences**

We can estimate a sum (or difference) by rounding each number to the place of the **leftmost nonzero digit** and then adding (or subtracting) these rounded numbers. This technique of estimating answers is especially helpful when working with decimal numbers, where the placement of the decimal point is so important.

Find each difference.

5. $9.262 - 3.07$

6. $170.44 - 28.973$

7. Estimate and then find the sum:

$$6.68 + 103.5 + 21.94$$

Example 7

Estimating Sums and Differences

a. Estimate the sum: $74 + 3.529 + 52.61$.

Solution

Estimate by adding rounded numbers.

74	rounds to →	70
3.529	rounds to →	4
52.61	rounds to →	+ 50
		124 ← estimate

b. Find the actual sum.

Solution

Find the actual sum.

$$
\begin{array}{r}
74.000 \\
3.529 \\
+ 52.610 \\
\hline
130.139 \end{array} \longleftarrow \text{actual sum}
$$

Now work margin exercise 7.

Practice Problems

Find each indicated sum or difference.

1. $46.2 + 3.07 + 2.6$ **2.** $9 + 5.6 + 0.58$

3. $6.4 - 3.7$ **4.** $18 - 0.4384$

5.	**6.**	**7.**	**8.**
1.1	0.09	51.3	78.1
0.32	65.1	− 6.29	− 69.3
2.4	9.7		
+ 6.01	+ 2.55		

Practice Problem Answers

1. 51.87   **2.** 15.18   **3.** 2.7   **4.** 17.5616
5. 9.83   **6.** 77.44   **7.** 45.01   **8.** 8.8

Exercises 3.2

Find each of the indicated sums. Estimate your answers, either mentally or on paper, before doing the actual calculations. Check to see that your sums are close to the estimated values. See Examples 1 through 4 and 7.

1. $0.6 + 0.4 + 1.3$

2. $5 + 6.1 + 0.4$

3. $0.59 + 6.91 + 0.05$

4. $3.48 + 16.6 + 25.02$

5.
```
  37.02
  25.0
   6.4
+  3.89
```

6.
```
  4.0086
  0.034
  0.6
+ 0.05
```

7.
```
   43.766
    9.33
   17.0
+ 206.0
```

8.
```
  52.3
   6.0
  21.01
+  4.005
```

9.
```
   2.051
   0.2006
   5.4
+ 37.0
```

10.
```
    5
    2.37
  463.0
+  10.88
```

11.
```
  47.3
  42.03
+ 29.003
```

12.
```
   1.007
  20.063
+  0.49
```

13.
```
  4.128
  0.02
+ 3.0
```

14.
```
  5.0015
  2.443
+ 0.0469
```

15.
```
  75.2
   3.682
+ 14.995
```

16.
```
  107.39
    5.061
   23.54
+  64.9801
```

17.
```
  34.967
  50.6
   8.562
+  9.3
```

18.
```
   4.156
   3.7
  25.682
+ 13.405
```

19.
```
  74.0
   3.529
  52.62
+  7.001
```

20.
```
  983.4
   47.518
  805.411
+ 300.766
```

Find each of the indicated differences. First estimate the differences mentally. See Examples 5 and 6.

21. $5.2 - 3.76$

22. $17.83 - 8.9$

23. $29.5 - 13.61$

24. $1.0057 - 0.03$

25.
```
  78.015
- 13.068
```

26.
```
  22.418
- 17.523
```

27.
```
  4.8
- 0.0026
```

28.
```
  31.009
- 0.534
```

29.
```
  4.0
- 1.0566
```

30.
```
  40.718
-  6.532
```

31.
```
  16.1
-  2.59
```

32.
```
  9.028
- 7.135
```

33.
```
  52.1
-  6.952
```

34.
```
  81.72
- 33.065
```

35.
```
  46.321
-  9.174
```

36.
```
  23.006
-  5.999
```

Find the estimated sum using rounded numbers. Then find the actual sum. See Example 7.

37. 58.2
      + 61.02

38. 29.63
      + 3.79

39. 51.07
       35.2
      + 6.19

40.   4.22
       71.6
      + 36.75

41. 121.6
       55.9
        8.32
      + 21.63

42. 44.4
       3.211
       0.19
      + 5.6

Find the estimated difference using rounded numbers. Then find the actual difference. See Example 7.

43. 51.21
      − 25.13

44. 1.345
      − 0.0691

45. 22
      − 12.91

46. 204
      − 38.08

47. 3.21
      − 0.589

48. 7521.22
      − 973.16

Find the perimeter of each figure.

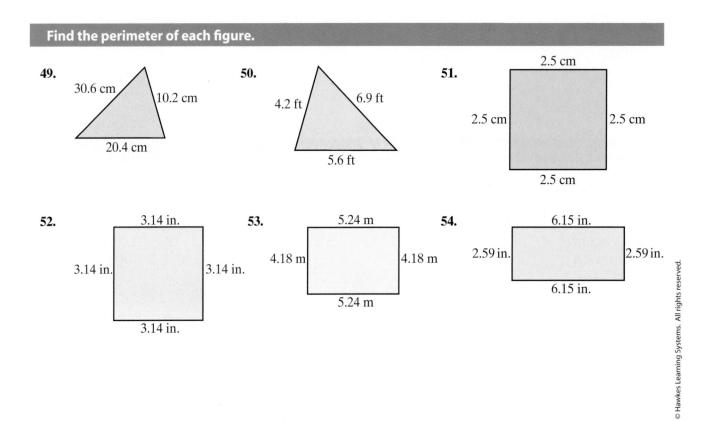

49. 30.6 cm, 10.2 cm, 20.4 cm

50. 4.2 ft, 6.9 ft, 5.6 ft

51. 2.5 cm, 2.5 cm, 2.5 cm, 2.5 cm

52. 3.14 in., 3.14 in., 3.14 in., 3.14 in.

53. 5.24 m, 4.18 m, 4.18 m, 5.24 m

54. 6.15 in., 2.59 in., 2.59 in., 6.15 in.

55. Goods and services: Theresa got a haircut for $30.00 and a manicure for $10.50. If she tipped the stylist $5, how much change did she receive from a $50 bill?

56. Goods and services: Mr. Johnson bought the following items at a department store: slacks, $32.50; shoes, $43.75; shirt, $18.60.

    a. How much did he spend?

    b. What was his change if he gave the clerk a $100 bill? (Tax was included in the prices.)

57. Construction: An architect's scale drawing shows a rectangular lot 2.38 inches on one side and 3.76 inches on the other side. What is the perimeter of (distance around) the rectangle in the drawing below?

2.38 in.

3.76 in.

58. Shipping: Jim is packing three books in a box for shipping. The weights of the books are 5.63 pounds, 12.4 pounds , and 3 pounds. If the shipping materials weigh 1.74 pounds, what is the total weight of the filled box?

    a. Estimate the total weight.

    b. Find the actual weight.

59. Carpentry: Three pieces of wood are to be cut off of a board which is 9.7 feet long. The respective lengths being cut off are 2 feet, 3.12 feet, and 1.85 feet. Assuming that no wood is lost in the cutting process, how much is left over from the original board? This must be done in two steps:

    a. Add the total length of the three boards which were cut off.

    b. Subtract this length from the original length of the board to find out what remains.

60. Gardening: David is preparing a four-sided garden plot with unequal sides of 7.5 feet, 26.34 feet, 36.92 feet, and 12.07 feet, respectively. How many feet of edging material must he use? (This is the same as finding the perimeter of the plot.)

    a. Estimate the total length.

    b. Compute the actual total length (or perimeter).

61. Driving Time: A road trip is being planned from New York City, NY to Lawrence, KS. The first leg of the trip will be to Cleveland, OH, which will take 2.817 hours driving time; then on to Des Moines, IA, which will take 11.15 hours; followed by Kansas City, MO, which will take exactly 3 hours; and the final leg to Lawrence, KS, which will take 0.7 hour.

Source: 2007 AAA Road Atlas

    a. Estimate the driving time.

    b. Compute the actual driving time.

62. Payroll: Sally is an hourly worker who works a different number of hours each week. For a given month she receives four paychecks as follows: $325.27, $450.83, $273.30, and $510.98. Each month she has to pay $352.78 for car payments, $650 for rent, and $55.25 for insurance. How much does Sally have remaining for all of her other expenses? To find the answer, do the following three steps:

    a. Compute her monthly pay.

    b. Compute her total of the three listed monthly expenses.

    c. Subtract her expenses from her monthly pay.

63. Boxes: Five boxes of unequal size are placed side-by-side along a wall, where the first box is 2.36 feet wide, the second is 1.76 feet wide, the third is 3.8 feet wide, the fourth is 0.94 feet wide, and the fifth is 6.17 feet wide. How much wall space is taken up?

a. Estimate this length.

b. Compute the actual amount of wall space.

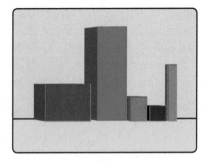

64. Milk: A group of people own three dairy cows to supply their own milk needs. During one milking, the first cow produced 2.79 gallons of milk, the second produced 5.34 gallons, while the third added 4.02 gallons. All of this milk was put into a 15.4 gallon holding bucket. **Source:** http://www.raw-milk-facts.com/dairy_cow_breeds.html

a. How much milk did the three cows produce during this milking?

b. How much more milk can the bucket hold?

65. Baking: An oven with very precise temperature control is 400.2°F. If the temperature is reduced by 15.327 degrees, what will be the new temperature?

66. Book: The total thickness of the pages of a book is 2.54 inches. If each of the covers is 0.187 inch, what is the total thickness of the book? To determine this, you must add the thickness of each of the two covers to the thickness of the pages.

67. Car Purchase: Carol has saved $19,273.22 towards the purchase of a new car, and she will take out a loan for any extra cost. The agreed upon price of the car with its accessories is $23,925.50, and the taxes, registration fees, and insurance add up to $1,437.62. How much must Carol borrow? (**Hint:** To solve this, first add up the total cost for the car and taxes, etc. Next, subtract Carol's savings to determine how much must be borrowed.)

68. Overtime: A certain generous company has the policy of paying overtime for all hours over 37.75 worked during a week. The daily hours John worked this week were, Day 1: 8.33 hours; Day 2: 6.2 hours; Day 3: 9.87 hours; Day 4: 8.5 hours; Day 5: 10.12 hours. How many hours of overtime will John receive this week? (**Hint:** First add the total number of hours worked during the week, and then subtract the threshold of 37.75 hours. The remainder will be the overtime hours.)

69. Steel Alloy: A special steel alloy for knife blades consists of iron, along with a mixture of other chemical elements. What is the total weight of the "ingredients" if 10 tons of iron, is combined with 0.28 ton of carbon, 1.58 ton of chromium, 0.269 ton of vanadium, and 0.0488 ton of manganese? **Source:** http://zknives.com/knives/steels/steelchart.ph

70. Grocery Purchase: Kevin made the following purchases at the local grocery store: steak for $11.27, lettuce for $2.85, one tomato for $0.63, plus $0.44 for sales tax.

a. Estimate the cost of the market basket.

b. Compute the actual cost.

c. If Kevin pays with a $20 bill, how much change will he receive?

71. College Supplies: A college student is purchasing the necessary textbooks and supplies for an education course, which consist of the textbook for $144.37, the accompanying lesson plan guide book for $38.17, and a package of 4 by 6 inch index cards for $1.79. In addition, she is purchasing a candy bar for $0.71 and a piece of bubble gum for $0.06.

    a. Estimate the total purchase cost.

    b. Compute the actual total purchase cost.

72. Change: A young child has a bunch of change in his pocket, and he wants to purchase a comic book which costs $1.73. If he has $0.75 in quarters, $0.40 in dimes, $0.35 in nickels, and $0.04 in pennies, does he have enough money to buy the comic book? If not, how much extra must he get so that he can buy it? (**Hint:** Add up the value of his coin collection. If it exceeds the cost, he is in good shape. If not, subtract the cash on hand from the cost of the comic book.)

73. Timer: A tightly-controlled timer is set at 759.99991 seconds.

    a. What will its setting be if the time is decreased by 1.00082 seconds?

    b. What will its setting be if the original time is increased by 1.00082 seconds?

74. Lemon Juice: Four unequal lemons were squeezed for juice, and they put out 0.0043 liter, 0.00056 liter, 0.01002 liter, and 0.0097 liter respectively.

    a. Estimate the total amount of juice which was collected.

    b. Compute the actual amount of juice collected.

75. Financial Aid: A college student is receiving assistance from two sources which will be applied to the following expenses: $17,993.74 for tuition and fees, $7,248.39 for room and board, and $1,537.71 for books and supplies. If one of the sources of assistance supplies $9438.72 and the other $8300, how much must the student cover? (**Hint:** Add the total costs, add up the total assistance, and subtract this from the total cost to find how much the student must cover.)

76. Dinner Bill: Tim eats dinner at an upscale restaurant. The appetizer costs $8.23, the entrée costs $21.78, the beverage costs $2.63, and the dessert costs $6.47 (the tax is included in these costs). If he gives a $7.50 tip, what is the total cost of the meal?

Writing & Thinking

77. Suppose that you are given two decimal numbers with 0 as their whole number parts.

    a. Explain how the sum might be more than 1.

    b. Explain why the sum cannot be more than 2.

78. Suppose that you are given two decimal numbers with 0 as their whole number parts.

    a. Explain why the difference would be less than 1.

    b. Explain how the difference might be 0.

Objectives

A Multiply decimal numbers.

B Multiply by powers of 10.

C Estimate products.

D Work applications using decimal numbers.

3.3 Multiplication with Decimal Numbers

Objective A **Multiplication with Decimal Numbers**

Multiplication with decimal numbers is similar to multiplication with whole numbers. The difference is in determining where to place the decimal point in the product. Two examples are shown here, in both fraction form and decimal form, to illustrate how the decimal point is to be placed in the product.

Products in Fraction Form

$$\frac{4}{10} \cdot \frac{6}{100} = \frac{24}{1000}$$

$$\frac{5}{1000} \cdot \frac{7}{100} = \frac{35}{100,000}$$

Products in Decimal Form

0.4 ← 1 decimal place in the factor

$\times\ \ 0.06$ ← 2 decimal places in the factor

0.024 ← 3 decimal places in the product

0.005 ← 3 decimal places in the factor

$\times\ \ 0.07$ ← 2 decimal places in the factor

0.00035 ← 5 decimal places in the product

As these examples indicate, there is **no need to keep the decimal points lined up for multiplication**.

To Multiply Decimal Numbers

1. Multiply the two numbers as if they were whole numbers.
2. Count the total number of places to the right of the decimal points in both numbers being multiplied.
3. Place the decimal point in the product so that the number of decimal places is the same as that found in Step 2.

Example 1

Multiplication with Decimal Numbers

Multiply: 2.432×5.1.

Solution

2.432 ← 3 decimal places ⎫
$\times\ \ \ \ 5.1$ ← 1 decimal place ⎬ total of 4 decimal places
⎭

2432

$12\ 160$
_____

12.4032 ← 4 decimal places in the product

Example 2

Multiplication with Decimal Numbers

Multiply: 4.35×12.6

Solution

$$
\begin{array}{r}
4.35 \quad \leftarrow \text{2 decimal places} \\
\times \ 12.6 \quad \leftarrow \text{1 decimal place} \\
\hline
2\ 610 \\
8\ 70 \\
43\ 5 \\
\hline
54.810 \quad \leftarrow \text{3 decimal places in the product}
\end{array}
$$

total of 3 decimal places

Example 3

Multiplication with Decimal Numbers

Multiply: $(0.046)(0.007)$

Solution

$$
\begin{array}{r}
0.046 \quad \leftarrow \text{3 decimal places} \\
\times \quad 0.007 \quad \leftarrow \text{3 decimal places} \\
\hline
0.000322 \quad \leftarrow \text{6 decimal places in the product}
\end{array}
$$

total of 6 decimal places

This means that three 0s had to be inserted between the 3 and the decimal point.

Completion Example 4

Multiplication with Decimal Numbers

Multiply: 3.4×5.8

Solution

$$
\begin{array}{r}
3.4 \quad \leftarrow \underline{} \text{ decimal place(s)} \\
\times \ 5.8 \quad \leftarrow \underline{} \text{ decimal place(s)} \\
\hline
272 \\
170 \\
\hline
\underline{} \quad \underline{} \text{ decimal place(s) in the product}
\end{array}
$$

total of ___ decimal place(s)

Find each product.

1. $(0.8)(0.2)$

2. 5.6×0.04

3.
$$
\begin{array}{r}
3.781 \\
\times \ 3.01 \\
\end{array}
$$

4. $(0.29)(11.4)$

Now work margin exercises 1 through 4.

Completion Example Answers

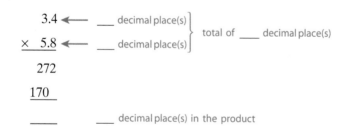

4.
$$
\begin{array}{r}
3.4 \quad \text{1 decimal place(s)} \\
\times \ 5.8 \quad \text{1 decimal place(s)} \\
\hline
2\ 72 \\
17\ 0 \\
\hline
19.72 \quad \text{2 decimal place(s) in the product.}
\end{array}
$$

total of 2 decimal place(s)

Objective B **Multiplication by Powers of 10**

The **powers of 10** are

$$1, \quad 10, \quad 100, \quad 1000, \quad 10{,}000, \quad 100{,}000, \quad \text{and so on.}$$

Multiplication by powers of 10 can be accomplished mentally by using the following general guidelines. (Remember that with whole numbers the decimal point is understood to be just to the right of the ones digit.)

To Multiply a Decimal Number by a Power of 10

1. Count the number of 0s in the power of 10.
2. Move the decimal point to the right the number of places equal to the number of 0s.

Multiplication by **10** moves the decimal point **one** place **to the right**.

Multiplication by **100** moves the decimal point **two** places **to the right**.

Multiplication by **1000** moves the decimal point **three** places **to the right**.

And so on.

5. Find each product.

a. $10(0.13)$

b. $1000(8.78)$

Example 5

Multiplication of a Decimal Number by a Power of 10

The following products illustrate multiplication by powers of 10.

a. $10(1.59) = 15.9$ Move the decimal point 1 place to the right.

b. $100(2.68) = 268. = 268$ Move the decimal point 2 places to the right.

c. $1000(0.9653) = 965.3$ Move the decimal point 3 places to the right.

d. $1000(7.2) = 7200. = 7200$ Move the decimal point 3 places to the right.

e. $10^2(3.5149) = 351.49$ Move the decimal point 2 places to the right. The exponent tells how many places to move the decimal point.

Now work margin exercise 5.

Objective C **Estimating Products**

Estimating products can be done by rounding each number to the place of the last nonzero digit on the left and multiplying these rounded numbers. This technique is particularly helpful in correctly placing the decimal point in the actual product.

Example 6

Estimating a Product

a. Estimate the product $(0.356)(6.1)$.

Solution

Estimate by multiplying rounded numbers.

$$
\begin{array}{rl}
0.4 & \text{(0.356 rounded)} \\
\times\ 6 & \text{(6.1 rounded)} \\
\hline
2.4 & \leftarrow \text{estimate}
\end{array}
$$

b. Find the product and use the estimation to help place the decimal point.

Solution

Find the actual product.

$$
\begin{array}{r}
0.356 \\
\times\ 6.1 \\
\hline
356 \\
2136 \\
\hline
2.1716 \quad \leftarrow \quad \text{actual product}
\end{array}
$$

The estimated product 2.4 helps place the decimal point correctly in the product 2.1716.

Now work margin exercise 6.

6. Estimate and then find the product.

$$11.287 \times 0.445$$

Objective D   Applications

Some applications may involve several operations. The problems do not usually say directly to add, subtract, or multiply. Experience and reasoning abilities are needed to decide which operations to perform with the given numbers. Example 7 illustrates a problem involving several steps and how answers can be estimated.

Example 7

Buying a Car

You can buy a car for $15,000 cash or you can make a down payment of $3750 and then pay $1093.33 each month for 12 months. How much can you save by paying cash?

Solution

a. Find the amount paid in monthly payments by multiplying the amount of each payment by 12. In this case, judgment dictates that we use 12 and do not round to 10, since we do not want to lose two full monthly payments in our estimate.

7. How much will you save on a new car by paying $500 a month for 24 months as opposed to $1050 a month for 12 months? Assume the down payment is the same.

Estimate	Actual Amount
$1000	$1093.33
× 12	× 12
2000	2186 66
10,000	10,933 30
$12,000 estimated monthly payments	$13,119.96 paid in monthly payments

b. Find the total amount paid by adding the down payment to the answer in part **a.**

Estimate	Actual Amount
$4 000 down payment	$3 750.00 down payment
+12,000 monthly payments	+13,119.96 monthly payments
$16,000 estimated total	$16,869.96 total paid

c. Find the savings by subtracting $15,000 (the cash price) from the answer to part **b.**

Estimate	Actual Amount
$16,000 estimated total	$16,869.96 total paid
−15,000 cash price	−15,000.00 cash price
$ 1 000 estimated savings	$ 1 869.96 savings by paying cash

The estimated $1000 saved by paying cash is reasonably close to the actual savings of $1869.96.

Now work margin exercise 7.

Practice Problems

Find the indicated products.

1.   1.23
   × 0.7

2.   6.884
   × 9.5

3.   0.08
   × 0.542

4. 1000(0.0079)

5. How much will you save by paying $950 cash for a new bicycle rather than payments of $55 a month for 24 months?

Practice Problem Answers
1. 0.861   **2.** 65.3980   **3.** 0.04336
4. 7.9   **5.** $370

228   Chapter 3   *Decimal Numbers*

© Hawkes Learning Systems. All rights reserved.

Exercises 3.3

Match each indicated product with the best estimate of that product. See Example 6.

1. _____ (i) 0.71×0.8 **A.** 0.06
   _____ (ii) 34.5×0.11 **B.** 0.56
   _____ (iii) 0.63×9.81 **C.** 3.0
   _____ (iv) 0.34×0.18 **D.** 5.0
   _____ (v) 4.6×1.2 **E.** 6.0

2. _____ (i) 1.75×0.04 **A.** 0.008
   _____ (ii) 1.75×0.004 **B.** 0.08
   _____ (iii) 17.5×0.04 **C.** 0.8
   _____ (iv) 1.75×4 **D.** 8.0

Find each of the indicated products. See Examples 1 through 4.

3. 0.6×0.7 4. 0.3×0.8 5. 0.2×0.2 6. 0.3×0.3

7. $8(2.7)$ 8. $4(9.6)$ 9. $(1.4)(0.3)$ 10. $(1.5)(0.6)$

11. $(0.2)(0.02)$ 12. $(0.3)(0.03)$ 13. 5.4×0.02 14. 7.3×0.01

15. 0.23×0.12 16. 0.15×0.15 17. $(8.1)(0.006)$ 18. $(7.1)(0.008)$

19. 0.06×0.01 20. 0.25×0.01 21. 3×0.125 22. 4×0.375

23. $(1.6)(0.875)$ 24. $(5.3)(0.75)$ 25. $(6.9)(0.25)$ 26. $(4.8)(0.25)$

27. 0.83×6.1 28. 0.27×0.24 29. 0.16×0.5 30. 0.28×0.5

Find each of the following products mentally by using your knowledge of multiplication by powers of ten. See Example 5.

31. 100×3.46 32. 100×20.57 33. 100×7.82

34. 100×6.93 35. $100(16.1)$ 36. $100(38.2)$

37. $10(0.435)$ 38. $10(0.719)$ 39. 10×1.86

40. 1000×4.1782 41. $(1000)(0.38)$ 42. $(1000)(0.47)$

43. $(10{,}000)(0.005)$ **44.** $10{,}000 \times 0.00615$ **45.** $10{,}000 \times 7.4$

Find the estimated product using rounded numbers. Then find the actual product. See Example 6.

46.
$$\begin{array}{r} 0.106 \\ \times\ 0.09 \\ \hline \end{array}$$

47.
$$\begin{array}{r} 1.07 \\ \times\ 0.5 \\ \hline \end{array}$$

48.
$$\begin{array}{r} 5.08 \\ \times\ 0.4 \\ \hline \end{array}$$

49.
$$\begin{array}{r} 0.0106 \\ \times\ 0.087 \\ \hline \end{array}$$

50.
$$\begin{array}{r} 0.0213 \\ \times\ 0.065 \\ \hline \end{array}$$

51.
$$\begin{array}{r} 83.105 \\ \times\ 0.111 \\ \hline \end{array}$$

52.
$$\begin{array}{r} 17.002 \\ \times\ 0.101 \\ \hline \end{array}$$

53.
$$\begin{array}{r} 86.1 \\ \times\ 0.057 \\ \hline \end{array}$$

54.
$$\begin{array}{r} 7.83 \\ \times\ 0.18 \\ \hline \end{array}$$

55.
$$\begin{array}{r} 95.62 \\ \times\ 0.57 \\ \hline \end{array}$$

56.
$$\begin{array}{r} 6.02 \\ \times\ 0.57 \\ \hline \end{array}$$

57.
$$\begin{array}{r} 8.034 \\ \times\ 0.29 \\ \hline \end{array}$$

Solve the following word problems. See Example 7.

58. Car payments: To buy a car, you can pay $2036.50 in cash, or you can put down $400 and make 18 monthly payments of $104.30. How much would you save by paying cash?

59. Stockroom: Suppose Ted is ordering more refrigerators for his stockroom after an applicance sale. He orders 12 new refrigerators at the whole sale price of $496.65.

   a. Estimate the total price Ted paid.

   b. What is the exact price paid?

60. Granite Countertops: If the price to buy and install granite countertops is $64.85 per square foot.

   a. Estimate the price of a new 11 square foot granite countertop that was installed?

   b. What is the exact amount paid for the granite countertops.?

61. Overtime pay: If you were paid a salary of $350 per week and $13.75 for each hour you worked over 40 hours in a week, how much would you make if you worked 45 hours in one week?

62. Geometry: Find the perimeter and area of a square with sides 3.2 ft long.

63. Geometry: Find the perimeter and area of a rectangle with length 10.5 cm and width 2.8 cm.

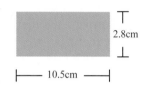

64. Geometry: Find the perimeter and area of a rectangle with length of 18.04 m and width 6.25 m.

65. Geometry: Find the perimeter and area of a square with sides 4.7 mm long.

66. Library funding: In 2003, Ohio led the nation in per capita funding for its public libraries with $39.87 spent for each person. How much funding would have been received that year by a library that served a town of 23,500 people?

67. Agriculture: In July of 2010 the average price per pound paid by the U.S. for beef imported from Australia and New Zealand was $1.45. At this price, what would be the value of 42,500 pounds of beef? **Source:** www.indexmundi. com

68. Automobile Prices: The list price for a particular automobile with accessories is $35,487. If the customer will pay cash with no trade-in, the dealer will give a discount of 0.18 times the list price. What will the car cost as a straight cash transaction? This cash value price will be the list price minus the discount.

69. Shopping: The total cost of a purchase is $583.47. If the sales tax is 0.065 of the purchase price, how much will the customer have to pay? (**Hint:** To calculate the sales tax, multiply the purchase price by the tax rate.) Round your answer to the nearest hundredth.

70. Gas Mileage: A family takes a long drive where the gasoline tank of the car must be filled three times, and the price of gasoline is different at each location. At the first stop the car took 12.537 gallons of gas at $3.129/gallon; at the second stop the car took 15.281 gallons at $2.449/gallon; and at the final stop the car took 6.785 gallons at $2.489/gallon. How much was spent on gasoline for the trip? Round your final answer to the nearest hundredth. (**Hint:** compute the cost for each leg of the trip, and then add these three costs)

71. Commision: A sales person at an appliance store receives a 0.12 commission on all items she sells. Assume a customer purchases a front loading washing machine for $996.45, a matching dryer for $891.68, and two matching pedestals for $282.17 each. **Source:** http://www.whirlpool.com/catalog/laundry_gallery.jsp#frontloadhe

a. Estimate what the commission will be. (**Hint:** Add all the prices before computing the commission. Commission = total sales times commission rate.)

b. What is the actual commission on the sale? Round your answer to the nearest hundredth.

72. TV Set: A rent-to-own store will rent a 50-inch plasma TV which is worth $1099 for $28.47 a week. After two years, the TV set belongs to the renter.

a. Assuming that there are 52 weeks in a year, how much will the customer have paid at the end of two years? (**Hint:** To get the total number of weeks, multiply the number of weeks in a year by the number of years.)

b. How much extra did the renter pay for the TV set by renting it over that 2 year period?

73. Down Payment: Each month, Carol is setting aside 0.26 times her take-home pay of $3428.84 for the down payment on a condominium.

a. Estimate how much is being placed in savings each month.

b. Compute the exact value of the savings each month. Round your answer to the nearest hundredth.

c. How much money is being saved each year? Use the monthly saving found in part **b.** for the computation.

74. Overtime Pay: Geoffrey works at a job with a regular pay of $12.50 an hour, up to 40 hours in a week. Any work which exceeds 40 hours is paid $18.25 an hour. If he works for 53.4 hours, what will his total pay be? (**Hint:** First compute the pay for the first 40 hours; second, compute how many hours overtime worked; third, compute the overtime pay; and fourth, add these regular and overtime pay.)

75. Commision: The salesperson at a high-end clothing store receives a salary of $375.50 plus a weekly commission of 0.18 times all sales over $850 for the week. How much will that person earn if he sells $3945 that week? (**Hint:** compute how much over the threshold of $850 the total sales were, and then multiply it by the rate of commission. Then add his salary to that value.)

76. Service Industry: It is common practice for restaurants to pay the servers less than minimum wage because their tips are considered to be part of their pay. One restaurant pays $2.75 an hour. If during a 9.4 hour day, the total value of the meals sold was $1044.83, how much would the server make that day if the average tip was 0.11 of the total bill? Round your answer to the nearest hundredth.

77. Textbook: A hardbound book has 257 pages, each of which is 0.0098 inch thick. The front and back cover are each 0.187 inch thick. How thick is the book?

78. Deck: A deck which is 7.2 meters long and 4.6 meters wide is to be painted.

a. Compute the area of the deck.

b. If a painter charges $3.75 a square meter to paint the deck, how much will the job cost?

3.4 Division with Decimal Numbers

Objective A **Division with Decimal Numbers**

The process of division (called the **division algorithm**) with decimal numbers is, in effect, the same as division with whole numbers. With whole numbers we are concerned with the remainder. But, with decimal numbers, we are concerned with the placement of the decimal point in the quotient. For example consider $950 \div 40$ as shown here.

$$
\begin{array}{r}
23. \quad \leftarrow \text{quotient} \\
\text{divisor} \longrightarrow 40\overline{)950.} \quad \leftarrow \text{dividend} \\
\underline{80} \\
150 \\
\underline{120} \\
30 \quad \leftarrow \text{remainder}
\end{array}
$$

By adding 0s onto the dividend, we can continue to divide and get a decimal quotient other than a whole number.

$$
\begin{array}{r}
23.75 \quad \leftarrow \text{quotient is a decimal number} \\
\text{divisor} \longrightarrow 40\overline{)950.00} \quad \leftarrow \text{0s added on} \\
\underline{80} \\
150 \\
\underline{120} \\
300 \\
\underline{280} \\
200 \\
\underline{200} \\
0
\end{array}
$$

If the divisor is a decimal number rather than a whole number, multiply both the divisor and dividend by a power of 10 so that the new divisor is a whole number. For example, we can write

$$
6.2\overline{)63.86} \quad \text{as} \quad \frac{63.86}{6.2} \cdot \frac{10}{10} = \frac{638.6}{62}.
$$

This means that

$$
6.2\overline{)63.86} \quad \text{is the same as} \quad 62\overline{)638.6}.
$$

Similarly, we can write

$$
1.23\overline{)4.6125} \quad \text{as} \quad \frac{4.6125}{1.23} \cdot \frac{100}{100} = \frac{461.25}{123} \quad \text{or} \quad 123\overline{)461.25}.
$$

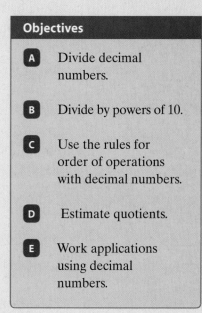

Objectives

A Divide decimal numbers.

B Divide by powers of 10.

C Use the rules for order of operations with decimal numbers.

D Estimate quotients.

E Work applications using decimal numbers.

To Divide Decimal Numbers

1. Move the decimal point in the divisor to the right so that the divisor is a whole number.
2. Move the decimal point in the dividend the same number of places to the right.
3. Place the decimal point in the quotient directly above the new decimal point in the dividend.
4. Divide just as with whole numbers. (0s may be added as needed to the dividend.)

notes

1. In moving the decimal point, you are multiplying both the divisor and dividend by a power of 10.
2. Be sure to place the decimal point in the quotient **before actually dividing**.
3. There must be a digit in the quotient above every digit to the right of the decimal point in the dividend.

Example 1

Division with Decimal Numbers

Find the quotient: $63.86 \div 6.2$

Solution

Step 1: Write down the numbers.

$$6.2\overline{)63.86}$$

Steps 2&3: Move both decimal points one place to the right so that the divisor becomes a whole number. Then place the decimal point in the quotient.

$$6.2.\overline{)63.8.6} \quad \longleftarrow \text{ decimal point in quotient}$$

Step 4: Proceed to divide as with whole numbers.

$$
\begin{array}{r}
10.3 \\
62.\overline{)638.6} \\
\underline{62} \\
18 \\
\underline{0} \\
186 \\
\underline{186} \\
0
\end{array}
$$

Completion Example 2

Division with Decimal Numbers

Divide: $24.225 \div 4.25$

Solution

$$
4.25.\overline{\smash{)}24.22.5}^{5.}
$$

Now work margin exercises 1 and 2.

In each of the preceding examples the remainder was eventually 0. This will not always be the case. We generally agree to some place of accuracy for the quotient before the division is performed. If the remainder is not 0 by the time this place of accuracy is reached in the quotient, then we divide one more place and round the quotient.

When the Remainder is Not 0

1. Decide first how many decimal places are to be in the quotient.
2. Divide until the quotient is **one digit past the place of desired accuracy**.
3. Using this last digit, round the quotient to the desired place of accuracy.

notes

If the remainder is eventually 0, the decimal number is said to be **terminating**.

For example, 14.2 and 0.3425 are terminating decimal numbers.

If the remainder is not eventually 0, the decimal number is said to be **nonterminating**.

For example, 1.33333… is a nonterminating decimal number.

Completion Example Answers

2.
$$
4.25.\overline{\smash{)}24.22.5}^{5.7}
$$
$$
\underline{2125}
$$
$$
2975
$$
$$
\underline{2975}
$$
$$
0
$$

Find each quotient.

1. $936 \div 0.5$

2. $14.872 \div 2.86$

Example 3

Division with Decimal Numbers

Find the quotient $82.3 \div 2.9$ to the nearest tenth.

Solution

Divide until the quotient is in hundredths (one place more than tenths); then round to tenths.

```
                        hundredths
                              read approximately equals

            2 8 . 3 7  ≈ 2 8 . 4
   2.9. ) 8 2 . 3 . 0 0      ←  Add 0s as needed.
          5 8
          2 4  3
          2 3  2
            1 1  0
              8  7
              2  30
              2  03
                 27
```

$82.3 \div 2.9 \approx 28.4$ accurate to the nearest tenth

Find the quotient to the indicated place of accuracy.

3. $83.541 \div 5.6$ to the nearest tenth.

4. $14.393 \div 9.3$ to the nearest hundredth.

Example 4

Division with Decimal Numbers

Find $1.83 \div 4.1$ to the nearest hundredth.

Solution

Divide until the quotient is in thousandths (one more place than hundredths); then round to hundredths.

```
                           read approximately equals

            0 . 4 4 6  ≈ 0 . 4 5
   4 .1. ) 1 . 8 . 3 0 0      ←  Add as many 0s as needed.
            1 6  4
            1 9 0
            1 6 4
              2 6 0
              2 4 6
                1 4
```

$1.83 \div 4.1 \approx 0.45$ accurate to the nearest hundredth

Now work margin exercises 3 and 4.

Objective B **Division by Powers of 10**

In section 3.3, we found that multiplication by powers of 10 can be accomplished by moving the decimal point to the right. Division by powers of 10 can be accomplished by moving the decimal point to the left.

> ### To Divide a Decimal Number by a Power of 10
>
> 1. Count the number of 0s in the power of 10.
> 2. Move the decimal point to the left the number of places equal to the number of 0s.
>
> Division by **10** moves the decimal point **one** place **to the left**.
>
> Division by **100** moves the decimal point **two** places **to the left**.
>
> Division by **1000** moves the decimal point **three** places **to the left**.
>
> And so on.

Two general guidelines will help you to understand how to work with powers of 10:

1. Multiplication by a power of 10 will make a number larger, so move the decimal point to the right.

2. Division by a power of 10 will make a number smaller, so move the decimal point to the left.

Example 5

Division by Powers of 10

The following quotients illustrate division by powers of 10.

a. $5.23 \div 100 = \dfrac{5.23}{100} = 0.0523$ Move decimal point 2 places to the left.

b. $817 \div 10 = \dfrac{817}{10} = 81.7$ Move decimal point 1 place to the left.

c. $495.6 \div 10^3 = 0.4956$ Move decimal point 3 places to the left. The exponent tells how many places to move the decimal point.

d. $286.5 \div 10^2 = 2.865$ Move decimal point 2 places to the left.

Now work margin exercise 5.

5. Find each quotient by performing the operation mentally.

  a. $\dfrac{16}{10}$

  b. $\dfrac{83.46}{1000}$

  c. $73.2 \div 10^2$

Order of Operations with Decimal Numbers

Just as with whole numbers, fractions, and mixed numbers, expressions with decimal numbers may involve more than one operation. To evaluate such an expression, use the rules for order of operations just as they were discussed in Sections 1.6 and 2.6.

Example 6

Order of Operations with Decimal Numbers

$3.1^2 + 7.05 \div 1.5$

Solution

$3.1^2 + 7.05 \div 1.5$	Evaluate the exponent.
$= 9.61 + 7.05 \div 1.5$	Divide.
$= 9.61 + 4.7$	Add.
$= 14.31$	

6. $2.4 \cdot 11.1 - 4.4^2$

7. $7.3 + \left(0.5^2 + 1.1\right)$

Example 7

Order of Operations with Decimal Numbers

$2.1\left(45.2 - 10.8\right) - 15.38$

Solution

$2.1\left(45.2 - 10.8\right) - 15.38$	Subtract inside the parentheses.
$= 2.1\left(34.4\right) - 15.38$	Multiply.
$= 72.24 - 15.38$	Subtract.
$= 56.86$	

Now work margin exercises 6 and 7.

Estimating Quotients

As with addition, subtraction, and multiplication, we can use estimating with division to help in placing the decimal point in the quotient and to verify the reasonableness of the quotient. In order to estimate with division, round both the divisor and the dividend to the place of the last nonzero digit on the left and then divide with these rounded values.

Example 8

Division with Decimal Numbers

a. Estimate the quotient $6.2 \div 0.302$.

Solution

Estimate the quotient using $6.2 \approx 6$ and $0.302 \approx 0.3$.

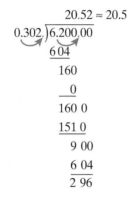

$$
\begin{array}{r}
20.\\
0.3.\overline{)6.0.}\\
\underline{6}\\
0\,0\\
\underline{0}\\
0
\end{array}
$$

b. Find the quotient to the nearest tenth.

Solution

Find the quotient to the nearest tenth.

$$
\begin{array}{r}
20.52 \approx 20.5\\
0.302.\overline{)6.200.00}\\
\underline{6\,04}\\
160\\
\underline{0}\\
160\,0\\
\underline{151\,0}\\
9\,00\\
\underline{6\,04}\\
2\,96
\end{array}
$$

The estimated value 20 is very close to the rounded quotient 20.5.

Now work margin exercise 8.

8. a. Estimate the quotient: $13.18 \div 2.41$

b. Find the quotient to the nearest tenth.

Objective E Applications

We know from Section 1.7 that the **average** of a set of numbers can be found by adding the numbers, then dividing the sum by the number of addends. The term **average** is used in phrases such as "average speed of 43 miles per hour" or "the average price of a pair of shoes." These averages can also be found by division. If the total amount of a quantity (distance, dollars, gallons of gas, etc.) and a number of units (time, items bought, miles, etc.) are known, then we can find the **average amount per unit** by dividing the total amount by the number of units.

9. If a sprinter runs 100 meters in 10.13 seconds, what is his approximate average speed in meters per second?

Example 9

Gas Mileage

The gas tank of a car holds 17 gallons of gasoline. Estimate how many miles per gallon the car averages if it will go 470 miles on one tank of gas.

Solution

To find an estimate we use rounded values:

$$17 \approx 20 \text{ gallons} \quad \text{and} \quad 470 \approx 500 \text{ miles}$$

Now divide to estimate the average number of miles per gallon:

$$
\begin{array}{r}
25 \text{ miles per gallon} \\
20\overline{)500} \\
\underline{40} \\
100 \\
\underline{100} \\
0
\end{array}
$$

The car averages about 25 miles per gallon.

Now work margin exercise 9.

If an average amount per unit is known, then a corresponding total amount can be found by multiplying by the number of units. For example, if you ride your bicycle at an average speed of 15.2 miles per hour, then the distance you travel can be found by multiplying your average speed by the time you spend riding in hours.

Example 10

Bicycle Speed

If you ride your bicycle at an average speed of 15.2 miles per hour, how far will you ride in 3.5 hours?

Solution

Multiply the average speed by the number of hours.

$$
\begin{array}{r}
15.2 \text{ miles per hour} \\
\times \quad 3.5 \text{ hours} \\
\hline
760 \\
\underline{456} \\
53.20 \text{ miles}
\end{array}
$$

You will ride 53.2 miles in 3.5 hours.

Now work margin exercise 10.

10. On average, you can store about 277.75 songs per gigabyte of memory. How many songs could you save on an iPod with 8 gigabytes of memory?

Example 11

Gasoline Prices

The price of a gallon of gas at the pump is $3.15. If the taxes you pay on each gallon of gas is 0.45 times the original price of a gallon of gas, what is the price of a gallon of gas before taxes (to the nearest penny)? (**Hint:** To find the price of a gallon of gas before taxes, divide the total price by 1.45.)

Solution

Performing the division gives the following result.

$$
\begin{array}{r}
2.172 \\
1.45\overline{\smash{)}3.15.000} \\
\underline{2\,90} \\
25\,0 \\
\underline{14\,5} \\
10\,50 \\
\underline{10\,15} \\
350 \\
\underline{290} \\
60
\end{array}
$$

So the cost of the gas is about $2.17 per gallon before taxes.

Now work margin exercise 11.

11. If a hotdog costs $2.39 (with taxes) at a little league hockey game and taxes were 0.06 times the actual price, how much (to the nearest cent) does the hotdog cost? (**Hint:** To find the price before taxes divide the total price by 1.06.)

Practice Problems

Find the indicated quotients.

1. $4.484 \div 1.9$

2. $2.5\overline{\smash{)}16.35}$

3. $4\overline{\smash{)}1.83}$ (Do not round your answer).

4. $0.06\overline{\smash{)}43.721}$ (to the nearest thousandth)

5. $\dfrac{42.31}{10}$

6. $87.96 \div 100$

7. Evaluate the following expression by following the rules for order of operations: $1.7(4.1+3.3)-6.2$

8. If you jog an average speed of 4.8 miles per hour, how far will you jog in 3.2 hours?

Practice Problem Answers

1. 2.36   **2.** 6.54   **3.** 0.4575   **4.** 728.683
5. 4.231   **6.** 0.08796   **7.** 6.38   **8.** 15.36

Exercises 3.4

Match the indicated quotient with the best estimate of that quotient. See Example 8.

1. _____ (i) $3.1\overline{)6.386}$ **A.** 0.02   **2.** _____ (i) $27.58 \div 0.003$ **A.** 100

_____ (ii) $0.1\overline{)216.5}$ **B.** 2   _____ (ii) $27.58 \div 0.03$ **B.** 10

_____ (iii) $3.7\overline{)281.6}$ **C.** 5   _____ (iii) $27.58 \div 0.3$ **C.** 10,000

_____ (iv) $18.5\overline{)127.9}$ **D.** 75   _____ (iv) $27.58 \div 3$ **D.** 1000

_____ (v) $4.1\overline{)0.0884}$ **E.** 2000

Divide. See Examples 1 and 2.

3. $4.68 \div 2$ **4.** $1.71 \div 3$ **5.** $4.95 \div 5$ **6.** $1.62 \div 9$

7. $0.064 \div 0.8$ **8.** $0.63 \div 0.7$ **9.** $82.24 \div 0.04$ **10.** $16.02 \div 0.03$

11. $48 \div 2.4$ **12.** $28 \div 5.6$

Find each quotient to the nearest tenth. See Example 3.

13. $8\overline{)455}$ **14.** $4\overline{)263}$ **15.** $9.4\overline{)6.538}$ **16.** $4.6\overline{)5}$

17. $7.05\overline{)0.4977}$ **18.** $0.37\overline{)4.683}$ **19.** $1.62\overline{)34}$ **20.** $1.33\overline{)75}$

Find each quotient to the nearest hundredth. See Example 4.

21. $24\overline{)0.1463}$ **22.** $1.23\overline{)14.911}$ **23.** $0.075\overline{)0.42753}$

24. $2.7\overline{)2.583}$ **25.** $23\overline{)62.949}$ **26.** $9\overline{)2}$

27. $13\overline{)65.476}$ **28.** $3.181\overline{)6}$

Find each quotient mentally by using your knowledge of division by powers of 10. See Example 5.

29. $78.4 \div 100$ **30.** $16.4963 \div 100$ **31.** $50.36 \div 100$

32. $45.621 \div 1000$ **33.** $73.85 \div 1000$ **34.** $18.6 \div 1000$

35. $\dfrac{167}{10}$ **36.** $\dfrac{138.1}{10}$ **37.** $\dfrac{7.85}{10}$

38. $\dfrac{1.54}{10{,}000}$ **39.** $\dfrac{169.9}{10{,}000}$ **40.** $\dfrac{10.413}{10{,}000}$

Evaluate each expression by using the rules for order of operations. See Examples 6 and 7.

41. $12.6 - 5.88 + 13.9 \cdot 6.5$ **42.** $8.6 \div 2.15 + 3.6 \cdot 20.3$ **43.** $1.5^2 - 0.56 + 2.2 \cdot 6.5$

44. $43.5 - 23.78 \div 5.8 + 4.7^2$ **45.** $(1.3 + 5.9) \cdot 2.6 + 8.16$ **46.** $4.2 + 23.79 \div (7.4 - 3.5)$

47. $(4.2 + 4.8) \cdot (3.3^2 - 9.9) + 7.2$ **48.** $(5.7 + 2.9)^2 - (2.1 + 4.1) \cdot 3.4$

Find the estimated quotients using rounded numbers. Then find the actual quotients to the nearest thousandth. See Example 8.

49. $23\overline{)71}$ **50.** $69\overline{)293}$ **51.** $85.3\overline{)24.31}$

52. $2.57\overline{)0.4961}$ **53.** $16.2\overline{)0.11623}$ **54.** $25.7\overline{)6.27}$

Solve the following word problems. See Examples 9 through 11.

55. Fuel Economy:
  a. If a car averages 24.6 miles per gallon, estimate how far it will go on 18 gallons of gas?
  b. Exactly how many miles will it go on 18 gallons of gas?

56. Bicycling:
  a. If Alberto Contador rode 145.09 miles in 5.35 hours, estimate how fast he was riding per hour.
  b. What was his average speed in miles per hour (to the nearest hundredth)?

57. Wholesale Purchasing: A quarter section of beef can be bought cheaper than the same amount of meat purchased a few pounds at a time.
  a. Estimate the cost per pound if 150 pounds costs $187.50.
  b. What is the cost per pound?

58. Automobile Travel:
  a. If you drive 9.5 hours at an average speed of 52.2 miles per hour, about how far will you drive?
  b. Exactly how far will you drive?

59. Custom Rims: If four new tires with custom rims cost $958.24, what did each individual tire with rim cost?

60. Bookstore: If you bought 6 books for a total price of $142.98, what average amount did you pay per book, including tax?

61. Sales Tax: If the total price of a stereo was $312.70 including a tax of 0.06 times the list price, you can find the list price by dividing the total price by 1.06. What was the list price? (**Note:** 1.06 represents the list price plus 0.06 times the list price.)

62. Mortgages: Suppose that the total interest paid on a 30-year mortgage for a home loan of $60,000 will be $189,570. What will be the payment each month if the payments are to pay off both the loan and the interest?

63. Baseball: In 2010 the New York Mets had a team batting average of 0.249 and had 1361 base hits. Find the number of team at bats, to the nearest whole number, by dividing base hits by batting average. **Source:** espn.go.com/mlb/

64. Baseball: In 2010, Adam Wainwright of the St. Louis Cardinals led his team with 20 wins and a 0.645 winning percentage. Find Wainwright's total number of pitching decisions (games won or lost), to the nearest whole number, by dividing wins by winning percentage. **Source:** espn.go.com/mlb/

65. Basketball: In 2009-2010, Steve Nash of the Phoenix Suns basketball team led the NBA with a 0.938 free-throw percentage. He successfully made 211 free throws. Find his number of attempted free throws, to the nearest whole number, by dividing free throws made by free-throw percentage. **Source:** NBA.com

66. Baseball: At Turner Field in Atlanta, home of the Braves baseball team, the distance down the right-field foul line is 330 feet. Convert this distance to meters, to the nearest tenth of a meter. (Use 1 m = 3.28 ft. See Appendix A.3 for a review of Metric to U.S. conversions.)

67. Running: A marathon is 26.219 miles in length. Convert this distance to kilometers. Round to the nearest tenth. (Use 1 km = 0.621 mi. See Appendix A.3 for a review of Metric to U.S. conversions.)

68. Mt. Everest: According to the latest measurements, Mt. Everest is 29,035 feet above sea level. If there are 5280 feet in a mile: **Source:** http://www.mnteverest.net/history.html

   a. Estimate how high Mt. Everest is in miles.

   b. What is the actual elevation in miles? Round your answer to the nearest tenth of a mile.

69. Taxes: After taxes, the cost of a car is $45,783.50. If the tax rate is 0.15 of the total cost, what is the actual price of the car before taxes? Round your answer to the nearest hundredth of a dollar. (**Hint:** Since the final cost is $1 + 0.15 = 1.15$ times the actual price, divide the final cost by 1.15 to get the pre-tax price of the car.)

70. Gasoline: Betty starts off the day with a full tank of gasoline, and she will get gasoline three times before reaching her destination. On her first fuel stop, the car took 18.32 gallons of gasoline and it went 398.4 miles. At the second stop the car took 20.48 gallons, after covering 423.7 miles. At the final fuel stop she made a point to completely fill the tank, which took 14.27 gallons after going 258.4 miles. What was the fuel consumption in miles per gallon for the entire trip? Round your answer to the nearest tenth. (**Hint:** Add the total number of gallons consumed, add the number miles covered, and divide the total number miles of gallons by the total number of gallons.)

71. Test Average: A professor has graded a test of five students, and their scores were 76.4, 100, 84.7, 10.2, and 68.3. What is the average of these five scores?

72. Retirement: Edward, who has retired, needs $875 to cover an expense, and he plans to take it out of his IRA. Because what was contributed was not taxed, 0.20 of the withdrawal will be withheld for federal and state income tax. How much should he withdraw so that he will receive his needed $875? (**Hint:** Since 0.20 is withheld, he will have $1 - 0.20 = 0.80$ of the withdrawal remaining. Divide the desired amount by 0.80 to find the total amount to be withdrawn.)

Writing & Thinking

73. Explain briefly how and why the decimal point is moved when dividing with decimal numbers.

74. Cycling: In 2005, Lance Armstrong won the Tour de France for a record seventh consecutive time. The race covered 2242 miles and his winning time was 86 hours 15 minutes 2 seconds. This time was 4 minutes 40 seconds better than the second place finisher, Ivan Basso of Italy, and 6 minutes 21 seconds better than the third place finisher, Jan Ullrich of Germany. To find the average speed of these three cyclists you need to first change each of their times into decimal form (minutes and seconds need to be expressed as parts of an hour) and then divide each time into 2242 to find each average speed in miles per hour.

(Remember that there are 60 seconds in a minute and 60 minutes in an hour.)

a. What was the average speed of Lance Armstrong?

b. What was the average speed of Ivan Basso?

c. What was the average speed of Jan Ullrich?

75. Cycling: During Lance Armstrong's Tour de France win in 2004, he traveled a total distance of 2107 miles in 83 hours 36 minutes 2 seconds. The second place finisher, Andreas Kloden, was 6 minutes 19 seconds behind.

a. What was Lance Armstrong's average speed in miles per hour?

b. What was Andreas Kloden's average speed in miles per hour?

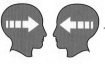

76. Do you know how to find the gas mileage (miles per gallon) that your car is getting? The gas mileage for your car can be calculated by following the steps given below. Find your gas mileage and compare it against the gas mileage of cars of other students in your class. (Your car might need some work if the mileage is not consistent or it is much worse than other similar sized cars.)

Step 1: Fill up your gas tank and write down the mileage indicated on the odometer.

Step 2: Drive the car for a few days.

Step 3: Fill up your gas tank again and write down the number of gallons needed to fill the tank and the new mileage indicated on the odometer.

Step 4: Find the number of miles that you drove by subtracting the new and old numbers indicated on the odometer.

Step 5: Divide the number of miles driven by the number of gallons needed to fill the tank. This number is your gas mileage (miles per gallon).

3.5 Decimal Numbers and Fractions

Objective A   **Changing Decimal Numbers to Fractions**

To understand how decimal numbers and fractions are related, we show how to change from one form to the other. From Section 3.1, we know that terminating decimal numbers can be written in fraction form with denominators that are powers of 10. For example,

$$0.35 = \frac{35}{100} \qquad \text{and} \qquad 0.017 = \frac{17}{1000}$$

The rightmost digit 5 is in the hundredths position, so the fraction has 100 in the denominator.

The rightmost digit 7 is in the thousandths position, so the fraction has 1000 in the denominator.

In each case, the denominator is the value of the position of the right most digit. Thus we proceed as follows to change a decimal number to fraction form.

To Change from Decimal Numbers to Fractions

A terminating decimal number can be written in fraction form by writing a fraction with the following:

1. a **numerator** that consists of the whole number formed by all the digits of the decimal number, and
2. a **denominator** that is the power of ten that names the position of the rightmost digit.

In Examples 1 – 5, each decimal number is changed to fraction form and then reduced, if possible, by using the factoring techniques discussed in Chapter 2 for reducing fractions.

Example 1

Changing Decimal Numbers to Fractions

$$0.25 = \frac{25}{100} = \frac{\cancel{5} \cdot \cancel{5} \cdot 1}{2 \cdot \cancel{5} \cdot 2 \cdot \cancel{5}} = \frac{1}{4}$$

hundredths

Convert to fraction form and then reduce:

1. 0.48

2. 0.95

3. 0.762

4. 0.03

5. 15.8

Example 2

Changing Decimal Numbers to Fractions

$$0.32 = \frac{32}{100} = \frac{\cancel{4} \cdot 8}{\cancel{4} \cdot 25} = \frac{8}{25}$$

hundredths

Example 3

Changing Decimal Numbers to Fractions

$$0.131 = \frac{131}{1000}$$

thousandths

Example 4

Changing Decimal Numbers to Fractions

$$0.075 = \frac{75}{1000} = \frac{\cancel{25} \cdot 3}{\cancel{25} \cdot 40} = \frac{3}{40}$$

thousandths

Example 5

Changing Decimal Numbers to Fractions

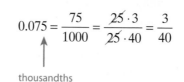

$$2.6 = \frac{26}{10} = \frac{\cancel{2} \cdot 13}{\cancel{2} \cdot 5} = \frac{13}{5}$$

tenths

or, as a mixed number,

$$2.6 = 2\frac{6}{10} = 2\frac{\cancel{2} \cdot 3}{\cancel{2} \cdot 5} = 2\frac{3}{5}$$

tenths

***Now work margin exercises* 1 through 5.**

Changing Fractions to Decimal Numbers

To change a fraction to decimal form, we use the meaning that the numerator is to be divided by the denominator. As we noted in Section 3.4:

If the remainder is eventually 0, the decimal number is said to be **terminating**.

If the remainder is never 0, the decimal number is said to be **nonterminating**.

The following examples illustrate fractions that convert to terminating decimal numbers.

Example 6

Changing Fractions to Decimal Numbers

$$\frac{3}{8} \longrightarrow 8\overline{)3.000} \longrightarrow \frac{3}{8} = 0.375$$

with long division:
```
      0.375
  8)3.000
    2 4
    ───
      60
      56
      ──
      40
      40
      ──
       0
```

Example 7

Changing Fractions to Decimal Numbers

$$\frac{5}{4} \longrightarrow 4\overline{)5.00} \longrightarrow \frac{5}{4} = 1.25$$

with long division:
```
      1.25
  4)5.00
    4
    ─
    1 0
      8
    ───
     20
     20
     ──
      0
```

Now work margin exercises 6 and 7.

Using a Calculator to Change Fractions to Decimal Numbers

To change a fraction to a decimal number, you will need to find the division key, $\div$, and the equal sign or enter key, $=$.

To change the fraction given above in example 7, press the keys 5 $\div$ 4. Then press $=$.

The display will read 1.25.

Nonterminating decimal numbers can be **repeating** or **nonrepeating**. A **nonterminating repeating decimal** (also called an **infinite repeating decimal number**) has a repeating pattern to its digits. Every fraction with a whole number numerator and nonzero denominator is either terminating or repeating. Such numbers are called **rational numbers**.

Convert to decimal notation:

6. $\frac{2}{5}$

7. $\frac{13}{20}$

The following examples illustrate how some fractions (using division) convert to repeating decimal numbers.

Example 8

Changing Fraction to Decimal Number

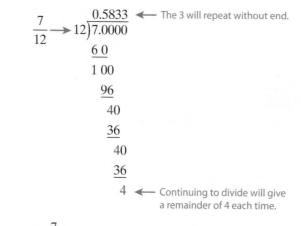

$$\frac{7}{12} \longrightarrow 12\overline{)7.0000}$$ with quotient 0.5833 ← The 3 will repeat without end.

$$\begin{array}{r} 0.5833 \\ 12\overline{)7.0000} \\ \underline{6\,0} \\ 1\,00 \\ \underline{96} \\ 40 \\ \underline{36} \\ 40 \\ \underline{36} \\ 4 \end{array}$$ ← Continuing to divide will give a remainder of 4 each time.

We write $\frac{7}{12} = 0.58333...$ where the three dots (called an ellipsis) mean "and so on" or to continue the pattern without stopping.

Or, if we agree to round, we can write $\frac{7}{12} = 0.583$ (to the nearest thousandth).

Convert to decimal notation:

8. $\frac{2}{15}$

9. $\frac{3}{13}$

Example 9

Changing Fraction to Decimal Number

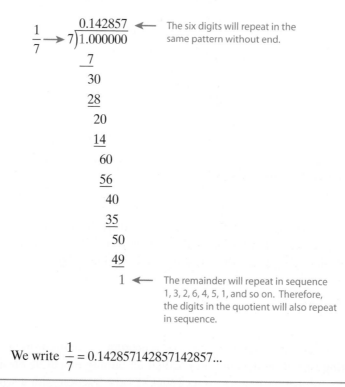

$$\frac{1}{7} \longrightarrow 7\overline{)1.000000}$$ with quotient 0.142857 ← The six digits will repeat in the same pattern without end.

$$\begin{array}{r} 0.142857 \\ 7\overline{)1.000000} \\ \underline{7} \\ 30 \\ \underline{28} \\ 20 \\ \underline{14} \\ 60 \\ \underline{56} \\ 40 \\ \underline{35} \\ 50 \\ \underline{49} \\ 1 \end{array}$$ ← The remainder will repeat in sequence 1, 3, 2, 6, 4, 5, 1, and so on. Therefore, the digits in the quotient will also repeat in sequence.

We write $\frac{1}{7} = 0.142857142857142857...$

Now work margin exercises 8 and 9.

Another way of writing repeating decimal numbers is to write a **bar** over the repeating digits. Thus we can write

$$\frac{1}{7} = 0.\overline{142857} \quad \text{and} \quad \frac{7}{12} = 0.58\overline{3}$$

Objective C **Operating with Both Fractions and Decimal Numbers**

As the following examples illustrate, we can perform operations and comparisons with both fractions and decimal numbers by changing the fractions to decimal form.

---

notes

- In some cases changing fractions to decimal form may involve rounding the decimal form of a number and settling for an approximate answer.
- To have a more accurate answer, we may need to change the decimals to fraction form and then perform the operations.

Example 10

Operating with Both Fractions and Decimal Numbers

Find the sum $10\frac{1}{2} + 7.32 + 5\frac{3}{5}$ in decimal form.

Solution

$$10\frac{1}{2} = 10.50 \qquad \left(\frac{1}{2} = 0.50\right)$$

$$7.32 = 7.32$$

$$+5\frac{3}{5} = 5.60 \qquad \left(\frac{3}{5} = 0.60\right)$$

$$\overline{\phantom{+5\frac{3}{5} = }23.42}$$

Now work margin exercise 10.

10. Find the sum

$$2.88 + \frac{1}{4} + 13.9$$

in decimal form.

11. Determine whether $\frac{9}{16}$ is larger than 0.52 by changing $\frac{9}{16}$ to decimal form and then comparing the two numbers. Find the difference.

Example 11

Operating with Both Fractions and Decimal Numbers

Determine whether $\frac{3}{16}$ is larger than 0.18 by changing $\frac{3}{16}$ to decimal form and then comparing the two numbers. Find the difference.

Solution

Divide first.

$$\begin{array}{r} 0.1875 \\ 16\overline{)3.0000} \\ \underline{1\,6} \\ 1\,40 \\ \underline{1\,28} \\ 120 \\ \underline{112} \\ 80 \\ \underline{80} \\ 0 \end{array} \qquad \text{So } \frac{3}{16} = 0.1875.$$

Now subtract.

$$\begin{array}{r} 0.1875 \\ -\,0.1800 \\ \hline 0.0075 \quad \text{difference} \end{array}$$

Thus $\frac{3}{16}$ is larger than 0.18 and their difference is 0.0075.

Now work margin exercise **11.**

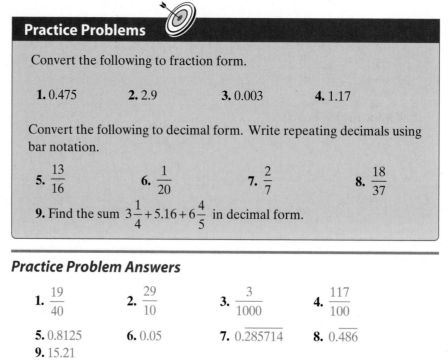

Practice Problems

Convert the following to fraction form.

1. 0.475   **2.** 2.9   **3.** 0.003   **4.** 1.17

Convert the following to decimal form. Write repeating decimals using bar notation.

5. $\frac{13}{16}$ **6.** $\frac{1}{20}$ **7.** $\frac{2}{7}$ **8.** $\frac{18}{37}$

9. Find the sum $3\frac{1}{4} + 5.16 + 6\frac{4}{5}$ in decimal form.

Practice Problem Answers

1. $\frac{19}{40}$ **2.** $\frac{29}{10}$ **3.** $\frac{3}{1000}$ **4.** $\frac{117}{100}$

5. 0.8125   **6.** 0.05   **7.** $0.\overline{285714}$ **8.** $0.\overline{486}$

9. 15.21

Exercises 3.5

Change each decimal number to fraction form. Do not reduce. See Examples 1 through 5.

1. 0.9

2. 0.3

3. 0.5

4. 0.8

5. 0.62

6. 0.38

7. 0.57

8. 0.41

9. 0.526

10. 0.625

11. 0.016

12. 0.012

13. 5.1

14. 7.2

15. 8.15

16. 6.35

Change each decimal number to fraction form or mixed number form and reduce if possible. See Examples 1 through 5.

17. 0.125

18. 0.36

19. 0.18

20. 0.375

21. 0.225

22. 0.455

23. 0.17

24. 0.029

25. 3.2

26. 1.25

27. 6.25

28. 2.75

Change each fraction to decimal form. If the decimal is nonterminating, write it using the bar notation over the repeating pattern of digits. See Examples 6 through 9.

29. $\dfrac{2}{3}$

30. $\dfrac{5}{16}$

31. $\dfrac{7}{11}$

32. $\dfrac{3}{11}$

33. $\dfrac{11}{16}$

34. $\dfrac{9}{16}$

35. $\dfrac{3}{7}$

36. $\dfrac{5}{7}$

37. $\dfrac{1}{6}$

38. $\dfrac{5}{18}$

39. $\dfrac{5}{9}$

40. $\dfrac{2}{9}$

Change each fraction to decimal form rounded to the nearest thousandth.

41. $\dfrac{7}{24}$

42. $\dfrac{16}{33}$

43. $\dfrac{5}{12}$

44. $\dfrac{13}{16}$

45. $\dfrac{1}{32}$

46. $\dfrac{1}{14}$

47. $\dfrac{16}{13}$

48. $\dfrac{20}{9}$

49. $\dfrac{30}{21}$

50. $\dfrac{40}{3}$

Perform the indicated operations by writing all the numbers in decimal form. Round to the nearest thousandth, if necessary. See Example 10.

51. $\dfrac{1}{4} + 0.25 + \dfrac{1}{5}$

52. $\dfrac{3}{4} + \dfrac{1}{10} + 3.55$

53. $\dfrac{5}{8} + \dfrac{3}{5} + 0.41$

54. $6 + 2\dfrac{37}{100} + 3\dfrac{11}{50}$

55. $2\dfrac{53}{100} + 5\dfrac{1}{10} + 7.35$

56. $37.02 + 25 + 6\dfrac{2}{5} + 3\dfrac{89}{100}$

57. $1\dfrac{1}{4} - 0.125$

58. $2\dfrac{1}{2} - 1.75$

59. $36.71 - 23\dfrac{1}{5}$

60. $3.1 - 2\dfrac{1}{100}$

61. $\left(\dfrac{35}{100}\right)^2 (0.73)$

62. $\left(5\dfrac{1}{10}\right)^2 (2.25)$

63. $\left(1\dfrac{3}{8}\right)(3.1)(2.6)$

64. $\left(1\dfrac{3}{4}\right)\left(2\dfrac{1}{2}\right)(5.35)$

65. $5\dfrac{54}{100} \div 2.1$

66. $72.16 \div \dfrac{2}{5}$

67. $13.65 \div \dfrac{1}{2}$

68. $91.7 \div \dfrac{1}{4}$

Using a calculator, change each of the following fractions to decimal form; then determine which number is larger and find the difference between the two numbers. (Use the bar notation over repeating digits, if necessary.) See Example 11.

69. $2\dfrac{1}{4}$, 2.3

70. $\dfrac{7}{8}$, 0.878

71. 0.28, $\dfrac{3}{11}$

72. $\dfrac{1}{3}$, 0.3　　　**73.** $\dfrac{22}{7}$, 3.3　　　**74.** $\dfrac{4}{9}$, 0.5

75. 3.5, $3\dfrac{2}{3}$　　　**76.** $5\dfrac{3}{4}$, 5.5

Write the three numbers in order, smallest to largest.　See Example 11.

77. 0.76, $\dfrac{3}{4}$, $\dfrac{7}{10}$　　**78.** 0.63, $\dfrac{5}{8}$, 0.64　　**79.** $\dfrac{5}{16}$, 0.3126, 0.314　　**80.** 0.083, $\dfrac{41}{500}$, $\dfrac{2}{25}$

Change any decimal number (not a whole number) into fraction or mixed number form.　See Examples 1 through 5.

81. By 2005 census estimates, there were 83.8 people per square mile in the United States.

82. The average weight for a one-year-old girl is 9.1 kg.

83. The median age for men at the beginning of their first marriage is 26.3 years.　The median age for women at the beginning of their first marriage is 24.1 years.

84. The maximum speed of a giant tortoise on land is about 0.17 mph.

85. There are about 21.5 students per teacher in California public schools.

86. The surface gravity on Mars is about 0.38 times the gravity on Earth.　The atmospheric pressure on Mars is about 0.01 times the atmospheric pressure on Earth.

Change each fraction or mixed number (not a whole number) into decimal form.　Round each number to the nearest hundredth.　See Examples 6 through 9.

87. In 2000 it was estimated that $\dfrac{7}{50}$ of U.S. households viewed 26 or more TV stations regularly.

88. In a recent year about $\dfrac{8}{57}$ of the advertising budget in the automotive industry was spent on newspaper ads.

89. In 2005 the average price of unleaded gasoline in California was $6\dfrac{6}{7}$ times the price it was in 1970.

90. In the 2004 presidential election, George W. Bush received $\dfrac{143}{269}$ of the popular vote.

Use a calculator to perform the indicated operations and write the answers in decimal form. Change any mixed numbers into improper fractions. If necessary, round decimal values to the nearest ten-thousandth.

91. $\dfrac{7}{10} + 2.1 + \dfrac{5}{10}$

92. $7 + 5\dfrac{1}{10} + 0.9$

93. $3.577 + 16.892 + 43\dfrac{2}{5}$

94. $4.0085 + 20\dfrac{1}{10} + 0.054 + 13$

95. $5\dfrac{1}{2} - 3.25$

96. $\dfrac{9}{14} - \dfrac{1}{3}$

97. $\dfrac{1}{3} + \dfrac{1}{3}$

98. $0.3 - \dfrac{1}{15}$

99. $2\dfrac{2}{25} - 1\dfrac{1}{5}$

100. $22\dfrac{7}{10} - 10\dfrac{1}{3}$

101. $6\dfrac{1}{8} \times 7\dfrac{1}{4}$

102. $\dfrac{5}{6} \times \dfrac{3}{7}$

Estimating Sums and Differences   page 217

    Estimate a sum (or difference) by rounding each number to the place of the **leftmost nonzero digit** and then adding (or subtracting) these rounded numbers.

Section 3.3: Multiplication with Decimal Numbers

To Multiply Decimal Numbers   page 224

1. Multiply the two numbers as if they were whole numbers.
2. Count the total number of places to the right of the decimal points in both numbers being multiplied.
3. Place the decimal point in the product so that the number of decimal places is the same as that found in Step 2.

To Multiply a Decimal Number by a Power of 10   page 226

1. Count the number of 0s in the power of 10.
2. Move the decimal point to the right the number of places equal to the number of 0s..

Estimating Products   page 226

    Estimate products by rounding each number to the place of the last nonzero digit on the left and multiplying these rounded numbers.

Section 3.4: Division with Decimal Numbers

To Divide Decimal Numbers   page 234

1. Move the decimal point in the divisor to the right so that the divisor is a whole number.
2. Move the decimal point in the dividend the same number of places to the right.
3. Place the decimal point in the quotient directly above the new decimal point in the dividend. (**Note:** Be sure to do this before dividing.)
4. Divide just as you would with whole numbers. (**Note:** 0s may be added in the dividend as needed to be able to continue the division process.)

When the Remainder is not 0   page 235

1. Decide first how many decimal places are to be in the quotient.
2. Divide until the quotient is **one digit past the place of desired accuracy**.
3. Using this last digit, round the quotient to the desired place of accuracy.

To Divide a Decimal Number by a Power of 10   page 237
   1. Count the number of 0s in the power of 10.
   2. Move the decimal point to the left the number of places equal to
      the number of 0s..

Order of Operations with Decimal Numbers   page 238

Estimating Quotients   page 238
   Estimate quotients by rounding both the divisor and the dividend to
   the place of the last nonzero digit on the left and then dividing with
   these rounded values.

Section 3.5: Decimal Numbers and Fractions

To Change from Decimal Numbers to Fractions   page 247
   A finite decimal number can be written in fraction form by writing a
   fraction with the following:
   1. a **numerator** that consists of the whole number formed by all the
      digits of the decimal number, and
   2. a **denominator** that is the power of ten that names the position of
      the rightmost digit.

To Change from Fractions to Decimal Numbers   page 249
   To change a fraction to decimal form, we use the meaning that the
   numerator is to be divided by the denominator.
   1. If the remainder is eventually 0, the decimal is said to be **terminating**.
   2. If the remainder is never 0, the decimal is said to be **nonterminating**.

Repeating Decimal Numbers   page 251
   One way to write repeating decimals is to write a bar over the
   repeating digits.

Chapter 3: Review

Section 3.1: Introduction to Decimal Numbers

Write the following numbers in decimal form.

1. $62\dfrac{9}{100}$

2. $15\dfrac{357}{1000}$

Write the following decimal numbers in mixed number form and simplify, if possible.

3. 81.47

4. 200.5

Write the following numbers in words.

5. 7.08

6. 12.137

Write the following numbers in decimal notation.

7. eight-four and seventy-five thousandths

8. three thousand and three thousandths

Find the larger number in each pair.

9. 0.0081, 0.081

10. 2.358, 2.349

Arrange the numbers in order, from smallest to largest.

11. 0.033, 0.03, 0.0303

12. 2.412, 2.41, 2.42, 2.4112

Round as indicated.

13. 5863 (to the nearest hundred)

14. 7.649 (to the nearest tenth)

15. 0.0385 (to the nearest thousandth)

16. 2.069896 (to the nearest hundred-thousandth)

Section 3.2: Addition and Subtraction with Decimal Numbers

Add or subtract as indicated.

17. $5.42 + 21.4$

18.
$$88.292$$
$$+ 15.98$$

19.
$$84$$
$$15.8$$
$$+ 24.63$$

20.
$$93.78$$
$$34.9$$
$$+ 84.1$$

21.
$$34.967$$
$$40.8$$
$$9.451$$
$$+ 8.2$$

22. $42.008 - 19.3$

23.
$$16.92$$
$$- 7.9$$

24.
$$5$$
$$- 1.0377$$

25.
$$107.83$$
$$- 79.99$$

26.
$$789.35$$
$$- 650.0$$

Solve the following word problem.

27. Find the perimeter of the triangle with sides 14.2 in., 16.3 in., and 24.8 in.

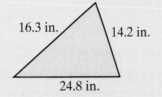

Section 3.3: Multiplication with Decimal Numbers

Find each of the indicated products.

28. $(0.8)(0.9)$

29. $(0.02)(0.32)$

30.
$$1.08$$
$$\times 1.6$$

31.
$$36.5$$
$$\times 4.7$$

32.
$$2.801$$
$$\times 0.19$$

33.
$$41.2$$
$$\times 0.031$$

Find each of the products mentally by using your knowledge of multiplication by powers of 10.

34. $100(2.35)$

35. $10(0.17632)$

36. $1000(5.9641)$

Solve the following word problem.

37. Find the area of a rectangle with length 48.6 feet and width 24.6 feet.

Section 3.4: Division with Decimal Numbers

Divide.

38. $14.68 \div 2$ **39.** $2.73 \div 3$ **40.** $16.05 \div 0.5$ **41.** $24 \div 2.4$

Find each quotient to the nearest hundredth.

42. $4\overline{)28.3}$ **43.** $0.5\overline{)16.923}$ **44.** $0.06\overline{)52.832}$ **45.** $1.4\overline{)26.7}$

Divide by using your knowledge of division by powers of 10.

46. $\dfrac{19.6}{10}$ **47.** $\dfrac{38.9}{100}$ **48.** $\dfrac{5.67}{1000}$

Solve the following problem.

49. You bought 9 items at the grocery store for $83.50 and paid tax figured at 0.08 times the price. What average amount did you pay per item, including tax?

Section 3.5: Decimal Numbers and Fractions

Change each decimal number to fraction or mixed number form. Reduce if possible.

50. 0.07 **51.** 2.025 **52.** 0.015

Change each fraction to decimal form. If the decimal number is nonterminating, write it using the bar notation over the repeating pattern of digits.

53. $\dfrac{1}{3}$ **54.** $\dfrac{5}{8}$ **55.** $2\dfrac{4}{9}$

Change each fraction to decimal form rounded to the nearest thousandth.

56. $\dfrac{15}{17}$ **57.** $\dfrac{99}{101}$

Solve the following word problems.

58. The buyer for a company purchased 17 cars at a price of $33,450 each. How much did he pay for the cars?

59. You are going to make a down payment of $275 on a new computer and ten equal monthly payments of $62.50. How much will you pay for the computer?

Use a calculator to perform the indicated operations. Change any mixed numbers into improper fractions. Write the answers in decimal form.

60. $0.96 + 1.73 + 19.2 + 18$

61. $\dfrac{1}{4} + \dfrac{7}{8}$

62. $3\dfrac{1}{2} \times 4\dfrac{1}{2}$

Chapter 3: Test

Solve the following problems.

1. Write 30.657
   a. in words   **b.** in mixed number form

2. Write the number two and thirty-two thousandths in decimal form.

3. Change $\dfrac{5}{16}$ to decimal form.

Write the following numbers in order from smallest to largest

4. 0.619, 0.72, 0.626   **5.** $\dfrac{1}{4}, \dfrac{1}{5}, 0.275$

Round as indicated.

		Nearest Tenth	Nearest Hundredth	Nearest Thousandth	Nearest Ten
6.	216.7049	**a.** _____	**b.** _____	**c.** _____	**d.** _____

		Nearest Tenth	Nearest Hundredth	Nearest Thousandth	Nearest Ten
7.	73.01485	**a.** _____	**b.** _____	**c.** _____	**d.** _____

Perform the indicated operations. Write each answer in decimal form. Round your answer to the nearest hundreth, if necessary.

8. $\begin{array}{r} 85.815 \\ + 17.943 \\ \hline \end{array}$ **9.** $\begin{array}{r} 0.35 \\ \times 0.84 \\ \hline \end{array}$ **10.** $\begin{array}{r} 95.6 \\ - 93.712 \\ \hline \end{array}$ **11.** $\begin{array}{r} 100.64 \\ - 82.495 \\ \hline \end{array}$

12. $82 \div 4.6$ **13.** $\begin{array}{r} 84 \\ 12.91 \\ + 25.2 \\ \hline \end{array}$ **14.** $\begin{array}{r} 16.31 \\ \times 0.785 \\ \hline \end{array}$ **15.** $\begin{array}{r} 1.92 \\ \times 1000 \\ \hline \end{array}$

16. $\dfrac{3.614}{100}$ **17.** $13\dfrac{2}{5} + 6 + 17.913$ **18.** $1\dfrac{1}{1000} - 0.09705$ **19.** $\left(4\dfrac{3}{4}\right)(0.6)\left(\dfrac{1}{2}\right)$

Find the estimated quotient using rounded numbers. Then find the actual quotient to the nearest hundredth.

20. $0.13\overline{)8.617}$

Solve the following word problems.

21. Gas mileage: An automobile gets 18.3 miles per gallon and the tank holds 21.4 gallons. How far (to the nearest mile) can the car travel on a full tank?

22. Coins: The modern U. S. quarter weighs 5.68 grams. Express this as a mixed number.

23. Volunteering: To meet his required community service hours at Tougaloo College, John is helping out at the local homeless shelter. This month he worked three unequal shifts of 5.34 hours, 2.7 hours, and 3.75 hours. (The school requires a total of 60 hours to graduate.)

   a. How many hours did John donate this month?

   b. If at the beginning of the month John had 31.54 hours remaining, how many more hours must he complete after this month?

24. Manufacturing: How many 1.57 ounce chocolate candy bars will a vat containing 289.5 ounces of molten chocolate produce? Round your answer to the nearest whole number.

Cumulative Review: Chapters 1 - 3

Write the following numbers in their English word equivalents.

1. **a.** 5,612,009

    b. 5612.009

Name the properties illustrated.

2. **a.** $15 + 6 = 6 + 15$

    b. $2(6 \cdot 7) = (2 \cdot 6)7$

    c. $51 + 0 = 51$

Round as indicated.

	Nearest Hundredth	Nearest Tenth	Nearest Whole Number	Nearest Hundred
3. 4591.057	**a.** _____	**b.** _____	**c.** _____	**d.** _____

Solve the following problems.

4. Change the following mixed number to a fraction: $12\dfrac{4}{5}$

5. Change the following fraction to a mixed number: $\dfrac{35}{8}$

6. Change each fraction to decimal form. If the decimal number is nonterminating, write it using the bar notation over the repeating pattern of digits.

    a. $\dfrac{7}{8}$

    b. $\dfrac{3}{11}$

7. Find the prime factorization of 420.

8. Find the LCM for each set of numbers.

    a. 28, 70

    b. 8, 16, 32, 64

Perform the indicated operations.

9. $3 + 15 + 9 + 7$

10. $\begin{array}{r} 2023 \\ -\ 579 \\ \hline \end{array}$

11. $612 \div 24$

12. $\begin{array}{r} 45 \\ \times\ 27 \\ \hline \end{array}$

13. $\dfrac{14}{15} - \dfrac{9}{10}$

14. $\begin{array}{r} 4\dfrac{5}{8} \\ +3\dfrac{9}{10} \\ \hline \end{array}$

15. $\begin{array}{r} 306\dfrac{3}{8} \\ -250\dfrac{13}{20} \\ \hline \end{array}$

16. $10\dfrac{1}{2} + 5\dfrac{3}{4} + 16\dfrac{1}{10}$

17. $\left(\dfrac{5}{6}\right)\left(\dfrac{7}{30}\right)\left(\dfrac{18}{11}\right)$

18. $\dfrac{\dfrac{1}{2} + \dfrac{5}{6}}{1 - \dfrac{5}{8}}$

19. $\begin{array}{r} 2.625 \\ 23.1 \\ +\ 4.66 \\ \hline \end{array}$

20. $15.2222 + 21.7779$

21. $3.0139 - 0.1187$

22. $\begin{array}{r} 54 \\ -16.23 \\ \hline \end{array}$

23. $(2.53)(0.62)$

24. $1000(2.899)$

25. $0.3\overline{\smash{)}10.74}$

26. $\dfrac{21.34}{100}$

Evaluate the expressions using the rules for order of operations.

27. $2^2 + 3^2$

28. $18 + 5\left(2 + 3^2\right) \div 11 \cdot 2$

29. $36 \div 4 + 9 \cdot 2^2$

30. $\left(\dfrac{1}{2}\right)^2\left(\dfrac{14}{15}\right) + \dfrac{4}{15} \div 2$

31. $4.5 \div 0.5 + (6.1 - 3.1)(4.8)$

32. $\dfrac{1}{5} + 0.3(6.1 - 5.9) + \left(\dfrac{1}{2}\right)^2$

Solve the following problems.

33. The difference between 5454 and 2310 is added to 697. What is the sum?

34. Find $\dfrac{3}{4}$ of 96.

35. Find the quotient $32.1 \div 1.7$ (to the nearest tenth).

36. Find the product of 21.6 and 0.35 and subtract the sum of 1.375 and 4.

Solve the following word problems.

37. Transportation: The business office at a university purchased five vans at a price of $41,540 each, including tax.

   a. Estimate how much was paid for the vans.

   b. Now calculate exactly how much was paid for the vans.

38. Buying a Car: Your friend plans to buy a used car for $16,000. He can either save up and pay cash or make a down payment of $1600 and 48 monthly payments of $420. What would you recommend that he do? Why?

39. **Dinner Bill:** You and three friends have dinner at a restaurant, and the total bill comes to $47.12. How much will each person pay, if each person pays an equal amount?

40. **Books:** A college student with $520.38 in her checking account purchases three books whose prices (including sales tax) are $35.67, $63.94, and $120.13 respectively. How much money will she have left over?

41. **Cross-Country Running:** Three people who competed in a cross-country race completed the course in the following times: $1\frac{3}{5}$ hours, $2\frac{1}{2}$ hours, and $3\frac{1}{4}$ hours respectively. What is the average time for the three runners?

42. **Real Estate:** The property line of a person's home consists of four sides of uneven length as shown below, where the lengths are 60.79 meters, 29.15 meters, 74.2 meters, and 35.86 meters, respectively. How long must a fence be (including any gates) to completely surround the property?

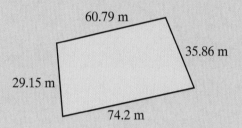

Ratios and Proportions, Percent, and Applications

Mathematics at Work!

Homeowners are familiar with trips to the store to buy such things as paint, fertilizer, and grass seed for repairs and maintenance. In many cases a homeowner must purchase goods in quantities greater than what is needed at one time for one particular job. The manufacturer's or distributor's packaging units will likely vary from their exact needs, and the store will not sell part of a can of paint or part of a bag of fertilizer. A little application of mathematics can help stretch the homeowner's dollar by minimizing any excess amounts that must be bought. Consider the following situation.

One bag of Weed Killer & Fertilizer contains 18 pounds of fertilizer and weed treatment with a recommended coverage of 5000 square feet. If your lawn is in the shape of a rectangle 150 feet by 220 feet, how many pounds of Weed Killer & Fertilizer do you need to cover the lawn? (See Exercises 44 and 45, Section 4.2).

4.1 Ratios and Proportions

Objective A   **Ratios**

We know two meanings for fractions.

1. To indicate a part of a whole.

$$\frac{3}{8} \quad \text{means} \quad \frac{3 \text{ pieces of pie}}{8 \text{ pieces in the whole pie}}$$

2. To indicate division.

$$\frac{3}{8} \quad \text{means} \quad 3 \div 8 \quad \text{or} \quad 8\overline{)3.000}$$

$$
\begin{array}{r}
0.375 \\
8\overline{)3.000} \\
\underline{24} \\
60 \\
\underline{56} \\
40 \\
\underline{40} \\
0
\end{array}
$$

A third use for fractions is to compare two quantities. Such a comparison is called a **ratio**. For example, the **ratio** $\frac{3}{4}$ might mean $\frac{3 \text{ dollars}}{4 \text{ dollars}}$ or $\frac{3 \text{ hours}}{4 \text{ hours}}$.

As another example, on the back of a box of pancake mix the ratio of pancake mix to milk is 2 to $1\frac{1}{2}$. We can write

$$\frac{2 \text{ cups mix}}{1\frac{1}{2} \text{ cups milk}} \qquad \text{or, with a colon,} \qquad 2 \text{ cups mix} : 1\frac{1}{2} \text{ cups milk.}$$

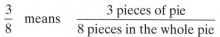

Ratio

A **ratio** is a comparison of two quantities by division. The ratio of a to b can be written as

$$\frac{a}{b} \qquad \text{or} \qquad a : b \qquad \text{or} \qquad a \text{ to } b.$$

Note: Generally, we will use the fraction notation for ratios.

Ratios have the following characteristics:
1. Ratios can be reduced, just as fractions can be reduced.
2. Whenever the units of the numbers in a ratio are the same, then the ratio has no units. We say the ratio is an **abstract number**.
3. When the numbers in a ratio have different units, then the numbers must be labeled to clarify what is being compared. Such a ratio is called a rate.

For example, the ratio of $\dfrac{35 \text{ miles}}{1 \text{ hour}}$ is a **rate** of 35 miles per hour (35 mph).

Example 1

Comparing Two Quantities as a Ratio

Compare the quantities 40 desks and 30 students as a ratio.

Solution

The ratio is $\dfrac{40 \text{ desks}}{30 \text{ students}}$ or in reduced form $\dfrac{4 \text{ desks}}{3 \text{ students}}$.

This means that there are 4 desks for every 3 students.
(What is the ratio of desks to students in your classroom?)

Now work margin exercise 1.

1. Compare 12 apples and 15 oranges as a ratio. Reduce your answer to lowest terms.

Example 2

Comparing Two Quantities as a Ratio

During baseball season, major league players' batting averages are published in the newspapers. If a player has a batting average of 0.250, what does this indicate?

Solution

A batting average is a ratio (or rate) of hits to times at bat. Thus a batting average of 0.250 means $\dfrac{250 \text{ hits}}{1000 \text{ at-bats}}$. Reducing gives

$$\dfrac{250 \text{ hits}}{1000 \text{ at-bats}} = \dfrac{250 \cdot 1 \text{ hits}}{250 \cdot 4 \text{ at-bats}} = \dfrac{1 \text{ hit}}{4 \text{ at-bats}}.$$

This means we can expect this player to get 1 hit for every 4 times at bat.

Now work margin exercise 2.

2. Inventory shows 500 washers and 4000 bolts. What is the ratio of washers to bolts? Reduce to lowest terms.

Example 3

Comparing Two Quantities as a Ratio

Write the comparison of 2 feet to 3 yards as a ratio.

Solution

a. We can write the ratio as $\dfrac{2 \text{ feet}}{3 \text{ yards}}$.

b. Another procedure is to change to common units. Because 1 yd = 3 ft, we have 3 yd = 9 ft and we can write the ratio as an abstract number:

$$\dfrac{2 \text{ feet}}{3 \text{ yards}} = \dfrac{2 \text{ feet}}{9 \text{ feet}} = \dfrac{2}{9} \quad \text{or} \quad 2:9 \quad \text{or} \quad 2 \text{ to } 9.$$

3. Write 3 quarters to 1 dollar in ratio form:

   a. using the given units.

   b. using common units.

Now work margin exercise 3.

Proportions

One way to compare the two fractions $\frac{3}{6}$ and $\frac{4}{8}$ is to find a common denominator and then compare the numerators. For example,

$$\frac{3}{6} = \frac{3}{6} \cdot \frac{4}{4} = \frac{12}{24} \quad \text{and} \quad \frac{4}{8} = \frac{4}{8} \cdot \frac{3}{3} = \frac{12}{24}.$$

We see that, because the resulting numerators, 12, are equal, the original fractions are equal.

Another comparison technique is to reduce both fractions:

$$\frac{3}{6} = \frac{3 \cdot 1}{3 \cdot 2} = \frac{1}{2} \quad \text{and} \quad \frac{4}{8} = \frac{4 \cdot 1}{4 \cdot 2} = \frac{1}{2}.$$

Because the original fractions reduce to the same fraction, we see that the original fractions are equal.

A third method is to set up an equation, called a **proportion**, which states that the two ratios are equal. If this statement is true, then the **cross products** must be equal.

$$\frac{3}{6} \bowtie \frac{4}{8} \text{ is true because } 3 \cdot 8 = 6 \cdot 4 \longleftarrow \text{cross products}$$

Proportions

A **proportion** is a statement that two ratios are equal.

In symbols, $\frac{a}{b} = \frac{c}{d}$ is a proportion ($b, d \neq 0$).

A proportion is true if the **cross products**, $a \cdot d$ and $b \cdot c$, are equal.

Example 4

Verifying the Proportion

Use the cross product technique to determine whether each proportion is true or false.

a. $\frac{6}{8} = \frac{15}{20}$

Solution

$$6 \cdot 20 = 120 \quad \text{and} \quad 8 \cdot 15 = 120$$

Therefore the cross products are equal and the proportion $\frac{6}{8} = \frac{15}{20}$ is true.

b. $\frac{5}{8} = \frac{7}{10}$

Solution

$5 \cdot 10 = 50$ and $8 \cdot 7 = 56$

Because the cross products are not equal ($50 \neq 56$), the proportion is false.

c. $\dfrac{9}{13} = \dfrac{4.5}{6.5}$

Solution

$$
\begin{array}{cc}
6.5 \quad \text{and} \quad 4.5 \\
\underline{\times\, 9} \qquad\quad \underline{\times\, 13} \\
58.5 \qquad\quad 13\,5 \\
\qquad\qquad \underline{45} \\
\qquad\qquad 58.5
\end{array}
$$

Because the cross products are equal, the proportion $\dfrac{9}{13} = \dfrac{4.5}{6.5}$ is true.

Determine whether the following proportions are true or false.

4. $\dfrac{3.2}{5} = \dfrac{4}{6.25}$

5. $\dfrac{\frac{4}{5}}{\frac{1}{3}} = \dfrac{1\frac{1}{3}}{1}$

Completion Example 5

Verifying the Proportion

Determine whether the proportion $\dfrac{2\frac{1}{3}}{7} = \dfrac{3\frac{1}{4}}{9\frac{3}{4}}$ is true or false.

Solution

$$2\frac{1}{3} \cdot 9\frac{3}{4} = \frac{7}{3} \cdot \frac{39}{4} = \underline{\hspace{1cm}} = \underline{\hspace{1cm}}$$

$$7 \cdot 3\frac{1}{4} = \frac{7}{1} \cdot \frac{13}{4} = \underline{\hspace{1cm}}$$

Because the cross products are _____ , the proportion is _____ .

Now work margin exercises 4 and 5.

Completion Example Answers

5. $2\frac{1}{3} \cdot 9\frac{3}{4} = \frac{7}{3} \cdot \frac{39}{4} = \frac{273}{12} = \frac{91}{4}$; $\qquad 7 \cdot 3\frac{1}{4} = \frac{7}{1} \cdot \frac{13}{4} = \frac{91}{4}$

Because the cross products are equal, the proportion is true.

Practice Problems

Write the following comparisons as ratios reduced to lowest terms.
(**Note:** There may be more than one form for a correct answer.)

1. 2 teachers to 24 students   **2.** 90 wins to 72 losses

3. 147 miles to 3 hours   **4.** 5 minutes to 200 seconds

Determine whether the following proportions are true or false.

5. $\dfrac{3.7}{11.1} = \dfrac{5.6}{16.8}$ **6.** $\dfrac{9}{8} = \dfrac{70}{64}$ **7.** $\dfrac{12}{19} = \dfrac{3}{4.75}$

Practice Problem Answers

1. $\dfrac{1 \text{ teacher}}{12 \text{ students}}$ **2.** $\dfrac{5 \text{ wins}}{4 \text{ losses}}$ **3.** $\dfrac{49 \text{ miles}}{1 \text{ hour}}$ or 49 mph

4. $\dfrac{1 \text{ minute}}{40 \text{ seconds}}$ or $\dfrac{3}{2}$ **5.** true   **6.** false

7. true

Exercises 4.1

Write the following comparisons as ratios reduced to lowest terms. Change to common units in the numerator and denominator whenever possible. See Examples 1 through 3.

1. 1 dime to 4 nickels

2. 5 nickels to 3 quarters

3. 5 dollars to 5 quarters

4. 6 dollars to 50 dimes

5. 250 miles to 5 hours

6. 270 miles to 4.5 hours

7. 50 miles to 2 gallons of gas

8. 60 miles to 5 gallons of gas

9. 30 chairs to 25 people

10. 25 people to 30 chairs

11. 18 inches to 2 feet

12. 36 inches to 2 feet

13. 8 days to 1 week

14. 21 days to 4 weeks

15. $200 in profit to $500 invested

16. $200 in profit to $1000 invested

17. 100 centimeters to 1 meter

18. 10 centimeters to 1 millimeter

19. 125 hits to 500 times at bat

20. 100 hits to 500 times at bat

Solve the following word problems. Change to common units in the numerator and denominator whenever possible. See Examples 1 through 3.

21. **Blood Types:** About 28 out of every 100 African-Americans have type-A blood. Express this fact as a ratio in lowest terms.

22. **Nutrition:** A serving of four home-baked chocolate chip cookies weighs 40 grams and contains 12 grams of fat. What is the ratio, in lowest terms, of fat grams to total grams?

23. **Standardized Testing:** In recent years, 18 out of every 100 students taking the SAT (Scholastic Aptitude Test) have scored 600 or above on the mathematics portion of the test. Write the ratio, in lowest terms, of the number of scores 600 or above to the number of scores below 600.

24. **Weather:** In a recent year, Albany, NY reported a total of 60 clear days, the rest being cloudy or partly cloudy. For a 365-day year, write the ratio, in lowest terms, of clear days to cloudy or partly cloudy days.

25. College Enrollment: In the fall semester of 2010, Bluefield College had 248 men enrolled full time, and 344 full time women. What is the ratio of men to women at the college?

Source: http://www.cappex.com/colleges/Bluefield-College-231554

26. Books: Different sellers of a given online book broker sell the same book title for a wide variation of prices. For a given title, the highest selling price (mint condition) is $35, while the lowest selling price is $10. What is the ratio of the highest price to the lowest price?

27. Cargo: One truck can carry 9.2 tons of cargo, while a second truck can haul 11.5 tons. What is the ratio of their capacities? (**Hint:** Multiply both the numerator and denominator by 10 to remove the decimal point.)

28. Salaries: If the CEO of a major corporation makes $4,000,000 a year, and the average annual salary of the workers is $30,000 a year, what is the ratio of the CEO's salary in comparison to that of the regular workers?

29. College Acceptances: A certain selective college has 18,200 applicants for 400 openings. What is the ratio of applicants to openings?

30. Coins: John has collected 7 quarters, while Walt has collected 25 dimes. How much money does Walt have compared with John's accumulation? (**Hint:** When converting the coin collections to their respective monetary values, express them in cents.)

31. Gas Consumption: If a car consumes 9 gallons of gasoline after being driven 225 miles, what is the ratio of the distance driven to the gasoline consumed?

32. Job Openings: An employer who has 6 job openings, is deluged with 444 applicants. What are the odds that a given applicant will be hired? (**Hint:** Express the ratio as job openings to applicants.)

33. Recipe: A recipe calls for 6 cups of milk and 4 cups bread crumbs. What is the ratio of milk to bread?

34. Farming: A farmer has a long narrow rectangular piece of land. If the length is 9504 feet, and the width is 1320 feet, what is the ratio of the length to the width?

Use the cross product technique to determine whether each proportion is true or false. See Examples 4 and 5.

35. $\dfrac{5}{6} = \dfrac{10}{12}$

36. $\dfrac{2}{7} = \dfrac{5}{17}$

37. $\dfrac{7}{21} = \dfrac{4}{12}$

38. $\dfrac{6}{15} = \dfrac{2}{5}$

39. $\dfrac{5}{8} = \dfrac{12}{17}$

40. $\dfrac{12}{15} = \dfrac{20}{25}$

41. $\dfrac{5}{3} = \dfrac{15}{9}$

42. $\dfrac{6}{8} = \dfrac{15}{20}$

43. $\dfrac{2}{5} = \dfrac{4}{10}$

44. $\dfrac{3}{5} = \dfrac{6}{10}$

45. $\dfrac{125}{1000} = \dfrac{1}{8}$

46. $\dfrac{3}{8} = \dfrac{375}{1000}$

47. $\dfrac{1}{4} = \dfrac{25}{100}$ **48.** $\dfrac{7}{8} = \dfrac{875}{1000}$ **49.** $\dfrac{3}{16} = \dfrac{9}{48}$ **50.** $\dfrac{2}{3} = \dfrac{66}{100}$

51. $\dfrac{1}{3} = \dfrac{33}{100}$ **52.** $\dfrac{14}{6} = \dfrac{21}{8}$ **53.** $\dfrac{4}{9} = \dfrac{7}{12}$ **54.** $\dfrac{19}{16} = \dfrac{20}{17}$

55. $\dfrac{3}{6} = \dfrac{4}{8}$ **56.** $\dfrac{12}{18} = \dfrac{14}{21}$ **57.** $\dfrac{7.5}{10} = \dfrac{3}{4}$ **58.** $\dfrac{6.2}{3.1} = \dfrac{10.2}{5.1}$

59. $\dfrac{8\frac{1}{2}}{2\frac{1}{3}} = \dfrac{4\frac{1}{4}}{1\frac{1}{6}}$ **60.** $\dfrac{6\frac{1}{5}}{1\frac{1}{7}} = \dfrac{3\frac{1}{10}}{\frac{8}{14}}$ **61.** $\dfrac{6}{24} = \dfrac{10}{48}$ **62.** $\dfrac{7}{16} = \dfrac{3\frac{1}{2}}{8}$

63. $\dfrac{10}{17} = \dfrac{5}{8\frac{1}{2}}$ **64.** $\dfrac{210}{7} = \dfrac{20}{2\frac{2}{3}}$ **65.** $\dfrac{8.5}{6.5} = \dfrac{4.5}{3.5}$ **66.** $\dfrac{12}{1.09} = \dfrac{36}{3.27}$

67. $\dfrac{6}{1.56} = \dfrac{2}{0.52}$ **68.** $\dfrac{3.75}{3} = \dfrac{7.5}{6}$ **69.** $\dfrac{3\frac{1}{5}}{1} = \dfrac{3\frac{3}{5}}{1\frac{2}{5}}$ **70.** $\dfrac{1\frac{1}{4}}{1\frac{1}{2}} = \dfrac{\frac{1}{4}}{\frac{1}{2}}$

Solve the following word problems.

71. Concrete: The quality of concrete is based on the ratio of bags of cement to a cubic yard of gravel. One batch of concrete consisted of 27 bags of cement into 9 cubic yards of gravel, while a second had 15 bags of cement mixed with 5 cubic yards of gravel. Use the cross product technique to determine whether both batches of concrete are equivalent.

72. Lemonade: Two children were selling lemonade at different stands. At the first stand the child mixed 2 cups of lemon juice with 8 cups of water, while at the second stand 3 cups of lemon juice were mixed with 13 cups of water. Use the cross product technique to determine if the two lemonades are of the same strength.

73. **Propane Heaters:** One propane heater will operate for $2\frac{2}{5}$ hours on $4\frac{2}{3}$ pounds of propane, while a second heater will operate for $4\frac{4}{5}$ hours with $9\frac{1}{3}$ pounds of propane. Use the cross product technique to determine if they operate at the same efficiency.

74. **Olive Oil:** One olive press will extract 3.2 cups of oil from 15.6 pounds of olives, while the second press will extract 9.5 cups of oil from 48.8 pounds of olives. Assuming that the olives came from the same batch, use the cross product technique to determine if the two presses operate at the same efficiency.

Writing & Thinking

75. Consider the proportion $\frac{16}{5} = \frac{22.4}{7}$. Show that multiplying both sides of the equation by 35 gives the same results as cross multiplication. Do you think that this method (of multiplying both sides of the equation by the least common denominator) will work for testing all proportions? Why or why not?

4.2 Solving Proportions

Solving Proportions

If one of the numbers in a proportion is unknown, it can be replaced with a variable, such as x. For example,

$$\frac{x}{12} = \frac{2}{3}.$$

We want to find the value of this unknown number. That is, we want to **solve the proportion**. This can be done as follows:

$$\frac{x}{12} = \frac{2}{3}$$

$3 \cdot x = 12 \cdot 2$ Find the cross products. (If the proportion is true, the cross products are equal.)

$\dfrac{\cancel{3} \cdot x}{\cancel{3}} = \dfrac{24}{3}$ Divide both sides by 3. 3 is the coefficient of x.

$x = 8$ Simplify.

To Solve a Proportion

1. Find the **cross products** (or **cross multiply**).
2. Divide both sides of the equation by the coefficient of the variable.
3. Simplify.

Example 1

Solving Proportions

Solve the proportion $\dfrac{3}{6} = \dfrac{5}{x}$.

Solution

$\dfrac{3}{6} = \dfrac{5}{x}$ Write the proportion.

Step 1: $3 \cdot x = 6 \cdot 5$ Find the cross products.

Step 2: $\dfrac{\cancel{3} \cdot x}{\cancel{3}} = \dfrac{30}{3}$ Divide both sides by 3, the coefficient of x.

Step 3: $x = 10$ Simplify.

Now work margin exercise 1.

1. Find x if $\dfrac{12}{x} = \dfrac{9}{15}$.

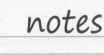

notes

■ As you work through solving a proportion, write each new equation below the previous equation. Keep the = signs aligned (as shown in Example 1). This will help you organize the steps and avoid simple errors.

■

Example 2

Solving Proportions

Find the value of y if $\dfrac{3}{8} = \dfrac{y}{2.4}$.

Solution

$\dfrac{3}{8} = \dfrac{y}{2.4}$ Write the proportion.

Step 1: $3 \cdot 2.4 = 8 \cdot y$ Find the cross products.

Step 2: $\dfrac{7.2}{8} = \dfrac{\cancel{8} \cdot y}{\cancel{8}}$ Divide both sides by 8.

Step 3: $0.9 = y$ Simplify.

Solve each proportion for the unknown term:

2. $\dfrac{3}{10} = \dfrac{R}{100}$

3. $\dfrac{\frac{1}{2}}{6} = \dfrac{5}{x}$

Example 3

Solving Proportions

Find A if $\dfrac{A}{7} = \dfrac{20}{\frac{2}{3}}$.

Solution

$\dfrac{A}{7} = \dfrac{20}{\frac{2}{3}}$ Write the proportion.

Step 1: $\dfrac{2}{3} \cdot A = 7 \cdot 20$ Find the cross products.

Step 2: $\dfrac{\frac{\cancel{2}}{\cancel{3}} \cdot A}{\frac{\cancel{2}}{\cancel{3}}} = \dfrac{140}{\frac{2}{3}}$ Divide both sides by $\frac{2}{3}$.

Step 3: $A = \dfrac{140}{1} \cdot \dfrac{3}{2}$ Multiply by the reciprocal and simplify.

$A = 210$

Now work margin exercises 2 and 3.

Completion Example 4

Solving Proportions

Solve the proportion $\dfrac{2\frac{1}{2}}{6} = \dfrac{3}{y}$.

Solution

Here we first change the mixed number $2\frac{1}{2}$ to the improper fraction $\dfrac{5}{2}$.

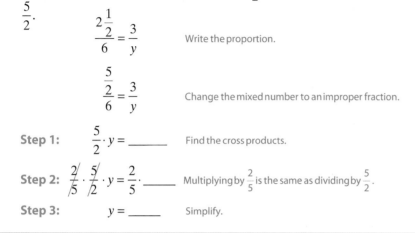

$$\dfrac{2\frac{1}{2}}{6} = \dfrac{3}{y} \qquad \text{Write the proportion.}$$

$$\dfrac{\frac{5}{2}}{6} = \dfrac{3}{y} \qquad \text{Change the mixed number to an improper fraction.}$$

Step 1: $\dfrac{5}{2} \cdot y = \underline{\hspace{1cm}}$ Find the cross products.

Step 2: $\dfrac{2}{5} \cdot \dfrac{5}{2} \cdot y = \dfrac{2}{5} \cdot \underline{\hspace{1cm}}$ Multiplying by $\frac{2}{5}$ is the same as dividing by $\frac{5}{2}$.

Step 3: $\qquad y = \underline{\hspace{1cm}}$ Simplify.

Now work margin exercise 4.

Objective B **Problem Solving with Proportions**

Problems that involve two ratios are the type that can be solved by using proportions.

To Solve a Word Problem by Using a Proportion

1. Identify the unknown quantity and use a variable to represent this quantity.
2. Set up a proportion in which the units are compared in the same order. (Make sure that the units are labeled so they can be seen to be in the right order.)
3. Solve the proportion.

Completion Example Answers

4.

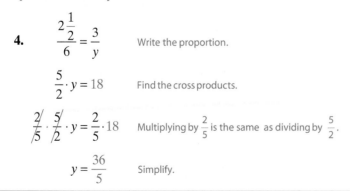

$$\dfrac{2\frac{1}{2}}{6} = \dfrac{3}{y} \qquad \text{Write the proportion.}$$

$$\dfrac{5}{2} \cdot y = 18 \qquad \text{Find the cross products.}$$

$$\dfrac{2}{5} \cdot \dfrac{5}{2} \cdot y = \dfrac{2}{5} \cdot 18 \qquad \text{Multiplying by } \frac{2}{5} \text{ is the same as dividing by } \frac{5}{2}.$$

$$y = \dfrac{36}{5} \qquad \text{Simplify.}$$

4. $\dfrac{\frac{8}{3}}{5} = \dfrac{\frac{1}{5}}{z}$

Example 5

Problems with Proportions

A motorcycle will travel 352 miles on 11 gallons of gas. How many miles will this motorcycle travel on 15 gallons of gas?

Solution

Assign the variable: Let x = unknown number of miles.

Set up the proportion: $\dfrac{352 \text{ miles}}{11 \text{ gallons}} = \dfrac{x \text{ miles}}{15 \text{ gallons}}$ The units are in the same order (miles to gallons) in each ratio.

Solve the proportion: $15 \cdot 352 = 11 \cdot x$

$$\frac{5280}{11} = \frac{\cancel{11} \cdot x}{\cancel{11}}$$

$$480 = x$$

The motorcycle will travel 480 miles on 15 gallons of gas.

5. One pound of candy costs $4.75. How many pounds of candy can be bought for $28.50?

6. Precinct 1 has 520 registered voters and Precinct 2 has 630 registered voters. If the ratio of actual voters to registered voters was the same in both precincts, how many voted in Precinct 2 in the last election if 104 voted in Precinct 1?

Example 6

Problems with Proportions

An architect draws the plans for a building by using a scale of $\dfrac{1}{2}$ inch to represent 10 feet. How many feet does 6 inches represent?

Solution

Assign the variable: Let y = unknown number of feet.

Set up the proportion: $\dfrac{\frac{1}{2} \text{ inch}}{10 \text{ feet}} = \dfrac{6 \text{ inches}}{y \text{ feet}}$

Solve the proportion: $\dfrac{1}{2} \cdot y = 10 \cdot 6$

$$\frac{\cancel{2}}{\cancel{1}} \cdot \frac{\cancel{1}}{\cancel{2}} \cdot y = \frac{2}{1} \cdot 60$$

$$y = 120$$

On these plans, 6 inches represents 120 feet.

Now work margin exercises 5 *and* 6.

Completion Example 7

Problems with Proportions

A recommended mixture of weed killer is 3 capfuls for 2 gallons of water. How many capfuls should be mixed with 5 gallons of water?

Solution

Assign the variable: Let x = unknown number of capfuls of weed killer.

Set up the proportion: $\dfrac{x \text{ capfuls}}{5 \text{ gallons}} = \dfrac{3 \text{ capfuls}}{2 \text{ gallons}}$

Solve the proportion:

$$\underline{\quad} \cdot x = 5 \cdot \underline{\quad}$$

$$\frac{\underline{\quad} \cdot x}{\underline{\quad}} = \frac{15}{\underline{\quad}}$$

$$x = \underline{\quad}$$

_____ capfuls of weed killer should be mixed with 5 gallons of water.

Now work margin exercise 7.

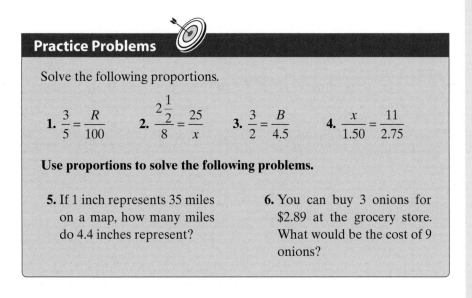

Practice Problems

Solve the following proportions.

1. $\dfrac{3}{5} = \dfrac{R}{100}$ **2.** $\dfrac{2\frac{1}{2}}{8} = \dfrac{25}{x}$ **3.** $\dfrac{3}{2} = \dfrac{B}{4.5}$ **4.** $\dfrac{x}{1.50} = \dfrac{11}{2.75}$

Use proportions to solve the following problems.

5. If 1 inch represents 35 miles on a map, how many miles do 4.4 inches represent?

6. You can buy 3 onions for $2.89 at the grocery store. What would be the cost of 9 onions?

Completion Example Answer

7. $2 \cdot x = 5 \cdot 3$

$$\frac{\cancel{2} \cdot x}{\cancel{2}} = \frac{15}{2}$$

$$x = 7.5$$

7.5 capfuls of weed killer should be mixed with 5 gallons of water.

Practice Problem Answers

1. $R = 60$ **2.** $x = 80$ **3.** $B = 6.75$ **4.** $x = 6$

5. 154 miles   **6.** $8.67

Exercises 4.2

1. $\dfrac{3}{6} = \dfrac{6}{x}$

2. $\dfrac{5}{7} = \dfrac{x}{28}$

3. $\dfrac{8}{B} = \dfrac{6}{30}$

4. $\dfrac{1}{2} = \dfrac{x}{100}$

5. $\dfrac{A}{3} = \dfrac{7}{2}$

6. $\dfrac{3}{5} = \dfrac{60}{D}$

7. $\dfrac{\frac{1}{2}}{x} = \dfrac{5}{10}$

8. $\dfrac{\frac{1}{3}}{x} = \dfrac{5}{9}$

9. $\dfrac{\frac{1}{8}}{6} = \dfrac{\frac{1}{2}}{w}$

10. $\dfrac{1}{4} = \dfrac{1\frac{1}{2}}{y}$

11. $\dfrac{1}{5} = \dfrac{x}{7\frac{1}{2}}$

12. $\dfrac{3}{5} = \dfrac{R}{200}$

13. $\dfrac{A}{4} = \dfrac{50}{100}$

14. $\dfrac{30}{B} = \dfrac{25}{100}$

15. $\dfrac{1}{3} = \dfrac{R}{100}$

16. $\dfrac{9}{x} = \dfrac{4\frac{1}{2}}{11}$

17. $\dfrac{x}{4} = \dfrac{1\frac{1}{4}}{5}$

18. $\dfrac{x}{3} = \dfrac{16}{3\frac{1}{5}}$

19. $\dfrac{3.5}{2.6} = \dfrac{10.5}{B}$

20. $\dfrac{7.8}{1.3} = \dfrac{x}{0.26}$

21. $\dfrac{150}{300} = \dfrac{R}{100}$

22. $\dfrac{12}{B} = \dfrac{25}{100}$

23. $\dfrac{A}{42} = \dfrac{65}{100}$

24. $\dfrac{A}{850} = \dfrac{30}{100}$

25. $\dfrac{5684}{B} = \dfrac{98}{100}$

Use proportions to solve the following word problems. See Examples 5 through 7.

26. **Cartography:** A mapmaker uses a scale of 2 inches to represent 30 miles. How many miles are represented by 3 inches?

27. **Gas prices:** If gasoline sells for $2.49 per gallon, how many gallons can be bought with $17.43?

28. **Investing:** An investor thinks she should make $12 for every $100 she invests. How much would she expect to make on a $1500 investment?

29. **Fabric:** The price of a certain fabric is $1.75 per yard. How many yards can be bought with $35 (not including tax)?

30. **Chemistry:** Two units of a certain gas weigh 175 grams. What is the weight of 5 units of this gas?

31. **Retail profit:** A store owner expects to make a profit of $2 on an item that sells for $10. How much profit will he expect to make on a larger but similar item that sells for $60?

32. **Taxes:** If property taxes are figured at $1.50 for every $100 in evaluation, what taxes will be paid on a home valued at $85,000?

33. **Taxes:** Sales tax is figured at 6¢ for every $1.00 of merchandise purchased. What was the purchase price on an item that had a sales tax of $2.04?

34. Shadows: A building 14 stories high casts a shadow 30 feet long at a certain time of day. What is the length of the shadow of a 20-story building at the same time of day in the same city?

35. Acceleration: A car is traveling at 45 miles per hour. Its speed is increased by 3 miles per hour every 2 seconds. By how much will its speed increase in 5 seconds? How fast will the car be traveling?

36. Travel: A salesman figured he drove 560 miles every two weeks. How far would he drive in three months (12 weeks)?

37. Travel: If you can drive 286 miles in $5\frac{1}{2}$ hours, how long will it take you to drive 468 miles at the same rate of speed?

38. Gas prices: If diesel fuel costs $2.27 per gallon, how much diesel fuel will $24.53 buy?

39. Investing: An investor made $144 in one year on a $1000 investment. What would she have earned if her investment had been $4500?

40. Maps: On a map, $1\frac{1}{2}$ inches represent 40 miles. How many inches represent 50 miles?

41. Lawn care: If 40 pounds of fertilizer are used on 2400 square feet of lawn, how many pounds of fertilizer are needed for a lawn of 5400 square feet?

42. Cooking: If 2 cups of flour are needed to make 12 biscuits, how much flour will be needed to make 9 of the same kind of biscuits?

43. Measurement: There are one thousand grams in one kilogram. How many grams are there in four and seven tenths kilograms?

44. Lawn care: One bag of Weed Killer & Fertilizer contains 18 pounds of fertilizer and weed treatment with a recommended coverage of 5000 square feet. If your lawn is in the shape of a rectangle 150 feet by 220 feet, how many pounds of Weed Killer & Fertilizer do you need to cover the lawn?

45. Lawn care: One bag of dichondra lawn food contains 20 pounds of fertilizer and its recommended coverage is 4000 square feet. If you want to cover a lawn that is in the shape of a rectangle 120 feet by 160 feet, how many pounds of lawn food do you need?

Collaborative Learning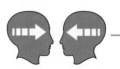

With the class separated into teams of two to four students, each team is to analyze the following two exercises and decide on the best of the given choices for the answers. The team leader is to discuss the team's choice and the reason for this choice with the class.

46. Killer Whales: Marine biologists at a killer whale feeding ground photograph and identify 35 whales during a research trip. The next year they return to the same location and identify 40 whales, 8 of which were ones they had identified the year before. About how many whales would you estimate are in the group that these biologists are studying?

    a. 40 **b.** 75 **c.** 100 **d.** 175

47. Wildlife management: Forest rangers are concerned about the population of deer in a certain region. To get a good estimate of the deer population, they find, tranquilize, and tag 50 deer. One month later, they again locate 50 deer and of these 50, 5 are deer that were previously tagged. On the basis of these procedures and results, what do they estimate to be the deer population of that region?

    a. 100 **b.** 150 **c.** 500 **d.** 1000

A   Understand that percent means hundredths.

B   Change decimal numbers to percents.

C   Change percents to decimal numbers.

4.3 Decimal Numbers and Percents

Objective A   **Percent Means Hundredths**

Income tax is a **percent** of your income.
Property tax is a **percent** of the value of your home.
Sales tax is a **percent** of the value of your purchase.
Your take-home income is a **percent** of your actual earnings.
Profit is a **percent** of the amount invested.
What **percent** of your income do you spend on food each month?

In other words, **percent** is an important part of your daily life. You need to understand percents.

The word percent comes from the Latin *per centum*, meaning **per hundred**. So **percent means hundredths**, or the **ratio of a number to 100**. The symbol % is called the **percent sign**. This sign has the same meaning as the fraction $\frac{1}{100}$. For example,

$$\frac{35}{100} = 35\left(\frac{1}{100}\right) = 35\% \quad \text{and} \quad \frac{60}{100} = 60\left(\frac{1}{100}\right) = 60\%.$$

In Figure 1, the large square is partitioned into 100 small squares. Each small square represents 1% or $\frac{1}{100}$ of the large square. Thus the shaded portion is

$$\frac{40}{100} = 40 \cdot \frac{1}{100} = 40\% \text{ of the large square.}$$

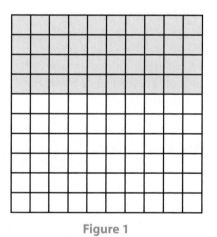

Figure 1

As will be illustrated in Example 1, a fraction with denominator 100 is the same as percent. The numerator is not changed regardless of whether it is a whole number, decimal number, or mixed number. The denominator is replaced by the % sign.

Example 1

Changing Fractions with Denominators of 100 to Percents

Each fraction is changed to a percent.

a. $\dfrac{7}{100} = 7\%$ Remember that **percent** means **hundredths**.

b. $\dfrac{83}{100} = 83\%$

c. $\dfrac{6.4}{100} = 6.4\%$ Note that the decimal point is not moved. The numerator is unchanged.

d. $\dfrac{100}{100} = 1 = 100\%$

e. $\dfrac{240}{100} = 240\%$ 240 is greater than 100 and the % is more than 100%.

Now work margin exercise **1**.

1. Change each fraction to a percent.

a. $\dfrac{9}{100}$

b. $\dfrac{1.25}{100}$

c. $\dfrac{125}{100}$

d. $\dfrac{6\frac{1}{4}}{100}$

Objective B **Changing Decimal Numbers to Percents**

One way to change a decimal number to a percent is to
a. First change the decimal number to fraction form with denominator 100.
b. Then change the fraction to percent form.

For example,

Decimal Form		Fraction Form		Percent Form
0.47	=	$\dfrac{47}{100}$	=	47%
0.93	=	$\dfrac{93}{100}$	=	93%
0.325	=	$\dfrac{32.5}{100}$	=	32.5%

By studying these examples carefully, we can see that we do not need the fraction form. To change from the decimal form we can go directly to the percent form simply by moving the decimal point two places to the right and writing the % sign.

To Change a Decimal Number to a Percent

Step 1: Move the decimal point two places to the right.

Step 2: Write the % sign.

2. Change each decimal number to a percent.

a. 0.34

b. 0.0082

c. 1

d. 5.799

Example 2

Changing Decimal Numbers to Percents

Change each decimal number to percent form by moving each decimal point two places to the right and writing the % sign.

a.   0.253   =   25.3%

decimal point moved   % sign added
two places to the right

b.   0.905 = 90.5%

c.   2.65 = 265%   Note that a number larger than 1 is more than 100%.

d.   0.01 = 1%   One hundredth is 1%.

e.   0.002 = 0.2%   Less than 1%.

Now work margin exercise **2.**

| Objective C | **Changing Percents to Decimal Numbers**
| --- |

To change percents to decimal numbers, we reverse the procedure of changing decimal numbers to percents. For example,

The decimal point is moved two places
left and the % sign is dropped.

$$38.\% = \frac{38}{100} = 0.38$$

understood decimal point

To Change a Percent to a Decimal Number

Step 1:   Move the decimal point two places to the left.

Step 2:   Delete the % sign.

Example 3

Changing Percents to Decimal Numbers

Change each percent to an equivalent decimal number by moving the decimal point two places to the left and deleting the % sign.

a. 76.% = 0.76 ◄—— % sign deleted

↑   ↑
understood   decimal point moved
decimal point   two places left

b. $18.5\% = 0.185$

c. $50\% = 0.50$

d. $100\% = 1.00$

e. $0.25\% = 0.0025$ Note that 0.25% is less than 1%.

***Now work margin exercise* 3.**

3. Change each percent to a decimal number.

a. 40%

b. 211%

c. 0.6%

d. 29.37%

The following relationships between decimal numbers and percents provide guidelines in changing from one form to the other.

1. A decimal number less than 0.01 is less than 1%.
2. A decimal number between 0.01 and 0.10 is between 1% and 10%.
3. A decimal number more than 1 is more than 100%.

Practice Problems

Change each fraction to a percent.

1. $\dfrac{73}{100}$ **2.** $\dfrac{9}{100}$ **3.** $\dfrac{14.2}{100}$

Change each decimal number to a percent.
4. 0.368   **5.** 0.04   **6.** 25

Change each percent to a decimal number.
7. 25%   **8.** 2.5%   **9.** 137%

Practice Problem Answers

1. 73%   **2.** 9%   **3.** 14.2%
4. 36.8%   **5.** 4%   **6.** 2500%
7. 0.25   **8.** 0.025   **9.** 1.37

Exercises 4.3

Find the percent of each square that is shaded. See Figure 1.

1.

2.

3.

4.

5.

6.

Change the following fractions to percents. See Example 1.

7. $\dfrac{20}{100}$

8. $\dfrac{9}{100}$

9. $\dfrac{15}{100}$

10. $\dfrac{62}{100}$

11. $\dfrac{53}{100}$

12. $\dfrac{68}{100}$

13. $\dfrac{125}{100}$

14. $\dfrac{200}{100}$

15. $\dfrac{336}{100}$

16. $\dfrac{13.4}{100}$

17. $\dfrac{0.48}{100}$

18. $\dfrac{0.5}{100}$

19. $\dfrac{2.14}{100}$

20. $\dfrac{1.62}{100}$

Change the following decimal numbers to percents. See Example 2.

21. 0.02

22. 0.09

23. 0.1

24. 0.7

25. 0.36

26. 0.52

27. 0.40

28. 0.65

29. 0.025

30. 0.035

31. 0.055

32. 0.004

33. 1.10

34. 1.75

35. 2

36. 2.3

37. 2%   **38.** 7%   **39.** 18%   **40.** 20%

41. 30%   **42.** 80%   **43.** 0.26%   **44.** 0.52%

45. 125%   **46.** 120%   **47.** 232%   **48.** 215%

49. 17.3%   **50.** 10.1%   **51.** 13.2%   **52.** 6.5%

Change percents to decimal numbers and decimal numbers to percents as directed in each problem.

53. Commission: In calculating his sales commission, Mr. Howard multiplies his total sales by the decimal number 0.12. Change 0.12 to a percent.

54. Sales Tax: Suppose that sales tax is figured at 0.0725 of the selling price. Change 0.0725 to a percent.

55. Mortgages: To calculate what your maximum house payment should be, a banker multiplied your income by 0.28. Change 0.28 to a percent.

56. Charity: A person regularly donates 0.2 of their income to charity. Write 0.2 as a percent.

57. Interest: The interest rate on a loan is 6.4%. Change 6.4% to a decimal number.

58. Commission: The sales commission for the clerk in a retail store is figured at 8.5%. Change 8.5% to a decimal number.

59. Discount: The discount during a special sale on dresses is 30%. Change 30% to a decimal number.

60. Health Insurance: With 27.8% of its residents under 65 years of age having no health coverage, Texas was recently ranked as the state with the highest percentage of health care uninsured. Write 27.8% as a decimal number.

61. License Fee: Suppose the state license fee is figured by multiplying the cost of your car by 0.065. Change 0.065 to a percent.

62. Exam Grades: When the professor graded a certain student exam, the quotient of the number of points the student received divided by the total number of possible points was 0.873. Write 0.873 as a percent.

63. Property Taxes: As of 2005, per capita property taxes in New Jersey were 476% higher than those in Alabama. Write 476% as a decimal number.

64. Recycling: According to the Aluminum Association, the percentage of aluminum cans recycled in 2003 was 50%, and this percentage rose to 51.2% in 2004. Write 50% and 51.2% as decimal numbers.

4.4 Fractions and Percents

Objective A **Changing Fractions and Mixed Numbers to Percents**

If a fraction has denominator 100, it can be changed to a percent by writing the numerator and adding the % sign. If the denominator is a factor of 100 (2, 4, 5, 10, 20, 25, or 50), the fraction can be changed to an equivalent fraction with denominator 100 and then changed to a percent. For example,

$$\frac{1}{4} = \frac{1}{4} \cdot \frac{25}{25} = \frac{25}{100} = 25\%$$

$$\frac{3}{5} = \frac{3}{5} \cdot \frac{20}{20} = \frac{60}{100} = 60\%$$

$$\frac{17}{50} = \frac{17}{50} \cdot \frac{2}{2} = \frac{34}{100} = 34\%$$

However, most fractions do not have factors of 100 as denominators. So, another approach (easily applied with calculators) is to change the fraction to decimal form by dividing. The resulting quotient will be a decimal number. The decimal number can be changed to a percent by moving the decimal point two places to the right and writing the % sign.

1. Change $\frac{13}{16}$ to a percent.

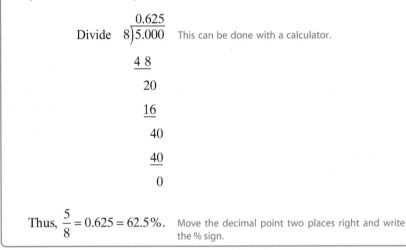

Example 1

Changing Fractions to Percents

Change $\frac{5}{8}$ to a percent.

Solution

Note that 8 is not a factor of 100 so we divide by using long division (or use a calculator).

$$\text{Divide} \quad 8)\overline{5.000} \quad \text{This can be done with a calculator.}$$

$$\begin{array}{r} 0.625 \\ \underline{4\ 8} \\ 20 \\ \underline{16} \\ 40 \\ \underline{40} \\ 0 \end{array}$$

Thus, $\frac{5}{8} = 0.625 = 62.5\%$. Move the decimal point two places right and write the % sign.

Now work margin exercise 1.

Example 2

Changing Mixed Numbers to Percents

Change $2\frac{1}{4}$ to a percent.

Solution

There are three ways to do this problem. All give the same answer.

Method 1: Change $2\frac{1}{4}$ to decimal form directly: $2\frac{1}{4} = 2.25 = 225\%$

Method 2: Change $2\frac{1}{4}$ to an improper fraction and divide to find the decimal form: $2\frac{1}{4} = \frac{9}{4}$ and $4\overline{)9.00}^{2.25}$

Thus, $2\frac{1}{4} = \frac{9}{4} = 2.25 = 225\%$.

Method 3: Change $2\frac{1}{4}$ to an improper fraction and then change the fraction to one with denominator 100. (Note that 4 is a factor of 100.)

$$2\frac{1}{4} = \frac{9}{4} = \frac{9}{4} \cdot \frac{25}{25} = \frac{225}{100} = 225\%$$

Now work margin exercise 2.

Example 3

Changing Fractions to Percents

Change $\frac{1}{3}$ to a percent.

Solution

a. Using a calculator, $\frac{1}{3} = 0.3333333...$

Rounding the decimal quotient to the nearest thousandth:

$\frac{1}{3} \approx 0.333 = 33.3\%$ The answer is rounded and thus not exact.

b. Without a calculator, we can divide and use fractions:

$$\frac{1}{3} = 0.33\overline{3} = 33\frac{1}{3}\% \text{ or } 33.\overline{3}\%$$

$$\begin{array}{r} 0.33\frac{1}{3} \\ 3\overline{)1.00} \\ \underline{9} \\ 10 \\ \underline{9} \\ 1 \end{array}$$

$33\frac{1}{3}\%$ and $33.\overline{3}\%$ are exact, and 33.3% is rounded.

Now work margin exercise 3.

2. Change $1\frac{1}{2}$ to a percent.

3. Change $\frac{3}{11}$ to a percent. Round this answer to the nearest thousandth.

4. The Braves little league baseball team won 14 of their 15 games this season. Find the percentage of games won this season.

Example 4

Changing Fractions to Percents

During the years 1921 to 2005, the New York Yankees baseball team played in 39 World Series Championships and won 26 of them. What percent of these championships did the Yankees win?

Solution

a. The percent won can be found by using a calculator and changing the fraction $\frac{26}{39}$ to decimal form and then changing the decimal to a percent. Using a calculator,

$$\frac{26}{39} = 0.6666666... \approx 0.667 = 66.7\%. \quad \text{(rounded)}$$

b. Without a calculator, we can divide and use fractions:

$$\frac{26}{39} = \frac{13 \cdot 2}{13 \cdot 3} = \frac{2}{3} = 0.6\overline{6} = 66\frac{2}{3}\% \text{ or } 66.\overline{6}\%$$

$$\begin{array}{r} 0.66\frac{2}{3} \\ 3\overline{)2.00} \\ \underline{18} \\ 20 \\ \underline{18} \\ 2 \end{array}$$

The Yankees won $66\frac{2}{3}\%$ or 66.7% of these championships.

Now work margin exercise 4.

Objective B **Changing Percents to Fractions and Mixed Numbers**

To Change a Percent to a Fraction or a Mixed Number

Step 1: Write the percent as a fraction with 100 as the denominator and drop the % sign.

Step 2: Reduce the fraction, if possible.

Change the percent to a fraction.

5. 80%

Example 5

Changing Percents to Fractions

Change 60% to a fraction and reduce, if possible.

Solution

$$60\% = \frac{60}{100} = \frac{3 \cdot 20}{5 \cdot 20} = \frac{3}{5}$$

Now work margin exercise 5.

Example 6

Changing Percents to Fractions

Change 130% to a mixed number.

Solution

$$130\% = \frac{130}{100} = \frac{13 \cdot 10}{10 \cdot 10} = \frac{13}{10} = 1\frac{3}{10}$$

Now work margin exercise 6.

6. Change the percent to a mixed number.

235%

Common Misunderstanding Concerning Percents

The fractions $\frac{1}{4}$ and $\frac{1}{2}$ are often confused with the percents $\frac{1}{4}\%$ and $\frac{1}{2}\%$. The differences can be clarified by using decimal numbers.

PERCENT	DECIMAL	FRACTION
$\frac{1}{4}\%$ (or 0.25%)	0.0025	$\frac{1}{400}$
$\frac{1}{2}\%$ (or 0.5%)	0.005	$\frac{1}{200}$
25%	0.25	$\frac{1}{4}$
50%	0.50	$\frac{1}{2}$

Thus

$$\frac{1}{4} = 0.25 \quad \text{and} \quad \frac{1}{4}\% = 0.0025$$

$$\mathbf{0.25 \neq 0.0025}$$

Similarly,

$$\frac{1}{2} = 0.50 \quad \text{and} \quad \frac{1}{2}\% = 0.005$$

$$\mathbf{0.50 \neq 0.005}$$

You can think of $\frac{1}{4}$ as being one-fourth of a dollar (a quarter) and $\frac{1}{4}\%$ as being one-fourth of a penny. $\frac{1}{2}$ can be thought of as one-half of a dollar and $\frac{1}{2}\%$ as one-half of a penny.

You should memorize the equivalent percent, decimal number, and fraction values in the following box. Look for patterns related to the fractions. Many calculations with these expressions can be done mentally.

Common Percent-Decimal-Fraction Equivalents

$$1\% = 0.01 = \frac{1}{100} \qquad 33\frac{1}{3}\% = 0.33\bar{3} = \frac{1}{3} \qquad 12\frac{1}{2}\% = 0.125 = \frac{1}{8}$$

$$25\% = 0.25 = \frac{1}{4} \qquad 66\frac{2}{3}\% = 0.66\bar{6} = \frac{2}{3} \qquad 37\frac{1}{2}\% = 0.375 = \frac{3}{8}$$

$$50\% = 0.50 = \frac{1}{2} \qquad\qquad\qquad\qquad\qquad 62\frac{1}{2}\% = 0.625 = \frac{5}{8}$$

$$75\% = 0.75 = \frac{3}{4} \qquad\qquad\qquad\qquad\qquad 87\frac{1}{2}\% = 0.875 = \frac{7}{8}$$

$$100\% = 1.00 = 1$$

Practice Problems

Change each fraction or mixed number to a percent.

1. $\dfrac{6}{25}$ **2.** $\dfrac{56}{64}$ **3.** $\dfrac{1}{18}$

4. $3\dfrac{1}{2}$ **5.** $5\dfrac{1}{10}$

Change each percent to a fraction or mixed number. Reduce the fractions.

6. 43% **7.** 292% **8.** 2.5%

9. 0.3% **10.** $37\dfrac{1}{2}\%$

Practice Problem Answers

1. 24% **2.** 87.5% **3.** 5.6% **4.** 350%

5. 510% **6.** $\dfrac{43}{100}$ **7.** $\dfrac{73}{25}$ **8.** $\dfrac{1}{40}$

9. $\dfrac{3}{1000}$ **10.** $\dfrac{3}{8}$

Exercises 4.4

Change the following fractions or mixed numbers to percents. See Examples 1 through 4.

1. $\dfrac{3}{100}$ 2. $\dfrac{16}{100}$ 3. $\dfrac{7}{100}$ 4. $\dfrac{29}{100}$ 5. $\dfrac{1}{2}$

6. $\dfrac{3}{4}$ 7. $\dfrac{1}{4}$ 8. $\dfrac{1}{20}$ 9. $\dfrac{11}{20}$ 10. $\dfrac{7}{10}$

11. $\dfrac{3}{10}$ 12. $\dfrac{3}{5}$ 13. $\dfrac{1}{5}$ 14. $\dfrac{2}{5}$ 15. $\dfrac{4}{5}$

16. $\dfrac{1}{50}$ 17. $\dfrac{13}{50}$ 18. $\dfrac{1}{25}$ 19. $\dfrac{12}{25}$ 20. $\dfrac{24}{25}$

21. $\dfrac{1}{8}$ 22. $\dfrac{5}{8}$ 23. $\dfrac{7}{8}$ 24. $\dfrac{7}{12}$ 25. $\dfrac{1}{12}$

26. $\dfrac{39}{50}$ 27. $\dfrac{17}{20}$ 28. $\dfrac{5}{6}$ 29. $1\dfrac{5}{12}$ 30. $1\dfrac{11}{12}$

31. $1\dfrac{1}{6}$ 32. $1\dfrac{3}{8}$ 33. $1\dfrac{1}{20}$ 34. $1\dfrac{1}{4}$ 35. $1\dfrac{3}{4}$

36. $1\dfrac{1}{5}$ 37. $1\dfrac{3}{8}$ 38. $2\dfrac{1}{2}$ 39. $2\dfrac{1}{10}$ 40. $2\dfrac{1}{15}$

Change the following percents to fractions or mixed numbers. See Examples 5 and 6.

41. 10% 42. 5% 43. 15% 44. 17% 45. 25%

46. 30% 47. 50% 48. $12\dfrac{1}{2}\%$ 49. $37\dfrac{1}{2}\%$ 50. $16\dfrac{2}{3}\%$

51. $33\dfrac{1}{3}\%$ 52. $66\dfrac{2}{3}\%$ 53. 33% 54. $\dfrac{1}{2}\%$ 55. $\dfrac{1}{4}\%$

56. 1%   **57.** 100%   **58.** 125%   **59.** 120%   **60.** 150%

61. 0.3%   **62.** 2.5%   **63.** 62.5%   **64.** 0.2%   **65.** 0.75%

Find the missing forms of each number. See Examples 1 through 6.

	Fraction form	Decimal form	Percent form
66.	$\dfrac{5}{8}$	_____	_____
67.	$\dfrac{11}{20}$	_____	_____
68.	_____	0.09	_____
69.	_____	1.75	_____
70.	_____	_____	36%
71.	_____	_____	10.5%

Change percents to fractions and fractions to percents as directed in each problem. See Example 4.

72. Discounts: A department store offers a 30% discount during a special sale on men's suits. Change 30% to a fraction reduced to lowest terms.

73. Investing: The portfolio of an investor increased by 385%. Express this as a mixed number reduced to lowest terms.

74. Income Tax: Betty pays income tax of 24% on her pay from working as a labor and delivery nurse. Express 24% as fraction reduced to lowest terms.

75. Voltage: A certain very precise voltmeter will measure the voltage with no greater than 0.035% error. This means that the difference between the actual voltage and the indicated voltage is no greater than 0.035% of the measured voltage; this is expressed as ±0.035%. Express this maximum error as a fraction reduced to lowest terms.

76. Travel: Malcolm planned to drive 360 miles on a trip. After 24 miles, what percent of the trip had he driven?

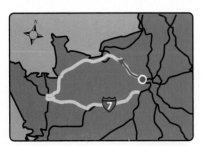

77. Measurements:

   a. There are 12 inches in a foot. What percent of a foot is an inch?

   b. There are 4 quarts in a gallon. What percent of a gallon is a quart?

   c. There are 16 ounces in a pound. What percent of a pound is an ounce?

78. Historic Items: A desk once belonging to George Washington was recently sold at auction. It was expected to bring $80,000 but actually sold for $165,000. Write the actual selling price as a percent of the expected price.

79. Commission: A salesperson received a $2275 commission for selling a $35,000 car. What percent of the selling price was the commission?

80. Student Government: In a sophomore class of 250 students, 10 represent the sophomore class on the student council. What percent of the class is on the student council?

81. Exam Grades: Out of a possible total of 240 points on an exam, David received 204 points. What percent of the exam did David get correct?

82. College Degrees: To receive a Bachelor of Science (BS) degree at Bluefield State College, the student must complete a total of 128 credit hours, of which 41 of these credits must be general education Core Skills courses. What percent of the total curriculum is dedicated to general education courses?
Source: 2010-2011 Bluefield, WV State College Catalogue, p. 76

83. Drinking Water: 3 gallons of a fluoride are in a 4,000,000 gallon supply of drinking water. What percent of the drinking water is flouride?

84. Blood Alcohol Level: According to the laws in the United States, a person can be arrested for driving under the influence of alcohol if the blood alcohol concentration (BAC) is 0.08% or greater, where the BAC in decimal form is defined as the number of grams of alcohol in 1 milliliter of blood. **Source:** http://www.ohsinc.com/drunk_driving_laws_blood_breath%20_alcohol_limits_CHART.htm
  a. Assume that 2 milliliters of blood has 0.0022 grams of alcohol, what is the BAC in decimal units? (**Hint:** Divide the amount of alcohol by the amount of blood.)
  b. Express this as a percent.
  c. Does this exceed the legal limit of 0.08%

85. Manufacturing: In a factory with antiquated machinery an employee could produce 25 items an hour. After the factory updated its equipment with state-of-the-art machines, the same employee could produce 350 items in an hour. This is an improvement of 325 items an hour. By what percent did production increase with the new state-of-the-art machines? (**Hint:** Divide the improvement of produced items by the number of items produced on the old machine.)

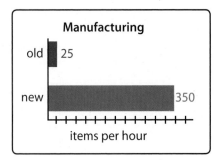

Manufacturing

old 25

new 350

items per hour

Writing & Thinking

86. Discuss a method you might use to memorize the percent equivalents for the fractions $\frac{1}{8}, \frac{2}{8}, \frac{3}{8}, \frac{4}{8}, \frac{5}{8}, \frac{6}{8}, \frac{7}{8}$, and $\frac{8}{8}$. Share these ideas in a discussion with your friends and classmates.

Objectives

A   Understand the proportion $\dfrac{P}{100} = \dfrac{A}{B}$.

B   Use the proportion $\dfrac{P}{100} = \dfrac{A}{B}$ to solve percent problems.

4.5 Solving Percent Problems by Using the Proportion: $\dfrac{P}{100} = \dfrac{A}{B}$

Objective A   **The Proportion** $\dfrac{P}{100} = \dfrac{A}{B}$

We know that **percent means hundredths** and percent is the ratio of a number to 100. For example, 25% can be written in the ratio form $\dfrac{25}{100}$. Using this ratio concept

"**25% of 60 is 15.**" can be written as the proportion $\dfrac{25}{100} = \dfrac{15}{60}$

$\quad\quad\downarrow\quad\quad\quad\downarrow\quad\quad\quad\downarrow$

    Percent   Base   Amount

In general,

"**P% of B is A.**" can be written as the proportion $\dfrac{P}{100} = \dfrac{A}{B}$.

$\quad\quad\downarrow\quad\quad\quad\downarrow\quad\quad\quad\downarrow$

    Percent   Base   Amount

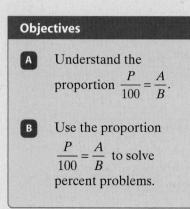

For the Proportion $\dfrac{P}{100} = \dfrac{A}{B}$

$P\% = $ **percent** (written as the ratio $\dfrac{P}{100}$)

$\quad B = $ **base** (number that we are finding the percent of)

$\quad A = $ **amount** (a part of the base)

Objective B   **Problem Solving with the Proportion** $\dfrac{P}{100} = \dfrac{A}{B}$

There are three basic types of percent problems. Each of these three types can be solved by using a proportion of the form $\dfrac{P}{100} = \dfrac{A}{B}$. If any two of the values P, A, and B are known, then the third can be found by solving the resulting proportion.

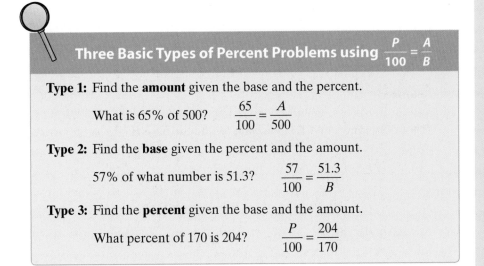

Three Basic Types of Percent Problems using $\frac{P}{100} = \frac{A}{B}$

Type 1: Find the **amount** given the base and the percent.

What is 65% of 500? $\frac{65}{100} = \frac{A}{500}$

Type 2: Find the **base** given the percent and the amount.

57% of what number is 51.3? $\frac{57}{100} = \frac{51.3}{B}$

Type 3: Find the **percent** given the base and the amount.

What percent of 170 is 204? $\frac{P}{100} = \frac{204}{170}$

The following examples illustrate how to substitute into the proportion $\frac{P}{100} = \frac{A}{B}$ and to solve the resulting proportion for the unknown number.

Example 1

Finding the Amount

What is 65% of 500?

Solution

$P\% = 65\%$ and $B = 500$. We want to find the **amount** A.

$$\frac{65}{100} = \frac{A}{500}$$ Substitute $P = 65$ and $B = 500$.

$$500 \cdot 65 = 100 \cdot A$$ Find the cross products.

$$\frac{\overset{5}{\cancel{500}} \cdot 65}{\cancel{100}} = \frac{\cancel{100} \cdot A}{\cancel{100}}$$ Divide both sides by 100.

$$325 = A$$ Simplify.

So 65% of 500 is **325**.

Now work margin exercise 1.

1. What is 15% of 80?

2. 86% of _____ is 430.

Example 2

Finding the Base

57% of _____ is 51.3?

Solution

$P\% = 57\%$ and $A = 51.3$. We want to find the **base** B.

$$\frac{57}{100} = \frac{51.3}{B}$$ Substitute $P = 57$ and $A = 51.3$.

$$57 \cdot B = 100 \cdot 51.3$$ Find the cross products.

$$\frac{\cancel{57} \cdot B}{\cancel{57}} = \frac{5130}{57}$$ Divide both sides by 57.

$$B = 90$$ Simplify.

So 57% of **90** is 51.3.

Now work margin exercise 2.

3. What percent of 90 is 19.8?

Example 3

Finding the Percent

What percent of 170 is 204?

Solution

$B = 170$ and $A = 204$. We want to find the **percent** P.

$$\frac{P}{100} = \frac{204}{170}$$ Subtitute $B = 170$ and $A = 204$.

$$170 \cdot P = 100 \cdot 204$$ Find the cross products.

$$\frac{\cancel{170} \cdot P}{\cancel{170}} = \frac{20,400}{170}$$ Divide both sides by 170.

$$P = 120$$ Simplify.

So **120**% of 170 is 204.

Now work margin exercise 3.

Example 4

Finding the Amount

Many food product labels now list total calories and number of calories from fat. Dietary experts believe that a healthy diet has at most 30% of its calories derived from fat. Following this guideline, if an adult consumes 2500 calories per day, how many calories should be from fat?

Solution

In this problem, $P\% = 30\%$ and $B = 2500$. We want to find the value of A. Substitution in the proportion $\dfrac{P}{100} = \dfrac{A}{B}$ gives

$$\frac{30}{100} = \frac{A}{2500}$$

$$30 \cdot 2500 = 100 \cdot A$$

$$\frac{75{,}000}{100} = \frac{\cancel{100} \cdot A}{\cancel{100}}$$

$$750 = A$$

So in a healthy diet of 2500 calories, no more than 750 calories should be derived from fat.

Now work margin exercise 4.

Practice Problems

Solve for the unknown quantities.

1. What is 85% of 60?

2. What percent of 14 is 4.2?

3. 63% of _____ is 504.

4. In one tennis match Andy served an "ace" on 15% of his serves. If he served 180 times, how many aces did he serve?

4. If you eat a meal which contains 300 calories from fat, what percentage of the recommended maximum of 750 is that amount?

Practice Problem Answers

1. 51   **2.** 30%   **3.** 800   **4.** 27

Exercises 4.5

Use the proportion $\dfrac{P}{100} = \dfrac{A}{B}$ to solve for the unknown quantity. Round your answer to the nearest hundreth, if necessary. See Examples 1 through 3.

1. Find 100% of 47.

2. Find 15% of 50.

3. 25% of 60 is _____ .

4. 80% of 80 is _____ .

5. 27 is 3% of what number?

6. 30% of _____ is 45.

7. 100% of _____ is 62.

8. 5 is 2% of what number?

9. _____ % of 44 is 66.

10. _____ % of 70 is 21.

11. What percent of 50 is 10?

12. What percent of 50 is 75?

13. Find 5% of 72.

14. 10% of 90 is _____ .

15. 14 is 20% of what number?

16. 440 is 110% of what number?

17. _____ % of 120 is 48.

18. 50% of _____ is 35.

19. 150% of _____ is 69.

20. What percent of 15 is 5?

21. 75% of 32 is _____ .

22. Find 60% of 40.

23. What percent of 54 is 18?

24. _____ % of 100 is 38.

25. What number is 50% of 35?

26. What number is 31% of 85?

27. 32 is 20% of what number?

28. 79 is 100% of what number?

29. 23 is _____ % of 10.

30. 18 is _____ % of 10.

31. 40 is $33\frac{1}{3}$% of _____ .

32. 76.8 is 15% of what number?

33. 142.6 is 23% of what number?

34. 20.25 is 25% of _____ .

35. 48 is _____ % of 24.

36. 8 is _____ % of 12.

37. _____ is 96% of 35.

38. _____ is 84% of 52.

39. _____ is 18% of 425.

40. _____ is 28% of 640. **41.** Find 18% of 345. **42.** Find 13.5% of 95.

43. What percent of 100 is 66.5? **44.** What percent of 32 is 24? **45.** What percent of 96 is 16?

46. What percent of 160 is 200? **47.** 200 is 125% of what number? **48.** 450 is 150% of what number?

49. 150% of 70 is _____ . **50.** 86.5% of 100 is _____ .

Answer the following word problems. See example 4.

51. Baseball Attendance: In 2010 the New York Yankees led the major leagues in home attendance, drawing an average of 46,491 fans to their home games. This figure represented 88.85% of the capacity of Yankee Stadium. Estimate how many fans the stadium can hold (to the nearest ten) when it is filled to capacity. **Source:** espn.go.com/mlb/attendance

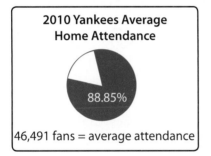

2010 Yankees Average Home Attendance

88.85%

46,491 fans = average attendance

52. College Enrollment: Baldwin-Wallace College, a small liberal arts college in Ohio, had a total enrollment of 4383 students in the 2008-2009 school year. Of that number, 79% of the students were undergraduates. How many undergraduates were enrolled at Baldwin-Wallace College in 2008-2009? Round your answer to the nearest whole number. **Source:** http://www.braintrack.com/college/u/baldwin-wallace-college

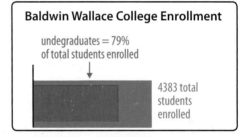

Baldwin Wallace College Enrollment

undegraduates = 79% of total students enrolled

4383 total students enrolled

53. Traveling by Car: Within the borders of West Virginia, the length of Interstate 77 is 186 miles. Of this, 88 miles is a toll road. What percentage of Interstate 77 in West Virginia is toll? Round your answer to the nearest hundredth of a percent.

54. Losing Weight While Ashley Johnston was on the TV show "The Biggest Loser" she dropped her weight to 191 pounds, which was 51.45% of her starting weight. What was her starting weight? Round your answer to the nearest pound.

Source: http://www.msnbc.msn.com/id/35487357/ns/health-fitness/

55. Real Estate: You want to purchase a new home for $122,000. The bank will loan you 80% of the purchase price. How much will the bank loan you? (This amount is called your mortgage and you will pay it off over several years with interest. For example, a 30-year loan will probably cost you a total of more than 3 times the original loan amount.)

56. Dining Out: Two people dined at the La Fiesta Mexican Grill where the food cost was $34.18. To that, a tax of $3.43 was added. What is the percentage of this tax? Round your answer to the nearest tenth of a percent.

57. Football: A kicker on a professional football team made 45 of 48 field goal attempts.

a. What percent of his attempts did he make?

b. What percent did he miss? Would you keep this player on your team or trade for a new kicker?

58. Basketball: In one season a basketball player missed 15% of her free throws. How many free throws did she attempt if she made 136 free throws?

59. Diversity: The University of Southern California has a total of 16,897 undergraduate students. Find the number of students in each of six groups by using the percents shown in the graph below. Round each answer to the nearest whole number. **Source:** http://education-portal.com/directory/school/University_of_Southern_California.html

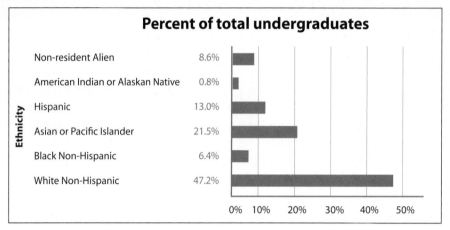

60. Women in the Military: Find the percent of women in each of the 5 branches of the military from the data shown in the following table. **Source:** http://en.wikipedia.org/wiki/United_States_Armed_Forces

Component	Military	Female
Army	548,000	74,411
Marine Corps	203,095	12,290
Navy	332,000	51,029
Air Force	323,000	64,137
Coast Guard	41,000	4,965

4.6 Solving Percent Problems by Using the Equation: $R \cdot B = A$

Objectives

A Understand the equation $R \cdot B = A$.

B Use the equation $R \cdot B = A$ to solve percent problems.

Objective A **The Equation $R \cdot B = A$**

In Section 4.5 we discussed how to solve the three types of percent problems by using the proportion

$$\frac{P}{100} = \frac{A}{B}.$$

If we let $P\% = R$ **where R is in decimal form**, the proportion takes the form $R = \dfrac{A}{B}$ **and cross multiplication gives** $R \cdot B = A$.

We call this last equation, $R \cdot B = A$, the basic equation for solving percent problems.

Now, consider the statement

$$16\% \text{ of } 50 \text{ is } 8.$$

This statement can be translated into an equation in the following way:

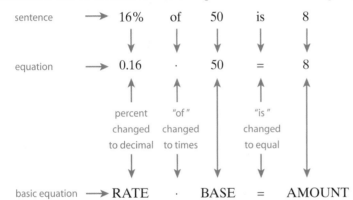

The terms represented in the basic equation are those discussed earlier and are explained in detail in the following box.

Terms Related to the Basic Equation $R \cdot B = A$

$R =$ **rate** or percent (as a decimal number or fraction)

$B =$ **base** (number we are finding the percent of)

$A =$ **amount** (a part of the base)

"of" means to multiply (the raised dot, $\cdot$, is used in the equation)

"is" means equals ($=$)

Objective B **Problem Solving with the Equation $R \cdot B = A$**

Many people have a difficult time with percent problems trying to decide whether to multiply or divide. By using the basic equation $R \cdot B = A$ (or the proportion discussed earlier), these difficulties can be avoided. The decision to multiply or divide to find the unknown quantity is determined when the given numbers are substituted into the equation.

notes

■ The basic equation can be written in the form $A = R \cdot B$. This form is convenient when solving for the amount, A.

■

Three Basic Types of Percent Problems Using $R \cdot B = A$

Type 1: Find the **amount** given the base and the percent.
What is 65% of 800? $A = 0.65 \cdot 800$

Type 2: Find the **base** given the percent and the amount.
42% of what number is 157.5? $0.42 \cdot B = 157.5$

Type 3: Find the **percent** given the base and the amount.
What percent of 92 is 115? $R \cdot 92 = 115$

1. What is 10% of 137?

Example 1

Finding the Amount

What is 65% of 800?

Solution

In this case, $R = 65\% = 0.65$ and $B = 800$. We want to find the **amount** A.

$$A = R \cdot B \qquad\qquad 0.65$$
$$A = 0.65 \cdot 800 \qquad \underline{\times \;\; 800}$$
$$A = 520 \qquad\qquad 520.00$$

So 65% of 800 is **520**.

Now work margin exercise **1.**

notes

■ The operations in these examples can be performed with a calculator or by hand, as shown in Example 1. In either case, the equations should be written so that the = signs are aligned one under the other.

■ Also, **writing the equations and the calculated values will help you remember whether you are multiplying or dividing.**

Example 2

Finding the Base

42% of what number is 157.5?

Solution

Here, $R = 42\% = 0.42$ and $A = 157.5$. We want to find the **base** B.

$$R \cdot B = A$$
$$0.42 \cdot B = 157.5$$
$$\frac{0.42 \cdot B}{0.42} = \frac{157.5}{0.42} \qquad \text{Divide both sides by 0.42.}$$
$$B = 375$$

So 42% of **375** is 157.5.

Now work margin exercise 2.

Example 3

Finding the Percent

_____% of 92 is 115.

Solution

For this problem, $B = 92$ and $A = 115$. We want to find the **percent** R.

$$R \cdot B = A$$
$$R \cdot 92 = 115$$
$$\frac{R \cdot 92}{92} = \frac{115}{92} \qquad \text{Divide both sides by 92.}$$
$$R = 1.25$$
$$R = 125\%$$

We have, **125%** of 92 is 115.

Now work margin exercise 3.

Percents can be changed to fraction form, and in some cases, the fraction form will simplify the work. For example, we know that

$$75\% = \frac{3}{4}, \qquad 33\frac{1}{3}\% = \frac{1}{3}, \qquad \text{and} \qquad 12\frac{1}{2}\% = \frac{1}{8}.$$

The following examples illustrate the use of fractions.

2. 8% of what number is 2?

3. _____% of 56 is 67.2.

Example 4

Finding the Amount

Find 75% of 56.

Solution

Here, $R = 75\% = \dfrac{3}{4}$ and $B = 56$. We want to find the **amount** A.

$$A = R \cdot B$$

$$A = \frac{3}{\overset{}{\cancel{4}}} \cdot \overset{14}{\cancel{56}} = 3 \cdot 14 = 42$$

So 75% of 56 is **42**.

4. Find 150% of 60.

5. 14 is $87\dfrac{1}{2}\%$ of _____ .

Example 5

Finding the Base

250 is 62.5% of _____ .

Solution

For this problem, $R = 62.5\% = \dfrac{5}{8}$ and $A = 250$. We want to find the **base** B.

$$R \cdot B = A$$

$$\frac{5}{8} \cdot B = 250$$

$$\frac{\cancel{8}}{\cancel{5}} \cdot \frac{\cancel{5}}{\cancel{8}} \cdot B = \frac{8}{\cancel{5}} \cdot \overset{50}{\cancel{250}} \qquad \text{Multiply both sides by } \frac{8}{5}, \text{ or divide by } \frac{5}{8}.$$

$$B = 8 \cdot 50$$

$$B = 400$$

Thus the amount 250 is 62.5% of **400**.

***Now work margin exercises* 4 and 5.**

The following two comments are helpful in understanding percents and the relative sizes of the bases and the amounts.

1. A percent is just another form of a fraction: $R = \dfrac{A}{B}$.

2. When you find a percent of a given number, the amount will be:
   a. smaller than the given number if the percent is less than 100%.
   b. larger than the given number if the percent is more than 100%.

Practice Problems

Use the basic equation $R \cdot B = A$ (or $A = R \cdot B$) to answer the following problems.

1. What is 40% of 73?

2. What percent of 88 is 55?

3. 87.5% of what number is 21?

4. _____ % of 4 is 7?

5. 12% of _____ is 15?

6. 150% of 12.6 is _____.

Practice Problem Answers

1. 29.2   2. 62.5%   3. 24

4. 175%   5. 125   6. 18.9

Exercises 4.6

1. 10% of 70 is what number?

2. 5% of 62 is what number?

3. Find 75% of 12.

4. Find 60% of 30.

5. 150% of _____ is 63.

6. 110% of _____ is 330.

7. 3% of _____ is 21.

8. 50% of _____ is 42.

9. _____% of 60 is 90.

10. _____% of 150 is 60.

11. What percent of 75 is 15?

12. What percent of 12 is 4?

13. 3 is 2% of what number?

14. _____% of 34 is 17.

15. _____% of 30 is 6.

16. 17 is 20% of what number?

17. 21 is 30% of what number?

18. 75 is 100% of what number?

19. 100% of 36 is _____.

20. 15% of 60 is _____.

21. 25% of 72 is _____.

22. 80% of 50 is _____.

23. What percent of 48 is 16?

24. What percent of 100 is 35?

25. What number is 50% of 25?

26. What number is 31% of 76?

27. 22 is 20% of _____.

28. 86 is 100% of _____.

29. 18 is what percent of 10?

30. 15 is what percent of 10?

31. 24 is $33\frac{1}{3}$% of what number?

32. 92.1 is 15% of what number?

33. 119.6 is 23% of _____.

34. 9.5 is 25% of _____.

35. 36 is _____% of 18.

36. 60 is _____% of 40.

37. _____ is 96% of 17.

38. _____ is 84% of 32.

39. 18% of 325 is what number?

40. 28% of 460 is what number?

41. Find 15.2% of 75.

42. Find 120% of 60.

43. What percent of 32 is 8?

44. What percent of 240 is 76.8?

45. 64.8 is _____ % of 180.

46. 27 is _____ % of 60.

47. 100 is 125% of what number?

48. 45 is 60% of what number?

Use fractions to solve the following percent problems. See Examples 4 and 5.

49. Find 50% of 32.

50. Find $66\frac{2}{3}$% of 60.

51. What is $12\frac{1}{2}$% of 80?

52. What is $62\frac{1}{2}$% of 16?

53. $33\frac{1}{3}$% of 75 is what number?

54. 25% of 150 is what number?

55. 75% of what number is 21?

56. 50% of what number is 35?

57. $37\frac{1}{2}$% of what number is 61.2?

58. 100% of what number is 76.3?

Solve the following word problems by using the basic equation $R \cdot B = A$.

59. **Casualties of War:** Only 27% of American deaths in the Revolutionary War occurred in battle. All other mortalities were the result of things such as exposure, disease, and starvation. Of the estimated 25,300 deaths, how many men died in battle (to the nearest hundred)?

60. **Compensation:** In 2010, Viacom Inc. paid its CEO $84.5 million in compensation (salary, bonus, and stock options). This was 250% of his total compensation in 2009. What was his total compensation package in 2009? (Round to the nearest million.) **Source:** www.cnbc.com

61. **Presidential Vetoes:** During his presidency, from 1945 to 1953, Harry Truman vetoed 250 congressional bills, and 12 of those vetoes were overridden. What percent of Truman's vetoes were overridden?

62. **Supreme Court Justices:** William O. Douglas served as an associate justice on the Supreme Court for 36 years, from 1939 to 1975, the longest tenure of any Supreme Court justice in history. He lived to be 82. What percent of his life (to the nearest one percent) did he spend on the Supreme Court?

63. **Gold:** 14 Karat gold is an alloy which consists of 58.33% gold by weight, while the remainder is alloy material. If a goldsmith starts with 20 grams of pure gold, how much will the 14 karat gold alloy weigh? Round your answer to the nearest hundredth of a gram. **Source:** www. gottrocks.com/chat-karat.htm

64. **Financing:** The minimum down payment to obtain the best financing rate on a house is 20%. Assuming that John has set aside $35,000 and wants to take advantage of the best financing rate, what is the most expensive house he can purchase? **Source:** http://articles. moneycentral.msn.com/Banking/HomeFinancing/ WhyYouNeedAHomeDownPayment.aspx

65. Test Grades: A student missed 6 problems on a mathematics test and received a grade of 85%. If all the problems were of equal value, how many problems were on the test?

66. Magazine Subscriptions: Magazine publishers will often give significant discounts to subscribers. The cover price of *Newsweek* is $133.65 for 6 months, but a 6-month subscription costs only $24! What is the percent savings (to the nearest tenth) for someone who subscribes to *Newsweek*?

67. Baseball: In 2010, the final standings of the top two teams in each division of the American and National Leagues in baseball were as listed here. With this information, find the percentage (accurate to three decimal places) of wins for each team in this list.

American League	Won	Lost	Pct.	National League	Won	Lost	Pct.
West				West			
Texas	90	72	_____	San Francisco	92	70	_____
Oakland	81	81	_____	San Diego	90	72	
Central				Central			
Minnesota	94	68	_____	Cincinnati	91	71	_____
Chicago	88	74	_____	St. Louis	86	76	_____
East				East			
Tampa Bay	96	66	_____	Philadelphia	97	65	_____
New York	95	67	_____	Atlanta	91	71	_____

68. Ideas: The Service Employees International Union asked employees for ideas and received 15,000 ideas related to ways to create and sustain good-paying jobs in its suggestion box. The results were published on December 2, 2005. The breakdown by topic is shown in the following bar graph. Find the number of suggestions related to each topic.

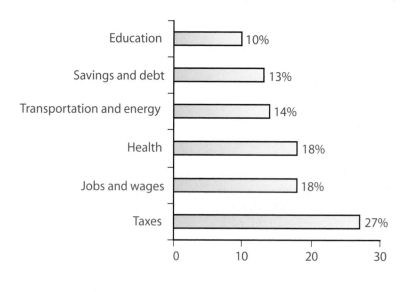

4.7 Applications: Discount, Sales Tax, Commission, and Percent Increase/Decrease

Objective A **The Problem Solving Process**

George Pólya (1887–1985), a famous professor at Stanford University, studied the process of discovery learning. Among his many accomplishments, he developed the following four-step process as an approach to problem solving:

1. Understand the problem.

2. Devise a plan.

3. Carry out the plan.

4. Look back over the results.

There are a variety of types of applications discussed throughout this text and in subsequent courses in mathematics, and you will find these four steps helpful as guidelines for understanding and solving all of them. Applying the necessary skills to solve exercises, such as adding fractions or solving equations, is not the same as accumulating the knowledge to solve problems. **Problem solving can involve careful reading, reflection, and some original or independent thought**.

Basic Steps for Solving Word Problems

1. Understand the problem. For example,
   a. Read the problem.
   b. Understand all the words.
   c. If it helps, restate the problem in your own words.
   d. Be sure that there is enough information.

2. Devise a plan using, for example, one or all of the following:
   a. Guess, estimate, or make a list of possibilities.
   b. Draw a picture or diagram.
   c. Use a variable and form an equation.

3. Carry out the plan. For example,
   a. Try all the possibilities you have listed.
   b. Study your picture or diagram for insight into the solution.
   c. Solve any equation that you may have set up.

4. Look back over the results. For example,
   a. Can you see an easier way to solve the problem?
   b. Does your solution actually work? Does it make sense in terms of the wording of the problem? Is it reasonable?
   c. If there is an equation, check your answer in the equation.

In this section, we will discuss applications that involve percent and the types of percent problems discussed in Sections 4.5 and 4.6. Your instructor may allow the use of calculators. If so, you should keep in mind that **a calculator is a tool to enhance, not replace, the necessary skills and abilities related to problem solving**. Your personal experience, knowledge, and general understanding of problem-solving situations will determine the ease with which you grasp the problems presented. Study the following examples carefully.

Objective B · Calculating Discounts

To attract customers or to sell goods that have been in stock for some time, retailers and manufacturers offer a **discount**. A discount is a reduction in the **original selling price** (or **marked price**) of an item. The new, reduced price is called the **sale price**. The **discount** is the difference between the original price and the sale price. The **rate of discount** (or **percent of discount**) is a percent of the original price.

As you study the examples and work through the exercises, be sure to **label each amount of money as to what it represents**. This will help in organizing the results and determining what operations to perform.

Example 1

Solving Discount Problems

A refrigerator that regularly sells for $1200 is on sale at a 20% discount.

a. What is the amount of the discount?

Solution

To find the discount, we find 20% of $1200.

$$20\% \text{ of } \$1200 \text{ is } \underline{\hspace{1cm}}$$

$$\frac{P}{100} = \frac{A}{B} \qquad \text{OR} \qquad R \cdot B = A$$

$$\frac{20}{100} = \frac{A}{1200} \qquad\qquad\qquad 0.20 \cdot 1200 = A$$

$$20 \cdot 1200 = 100 \cdot A \qquad\qquad\qquad 240 = A$$

$$\frac{24{,}000}{100} = \frac{\cancel{100} \cdot A}{\cancel{100}}$$

$$240 = A$$

The discount is **$240**.

b. What is the sale price?

Solution

Find the sale price by subtracting the discount from the original price.

$$
\begin{array}{ll}
\$1200.00 & \text{original price} \\
-\ 240.00 & \text{discount} \\
\hline
\$\ 960.00 & \text{sale price}
\end{array}
$$

Now work margin exercise 1.

Example 2

Solving Discount Problems

Large fluffy towels were on sale at a discount of 30%. If the sale price was $8.40, what was the original price?

Solution

In this case, we already know the sale price. Now, we need to reason that because the discount was 30%, the sale price represents 70% of the original price. (100% − 30% = 70%) Also, note that the original price will be more than the sale price of $8.40.

70% of _____ is $8.40.

$$
\frac{P}{100} = \frac{A}{B}
$$

$$
\frac{70}{100} = \frac{8.40}{B}
$$

$$
70 \cdot B = 100 \cdot 8.40
$$

$$
\frac{\cancel{70} \cdot B}{\cancel{70}} = \frac{840}{70}
$$

$$
B = 12
$$

OR

$$
R \cdot B = A
$$

$$
0.70 \cdot B = 8.40
$$

$$
\frac{\cancel{0.70}\,B}{\cancel{0.70}} = \frac{8.40}{0.70}
$$

$$
B = 12
$$

The original price of the towels was **$12.00** each.

Now work margin exercise 2.

Objective C **Calculating Sales Tax**

Sales tax is a tax charged on the actual selling price of goods sold by retailers. The **rate of the sales tax** (a percent of the actual selling price) varies from state to state (or even city to city in some cases). States and cities use sales tax to pay for public services.

Margin exercises:

1. A pair of shoes originally priced at $52 is now discounted 25%.
  a. What is the amount of the discount?

  b. What is the sale price?

2. You buy a watch for $90, which is a 40% discount off the original price. What did the watch cost originally?

3. Assuming a 7% sales tax rate, what would be the final cost of a discounted pair of shoes priced at $39?

Example 3

Solving Sales Tax Problems

If the sales tax rate is 6%, what would be the final cost of a laptop computer priced at $899?

Solution

First, find the sales tax by taking 6% of $899.

6% of 899 is _____

$$\frac{P}{100} = \frac{A}{B} \qquad \text{OR} \qquad R \cdot B = A$$

$$\frac{6}{100} = \frac{A}{899} \qquad\qquad\qquad 0.06 \cdot 899 = A$$

$$6 \cdot 899 = 100 \cdot A \qquad\qquad\qquad 53.94 = A$$

$$\frac{5394}{100} = \frac{\cancel{100} \cdot A}{\cancel{100}}$$

$$53.94 = A$$

Next, find the final cost by adding the sales tax to the original price.

$ 899.00	original price
+ 53.94	sales tax
$952.94	final cost

The final cost of the laptop would be **$952.94**.

Now work margin exercise 3.

Objective D Calculating Commission

A **commission** is a fee paid to an agent or salesperson for a service. Commissions are usually a percent of a negotiated contract or a percent of sales. In some cases, salespeople earn a straight commission on what they sell. In other cases, the salesperson earns a base salary plus a commission on sales above a certain amount.

Example 4

Solving Commission Problems

Susan sells women's shoes. She earns a salary of $2000 a month plus a commission of 8% on what she sells over $8500. What did Susan earn the month she sold $22,500 worth of shoes?

Solution

First, find the base for her commission by subtracting $8500 from her sales.

$22,500   total sales

$- 8\ 500$

$14,000   base for commission

Next, find the amount of the commission by taking 8% of $14,000.

$$\frac{P}{100} = \frac{A}{B} \qquad \text{OR} \qquad A = R \cdot B$$

$$\frac{8}{100} = \frac{A}{14,000} \qquad\qquad A = 0.08 \cdot 14,000$$

$$8 \cdot 14,000 = 100 \cdot A \qquad\qquad A = 1120 \quad \text{amount of commission}$$

$$\frac{112,000}{100} = \frac{100 \cdot A}{100}$$

$$1120 = A \quad \text{amount of commission}$$

Finally, add her salary and the amount of the commission to find what she earned.

$\$\ 2000$ salary

$+\ 1120$ commission

$\$\ 3120$ total pay for the month

Susan earned **$3120** for the month.

Now work margin exercise 4.

4. Lynsay earns a salary of $1250 a month plus a commission of 5% on all electronics she sells at her job at the local computer store. What did she earn the month she sold $28,640 in electronics?

Objective E   Calculating Percent Increase and Percent Decrease

Over a period of time values of property, such as a home, can increase or decrease. Similarly, values of stocks in the stock market, populations of cities, and countries, and quantities of products sold in a year can, and do, increase or decrease. At times it is helpful to know by what percent the value changed. This is called finding the **percent of increase** (or the **percent of decrease**). Examples 5 and 6 illustrate these ideas. (**Note:** In many cases the increase in value is called **appreciation**, and the decrease in value is call **depreciation**.)

5. In November, a hockey player scored 12 times. The following month, he scored 15 times. What was his percent increase in scoring from November to December?

Example 5

Finding the Percent Increase

Ben's last two exam scores in algebra were 80 and 88. What was the percent increase from the first exam to the second?

Solution

First, find the increase in score from the first exam to the second.

$$\begin{array}{rl} 88 & \text{2nd exam} \\ -\ 80 & \text{1st exam} \\ \hline 8 & \text{amount of change} \end{array}$$

Now determine the percent increase by finding what percent 8 is of 80. Use 80 as the base because it is the number that was increased.

$$\frac{P}{100} = \frac{A}{B}$$ OR $$R \cdot B = A$$

$$\frac{P}{100} = \frac{8}{80}$$ $$R \cdot 80 = 8$$

$$80 \cdot P = 8 \cdot 100$$ $$\frac{R \cdot \cancel{80}}{\cancel{80}} = \frac{8}{80}$$

$$\frac{\cancel{80} \cdot P}{\cancel{80}} = \frac{800}{80}$$ $$R = 0.1$$

$$P = 10$$

So Ben's score increased by 10%.

Now work margin exercise 5.

Example 6

Finding the Percent Decrease

Three years ago you bought a new car for $25,000. Your business is doing well and you are now looking for another new car and want to trade in your first car. The dealer has told you that the trade in value of your car is now $17,500. What is the percent decrease in the value of your car?

Solution

First find the actual decrease in the value:

$$\begin{array}{rl} \$25,000 & \text{new car value} \\ -\ \$17,500 & \text{trade in value} \\ \hline \$7\ 500 & \text{amount decreased} \end{array}$$

Now we find the percent decrease by finding what percent $7500 (the decrease) is of $25,000 (the base price).

$$\frac{P}{100} = \frac{A}{B} \qquad \text{OR} \qquad R \cdot B = A$$

$$\frac{P}{100} = \frac{7500}{25,000} \qquad\qquad R \cdot 25,000 = 7500$$

$$25,000 \cdot P = 100 \cdot 7500 \qquad \frac{R \cdot \cancel{25,000}}{\cancel{25,000}} = \frac{7500}{25,000}$$

$$\frac{\cancel{25,000} \cdot P}{\cancel{25,000}} = \frac{750,000}{25,000} \qquad\qquad R = 0.3$$

$$P = 30$$

So the percent decrease in the value of your car is **30%**.

Now work margin exercise 6.

Practice Problems

1. A television which normally sells for $300 is priced at a 10% discount. Find **a.** the amount of the discount and **b.** the sale price.

2. If the sales tax rate is 6.5%, what is the tax on an $800 purchase?

3. A realtor works on a 5% commision. What would be her commission on a house that sold for $485,000?

4. Central Valley Community College had 48 teams compete at their 1st annual corn hole tournament. The following year they had 54 teams compete. What was the percent increase in competing teams?

6. Tom is looking to sell his house. Unfortunately the housing market in his area has not done so well, and he has to sell his house for less than he bought it for. If Tom's purchasing price of the house was $200,000 and he plans to sell it for $194,000, what is the percent of decrease in the value of Tom's house?

Practice Problem Answers

1. a. $30   **b.** $270
2. $52
3. $24,250
4. 12.5%

Exercises 4.7

Solve the following word problems. Note that the problems may involve several calculations, so follow Pólya's four-step problem-solving process as closely as you can. See Examples 1 through 6.

1. **Purchasing:** A store owner received a 3% discount from the manufacturer when she bought $15,500 worth of dresses.
   a. What was the amount of the discount?
   b. What did she pay for the dresses?

2. **Office Supplies:** A new briefcase was priced at $275. If it was to be marked down 30%:
   a. What was the amount of the discount?
   b. What would be the new price?

3. **Sales Tax:** If the sales tax in a certain state is figured at 6%:
   a. How much tax is there on a purchase of $30.20?
   b. What is the total amount paid for the purchase?

4. **Sales Tax:** If sales tax was figured at 6%:
   a. How much tax was paid on the purchase of three textbooks priced at $55.00, $25.50, and $34.95?
   b. What would be the total cost of all three books?

5. **Shopping:** The discount on a fur coat was $150. This was a 20% discount.
   a. What was the original selling price of the coat?
   b. What was the sale price?
   c. What was the total amount paid for the coat if a 6% sales tax was added to the sale price?

6. **Shopping:** The discount on men's suits was $50, and they were on sale for $200.
   a. What was the original selling price?
   b. What was the rate of discount?
   c. What was the total amount paid for the suit if an 8% sales tax was added to the sale price?

7. **Selling Books:** In order to get more subscribers, a book club offered three books for a total price of $7.02. The total selling price was originally $17.55 for all three books.
   a. What was the amount of the discount?
   b. Based on the original selling price, what was the rate of the discount on these three books?

8. **Household Goods:** Sheets are marked $22.50 and pillowcases $7.50. What is the sale price of each item if each item is discounted 25% off the marked price?

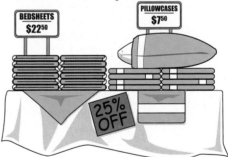

9. **Household Goods:** Towels were on sale at a discount of 30%. If the sale price was $3.01, what was the original price?

10. **Electronics:** Computer disks were on sale for $5.24 per box. What was the original price per box if the sale price represents a discount of 20%?

11. **Car Repair:** An auto supply store received a shipment of auto parts and a bill for $845.30. Some of the parts were not as ordered, and they were returned immediately. The value of the parts returned was $175.50. The terms of the billing provided the store with a 2% discount if it paid cash (for the parts it kept) within two weeks. What did the store pay for the parts it kept if it paid cash within two weeks?

12. **School Supplies:** Linda is enrolled in a calculus course. She has the choice of buying the text in hardback form for $60.00 or in paperback form for $46.50. Tax is figured at 5% of the selling price. The bookstore buys back hardback books for 40% of the selling price and paperback books for 30% of the selling price.
   a. Which book is the more economical buy for Linda if she sells her book back to the bookstore at the end of the semester?
   b. How much does she save?

13. **Roofing:** In the roofing business, shingles are sold by the "square," which is enough material to cover a 10 ft by 10 ft square (or 100 square feet). A roofing supplier has a closeout on shingles at a 30% discount.
   a. If the original price was $230 per square, what is the sale price per square?
   b. How much would a roofer pay for 34 squares?

14. **Auto Repair:** An auto dealer paid $8730 for a large order of special parts. This was not the original price. The amount paid reflects a 3% discount off the original price because the dealer paid cash. What was the original price of the parts?

15. **Property Tax:** The property taxes on a house were $1050. What was the tax rate if the house was valued at $70,000?

16. **Sales Tax:** If sales tax is figured at 7.25%, how much tax will be added to the total purchase price of three textbooks priced at $25.00, $35.00, and $52.00?

17. **Real Estate:** A realtor works on a 6% commission. What is his commission on a house he sold for $195,000?

18. **Real Estate:** A realtor selling commercial property works on a 4% commission. What is her commission on a building she sold for $875,000?

19. **Real Estate:** A realtor works on 6% commission. What is his commission on a house he sold for $125,000?

20. **Car Sales:** A car saleswoman earns a commission of 7% on each car she sells. How much did she earn on the sale of a car for $12,500?

21. **Sales:** If a salesman works on a 10% commission only (no monthly salary), how much merchandise will he have to sell to earn $2800 in one month?

22. **Sales:** A sales clerk receives a monthly salary of $500 plus a commission of 6% on all sales over $3500. What did the clerk earn the month she sold $8000 in merchandise?

23. Sales: A sales clerk receives a monthly salary of $1295 plus a commission of 7% on all sales over $2500. What did the clerk earn the month she sold $16,000 in merchandise?

24. Shoe Sales: A shoe saleswoman works on a fixed salary of $940 per month plus a 5% commission. How much did she make during the month in which she sold $7500 worth of shoes?

25. Real Estate: Suppose you sell your home for $180,000 and you owe the Savings and Loan $60,000 on the first trust deed. You pay a real estate agent a commission of 6% of the selling price and other fees and taxes totaling $1200. How much cash do you have from the sale?

26. Real Estate: Scott is buying a beach house off the coast of South Carolina for $260,000. The bank will loan him $225,000. If he must also pay a 4% commission to the real estate agent and $3800 in taxes and other fees. How much cash does he need?

27. Real Estate: Suppose you sell your three bedroom home for $180,000 and you owe the balance of the mortgage of $55,000 to the bank. You pay a real estate agent a fee of 6% of the selling price and other fees and taxes that total $1500. How much cash do you have after the sale? (You may have to pay income taxes later.)

28. Bonus: A computer programmer was told that he would be given a bonus of 5% of any money his programs could save the company. How much would he have to save the company to earn a bonus of $500?

29. Cereal Sales: Due to the increasing cost of breakfast cereals, more and more people are buying private-label brands rather than national brands. In a recent year, the sale of private-label cereals rose from 170 million boxes to 180 million boxes. What was the percent increase in sales (to the nearest tenth of a percent)?

30. School Teachers: The decade from 1960 to 1970 saw the largest 10-year increase in the number of male elementary and high school teachers in our nation's history. There was an increase-from 402,000 to 691,000. What was the percent increase from 1960 to 1970 (to the nearest one percent)?

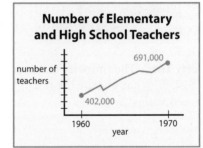

31. Baseball Attendance: The average attendance to a Yankees game in 2009 was 45,364 fans. In 2010 the average attendance grew to 46,491 fans. Find the percent increase in attendance. Round your answer to the nearest thousandth.

32. Population: According to U.S. Census information from 1990 and 2000, Las Vegas, Nevada had the largest population growth among the nation's 100 most populated cities. In 1990 the population of Las Vegas was 852,854. The percent increase between 1990 and 2000 was 83.3%. What was the population of Las Vegas in 2000 (to the nearest whole number)?

33. Retail Sales: In February 2011, Radio Shack sold a 17.3" HP laptop computer for the reduced price of $429.97. The original price was $599.99.

   a. How much was the price reduced in terms of dollars?

   b. Find the percent decrease or reduction. Round your answer to the nearest hundredth of a percent.

34. Population: The 2000 population in Welch, WV was 5,343, while the 2009 population dropped to 4,426. What was the percent decrease of population in that nine year period? Round your answer to the nearest tenth of a percent. **Source:** http://www.city-data.com/zips/24801.htm

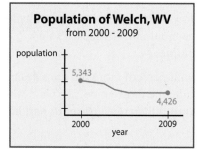

35. Population: The population of Zimbabwe was 11,651,858 in 2010 with an expected percent increase of 2.95% per year. At this rate of increase, what was the expected population of Zimbabwe for the year 2012? Round your answer to the nearest whole number. (**Hint:** First find the expected population for 2011. Then use this answer to find the expected population for 2012.) **Source:** CIA World Factbook

36. Enrollment: In 1966 the student enrollment at California Polytechnic State University in San Luis Obispo, CA was 7,740. In 1977 the university had 15,502 students. Since that time the enrollment growth has slowed. What is the percent increase of student enrollment during that eleven year period? Round your answer to the nearest tenth of a percent. **Source:** http://lib.calpoly.edu/universityarchives/history/timeline/

37. Circulation: The circulation of the *Washington Post* newspaper was approximately 730,000 in 2000, and it dropped to 570,000 in 2009. What was the percent decrease in circulation? Round your answer to the nearest percent. **Source:** http://www.theawl.com/2009/10/a-graphic-history-of-newspaper-circulation-over-the-last-two-decades

38. Stock Market: The Dow Jones Industrial Index had a peak of 13,930 in October of 2007, but dropped to a minimum of 7,063 in February of 2009. Fortunately this dip was short lived, and the market started increasing again. What was the percent decrease in the stock market drop according to the Dow Jones Industrial Index during this sixteen month interval? Round your answer to the nearest tenth of a percent. **Source:** http://stockcharts.com/charts/historical/djia2000.html

39. One shoe salesman worked on a straight 9% commission. His friend worked on a salary of $400 plus a 5% commission.

    a. How much did each salesman make during the month in which each sold $7500 worth of shoes?

    b. What percent more did the salesman who made the most make? Explain why there is more than one answer to part **b.**

40. A man weighed 200 pounds. He lost 20 pounds in 3 months. Then he gained back 20 pounds 2 months later.

    a. What percent of his weight did he lose in the first 3 months?

    b. What percent of his weight did he gain back? The loss and the gain are the same, but the two percents are different. Explain why.

Collaborative Learning

41. With the class separated into teams of two to four students, each team is to analyze the following problem and decide how to answer the related questions. Then each team leader is to present the team's answers and related ideas to the class for general discussion.

Jerry works in a bookstore and gets a salary of $500 per month plus a commission of 3% on whatever he sells over $2000. Wilma works in the same store, but she has decided to work on a straight 8% commission.

    a. At what amount of sales will Jerry and Wilma make the same amount of money?

    b. Up to that point, who would be making more?

    c. After that point, who would be making more? Explain briefly. (If you were offered a job at this bookstore, which method of payment would you choose?)

4.8 Applications: Profit, Simple Interest, and Compound Interest

Objectives

A   Understand percent of profit.

B   Calculate simple interest.

C   Calculate compound interest.

Objective A   **Percent of Profit**

Manufacturers and retailers are concerned with the **profit on each item** produced or sold. In this sense, the profit on an item is the difference between the selling price to the customer and the cost to the company. For example, suppose that a department store can buy a certain light fixture from a manufacturer for $80 (the store's cost) and sell the fixture for $100 (the selling price). The profit for the store is the difference between the selling price and the cost of the fixture. That is, in this example,

$$\text{Profit} = \$100 - \$80 = \$20.$$

The **percent of profit** is a ratio. And, as the following discussion indicates, this ratio can be based either on cost or on selling price.

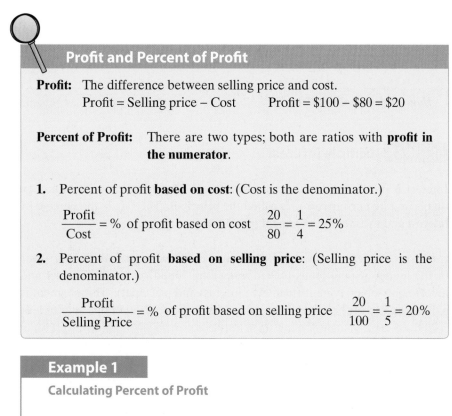

Profit and Percent of Profit

Profit: The difference between selling price and cost.

     Profit = Selling price − Cost   Profit = $100 − $80 = $20

Percent of Profit: There are two types; both are ratios with **profit in the numerator**.

1. Percent of profit **based on cost**: (Cost is the denominator.)

$$\frac{\text{Profit}}{\text{Cost}} = \% \text{ of profit based on cost} \qquad \frac{20}{80} = \frac{1}{4} = 25\%$$

2. Percent of profit **based on selling price**: (Selling price is the denominator.)

$$\frac{\text{Profit}}{\text{Selling Price}} = \% \text{ of profit based on selling price} \qquad \frac{20}{100} = \frac{1}{5} = 20\%$$

Example 1

Calculating Percent of Profit

A retail store markets calculators that cost the store $45 each and are sold to customers for $60 each.

a. What is the profit on each calculator?

Solution

First find the profit.

$60.00	selling price
− 45.00	cost
$15.00	profit

The profit is **$15** per calculator.

1. A music store sells certain CD's for $15 each when the CD's actually cost the store $6.

   a. What is the profit on each CD?

   b. What is the percent of profit based on cost?

   c. What is the percent of profit based on selling price?

For **b.** and **c.**, use a ratio and then change the fraction to a percent to find each percent of profit.

b. What is the percent of profit based on cost?

Solution

For profit based on cost, remember that cost is in the denominator.

$$\frac{\$15 \text{ profit}}{\$45 \text{ cost}} = \frac{1}{3} = 33\frac{1}{3}\% \qquad \text{profit based on cost}$$

c. What is the percent of profit based on selling price?

Solution

For profit based on selling price, remember that selling price is in the denominator.

$$\frac{\$15 \text{ profit}}{\$60 \text{ selling price}} = \frac{1}{4} = 25\% \qquad \text{profit based on selling price}$$

Now work margin exercise 1.

Objective B   Simple Interest

Interest is money paid for the use of money. The initial amount of money that is invested or borrowed is called the **principal**. The rate is the **percent of interest** and is stated as an **annual (yearly) rate**.

Interest is either paid or earned, depending on whether you are the borrower or the lender. Some loans (called **notes**) are based on **simple interest** and involve only one payment (including interest and principal). This payment is paid at the end of the term of the loan, usually for a period of one year or less.

A **formula** is a general statement (usually an equation) that relates two or more variables. The following formula is used to calculate simple interest.

Formula for Calculating Simple Interest

$$I = P \cdot r \cdot t,$$

where
  I = interest (earned or paid)
  P = principal (the amount invested or borrowed)
  r = rate of interest (stated as an annual rate) in decimal number or fraction form
  t = time (one year or fraction of a year)

Note: For calculation purposes, we will use 360 days in one year and 30 days in a month. However, many lending institutions now (with the advent of computers) base their calculations on 365 days per year and pay or charge interest on a daily basis.

Example 2

Calculating Simple Interest

You want to borrow $2000 from your bank for one year. If the interest rate is 5.5%, how much interest would you pay?

Solution

Use the formula for simple interest: $I = P \cdot r \cdot t$, with

$$P = \$2000, \quad r = 5.5\% = 0.055, \quad \text{and} \quad t = 1 \text{ year}$$

$$I = 2000 \cdot 0.055 \cdot 1 = 110$$

You pay $110 in interest.

Now work margin exercise 2.

Completion Example 3

Calculating Simple Interest

Sylvia borrowed $2400 at 5% interest for 90 days (3 months). How much interest did she have to pay?

Solution

$$P = \$2400 \quad r = 5\% = \underline{\quad} \quad \text{and} \quad t = 90 \text{ days} = \frac{90}{360} \text{ yr} = \underline{\quad} \text{ yr}$$

$$I = \underline{\quad} \cdot \underline{\quad} \cdot \underline{\quad} = \underline{\quad}$$

Sylvia had to pay _____ in interest.

If you know the values of any three of the variables in the formula $I = P \cdot r \cdot t$, you can find the unknown value by substituting into the formula and solving for the unknown. This procedure is illustrated in Examples 4 and 5.

Now work margin exercise 3.

Example 4

Calculating Principal using Simple Interest

What principal would you need to invest at a rate of 6% to earn $450 in 6 months?

Solution

Here the principal P is unknown. We do know

$$I = \$450, \quad r = 6\% = 0.06, \quad \text{and} \quad t = \frac{6}{12} = \frac{1}{2} \text{yr}.$$

Completion Example Answers

3. $P = \$2400, r = 5\% = 0.05, t = 90 \text{ days} = \dfrac{90}{360} \text{ yr} = \dfrac{1}{4} \text{ yr}$

$$I = 2400 \cdot 0.05 \cdot \frac{1}{4} = \$30$$

Sylvia had to pay $30 in interest.

2. If you were to borrow $1500 at 10% for one year, how much interest would you pay?

3. Ralph borrowed $3000 at 7% for sixty days. How much interest did he pay?

4. How much (what principal) would you need to invest if your investment returned 8% interest and you wanted to make $500 in interest in 90 days?

Substituting and solving for P we have

$$450 = P \cdot 0.06 \cdot \frac{1}{2}$$

$$450 = P \cdot 0.03$$

$$\frac{450}{0.03} = \frac{P \cdot 0.03}{0.03} \qquad \text{Divide both sides by 0.03.}$$

$$15{,}000 = P$$

You would need to invest $15,000 to earn $450 in 6 months at a rate of 6%.

Now work margin exercise 4.

5. Rick wants to borrow $1000 at 12% from his uncle and is willing to pay $100 in simple interest. How long can he keep the money?

Example 5

Calculating Time using Simple Interest

Stuart wants to borrow $1500 from his father and is willing to pay $15 in interest. His father told Stuart that he would want interest at 4%. How long can Stuart keep the money?

Solution

In this case, the unknown is time t. We do know

$$I = \$15, \qquad r = 4\% = 0.04, \qquad \text{and} \qquad P = \$1500$$

Substituting and solving for t gives

$$15 = 1500 \cdot 0.04 \cdot t$$

$$15 = 60 \cdot t$$

$$\frac{15}{60} = \frac{60 \cdot t}{60} \qquad \text{Divide both sides by 60.}$$

$$\frac{1}{4} = t$$

Stuart can keep the money for $\frac{1}{4}$ yr (or 3 months).

Now work margin exercise 5.

Objective C **Compound Interest**

Interest paid on interest earned is called **compound interest**. To calculate compound interest, we can calculate the simple interest for each period of time that interest is compounded. **A new principal is used for each calculation.** The new principal is the previous principal plus the earned interest. The calculations can be performed in a step-by-step manner, as indicated in the following outline.

To Calculate Compound Interest

Step 1: Use the formula $I = P \cdot r \cdot t$, to calculate simple interest.

Let $t = \dfrac{1}{n}$ where n is the number of periods per year for compounding.

For example:

for compounding **annually**, $n = 1$ and $t = \dfrac{1}{1} = 1$.

for compounding **semiannually**, $n = 2$ and $t = \dfrac{1}{2}$

for compounding **quarterly**, $n = 4$ and $t = \dfrac{1}{4}$

for compounding **bi-monthly**, $n = 6$ and $t = \dfrac{1}{6}$

for compounding **monthly**, $n = 12$ and $t = \dfrac{1}{12}$

for compounding **daily**, $n = 360$ and $t = \dfrac{1}{360}$.

Step 2: Add this interest to the principal to create a new value for the principal.

Step 3: Repeat steps 1 and 2 however many times the interest is to be compounded.

Example 6

Calculating Compound Interest

If a savings account of $1200 is compounded annually (once a year) at 5%, how much interest will be earned in three years?

Solution

The account is compounded annually, so $n = 1$ and $t = \dfrac{1}{1} = 1$.

Use the formula for simple interest, $I = P \cdot r \cdot t$, with $r = 5\% = 0.05$ and $t = 1$.

The principal will change each year.

a. First year: the principal is $P = \$1200$.
$I = \$1200 \cdot 0.05 \cdot 1 = \$60$ interest for the first year.

b. Second year: the new principal is $P = \$1200 + \$60 = \$1260$
$I = \$1260 \cdot 0.05 \cdot 1 = \$63$ interest for the second year.

c. Third year: the new principal is $P = \$1260 + \$63 = \$1323$
$I = \$1323 \cdot 0.05 \cdot 1 = \$66.15$ interest for the third year.

The total interest earned in three years will be

6. George deposits $500 in a savings account that pays 6% interest compounded quarterly. How much interest will he earn in 9 months?

7. If a principal of $3000 is compounded bi-monthly (6 times a year) at 7%, what will be the balance in the account at the end of one year?

$$\begin{array}{r} \$\ 60.00 \\ 63.00 \\ +\ 66.15 \\ \hline \$189.15 \end{array}$$

(Note that, because the principal is larger each year, the interest earned increases each year.)

Now work margin exercise 6.

Now work margin exercise 6.

Example 7

Calculating Compound Interest

If an account of $5000 is compounded monthly (12 times a year) at 6%, what will be the balance in the account at the end of four months?

Solution

The account is compounded monthly, so $n = 12$ and $t = \dfrac{1}{12}$.

Use the formula for simple interest, $I = P \cdot r \cdot t$, with $r = 6\% = 0.06$ and $t = \dfrac{1}{12}$.

The principal will change each month.

a. First month: the principal is $P = \$5000$.

$I = \$5000 \cdot 0.06 \cdot \dfrac{1}{12} = \$25.00$ interest for the first month

b. Second month: the new principal is $P = \$5000 + \$25 = \$5025.00$.

$I = \$5025 \cdot 0.06 \cdot \dfrac{1}{12} \approx \$25.13$ interest for the second month

c. Third month: the new principal is $P = \$5025 + \$25.13 = \$5050.13$.

$I = \$5050.13 \cdot 0.06 \cdot \dfrac{1}{12} \approx \$25.25$ interest for the third month

d. Fourth month: the new principal is $P = \$5050.13 + \$25.25 = \$5075.38$.

$I = \$5075.38 \cdot 0.06 \cdot \dfrac{1}{12} \approx \$25.38$ interest for the fourth month

The total interest earned in four months will be

$$\begin{array}{r} \$\ 25.00 \\ 25.13 \\ 25.25 \\ +\ 25.38 \\ \hline \$100.76 \end{array}$$

The balance in the account will be $\$5000.00 + \$100.76 = \$5100.76$.

Now work margin exercise 7.

Now work margin exercise 7.

notes

Loans in savings accounts, house payments, and credit card debts are based on compound interest, compounded daily over periods of years. The calculations, including monthly earnings or payments, are generally performed with computers. These calculations are related to the compound interest formula $A = P\left(1 + \dfrac{r}{n}\right)^{nt}$.

Practice Problems

1. A company manufactures and sells plastic boxes that cost $21 each to produce, and that sell for $28 each.
a. How much profit does the company make on each box?
b. What is the percent of profit based on cost?
c. What is the percent of profit based on selling price?

2. If you were to borrow $1000 at 5% for nine months, how much interest would you pay?

3. What interest rate would you be paying if you borrowed $1000 for 6 months and paid $60 in interest?

4. You deposit $1500 at 4% to be compounded semiannually. How much interest will you earn in 3 years?

5. A principal of $2500 is deposited at 6% to be compounded monthly. How much will the account be worth in 6 months?

Practice Problem Answers

1. a. $7   **b.** $33\dfrac{1}{3}\%$ **c.** 25%

2. $37.50   **3.** 12%

4. $189.24   **5.** $2575.94

Exercises 4.8

1. **Art:** An art gallery sells paintings by a well-known artist for $2500 each. The gallery owner has agreed to pay the artist $2000 for each painting of a certain size.
   a. What is the profit on each painting?
   b. What is the percent of profit based on cost?
   c. What is the percent of profit based on selling price?

2. **Golf:** The Golf Pro Shop had a set of 10 golf clubs that were marked on sale for $860. This was a discount of 20% off the original selling price.
   a. What was the original selling price?
   b. If the clubs cost the Golf Pro Shop $602, what was its profit?
   c. What was the shop's percent of profit based on the original selling price?
   d. What was the percent of profit based on the sale price?

3. **Electronics:** The cost of a 20-inch television set to a store owner was $450, and she sold the set for $630.
   a. What was her profit?
   b. What was her percent of profit based on cost?
   c. What was her percent of profit based on selling price?

4. **Cars:** A car dealer bought a five year old used car for $2500. He marked up the price so that he would make a profit of 25% based on his cost.
   a. What was the selling price?
   b. If the customer paid 8% of the selling price in taxes and fees, what was the customer's total cost for the car?

Solve the following problems related to simple interest ($I = P \cdot r \cdot t$). Round your answer to the nearest cent, if necessary. See Examples 2 through 5.

5. What is the simple interest paid on $500 at 6% for one year?

6. What is the simple interest paid on $2000 at 8% for one year?

7. How much interest would be paid on a loan of $5000 at 8% for 6 months?

8. How much interest would be paid on a loan of $3000 at 5% for 9 months?

9. What principal will earn $50 in interest if it is invested at 6% for one year?

10. What principal will earn $75 in interest if it invested at 5% for 6 months?

11. How long will it take for $1000 invested at 5% to earn $50 in simple interest?

12. What length of time will it take to earn $70 in simple interest if $2000 is invested at 7%?

13. If interest is paid at 6% for one year, what will a principal of $1800 earn?

14. If a principal of $900 is invested at a rate of 4% for 90 days, what will be the interest earned?

15. If you borrow $750 for 30 days at 9%, how much interest will you pay?

16. How much interest would be paid on a 60-day loan of $500 at 4%?

17. If you charge $1000 worth of merchandise at a local department store at 18% interest, how much will you owe at the end of 60 days?

18. A friend wants to borrow $500 from you for 8 months and is willing to pay you interest at 6%. How much would he owe you at the end of the 8 months?

19. What rate of interest is charged if a loan of $2500 for 90 days is paid off with $2562.50? (**Note:** The payoff is principal plus interest.)

20. A bank decides to loan $5 million to a contractor to build new homes. How much interest will the bank earn in one year if the interest rate is 9.2%?

21. Determine the missing item in each row.

Principal	Rate	Time	Interest
$400	16%	90 days	
	15%	120 days	$5.00
$560	12%		$5.60
$2700		40 days	$25.50

22. Determine the missing item in each row.

Principal	Rate	Time	Interest
$600	15%	30 days	
$500	18%		$15.00
$450		90 days	$22.50
	10%	30 days	$1.50

Solve the following compound interest problems. (Use the formula for simple interest $I = P \cdot r \cdot t$, repeatedly.) See Examples 6 and 7.

23. You loan your cousin $2000 compounded annually at 5% for 3 years. How much interest will your cousin owe you?

24. Juan borrowed $5000 from his uncle compounded annually at 6% for 4 years. How much interest will he owe his uncle at the end of 4 years?

25. If $9000 is deposited in a savings account compounded monthly at 4%, what will be the balance in the account in 6 months?

26. Jeremy put $3500 in a savings account at 5.5% compounded quarterly (every 3 months) for 6 months. How much interest did he earn? (Round to the nearest cent.)

27. **a.** How much interest will be earned in 1 year on a loan of $4000 compounded quarterly at 4%?
   b. How much will be owed by the borrower?

28. **a.** How much interest will be earned on a savings account of $3000 in two years if interest is compounded annually at 5.5%?
   b. If interest is compounded semiannually (twice a year)?

29. Calculate the interest earned in six months on $20,000 compounded monthly at 8%.

30. If interest is calculated at 10% compounded quarterly, what will be the value of $15,000 in 9 months?

For the Proportion $\dfrac{P}{100} = \dfrac{A}{B}$

page 300

$P\%$ = **percent** (written as the ratio $\dfrac{P}{100}$)

B = **base** (number that we are finding the percent of)

A = **amount** (a part of the base)

The Three Basic Types of Percent Problems Using $\dfrac{P}{100} = \dfrac{A}{B}$

page 301

1. Find the **amount** given the base and the percent.

   What is 65% of 500? $\dfrac{65}{100} = \dfrac{A}{500}$

2. Find the **base** given the percent and the amount.

   57% of what number is 51.3? $\dfrac{57}{100} = \dfrac{51.3}{B}$

3. Find the **percent** given the base and the amount.

   What percent of 170 is 204? $\dfrac{P}{100} = \dfrac{204}{170}$

Section 4.6: Solving Percent Problems by Using the Equation: $R \cdot B = A$

Terms Related to the Basic Equation $R \cdot B = A$

page 307

R = **rate** or percent (as a decimal number or fraction)

B = **base** (number we are finding the percent of)

A = **amount** (a part of the base)

"of" means to multiply (the raised dot, $\cdot$, is used in the equation)

"is" means equals ($=$)

The Three Basic Types of Percent Problems: $R \cdot B = A$

page 308

1. Find the **amount** given the base and the percent.

   What is 65% of 800? $A = 0.65 \cdot 800$

2. Find the **base** given the percent and the amount.

   42% of what number is 157.5? $0.42 \cdot B = 157.5$

3. Find the **percent** given the base and the amount.

   What percent of 92 is 115? $R \cdot 92 = 115$

Basic Steps for Solving Word Problems   page 315
1. Understand the problem.
2. Devise a plan.
3. Carry out the plan.
4. Look back over the results.

Discount   page 316
A **discount** is a reduction in the original selling price of an item.

Sales Tax   page 317
Sales tax is a tax charged on the actual selling price of goods sold by retailers.

Commission   page 318
A **commission** is a fee paid to an agent or salesperson for a service. Commissions are usually a percent of a negotiated contract or a percent of sales.

Percent Increase/Decrease   page 319
The percent a value has changed is the **percent increase** (or **decrease**).

Terms Related to Profit and Percent of Profit   page 327
Profit: The difference between selling price and cost.
  Profit = Selling price − Cost
Percent of Profit: There are two types; both are ratios with **profit in the numerator**.
1. Percent of profit **based on cost**: (Cost is the denominator.)

$$\frac{\text{Profit}}{\text{Cost}} = \% \text{ of profit based on cost}$$

2. Percent of profit **based on selling price**: (Selling price is the denominator.)

$$\frac{\text{Profit}}{\text{Selling Price}} = \% \text{ of profit based on selling price}$$

Formula for Calculating Simple Interest: page 328
$$I = P \cdot r \cdot t,$$
where

I = interest (earned or paid)

P = principal (the amount invested or borrowed)

r = rate of interest (stated as an annual rate) in decimal number or fraction form

t = time (one year or fraction of a year)

Note: For calculation purposes, we will use 360 days in one year and 30 days in a month. However, many lending institutions now (with the advent of computers) base their calculations on 365 days per year and pay or charge interest on a daily basis.

To Calculate Compound Interest: page 331

1. Use the formula $I = P \cdot r \cdot t$, to calculate simple interest.

   Let $t = \dfrac{1}{n}$ where n is the number of periods per year for compounding.

   For example:

   for compounding **annually**, $n = 1$ and $t = \dfrac{1}{1} = 1$.

   for compounding **semiannually**, $n = 2$ and $t = \dfrac{1}{2}$

   for compounding **quarterly**, $n = 4$ and $t = \dfrac{1}{4}$

   for compounding **bi-monthly**, $n = 6$ and $t = \dfrac{1}{6}$

   for compounding **monthly**, $n = 12$ and $t = \dfrac{1}{12}$

   for compounding **daily**, $n = 360$ and $t = \dfrac{1}{360}$.

2. Add this interest to the principal to create a new value for the principal.

3. Repeat steps 1 and 2 however many times the interest is to be compounded.

Chapter 4: Review

Section 4.1: Ratios and Proportions

Write the following comparisons as ratios reduced to lowest terms. Use common units in the numerator and denominator whenever possible.

1. 34 students to 40 desks

2. 134 miles to 2 hours

3. 6 minutes to 1 hour

4. 10 cars to 40 tires

5. 7 days to 2 weeks

6. 24 inches to 3 feet

Use the cross product technique to determine whether each proportion is true or false.

7. $\dfrac{3}{8} = \dfrac{325}{1000}$

8. $\dfrac{7.5}{4.5} = \dfrac{9.5}{6.5}$

9. $\dfrac{2\frac{1}{2}}{\frac{3}{4}} = \dfrac{1\frac{1}{2}}{\frac{1}{4}}$

10. $\dfrac{6}{1.56} = \dfrac{1}{0.26}$

Section 4.2: Solving Proportions

Solve the following proportions.

11. $\dfrac{7}{21} = \dfrac{y}{6}$

12. $\dfrac{7}{B} = \dfrac{5}{15}$

13. $\dfrac{3}{16} = \dfrac{9}{x}$

14. $\dfrac{\frac{2}{3}}{3} = \dfrac{y}{126}$

15. $\dfrac{1}{5} = \dfrac{x}{2\frac{1}{2}}$

16. $\dfrac{7.2}{y} = \dfrac{4.8}{14.4}$

17. $\dfrac{13.5}{B} = \dfrac{15}{100}$

18. $\dfrac{A}{595} = \dfrac{6}{100}$

Use proportions to solve the following word problems.

19. A baseball team bought 8 bats for $96. What would they pay for 10 bats?

20. An architect drew plans for a city park using a scale of $\dfrac{1}{4}$ inch to represent 25 feet. How many feet would 2 inches represent?

21. Two numbers are in the ratio of 4 to 3. The number 10 is in that same ratio to a fourth number. What is the fourth number?

22. There are 2.54 centimeters in 1 inch. How many centimeters are there in 1 foot? (Remember, there are 12 inches in 1 foot.)

Section 4.3: Decimal Numbers and Percents

Change the following fractions to percents.

23. $\dfrac{27}{100}$ **24.** $\dfrac{7}{100}$ **25.** $\dfrac{0.6}{100}$ **26.** $\dfrac{425}{100}$

Change the following decimal numbers to percents.

27. 0.03   **28.** 0.36   **29.** 0.005   **30.** 1.63

Change the following percents to a decimal numbers.

31. 8%   **32.** 0.42%   **33.** 183%   **34.** 15.8%

Section 4.4: Fractions and Percents

Change the following fractions or mixed numbers to percents.

35. $\dfrac{4}{25}$ **36.** $\dfrac{13}{20}$ **37.** $2\dfrac{3}{10}$ **38.** $1\dfrac{5}{8}$

Change the following percents to fractions or mixed numbers.

39. 19%   **40.** 60%   **41.** 130%   **42.** 0.7%

43. $33\dfrac{1}{3}\%$ **44.** $\dfrac{1}{10}\%$

Section 4.5: Solving Percent Problems by Using the Proportions: $\dfrac{P}{100} = \dfrac{A}{B}$

Use the proportion $\dfrac{P}{100} = \dfrac{A}{B}$ to solve for the unknown quantity.

45. 20% of 60 is _____.   **46.** 30% of _____ is 15.

47. _____% of 25 is 10.   **48.** _____% of 120 is 60.

49. 110% of _____ is 77.   **50.** Find 22.6% of 90.

51. What percent of 64 is 48?   **52.** 600 is 150% of what number?

Section 4.6: Solving Percent Problems by Using the Equation: $R \cdot B = A$

Use the basic equation $R \cdot B = A$ to solve for the unknown quantity.

53. 2% of ____ is 45.

54. 80% of 10 is ___.

55. ____% of 150 is 60.

56. ____% of 100 is 27.

57. What is $12\frac{1}{2}\%$ of 160?

58. 75% of what number is 42?

59. Find 16.2% of 80.

60. What percent of 55 is 110?

Solve the following word problems by using the basic equation $R \cdot B = A$.

61. Gasoline Prices: In the San Francisco area between 2000 and 2010, the price of gasoline was very volatile. The average price of a gallon of gasoline in January 2000 in San Francisco was $1.49 per gallon. By January 2010, the average price of a gallon of gasoline had risen to $3.03. What was the percent increase for the price of a gallon of gasoline from 2000 to 2010 in the San Francisco area?

62. Nursing: According to the Department of Health and Human Services, there were 2,909,357 licensed registered nurses in the United States in 2009. Only 5.78% of these registered nurses were men. How many male licensed registered nurses were there in the United States in 2009?

Section 4.7: Applications: Discount, Sales Tax, Commission, and Percent Increase/Decrease

Solve the following word problems. Note that the problems may involve several calculations, so follow Polya's four-step problem-solving process as closely as you can.

63. Discounts: A new suitcase was priced at $325. If it was to be marked down 30%,

   a. What would be the discount?

   b. What would be the new price?

64. Discounts: The discount on rain jackets was $25, and they were on sale for $75,

   a. What was the original selling price?

   b. What was the rate of discount?

65. Commissions: A realtor works on a 5% commission. What was the commission on a property that sold for $925,000?

66. Automobile: An automobile dealer buys used cars and then marks up the price so he will make a profit of 30% of his cost. He bought one car for $2000.

   a. What was the selling price for this car?

   b. If the customer paid 6% of the selling price for taxes and other fees, what did the customer pay for this car?

67. Property tax: The property taxes on a home were $1500. What was the tax rate if the property was valued at $125,000 for tax purposes?

68. Commissions: Lisa worked in a furniture store for a salary of $1750 a month plus a commission of 8% on any sales she had over $20,000. What did she earn the month she sold $45,000 worth of furniture?

Section 4.8: Applications: Profit, Simple Interest, and Compound Interest

Solve the following word problems. Note that the problems may involve several calculations, so follow Polya's four-step problem-solving process as closely as you can.

69. Profit: Women's coats were on sale for $250.

  a. If the coats cost the store owner $200, what was his percent of profit based on cost?

  b. What was his percent of profit based on selling price?

70. Sports Goods: The Tennis Shop had a sale on new tennis rackets. The rackets were marked on sale for $300 which was 20% off the original price.

  a. What was the original price?

  b. If the rackets cost the tennis pro $250 each, what was his profit?

  c. What was his percent of profit based on cost?

  d What was his percent of profit based on selling price?

71. Interest: Renee wants to borrow $2500 from her friend and is willing to pay her $100 in interest. If her friend is asking for 4% simple interest, how long can Renee keep the money?

72. Interest: How much money will you need to invest if your investment will earn 5.5% interest and you would like to make $275 in 6 months?

73. Interest: Ellen loaned Kathleen, her sister-in-law, $5000 for one year and asked for only $100 in interest.

  a. What interest rate was Ellen asking?

  b. How much did Kathleen pay Ellen at the end of the year?

74. Interest: If an account is compounded annually (once a year) at 4.5%, how much interest will a principal of $2500 earn in three years?

75. Interest: A principal of $10,000 is to be compounded monthly at 8% for 4 months.

  a. How much interest will be earned in the 4 months?

  b. What will be the balance in the account at the end of 4 months?

Chapter 4: Test

Write the following comparisons as ratios reduced to lowest terms. Change to common units in the numerator and denominator whenever possible.

1. 3 weeks to 35 days

2. 6 nickels to 3 quarters

3. 220 miles to 4 hours

Solve the following proportions.

4. $\dfrac{9}{17} = \dfrac{x}{51}$

5. $\dfrac{\frac{3}{8}}{x} = \dfrac{9}{20}$

Solve the following word problems.

6. On a certain map, 2 inches represents 15 miles. How far apart are two towns that are 3.2 inches apart on the map?

7. If you can buy 4 tires for $236, what will be the cost of 5 of the same type of tires?

Change the following percents to decimal numbers.

8. 35%

9. 7.1%

10. 132%

Change the following fractions or mixed numbers to percents.

11. $\dfrac{85}{100}$

12. $\dfrac{4}{25}$

13. $1\dfrac{3}{5}$

Change the following percents to fractions or mixed numbers.

14. 14%

15. 400%

16. $12\dfrac{1}{2}\%$

Solve each problem for the unknown quantity. Round your answer to the nearest hundreth, if necessary.

17. 30% of 52 is _____.

18. 42% of _____ is 18.

19. What is 36% of 250?

20. Find 65% of 130.

21. 25% of what number is 48?

22. What percent of 104 is 39?

23. 62 is _____% of 31.

24. 0.975 is $6\dfrac{1}{2}\%$ of _____.

25. **Discounts:** A shirt was marked 25% off. What would you pay for the shirt if the original price was $15 and you had to pay 6% sales tax?

26. **Profit:** Men's topcoats were on sale for $180.

    a. If the store owner paid $120 for the coats, what was his percent of profit based on his cost?

    b. What was his percent of profit based on the selling price?

27. **Commissions:** A saleswoman works on a 9% commission on her sales over $10,000 each month, plus a base salary of $600 per month. How much did she make the month she sold $25,000 in merchandise?

28. **Interest:** How much simple interest will be earned in one year on a savings account of $1500 if the bank pays 4.5% interest?

29. **Interest:** You made $42 in interest on an investment at 7% for 9 months. What principal did you invest?

30. **Interest:** In an investment of $7500 earns simple interest of $131.25 in 90 days, what is the rate of interest?

Cumulative Review: Chapters 1 - 4

Solve the following problems.

1. Write the following in standard notation: two hundred thousand, sixteen.

2. Write the following in decimal notation: three hundred and four thousandths.

3. Round 16.996 to the nearest hundredth.

4. Write $\dfrac{31}{8}$ as a mixed number.

5. Find the decimal equivalent to $\dfrac{14}{35}$.

6. Find the decimal equivalent to $\dfrac{21}{40}$.

7. Write $\dfrac{9}{5}$ as a percent.

8. Write $1\dfrac{1}{2}\%$ as a decimal.

Perform the indicated operations.

9. $\begin{array}{r} 4591 \\ + 5568 \\ \hline \end{array}$

10. $\dfrac{2}{15}+\dfrac{11}{15}+\dfrac{7}{15}$

11. $4-\dfrac{3}{11}$

12. $70\dfrac{1}{4}-23\dfrac{5}{6}$

13. $2\dfrac{4}{15}+3\dfrac{1}{6}+4\dfrac{7}{10}$

14. $403-4.012$

15. $71+0.354+4.39$

16. $\dfrac{2}{5}\cdot\dfrac{1}{3}\cdot\dfrac{4}{7}$

17. $(0.27)(0.043)$

18. $83\overline{)4482}$

19. $4\dfrac{5}{7}\cdot 2\dfrac{6}{11}$

20. $6\div 3\dfrac{1}{3}$

21. $(700)(8000)$

22. $27.404\div 0.34$

Perform the indicated operations.

23. $\left(36\div 3^{2}\cdot 2\right)+12\div 4-2^{2}$

24. $\dfrac{5}{6}+6\left(\dfrac{5}{3}-\dfrac{1}{3}\right)^{2}-\dfrac{7}{8}\div\dfrac{1}{4}$

Solve the following problems.

25. Write the following expression in exponential form.
$3\cdot 3\cdot 3\cdot 3\cdot 5\cdot 5\cdot 7$

26. Use the tests for divisibility to determine if 732 can be divided exactly by 2, 3, 4, 5, 6, 9, and 10.

27. Find the prime factorization of 396.

28. Find the LCM of 14, 21, and 30.

29. Division by _____ is undefined.

30. Determine whether the proportion $\dfrac{7}{8} = \dfrac{3.5}{4}$ is true or false.

31. 15% of _____ is 7.5.

32. $9\dfrac{1}{4}\%$ of 200 is _____.

33. 65 is _____ % of 26.

34. Solve for x: $\dfrac{1\frac{2}{3}}{x} = \dfrac{10}{2\frac{1}{4}}$

Solve the following word problems.

35. Gender: In a certain company, three out of every five employees are male. How many female employees are there out of the 490 people working for this company?

36. Number Problem: The sum of two numbers is 521. If one of the numbers is 196, what is the other number?

37. Pie: Margaret has $\dfrac{3}{8}$ of a pie left after her boss and his wife came over for dinner.

a. If the next day she told her three children they could evenly split the leftover pie, what portion of the pie does each child get?

b. If the whole pie was 2720 calories, how many calories did each child consume?

38. Rectangles: Refer to the figure below to find:

a. the perimeter

b. the area

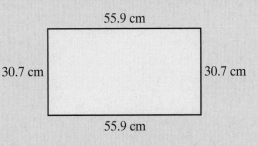

55.9 cm

30.7 cm   30.7 cm

55.9 cm

39. Math Scores: James had the following scores on his first four math tests: 75, 87, 79, and 79. What was his average score?

40. Diet: An Amy's Black Bean Enchilada frozen dinner contains 740 milligrams of sodium, which represents 30.83% of the average daily recommended sodium intake. How many milligrams of sodium are allowed for the day? Round your answer to the nearest whole number.

41. Sales Tax: A discounted room at the Red Roof Inn in Chapel Hill, NC costs $53.99 for one night, and the total occupancy tax rate is 13.74%. What is the total cost of the night's stay? Round your answer to the nearest cent.

42. Percent Decrease: On March 15, 2011, the Nikkei Index in Japan dropped from 9625.45 at opening to 8605.15 at closing. What was the percent decrease in the Index? Round your answer to the nearest tenth of a percent. **Source:** http://finance.yahoo.com/news/ Hedge-funds-hammer-Japan-rb-3702155011.html?x=0&. v=14

43. Interest: An investment pays $6\dfrac{1}{4}\%$ simple interest. What is the interest on $4800 invested for 9 months?

44. Interest: How much is put into a savings account that pays 5.5% simple interest if after 6 months, the interest earned is $123.75?

Geometry

Mathematics at Work!

Geometric figures and their properties are part and parcel of our daily lives. **Geometry** is an integral part of art, employing concepts such as a vanishing point and ways of creating three-dimensional perspective on a two-dimensional canvas. Architects and engineers use geometric shapes in designing buildings and other infrastructures for beauty and structural strength. Geometric patterns are evident throughout nature in snow flakes, leaves, pine cones, butterflies, trees, crystal formations, camouflage for deer, fish, and tigers, and so on.

As an example of geometry in the human-created world, consider sewer covers in streets that you drive over every day. Why are these covers circular in shape? Why are they not in the shape of a square or a rectangle? One reason might be that a circle cannot "fall through" a slightly smaller circle, whereas a square or rectangle can be turned so that either will "fall through" a slightly smaller version of itself. Thus, circular sewer covers are much safer than other shapes because they avoid the risk of the cover accidentally falling down into the hole.

To understand this idea, draw a square, a rectangle, a triangle, and a hexagon (six-sided figure) on a piece of paper. Cut out these figures, and show how easily they can be made to pass through the holes made by the corresponding cutouts. Then follow the same procedure with a circle. The difference will be clear.

Objectives

A   Understand the terms point, line, and plane.

B   Know the definition of an angle and how to measure an angle.

C   Be able to classify an angle by its measure.

D   Recognize complementary angles and supplementary angles.

E   Recognize congruent angles, vertical angles, and adjacent angles.

F   Know when lines are parallel and perpendicular.

5.1 Angles

Objective A   **Introduction to Geometry**

Plane geometry is the study of the properties of figures in a plane. The three most basic ideas in plane geometry are **point**, **line**, and **plane**. These terms are considered so fundamental that they are simply called **undefined terms**. These undefined terms provide the foundation for the study of geometry and the definitions of other geometric figures such as **line segment**, **ray**, **angle**, **triangle**, **polygon**, and so on.

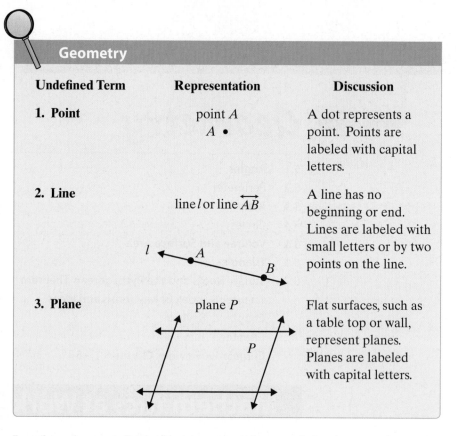

Geometry		
Undefined Term	**Representation**	**Discussion**
1. Point	point A $A \bullet$	A dot represents a point. Points are labeled with capital letters.
2. Line	line l or line $\overleftrightarrow{AB}$	A line has no beginning or end. Lines are labeled with small letters or by two points on the line.
3. Plane	plane P	Flat surfaces, such as a table top or wall, represent planes. Planes are labeled with capital letters.

In a formal approach to plane geometry these undefined terms along with certain assumptions known as axioms, statements called theorems, and a formal system of logic are studied in detail. This approach to the study of geometry is credited to Euclid (about 300 B.C.) which is why the plane geometry courses generally given in high school are known as courses in Euclidean geometry.

Objective B   **Angles and Measures of Angles**

We begin the discussion of angles with the definitions of a **ray** and an **angle** by using the undefined terms **point** and **line**.

Ray and Angle

Term	Definition	Illustrations with Notation
Ray	A **ray** consists of a point (called the **endpoint**) and all the points on a line on one side of that point.	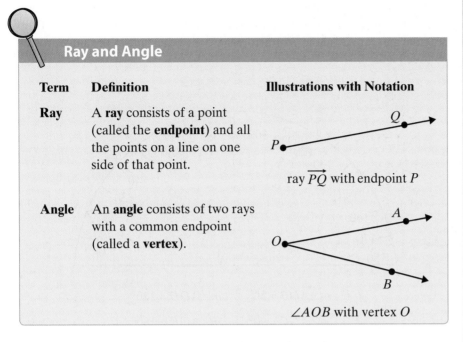 ray $\overrightarrow{PQ}$ with endpoint P
Angle	An **angle** consists of two rays with a common endpoint (called a **vertex**).	$\angle AOB$ with vertex O

In an angle, the two rays are called the **sides** of the angle.

Every angle has a **measurement** or **measure** associated with it. Suppose that a circle is divided into 360 equal arcs. If two rays are drawn from the center of the circle through two consecutive points of division on the circle, then that angle is said to **measure one degree** (symbolized $1°$). For example, in Figure 1, a device called a protractor shows that the measure of $\angle AOB$ is 60 degrees. (We write $m\angle AOB = 60°$.)

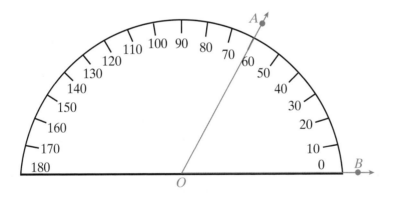

The protractor shows $m\angle AOB = 60°$

Figure 1

To measure an angle with a protractor, lay the bottom edge of the protractor along one side of the angle with the vertex at the marked center point. Then read the measure from the protractor where the other side of the angle crosses it. (See Figure 2.)

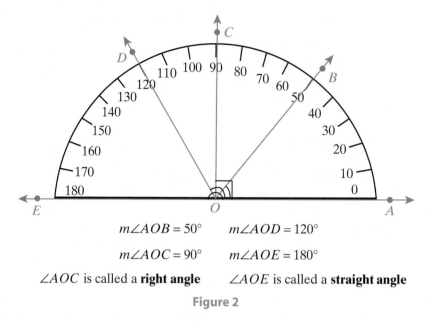

$$m\angle AOB = 50° \qquad m\angle AOD = 120°$$

$$m\angle AOC = 90° \qquad m\angle AOE = 180°$$

$\angle AOC$ is called a **right angle** $\qquad \angle AOE$ is called a **straight angle**

Figure 2

Labeling Angles

Three common ways of labeling angles (Figure 3) are:

a. Using three capital letters with the vertex as the middle letter.
b. Using single numbers such as 1, 2, 3.
c. Using the single capital letter at the vertex when the meaning is clear.

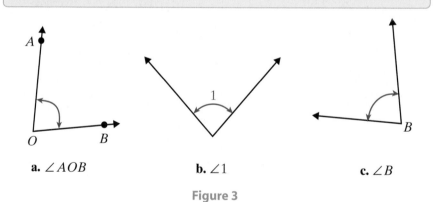

a. $\angle AOB$ \qquad b. $\angle 1$ \qquad c. $\angle B$

Figure 3

Objective C **Classifying Angles by their Measures**

Angles can be classified (or named) according to their measures.

notes

We will use the two inequality symbols

< (read "is less than")
> (read "is greater than")

to indicate the relative sizes of the measures of angles.

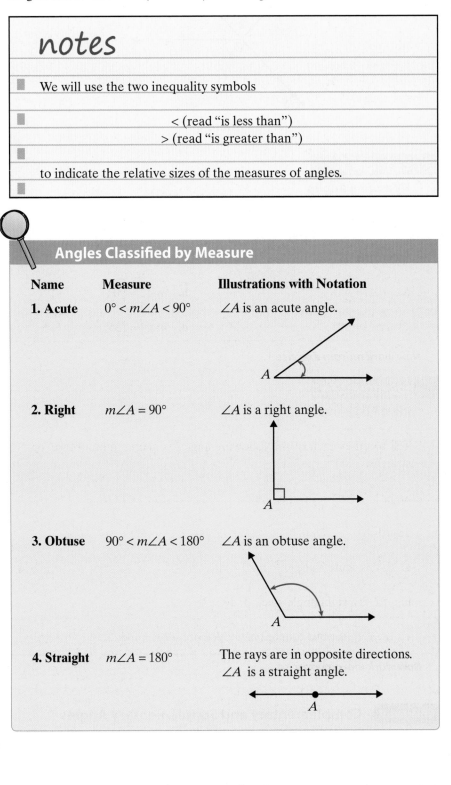

Angles Classified by Measure

Name	Measure	Illustrations with Notation
1. Acute	$0° < m\angle A < 90°$	$\angle A$ is an acute angle.
2. Right	$m\angle A = 90°$	$\angle A$ is a right angle.
3. Obtuse	$90° < m\angle A < 180°$	$\angle A$ is an obtuse angle.
4. Straight	$m\angle A = 180°$	The rays are in opposite directions. $\angle A$ is a straight angle.

The following figure is used for Examples 1 and 2.

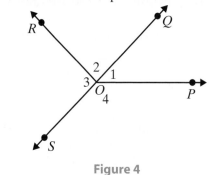

Figure 4

1. Check the measures of $\angle 1$ and $\angle 2$ below using a protractor.

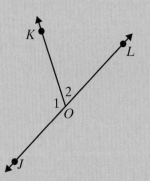

2. Identify the following (from Examples 1 and 2) as acute, right, obtuse, or straight:

a. $\angle 3$
b. $\angle 4$
c. $\angle SOQ$

Example 1

Measuring Angles

Use a protractor to check the measures of the above angles.

Solutions

a. $m\angle 1 = 45°$

b. $m\angle 2 = 90°$

c. $m\angle 3 = 90°$

d. $m\angle 4 = 135°$

Now work margin exercise 1.

Example 2

Identifying Angles

Tell whether each of the following angles is acute, right, obtuse, or straight.

a. $\angle 1$

b. $\angle 2$

c. $\angle POR$

Solutions

a. $\angle 1$ is acute since $0° < m\angle 1 < 90°$.

b. $\angle 2$ is a right angle since $m\angle 2 = 90°$.

c. $\angle POR$ is obtuse since $m\angle POR = 45° + 90° = 135° > 90°$.

Now work margin exercise 2.

Objective D   **Complementary and Supplementary Angles**

Two Angles are

1. **Complementary** if the sum of their measures is 90°.
2. **Supplementary** if the sum of their measures is 180°.

Example 3

Identifying Angles

In the figure shown,

a. $\angle 1$ and $\angle 2$ are complementary since $m\angle 1 + m\angle 2 = 90°$.

b. $\angle COD$ and $\angle COA$ are supplementary since $m\angle COD + m\angle COA = 70° + 110° = 180°$.

c. $\angle AOD$ is a straight angle since $m\angle AOD = 180°$.

d. $\angle BOA$ and $\angle BOD$ are supplementary; and in this case $m\angle BOD = 90°$.

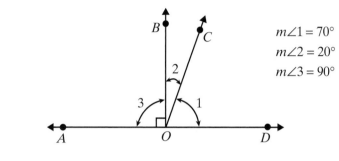

$$m\angle 1 = 70°$$
$$m\angle 2 = 20°$$
$$m\angle 3 = 90°$$

Now work margin exercise 3.

3. In the figure shown in Example 3, are $\angle 2$ and $\angle 3$ supplementary, complementary, or neither? Explain your answer.

Example 4

Measuring Angles

In the figure below, $\overleftrightarrow{PS}$ is a straight line and $m\angle QOP = 30°$. Find the measures of

a. $\angle QOS$ and   **b.** $\angle SOP$

c. Are any pairs supplementary?

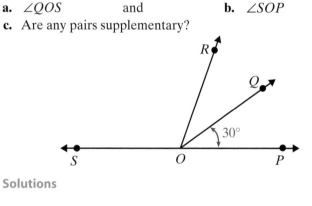

Solutions

a. $m\angle QOS = 150°$.

b. $m\angle SOP = 180°$.

c. Yes, $\angle QOP$ and $\angle QOS$ are supplementary and $\angle ROP$ and $\angle ROS$ are supplementary.

Now work margin exercise 4.

4. In the figure shown in Example 4, assume $m\angle ROQ = 40°$ and answer the following questions:

a. What is the measure of $\angle SOR$?

b. Are $\angle SOR$ and $\angle ROP$ equal?

Congruent, Vertical, and Adjacent Angles

If two angles have the same measure, they are said to be **congruent angles** (symbolized as ≅). As shown in the following figure, $m\angle A = m\angle B = 30°$ and $\angle A \cong \angle B$ (read "angle *A* **is congruent to** angle *B*.")

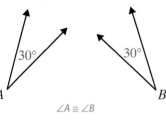

$\angle A \cong \angle B$

Figure 5

5. Identify the congruent angles in the figure below.

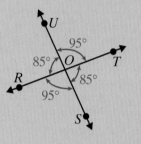

Example 5

Identifying Congruent Angles

In the figure shown below, identify the congruent angles.

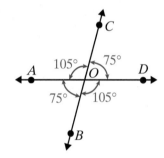

Solution

$\angle AOB \cong \angle COD$ and $\angle COA \cong \angle DOB$

Now work margin exercise **5.**

Two lines **intersect** if there is one point on both lines. If two lines intersect, then two pairs of **vertical angles** are formed. (See Figure 6.)

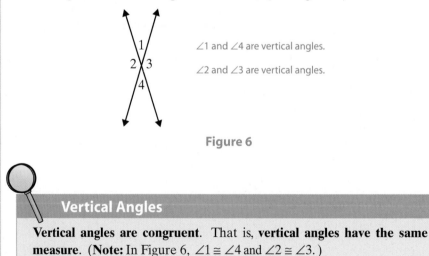

∠1 and ∠4 are vertical angles.

∠2 and ∠3 are vertical angles.

Figure 6

Vertical Angles

Vertical angles are congruent. That is, **vertical angles have the same measure**. (**Note:** In Figure 6, $\angle 1 \cong \angle 4$ and $\angle 2 \cong \angle 3$.)

Example 6

Measures of Vertical Angles

In the figure shown, three lines intersect at the point O.
Find the measures of the following angles:

a. $\angle TOU$ **b.** $\angle ROS$ **c.** $\angle POQ$ **d.** $\angle SOT$

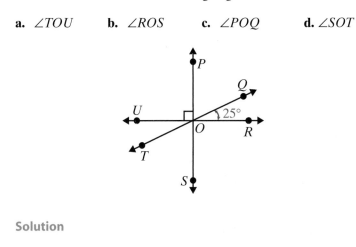

Solution

a. $\angle QOR$ and $\angle TOU$ are vertical angles and are congruent. They have the same measure, so $m\angle TOU = 25°$.

b. $\angle POU$ and $\angle ROS$ are vertical angles and are congruent. They have the same measure, so $m\angle ROS = 90°$.

c. Now, because $\overleftrightarrow{RU}$ is a straight line, we have
$m\angle POQ = 180° - (90° + 25°) = 180° - 115° = 65°$.

d. $\angle POQ$ and $\angle SOT$ are vertical angles and are congruent. They have the same measure, so $m\angle SOT = 65°$.

Now work margin exercise 6.

Adjacent Angles

Two angles are **adjacent** if they have a common side. (See Figure 7.)

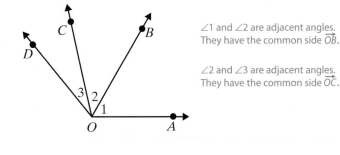

$\angle 1$ and $\angle 2$ are adjacent angles. They have the common side $\overrightarrow{OB}$.

$\angle 2$ and $\angle 3$ are adjacent angles. They have the common side $\overrightarrow{OC}$.

Figure 7

6. Find the measures of the following angles in the figure below.

a. $\angle COD$

b. $\angle DOE$

c. $\angle EOF$

d. $\angle BOC$

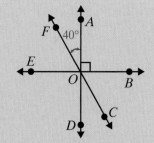

7. In the figure below, $\overleftrightarrow{WY}$ and $\overleftrightarrow{VX}$ are straight lines.

a. Name an angle adjacent to $\angle XQZ$.

b. What is $m\angle WQV$?

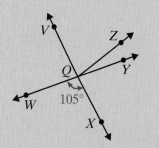

Example 7

Finding Adjacent Angles

In the figure below, $\overleftrightarrow{AC}$ **and** $\overleftrightarrow{BD}$ are straight lines.

a. Name an angle adjacent to $\angle EOD$.

b. What is $m\angle AOD$?

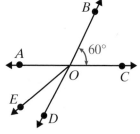

Solution

a. Three angles adjacent to $\angle EOD$ are:
$$\angle AOE, \angle BOE, \text{ and } \angle COD.$$

b. Since $\angle BOC$ and $\angle AOD$ are vertical angles, they have the same measure. So $m\angle AOD = 60°$.

Now work margin exercise 7.

Objective F **Parallel and Perpendicular Lines**

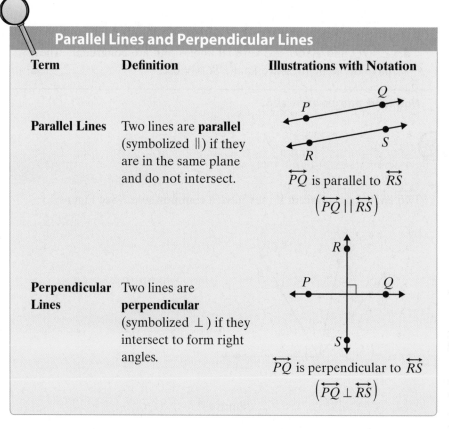

Parallel Lines and Perpendicular Lines		
Term	**Definition**	**Illustrations with Notation**
Parallel Lines	Two lines are **parallel** (symbolized ∥) if they are in the same plane and do not intersect.	$\overleftrightarrow{PQ}$ is parallel to $\overleftrightarrow{RS}$ $\left(\overleftrightarrow{PQ} \parallel \overleftrightarrow{RS}\right)$
Perpendicular Lines	Two lines are **perpendicular** (symbolized ⊥) if they intersect to form right angles.	$\overleftrightarrow{PQ}$ is perpendicular to $\overleftrightarrow{RS}$ $\left(\overleftrightarrow{PQ} \perp \overleftrightarrow{RS}\right)$

A **transversal** is a line in a plane that intersects two or more lines in that plane in different points. As shown in Figure 8, eight angles are formed when a transversal intersects two lines.

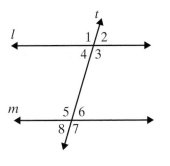

Figure 8

Some of the eight angles formed are named as follows:

Corresponding angles: $\angle 1$ and $\angle 5$, $\angle 2$ and $\angle 6$, $\angle 3$ and $\angle 7$, $\angle 4$ and $\angle 8$

Alternate interior angles: $\angle 3$ and $\angle 5$, $\angle 4$ and $\angle 6$

If two parallel lines are cut by a transversal, then the following two statements are true:

1. **Corresponding angles are congruent.**
2. **Alternate interior angles are congruent.**

Example 8

Finding Measures of Angles

Lines l and m are parallel, t is a transversal, and $m\angle 1 = 50°$. Find the measures of the other 7 angles.

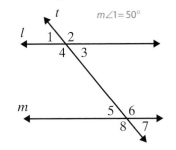

Solution

One way of reasoning (among several) is as follows:

$\angle 1$ and $\angle 3$ are vertical angles so $m\angle 1 = m\angle 3 = 50°$.

$\angle 1$ and $\angle 2$ are supplementary angles so $m\angle 2 = 130°$.

$\angle 2$ and $\angle 4$ are vertical angles so $m\angle 4 = 130°$.

Now, because $\angle 1$ and $\angle 5$ are corresponding angles, they have the same measures and $m\angle 5 = 50°$.

Because $\angle 4$ and $\angle 6$ are alternate interior angles, they have the same measures and $m\angle 6 = 130°$.

Again, with vertical angles $m\angle 5 = m\angle 7 = 50°$ and $m\angle 6 = m\angle 8 = 130°$.

Now work margin exercise 8.

8. Given that lines l and m are parallel, t is a transversal, and $m\angle 2 = 80°$, determine $m\angle 4, m\angle 6$, and $m\angle 5$.

Practice Problems

Identify each given angle as acute, right, obtuse, or straight.

1. $m\angle A = 83°$ **2.** $m\angle B = 149°$ **3.** $m\angle C = 180°$ **4.** $m\angle 1 = 7°$

5. $\angle S$ is a straight angle. Sketch a drawing of the angle.

6. Two straight lines intersect and angles 5, 6, 7, and 8 are formed. $\angle 5$ and $\angle 7$ are vertical. $\angle 6$ and $\angle 8$ are vertical.

   a. Which angles are supplementary?
   b. Which angles are congruent?
   c. Which angles are acute?
   d. Which angles are obtuse?
   e. Name three pairs of adjacent angles.

Practice Problem Answers

1. acute   **2.** obtuse   **3.** straight
4. acute

5.

6. a. $\angle 5$ and $\angle 6$; $\angle 7$ and $\angle 8$; $\angle 5$ and $\angle 8$; $\angle 6$ and $\angle 7$
   b. $\angle 5$ and $\angle 7$; $\angle 6$ and $\angle 8$
   c. $\angle 5$ and $\angle 7$
   d. $\angle 6$ and $\angle 8$
   e. $\angle 5$ and $\angle 8$; $\angle 6$ and $\angle 7$; $\angle 5$ and $\angle 6$; $\angle 7$ and $\angle 8$

Exercises 5.1

1. right angle

2. parallel lines

3. perpendicular lines

4. transversal

Use a protractor to measure all the angles in each figure. Each line segment may be extended as a ray to form the side of an angle. See Example 1.

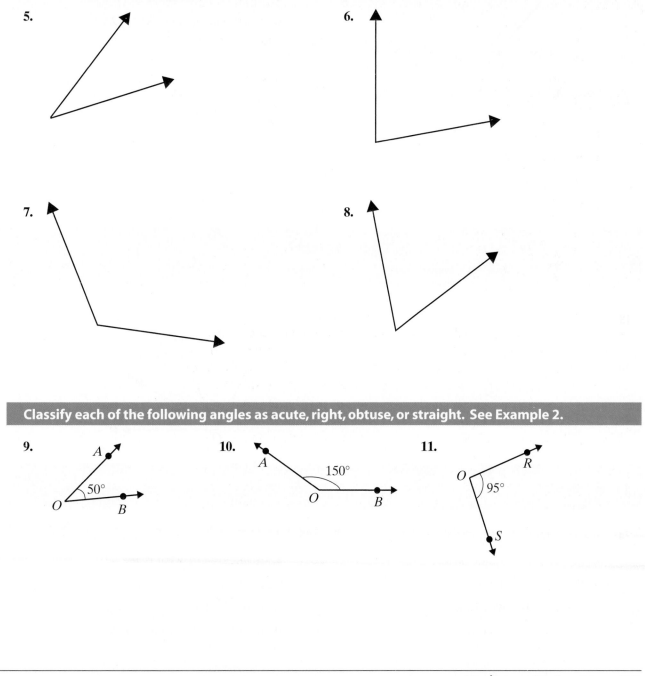

5.

6.

7.

8.

Classify each of the following angles as acute, right, obtuse, or straight. See Example 2.

9.
A
50°
O
B

10.
A
150°
O
B

11.
O
R
95°
S

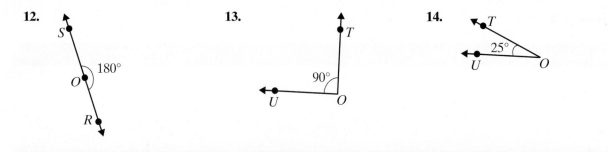

12. 180°

13. 90°

14. 25°

15. In the figure shown below, $\overleftrightarrow{DC}$ is a straight line and $m\angle BOA = 90°$.

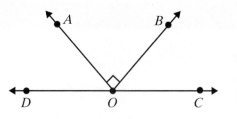

    a. What type of angle is $\angle AOC$?

    b. What type of angle is $\angle BOC$?

    c. What type of angle is $\angle BOA$?

16. Name the type of angle formed by the hands on a clock.

    a. at six o'clock

    b. at three o'clock

    c. at one o'clock

    d. at five o'clock

17. What is the measure of each angle formed by the hands of the clock in Exercise 16?

18. $\angle AOX$

19. $\angle BOX$

20. $\angle COX$

21. $\angle BOY$

22. $\angle AOY$

23. $\angle AOC$

24. Assume that $\angle 1$ and $\angle 2$ are complementary.

    a. If $m\angle 1 = 15°$, what is $m\angle 2$?

    b. If $m\angle 1 = 3°$, what is $m\angle 2$?

    c. If $m\angle 1 = 45°$, what is $m\angle 2$?

    d If $m\angle 1 = 75°$, what is $m\angle 2$?

25. Assume that $\angle 3$ and $\angle 4$ are supplementary.

    a. If $m\angle 3 = 45°$, what is $m\angle 4$?

    b. If $m\angle 3 = 90°$, what is $m\angle 4$?

    c. If $m\angle 3 = 110°$, what is $m\angle 4$?

    d. If $m\angle 3 = 135°$, what is $m\angle 4$?

26. **a.** What is the supplement of a right angle?

    b. What is the supplement of an obtuse angle?

    c. What is the supplement of an acute angle?

27. In the figure shown below:

    a. Name all the pairs of supplementary angles.

    b. Name all the pairs of complementary angles.

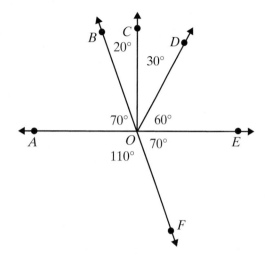

Use the definitions of adjacent and vertical angles to answer the following questions. See Examples 4 and 5.

28. The figure below shows two intersecting lines.

    a. If $m\angle 1 = 30°$, what is $m\angle 2$?

    b. Is $m\angle 3 = 30°$? Give a reason for your answer other than the fact that $\angle 1$ and $\angle 3$ are vertical angles.

    c. Name four pairs of adjacent angles.

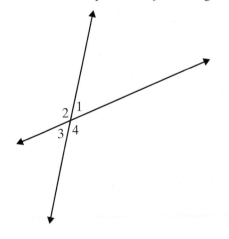

29. In the figure shown below, $\overleftrightarrow{AB}$ is a straight line.

    a. Name two pairs of adjacent angles.

    b. Name two vertical angles if there are any.

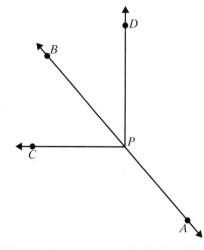

30. Given that $m\angle 1 = 42°$ in the figure shown below, find the measures of the other three angles.

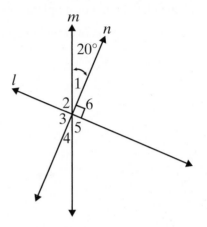

31. In the figure shown below, l, m, and n are straight lines with $m\angle 1 = 20°$ and $m\angle 6 = 90°$.

a. Find the measures of the other four angles.

b. Which angle is supplementary to $\angle 6$?

c. Which angles are complementary to $\angle 1$?

32. In the figure shown below, $m\angle 2 = m\angle 3 = 40°$. Find all other pairs of angles that have equal measures.

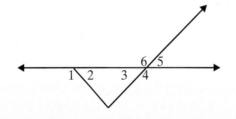

Use the figures to answer Exercises 33 through 37.

33. If $m\angle 1 = 125°$, then $m\angle 3 =$ _____.
Explain your reasoning.

34. If $m\angle 8 = 55°$, then $m\angle 6 =$ _____.
Explain your reasoning.

35. What is $m\angle 7$?

Explain your reasoning.

36. Does $m\angle 2 = m\angle 6$?

Explain your reasoning.

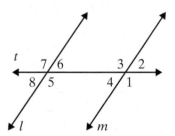

37. Line $\overleftrightarrow{AB}$ and ray $\overrightarrow{PQ}$ are perpendicular.

a. Which angles are acute?

b Which angles are obtuse?

c. Which angles are right angles?

d. Which pairs of angles are vertical angles?

e. Which pairs of angles are complementary?

f. Which pairs of angles are supplementary?

g. Which pairs of angles are adjacent?

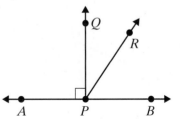

5.2 Perimeter

Objective A **Polygons**

A **line segment** consists of two points on a line and all the points between them.

line segment $\overline{AB}$

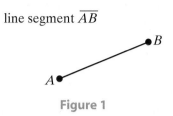

Figure 1

Objectives

A Know what types of geometric figures are polygons.

B Find the perimeters of polygons.

Familiar geometric figures called **polygons** (such as triangles, rectangles and squares) are defined in terms of line segments.

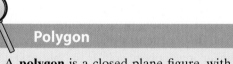

Polygon

A **polygon** is a closed plane figure, with three or more sides, in which each side is a line segment.

Each point where two sides meet is called a **vertex**.

Note: A **closed figure** begins and ends at the same point.

A **triangle** is a polygon with three sides.

A **rectangle** is a polygon with four sides in which adjacent sides are perpendicular (meet at a 90° angle).

A **square** is a rectangle in which all four sides are the same length.

A **trapezoid** is a four-sided polygon with one pair of opposite sides that are parallel.

A **parallelogram** is a four-sided polygon with both pairs of opposite sides parallel.

An illustration of each of these polygons is shown in Figure 2.

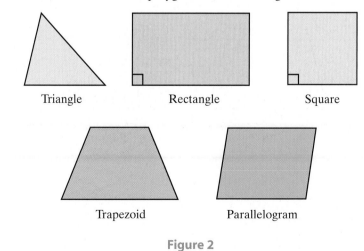

Triangle   Rectangle   Square

Trapezoid   Parallelogram

Figure 2

Formulas for Perimeter

As stated in Chapter 4, a **formula** is a general statement (usually an equation) that relates two or more variables. For example, we used the formula for simple interest, $I = Prt$ in Section 4.8. Note that multiplication is indicated when variables or constants and variables are written next to each other. The formulas used in this section are used to calculate the **perimeters** of polygons.

> ### Perimeter
>
> The **perimeter** P of a polygon is the sum of the lengths of its sides.

The formulas for the perimeters of several geometric figures are given here. In the rectangle l represents length and w represents width. (Customarily, the width is the shorter of the two.) In the square s is the length of one side, and all four sides have the same length. In using these formulas you should be familiar with the following basic units of length and the corresponding abbreviations. (**Note:** Tables of Conversion between units in the metric system and the U.S. customary system are in the Appendix A.3.)

From the metric system	**From the U.S. customary system**
millimeter (mm)	inch (in.)
centimeter (cm)	foot (ft)
meter (m)	yard (yd)
kilometer (km)	mile (mi)

Be sure to label your answers with the correct units.

Formulas for the Perimeters of Five Polygons

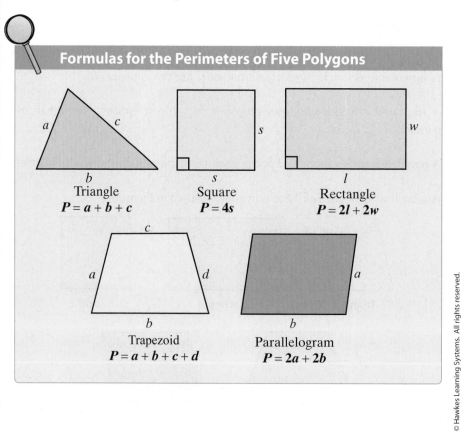

Triangle
$P = a + b + c$

Square
$P = 4s$

Rectangle
$P = 2l + 2w$

Trapezoid
$P = a + b + c + d$

Parallelogram
$P = 2a + 2b$

Example 1

Finding the Perimeter of a Square

Find the perimeter of a square with sides of length 16 inches.

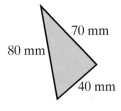

16 in.

Solution

Using the formula for the perimeter of a square, we have

$$P = 4s$$

$$P = 4 \cdot 16 = 64 \text{ in.}$$

The perimeter of the square is 64 inches.

Now work margin exercise 1.

Example 2

Finding the Perimeter of a Triangle

Find the perimeter of a triangle with sides of length 40 mm, 70 mm, and 80 mm.

70 mm

80 mm

40 mm

Solution

$$P = a + b + c$$

$$P = 40 + 70 + 80$$

$$= 190 \text{ mm}$$

The perimeter of the triangle is 190 mm.

Now work margin exercise 2.

1. Find the perimeter of a square with one side of length 8 ft.

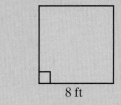

8 ft

2. Find the perimeter of the figure shown below.

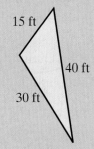

15 ft

40 ft

30 ft

3. Find the perimeter of the following rectangle.

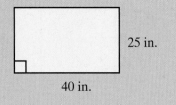

25 in.

40 in.

Example 3

Finding the Perimeter of a Rectangle

Find the perimeter of the rectangle shown.

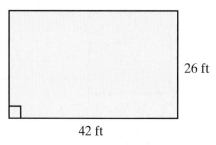

26 ft

42 ft

Solution

$$P = 2l + 2w$$
$$P = 2 \cdot 42 + 2 \cdot 26$$
$$= 84 + 52$$
$$= 136 \text{ ft}$$

The perimeter of the rectangle is 136 ft.

Now work margin exercise 3.

4. Find the perimeter of the figure shown below.

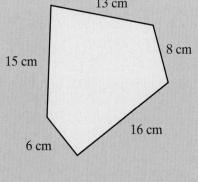

13 cm

8 cm

15 cm

6 cm

16 cm

Example 4

Finding the Perimeter of a Polygon

Find the perimeter of the polygon shown.
(**Note:** A 5-sided polygon is called a **pentagon**.)

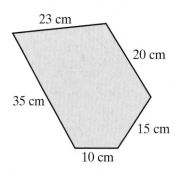

23 cm

20 cm

35 cm

15 cm

10 cm

Solution

The perimeter of any polygon is found by adding the lengths of the sides. Thus, for this pentagon we have

$$P = 10 + 15 + 20 + 23 + 35 = 103 \text{ cm}$$

Now work margin exercise 4.

Example 5

Finding the Cost of a Fencing Project

For security, a chain link fence is to be built around a new warehouse building. The property is "L" shaped as shown.

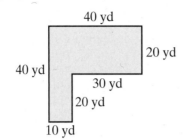

a. How many yards of fencing will be needed?
b. What will be the cost of the entire fencing project if the price is $12.50 per yard?

Solution

a. The perimeter of the property is

$$P = 40 + 40 + 20 + 30 + 20 + 10 = 160 \text{ yd.}$$

b. The cost will be

$$\text{Cost} = \$12.50 \cdot 160 = \$2000.$$

Now work margin exercise 5.

Example 6

Finding the Perimeter of a Polygon

If a rectangle has length 15 ft and width 12 ft, then its perimeter is:

$$P = 2 \cdot 15 + 2 \cdot 12$$
$$= 30 + 24$$
$$= 54 \text{ ft}$$

12 ft

15 ft

Now, if a small rectangle is cut from one corner of the original rectangle, then a new shape is formed. An interesting fact is that, regardless of the size of the cut out rectangle, this new shape will have the **same perimeter** as the original rectangle. This is because the length and width of the segments of the new shape are the same as those of the cut out. Note the colored edges.

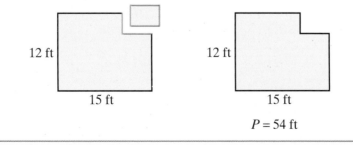

12 ft

15 ft

12 ft

15 ft

$$P = 54 \text{ ft}$$

Now work margin exercise 6.

5. Brandy is adding plastic trimming around the edge of her new swimming pool shown below:

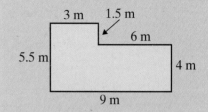

a. How many meters of trimming will she need?

b. What will be the total cost of the trimming if she can buy it for $7.50 a meter?

6. Find the perimeter of the figure shown below.

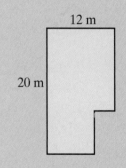

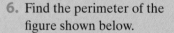

Practice Problems

Sketch a figure as an aid for each problem.

1. Find the perimeter of a square with sides of length 15 inches.

2. Find the perimeter of a parallelogram with sides of length 20 m and 25 m.

3. What is the perimeter of a rectangle with width 8 yards and length 12 yards?

4. Find the perimeter of the polygon shown:

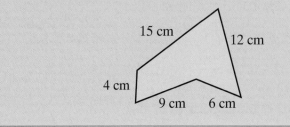

15 cm   12 cm

4 cm   9 cm   6 cm

Practice Problem Answers

1. 60 inches   **2.** 90 m   **3.** 40 yards   **4.** 46 cm

Exercises 5.2

Use the definitions and formulas from this section and answer the questions below.

1. Write the definition of a **polygon**.

2. Match each formula for perimeter to its corresponding geometric figure.

   _____ **a.** square   **A.** $P = 2l + 2w$

   _____ **b.** parallelogram   **B.** $P = 4s$

   _____ **c.** rectangle   **C.** $P = 2b + 2a$

   _____ **d.** trapezoid   **D.** $P = a + b + c$

   _____ **e.** triangle   **E.** $P = a + b + c + d$

3. True or false (If a statement is false, explain why.)

   a. Every square is a rectangle.

   b. Every rectangle is a square.

   c. Every parallelogram is a rectangle.

   d. Every rectangle is a parallelogram.

Find the perimeter of each figure described below. See Examples 1 and 2.

4. A parallelogram with sides of length 15 cm and 7 cm.

5. A square with sides of length 11 m.

6. A triangle with sides of length 21 in., 67 in., and 55 in.

7. A parallelogram with sides of length 42 mm and 34 mm.

8. A rectangle with sides of length 24 cm and 34 cm.

9. A trapezoid with sides of length 31 ft, 39 ft, 45 ft, and 51 ft.

10. A trapezoid with sides of length 14.2 yd, 10.1 yd, 8 yd, and 15.8 yd.

11. A triangle with side of length 7.5 in., 17 in., and 13.6 in.

12. A square with sides of length $4\frac{1}{2}$ km.

13. A rectangle with sides of length $3\frac{1}{4}$ ft and $2\frac{5}{6}$ ft.

Find the perimeter of each figure. See Examples 1 through 6.

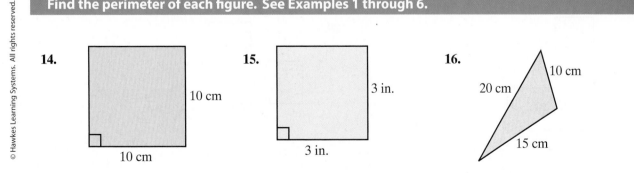

14. 10 cm, 10 cm

15. 3 in., 3 in.

16. 10 cm, 20 cm, 15 cm

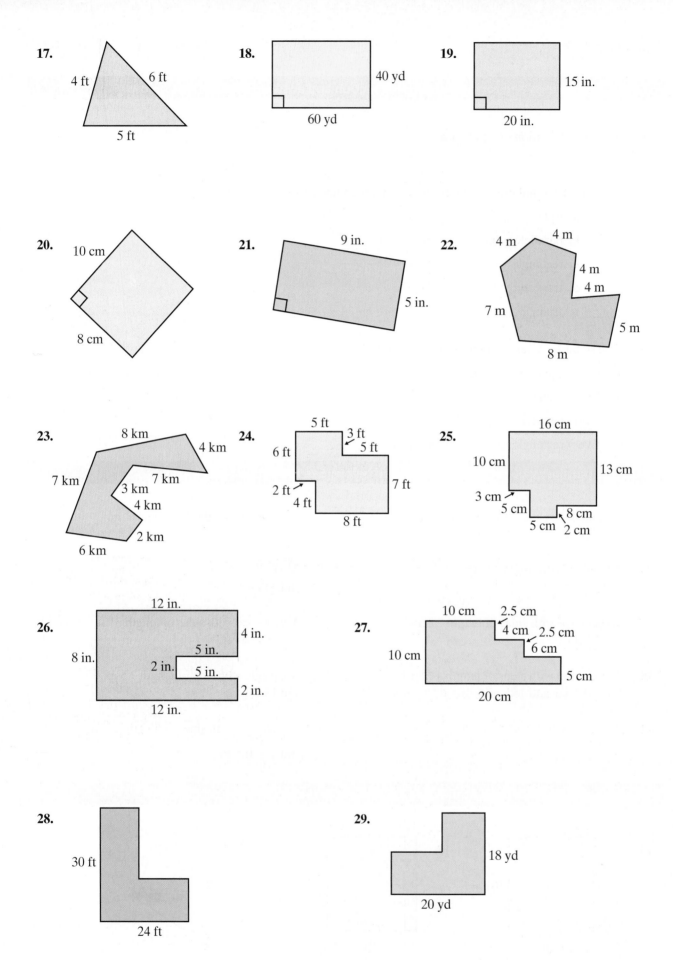

17. 4 ft, 6 ft, 5 ft

18. 40 yd, 60 yd

19. 15 in., 20 in.

20. 10 cm, 8 cm

21. 9 in., 5 in.

22. 4 m, 4 m, 4 m, 4 m, 7 m, 5 m, 8 m

23. 8 km, 4 km, 7 km, 7 km, 3 km, 4 km, 2 km, 6 km

24. 5 ft, 3 ft, 5 ft, 6 ft, 2 ft, 4 ft, 7 ft, 8 ft

25. 16 cm, 10 cm, 13 cm, 3 cm, 5 cm, 5 cm, 8 cm, 2 cm

26. 12 in., 4 in., 5 in., 8 in., 2 in., 5 in., 2 in., 12 in.

27. 10 cm, 2.5 cm, 4 cm, 2.5 cm, 6 cm, 10 cm, 5 cm, 20 cm

28. 30 ft, 24 ft

29. 18 yd, 20 yd

30. A **regular hexagon** is a six-sided figure with all six sides equal and all six angles equal. Find the perimeter of a regular hexagon with one side measuring 19 centimeters.

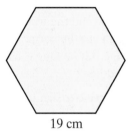

19 cm

31. A **regular octagon** is an eight-sided figure with all eight sides equal and all eight angles equal. Find the perimeter of a regular octagon if one side measures 12 inches.

(**Note:** Where do you see regular octagons on a regular basis?)

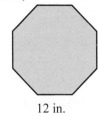

12 in.

32. **Gardening:** A five-pointed star-shaped flower plot, where each edge of the star is 6.5 feet, is placed in the middle of a lawn.

   a. What is the perimeter of the plot?

   b. If edging material costs $2.40 per foot, how much will it cost to fully enclose the star?

6.5 ft

33. **Military:** The Pentagon near Washington, D.C. is a five-sided building where each outside wall is 921 feet. (**Source:** http://www.infoplease.com/spot/pentagon1.html)

   a. What is the perimeter of the building?

   b. If it takes a person 0.00341 minute to walk 1 foot, how long will it take the person to walk completely around the building? Round your answer to the nearest tenth of a minute.

34. **Decorating:** A rectangular picture frame is 10.5 inches high and 8.7 inches wide. How much picture framing material must be used to frame the picture?

(**Hint:** This is the same as the perimeter of the outer edge.)

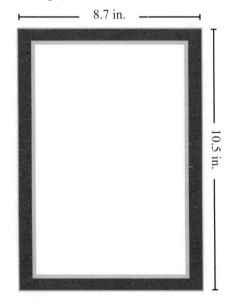

8.7 in.

10.5 in.

35. **Architecture:** A public building consists of a rectangular portion used for office space that is 85 feet long and 27 feet wide connected to an octagonal shaped auditorium where each of the eight sides is the same length as the width of the rectangular portion. The roof is constructed so that there are rain gutters on all sides of the building. How much guttering must be purchased? (**Hint:** One side of the rectangle and one side of the octagon will not be included in computing the perimeter.)

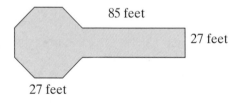

85 feet

27 feet

27 feet

36. Real estate: A property owner has an odd-shaped lot as shown in the drawing below.

   a. What is the perimeter of the lot?

   b. If the cost of constructing a fence is $15.00 a foot, how much will it cost to construct the fence?

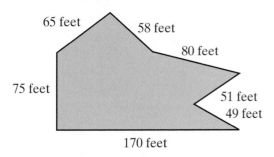

65 feet   58 feet

80 feet

75 feet

51 feet
49 feet

170 feet

37. Walking: For exercise, John will walk along the path which is indicated by the solid line in the drawing below. Note that he cuts a corner, where both sides are the same length, as shown.

   a. How many meters did John walk?

   b. How long would the walk have been if John didn't cut the corner, but rather walked the full rectangle, as indicated by; the dotted line?

   c. How do these two distances compare?

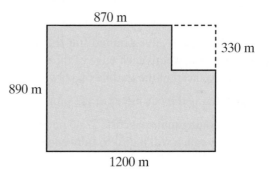

870 m

330 m

890 m

1200 m

Writing & Thinking

38. Perimeter: The perimeter of a standard sheet of paper ($8\frac{1}{2}$ inches wide and 11 inches long) is

$P = 2 \cdot 8\frac{1}{2} + 2 \cdot 11 = 17 + 22 = 39$ inches. Use a pair of scissors to cut a rectangle from one corner of

a standard sheet of paper and measure the perimeter of the new figure. Repeat this process several times (in some cases cut more than one rectangle from a different corner and measure the perimeter each time). Give a brief explanation of the results in each case.

5.3 Area

Objective A **The Concept of Area**

Area is a measure of the interior of (or surface enclosed by) a plane figure and is measured in **square units**. The concept of area was discussed in Chapter 1 and is illustrated again in Figure 1. (**Note:** in.² is read "inches squared" or "square inches.")

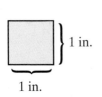
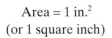

} 1 in.

1 in.

Area = 1 in.²
(or 1 square inch)

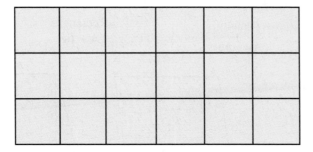

Area = 6 in · 3 in = 18 in.²
There are 18 squares that are each 1 in.²
for a total of 18 in.²

Figure 1

You should be familiar with the following basic units of area and the corresponding abbreviations.

From the metric system

square millimeters $\left(mm^2\right)$

square centimeters $\left(cm^2\right)$

square kilometers $\left(km^2\right)$

square meters $\left(m^2\right)$

From the U.S. customary system

square inches $\left(in.^2\right)$

square feet $\left(ft^2\right)$

square yards $\left(yd^2\right)$

square miles $\left(mi^2\right)$

Objective B **Formulas for Area**

The formulas for the areas of the geometric figures shown here are independent of the units of measure used. **Be sure to label your answers with the correct units**.

notes

- In triangles and other figures, we have used the letter h to represent the **height** of the figure. The height is also called the **altitude** and is perpendicular to the base.

Objectives

A Understand the concept of area.

B Know the formulas for finding the area of five polygons.

Formulas for the Area of Five Polygons

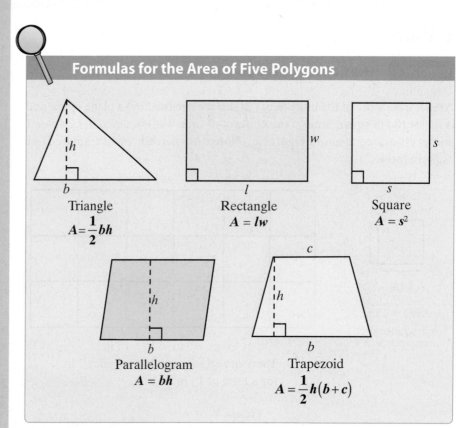

Triangle
$$A = \frac{1}{2}bh$$

Rectangle
$$A = lw$$

Square
$$A = s^2$$

Parallelogram
$$A = bh$$

Trapezoid
$$A = \frac{1}{2}h(b+c)$$

1. Find the area of the given triangle.

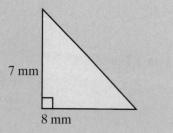

7 mm

8 mm

Example 1

Finding the Area of a Triangle

Find the area of a triangle with height 4 in. and base 10 in.

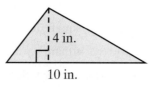

4 in.

10 in.

Solution

To find the area of a triangle, multiply $\frac{1}{2}$ times the base times the height. (Be sure to label the answer in square inches.)

$$A = \frac{1}{2}bh$$

$$A = \frac{1}{2} \cdot 10 \cdot 4 = 20 \text{ in.}^2$$

The area of the triangle is 20 in.^2

Now work margin exercise 1.

Example 2

Finding the Area of a Trapezoid

Find the area of a trapezoid with altitude 6 in. and parallel sides of length 12 in. and 24 in.

Solution

First draw a figure and label the lengths of the known parts.

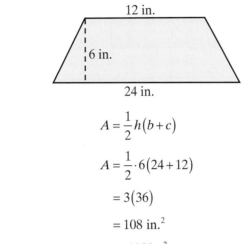

$$A = \frac{1}{2}h(b+c)$$

$$A = \frac{1}{2} \cdot 6(24 + 12)$$

$$= 3(36)$$

$$= 108 \text{ in.}^2$$

The area of the trapezoid is 108 in.^2

Now work margin exercise 2.

Example 3

Finding the Area of a Composite Figure

Find the area of the figure shown here with the indicated dimensions.

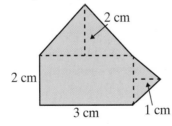

Solution

To find the area of this figure, find the area of each part and then add the three areas. The figure is made up of two triangles and one rectangle.

Rectangle	**Larger Triangle**	**Smaller Triangle**
$A = lw$	$A = \frac{1}{2}bh$	$A = \frac{1}{2}bh$
$A = 2 \cdot 3 = 6 \text{ cm}^2$	$A = \frac{1}{2} \cdot 3 \cdot 2 = 3 \text{ cm}^2$	$A = \frac{1}{2} \cdot 2 \cdot 1 = 1 \text{ cm}^2$

$$\text{total area} = 6 \text{ cm}^2 + 3 \text{ cm}^2 + 1 \text{ cm}^2$$

$$= 10 \text{ cm}^2$$

Now work margin exercise 3.

2. Find the area of a trapezoid with altitude 3 cm and parallel sides of length 9 cm and 15 cm.

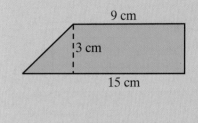

3. Find the area of the figure shown here with the indicated dimensions.

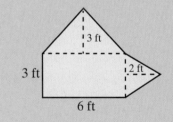

4. A rectangle is cut out of a rectangle as shown. Find the area of the yellow shaded region.

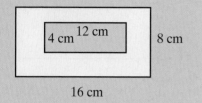

16 cm

Example 4

Finding the Area of a Rectangle

A square is cut out of a rectangle as shown. Find the area of the orange shaded region.

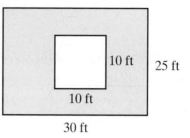

Solution

There are three steps in finding the area of the shaded region. Find the area of the outer figure. Find the area of the inner figure. Find the difference between the areas.

Step 1: Find the area of the rectangle.

$$A = lw$$

$$A = 30 \cdot 25 = 750 \text{ ft}^2$$

Step 2: Find the area of the square.

$$A = s^2$$

$$A = 10^2 = 100 \text{ ft}^2$$

Step 3: Find the difference between the two areas.

$$\text{Area of shaded region} = 750 - 100 = 650 \text{ ft}^2$$

The area of the shaded region is 650 ft^2.

Now work margin exercise 4.

Example 5

Finding the Area of a Polygon

The polygon shown here is a rectangle with a rectangular piece missing. Find the area of the polygon.

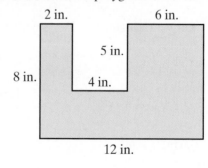

Solution

There are several ways of finding the area of this figure. One way is to find the area of each of the three parts as illustrated here and then adding the three areas.

$A_1 = 2 \cdot 5 = 10$ in.^{2}

$A_2 = 5 \cdot 6 = 30$ in.^{2}

$A_3 = 12 \cdot 3 = 36$ in.^{2}

Note that the height of the bottom rectangle is 3.

$A_{\text{Total}} = 10 + 30 + 36 = 76$ in.^{2}

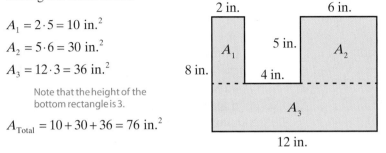

Now work margin exercise 5.

Example 6

Finding the Perimeter and Area

A baseball infield is in the shape of a square 90 feet on each side.
a. What is the perimeter of the infield?
b. What is the area of the infield?

Solutions

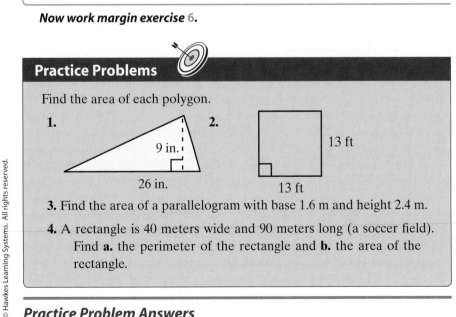

a. $P = 4s$

$P = 4 \cdot 90 = 360$ ft

b. $A = s^2$

$A = 90^2 = 8100$ ft^2

The perimeter of the infield is 360 feet and the area is 8100 square feet.

Now work margin exercise 6.

Practice Problems

Find the area of each polygon.

1. [triangle: 9 in. height, 26 in. base]

2. [square: 13 ft by 13 ft]

3. Find the area of a parallelogram with base 1.6 m and height 2.4 m.

4. A rectangle is 40 meters wide and 90 meters long (a soccer field). Find **a.** the perimeter of the rectangle and **b.** the area of the rectangle.

Practice Problem Answers

1. 117 in.^{2} **2** 169 ft^2 **3.** 3.84 m^2 **4. a.** 260 m **b.** 3600 m^2

---

5. Find the area of the polygon.

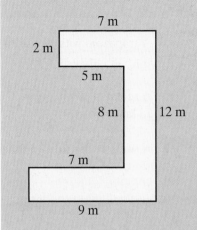

6. A park is in the shape of a square 30 yd on each side.

a. What is the perimeter of the park?

b. What is the area of the park?

Exercises 5.3

Find the area of each of the figures described below. See Examples 1 and 2.

1. A trapezoid with height 10 cm and parallel sides of length 15 cm and 18 cm.

2. A square with sides of length 6 in.

3. A rectangle with length 21 km and width 25 km.

4. A parallelogram with height 5 m and base 12 m.

5. A square with sides of length 9 ft.

6. A trapezoid with height 30 mm and parallel sides of length 45 mm and 50 mm.

7. A triangle with height $\frac{8}{9}$ in. and base $\frac{5}{12}$ in.

8. A rectangle with length $1\frac{1}{4}$ mi and width $2\frac{1}{2}$ mi.

9. A paralleogram with height 2.3 ft and base 11.9 ft.

10. A triangle with height 16.4 cm and base 8.2 cm.

Find the area of each figure. See Examples 1 through 5.

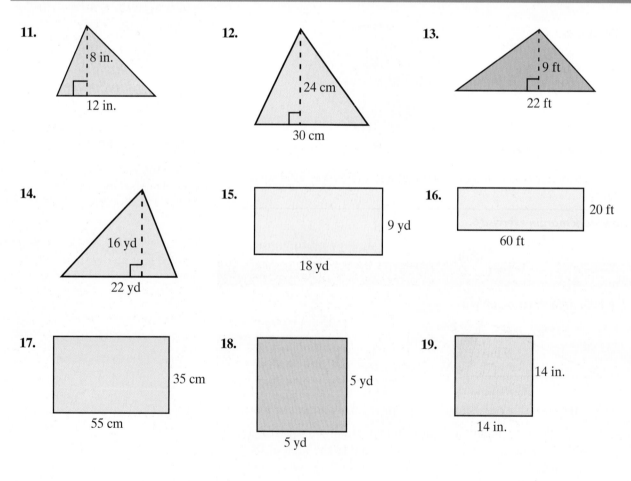

11. 8 in.   12 in.

12. 24 cm   30 cm

13. 9 ft   22 ft

14. 16 yd   22 yd

15. 9 yd   18 yd

16. 20 ft   60 ft

17. 35 cm   55 cm

18. 5 yd   5 yd

19. 14 in.   14 in.

20. 1 ft, 1 ft

21. 6 cm, 8 cm

22. 8 ft, 12 ft

23. 14 in., 10 in., 18 in.

24. 15 yd, 12 yd, 12 yd

25. 4 cm, 7 cm, 10 cm

26. 3 in., 10 in., 3 in., 10 in.

27. 11 m, 9 m, 4 m, 13 m

28. 4 yd, 14 yd, 7 yd, 9 yd

29. 4 in., 7 in., 10 in., 3 in., 3 in., 12 in.

30. 5 cm, 3.5 cm, 3 cm, 2.5 cm, 15 cm, 9 cm

31. 6 mm, 15 mm, 10 mm, 20 mm

32. 5 m, 5 m, 4 m, 10 m

33. 8 cm, 10 cm, 3 cm, 6 cm

34. 5 yd, 12 yd, 5 yd, 18 yd

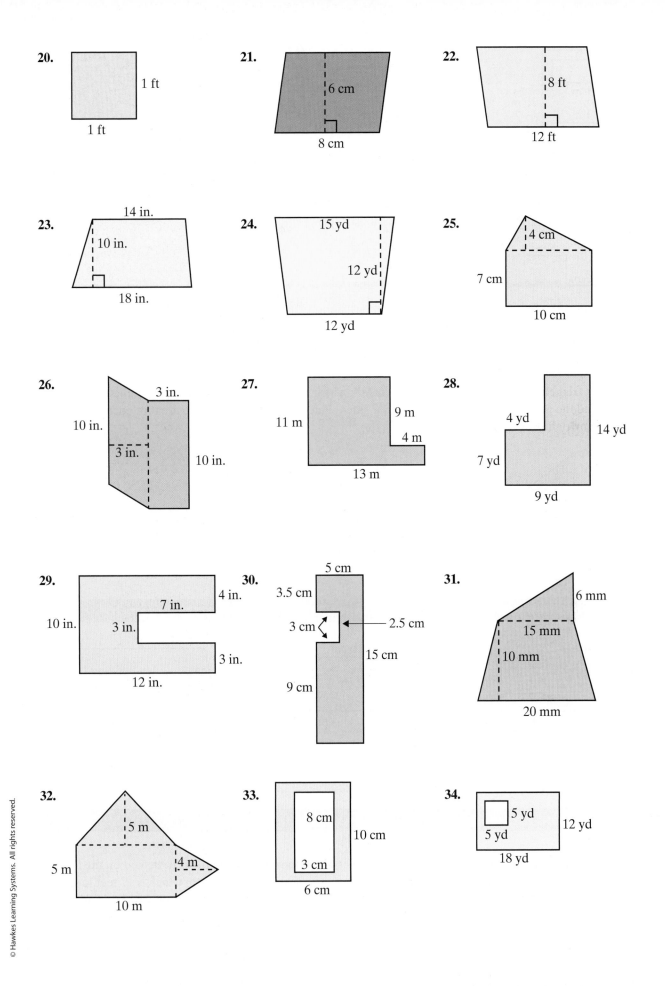

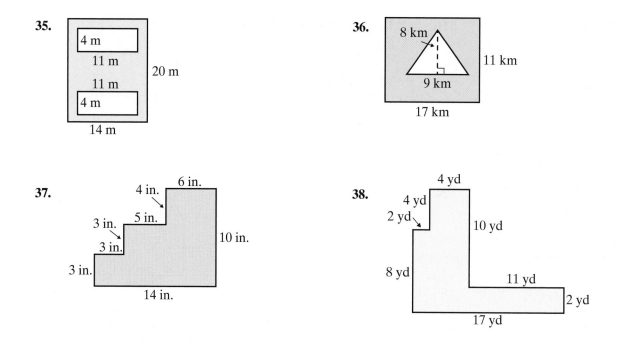

35.

4 m
11 m
11 m
4 m
20 m
14 m

36.

8 km
11 km
9 km
17 km

37.

6 in.
4 in.
5 in.
3 in.
3 in.
3 in.
10 in.
14 in.

38.

4 yd
4 yd
2 yd
10 yd
8 yd
11 yd
2 yd
17 yd

39. A **right triangle** is a triangle with one angle of 90°, which means that two sides are perpendicular. The base and the height are the two perpendicular sides. Find the perimeter and the area for each of the following right triangles.

a.

3 ft
5 ft
4 ft

b.

12 cm
13 cm
5 cm

c.

12 in.
20 in.
16 in.

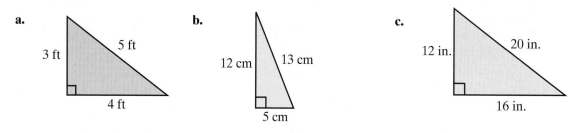

Find a. the perimeter and b. the area of each figure.   See Example 6.

40.

12 ft
16 ft

41.

10 cm
13 cm
11 cm
17 cm
30 cm

42.

5 ft
6 ft
10 ft

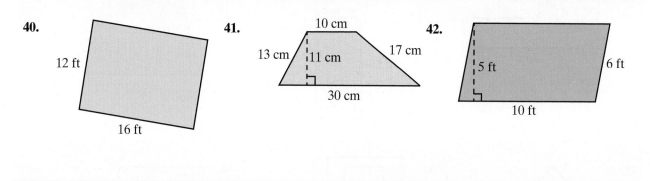

Find the solutions to the following word problems.

43. Cartography: The boundaries of a certain small town form a parallelogram with a length of 4.5 miles, and a height of 2.6 miles. What is the area within the town limits?

44. Flooring: Vinyl tile is to be laid on the floor of a rectangular room which is 17 feet long and 12 feet wide. How many square feet of tile must be put down?

45. Sailing: A sailboat has a triangular sail with the dimensions as shown in the drawing below. Note that the 12 foot measurement is the height of the triangle.

a. What is the area of the sail?

b. What is the perimeter of the sail?

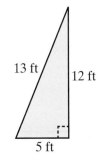

13 ft   12 ft

5 ft

46 Electronics: A square electronics circuit board is 18 centimeters on each side. On the center of one of the edges is a 8 by 1.5 centimeter rectangular lip for plugging in.

a. What is the total perimeter of the circuit board, including the lip?

b. What is the area of the circuit board?

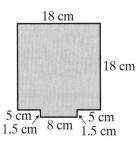

18 cm

18 cm

5 cm   5 cm
1.5 cm   8 cm   1.5 cm

47. Landscaping: A trapezoidal patio with a square opening for flowers is to be constructed. As shown on the drawing below, the ends of the trapezoid are 12 ft and 9 ft respectively, with a height of 15 ft. Each side of the square cutout is 3 ft.

a. What is the area of the concrete surface?

b. If the charge to pour and finish the concrete is $9.50 a square foot, what will it cost?

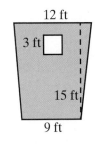

12 ft

3 ft

15 ft

9 ft

48. Landscaping: David is planting a five-sided lawn as shown in the figure below. The lawn consists of a 50 foot by 40 foot rectangle and an attached 14 foot high triangle.

a. What is the area of the lawn to be planted?

b. If one pound of grass seed will cover 200 square feet, how many pounds will be necessary to cover the entire lawn? (**Hint:** Divide the area by the number of square feet that one pound of seed will cover.)

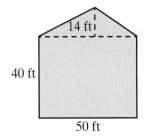

14 ft

40 ft

50 ft

49. Draw a rectangle and a **diagonal** (a segment from one vertex to an opposite vertex) forming two triangles. (See the figure.) Discuss the fact that the area of one of the triangles can be found by using the formula $A = (b \cdot h) \div 2$.

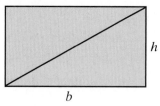

50. Draw a rectangle and choose any point on one side of the rectangle. Draw line segments to the vertices on the opposite side (forming three triangles). Now cut out the two triangles on each end. Place these triangles over the remaining triangle to show that the total of the two areas is equal to the area of the remaining triangle. Do this three different times choosing a different point each time. What fact does this illustrate about the area of a triangle?

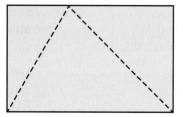

5.4 Circles

Objective A **Circles**

The concepts of perimeter and area were discussed in Sections 5.2 and 5.3. In this section we discuss circles and the formulas for circumference (perimeter) and area that are related to circles.

Important Terms and Definitions for Circles

Circle: The set of all points in a plane that are equidistant from a fixed point called the **center** of the circle.

Radius: The distance from the center of a circle to any point on the circle. (The letter r is used to represent the radius of a circle.)

Diameter: The distance from one point on a circle through the center to the point directly opposite it. (The letter d is used to represent the diameter of a circle.)

Circumference: The perimeter of a circle.

Center

Radius (r)

Diameter
($d = 2r$)

Objective B **Formulas for Circumference and Area of a Circle**

Note carefully that a diameter is twice as long as a radius. That is, $d = 2r$. This relationship leads to two formulas for circumference, one with the diameter and the other with the radius.

Formulas for Circles

For **circumference**: $C = 2\pi r$ and $C = \pi d$

For **area**: $A = \pi r^2$

notes

The Greek letter π (Pi) is the symbol used for the constant 3.1415926535…. This number is an infinite nonrepeating decimal. For our purposes, we will use π = 3.14 (accurate to hundredths). However, you should always be aware that 3.14 is only an approximation for π and that related answers are only approximations.

1. Find the circumference and area of a circle with a radius of 11 meters.

Example 1

Finding the Circumference and Area of a Circle

Find **a.** the circumference and **b.** the area of a circle with a radius of 6 ft.

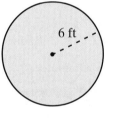

6 ft

Solutions

a. Using the formula for circumference:

$$C = 2\pi r$$
$$C = 2 \cdot 3.14 \cdot 6 = 37.68$$

The circumference is 37.68 ft.

b. Using the formula for area:

$$A = \pi r^2$$
$$A = 3.14 \cdot 6^2 = 3.14 \cdot 36 = 113.04$$

The area is 113.04 ft².

Now work margin exercise **1.**

Example 2

Finding the Circumference and Area of a Circle

Find **a.** the circumference and **b.** the area of a circle with a diameter of 5.2 in.

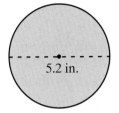

5.2 in.

2. Find the circumference and area of a circle with a diameter of 15 feet.

Solutions

a. Using the formula for circumference:
$$C = \pi d$$
$$C = \pi d = 3.14 \cdot 5.2 = 16.328.$$

The circumference is 16.328 in.

b. Using the formula for area:

$$A = \pi r^2 \quad \left(\text{In this case } r = \frac{1}{2} \cdot 5.2 = 2.6 \right).$$

$$A = 3.14 \cdot (2.6)^2 = 3.14 \cdot 6.76 = 21.2264$$

The area is 21.2264 in.^{2}

Now work margin exercise 2.

Example 3

Finding the Perimeter of a Semicircle

Find the perimeter of a figure that is a **semicircle** (half of a circle) including its diameter. The diameter is 20 cm long.

Solution

3. Find the perimeter of a semicircle which has a diameter of 26 inches.

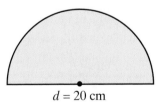

$d = 20$ cm

Now we find the perimeter of the figure by adding the length of the semicircle (which is half of the circumference of a circle) to the length of the diameter.

Length of semicircle: $\frac{1}{2}C = \frac{1}{2}\pi d = \frac{1}{2} \cdot 3.14 \cdot 20 = 3.14 \cdot 10 = 31.4$ cm

Perimeter of figure: $P = 31.4 + 20 = 51.4$ cm

Now work margin exercise 3.

4. Find the area of the shaded portion of the figure below. Round to the nearest hundreth.

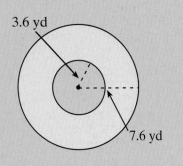

3.6 yd

7.6 yd

Example 4

Finding the Area of a Washer

Find the area of the washer (shaded portion) with dimensions as shown in the figure.

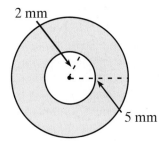

2 mm

5 mm

Solution

Subtract the area of the inside (smaller) circle from the area of the outside (larger) circle.

Larger Circle	**Smaller Circle**
$A = \pi r^2$	$A = \pi r^2$
$A = 3.14\left(5^2\right)$	$A = 3.14\left(2^2\right)$
$\quad = 3.14(25)$	$\quad = 3.14(4)$
$\quad = 78.50\,\text{mm}^2$	$\quad = 12.56\,\text{mm}^2$

Washer

$78.50\,\text{mm}^2$

$-12.56\,\text{mm}^2$

$\overline{\quad 65.94\,\text{mm}^2}$ area of washer

Now work margin exercise 4.

Example 5

Finding the Perimeter and Area

Find **a.** the perimeter and **b.** the area of the figure shown here with a square base and a semicircle cut out of one side. One side of the square is 10 inches long.

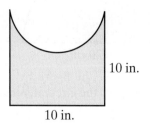

10 in.

10 in.

Solutions

a. The perimeter of the figure is the sum of the lengths of three sides of the square and the length of the semicircle. (Note that the diameter of the semicircle is 10 in.)

Three sides of the square = 10 + 10 + 10 = 30 in.

Length of semicircle = $\dfrac{1}{2}C = \dfrac{1}{2}\pi d = \dfrac{1}{2} \cdot 3.14 \cdot 10 = 15.70$ in.

Perimeter of the figure = 30 + 15.70 = 45.70 in.

b. Find the area of the square then subtract the area of the semicircle.

Area of square = $s^2 = 10^2 = 100$ in.^{2}

Area of semicircle = $\dfrac{1}{2}A = \dfrac{1}{2} \cdot \pi r^2 = \dfrac{1}{2} \cdot 3.14 \cdot 5^2 = 39.25$ in.^{2}

Area of the figure = 100 – 39.25 = 60.75 in.^{2}

Now work margin exercise 5.

Practice Problems

1. Find the circumference and area of a circle with a radius of 3 inches.

2. Find the circumference and area of a circle with a diameter of 14 centimeters.

3. Find the perimeter and area of a semicircle with a diameter of 8 feet.

5. Find **a.** the perimeter and **b.** the area of the figure shown here with a square base and a semicircle cut out of one side. One side of the square is 6 centimeters long.

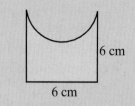

6 cm

6 cm

Practice Problem Answers

1. 18.84 in.; 28.26 in.^{2} **2.** 43.96 cm; 153.86 cm^2 **3.** 20.56 ft; 25.12 ft^2

Exercises 5.4

Find the circumference and area of each circle. (Use π = 3.14.) See Examples 1 and 2.

1. A circle with radius 5 ft.

2. A circle of diameter 1 ft.

3. A circle with radius 70 cm.

4. A circle with diameter 2.4 inches.

5. A circle with diameter 6.2 yd.

6. A circle with radius 0.5 meters.

7. A circle with diameter 14 m.

8. A circle of radius 1 ft.

9. A circle with radius $\frac{3}{4}$ feet.

10. A circle with diameter $12\frac{1}{5}$ miles.

Find the perimeter and area of each figure. (Use π = 3.14.) See Examples 3 through 5.

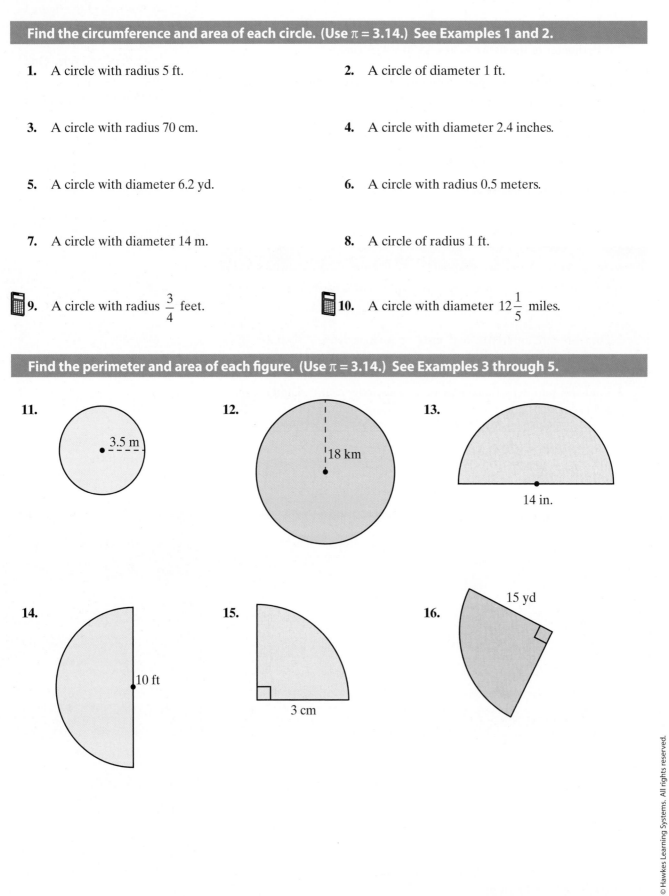

11. 3.5 m

12. 18 km

13. 14 in.

14. 10 ft

15. 3 cm

16. 15 yd

17.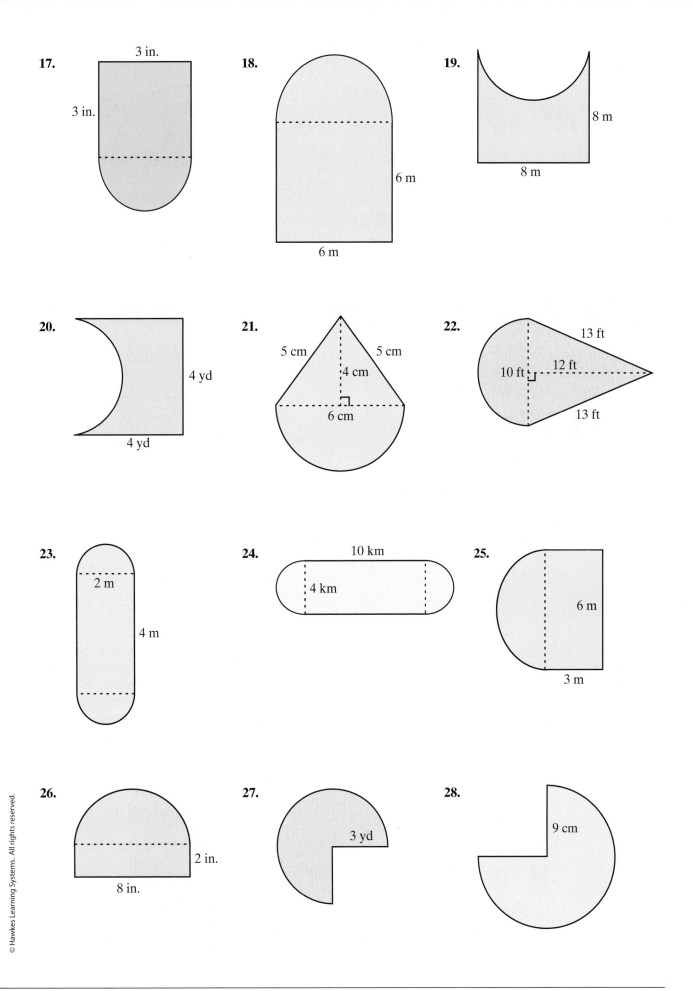

3 in.

3 in.

18.

6 m

6 m

19.

8 m

8 m

20.

4 yd

4 yd

21.

5 cm   5 cm

4 cm

6 cm

22.

13 ft

10 ft   12 ft

13 ft

23.

2 m

4 m

24.

10 km

4 km

25.

6 m

3 m

26.

2 in.

8 in.

27.

3 yd

28.

9 cm

29. Find the area of the shaded portion in the figure.

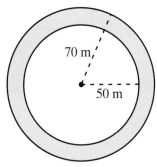

30. Find the area of the shaded portion in the figure.

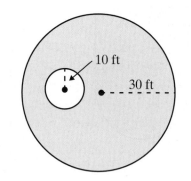

Find the answers to the following word problems. (Use $\pi = 3.14$.)

31. Architecture: An entrance door has a semicircular window which is 24 in. across the bottom, at near the top of the door.

a. Find the perimeter of the window.

b. Find the area of the window.

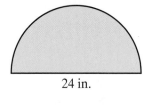

24 in.

32. Pizza: A large 16 in. pizza is cut into eight pieces.

a. What is the perimeter of a single piece?

b. What is the area of this piece of pizza?

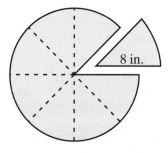

8 in.

33 Traffic: Some cities use traffic circles rather than traffic lights at major intersections. A traffic circle is a circular road at the intersection, where the driver will enter, drive counterclockwise, and then leave on the desired road. If the outer diameter of the circle is 220 feet, and the road is 25 feet wide, what is the surface area of the traffic circle?

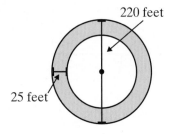

220 feet

25 feet

34. Architecture: An entry door is in the shape of a rectangle, but with the top part being a semicircle as shown below. Compute its area. (**Hint:** To find the height of the rectangular portion of the door, subtract the height (radius) of the semicircle from the total height of the door.)

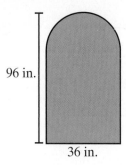

96 in.

36 in.

35. Furniture: One common design for speaker cabinets in high quality stereo systems is the bass reflex cabinet, where there is a port (either rectangular or circular) in the front of the cabinet in addition to the speaker itself. If the front panel is 26 inches high and 18 inches wide, what will the area of the panel be if a 12 inch hole is cut out for the speaker, and a 10 inch by 4 inch rectangular port is cutout as shown below.

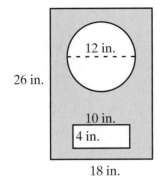

26 in.

12 in.

10 in.

4 in.

18 in.

36. Mechanics: A 10 inch flywheel has a square opening in the center to connect it to the driving shaft. Assuming that this square hole is 1 inch on the side, what is the area of the flywheel?

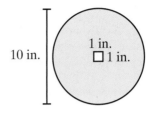

10 in.

1 in.

1 in.

Writing & Thinking

37. With a piece of string and a ruler, carefully measure the circumference and the diameter of several circular figures in your home. Divide each circumference by the length of the corresponding diameter. What result do you observe? Explain this result in terms of a formula.

5.5 Volume and Surface Area

Objectives

A Understand the concept of volume.

B Know the formulas for finding the volume of five geometric solids.

C Understand the concept of surface area.

D Know the formulas for finding the surface area of three geometric solids.

Objective A The Concept of Volume

Volume is the measure of the space enclosed by a three-dimensional figure and is measured in **cubic units**. The concept of volume is illustrated in Figure 1 in terms of cubic inches.

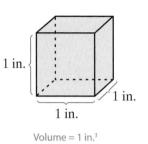

1 in.
1 in.
1 in.
Volume = 1 in.³
(or 1 cubic inch)

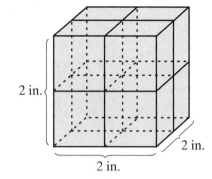

2 in.
2 in.
2 in.

Volume = length × height × width = 2 × 2 × 2 = 8 in.³
There are a total of 8 cubes that are each 1 in.³ for a total of 8 in.³

Figure 1

Some of the basic units of volume and the corresponding abbreviations:

From the metric system

cubic millimeters (mm³)
cubic centimeters (cm³)
cubic meters (m³)

From the U.S. customary system

cubic inches (in.³)
cubic feet (ft³)
cubic yards (yd³)

Objective B Formulas for Volume

Five common geometric solids and the corresponding formulas for calculating the volume of each type of solid are shown here. **Be sure to label your answers with the correct units.**

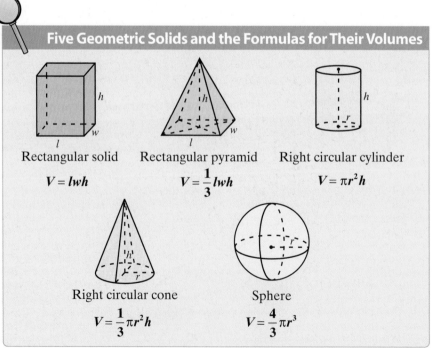

Five Geometric Solids and the Formulas for Their Volumes

Rectangular solid
$V = lwh$

Rectangular pyramid
$V = \dfrac{1}{3}lwh$

Right circular cylinder
$V = \pi r^2 h$

Right circular cone
$V = \dfrac{1}{3}\pi r^2 h$

Sphere
$V = \dfrac{4}{3}\pi r^3$

Example 1

Finding the Volume of a Rectangular Solid

Find the volume of the rectangular solid with length 8 in., width 4 in., and height 1 ft. Write your answer in cubic inches and in cubic feet.

1. Find the volume of the solid below.

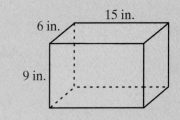

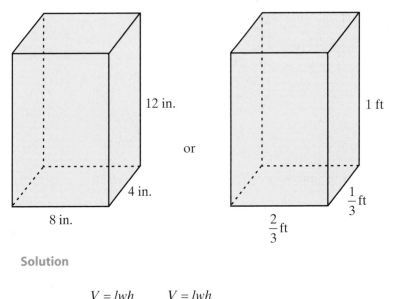

or

Solution

$$V = lwh \qquad V = lwh$$

$$V = 8 \cdot 4 \cdot 12 \qquad V = \frac{2}{3} \cdot \frac{1}{3} \cdot 1$$

$$= 384 \text{ in.}^3 \qquad = \frac{2}{9} \text{ ft}^3$$

Now work margin exercise 1.

Example 2

Finding the Volume of a Sphere

Find the volume of a sphere with radius 9 cm.

2. Find the volume of the sphere below.

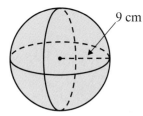

Solution

Using the formula for the volume of a sphere:

$$V = \frac{4}{3} \pi r^3$$

$$V = \frac{4}{3} \cdot 3.14 \cdot 9^3 = \frac{4}{3} \cdot 3.14 \cdot \overset{3}{\cancel{9}} \cdot 9 \cdot 9 = 3052.08 \text{ cm}^3$$

The volume of the sphere is 3052.08 cubic centimeters.

Now work margin exercise 2.

3. Find the volume of the cylinder below.

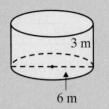

3 m

6 m

Example 3

Finding the Volume of a Cylinder

What is the volume of a cylinder with a height of 10 mm and a circular base with a diameter of 8 mm?

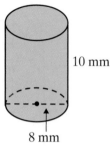

10 mm

8 mm

Solution

We know the diameter of 8 mm but we need the radius.

The radius is half of the diameter: $r = \dfrac{1}{2} \cdot 8 = 4$.

So applying the formula for the volume of a cylinder, we have:

$$V = \pi r^2 h$$

$$V = 3.14 \cdot 4^2 \cdot 10 = 3.14 \cdot 160 = 502.4 \text{ mm}^3$$

The volume of the cylinder is 502.4 cubic millimeters.

Now work margin exercise **3.**

Example 4

Finding the Volume of a Solid

Find the volume of a solid with the dimensions indicated. (Use $\pi = 3.14$.)

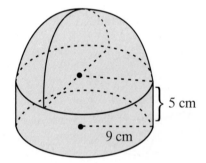

5 cm

9 cm

Solution

On top of the cylinder is a **hemisphere** (one-half of a sphere). Find the volume of the cylinder and hemisphere and add the results.

Cylinder   **Hemisphere**

$V = \pi r^2 h$ $V = \dfrac{1}{2} \cdot \dfrac{4}{3} \pi r^3$ half the volume of a sphere

$V = 3.14(9^2)(5)$ $V = \dfrac{2}{3}(3.14)(9^3)$

$= 1271.7 \text{ cm}^3$ $= 1526.04 \text{ cm}^3$

Total Volume

$$1271.70 \text{ cm}^3$$
$$+1526.04 \text{ cm}^3$$
$$\overline{2797.74 \text{ cm}^3} \quad \text{total volume}$$

Now work margin exercise 4.

4. Find the volume of the solid below.

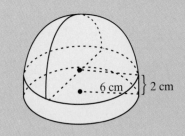

A **cube** is a rectangular solid in which the length, width, and height are all equal. If s is the length of one edge, the formula for the volume of the cube is $V = s^3$.

Example 5

Finding the Volume of a Cube

Find the volume of a cube with $s = 3$ yards in both cubic yards and cubic feet. (1 yard = 3 feet).

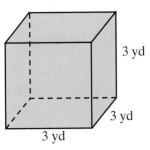

3 yd

3 yd

3 yd

Solution

We apply the formula $V = s^3$ twice; once by using $s = 3$ yards and once by using $s = 9$ feet.

$$V = s^3 \qquad\qquad V = s^3$$
$$V = 3^3 = 27 \text{ yd}^3 \quad V = 9^3 = 729 \text{ ft}^3$$

Notice that because feet are smaller than yards, there are many more cubic feet than cubic yards in the cube.

Now work margin exercise 5.

5. Find the volume of the cube below.

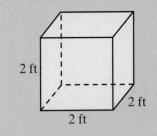

2 ft

2 ft

2 ft

Objective C **The Concept of Surface Area**

The **surface area** (*SA*) of a geometric solid is a measure of the area of the outside surface in square units. Surface area is particularly important in fields related to building and construction. The surface area of a rectangular solid is found by adding the areas of the six sides. Each side is a rectangle. The front and back are the same. The left and right sides are the same. The top and bottom are the same. If the edges are labeled as *l*, *w*, and *h*, we see that the total surface area can be calculated by using the following formula:

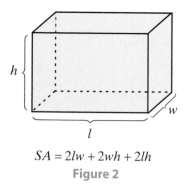

$$SA = 2lw + 2wh + 2lh$$
Figure 2

Objective D **Formulas for Surface Area**

Cylinders and spheres are two other solids where surface area can be an important consideration. The cost of building and maintaining the surface area of such figures can be a determining factor for an architect in using a particular shape in designing structures. The formulas for surface areas of three solids are given here.

Be sure to label your answers with square units.

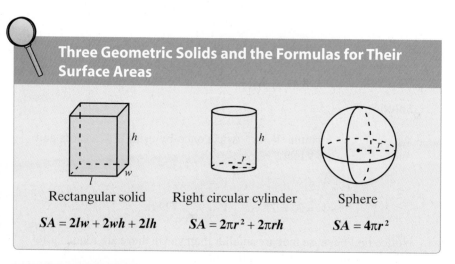

Three Geometric Solids and the Formulas for Their Surface Areas

Rectangular solid	Right circular cylinder	Sphere
$SA = 2lw + 2wh + 2lh$	$SA = 2\pi r^2 + 2\pi rh$	$SA = 4\pi r^2$

Example 6

Finding the Surface Area of a Rectangular Solid

A cereal box in the shape of a rectangular solid has the following dimensions:

$$l = 30 \text{ cm}, \qquad w = 10 \text{ cm}, \qquad h = 40 \text{ cm}$$

Find the surface area of the box.

Solution

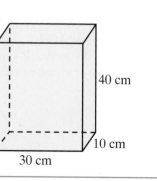

$$SA = 2lw + 2wh + 2lh$$
$$SA = 2 \cdot 30 \cdot 10 + 2 \cdot 10 \cdot 40 + 2 \cdot 30 \cdot 40$$
$$= 600 + 800 + 2400$$
$$= 3800 \text{ cm}^2$$

Example 7

Finding the Surface Area of a Right Circular Cylinder

A coffee can in the shape of a cylinder has the following dimensions:

$$r = 2 \text{ in.}, \qquad h = 5 \text{ in.}$$

Find the surface area of the can. (Use $\pi = 3.14$.)

Solution

$$SA = 2\pi r^2 + 2\pi rh$$

$$SA = 2 \cdot 3.14 \cdot 2^2 + 2 \cdot 3.14 \cdot 2 \cdot 5$$

$$= 25.12 + 62.8$$

$$= 87.92 \text{ in.}^2$$

5 in.

2 in.

Now work margin exercises 6 and 7.

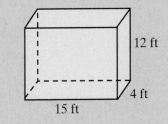

6. Find the surface area of the rectangular solid below.

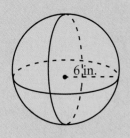

12 ft

4 ft

15 ft

7. Find the surface area of a sphere with radius 6 in.

6 in.

Practice Problems

1. Find the volume of a rectangular pyramid with length 4.5 cm, width 3.2 cm, and height 1.6 cm.

2. A right circular cone has a height of 12 ft and a circular base with radius 6 ft. What is the volume of the cone?

3. A ball in the shape of a sphere has a diameter of 18 in. Air is blown into the ball until it has a new diameter of 20 in. What is the change in the volume of the ball? Round your answer to the hundredth place.

Practice Problem Answers

1. 7.68 cm^3 **2.** 452.16 ft^3 **3.** 1134.59 in.^{3}

Exercises 5.5

Match each of the following formulas with its corresponding geometric figure.

1. _____ **a.** rectangular solid   **A.** $V = \dfrac{4}{3}\pi r^3$

   _____ **b.** rectangular pyramid   **B.** $V = \dfrac{1}{3}\pi r^2 h$

   _____ **c.** right circular cylinder   **C.** $V = lwh$

   _____ **d.** right circular cone   **D.** $V = \pi r^2 h$

   _____ **e.** sphere   **E.** $V = \dfrac{1}{3}lwh$

Find the volume of each of the following solids in a convenient unit. See Examples 1 through 5. (Use $\pi = 3.14$.)

2. A rectangular solid with length 5 in., width 2 in., and height 7 in.

3. A right circular cylinder 15 in. high and 1 ft in diameter.

4. A sphere with radius 4.5 cm.

5. A sphere with diameter 12 ft.

6. A right circular cone 3 mm high with a 2 mm radius.

7. A rectangular pyramid with length 8 cm, width 1 cm, and height 30 cm.

8.
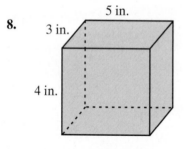
5 in.
3 in.
4 in.

9.
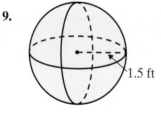
1.5 ft

10.
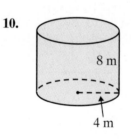
8 m
4 m

11.
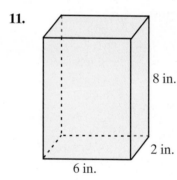
8 in.
2 in.
6 in.

12.
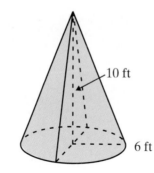
10 ft
6 ft

13.
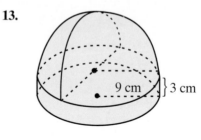
9 cm
3 cm

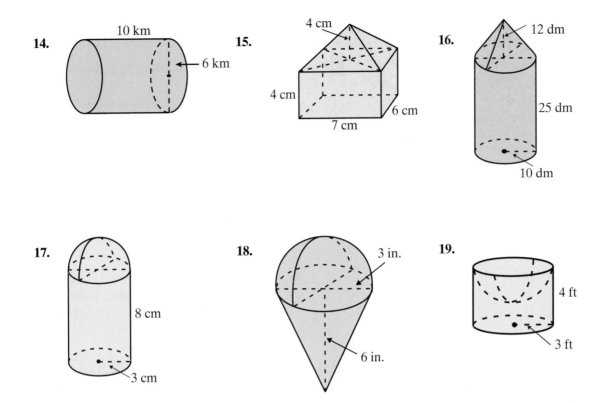

14. 10 km, 6 km

15. 4 cm, 4 cm, 7 cm, 6 cm

16. 12 dm, 25 dm, 10 dm

17. 8 cm, 3 cm

18. 3 in., 6 in.

19. 4 ft, 3 ft

Find the surface area of each of the following solids in a convenient unit. See Examples 6 and 7. (Use π = 3.14.)

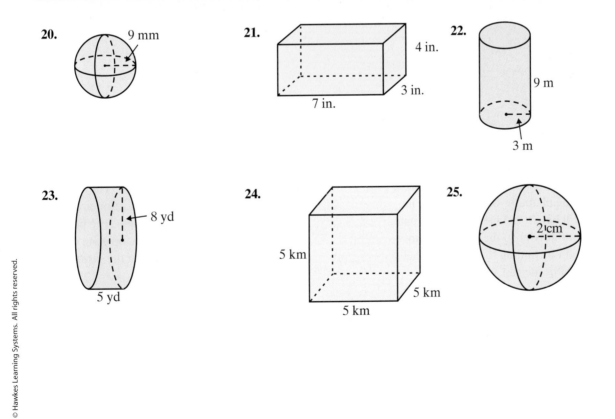

20. 9 mm

21. 4 in., 3 in., 7 in.

22. 9 m, 3 m

23. 8 yd, 5 yd

24. 5 km, 5 km, 5 km

25. 2 cm

26. Pyramids: The Great Pyramid of Egypt, which is located in Giza, has a square base of 231 m on each side, and its height is 146 m. What is its volume? **Source:** http://www.timstouse. com/EarthHistory/Egypt/GreatPyramid/interestingfacts. html

27. Manufacturing: A standard 55 gallon round steel drum is about 23 in. in diameter and 34.5 in. high. Assuming that the drum is totally enclosed, what is its surface area? **Source:** http://www.yankeecontainers.com/Steel-Drums-Closed-Top.html

28. Storage: A specially lined storage chest has the inside dimensions of 5 ft long, 3 ft wide, and 2 ft high.

  a. What is its volume?

  b. What is the surface area of the interior?

29. Furniture: A cubic hassock is 1.5 ft long in each direction.

  a. What is the volume of the hassock?

  b. How many square feet of material are necessary to cover the hassock? Assume the bottom is also being covered in the material.

30. Going to the Beach: A group of college students went to the beach and inflated their 2-ft spherical beach ball, whose radius is 1 ft.

  a. What is the volume of the ball? Round your answer to the nearest hundredth.

  b. What is the surface area of the ball?

31. Ice Cream: A 6 in. tall ice cream cone is filled solid with ice cream where the final scoop of ice cream forms a perfect hemisphere above the top of the cone. What is the total volume of ice cream in the cone if the top of the cone has a 2.2 in. opening? Round your answer to the nearest hundredth.

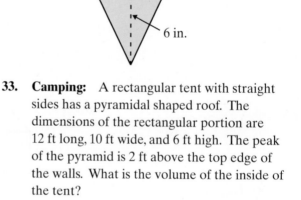

32. Trash Cans: A cylindrical trash can has a hemispherical top (with a trap door for the trash). If the diameter of the can is 16 in., and its total height is 38 in., find its volume. Round your answer to the nearest tenth. (**Hint:** Begin by finding the height of the straight part of the can.)

33. Camping: A rectangular tent with straight sides has a pyramidal shaped roof. The dimensions of the rectangular portion are 12 ft long, 10 ft wide, and 6 ft high. The peak of the pyramid is 2 ft above the top edge of the walls. What is the volume of the inside of the tent?

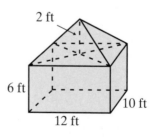

5.6 Triangles

Objective A **Triangles Classified by Sides**

A **triangle** consists of the three line segments that join three points that do not lie on a straight line (not collinear). The line segments are called the **sides** of the triangle and the points are called the **vertices** of the triangle. If the points are labeled A, B, and C, the triangle is symbolized $\triangle ABC$. See Figure 1.)

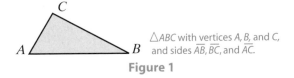

△ABC with vertices A, B, and C, and sides $\overline{AB}$, $\overline{BC}$, and $\overline{AC}$.

Figure 1

The sides of a triangle are said to determine three angles, and these angles are labeled by the vertices. Thus the angles of $\triangle ABC$ are $\angle A$, $\angle B$, and $\angle C$. (Since the definition of an angle involves rays, we can think of the sides of the triangles extended as rays that form these angles.)

Triangles are classified according to the lengths of their sides and according to the measures of their angles.

Note: The line segment with endpoints A and B is indicated by placing a bar over the letters, as in $\overline{AB}$. The length of the segment is indicated by writing only the letters, as in AB.

Triangles Classified by Sides

(**Note:** In the figures, sides with equal length are indicated by the same number of slash marks.)

Name	Property	Example
1. Scalene	No two sides are equal. 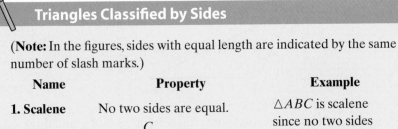	$\triangle ABC$ is scalene since no two sides have equal lengths.
2. Isosceles	At least two sides have equal lengths. 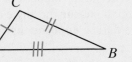	$\triangle PQR$ is isosceles since $PR = QR$.
3. Equilateral	All three sides have equal lengths. 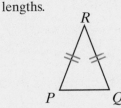	$\triangle XYZ$ is equilateral since $XY = XZ = YZ$

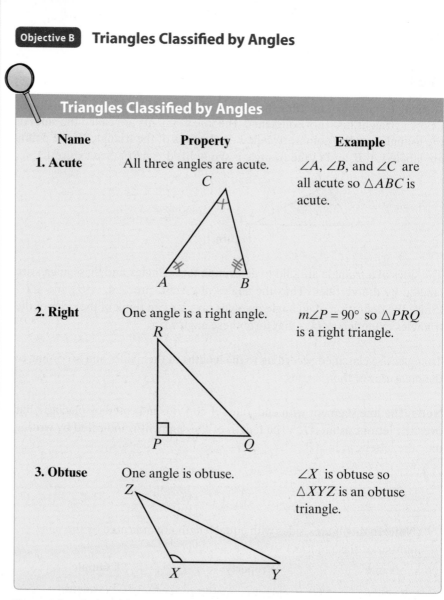

Name	Property	Example
1. Acute	All three angles are acute.	$\angle A$, $\angle B$, and $\angle C$ are all acute so $\triangle ABC$ is acute.
2. Right	One angle is a right angle.	$m\angle P = 90°$ so $\triangle PRQ$ is a right triangle.
3. Obtuse	One angle is obtuse.	$\angle X$ is obtuse so $\triangle XYZ$ is an obtuse triangle.

Every triangle is said to have six parts – namely, three angles and three sides. Two sides of a triangle are said to **include** the angle at their common endpoint or vertex. The third side is said to be **opposite** this angle.

The sides in a right triangle have special names. The longest side, opposite the right angle, is called the **hypotenuse**, and the other two sides are called **legs**. (See Figure 2.)

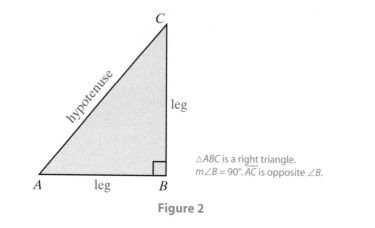

$\triangle ABC$ is a right triangle.
$m\angle B = 90°$. $\overline{AC}$ is opposite $\angle B$.

Figure 2

Three Important Statements about any Triangle

1. The sum of the measures of the angles is 180°.
2. The sum of the lengths of any two sides must be greater than the length of the third side.
3. Longer sides are opposite angles with larger measures.

Example 1

Classifications of Triangles

In △ABC below, AB = AC. What kind of triangle is △ABC ?

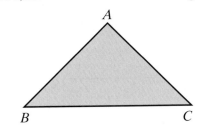

Solution

△ABC is isosceles because two sides have equal lengths.

Now work margin exercise 1.

Example 2

Possible Triangles

△PQR was drawn and then the lengths of the sides were labeled as shown in the figure. Explain why the labels cannot be correct.

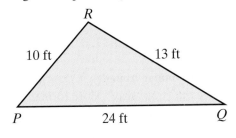

Solution

The labels cannot be correct because PR + QR = 10 ft + 13 ft = 23 ft and PQ = 24 ft, which is greater than the sum of the other two sides. In a triangle, the sum of the lengths of any two sides must be greater than the length of the third side.

Now work margin exercise 2.

1. In △DEF, DE = 5, EF = 7, and DF = 8. What kind of triangle is △DEF ?

2. Is it possible to have a triangle with sides of length 15 in., 37 in. and 51 in.?

3. Now assume that in
$\triangle BOR$, $m\angle B = 35°$ and
$m\angle O = 55°$. Answer the
questions **a. – e.** as given
in Example 3.

Example 3

Finding Angles and Sides

In $\triangle BOR$ below, $m\angle B = 50°$ and $m\angle O = 70°$.

a. What is $m\angle R$?

b. What kind of triangle is $\triangle BOR$?

c. Which side is opposite $\angle R$?

d. Which sides include $\angle R$?

e. Is $\triangle BOR$ a right triangle? Why or why not?

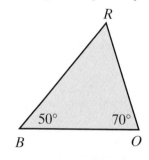

Solutions

a. The sum of the measures of the angles must be 180°.
Since $50° + 70° = 120°$, $m\angle R = 180° - 120° = 60°$

b. $\triangle BOR$ is an acute triangle since all the angles are acute. Also,
$\triangle BOR$ is scalene because no two sides are equal.

c. $\overline{BO}$ is opposite $\angle R$.

d. $\overline{RB}$ and $\overline{RO}$ include $\angle R$.

e. $\triangle BOR$ is not a right triangle because none of the angles are right
angles.

Now work margin exercise 3.

Objective C **Similar Triangles**

Two triangles are said to be **similar triangles** if they have the same "shape."
They may or may not have the same "size." More formally, if two triangles are
similar, they have the following two properties.

Similar Triangles

1. In similar triangles, the **corresponding angles have the same measure**.
    (We say that the corresponding angles are congruent.)
2. In similar triangles, the **lengths of the corresponding sides are
    proportional**. (See Figure 3.)

In similar triangles, **corresponding sides** are those sides opposite the congruent angles (angles with the same measure) in the respective triangles. (See Figure 3 and the following discussion.)

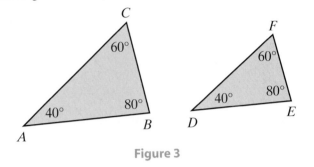

Figure 3

For the similar triangles in Figure 3, the following relationships are true.

$\angle A$ corresponds to $\angle D$; $m\angle A = m\angle D$.

$\angle B$ corresponds to $\angle E$; $m\angle B = m\angle E$.

$\angle C$ corresponds to $\angle F$; $m\angle C = m\angle F$.

The corresponding angles are congruent.

$\overline{AB}$ corresponds to $\overline{DE}$.

$\overline{BC}$ corresponds to $\overline{EF}$.

$\overline{AC}$ corresponds to $\overline{DF}$.

$$\frac{AB}{DE} = \frac{BC}{EF} = \frac{AC}{DF}$$

The lengths of corresponding sides are proportional.

We write $\triangle ABC \sim \triangle DEF$. **(~ is read "is similar to")**

Note: The notation for similar triangles indicates the respective correspondences of angles and sides by the order in which the vertices of each triangle are identified. For example, we could have written $\triangle BCA \sim \triangle EFD$ or $\triangle CAB \sim \triangle FDE$. Be sure to follow this pattern when indicating similar triangles.

Example 4

Similarity of Triangles

Given the two triangles $\triangle ABC$ and $\triangle AXY$ with

$m\angle ABC = m\angle AXY = 90°$ (as shown in the figure)

determine whether or not $\triangle ABC \sim \triangle AXY$.

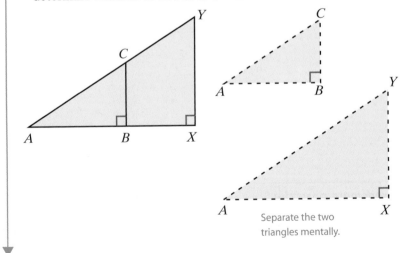

Separate the two triangles mentally.

4. Which of the following statements is correct in reference to the figure given in Example 4?

a. $\triangle BAC \sim \triangle XYA$

b. $\triangle CBA \sim \triangle YXA$

Solution

We can show that the corresponding angles are congruent as follows:

$m\angle CAB = m\angle YAX$ because they are the same angle.

$m\angle CBA = m\angle YXA$ because both are right angles (90°).

$m\angle BCA = m\angle XYA$ because the sum of the measures of the angles in each triangle must be 180°.

Therefore the corresponding angles are congruent, and the triangles are similar.

Now work margin exercise 4.

5. Using the information given in Examples 4 and 5, and given that $AC = 5$ cm, find CY.

Example 5

Finding the Length of a Side of a Triangle

Refer to the figure below. If $AB = 4$ centimeters, $BX = 2$ centimeters, and $BC = 3$ centimeters, find XY.

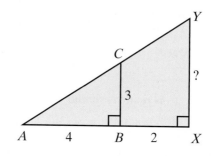

Solution

From Example 4 we know that the two triangles are similar; therefore, the lengths of their corresponding sides are proportional. Since $\overline{AB}$ and $\overline{AX}$ are corresponding sides (they are opposite congruent angles) and $\overline{BC}$ and $\overline{XY}$ are corresponding sides (they are opposite congruent angles), the following proportion is true:

$$\frac{AB}{AX} = \frac{BC}{XY}$$

Note that, $AX = AB + BX = 4 + 2 = 6$ cm.

Thus,

$$\frac{4 \text{ cm}}{6 \text{ cm}} = \frac{3 \text{ cm}}{XY}$$

$$4 \cdot XY = 3 \cdot 6$$

$$\frac{\cancel{4} \cdot XY}{\cancel{4}} = \frac{18}{4}$$

$$XY = 4.5 \text{ cm}$$

Now work margin exercise 5.

Congruent Triangles

If two triangles have the same "shape" and the same "size", they are congruent. Two congruent triangles have the following two properties.

Two Triangles are Congruent If:

1. The corresponding angles have the same measure.
2. The lengths of corresponding sides are equal.

In Figure 4 triangles *ABC* and *DEF* are congruent.

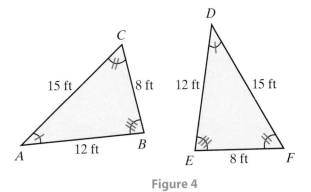

Figure 4

We write $\triangle ABC \cong \triangle DEF$. (Remember $\cong$ is read **"is congruent to"**)

Note that the corresponding angles are congruent. That is, $m\angle A = m\angle D$, $m\angle B = m\angle E$, and $m\angle C = m\angle F$. And, corresponding sides have the same lengths.

The following three properties can be used to determine whether two triangles are congruent.

Determining Congruent Triangles

1. **Side-Side-Side (SSS)**
    If two triangles are such that the lengths of the three sides of one triangle are equal to the lengths of corresponding sides of the other triangle, then the two triangles are congruent.

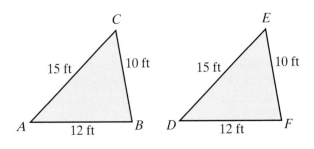

These two triangles are congruent by SSS.

2. **Side-Angle-Side (SAS)**

   If two triangles are such that the lengths of two sides of one triangle equal the lengths of corresponding sides of the other triangle and the angles included between the sides are congruent, then the two triangles are congruent.

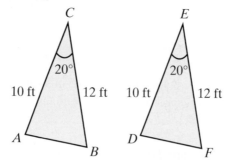

   These two triangles are congruent by SAS.

3. **Angle-Side-Angle (ASA)**

   If two triangles are such that two angles in one triangle are congruent to two angles in the other triangle and the lengths of the included sides are equal, then the two triangles are congruent.

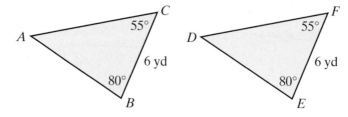

   These two triangles are congruent by ASA.

6. Determine whether triangles *JKL* and *MNO* are congruent. Explain.

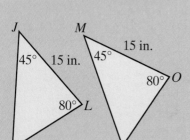

Example 6

Determining Congruent Triangles

Determine whether triangles *PQR* and *MNO* are congruent.

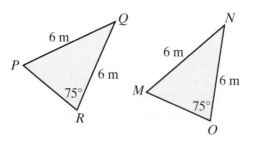

Solution

Because the lengths of two sides of one triangle equal the lengths of two sides of the other triangle and the angles included between the two sides are congruent, the two triangles are congruent by SAS (Side-Angle-Side).

Now work margin exercise 6.

Practice Problems

Determine whether it is possible for a triangle to have the following side lengths. If such a triangle exists, classify it as scalene, isosceles, or equilateral.

1. 8 in., 13 in., 21 in.

2. 32 mm, 5 mm, 32 mm

3. 3 ft, 4 ft, 5 ft

4. 16 cm, 16 cm, 16 cm

5. Given $\triangle ABC \sim \triangle JKL$. Find the value of x.

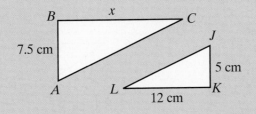

6. Determine whether the two triangles are congruent. Explain.

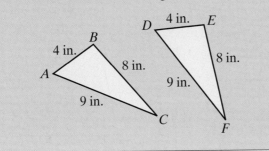

Practice Problem Answers

1. no

2. yes; isosceles

3. yes; scalene

4. yes; equilateral

5. 18

6. They are congruent because of SSS.

Exercises 5.6

Name each of the following triangles in the most precise way possible, given the indicated measures of angles and lengths of sides. See Examples 1 through 3.

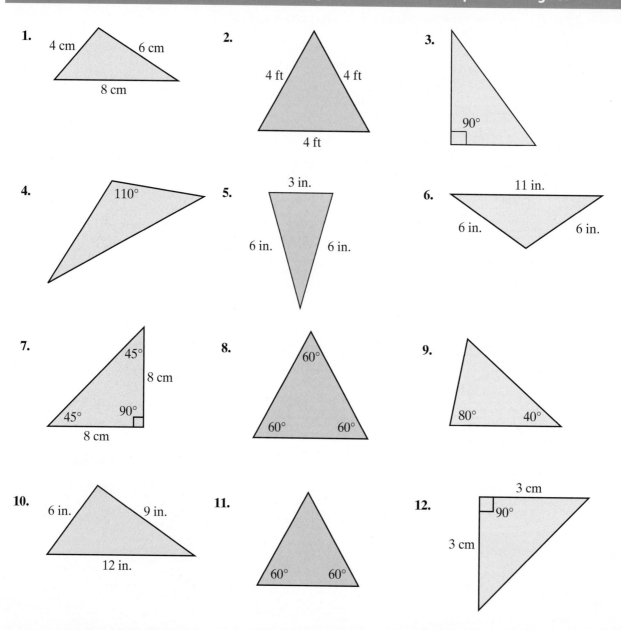

1.
4 cm
6 cm
8 cm

2.
4 ft
4 ft
4 ft

3.
90°

4.
110°

5.
3 in.
6 in.
6 in.

6.
11 in.
6 in.
6 in.

7.
45°
8 cm
45°
90°
8 cm

8.
60°
60°
60°

9.
80°
40°

10.
6 in.
9 in.
12 in.

11.
60°
60°

12.
3 cm
90°
3 cm

Determine whether each pair of triangles is similar. If the triangles are similar, explain why and indicate the similarity by using the ~ symbol. See Example 4.

13.
A
4
3
B   2   C

X
4   4
Y   2   Z

14.
F
5   4
H   3   G

L
3.5   3
N   2   M

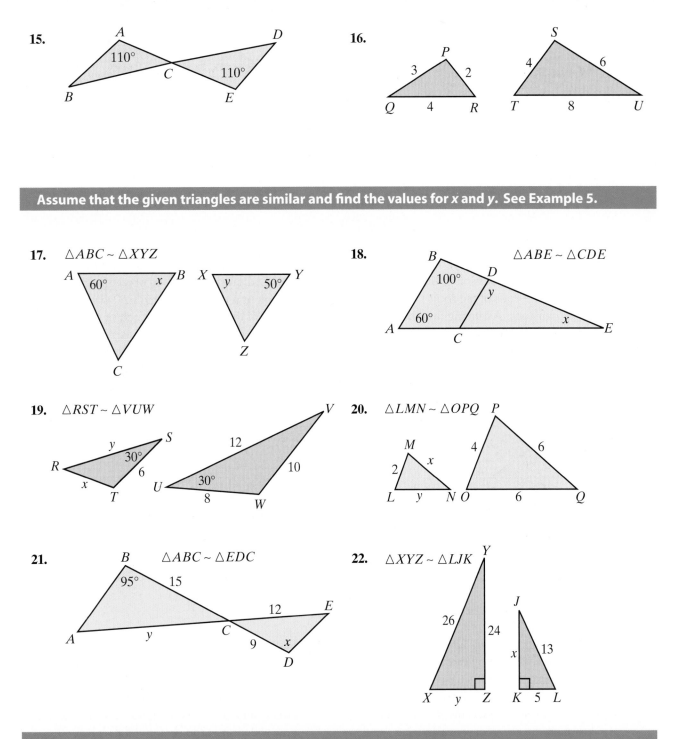

15.

16.

Assume that the given triangles are similar and find the values for *x* and *y*. See Example 5.

17. $\triangle ABC \sim \triangle XYZ$

18. $\triangle ABE \sim \triangle CDE$

19. $\triangle RST \sim \triangle VUW$

20. $\triangle LMN \sim \triangle OPQ$

21. $\triangle ABC \sim \triangle EDC$

22. $\triangle XYZ \sim \triangle LJK$

Determine whether each pair of triangles congruent. If they are congruent, state the property that makes them congruent. See Example 6.

23.

24.

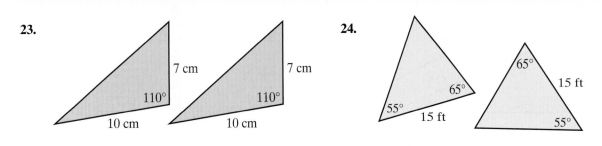

25.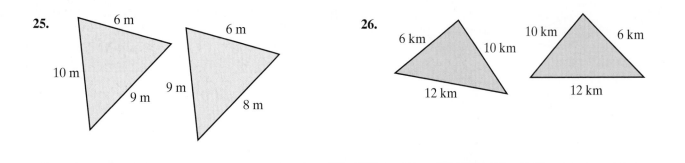

26.

Solve the following word problems.

27. Construction: A child's playhouse is a built to look like a smaller version of the family house, where the ends of the roofs have similar proportions. Assuming that the width of the main house (*AB*) is 32 feet, and the sloped height of the roof for one of the sides is 20 feet, what is the sloped height of the side (*DE*) of the playhouse roof, if the width of the playhouse (*DF*) is 12 feet?

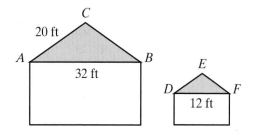

28. Cameras: A camera uses a lens which will look at a properly focused object (such as a person or a tree), and then display an inverted image of this object on a screen or film which is on the opposite of the lens as shown in the figure below. Notice that this diagram forms two similar right triangles. If a picture of a 50 foot tall building (*AB*), which is 150 feet from the lens (*AC*) is photographed, how high is the image (*DE*) if the film on the opposite side (*CD*) is 0.3 inch from the lens?

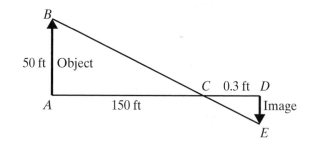

29. Measuring Height: A hill has a straight sloping surface which is 500 meters long before it levels out. If the surface has risen 8 meters when you are 25 meters from the base of the hill, how high is the peak? **Hint:** As shown on the accompanying drawing, there are two similar triangles, △*ABC*, and △*AED*. Solve for *DE*.

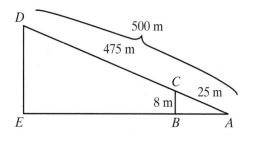

30. Sailing: A sloop is a sailboat which has two similar triangle sails on a single mast. If the smaller sail is 12 feet along the mast (*CB*), and 5 feet along its bottom (*AC*), and the larger sail is 16.5 feet along the mast (*ZY*), how wide is it at the bottom (*XZ*)? Round your answer to the nearest tenth.

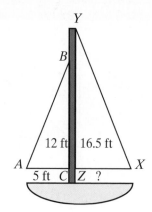

31. In $\triangle XYZ$ and $\triangle UVW$ below, $m\angle Z = 30°$ and $m\angle W = 30°$.

    a. If both triangles are isosceles, what are the measures of the other four angles? (In an isosceles triangle, the angles opposite the equal sides must be congruent.)

    b. Are the triangles similar? Explain.

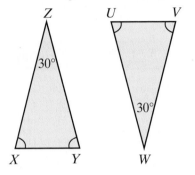

32. Given $\triangle ABC$ and $\triangle DEF$ shown below

    a. Are the triangles similar? Explain.

    b. If so, which angles are congruent?°

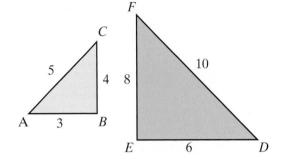

Writing & Thinking

33.   **a.** Is it possible to form a triangle, $\triangle STV$, if $ST = 12$ centimeters, $TV = 9$ centimeters, and $SV = 15$ centimeters? Explain your reasoning.

    b. If so, what kind of triangle is $\triangle STV$?

    c. Use a ruler to draw this triangle, if possible.

34.   **a.** Is it possible to form a triangle, $\triangle ABC$ if $AB = 6$ inches, $BC = 8$ inches, and $CA = 14$ inches? Explain your reasoning.

    b. Use a ruler to draw this triangle, if possible.

5.7 Square Roots and the Pythagorean Theorem

A Understand and calculate square roots.

B Understand the Pythagorean Theorem.

Objective A Square Roots

A number is **squared** when it is multiplied by itself. For example,

$$7^2 = 7 \cdot 7 = 49$$

$$\text{and } 10^2 = 10 \cdot 10 = 100.$$

The result is called a **perfect square**. Thus, 49 and 100 are perfect squares. Table 1 shows the perfect squares found by squaring the whole numbers from 1 to 20.

Squares of Whole Numbers from 1 to 20				
$1^2 = 1$	$5^2 = 25$	$9^2 = 81$	$13^2 = 169$	$17^2 = 289$
$2^2 = 4$	$6^2 = 36$	$10^2 = 100$	$14^2 = 196$	$18^2 = 324$
$3^2 = 9$	$7^2 = 49$	$11^2 = 121$	$15^2 = 225$	$19^2 = 361$
$4^2 = 16$	$8^2 = 64$	$12^2 = 144$	$16^2 = 256$	$20^2 = 400$

Table 1

The square of 5 is $5^2 = 25$, and the number 5 is called the **square root** of 25. We write

$$\sqrt{25} = 5. \quad \text{(Read "the square root of 25 is 5.")}$$

Similarly,

$$\sqrt{49} = 7, \quad \sqrt{100} = 10, \quad \sqrt{1} = 1, \quad \text{and} \quad \sqrt{0} = 0.$$

Terminology of Radicals

The symbol $\sqrt{}$ is called a **radical sign**.

The number under the radical sign is called the **radicand**.

The complete expression, such as $\sqrt{49}$, is called a **radical** or **radical expression**.

Table 2 contains the square roots of the perfect square numbers from 1 to 400 (the squares of the numbers from 1 to 20). Note that this table is just another way of looking at Table 1.

Square Roots of Perfect Squares from 1 to 400				
$\sqrt{1} = 1$	$\sqrt{25} = 5$	$\sqrt{81} = 9$	$\sqrt{169} = 13$	$\sqrt{289} = 17$
$\sqrt{4} = 2$	$\sqrt{36} = 6$	$\sqrt{100} = 10$	$\sqrt{196} = 14$	$\sqrt{324} = 18$
$\sqrt{9} = 3$	$\sqrt{49} = 7$	$\sqrt{121} = 11$	$\sqrt{225} = 15$	$\sqrt{361} = 19$
$\sqrt{16} = 4$	$\sqrt{64} = 8$	$\sqrt{144} = 12$	$\sqrt{256} = 16$	$\sqrt{400} = 20$

Table 2

Example 1

Calculating Squares and Square Roots

Use your memory of the results in Tables 1 and 2 to answer the following.

a. 15^2 **b.** 11^2 **c.** $\sqrt{256}$ **d.** $\sqrt{81}$

Solutions

a. $15^2 = 225$ **b.** $11^2 = 121$ **c.** $\sqrt{256} = 16$ **d.** $\sqrt{81} = 9$

Now work margin exercise 1.

1. Find the following.

  a. 18^2

  b. $\sqrt{169}$

Most numbers are not perfect squares and the square roots of these numbers are not found as easily as those in Example 1 and in Table 2. In fact, most square roots are **irrational numbers (infinite nonrepeating decimals)**; that is, most square roots can be only approximated with decimals.

Decimal approximations of $\sqrt{2}$ are shown here. Note that we are looking for a decimal number whose square is 2.

$$
\begin{array}{llll}
1.4 & 1.414 & 1.41421 & 1.41422 \\
\times\ 1.4 & \times\ 1.414 & \times\ 1.41421 & \times\ 1.41422 \\
\hline
56 & 5656 & 141421 & 282844 \\
1\,4 & 1414 & 282842 & 282844 \\
\hline
1.96 & 5656 & 565684 & 565688 \\
& 1\,414 & 141421 & 141222 \\
\hline
& 1.999396 & 565684 & 565688 \\
& & 1\,41421 & 1\,41422 \\
\hline
& & 1.9999899241 & 2.0000182084 \\
\end{array}
$$

So, we see that $\sqrt{2}$ is between 1.41421 and 1.41422.

Examples 2 and 3 show how to find square roots (or approximate square roots) by using a calculator.

Example 2

Calculating Square Roots Using a Calculator

Find $\sqrt{2}$ rounded to the nearest thousandth by using a calculator.

Solution

To find the square root of a number, you will need to find the square root key, $\boxed{\sqrt{\ }}$, and the equal sign key, $\boxed{=}$.
Therefore, press $\boxed{\sqrt{\ }}$ followed by $\boxed{2}$. Then press $\boxed{=}$.
The display will read 1.414213562.

Rounding to the nearest thousandth, we will have 1.414.

Use a calculator to find the following square roots accurate to four decimal places.

2. $\sqrt{5}$

3. $\sqrt{7.5}$

Example 3

Calculating Square Roots using a Calculator

Find $\sqrt{18}$ rounded to the nearest thousandth by using a calculator.

Solution

Following the steps from Example 2, press $\boxed{\sqrt{}}$ followed by $\boxed{1}$ $\boxed{8}$. Then press $\boxed{=}$. The display will read 4.242640687.

Rounding to the nearest thousandth, we will have 4.243.

Now work margin exercises 2 and 3.

Objective B The Pythagorean Theorem

The following discussion involving right triangles serves as an application of squares and square roots and uses the terms defined below.

Right triangle: A triangle containing a right angle $(90°)$.

Hypotenuse: The longest side of a right triangle; the side opposite the right angle.

Leg: Each of the other two sides of a right triangle.

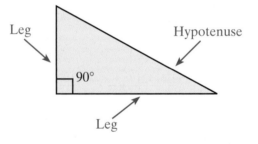

Pythagoras (c. 585 – 501 B.C.), a famous Greek mathematician, is given credit for discovering the following theorem (although historians have found that the facts of the theorem were known before the time of Pythagoras).

The Pythagorean Theorem

In a right triangle, the square of the length of the hypotenuse is equal to the sum of the squares of the lengths of the two legs:

$$c^2 = a^2 + b^2.$$

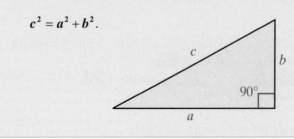

Example 4

Right Triangles

Show that a triangle with sides of lengths 3 inches, 4 inches, and 5 inches must be a right triangle.

Solution

If the triangle is a right triangle, then its three sides must satisfy the property stated in the Pythagorean Theorem: $c^2 = a^2 + b^2$. Or, in this case, $5^2 = 3^2 + 4^2$. Since $25 = 9 + 16$ is a true statement, the triangle is a right triangle.

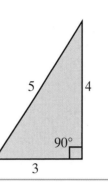

Now work margin exercise 4.

4. Is a triangle with sides of length 4 in., 7 in. and 8 in. a right triangle?

Example 5

Finding the Length of the Hypotenuse

Find the length of the hypotenuse of a right triangle with legs of length 12 cm and 5 cm.

Solution

Let c = the length of the hypotenuse.
Now, by the Pythagorean Theorem,

$$c^2 = 12^2 + 5^2$$
$$c^2 = 144 + 25$$
$$c^2 = 169$$
$$c = \sqrt{169} = 13$$

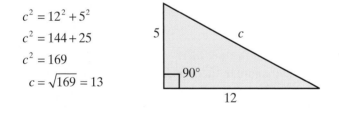

The length of the hypotenuse is 13 centimeters.

Now work margin exercise 5.

5. Find the length of the hypotenuse of a right triangle with legs of length 8 cm and 15 cm.

Rarely do all three sides of a right triangle have whole number values. Examples 6 and 7 illustrate right triangles with irrational numbers as the lengths of the hypotenuses.

6. If the legs of a right triangle measure 8 yards and 12 yards, find the length of the hypotenuse.

Example 6

Finding the Length of the Hypotenuse

Find the length of the hypotenuse of a right triangle in which both legs have a length of 1 meter.

Solution

$$c^2 = 1^2 + 1^2$$
$$c^2 = 1 + 1 = 2$$
$$c = \sqrt{2}$$

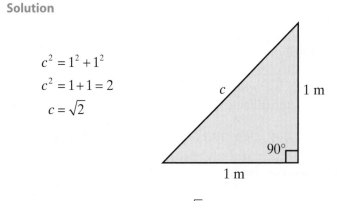

The length of the hypotenuse is $\sqrt{2}$ meters (or about 1.41 meters).

Now work margin exercise 6.

7. A 35-foot-tall tree casts a shadow which is 30 ft long. Find the distance from the top of the tree to the end of its shadow.

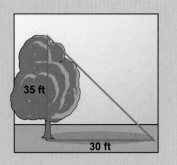

Example 7

Finding the Length of the Hypotenuse

A guy wire is attached to the top of a telephone pole and anchored to the ground 10 feet from the base of the pole. If the pole is 20 feet high, what is the length of the guy wire?

Solution

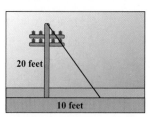

Let x = the length of the guy wire. Then, by the Pythagorean theorem,

$$x^2 = 10^2 + 20^2$$
$$x^2 = 100 + 400$$
$$x^2 = 500$$
$$x = \sqrt{500} \approx 22.36$$

The guy wire is about 22.36 feet long.

Now work margin exercise 7.

Practice Problems

Find the following values using your memory of Tables 1 and 2.

1. $\sqrt{324}$ **2.** 8^2

3. 19^2 **4.** $\sqrt{121}$

Use a calculator to find the following values accurate to four decimal places.

5. $\sqrt{89}$ **6.** $\sqrt{17}$

Determine whether triangles with the following side lengths are right triangles.

7. 8 cm, 10 cm, 13 cm   **8.** 40 in., 41 in., 9 in.

Practice Problem Answers

1. 18   **2.** 64
3. 361   **4.** 11
5. 9.4340   **6.** 4.1231
7. no; $13^2 \neq 8^2 + 10^2$ **8.** yes; $41^2 = 40^2 + 9^2$

Exercises 5.7

Use your memory of the results in Tables 1 and 2 to find the values of the following expressions. See Example 1.

1. 12^2

2. 17^2

3. $\sqrt{225}$

4. $\sqrt{361}$

5. $\sqrt{36}$

6. $\sqrt{81}$

7. 20^2

8. 8^2

9. $\sqrt{169}$

10. $\sqrt{196}$

Solve the following problems.

11. Show by squaring 1.732 and 1.733 that $\sqrt{3}$ is between these two numbers.

12. Show by squaring 2.236 and 2.237 that $\sqrt{5}$ is between these two numbers.

Use a calculator to find the value of each square root accurate to four decimal places. See Examples 2 and 3.

13. $\sqrt{12}$

14. $\sqrt{28}$

15. $\sqrt{24}$

16. $\sqrt{32}$

17. $\sqrt{48}$

18. $\sqrt{288}$

19. $\sqrt{363}$

20. $\sqrt{242}$

21. $\sqrt{0.25}$

22. $\sqrt{0.36}$

23. $\sqrt{0.81}$

24. $\sqrt{1.21}$

25. $\sqrt{1.44}$

26. $\sqrt{3.61}$

27. $\sqrt{2.25}$

28. $\sqrt{1.96}$

29. $\sqrt{8100}$

30. $\sqrt{3600}$

31. $\sqrt{900}$

32. $\sqrt{1600}$

33. $\sqrt{0.0025}$

34. $\sqrt{0.0016}$

35. $\sqrt{8}$

36. $\sqrt{34}$

37. $\sqrt{10}$

38. $\sqrt{19}$

39. $\sqrt{45}$

40. $\sqrt{5}$

41. $\sqrt{1.5129}$

42. $\sqrt{4.6225}$

43. $\sqrt{9.0601}$

44. $\sqrt{1030.41}$

45. $\sqrt{800}$

46. $\sqrt{500}$

47. $\sqrt{0.003}$

48. $\sqrt{0.004}$

49. $\sqrt{0.0009}$

50. $\sqrt{0.000025}$

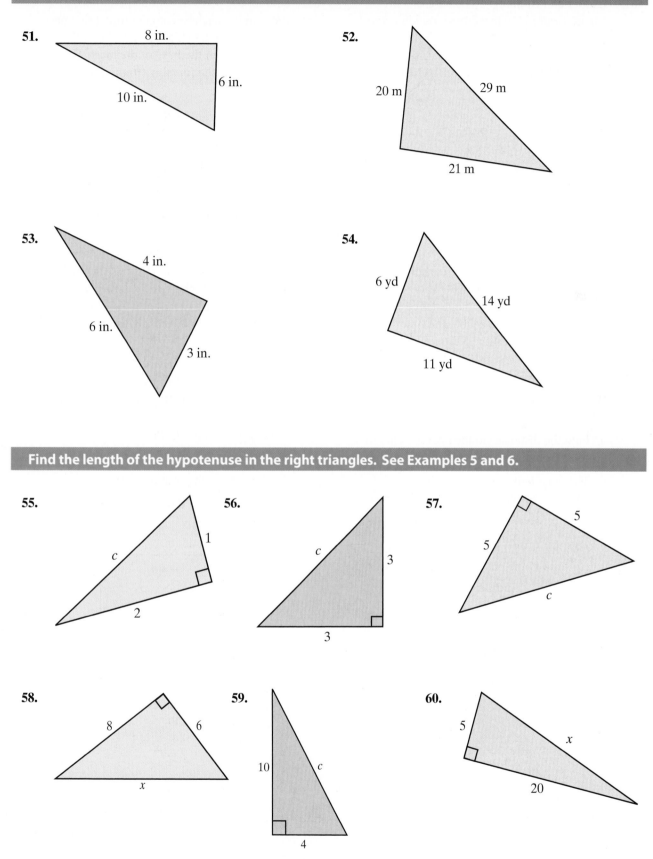

51. 8 in. 6 in. 10 in.

52. 20 m 29 m 21 m

53. 4 in. 6 in. 3 in.

54. 6 yd 14 yd 11 yd

Find the length of the hypotenuse in the right triangles. See Examples 5 and 6.

55. *c* 1 2

56. *c* 3 3

57. 5 5 *c*

58. 8 6 *x*

59. 10 *c* 4

60. 5 *x* 20

61. Ladder and Building: The base of a fire-engine ladder is 20 feet from a building and reaches to a fourth floor window 60 feet above the ground level. How far is the ladder extended (to the nearest tenth of a foot)?

62. Sailing: A forestay that helps support a ship's mast reaches from the top of the mast, which is 20 meters high, to a point on the deck 10 meters from the base of the mast. What is the length of the forestay (to the nearest tenth of a meter)?

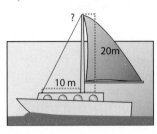

63. Architecture: The Xerox Center building in Chicago is 500 feet tall. At a certain time of day, it casts a shadow that is 150 feet long. At that time of day, what is the distance (to the nearest tenth of a foot) from the tip of the shadow to the top of the Xerox building?

64. Airplane distance: If an airplane passes directly over your head at an altitude of 1 mile, how far is the airplane from your position (to the nearest tenth of a mile) after it has flown 4 miles farther at the same altitude?

65. Baseball diamond: The shape of a baseball infield is a square with sides 90 feet long.

  a. Find the distance (to the nearest tenth of a foot) from home plate to second base.

  b. The diagonals of the square intersect halfway between home plate and second base. If the pitcher's mound is $60\frac{1}{2}$ feet from home plate, is the pitcher's mound closer to home plate or to second base?

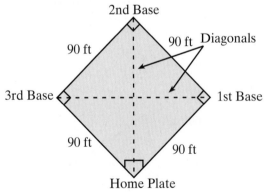

66. Geometry: A diagonal of a square is the line segment from one corner to an opposite corner of the square. Find

  a. the perimeter of the square below

  b. the area of the square below

  c. the length of a diagonal of the square below.

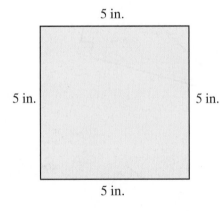

67. Painting: Before painting a picture on canvas, an artist must stretch the canvas on a rectangular wooden frame. To be sure that the corners of the canvas are true right angles, the artist can measure the diagonals of the stretched canvas. What should be the diagonal measure, to the nearest tenth of an inch, of a canvas whose sides are 24 inches and 38 inches in length?

68. Construction: While installing windows in a new home, a builder measures the diagonals of rectangular window casements to verify that their corners are true right angles. What should be the diagonal measure, to the nearest tenth of an inch, of a window casement with dimensions 36 inches by 54 inches?

69. Quilting: To create a square inside a square, a quilting pattern requires four triangular pieces like the one shaded in the figure shown here. If the square in the center measures 10 cm on a side, and the two legs of each triangle are of equal length, how long are the legs of each triangle, to the nearest tenth of a centimeter?

70. Sports: The shape of home plate in the game of baseball can be created by cutting off two triangular pieces at the corners of a square, as shown in the figure. If each of the triangular pieces has a hypotenuse of 12 inches and legs of equal length, what is the length of one side of the original square, to the nearest tenth of an inch?

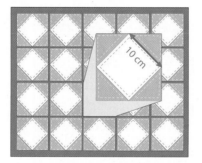

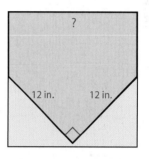

71. Edging: A square plot with an area of 518 square feet must be bordered by edging material.
  a. How long is each side of the plot? Since the area of a square is found by squaring length of one of the sides, the length must be the square root of the area. Round your answer to the nearest hundredth of a foot.
  b. How much edging material must be obtained? Remember that a square has four sides of equal length.

72. Pool: A circular children's wading pool has an area of 293 square feet.
  a. Find the radius of the pool. The area of a circle is approximately 3.14 times the square of the radius. Therefore the radius is the square root of the area divided by the square root of 3.14. Round your answer to the nearest hundredth of a foot.
  b. Find the inside diameter of the pool. The diameter is twice the radius.

73. Planting Trees: The city needs to replant an 18-foot tree. To ensure the tree does not fall over, three wires are to be attached to the tree 3 feet from the top. If the wires will extend 8 feet from the base of the tree:
  a. How long will each wire be? Round to the nearest hundredth.
  b. What will be the total length of the three wires?

74. Hiking: A hiker hikes 9 kilometers north, and then turns left and hikes 11 kilometers west. If she takes the shortest path, how long will she have to walk to get back? Assume that the terrain is flat with no obstructions. Round the answer to the closest tenth.

75. **Line Length:** A 20 meter long line is connected to the top of a cliff, and is stretched out 15 meters on the level ground below the cliff. How high is the cliff? Round to the nearest hundredth of a meter. (**Hint:** Modifying the Pythagorean Theorem, the formula becomes: $b^2 = c^2 - a^2$.)

Writing & Thinking

76. If three whole numbers satisfy the Pythagorean Theorem, these three numbers are called a Pythagorean triple. For example, 3, 4, and 5 are a Pythagorean triple because

$$3^2 + 4^2 = 5^2 \quad \left(\text{or } 9 + 16 = 25\right).$$

Another Pythagorean triple is 5, 12, and 13 because

$$5^2 + 12^2 = 13^2 \quad \left(\text{or } 25 + 144 = 169\right).$$

Complete the following table by finding a, b, and c, and tell which sets of these three numbers (if any) constitute a Pythagorean triple. The first one is done for you.

m	n	$a = 2mn$	$b = m^2 - n^2$	$c = m^2 + n^2$	Pythagorean triple?
5	1	**10**	**24**	**26**	**yes:** $10^2 + 24^2 = 26^2$
7	1				
3	2				
7	2				
5	3				
11	3				
13	7				

Extension: Choose some of your own numbers for m and n. Are your results Pythagorean triples? (**Note:** m must be larger than n so that $m^2 - n^2$ will be positive.)

Section 5.1: Angles

Basic Geometric Terms   page 350
   Point, line, and plane

Rays   page 351
   A **ray** consists of a point (called the **endpoint**) and all the points on a
   line on one side of that point.

Angles   page 351
   An **angle** consists of two rays with a common endpoint (called a **vertex**).

Three Common Ways to Label Angles   page 352
   1. Using three capital letters with the vertex as the middle letter.
   2. Using single numbers such as 1, 2, 3.
   3. Using the single capital letter at the vertex if the meaning is clear.

Angles Classified by Measure   page 353
   1. If $\angle A$ is an **acute** angle, then $0° < m\angle A < 90°$.
   2. If $\angle A$ is a **right** angle, then $m\angle A = 90°$.
   3. If $\angle A$ is an **obtuse** angle, then $90° < m\angle A < 180°$.
   4. If $\angle A$ is a **straight** angle, then $m\angle A = 180°$.

More Classification of Angles   pages 354, 356
   Two angles are:
   1. **Complementary** if the sum of their measures is 90°.
   2. **Supplementary** if the sum of their measures is 180°.
   3. **Congruent** if they have the same measure.

Vertical Angles   page 356
   Vertical angles are congruent. That is, **vertical angles have the same
   measure.**

Adjacent Angles   page 357
   Two angles are **adjacent** if they have a common side.

Parallel Lines   page 358
   Two lines are **parallel** (symbolized ∥) if they are in the same plane and
   do not intersect.

Perpendicular Lines   page 358
   Two lines are **perpendicular** (symbolized ⊥) if they intersect to form
   right angles.

Transversal   page 358
   A line in a plane that intersects two or more lines in that plane in
   different points is called a **transversal**. If two parallel lines are cut by a
   transversal, then
   1. **Corresponding angles are equal.**
   2. **Alternate interior angles are equal.**

Polygon page 365

A **polygon** is a closed plane figure, with three or more sides, in which each side is a line segment.

Each point where two sides meet is called a **vertex**.

(**Note:** A **closed figure** begins and ends at the same point.)

Types of Polygons page 365

Triangle – A polygon with three sides.

Rectangle – A polygon with four sides in which adjacent sides are perpendicular.

Square – A rectangle in which all four sides are the same length.

Trapezoid – A four-sided polygon with one pair of opposite sides that are parallel.

Parallelogram – A four-sided polygon with both pairs of opposite sides parallel.

Formula page 366

A **formula** is a general statement (usually an equation) that relates two or more variables.

Perimeter page 366

The **perimeter** P of a polygon is the sum of the lengths of its sides.

Formulas for Perimeter page 366

Triangle: $P = a + b + c$

Square: $P = 4s$

Rectangle: $P = 2l + 2w$

Trapezoid: $P = a + b + c + d$

Parallelogram: $P = 2a + 2b$

Area page 375

Area is a measure of the interior of (or surface enclosed by) a figure in a plane.

Height of a Figure page 375

In triangles and other figures, we have used the letter h to represent the **height** of the figure. The height is also called the **altitude** and is perpendicular to the base.

Formulas for Area page 376

Triangle: $A = \dfrac{1}{2}bh$

Rectangle: $A = lw$

Square: $A = s^2$

Parallelogram: $A = bh$

Trapezoid: $A = \dfrac{1}{2}h(b + c)$

Section 5.4: Circles

Circle   page 385

A **circle** is the set of all points in a plane that are some fixed distance from a fixed point called the **center** of the circle.

Radius   page 385

The **radius** is the fixed distance from the center of a circle to any point on the circle. (The letter r is used to represent the radius of a circle.)

Diameter   page 385

The **diameter** is the distance from one point on a circle through the center to the point directly opposite it. (The letter d is used to represent the diameter of a circle.)

Circumference   page 385

The **circumference** is the perimeter of a circle.

Formulas for Circles   page 386

Circumference: $C = 2\pi r$ and $C = \pi d$
Area: $A = \pi r^2$

Section 5.5: Volume and Surface Area

Volume   page 394

Volume is the measure of the space enclosed by a three-dimensional figure and is measured in cubic units.

Formulas for Volume   page 394

Rectangular solid: $v = lwh$

Rectangular pyramid: $v = \frac{1}{3}lwh$

Right circular cylinder: $v = \pi r^2 h$

Right circular cone: $v = \frac{1}{3}\pi r^2 h$

Sphere: $v = \frac{4}{3}\pi r^3$

Surface Area   page 397

The **surface area** (SA) of a geometric solid is a measure of the outside surface in square units.

Formulas for Surface Area   page 398

Rectangular solid: $SA = 2lw + 2wh + 2lh$
Right circular cylinder: $SA = 2\pi r^2 + 2\pi rh$
Sphere: $SA = 4\pi r^2$

© Hawkes Learning Systems. All rights reserved.

Key Terms and Ideas   Chapter 5   429

Square Roots of Perfect Squares from 1 to 400   page 416
(See Table 2).

Right Triangle   page 418
A **right triangle** is a triangle containing a right angle (90°).

Hypotenuse   page 418
The **hypotenuse** is the longest side of a right triangle; the side opposite the right angle.

Leg   page 418
A **leg** of a triangle is either of the other two sides of a right triangle (the sides that are not the hypotenuse).

The Pythagorean Theorem   page 418
In a right triangle, the square of the length of the hypotenuse is equal to the sum of the squares of the lengths of the two legs: $c^2 = a^2 + b^2$.

Chapter 5: Review

Section 5.1: Angles

Use a protractor to measure all the angles in each figure. Each line segment may be extended as a ray to form the side of an angle.

1.

2.

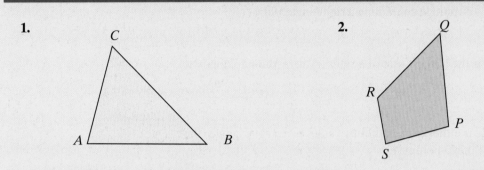

Classify each of the following angles as acute, right, obtuse, or straight.

3.

4.

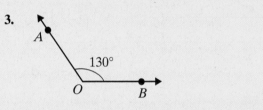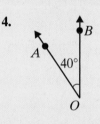

In the figure shown below, $m\angle POQ = 90°$, $m\angle QOR = 60°$ and $\overline{OA}$ and $\overline{OB}$ are angle bisectors. Find the measures of the following angles.

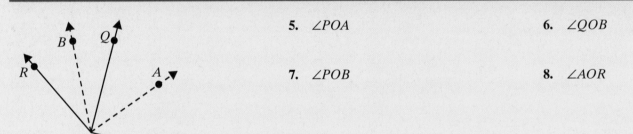

5. ∠*POA*

6. ∠*QOB*

7. ∠*POB*

8. ∠*AOR*

Use the definition of complementary, supplementary, and straight angles to answer the following questions.

9. Assume that ∠1 and ∠2 are complementary.

   a. If $m\angle 1 = 20°$, what is $m\angle 2$?

   b. If $m\angle 1 = 13°$, what is $m\angle 2$?

10. Assume that ∠3 and ∠4 are supplementary.

   a. If $m\angle 3 = 90°$, what is $m\angle 4$?

   b. If $m\angle 3 = 150°$, what is $m\angle 4$?

11. Given that $m\angle 1 = 57°$ in the figure shown to the right, find:

    a. The measures of the other three angles.

    b. Two pairs of adjacent angles.

    c. Two pairs of vertical angles.

Section 5.2: Perimeter

12. Write three formulas for perimeter that you learned in this section.

13. A triangle with sides of length 6 m, $2\frac{1}{2}$ m, and $6\frac{1}{3}$ m.

14. A parallelogram with sides of length 32 cm and 23 cm

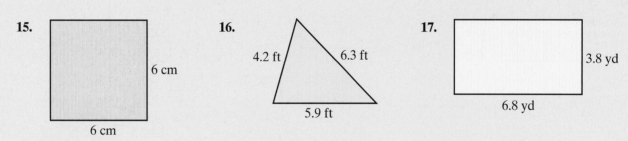

15. 6 cm, 6 cm

16. 4.2 ft, 6.3 ft, 5.9 ft

17. 3.8 yd, 6.8 yd

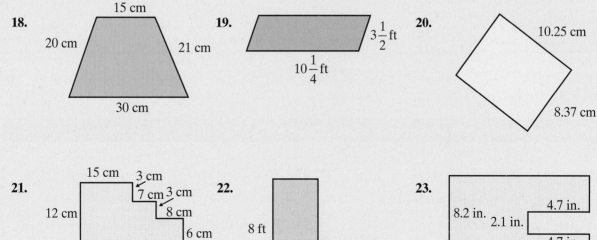

18. 15 cm, 20 cm, 21 cm, 30 cm

19. $3\frac{1}{2}$ ft, $10\frac{1}{4}$ ft

20. 10.25 cm, 8.37 cm

21. 15 cm, 3 cm, 7 cm, 3 cm, 8 cm, 12 cm, 6 cm, 30 cm

22. 8 ft, 7 ft

23. 4.7 in., 8.2 in., 2.1 in., 4.7 in., 12.5 in.

Section 5.3: Area

Find the area of each figure.

24. A square with sides of length 16 ft

25. A trapezoid with height 10 yd and parallel sides of 8 yd and 11 yd

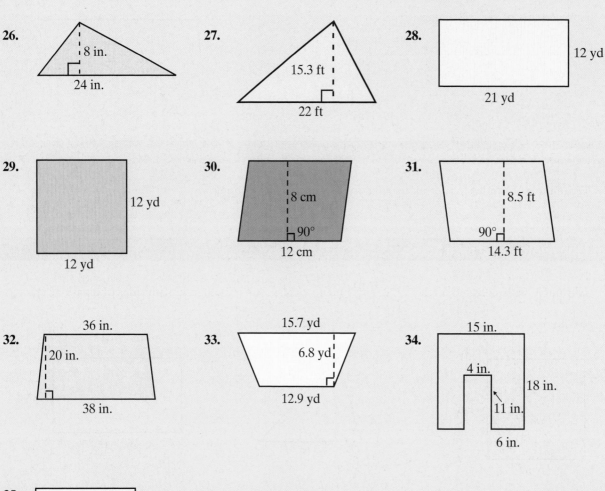

26. 8 in. 24 in.

27. 15.3 ft 22 ft

28. 12 yd 21 yd

29. 12 yd 12 yd

30. 8 cm 90° 12 cm

31. 8.5 ft 90° 14.3 ft

32. 36 in. 20 in. 38 in.

33. 15.7 yd 6.8 yd 12.9 yd

34. 15 in. 4 in. 18 in. 11 in. 6 in.

35. 5 yd 3 yd 7 yd 11 yd

Section 5.3: Circles

Find the circumference and area of each circle. (Use π = 3.14).

36. A circle with diameter 16.2 yd.

37. A circle with radius 8 ft.

38. A circle of radius 6 ft.

39. A circle of diameter 6 ft.

Find the perimeter and area of each figure. (Use π = 3.14).

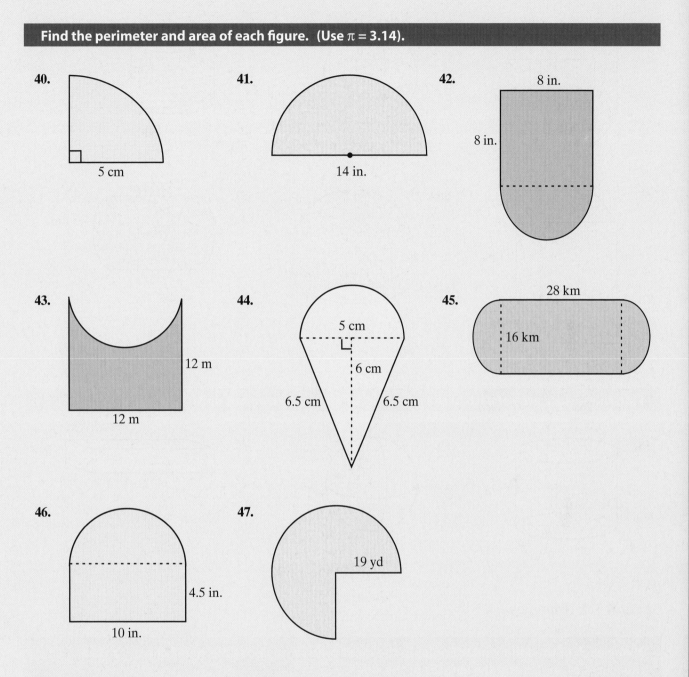

40.
5 cm

41.
14 in.

42.
8 in.
8 in.

43.
12 m
12 m

44.
5 cm
6 cm
6.5 cm   6.5 cm

45.
28 km
16 km

46.
4.5 in.
10 in.

47.
19 yd

Section 5.4: Volume and Surface Area

Find the volume of each of the following solids with the dimensions indicated.

48. A rectangular solid with length 6 in., width 3 in., and height 8 in.

49. A sphere with diameter 24 ft.

50. A right circular cone 9 dm high with a 4 dm radius.

51. A rectangular pyramid with length 10 cm, width 12 cm, and height 5 cm.

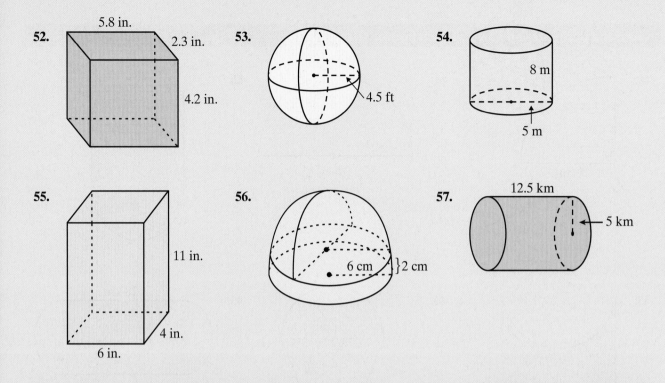

52. 5.8 in. 2.3 in. 4.2 in.

53. 4.5 ft

54. 8 m, 5 m

55. 11 in. 4 in. 6 in.

56. 6 cm }2 cm

57. 12.5 km, 5 km

Find the surface area of each of the following solids in a convenient unit. (Use π = 3.14.)

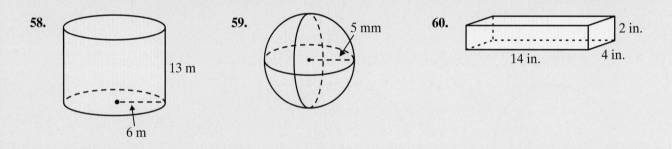

58. 13 m, 6 m

59. 5 mm

60. 2 in. 14 in. 4 in.

Section 5.6: Triangles

Name each of the following triangles in the most precise way possible, given the indicated measures of angles and the lengths of the sides.

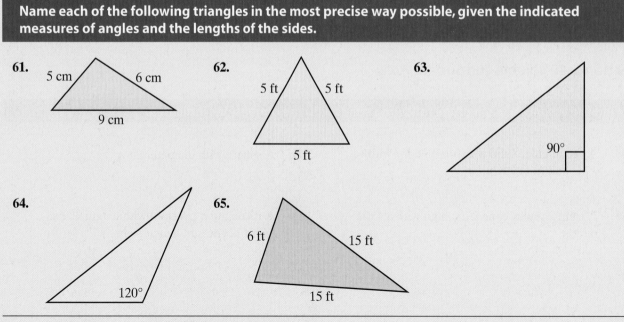

61. 5 cm, 6 cm, 9 cm

62. 5 ft, 5 ft, 5 ft

63. 90°

64. 120°

65. 6 ft, 15 ft, 15 ft

Assume that the given triangles are similar and find the values for *x* and *y*.

66. △*ABD* ~ △*EFD*

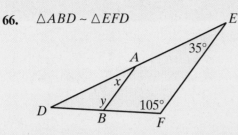

67. △*LMN* ~ △*QPO*

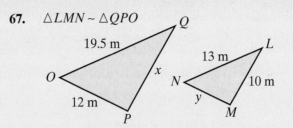

Determine whether each pair of triangles is congruent. If they are congruent, state the property that makes them congruent.

68.

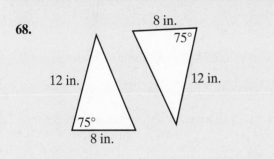

69.

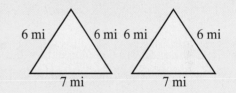

70.

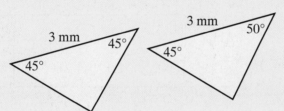

Section 5.7: Square Roots and the Pythagorean Theorem

Find the following squares and square roots.

71. 11^2

72. $\sqrt{81}$

Use your calculator to find the square roots accurate to four decimal places.

73. $\sqrt{60}$

74. $\sqrt{10.24}$

75. $\sqrt{92}$

76. $\sqrt{0.3969}$

Solve the following word problems.

77. Determine whether or not a triangle with sides of length 10 in., 24 in., and 26 in. is a right triangle.

78. Find the length (accurate to four decimal places) of the hypotenuse of a right triangle with two equal legs of length 4 m.

Chapter 5: Test

Solve the following problems.

1. In the figure shown to the right, $\overleftrightarrow{AD}$ and $\overleftrightarrow{BE}$ are straight lines.

   a. What type of angle is $\angle BOC$?

   b. What type of angle is $\angle AOE$?

   c. Name a pair of vertical angles.

   d. Name two pairs of supplementary angles.

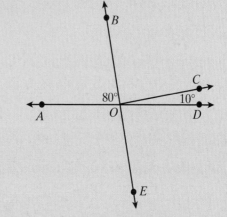

2. a. If $\angle 1$ and $\angle 2$ are complementary and $m\angle 1 = 35°$, what is $m\angle 2$?

   b. If $\angle 3$ and $\angle 4$ are supplementary and $m\angle 3 = 15°$, what is $m\angle 4$?

3. Find **a.** the circumference and **b.** the area of the following circle. (Use $\pi = 3.14$.)

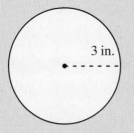

4. Find **a.** the perimeter and **b.** the area of the trapezoid with dimensions shown here.

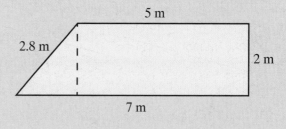

5. a. Find the perimeter of a rectangle that is $8\frac{1}{2}$ inches wide and $11\frac{2}{3}$ inches long.

   b. Find the area of the rectangle.

6. Find the area of a square that has sides the length of 14 inches.

7. Find the area of a right triangle if its legs are 4 centimeters and 5 centimeters long.

Find the volume of each of the following solids with the dimensions indicated.

8. Find the volume of the cylinder shown here with the given dimensions. (Use $\pi = 3.14$.)

9. Find the volume of the rectangular solid shown here with the given dimensions.

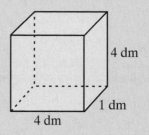

10. Name the type of each of the following triangles based on the measures and shapes shown.

a.
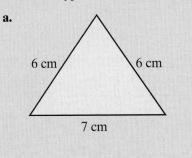

b.
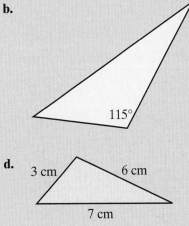

c.
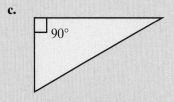

d.
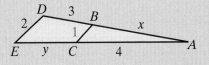

11. a. Find the value of x.

    b. What kind of triangle is $\triangle RST$?

    c. Which side is opposite $\angle S$?

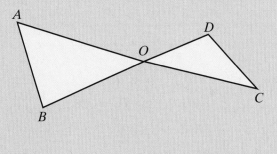

12. Find the value of x given $\triangle ABC \sim \triangle ADE$.

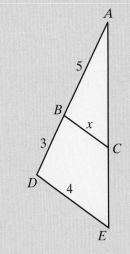

13. Given $\angle B \cong \angle D$, determine whether or not $\triangle AOB \sim \triangle COD$. Explain your reasoning.

14. Given that $\triangle ABC \sim \triangle ADE$ find the value x and the value of y.

15. Find **a.** the perimeter and **b.** the area of the figure shown.

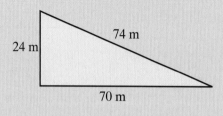

20 ft

10 ft

16. Find **a.** the perimeter and **b.** the area of the figure shown.

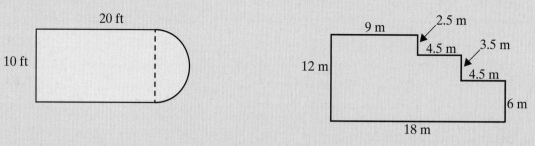

9 m   2.5 m
          4.5 m   3.5 m
12 m   4.5 m
                        6 m
      18 m

17. Is the figure with dimensions as shown a right triangle? Why or why not? If the figure is a right triangle, find **a.** the perimeter and **b.** the area of the figure shown right.

74 m

24 m

70 m

18. Find the volume of a circular cylinder with radius 10 ft and height 4 ft.

19. Find the volume of a circular cone with diameter 8 cm and height 15 cm.

20. Geometry: The lengths of the three sides of a triangle are $5\frac{4}{5}$ centimeters, $3\frac{3}{5}$ centimeters, and $7\frac{1}{5}$ centimeters. The height of the triangle is $2\frac{3}{5}$ cm.

a. Find the perimeter of the triangle.

b. Find the area of the triangle.

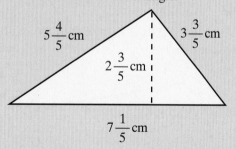

$5\frac{4}{5}$ cm $3\frac{3}{5}$ cm

$2\frac{3}{5}$ cm

$7\frac{1}{5}$ cm

21. Geometry:

a. Find the perimeter of a rectangle that is 8 inches wide and $11\frac{1}{2}$ inches long.

b. Find the area of the rectangle.

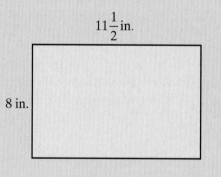

$11\frac{1}{2}$ in.

8 in.

22. Geometry: A triangle has dimensions as shown in the figure. Find **a.** the perimeter and **b.** the area of the triangle.

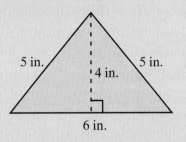

5 in.   5 in.

4 in.

6 in.

23. Geometry: A right triangle has sides of length 10 feet, 24 feet, and 26 feet. **a.** Draw a sketch and label the sides of the triangle. Find **b.** the perimeter and **c.** the area of the triangle.

24. a. What theorem do you use to determine whether or not a triangle is a right triangle?

b. If a triangle has sides of 17 inches, 15 inches and 8 inches, is it a right triangle? Why or why not?

25. Football A football field is rectangular in shape—100 yards long and 40 yards wide.

a. What is the perimeter of the field?

b. What is the area of the field?

c. What is the distance (to the nearest tenth of a yard) from one corner of the field to another corner? (That is, what is the length of a diagonal of the rectangle?)

100 yards

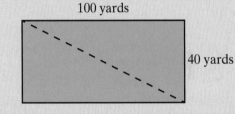

40 yards

Cumulative Review: Chapters 1 - 5

Solve the following problems.

1. Find the LCM of 12, 15, and 40.

2. Determine the prime factorization for 76.

3. Write $\dfrac{25}{6}$ as a mixed number.

4. Write $\dfrac{11}{5}$ as a percent.

5. Write 150% as a mixed number with the fraction part reduced.

6. Write $2\dfrac{1}{3}$ as a percent.

7. Find 100% of 75.

8. Write the number 423.85 in its English word equivalent.

9. Round 166.075 to the nearest hundred.

Perform the indicated operation and reduce all fractions.

10. $\dfrac{16}{0}$

11. $\dfrac{0}{3}$

12. $(500)(7000)$

13. $(2.4)(6.1)$

14. $\dfrac{5}{8}+\dfrac{3}{4}+\dfrac{11}{40}$

In the following exercises, a. estimate the answer, then b. find the actual answer.

15.
$$\begin{array}{r} 75.63 \\ 81.45 \\ +\,146.98 \end{array}$$

16.
$$\begin{array}{r} 8000.0 \\ -\,6476.9 \end{array}$$

17.
$$\begin{array}{r} 43.8 \\ \times\,2.7 \end{array}$$

Perform the indicated operation and reduce all fractions.

18. $(17-15)(32-21)$

19. $7^2 \cdot 5 - 2.1(6) \div \dfrac{1}{2}$

20. $\dfrac{1}{2}-\dfrac{2}{3}\cdot\left(\dfrac{1}{2}\right)^2+\dfrac{9}{10}$

Use your calculator to find the following square roots accurate to four decimal places.

21. a. $\sqrt{11}$

   b. $\sqrt{441}$

   c. $\sqrt{2000}$

   d. $\sqrt{3.56}$

22. In the figure shown to the right, $\overleftrightarrow{AD}$, $\overleftrightarrow{BE}$, and $\overleftrightarrow{CF}$ are straight lines.

    a. What type of angle is $\angle AOC$?

    b. What type of angle is $\angle FOE$?

    c. What angles are supplementary to $\angle COD$?

    d. Is $m\angle AOB$ equal to $m\angle DOE$? Explain your reasoning.

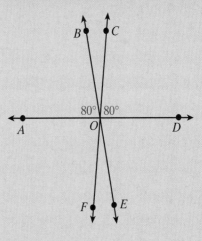

23. Name the type of each of the following triangles based on the measures and shapes shown.

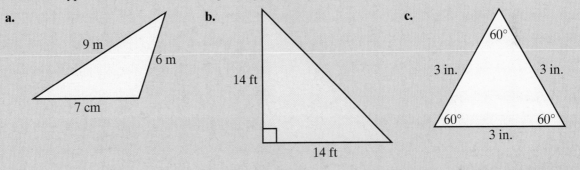

a.

9 m
6 m
7 cm

b.

14 ft
14 ft

c.

60°
3 in.   3 in.
60°   60°
3 in.

24. **Temperature:** One week in March 2011. San Diego, CA had the following high temperatures: 63°F, 62°F, 66°F, 77°F, 72°F, 68°F, 63°F. Find the average high temperature for the week. (Round your answer to the nearest tenth.)

25. **Cell Phones:** Melinda's cell phone plan comes with 800 minutes a month. So far this month she has only used 560 minutes.

    a. How many more minutes does she have to use this month?

    b. What fraction of her monthly minutes does she have left?

26. **Thai Food:** Mandy and three friends decided to go out for Thai food after work. When their appetizers were brought out they saw that there were 6 crab wontons and 5 spring rolls. If they plan to share everything evenly, how many wontons and how many spring rolls did each person get? Write your answers as mixed numbers.

27. **Eating Out:** Matt took his girlfriend out to a local Mexican restaurant for her birthday. The bill for dinner came to 28.65 (including tax). If he plans to leave the waiter a tip of 20% of the bill (including tax), how much did he pay in total for dinner?

28. **Cartography:** The scale on map indicates that $1\frac{3}{4}$ inches represents 50 miles. How many miles apart are two cities marked 2 inches apart on the map?

29. **School supplies:** A student decided to buy a new laptop. She paid $630.50. This was not the original price. She received a 3% discount for paying cash. What was the original price?

30. **Furniture:** The sale price of a new sofa was $752. This was $\frac{4}{5}$ of the original price. What was the original price of the sofa?

31. **Interest:** What is the rate of interest if the simple interest earned on $5000 for 90 days is $156.25?

32. **Triangles:** Find the area of a triangle with base 20 cm and height 14 cm.

33. **Circles:** What is the circumference of a circle with diameter 28 in.? (Use $\pi = 3.14$.)

34. **Pyramids:** Find the volume of a rectangular pyramid with length 12 mm, width 9 mm, and height 5 mm.

35. **Cylinders:** Find the surface area of a right circular cylinder with a radius of 4 in. and a height of 7 in. (Use $\pi = 3.14$.)

36. **Trapezoids:** Find **a.** the perimeter and **b.** the area of the trapezoid shown.

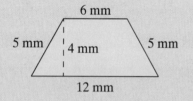

37. **Polygons:** Find **a.** the perimeter and **b.** the area of the figure shown.

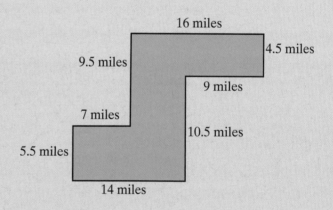

38. **Triangles:** Find the length of the hypotenuse of a right triangle with both legs of length 15 feet. (Write the answer in both radical form and decimal form rounded to the nearest hundredth).

39. **Triangles:** Use the Pythagorean Theorem to show that a triangle with sides of 12 cm, 16 cm, and 20 cm is a right triangle.

Statistics, Graphs, and Probability

Mathematics at Work!

Assessing the likelihood of certain situations is a useful tool in everyday life. For example, say you were playing cards. How would you wager differently if you knew you had only a 3% chance of winning a hand versus if you knew you had a 53% chance of winning? In both cases, you need to be able to count the number of possible outcomes and then calculate the likelihood of the one situation you have.

A standard deck of cards is 52 cards with four suits (hearts, diamonds, spades, and clubs) and 13 cards in each suit. The cards are ace, king, queen, jack, 10, 9, 8, 7, 6, 5, 4, 3, and 2. If one card is drawn from a deck of cards, find the probability of each of the following events. (See Exercises 29 – 36 in Section 6.4).

1. The card is an ace.
2. The card is the 3 of diamonds.
3. The card is a club.
4. The card is a queen or a jack.
5. The card is red.
6. The card is a 10, 9, or 8.
7. The card is the king of hearts.
8. The card is a 1.

6.1 Statistics: Mean, Median, Mode, and Range

Objective A Statistical Terms

Statistics is the study of how to gather, organize, analyze, and interpret numerical information. A **statistic** is a particular measure or characteristic of a part, or **sample**, of a larger collection of items called the **population** of interest. The population can be a collection of people, animals, objects, or numbers related to information of interest.

In this section we will study four measures that are easily found or calculated: **mean**, **median**, **mode**, and **range**. The mean, median, and mode are measures that describe the "average" or "middle" of a set of data. The range is a measure that describes how "spread out" the data is. (**Note:** The concept of **average** was introduced in Section 1.7).

Terms Used in the Study of Statistics

Data: Value(s) measuring some characteristic of interest such as income, height, weight, grade point averages, scores on tests, and so on. (We will consider only numerical data.)

Mean: The sum of all the data divided by the number of data items. (Also called the **arithmetic average**.)

Median: The middle data item. (Arrange the data in order and pick out the middle item.)

Mode: The single data item that appears the most number of times. (Some data may have more than one mode. We will leave the discussion of such a situation to a course in statistics. In this text, if the data have a mode, there will be only one mode.)

Range: The difference between the largest and smallest data items.

Objective B Calculating Mean, Median, Mode, and Range

The following two sets of data are used in answering the questions in Examples 1, 2, and 3.

Group A: Annual Income for 8 Families			
$28,000	$22,000	$25,000	$27,000
$45,000	$80,000	$25,000	$30,000

Group B: Grade Point Average (GPA) for 11 Students					
2.0	2.0	1.9	3.1	3.5	2.9
2.5	3.6	2.0	2.4	3.4	

Example 1

Finding the Mean

Find the mean income for the families in Group A.

Solution

Find the sum of the 8 incomes and divide by 8.

Add	**Divide**

$ 28,000

22,000

25,000

27,000

45,000

80,000

25,000

+ 30,000

$282,000

$$\begin{array}{r} 35{,}250 \\ 8\overline{)282{,}000} \\ \underline{24} \\ 42 \\ \underline{40} \\ 20 \\ \underline{16} \\ 40 \\ \underline{40} \\ 00 \\ \underline{0} \\ 0 \end{array}$$

For Group A: The mean annual income is $35,250. (You may want to use a calculator to do this arithmetic).

Now work margin exercise 1.

To Find the Median

1. Arrange the data in order.
2. If there is an **odd** number of items, the median is the middle item.
3. If there is an **even** number of items, the median is the average of the two middle items.

1. Find the mean grade point average for the students in Group B. Round your answer to the nearest hundredth.

2. Find the median age for the following set of data.

Ages of 8 People			
35	14	37	51
28	19	35	45

Example 2

Finding the Median

Find the median income for Group A and the median GPA for Group B.

Solution

The data are not in order, so we arrange both sets to be in order:

Group A (in order):

$22,000; $25,000; $25,000;   **$27,000**;   **$28,000**;

$30,000; $45,000; $80,000

There are 8 items (an **even** number) so we find the middle two items and average them: These items are the 4th and 5th items. (Count 4 from the left and 4 from the right.)
The data are $27,000 and $28,000:

$$\text{median} = \frac{27,000 + 28,000}{2} = \frac{55,000}{2} = \$27,500.$$

Group B (in order):

1.9;   2.0;   2.0;   2.0;   2.4;   **2.5**;   2.9;   3.1;   3.4;   3.5;   3.6

For Group B, there are 11 items (an **odd** number) and the median is the 6th item. So the median is 2.5.

Now work margin exercise 2.

3. Find the mode and the range for the following set of data.

Ages of 8 People			
35	14	37	51
28	19	35	45

Example 3

Finding the Mode and Range

Find the mode and the range for both Group A and Group B.

Solution

The mode is the most frequent item. From the arranged data in Example 2, we can see that:

for Group A, the **mode** is $25,000.

for Group B, the **mode** is 2.0.

The range is the difference between the largest and smallest items:

Group A **range** = $80,000 − $22,000 = $58,000.

Group B **range** = 3.6 − 1.9 = 1.7.

Now work margin exercise 3.

Of the four statistics mentioned in this section, the mean and median are most commonly used. Many people feel that the mean (or arithmetic average) is relied on too much in reporting central tendencies. A few very large (or very small) data items can distort the mean as a picture of a central tendency. As you can see in the Group A data, the median of $27,500 is probably more representative of the data than the mean of $35,250. Note how the one income of $80,000 raises the mean considerably.

When you read an article in a magazine or newspaper that reports means or medians, you should now have a better understanding of the implications.

Practice Problems

Two sets of data, Group A and Group B, are given. Find the following statistics for each group.

a. mean   **b.** median   **c.** mode   **d.** range

Group A: Body Temperature (in Fahrenheit degrees) of 8 People							
96.4°	98.6°	98.7°	99.8°	99.2°	101.2°	98.6°	97.1°

Group B: The Time (in minutes) of 11 Movies					
100 min	90 min	113 min	110 min	88 min	90 min
155 min	88 min	105 min	93 min	90 min	

Practice Problem Answers

	Mean	Median	Mode	Range
Group A	98.7°F	98.65°F	98.6°F	4.8°F
Group B	102 min	93 min	90 min	67 min

Exercises 6.1

For each of the following problems, find a. the mean, b. the median, c. the mode, and d. the range of the given data. See Examples 1 through 3.

1. **Test Scores:** Ten math students had the following scores on a final exam.

75	83	93	65	85
85	88	90	55	71

2. **Exercising:** Joe did the following number of sit-ups each morning for a week.

25	52	48	42	38	58	52

3. **Studying:** Fifteen college students reported the following hours of sleep the night before an exam.

4	6	6	7	6.5	6.5	7.5	8.5
5	6	4.5	5.5	9	3	8	

4. **Basketball:** The local high school basketball team scored the following points per game during their 20-game season.

85	60	62	70	75	52	88
50	80	72	90	85	85	93
70	75	68	73	65	82	

5. **Car Repairs:** Stacey went to six different repair shops to get the following estimates to repair her car.

$425	$525	$325	$300	$500	$325

6. **Golf:** Mike kept track of his golf scores for twelve rounds of eighteen holes each. His scores were as follows. Round to the nearest tenth, if necessary.

85	90	82	85	87	80
78	82	88	82	86	81

7. **Weather:** The local weather station recorded the following daily high temperatures (in degrees Fahrenheit) for one month.

75°	76°	76°	78°	85°	82°	85°	88°
90°	90°	88°	95°	96°	92°	88°	88°
80°	80°	78°	80°	78°	76°	77°	75°
75°	74°	70°	70°	72°	73°		

8. **Public Services:** The Big City fire department reported the following mileage for tires used on their nine fire trucks.

14,000	14,000	11,000	15,000	9000
14,000	12,000	10,000	9000	

9. **Construction:** The city planning department issued the following numbers of building permits over a three-week period (15 business days).

17	19	18	35	30	29	23	14
18	16	20	18	18	25	30	

10. **Speed Detection:** Police radar measured the following speeds in miles per hour of 35 cars on one street.

28	24	22	38	40	25	24	35	25
23	22	50	31	37	45	28	30	30
30	25	35	32	45	52	24	26	18
20	30	32	33	48	58	30	25	

11. Fishing: On a one-day fishing trip, Mr. and Mrs. Milster recorded the following lengths of the ten fish they caught (measured in inches).

14.3	13.6	10.5	15.5	20.1
10.9	12.4	25.0	30.2	32.5

12. Weight: A girl's organization has fifteen 16-year-old girls with the following weights in pounds:

142	124	179	161	117
232	135	126	101	155
129	133	120	113	124

13. Test Scores: Dr. Wright recorded the following nine test scores for students in his statistics course.

95 82 85 71 65 85 62 77 98

14. Presidents: The ages of the first five U.S. presidents on the date of their inaugurations were as follows. (The presidents were Washington, Adams, Jefferson, Madison, and Monroe.)

57   61   57   57   58

15. Income: Family incomes in a survey of eight students are as follows.

$35,000	$28,000	$42,000	$71,000
$63,000	$36,000	$51,000	$63,000

16. College Tuition: Resident tuition charged by 10 colleges in North Carolina in 2010 are listed in the chart below. Round to the nearest tenth, if necessary. **Source:** collegeboard.com

$4088	$5175	$5076	$3639	$3756
$6529	$4479	$5922	$5138	$5124

17. College Tuition: Nonresident tuition charged by 10 colleges in North Carolina in 2010 are as follows. **Source:** collegeboard.com

$16,487	$13,234	$17,831	$13,276	$14,220
$19,064	$15,052	$24,736	$16,185	$14,721

18. College Students: The following list contains the ages of 20 students surveyed in a college chemistry class.

18	23	23	23	22	18	21	20	18	20
19	20	21	19	23	36	35	26	17	24

19. Travel: The distances from Chicago to selected cities are shown below.

Boston:	980 miles
Cleveland:	345 miles
Dallas:	930 miles
Denver:	1050 miles
Detroit:	280 miles
Indianapolis:	190 miles
Los Angeles:	2110 miles
San Francisco:	2210 miles
New Orleans:	950 miles
Miami:	1390 miles
Seattle:	2050 miles

20. **Air Travel:** Passengers (to the nearest thousand in 2009) in the world's 10 busiest airports:

Atlanta, Hartsfield: 88,032,000	Paris, Charles de Gaulle: 57,907,000
London, Heathrow: 66,038,000	Los Angeles, LAX: 56,521,000
Beijing, PEK: 65,372,000	Dallas/Ft Worth, DFW: 56,030,000
Chicago, O'Hare: 64,158,000	Frankfurt–Main: 50,933,000
Tokyo, Haneda: 61,904,000	Denver, DEN: 50,167,000

Source: Airports Council International

Writing & Thinking

21. Your grade point average (GPA) is a form of a **weighted average**. That is, 4 units of A counts more than 4 units of B. The most common weight for grades is A, 4 points; B, 3 points; C, 2 points; D, 1 point; and F, 0 points. To find a GPA,

1. Multiply the points for each grade by the number of units for the course.

2. Find the sum of these products.

3. Divide this sum by the total number of units taken.

Find the GPA (to the nearest tenth) for each of the following situations:

a. 3 units of A in astronomy, 4 units of B in geometry, 3 units of C in sociology, and 4 units of A in biology.

b. 5 units of B in history, 4 units of C in calculus, 3 units of A in computer science, and 4 units of D in geology.

c. Your own GPA for the last semester (or your anticipated GPA for this semester).

Collaborative Learning

22. With the class separated into teams of 2 to 4 students, each team is to go on campus and survey 50 students and ask each student how many minutes it takes him or her to drive to school from home.

a. Each team is to find the mean, median, mode, and range for the 50 responses.

b. Each team is to bring all the data to class, and the class is to pool the information and find the mean, median, mode, and range for the pooled data.

c. The class is to discuss the results of the individual teams and the pooled data and what use such information might have for the administration of the college.

6.2 Reading Graphs

Objective A   **Introduction to Graphs**

Graphs are pictures of numerical information. Graphs appear almost daily in newspapers and magazines and frequently in textbooks and corporate reports. Well-drawn graphs can organize and communicate information accurately, effectively, and fast. Most computers can be programmed to draw graphs, and anyone whose work involves a computer in any way will probably be expected to understand graphs and even to create graphs.

There are many different types of graphs, and each type is particularly well-suited to the display and clarification of certain types of information. In this section, we will discuss in detail the uses of bar graphs, circle graphs, line graphs, and histograms.

The Purposes of Four Types of Graphs

1. **Bar Graphs:** to emphasize comparative amounts.
2. **Circle Graphs:** to help in understanding percents or parts of a whole. Circle graphs are also called **pie charts.**
3. **Line Graphs:** to indicate tendencies or trends over a period of time.
4. **Histograms:** to indicate data in **classes** (a range or interval of numbers).

A common characteristic of all graphs is that they are intended to communicate information about numerical data quickly and easily. With this in mind, note the following three properties of all graphs:

1. They should be clearly labeled.
2. They should be easy to read.
3. They should have appropriate titles.

Objectives

A   Learn the purposes and properties of graphs.

B   Read bar graphs.

C   Read circle graphs.

D   Read line graphs.

E   Read histograms.

1. Use the bar graph in Example 1 to answer the following questions.

   a. What were the sales in May?

   b. Which month had sales of $50,000?

   c. What was the amount of increase in sales between April and June?

   d. What was the percent of increase in sales from April to June?

Example 1

Reading a Bar Graph

Examine the following bar graph. Note that the scale on the left (sales) and the categories at the bottom (months) are clearly labeled and the graph itself has a title.

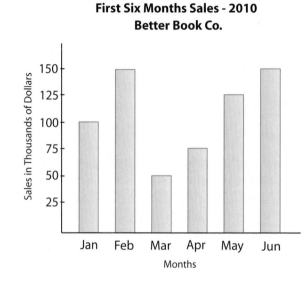

First Six Months Sales - 2010
Better Book Co.

Solution

The following questions can be answered by reading information directly from the graph.

a. What were the sales in January? (Note that the scale on the left of the graph says that sales are in thousands of dollars.) $100,000

b. During what month were sales lowest? March

c. During what month were sales highest?   February and June

d. What were the sales during each of the highest sales months? $150,000

e. What were the sales in April? $75,000

The following questions can be answered by reading information from the graph and making some calculations.

f. What was the amount of decrease in sales between February and March?

$$\begin{array}{rl} \$150,000 & \text{February sales} \\ -\ 50,000 & \text{March sales} \\ \hline \$100,000 & \text{decrease in sales} \end{array}$$

g. What was the percent of decrease in sales from February to March?

$$\frac{\text{decrease}}{\text{February sales}}\ \frac{100,000}{150,000} = \frac{2}{3} \approx 0.666 \approx 67\%$$

Now work margin exercise **1.**

Reading a Circle Graph

Example 2

Reading a Circle Graph

Examine the following circle graph. Percents budgeted for various items in a household for one year are given. Use the information in the graph to calculate what amount will be allocated to each item indicated if the family income was $45,000.

Household Budget for One Year

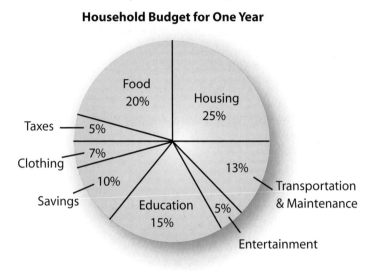

Solution

Item	Amount
Housing	$0.25 \times \$45,000 = \$11,250$
Food	$0.20 \times \$45,000 = \$9000$
Taxes	$0.05 \times \$45,000 = \$2250$
Clothing	$0.07 \times \$45,000 = \$3150$
Savings	$0.10 \times \$45,000 = \$4500$
Education	$0.15 \times \$45,000 = \$6750$
Entertainment	$0.05 \times \$45,000 = \$2250$
Tran. &Maintenance	$0.13 \times \$45,000 = \$5850$

Now work margin exercise 2.

2. Use the circle graph in Example 2 to answer the following question. If the family income increased from $45,000 to $55,000 and the percent allocated to each item does not change, how much will the family spend on each of the following items?

a. Housing

b. Savings

c. Clothing

3. Use the line graph in Example 3 to answer the following questions.

a. What was the high temperature for the week (highest daily high)?

b. What was the low temperature for the week (lowest daily low)?

c. Find the maximum daily difference between daily high and low temperatures in a single day for the week.

Objective D
 Reading a Line Graph

Example 3

Reading a Line Graph

Examine the following line graph. This graph shows the relationships between daily high and low temperatures. You can tell that temperatures tended to rise during the week but fell sharply on Saturday.

(Note that the temperature scale on the left does not start at 0. There is a break indicated in that scale).

Find the mean of the differences between the daily high and low temperatures for the week shown.

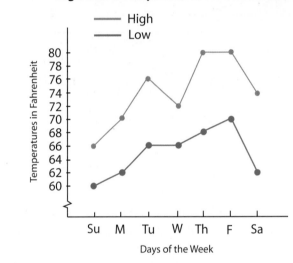

High & Low Temperatures for One Week

Solution

First, find the differences for each day. Then, find the mean of these differences.

Sunday:	66 – 60	= 6°
Monday:	70 – 62	= 8°
Tuesday:	76 – 66	= 10°
Wednesday:	72 – 66	= 6°
Thursday:	80 – 68	= 12°
Friday:	80 – 70	= 10°
Saturday:	74 – 62	= 12°

$$\begin{array}{r} 9.1° \quad \text{mean difference} \\ 7\overline{)64.0} \\ \underline{63} \\ 1\,0 \\ \underline{7} \\ 3 \end{array}$$

64°   total of differences

Now work margin exercise 3.

Reading a Histogram

For bar graphs (see Example 1), the base line is labeled for each bar with individual categories (such as people, months, days of the week, and states). Another type of bar graph called a **histogram**, uses a base line marked with numbers that indicate the boundaries of intervals (or ranges) of numbers. This range of numbers is called a **class** (each bar on a histogram represents a class). The bars are placed next to each other with no space between them.

Terms Related to Histograms

Class: an interval (or range) of numbers that contains data items.

Lower class limit: the smallest whole number that belongs to a class.

Upper class limit: the largest whole number that belongs to a class.

Class boundaries: numbers that are halfway between the upper limit of one class and the lower limit of the next class.

Class width: the difference between the class boundaries of a class (the width of each bar).

Frequency: the number of data items in a class.

Example 4

Reading a Histogram

Examine the following histogram. This histogram summarizes the scores of 50 students on an English placement test. Refer to the graph to answer the following questions.

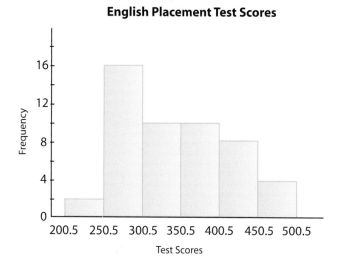

English Placement Test Scores

4. Use the histogram in Example 4 to answer the following questions.

   a. Which class has the least frequency?

   b. What percent of the scores are below 300.5?

   c. Which two classes appear to be equal?

Solution

a. How many classes are represented?   6
b. What are the class limits of the first class?   201 and 250
c. What are the class boundaries of the second class?   250.5 and 300.5

   d. What is the width of each class?   50
   e. Which class has the greatest frequency?   second class
   f. What is this frequency?   16

   g. What percent of the scores are between 200.5 and 250.5? $\dfrac{2}{50} = 4\%$

   h. What percent of the scores are above 400.5? $\dfrac{12}{50} = 24\%$

Now work margin exercise 4.

Practice Problems

1. Using the bar graph in Example 1, what was the amount of increase in sales between March and April?

2. Using the circle graph in Example 2, if the person's income decreases to $30,000 and the percent spent on each item does not change, what is the amount spent on entertainment over the year?

3. Using the line graph in Example 3, find the minimum difference between the high and low temperatures of a single day in the week shown.

4. Using the histogram in Example 4, what is the frequency of the class with the least frequency?

Practice Problem Answers
   1. $25,000   **2.** $1500   **3.** 6°   **4.** 2

Exercises 6.2

1. **Fields of Study:** The following bar graph shows the numbers of students in five fields of study at a university.

Declared College Majors at Downstate University

Field of Study:
- Math & Engineering
- Chemistry & Phys.
- Computer Science
- Humanities
- Social Science

Numbers in Hundreds: 1 2 3 4 5 6 7 8 9 10

a. Which field(s) of study has the largest number of declared majors?

b. Which field(s) of study has the smallest number of declared majors?

c. How many declared majors are indicated in the entire graph?

d. What percent are computer science majors? Round your answer to the nearest tenth of a percent.

2. **Traffic:** The following bar graph shows the number of vehicles that crossed one intersection during a two-week period.

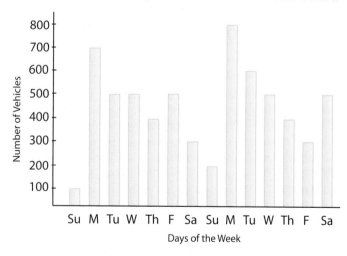

Traffic at One Intersection over a Two Week Period

Number of Vehicles: 100 200 300 400 500 600 700 800

Days of the Week: Su M Tu W Th F Sa Su M Tu W Th F Sa

a. On which day did the highest number of vehicles cross the intersection? How many crossed that day?

b. What was the mean number of vehicles that crossed the intersection on the two Sundays?

c. What was the total number of vehicles that crossed the intersection during the two weeks?

d. About what percent of the total traffic was counted on Saturdays? Round your answer to the nearest tenth of a percent.

3. **College Life:** The following bar graphs show the number of hours worked each week and the GPA's of five college students. When comparing the following two graphs, assume that all five students graduated with comparable grades from the same high school.

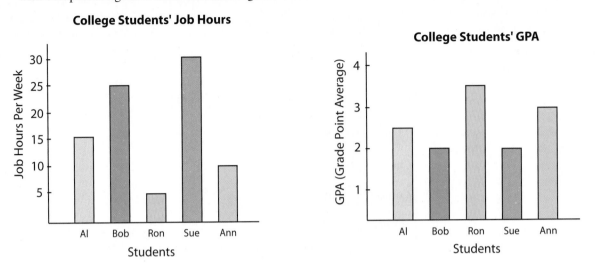

a. Who worked the most hours per week?

b. Who had the lowest GPA?

c. If Ron spent 30 hours per week studying for his classes, then length of his total work week is the sum of the time he spent studying and the time he spent working. What percent of his work week did he spend studying? Round your answer to the nearest tenth of a percent.

d. Which two students worked the most hours? Which two students had the lowest GPA's? Do you think that this is typical?

e. Do you think that the two graphs shown here could be set as one graph? If so, show how you might do this.

4. **Budgeting:** The following circle graph represents the various areas of spending for a school with a total budget of $34,500,000.

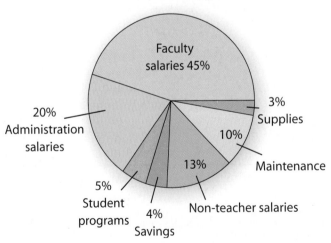

a. What amount will be allocated to each category?

b. What percent will be for expenditures other than salaries?

c. How much will be spent on maintenance and supplies?

d. How much more will be spent on teachers' salaries than on administration salaries?

5. **Television Programming:** The following circle graph represents the types of shows broadcast on television station KCBA. The station is off the air from 2 A.M. to 6 A.M., so there are only 20 hours of daily programming. Sports are not shown in the graph below because they are considered special events.

20-Hour TV Programming at Station KCBA

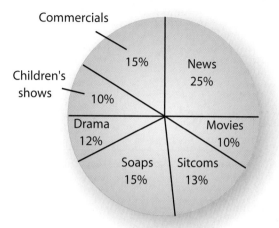

a. In the 20-hour period shown, how much time (in minutes) is devoted daily to each category?

b. What category has the most time devoted to it?

c. How much total time (in minutes) is devoted to drama, soaps, and sitcoms?

6. **Budgeting:** Mike just graduated from college and decided that he should try to live within a budget. The circle graph shows the categories he chose and the percents he allowed. His beginning take home salary is $24,000.

Mike's Budget

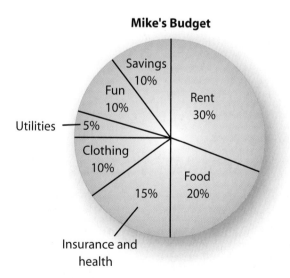

a. How much did he budget for each category?

b. What category was smallest in his budget?

c. What total amount did he budget for food, clothing, and rent?

7. Rainfall: The following line graph shows the total monthly rainfall in Tampa, Florida for the first 6 months of 2010. **Source:** weather.gov

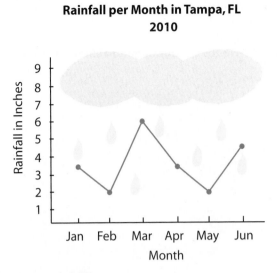

Rainfall per Month in Tampa, FL
2010

a. Which months had the least rainfall?

b. What was the most rainfall in a month?

c. What month had the most rainfall?

d. What was the mean rainfall over the six-month period (to the nearest hundredth)?

8. Mortgage Rates: The following line graph shows the average monthly mortgage rates for June and December for 2006-2010. **Source:** http://www.mortgage-x.com/trends.htm/

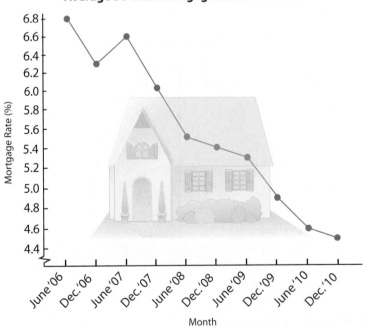

Average 30-Year Mortgage Rates in 2006-2010

a. During what month or months were mortgage rates highest?

b. Lowest?

c. What was the mean of the interest rates given over the 5 year period? Round each value to the nearest tenth of a percent.

9. Baseball: The following line graph shows the number of home runs hit each month of the 2010 Season by José Bautista and Albert Pujols. **Source:** mlb.com

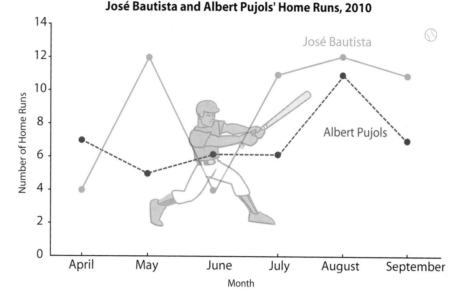

José Bautista and Albert Pujols' Home Runs, 2010

a. During which month did Albert hit the most home runs?

b. How much higher was his total for that month than for the lowest month?

c. In what months did José hit less home runs than Albert?

d. What was the difference between José and Albert's home runs in July?

e. What percent of José's total home runs did he hit in May?

f. What percent of Albert's home runs did he hit in April? Round your answer to the nearest percent.

10. Investing: The following line graphs shows the stock market prices for oil, steel, and wheat over the course of a week. Assume that on Monday morning you had 100 shares of each of the three stocks shown.

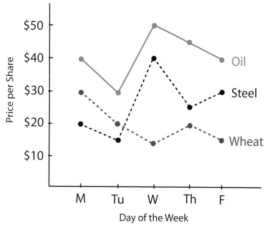

Stock Market Price for One Week

a. If you held the stock all week, on which stock would you have lost money?

b. How much would you have lost?

c. On which stock would you have gained money?

d. How much would you have gained?

e. On which stock could you have made the most money if you had sold at the best time?

f. How much could you have made?

11. Population: The following line graph shows the change in the percent of the U.S. population living in each of four major regions over the last century. **Source:** U.S. Census Bureau, decennial census of population, 1900 to 2000.

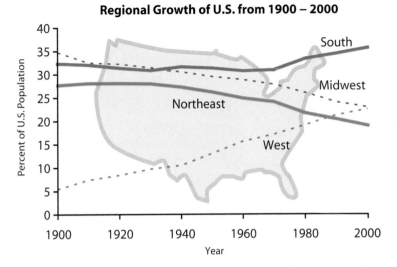

Regional Growth of U.S. from 1900 – 2000

a. Approximately what percent of the population was in each of the four regions in 1900?

b. In 2000?

c. Which region seems to have had the most stable percent of the population between 1900 and 2000?

d. What is the difference between the highest and lowest percents for this region?

e. Which region has had the most growth?

f. What was its lowest percent and when?

g. What was its highest percent and when?

h. Which region has had the most decline?

12. Tires: The following histogram summarizes the tread life for 100 types of new tires.

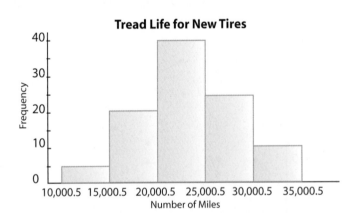

Tread Life for New Tires

a. How many classes are represented?

b. What is the width of each class?

c. Which class has the highest frequency?

d. What is this frequency?

e. What are the class boundaries of the second class?

f. How many tires were tested?

g. What percent of the tires were in the first class?

h. What percent of the tires lasted more than 25,000 miles?

13. **Fuel Efficiency:** A certain number of new cars were evaluated to find how many miles per gallon could be driven with a gallon of gas. The data is summarized in the following histogram.

Miles per Gallon for New Cars Tested

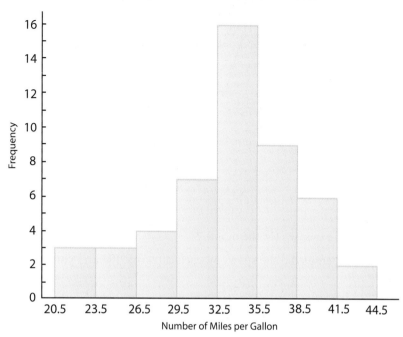

Number of Miles per Gallon

a. How many classes are represented?

b. What is the class width?

c. Which class has the smallest frequency?

d. What is this frequency?

e. What are the class limits for the third class?

f. How many cars were tested?

g. How many cars tested below 30 miles per gallon?

h. What percent of the cars tested above 38 miles per gallon?

14. **Government:** The following circle graph represents the various sources of income for a city government with a total income of $100,000,000.

Sources of City Revenues

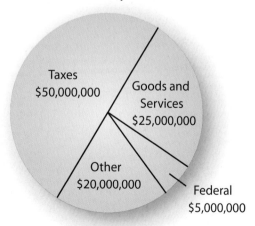

a. What is the city's largest source of income?

b. What percent of income comes from goods and services?

c. What is the ratio of income from taxes to the total income?

15. Cars: The following circle graph shows Sally's car expenses for the month of June.

Monthly Car Expenses

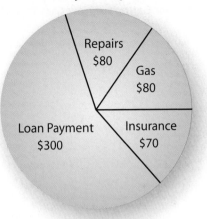

a. What were her total car expenses for the month?

b. What percent of her expenses did she spend on each category? Round your answer to the nearest tenth of a percent.

c. What was the ratio of her insurance expenses to her gas expenses?

6.3 Constructing Graphs from Databases

Objective A   **Constructing a Bar Graph**

In this section, you will be given a table of data and asked to construct your own graph, either a **bar graph** or a **circle graph**.

A   Organize and represent given data in the form of a bar graph.

B   Organize and represent given data in the form of a circle graph.

---

notes

◼ All of the graphs discussed here can be done with a computer and a spreadsheet program such as Excel. If you have access to a computer, your instructor may choose to have you work the problems in this section with a spreadsheet program.

◼

As illustrated in Section 6.2, a bar graph may have either horizontal or vertical bars. In both cases, the length of each bar represents the frequency of the data in a category being graphed. For consistency and to simplify the directions, we will discuss the construction of vertical bar graphs only.

Steps to Follow in Constructing a Vertical Bar Graph

1.   Draw a vertical axis and a horizontal axis.
2.   Mark an appropriate scale on the vertical axis to represent the frequency of each category. (The scale must be uniform. That is, the distance between consecutive marks must represent the same amount.)
3.   Mark the categories of data along the horizontal axis.
4.   Draw the vertical bar for each category so that the height of the bar reaches the frequency of the data in that category.
5.   The bars have the same width and do not touch each other.

Example 1

Constructing a Bar Graph

Construct a bar graph that represents the following population data.

1.	New York, NY	8,392,000
2.	Los Angeles, CA	3,832,000
3.	Chicago, IL	2,851,000
4.	Houston, TX	2,258,000
6.	Phoenix, AZ	1,594,000
5.	Philadelphia, PA	1,547,000
7.	San Antonio, TX	1,374,000
8.	San Diego, CA	1,306,000

1. Construct a bar graph using the following data of 5 college students' GPA at the end of the semester.

Ian 2.50
Marcus 3.25
Luke 3.00
Jenna 1.75
Denise 3.75

Steps 1 and 2: Draw the vertical axis and horizontal axis and mark a scale on the vertical axis that will encompass the numbers from 0 to 8.392 million people. (On this graph, we have chosen to mark the numbers from 0 to 9.0 in a scale of 1 unit.)

Steps 3 and 4: The horizontal axis marks are labeled with the names of the cities represented. The height of each vertical bar corresponds to the population (in millions) of each city as given.

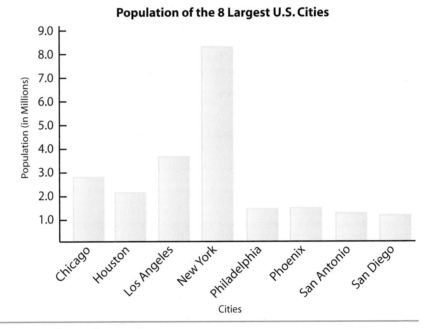

Population of the 8 Largest U.S. Cities

Now work margin exercise 1.

Objective B **Constructing a Circle Graph**

A **circle graph** is a circle that is marked in **sectors** (pie-shaped wedges) that correspond to percentages of data in each category represented. Refer to Section 5.1 for information on angles and protractors and Section 5.4 for information on circles.

If a radius of a circle is rotated completely around the circle (like the motion of the minute hand of a clock in 1 hour), the radius is said to rotate 360°. Therefore, to find the central angle needed in the graph to represent a percent of the data, multiply that percent (in decimal form) by 360°.

Steps to Follow in Constructing a Circle Graph

1. Find the central angle (angle at the center of the circle) for each category by multiplying the corresponding percent (in decimal form) by 360°.
2. Draw a circle.
3. Draw each central angle (use a protractor), and label each sector with the name and corresponding percent of each category.

Example 2

Constructing a Circle Graph

Construct a circle graph that represents the following data.

Ethnic Breakdown of Students Who Took the SAT (Scholastic Assessment Test), Nationwide, 2010

Ethnicity	Percent
African-American	13%
Asian-American	11%
Hispanic	14%
Native American	1%
White	54%
Other	3%
No Response	4%

Source: CollegeBoard.com

Step 1: Find each percent of 360°.

African-American: 13% of 360° = $0.13 \times 360° = 46.8°$

Asian-American: 11% of 360° = $0.11 \times 360° = 39.6°$

Hispanic: 14% of 360° = $0.14 \times 360° = 50.4°$

Native American: 1% of 360° = $0.01 \times 360° = 3.6°$

White: 54% of 360° = $0.54 \times 360° = 194.4°$

Other: 3% of 360° = $0.03 \times 360° = 10.8°$

No response: 4% of 360° = $0.04 \times 360° = 14.4°$

Steps 2 and 3: Draw a circle, mark the central angles as close to the actual degrees as is practical, and label each sector. Note that the order of the sectors (pie slices) is not important.

Ethnic Breakdown for SAT, Nationwide, 2010

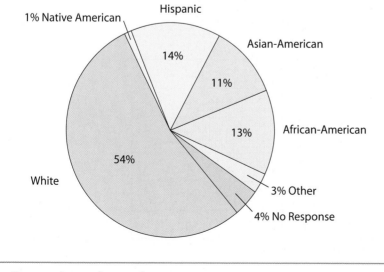

Now work margin exercise 2.

2. Construct a circle graph that represents the following data based on the 2010 school budget for Richland County.

Faculty Salaries	40%
Administration	20%
Student programs	15%
Non-teacher Salaries	12%
Maintenance	10%
Supplies	3%

Exercises 6.3

1. **Geography:**   Construct a bar graph that represents the following data.

Largest Islands of the World

Island	Area in Square Miles (nearest ten thousand)
Greenland	840,000
New Guinea	310,000
Borneo	290,000
Madagascar	230,000
Baffin	200,000
Sumatra	180,000
Honshu	90,000
Great Britain	90,000

Source: Infoplease

2. **Movies:**   Construct a bar graph that represents the following data.

Top 10 Films by U.S./Canada Box Office Earnings, 2010

Motion Picture	Box Office (in millions of dollars)
Avatar	$476.9
Toy Story 3	$415.0
Alice in Wonderland	$334.2
Iron Man 2	$312.1
The Twilight Saga: Eclipse	$300.5
Inception	$292.6
Harry Potter and the Deathly Hallows, Part 1	$280.2
Despicable Me	$251.1
Shrek Forever After	$238.4
How to Train Your Dragon	$217.6

Source: Motion Picture Association of America

3. **Architecture:** Construct a bar graph that represents the following data.

Well-Known U.S. Skyscrapers

Building and City	Height (in Stories)
AMOCO (Chicago)	80
Chrysler (NY)	77
Empire State (NY)	102
First Interstate World Center (LA)	73
John Hancock Center (Chicago)	100
Willis Tower (Chicago)	110
Texas (Houston)	75
One Liberty Place (Philadelphia)	61

4. **Anatomy:** Construct a circle graph that represents the following data.

Percent of Population with Particular Blood Types

Type of Blood	Percent of Population
O positive $\left(O^+\right)$	38%
O negative $\left(O^-\right)$	7%
A positive $\left(A^+\right)$	34%
A negative $\left(A^-\right)$	6%
B positive $\left(B^+\right)$	9%
B negative $\left(B^-\right)$	2%
AB positive $\left(AB^+\right)$	3%
AB negative $\left(AB^-\right)$	1%

Source: AABB.org

5. Population: Construct a bar graph that represents the following data.

Population of Countries that Begin with the Letter L (Mid 2011 Estimates)

Country	Population
Laos	6,477,000
Latvia	2,205,000
Lebanon	4,143,000
Lesotho	1,925,000
Liberia	3,787,000
Libya	6,598,000
Liechtenstein	35,000
Lithuania	3,536,000
Luxembourg	503,000

Source: CIA World Factbook

6. Civil Engineering: Construct a bar graph that represents the following data.

Notable Modern Bridges

Bridge and Location	Span (in meters)
Brooklyn (New York)	500
Bosporus (Istanbul)	1100
Fourth Road (Scotland)	1000
George Washington (New York)	1100
Golden Gate (San Francisco)	1300
Humber (Hull, Britain)	1400
Mackinac Straits (Michigan)	1200
Newport (Rhode Island)	500
Ponte 25 de Abril (Lisbon)	1000
Verazano-Narrows (New York)	1300

7. **Energy:** Construct a circle graph that represents the following data.

World Sources of Energy for 2007

Source of Energy	Percent
Liquids	35%
Natural Gas	23%
Coal	27%
Nuclear	5%
Renewables	10%

Source: U. S. Energy Information Administration

8. **Olympics:** Construct a circle graph that represents the following data. Divide each number of medals by the total number to get corresponding percent for each country.

Total Medals Awarded to Top 7 Countries, 2010 Winter Olympic Games (Vancouver)

Country	Number of Medals
United States	37
Germany	30
Canada	26
Norway	23
Austria	16
Russia	15
South Korea	14
Total	161

9. Health: Construct a circle graph that represents the levels of cholesterol from a sample of 100 students reporting for a physical examination.

Cholesterol Levels of 100 Students

Cholesterol Level	Number of Students
Recommended	35
Borderline	15
Moderate Risk	40
High Risk	10

10. Employment: Construct a circle graph that represents the following data.

2010 Distribution of Employment by Industry

Industry	Percent
Professional and Business Services	11%
Education and Health Services	23%
Manufacturing	10%
Wholesale and Retail Trade	14%
Construction	7%
Agriculture, Forestry, Fishing, and Hunting	2%
Other	33%

Source: U.S. Department of Labor

11. Technology: Construct a circle graph that represents the following data.

2009 World Broadband Subscribers by Region

Region	Percent
North America	21%
South and East Asia	25%
Western Europe	24%
Asia and the Pacific	14%
Other	16%

Source: Point Topic World Broadband Statistics Report

12. **Population:** Construct a bar graph that represents the following data.

World Population by Decade

Year	Total World Population
1950	2,556,000,053
1960	3,039,451,023
1970	3,706,618,163
1980	4,453,831,714
1990	5,278,639,789
2000	6,082,966,429
2010	6,852,472,823

Source: US Census Bureau

13. **Music:** Construct a bar graph that represents the following data.

Top 10 Selling Musical Artists in 2010

Artist	Units Sold
Taylor Swift	4,470,000
Eminem	4,317,000
Lady Antebellum	3,848,000
Justin Bieber	3,728,000
Glee Cast	3,603,000
Susan Boyle	2,711,000
Lady Gaga	2,591,000
Michael Jackson	2,118,000
Zac Brown Band	1,824,000
The Beatles	1,697,000

Source: Nielsen Soundscan

Collaborative Learning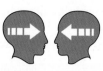

14. **Agriculture:** With the class separated into teams of two to four students, find access to a computer and a computer spreadsheet program, such as Excel, and enter the following data related to crop production on farms. Have the program find the mean and median for each category of crop for the years listed here. Also, have the program draw a bar graph for each category of crop by using the years listed here on the base line. Each team leader is to discuss how the team learned to use the program and bring printouts of the graphs. A general classroom discussion related to the power and efficiency of the computer should follow.

Crop Production in the U.S. (in million bushels)

Year	Corn	Sorghum	Barley	Oats	Rye
2003	10,087	411	278	144	8.63
2004	11,806	454	280	116	8.26
2005	11,112	393	212	115	7.54
2006	10,531	277	180	94	7.19
2007	13,038	497	210	90	6.31
2008	12,092	472	240	89	7.98
2009	13,092	383	227	93	6.99
2010	12,447	345	180	81	7.43

Source: U.S. Department of Agriculture, National Agricultural Statistics Survey

6.4 Probability

Objective A **Probability**

Activities involving chance such as tossing a coin, rolling a die, spinning a wheel in a game, and predicting weather are called **experiments**. The likelihood of a particular result is called its **probability**. For example,

The weather person might predict rain with a probability of 30%.

In tossing a fair coin, the probability of heads (H) is $\frac{1}{2}$.

Terms Related to Probability

Experiment: an activity in which the result is random in nature.

Outcome: an individual result of an experiment.

Sample Space: the set of all possible outcomes of an experiment.

Event: some (or all) of the outcomes from the sample space.

A **tree diagram** can be used to "picture" the possible outcomes of an experiment. Each branch of the tree diagram shows a separate outcome. For example, the tree diagram shown here illustrates the possible outcomes of tossing a coin.

H

T

Tree Diagram

Example 1

Drawing a Tree Diagram

A coin is tossed twice. Draw a tree diagram illustrating this experiment and list the possible outcomes in the sample space.

Solution

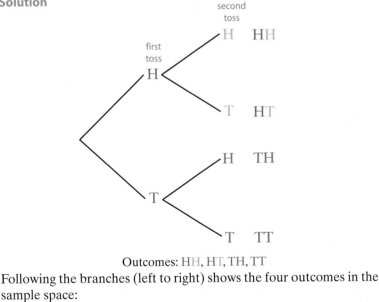

Outcomes: HH, HT, TH, TT

Following the branches (left to right) shows the four outcomes in the sample space:

$$S = \{HH, HT, TH, TT\}.$$

Now work margin exercise 1.

1. A coin is tossed and then one of the numbers (1, 2, and 3) is chosen at random from a box. Draw a tree diagram illustrating the possible outcomes of the experiment and list the outcomes in the sample space.

2. A coin is tossed and then a six-sided die is rolled. Draw a tree diagram illustrating the outcomes of the experiment and list the outcomes in the sample space.

3. A 4-sided die is rolled twice. Draw a tree diagram illustrating the outcomes of the experiment and list the outcomes in the sample space.

Example 2

Drawing a Tree Diagram

A coin is tossed three times. Draw a tree diagram illustrating this experiment and list the outcomes in the sample space.

Solution

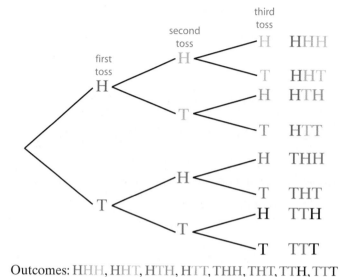

Outcomes: HHH, HHT, HTH, HTT, THH, THT, TTH, TTT

There are eight outcomes in the sample space.
$$S = \{HHH, HHT, HTH, HTT, THH, THT, TTH, TTT\}.$$

Example 3

Drawing a Tree Diagram

A 6-sided die is rolled. Draw a tree diagram illustrating this experiment.

Solution

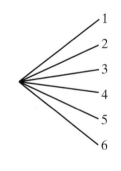

Now work margin exercises **2** *and* **3**.

Objective B **Finding the Probability of an Event**

If a coin is tossed three times, there are eight outcomes (as shown in Example 2).

$$S = \{HHH, HHT, HTH, HTT, THH, THT, TTH, TTT\}$$

Choosing only those outcomes with two heads is an **event**. If we label this event A, then
$$A = \{\text{HHT}, \text{HTH}, \text{THH}\}.$$

The probability of an event is the likelihood of it occurring. Because there are eight possible outcomes in the sample space and three in event A, we have probability of $A = \dfrac{3}{8}$.

Probability of an Event

$$\text{probability of an event} = \frac{\text{number of outcomes in event}}{\text{number of outcomes in sample space}}$$

Probabilities have the following two basic characteristics.
1. Probabilities are between 0 and 1, inclusive.
   If an event can never occur, its probability is 0.
   If an event will always occur, its probability is 1.
2. The sum of the probabilities of the outcomes in a sample space is 1.

Example 4

Finding the Probability

A coin is tossed twice. Find the probability that both tosses are tails, TT.

Solution

There are four outcomes in the sample space: $S = \{\text{HH}, \text{HT}, \text{TH}, \text{TT}\}$. One of these outcomes is the event TT.
So,
$$\text{probability of TT} = \frac{1}{4}.$$

Example 5

Finding the Probability

A die is rolled once. Find the probability of rolling an even number.

Solution

There are six possible outcomes in rolling a die. The sample space is
$$S = \{1, 2, 3, 4, 5, 6\}.$$

Of these six, three are even numbers: $2, 4, 6$.
So,
$$\text{probability of an even number} = \frac{3}{6} = \frac{1}{2}.$$

4. There are three M&M candies in a box: one red, one yellow, and one green. Two pieces are chosen. What is the probability that these pieces are the red one and the green one?

5. If a coin is tossed three times, what is the probability that the tosses are all heads or all tails?

6. One card is drawn from a standard deck of cards. Find the probability of each of the following events:

   a. The card is the queen of clubs.

   b. The card is a king.

   c. The card is a 3, 4, or 5.

Example 6

Finding the Probability

A standard deck of cards is 52 cards with four suits (hearts, diamonds, spades, and clubs) and 13 cards in each suit. The cards are ace, king, queen, jack, 10, 9, 8, 7, 6, 5, 4, 3, and 2. If one card is drawn from a deck of cards, find the probability of each of the following events.

a. The card is a 10.

Solution

There are four 10s in a deck (one 10 in each suit). So,

$$\text{probability of a 10} = \frac{4}{52} = \frac{\cancel{4} \cdot 1}{\cancel{4} \cdot 13} = \frac{1}{13}.$$

b. The card is a diamond.

Solution

There are 13 diamonds in a deck (diamonds is one of the suits). So,

$$\text{probability of a diamond} = \frac{13}{52} = \frac{\cancel{13} \cdot 1}{4 \cdot \cancel{13}} = \frac{1}{4}.$$

c. The card is a 2 or a 3.

Solution

There are four 2s and four 3s in a deck (one of each in each suit) for a total of 8. So,

$$\text{probability of a 2 or a 3} = \frac{8}{52} = \frac{\cancel{4} \cdot 2}{\cancel{4} \cdot 13} = \frac{2}{13}.$$

Now work margin exercises **4 through 6 .**

Draw a tree diagram for each experiment and list the possible outcomes in the sample space. See Examples 1 through 3.

1. **Games:** A coin is tossed followed by choosing a number from 1 to 4.

2. **Games:** Four marbles are in a box one red, one white, one blue, and one purple. One is chosen.

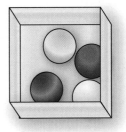

3. **Pizza:** In ordering a pizza, a crust is chosen (crispy or thick) followed by choosing a topping (cheese, pepperoni, or sausage).

4. **Games:** A game is played by a coin toss followed by spinning a spinner with three colors: yellow, blue, and orange.

5. **Fashion:** You have two pairs of pants: jeans and khaki pants. You also have three shirts: a blue one, a green one, and a black one. How many outfits can you make?

6. **Coins:** A coin is tossed four times.

7. **Games:** A spinner with two colors (yellow and blue) is spun followed by rolling a die.

8. **Games:** Two digits from 0 to 9 are chosen at random. The first one is even and the second one is odd.

9. **Marbles:** A box contains 5 marbles: two red, one white, two blue. What is the probability of choosing a blue marble from the box?

10. **Gumball Machines:** A machine contains only 5 gumballs: 3 yellow, one green, one white. What is the probability of getting a yellow gumball when you put a coin in the machine?

11. **Education:** Your English professor chooses students randomly to answer questions. If the class has 20 students, what is the probability that you will be selected to answer the next question?

12. **Clothes:** There are four socks in a drawer (two black and two blue). Two socks are chosen at random. What is the probability that the chosen socks are a matching pair?

Not all experiments are easily pictured with tree diagrams. If two dice are rolled there are 36 possible outcomes and these outcomes can be represented in table form as shown here. In this table, we have illustrated the two dice in different colors, blue and red. Use this table to find the probabilities of the events described. See Example 5.

	1	2	3	4	5	6
1	1, 1	1, 2	1, 3	1, 4	1, 5	1, 6
2	2, 1	2, 2	2, 3	2, 4	2, 5	2, 6
3	3, 1	3, 2	3, 3	3, 4	3, 5	3, 6
4	4, 1	4, 2	4, 3	4, 4	4, 5	4, 6
5	5, 1	5, 2	5, 3	5, 4	5, 5	5, 6
6	6, 1	6, 2	6, 3	6, 4	6, 5	6, 6

13. a sum of 7

14. a sum of 6

15. a sum of 15

16. a sum of 4

17. a sum of 2

18. a sum of 8

19. neither die is even

20. at least one die is 6

A box contains four pieces of paper with the numbers 1, 2, 3, and 4 written on them. A piece of paper is drawn at random. Find the probability of each event.

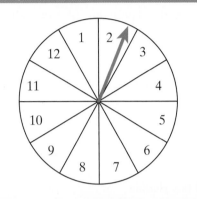

21. The number is 4.

22. The number is odd.

23. The number is not 3.

24. The number is less than 3.

A spinner (shown) with the numbers from 1 to 12 is spun once. Find the probability of each event.

25. The number is an even number?

26. The number is a prime number?

27. The number is a multiple of 5?

28. The number is less than 13?

A standard deck of cards is 52 cards with four suits (hearts, diamonds, spades, and clubs) and 13 cards in each suit. The cards are ace, king, queen, jack, 10, 9, 8, 7, 6, 5, 4, 3, and 2. If one card is drawn from a deck of cards, find the probability of each of the following events. See Example 6.

29. The card is an ace.

30. The card is the 3 of diamonds.

31. The card is a club.

32. The card is a queen or a jack.

33. The card is red.

34. The card is a 10, 9, or 8.

35. The card is the king of hearts.

36. The card is a 1.

Writing & Thinking

37. Give an example of an event that you think has a probability of 1.

38. Explain, in your own words, why the probability of an event cannot be greater than 1.

Section 6.1: Statistics: Mean, Median, Mode, and Range

Data — page 446
Value(s) measuring some characteristic of interest such as income, height, weight, grade point averages, scores on tests, and so on. (We will consider only numerical data).

Mean — page 446
The sum of all the data divided by the number of data items. (Also called the **arithmetic average**.)

Median — page 446
The middle data item. (Arrange the data in order and pick out the middle item).

Mode — page 446
The single data item that appears the most number of times. (Some data may have more than one mode).

Range — page 446
The difference between the largest and smallest data items.

To Find the Median — page 447
1. Arrange the data in order.
2. If there is an **odd** number of items, the median is the middle item.
3. If there is an **even** number of items, the median is the average of the two middle items.

Section 6.2: Reading Graphs

Types of Graphs — page 453
1. **Bar Graphs:** to emphasize comparative amounts.
2. **Circle Graphs:** to help in understanding percents or parts of a whole.
3. **Line Graphs:** to indicate tendencies or trends over a period of time.
4. **Histograms:** to indicate data in classes (a range or interval of numbers).

Terms Related to Histograms — page 457
Class: an interval (or range) of numbers that contains data items.
Lower class limit: the smallest whole number that belongs to a class.
Upper class limit: the largest whole number that belongs to a class.
Class boundaries: numbers that are halfway between the upper limit of one class and the lower limit of the next class.
Class width: the difference between the class boundaries of a class (the width of each bar).
Frequency: the number of data items in a class.

Steps to Follow in Constructing a Vertical Bar Graph   page 467
1. Draw a vertical axis and a horizontal axis.
2. Mark an appropriate scale on the vertical axis to represent the frequency of each category. (The scale must be uniform. That is, the distance between consecutive marks must represent the same amount).
3. Mark the categories of data along the horizontal axis.
4. Draw the vertical bar for each category so that the height of the bar reaches the frequency of the data in that category.
5. The bars have the same width and do not touch each other.

Steps to Follow in Constructing a Circle Graph   page 468
1. Find the central angle (angle at the center of the circle) for each category by multiplying the corresponding percent (in decimal form) by 360°.
2. Draw a circle.
3. Draw each central angle (use a protractor), and label each sector with the name and corresponding percent of each category.

Section 6.4: Probability

Terms Related to Probability   page 477
Experiment: an activity in which the result is random in nature.
Outcome: an individual result of an experiment.
Sample Space: the set of all possible outcomes of an experiment.
Event: some (or all) of the outcomes from the sample space.

Probability of an Event   page 479

$$\text{probability of an event} = \frac{\text{number of outcomes in event}}{\text{number of outcomes in sample space}}$$

Chapter 6: Review

Section 6.1: Statistics: Mean, Median, Mode, and Range

Find the following statistics for the given data. a. the mean, b. the median, c. the mode, and d. the range.

1. 27   36   45   72   63   36   27   18   36   90

2. **Engineering:** The capacity (in thousands of cubic meters) of the world's 5 largest dam areas follows:

Dams	Capacity (in thousands of cubic meters)
New Cornelia Tailings, Arizona	209,500
Pati, Argentina	238,200
Tarbela, Pakistan	121,720
Syncrude Tailings, Canada	540,000
Chapeton, Argentina	396,200

   a. Mean = _____
   b. Median = _____
   c. Mode = _____
   d. Range = _____

3. **Libraries:** The volumes (to nearest thousand) in top college libraries in the United States are as follows:

College Library	Volume (to nearest thousand)
Harvard	13,617,000
Yale	9,932,000
U of Illinois	9,024,000
UC Berkeley	8,628,000
U of Texas	7,495,000
U of Michigan	6,973,000
UCLA	7,010,000
Columbia	6,906,000

   a. Mean = _____
   b. Median = _____
   c. Mode = _____
   d. Range = _____

4. **Football:** The winning margin for each of the first 40 Super Bowls are as follows. Round your answers to the nearest tenth, if necessary.

   25   19   9   16   3   21   7   17   10   4   18   17   4   12
   17   5   10   29   22   36   19   32   4   45   1   13   35   17
   23   10   14   7   15   7   27   3   27   3   3   11

   a. Mean = _____   b. Median = _____
   c. Mode = _____   d. Range = _____

5. **Military:** The Medal of Honor is the nation's highest military award for uncommon valor by men and women in battle. The medals awarded by conflict are as follows. Round your answers to the nearest tenth, if necessary. **Source:** Congressional Medal of Honor Society

Civil War:	1520	Indian Wars (1861-1898):	428
Korean Expedition (1871):	15	Spanish-American War:	109
Philippines/Samoa (1899-1913):	91	Boxer Rebellion (1900):	59
Dominican Republic (1904):	3	Nicaragua (1911):	2
Mexico (Veracruz) (1914):	55	Haiti (1915):	6
Miscellaneous (1861-1920):	166	World War I:	124
Haitian Action (1919-20):	2	Korean War:	131
World War II:	433	Somalia:	2
Vietnam War:	238		

a. Mean = _____   b. Median = _____

c. Mode = _____   d. Range = _____

Section 6.2: Reading Graphs

Answer the questions related to each of the graphs. Some questions can be answered directly from the graphs; others may require some calculations.

6. **Cell Phones:** The circle graph below indicates the percent of the smartphone market that used certain operating systems (OS) in the first quarter of 2010. Use this information to answer the following questions.

a. What percent of the market used the Android OS?

b. Which OS was used by only 4% of the market share in the first quarter of 2010?

c. What is the difference between the percent of the market using RIM Blackberry OS and the percent of the market using Microsoft Windows Mobile?

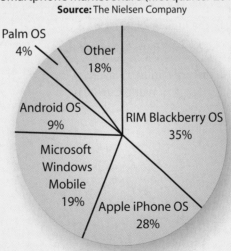

Smartphone Market Share (first quarter 2010)
Source: The Nielsen Company

7. **Television:** Use the information in the bar graph shown to answer the following questions.

   a. Which TV show was watched by 8,446,000 people on September 19, 2010?

   b. Which TV Program was watched by the most people?

   c. How many more people watched America's Got Talent than watched Jersey Shore?

   d. How many total people were tuned in to Parenthood and The Closer?

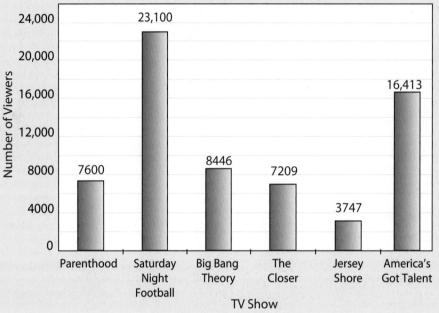

Primetime Broadcast Total Viewing for September 19, 2010
Source: The Nielsen Company

8. **Soccer:** The histogram shown below gives the total number of minutes played by the top 49 players in the 2010 World Cup. Round your answer to the nearest tenth, if necessary.

   a. How many classes are represented?

   b. What is the width of each class?

   c. Which class has the highest frequency?

   d. What is this frequency?

   e. What are the class boundaries of the third class?

   f. What percent of players were in the highest class?

   g. What percent of the players played more than 549 minutes?

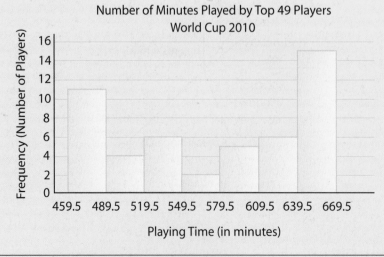

Number of Minutes Played by Top 49 Players
World Cup 2010

9. **Unemployment:** Use the information in the line graph below to answer the following questions.

   a. Which month saw the lowest percent of unemployment?

   b. Which month saw the highest percent of unemployment?

   c. What was the highest percent of unemployment?

   d. What is the mean percent of unemployment between January of 2010 and September of 2010? Round your answer to the nearest hundreth of a percent.

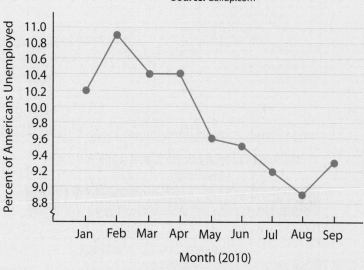

United States Unemployment Rate 2010

Source: Gallup.com

Section 6.3: Constructing Graphs from Databases

Construct a graph of the indicated type that represents the given data.

10. **Spending:** Construct a circle graph that represents the following information:

2008 Average Annual Expenditures

Source: Bureau of Labor Statistics

Expenditure	Percent of Annual Income Spent
Food	13%
Housing	34%
Transportation	17%
Healthcare	6%
Entertainment	5%
Other	25%

11. Social Networking: Construct a circle graph that represents the following information:

Most Visited Social Networking Sites
(week of September 25, 2010)
Source: Experian Hitwise

Social Networking Site	Percent of Total Hits
Facebook	62%
YouTube	17%
MySpace	6%
Twitter	1%
Other	14%

12. Income: Construct a bar graph that represents the following information:

Median Income by State (2008-2009 Average)
Source: U.S. Census Bureau

State	Median Income
Alaska	$62,675
Idaho	$47,009
South Carolina	$41,548
Hawaii	$58,469
Mississippi	$35,693
Iowa	$50,337

13. Salaries: Construct a bar graph that represents the following information:

2009 Average Starting Salaries for New College Graduates
Source: University Language Services

Discipline or Major	Average Starting Salary
Computer Science	$61,407
Marketing	$43,325
Psychology	$34,284
Economics	$49,829
Liberal Arts	$36,175
Civil Engineering	$52,048
History	$37,861

Section 6.4: Probability

14. What is the term for the individual results of an experiment?

15. Explain, in your own words, what an event is.

16. Games: Draw a tree diagram for the experiment of spinning a spinner with two colors (red and yellow) twice.

17. Games: Draw a tree diagram for the experiment of drawing a number from 1 to 5 at random out of a hat.

18. Games: Draw a tree diagram for rolling a die and then tossing a coin twice.

19. Cards: What is the probability of drawing an ace of hearts from a standard deck of cards?

20. Cards: What is the probability of drawing the number 11 from a standard deck of cards?

21. Marbles: Box 1 contains marbles numbered 1, 2, and 3. Box 2 contains marbles numbered 4 and 5. A marble is drawn from each box. What is the probability that

a. both marbles are numbered with prime numbers?

b. both marbles are numbered with even numbers?

c. the sum of the numbers is 9?

d. the sum of the numbers is a multiple of 3?

1. **Miles Per Gallon:** Ten cars had the following miles-per-gallon ratings:

   26, 21, 20, 34, 20, 30, 25, 25, 25, 28

   Find the following statistics for these ten data items.

   a. Mean = _____   **b.** Median = _____

   c. Mode = _____   **d.** Range = _____

2. **Banking:** Financial institutions sometimes fail. Most are insured by the FDIC (Federal Deposit Insurance Corporation). In the years from 2001 to 2010 the following numbers of financial institutions have failed.

Year	2001	2002	2003	2004	2005	2006	2007	2008	2009	2010
Number Failed	4	11	3	4	0	0	3	25	140	157

   Find the following statistics for closures over these ten years:

   a. Mean = _____   **b.** Median = _____

   c. Mode = _____   **d.** Range = _____

3. **Football:** The bar graph below shows the team total offensive yards of six NFL teams. Use this information to answer the following questions.

   a. Which team had the greatest number of yards in 2009?

   b. Which team had the fewest number of yards?

   c. How many yards were totaled by the team found in part **b.**?

   d. How many more yards did the offense of the Minnesota Vikings total than the offense of the New York Jets?

   e. Of the six teams shown, what percent of yards were gained by the Indianapolis Colts? Round your answer to the nearest tenth of a percent.

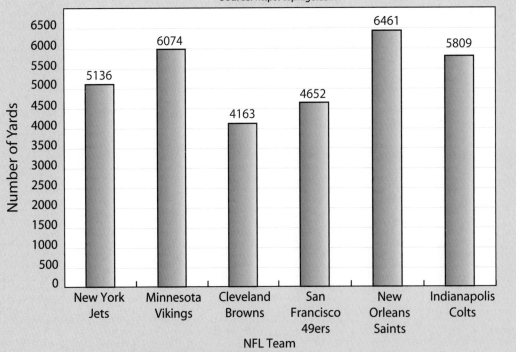

NFL Team Total Offensive Yards for 2009 Season
Source: http://espn.go.com

4. **Video Games:** Use the circle graph shown to the right to answer the following questions.

   a. Which video game console has the highest sales?

   b. What percent of people purchased a PlayStation3?

   c. What total percent of video game consoles purchased were made by Nintendo?

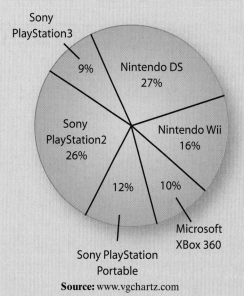

2010 Sales for Top 6 Video Game Consoles

Sony PlayStation3 9%
Nintendo DS 27%
Sony PlayStation2 26%
Nintendo Wii 16%
12%
10%
Sony PlayStation Portable
Microsoft XBox 360

Source: www.vgchartz.com

5. **Housing:** Construct a circle graph that represents the following information:

Year Owner-Occupied Housing Units were Built
Source: U.S. Census Bureau, 2005-2009
American Community Survey

Year House Built	Percent of Total Owner-Occupied Houses
2000 or later	12%
1980 to 1999	30%
1960 to 1979	27%
1940 to 1959	18%
1939 or earlier	13%

6. **Games:** List the outcomes in the sample space of the experiment of tossing a coin and then randomly selecting the number 1 or 2. Draw a tree diagram for this experiment.

7. **Coins:** What is the probability of getting two heads when tossing a coin twice?

8. **Games:** If two dice are rolled, what is the probability that the sum of the dice is 12?

9. **Cards:** What is the probability of drawing a black card from a standard deck of cards?

10. **Clothes:** A drawer has three socks in it: two blue and one brown.

   a. In a single random selection what is the probability that the sock is blue?

   b. In selecting two socks, what is the probability that they are the blue pair?

Cumulative Review: Chapters 1 - 6

Perform the indicated operations.

1. $48 \div 12 \div 4 - 1 + 6$

2. $4(7-2) \div 10 + 5$

3. $18 + 18 \div 2 \div 3 - 3 \cdot 1$

4. $1\frac{2}{3} \cdot 1\frac{12}{15}$

5. $\frac{6}{13} \div \frac{5}{6}$

6. $\frac{2}{35} + \frac{3}{7}$

7. $6\frac{1}{6} - 4\frac{2}{7}$

8. $\frac{4}{5} \div 1\frac{1}{5} + \frac{2}{15} \cdot 1\frac{3}{7}$

9. $85.815 - 17.943$

10. $13.4 + 6 + 17.913$

11. $(1.92)(1000)$

12. $32.07 + 14 \div 2.5 \cdot 1.334$

Using the tests for divisibility, determine which of the numbers 2, 3, 4, 5, 6, 9, and 10 will divide exactly into each of the following numbers.

13. 372

14. 375

Decide whether each of the following numbers is prime or composite. If the number is composite, find at least three factors for the number.

15. 19

16. 63

Find the LCM for each set of numbers.

17. 36, 54, 72

18. 15, 25, 30, 40

Change each decimal to fraction form.

19. 0.875

20. 4.25

Change each fraction to decimal form.

21. $\frac{7}{16}$

22. $\frac{9}{15}$

Change each decimal to a percent.

23. 0.25

24. 1.384

Change each percent to a decimal.

25. 38.2%

26. 259%

Solve the following proportions. Reduce all fractions.

27. $\dfrac{10}{12} = \dfrac{x}{6}$

28. $\dfrac{1.7}{5.1} = \dfrac{100}{y}$

Find the perimeter and area of the following figures.

29.

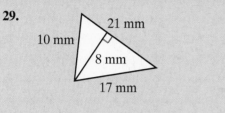

30.

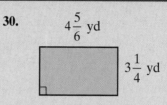

31.

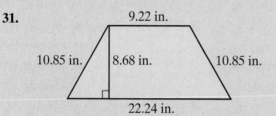

32. Assume that the given triangles are similar and find the values for x and y.

a. $\triangle ABC \sim \triangle DEC$

b. $\triangle LMN \sim \triangle OPQ$

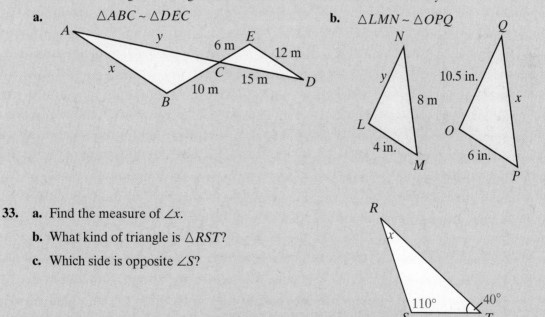

33. **a.** Find the measure of $\angle x$.

b. What kind of triangle is $\triangle RST$?

c. Which side is opposite $\angle S$?

Solve the following word problems.

34. **Grades:** In a calculus class, the scores on the first exam were as follows: five students made a 94, two students made a 93, seven students made an 89, one student made an 88, six students made a 72 and three students made a 50. Determine the mean grade for this first exam, rounded to the nearest tenth of a percent.

35. **Simple Interest:** Sally needs to borrow $1200 to help pay off her student loans. Her parents offer to loan her the money if she pays them an easy interest rate of 0.025 and promises to pay them back in 9 months. How much interest will she pay her parents? Use the formula $I = Prt$.

36. Pythagorean Theorem: Mike lives in the same neighborhood as his best friend Daniel. To walk to Daniel's house, Mike walks down one road for 352 yards, turns and walks down another road 264 yards. What is the shortest distance between Mike and Daniel's houses?

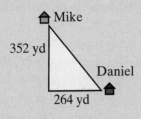

37. Similar Triangles: In the figure below, $\triangle ABC \sim \triangle DEF$. Determine the values of x and y.

38. Military: The armed forces of countries are measured in several ways: active troops, reserve troops, tanks, navy (carriers, cruisers, frigates, destroyers, submarines), and combat aircraft. For 2009, the following countries had active troops in the following numbers:

Afghanistan: 93,800   Egypt: 468,500   China: 2,285,000

France: 352,771   U.S.:   1,580,255   Chile: 60,560

a. What was the mean number of active troops for these countries? Round your answer to the nearest tenth.

b. What was the median number of active troops for these countries?

39. Government: Each state pays its governor a salary. As of 2010, the governor's salaries from ten states were as follows:

New York: $179,000   Michigan: $177,000   California: $206,500

New Jersey: $175,000   Virginia: $166,000   Vermont: $142,542

Illinois: $177,500   Connecticut: $150,000   Washington: $166,891

Maryland: $150,000

Find the following statistics related to these ten salaries:

a. Mean = _____   **b.** Median = _____

c. Mode = _____   **d.** Range = _____

Family Budget

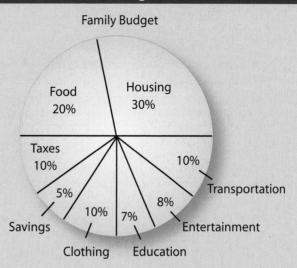

40. The circle graph shows a home budget. What amount will be spent on each category if the family income is $35,000?

41. How much more will the family spend for food than for clothing if their income is increased to $40,000 (to the nearest dollar)?

42. What fractional part of the family income is spent for food, housing, and taxes combined?

43. How much will the family spend for food, housing, and transportation combined if their income is reduced to $30,000 (to the nearest dollar)?

44. Construct a bar graph that represents the following data.

First Six Months Sales - 2010
Better Book Co.

Months	Sales in Thousands of Dollars
January	100
February	150
March	50
April	75
May	125
June	150

45. **Circle Graph:** Construct a circle graph that represents the following data.

20-Hour TV Programming at Station WFMJ

Category	Percent
Commercials	10%
News	30%
Movies	5%
Sitcoms	20%
Soaps	16%
Drama	14%
Children's Shows	5%

46. **Probability:** List the outcomes in the sample space of the experiment of tossing a coin and then randomly choosing one of the numbers 1, 3, 5, and 7. Draw a tree diagram for this experiment.

47. **Probability:** Two dice are rolled. (See the table on page 482.)

   a. What is the probability that the sum of the dice is 10?

   b. What is the probability that both of the dice are even?

48. **Probability:** One card is drawn from a standard deck of cards. What is the probability that the card is

   a. the queen of hearts?

   b. a diamond?

   c. a 9?

Introduction to Algebra

Mathematics at Work!

Algebraic expressions can be used to create mathematical models that generate interesting data about the world around us. Knowing what an algebraic expression means and how to evaluate it at different points can help you extract useful information and make better decisions. Consider the following situation.

You want to visit a local amusement park and are deciding which day is the best to go. You find out that the number of people who attend the amusement park during the summer is equal to $-1.25x^2 + 100x + 3000$, where x is the number of days after June 1. You are free to go to the park on either June 11 (10 days after June 1) or on July 21 (51 days after June 1). If you want to go to the park on the day with fewer people, which day should you plan to go? (See Section 7.7).

<table>
<tr><td rowspan="6" valign="top">

Objectives

A Graph numbers on number lines.

B Identify types of numbers.

C Understand inequality symbols such as < and >.

D Know the meaning of absolute value.

E Graph absolute value inequalities.

</td></tr>
</table>

7.1 The Real Number Line and Absolute Value

Numbers and number concepts form the foundation for the study of algebra. In this chapter, you will learn about positive and negative numbers and how to operate with these numbers. The following list indicates ways that positive and negative numbers occur frequently in our daily lives.

	Negative	**Zero**	**Positive**
Temperatures are recorded as:	below zero	zero	above zero
The stock market will show:	a loss	no change	a gain
Altitude can be measured as:	below sea level	sea level	above sea level
Businesses will report:	losses	break even	profits

Objective A **Number Lines**

To help in understanding different types of numbers and their relationships to each other, we begin with a "picture" called a **number line**. For example, choose some point on a horizontal line and label it with the number 0 (Figure 1).

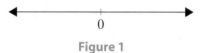

Figure 1

Now choose another point on the line to the right of 0 and label it with the number 1 (Figure 2). This point is arbitrary and once it is chosen, a units scale is determined for the remainder of the line.

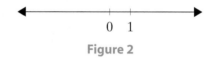

Figure 2

We now have a number line. Points corresponding to all the whole numbers are determined. The point corresponding to 2 is the same distance from 1 as 1 is from 0, and so on (Figure 3).

Figure 3

The **graph** of a number is the point that corresponds to the number and the number is called the **coordinate** of the point. We will follow the convention of using the terms "number" and "point" interchangeably. For example, one point might be "seven" and another point "two." The graph of 7 is indicated by marking the point corresponding to 7 with a large dot (Figure 4).

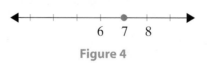

Figure 4

The graph of the set $A = \{2, 4, 6\}$ is shown in Figure 5.

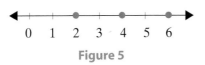

Figure 5

On a horizontal number line, the point one unit to the left of 0 is the **opposite** of 1. It is called **negative** 1 and is symbolized –1. Similarly, the point two units to the left of 0 is the opposite of 2, called negative 2, and symbolized –2, and so on (Figure 6).

The opposite of 1 is –1;   The opposite of –1 is $-(-1) = +1$;

The opposite of 2 is –2;   The opposite of –2 is $-(-2) = +2$;

The opposite of 3 is –3;   The opposite of –3 is $-(-3) = +3$;
and so on.   and so on.

---

notes

- The negative sign $(-)$ indicates the opposite of a number as well as a negative number. It is also used, as we will see in Section 7.3, to
- indicate subtraction. (Note on your calculator the subtraction key on the right side and the negative key $(-)$ at the bottom of the key pad).

Numbers and Their Opposites

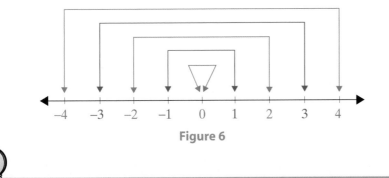

Figure 6

Integers

The set of numbers consisting of the whole numbers and their opposites is called the set of **integers**: $\mathbb{Z} = \{..., -3, -2, -1, 0, 1, 2, 3, ...\}$.

The natural numbers are also called **positive integers**. Their opposites are called **negative integers**. **Zero is its own opposite and is neither positive nor negative** $(0 = -0)$ (Figure 7). Note that the opposite of a positive integer is a negative integer, and the opposite of a negative integer is a positive integer.

Integers: $\{..., -3, -2, -1, 0, 1, 2, 3, ...\}$

Positive integers: $\{1, 2, 3, 4, 5, ...\}$

Negative integers: $\{... -4, -3, -2, -1\}$

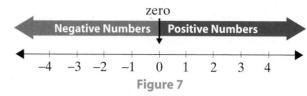

Figure 7

1. State the opposite of

   a. – 10

   b. – 8

   c. + 17

Example 1

Opposites

a. State the opposite of 7.

Solution

-7

b. State the opposite of -4.

Solution

$-(-4)$ or $+4$

In words, the opposite of -4 is $+4$.

Now work margin exercise **1.**

---

2. a. Graph the set of integers $\{-2,-1,0\}$ on a number line.

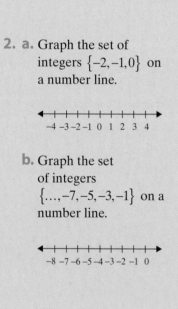

b. Graph the set of integers $\{...,-7,-5,-3,-1\}$ on a number line.

Example 2

Number Line

a. Graph the set of integers $\{-3, -1, 1, 3\}$.

Solution

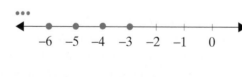

b. Graph the set of integers $\{...,-5,-4,-3\}$.

Solution

The three dots above the number line indicate that the pattern in the graph continues without end.

Now work margin exercise **2.**

Objective B **Types of Numbers**

The integers are not the only numbers that can be represented on a number line. Fractions and decimal numbers such as $\frac{1}{2}, -\frac{4}{3}, \frac{3}{4}$, and -2.3, as well as numbers such as π and $\sqrt{2}$, can also be represented (Figure 8).

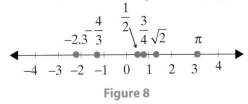

Figure 8

Numbers that can be written as fractions and whose numerators and denominators are integers have the technical name **rational numbers** ($\mathbb{Q}$). Positive and negative decimal numbers and the integers themselves can also be classified as rational numbers. For example, the following numbers are all rational numbers:

$$1.3 = \frac{13}{10}, \quad 5 = \frac{5}{1}, \quad -4 = \frac{-4}{1}, \quad \frac{3}{8}, \quad \text{and} \quad \frac{17}{6}.$$

The variables a and b are used to represent integers in the following definition of a **rational number**.

Rational Numbers

A **rational number** is a number that can be written in the form of $\frac{a}{b}$ where a and b are integers and $b \neq 0$. ($\neq$ is read "is not equal to").

OR

A **rational number** is a number that can be written in decimal form as a terminating decimal or as an infinite repeating decimal.

Other numbers on a number line, such as $\sqrt{2}, \sqrt{3}, \pi,$ and $\sqrt[3]{5}$, are called **irrational numbers**. These numbers can be written as **infinite nonrepeating decimal numbers**. All rational numbers and irrational numbers are classified as **real numbers** ($\mathbb{R}$) and can be written in some decimal form. The number line is called the **real number line**.

The following decimal values and approximations can be found with a calculator.

Examples of Rational Numbers:

$\dfrac{3}{4} = 0.75$　　　　　This decimal number is terminating.

$\dfrac{1}{3} = 0.33333333\ldots$　　　　There is an infinite number of 3s in this repeating pattern.

$\dfrac{3}{11} = 0.27272727\ldots$　　　　This repeating pattern shows that there may be more than one digit in the pattern.

Examples of Irrational Numbers:

$\sqrt{2} = 1.414213562\ldots$　　　　There is an infinite number of digits with no repeating pattern.

$\pi = 3.141592653\ldots$　　　　There is an infinite number of digits with no repeating pattern.

$1.41441444144441\ldots$　　　　There is an infinite number of digits and a pattern of sorts. However, the pattern is nonrepeating.

The following diagram (Figure 9) illustrates the relationships among the various categories of real numbers.

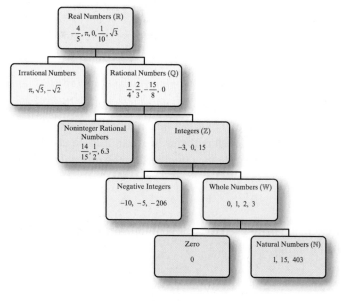

Figure 9

3. List the numbers in the set
$$S = \left\{-6, -\frac{1}{7}, 1, \sqrt{11}, 20\right\}$$
that are:

a. whole numbers

b. integers

c. rational numbers

d. real numbers

Example 3

Identifying Types of Numbers

List the numbers in the set $S = \left\{-5, -\dfrac{3}{4}, 0, \sqrt{2}, 17\right\}$ that are:

a. Whole numbers

Solution

0 and 17 are whole numbers.

b. Integers

Solution

$-5, 0$, and 17 are integers.

c. Rational numbers

Solution

$-5, -\dfrac{3}{4}, 0$, and 17 are rational numbers.

d. Real numbers

Solution

All numbers in S are real numbers.

Now work margin exercise 3.

Objective C   Inequality Symbols

On a horizontal number line, **smaller numbers are always to the left of larger numbers**. Each number is smaller than any number to its right and larger than any number to its left. Two symbols used to indicate order are

$$<\qquad \text{read "is less than"}$$

$$\text{and}\quad >\qquad \text{read "is greater than".}$$

Using the real number line in Figure 10, you can see the following relationships.

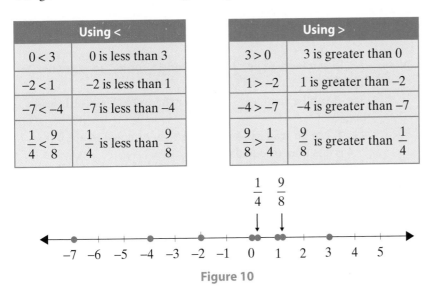

Using <	
$0 < 3$	0 is less than 3
$-2 < 1$	-2 is less than 1
$-7 < -4$	-7 is less than -4
$\dfrac{1}{4} < \dfrac{9}{8}$	$\dfrac{1}{4}$ is less than $\dfrac{9}{8}$

Using >	
$3 > 0$	3 is greater than 0
$1 > -2$	1 is greater than -2
$-4 > -7$	-4 is greater than -7
$\dfrac{9}{8} > \dfrac{1}{4}$	$\dfrac{9}{8}$ is greater than $\dfrac{1}{4}$

Figure 10

Two other symbols commonly used are

$$\le, \qquad \text{read "is less than or equal to"}$$

$$\text{and}\quad \ge, \qquad \text{read "is greater than or equal to".}$$

For example, $5 \ge -10$ is true since 5 is greater than -10. Also, $5 \ge 5$ is true since 5 does equal 5.

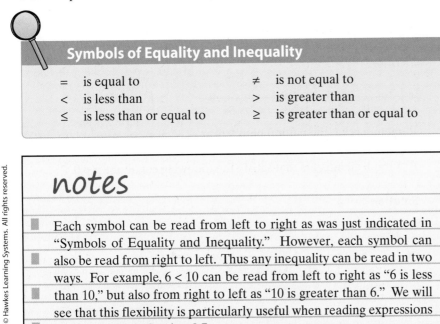

Symbols of Equality and Inequality

$=$	is equal to	$\ne$	is not equal to
$<$	is less than	$>$	is greater than
$\le$	is less than or equal to	$\ge$	is greater than or equal to

notes

Each symbol can be read from left to right as was just indicated in "Symbols of Equality and Inequality." However, each symbol can also be read from right to left. Thus any inequality can be read in two ways. For example, $6 < 10$ can be read from left to right as "6 is less than 10," but also from right to left as "10 is greater than 6." We will see that this flexibility is particularly useful when reading expressions with variables in Section 8.7.

4. Determine whether each of the following statements is true or false.

a. $2 < 10$

b. $6 < -5$

c. $9 \geq -9$

d. $8.5 \leq 8.5$

e. $-3 < -7$

Example 4

Inequalities

Determine whether each of the following statements is true or false.

$7 < 15$ True, since 7 is less than 15. 7 is to the left of 15 on the number line.

$3 > -1$ True, since 3 is greater than −1. 3 is to the right of −1 on the number line.

$4 \geq -4$ True, since 4 is greater than −4. 4 is to the right of −4 on the number line.

$2.7 \geq 2.7$ True, since 2.7 is equal to 2.7.

$-5 < -6$ False, since −5 is greater than −6.

Note: $7 < 15$ can be read as "7 is less than 15" or as "15 is greater than 7."

$3 > -1$ can be read as "3 is greater than −1" or as "−1 is less than 3."

$4 \geq -4$ can be read as "4 is greater than or equal to −4" or as "−4 is less than or equal to 4."

Now work margin exercise 4.

5. a. Graph the following set of real numbers.

$$\left\{-2.5, -1, 0, \frac{5}{4}, 4\right\}$$

b. Graph all natural numbers less than or equal to 6.

c. Graph all integers less than 2.

Example 5

Graphing Sets of Numbers

a. Graph the set of **real numbers** $\left\{-\frac{3}{4}, 0, 1, 1.5, 3\right\}$.

Solution

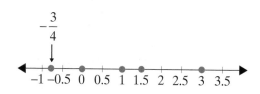

b. Graph all **natural numbers** less than or equal to 3.

Solution

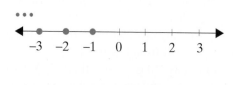

Remember that the natural numbers are $1, 2, 3, 4, \ldots$.

c. Graph all **integers** that satisfy the following condition: $a < 0$.

Solution

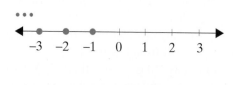

Remember, the three dots above the number line indicate that the pattern in the graph continues without end.

Now work margin exercise 5.

Objective D   **Absolute Value**

In working with the real number line, you may have noticed that any integer and its opposite lie the same number of units from 0 on the number line. For example, both +7 and −7 are seven units from 0 (Figure 11). The + and − signs indicate direction and the 7 indicates distance.

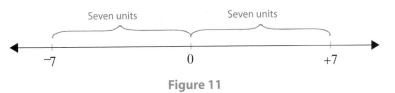

Figure 11

The **distance a number is from 0 on a number line** is called its **absolute value** and is symbolized by two vertical bars, $|\;\;|$. Thus $|+7| = 7$ and $|-7| = 7$. Similarly,

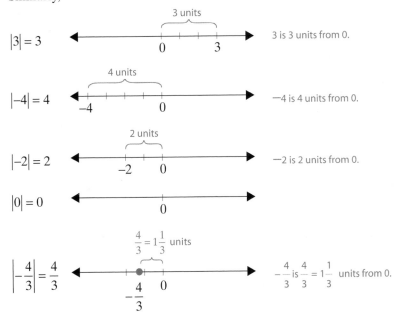

Since distance (similar to length) is never negative, the absolute value of a number is never negative. Or, the absolute value of a nonzero number is always positive.

Absolute Value

The **absolute value** of a real number is its distance from 0. Note that the absolute value of a real number is never negative.

$$|a| = a \quad \text{if } a \text{ is a positive number or } 0.$$

$$|a| = -a \quad \text{if } a \text{ is a negative number.}$$

notes

The symbol $-a$ should be thought of as the "opposite of a." Since a is a variable, a might represent a positive number, a negative number, or 0. This use of symbols can make the definition of absolute value difficult to understand at first. As an aid to understanding the use of the negative sign, consider the following examples.

$$\text{If } a = -6, \text{ then } -a = -(-6) = 6.$$

Similarly,

$$\text{If } x = -1, \text{ then } -x = -(-1) = 1.$$
$$\text{If } y = -10, \text{ then } -y = -(-10) = 10.$$

Remember that $-a$ (the opposite of a) represents a positive number whenever a represents a negative number.

6. a. Find the absolute value:

$$|+4|$$

b. Find the absolute value of $-|-7.4|$

7. True or false:

$$|-20| \leq 20$$

Example 6

Absolute Value

a. $|6.3| = 6.3$

The number 6.3 is 6.3 units from 0. Also, 6.3 is positive so its absolute value is the same as the number itself.

b. $-|-2.9| = -(2.9) = -2.9$

The opposite of the absolute value of -2.9.

Now work margin exercise 6.

Example 7

Absolute Value

True or false: $|-4| \geq 4$?

Solution

True, since $|-4| = 4$ and $4 \geq 4$.

Now work margin exercise 7.

Example 8

Absolute Value

a. If $|x| = 7$, what are the possible values for x?

Solution

$x = 7$ or $x = -7$ since $|7| = 7$ and $|-7| = 7$.

b. If $|x| = -3$, what are the possible values for x?

Solution

There are no values of x for which $|x| = -3$. The absolute value can never be negative. There is **no solution**.

Now work margin exercise 8.

8. a. If $|z| = 3$, what are the possible values of z?

b. If $|y| = -19$, what are the possible values of y?

Objective E ## Graphing Absolute Value Inequalities

The solution sets and the corresponding graphs for absolute value inequalities can be finite or infinite, depending on the nature of the inequality. The following examples illustrate both situations.

Example 9

Graphing Absolute Value Inequalities

If $|x| \geq 4$, what are the possible integer values for x? Graph these integers on a number line.

Solution

There are an infinite number of integers 4 or more units from 0, both negative and positive. These integers are $\{\ldots -7, -6, -5, -4, 4, 5, 6, 7, \ldots\}$.

Now work margin exercise 9.

9. If $|x| \geq 1$, what are the possible integer values for x? Graph these integers on a number line.

10. If $|x| < 3$, what are the possible integer values for x? Graph these integers on a number line.

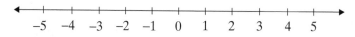

-4 –3 –2 –1 0 1 2 3 4

Completion Example 10

Graphing Absolute Value Inequalities

If $|x| < 4$, what are the possible integer values for x? Graph these integers on a number line.

Solution

The integers that are less than 4 units from 0 have absolute values less than 4. These integers are _____.

$$\xleftarrow{\hspace{0.5cm}} \overset{-5}{|} \quad \overset{-4}{|} \quad \overset{-3}{|} \quad \overset{-2}{|} \quad \overset{-1}{|} \quad \overset{0}{|} \quad \overset{1}{|} \quad \overset{2}{|} \quad \overset{3}{|} \quad \overset{4}{|} \quad \overset{5}{|} \xrightarrow{\hspace{0.5cm}}$$

Now work margin exercise 10.

Practice Problems

Fill in the blank with the appropriate symbol: $<, >$, or $=$.

1. -2 ____ 1 **2.** $1\frac{6}{10}$ ____ 1.6 **3.** $-(-4.1)$ ____ -7.2

4. Graph the set of all negative integers on a number line.

5. True or false: $3.6 \le |-3.6|$

6. If $|x| = 8$, what are the possible values for x?

7. If $|x| = -6$, what are the possible values for x?

8. If $|x| > 2$, what are the possible values for x? Graph these integers on a number line.

Completion Example Answers

10. These integers are $\{-3, -2, -1, 0, 1, 2, 3\}$.

$$\xleftarrow{\hspace{0.5cm}} \overset{-4}{|} \quad \overset{-3}{\bullet} \quad \overset{-2}{\bullet} \quad \overset{-1}{\bullet} \quad \overset{0}{\bullet} \quad \overset{1}{\bullet} \quad \overset{2}{\bullet} \quad \overset{3}{\bullet} \quad \overset{4}{|} \xrightarrow{\hspace{0.5cm}}$$

Practice Problem Answers

1. $<$ **2.** $=$ **3.** $>$

4. 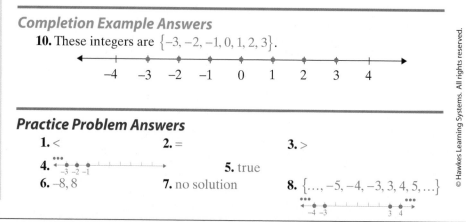   **5.** true

6. $-8, 8$ **7.** no solution   **8.** $\{\dots, -5, -4, -3, 3, 4, 5, \dots\}$

Exercises 7.1

List the numbers in the set $A = \left\{-7, -\sqrt{6}, -2, -\dfrac{5}{3}, -1.4, 0, \dfrac{3}{5}, \sqrt{5}, \sqrt{11}, 4, 5.9, 8\right\}$ **that are described in each exercise. See Example 3.**

1. Natural numbers

2. Whole numbers

3. Integers

4. Irrational numbers

5. Rational numbers

6. Real numbers

Graph each set of real numbers on a real number line. See Examples 2 and 5.

7. $\{1, 2, 5, 6\}$

8. $\{-3, -2, 0, 1\}$

9. $\{2, -3, 0, -1\}$

10. $\{-2, -1, 4, -3\}$

11. $\left\{0, -1, \dfrac{7}{4}, 3, 1\right\}$

12. $\left\{-2, -1, -\dfrac{1}{3}, 2\right\}$

13. $\left\{-\dfrac{3}{4}, 0, 2, 3.6\right\}$

14. $\left\{-3.4, -2, -0.5, 1, \dfrac{5}{2}\right\}$

15. $\left\{-\dfrac{7}{2}, -1.5, 1, \dfrac{4}{3}, 2\right\}$

16. $\left\{-4, -\dfrac{7}{3}, -1, 0.2, \dfrac{5}{2}\right\}$

17. all whole numbers less than 4

18. all integers less than 4

19. all integers greater than or equal to -2

20. all negative integers greater than -4

21. all whole numbers less than 0

22. all integers less than or equal to 0

Fill in each blank with the appropriate symbol that will make the statement true: <, >, or =. See Example 4.

23. $4 \underline{\hspace{1cm}} 6$

24. $-3 \underline{\hspace{1cm}} 1$

25. $-2 \underline{\hspace{1cm}} -4$

26. $-8 \underline{\hspace{1cm}} 0$

27. $-20 \underline{\hspace{1cm}} -19$

28. $-67 \underline{\hspace{1cm}} -50$

29. $-(-4.3) \underline{\hspace{1cm}} 4.3$

30. $5.6 \underline{\hspace{1cm}} -(-8.7)$

31. $-\dfrac{3}{4} \underline{\hspace{1cm}} -1$

32. $-2.3 \_\_\_\_ -2\frac{3}{10}$

33. $\frac{1}{3} \_\_\_\_ \frac{1}{2}$

34. $-\frac{1}{2} \_\_\_\_ -\frac{1}{3}$

35. $|-4| \_\_\_\_ 4$

36. $|7| \_\_\_\_ -7$

37. $|-8| \_\_\_\_ -8$

38. $-15 \_\_\_\_ |-15|$

> **Determine whether each statement is true or false. If a statement is false, rewrite it in a form that is a true statement. (There may be more than one way to correct a statement.) See Examples 4 and 7.**

39. $0 = -0$

40. $-22 < -16$

41. $-9 > -8.5$

42. $-17 \leq 17$

43. $4.7 \geq 3.5$

44. $|-5| = 5$

45. $-|-7| \geq -|7|$

46. $|-8| \geq 4$

47. $-|-3| < -|4|$

48. $\left|-\frac{5}{2}\right| < 2$

> **List the possible values for x for each statement. See Example 8.**

49. $|x| = 5$

50. $|x| = 8$

51. $|x| = 2$

52. $|x| = 0$

53. $|x| = -6$

54. $|x| = -1$

55. $|x| = 23$

56. $|x| = 105$

> **Determine the integer values for x that satisfy the conditions given in each statement. Graph the integers on a number line. See Examples 9 and 10.**

57. $a > 3$

58. $x \leq 2$

59. $|x| > 5$

60. $|a| > 6$

61. $|x| \leq 2$

62. $|y| \geq 5$

63. $|a|$ is (never, sometimes, always) equal to a.

64. $|x|$ is (never, sometimes, always) equal to $-x$.

65. $|y|$ is (never, sometimes, always) equal to a positive integer.

66. $|x|$ is (never, sometimes, always) greater than x.

Writing & Thinking

67. Explain, in your own words, how an expression such as $-y$ might represent a positive number.

68. Explain, in your own words, the meaning of absolute value.

7.2 Addition with Real Numbers

Objectives

A Add real numbers.

B Add real numbers in a vertical format.

C Determine if a given real number is a solution to an equation.

Objective A Addition with Real Numbers

Picture a straight line in an open field and numbers marked on a number line. An archer stands at 0 and shoots an arrow to +3, then stands at 3 and shoots the arrow 5 more units in the positive direction (to the right). Where will the arrow land (Figure 1)?

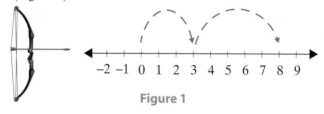

Figure 1

Naturally, you have figured out that the answer is +8. What you have done is added the two positive integers, +3 and +5.

$$(+3)+(+5)=+8 \quad \text{or} \quad 3+5=8$$

Suppose another archer shoots an arrow in the same manner as the first but in the opposite direction. Where would his arrow land? The arrow lands at –8. You have just added –3 and –5 (Figure 2).

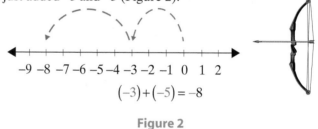

$$(-3)+(-5)=-8$$

Figure 2

If the archer stands at 0 and shoots an arrow to +3 and then the archer goes to +3 and turns around and shoots an arrow 5 units in the opposite direction, where will the arrow stick? Would you believe at –2 (Figure 3)?

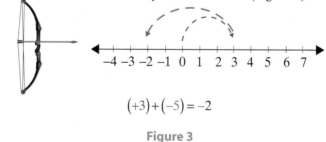

$$(+3)+(-5)=-2$$

Figure 3

For our final archer, the first shot is to –3. Then after going to –3, he turns around and shoots 5 units in the opposite direction. Where is the arrow? It is at +2 (Figure 4).

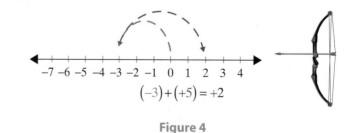

$$(-3)+(+5)=+2$$

Figure 4

In summary:

1. The sum of two positive real numbers is positive.

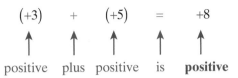

$$(+3) \quad + \quad (+5) \quad = \quad +8$$

positive   plus   positive   is   **positive**

2. The sum of two negative real numbers is negative.

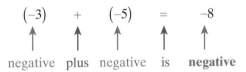

$$(-3) \quad + \quad (-5) \quad = \quad -8$$

negative   plus   negative   is   **negative**

3. The sum of a positive real number and a negative real number may be negative or positive (or zero) depending on which number is farther from 0.

$$(+3) \quad + \quad (-5) \quad = \quad -2 \qquad (+5) \quad + \quad (-3) \quad = \quad +2$$

positive plus negative is **negative**   positive plus negative is **positive**

The rules for addition with real numbers can be summarized in the following manner.

Rules for Addition with Real Numbers

1. To add two real numbers with **like signs**,
   a. add their absolute values and
   b. use the common sign.

2. To add two real numbers with **unlike signs**,
   a. subtract their absolute values (the smaller from the larger), and
   b. use the sign of the number with the larger absolute value.

Study the following examples carefully to help in understanding the rules for addition.

Example 1

Addition with Like Signs

a. $(+10)+(+3)$

$= +\left(|+10|+|+3|\right)$

$= +\left(10+3\right)$

$= 13$

1. Find each of the following sums:

a. $(-15)+(-5)$

b. $(+4)+(+11)$

c. $(+3.1)+(+2.8)$

d. $\left(-\dfrac{2}{5}\right)+\left(-\dfrac{1}{3}\right)$

b. $(-10)+(-3)$

$\quad = -\big(|-10|+|-3|\big)$

$\quad = -(10+3)$

$\quad = -13$

c. $(-1.4)+(-2.5) = -\big(|-1.4|+|-2.5|\big)$

$\qquad\qquad\qquad = -(1.4+2.5)$

$\qquad\qquad\qquad = -3.9$

d. $\left(+\dfrac{3}{4}\right)+\left(+\dfrac{1}{3}\right) = +\left(\left|+\dfrac{3}{4}\right|+\left|+\dfrac{1}{3}\right|\right)$

$\qquad\qquad\qquad = +\left(\dfrac{3}{4}+\dfrac{1}{3}\right)$

$\qquad\qquad\qquad = +\left(\dfrac{3}{4}\cdot\dfrac{3}{3}+\dfrac{1}{3}\cdot\dfrac{4}{4}\right)$

$\qquad\qquad\qquad = +\left(\dfrac{9}{12}+\dfrac{4}{12}\right)$

$\qquad\qquad\qquad = \dfrac{13}{12}$

Now work margin exercise **1.**

Example 2

Addition with Unlike Signs

a. $(-10)+(+3)$

$\quad = -\big(|-10|-|+3|\big)$

$\quad = -(10-3)$

$\quad = -7$

b. $(+10)+(-3)$

$\quad = +\big(|+10|-|-3|\big)$

$\quad = +(10-3)$

$\quad = 7$

c. $(-9.5)+(8.7) = -\big(|-9.5|-|8.7|\big)$

$\qquad\qquad\qquad = -(9.5-8.7)$

$\qquad\qquad\qquad = -0.8$

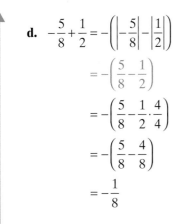

d. $-\dfrac{5}{8}+\dfrac{1}{2}=-\left(\left|-\dfrac{5}{8}\right|-\left|\dfrac{1}{2}\right|\right)$

$\qquad = -\left(\dfrac{5}{8}-\dfrac{1}{2}\right)$

$\qquad = -\left(\dfrac{5}{8}-\dfrac{1}{2}\cdot\dfrac{4}{4}\right)$

$\qquad = -\left(\dfrac{5}{8}-\dfrac{4}{8}\right)$

$\qquad = -\dfrac{1}{8}$

Now work margin exercise 2.

Example 3

Adding More Than Two Real Numbers

One technique to add more than two real numbers is to add left to right.

a. $-3+2+(-5)=-1+(-5)=-6$

b. $6.0+(-4.3)+(-1.5)=1.7+(-1.5)=0.2$

Now work margin exercise 3.

Objective B **Addition in a Vertical Format**

An **equation** is a statement that two expressions are equal. Equations in algebra are almost always written horizontally, so addition (and subtraction) with real numbers is done much of the time in the horizontal format. However, there are situations (such as in long division) where sums (and differences) are written in a vertical format as illustrated in Example 4. If more than two numbers are being added, then (at this time) add two at a time starting from the top. (Later, other methods will be discussed).

notes

The positive sign (+) may be omitted when writing positive numbers, but the negative sign (–) must always be written for negative numbers. Thus, if there is no sign in front of an real number, the real number is understood to be positive.

2. Find each of the following sums:

a. $(+10)+(-2)$

b. $(-12)+(+5)$

c. $(+1.9)+(-4.2)$

d. $\left(-\dfrac{1}{3}\right)+\left(+\dfrac{7}{9}\right)$

3. Find each of the following sums:

a. $-7+5+(-3)$

b. $-3.2+(-6.1)+5.7$

Addition with Real Numbers   **Section 7.2**   **517**

4. Find each of the following sums:

a. $\begin{array}{r} -13 \\ +\ 6 \\ \hline \end{array}$

b. $\begin{array}{r} -1.5 \\ -2.8 \\ \hline \end{array}$

5. Find each of the following sums:

a. $\begin{array}{r} -15 \\ 6 \\ -25 \\ \hline \end{array}$

b. $\begin{array}{r} 56 \\ -12 \\ -5 \\ \hline \end{array}$

Example 4

Addition in a Vertical Format

Find each sum.

a. $\begin{array}{r} -10 \\ +\ 7 \\ \hline -3 \end{array}$

b. $\begin{array}{r} -4.5 \\ -9.1 \\ \hline -13.6 \end{array}$

Now work margin exercise 4.

Example 5

Addition in a Vertical Format

Find the sums.

a. $\begin{array}{r} -20 \\ 7 \\ -30 \\ \hline \end{array}$

Solution

One technique for adding several real numbers is to mentally add the positive and negative real numbers separately and then add these results. (We are, in effect, using the commutative and associative properties of addition).

$\begin{array}{r} -20 \\ 7 \\ -30 \\ \hline -43 \end{array}$ or $\begin{array}{r} -50 \\ 7 \\ \hline -43 \end{array}$

b. $\begin{array}{r} 42 \\ -10 \\ -\ 3 \\ \hline \end{array}$

Solution

$\begin{array}{r} 42 \\ -10 \\ -\ 3 \\ \hline 29 \end{array}$ or $\begin{array}{r} 42 \\ -13 \\ \hline 29 \end{array}$

Now work margin exercise 5.

Objective C Real Number Solutions to Equations

Now that we know how to add positive and negative real numbers, we can determine whether or not a particular real number satisfies an equation that contains a variable. A number is said to be a **solution** or to **satisfy an equation** if it gives a true statement when substituted for the variable.

Example 6

Determining Possible Solutions

Determine whether or not the given real number is a solution to the given equation by substituting for the variable and adding.

a. $x + 5 = -2$ given that $x = -7$

Solution

$(-7) + 5 = -2$ is true, so -7 is a solution.

b. $1.8 + z = -3.9$ given that $z = 2.1$

Solution

$1.8 + 2.1 = -3.9$ is false since $1.8 + 2.1 = +3.9$
So, 2.1 is not a solution.

c. $|x| + (-7) = -3$ given that $x = -4$

Solution

$|-4| + (-7) = 4 + (-7) = -3$ is true, so -4 is a solution.

d. $x + \left(-\dfrac{1}{5}\right) = -\dfrac{1}{4}$ given that $x = -\dfrac{1}{20}$

Solution

$$\left(-\frac{1}{20}\right) + \left(-\frac{1}{5}\right) = -\frac{1}{20} + \left(-\frac{1}{5} \cdot \frac{4}{4}\right) = -\frac{1}{20} + \left(-\frac{4}{20}\right) = -\frac{5}{20} = -\frac{1}{4}$$

is true, so $-\dfrac{1}{20}$ is a solution.

Now work margin exercise 6.

6. Determine whether $x = -5$ is a solution to each equation.

a. $x + 14 = 9$

b. $-2.8 + x = 2.2$

c. $x + (-25) = -20$

d. $x - \dfrac{10}{3} = \dfrac{13}{6}$

Practice Problems

Find each sum. Add from left to right if there are more than two numbers.

1. $+16 + (-10)$

2. $-12 + (8)$

3. $-\dfrac{1}{6} + \left(-\dfrac{2}{3}\right)$

4. $-1.3 + (1.3)$

5. $-\dfrac{3}{10} + \left(-\dfrac{1}{15}\right)$

6. $+6 + (-7) + (-1)$

7. $+100 + (-100) + (+10)$

Find the indicated sum

8. $\begin{array}{r} -42 \\ +\ \ 9 \\ \hline \end{array}$

9. $\begin{array}{r} 16 \\ -10 \\ \underline{-4} \end{array}$

10. Determine whether or not $x = -3$ is a solution to the equation $x + (-6) = -9$.

Practice Problem Answers

1. 6   **2.** -4 **3.** $-\dfrac{5}{6}$ **4.** 0

5. $-\dfrac{11}{30}$ **6.** -2 **7.** 10   **8.** -33

9. 2   **10.** yes

Exercises 7.2

Find the sum. Reduce any fractions to lowest terms. See Examples 1 through 3.

1. $4 + 9$

2. $8 + (-3)$

3. $(-9) + 5$

4. $(-7) + (-3)$

5. $(-9) + 9$

6. $2 + (-8)$

7. $11 + (-6)$

8. $(-12) + 3$

9. $-18 + 5$

10. $26 + (-26)$

11. $-5 + (-3)$

12. $11 + (-2)$

13. $(-2) + (-8)$

14. $10 + (-3)$

15. $17 + (-17)$

16. $(-7) + 20$

17. $21 + (-4)$

18. $2.1 + (-4.6)$

19. $-1.5 + (-3.1)$

20. $(-15) + (-3)$

21. $(-12) + (-17)$

22. $24 + (-16)$

23. $-4 + (-5)$

24. $-4.3 + (-5.8)$

25. $-6.9 + (-8.5)$

26. $(-6) + (-8)$

27. $9 + (-12)$

28. $-12 + 9$

29. $(-33) + (-21)$

30. $(-21) + 18$

31. $9.7 + (-12.2)$

32. $-19.6 + 4.1$

33. $\dfrac{3}{14} + \dfrac{3}{14}$

34. $\dfrac{1}{10} + \dfrac{3}{10}$

35. $\dfrac{3}{4} + \left(-\dfrac{1}{8}\right)$

36. $\dfrac{5}{17} + \left(-\dfrac{15}{34}\right)$

37. $-\dfrac{5}{2} + \dfrac{3}{4}$

38. $-\dfrac{1}{6} + \dfrac{7}{15}$

39. $-3 + 4 + (-8)$

40. $(-9) + (-6) + 5$

Find each of the indicated sums in a vertical format. See Examples 4 and 5.

41.
$$\begin{array}{r} -32 \\ +\ \ 8 \\ \hline \end{array}$$

42.
$$\begin{array}{r} -53 \\ +\ 19 \\ \hline \end{array}$$

43.
$$\begin{array}{r} 3.7 \\ -0.6 \\ \hline \end{array}$$

44.
$$\begin{array}{r} -7.5 \\ -1.6 \\ \hline \end{array}$$

45.
$$\begin{array}{r} 102 \\ -21 \\ -5 \\ \hline \end{array}$$

46.
$$\begin{array}{r} 130 \\ -45 \\ -32 \\ \hline \end{array}$$

47.
$$\begin{array}{r} -210 \\ -200 \\ +\ 100 \\ \hline \end{array}$$

48.
$$\begin{array}{r} -108 \\ -105 \\ -330 \\ \hline \end{array}$$

49.
$$\begin{array}{r} 35 \\ 2 \\ -5 \\ -5 \\ \hline \end{array}$$

50.
$$\begin{array}{r} -56 \\ -3 \\ -1 \\ +\ 3 \\ \hline \end{array}$$

Find each of the indicated sums. Be sure to find the absolute values first.

51. $13 + |-5|$

52. $|-2| + (-5)$

53. $|-10| + |-4|$

54. $|-7| + (+7)$

55. $|-18| + |+17|$

56. $|-14| + |-6|$

Determine whether or not the given number is a solution to the given equation by substituting and then evaluating. See Example 6.

57. Determine whether or not $x = -2$ is a solution of $x + 4 = 2$.

58. Determine whether or not $x = -3$ is a solution of $x + (-7) = 10$.

59. Is $x = -4$ a solution of $-10 + x = -14$?

60. Is $y = -2$ a solution of $y + 9 = 7$?

61. Is $y = -0.6$ a solution of $1.7 + y = 1.1$?

62. Determine whether or not $z = 2.3$ is a solution of $z + (-3.1) = 5.4$.

63. Determine whether or not $z = 18$ is a solution of $z + (-12) = 6$.

64. Determine whether or not $x = -7$ is a solution of $x + 3 = -10$.

65. Is $y = -2$ a solution of $|y| + (-10) = -12$?

66. Is $x = -2$ a solution of $|x| + 15 = 17$?

67. Is $x = \dfrac{5}{3}$ a solution of $x + \left(-\dfrac{5}{3}\right) = -\dfrac{5}{6}$?

68. Is $x = -\dfrac{4}{3}$ a solution of $\dfrac{3}{4} + x = -\dfrac{7}{12}$?

69. Determine whether or not $z = -72$ is a solution of $42 + |z| = -30$.

70. Determine whether or not $x = -4$ is a solution of $|x| + (-5) = -1$.

Choose the response that correctly completes each statement. In each problem, give two examples that illustrate your reasoning.

71. If x and y are real numbers, then $x + y$ is (never, sometimes, always) equal to 0.

72. If x and y are real numbers, then $x + y$ is (never, sometimes, always) negative.

73. If x and y are real numbers, then $x + y$ is (never, sometimes, always) positive.

74. If x is a positive real number and y is a negative real number, then $x + y$ is (never, sometimes, always) equal to 0.

75. If x and y are positive real numbers, then $x + y$ is (never, sometimes, always) equal to 0.

76. If x and y are both negative real numbers, then $x + y$ is (never, sometimes, always) equal to 0.

77. If x is a negative real number, then $-x$ is (never, sometimes, always) negative.

78. If x is a positive real number, then $-x$ is (never, sometimes, always) negative.

Use your calculator to find the value of each of the following expressions.

79. $47 + (-29) + 66$

80. $56 + (-41) + (-28)$

81. $2932 + 4751 + (-3876)$

82. $(-8154) + 2147 + (-136)$

83. $(-16,945) + (-27,302) + (-53,467)$

84. $(-12,299) + 15,631 + (-47,558)$

Writing & Thinking

85. Describe, in your own words, how the sum of the absolute values of two numbers might be 0. (Is this even possible?)

86. Describe in your own words the conditions under which the sum of two integers will be 0.

Objectives

A Subtract real numbers.

B Subtract real numbers in a vertical format.

C Find the change in value between two real numbers.

D Determine if a given real number is a solution to an equation.

1. Find the additive inverse (opposite) of each number:

a. 14

b. –9

Objective A **Subtraction with Real Numbers**

Before developing methods of **subtraction** with real numbers, we state and illustrate an important relationship between any real number and its opposite.

> ### Additive Inverse
>
> The **opposite** of a real number is called its **additive inverse**. The sum of a number and its additive inverse is zero. Symbolically, for any real number a,
> $$a + (-a) = 0.$$

Example 1

Additive Inverse

a. Find the additive inverse (opposite) of 3.
   Solution: The additive inverse of 3 is –3. $3 + (-3) = 0$

b. Find the additive inverse (opposite) of –7.3.
   Solution: The additive inverse of –7.3
   is $-(-7.3) = +7.3$. $(-7.3) + (+7.3) = 0$

c. Find the additive inverse (opposite) of 0.
   Solution: The additive inverse of 0 is $-0 = 0$.
   That is, 0 is its own opposite. $(0) + (-0) = 0 + 0 = 0$

Now work margin exercise 1.

Intuitively, we can think of addition of numbers (positive or negative or both) as "*piling on*" or "*accumulating*" numbers. For example, when we add positive numbers (or negative numbers), the sum is a number more positive (or more negative). Figure 1 illustrates these ideas.

a. Adding positive numbers "piles on" or "accumulates" the numbers in a positive direction.

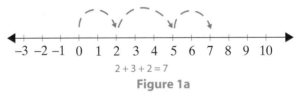

$$2 + 3 + 2 = 7$$
Figure 1a

b. Adding negative numbers "piles on" or "accumulates" the numbers in a negative direction.

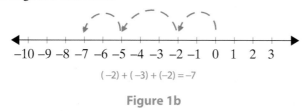

$$(-2) + (-3) + (-2) = -7$$

Figure 1b

In **subtraction** we want to find the "*difference between*" two numbers. On a number line this translates as the "*distance between*" the two numbers **with direction considered**. As illustrated in Figure 2, the *distance between* 6 and 1 is five units. To find this *distance*, we subtract: $6 - 1 = 6 + (-1) = 5$. Note that to subtract 1, we add the opposite of 1 (which is –1).

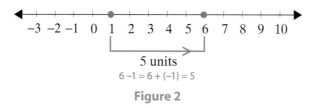

5 units

$$6 - 1 = 6 + (-1) = 5$$

Figure 2

In Figure 3 we see that the distance between 6 and –4 is ten units. Again, to find this distance, we subtract $6 - (-4)$. But we know that we must have 10 as an answer. To get 10, we add the opposite of –4 (which is +4) as follows: $6 - (-4) = 6 + (+4) = 10$.

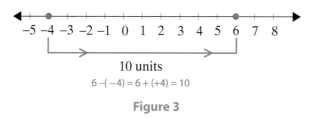

10 units

$$6 - (-4) = 6 + (+4) = 10$$

Figure 3

To understand the "direction" in subtraction, think of subtraction on the number line as:

$$(\text{end value}) - (\text{beginning value}).$$

This means that for $6 - (-4) = 6 + (+4) = 10$,

we have   6   –   (–4)   =   6 + (+4) = 10

↑   ↑

(end value) – (beginning value)

and, as illustrated in Figure 3, we would start at –4 and move 10 units in the **positive direction** to end at 6.

As illustrated in Figures 1, 2, and 3, **subtraction is defined in terms of addition**.

> **To subtract, add the opposite of the number being subtracted.**

If we reverse the order of subtraction, then the answer must indicate the opposite direction.

This means that for $-4-6=-4+(-6)=-10$, we have

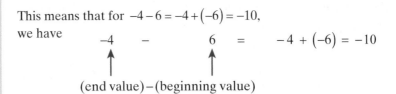

(end value) − (beginning value)

and, as illustrated in Figure 4, we would start at 6 and move 10 units in the **negative direction** to end at −4. We see that −4 and 6 are still ten units apart but subtraction now indicates a negative direction.

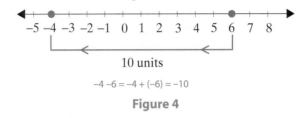

10 units

$-4-6=-4+(-6)=-10$

Figure 4

The formal definition of subtraction is as follows.

Subtraction

For any real numbers a and b,

$$a-b=a+(-b).$$

In words, to subtract b from a, **add** the **opposite** of b to a.

We change the sign of the second number and follow the rules for addition.

$$(+1)-(-4)=(+1)+\left[-(-4)\right]=(+1)+(+4)=+5$$

add **opposite**

$$(-10)-(-3)=(-10)+\left[-(-3)\right]=(-10)+(+3)=-7$$

add **opposite**

Example 2

Subtraction

Find the following differences.

a. $(-1)-(-4)=(-1)+(+4)=+3$

b. $(-1)-(-8)=(-1)+(+8)=7$

c. $(10.3)-(-2.3)=10.3+(+2.3)=10.3+2.3=12.6$

d. $\dfrac{3}{16}-\left(-\dfrac{1}{16}\right)=\dfrac{3}{16}+\left(+\dfrac{1}{16}\right)=\dfrac{4}{16}=\dfrac{\cancel{4}\cdot 1}{\cancel{4}\cdot 4}=\dfrac{1}{4}$

e. $-\dfrac{2}{3}-\left(-\dfrac{1}{6}\right)=-\dfrac{2}{3}+\left(+\dfrac{1}{6}\right)=\left(-\dfrac{2}{3}\cdot\dfrac{2}{2}\right)+\dfrac{1}{6}=-\dfrac{4}{6}+\dfrac{1}{6}=-\dfrac{3}{6}=-\dfrac{1}{2}$

Now work margin exercise 2.

In practice, the notation $a-b$ is thought of as addition of signed numbers. That is, because $a-b=a+(-b)$, we think of the plus sign, +, as being present in $a-b$. In fact, an expression such as $4-19$ can be thought of as "four plus negative nineteen." We have

$$4-19=4+(-19)=-15$$

$$-25-30=-25+(-30)=-55$$

$$-3-(-17)=-3+(+17)=14$$

$$24-11-6=24+(-11)+(-6)=7.$$

Generally, the second step is omitted and we go directly to the answer by computing the sum mentally.

$$4-19=-15$$

$$-25-30=-55$$

$$-3-(-17)=14$$

$$24-11-6=7$$

2. Find the following differences.

a. $(-4)-(-4)$

b. $(-3)-(+8)$

c. $-1.5-(-5.3)$

d. $-\dfrac{1}{8}-\left(+\dfrac{5}{8}\right)$

e. $\dfrac{4}{5}-\left(-\dfrac{1}{2}\right)$

Objective B — Subtraction in a Vertical Format

As with addition, real numbers can be written vertically in subtraction. One number is written underneath the other, and the sign of the real number being subtracted (the bottom number) is changed and addition is performed.

3. Subtract.

a.
$$
\begin{array}{r}
-27 \\
-(+19) \\
\hline
\end{array}
$$

b.
$$
\begin{array}{r}
-20 \\
-(+87) \\
\hline
\end{array}
$$

c.
$$
\begin{array}{r}
-56 \\
-(-56) \\
\hline
\end{array}
$$

d.
$$
\begin{array}{r}
6.8 \\
-(-5.1) \\
\hline
\end{array}
$$

Example 3

Subtraction

a.
Subtract		**Add**
43		43
$-(-25)$	$\xrightarrow[\text{change}]{\text{sign}}$	$+25$
		68

b.
Subtract		**Add**
-38		-38
$-(+11)$	$\xrightarrow[\text{change}]{\text{sign}}$	-11
		-49

c.
Subtract		**Add**
-7.3		-7.3
$-(-3.2)$	$\xrightarrow[\text{change}]{\text{sign}}$	$+3.2$
		-4.1

d.
Subtract		**Add**
17		17
$-(+69)$	$\xrightarrow[\text{change}]{\text{sign}}$	-69
		-52

Now work margin exercise 3.

Change in Value

To find the **change in value** between two numbers, take the end value and subtract the beginning value. Symbolically,

$$\text{change in value} = (\text{end value}) - (\text{beginning value}).$$

Example 4

Change in Value

a. At noon on Tuesday the temperature was 34°F. By noon on Thursday the temperature had changed to –5°F. How much did the temperature change between Tuesday and Thursday?

12pm Tuesday   12pm Thursday

Solution

For change in value:

$$(\text{end value}) - (\text{beginning value}) = (-5) - (+34)$$
$$= -5 + (-34)$$
$$= -39$$

Between Tuesday and Thursday the temperature changed –39°F (or dropped 39°F).

b. A jet pilot flew her plane from an altitude of 30,000 ft to an altitude of 12,000 ft. What was the change in altitude?

Solution

end altitude – beginning altitude = change in altitude

12,000 – 30,000 = –18,000 ft

(This means that the plane *descended* 18,000 ft.)

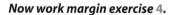

Now work margin exercise 4.

4. a. At 3 P.M. on Friday the temperature was 63°F. By midnight the temperature had changed to 44°F. How much did the temperature change between 3 P.M. and midnight?

b. A drone plane flew from an altitude of 25,000 ft to an altitude of 14,000 ft. What was the change in altitude?

5. a. If Andrea's sales for last week were as follows, what were her total net sales?

Day	Net Sales
Monday	3
Tuesday	8
Wednesday	3
Thursday	6
Friday	−1

b. Shawn weighed 236 lb when he started to diet. The first month he lost 8 lb, the second month he gained 3 lb, and the third month he lost 10 lb. What was his weight after 3 months of dieting?

The **net change** in a measure is the algebraic sum of several numbers. Example 5 illustrates how positive and negative numbers can be used to find the net change of weight (gain or loss) and sales over a period of time.

Example 5

Net Change

a. Susan is a salesperson for a shoe store. Last week her sales of pairs of shoes were as follows:

Day	Sales	Returns	Daily Net Sales
Monday	7	1	6
Tuesday	3	0	3
Wednesday	2	4	−2
Thursday	6	1	5
Friday	8	3	5

What were Susan's net sales for last week?

Solution

$$6 + 3 + (-2) + 5 + 5 = 17$$

Susan's net sales for the week were 17 pairs of shoes.

b. Robert weighed 230 lb when he started to diet. The first month he lost 7 lb, the second month he gained 2 lb, and the third month he lost 5 lb. What was his weight after 3 months of dieting?

Solution

$$230 + (-7) + (+2) + (-5)$$
$$= 223 + (+2) + (-5)$$
$$= 225 + (-5)$$
$$= 220 \, \text{lb}$$

Now work margin exercise 5.

Now both addition and subtraction can be used to help determine whether or not a particular real number is a solution to an equation. (This skill will prove useful in evaluating formulas and solving equations.)

Example 6

Evaluating Possible Solutions

Determine whether or not the given real number is a solution to the given equation by substituting for the variable and then evaluating.

a. $x - (-6) = -10$ given that $x = -14$

Solution

$$x - (-6) = -10$$
$$(-14) - (-6) \overset{?}{=} -10$$
$$-14 + (+6) \overset{?}{=} -10$$
$$-8 \neq -10$$

-14 **is not** a solution.

b. $7 - y = -1$ given that $y = 8$

Solution

$$7 - y = -1$$
$$7 - (8) \overset{?}{=} -1$$
$$-1 = -1$$

8 **is** a solution.

c. $a - \dfrac{7}{12} = -\dfrac{1}{3}$ given that $a = \dfrac{1}{4}$

Solution

$$a - \frac{7}{12} = -\frac{1}{3}$$
$$\frac{1}{4} - \frac{7}{12} \overset{?}{=} -\frac{1}{3}$$
$$-\frac{4}{12} \overset{?}{=} -\frac{1}{3}$$
$$-\frac{1}{3} = -\frac{1}{3}$$

$\dfrac{1}{4}$ **is** a solution.

Now work margin exercise 6.

6. Determine whether or not the given real number is a solution to the given equation by substituting for the variable and then evaluating.

a. $x - (-2) = 5$ given that $x = 3$

b. $2 - y = -13$ given that $y = 15$

c. $a - \dfrac{1}{6} = -\dfrac{1}{4}$ given that $a = -\dfrac{1}{12}$

Practice Problems

1. What is the additive inverse of 85?
2. Find the difference: $-6-(-5)$
3. Simplify: $-6-4-(-2)$
4. Perform the indicated subtraction:
$$\begin{array}{r} 63 \\ -(-27) \\ \hline \end{array}$$

5. Perform the operations on each side of the inequality symbol to determine whether the expression is true or false: $-5+(-3) < -5-(-3)$

6. Determine whether or not $x = -4$ is a solution to the equation $x-(-5)=1$.

Practice Problem Answers

1. -85	**2.** -1	**3.** -8
4. 90	**5.** true	**6.** -4 is a solution

Exercises 7.3

1. 11   **2.** 17   **3.** −6   **4.** −2   **5.** 4.7

6. −3.4   **7.** 0   **8.** $\dfrac{9}{16}$ **9.** $-\dfrac{5}{7}$ **10.** −257

Simplify the expressions. Reduce any fractions to lowest terms. See Example 2.

11. $8-3$ **12.** $5-7$ **13.** $-4-6$ **14.** $-18-17$

15. $3-(-4)$ **16.** $5-(-7)$ **17.** $-8-(-11)$ **18.** $-14-2$

19. $0-(-12)$ **20.** $-8-7$ **21.** $16-(-8)$ **22.** $15-23$

23. $2.8-(-3.1)$ **24.** $5.3-(-1.7)$ **25.** $-1.4-2.6$ **26.** $-8.5-7.1$

27. $1.6-(-8.4)$ **28.** $1.5-2.3$ **29.** $\dfrac{2}{5}-\dfrac{3}{4}$ **30.** $\dfrac{7}{15}-\dfrac{2}{15}$

31. $\dfrac{5}{16}-\dfrac{9}{16}$ **32.** $\dfrac{5}{6}-\dfrac{7}{10}$ **33.** $\dfrac{9}{20}-\dfrac{3}{8}$ **34.** $\dfrac{3}{14}-\dfrac{5}{6}$

Perform the indicated subtraction. See Example 3.

35. $\begin{array}{r} 27 \\ -(+42) \\ \hline \end{array}$ **36.** $\begin{array}{r} 19 \\ -(+26) \\ \hline \end{array}$ **37.** $\begin{array}{r} -23 \\ -(-7) \\ \hline \end{array}$ **38.** $\begin{array}{r} -41 \\ -(-8) \\ \hline \end{array}$

39. $\begin{array}{r} -21 \\ -(+36) \\ \hline \end{array}$ **40.** $\begin{array}{r} -27 \\ -(+27) \\ \hline \end{array}$ **41.** $\begin{array}{r} -4.7 \\ -(+1.3) \\ \hline \end{array}$ **42.** $\begin{array}{r} -1.9 \\ -(-2.6) \\ \hline \end{array}$

Solve the following word problems.

43. Find the difference between −5 and −6. (**Hint:** Subtract the numbers in the order given.)

44. Find the difference between 30 and −12. (**Hint:** Subtract the numbers in the order given.)

45. Subtract −3 from −10.

46. Subtract −2 from 6.

47. Subtract 13 from −13.

48. Subtract 20 from −20.

49. Find the sum of −12 and 6. Then subtract −17.

50. Find the sum of 11 and −13. Then subtract 25.

Find the net change in value of each expression by performing the indicated operations. See Example 5.

51. $-6+(-4)-5$ **52.** $-2-2+11$ **53.** $6+(-3)+(-4)$ **54.** $-3+(-7)+2$

55. $-5-2-(-4)$ **56.** $-8-5-(-3)$ **57.** $-7-(-2)+6$ **58.** $-3-(-3)+(-6)$

59. $9.7-1.6-(8.1)$ **60.** $-11.3+5.3-7.9$

Perform the operations on each side of the blank and then fill in the blank with the proper symbol: <, >, or =.

61. $-6+(-2)$ _____ $3+(-8)$ **62.** $-4-(-3)$ _____ $-4+(-3)$

63. $5.1-8.2$ _____ $8.2-5.1$ **64.** $7-(-3)$ _____ $-3-7$

65. $\dfrac{11}{2}+\left(-\dfrac{3}{4}\right)$ _____ $\dfrac{11}{2}-\dfrac{3}{4}$ **66.** $0-6$ _____ $0-(-6)$

67. $-8-(-8)$ _____ $-14-13$ **68.** $-\dfrac{7}{4}-\left(-\dfrac{3}{8}\right)$ _____ $\dfrac{1}{2}-\dfrac{7}{3}$

69. $-151-86$ _____ $-107-141$ **70.** $2.5-6.2$ _____ $-1.1-2.3$

Solve the following word problems. See Example 4.

71. Temperature: At 2 P.M. the temperature was 76°F. At 8 P.M. the temperature was 58°F. What was the change in temperature?

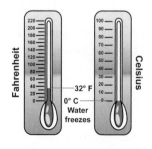

72. Astronomy: According to NASA, the temperature on the earth's moon during the day is 260°F and during the night it is −280°F. What is the daily temperature change on the moon? **Source**: NASA.gov

73. Stock Market: On Monday, February 8, 2010 Nike stock opened at $61 per share. A month later, on Monday, March 8, 2010, the stock opened at $69 per share. Find the change in price of the stock. **Source:** NYSE

74. Used Cars: Mr. Meade is having a hard time selling his old car. He just slashed the price from $3500 to $2750. By how much did he change the price?

75. Travel: If you travel from the top of Mt. Whitney, elevation 14,495 ft, to the floor of Death Valley, elevation 282 ft below sea level, what is the change in elevation?

76. Military: A submarine submerged 280 ft below the surface of the sea and fired a rocket that reached an altitude of 30,000 ft. What was the change in altitude of the rocket?

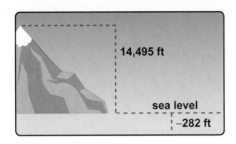

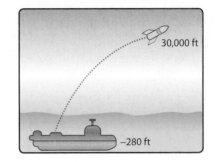

77. History: The great English mathematician and scientist Isaac Newton was born in 1642 and died in 1727. How old was he when he died? (Assume he lived past his birthday.)

78. History: The famous pop culture icon Michael Jackson was born in August 1958 and died in June 2009. How old was he when he died? (Notice that he did not reach his birthday in 2009.)

Determine whether or not the given number is a solution to the given equation by substituting and then evaluating. See Example 6.

79. Is $x = -8$ a solution of $x + 5 = -3$?

80. Is $x = -3$ a solution of $x - 6 = -9$?

81. Is $y = -2$ a solution of $15 - y = 17$?

82. Is $x = 3$ a solution of $11 - x = 8$?

83. Determine whether or not $x = -\dfrac{2}{5}$ is a solution of $x - \dfrac{3}{10} = -\dfrac{7}{10}$.

84. Determine whether or not $y = \dfrac{1}{6}$ is a solution of $y - \dfrac{1}{2} = -\dfrac{1}{3}$.

85. Determine whether or not $x = 0.7$ is a solution of $-2.1 - x = 1.4$.

86. Is $x = -7.8$ a solution of $x + 10.3 = 3.5$?

87. Is $x = 1$ a solution of $x - 2 = -3$?

88. Determine whether $y = 12$ is a solution of $-18 - y = -30$.

89. Determine whether or not $x = -4$ is a solution of $|x| - (-10) = 14$.

90. Determine whether or not $y = -9$ is a solution of $|y| - 12 = -3$.

91. Determine whether or not $z = -5$ is a solution of $|z| - 5 = 0$.

92. Determine whether or not $x = 16$ is a solution of $-16 - |x| = 0$.

93. Is $x = -28$ a solution of $|x| - |-3| = 25$?

94. Is $x = -11$ a solution of $|-2| + |x| = 13$?

Use your calculator to find the value indicated on each side of the blank and then fill in the blank with the proper symbol: <, >, or =.

95. $648 - (-396)$ _____ $124 - 163$

96. $-19,824 - 23,417$ _____ $12,793 - (-14,387)$

97. $-43,931 - (-28,677)$ _____ $-(13,665 + 21,425)$

98. $-(24,295 + 13,107)$ _____ $-48,261 - (-16,276)$

Answer the following word problems. Temperatures above 0° and below 0° as well as gains and losses can be thought of in terms of positive and negative numbers. Use positive and negative numbers to answer the following problems. See Example 5.

99. Football: At the 2010 Rose Bowl, during one possession for the Ohio State Buckeyes, the team gained 28 yards, gained 9 yards, lost 2 yards, gained 5 yards, gained 1 yard, gained 10 yards, lost 5 yards on a penalty, and lost 11 yards. What was the team's net yardage on this possession of the football? Source ESPN

100. Temperature: Beginning at a temperature of 10°C, the temperature in a scientific experiment was measured hourly for four hours. It dropped 5°, dropped 8°, dropped 6°, then rose 3°. What was the final temperature recorded?

101. Dieting: Harry and his wife went on a diet for 5 weeks. During those 5 weeks, Harry lost 5 pounds, gained 3 pounds, lost 2 pounds, lost 4 pounds, and gained 1 pound. What was his total loss (or gain) for the 5 weeks? If he weighed 210 pounds when he started the diet plan, what did he weigh at the end of the 5-week period? (During the same time, his wife lost 10 pounds).

102. Stock Market: In a 5-day week the NASDAQ posted a gain of 38 points, a loss of 65 points, a loss of 32 points, a gain of 10 points, and a gain of 15 points. If the NASDAQ started the week at 2350 points, what was the market at the end of the week?

103. Banking: Many personal securities accounts have a cash fund to receive dividends. The investor is free to write checks on that account, as well as make deposits. Unlike regular checking accounts, there is no overdraft penalty when the balance is less than zero dollars; instead, a nominal interest rate is charged until the balance becomes positive. In addition, this type of account will accrue interest on positive balances. At the beginning of an account period the balance is $3520, and five transactions are made before the next statement is posted as follows: $300 deposit, $2500 withdrawal, $1800 withdrawal, $253 deposit, and $450 withdrawal. What is the balance at the close of this accounting period?

104. Rainwater Collection: A family has a rain cistern to collect rainwater runoff from the roof to water their lawn and garden. When it rains, the garden does not need to be watered, and the cistern just collects additional water for future use. When it doesn't rain, the stored water is needed to water the plants. At the beginning of a five week period the cistern has 873 gallons of water, and the net intake/outflow was noted at the end of each week. First it rose by 461 gallons, then it dropped by 349 gallons, then it dropped by 217 gallons, followed by a 275 gallon increase, and a 177 gallon increase. How many gallons does the cistern contain after this five week period?

105. Gambling: John goes to a casino in Las Vegas and plays six games. His wins and losses for each respective game was $15 gained, $50 lost, $20 lost, $80 gained, $100 lost, and $40 lost. What was John's net gain or loss?

106. Population: The population of San Bernardino County, California in 2003 was 197,126. The population shift for the next five years is shown on the chart below. **Source:** http://www.idcide.com/citydata/ca/san-bernardino.htm

Year	Population Shift
2004	+2696
2005	−554
2006	+162
2007	−955
2008	+105

a. What was the change in population in that five year period?

b. What is the population in 2008?

Writing & Thinking

107. Explain, in your own words, how to find the difference between a positive and a negative number.

108. What is the additive inverse of 0? Why?

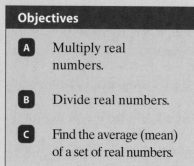

7.4 Multiplication and Division with Real Numbers

Objective A **Multiplication with Real Numbers**

Multiplication is shorthand for repeated addition. That is,

$$7+7+7+7+7 = 5\cdot 7 = 35$$

and

$$(-6)+(-6)+(-6) = 3(-6) = -18.$$

Similarly,

$$(-2)+(-2)+(-2)+(-2)+(-2) = 5(-2) = -10.$$

Repeated addition with a negative number results in a product of a positive number and a negative number. Since the sum of negative numbers is negative, we have the following general rule.

The product of a positive real number and a negative real number is negative.

Objectives

A Multiply real numbers.

B Divide real numbers.

C Find the average (mean) of a set of real numbers.

---

1. Find each product.

a. $4(-7)$

b. $-8(4)$

c. $-1(3)$

d. $4(-1.6)$

e. $\dfrac{5}{3}\cdot\left(-\dfrac{21}{10}\right)$

Example 1

Products of Positive and Negative Real Numbers

a. $5(-3) = (-3)+(-3)+(-3)+(-3)+(-3) = -15$

b. $7(-10) = -70$

c. $42(-1) = -42$

d. $3(-5.2) = -15.6$

e. $\dfrac{3}{7}\cdot\left(-\dfrac{14}{9}\right) = -\dfrac{\cancel{3}\cdot 2\cdot\cancel{7}}{\cancel{7}\cdot\cancel{3}\cdot 3} = -\dfrac{2}{3}$

Now work margin exercise 1.

The product of two negative real numbers can be explained in terms of opposites. We also need the fact that for any real number a, we can think of the **opposite of a**, $(-a)$, as the product of -1 and a. That is, we have $-a = -1\cdot a$. Thus, $-4 = -1\cdot 4 = -1(4)$. And in the product $-4(-7)$ we have,

$$-4(-7) = -1(4)(-7) = -1\big[(4)(-7)\big] = -1(-28) = -(-28) = 28.$$

Although one example does not prove a rule, this process can be used in general to arrive at the following correct conclusion:

The product of two negative real numbers is positive.

Example 2

Products of Negative Real Numbers

a. $(-4)(-9) = +36$

b. $-\dfrac{3}{2}\left(-\dfrac{2}{9}\right) = +\dfrac{\cancel{3}\cdot\cancel{2}}{\cancel{2}\cdot\cancel{3}\cdot 3} = +\dfrac{1}{3}$

c. $-2(-6.7) = +13.4$

d. $(-1)(-5)(-3)(-2) = 5(-3)(-2) = -15(-2) = +30$

Now work margin exercise 2.

What happens if a number is multiplied by 0? For example, $3(0) = 0+0+0 = 0$. In fact, **multiplication by 0 always gives a product of 0**.

Example 3

Multiplication by 0

a. $6 \cdot 0 = 0$

b. $-13 \cdot 0 = 0$

Now work margin exercise 3.

The rules for multiplication with real numbers can be summarized as follows. Remember that these rules apply to integers, decimals, and fractions.

Rules for Multiplication with Real Numbers

If a and b are positive real numbers, then

1. The product of two positive numbers is positive: $a \cdot b = +ab$.
2. The product of two negative numbers is positive: $(-a)(-b) = +ab$.
3. The product of a positive number and a negative number is negative: $a(-b) = -ab$.
4. The product of 0 and any number is 0: $a \cdot 0 = 0$ and $(-a)\cdot 0 = 0$.

Objective B **Division with Real Numbers**

The rules for multiplication lead directly to the rules for division because division is defined in terms of multiplication. For convenience, division is indicated in fraction form.

2. Find each product.

a. $-6(-4)$

b. $-\dfrac{4}{5}\left(-\dfrac{11}{6}\right)$

c. $-5(-3.5)$

d. $(-2)(-3)(-4)(-2)$

3. Find the product.

$-19(0)$

Division with Real Numbers

For real numbers a, b, and x (where $b \neq 0$),

$$\frac{a}{b} = x \text{ means that } a = b \cdot x.$$

For real numbers a and b (where $b \neq 0$),

$$\frac{a}{0} \text{ is } \mathbf{undefined}, \text{ but } \frac{0}{b} = 0.$$

Note: See Section 1.4 for a thorough discussion on division by 0.

4. Perform each division.

a. $\dfrac{42}{7}$

b. $\dfrac{-42}{7}$

c. $\dfrac{42}{-7}$

d. $\dfrac{-42}{-7}$

e. $\dfrac{-3}{0}$

f. $\dfrac{0}{-3}$

Example 4

Division with Real Numbers

a. $\dfrac{36}{9} = 4$ because $36 = 9 \cdot 4$.

b. $\dfrac{-36}{9} = -4$ because $-36 = 9(-4)$.

c. $\dfrac{36}{-9} = -4$ because $36 = -9(-4)$.

d. $\dfrac{-36}{-9} = 4$ because $-36 = -9(4)$.

e. $\dfrac{8}{0}$ is undefined.

f. $\dfrac{0}{-6} = 0$ because $-6 \cdot 0 = 0$.

Now work margin exercise 4.

The rules for division can be stated as follows.

Rules for Division with Real Numbers

If a and b are positive real numbers (where $b \neq 0$),

1. The quotient of two positive numbers is positive: $\dfrac{a}{b} = +\dfrac{a}{b}$.

2. The quotient of two negative numbers is positive: $\dfrac{-a}{-b} = +\dfrac{a}{b}$.

3. The quotient of a positive number and a negative number is negative:

$$\frac{-a}{b} = -\frac{a}{b} \text{ and } \frac{a}{-b} = -\frac{a}{b}.$$

Example 5

Division with Fractions and Decimals

Perform each division. With fractions, reduce quotients to lowest terms. With decimals, round quotients to the nearest tenth.

a. $-\dfrac{16}{7} \div \left(-\dfrac{2}{21}\right) = -\dfrac{16}{7} \cdot \left(-\dfrac{21}{2}\right) = +\dfrac{2 \cdot 8 \cdot 3 \cdot \cancel{7}}{\cancel{7} \cdot \cancel{2}} = \dfrac{24}{1} = +24$

b. $-\dfrac{20}{12} \div \dfrac{15}{2} = -\dfrac{20}{12} \cdot \dfrac{2}{15} = -\dfrac{\cancel{4} \cdot \cancel{5} \cdot 2}{\cancel{4} \cdot 3 \cdot 3 \cdot \cancel{5}} = -\dfrac{2}{9}$

c. $-5.7 \div 4.2 \approx -1.4$ (to the nearest tenth)

d. $-16.54 \div (-5.1) \approx +3.2$ (to the nearest tenth)

Now work margin exercise 5.

---

notes

▪ The following common rules about multiplication and division with two nonzero real numbers are helpful in remembering the signs of answers.

▪

   a. If the numbers have the same sign, both the product and quotient
▪   will be positive.

   b. If the numbers have different signs, both the product and quotient
▪   will be negative.

Objective C   **Average (or Mean)**

Average (or **mean**) has already been discussed with whole numbers, fractions, and decimal numbers. Now we apply this discussion to negative numbers. Newspapers and magazines report average income, average sales, average attendance at sporting events, and so on. The mean of a set of numbers is particularly important in the study of statistics as was discussed in Section 6.1. For example, scientists might be interested in the mean IQ of the students attending a certain university or the mean height of students in the fourth grade.

Average

The **average** (or **mean**) of a set of numbers is the value found by adding the numbers in the set and then dividing the sum by the number of numbers in the set.

---

5. Perform each division. With fractions, reduce quotients to lowest terms. With decimals, round quotients to the nearest tenth.

a. $\dfrac{-14}{5} \div \left(-\dfrac{7}{30}\right)$

b. $-\dfrac{21}{10} \div \dfrac{6}{5}$

c. $-3.4 \div 1.6$

d. $-21.32 \div (-7.4)$

6.
a. At noon on five consecutive days in Madison Heights, Michigan the temperatures were –3°, 5°, 8°, –4°, and 14°(in degrees Fahrenheit). (Negative numbers represent temperatures below zero). Find the average of these noonday temperatures.

b. In a placement exam for biology, a group of ten students had the following scores: 4 students scored 72, 2 students scored 78, 3 students scored 85, and 1 student scored 91. What was the mean score for this group of students?

c. The following speeds (in miles per hour) of twenty cars were recorded at a certain point on a busy road.

45	55	43
47	48	51
54	52	44
46	41	49
50	48	44
45	43	47
53	49	

Find the average speed of these cars. Round your answer to the nearest tenth.

Example 6

Average

a. At noon on five consecutive days in Aspen, Colorado the temperatures were –5°, 7°, 6°, –7°, and 14° (in degrees Fahrenheit). (Negative numbers represent temperatures below zero). Find the average of these noonday temperatures.

Solution

First, add the five temperatures. $(-5)+7+6+(-7)+14 = 15$

Now divide the sum, 15, by the number of temperatures, 5.
$$\frac{15}{5} = 3$$
The average noon temperature was 3° F.

b. In a placement exam for mathematics, a group of ten students had the following scores: 3 students scored 75, 2 students scored 80, 1 student scored 82, 3 students scored 85, and 1 student scored 88. What was the mean score for this group of students?

Solution

To find the total of all the scores, we multiply and then add. This is more efficient than adding all ten scores.

$$
\left.
\begin{array}{l}
75 \cdot 3 = 225 \\
80 \cdot 2 = 160 \\
82 \cdot 1 = 82 \\
85 \cdot 3 = 255 \\
88 \cdot 1 = 88
\end{array}
\right\} \text{Multiply.}
$$

$225 + 160 + 82 + 255 + 88 = 810$ Add.

$810 \div 10 = 81$ Divide by the number of scores.

The mean score on the placement test for this group of students was 81.

c. The following speeds (in miles per hour) of fifteen cars were recorded at a certain point on a freeway.

70	75	65	60	61	64	68	72
59	68	82	76	70	68	50	

Find the average speed of these cars. (One car received a speeding ticket, while another had a broken muffler).

Solution

Using a calculator, the sum of the speeds is 1008 mph.
Dividing by 15 gives the average speed: $1008 \div 15 = 67.2$ mph

Now work margin exercise 6.

Practice Problems

Find the following products.

1. $5(-3)$

2. $-6(-4)$

3. $-8(4)$

4. $-12(0)$

5. $-9(-2)(-1)$

6. $3(-20)(5)$

7. $5(-3.7)$

8. $-4.1(-4.5)$

9. $\left(-\dfrac{1}{6}\right)\left(-\dfrac{2}{3}\right)$

10. $\dfrac{3}{5} \cdot \dfrac{-5}{4}$

Find the quotients.

11. $\dfrac{-30}{10}$

12. $\dfrac{40}{-10}$

13. $\dfrac{-20}{-10}$

14. $\dfrac{-7}{0}$

15. $\dfrac{0}{13}$

16. $\dfrac{7.5}{-3}$

17. $\dfrac{-4.32}{-4}$

18. $\dfrac{-5.2}{-2.6}$

19. Find the mean of the set of integers -16, 20, 32, and 92.

Practice Problem Answers

1. -15 **2.** 24 **3.** -32 **4.** 0

5. -18 **6.** -300 **7.** -18.5 **8.** 18.45

9. $\dfrac{1}{9}$ **10.** $-\dfrac{3}{4}$ **11.** -3 **12.** -4

13. 2 **14.** undefined   **15.** 0 **16.** -2.5

17. 1.08 **18.** 2 **19.** 32

Exercises 7.4

1. $4(-3)$

2. $6(-5)$

3. $12 \cdot 4$

4. $19 \cdot 3$

5. $(-8)(-7)$

6. $(-11)(-2)$

7. $-3 \cdot 7$

8. $-7 \cdot 5$

9. $(-14)(-4)$

10. $(-11)(-6)$

11. $(-13)(-2)$

12. $(-8)(-9)$

13. $10(-7)$

14. $(-5)(12)$

15. $(-2)(-3)(-4)$

16. $(-6)(-3)(-9)$

17. $-8 \cdot 4 \cdot 9$

18. $(-3)(2)(-3)$

19. $(-7)(-16)(0)$

20. $-9 \cdot 0 \cdot 4$

21. $(-2)(4.5)$

22. $(-5)(-3.8)$

23. $4.3(-1.7)$

24. $(-2.6)(-0.2)$

25. $-\dfrac{3}{8} \cdot \dfrac{4}{9}$

26. $\dfrac{4}{5} \cdot \dfrac{-3}{14}$

27. $\dfrac{-4}{5} \cdot \dfrac{-9}{2}$

28. $-\dfrac{3}{4} \cdot -\dfrac{6}{7}$

29. $-4 \cdot \dfrac{3}{5}$

30. $-7 \cdot \dfrac{5}{6}$

31. $\dfrac{5}{2} \cdot \dfrac{-15}{10} \cdot \dfrac{6}{5}$

32. $\dfrac{-4}{7} \cdot \dfrac{2}{5} \cdot \dfrac{-2}{13}$

Find the indicted quotients and reduce any fractions to lowest terms. Round answers with decimals to the nearest tenth. See Examples 4 and 5.

33. $\dfrac{-8}{-2}$

34. $\dfrac{-20}{-10}$

35. $\dfrac{-30}{5}$

36. $\dfrac{-51}{3}$

37. $\dfrac{-26}{-13}$

38. $\dfrac{-91}{-7}$

39. $\dfrac{0}{6}$

40. $\dfrac{16}{0}$

41. $\dfrac{39}{-13}$

42. $\dfrac{44}{-4}$

43. $\dfrac{-34}{2}$

44. $\dfrac{-36}{9}$

45. $\dfrac{-3}{0}$

46. $\dfrac{0}{-7}$

47. $\dfrac{-60}{-12}$

48. $\dfrac{-48}{-16}$

49. $\dfrac{-4.8}{8}$

50. $\dfrac{-5.6}{7}$

51. $\dfrac{-4}{-0.2}$

52. $\dfrac{-3}{-8}$

53. $\dfrac{2.99}{-1.3}$

54. $\dfrac{2.8}{-1.4}$

55. $\dfrac{-2}{15} \div \dfrac{8}{5}$

56. $\dfrac{-3}{5} \div \dfrac{-9}{10}$

57. $\dfrac{6}{11} \div \dfrac{4}{3}$

58. $\dfrac{-10}{3} \div \dfrac{-7}{5}$

59. $\dfrac{-9}{14} \div \dfrac{54}{35}$

60. $\dfrac{45}{8} \div \dfrac{35}{12}$

Determine whether each statement is true or false. If a statement is false, rewrite it in a form that is true. (There may be more than one correct new form).

61. $(-4)(6) \geq 3 \cdot 8$

62. $(-7)(-9) \leq 3 \cdot 21$

63. $-\dfrac{3}{4} \cdot \dfrac{5}{8} = \dfrac{15}{16}\left(-\dfrac{1}{2}\right)$

64. $(-6.0)(9.1) = (6.1)(-9.0)$

65. $6(-3) > (-14) + (-4)$

66. $7 + 8 > (-10) + (-5)$

67. $-\dfrac{2}{3} + \dfrac{1}{2} \leq \dfrac{1}{2} - \dfrac{1}{4}$

68. $1.7 + (-3.9) < (-1.4) + (-4.2)$

69. $(-4)(9) = (-24) + (-12)$

70. $14 + 6 \leq (-2)(-10)$

Solve the following word problems. See Example 6.

71. Find the mean of the following set of integers: $-10, 15, 16, -17, -34,$ and -42.

72. Find the mean of the following set of integers: $-72, -100, -54, 82,$ and -96.

73. **Airline Travel:** The costs of a one-way flight from Baltimore, MD to Orlando, FL on seven different airlines are as follows: $189, $134, $131, $231, $134, $213, $109. What is the average cost of a flight from Baltimore to Orlando? **Source:** Expedia.com

74. **Car Accidents:** The numbers of fatal motor-vehicle accidents in the United States each year from 1998 to 2008 are as follows: 37,107; 37,140; 37,526; 37,862; 38,491; 38,477; 38,444; 39,252; 38,648; 37,435; and 34,017. What was the average number of fatal accidents in the United States per year from 1998 to 2008? (Round your answer to the nearest tenth). **Source:** NHTSA

75. **Business:** Twenty business executives made the following numbers of telephone calls during one week. Find the mean number of calls (to the nearest tenth) made by these executives.

20	16	14	11	51
40	36	28	52	25
18	16	42	49	12
18	22	33	9	19

76. **Health:** The blood calcium level (in milligrams per deciliter) for 20 patients was reported as follows:

8.2	10.2	9.3	8.5	7.3
9.7	9.6	8.3	9.8	9.1
9.4	11.1	10.0	8.5	9.9
8.6	10.2	9.4	9.1	9.2

Find the mean blood calcium level (to the nearest hundredth) for these patients.

77. **Exam Scores:** Fifteen students scored the following scores on an exam in accounting: 1 scored 67, 4 scored 73, 3 scored 77, 2 scored 80, 3 scored 88, and 2 scored 93. What was the average score for these students?

78. **Exam Scores:** On an exam in history, a class of twenty-one students had the following test scores: 4 scored 65, 3 scored 70, 6 scored 78, 2 scored 82, 1 scored 85, 3 scored 91, and 2 scored 95. What was the mean score (to the nearest tenth) on this test for the class?

The frequency of a number is a count of how many times that number appears. In statistics, data is commonly given in the table form of a frequency distribution as illustrated. To find the mean, multiply each number by its frequency, add these products, and divide the sum by the sum of the frequencies.

---

79. **Height:** The heights of the top 30 NBA scorers for the 2009–2010 season are listed in the frequency table. Find the mean height for these men. **Source:** NBA

Height (in inches)	72	73	74	75	76	77	78	79	80	81	82	83	84
Frequency	1	1	1	5	1	0	5	4	3	3	1	3	2

80. **Books Read:** The students in a psychology class were asked the number of books that they had read in the last month. The following frequency distribution indicates the results. Find the mean number of books read (to the nearest tenth) by these students.

Number of Books	0	1	2	3	4	5
Frequency	3	2	6	4	2	1

81. **Watching TV:** The following bar graph shows the approximate amounts of time per week spent watching TV for six groups (by age and sex) of people 18 years of age and older. What is the average amount of time per week people over the age of 18 spend watching TV? (Assume each group has the same number of people.)

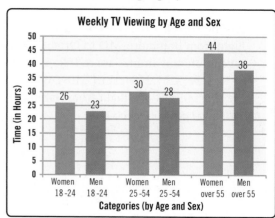

82. Phone Calls: The following pictograph shows the number of phone calls received by a 1-hour radio talk show in Seattle during one week. (Not all calls actually get on the air). What was the mean number of calls per show received that week?

Phone Calls Received by a Seattle Talk Show

Monday
Tuesday
Wednesday
Thursday
Friday

Each  represents 10 phone calls

83. Geography: The following bar graph shows the area of each of the five Great Lakes. Lake Superior, with an area of about 31,800 square miles, is the world's largest fresh water lake. What is the mean size of these lakes?

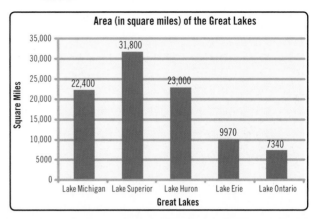

Area (in square miles) of the Great Lakes

Use a graphing calculator to find the value of each expression. Round quotients to the nearest hundredth, if necessary. Remember the negative sign **(–)** is next to ENTER .

For example, $-14.8 \div (-5)$ would appear as follows:

```
-14.8/(-5)
            2.96
```

84. $(27)(-24)(-180)$ **85.** $(-461)(-45)(-17)$ **86.** $(54)(-17)(-24)$

87. $(-77,459) \div 29$ **88.** $(-62,234) \div (-37)$ **89.** $(-35 - 45 - 56) \div 3$

90. $(-52 - 30 - 40 - 60) \div 4$ **91.** $72 \div (15 - 22)$ **92.** $95 \div (-3 - 7)$

93. $-15.3 \div (-5.4)$ **94.** $(-13.4)(-2.5)(-1.63)$ **95.** $(-2.5)(-3.41)(-10.6)$

Writing & Thinking

96. Explain the conditions under which the quotient of two numbers is 0.

97. Explain, in your own words, why division by 0 is not a valid arithmetic operation.

Objectives

A Use the rules for order of operations to evaluate expressions.

7.5 Order of Operations with Real Numbers

Objective A **Using the Rules for Order of Operations**

The **rules for order of operations** were discussed in Section 1.6 in evaluating expressions with whole numbers. For example, the expression with whole numbers

$$45 \div 9 + 2 \cdot 3^2$$

can be evaluated as follows:

$$45 \div 9 + 2 \cdot 3^2 = 45 \div 9 + 2 \cdot 9$$
$$= 5 + 2 \cdot 9$$
$$= 5 + 18$$
$$= 23$$

In this section we will see how to apply these rules with other real numbers, including integers, decimal numbers, and fractions.

Rules for Order of Operations

1. Simplify within grouping symbols, such as parentheses ( ), brackets [ ], and braces { }, working from the innermost grouping outward.

2. Find any powers indicated by exponents.

3. Moving from **left to right**, perform any multiplications **or** divisions **in the order they appear**.

4. Moving from **left to right**, perform any additions **or** subtractions **in the order they appear**.

notes

■ Other grouping symbols are the absolute value bars (such as $|3+5|$),

■ the fraction bar (as in $\dfrac{4+7}{10+1}$), and the square root symbol

■ (such as $\sqrt{5+11}$).

These rules are very explicit and should be studied carefully. Note that in rule 3, neither multiplication nor division has priority over the other. Whichever of these operations occurs first, **moving left to right**, is done first. In rule 4, addition and subtraction are handled in the same way. Unless they occur within grouping symbols, **addition and subtraction are the last operations to be performed**.

A well-known mnemonic device for remembering the rules for order of operations is the following:

Please	Excuse	My	Dear	Aunt	Sally
↓	↓	↓	↓	↓	↓

Parentheses Exponents Multiplication Division Addition Subtraction

notes

Even though the mnemonic **PEMDAS** is helpful, remember that multiplication and division are performed as they appear, left to right. Also, addition and subtraction are performed as they appear, left to right.

For example

$$12 \div 3 \cdot 4 = 4 \cdot 4 = 16,$$

but

$$12 \cdot 3 \div 4 = 36 \div 4 = 9.$$

A negative sign in front of a variable indicates that the variable is to be multiplied by –1. For example,

$$-x^2 = -1 \cdot x^2.$$

This is consistent with the rules for order of operations (indicating that exponents come before multiplication) and is particularly useful in determining the values of expressions involving negative numbers and exponents. For example, each of the expressions

$$-7^2 \quad \text{and} \quad (-7)^2$$

has a different value. By the order of operations,

$$-7^2 = -1 \cdot 7^2 = -1 \cdot 49 = -49,$$

but

$$(-7)^2 = (-7)(-7) = 49.$$

In the second expression the base is a negative number. Remember that if the base is a negative number, then the negative number must be placed in parentheses.

The following examples show how to apply the rules. In some cases, more than one step can be performed at the same time. This is possible when parts are separated by + or − signs or are within separate symbols of inclusion. **Work through each of the following examples step by step, and rewrite the examples on a separate sheet of paper.**

1. Use the rules for order of operations to evaluate the following expressions.

   a. $56 \div 7 - 5 \cdot 3^2$

   b. $2(5^2 - 9) - 2 \cdot 3^3$

   c. $7 - \left[(4 \cdot 6 - 3^2) \div 5 + 4^2 \right]$

Example 1

Order of Operations with Integers

Use the rules for order of operations to evaluate each of the following expressions.

a. $36 \div 4 - 6 \cdot 2^2$

Solution

$36 \div 4 - 6 \cdot 2^2$

$\begin{aligned}
&= 36 \div 4 - 6 \cdot 4 && \text{exponents} \\
&= 9 - 24 && \text{Divide and multiply, left to right.} \\
&= -15 && \text{Subtract.}
\end{aligned}$

b. $2(3^2 - 1) - 3 \cdot 2^3$

Solution

$2(3^2 - 1) - 3 \cdot 2^3$

$\begin{aligned}
&= 2(9 - 1) - 3 \cdot 8 && \text{exponents} \\
&= 2(8) - 3 \cdot 8 && \text{Subtract inside the parentheses.} \\
&= 16 - 24 && \text{Multiply.} \\
&= -8 && \text{Subtract (or add algebraically).}
\end{aligned}$

c. $9 - 2\left[(3 \cdot 5 - 7^2) \div 2 + 2^2 \right]$

Solution

$9 - 2\left[(3 \cdot 5 - 7^2) \div 2 + 2^2 \right]$

$\begin{aligned}
&= 9 - 2\left[(3 \cdot 5 - 49) \div 2 + 4 \right] && \text{exponents} \\
&= 9 - 2\left[(15 - 49) \div 2 + 4 \right] && \text{Multiply inside the parentheses.} \\
&= 9 - 2\left[(-34) \div 2 + 4 \right] && \text{Subtract inside the parentheses.} \\
&= 9 - 2\left[-17 + 4 \right] && \text{Divide inside the brackets.} \\
&= 9 - 2\left[-13 \right] && \text{Add inside the brackets.} \\
&= 9 + 26 && \text{Multiply.} \\
&= 35 && \text{Add.}
\end{aligned}$

Note: Because of the rules for order of operations at no time did we even subtract 9 − 2.

Now work margin exercise 1.

Completion Example 2

Order of Operations with Integers

Evaluate the expression $6\left(5^2 - 4^2\right) - 2 \cdot 3^3$.

Solution

$6\left(5^2 - 4^2\right) - 2 \cdot 3^3$

$= 6(\underline{\quad} - \underline{\quad}) - 2 \cdot \underline{\quad}$ exponents

$= 6(\underline{\quad}) - 2 \cdot 27$ Simplify within parentheses.

$= \underline{\quad} - \underline{\quad}$ Multiply.

$= \underline{\quad}$ Subtract.

Now work margin exercise 2.

Example 3

Order of Operations with Fractions

Use the rules for order of operations to evaluate each of the following expressions.

a. $2\dfrac{1}{3} \div \left(\dfrac{1}{4} + \dfrac{1}{3}\right)$

Solution

$2\dfrac{1}{3} \div \left(\dfrac{1}{4} + \dfrac{1}{3}\right)$

$= \dfrac{7}{3} \div \left(\dfrac{3}{12} + \dfrac{4}{12}\right)$ Change the mixed number to an improper fraction and find the LCD in the parentheses.

$= \dfrac{7}{3} \div \left(\dfrac{7}{12}\right)$ Add within parentheses.

$= \dfrac{\cancel{7}}{\cancel{3}} \cdot \dfrac{\overset{4}{\cancel{12}}}{\cancel{7}}$ Divide and reduce.

$= 4$

2. Evaluate the expression.

$$-3 + (3+2)^2 + \left(4 - 3^2\right)$$

Completion Example Answer

2. $6\left(5^2 - 4^2\right) - 2 \cdot 3^3$

$= 6\left(25 - 16\right) - 2 \cdot 27$

$= 6\left(9\right) - 2 \cdot 27$

$= 54 - 54$

$= 0$

3. Evaluate the expression.

$$\frac{1}{10} \div \frac{3}{4} \cdot \frac{1}{2} - \frac{3}{5} \cdot \frac{7}{30}$$

b. $\dfrac{3}{5} \cdot \dfrac{5}{6} + \dfrac{1}{4} \div \left(\dfrac{5}{2}\right)^2$

Solution

$$\frac{3}{5} \cdot \frac{5}{6} + \frac{1}{4} \div \left(\frac{5}{2}\right)^2$$

$$= \frac{3}{5} \cdot \frac{5}{6} + \frac{1}{4} \div \left(\frac{25}{4}\right) \qquad \text{exponents}$$

$$= \frac{\cancel{3}}{\cancel{5}} \cdot \frac{\cancel{5}}{\cancel{6}_2} + \frac{1}{\cancel{4}} \cdot \frac{\cancel{4}}{25} \qquad \text{Divide, then multiply and reduce.}$$

$$= \frac{1}{2} + \frac{1}{25}$$

$$= \frac{25}{50} + \frac{2}{50} \qquad \text{Find the LCD.}$$

$$= \frac{27}{50} \qquad \text{Add.}$$

Now work margin exercise 3.

Order of Operations with a Calculator

Some calculators are programmed to follow the rules for order of operations. To test for if your calculator is programmed in this manner, let's evaluate $7 + 4 \cdot 2$. Press the keys

[7] [+] [4] [×] [2] and then [=].

If the answer on your calculator screen shows 15, then you do have a built in order of operations. This means that most of the time, you can type in the problem as it is written with the calculator performing the order of operations in a proper manner. If the calculator gave you an answer of 22, then you will need to be careful when entering problems.

If your calculator has parentheses buttons, (and), you can evaluate the expression $(6 + 2) + (8 + 1) \div 9$ by pressing the keys

[( ] [6] [+] [2] [ )] [+] [( ] [8] [+] [1] [ )] [÷] [9].
Then press [=].
The display will read 9.

Practice Problems

Use the rules for order of operations to evaluate each expression.

1. $36 + 2 \cdot 6$

2. $8 \cdot 2 + 4(-2) - 4^2$

3. $9 \cdot 3 \div (2^2 - 5)$

4. $\left(\dfrac{1}{6}\right)^2 \div \dfrac{7}{12} - \dfrac{5}{8}$

5. $\left(-\dfrac{3}{4}\right) + \dfrac{3}{8} \cdot \dfrac{4}{5} \div \dfrac{2}{5} + \dfrac{2}{3}$

6. $-15 \div \left(\dfrac{1}{3} + \dfrac{1}{10}\right)$

Simplify.

7. $\dfrac{2}{5a} \div \dfrac{16}{3a} - \dfrac{2b}{7} \cdot \dfrac{7}{5b}$

Practice Problem Answers

1. 38.6

2. −8

3. −27

4. $-\dfrac{97}{168}$

5. $\dfrac{2}{3}$

6. $-\dfrac{450}{13}$

7. $-\dfrac{13}{40}$

Exercises 7.5

Use the rules for order of operations to evaluate the expressions. Change mixed numbers to improper fractions before evaluating. See Examples 1 through 3.

1. **a.** $24 \div 4 \cdot 6$
    b. $24 \cdot 4 \div 6$

2. **a.** $20 \div 5 \cdot 2$
    b. $20 \cdot 5 \div 2$

3. $15 \div (-3) \cdot 3 - 10$

4. $20 \cdot 2 \div 2^2 + 5(-2)$

5. $3^2 \div (-9) \cdot \left(4 - 2^2\right) + 5(-2)$

6. $4^2 \div (-8)(-2) + 3\left(2^2 - 5^2\right)$

7. $14 \cdot 3 \div (-2) - 6(4)$

8. $6(13 - 15)^2 \cdot 8 \div 2^2 + 3(-1)$

9. $-10 + 15 \div (-5) \cdot 3^2 - 10^2$

10. $16 \cdot 3 \div \left(2^2 - 5\right)$

11. $2 - 5\left[(-20) \div (-4) \cdot 2 - 40\right]$

12. $9 - 6\left[(-21) \div 7 \cdot 2 - (-8)\right]$

13. $(7 - 10)\left[49 \div (-7) + 20 \cdot 3 - (-10)\right]$

14. $(9 - 11)\left[(-10)^2 \cdot 2 + 6(-5)^2 - 10^2 + 3 \cdot 5\right]$

15. $8 - 9\left[(-39) \div (-13) + 7(-2) - (-2)^2\right]$

16. $6 - 20\left[(-15) \div 3 \cdot 5 + 6 \cdot 2 \div 3\right]$

17. $|16 - 20|\left[32 \div |3 - 5| - 5^2\right]$

18. $|10 - 30|\left[4^2 \cdot |5 - 8| \div (-2)^2 + |17 - 18|\right]$

19. $(-10) + (-2) + |2 - 4|$

20. $|16 - 20| + (-10)^2 + 5^2$

21. $\dfrac{3}{8} \cdot \dfrac{4}{5} + \dfrac{1}{15}$

22. $\dfrac{1}{4} \cdot \dfrac{12}{15} + \dfrac{2}{7}$

23. $\dfrac{1}{3} \div \dfrac{1}{2} - \dfrac{5}{6} \cdot \dfrac{3}{4}$

24. $\dfrac{2}{9} \div \dfrac{14}{3} - \dfrac{1}{6} \cdot \dfrac{4}{7}$

25. $\left(\dfrac{5}{6}\right)^2 \div \dfrac{5}{12} - \dfrac{3}{8}$

26. $\left(\dfrac{2}{5}\right)^2 \cdot \dfrac{3}{8} + \dfrac{1}{5} \div \dfrac{3}{4}$

27. $\dfrac{7}{6} \cdot 2^2 - \dfrac{2}{3} \div 3\dfrac{1}{5}$

28. $\dfrac{3}{4} \div 3^2 - 4\left(\dfrac{3}{2}\right)^2$

29. $\left(-\dfrac{3}{4}\right) \div \left(-\dfrac{3}{5}\right) \cdot \dfrac{7}{8} + \dfrac{3}{16}$

30. $\left(-\dfrac{2}{3}\right) \div \dfrac{7}{12} - \dfrac{2}{7} + \left(-\dfrac{1}{2}\right)^2$

31. $\left(-\dfrac{9}{10}\right) + \dfrac{5}{8} \cdot \dfrac{4}{5} \div \dfrac{6}{10} + \dfrac{2}{3}$

32. $\dfrac{5}{8} \div \dfrac{1}{10} + \left(-\dfrac{1}{3}\right)^2 \cdot \dfrac{3}{5}$

33. $-2\dfrac{1}{2}\cdot 3\dfrac{1}{5}\div\dfrac{3}{4}-\dfrac{7}{10}$

34. $-\dfrac{5}{9}-\dfrac{1}{3}\cdot\dfrac{2}{3}-3\dfrac{1}{3}$

35. $\left(\dfrac{1}{2}-1\dfrac{3}{4}\right)\div\left(\dfrac{2}{3}+\dfrac{3}{4}\right)$

36. $\left(\dfrac{5}{8}+2\dfrac{1}{4}\right)\div\left(1-\dfrac{1}{4}\right)$

37. $-15\div\left(\dfrac{1}{4}-\dfrac{7}{8}\right)$

38. $-12\div\left(\dfrac{1}{2}+\dfrac{1}{10}\right)$

39. $\left(1\dfrac{1}{10}-3\dfrac{1}{5}\right)\div 3\dfrac{1}{2}+1\dfrac{3}{10}$

40. $4\left(\dfrac{1}{2}\right)^{2}+\dfrac{7}{10}\div\dfrac{3}{5}\cdot\dfrac{5}{18}-\dfrac{9}{10}$

41. Find the average of the three numbers: $3\dfrac{2}{5}$, $5\dfrac{3}{4}$, and $6\dfrac{1}{10}$.

42. Find the average of the four numbers: $1\dfrac{1}{8}$, $2\dfrac{1}{2}$, $-5\dfrac{1}{2}$, and $-\dfrac{5}{8}$.

43. If the square of $\dfrac{7}{8}$ is subtracted from the square of $\dfrac{3}{4}$, what is the difference?

44. Find the quotient if the sum of $\dfrac{4}{5}$ and $\dfrac{2}{15}$ is divided by the difference between $\dfrac{7}{8}$ and $2\dfrac{1}{4}$.

Use a graphing calculator to evaluate each of the expressions.

45. $3.4\div 4+5\cdot 8.32$

46. $8.1\div 5+16.3\cdot 7$

47. $0.75\div 1.5+7\cdot 3.1^{2}$

48. $1.05\div(-3)\cdot 3.7-1.1^{2}$

49. $6.32\cdot 8.4\div 16.8+3.5^{2}$

50. $(82.7+16.2)\div(14.83-19.83)^{2}$

Writing & Thinking

51. Explain, in your own words, why the following expression cannot be evaluated.

$$\left(24-2^{4}\right)+6(3-5)\div\left(3^{2}-9\right)$$

52. Consider any number between 0 and 1. If you square this number, will the result be larger or smaller than the original number? Is this always the case? Explain.

53. Consider any number between −1 and 0. If you square this number, will the result be larger or smaller than the original number? Is this always the case? Explain.

7.6 Properties of Real Numbers

Objective A   **Applying the Properties of Real Numbers**

Operations with whole numbers, decimals, and fractions were covered in Chapters 1, 2, and 3. In Chapter 7 we have introduced negative numbers and operations with negative numbers. As we continue to work with all types of real numbers and algebraic expressions we will need to understand that certain properties are true for addition and multiplication. However, these same properties are not true for subtraction and division. For example,

the order of the numbers in addition **does not** change the result:

$$17 + 5 = 5 + 17 = 22$$

but, the order of the numbers in subtraction **does** change the result:

$$17 - 5 = 12 \quad \text{but} \quad 5 - 17 = -12.$$

We say that addition is **commutative** while subtraction is **not commutative**. The various properties of real numbers under the operations of addition and multiplication are summarized here. These properties are used throughout algebra and mathematics in developing formulas and general concepts.

Properties of Addition and Multiplication

In this table a, b, and c are real numbers.

Name of Property

For Addition		For Multiplication
$a + b = b + a$	Commutative property	$ab = ba$
$3 + 6 = 6 + 3$		$4 \cdot 9 = 9 \cdot 4$
$(a + b) + c = a + (b + c)$	Associative property	$a(bc) = (ab)c$
$(2 + 5) + 4 = 2 + (5 + 4)$		$6 \cdot (2 \cdot 7) = (6 \cdot 2) \cdot 7$
$a + 0 = 0 + a = a$	Identity	$a \cdot 1 = 1 \cdot a = a$
$20 + 0 = 0 + 20 = 20$		$-2 \cdot 1 = 1 \cdot (-2) = -2$
$a + (-a) = 0$	Inverse	$a \cdot \dfrac{1}{a} = 1 \; \left(\text{for } a \neq 0\right)$
$10 + (-10) = 0$		$3 \cdot \dfrac{1}{3} = 1$

Zero-Factor Law

$$a \cdot 0 = 0 \cdot a = 0 \qquad\qquad -5 \cdot 0 = 0 \cdot (-5) = 0$$

Distributive Property of Multiplication over Addition

$$a(b + c) = ab + ac \qquad\qquad 3(x + 5) = 3 \cdot x + 3 \cdot 5$$

notes

The number **0** is called the **additive identity** because when 0 is added to a number the result is the same number. Likewise, the number **1** is called the **multiplicative identity** because when a number is multiplied by 1 the result is the same number. Also, the **additive inverse** of a number is its **opposite** and the **multiplicative inverse** of a number is its **reciprocal**.

The raised dot is optional when indicating multiplication between two variables or a number and a variable. For example,

$$x \cdot y = xy \quad \text{and} \quad 5 \cdot x = 5x.$$

In the case of a number and a variable, the number is called the **coefficient** of the variable. So in

$$5x \text{ the number } 5 \text{ is the } \textbf{coefficient} \text{ of } x.$$

Thus, in an illustration of the distributive property involving a variable, we can write

$$5(x - 6) = 5 \cdot x - 5 \cdot 6 = 5x - 30.$$

Example 1

Properties of Addition and Multiplication

State the name of each property being illustrated.

a. $(-7) + 13 = 13 + (-7)$

Solution   commutative property of addition

b. $8 + (9 + 1) = (8 + 9) + 1$

Solution   associative property of addition

c. $(-25) \cdot 1 = -25$

Solution   multiplicative identity

d. $3(x + y) = 3x + 3y$

Solution   distributive property

e. $0 \cdot 14 = 0$

Solution   zero-factor law

Now work margin exercise 1.

1. State the name of each property being illustrated.

a. $(-4) + 16 = 16 + (-4)$

b. $14 + (2 + 5) = (14 + 2) + 5$

c. $(-37) \cdot 0 = 0$

d. $5(z + w) = 5z + 5w$

e. $7 \cdot (3 \cdot 8) = (7 \cdot 3) \cdot 8$

Objective B   Identifying Properties

2. State the property illustrated and show that the statement is true for the value given for the variable.

a. $x + 21 = 21 + x$

given that $x = -7$

b. $(5 \cdot 4)x = 5(4x)$

given that $x = 2$

c. $11(y + 3) = 11y + 33$

given that $y = -4$

Example 2

Properties of Addition and Multiplication

In each of the following equations, state the property illustrated and show that the statement is true for the value given for the variable by substituting the value in the equation and evaluating.

a. $x + 14 = 14 + x$ given that $x = -4$

Solution

The commutative property of addition is illustrated.

$$(-4) + 14 = 10 \quad \text{and} \quad 14 + (-4) = 10$$

b. $(3 \cdot 6)x = 3(6x)$ given that $x = 5$

Solution

The associative property of multiplication is illustrated.

$$(3 \cdot 6) \cdot 5 = 18 \cdot 5 = 90 \quad \text{and} \quad 3 \cdot (6 \cdot 5) = 3 \cdot 30 = 90$$

c. $12(y + 3) = 12y + 36$ given that $y = -2$

Solution

The distributive property is illustrated.

$$12(-2 + 3) = 12(1) = 12 \quad \text{and} \quad 12(-2) + 36 = -24 + 36 = 12$$

Now work margin exercise **2.**

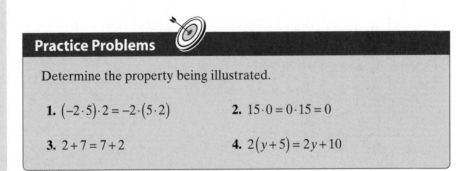

Practice Problems

Determine the property being illustrated.

1. $(-2 \cdot 5) \cdot 2 = -2 \cdot (5 \cdot 2)$ **2.** $15 \cdot 0 = 0 \cdot 15 = 0$

3. $2 + 7 = 7 + 2$ **4.** $2(y + 5) = 2y + 10$

Practice Problem Answers

1. associative property of multiplication

2. zero-factor law

3. commutative property of addition

4. distributive property

Exercises 7.6

Complete the expressions using the given property. Do not simplify.

1. $7 + 3 =$ _____   commutative property of addition

2. $(6 \cdot 9) \cdot 3 =$ _____   associative property of multiplication

3. $19 \cdot 4 =$ _____   commutative property of multiplication

4. $18 + 5 =$ _____   commutative property of addition

5. $6(5 + 8) =$ _____   distributive property

6. $16 + (9 + 11) =$ _____   associative property of addition

7. $2 \cdot (3x) =$ _____   associative property of multiplication

8. $3(x + 5) =$ _____   distributive property

9. $3 + (x + 7) =$ _____   associative property of addition

10. $9(x + 5) =$ _____   distributive property

11. $6 \cdot 0 =$ _____   zero-factor law

12. $6 \cdot 1 =$ _____   multiplicative identity

13. $0 + (x + 7) =$ _____   additive identity

14. $0 \cdot (-13) =$ _____   zero-factor law

15. $2(x - 12) =$ _____   distributive property

16. $(-5) + 5 =$ _____   additive inverse

17. $6.3 + (-6.3) =$ _____   additive inverse

18. $3 \cdot \dfrac{1}{3} =$ _____   multiplicative inverse

Name the property of real numbers illustrated. See Example 1.

19. $5 + 16 = 16 + 5$

20. $5 \cdot 16 = 16 \cdot 5$

21. $32 \cdot 1 = 32$

22. $32 + 0 = 32$

23. $5 + (3 + 1) = (5 + 3) + 1$

24. $5 + (3 + 1) = (3 + 1) + 5$

25. $13(y+2) = (y+2) \cdot 13$

26. $13(y+2) = 13y + 26$

27. $6(2 \cdot 9) = (2 \cdot 9) \cdot 6$

28. $6(2 \cdot 9) = (6 \cdot 2) \cdot 9$

29. $5 \cdot \dfrac{1}{5} = 1$

30. $14 \cdot \dfrac{1}{14} = 1$

31. $7.1 + (-7.1) = 0$

32. $(-9) + 9 = 0$

33. $1 \cdot 14.2 = 14.2$

34. $(5 \cdot 3) \cdot -7 = 5(3 \cdot -7)$

35. $5.68 \cdot 0 = 0 \cdot 5.68 = 0$

36. $0 + 5.68 = 5.68$

37. $2 + (x+6) = (2+x) + 6$

38. $2(x+6) = 2x + 12$

First evaluate each expression using the rules for order of operations and then use the distributive property to evaluate the same expression. The value must be the same.

39. $6(3+8)$

40. $7(8-5)$

41. $10(2-9)$

42. $13(5+3)$

In each of the following equations, state the property illustrated and show that the statement is true for the value of $x = 4$, $y = -2$, or $z = 3$ by substituting the corresponding value in the equation and evaluating. See Example 2.

43. $6 \cdot x = x \cdot 6$

44. $19 + z = z + 19$

45. $8 + (5+y) = (8+5) + y$

46. $(2 \cdot 7) \cdot x = 2 \cdot (7x)$

47. $5(x+18) = 5x + 90$

48. $(2z+14) + 3 = 2z + (14+3)$

49. $(6 \cdot y) \cdot 9 = 6 \cdot (y \cdot 9)$

50. $11 \cdot x = x \cdot 11$

51. $z + (-34) = -34 + z$

52. $3(y+15) = 3y+45$ **53.** $2(3+x) = 2(x+3)$ **54.** $(y+2)(y-4) = (y-4)(y+2)$

55. $5+(x-15) = (x-15)+5$ **56.** $z+(4+x) = (4+x)+z$ **57.** $(3x)\cdot 5 = 3\cdot(x\cdot 5)$

58. $(x+y)+z = x+(y+z)$

Writing & Thinking

59. **a.** The distributive property illustrated as $a(b+c) = ab+ac$ is said to "distribute multiplication over addition." Explain, in your own words, the meaning of this phrase.

 b. What would an expression that "distributes addition over multiplication" look like? Explain why this would or would not make sense.

Objectives

A   Identify like terms.

B   Simplify algebraic expressions by combining like terms.

C   Evaluate expressions for given values of the variables.

7.7 Simplifying and Evaluating Algebraic Expressions

Objective A   **Identifying Like Terms**

A single number is called a **constant**. Any constant or variable or the indicated product and/or quotient of constants and variables is called a **term**. Examples of terms are

$$16, \quad 3x, \quad -5.2, \quad 1.3xy, \quad -5x^2, \quad 14b^3, \quad \text{and} \quad -\frac{x}{y}.$$

Note: As discussed in Section 1.6, an expression with 2 as the exponent is read "squared" and an expression with 3 as the exponent is read "cubed". Thus $7x^2$ is read "seven x squared" and $-4y^3$ is read "negative four y cubed".

The number written next to a variable is called the **coefficient** (or the **numerical coefficient**) of the variable. For example, in

$$5x^2, 5 \text{ is the coefficient of } x^2.$$

Similarly, in

$$1.3xy, 1.3 \text{ is the coefficient of } xy.$$

notes

- If no number is written next to a variable, the coefficient is understood to be 1. If a negative sign $(-)$ is next to a variable, the coefficient is understood to be -1. For example,

$$x = 1 \cdot x, \quad a^3 = 1 \cdot a^3, \quad -x = -1 \cdot x, \quad \text{and} \quad -y^5 = -1 \cdot y^5.$$

Like Terms

Like terms (or **similar terms**) are terms that are constants or terms that contain the same variables raised to the same powers.

Note: If there is more than one variable in a term, then the sum of the exponents on the variables is the **degree** of the term.

Like Terms

$-6,\ 1.84,\ 145,\ \dfrac{3}{4}$	are like terms because each term is a constant.
$-3a,\ 15a,\ 2.6a,\ \dfrac{2}{3}a$	are like terms because each term contains the same variable a, raised to the same exponent, 1. (Remember that $a = a^1$). These terms are first-degree in a.
$5xy^2$ and $-3.2xy^2$	are like terms because each term contains the same two variables, x and y, with x first-degree in both terms and y second-degree in both terms.

Unlike Terms

$8x$ and $-9x^2$ are unlike terms (**not** like terms) because the variable x is not of the same power in both terms. $8x$ is first-degree in x and $-9x^2$ is second-degree in x.

Example 1

Like Terms

From the following list of terms, pick out the like terms.

$$-7, 2x, 4.1, -x, 3x^2y, 5x, -6x^2y, \text{ and } 0$$

Solution

$-7, 4.1,$ and 0 are like terms. (All are constants).

$2x, -x,$ and $5x$ are like terms.

$3x^2y$ and $-6x^2y$ are like terms.

Now work margin exercise 1.

1. From the following list of terms, pick out the like terms.

$$-5.3, 4y, 9, -2y, 0,$$
$$10xy^2, -xy^2, \text{ and } 7y$$

Objective B **Simplifying Algebraic Expressions**

An **algebraic expression** is a combination of variables and numbers using any of the operations of addition, subtraction, multiplication, or division, as well as exponents. Examples of algebraic expressions are

$$x^2 - 14, \quad \frac{2xy}{3z^3} + y^2, \quad \frac{12x - 9}{3x - 4}, \quad \text{and} \quad 10x^3 - 4x^2 - 11x + 1.$$

To simplify expressions that contain like terms we want to **combine like terms**.

Combining Like Terms

To **combine like terms**, add (or subtract) the coefficients and keep the common variable expression.

The procedure for combining like terms uses the distributive property in the form

$$ba + ca = (b + c)a.$$

In particular, with numerical coefficients, we can combine like terms as follows.

$$9x + 6x = (9 + 6)x \qquad \text{by the distributive property}$$

$$= 15x \qquad \text{Add the coefficients algebraically.}$$

$$3x^2 - 5x^2 = (3 - 5)x^2 \qquad \text{by the distributive property}$$

$$= -2x^2 \qquad \text{Add the coefficients algebraically.}$$

$$xy - 1.6xy = (1.0 - 1.6)xy \qquad \text{by the distributive property}$$

Note: $xy = 1xy = 1.0xy$

$$= -0.6xy \qquad \text{Add the coefficients algebraically.}$$

Example 2

Combining Like Terms

Combine like terms whenever possible.

a. $8x + 10x$

Solution

$$8x + 10x = (8 + 10)x \qquad \text{by the distributive property}$$

$$= 18x$$

b. $6.5y - 2.3y$

Solution

$$6.5y - 2.3y = (6.5 - 2.3)y \qquad \text{by the distributive property}$$

$$= 4.2y$$

c. $2x^2 + 3a + x^2 - a$

Solution

$$2x^2 + 3a + x^2 - a$$

$$= 2x^2 + x^2 + 3a - a \qquad \text{Use the commutative property of addition.}$$

$$= (2 + 1)x^2 + (3 - 1)a \qquad \text{Note: } +x^2 = +1x^2 \text{ and } -a = -1a$$

$$= 3x^2 + 2a$$

d. $4(n - 7) + 5(n + 1)$

Solution

$$4(n - 7) + 5(n + 1)$$

$$= 4n - 28 + 5n + 5 \qquad \text{Simplify by using the distributive property twice.}$$

$$= 4n + 5n - 28 + 5 \qquad \text{Use the commutative property of addition.}$$

$$= (4 + 5)n + (-28 + 5) \qquad \text{by the distributive property}$$

$$= 9n - 23 \qquad \text{Combine like terms.}$$

e. $\dfrac{x+3x}{2}+5x$

Solution

A fraction bar is a grouping symbol, similar to parentheses. So combine like terms in the numerator first.

$$\dfrac{x+3x}{2}+5x = \dfrac{4x}{2}+5x$$

$$= \dfrac{4}{2}\cdot x+5x$$

$$= 2x+5x \qquad \text{Reduce the fraction.}$$

$$= 7x \qquad \text{Combine like terms.}$$

Now work margin exercise 2.

2. Combine like terms whenever possible.

a. $7x+2x$

b. $4.2z-3.1z$

c. $4x^2+7y+3x^2-2y$

d. $6(a-2)+4(a+5)$

e. $\dfrac{x+5x}{3}+6x$

Objective C **Evaluating Algebraic Expressions**

In most cases, if an expression is to be evaluated, like terms should be combined first and then the resulting expression evaluated by following the rules for order of operations.

Parentheses must be used around negative numbers when substituting.

Without parentheses, an evaluation can be dramatically changed and lead to wrong answers, particularly when even exponents are involved. We analyze with the exponent 2 as follows.

In general, except for $x = 0$,

1. $-x^2$ **is negative** $\left[-6^2 = -1\cdot 6^2 = -1\cdot 36 = -36 \right]$

2. $(-x)^2$ **is positive** $\left[(-6)^2 = (-6)(-6) = 36 \right]$

3. $-x^2 \neq (-x)^2$ $\left[-36 \neq 36 \right]$

To Evaluate an Algebraic Expression

1. Combine like terms, if possible.
2. Substitute the values given for any variables.
3. Follow the rules for order of operations.

3. a. Evaluate $2x^2$ for $x = 4$ and for $x = 5$.

b. Evaluate $-2x^2$ for $x = 4$ and for $x = 5$.

Example 3

Evaluating Algebraic Expressions

a. Evaluate x^2 for $x = 3$ and for $x = -4$.

Solution

For $x = 3$, $x^2 = (3)^2 = 9$.

For $x = -4$, $x^2 = (-4)^2 = 16$.

b. Evaluate $-x^2$ for $x = 3$ and for $x = -4$.

Solution

For $x = 3$, $-x^2 = -(3)^2 = -1(9) = -9$.

For $x = -4$, $-x^2 = -(-4)^2 = -1(16) = -16$.

Now work margin exercise 3.

Example 4

Simplifying and Evaluating Algebraic Expressions

Simplify each expression below by combining like terms. Then evaluate the resulting expression using the given values for the variables.

a. Simplify and evaluate $2x + 5 + 7x$ for $x = -3$.

Solution

Simplify first.
$$2x + 5 + 7x = 2x + 7x + 5$$
$$= 9x + 5$$

Now evaluate.
$$9x + 5 = 9(-3) + 5$$
$$= -27 + 5$$
$$= -22$$

b. Simplify and evaluate $3ab - 4ab + 6a - a$ for $a = 2, b = -1$.

Solution

Simplify first.
$$3ab - 4ab + 6a - a = -ab + 5a$$
Now evaluate.
$$-ab + 5a = -1(2)(-1) + 5(2) \quad \textbf{Note:} \ -ab = -1ab$$
$$= 2 + 10$$
$$= 12$$

c. Simplify and evaluate $\dfrac{5x+3x}{4}+2(x+1)$ for $x = 5$.

Solution

Simplify first.

$$\dfrac{5x+3x}{4}+2(x+1) = \dfrac{8x}{4}+2x+2$$
$$= 2x+2x+2$$
$$= 4x+2$$

Now evaluate.

$$4x+2 = 4(5)+2$$
$$= 20+2$$
$$= 22$$

Now work margin exercise 4.

4. Simplify and evaluate the following:

a. $4x+6+3x$ for $x=-4$

b. $5ab-8ab+2a-3a$

for $a=-3, b=1$

c. $\dfrac{7y+2y}{3}+2(y-8)$

for $y=4$

Practice Problems

Simplify the following expressions by combining like terms.

1. $-2x-5x$ **2.** $12y+6-y+10$

3. $5(x-1)+4x$ **4.** $2b^2 -a+b^2 +a$

Simplify the expression. Then evaluate the resulting expression for $x = 3$ and $y = -2$.

5. $2(x+3y)+4(x-y)$

Practice Problem Answers

1. $-7x$ **2.** $11y+16$ **3.** $9x-5$

4. $3b^2$ **5.** $6x+2y; 14$

Pick out the like terms in each list of terms. See Example 1.

1. $-5, \dfrac{1}{6}, 7x, 8, 9x, 3y$

2. $-2x^2, -13x^3, 5x^2, 14x^2, 10x^3$

3. $5xy, -x^2, -6xy, 3x^2y, 5x^2y, 2x^2$

4. $3ab^2, -ab^2, 8ab, 9a^2b, -10a^2b, ab, 12a^2$

5. $24, 8.3, 1.5xyz, -1.4xyz, -6, xyz, 5xy^2z, 2xyz^2$

6. $-35y, 1.62, -y^2, -y, 3y^2, \dfrac{1}{2}, 75y, 2.5y^2$

Find the value of each numerical expression.

7. $(-8)^2$

8. -8^2

9. -11^2

10. $(-6)^2$

Simplify each expression by combining like terms. See Example 2.

11. $8x + 7x$

12. $3y + 8y$

13. $5x + (-2x)$

14. $7x + (-3x)$

15. $-n - n$

16. $-x - x$

17. $6y^2 - y^2$

18. $16z^2 - 5z^2$

19. $23x^2 + 11x^2$

20. $18x^3 + 7x^3$

21. $4x + 2 + 3x$

22. $3x - 1 + x$

23. $2x - 3y - x - y$

24. $x + y + x - 2y$

25. $2x^2 - 2y + 5x^2 + 6x^2$

26. $4a + 2a - 3b - a$

27. $3(n+1) - n$

28. $2(n-4) + n + 1$

29. $5(a-b) + 2a - 3b$

30. $4a - 3b + 2(a+2b)$

31. $3(2x+y) + 2(x-y)$

32. $4(x+5y)+3(2x-7y)$

33. $2x+3x^2-3x-x^2$

34. $2y^2+4y-y^2-3y$

35. $2n^2-6n+1-4n^2+8n-3$

36. $3n^2+2n-5-n^2+n-4$

37. $3x^2+4xy-5xy+y^2$

38. $2x^2-5xy+11xy+3$

39. $y-\dfrac{4y+5y}{3}$

40. $z-\dfrac{3z+5z}{4}$

41. $\dfrac{2x+3x}{3}+x$

42. $\dfrac{2y+4y}{5}-2y$

For the following expressions, a. simplify, and b. evaluate the simplified expression for $x = 4$, $y = 3$, $a = -2$, and $b = -1$. See Examples 2 through 4.

43. $5x+4-2x$

44. $7x-17-x$

45. $x-10-3x+2$

46. $6a+5a-a+13$

47. $3(y-1)+2(y+2)$

48. $4(y+3)+5(y-2)$

49. $-5(x+y)+2(x-y)$

50. $-2(a+b)+3(b-a)$

51. $8.3x^2-5.7x^2+x^2+2$

52. $3.1a^2-0.9a^2+4a-5.3a^2$

53. $5ab+b^2-2ab+b^3$

54. $5a+ab^2-2ab^2+3a$

55. $2.4(x+1)+1.3(x-1)$

56. $1.3(y+2)-2.6(8-y)$

57. $\dfrac{3a+5a}{-2}+12a$

58. $8a+\dfrac{5a+4a}{9}$

59. $\dfrac{-4b-2b}{-3}+\dfrac{2b+5b}{7}$

60. $\dfrac{5b+3b}{4}+\dfrac{-4b-b}{-5}$

61. $2x+3\left[x-2(9+x)\right]$

62. $5x-2\left[x+5(x-3)\right]$

63. Explain the difference between -5^2 and $(-5)^2$.

64. The text recommends simplifying an expression (combining like terms) before evaluating. Do you think this is necessary?

Evaluate the expression $4x^2 - 5(x+2) + 3x + 10 + 2x$ for $x = 3$:

a. by substituting and then evaluating.

b. by first simplifying and then evaluating.

Which method would you recommend? Why?

7.8 Translating English Phrases and Algebraic Expressions

Objectives

A Translate English phrases into algebraic expressions.

B Translate algebraic expressions into English phrases.

Objective A Translating English Phrases into Algebraic Expressions

Algebra is a language of mathematicians, and to understand mathematics, you must understand the language. We want to be able to change English phrases into their "algebraic" equivalents and vice versa. So if a problem is stated in English, we can translate the phrases into algebraic symbols and proceed to solve the problem according to the rules and methods developed for algebra.

Certain words are the keys to the basic operations. Some of these words are listed here and highlighted in boldface in Example 1.

Key Words To Look For When Translating Phrases				
Addition	**Subtraction**	**Multiplication**	**Division**	**Exponent (Powers)**
add	subtract (from)	multiply	divide	square of
sum	difference	product	quotient	cube of
plus	minus	times		
more than	less than	twice		
increased by	decreased by	of (with fractions and percent)		
	less			

The following examples illustrate how these key words used in English phrases can be translated into algebraic expressions. **Note that in each case "a number" or "the number" implies the use of a variable (an unknown quantity).**

Example 1

Translating English Phrases into Algebraic Expressions

English Phrase — Algebraic Expression

a. the **product** of 3 and x

    3 **times** x $3x$

    3 **multiplied by** the number represented by x

b. 3 **added to** a number

    the **sum** of z and 3

    z **plus** 3 $z + 3$

    3 **more than** z

    z **increased by** 3

1. Translate the English phrases into algebraic expressions.

a. the product of 7 and x

b. 5 increased by n

c. 4 times the quantity found by adding a number to 2

d. 3 more than the product of 2 and a number

e. 9 multiplied by a number, less 4

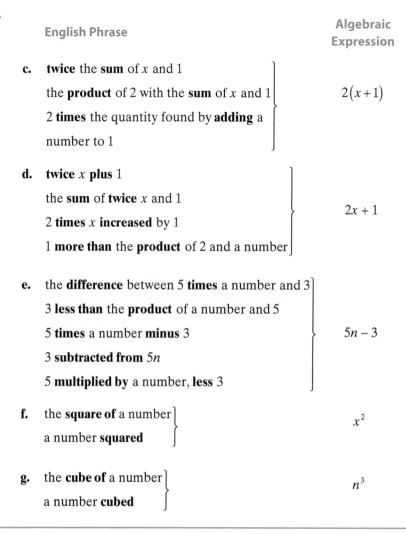

English Phrase	Algebraic Expression
c. twice the **sum** of x and 1 the **product** of 2 with the **sum** of x and 1 2 **times** the quantity found by **adding** a number to 1	$2(x+1)$
d. twice x **plus** 1 the **sum** of **twice** x and 1 2 **times** x **increased** by 1 1 **more than** the **product** of 2 and a number	$2x + 1$
e. the **difference** between 5 **times** a number and 3 3 **less than** the **product** of a number and 5 5 **times** a number **minus** 3 3 **subtracted from** $5n$ 5 **multiplied by** a number, **less** 3	$5n - 3$
f. the **square of** a number a number **squared**	x^2
g. the **cube of** a number a number **cubed**	n^3

Now work margin exercise **1.**

notes

In Example 1b, the phrase "the sum of z and 3" was translated as $z + 3$. If the expression had been translated as $3 + z$, there would have been no mathematical error because addition is commutative. That is, $z + 3 = 3 + z$. However, in part e, the phrase "3 less than the product of a number and 5" must be translated as it was because subtraction is **not** commutative. Thus

"3 less than 5 times a number" means $5n - 3$
while "5 times a number less than 3" means $3 - 5n$
and "3 less 5 times a number" means $3 - 5n$.

Therefore, be very careful when writing and/or interpreting expressions indicating subtraction. Be sure that the subtraction is in the order indicated by the wording in the problem. The same is true with expressions involving division.

The words **quotient** and **difference** deserve special mention because their use implies that the numbers given are to be operated on in the order given. That is, division and subtraction are done with the values in the same order that they are given in the problem. For example:

the quotient of y and 5 $\longrightarrow$ $\dfrac{y}{5}$

the quotient of 5 and y $\longrightarrow$ $\dfrac{5}{y}$

the difference between 6 and x $\longrightarrow$ $6 - x$

the difference between x and 6 $\longrightarrow$ $x - 6$

If we did not have these agreements concerning subtraction and division, then the phrases just illustrated might have more than one interpretation and be considered **ambiguous**.

An **ambiguous phrase** is one whose meaning is not clear or for which there may be two or more interpretations. This is a common occurrence in ordinary everyday language, and misunderstandings occur frequently. Imagine the difficulties diplomats have in communicating ideas from one language to another trying to avoid ambiguities. Even the order of subjects, verbs, and adjectives may not be the same from one language to another. Translating grammatical phrases in any language into mathematical expressions is quite similar. To avoid ambiguous phrases in mathematics, we try to be precise in the use of terminology, to be careful with grammatical construction, and to follow the rules for order of operations.

Objective B **Translating Algebraic Expressions into English Phrases**

Consider the three expressions to be translated into English:

$$7(n+1),\ 6(n-3),\ \text{and}\ 7n+1.$$

In the first two expressions, we indicate the parentheses with a phrase such as "the quantity" or "the sum of" or "the difference between." **Without the parentheses, we agree that the operations used in the expression are to be indicated in the order given.** Thus

$7(n+1)$ can be translated as "seven times the sum of a number and 1,"

$6(n-3)$ can be translated as "six times the difference between a number and 3,"

while $7n+1$ can be translated as "seven times a number plus 1".

2. Translate the algebraic expressions to English phrases.

   a. $10x$

   b. $4a + 7$

   c. $7(z - 5)$

3. Translate the English phrases to algebraic expressions.

   a. the quotient of a number and –9

   b. 3 less than 4 times a number

   c. twice the sum of 8 and a number

   d. the number of inches in f feet

   e. The cost of renting a trailer for one day and driving x miles if the rate is $25 per day plus $0.33 per mile.

Example 2

Translating Algebraic Expressions to English Phrases

Write an English phrase that indicates the meaning of each algebraic expression.

Algebraic Expression	Possible English Phrase
a. $5x$	the product of 5 and a number
b. $2n + 8$	twice a number increased by 8
c. $3(a-2)$	three times the difference between a number and 2

Now work margin exercise 2.

Example 3

Translating English Phrases to Algebraic Expressions

Change each phrase into an equivalent algebraic expression.

English Phrase	Algebraic Expression
a. the quotient of a number and –4	$\dfrac{x}{-4}$
b. 6 less than 5 times a number	$5y - 6$
c. twice the sum of 3 and a number	$2(3+n)$
d. the number of minutes in h hours	$60h$
e. the cost of renting a truck for one day and driving x miles if the rate is $30 per day plus $0.25 per mile	$30 + 0.25x$

Now work margin exercise 3.

Practice Problems

Change the following phrases to algebraic expressions.

  1. 7 less than a number   **2.** the quotient of y and 5

  3. 14 more than 3 times a number

Change the following algebraic expressions into English phrases. (There may be more than one correct translation.)

  4. $10 - x$ **5.** $2(y-3)$ **6.** $5n + 3n$

Practice Problem Answers

  1. $x - 7$ **2.** $\dfrac{y}{5}$ **3.** $3y + 14$

  4. 10 decreased by a number

  5. twice the difference between a number and 3

  6. 5 times a number plus 3 times the same number

Exercises 7.8

Translate each algebraic expression into an equivalent English phrase. There may be more than one correct translation. See Example 2.

1. $4x$

2. $-9x$

3. $x + 5$

4. $4x - 7$

5. $7(x + 1.1)$

6. $3.2(x + 2.5)$

7. $-2(x - 8)$

8. $10(x + 4)$

9. $\dfrac{6}{(x-1)}$

10. $\dfrac{9}{(x+3)}$

11. $5(2x + 3)$

12. $3(4x - 5)$

Write each pair of expressions in words. Notice the differences between the algebraic expressions and the corresponding English phrases.

13. $3x + 7;\ 3(x + 7)$

14. $4x - 1;\ 4(x - 1)$

15. $7x - 3;\ 7(x - 3)$

16. $5(x + 6);\ 5x + 6$

Write the algebraic expressions described by the English phrases. Choose your own variable. See Example 1.

17. 6 added to a number

18. 7 more than a number

19. 4 less than a number

20. a number decreased by 13

21. the quotient of twice a number and 10

22. the difference between a number and 3, all divided by 7

23. 5 subtracted from three times a number

24. the sum of twice a number and four times the number

25. 8 minus twice a number

26. the sum of a number and 9 times the number

27. twenty decreased by 4.8 times a number

28. the difference between three times a number and five times the same number

29. 9 times the sum of a number and 2

30. 3 times the difference between a number and 8

31. 13 less than the product of 4 and the sum of a number and 1

32. 4 more than the product of 8 with the difference between a number and 6

33. eight more than the product of 3 and the sum of a number and 6

34. six less than twice the difference between a number and 7

35. four less than 3 times the difference between 7 and a number

36. nine more than twice the sum of 17 and a number

37. **a.** 6 less than a number
    b. 6 less a number

38. **a.** 5 less than 3 times a number
    b. 5 less 3 times a number

39. **a.** 20 less than a number
    b. 20 less a number

40. **a.** 6 less than 4 times a number
    b. 6 less 4 times a number

Write the algebraic expressions described by the English phrases. See Example 3.

41. **Time:** the number of hours in d days

42. **Graphing Calculators:** the cost of x graphing calculators if one calculator costs $115

43. **Gas Prices:** the cost of x gallons of gasoline if the cost of one gallon is $3.15

44. **Time:** the number of seconds in m minutes

45. **Time:** the number of days in y years. (Assume 365 days in a year.)

46. **Candy:** the cost of x pounds of candy at $4.95 a pound

47. **Time:** the number of days in t weeks and 3 days

48. **Time:** the number of minutes in h hours and 20 minutes

49. **Football:** the points scored by a football team on t touchdowns (7 points) and 1 field goal (3 points)

50. **Vacation Time:** the amount of vacation days an employee has after w weeks if she gets 0.2 vacation days for every week she works

51. **Car Rentals:** the cost of renting a car for one day and driving m miles if the rate is $20 per day plus 15 cents per mile

52. **Fishing:** the cost of purchasing a fishing rod and reel if the rod costs x dollars and the reel costs $8 more than twice the cost of the rod

53. **Rectangles:** the perimeter of a rectangle if the width is *w* centimeters and the length is 3 cm less than twice the width

3 cm less than twice the width

54. **Squares:** the area of a square with side *c* centimeters

Writing & Thinking

55. Discuss the meaning of the term "ambiguous phrase."

Addition with Real Numbers page 515
1. To add two real numbers with **like signs**,
   a. add their absolute values and
   b. use the common sign.
2. To add two real numbers with **unlike signs**,
   a. subtract their absolute values (the smaller from the larger), and
   b. use the sign of the number with the larger absolute value.

Section 7.3: Subtraction with Real Numbers

Additive Inverse page 524

The **opposite** of a real number is called its **additive inverse**.
The sum of a number and its additive inverse is zero.
Symbolically, for any real number a, $a + (-a) = 0$.

Subtraction with Real Numbers page 526

For any real numbers a and b, $a - b = a + (-b)$.
In words, to subtract b from a, **add** the **opposite** of b to a.

Change in Value and Net Change pages 529, 530

To find the **change in value** between two numbers, take the end value and subtract the beginning value. The **net change** in a measure is the algebraic sum of several numbers.

Section 7.4: Multiplication and Division with Real Numbers

Rules for Multiplication with Positive and Negative Real Numbers page 539

For positive real numbers a and b,
1. The product of two positives is positive:
$$(a)(b) = +ab$$
2. The product of two negatives is positive:
$$(-a)(-b) = +ab$$
3. The product of a positive and a negative is negative:
$$a(-b) = -ab$$
4. The product of 0 and any number is 0:
$$a \cdot 0 = 0 \text{ and } (-a) \cdot 0 = 0$$

Division with Real Numbers page 540

For real numbers a, b, and x (where $b \neq 0$),

$$\frac{a}{b} = x \text{ means that } a = b \cdot x.$$

For real numbers a and b (where $b \neq 0$),

$$\frac{a}{0} \text{ is undefined and } \frac{0}{b} = 0.$$

Division by 0 is Undefined   page 540

Rules for Division with Positive and Negative Real Numbers   page 540

If a and b are positive real numbers (where $b \neq 0$),

1. The quotient of two positive numbers is positive:
$$\frac{a}{b} = +\frac{a}{b}$$

2. The quotient of two negative numbers is positive:
$$\frac{-a}{-b} = +\frac{a}{b}$$

3. The quotient of a positive number and a negative number is negative: $\dfrac{-a}{b} = -\dfrac{a}{b}$ and $\dfrac{a}{-b} = -\dfrac{a}{b}$

Average   page 541

The **average** (or **mean**) of a set of numbers is the value found by adding the numbers in the set and then dividing the sum by the number of numbers in the set.

Rules for Order of Operations   page 548

1. Simplify within grouping symbols, such as parentheses ( ), brackets [ ], and braces { }, working from the **innermost** grouping outward.
2. Find any powers indicated by exponents.
3. Moving from **left to right**, perform any multiplications **or** divisions **in the order they appear**.
4. Moving from **left to right**, perform any additions **or** subtractions **in the order they appear**.

Properties of Addition and Multiplication   page 556

For Addition	Name of Property	For Multiplication
$a + b = b + a$	Commutative property	$ab = ba$
$(a+b)+c = a+(b+c)$	Associative property	$a(bc) = (ab)c$
$a + 0 = 0 + a = a$	Identity	$a \cdot 1 = 1 \cdot a = a$
$a + (-a) = 0$	Inverse	$a \cdot \dfrac{1}{a} = 1$ (for $a \neq 0$)

Zero-Factor Law   page 556

$$a \cdot 0 = 0 \cdot a = 0$$

Distributive Property of Multiplication over Addition   page 556

$$a(b+c) = ab + ac$$

Section 7.7: Simplifying and Evaluating Algebraic Expressions

Algebraic Vocabulary   page 562

    Constant: a single number.

    Term: any constant or variable or the indicated product and/or quotient of constants and variables.

    Numerical coefficient: the number written next to a variable.

    Algebraic expression: a combination of variables and numbers using any of the operations of addition, subtraction, multiplication, or division, as well as exponents.

Like Terms   page 562

    Like terms (or **similar terms**) are terms that are constants or terms that contain the same variables that are raised to the same powers.

Combining Like Terms   page 563

    To **combine like terms**, add (or subtract) the coefficients and keep the common variable expression.

Evaluating Algebraic Expressions   page 565

    1. Combine like terms, if possible.

    2. Substitute the values given for any variables.

    3. Follow the rules for order of operations.

Section 7.8: Translating English Phrases and Algebraic Expressions

Key Words to Look for When Translating Phrases   page 571

    Addition: add, sum, plus, more than, increased by

    Subtraction: subtract (from), difference, minus, less than, decreased by, less

    Multiplication: multiply, product, times, twice, of

    Division: divide, quotient

    Exponent (Powers): square of, cube of

Translating English Phrases into Algebraic Expressions page 571

Translating Algebraic Expressions into English Phrases page 573

Section 7.1: The Real Number Line and Absolute Value

List the numbers described below.

1. Given the set of numbers $\left\{-9.3, -2, -1.343434..., -\dfrac{3}{4}, 0, \sqrt{2}, \dfrac{2}{1}, \dfrac{10}{3}\right\}$, tell which numbers are

   a. natural numbers   b. integers   c. rational numbers   d. irrational numbers

Graph each set of real numbers on a real number line.

2. $\{-3, 0, 3, 4\}$

3. $\left\{-4, -1, -\dfrac{1}{4}, 2.5\right\}$

4. All positive integers less than or equal to 5.

Fill in each blank with the appropriate symbol: <, >, or =.

5. -5 _____ -3

6. $2\dfrac{9}{10}$ _____ 2.09

Determine whether each statement is true or false. If a statement is false, rewrite it in a form that is a true statement. (There may be more than one way to correct a statement).

7. $-14 \leq 14$

8. $|-2.3| < 0$

List the possible values for x for each statement.

9. $|x| = 5$

10. $|x| = -5$

Determine the integer values for x that satisfy the conditions given in each statement. Graph the integers on a number line.

11. $x \leq 3$

12. $|x| > 3$

Section 7.2: Addition with Real Numbers

Find the following sums. Reduce any fractions to lowest terms.

13. $6 + 9$

14. $8 + (-3)$

15. $18 + (-18)$

16. $-2.3 + (-10.6)$

17. $-12 + (-1) + (-5)$

18. $-\dfrac{5}{12} + \dfrac{1}{3} - \dfrac{5}{6}$

19.
$$\begin{array}{r} -8 \\ 23 \\ \underline{-7} \end{array}$$

20.
$$\begin{array}{r} -22 \\ -10 \\ \underline{-13} \end{array}$$

Determine whether or not the given number is a solution to the given equation by substituting and then evaluating.

21. $x + (-8.2) = -10.3$ given that $x = 2.1$

22. $|y| + (-4) = -1$ given that $y = 3$

Section 7.3: Subtraction with Real Numbers

Simplify the expressions. Reduce any fractions to lowest terms.

23. $-4 - 3$

24. $-\dfrac{2}{3} - \dfrac{1}{15}$

25. $-7 + (-2) - (-5)$

Perform the indicated subtraction.

26.
$$\begin{array}{r} 3.2 \\ \underline{-(4.1)} \end{array}$$

27.
$$\begin{array}{r} -24 \\ \underline{-(-7)} \end{array}$$

28. Find the difference between 17 and –20.

Solve the following word problems.

29. Airplanes: An airplane drops in altitude from 30,000 feet to 24,000 feet. What is the change in altitude in terms of signed numbers?

30. Temperature: Find the change in temperature if the temperature drops from 30°F at noon to –3°F at midnight.

Determine whether or not the given number is a solution to the given equation by substituting and then evaluating.

31. $x - 3 = -4$ given that $x = -7$

32. $1.2 - x = -5.9$ given that $x = -7.1$

Section 7.4: Multiplication and Division with Real Numbers

Find the indicated products and quotients. Reduce any fractions to lowest terms.

33. $8(-3.2)$

34. $(-9)(-10)$

35. $\dfrac{2}{7} \cdot \dfrac{-21}{8}$

36. $-4 \cdot 3 \cdot (-2)$

37. $\dfrac{-10}{-2}$

38. $\dfrac{-16}{0}$

39. $\dfrac{0}{-16}$

40. $\dfrac{3.2}{-8}$

41. $\dfrac{-32}{25} \div \dfrac{16}{15}$

Solve the following word problems.

42. Find the mean of the following set of integers: $30, 25, 70, 85,$ and 100.

43. **Height:** The heights of twenty women were recorded as shown in the following frequency distribution. Find the mean height (to the nearest tenth of an inch) for these women.

Height (in inches)	60	61	63	66	70	71
Frequency	2	1	6	4	4	3

44. **Hurricanes:** The bar graph shows the number of hurricanes during the Atlantic Hurricane Seasons from 2004 – 2009. What was the average number of hurricanes per season in these years? (Round your answer to the nearest tenth.)

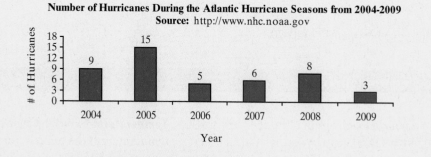

Number of Hurricanes During the Atlantic Hurricane Seasons from 2004-2009
Source: http://www.nhc.noaa.gov

Section 7.5 Order of Operations with Real Numbers

Use the rules for order of operations to evaluate each expression.

45. $16 \div 2 \cdot 8 + 24$

46. $-36 \div (-2)^2 + 25 - 2(16 - 17)$

47. $9(-1 + 2^2) - 7 - 6 \cdot 2^2$

48. $2\left(-1+5^2\right)-4-2\cdot 3^2$

49. $10-2\left[(19-14)\div 5+3(-4)\right]$

50. $\left(\dfrac{1}{15}-\dfrac{1}{12}\right)\div 6$

51. $\dfrac{2}{3}\div\dfrac{7}{12}-\dfrac{2}{7}+\left(\dfrac{1}{2}\right)^2$

52. If the product of $\dfrac{3}{5}$ and $\dfrac{3}{4}$ is divided by the product of $\dfrac{4}{5}$ and $-\dfrac{1}{2}$, what is the quotient?

Section 7.6: Properties of Real Numbers

Name the property of real numbers illustrated.

53. $9+15=15+9$

54. $8+(3+2)=(8+3)+2$

55. $7+0=7$

56. $6+(-6)=0$

57. $8(-7)=-7(8)$

58. $9.2\cdot 1=9.2$

59. $19\cdot\dfrac{1}{19}=1$

60. $3(2+x)=6+3x$

Section 7.7: Simplifying and Evaluating Algebraic Expressions

Simplify each expression by combining like terms.

61. $3y+4y-y$

62. $-a^2-5a^2+3a^2$

63. $4x+2-7x$

64. $4(n+1)-n$

65. $2y^2+5y-y^2+2y$

66. $12a+3a^2-8a+4a^2$

For the following expressions, a. simplify, and b. evaluate the simplified expression for $x=-1$, $y=4$, and $a=-2$.

67. $5a^2+3a-6a^2+a+7$

68. $20-17+13x-20+4x$

69. $-4(y+6)+2(y-1)$

70. $3y+\dfrac{10y-2y}{2}$

Section 7.8: Translating English Phrases and Algebraic Expressions

Translate each algebraic expression into an equivalent English phrase. (There may be more than one correct translation).

71. $5n + 3$

72. $-3(x + 2)$

73. $5(x - 3)$

74. $\dfrac{7n}{33}$

75. $\dfrac{50}{6y}$

Write the algebraic expressions described by the English phrases. Choose your own variable.

76. 28 decreased by 6 times a number

77. 72 plus 8 times the sum of a number and 2

78. 32 less than the product of a number and 10

79. Time: the number of hours in x days and 5 hours.

80. Frozen Yogurt: the cost of y ounces of yogurt at 49¢ an ounce.

Chapter 7: Test

Solve the following problems.

1. Given the set of numbers $\left\{-5, -\pi, -1, -\dfrac{1}{3}, 0, \dfrac{1}{2}, 3\dfrac{1}{4}, 7.121212...\right\}$, tell which numbers are

   a. integers   b. rational numbers   c. irrational numbers   d. real numbers

2. What two numbers are 13 units from -10 on a number line? Show a number line graph and mark the numbers.

3. Fill in the blanks with the proper symbol: $<$, $>$, or $=$.

   a. $-4 \underline{\hspace{1cm}} -2$ b. $-2(-2) \underline{\hspace{1cm}} 0$ c. $|-9| \underline{\hspace{1cm}} |9|$

4. List the **integers** that satisfy the following inequalities:

   a. $|x| < 3$ b. $|x| \geq 9$

Perform the indicated operations.

5. $13 - (-16) + (-6)$ 6. $1.5 - 4.5 + 1.7$ 7. $(-7)(-16)$ 8. $41(-5)(-3)(0)$

9. $\dfrac{-57}{19}$ 10. $\dfrac{4.5}{-15}$ 11. $-\dfrac{5}{6} \cdot \dfrac{8}{15}$ 12. $\dfrac{-2}{9} \div \dfrac{-4}{7}$

Use the rules for order of operations to evaluate (or simplify) each expression.

13. $36 \div (-9) \cdot 2 - 16 + 4(-5)$ 14. $4^2 + \left[24 + 3(4 - 5)\right] \div 7$

15. $\dfrac{7}{8} - \dfrac{1}{3} \div \dfrac{5}{6} + \dfrac{1}{4}$ 16. $\dfrac{7}{12} \cdot \dfrac{9}{56} \div \dfrac{5}{16} - \left(\dfrac{1}{3}\right)^2$

Name each property of real numbers illustrated.

17. $-3 + 0 = -3$ 18. $8 \cdot 2 = 2 \cdot 8$

19. $13 + (9 + 1) = (13 + 9) + 1$ 20. $-5 \cdot 0 = 0$

Simplify each expression by combining like terms, and then evaluate each simplified expression for $x = -2$ and $y = 3$.

21. $7x + 8x^2 - 3x^2$

22. $5y - y - 6 + 2y$

Translate each expression into an equivalent English phrase. (There may be more than one correct translation).

23. $5n + 18$

24. $3(x + 6)$

Write an algebraic expression described by each English phrase.

25. the product of a number and 6 decreased by 3

26. 4 less than twice the sum of a number and 15

Answer the following word problems.

27. Find the sum of $\frac{3}{4}$ and $\frac{9}{10}$. Then multiply by $-\frac{5}{4}$. What is the product?

28. Find the average of the following set of numbers: $\{-12, -15, -32, -31\}$.

29. **Gasoline:** You know that the gas tank in your car holds 20 gallons of gas and the gauge reads $\frac{1}{4}$ full.

   a. How many gallons will be needed to fill the tank?

   b. If the price of gas is $3.10 per gallon and you have $40, do you have enough cash to fill the tank? How much extra money do you have or how much are you short?

30. **Stock Market:** In a 5-day week, the Dow Jones stock market average quote showed a gain of 32 points, a gain of 140 points, a loss of 30 points, a loss of 53 points, and a gain of 63 points. What was the net change in the stock market for the week?

31. **Temperature:** The temperature at midnight in the mountain resort of Vail, Colorado was $-10°F$ ($10°$ below 0). At 6 A.M. the next morning the temperature was $5°F$. What was the change in temperature?

Cumulative Review: Chapters 1 - 7

Evaluate each of the following expressions.

1. $6 \cdot 3 + 15 \div 3 - 2^2$

2. $20 - (4 + 2 \cdot 5) \div 7 + 7$

3. $-12(-2)(-3)$

4. $4 - (-18) + (-6)$

5. $2 - (-8 \div 4 - 5) + 10(-1)$

6. $3 + (14 - 3^2 + 8 \cdot 3)$

7. $17 - |8 - 14 + 1| \cdot 2 + (-3)^2$

Using the tests for divisibility, decide which of the numbers 2, 3, 4, 5, 6, 9, and 10 (if any) will divide each of the following numbers.

8. 648

9. 227

Answer the following questions.

10. Find the prime factorization of 264.

11. State which property is illustrated: $-15 + (6 + 21) = (-15 + 6) + 21$.

12. Reduce each fraction to lowest terms.

    a. $\dfrac{36}{48}$

    b. $\dfrac{140}{210}$

13. Change $8\dfrac{5}{6}$ to an improper fraction.

14. Find the LCM of 12, 30 and 40.

15. Arrange the numbers $\dfrac{6}{7}, \dfrac{3}{4}$, and $\dfrac{13}{16}$ in order, from smallest to largest.

Perform the indicated operations. Reduce all fractions to lowest terms.

16. $\dfrac{11}{12} - \dfrac{5}{12} + \dfrac{1}{12}$

17. $-\dfrac{8}{9} \div 6$

18. $2\dfrac{1}{7} \cdot 1\dfrac{1}{3} - \dfrac{1}{2}$

19. $\left(3 - 1\dfrac{2}{5}\right) \cdot \left(\dfrac{1}{2}\right)^2 + 2\dfrac{1}{10}$

20. $-\dfrac{3}{4} - \dfrac{1}{3} \cdot \dfrac{2}{3} + \dfrac{5}{9}$

21. $(1000)(8.71943)$

22. $2.87\overline{)18.368}$

23. $5.13 + 15.75 \div 1.5^2 - 7.62$

24. Round 5604.6894 to the nearest hundredth.

25. Change $\dfrac{2}{3}$ to decimal form.

Simplify each radical expression.

26. $\sqrt{\dfrac{4}{81}}$

27. $\sqrt{0.0169}$

Solve the following problems.

28. Find 32% of 49.

29. Solve the proportion: $\dfrac{\frac{2}{5}}{3} = \dfrac{4}{x}$.

30. Use the bar graph to answer the following questions.

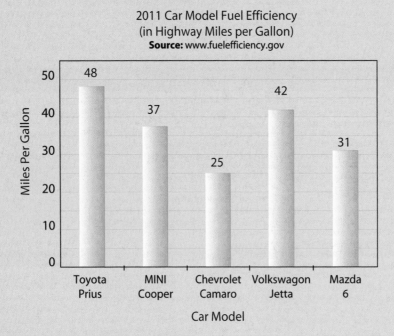

2011 Car Model Fuel Efficiency
(in Highway Miles per Gallon)
Source: www.fuelefficiency.gov

a. Which car model is the most fuel efficient?

b. Which car model is the least fuel efficient?

c. How much more fuel efficient is the Volkswagon Jetta than the Mazda 6?

31. Assuming the triangles below are similar, find values for m and n.

$$\triangle XYZ \sim \triangle PQR$$

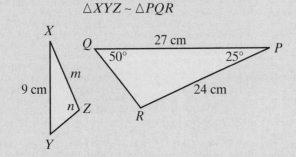

32. The following table contains running times for 10 movies from 2010. Find the mean, median, mode, and range of the data.

Movie	Running Time in Minutes
Toy Story 3	103
The Social Network	120
Inception	148
Black Swan	108
True Grit	110
The King's Speech	118
How to Train Your Dragon	98
The Fighter	115
Harry Potter and the Deathly Hallows, Part 1	146
Alice in Wonderland	108

Mean= _____

Median= _____

Mode= _____

Range= _____

Graph each set of numbers on a number line.

33. $\left\{ \dfrac{9}{2}, -3, 1, -0.5 \right\}$

34. All odd integers between -7 and 7.

For the following expressions, a. simplify, and b. evaluate the simplified expression for $x = -3$ and $y = -2$.

35. $2(x - 5) - 4x$

36. $y^2 - 3y + 2y^2 + 9y$

Write an algebraic phrase for the following English phrases.

37. 5 more than twice the difference of a number and 8

38. the product of 12 and the sum of a number and 7

Solve the following problems.

39. **Average:** Find the average of the following set of numbers: $\{-22, -17, -38, -19\}$.

40. **Number Problem:** If the quotient of -84 and 12 is subtracted from the product of -16 and 5, what is the difference?

41. **Right Triangle:** Find the length of the hypotenuse (to the nearest tenth of an inch) of a right triangle with legs of length 8 in. and 12 in.

42. **Interest:** If $2500 is invested at 4% interest. How much simple interest will be earned in two years for this investment?

43. **Shoes:** A pair of shoes is marked 20% off. What would Alicia pay for the shoes if the original price of the shoes was $45 and she had to pay 7% sales tax?

44. **Drawing:** Michael plans to make a drawing of a house using a scale of 2 inches to 25 feet. How many feet are represented by 3.2 inches?

45. **Playground:** A rectangular playground is 100.5 yards long and 62.3 yards wide. Find

    a. the perimeter and

    b. the area.

46. **Cube:** Find the surface area of a cube with a side that measures 18 cm.

47. **Cards:** In a deck of 52 cards, find the following probabilities:

    a. the probability of drawing a king.

    b. the probability of drawing a red card.

48. **Temperature:** One day in September, at noon, the temperature was 68° F. What was the change in the temperature, if at 9 pm it was recorded 44° F?

Choose the response that correctly completes each statement.

49. "If x and y are both negative integers, then $x + y$ is (never, sometimes, always) positive".

50. "If x and y are both positive integers, then $x - y$ is (never, sometimes, always) positive".

8

Solving Linear Equations and Inequalities

Mathematics at Work!

Suppose that you are a traveling salesperson and your company has allowed you $65 per day for a car rental. You rent a car for $45 per day plus $0.05 per mile. How many miles can you drive for the $65 allowed?

Plan: Set up an equation relating the amount of money spent and the money allowed in your budget and solve the equation.

Solution: Let x = the number of miles driven.

$$\overbrace{\text{Money spent}}^{} = \text{Money budgeted}$$

$$45 + 0.05x = 65$$

$$45 + 0.05x - 45 = 65 - 45$$

$$0.05x = 20$$

$$\frac{0.05x}{0.05} = \frac{20}{0.05}$$

$$x = 400 \text{ miles}$$

You would like to get the most miles for your money. The car rental agency down the street charges only $39 per day and $0.08 per mile. How many miles would you get for your $65 with this price structure? Is this a better deal for you? (See Section 8.2.)

8.1 Solving Linear Equations: $x + b = c$ and $ax = c$

Objective A **Linear Equations**

In this section, we will discuss solving linear equations in the following two forms:

$$x + b = c \quad \text{and} \quad ax = c.$$

In these equations we treat a, b, and c as constants and the variable x as the unknown quantity.

In the following sections, we will combine the techniques developed here and discuss solving linear equations in the forms:

$$ax + b = c \quad \text{and} \quad ax + b = cx + d.$$

An **equation** is a statement that two algebraic expressions are equal. That is, both expressions represent the same number. If an equation contains a variable, any number that gives a true statement when substituted for the variable is called a **solution** to the equation. The solutions to an equation form a **solution set**. The process of finding the solution set is called **solving the equation**.

Linear Equation in x

If a, b, and c are **constants** and $a \neq 0$, then a **linear equation in x** is an equation that can be written in the form

$$ax + b = c.$$

Note: A linear equation in x is also called a **first-degree equation in x** because the variable x can be written with the exponent 1. That is, $x = x^1$.

Objective B **Solving Equations of the Form $x + b = c$**

To begin, we need the **addition principle of equality** given below.

Addition Principle of Equality

If the same algebraic expression is added to both sides of an equation, the new equation has the same solutions as the original equation. Symbolically, if A, B, and C are algebraic expressions, then the equations

$$A = B$$

and

$$A + C = B + C$$

have the same solutions.

Equations with the same solutions are said to be **equivalent equations**.

The objective of solving linear (or first–degree) equations is to get the variable by itself (with a coefficient of +1) on one side of the equation and any constants on the other side. The following procedure will help in solving linear equations such as

$$x - 3 = 7, \quad -11 = y + 5, \quad 3z - 2z + 2 = 3 + 8, \quad \text{and} \quad x - \frac{2}{5} = \frac{3}{10}.$$

Procedure for Solving Linear Equations that Simplify to the Form $x + b = c$

1. Combine any like terms on each side of the equation.

2. Use the **addition principle of equality** and add the opposite of the constant b to both sides. The objective is to isolate the variable on one side of the equation (either the left side or the right side) with a coefficient of +1.

3. Check your answer by substituting it for the variable in the original equation.

Every linear equation has exactly one solution. This means that once a solution has been found, there is no need to search for another solution.

Example 1

Solving $x + b = c$

Solve the equation $x - 3 = 7$

Solution

$x - 3 = 7$	Write the equation.
$x - 3 + 3 = 7 + 3$	Add 3 (the opposite of -3) to both sides.
$x = 10$	Simplify.

Check:

$$x - 3 = 7$$

$$(10) - 3 \overset{?}{=} 7 \qquad \text{Substitute } x = 10.$$

$$7 = 7 \qquad \text{true statement}$$

Example 2

Solving $x + b = c$

Solve the equation $-11 = y + 5$

Solution

$$-11 = y + 5$$ Write the equation. Note that the variable can be on the right side.

$$-11 - 5 = y + 5 - 5$$ Add -5 (the opposite of $+5$) to both sides.

$$-16 = y$$ Simplify.

Check:

$$-11 = y + 5$$

$$-11 \overset{?}{=} (-16) + 5$$ Substitute $y = -16$.

$$-11 = -11$$ true statement

Solve the following equations.

1. $x - 5 = 12$

2. $-8 = x + 4$

3. $x - 3.7 = 0$

Example 3

Solving $x + b = c$

Solve the equation $-4.1 + x = 0$

Solution

$$-4.1 + x = 0$$ Write the equation.

$$-4.1 + x + 4.1 = 0 + 4.1$$ Add 4.1 (the opposite of -4.1) to both sides.

$$x = 4.1$$ Simplify.

Check:

$$-4.1 + x = 0$$

$$-4.1 + (4.1) \overset{?}{=} 0$$ Substitute $x = 4.1$.

$$0 = 0$$ true statement

Now work margin exercises **1 through 3.**

Example 4

Solving $x + b = c$

Solve the equation $x - \dfrac{2}{5} = \dfrac{3}{10}$

Solution

$$x - \dfrac{2}{5} = \dfrac{3}{10} \qquad \text{Write the equation.}$$

$$x - \dfrac{2}{5} + \dfrac{2}{5} = \dfrac{3}{10} + \dfrac{2}{5} \qquad \text{Add } \dfrac{2}{5}, \left(\text{the opposite of } -\dfrac{2}{5} \right) \text{to both sides.}$$

$$x = \dfrac{3}{10} + \dfrac{4}{10} \qquad \text{Simplify. (The common denominator is 10.)}$$

$$x = \dfrac{7}{10} \qquad \text{Simplify.}$$

Check:

$$x - \dfrac{2}{5} = \dfrac{3}{10}$$

$$\left(\dfrac{7}{10} \right) - \dfrac{2}{5} \stackrel{?}{=} \dfrac{3}{10} \qquad \text{Substitute } x = \dfrac{7}{10}.$$

$$\dfrac{7}{10} - \dfrac{4}{10} \stackrel{?}{=} \dfrac{3}{10} \qquad \text{The common denominator is 10.}$$

$$\dfrac{3}{10} = \dfrac{3}{10} \qquad \text{true statement}$$

Now work margin exercise 4.

Example 5

Simplfying and Solving Equations

Simplify and solve the equation $3z - 2z + 2 = 3 + 8$

Solution

$$3z - 2z + 2 = 3 + 8 \qquad \text{Write the equation.}$$

$$z + 2 = 11 \qquad \text{Combine like terms on both sides of the equation.}$$

$$z + 2 - 2 = 11 - 2 \qquad \text{Add } -2 \left(\text{the opposite of } +2 \right) \text{to both sides.}$$

$$z = 9 \qquad \text{Simplify.}$$

Check:

$$3z - 2z + 2 = 3 + 8$$

$$3(9) - 2(9) + 2 \stackrel{?}{=} 3 + 8 \qquad \text{Substitute } z = 9.$$

$$27 - 18 + 2 \stackrel{?}{=} 3 + 8 \qquad \text{Simplify.}$$

$$11 = 11 \qquad \text{true statement}$$

Now work margin exercise 5.

4. Solve the following equation.

$$x - \dfrac{3}{8} = \dfrac{3}{4}$$

5. Simplify and solve the following equation.

$$5z - 4z + 3 = 5 - 7$$

Note that, as illustrated in Examples 1-5, variables other than x may be used.

6. Simplify and solve the following equation.

$$4z + 0.8 - 3z = 3.1 + 1.9$$

Completion Example 6

Simplifying and Solving Equations

Supply the reasons for each step in solving the equation.

$$5x - 4x - 1.5 = 6.3 + 4.0$$

Solution

$5x - 4x - 1.5 = 6.3 + 4.0$	_____
$x - 1.5 = 10.3$	_____
$x - 1.5 + 1.5 = 10.3 + 1.5$	_____
$x = 11.8$	_____

Now work margin exercise 6.

Objective C **Solving Equations of the Form $ax = c$**

To solve equations of the form $ax = c$, where $a \neq 0$, we can use the idea of the reciprocal of the coefficient a. For example, as we studied in Chapter 2, the reciprocal of $\frac{3}{4}$ is $\frac{4}{3}$ and $\frac{3}{4} \cdot \frac{4}{3} = 1$. Also, we need the **multiplication** (or **division**) **principle of equality** as stated below.

Multiplication (or Division) Principle of Equality

If both sides of an equation are multiplied by (or divided by) the same nonzero constant, the new equation has the same solutions as the original equation. Symbolically, if A and B are algebraic expressions and C is any nonzero constant, then the equations

$$A = B$$

and $\quad AC = BC \quad$ where $C \neq 0$

and $\quad \dfrac{A}{C} = \dfrac{B}{C} \quad$ where $C \neq 0$

have the same solutions. We say that the equations are equivalent.

Completion Example Answers

6.

$5x - 4x - 1.5 = 6.3 + 4.0$	Write the equation.
$x - 1.5 = 10.3$	Combine like terms on both sides of the equation.
$x - 1.5 + 1.5 = 10.3 + 1.5$	Add 1.5 (the opposite of −1.5) to both sides of the equation.
$x = 11.8$	Simplify.

Remember that the objective is to get the variable by itself on one side of the equation. That is, we want the variable to have +1 as its coefficient. The following procedure will accomplish this.

Procedure for Solving Linear Equations that Simplify to the Form $ax = c$

1. Combine any like terms on each side of the equation.
2. Use the **multiplication** (or **division**) **principle of equality** and multiply both sides of the equation by the reciprocal of the coefficient of the variable. (**Note:** This is the same as dividing both sides of the equation by the coefficient.) Thus the coefficient of the variable will become +1.
3. Check your answer by substituting it for the variable in the original equation.

Example 7

Solving $ax = c$

Solve the equation $5x = 20$

Solution

$$5x = 20$$ Write the equation.

$$\frac{1}{5} \cdot (5x) = \frac{1}{5} \cdot 20$$ Multiply by $\frac{1}{5}$, the reciprocal of 5.

$$\left(\frac{1}{5} \cdot 5\right) x = \frac{1}{5} \cdot \frac{20}{1}$$ Use the associative property of multiplication.

$$1 \cdot x = 4$$ Simplify.

$$x = 4$$

Check:

$$5x = 20$$

$$5 \cdot (4) \overset{?}{=} 20$$ Substitute $x = 4$.

$$20 = 20$$ true statement

Multiplying by the reciprocal of the coefficient is the same as **dividing** by the coefficient itself. So, we can multiply both sides by $\frac{1}{5}$, as we did, or we can divide both sides by 5. In either case, the coefficient of x becomes +1.

$$5x = 20$$

$$\frac{5x}{5} = \frac{20}{5}$$ Divide both sides by 5.

$$x = 4$$ Simplify.

Example 8

Solving $ax = c$

Solve the following equation.

$$1.1x + 0.2x = 12.2 - 3.1$$

Solution

When decimal coefficients or constants are involved, you might want to use a calculator to perform some of the arithmetic.

$1.1x + 0.2x = 12.2 - 3.1$	Write the equation.
$1.3x = 9.1$	Combine like terms.
$\dfrac{1.3x}{1.3} = \dfrac{9.1}{1.3}$	Use a calculator or pencil and paper to divide.
$x = 7.0$	

Check:

$1.1x + 0.2x = 12.2 - 3.1$	
$1.1(7) + 0.2(7) \overset{?}{=} 12.2 - 3.1$	Substitute $x = 7$.
$7.7 + 1.4 \overset{?}{=} 9.1$	Simplify.
$9.1 = 9.1$	true statement

Solve the following equations.

7. $3x = 33$

8. $2.5x + 0.4x = 15.2 - 3.6$

9. $-x = -15$

Example 9

Solving $ax = c$

Solve the following equation.

$$-x = 4$$

Solution

$-x = 4$	Write the equation.
$-1x = 4$	-1 is the coefficient of x.
$\dfrac{-1x}{-1} = \dfrac{4}{-1}$	Divide by -1 so that the coefficient will become $+1$.
$x = -4$	

Check:

$-x = 4$	
$-(-4) \overset{?}{=} 4$	Substitute $x = -4$.
$4 = 4$	true statement

Now work margin exercises 7 through 9.

Completion Example 10

Solving $ax = c$

Supply the reasons for each step in solving the following equation.

$$\frac{4x}{5} = \frac{3}{10}$$ (This could be written $\frac{4}{5}x = \frac{3}{10}$ because $\frac{4}{5}x$ is the same as $\frac{4x}{5}$.)

Solution

$$\frac{4x}{5} = \frac{3}{10}$$ _____

$$\frac{5}{4} \cdot \frac{4}{5}x = \frac{5}{4} \cdot \frac{3}{10}$$ _____

$$1 \cdot x = \frac{1 \cdot \cancel{5}}{4} \cdot \frac{3}{2 \cdot \cancel{5}}$$ _____

$$x = \frac{3}{8}$$

Now work margin exercise 10.

Example 11

Application

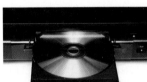

Mark had a coupon for $50 off the Blu-Ray player he just purchased. With the coupon the price was $165.50. Solve the equation $y - 50 = 165.50$ to determine the original price of the Blu-Ray player.

Solution

$y - 50 = 165.50$
$y - 50 + 50 = 165.50 + 50$ Use the addition principle by adding 50 to both sides.
$y = 215.50$ Simplify.

The original price of the Blu-Ray player was $215.50.

Now work margin exercise 11.

---

10. Solve the following equation.

$$\frac{2x}{3} = \frac{4}{5}$$

11. Mary has a coupon for $20 off at a local department store. If she uses the coupon to buy a wool coat for $84.99, solve the equation $y - 20 = 84.99$ to determine the original price of the coat.

Completion Example Answers

10. $$\frac{4x}{5} = \frac{3}{10}$$ <u>Write the equation.</u>

$$\frac{5}{4} \cdot \frac{4}{5}x = \frac{5}{4} \cdot \frac{3}{10}$$ <u>Multiply both sides by $\frac{5}{4}$.</u>

$$1 \cdot x = \frac{1 \cdot \cancel{5}}{4} \cdot \frac{3}{2 \cdot \cancel{5}}$$ <u>Simplify.</u>

$$x = \frac{3}{8}$$

Practice Problem Answers

1. $x = -21$ **2.** $y = 1.5$ **3.** $x = -5$

4. $y = 55$ **5.** $z = 2$ **6.** $x = 8$

7. $x = \dfrac{10}{3}$ **8.** $x = -4.3$ **9.** $\dfrac{4}{15} = x$

Exercises 8.1

Solve each of the following linear equations. (Remember that the objective is to isolate the variable on one side of the equation with a coefficient of +1.) See Examples 1 through 10.

1. $x - 6 = 1$ **2.** $x - 10 = 9$ **3.** $y + 7 = 3$ **4.** $y + 12 = 5$ **5.** $x + 15 = -4$

6. $x + 17 = -10$ **7.** $22 = n - 15$ **8.** $36 = n - 20$ **9.** $6 = z + 12$ **10.** $18 = z + 1$

11. $x - 20 = -15$ **12.** $x - 10 = -11$ **13.** $y + 3.4 = -2.5$ **14.** $y + 1.6 = -3.7$ **15.** $x + 3.6 = 2.4$

16. $x + 2.7 = 3.8$ **17.** $x + \dfrac{1}{20} = \dfrac{3}{5}$ **18.** $n - \dfrac{2}{7} = \dfrac{3}{14}$ **19.** $5x = 45$ **20.** $9x = 108$

21. $32 = 4y$ **22.** $51 = 17y$ **23.** $\dfrac{3x}{4} = 15$ **24.** $\dfrac{5x}{7} = 65$ **25.** $4 = \dfrac{2y}{5}$

26. $46 = \dfrac{46y}{5}$ **27.** $4x - 3x = 10 - 12$ **28.** $7x - 6x = 13 + 15$ **29.** $7x - 8x = 13 - 25$

30. $10n - 11n = 20 - 14$ **31.** $3n - 2n + 6 = 14$ **32.** $7n - 6n + 13 = 22$ **33.** $1.7y + 1.3y = 6.3$

34. $2.5y + 7.5y = 4.2$ **35.** $\dfrac{3}{4}x = \dfrac{5}{3}$ **36.** $\dfrac{5}{6}x = \dfrac{5}{3}$ **37.** $7.5x = -99.75$

38. $-14 = 0.7x$ **39.** $1.5y - 0.5y + 6.7 = -5.3$ **40.** $2.6y - 1.6y - 5.1 = -2.9$

41. $10x - 9x - \dfrac{1}{2} = -\dfrac{9}{10}$ **42.** $6x - 5x + \dfrac{3}{4} = -\dfrac{1}{12}$ **43.** $1.4x - 0.4x + 2.7 = -1.3$

44. $3.5y - 2.5y - 6.3 = -1.0 - 2.5$ **45.** $\dfrac{7x}{4} - \dfrac{3x}{4} + \dfrac{7}{8} = \dfrac{3}{2}$ **46.** $\dfrac{5n}{2} - \dfrac{3n}{2} + \dfrac{4}{5} = \dfrac{7}{5} - \dfrac{1}{10}$

47. $6.2 = -3.5 + 7n - 6n$ **48.** $-7.2 = 1.3n - 0.3n - 1.0$ **49.** $1.7x = -5.1 - 1.7$

50. $3.2x = 2.8 - 9.2$

51. **World Languages:** The Japanese writing system consists of three sets of characters, two with 81 characters (which all Japanese students must know), and a third, kanji, with over 50,000 characters (of which only some are used in everyday writing). If a Japanese student knows 2107 total characters, solve the equation $x + 2(81) = 2107$ to determine the number of kanji characters the student knows.

52. **Astronomy:** The diameter of the Milky Way, the galaxy our solar system is in, is approximately 23,585 times the distance from the sun to the nearest star, Proxima Centauri. Considering that the Milky Way is roughly 100,000 light years across, solve the equation $23,585x = 100,000$ to find the number of light years from the sun to this star. (Round your answer to the nearest hundreth.)

53. **Sculpture:** A sculptor has decided to begin a project to make scale models of famous landmarks out of stone. His first model will be of one of the moai, giant human figures carved from stone on Easter Island. If his model is to be 1/12 scale and the original moai weighs 75 tons, solve the equation $12x = 75$ to determine how many tons his completed sculpture will weigh.

54. **Writing:** An author is determined to have his first novel published by the publisher of George Orwell's *1984*, his favorite book. However, his contract with the publisher requires his novel to be at least 75,000 words, and he has only written 63,500. Solve the following equation to determine how many more words he must write.
$63,500 + x = 75,000$

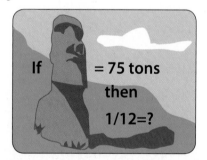

If   = 75 tons
then
1/12=?

Use a calculator to help solve the following equations.

55. $y + 32.861 = -17.892$

56. $x - 41.625 = 59.354$

57. $17.61x - 16.61x + 27.059 = 9.845$

58. $14.83y - 8.65 - 13.83y = 17.437 + 1.0$

59. $2.637x = 648.702$

60. $-0.3057y = 316.7052$

61. $-x = 145.6 + 17.89 - 10.32$

62. $-y = 143.5 + 178.462 - 200$

Writing & Thinking

63. **a.** Is the expression $6 + 3 = 9$ an equation? Explain.

  b. Is 4 a solution to the equation $5 + x = 10$? Explain.

8.2 Solving Linear Equations: $ax + b = c$

Objective A **Solving Equations of the Form $ax + b = c$**

In Section 8.1 we learned the equations to be solved were in one of two forms: $x + b = c$ or $ax = c$. In the first type the coefficient of the variable x was always +1 and we had to add or subtract constants to solve the equation. (We used the addition principle.) In the second type, we had to multiply both sides by the reciprocal of the coefficient of the variable (or divide both sides by the coefficient itself). Now we will apply both of these techniques in solving the same equation. That is, in the form $ax + b = c$ the coefficient a may be a number other than 1.

The **general procedure for solving linear equations** is now a combination of the procedures stated in Section 8.1.

> ### Procedure for Solving Linear Equations that Simplify to the Form $ax + b = c$
>
> 1. Combine like terms on both sides of the equation.
> 2. Use the **addition principle of equality** and add the opposite of the constant b to both sides.
> 3. Use the **multiplication** (or **division**) **principle of equality** to multiply both sides by the reciprocal of the coefficient of the variable (or divide both sides by the coefficient itself). The coefficient of the variable will become +1.
> 4. Check your answer by substituting it for the variable in the original equation.

Example 1

Solving Linear Equations of the Form $ax + b = c$

Solve the equation $3x + 3 = -18$

Solution

$3x + 3 = -18$	Write the equation.
$3x + 3 - 3 = -18 - 3$	Add –3 to both sides.
$3x = -21$	Simplify.
$\dfrac{3x}{3} = \dfrac{-21}{3}$	Divide both sides by 3.
$x = -7$	Simplify.

Check:

$3x + 3 = -18$

$3(-7) + 3 \overset{?}{=} -18$ 　　　Substitute $x = -7$.

$-21 + 3 \overset{?}{=} -18$ 　　　Simplify.

$-18 = -18$ 　　　true statement

Example 2

Solving Linear Equations of the Form $ax + b = c$

Solve the equation $-26 = 2y - 14 - 4y$

Solution

$-26 = 2y - 14 - 4y$	Write the equation.
$-26 = -2y - 14$	Combine like terms.
$-26 + 14 = -2y - 14 + 14$	Add 14 to both sides.
$-12 = -2y$	Simplify.
$\dfrac{-12}{-2} = \dfrac{-2y}{-2}$	Divide both sides by –2.
$6 = y$	Simplify.

Check:

$-26 = 2y - 14 - 4y$	
$-26 \overset{?}{=} 2(6) - 14 - 4(6)$	Substitute $y = 6$.
$-26 \overset{?}{=} 12 - 14 - 24$	Simplify.
$-26 = -26$	true statement

Now work margin exercises **1** *and* **2.**

Examples 3 and 4 illustrate solving equations with decimal coefficients. You may choose to work with the decimal coefficients as they are. However, another approach, shown here, is to multiply both sides in such a way to give integer coefficients. Generally, integers are easier to work with than decimals.

Example 3

Solving Linear Equations Involving Decimals

Solve the equation $16.53 - 18.2z - 7.43 = 0$

Solution

$16.53 - 18.2z - 7.43 = 0$	Write the equation.
$100(16.53 - 18.2z - 7.43) = 100(0)$	Multiply both sides by 100. (This results in integer coefficients.)
$1653 - 1820z - 743 = 0$	Simplify.
$910 - 1820z = 0$	Combine like terms.
$910 - 1820z - 910 = 0 - 910$	Add –910 to both sides. Simplify.
$\dfrac{-1820z}{-1820} = \dfrac{-910}{-1820}$	Divide both sides by –1820.
$z = 0.5 \left(\text{or } z = \dfrac{1}{2} \right)$	Simplify.

Check:

$$16.53 - 18.2z - 7.43 = 0$$

$$16.53 - 18.2(0.5) - 7.43 \overset{?}{=} 0 \qquad \text{Substitute } z = 0.5.$$

$$16.53 - 9.10 - 7.43 \overset{?}{=} 0 \qquad \text{Simplify.}$$

$$0 = 0 \qquad \text{true statement}$$

Solve the following equations.

3. $3.69 + 12.8z - 6.25 = 0$

4. $0.25x + 0.5x - 2.3 = 1.45$

Example 4

Solving Linear Equations Involving Decimals

Solve the equation $5.1x + 7.4 - 1.8x = -9.1$

Solution

Because the decimal numbers are accurate to tenths we could multiply both sides by 10 to get integer coefficients. However, we can also work with the decimal numbers directly as shown here:

$$5.1x + 7.4 - 1.8x = -9.1 \qquad \text{Write the equation.}$$

$$3.3x + 7.4 = -9.1 \qquad \text{Combine like terms.}$$

$$3.3x + 7.4 - 7.4 = -9.1 - 7.4 \qquad \text{Add } -7.4 \text{ to both sides.}$$

$$3.3x = -16.5 \qquad \text{Simplify.}$$

$$\frac{3.3x}{3.3} = \frac{-16.5}{3.3} \qquad \text{Divide both sides by the coefficient 3.3.}$$

$$x = -5 \qquad \text{Simplify.}$$

Check:

$$5.1x + 7.4 - 1.8x = -9.1$$

$$5.1(-5) + 7.4 - 1.8(-5) \overset{?}{=} -9.1 \qquad \text{Substitute } z = -5.$$

$$-25.5 + 7.4 + 9 \overset{?}{=} -9.1 \qquad \text{Simplify.}$$

$$-9.1 = -9.1 \qquad \text{true statement}$$

Now work margin exercises 3 *and* 4.

Examples 5 and 6 illustrate solving equations with coefficients that are fractions. You may choose to work with the the coefficients as they are. However, another approach, shown here, is to multiply both sides by the LCM of the denominators which will give integer coefficients. Generally, integers are easier to work with than fractions.

Example 5

Solving Linear Equations with Fractional Coefficients

Solve the equation $\dfrac{5}{6}x - \dfrac{5}{2} = -\dfrac{10}{9}$

Solution

$$\frac{5}{6}x - \frac{5}{2} = -\frac{10}{9}$$
Write the equation.

$$18\left(\frac{5}{6}x - \frac{5}{2}\right) = 18\left(-\frac{10}{9}\right)$$
Multiply both sides by 18 (the LCM of the denominators).

$$18\left(\frac{5}{6}x\right) - 18\left(\frac{5}{2}\right) = 18\left(-\frac{10}{9}\right)$$
Apply the distributive property.

$$\frac{\overset{3}{\cancel{18}}}{1} \cdot \frac{5}{\cancel{6}}x - \frac{\overset{9}{\cancel{18}}}{1} \cdot \frac{5}{\cancel{2}} = \frac{\overset{2}{\cancel{18}}}{1} \cdot \left(-\frac{10}{\cancel{9}}\right)$$
Multiply.

$$15x - 45 = -20$$
Simplify.

$$15x - 45 + 45 = -20 + 45$$
Add 45 to both sides.

$$15x = 25$$
Simplify.

$$\frac{15x}{15} = \frac{25}{15}$$
Divide both sides by 15.

$$x = \frac{5}{3}$$
Simplify.

Check:

$$\frac{5}{6}x - \frac{5}{2} = -\frac{10}{9}$$

$$\frac{5}{6}\left(\frac{5}{3}\right) - \frac{5}{2} \overset{?}{=} -\frac{10}{9}$$
Substitute $x = \dfrac{5}{3}$.

$$\frac{25}{18} - \frac{45}{18} \overset{?}{=} -\frac{20}{18}$$
Simplify.

$$-\frac{20}{18} = -\frac{20}{18}$$
true statement

notes

About Checking: Checking can be quite time-consuming and need not be done for every problem. This is particularly important on exams. You should check only if you have time after the entire exam is completed.

Example 6

Solving Linear Equations with Fractional Coefficients

Solve the equation $\frac{1}{2}x + \frac{3}{4}x + \frac{7}{2} - \frac{2}{3}x = 0$.

Solution

$$\frac{1}{2}x + \frac{3}{4}x + \frac{7}{2} - \frac{2}{3}x = 0 \qquad \text{Write the equation.}$$

$$12\left(\frac{1}{2}x + \frac{3}{4}x + \frac{7}{2} - \frac{2}{3}x\right) = 12(0) \qquad \text{Multiply both sides by 12 (the LCM of the denominators).}$$

$$12\left(\frac{1}{2}x\right) + 12\left(\frac{3}{4}x\right) + 12\left(\frac{7}{2}\right) - 12\left(\frac{2}{3}x\right) = 12(0) \qquad \text{Apply the distributive property.}$$

$$6x + 9x + 42 - 8x = 0 \qquad \text{Simplify.}$$

$$7x + 42 = 0 \qquad \text{Combine like terms.}$$

$$7x + 42 - 42 = 0 - 42 \qquad \text{Add } -42 \text{ to both sides.}$$

$$7x = -42 \qquad \text{Simplify.}$$

$$\frac{7x}{7} = \frac{-42}{7} \qquad \text{Divide both sides by 7.}$$

$$x = -6 \qquad \text{Simplify. Checking will show that } -6 \text{ is the solution.}$$

Solve the following equations.

5. $\frac{3}{5}x - \frac{7}{3} = \frac{-17}{15}$

6. $\frac{1}{3}x + \frac{3}{8}x + \frac{9}{2} - \frac{3}{4}x = 0$

7. $\frac{2}{3}x - \frac{5}{9} = \frac{1}{18}$

Completion Example 7

Solving Linear Equations with Fractional Coefficients

Solve the equation $\frac{3}{4}y + \frac{1}{2} = \frac{5}{8}$.

Solution

$$\frac{3}{4}y + \frac{1}{2} = \frac{5}{8} \qquad \text{Write the equation.}$$

$$\underline{\quad}\left(\frac{3}{4}y + \frac{1}{2}\right) = \underline{\quad}\left(\frac{5}{8}\right) \qquad \text{Multiply both sides by } \underline{\quad}.$$

$$\underline{\quad}\left(\frac{3}{4}\right)y + \underline{\quad}\left(\frac{1}{2}\right) = \underline{\quad}\left(\frac{5}{8}\right) \qquad \text{Apply the distributive property.}$$

$$\underline{\quad}y + \underline{\quad} = \underline{\quad} \qquad \text{Simplify.}$$

$$\underline{\quad}y + 4 - \underline{\quad} = 5 - \underline{\quad} \qquad \text{Subtract } \underline{\quad} \text{ from both sides.}$$

$$6y = \underline{\quad} \qquad \text{Simplify.}$$

$$\frac{6y}{\underline{\quad}} = \frac{1}{\underline{\quad}} \qquad \text{Divide both sides by } \underline{\quad}.$$

$$y = \underline{\quad} \qquad \text{Simplify.}$$

Now work margin exercises 5 through 7.

Practice Problems

Solve the following linear equations.

1. $x + 14 - 8x = -7$

2. $2.4 = 2.6y - 5.9y - 0.9$

3. $n - \dfrac{2n}{3} - \dfrac{1}{2} = \dfrac{1}{6}$

4. $\dfrac{3x}{14} + \dfrac{1}{2} - \dfrac{x}{7} = 0$

Completion Example Answers

7.

$$\frac{3}{4}y + \frac{1}{2} = \frac{5}{8}$$ 　Write the equation.

$$\underline{8}\left(\frac{3}{4}y + \frac{1}{2}\right) = \underline{8}\left(\frac{5}{8}\right)$$ 　Multiply both sides by $\underline{8}$.

$$\underline{8}\left(\frac{3}{4}\right)y + \underline{8}\left(\frac{1}{2}\right) = \underline{8}\left(\frac{5}{8}\right)$$ 　Apply the distributive property.

$$\underline{6}\,y + \underline{4} = \underline{5}$$ 　Simplify.

$$6y + 4 - \underline{4} = 5 - \underline{4}$$ 　Subtract $\underline{4}$ from both sides.

$$6y = \underline{1}$$ 　Simplify.

$$\frac{6y}{\underline{6}} = \frac{\underline{1}}{\underline{6}}$$ 　Divide both sides by $\underline{6}$.

$$y = \frac{\underline{1}}{\underline{6}}$$ 　Simplify.

Practice Problem Answers

1. $x = 3$ 　　**2.** $y = -1$ 　　**3.** $n = 2$ 　　**4.** $x = -7$

Exercises 8.2

Solve each of the following linear equations. See Examples 1 through 7.

1. $3x + 11 = 2$

2. $3x + 10 = -5$

3. $5x - 4 = 6$

4. $4y - 8 = -12$

5. $6x + 10 = 22$

6. $3n + 7 = 19$

7. $9x - 5 = 13$

8. $2x - 4 = 12$

9. $1 - 3y = 4$

10. $5 - 2x = 9$

11. $14 + 9t = 5$

12. $5 + 2x = -7$

13. $-5x + 2.9 = 3.5$

14. $3x + 2.7 = -2.7$

15. $10 + 3x - 4 = 18$

16. $5 + 5x - 6 = 9$

17. $15 = 7x + 7 + 8$

18. $14 = 9x + 5 + 8$

19. $5y - 3y + 2 = 2$

20. $6y + 8y - 7 = -7$

21. $x - 4x + 25 = 31$

22. $3y + 9y - 13 = 11$

23. $-20 = 7y - 3y + 4$

24. $-20 = 5y + y + 16$

25. $4n - 10n + 35 = 1 - 2$

26. $-5n - 3n + 2 = 34$

27. $3n - 15 - n = 1$

28. $2n + 12 + n = 0$

29. $5.4x - 0.2x = 0$

30. $0 = 5.1x + 0.3x$

31. $\dfrac{1}{2}x + 7 = \dfrac{7}{2}$

32. $\dfrac{3}{5}x + 4 = \dfrac{9}{5}$

33. $\dfrac{1}{2} - \dfrac{8}{3}x = \dfrac{5}{6}$

34. $\dfrac{2}{5} - \dfrac{1}{2}x = \dfrac{7}{4}$

35. $\dfrac{3}{2} = \dfrac{1}{3}x + \dfrac{11}{3}$

36. $\dfrac{11}{8} = \dfrac{1}{5}x + \dfrac{4}{5}$

37. $\dfrac{7}{2} - 5 - \dfrac{5}{2}x = 9$

38. $\dfrac{8}{3} + 2 - \dfrac{7}{3}x = 6$

39. $\dfrac{5}{8}x - \dfrac{1}{4}x + \dfrac{1}{2} = \dfrac{3}{10}$

40. $\dfrac{1}{2}x + \dfrac{3}{4}x - \dfrac{5}{3} = \dfrac{5}{6}$

41. $\dfrac{y}{2} + \dfrac{1}{5} = 3$

42. $\dfrac{y}{3} - \dfrac{2}{3} = 7$

43. $\dfrac{7}{8} = \dfrac{3}{4}x - \dfrac{5}{8}$

44. $\dfrac{1}{10} = \dfrac{4}{5}x + \dfrac{3}{10}$

45. $\dfrac{y}{7} + \dfrac{y}{28} + \dfrac{1}{2} = \dfrac{3}{4}$

46. $\dfrac{5y}{6} - \dfrac{7y}{8} - \dfrac{1}{12} = \dfrac{1}{3}$

47. $x + 1.2x + 6.9 = -3.0$

48. $3x - 0.75x - 1.72 = 3.23$

49. $10 = x - 0.5x + 32$

50. $33 = y + 3 - 0.4y$

51. $2.5x + 0.5x - 3.5 = 2.5$

52. $4.7 - 0.5x - 0.3x = -0.1$

53. $6.4 + 1.2x + 0.3x = 0.4$

54. $5.2 - 1.3x - 1.5x = -0.4$

55. $-12.13 = 2.42y + 0.6y - 13.64$

56. $-7.01 = 1.75x + 3.05x - 8.45$

57. $-0.4x + x + 17.2 = 18.1$

58. $y - 0.75y + 13.76 = 14.66$

59. $0 = 17.3x - 15.02x - 0.456$

60. $0 = 20.5x - 16.35x + 0.1245$

Solve the following word problems.

61. Music: The tickets for a concert featuring the new hit band, Flying Sailor, sold out in 2.5 hours. If there were 35,000 tickets sold, solve the equation $35,000 - 2.5x = 0$ to find the number of tickets sold per hour.

62. Parking Lots: A rectangular-shaped parking lot is to have a perimeter of 450 yards. If the width must be 90 yards because of a building code, solve the equation $2l + 2(90) = 450$ to determine the length of the parking lot.

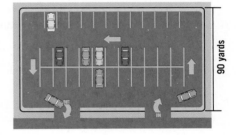

63. Temperature: Jeff, who lives in England, is reading a letter from his pen pal in the United States. His pen pal says that the temperature in his city was 97.7° Fahrenheit one day, so it was too hot to play soccer outside. Jeff doesn't know how hot this is, because he is used to temperatures in Celsius. Help Jeff solve the equation below to determine the temperature in degrees Celsius.

$$1.8C + 32 = 97.7$$

64. Height: The tallest man-made structure in the world is the Burj Khalifa in Dubai, which stands at 2717 feet tall. The tallest tree in the world is a Mendocino tree in California. If 7 of these trees were stacked on top of each other, they would still be 144.5 feet shorter than the Burj Khalifa. Solve the equation below to determine the height of the tree.

$$7x + 144.5 = 2717$$

Use a calculator to help solve the linear equations.

65. $0.15x + 5.23x - 17.815 = 15.003$

66. $15.97y - 12.34y + 16.95 = 8.601$

67. $13.45x - 20x - 17.36 = -24.696$

68. $26.75y - 30y + 23.28 = 4.4625$

8.3 Solving Linear Equations: $ax + b = cx + d$

Objective A **Solving Equations of the Form $ax + b = cx + d$**

Now we are ready to solve linear equations of the most general form $ax + b = cx + d$ where constants and variables may be on both sides. There may also be parentheses or other symbols of inclusion. **Remember that the objective is to get the variable on one side of the equation by itself with a coefficient of $+1$.**

General Procedure for Solving Linear Equations that Simplify to the Form $ax + b = cx + d$

1. Simplify by removing any grouping symbols and combining like terms on each side of the equation.

2. Use the **addition principle of equality** and add the opposite of a constant term and/or variable term to both sides so that variables are on one side and constants are on the other side.

3. Use the **multiplication** (or **division**) **principle of equality** to multiply both sides by the reciprocal of the coefficient of the variable (or divide both sides by the coefficient itself). The coefficient of the variable will become $+1$.

4. Check your answer by substituting it for the variable in the original equation.

Example 1

Solving Equations of the Form $ax + b = cx + d$

Solve the following equation.

$$5x + 3 = 2x - 18$$

Solution

$5x + 3 = 2x - 18$	Write the equation.
$5x + 3 - 3 = 2x - 18 - 3$	Add -3 to both sides.
$5x = 2x - 21$	Simplify.
$5x - 2x = 2x - 21 - 2x$	Add $-2x$ to both sides.
$3x = -21$	Simplify
$\dfrac{3x}{3} = \dfrac{-21}{3}$	Divide both sides by 3.
$x = -7$	Simplify.

Check:

$$5x + 3 = 2x - 18$$

$$5(-7) + 3 \overset{?}{=} 2(-7) - 18 \qquad \text{Substitute } x = -7.$$

$$-35 + 3 \overset{?}{=} -14 - 18 \qquad \text{Simplify.}$$

$$-32 = -32 \qquad \text{true statement}$$

Example 2

Solving Equations of the Form $ax + b = cx + d$

Solve the following equation.

$$4x + 1 - x = 2x - 13 + 5$$

Solution

$4x + 1 - x = 2x - 13 + 5$	Write the equation.
$3x + 1 = 2x - 8$	Combine like terms.
$3x + 1 - 1 = 2x - 8 - 1$	Add −1 to both sides.
$3x = 2x - 9$	Simplify.
$3x - 2x = 2x - 9 - 2x$	Add −2x to both sides.
$x = -9$	Simplify.

Check:

$$4x + 1 - x = 2x - 13 + 5$$

$$4(-9) + 1 - (-9) \overset{?}{=} 2(-9) - 13 + 5 \qquad \text{Substitute } x = -9.$$

$$-36 + 1 + 9 \overset{?}{=} -18 - 13 + 5 \qquad \text{Simplify.}$$

$$-26 = -26 \qquad \text{true statement}$$

***Now work margin exercises* 1 and 2.**

Solve the following equations.

1. $7x + 4 = 3x - 20$

2. $6x + 3 - 2x = 2x - 15 + 4$

Example 3

Solving Linear Equations Involving Decimals

Solve the following equation.

$$6y + 2.5 = 7y - 3.6$$

Solution

$6y + 2.5 = 7y - 3.6$	Write the equation.
$6y + 2.5 + 3.6 = 7y - 3.6 + 3.6$	Add 3.6 to both sides.
$6y + 6.1 = 7y$	Simplify.
$6y + 6.1 - 6y = 7y - 6y$	Add −6y to both sides.
$6.1 = y$	Simplify.

3. Solve the following equation.

$$5y + 1.4 = 7y - 2.8$$

▲ **Check:**

$$6y + 2.5 = 7y - 3.6$$

$$6(6.1) + 2.5 \overset{?}{=} 7(6.1) - 3.6 \qquad \text{Substitute } y = 6.1.$$

$$36.6 + 2.5 \overset{?}{=} 42.7 - 3.6 \qquad \text{Simplify.}$$

$$39.1 = 39.1 \qquad \text{true statement}$$

Now work margin exercise 3.

---

4. Solve the following equation.

$$\frac{1}{4}x + \frac{2}{5} = \frac{1}{2}x - \frac{6}{10}$$

Example 4

Solving Linear Equations with Fractional Coefficients

Solve the following equation.

$$\frac{1}{3}x + \frac{1}{6} = \frac{2}{5}x - \frac{7}{10}$$

Solution

$$\frac{1}{3}x + \frac{1}{6} = \frac{2}{5}x - \frac{7}{10} \qquad \text{Write the equation.}$$

$$30\left(\frac{1}{3}x + \frac{1}{6}\right) = 30\left(\frac{2}{5}x - \frac{7}{10}\right) \qquad \begin{array}{l}\text{Multiply both sides by the}\\\text{LCM, 30.}\end{array}$$

$$30\left(\frac{1}{3}x\right) + 30\left(\frac{1}{6}\right) = 30\left(\frac{2}{5}x\right) - 30\left(\frac{7}{10}\right) \qquad \text{Apply the distributive property.}$$

$$10x + 5 = 12x - 21 \qquad \text{Simplify.}$$

$$10x + 5 - 5 = 12x - 21 - 5 \qquad \text{Add } -5 \text{ to both sides.}$$

$$10x = 12x - 26 \qquad \text{Simplify.}$$

$$10x - 12x = 12x - 26 - 12x \qquad \text{Add } -12x \text{ to both sides.}$$

$$-2x = -26 \qquad \text{Simplify.}$$

$$\frac{-2x}{-2} = \frac{-26}{-2} \qquad \text{Divide both sides by } -2.$$

$$x = 13 \qquad \text{Simplify.}$$

Now work margin exercise 4.

Example 5

Solving Equations with Parentheses

Solve the following equation.

$$2(y-7) = 4(y+1) - 26$$

Solution

$2(y-7) = 4(y+1) - 26$	Write the equation.
$2y - 14 = 4y + 4 - 26$	Use the distributive property.
$2y - 14 = 4y - 22$	Combine like terms.
$2y - 14 + 22 = 4y - 22 + 22$	Add 22 to both sides. Here we will put the variable on the right side to get a positive coefficient of y.
$2y + 8 = 4y$	Simplify.
$2y + 8 - 2y = 4y - 2y$	Add $-2y$ to both sides.
$8 = 2y$	Simplify.
$\dfrac{8}{2} = \dfrac{2y}{2}$	Divide both sides by 2.
$4 = y$	Simplify.

Now work margin exercise 5.

5. Solve the following equation.

$$3(y-5) = 8(y+1) - 3$$

Example 6

Solving Equations with Parentheses

Solve the following equation.

$$-2(5x+13) - 2 = -6(3x-2) - 41$$

Solution

$-2(5x+13) - 2 = -6(3x-2) - 41$	Write the equation.
$-10x - 26 - 2 = -18x + 12 - 41$	Use the distributive property. Be careful with the signs.
$-10x - 28 = -18x - 29$	Combine like terms.
$-10x - 28 + 18x = -18x - 29 + 18x$	Add $18x$ to both sides.
$8x - 28 = -29$	Simplify.
$8x - 28 + 28 = -29 + 28$	Add 28 to both sides.
$8x = -1$	Simplify.
$\dfrac{8x}{8} = \dfrac{-1}{8}$	Divide both sides by 8.
$x = -\dfrac{1}{8}$	Simplify.

Now work margin exercise 6.

6. Solve the following equation.

$$-5(3x+7) - 11 = -7(x-1) - 5$$

7. Solve the following equation.

$$2 - 2(x + 5) = 3(x - 6)$$

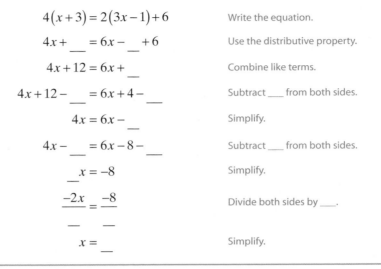

Completion Example 7

Solving Equations with Parentheses

Solve the equation $4(x + 3) = 2(3x - 1) + 6$

Solution

$4(x + 3) = 2(3x - 1) + 6$	Write the equation.
$4x + \underline{} = 6x - \underline{} + 6$	Use the distributive property.
$4x + 12 = 6x + \underline{}$	Combine like terms.
$4x + 12 - \underline{} = 6x + 4 - \underline{}$	Subtract ___ from both sides.
$4x = 6x - \underline{}$	Simplify.
$4x - \underline{} = 6x - 8 - \underline{}$	Subtract ___ from both sides.
$\underline{}x = -8$	Simplify.
$\dfrac{-2x}{\underline{}} = \dfrac{-8}{\underline{}}$	Divide both sides by ___.
$x = \underline{}$	Simplify.

Now work margin exercise 7.

Objective B **Conditional Equations, Identities and Contradictions**

When solving equations, there are times that we are concerned with the number of solutions that an equation has. If an equation has a finite number of solutions (the number of solutions is a countable number), the equation is said to be a **conditional equation**. As stated earlier, every linear equation has exactly one solution. Thus, **every linear equation is a conditional equation**. However, in some cases, simplifying an equation will lead to a statement that is always true, such as $0 = 0$. In these cases the original equation is called an

---

Completion Example Answers

7.

$4(x + 3) = 2(3x - 1) + 6$	Write the equation.
$4x + \underline{12} = 6x - \underline{2} + 6$	Use the distributive property.
$4x + 12 = 6x + \underline{4}$	Combine like terms.
$4x + 12 - \underline{12} = 6x + 4 - \underline{12}$	Subtract $\underline{12}$ from both sides.
$4x = 6x - \underline{8}$	Simplify.
$4x - \underline{6x} = 6x - 8 - \underline{6x}$	Subtract $\underline{6x}$ from both sides.
$\underline{-2}x = -8$	Simplify.
$\dfrac{-2x}{\underline{-2}} = \dfrac{-8}{\underline{-2}}$	Divide both sides by $\underline{-2}$.
$x = \underline{4}$	Simplify.

identity and has an infinite number of solutions which can be written as all real numbers or $\mathbb{R}$. If the equation simplifies to a statement that is never true, such as $0 = 2$, then the original equation is called a **contradiction** and there is no solution. Table 1 summarizes these ideas.

Type of Equation	Number of Solutions
Conditional	Finite number of solutions
Identity	Infinite number of solutions
Contradiction	No solution

Table 1

Example 8

Solutions of Equations

Determine whether the equation $3(x+5)+1 = -11$ is a conditional equation, an identity, or a contradiction.

Solution

$3(x+5)+1 = -11$	Write the equation.
$3x+15+1 = -11$	Use the distributive property.
$3x+16 = -11$	Combine like terms.
$3x+16-16 = -11-16$	Add -16 to both sides.
$3x = -27$	Simplify.
$\dfrac{3x}{3} = \dfrac{-27}{3}$	Divide both sides by 3.
$x = -9$	Simplify.

The equation has one solution. Therefore it is a conditional equation.

Example 9

Solutions of Equations

Determine whether the equation $3(x-25)+3x = 6(x+10)$ is a conditional equation, an identity, or a contradiction.

Solution

$3(x-25)+3x = 6(x+10)$	Write the equation.
$3x-75+3x = 6x+60$	Use the distributive property.
$6x-75 = 6x+60$	Combine like terms.
$6x-75-6x = 6x+60-6x$	Add $-6x$ to both sides.
$-75 = 60$	Simplify.

The last equation is never true. Therefore, the original equation is a contradiction and has no solution.

Determine whether each of the following equations is a conditional equation, an identity, or a contradiction.

8. $4(x-17)+3x=7x+49$

9. $-3(x-4)=12-2x-x$

10. $5x+9=-13+3x$

Example 10

Solutions of Equations

Determine whether the following equation is a conditional equation, an identity, or a contradiction.

$$-2(x-7)+x=14-x$$

Solution

$-2(x-7)+x=14-x$	Write the equation.
$-2x+14+x=14-x$	Use the distributive property.
$14-x=14-x$	Combine like terms.
$14-x-14=14-x-14$	Add -14 to both sides.
$-x=-x$	
$-x+x=-x+x$	Add x to both sides.
$0=0$	Simplify.

The last equation is always true. Therefore, the original equation is an identity and has an infinite number of solutions. Every real number is a solution.

Now work margin exercises 8 through 10.

Practice Problems

Solve the following linear equations.

1. $x+14-6x=2x-7$ **2.** $6.4x+2.1=3.1x-1.2$

3. $\dfrac{2x}{3}-\dfrac{1}{2}=x+\dfrac{1}{6}$ **4.** $\dfrac{3}{14}n+\dfrac{1}{4}=\dfrac{1}{7}n-\dfrac{1}{4}$

5. $5-(y-3)=14-4(y+2)$

Determine whether each of the following equations is a conditional equation, an identity, or a contradiction.

6. $7(x-3)+42=7x+21$ **7.** $-2x+14=-2(x+1)+10$

Practice Problem Answers

1. $x=3$ **2.** $x=-1$ **3.** $x=-2$ **4.** $n=-7$

5. $y=-\dfrac{2}{3}$ **6.** identity   **7.** contradiction

Exercises 8.3

Solve each of the following linear equations. See Examples 1 through 7.

1. $3x + 2 = x - 8$

2. $5x + 1 = 2x - 5$

3. $4n - 3 = n + 6$

4. $6y + 3 = y - 7$

5. $3y + 18 = 7y - 6$

6. $2y + 5 = 8y + 10$

7. $3x + 11 = 8x - 4$

8. $9x + 3 = 5x - 9$

9. $14n = 3n$

10. $1.6x = 0.8x$

11. $6y - 2.1 = y - 2.1$

12. $13x + 5 = 2x + 5$

13. $2(z + 1) = 3z + 3$

14. $6x - 3 = 3(x + 2)$

15. $16y + 23y - 3 = 16y - 2y + 2$

16. $5x - 2x + 4 = 3x + x - 1$

17. $0.25 + 3x + 6.5 = 0.75x$

18. $0.9y + 3 = 0.4y + 1.5$

19. $6.5 + 1.2x = 0.5 - 0.3x$

20. $x - 0.1x + 0.8 = 0.2x + 0.1$

21. $\dfrac{2}{3}x + 1 = \dfrac{1}{3}x - 6$

22. $\dfrac{4}{5}n + 2 = \dfrac{2}{5}n - 4$

23. $\dfrac{y}{5} + \dfrac{3}{4} = \dfrac{y}{2} + \dfrac{3}{4}$

24. $\dfrac{5n}{6} + \dfrac{1}{9} = \dfrac{3n}{2} + \dfrac{1}{9}$

25. $\dfrac{3}{8}\left(y - \dfrac{1}{2}\right) = \dfrac{1}{8}\left(y + \dfrac{1}{2}\right)$

26. $\dfrac{1}{2}\left(\dfrac{x}{2} + 1\right) = \dfrac{1}{3}\left(\dfrac{x}{2} - 1\right)$

27. $\dfrac{2x}{3} + \dfrac{x}{3} = -\dfrac{3}{4} + \dfrac{x}{2}$

28. $\dfrac{3}{4}x + \dfrac{1}{5}x = \dfrac{1}{2}x - \dfrac{3}{10}$

29. $x + \dfrac{2}{3}x - 2x = \dfrac{x}{6} - \dfrac{1}{8}$

30. $3x + \dfrac{1}{2}x - \dfrac{2}{5}x = \dfrac{x}{10} + \dfrac{7}{20}$

31. $3(1 + 9x) = 6(2 - 4x)$

32. $4(5 - x) = 8(3x + 10)$

33. $3(4x - 1) = 4(2x - 3) + 8$

34. $7(2x - 1) = 5(x + 6) - 13$

35. $5 - 3(2x + 1) = 4(x - 5) + 6$

36. $-2(y + 5) - 4 = 6(y - 2) + 2$

37. $8 + 4(2x - 3) = 5 - (x + 3)$

38. $8(3x + 5) - 9 = 9(x - 2) + 14$

39. $4.7 - 0.3x = 0.5x - 0.1$

40. $5.8 - 0.1x = 0.2x - 0.2$

41. $0.2(x + 3) = 0.1(x - 5)$

42. $0.4(x + 3) = 0.3(x - 6)$

43. $\frac{1}{2}(4-8x)=\frac{1}{3}(4x+7)-3$

44. $3+\frac{1}{4}(x-4)=\frac{2}{5}(2+3x)$

45. $0.6x-22.9=1.5x-18.4$

46. $0.1y+3.8=5.72-0.3y$

47. $0.12n+0.25n-5.895=4.3n$

48. $0.15n+32n-21.0005=10.5n$

49. $0.7(x+14.1)=0.3(x+32.9)$

50. $0.8(x-6.21)=0.2(x-24.84)$

Determine whether each of the following equations is a conditional equation, an identity, or a contradiction. See Examples 8 through 10.

51. $2(3x-1)+5=3$

52. $-2x+13=-2(x-7)$

53. $5x+13=-2(x-7)+3$

54. $3x+9=-3(x-3)+6x$

55. $7(x-1)=-3(3-x)+4x$

56. $3(x-2)+4x=6(x-1)+x$

57. $5(x+1)=3(x+1)+2(x+1)$

58. $8x-20+x=-3(5-2x)+3(x-4)$

59. $2x+3x=5.2(3-x)$

60. $5.2x+3.4x=0.2(x-0.42)$

Solve each of the following word problems.

61. **Construction:**   A farmer is putting a shed on his property. He has two designs. One uses wood and would cost $40 per square foot plus an extra $12,000 in materials. The other design is metal and would cost $55 per square foot plus an additional $13,000. Both sheds are the same size, and the wood shed costs $\frac{3}{4}$ what the metal shed costs. Solve the following equation to determine how many square feet the shed will be.

$$40x+12,000=\frac{3}{4}(55x+13,000)$$

62. **Renting Buildings:**   Two rival shoe companies want to rent the same empty building for their office and shipping space. Schulster's Shoes needs 1000 square feet for offices, 600 for shipping, and another 6 square feet for every packaged box of shoes. Shoes, Shoes, Shoes! needs 750 square feet for offices, 400 for shipping, and 9 square feet per packaged box of shoes. If only one company will fit exactly into the empty building and they both plan to have the same amount of inventory, solve the following equation to determine how many boxes of shoes both companies hope to have at any given time.

$$1000+600+6x=750+400+9x$$

63. Ice Cream: An ice cream shop is having a special "Ice Cream Sunday" event in which they are giving away giant mixed sundaes of 3 scoops of vanilla ice cream and 2 scoops of chocolate. If they have 24 gallons of chocolate and 36 gallons of vanilla to start with, solve the given equation to determine how many sundaes they will have made when they run out of ice cream. (For this problem, we assume a gallon = 20 scoops.)

$$36 - \frac{1}{20}(3x) = 24 - \frac{1}{20}(2x)$$

64. Music: A guitarist and a drummer are getting ready for a gig. The length of the gig will depend on how much material they have prepared. For every hour of the show, the guitarist must practice for 5 days and the drummer for 3 days. Since the guitarist knows some of the songs, he saves 3 days of practice time. The drummer hurts his hand and loses 3 days of practice. If they plan to start and finish practicing at the same time, solve the following equation to determine how long the show will be $5x - 3 = 3x + 3$

Use a calculator to help solve the linear equations.

65. $0.17x - 23.0138 = 1.35x + 36.234$

66. $48.512 - 1.63x = 2.58x + 87.63553$

67. $0.32(x + 14.1) = 2.47x + 2.21795$

68. $1.6(9.3 + 2x) = 0.2(3x + 133.94)$

8.4 Applications: Number Problems and Consecutive Integers

Objective A   **Number Problems**

Previously, we have discussed translating English phrases into algebraic expressions, and learned the Pólya four-step process as an approach to problem solving. Now we will use those skills to read number problems and translate the sentences and phrases in the problem into a related equation. The solution of this equation will be the solution to the problem.

Example 1

Number Problems

If a number is decreased by 36 and the result is 76 less than twice the number, what is the number?

Solution

Let n = the unknown number.

a number is decreased by 36	the result is	76 less than twice the number
$n - 36$	$=$	$2n - 76$

$$n - 36 - n = 2n - 76 - n$$
$$-36 = n - 76$$
$$-36 + 76 = n - 76 + 76$$
$$40 = n$$

The number is 40.

Example 2

Number Problems

Three times the sum of a number and 5 is equal to twice the number plus 5. Find the number.

Solution

Let x = the unknown number.

3 times the sum of a number and 5	is equal to	twice the number plus 5
$3(x + 5)$	$=$	$2x + 5$

$$3x + 15 = 2x + 5$$
$$3x + 15 - 2x = 2x + 5 - 2x$$
$$x + 15 = 5$$
$$x + 15 - 15 = 5 - 15$$
$$x = -10$$

The number is −10.

Example 3

Number Problems

One integer is 4 more than three times a second integer. Their sum is 24. What are the two integers?

Solution

Let n = the second integer, then $3n + 4$ = the first integer.

$$(\text{1st integer}) + (\text{2nd integer}) = 24 \qquad \text{Their sum is 24.}$$
$$(3n + 4) + n = 24$$
$$4n + 4 = 24$$
$$4n + 4 - 4 = 24 - 4$$
$$4n = 20$$
$$\frac{4n}{4} = \frac{20}{4}$$
$$n = 5$$
$$3n + 4 = 19$$

The two integers are 5 and 19.

***Now work margin exercises* 1 through 3.**

Objective B Consecutive Integers

Remember that the set of **integers** consists of the whole numbers and their opposites.

$$\mathbb{Z} = \{\ldots, -4, -3, -2, -1, 0, 1, 2, 3, 4, \ldots\}$$

Even integers are integers that are divisible by 2. The even integers are

$$\{\ldots, -6, -4, -2, 0, 2, 4, 6, \ldots\}.$$

Odd integers are integers that are not even. If an odd integer is divided by 2 the remainder will be 1. The odd integers are

$$\{\ldots, -5, -3, -1, 1, 3, 5, \ldots\}.$$

notes

In this discussion we will be dealing only with integers. Therefore, if you get a result that has a fraction or decimal number (not an integer), you will know that an error has been made and you should correct some part of your work.

1. If a number is decreased by 28 and the result is 92 less than twice the number, what is the number?

2. Six times the sum of a number and 3 is equal to 3 times the number plus 3. Find the number.

3. One integer is 3 more than 4 times a second integer. Their sum is 28. What are the two integers?

The following terms and the ways of representing the integers must be understood before attempting the problems.

Consecutive Integers

Integers are **consecutive** if each is 1 more than the previous integer. Three consecutive integers can be represented as

$$n, \ n+1, \ \text{and} \ n+2$$

where n is an integer.

Consecutive Even Integers

Even integers are **consecutive** if each is 2 more than the previous even integer. Three consecutive even integers can be represented as

$$n, \ n+2, \ \text{and} \ n+4$$

where n is an **even** integer.

Consecutive Odd Integers

Odd integers are **consecutive** if each is 2 more than the previous odd integer. Three consecutive odd integers can be represented as

$$n, \ n+2, \ \text{and} \ n+4$$

where n is an **odd** integer.

Note that consecutive even and consecutive odd integers are represented in the same way:
$$n, \ n+2, \ \text{and} \ n+4.$$

The value of the first integer, n, determines whether the remaining integers are odd or even. For example,

$\underline{n \text{ is odd}}$		$\underline{n \text{ is even}}$
If $\quad n = 11$	**or**	If $\quad n = 36$
then $\ n+2 = 13$		then $\ n+2 = 38$
and $\ n+4 = 15$		and $\ n+4 = 40$

Example 4

Consecutive Integers

The sum of three consecutive **odd** integers is −3. What are the integers?

Solution

Let $n =$ the first odd integer,
then $n + 2 =$ the second odd integer
and $n + 4 =$ the third odd integer.
Set up and solve the related equation.

$$(\text{1st integer}) + (\text{2nd integer}) + (\text{3rd integer}) = -3$$

$$n + (n + 2) + (n + 4) = -3$$

$$3n + 6 = -3$$

$$3n + 6 - 6 = -3 - 6$$

$$3n = -9$$

$$\frac{3n}{3} = \frac{-9}{3}$$

$$n = -3$$

$$n + 2 = -1$$

$$n + 4 = 1$$

The three consecutive odd integers are −3, −1, and 1.

Now work margin exercise 4.

4. The sum of three consecutive odd integers is −9. What are the integers?

Example 5

Consecutive Integers

Find three consecutive integers such that the sum of the first and third is 76 less than three times the second.

Solution

Let $n =$ the first integer,
then $n + 1 =$ the second integer
and $n + 2 =$ the third integer.
Set up and solve the related equation.

5. Find three consecutive even integers such that 6 less than three times the first is equal to the sum of the other two.

$$(\text{1st integer}) + (\text{3rd integer}) = 3(\text{2nd integer}) - 76$$
$$n + (n+2) = 3(n+1) - 76$$
$$2n + 2 = 3n + 3 - 76$$
$$2n + 2 = 3n - 73$$
$$2n + 2 + 73 = 3n - 73 + 73$$
$$2n + 75 = 3n$$
$$2n + 75 - 2n = 3n - 2n$$
$$75 = n$$
$$76 = n + 1$$
$$77 = n + 2$$

The three consecutive integers are 75, 76, and 77.

Check:

$$75 + 77 = 152 \quad \text{and} \quad 3(76) - 76 = 228 - 76 = 152$$

Now work margin exercise 5.

Objective C **Other Applications**

As you learn more abstract mathematical ideas, you will find that you will use these ideas and the related processes to solve a variety of everyday problems as well as problems in specialized fields of study. Generally, you may not even be aware of the fact that you are using your mathematical knowledge. However, these skills and ideas will be part of your thinking and problem solving techniques for the rest of your life.

6. Jim plans to budget $\frac{3}{7}$ of his monthly income to send his son Taylor to private school. If the school he'd like Taylor to attend costs $1200 a month, what monthly income does Jim need to be able to afford the school?

Example 6

Living Expenses

Joe wants to budget $\frac{2}{5}$ of his monthly income for rent. He found an apartment he likes for $800 a month. What monthly income does he need to be able to afford this apartment?

Solution

Let $x =$ Joe's monthly income, then $\frac{2}{5}x =$ rent.

$$\frac{2}{5}x = 800$$
$$\frac{5}{2} \cdot \frac{2}{5}x = \frac{5}{2} \cdot \frac{800}{1}$$
$$x = 2000$$

Joe's monthly income should be $2000.

Now work margin exercise 6.

Example 7

School Supplies

A student bought a calculator and a textbook for a total of $200.80 (including tax). The textbook cost $20.50 more than the calculator. He then challenged a friend to calculate the cost of each item.

Solution

His friend proceeded as follows.
Let x = cost of the calculator,
then $x + 20.50$ = cost of the textbook.
The equation to be solved is:

$$\underbrace{x + 20.50}_{\substack{\text{cost of} \\ \text{textbook}}} + \underbrace{x}_{\substack{\text{cost of} \\ \text{calculator}}} = \overbrace{200.80}^{\text{total spent}}$$

$$2x + 20.50 = 200.80$$

$$2x + 20.50 - 20.50 = 200.80 - 20.50$$

$$2x = 180.30$$

$$\frac{2x}{2} = \frac{180.30}{2}$$

$$x = 90.15 \qquad \text{cost of calculator}$$

$$x + 20.50 = 110.65 \qquad \text{cost of textbook}$$

The calculator costs $90.15 and the textbook costs $90.15 + $20.50 = $110.65, with tax included in each price.

Now work margin exercise 7.

7. Graeme bought bagpipes and a kilt for a total of $507.30 (including tax). If the bagpipes cost $70.40 more than the kilt, what was the cost of each item?

> **Read each problem carefully, translate the various phrases into algebraic expressions, set up an equation, and solve the equation. See Examples 1 through 7.**

1. Five less than a number is equal to 13 decreased by the number. Find the number.

2. Three less than twice a number is equal to the number. What is the number?

3. Thirty-six is 4 more than twice a certain number. Find the number.

4. Fifteen decreased by twice a number is 27. Find the number.

5. Seven times a certain number is equal to the sum of twice the number and 35. What is the number?

6. The difference between twice a number and 3 is equal to 6 decreased by the number. Find the number.

7. Fourteen more than three times a number is equal to 6 decreased by the number. Find the number.

8. Two added to the quotient of a number and 7 is equal to −3. What is the number?

9. The quotient of twice a number and 5 is equal to the number increased by 6. What is the number?

10. Three times the sum of a number and 4 is equal to −9. Find the number.

11. Four times the difference between a number and 5 is equal to the number increased by 4. What is the number?

12. When 17 is added to six times a number, the result is equal to 1 plus twice the number. What is the number?

13. If the sum of twice a number and 5 is divided by 11, the result is equal to the difference between 4 and the number. Find the number.

14. If the difference between a number and 21 is divided by 2, the result is 4 times the number. What is the number?

15. Twice a number increased by three times the number is equal to 4 times the sum of the number and 3. Find the number.

16. Twice the difference between a number and 10 is equal to 6 times the number plus 16. What is the number?

17. The sum of two consecutive odd integers is 60. What are the integers?

18. The sum of two consecutive even integers is 78. What are the integers?

19. Find three consecutive integers whose sum is 69.

20. Find three consecutive integers whose sum is 93.

21. The sum of four consecutive integers is 74. What are the integers?

22. Find four consecutive integers whose sum is 90.

23. 171 minus the first of three consecutive integers is equal to the sum of the second and third. What are the integers?

24. If the first of three consecutive integers is subtracted from 120, the result is the sum of the second and third. What are the integers?

25. Four consecutive integers are such that if 3 times the first is subtracted from 208, the result is 50 less than the sum of the other three. What are the integers?

26. Find two consecutive integers such that twice the first plus three times the second equals 83.

27. Find three consecutive even integers such that the first plus twice the second is 54 less than four times the third.

28. Find three consecutive odd integers such that 4 times the first is 44 more than the sum of the second and third.

29. Find three consecutive even integers such that if the first is subtracted from the sum of the second and third, the result is 66.

30. Find three consecutive even integers such that their sum is 168 more than the second.

31. Find three consecutive odd integers such that the sum of twice the first and three times the second is 7 more than twice the third.

32. Find three consecutive even integers such that the sum of three times the first and twice the third is twenty less than six times the second.

33. School Supplies: A mathematics student bought a calculator and a textbook for a course in statistics. If the text costs $49.50 more than the calculator, and the total cost for both was $125.74, what was the cost of each item?

34. Electronics: The total cost of a computer flash drive and a color printer was $225.50, including tax. If the cost of the flash drive was $170.70 less than the printer, what was the cost of each item?

35. Real Estate: A real estate agent says that the current value of a 25-year old home is $90,000 more than twice its value when it was new. If the current value is $310,000, what was the value of the home when it was new?

36. Guitars: On average the number of electric guitars sold in Texas each year is 91,399, which is seven times the average number of guitars sold each year in Wyoming. How many electric guitars, on average, are sold each year in Wyoming?

37. Classic Cars: A classic car is now selling for $1500 more than three times its original price. If the selling price is now $12,000, what was the car's original price?

38. Mail: On August 24, the Fernandez family received 19 pieces of mail, consisting of magazines, bills, letters, and ads. If they received the same number of magazines as letters, three more bills than letters, and five more ads than bills, how many magazines did they receive?

39. Class Size: A high school graduating class is made up of 542 students. If there are 56 more girls than boys, how many boys are in the class?

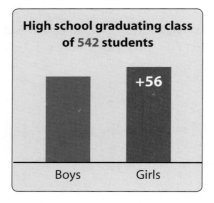

High school graduating class of **542** students

Boys   Girls

40. Golfing: Lucinda bought two buckets of golf balls for the driving range. She gave the pro-shop clerk a 50-dollar bill and received $10.50 in change. What was the cost of one bucket of golf balls? (Tax was included.)

41. Guitars: A guitar manufacturer spent $158 million on the production of acoustic and electric guitars last year. If the amount the company spent producing acoustic guitars was $68 million more than it spent on producing electric guitars, how much did the company spend producing electric guitars?

42. Text Messages: Sarah's text messaging plan costs $10 for the first 500 messages and 20¢ for each additional text message. If she owes $15.40 for text messaging in the month of May, how many text messages did she send that month?

43. Rental Cars: If the rental price on a car is a fixed price per day plus $0.28 per mile, what was the fixed price per day if the total paid was $140 and the car was driven for 250 miles in two days?

44. Rental Cars: The U-Drive Company charges $20 per day plus 22¢ per mile driven. For a one-day trip, Louis paid a rental fee of $66.20. How many miles did he drive?

45. Carpentry: A 29 foot board is cut into three pieces at a sawmill. The second piece is 2 feet longer than the first and the third piece is 4 feet longer than the second. What are the lengths of the three pieces?

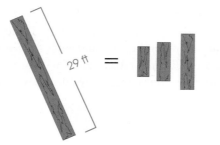

46. Triangles: The three sides of a triangle are n, $4n - 4$, $2n + 7$ (as shown in the figure below). If the perimeter of the triangle is 59 cm, what is the length of each side?

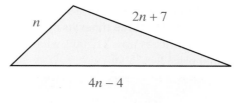

n $2n + 7$

$4n - 4$

47. Construction: Joe Johnson decided to buy a lot and build a house on the lot. He knew that the cost of constructing the house was going to be $25,000 more than the cost of the lot. He told a friend that the total cost was going to be $275,000. As a test to see if his friend remembered the algebra they had together in school, he challenged his friend to calculate what he paid for the lot and what he was going to pay for the house. What was the cost of the lot and the cost of the house?

48. Triangles: The three sides of a triangle are x, $3x - 1$, and $2x + 5$ (as shown in the figure below). If the perimeter of the triangle is 64 inches, what is the length of each side? (**Reminder:** The perimeter of a triangle is the sum of the lengths of the sides.)

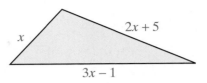

Make up your own word problem that might use the given equation in its solution. Be creative! Then solve the equation and check to see that the answer is reasonable.

49. $5x - x = 8$

50. $2x + 3 = 9$

51. $n + (n + 1) = 33$

52. $n + (n + 4) = 3(n + 2)$

53. $3(n + 1) = n + 53$

54. $2(n + 2) - 6 = n + 4 - n$

Writing & Thinking

55. **a** How would you represent four consecutive odd integers?

b. How would you represent four consecutive even integers?

c. Are these representations the same? Explain.

Objectives

A Understand the concept of a formula, and recognize several common formulas.

B Evaluate formulas for given values of the variables.

C Solve formulas for specified variables in terms of the other variables.

8.5 Working with Formulas

Objective A Formulas

Formulas are general rules or principles stated mathematically. There are many formulas in such fields of study as business, economics, medicine, physics, and chemistry as well as mathematics. Some of these formulas and their meanings are shown here.

notes

Be sure to use the letters just as they are given in the formulas. In mathematics, there is little or no flexibility between capital and small letters as they are used in formulas. In general, capital letters have special meanings that are different from corresponding small letters. For example, capital **A** may mean the area of a triangle and small **a** may mean the length of one side, two completely different ideas.

Listed here are a few formulas with a description of the meaning of each formula. There are of course many more formulas, some of which you will see in the exercises.

Formula	Meaning
$I = Prt$	The **simple interest** I earned by investing money is equal to the product of the principal P times the rate of interest r times the time t in one year or part of a year.
$C = \dfrac{5}{9}(F - 32)$	**Temperature** in degrees Celsius C equals $\dfrac{5}{9}$ times the difference between the Fahrenheit temperature F and 32.
$d = rt$	The **distance traveled** d equals the product of the rate of speed r and the time t.
$P = 2l + 2w$	The **perimeter** P **of a rectangle** is equal to twice the length l plus twice the width w.
$L = 2\pi rh$	The **lateral surface area** L (top and bottom not included) of a cylinder is equal to 2π times the radius r of the base times the height h.
$F = ma$	In physics, the **force** F acting on an object is equal to its mass m times its acceleration a.

Formula	Meaning
$\alpha + \beta + \gamma = 180°$	The **sum of the angles** ($\alpha, \beta,$ and γ) **of a triangle** is 180°. **Note:** $\alpha, \beta,$ and γ are the Greek lowercase letters alpha, beta, and gamma, respectively.
$P = a + b + c$	The **perimeter** P **of a triangle** is equal to the sum of the lengths of the three sides $a, b,$ and c.

Table 1

Objective B Evaluating Formulas

If you know values for all but one variable in a formula, you can substitute those values and find the value of the unknown variable using the techniques for solving equations discussed earlier in this chapter. This section presents a variety of formulas from real-life situations, with complete descriptions of the meanings of these formulas. Working with these formulas will help you become familiar with a wide range of applications of algebra and provide practice in solving equations.

Example 1

Evaluating Formulas

A **note** is a loan for a period of 1 year or less, and the interest earned (or paid) is called **simple interest**. (Simple interest was first introduced in Chapter 4). A note involves only one payment at the end of the term of the note and includes both principal and interest. The formula for calculating simple interest is:

$$I = Prt$$

where I = **interest** (earned or paid)

  P = **principal** (the amount invested or borrowed)

  r = **rate of interest** (stated as an annual or yearly rate in percent form)

  t = **time** (one year or part of a year)

Note: The rate of interest is usually given in percent form and converted to decimal or fraction form for calculations. **For the purpose of calculations, we will use 360 days in one year and 30 days in a month.** Before the use of computers, this was common practice in business and banking.

Maribel loaned $5000 to a friend for 90 days at an annual interest rate of 8%. How much will her friend pay her at the end of the 90 days?

Solution

Here, $P = \$5000,$

  $r = 8\% = 0.08,$

  $t = 90 \text{ days} = \dfrac{90}{360} \text{ year} = \dfrac{1}{4} \text{ year}.$

1. You get a loan for $2000 for 60 days at an annual interest rate of 6%. How much do you have to pay at the end of the loan period, 60 days?

Find the interest by substituting in the formula $I = Prt$ and evaluating.

$$I = 5000 \cdot \overset{0.02}{\cancel{0.08}} \cdot \frac{1}{\cancel{4}}$$

$$= 5000 \cdot 0.02$$

$$= \$100.00$$

The interest is $100 and the amount to be paid at the end of 90 days is principal + interest = $5000 + $100 = $5100.

Now work margin exercise 1.

Example 2

Evaluating Formulas

Given the formula $C = \frac{5}{9}(F - 32)$, first find C if $F = 212°$ (212 degrees Fahrenheit) and then find F if $C = 20°$ (20 degrees Celsius).

Solution

$F = 212°$, so substitute 212 for F in the formula.

$$C = \frac{5}{9}(212 - 32)$$

$$= \frac{5}{9}(180)$$

$$= 100$$

That is, 212°F is the same as 100°C. Water will boil at 212°F at sea level. This means that if the temperature is measured in degrees Celsius instead of degrees Fahrenheit, water will boil at 100°C at sea level.

2. Given $C = \frac{5}{9}(F - 32)$, find F if $C = 50°$.

$C = 20°$, so substitute 20 for C in the formula.

$$20 = \frac{5}{9}(F - 32) \qquad \text{Now solve for } F.$$

$$\frac{9}{5} \cdot 20 = \frac{9}{5} \cdot \frac{5}{9}(F - 32) \qquad \text{Multiply both sides by } \frac{9}{5}.$$

$$36 = F - 32 \qquad \text{Simplify.}$$

$$36 + 32 = F - 32 + 32 \qquad \text{Add 32 to both sides.}$$

$$68 = F \qquad \text{Simplify.}$$

That is, a temperature of 20°C is the same as a comfortable spring day temperature of 68°F.

Now work margin exercise 2.

Example 3

Evaluating Formulas

The lifting force F exerted on an airplane wing is found by multiplying some constant k by the area A of the wing's surface and by the square of the plane's velocity v. The formula is $F = kAv^2$. Find the force on a plane's wing during take-off if the area of the wing is 120 ft², k is $\frac{4}{3}$, and the plane is traveling 80 miles per hour during take off.

Solution

We know that $k = \frac{4}{3}$, $A = 120$, and $v = 80$. Substitution gives

$$F = \frac{4}{3} \cdot 120 \cdot 80^2$$
$$= \frac{4}{3} \cdot 120 \cdot 6400$$
$$= 160 \cdot 6400$$
$$F = 1{,}024{,}000 \text{ lb}$$

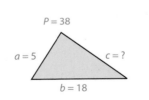 The force is measured in pounds.

Now work margin exercise 3.

Example 4

Evaluating Formulas

The perimeter of a triangle is 38 feet. One side is 5 feet long and a second side is 18 feet long. How long is the third side?

Solution

Using the formula $P = a + b + c$, substitute $P = 38$, $a = 5$, and $b = 18$. Then solve for the third side.

$$P = a + b + c$$
$$38 = 5 + 18 + c$$
$$38 = 23 + c$$
$$38 - 23 = 23 + c - 23$$
$$15 = c$$

$P = 38$

$a = 5$ $c = ?$

$b = 18$

The third side is 15 feet long.

Now work margin exercise 4.

3. Use the formula $F = kAv^2$ to find the force exerted on a plane's wing during take-off if the area of the wing is 100 ft², $k = \frac{5}{4}$, and $v = 90$ mph.

4. Two angles of a triangle measure 65° and 85°. What is the measure of the third angle of the triangle?

Objective C Solving Formulas for Different Variables

We say that the formula $d = rt$ is "solved for" d in terms of r and t. Similarly, the formula $A = \frac{1}{2}bh$ is solved for A in terms of b and h, and the formula $P = R - C$ (profit is equal to revenue minus cost) is solved for P in terms of R and C. Many times we want to use a certain formula in another form. We want the formula "solved for" some variable other than the one given in terms of the remaining variables. **Treat the variables just as you would constants in solving linear equations.** Study the following examples carefully.

5. Given $P = IV$, solve for I in terms of P and V.

Example 5

Solving for Different Variables

Given $d = rt$, solve for t in terms of d and r. We want to represent the time in terms of distance and rate. We will use this concept later in word problems.

Solution

$$d = rt \qquad \text{Treat } r \text{ and } d \text{ as if they were constants.}$$

$$\frac{d}{r} = \frac{rt}{r} \qquad \text{Divide both sides by } r.$$

$$\frac{d}{r} = t \qquad \text{Simplify.}$$

Now work margin exercise 5.

6. Given $P = \dfrac{I}{rt}$, solve for t in terms of I, r, and P.

Example 6

Solving for Different Variables

Given $V = \dfrac{k}{P}$, solve for P in terms of V and k.

Solution

$$V = \frac{k}{P}$$

$$P \cdot V = P \cdot \frac{k}{P} \qquad \text{Multiply both sides by } P.$$

$$PV = k \qquad \text{Simplify.}$$

$$\frac{PV}{V} = \frac{k}{V} \qquad \text{Divide both sides by } V.$$

$$P = \frac{k}{V} \qquad \text{Simplify.}$$

Now work margin exercise 6.

© Hawkes Learning Systems. All rights reserved.

Example 7

Solving for Different Variables

Given $C = \dfrac{5}{9}(F - 32)$ as in Example 2, solve for F in terms of C.
This would give a formula for finding Fahrenheit temperature given a Celsius temperature value.

Solution

$$C = \frac{5}{9}(F - 32)$$ Treat C as a constant.

$$\frac{9}{5} \cdot C = \frac{9}{5} \cdot \frac{5}{9}(F - 32)$$ Multiply both sides by $\dfrac{9}{5}$.

$$\frac{9}{5}C = F - 32$$ Simplify.

$$\frac{9}{5}C + 32 = F - 32 + 32$$ Add 32 to both sides.

$$\frac{9}{5}C + 32 = F$$ Simplify.

Thus $F = \dfrac{9}{5}C + 32$ is solved for F, and $C = \dfrac{5}{9}(F - 32)$ is solved for C.

These are two forms of the same formula.

Now work margin exercise 7.

Example 8

Solving for Different Variables

Given the equation $2x + 4y = 10$,
1. solve for x in terms of y, and then
2. solve for y in terms of x.

This equation is typical of the algebraic equations that we will discuss in later sections.

Solution

1. Solving for x yields

$$2x + 4y = 10$$ Treat $4y$ as a constant.

$$2x + 4y - 4y = 10 - 4y$$ Subtract $4y$ from both sides.
                   (This is the same as adding $-4y$.)

$$2x = 10 - 4y$$ Simplify.

$$\frac{2x}{2} = \frac{10 - 4y}{2}$$ Divide both sides by 2.

$$x = \frac{10}{2} - \frac{4y}{2}$$

$$x = 5 - 2y$$ Simplify.

7. Given $y = \dfrac{2}{5}(x + 3)$ solve for x.

8. Given the equation $16y + 25z = 400,$

    a. solve for y in terms of z, and then

    b. solve for z in terms of y.

9. Given $4x + 8y + 12z = 16$ solve for x.

---

2. Solving for y yields

$$2x + 4y = 10 \qquad \text{Treat } 2x \text{ as a constant.}$$

$$2x + 4y - 2x = 10 - 2x \qquad \text{Subtract } 2x \text{ from both sides.}$$

$$4y = 10 - 2x \qquad \text{Simplify.}$$

$$\frac{4y}{4} = \frac{10 - 2x}{4} \qquad \text{Divide both sides by 4.}$$

$$y = \frac{10}{4} - \frac{2x}{4}$$

$$y = \frac{5}{2} - \frac{x}{2} \qquad \text{Simplify.}$$

or we can write

$$y = \frac{5 - x}{2} \quad \text{or} \quad y = -\frac{1}{2}x + \frac{5}{2}. \qquad \text{All forms are correct.}$$

Now work margin exercise 8.

Completion Example 9

Solving for a Variable

Given $3x - y = 15$, solve for y in terms of x.

Solution

Solving for y gives

$$3x - y = 15$$

$$3x - y - 3x = 15 - 3x \qquad \underline{\hspace{5cm}}$$

$$-y = 15 - 3x$$

$$-1(-y) = -1(15 - 3x) \qquad \underline{\hspace{5cm}}$$

$$y = -15 + 3x \qquad \underline{\hspace{5cm}}$$

$$\text{or} \qquad y = 3x - 15$$

Now work margin exercise 9.

Completion Example Answers

9.

$$3x - y = 15$$

$$3x - y - 3x = 15 - 3x \qquad \underline{\text{Subtract } 3x \text{ from both sides.}}$$

$$-y = 15 - 3x$$

$$-1(-y) = -1(15 - 3x) \qquad \underline{\text{Multiply both sides by } -1 \text{ (or divide both sides by } -1).}$$

$$y = -15 + 3x \qquad \underline{\text{Simplify using the distributive property.}}$$

$$\text{or} \qquad y = 3x - 15$$

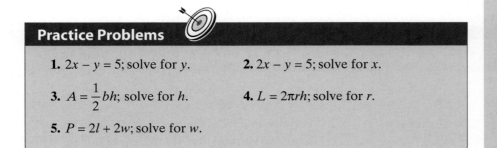

Practice Problems

1. $2x - y = 5$; solve for y. **2.** $2x - y = 5$; solve for x.

3. $A = \dfrac{1}{2}bh$; solve for h. **4.** $L = 2\pi rh$; solve for r.

5. $P = 2l + 2w$; solve for w.

Practice Problem Answers

1. $y = 2x - 5$ **2.** $x = \dfrac{y + 5}{2}$ **3.** $h = \dfrac{2A}{b}$

4. $r = \dfrac{L}{2\pi h}$ **5.** $w = \dfrac{P - 2l}{2}$

Exercises 8.5

Refer to Example 1 for information concerning simple interest and the related formula *I* = *Prt*. (Note: For ease of calculation, assume 360 days in a year and 30 days in each month.)

Simple Interest

1. You want to borrow $4000 at 12% for only 90 days. How much interest would you pay?

2. For how many days must you leave $1000 in a savings account at 5.5% to earn $11.00 in interest?

3. What principal would you need to invest to earn $450 in simple interest in 6 months if the interest rate was 9%?

4. After 30 days, Gustav received $25 in simple interest on his savings account of $12,000. What was the interest rate?

5. A savings account of $3500 is left for 9 months and draws simple interest at a rate of 7%.

   a. How much interest is earned?

   b. What is the balance in the account at the end of the 9 months?

6. Tim just deposited $2562.50 to pay off a 3 month loan of $2500.

   a. How much of what he deposited was interest on the loan?

   b. What rate of interest was he charged?

In the following application problems, read the descriptive information carefully and then substitute the values given in the problem for the corresponding variables in the formulas. Evaluate the resulting expression for the unknown variable.

Velocity

If an object is shot upward with an initial velocity v_0 in feet per second, the velocity v in feet per second is given by the formula $v = v_0 - 32t$, where t is time in seconds. (v_0 is read "v *sub zero*." The 0 is called a subscript.)

7. An object projected upward with an initial velocity of 106 feet per second has a velocity of 42 feet per second. How many seconds have passed?

8. Find the initial velocity of an object if the velocity after 4 seconds is 48 feet per second.

Medicine

In nursing, one procedure for determining the dosage for a child is

$$\text{child's dosage} = \frac{\text{age of child in years}}{\text{age of child} + 12} \cdot \text{adult dosage}$$

9. If the adult dosage of a drug is 20 milliliters, how much should a 3-year-old child receive?

10. If the adult dosage of a drug is 340 milligrams, how much should a 5-year-old child receive?

Investment

The total amount of money in an account with P dollars invested in it is given by the formula $A = P + Prt$, where r is the rate expressed as a decimal and t is time (one year or part of a year).

11. If $1000 is invested at 6% interest, find the total amount in the account after 6 months.

12. How long will it take an investment of $600, at an annual rate of 5%, to be worth $615?

Carpentry

The number N of rafters in a roof or studs in a wall can be found by the formula $N = \dfrac{L}{d} + 1$, where L is the length of the roof or wall and d is the center-to-center distance from one rafter or stud to the next. Note that L and d must be in the same units.

13. How many rafters will be needed to build a roof 26 ft long if they are placed 2 ft apart center-to-center?

14. A wall has studs placed 16 in. apart center-to-center. If the wall is 20 ft long, how many studs are in the wall?

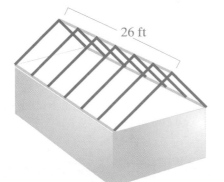

26 ft

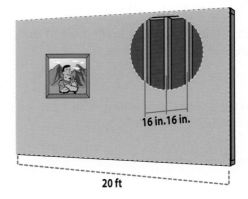

16 in. 16 in.

20 ft

15. How long is a wall if it requires 22 studs placed 16 in. apart center-to-center?

16. What should the center-to-center distance be if you are building a 33 ft long roof using 12 rafters?

Cost

The total cost C of producing x items can be found by the formula $C = ax + k$, where a is the cost per item and k is the fixed costs (rent, utilities, and so on).

17. Find the total cost of producing 30 items if each costs $15 and the fixed costs are $580.

18. The total cost to produce 80 dolls is $1097.50. If each doll costs $9.50 to produce, find the fixed costs.

19. It costs a company $3.60 to produce a calculator. Last week the total costs were $1308. If the fixed costs are $480 weekly, how many calculators were produced last week?

20. Each week an electronics company builds 60 mp3 players for a total cost of $5340. If the fixed costs for a week are $750, what is the cost to produce each mp3 player?

Profit

The profit P is given by the formula $P = R - C$, where R is the revenue and C is the cost.

21. Find the revenue (income) of a company that shows a profit of $3.2 million and costs of $1.8 million.

22. Find the revenue of a company that shows a profit of $3.2 million and costs of $5.7 million.

Depreciation

Many items decrease in value as time passes. This decrease in value is called depreciation. One type of depreciation is called linear depreciation. The value V of an item after t years is given by $V = C - Crt$, where C is the original cost and r is the rate of depreciation expressed as a decimal.

23. If you buy a car for $6000 and it depreciates linearly at a rate of 10% per year, what will be its value after 6 years?

24. A contractor buys a 4 year-old piece of heavy equipment valued at $20,000. If the original cost of this equipment was $25,000, find the rate of depreciation.

Distance, Rate, and Time

The distance traveled d is given by the formula $d = rt$, where r is the rate of speed and t is the time it takes.

25. How long will a truck driver take to travel 350 miles if he averages 50 mph?

26. What is the average rate of speed of a biker who bikes 21.92 miles in 68.5 min?

27. What is Jonathan's average rate of speed if he hikes 10.4 miles in 6.4 hours?

28. How long will it take a train traveling at 40 mph to go 140 miles?

Solve each formula for the indicated variable. See Examples 5 through 9.

29. $P = a + b + c$; solve for b.

30. $P = 3s$; solve for s.

31. $F = ma$; solve for m.

32. $C = \pi d$; solve for d.

33. $A = lw$; solve for w.

34. $P = R - C$; solve for C.

35. $R = np$; solve for n.

36. $v = k + gt$; solve for k.

37. $I = A - P$; solve for P.

38. $L = 2\pi rh$; solve for h.

39. $A = \dfrac{m+n}{2}$; solve for m.

40. $P = a + 2b$; solve for a.

41. $I = Prt$; solve for t.

42. $R = \dfrac{E}{I}$; solve for E.

43. $P = a + 2b$; solve for b.

44. $c^2 = a^2 + b^2$; solve for b^2.

45. $\alpha + \beta + \gamma = 180°$; solve for β.

46. $y = mx + b$; solve for x.

47. $V = lwh$; solve for h.

48. $v = -gt + v_0$; solve for t.

49. $A = \dfrac{1}{2}bh$; solve for b.

50. $R = \dfrac{E}{I}$; solve for I.

51. $A = \pi r^2$; solve for π.

52. $A = \dfrac{R}{2L}$; solve for L.

53. $K = \dfrac{mv^2}{2g}$; solve for g.

54. $x + 4y = 4$; solve for y.

55. $2x + 3y = 6$; solve for y.

56. $3x - y = 14$; solve for y.

57. $5x + 2y = 11$; solve for x.

58. $-2x + 2y = 5$; solve for x.

59. $A = \dfrac{1}{2}h(b+c)$; solve for b.

60. $A = P(1+rt)$; solve for r.

61. $R = \dfrac{3(x-12)}{8}$; solve for x.

62. $-2x - 5 = -3(x + y)$; solve for x. **63.** $3y - 2 = x + 4y + 10$; solve for y. **64.** $V = \dfrac{1}{3}\pi r^2 h$; solve for h.

Make up a formula for each of the following situations.

65. **Concert Tickets:** Each ticket for a concert costs $\$t$ per person and parking costs $\$9.00$. What is the total cost per car C if there are n people in a car?

66. **Car Rentals:** ABC Car Rental charges $\$25$ per day plus $\$0.12$ per mile. What would you pay per day for renting a car from ABC if you were to drive the car x miles in one day?

67. **Computers:** Top-of-The-Line computer company knows that the cost (labor and materials) of producing a computer is $\$325$ per computer per week and the fixed overhead costs (lighting, rent, etc.) are $\$5400$ per week. What are the company's weekly costs of producing n computers per week?

68. **Computers:** If the Top-of-The-Line computer company (see Exercise 67) sells its computers for $\$683$ each, what is its profit per week if it sells the same number n that it produces? (Remember that profit is equal to revenue minus costs, or $P = R - C$.)

69. The formula $z = \dfrac{x - \bar{x}}{s}$ is used extensively in statistics. In this formula, x represents one value in a set of data, $\bar{x}$ represents the average (or mean) of those numbers in the set, and s represents a value called the standard deviation of the numbers. (The standard deviation is a positive number and is a measure of how "spread out" the numbers are.) The values for z are called z-scores, and they measure the number of standard deviation units a number x is from the mean $\bar{x}$.

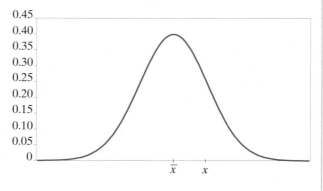

    a. If $\bar{x} = 70$, what will be the z-score for $x = 70$? Does this z-score depend on the value of s? Explain.

    b. For what values of x will the corresponding z-scores be negative?

    c. Calculate your z-score on each of the last two test scores in this class. (Your instructor will give you the mean and standard deviation for each test.) What do these scores tell you about your performance on the two exams?

70. Suppose that, for a particular set of exam scores, $\bar{x} = 72$ and $s = 6$. Find the z-score that corresponds to a score of

    a. 78   **b.** 66   **c.** 81   **d.** 60

8.6 Applications: Distance-Rate-Time, Interest, Average

Objectives

A Solve distance-rate-time problems by using linear equations.

B Solve simple interest problems by using linear equations.

C Solve average problems by using linear equations.

D Solve cost problems by using linear equations.

Objective A   Distance-Rate-Time

Word problems (or applications) are designed to teach you to read carefully, to organize, and to think clearly. Whether or not a particular problem is easy for you depends a great deal on your personal experiences and general reasoning abilities. The problems generally do not give specific directions to add, subtract, multiply, or divide. You must decide what relationships are indicated through careful analysis of the problem.

Problems involving distance usually make use of the relationship indicated by the formula $d = rt$, where d = distance, r = rate, and t = time. A chart or table showing the known and unknown values is quite helpful and is illustrated in the next example.

Example 1

Distance-Rate-Time

A motorist averaged 45 mph for the first part of a trip and 54 mph for the last part of the trip. If the total trip of 303 miles took 6 hours, what was the time for each part?

Solution

Analysis of strategy:

What is being asked for?
Total time minus time for 1st part of trip gives time for 2nd part of trip.

Let t = time for 1st part of trip
         $6 - t$ = time for 2nd part of trip

	rate	x	time	=	distance
1st Part	45		t		$45t$
2nd Part	54		$6-t$		$54(6-t)$

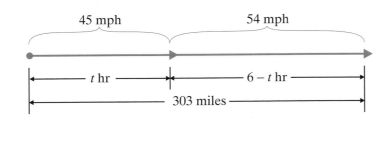

45 mph   54 mph

t hr $6 - t$ hr

303 miles

1. Liang drove 200 miles in 4 hours. His average speed for the first part of the trip was 60 mph and his average speed for the second part of the trip was 45 mph. How long did each part of the trip take him?

1st part distance	+	2nd part distance	=	total distance	Form an equation relating the given infromation.

$$45t \ + \ 54(6-t) \ = \ 303 \qquad \text{Solve the equation.}$$

$$45t \ + \ 324 - 54t \ = \ 303$$

$$324 - 9t \ = \ 303$$

$$-9t \ = \ -21$$

$$t \ = \frac{-21}{-9} = \frac{7}{3} \ \text{hr} \qquad \text{1st part of the trip}$$

$$6 - t \ = 6 - \frac{7}{3} = \frac{11}{3} \ \text{hr} \qquad \text{2nd part of the trip}$$

Check:

$$45 \cdot \frac{7}{3} = 15 \cdot 7 = 105 \text{ miles (1st part)}$$

$$54 \cdot \frac{11}{3} = 18 \cdot 11 = 198 \text{ miles (2nd part)}$$

$$105 + 198 = 303 \text{ miles in total}$$

The first part took $\frac{7}{3}$ hr or $2\frac{1}{3}$ hr.

The second part took $\frac{11}{3}$ hr or $3\frac{2}{3}$ hr.

***Now work margin exercise* 1.**

Objective B **Interest**

To solve problems related to interest on money invested for one year, you need to know the basic relationship among the principal P (amount invested), the annual rate of interest r, and the amount of interest I (money earned). This relationship is described in the formula $Pr = I$. (This is the formula for simple interest, $I = Prt$, with $t = 1$.) We use this relationship in Example 2.

Example 2

Interest

Kara has had $40,000 invested for one year, some in a savings account which paid 7% and the rest in a high-risk stock which yielded 12% for the year. If her interest income last year was $3550, how much did she have in the savings account and how much did she invest in the stock?

Solution

Let x = amount invested at 7%

$40,000 - x$ = amount invested at 12%

Total amount invested minus amount invested at 7% represents amount invested at 12%.

	principal	·	rate	=	interest
Savings Account	x		0.07		$0.07(x)$
Stock	$40{,}000 - x$		0.12		$0.12(40{,}000 - x)$

$$\underbrace{0.07(x)}_{\text{interest at 7\%}} + \underbrace{0.12(40{,}000-x)}_{\text{interest at 12\%}} = \underbrace{3550}_{\text{total interest}}$$

$$
\begin{aligned}
7x + 12(40{,}000 - x) &= 355{,}000 \\
7x + 480{,}000 - 12x &= 355{,}000 \\
-5x &= -125{,}000 \\
x &= 25{,}000 \\
40{,}000 - x &= 15{,}000
\end{aligned}
$$

Multiply both sides of the equation by 100 to eliminate the decimal.

amount invested at 7%

amount invested at 12%

Check:

$25{,}000(0.07) = 1750$ and $15{,}000(0.12) = 1800$
and $\$1750 + \$1800 = \$3550.$
Kara had \$25,000 in the savings account at 7% interest and invested \$15,000 in the stock at 12% interest.

Now work margin exercise 2.

2. Mark has had \$15,000 invested for 1 year, some in a low-risk stock yielding 4% and the rest in a high-risk stock yielding 15%. If his investment has earned \$1997 in interest, how much did he invest in the low-risk stock and how much in the high-risk stock?

Objective C **Average (or Mean)**

The **average** (or **mean**) of a set of numbers was defined and discussed previously. In this section we use the concept of average to find unknown numbers.

Example 3

Average (or Mean)

Suppose that you have scores of 85, 92, 82 and 88 on four exams in your English class. What score will you need on the fifth exam to have an average of 90?

Solution

Let $x =$ your score on the fifth exam.
The sum of all the scores, including the unknown fifth exam, divided by 5 must equal 90.

3. Suppose you earned the following four scores on the first four biology tests of the semester, 90, 95, 97, and 90, respectively. You can exempt the final if your average on the five tests is 85 or higher. What is the minimum score you can earn on the fifth test and still exempt the final?

$$\frac{85 + 92 + 82 + 88 + x}{5} = 90$$

$$\frac{347 + x}{5} = 90$$

$$5 \cdot \frac{347 + x}{5} = 5 \cdot 90$$

$$347 + x = 450$$

$$x = 103$$

Assuming that each exam is worth 100 points, you cannot attain an average of 90 on the five exams.

Now work margin exercise 3.

Objective D **Cost**

Example 4

Cost

A jeweler paid $350 for a ring. He wants to price the ring for sale so that he can give a 30% discount on the marked selling price and still make a profit of 20% on his cost. What should be the marked selling price of the ring?

Solution

Again, we make use of the relationship:

$S - C = P$ (selling price − cost = profit).
Let x = marked selling price,
then $x - 0.30x$ = actual selling price
and 350 = cost.

actual selling price	−	cost	=	profit	

The actual selling price is the marked selling price minus the 30% discount on the ring.

$$x - 0.30x \quad - \quad 350 \quad = \quad 0.20(350)$$
$$0.70x \quad - \quad 350 \quad = \quad 70$$
$$0.70x \quad = \quad 420$$
$$x \quad = \quad 600$$

The profit is 20% of what he paid originally.

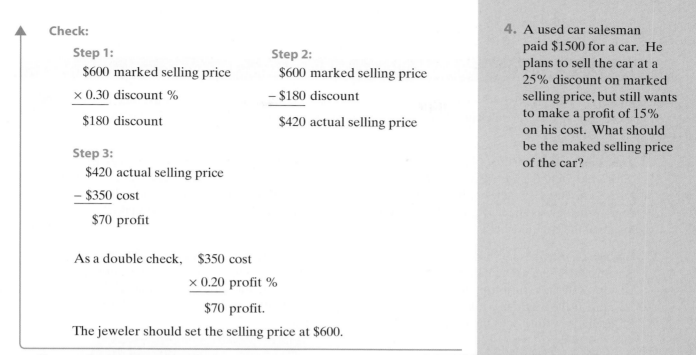

Check:

Step 1:

    $600 marked selling price

$\underline{\times\ 0.30}$ discount %

    $180 discount

Step 2:

    $600 marked selling price

$\underline{-\ \$180}$ discount

    $420 actual selling price

Step 3:

    $420 actual selling price

$\underline{-\ \$350}$ cost

     $70 profit

As a double check, $350 cost

                   $\underline{\times\ 0.20}$ profit %

                    $70 profit.

The jeweler should set the selling price at $600.

Now work margin exercise 4.

4. A used car salesman paid $1500 for a car. He plans to sell the car at a 25% discount on marked selling price, but still wants to make a profit of 15% on his cost. What should be the maked selling price of the car?

Exercises 8.6

1. **Hiking:** What is Nathan's average rate of speed if he hikes 12.6 miles in 7.5 hours?

2. **Biking:** What is Amy's average rate of speed if she bikes 39.6 miles in 2.2 hours?

3. **Traveling by Car:** Jamie plans to take the scenic route from Los Angeles to San Francisco. Her GPS tells her it is a 420 mile trip. If she figures she'll average 48 mph, how long will the trip take her?

4. **Traveling by Car:** Scott's averaging 60 mph on his drive from St. Louis, MO to Memphis, TN. If the total trip is 285 miles, how long should he expect the drive to take?

5. **Biking:** Jane rides her bike to Lake Junaluska. Going to the lake, she averages 12 mph. On the return trip, she averages 10 mph. If the round trip takes a total of 5.5 hours, how long does the return trip take?

	rate ·	time	= distance
Going	12	$5.5 - t$	
Returning	10	t	

6. **Plane Speeds:** Two planes which are 2475 miles apart fly toward each other. Their speeds differ by 75 mph. If they pass each other in 3 hours, what is the speed of each?

	rate ·	time	= distance
1st plane	r	3	
2nd plane	$r + 75$	3	

7. **Traveling by Car:** Marcus drives from Chicago to Detroit in 6 hours. On the return trip, his speed is increased by 10 mph and the trip takes 5 hours. Find his rate on the return trip. How far apart are the towns?

	rate ·	time	= distance
Going	r	6	
Returning	$r + 10$	5	

8. **Hiking:** Tim and Barb have 8 hours to spend on a mountain hike. They can walk up the trail at an average of 2 mph and can walk down at an average of 3 mph. How long should they plan to hike uphill before turning around?

	rate ·	time	= distance
Uphill	2	x	
Downhill	3	$8 - x$	

9. **Traveling by Car:** The Reeds are moving across Texas. Mr. Reed leaves $3\frac{1}{2}$ hours before Mrs. Reed. If he averages 40 mph and she averages 60 mph, how long will Mrs. Reed have to drive before she overtakes Mr. Reed?

10. **Traveling by Car:** After traveling for 40 minutes, Mr. Koole had to slow to $\frac{2}{3}$ his original speed for the rest of the trip due to heavy traffic. The total trip of 84 miles took 2 hours. Find his original speed.

11. **Train Speeds:** A train leaves Cincinnati at 2:00 PM. A second train leaves the same station in the same direction at 4:00 PM. The second train travels 24 mph faster than the first. If the second train overtakes the first at 7:00 PM, what is the speed of each of the two trains?

12. **Running:** Maria runs through the countryside at a rate of 10 mph. She returns along the same route at 6 mph. If the total trip took 1 hour 36 minutes, how far did she run in total?

13. **Traveling by Car:** The distance from Atlanta, Georgia to Washington, DC is 620 miles. Driving in the middle of the night it takes about 9 hours to get to Washington. Due to higher traffic volume, it takes 2 more hours to travel there during the day. What is the average rate of the driver during the day and during the night? (Round to the nearest whole number.)

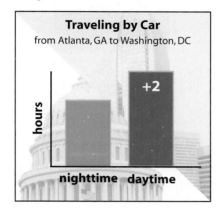

14. **Traveling by Car:** Mr. Kent drove to a conference. The first half of the trip took 3 hours due to traffic. Traffic let up for the second half of the trip and he was able to increase his speed by 20 mph to make sure he got there on time. Find his rates of speed if he traveled 2 hours at the second rate.

15. **Exercising:** Jayden walked to his friend's house at a rate of 4 mph to borrow his friend's bicycle. Coming back home, he rode the bicycle at an average rate of 12 mph. The total time for the round trip was 1 hour 30 minutes. How far away does Jayden's friend live?

16. **Exercising:** Once a week Felicia walks/runs for a total of 6 miles. Felicia spends twice as much time walking as she does running. If she walks at a rate of 4 mph and runs three times faster than she walks, what is the time for each part?

17. **Investing:** Amanda invests $25,000, part at 5% and the rest at 6%. The annual return on the 5% investment exceeds the annual return on the 6% investment by $40. How much did she invest at each rate?

18. **Investing:** Mr. Hill invests $10,000, part at 5.5% and part at 6%. The annual interest from the 5.5% investment exceeds the annual interest from the 6% investment by $251. How much did he invest at each rate?

19. **Investing:** The annual interest earned on a $6000 investment was $120 less than the interest earned on $10,000 invested at 1% less interest per year. What was the rate of interest on each amount?

20. **Investing:** Two investments totaling $16,000 produce an annual income of $1140. One investment yields 6% a year, while the other yields 8% per year. How much is invested at each rate?

21. **Investing:** The annual interest on a $4000 investment exceeds the interest earned on a $3000 investment by $80. The $4000 is invested at a 0.5% higher rate of interest than the $3000. What is the interest rate of each investment?

22. **Investing:** D'Andra makes two investments that total $12,000. One investment yields 8% per year and the other 10% per year. The total interest for one year is $1090. Find the amount invested at each rate.

23. **Investing:** A company invested $42,000 into two simple interest accounts. The annual interest rate on one of the accounts is 4.5% while the rate on the other is 6%. How much should the company invest in each account so that the two accounts will produce an equal annual interest income?

24. **Shopping:** A particular style of shoe costs the dealer $81 per pair. At what price should the dealer mark them so he can sell them at a 10% discount off the selling price and still make a 25% profit?

25. **Investing:** Sebastian would like both of his investments, a total of $12,000, to bring him the same annual interest income. One of his investments is at 5.5% annual interest rate and the other at 7%. Find the amount of money that Sebastian should invest in each account.

26. **Used Car Sales:** A car dealer paid $2850 for a used car. For his upcoming Labor Day sale, he wants to offer a 5% discount off the posted selling price, but would still like to make a 40% profit. What price should he advertise for that car?

27. **Investing:** Gabriella got some money from her grandparents as a graduation present. She decided to invest all of it. Part of the money was invested at a 2.5% interest rate, and the rest at a 4% interest rate. She invested $200 more in the 4% account than the 2.5% account. If her annual interest income was $47, how much did she invest at each rate?

28. **Investing:** Jordan is earning 1.5% interest from money invested in a savings account and 4% interest on a mutual bond fund. If the total of his investments is $18,000 and the annual interest from the savings account is less than the annual interest from the bond by $60, how much has Jordan invested at each rate?

29. **Investing:** A small company invested $20,500 such that a part of the money is in an account with a 4% interest rate and the rest at a 5% rate. The annual interest from the 5% account is $35 more than the interest earned from the 4% account. Find the amount of money the company invested at each rate.

30. **Shopping:** During the month of January, a department store would like to have a sale of 40% off of women's long boots. The store purchased the boots for $33 per pair. How much should the selling price be, if the manager of the store wants to make a 10% profit per pair?

31. **Shopping:** Carla plans on buying a new pair of sandals for the summer. They are on sale for 20% off of the original price. What was the original price of the sandals if she pays $34.83 with 7.5% sales tax?

32. **Investing:** Last year an individual invested some money at a 5% interest rate and $2200 less than that amount at a 6% interest rate. If his interest income was $880, how much did he invested at each rate?

33. Shopping: A store purchased a certain style of leather jacket at $70 per jacket. If the store wants to sell the jackets at a 20% discount and still make a profit of 30%, what should be the marked selling price for each jacket?

34. Electronics: After receiving a 10% off coupon in the mail, Mark decided that it was time to buy a Blu-Ray player. Using the coupon and paying 8% sales tax, the final price came to $213.84. What was the listed price of the Blu-Ray player?

35. Exam Scores: Marissa has five exam scores of 75, 82, 90, 85, and 77 in her chemistry class. What score does she need on the final exam to have an average of 80 (and thus earn a grade of B)? (All exams have a maximum of 100 points.)

36. Exam Scores: Gerald had scores of 80, 92, 89, and 95 on four exams in his algebra class. What score will he need on his fifth exam to have an overall average of 90? (All exams have a maximum of 100 points.)

37. Biking: While riding her bike to the park and back home five times, Stacey timed herself at 60 min, 62 min, 55 min (the wind was helping), 58 min, and 63 min. She had set a goal of averaging 60 minutes for her rides. How many minutes will she need on her sixth ride to attain her goal?

38. Cell Phones: For every 4 week period, Lauren wants to make an average of 6 phone calls per week from her prepaid cell phone. The first week she made 9 phone calls, the second week 6 phone calls, and the third week 5 phone calls. How many phone calls does Lauren need to make in the fourth week to make sure she stays on track with her goal?

39. TV watching: While growing up, Jason was allowed to watch TV an average of 3 hours a day over a one week period. One particular week he watched 1 hour, 2 hours, 1 hour, 3 hours, 3 hours, and 5 hours. How many hours could Jason watch the seventh and last day of the week and still obey his parents?

40. Budgeting: A college student realized that he was spending too much money on video games. For the remaining 5 months of the year his goal is to spend an average of $50 a month towards his hobby. How much can he spend in December, taking into consideration that the other 4 months he spent $70, $25, $105, and $30 respectively?

41. Exam Scores: Wade has scores of 59, 68, 76, 84 and 69 on the first five tests in his social studies class. He knows that the final exam counts as two tests. What score will he need on the final to have an average of 70? (All tests and exams have a maximum of 100 points.)

42. Exam Scores: A statistics student has grades of 86, 91, 95, and 76 on four hour-long exams. What score must he receive on the final exam to have an average of 90 if:

a. the final is equivalent to a single hour-long exam (100 points maximum)?

b. the final is equivalent to two hour-long exams (200 points maximum)?

Objectives

A   Understand and use set-builder notation.

B   Understand and use interval notation.

C   Solve linear inequalities.

D   Solve compound inequalities.

E   Learn how to apply inequalities to solve word problems.

8.7 Linear Inequalities

Objective A   **Sets and Set-Builder Notation**

A **set** is a collection of objects or numbers. The items in the set are called **elements**, and sets are indicated with braces $\{\ \}$, and named with capital letters. If the elements are listed within the braces the set is said to be in **roster form**. For example,

$$A = \left\{ \frac{3}{4}, 1.7, 2, 4, 6, \sqrt{89} \right\}$$

$$\mathbb{W} = \{0, 1, 2, 3, 4, 5, ...\}$$

$$\mathbb{Z} = \{..., -4, -3, -2, -1, 0, 1, 2, 3, 4, ...\}$$

are all in roster form.

The symbol $\in$ is read "is an element of" and is used to indicate that a particular number belongs to a set. For example, in the previous illustrations, $1.7 \in A$, $0 \in \mathbb{W}$, and $-3 \in \mathbb{Z}$.

If the elements in a set can be counted, as in A above, the set is said to be **finite**. If the elements cannot be counted, as in $\mathbb{W}$ and $\mathbb{Z}$ above, the set is said to be **infinite**. If a set has absolutely no elements, it is called the **empty set** or **null set** and is written in the form $\{\ \}$ or with the special symbol $\varnothing$. For example, the set of all people over 15 feet tall is the empty set, $\varnothing$.

The notation $\{x|\quad\}$ is read "the set of all x such that ..." and is called **set-builder notation**. The vertical bar $\left(\,|\,\right)$ is read "such that." A statement following the bar gives a condition (or restriction) for the variable x. For example,

$\{x | x$ is an even integer$\}$ is read "the set of all x such that x is an even integer."

The following note highlights the two important set concepts **union** and **intersection**.

Special Comments about Union and Intersection

The concepts of union and intersection are part of set theory which is very useful in a variety of courses including abstract algebra, probability, and statistics. These concepts are also used in analyzing inequalities and analyzing relationships among sets in general.

The **union** (symbolized $\cup$, as in A $\cup$ B) of two (or more) sets is the set of all elements that belong to either one set or the other set or to both sets. The **intersection** (symbolized $\cap$, as in A $\cap$ B) of two (or more) sets is the set of all elements that belong to both sets. The word **or** is used to indicate union and the word **and** is used to indicate intersection. For example, if $A = \{1, 2, 3\}$ and $B = \{2, 3, 4\}$, then the numbers that belong to A **or** B is the set $A \cup B = \{1, 2, 3, 4\}$. The set of numbers that belong to A **and** B is the set $A \cap B = \{2, 3\}$. These relationships can be illustrated using the following Venn diagram.

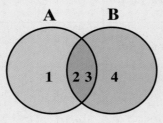

Similarly, union and intersection notation can be used for sets with inequalities.

For example, $\{x | x < a \text{ or } x > b\}$ can be written in the form

$$\{x | x < a\} \cup \{x | x > b\}.$$

Also, $\{x | x > a \text{ and } x < b\}$ can be written in the form

$$\{x | x > a\} \cap \{x | x < b\} \text{ or } \{x | a < x < b\}.$$

Objective B **Intervals of Real Numbers**

Suppose that a and b are two real numbers and that $a < b$. The set of all real numbers between a and b is called an **interval of real numbers**. Intervals of real numbers are used in relationship to everyday concepts such as the length of time you talk on your cell phone, the shelf life of chemicals or medicines, and the speed of an airplane.

As an aid in reading inequalities and graphing inequalities correctly, note that an inequality may be read either from left to right or right to left. Because we are concerned about which numbers satisfy an inequality, we **read the variable first**:

$$x > 7 \quad \text{is read from left to right as "}x\text{ is greater than 7."}$$
and $\quad 7 < x \quad$ is read from right to left as "x is greater than 7."

A compound interval such as $-3 < x < 6$ is read

"x is greater than -3 and x is less than 6."

Again, note that *the variable is read first*.

Various types of intervals and their corresponding **interval notations** are listed in Table 1. In this notation, brackets [and] indicate an endpoint is included and parentheses (and) indicate an endpoint is not included.

Type of Interval	Algebraic Notation	Interval Notation	Graph
Open Interval	$a < x < b$	(a, b)	
Closed Interval	$a \le x \le b$	$[a, b]$	
Half-open Interval	$\begin{cases} a \le x < b \\ a < x \le b \end{cases}$	$\begin{array}{l}[a, b) \\ (a, b]\end{array}$	
Open Interval	$\begin{cases} x > a \\ x < b \end{cases}$	$\begin{array}{l}(a, \infty) \\ (-\infty, b)\end{array}$	
Half-open Interval	$\begin{cases} x \ge a \\ x \le b \end{cases}$	$\begin{array}{l}[a, \infty) \\ (-\infty, b]\end{array}$	

Table 1

- In an **open interval**, neither endpoint is included.
- In a **closed interval**, both endpoints are included.
- In a **half-open interval**, only one end point is included.

notes

The symbol for infinity ∞ (or −∞) is not a number. It is used to indicate that the interval is to include all real numbers from some point on (either in the positive direction or the negative direction) without end.

Example 1

Graphing Intervals

Graph the open interval $(3, \infty)$.

Solution

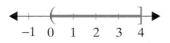

Now work margin exercise 1.

Example 2

Graphing Intervals

Graph the half-open interval $0 < x \le 4$.

Solution

Now work margin exercise 2.

Example 3

Graphing Intervals

Represent the following graph using algebraic notation, and state what kind of interval it is.

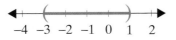

Solution

$x \ge 1$ is a half-open interval.

Now work margin exercise 3.

Example 4

Graphing Intervals

Represent the following graph using interval notation, and state what kind of interval it is.

Solution

$(-3, 1)$ is an open interval.

Now work margin exercise 4.

1. Graph the half-open interval $(-\infty, 6]$.

2. Graph the closed interval $-1 \le x \le 1$.

3. Represent the following graph using algebraic notation, and state what kind of interval it is.

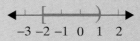

4. Represent the following graph using interval notation, and state what kind of interval it is.

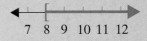

Example 5 and Example 6 illustrate how the concepts of **union** (indicated by **or**) and **intersection** (indicated by **and**) are related to the graphs of intervals of real numbers on a real number line.

5. Graph the set
$$\{t \mid t \le 8 \ \textbf{or} \ t > 10\}.$$

Example 5

Sets of Real Numbers Illustrating Union

Graph the set $\{x \mid x > 5 \ \textbf{or} \ x \le 4\}$. The word **or** implies those values of x that satisfy **at least one** of the inequalities.

Solution

$x > 5$

$x \le 4$

$x > 5$ **or** $x \le 4$

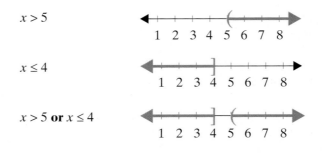

The solution graph shows the union $\cup$ of the first two graphs.

Now work margin exercise 5.

6. Graph the set
$$\{y \mid y \ge -1 \ \textbf{and} \ y < 3\}.$$

Example 6

Sets of Real Numbers Illustrating Intersection

Graph the set $\{x \mid x \le 2 \ \textbf{and} \ x \ge 0\}$. The word **and** implies those values of x that satisfy both inequalities.

Solution

$x \le 2$

$x \ge 0$

$x \le 2$ **and** $x \ge 0$

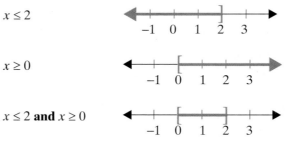

The solution graph shows the intersection $\cap$ of the first two graphs. In other words, the third graph shows the points in common between the first two graphs in this example.

This set can also be indicated as $\{x \mid 0 \le x \le 2\}$.

Now work margin exercise 6.

Objective C **Solving Linear Inequalities**

In this section we will solve **linear inequalities**, such as $6x + 5 \leq -7$, and write the solution in **interval notation**, such as $(-\infty, -2]$, to indicate all real numbers x less than or equal to -2. Note that this can also be written in set-builder notation as $\{x \mid x \in (-\infty, -2]\}$ or $\{x \mid x \leq -2\}$.

> ### Linear Inequalities
>
> Inequalities of the given form, where $a, b,$ and c are real numbers and $a \neq 0$,
>
> $$ax + b < c \quad \text{and} \quad ax + b \leq c$$
>
> $$ax + b > c \quad \text{and} \quad ax + b \geq c$$
>
> are called **linear inequalities**.
>
> The inequalities $c < ax + b < d$ and $c \leq ax + b \leq d$ are called **compound linear inequalities**. (This includes $c < ax + b \leq d$ and $c \leq ax + b < d$ as well.)

The solutions to linear inequalities are intervals of real numbers, and the methods for solving linear inequalities are similar to those used to solve linear equations. There is only one important exception.

Multiplying or dividing both sides of an inequality by a negative number causes the "sense" of the inequality to be reversed.

By the sense of the inequality, we mean "less than" or "greater than." Consider the following examples.

We know that $6 < 10$.

Add 5 to both sides	**Add –7 to both sides**	**Multiply both sides by 3**
$6 \;<\; 10$	$6 \;<\; 10$	$6 \;<\; 10$
$6+5 \;\;?\;\; 10+5$	$6+(-7) \;\;?\;\; 10+(-7)$	$3 \cdot 6 \;\;?\;\; 3 \cdot 10$
$11 \;<\; 15$	$-1 \;<\; 3$	$18 \;<\; 30$

In the three cases just illustrated, addition, subtraction, and multiplication by a positive number, the sense of the inequality stayed the same. It remained <. Now we will see that multiplying or dividing each side by a negative number will **reverse the sense** of the inequality, from < to > or from > to <. This concept also applies to ≤ and ≥.

Multiply both sides by –3	**Divide both sides by –2**
$6 \;<\; 10$	$6 \;<\; 10$
$-3 \cdot 6 \;\;?\;\; -3 \cdot 10$	$\dfrac{6}{-2} \;\;?\;\; \dfrac{10}{-2}$
$-18 \;>\; -30$	$-3 \;>\; -5$

In each of these last two examples, the sense of the inequality is changed from < to >. While two examples do not prove a rule to be true, this particular rule is true and is included in the following rules for solving inequalities.

Rules for Solving Linear Inequalities

1. Simplify each side of the inequality by removing any grouping symbols and combining like terms.
2. Use the addition property of equality to add the opposites of constants or variable expressions so that variable expressions are on one side of the inequality and constants are on the other.
3. Use the multiplication property of equality to multiply both sides by the reciprocal of the coefficient of the variable (that is, divide both sides by the coefficient) so that the new coefficient is 1. **If this coefficient is negative, reverse the sense of the inequality.**
4. A quick (and generally satisfactory) check is to select any one number in your solution and substitute it into the original inequality.

As with solving equations, the object of solving an inequality is to find equivalent inequalities of simpler form that have the same solution set. We want the variable with a coefficient of +1 on one side of the inequality and any constants on the other side. One key difference between solving linear equations and solving linear inequalities is that linear equations have only one solution while linear inequalities generally have an infinite number of solutions.

Example 7

Solving Linear Inequalities

Solve the linear inequality $6x + 5 \le -1$ and graph the solution set. Write the solution set using interval notation. Assume that x is a real number.

Solution

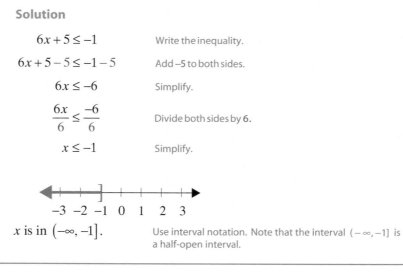

$6x + 5 \le -1$	Write the inequality.
$6x + 5 - 5 \le -1 - 5$	Add -5 to both sides.
$6x \le -6$	Simplify.
$\dfrac{6x}{6} \le \dfrac{-6}{6}$	Divide both sides by 6.
$x \le -1$	Simplify.

x is in $(-\infty, -1]$.   Use interval notation. Note that the interval $(-\infty, -1]$ is a half-open interval.

Example 8

Solving Linear Inequalities

Solve the linear inequality $x - 3 > 3x + 4$ and graph the solution set. Write the solution set using interval notation. Assume that x is a real number.

Solution

$$x - 3 > 3x + 4 \qquad \text{Write the inequality.}$$

$$x - 3 - x > 3x + 4 - x \qquad \text{Add } -x \text{ to both sides.}$$

$$-3 > 2x + 4 \qquad \text{Simplify.}$$

$$-3 - 4 > 2x + 4 - 4 \qquad \text{Add } -4 \text{ to both sides.}$$

$$-7 > 2x \qquad \text{Simplify.}$$

$$\frac{-7}{2} > \frac{2x}{2} \qquad \text{Divide both sides by 2.}$$

$$\frac{-7}{2} > x \qquad \text{Simplify.}$$

$$\text{or} \qquad x < -\frac{7}{2}$$

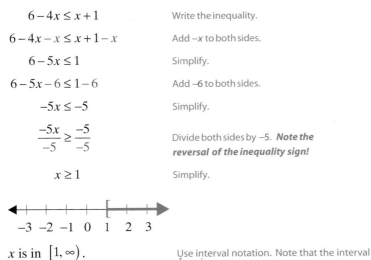

$$-4 \quad -\frac{7}{2} \quad -3$$

x is in $\left(-\infty, -\frac{7}{2} \right)$. Use interval notation. Note that the interval $\left(-\infty, -\frac{7}{2} \right)$ is an open interval

Example 9

Solving Linear Inequalities

Solve the linear inequality $6 - 4x \le x + 1$ and graph the solution set. Write the solution set using interval notation. Assume that x is a real number.

Solution

$$6 - 4x \le x + 1 \qquad \text{Write the inequality.}$$

$$6 - 4x - x \le x + 1 - x \qquad \text{Add } -x \text{ to both sides.}$$

$$6 - 5x \le 1 \qquad \text{Simplify.}$$

$$6 - 5x - 6 \le 1 - 6 \qquad \text{Add } -6 \text{ to both sides.}$$

$$-5x \le -5 \qquad \text{Simplify.}$$

$$\frac{-5x}{-5} \ge \frac{-5}{-5} \qquad \text{Divide both sides by } -5. \textbf{\textit{Note the}} \\ \textbf{\textit{reversal of the inequality sign!}}$$

$$x \ge 1 \qquad \text{Simplify.}$$

$$-3 \quad -2 \quad -1 \quad 0 \quad 1 \quad 2 \quad 3$$

x is in $[1, \infty)$. Use interval notation. Note that the interval $[1, \infty)$ is a half-open interval.

Solve the following linear inequalities and graph the solution sets. Write the solution sets using interval notation. Assume that x is a real number.

7. $3x + 8 > -10$

8. $7 - x \leq 9x - 1$

9. $5x - 9 \geq 2 - 4(x + 3)$

10. $6x - (9 - x) \leq 3x - 1$

Completion Example 10

Solving Linear Inequalities

Supply the reasons for each step in solving the linear inequality and graph the solution set. Use interval notation to write the solution set.

$$2x + 5 < 3x - (7 - x)$$

Solution

$2x + 5 < 3x - (7 - x)$	Write the inequality.
$2x + 5 < 3x - 7 + x$	Distribute the negative sign.
$2x + 5 < 4x - 7$	Combine like terms.
$2x + 5 - 2x < 4x - 7 - 2x$	_____
$5 < 2x - 7$	_____
$5 + 7 < 2x - 7 + 7$	_____
$12 < 2x$	_____
$\dfrac{12}{2} < \dfrac{2x}{2}$	_____
$6 < x$	_____

Graph the interval on a number line.

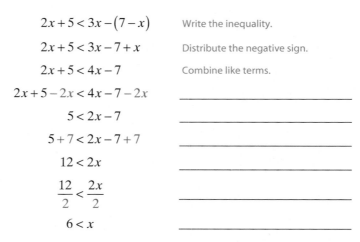

2   3   4   5   6   7   8

x is in _____   Use interval notation. Note that the interval is an open interval.

Now work margin exercises 7 through 10.

Completion Example Answers

10.

$2x + 5 < 3x - (7 - x)$	Write the inequality.
$2x + 5 < 3x - 7 + x$	Distribute the negative sign.
$2x + 5 < 4x - 7$	Combine like terms.
$2x + 5 - 2x < 4x - 7 - 2x$	Add $-2x$ to both sides.
$5 < 2x - 7$	Simplify.
$5 + 7 < 2x - 7 + 7$	Add 7 to both sides.
$12 < 2x$	Simplify.
$\dfrac{12}{2} < \dfrac{2x}{2}$	Divide both sides by 2.
$6 < x$	Simplify.

2   3   4   5   6   7   8

x is in $(6, \infty)$ Use interval notation. Note that the interval is an open interval.

Objective D **Solving Compound Inequalities**

Compound inequalities have three parts and can arise when a variable or variable expression is to be between two numbers. For example, the inequality

$$5 < x + 3 < 10$$

indicates that the values for the expression $x + 3$ are to be between 5 and 10. To solve this inequality, subtract 3 (or add -3) from each part.

$$5 < x + 3 < 10$$
$$5 - 3 < x + 3 - 3 < 10 - 3$$
$$2 < x < 7$$

Thus the variable x is isolated with coefficient $+1$, and we see that the solution set is the interval of real numbers $(2, 7)$. The graph of the solution set is the following.

   1 2 3 4 5 6 7 8

Example 11

Solving Compound Inequalities

Solve the compound inequality $-5 \leq 4x - 1 < 11$ and graph the solution set. Write the solution set using interval notation. Assume that x is a real number.

Solution

$-5 \leq \quad 4x - 1 \quad < 11$		Write the inequality.
$-5 + 1 \leq \quad 4x - 1 + 1 < 11 + 1$		Add 1 to each part.
$-4 \leq \quad 4x \quad < 12$		Simplify.
$\dfrac{-4}{4} \leq \quad \dfrac{4x}{4} \quad < \dfrac{12}{4}$		Divide each part by 4.
$-1 \leq \quad x \quad < 3$		Simplify.

  -2 -1 0 1 2 3 4 5

The solution set is the half-open interval $[-1, 3)$.

Example 12

Solving Compound Inequalities

Solve the compound inequality $5 \leq -3 - 2x \leq 13$ and graph the solution set. Write the solution set using interval notation. Assume that x is a real number.

Solution

$$5 \leq \quad -3 - 2x \quad \leq 13 \qquad \text{Write the inequality.}$$

$$5 + 3 \leq -3 - 2x + 3 \leq 13 + 3 \qquad \text{Add 3 to each part.}$$

$$8 \leq \quad -2x \quad \leq 16 \qquad \text{Simplify.}$$

$$\frac{8}{-2} \geq \quad \frac{-2x}{-2} \quad \geq \frac{16}{-2} \qquad \text{Divide each part by } -2. \text{ Note that}$$
the inequalities change sense.

$$-4 \geq \quad x \quad \geq -8 \qquad \text{Simplify.}$$

$$\left(\text{or} -8 \leq \quad x \quad \leq -4 \right)$$

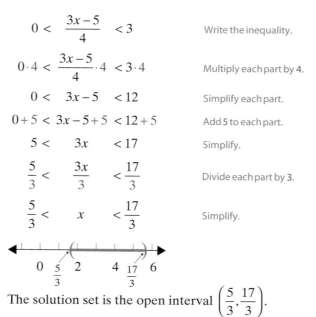

The solution set is the closed interval $[-8, -4]$.

Example 13

Solving Compound Inequalities

Solve the compound inequality $0 < \dfrac{3x - 5}{4} < 3$ and graph the solution set. Write the solution set using interval notation. Assume that x is a real number.

Solution

$$0 < \quad \frac{3x - 5}{4} \quad < 3 \qquad \text{Write the inequality.}$$

$$0 \cdot 4 < \quad \frac{3x - 5}{4} \cdot 4 \quad < 3 \cdot 4 \qquad \text{Multiply each part by 4.}$$

$$0 < \quad 3x - 5 \quad < 12 \qquad \text{Simplify each part.}$$

$$0 + 5 < 3x - 5 + 5 < 12 + 5 \qquad \text{Add 5 to each part.}$$

$$5 < \quad 3x \quad < 17 \qquad \text{Simplify.}$$

$$\frac{5}{3} < \quad \frac{3x}{3} \quad < \frac{17}{3} \qquad \text{Divide each part by 3.}$$

$$\frac{5}{3} < \quad x \quad < \frac{17}{3} \qquad \text{Simplify.}$$

The solution set is the open interval $\left(\dfrac{5}{3}, \dfrac{17}{3} \right)$.

Solve the following compound inequalities and graph the solution sets. Write the solution sets using interval notation. Assume that x is a real number.

11. $-12 < 5x + 3 \leq 8$

12. $-3 < 5 - 4x < 17$

13. $7 \leq \dfrac{5x - 1}{2} \leq 13$

Now work margin exercises 11 through 13.

Objective E   Applications of Linear Inequalities

In the following two examples, we show how inequalities can be related to real-world problems.

Example 14

Application with an Inequality

A math student has grades of 85, 98, 93, and 90 on four examinations. If he must average 90 or better to receive an A for the course, what scores can he receive on the final exam and earn an A? (Assume that the final exam counts the same as the other exams.)

Solution

Let x = score on final exam.
The average is found by adding the scores and dividing by 5.

$$\frac{85 + 98 + 93 + 90 + x}{5} \geq 90$$

$$\frac{366 + x}{5} \geq 90 \qquad \text{Simplify the numerator.}$$

$$5\left(\frac{366 + x}{5}\right) \geq 5 \cdot 90 \qquad \text{Multiply both sides by 5.}$$

$$366 + x \geq 450 \qquad \text{Simplify.}$$

$$366 + x - 366 \geq 450 - 366 \qquad \text{Add } -366 \text{ to each side.}$$

$$x \geq 84$$

If the student scores 84 or more on the final exam, he will average 90 or more and receive an A in math.

***Now work margin exercise** 14.*

14. A certain anesthetic must be administered in a way that requires the average dosage to remain under 450 milligrams per hour in order to guarantee patient safety. This anesthetic is being used during a 4-hour surgery. For the first three hours, the dosages were as follows: 500 milligrams the first hour, 380 milligrams during the second hour, and 520 milligrams the third hour. What is the maximum dosage that can be administered safely to the patient for the final dosage?

15. Ashley is planning her wedding and is ordering flowered centerpieces. She wants some rose centerpieces and some lily centerpieces. The rose centerpieces cost $225, and the lily centerpieces cost $175. If she needs a total of 20 centerpieces, and she has a budget of $3900 to spend, what is the maximum number of rose centerpieces she can buy?

Example 15

Application with an Inequality

Ellen is going to buy 30 stamps, some 28-cent and some 44-cent. If she has $9.68, what is the maximum number of 44-cent stamps she can buy?

Solution

Let x = number of 44-cent stamps,
    then $30 - x$ = number of 28-cent stamps.

Ellen cannot spend more than $9.68.

$$0.44x + 0.28(30 - x) \leq 9.68$$

$$0.44x + 8.40 - 0.28x \leq 9.68$$

$$0.16x + 8.40 \leq 9.68$$

$$0.16x + 8.40 - 8.40 \leq 9.68 - 8.40$$

$$0.16x \leq 1.28$$

$$\frac{0.16x}{0.16} \leq \frac{1.28}{0.16}$$

$$x \leq 8$$

Ellen can buy at most eight 44-cent stamps if she buys a total of 30 stamps.

Now work margin exercise **15.**

Practice Problems

Graph each set of real numbers on a real number line.

1. $\{x \mid x \leq 3 \text{ and } x > 0\}$ **2.** $\{x \mid x \geq -1.5\}$ **3.** $\{x \mid x > 2 \text{ or } x < -4\}$

Solve each of the following inequalities and graph the solution sets. Write each solution set in interval notation. Assume that x is a real number.

4. $7 + x < 3$ **5.** $-5 \leq 2x + 1 < 9$ **6.** $\dfrac{x}{2} + 1 \leq \dfrac{1}{3}$

Practice Problem Answers

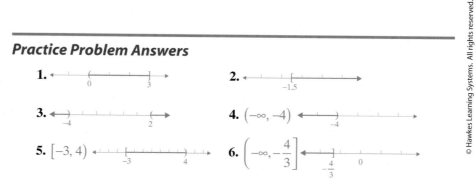

1.
2.
3.
4. $(-\infty, -4)$
5. $[-3, 4)$
6. $\left(-\infty, -\dfrac{4}{3}\right]$

Exercises 8.7

Graph each set of indicated numbers on a real number line. See Examples 1 through 4.

1. $\{x \mid x \text{ is a whole number less than } 3\}$

2. $\{x \mid x \text{ is an integer with } |x| \leq 3\}$

3. $\{x \mid x \text{ is a prime number less than } 20\}$

4. $\{x \mid x \text{ is a positive whole number divisible by } 0\}$

5. $\{x \mid x \text{ is a composite number between 2 and } 10\}$

6. $\{x \mid x \geq 4, x \text{ is an integer}\}$

7. $\{x \mid -8 < x < 0, x \text{ is a whole number}\}$

8. $\{x \mid -2 < x < 12, x \text{ is an integer}\}$

Use set-builder notation to indicate each set of numbers as described.

9. The set of all real numbers between 3 and 5, including 3

10. The set of all real numbers between −4 and 4

11. The set of all real numbers greater than or equal to −2.5

12. The set of all real numbers between −1.8 and 5, including both of these numbers

The graphs of sets of real numbers are given. a. Use set-builder notation to indicate the set of numbers shown in each graph. b. Use interval notation to represent the graph. c. Tell what type of interval is illustrated.

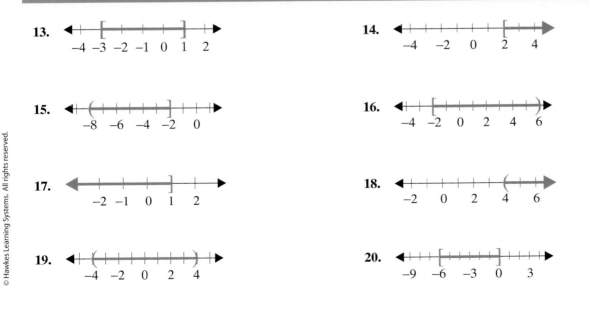

13.
14.
15.
16.
17.
18.
19.
20.

Graph each interval on a real number line and tell what type of interval it is. Assume that x is a real number.

21. $x \leq -3$

22. $x \geq -0.5$

23. $x > 4$

24. $x < -\dfrac{1}{10}$

25. $0 < x \leq 2.5$

26. $-1.5 \leq x < 3.2$

27. $-2 \leq x \leq 0$

28. $-1 \leq x \leq 1$

29. $4 > x \geq 2$

30. $0 > x \geq -5$

Solve the inequalities and graph the solution sets. Write each solution set using interval notation. Assume that x is a real number. See Examples 7 through 10.

31. $2x + 3 < 5$

32. $4x - 7 \geq 9$

33. $14 - 5x < 4$

34. $23 < 7x - 5$

35. $6x - 15 > 1$

36. $9 - 2x < 8$

37. $5.6 + 3x \geq 4.4$

38. $12x - 8.3 < 6.1$

39. $1.5x + 9.6 < 12.6$

40. $0.8x - 2.1 \geq 1.1$

41. $2 + 3x \geq x + 8$

42. $x - 6 \leq 4 - x$

43. $3x - 1 \leq 11 - 3x$

44. $5x + 6 \geq 2x - 2$

45. $4 - 2x < 5 + x$

46. $4 + x > 1 - x$

47. $x - 6 > 3x + 5$

48. $4 + 7x \leq 4x - 8$

49. $\dfrac{x}{2} - 1 \leq \dfrac{5x}{2} - 3$

50. $\dfrac{x}{4} + 1 \leq 5 - \dfrac{x}{4}$

51. $\dfrac{x}{3} - 2 > 1 - \dfrac{x}{3}$

52. $\dfrac{5x}{3} + 2 > \dfrac{x}{3} - 1$

53. $6x + 5.91 < 1.11 - 2x$

54. $4.3x + 21.5 \geq 1.7x + 0.7$

55. $6.2x - 5.9 > 4.8x + 3.2$

56. $0.9x - 11.3 < 3.1 - 0.7x$

57. $4(6 - x) < -2(3x + 1)$

58. $-3(2x - 5) \leq 3(x - 1)$

59. $-(3x + 8) \geq 2(3x + 1)$

60. $6(3x + 1) < 5(1 - 2x)$

61. $11x + 8 - 5x \geq 2x - (4 - x)$

62. $1 - (2x + 8) < (9 + x) - 4x$

63. $5 - 3(4 - x) + x \leq -2(3 - 2x) - x$

64. $x - 2(x + 3) \geq 7 - (4 - x) + 11$

65. $\dfrac{2(x - 1)}{3} < \dfrac{3(x + 1)}{4}$

66. $\dfrac{3(x - 2)}{2} \geq \dfrac{4(x - 1)}{3}$

67. $\dfrac{x - 2}{4} > \dfrac{x + 2}{2} + 6$

68. $\dfrac{x + 4}{9} \leq \dfrac{x}{3} - 2$

69. $\dfrac{2x + 7}{4} \leq \dfrac{x + 1}{3} - 1$

70. $\dfrac{4x}{7} - 3 > \dfrac{x - 6}{2} - 4$

Solve the compound inequalities and graph the solution sets. Write each solution set using interval notation. Assume that x is a real number. See Examples 11 through 13.

71. $-4 < x + 5 < 6$

72. $2 \leq -x + 2 \leq 6$

73. $3 \geq 4x - 3 \geq -1$

74. $13 > 3x + 4 > -2$

75. $1 \leq \dfrac{2}{3}x - 1 \leq 9$

76. $-2 \leq \dfrac{1}{2}x - 5 \leq -1$

77. $14 > -2x - 6 > 4$

78. $-11 \geq -3x + 2 > -20$

79. $-1.5 < 2x + 4.1 < 3.5$

80. $0.9 < 3x + 2.4 < 6.9$

81. Test Scores: A statistics student has grades of 82, 95, 93, and 78 on four hour-long exams. He must average 90 or higher to receive an A for the course. What scores can he receive on the final exam and earn an A if:

   a. The final is equivalent to a single hour-long exam (100 points maximum)?

   b. The final is equivalent to two hourly exams (200 points maximum)?

82. Test Scores: To receive a grade of B in a chemistry class, Melissa must average 80 or more but less than 90. If her five hour-long exam scores were 75, 82, 90, 85, and 77, what score does she need on the final exam (100 points maximum) to earn a grade of B?

83. Car Sales: A car salesman makes $1000 each day that he works and makes approximately $250 commission for each car he sells. If a car salesman wants to make at least $3500 in one day, how many cars does he need to sell?

84. Postage: Allison is going to the post office to buy 38¢ stamps and 2¢ adjustment stamps. Since the current postage rate is 44¢, she will need 3 times as many 2¢ adjustment stamps as 38¢ stamps. If she has $11 to spend, what is the largest number of 38¢ stamps she can buy?

Writing & Thinking

85. a. Write a list of three situations where inequalities might be used in daily life.

   b. Illustrate theses situations with algebraic inequalities and appropriate numbers.

Section 8.1: Solving Linear Equations: $x + b = c$ and $ax = c$

Solution — page 594

A **solution** to an equation is a number that gives a true statement when substituted for the variable in the equation.

Linear Equations in x — page 594

If a, b, and c are **constants** and $a \neq 0$, then a **linear equation in x** (or **first-degree equation in x**) is an equation that can be written in the form $ax + b = c$.

Addition Principle of Equality — page 594

If the same algebraic expression is added to both sides of an equation, the new equation has the same solutions as the original equation. Symbolically, if A, B, and C are algebraic expressions, then the equations
$$A = B \quad \text{and} \quad A + C = B + C$$
have the same solutions.

Equivalent Equations — page 594

Equations with the same solutions are said to be **equivalent equations**.

Procedure for Solving Linear Equations that Simplify to the Form $x + b = c$ — page 595

Multiplication (or Division) Principle of Equality — page 598

If both sides of an equation are multiplied by (or divided by) the same nonzero constant, the new equation has the same solutions as the original equation. Symbolically, if A and B are algebraic expressions and C is any nonzero constant, then the equations
$$A = B$$
$$\text{and} \quad AC = BC \qquad \text{where } C \neq 0$$
$$\text{and} \quad \frac{A}{C} = \frac{B}{C} \qquad \text{where } C \neq 0$$
have the same solutions.

Procedure for Solving Linear Equations that Simplify to the Form $ax = c$ — page 599

Procedure for Solving Linear Equations that Simplify to the Form $ax + b = c$ page 606
1. Combine like terms on both sides of the equation.
2. Use the **addition principle of equality** and add the opposite of the constant b to both sides.
3. Use the **multiplication** (or **division**) **principle of equality** to multiply both sides by the reciprocal of the coefficient of the variable (or divide both sides by the coefficient itself). The coefficient of the variable will become +1.
4. Check your answer by substituting it for the variable in the original equation.

Section 8.3 : Solving Linear Equations: $ax + b = cx + d$

Procedure for Solving Linear Equations that Simplify to the Form $ax + b = cx + d$
page 614

1. Simplify by removing any grouping symbols and combining like terms on each side of the equation.
2. Use the **addition principle of equality** and add the opposite of a constant term and/or variable term to both sides so that variables are on one side and constants are on the other side.
3. Use the **multiplication** (or **division**) **principle of equality** to multiply both sides by the reciprocal of the coefficient of the variable (or divide both sides by the coefficient itself). The coefficient of the variable will become +1.
4. Check your answer by substituting it for the variable in the original equation.

Types of Equations and their Solutions   page 619

Type of Equation	Number of Solutions
Conditional	Finite number of solutions
Identity	Infinite number of solutions
Contradiction	No solution

Section 8.4: Applications: Number Problems and Consecutive Integers

Number Problems   page 624

Consecutive Integers   page 626
Consecutive integers are two integers that differ by 1. Three consecutive integers can be represented as $n, n + 1$, and $n + 2$ where n is an integer.

Consecutive Even Integers   page 626
Consecutive even integers are two even integers that differ by 2. Two consecutive even integers can be represented as n and $n + 2$ where n is even.

Consecutive Odd Integers   page 626
Consecutive odd integers are two odd integers that differ by 2. Two consecutive odd integers can be represented as n and $n + 2$ where n is odd.

Lists of Common Formulas   page 634

Formula	Meaning
$I = Prt$	The **simple interest** I earned by investing money is equal to the product of the principal P times the rate of interest r times the time t in one year or part of a year.
$C = \dfrac{5}{9}(F - 32)$	**Temperature** in degrees Celsius C equals $\dfrac{5}{9}$ times the difference between the Fahrenheit temperature F and 32.
$d = rt$	The **distance traveled** d equals the product of the rate of speed r and the time t.
$P = 2l + 2w$	The **perimeter P of a rectangle** is equal to twice the length l plus twice the width w.
$L = 2\pi rh$	The **lateral surface area** L (top and bottom not included) of a cylinder is equal to 2π times the radius r of the base times the height h.
$F = ma$	In physics, the **force** F acting on an object is equal to its mass m times its acceleration a.
$\alpha + \beta + \gamma = 180°$	The **sum of the angles** ($\alpha, \beta,$ and γ) **of a triangle** is $180°$.
$P = a + b + c$	The **perimeter P of a triangle** is equal to the sum of the lengths of the three sides $a, b,$ and c.

Evaluating Formulas   page 635

If you know values for all but one variable in a formula, you can substitute those values and find the value of the unknown variable using the techniques for solving equations.

Solving Formulas for Different Variables   page 638

To solve a formula for a specific variable, treat the variables just as you would constants in solving linear equations.

Distance-Rate-Time Problems   page 647

The **distance traveled** d equals the product of the rate of speed r and the time t. This relationship can be written as $d = rt$.

Simple Interest Problems   page 648

The **simple interest** I earned by investing money is equal to the product of the principal P times the rate of interest r times the time t in one year or part of a year. This relationship can be written as $I = Prt$.

Average Problems   page 649

Use the concept of average to find unknown numbers.

Cost Problems   page 650

$S - C = P$ (selling price – cost = profit).

Terminology Regarding Sets page 656

    Set

    Elements

    Finite and infinite sets

    The empty set (or null set), $\varnothing$

    Set-builder notation

Union page 657

    The **union** (symbolized $\cup$, as in A $\cup$ B) of two (or more) sets is the set of all elements that belong to either one set or the other set or to both sets.

Intersection page 657

    The **intersection** (symbolized $\cap$, as in A $\cap$ B) of two (or more) sets is the set of all elements that belong to both sets.

Interval Notation page 658

Type of Interval	Interval Notation
Open	(a, b), (a, ∞), $(-\infty, b)$
Closed	$[a, b]$
Half-open	$[a, \infty)$, $(-\infty, b]$, $[a, b)$, $(a, b]$

Linear Inequalities page 661

    Inequalities of the given form where a, b, and c are real numbers and $a \neq 0$,

$$ax + b < c \quad \text{and} \quad ax + b \leq c$$
$$ax + b > c \quad \text{and} \quad ax + b \geq c$$

    are called linear inequalities.

Rules for Solving Linear Inequalities page 662

  1. Simplify each side of the inequality by removing any grouping symbols and combining like terms.

  2. Use the addition property of equality to add the opposites of constants or variable expressions so that variable expressions are on one side of the inequality and constants are on the other.

  3. Use the multiplication property of equality to multiply both sides by the reciprocal of the coefficient of the variable (that is, divide both sides by the coefficient) so that the new coefficient is 1. If this coefficient is negative, reverse the sense of the inequality.

  4. A quick check is to select any one number in your solution and substitute it into the original inequality.

Compound Inequalities page 665

    The inequalities $c < ax + b < d$ and $c \leq ax + b \leq d$ are called **compound linear inequalities**. (This includes $c < ax + b \leq d$ and $c \leq ax + b < d$ as well.)

Applications of Linear Inequalities page 667

Chapter 8: Chapter Review

Section 8.1: Solving Linear Equations: $x + b = c$ and $ax = c$

Solve each of the following linear equations.

1. $x - 6 = 1$

2. $x - 4 = 10$

3. $y + 5 = -6$

4. $0 = n + 8$

5. $7n = 42$

6. $\dfrac{2}{3}y = -12$

7. $\dfrac{5}{6}y = -25$

8. $3.2y - 2.2y = 1.3 - 0.9$

9. $2.5x - 1.5x + 3 = 10$

10. $6t = -54$

Section 8.2: Solving Linear Equations: $ax + b = c$

Solve each of the following linear equations.

11. $2x + 3 = -7$

12. $3x + 5 = -19$

13. $2y + 2y - 6 = 30$

14. $3y + 10 + 2y = -25$

15. $-30 = 6x - 4x + 20$

16. $5y + 6y + 17 = -16$

17. $\dfrac{3}{8}n + \dfrac{1}{4}n + \dfrac{1}{2} = \dfrac{3}{4}$

18. $\dfrac{2}{3}x - \dfrac{1}{2}x + \dfrac{5}{6} = 2$

19. $5.32 + 0.4x - 0.6x = 7.1$

20. $8.62 = 0.45x - 0.15x + 5.02$

Section 8.3: Solving Linear Equations: $ax + b = cx + d$

Solve each of the following linear equations.

21. $8y - 2.1 = y + 5.04$

22. $2.4n = 3.6n$

23. $2(x + 3) = 3(x - 7)$

24. $\dfrac{1}{3}n + 6 = \dfrac{1}{4}n + 5$

25. $6(2x - 1) = 4(x + 3) + 14$ =

26. $8 + 4(x - 4) = 5 - (x - 14)$

27. $-2(y + 2) - 4 = 6 - (y + 12)$

28. $\dfrac{5x}{6} + \dfrac{1}{18} = \dfrac{x}{3}$

29. $\dfrac{x}{6} + \dfrac{1}{15} - \dfrac{2x}{3} = 0$

30. $2(y + 8.3) + 1.2 = y - 16.6 - 1.2$

Determine whether each of the following equations is a conditional equation, an identity, or a contradiction.

31. $3(x - 1) = 3 - x + 4x$

32. $7 + 2(3x + 1) = 3(2x - 5) + 24$

Section 8.4: Applications: Number Problems and Consecutive Integers

Read each problem carefully, translate the various phrases into algebraic expressions, set up an equation, and solve the equation.

33. Forty-two is 4 more than twice a certain number. What is the number?

34. The difference between three times a number and 7 is equal to 17 decreased by the number. Find the number.

35. The quotient of twice a number and 9 is seven less than the number. What is the number?

36. Find three consecutive odd integers whose sum is 279.

37. Three consecutive even integers are such that if the second is added to twice the first the result is 18 more than the third. Find these integers.

38. Find four consecutive integers whose sum is 190.

39. **Real Estate:** A real estate agent says that the current value of a home is $150,000 more than twice its value when it was new. If the current value is $640,000, what was the value of the home when it was new?

40. **Buying a Car:** Karen wants to pay cash for a new car that costs $15,325. So far, she has saved $7621 for the car, and has determined that she can afford to put away $642 a month towards it. How many months will it take her to save up for the car?

41. **Candy:** Jeremy has devised a plan to get the most candy possible this Halloween. He intends to visit a neighborhood that gives away 5 pieces of candy per house and then finish at his neighbor's house where he will get 36 pieces at the end of the night. If he ends the night with 161 pieces of candy, how many houses did he visit in the other neighborhood?

42. **Collecting:** Anferny loves collecting pens of all types. His favorites are rare, old pens, but he also likes regular store-bought ones. His pen collection currently has 16 rare pens, and 82 regular ones. If his mother wants to bring his collection up to an even 200 pens for his birthday, how many packs of pens must she get him if each pack has 6 pens?

Section 8.5: Working with Formulas

Solve the following word problems.

43. **Temperature:** Given the formula $C = \frac{5}{9}(F - 32)$, find the value of C if $F = 32°$.

44. **Driving Speed:** Using the formula $d = rt$, find Karl's average rate of speed if he drove 190 miles in 4 hours.

45. **Soccer:** If the perimeter of a rectangular soccer field is planned to be 150 meters and because of building restrictions the width has to be 35 meters, what is the planned length of the field? $(P = 2l + 2w)$

46. **Pressure of a Gas:** Volume and pressure of gas in a container are known to be related by the formula $V = \frac{k}{P}$ where k is a constant that depends on the type of gas involved. Find P if $k = 3.2$ and $V = 20$ cubic centimeters.

47. $L = 2\pi rh$; solve for π.

48. $P = R - C$; solve for R.

49. $v = v_0 - gt$; solve for g.

50. $K = \dfrac{mv^2}{2g}$; solve for m.

51. $3x + y = 6$; solve for y.

52. $5x - 2y = 10$; solve for x.

Section 8.6: Applications: Distance-Rate-Time, Interest, Average

Solve the following word problems.

53. **Traveling by Car:** Alicia drove to her parents' home in 6 hours. She figured that she averaged 50 mph for the first part of the trip and 62 mph for the second part. The trip was 328 miles. As a math quiz for her sister, she asked her sister how much time she took for each part of the trip. Her sister, being a math whiz, told her what two numbers?

54. **Traveling by Car:** Jerry drove from St. Louis, MO to Kansas City, MO in 5.5 hours. On the return trip he decided to increase his speed by 10 mph and the return took 4.5 hours. Find the rate on the return trip. What is the distance from St. Louis to Kansas City?

55. **Traveling by Car:** Jacksonville and Tampa are about 210 miles apart. George leaves Tampa at the same time that Joshua leaves Jacksonville. They drive towards each other and meet in 2 hours. What was each person's speed, knowing that George traveled 5 mph faster than Joshua?

56. **Investing:** Reginald has had $40,000 invested for one year, some in a savings account which paid 4%, the rest in stocks which earned 6.5% for the year. If his interest income for the year was $1850, how much did he have in each investment?

57. **Investing:** Ms. Clark has two investments totaling $20,000. One investment yields 6% a year and the other yields 8%. If her annual income from these investments is $1520, how much did she invest at each rate?

58. **Shopping:** A men's store pays $50 per pair for a certain style of men's pants. The store wants to be able to advertise a sale of 15% off the marked price and still make a profit of 36%. What should be the marked price?

59. **Saving:** For the next six months Sarah is planning to put money in her savings account for an average of $200 a month. The first five months she managed to save the following amounts: $250, $100, $180, $260, $190 respectively. How much will she need to save the last month to keep up with her goal?

60. **Test Scores:** Sundar received scores of 71, 85, 83, 62, and 70 on his five Physics tests this semester. The final exam counts as two tests. What score will he need on the final to have an average of 75 for the semester? (All tests and exams have a maximum of 100 points.)

61. **Shopping:** Maria used a coupon for 20% off her purchase to buy a new digital camera. If sales tax is 8% and she paid $190.08, what was the original price of the camera?

Graph each set of indicated numbers on a real number line.

62. $\{x \mid x$ is a composite number between 2 and 20$\}$ **63.** $\left\{x \mid x$ is a whole number with $|x| \leq \dfrac{3}{4}\right\}$

Use set-builder notation to indicate the set of numbers as described.

64. The set of all real numbers between −1 and 6, including −1.

The graphs of sets of real numbers are given. a. Use set-builder notation to indicate the set of numbers shown in each graph. b. Use interval notation to represent the graph. c. Tell what type of interval is illustrated.

65.

```
◄──┼──(┼──┼──┼──┼──┼──)┼──┼──►
   -4  -2   0   2   4
```

66.

```
◄──┼──┼──┼──┼──┼──┼──[┼──┼──►
   -6     -3    0    3    6
```

Solve the inequalities and graph the solution sets. Write each solution set using interval notation. Assume that x is a real number.

67. $4x + 6 \geq 10$ **68.** $3x - 7 \leq 14$ **69.** $x - 6 > 3x + 4$

70. $\dfrac{1}{2} + \dfrac{x}{5} < 1 - \dfrac{x}{5}$ **71.** $-4 < 5x + 1 < 16$ **72.** $0 < -\dfrac{1}{2}x + 3 \leq 2$

Chapter 8: Test

Solve each of the following linear equations.

1. $0 = 2x + 8$

2. $y + 13 = -16$

3. $\dfrac{5}{3}x + 1 = -4$

4. $3 + 6x - 8x = 15$

5. $4x - 5 - x = 2x + 5 - x$

6. $8(3 + y) = -4(2y - 6)$

7. $\dfrac{3}{2}a + \dfrac{1}{2}a = -3 + \dfrac{3}{4}$

8. $6 + \dfrac{3}{2}x = 9 + \dfrac{1}{3}x - \dfrac{5}{6}x$

9. $0.7x + 2 = 0.4x + 8$

10. $4(2x - 1) + 3 = 2(x - 4) - 5$

Determine whether each of the following equations is a conditional equation, an identity, or a contradiction.

11. $9x + x - 4(x + 3) = 6(x - 2)$

12. $\dfrac{1}{4}(4x + 1) = -\dfrac{1}{2}(2x - 1)$

Set up an equation for each word problem and solve the equation.

13. One number is 5 more than twice another number. Their sum is –22. Find the numbers.

14. Find three consecutive odd integers such that three times the second is equal to 27 more than the sum of the first and the third.

15. Find two consecutive integers such that twice the first added to three times the second is equal to 83.

16. **Basketball:** Brandon scored 21 points in his basketball game last night. If he made two free throws (1 point each) and one 3-point shot, how many 2-point shots did he make?

Solve each formula for the indicated variable.

17. $N = mrt + p$; solve for m.

18. $5x + 3y - 7 = 0$; solve for y.

Solving the following application problems.

19. **Text Messaging:** The cost to send/receive t text messages in a given month on AT&T's Messaging 1500 plan is $C = 0.05(t - 1500) + 15.00$ when $t \geq 1500$. How much did Jena spend the month she sent/received 1780 text messages?

20. **Payment Plans:** When purchasing an item on an installment plan, the total cost C equals the down payment d plus the product of the monthly payment p and the number of months t. $(C = d + pt)$ Use this information to answer the following questions.

a. A refrigerator costs \$857.60 if purchased on the installment plan. If the monthly payments are \$42.50 and the down payment is \$92.60, how long will it take to pay for the refrigerator?

b. A used car will cost \$3250 if purchased on an installment plan. If the monthly payments are \$115 for 24 months, what will be the down payment?

21. Investing: George and Marie have investments in two accounts totaling $25,000. One investment yields 5% a year and the other yields 3.5%. If their annual income from the 5% account yields $145 more each year than the other account, what amount do they have in each account?

22. Traveling by Bus: A bus leaves Kansas City headed for Phoenix traveling at a rate of 48 mph. Thirty minutes later, a second bus follows, traveling at 54 mph. How long will it take the second bus to overtake the first?

23. Shopping: The manager of a jewelry store purchased a selection of watches from the manufacturer for $190 each. Next weekend he plans to have a 20% off sale. What price should he mark on each watch so that he will still make a profit of 40% on his cost?

24. Test Scores: In his calculus class a student has the following test scores: 77, 82, and 73. If his last test will have a maximum of 100 points, what score does he need to finish the class with an average of 80?

The graphs of sets of real numbers are given. a. Use set-builder notation to indicate the set of numbers shown in each graph. b. Use interval notation to represent the graph. c. Tell what type of interval is illustrated.

25.

$$-2\ -1\ 0\ 1\ 2\ 3\ 4\ 5\ 6$$

26.

$$-3\ -2\ -1\ 0\ 1\ 2\ 3\ 4\ 5$$

Solve the inequalities and graph the solution sets. Write each solution set in interval notation. Assume that x is a real number.

27. $3x + 7 < 4(x + 3)$

28. $\dfrac{2x + 5}{4} > x + 3$

29. $-1 < 3x + 2 < 17$

Cumulative Review: Chapters 1 - 8

1. Round 16.996 to the nearest hundredth.

2. Find the decimal equivalent to $\dfrac{14}{35}$.

3. Write $1\dfrac{1}{2}\%$ as a decimal.

4. Use the tests for divisibility to determine if 732 can be divided exactly by 2, 3, 4, 5, 9, and 10.

5. Find the prime factorization of 396.

6. Find the LCM of 14, 21, and 30.

7. Division by _____ is undefined.

8. In the proportion $\dfrac{7}{8} = \dfrac{140}{160}$, the extremes are _____ and _____.

9. 15% of _____ is 7.5.

10. Solve for x: $\dfrac{1\frac{2}{3}}{x} = \dfrac{10}{2\frac{1}{4}}$

Solve the following word problems.

11. **Employment:** In a certain company, three out of every five employees are male. How many female employees are there out of the 490 people working for this company?

12. **Test Scores:** James has the following scores on his first three math tests: 75, 87, and 79. To earn a B grade for the course, he must have at least an 80 average. What is the least score that he can get on his fourth test to maintain a B average?

13. **Investing:** An investment pays $6\dfrac{1}{4}\%$ simple interest. What is the interest on $4800 invested for 8 months?

14. **Architecture:** A scale on an architect's drawing indicates $2\dfrac{1}{2}$ inches represents 50 feet. How many feet apart are two points 6 inches apart on the drawing?

15. **Cereal Boxes:** A cereal box in the shape of a rectangular solid is 5 cm wide, 19 cm long, and 27 cm high. What is the volume of the box? What is the surface area of the box?

16. Polygons: Find **a.** the circumference of the inner circle, **b.** the circumference of the outer circle, and **c.** the area of the shaded figure. (Use $\pi = 3.14$.)

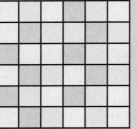

4 ft

10 ft

17. Triangles: Two similar triangles are shown. Find the values of x and y.

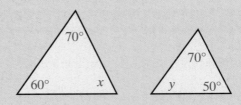

70°

60° x

70°

y 50°

18. Darts: A square dart board has small squares colored in yellow, pink, and green as shown in the figure here.

a. What is the probability that a dart that hits the board will hit a yellow square?

b. What is the probability that a dart that hits the board will not hit a green square?

19. Games: Two dice are rolled. The sum of the two numbers showing on top is calculated. (There are 36 possible pairs of numbers.) Find

a. the probability that the sum is 12.

b. the probability that the sum is 13.

c. the probability that the sum is more than 10.

Hint: You may want to list all 36 pairs.

Find the following statistics for the information indicated in the graph.

20. Mr. John kept the mileage records for 5 months as shown in the following bar graph.

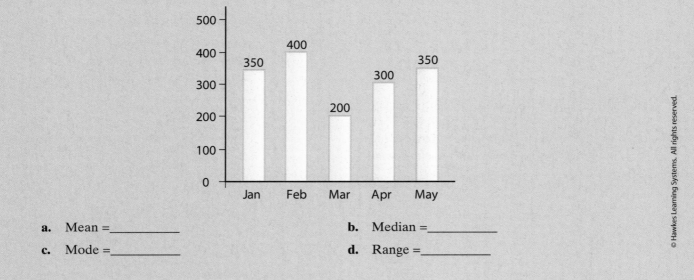

a. Mean = _____

b. Median = _____

c. Mode = _____

d. Range = _____

Solve the following word problem.

21. The circle graph shows a family budget for one year. What amount will be spent in each category if the family income is $45,000?

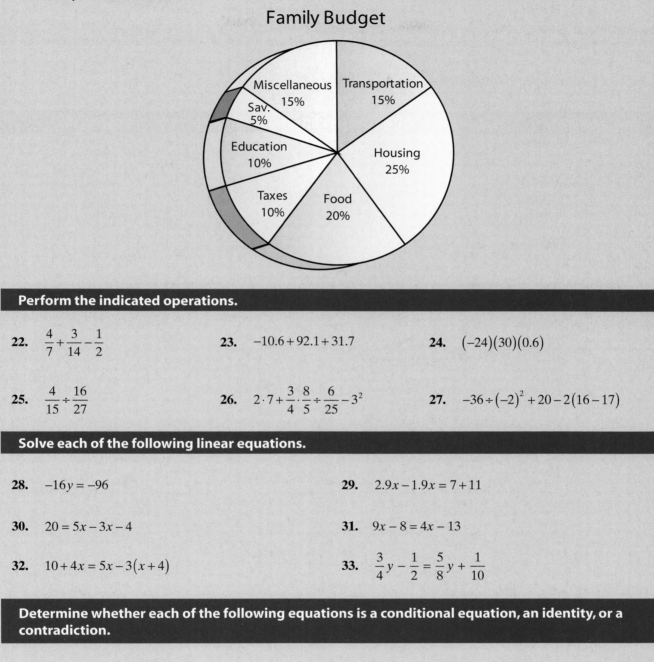

Family Budget

Perform the indicated operations.

22. $\dfrac{4}{7} + \dfrac{3}{14} - \dfrac{1}{2}$

23. $-10.6 + 92.1 + 31.7$

24. $(-24)(30)(0.6)$

25. $\dfrac{4}{15} \div \dfrac{16}{27}$

26. $2 \cdot 7 + \dfrac{3}{4} \cdot \dfrac{8}{5} \div \dfrac{6}{25} - 3^2$

27. $-36 \div (-2)^2 + 20 - 2(16 - 17)$

Solve each of the following linear equations.

28. $-16y = -96$

29. $2.9x - 1.9x = 7 + 11$

30. $20 = 5x - 3x - 4$

31. $9x - 8 = 4x - 13$

32. $10 + 4x = 5x - 3(x + 4)$

33. $\dfrac{3}{4}y - \dfrac{1}{2} = \dfrac{5}{8}y + \dfrac{1}{10}$

Determine whether each of the following equations is a conditional equation, an identity, or a contradiction.

34. $0.2x + 25 = -1.6 + 0.1x$

35. $3(x - 5) + 4 = x + 2(x - 5)$

Solve the following word problems.

36. Find three consecutive even integers such that the sum of the first and twice the second is equal to 14 more than twice the third.

37. **Electronics:** The total cost of a portable hard drive and a spare laptop battery was $130.75. If the hard drive cost $38.45 more than the battery, what was the cost of each item?

Use set-builder notation to indicate each set of numbers as described.

38. the set of all real numbers greater than or equal to −8.3

Graph each set of indicated numbers on a real number line.

39. $\{x \mid x \text{ is a prime number less than } 15\}$

The graphs of sets of real numbers are given. Use interval notation to represent the graph.

40.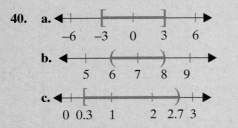

a.

b.

c.

Solve each of the inequalities. Graph each solution set on a real number line and write the solution set using interval notation.

41. $-3(7 - 2x) \geq 2 + 3x - 10$

42. $0 < \dfrac{1}{2}x + 3 < 4$

Linear Equations and Inequalities in Two Variables

Mathematics at Work!

Linear equations (or linear functions) can be used to represent many real-world applications. They play an especially large role in the business world, where they may be used to represent supply functions, demand functions, cost functions, or profit functions.

A pharmaceutical company is examining the demand of a certain drug. When the price was set at $55 per unit, sales were 1.2 million units. When the price was set at $40 per unit, sales increased to 1.8 million units. Using the price as the independent variable x and the quantity demanded (in millions of units) as the dependent variable y, find the linear equation that models this relationship. According to this equation, at what price will sales drop to 0? (See Section 9.4.)

9.1 The Cartesian Coordinate System

René Descartes (1596 – 1650), a famous French mathematician, developed a system for solving geometric problems using algebra. This system is called the **Cartesian coordinate system** in his honor. Descartes based his system on a relationship between points in a plane and **ordered pairs** of real numbers. This section begins by relating algebraic formulas with ordered pairs and then shows how these ideas can be related to geometry.

Objective A **Equations in Two Variables**

Equations such as $d = 60t$, $I = 0.05P$, and $y = 2x + 3$ represent relationships between pairs of variables. For example, in the first equation, if $t = 3$, then $d = 60 \cdot 3 = 180$. With the understanding that t is first and d is second, we can represent $t = 3$ and $d = 180$ in the form of an ordered pair $(3, 180)$. In general, if t is the first number and d is the second number, then solutions to the equation $d = 60t$ can be written in the form of ordered pairs (t, d). Thus we see that $(180, 3)$ is different from $(3, 180)$. **The order of the numbers in an ordered pair is critical**.

We say that $(3, 180)$ **is a solution of** (or **satisfies**) the equation $d = 60t$. Similarly, $(5, 300)$ represents $t = 5$ and $d = 300$ and satisfies the equation $d = 60t$. In the same way, $(100, 5)$ satisfies $I = 0.05P$ where $P = 100$ and $I = 0.05 \cdot 100 = 5$. In this equation, solutions are ordered pairs in the form (P, I).

For the equation $y = 2x + 3$, ordered pairs are in the form (x, y), and $(2, 7)$ satisfies the equation. If $x = 2$, then substituting in the equation gives $y = 2 \cdot 2 + 3 = 7$. In the ordered pair (x, y), x is called the **first coordinate** and y is called the **second coordinate**. To find ordered pairs that satisfy an equation in two variables, we can **choose any value** for one variable and find the corresponding value for the other variable by substituting into the equation. For example:

for the equation $y = 2x + 3$,

Choices for x:	Substitution:	Ordered Pairs:
$x = 1$	$y = 2(1) + 3 = 5$	$(1, 5)$
$x = -2$	$y = 2(-2) + 3 = -1$	$(-2, -1)$
$x = \dfrac{1}{2}$	$y = 2\left(\dfrac{1}{2}\right) + 3 = 4$	$\left(\dfrac{1}{2}, 4\right)$

Table 1

All the ordered pairs $(1, 5), (-2, -1),$ and $\left(\dfrac{1}{2}, 4\right)$ satisfy the equation $y = 2x + 3$.

There is an infinite number of such ordered pairs. Any real number could have been chosen for x and the corresponding value for y calculated.

Since the equation $y = 2x + 3$ is solved for y, we say that the value of y "depends" on the choice of x. Thus in an ordered pair of the form (x, y), the first coordinate x is called the **independent variable** and the second coordinate y is called the **dependent variable**.

In the following table the first variable in each case is the independent variable and the second variable is the dependent variable. Corresponding ordered pairs would be of the form (t, d) and (x, y). The choices for the values of the independent variables are arbitrary. There are an infinite number of other values that could have just as easily been chosen.

$d = 60t$			$y = 2x + 3$		
t	d	(t, d)	x	y	(x, y)
5	$60(5) = 300$	$(5, 300)$	-2	$2(-2) + 3 = -1$	$(-2, -1)$
10	$60(10) = 600$	$(10, 600)$	-1	$2(-1) + 3 = 1$	$(-1, 1)$
12	$60(12) = 720$	$(12, 720)$	0	$2(0) + 3 = 3$	$(0, 3)$
15	$60(15) = 900$	$(15, 900)$	3	$2(3) + 3 = 9$	$(3, 9)$

Table 2

Objective B **Graphing Ordered Pairs**

The Cartesian coordinate system relates algebraic equations and ordered pairs to geometry. In this system, two number lines intersect at right angles and separate the plane into four **quadrants**. The **origin**, designated by the ordered pair $(0, 0)$, is the point of intersection of the two lines. The horizontal number line is called the **horizontal axis** or **x-axis**. The vertical number line is called the **vertical axis** or **y-axis**. Points that lie on either axis are not in any quadrant. They are simply on an axis (Figure 1).

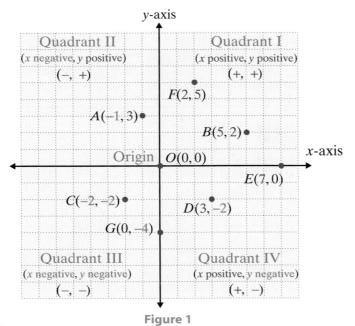

Figure 1

The following important relationship between ordered pairs of real numbers and points in a plane is the cornerstone of the Cartesian coordinate system.

One-to-One Correspondence

There is a **one-to-one correspondence** between points in a plane and ordered pairs of real numbers.

In other words, for each point in a plane there is one and only one corresponding ordered pair of real numbers, and for each ordered pair of real numbers there is one and only one corresponding point in the plane.

The **graphs of the points** $A(2, 1)$, $B(-2, 3)$, $C(-3, -2)$, $D(1, -2)$, and $E(3, 0)$ are shown in Figure 2. (**Note:** An ordered pair of real numbers and the corresponding point on the graph are frequently used to refer to each other. Thus, the ordered pair $(2, 1)$ and the point $(2, 1)$ are interchangeable ideas.)

Point	Quadrant
$A(2, 1)$	I
$B(-2, 3)$	II
$C(-3, -2)$	III
$D(1, -2)$	IV
$E(3, 0)$	x-axis

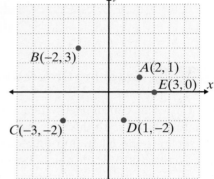

Figure 2

notes

■ Unless otherwise stated, assume that the grid lines (as shown here in Figure 2) are one unit apart.

■

Example 1

Graphing Ordered Pairs

Graph the sets of ordered pairs.

a. $\left\{ A(-2, 1), B(-1, -4), C(0, 2), D(1, 3), E(2, -3) \right\}$

Note: The listing of ordered pairs within the braces can be in any order.

Solution

To locate points: start at the **origin** $(0,0)$, move left or right for the x-coordinate and up or down for the y-coordinate.

For $A(-2,1)$, move 2 units left and 1 unit up.

For $B(-1,-4)$, move 1 unit left and 4 units down.

For $C(0,2)$, move no units left or right and 2 units up.

For $D(1,3)$, move 1 unit right and 3 units up.

For $E(2,-3)$, move 2 units right and 3 units down.

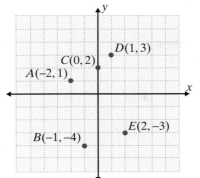

b. $\{A(-1,3), B(0,1), C(1,-1), D(2,-3), E(3,-5)\}$

Solution

To locate each point, start at the **origin**, and:

For $A(-1,3)$, move 1 unit left and 3 units up.

For $B(0,1)$, move no units left or right and 1 unit up.

For $C(1,-1)$, move 1 unit right and 1 unit down.

For $D(2,-3)$, move 2 units right and 3 units down.

For $E(3,-5)$, move 3 units right and 5 units down.

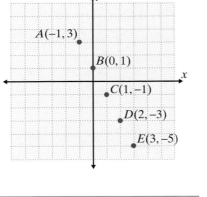

Now work margin exercise 1.

1. Graph the sets of ordered pairs.

a. $A(3,-1), B(-1,-4),$ $C(4,0), D(-3,2),$ $E(3,3)$

b. $A(0,2), B(-3,-1),$ $C(-2,0), D(4,-3),$ $E(3,1)$

Objective C   **Finding Ordered Pairs that Satisfy a Given Equation**

The points (ordered pairs) in Example 1b can be shown to satisfy the equation $y = -2x + 1$. For example, using $x = -1$ in the equation yields,

$$y = -2(-1) + 1 = 2 + 1 = 3$$

and the ordered pair $(-1, 3)$ satisfies the equation. Similarly, letting $y = 1$ gives,

$$1 = -2x + 1$$
$$0 = -2x$$
$$0 = x$$

and the ordered pair $(0, 1)$ satisfies the equation.

We can write all the ordered pairs in Example 1b in table form.

x	−2x + 1 = y	(x, y)
−1	$-2(-1)+1=3$	$(-1, 3)$
0	$-2(0)+1=1$	$(0, 1)$
1	$-2(1)+1=-1$	$(1, -1)$
2	$-2(2)+1=-3$	$(2, -3)$
3	$-2(3)+1=-5$	$(3, -5)$

Table 3

Example 2

Determining Ordered Pairs

a. Determine which, if any, of the ordered pairs $(0, -2)$, $\left(\frac{2}{3}, 0\right)$, and $(2, 5)$ satisfy the equation $y = 3x - 2$.

Solution

We will substitute 0, $\frac{2}{3}$, and 2 for x in the equation $y = 3x - 2$ and see if the corresponding y-values match those in the given ordered pairs.

$x = 0$:　　$y = 3(0) - 2 = -2$ so, $(0, -2)$ satisfies the equation.

$x = \frac{2}{3}$:　　$y = 3\left(\frac{2}{3}\right) - 2 = 0$ so, $\left(\frac{2}{3}, 0\right)$ satisfies the equation.

$x = 2$:　　$y = 3(2) - 2 = 4$ so, $(2, 4)$ satisfies the equation.

The point $(2, 5)$ does not satisfy the equation $y = 3x - 2$ because, as just illustrated, $y = 4$ when $x = 2$, not 5.

b. Determine the missing coordinate in each of the following ordered pairs so that the points will satisfy the equation $2x + 3y = 12$: $(0, \ )$, $(3, \ )$, $(\ , 0)$, $(\ , -2)$.

Solution

The missing values can be found by substituting the given values for x (or for y) into the equation $2x + 3y = 12$ and solving for the other variable.

For $(0, \ )$, let $x = 0$:

$$2(0) + 3y = 12$$
$$3y = 12$$
$$y = 4$$

For $(3, \ )$, let $x = 3$:

$$2(3) + 3y = 12$$
$$6 + 3y = 12$$
$$3y = 6$$
$$y = 2$$

The ordered pair is $(0, 4)$.

The ordered pair is $(3, 2)$.

For $(\ , 0)$, let $y = 0$:

$$2x + 3(0) = 12$$
$$2x = 12$$
$$x = 6$$

For $(\ , -2)$, let $y = -2$:

$$2x + 3(-2) = 12$$
$$2x - 6 = 12$$
$$2x = 18$$
$$x = 9$$

The ordered pair is $(6, 0)$.

The ordered pair is $(9, -2)$.

c. Complete the table below so that each ordered pair will satisfy the equation $y = 1 - 2x$.

x	$y = 1 - 2x$	y	(x,y)
0			
		3	
$\frac{1}{2}$			
5			

Solution

Substituting each given value for x or y into the equation $y = 1 - 2x$ gives the following table of ordered pairs.

x	$y = 1 - 2x$	y	(x,y)
0	$1 = 1 - 2(\mathbf{0})$	**1**	$(\mathbf{0}, \mathbf{1})$
−1	$3 = 1 - 2(-1)$	**3**	$(\mathbf{-1}, \mathbf{3})$
$\frac{1}{2}$	$0 = 1 - 2\left(\dfrac{1}{2}\right)$	**0**	$\left(\dfrac{1}{2}, 0\right)$
5	$-9 = 1 - 2(\mathbf{5})$	**−9**	$(\mathbf{5}, \mathbf{-9})$

For $x = \mathbf{0}$:

$$y = 1 - 2(\mathbf{0}) = 1$$

For $y = \mathbf{3}$:

$$(\mathbf{3}) = 1 - 2x$$
$$2 = -2x$$
$$-1 = x$$

For $x = \dfrac{\mathbf{1}}{\mathbf{2}}$:

$$y = 1 - 2\left(\dfrac{1}{2}\right) = 0$$

For $x = \mathbf{5}$:

$$y = 1 - 2(\mathbf{5}) = -9$$

Now work margin exercise 2.

2. a. Determine which, if any, of the ordered pairs $(0, -3), (3, 6), (2, 5)$ satisfy the equation $y = 4x - 3$.

b. Determine the missing coordinate in each of the following ordered pairs so that the points will satisfy the equation $x + 5y = 15$.

$(0, \ ), (5, \ )$
$(\ , 0), (\ , -3)$

c. Complete the table below so that each ordered pair will satisfy the equation $y = 2 - 3x$.

x	y	(x,y)
0		
	1	
−2		
$\frac{2}{3}$		

notes

- Although this discussion is related to ordered pairs of real numbers, most of the examples use ordered pairs of **integers**. This is because ordered pairs of integers are relatively easy to locate on a graph and relatively easy to read from a graph. Ordered pairs with fractions, decimals, or radicals must be located by estimating the positions of the points. The precise coordinates intended for such points can be difficult or impossible to read because large dots must be used so the points can be seen. **Even with these difficulties, you should understand that we are discussing ordered pairs of real numbers and that points with fractions, decimals, and radicals as coordinates do exist and should be plotted by estimating their positions.**

Objective D **Locating Points on a Line**

Example 3

Reading Points on a Graph

The graphs of two lines are given. Each line contains an infinite number of points. Use the grid to help you locate (or estimate) three points on each line.

a.

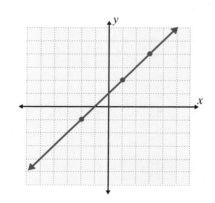

Solution

Three points on this graph are $(-2, -1)$, $(1, 2)$, and $(3, 4)$. (Of course there is more than one correct answer to this type of question. Use your own judgement.)

b.

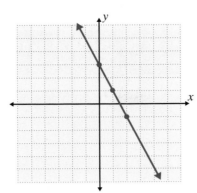

3. The graphs of two lines are given. Use the grid to locate three points on each line.

a.

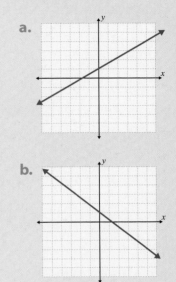

b.

Solution

Three points on this graph are $(0,3)$, $(1,1)$, and $(2,-1)$. (You may also estimate with fractions. For example, one point appears to be approximately $\left(\frac{1}{2}, 2\right)$.)

Now work margin exercise 3.

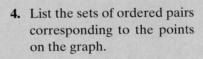

Practice Problems

1. Determine which ordered pairs satisfy the equation $3x + y = 14$.

   a. $(5, -1)$ **b.** $(4, 2)$ **c.** $(-1, 17)$

2. Given $3x + y = 5$, find the missing coordinate of each ordered pair so that it will satisfy the equation.

   a. $(0, \ )$ **b.** $\left(\frac{1}{3}, \ \right)$ **c.** $(\ , 2)$

3. Complete the table so that each ordered pair will satisfy the equation $y = \frac{2}{3}x + 1$.

x	y
0	
	-2
-3	
6	

4. List the sets of ordered pairs corresponding to the points on the graph.

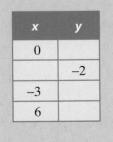

Practice Problem Answers

1. all satisfy the equation   2. **a.** $(0, 5)$ **b.** $\left(\frac{1}{3}, 4\right)$ **c.** $(1, 2)$

3. $(0, 1), \left(-\frac{9}{2}, -2\right), (-3, -1), (6, 5)$

4. $\{A(-2, -2), B(0, 0), C(0, 5), D(3, -1)\}$

Exercises 9.1

1.

2.

3.

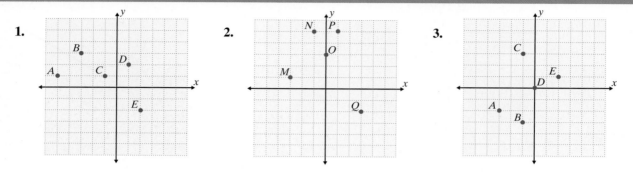

4.

5.

6.

7.

8.

9.

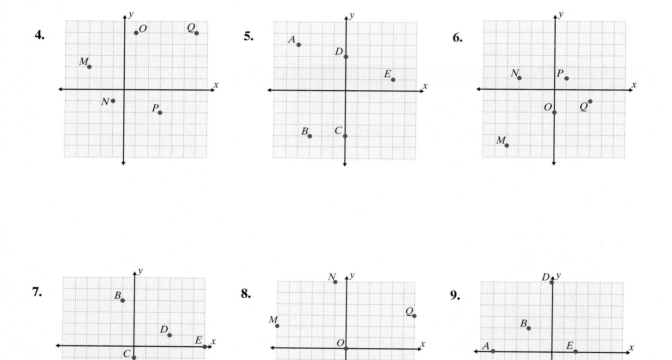

10.

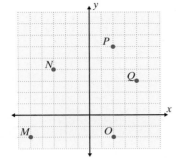

11. $\{A(4,-1), B(3,2), C(0,5), D(1,-1), E(1,4)\}$

12. $\{A(-1,-1), B(-3,-2), C(1,3), D(0,0), E(2,5)\}$

13. $\{A(1,2), B(0,2), C(-1,2), D(2,2), E(-3,2)\}$

14. $\{A(-1,4), B(0,-3), C(2,-1), D(4,1), E(-1,-1)\}$

15. $\{A(1,0), B(3,0), C(-2,1), D(-1,1), E(0,0)\}$

16. $\{A(-1,-1), B(0,1), C(1,3), D(2,5), E(3,10)\}$

17. $\{A(4,1), B(0,-3), C(1,-2), D(2,-1), E(-4,2)\}$

18. $\{A(0,1), B(1,0), C(2,-1), D(3,-2), E(4,-3)\}$

19. $\{A(1,4), B(-1,-2), C(0,1), D(2,7), E(-2,-5)\}$

20. $\{A(0,0), B(-1,3), C(3,-2), D(0,4), E(-7,0)\}$

21. $\left\{A(1,-3), B\left(-4,\frac{3}{4}\right), C\left(2,-2\frac{1}{2}\right), D\left(\frac{1}{2},4\right)\right\}$

22. $\left\{A\left(\frac{3}{4},\frac{1}{2}\right), B\left(2,-\frac{5}{4}\right), C\left(\frac{1}{3},-2\right), D\left(-\frac{5}{3},2\right)\right\}$

23. $\{A(1.6, -2), B(3, 2.5), C(-1, 1.5), D(0, -2.3)\}$ **24.** $\{A(-2, 2), B(-3, 1.6), C(3, 0.5), D(1.4, 0)\}$

Determine which, if any, of the ordered pairs satisfy the given equations.

25. $2x - y = 4$
a. $(1, 1)$
b. $(2, 0)$
c. $(1, -2)$
d. $(3, 2)$

26. $x + 2y = -1$
a. $(1, -1)$
b. $(1, 0)$
c. $(2, 1)$
d. $(3, -2)$

27. $4x + y = 5$
a. $\left(\dfrac{3}{4}, 2\right)$
b. $(4, 0)$
c. $(1, 1)$
d. $(0, 3)$

28. $2x - 3y = 7$
a. $(1, 3)$
b. $\left(\dfrac{1}{2}, -2\right)$
c. $\left(\dfrac{7}{2}, 0\right)$
d. $(2, 1)$

29. $2x + 5y = 8$
a. $(4, 0)$
b. $(2, 1)$
c. $(1, 1.2)$
d. $(1.5, 1)$

30. $3x + 4y = 10$
a. $(-2, 3)$
b. $(0, 2.5)$
c. $(4, -2)$
d. $(1.2, 1.6)$

Determine the missing coordinate in each of the ordered pairs so that the point will satisfy the equation given.

31. $x - y = 4$
a. $(0, \quad)$
b. $(2, \quad)$
c. $(\quad, 0)$
d. $(\quad, -3)$

32. $x + y = 7$
a. $(0, \quad)$
b. $(-1, \quad)$
c. $(\quad, 0)$
d. $(\quad, 3)$

33. $x + 2y = 6$
a. $(0, \quad)$
b. $(2, \quad)$
c. $(\quad, 0)$
d. $(\quad, 4)$

34. $3x + y = 9$
a. $(0, \quad)$
b. $(4, \quad)$
c. $(\quad, 0)$
d. $(\quad, 3)$

35. $4x - y = 8$
a. $(0, \quad)$
b. $(1, \quad)$
c. $(\quad, 0)$
d. $(\quad, 4)$

36. $x - 2y = 2$
a. $(0, \quad)$
b. $(4, \quad)$
c. $(\quad, 0)$
d. $(\quad, 3)$

37. $2x + 3y = 6$
a. $(0, \quad)$
b. $\left(-1, \quad\right)$
c. $(\quad, 0)$
d. $(\quad, -2)$

38. $5x + 3y = 15$
a. $(0, \quad)$
b. $\left(2, \quad\right)$
c. $(\quad, 0)$
d. $\left(\quad, 4\right)$

39. $3x - 4y = 7$
a. $\left(0, \quad\right)$
b. $(1, \quad)$
c. $\left(\quad, 0\right)$
d. $\left(\quad, \dfrac{1}{2}\right)$

40. $2x + 5y = 6$
a. $\left(0, \quad\right)$
b. $\left(\dfrac{1}{2}, \quad\right)$
c. $(\quad, 0)$
d. $(\quad, 2)$

Complete the tables so that each ordered pair will satisfy the given equation. Graph the resulting sets of ordered pairs.

41. $y = 3x$

x	y
0	
	-3
-2	
	6

42. $y = -2x$

x	y
0	
	4
3	
	-2

43. $y = 2x - 3$

x	y
0	
	-1
-2	
	3

44. $y = 3x + 5$

x	y
0	
	-4
-2	
	2

45. $y = 9 - 3x$

x	y
0	
	0
1	
	-3

46. $y = 6 - 2x$

x	y
0	
	0
-2	
	-2

47. $y = \dfrac{3}{4}x + 2$

x	y
0	
	5
-4	
	$\dfrac{5}{4}$

48. $y = \dfrac{3}{2}x - 1$

x	y
0	
	2
-2	
	$-\dfrac{5}{2}$

49. $3x - 5y = 9$

x	y
0	
	0
-2	
	-1

50. $4x + 3y = 6$

x	y
0	
	0
3	
	-1

51. $5x - 2y = 10$

x	y
0	
	0
-1	
	5

52. $3x - 2y = 12$

x	y
	0
0	
	-3
6	

53. $2x + 3.2y = 6.4$

x	y
0	
3.2	
	0.8
	-0.2

54. $3x + y = -2.4$

x	y
	0
0	
	0.6
1.6	

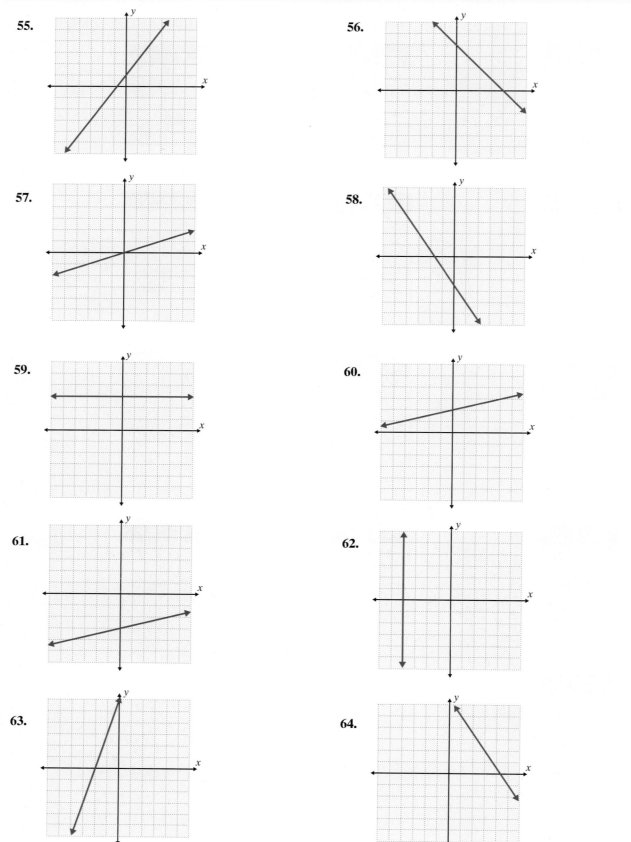

55.

56.

57.

58.

59.

60.

61.

62.

63.

64.

65. **Exchange Rate:** At one point in 2010, the current exchange rate from U.S. dollars to Euros was $E = 0.7802D$ where E is euros and D is dollars.

   a. Make a table of ordered pairs for the values of D and E if D has the values $100, $200, $300, $400, and $500.

   b. Graph the points corresponding to the ordered pairs.

D	E
100	
200	
300	
400	
500	

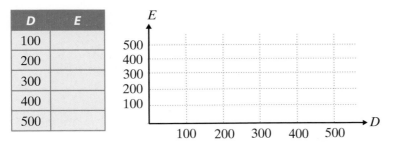

66. **Temperature:** Given the equation $F = \dfrac{9}{5}C + 32$ where C is temperature in degrees Celsius and F is the corresponding temperature in degrees Fahrenheit:

   a. Make a table of ordered pairs for the values of C and F if C has the values $-20°, -10°, -5°, 0°, 5°, 10°,$ and $15°$.

   b. Graph the points corresponding to the ordered pairs.

C	F
-20	
-10	
-5	
0	
5	
10	
15	

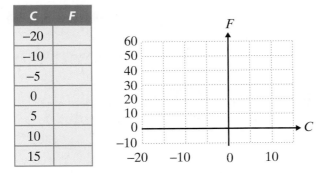

67. **Falling Objects:** Given the equation $d = 16t^2$, where d is the distance an object falls in feet and t is the time in seconds that the object falls:

   a. Make a table of ordered pairs for the values of t and d with the values of $1, 2, 3.5, 4, 4.5,$ and 5 for t seconds.

   b. Graph the points corresponding to the ordered pairs.

   c. These points do not lie on a straight line. What feature of the equation might indicate to you that the graph is not a straight line?

t	d
1	
2	
3.5	
4	
4.5	
5	

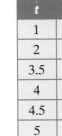

68. Volume: Given the equation $V = 9h$, where V is the volume (in cubic centimeters) of a box with a variable height h in centimeters and a fixed base of area 9 cm².

    a. Make a table of ordered pairs for the values of h and V with h as the values 2 cm, 3 cm, 5 cm, 8 cm, 9 cm, and 10 cm.

    b. Graph the points corresponding to the ordered pairs.

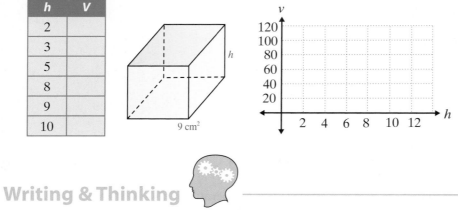

h	V
2	
3	
5	
8	
9	
10	

Writing & Thinking

In statistics, data is sometimes given in the form of ordered pairs where each ordered pair represents two pieces of information about one person. For example, ordered pairs might represent the height and weight of a person or the person's number of years of education and that person's annual income. The ordered pairs are plotted on a graph and the graph is called a **scatter diagram** (or **scatter plot**). Such scatter diagrams are used to see if there is any pattern to the data and, if there is, then the diagram is used to predict the value for one of the variables if the value of the other is known. For example, if you know that a person's height is 5 ft 6 in., then his or her weight might be predicted from information indicated in a scatter diagram that has several points of known information about height and weight.

69.   **a.** The following table of values indicates the number of push-ups and the number of sit-ups that ten students did in a physical education class. Plot these points in a scatter diagram.

Person	#1	#2	#3	#4	#5	#6	#7	#8	#9	#10
x (push-ups)	20	15	25	23	35	30	42	40	25	35
y (sit-ups)	25	20	20	30	32	36	40	45	18	40

    b. Does there seem to be a pattern in the relationship between push-ups and sit-ups? What is this pattern?

    c. Using the scatter diagram in part **a.**, predict the number of sit-ups that a student might be able to do if he or she has just done each of the following numbers of push-ups: 22, 32, 35, and 45. (**Note:** In each case, there is no one correct answer. The answers are only estimates based on the diagram.)

70. Ask ten friends or fellow students what their height and shoe size is. (You may want to ask all men or all women since the scale for men and womens' shoe sizes is different.) Organize the data in table form and then plot the corresponding scatter diagram. Knowing your own height, does the pattern indicated in the scatter diagram seem to predict your shoe size?

71. Ask ten friends or fellow students what their height and age is. Organize the data in table form and then plot the corresponding scatter diagram. Knowing your own height, does the pattern indicated in the scatter diagram seem to predict your age? Do you think that all scatter diagrams can be used to predict information related to the two variables graphed? Explain.

9.2 Graphing Linear Equations in Two Variables: $Ax + By = C$

Objective A   **The Standard Form: $Ax + By = C$**

In Section 9.1, we discussed ordered pairs of real numbers and graphed a few points (ordered pairs) that satisfied particular equations. Now suppose we want to graph all the points that satisfy an equation such as:

$$3 - 3x = y.$$

The **solution set** for equations of this type (in the two variables x and y) consists of an infinite set of ordered pairs in the form (x, y) that satisfy the equation.

To find some of the solutions of the equation $3 - 3x = y$, we form a table (as we did in Section 9.1) by:

1. choosing arbitrary values for x and
2. finding the corresponding values for y by substituting into the equation.

In Figure 1, we have found five ordered pairs that satisfy the equation and graphed the corresponding points.

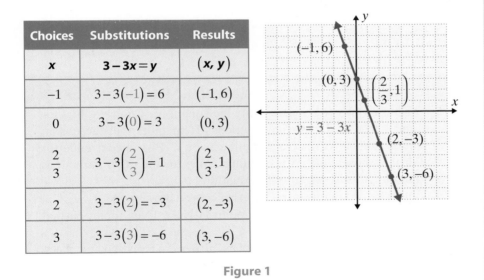

Choices	Substitutions	Results
x	$3 - 3x = y$	(x, y)
-1	$3 - 3(-1) = 6$	$(-1, 6)$
0	$3 - 3(0) = 3$	$(0, 3)$
$\dfrac{2}{3}$	$3 - 3\left(\dfrac{2}{3}\right) = 1$	$\left(\dfrac{2}{3}, 1\right)$
2	$3 - 3(2) = -3$	$(2, -3)$
3	$3 - 3(3) = -6$	$(3, -6)$

Figure 1

The five points in Figure 1 appear to lie on a line. They, in fact, do lie on a line, and any ordered pair that satisfies the equation $y = 3 - 3x$ will also lie on that same line.

Just as we use the terms **ordered pair** and **point** (the graph of an ordered pair) interchangeably, we use the terms **equation** and **graph of an equation** interchangeably. The equations

$$2x + 3y = 4, \quad y = -5, \quad x = 1.4, \quad \text{and} \quad y = 3x + 2$$

are called **linear equations**, and their graphs are lines on the Cartesian plane.

Standard Form of a Linear Equation

Any equation of the form

$$Ax + By = C,$$

where A, B, and C are real numbers and A and B are not both equal to 0, is called the **standard form** of a **linear equation**.

notes

▪ Note that in the standard form $Ax + By = C$, A and B may be positive, negative, or 0, but A and B cannot **both** equal 0.

▪

Objective B   **Graphing Linear Equations**

Every line corresponds to some linear equation, and the graph of every linear equation is a line. We know from geometry that **two points determine a line.** This means that the graph of a linear equation can be found by locating any two points that satisfy the equation.

To Graph a Linear Equation in Two Variables

1. Locate any two points that satisfy the equation. (Choose values for x and y that lead to simple solutions. Remember that there is an infinite number of choices for either x or y. But, once a value for x or y is chosen, the corresponding value for the other variable is found by substituting into the equation.)
2. Plot these two points on a Cartesian coordinate system.
3. Draw a line through these two points. (**Note:** Every point on that line will satisfy the equation.)
4. **To check:** Locate a third point that satisfies the equation and check to see that it does indeed lie on the line.

Example 1

Graphing a Linear Equation in Two Variables

Graph each of the following linear equations.

a. $2x + 3y = 6$

Solution

Make a table with headings x and y and, whenever possible, **choose values for x or y that lead to simple solutions for the other variable.** (Values chosen for x and y are shown in red.)

x	$2x + 3y = 6$	y
0	$2(0) + 3y = 6$	2
−3	$2(-3) + 3y = 6$	4
3	$2x + 3(0) = 6$	0
$\dfrac{5}{2}$	$2x + 3\left(\dfrac{1}{3}\right) = 6$	$\dfrac{1}{3}$

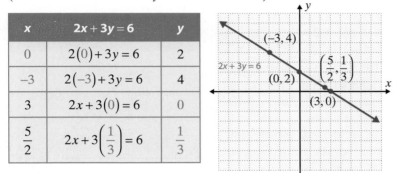

b. $x - 2y = 1$

Solution

Solve the equation for x ($x = 2y + 1$) and substitute 0, 1, and 2 for y.

Results	Substitutions	Choices
x	$x = 2y + 1$	y
1	$x = 2(0) + 1$	0
3	$x = 2(1) + 1$	1
5	$x = 2(2) + 1$	2

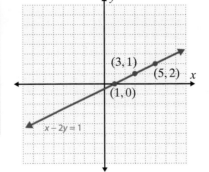

c. $y = 2x$

Solution

Substitute −1, 0, and 1 for x.

Choices	Substitutions	Results
x	$y = 2x$	y
−1	$y = 2(-1)$	−2
0	$y = 2(0)$	0
1	$y = 2(1)$	2

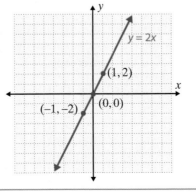

Now work margin exercise 1.

1. Graph each of the following linear equations.

a. $3x + 2y = 6$

b. $2x - y = 2$

c. $y = 3x$

Objective C   **Locating the *y*-intercept and *x*-intercept**

While the choice of the values for *x* or *y* can be arbitrary, letting $x = 0$ will locate the point on the graph where the line crosses (or intercepts) the *y*-axis. This point is called the **y-intercept** and is of the form $(0, y)$. The **x-intercept** is the point found by letting $y = 0$. This is the point where the line crosses (or intercepts) the *x*-axis and is of the form $(x, 0)$. These two points are generally easy to locate and are frequently used as the two points for drawing the graph of a linear equation. If the line passes through the point $(0, 0)$, then the *y*-intercept and the *x*-intercept are the same point, namely the origin. In this case you will need to locate some other point to draw the graph.

Intercepts

1. To find the **y-intercept** (where the line crosses the *y*-axis), substitute $x = 0$ and solve for *y*.
2. To find the **x-intercept** (where the line crosses the *x*-axis), substitute $y = 0$ and solve for *x*.

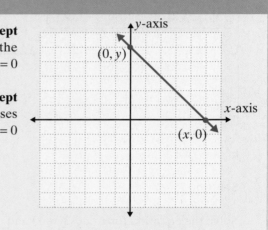

Example 2

x- and *y*-Intercepts

Graph the following linear equations by locating the *y*-intercept and the *x*-intercept.

a. $x + 3y = 9$

Solution

$x = 0 \rightarrow (0) + 3y = 9$
$\qquad\qquad\quad 3y = 9$
$\qquad\qquad\quad\ \ y = 3$

$(0, 3)$ is the *y*-intercept.

$y = 0 \rightarrow x + 3(0) = 9$
$\qquad\qquad\quad x = 9$
$(9, 0)$ is the *x*-intercept.
Plot the two intercepts and draw the line that contains them.

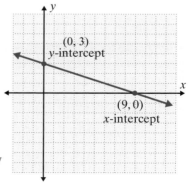

b. $3x - 2y = 12$

Solution

$x = 0 \rightarrow 3(0) - 2y = 12$

$-2y = 12$

$y = -6$

$(0, -6)$ is the y-intercept.

$y = 0 \rightarrow 3x - 2(0) = 12$

$3x = 12$

$x = 4$

$(4, 0)$ is the x-intercept.
Plot the two intercepts and
draw the line that contains them.

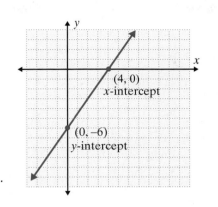

Graph the following linear
equations by locating
the x-intercept and the
y-intercept.

2. a. $x + 2y = 6$

   b. $5x - 3y = 15$

3. $4x - y = 8$

Completion Example 3

x- and *y*-Intercepts

Graph the following linear equation by locating the y-intercept and
the x-intercept.

$x - 5y = 5$

Solution

$x = 0 \rightarrow \underline{\quad} - 5y = 5$

$\underline{\qquad\qquad}$

$\underline{\qquad\qquad}$

$\underline{\qquad}$ is the y-intercept.

$y = 0 \rightarrow x - 5 \cdot \underline{\quad} = 5$

$\underline{\qquad\qquad}$

$\underline{\qquad\qquad}$

$\underline{\qquad}$ is the x-intercept.
Plot the two intercepts and
draw the line that contains them.

Now work margin exercises **2** *and* **3.**

Completion Example Answers

$x = 0 \;\rightarrow\; \underline{0} - 5y = 5$

$\qquad\qquad \dfrac{\;-5y = 5\;}{}$

$\qquad\qquad \dfrac{\;y = -1\;}{}$

$\underline{(0,-1)}$ is the *y*-intercept.

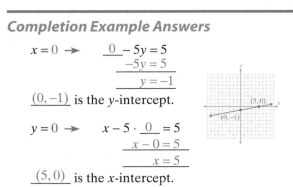

$y = 0 \;\rightarrow\; x - 5 \cdot \underline{0} = 5$

$\qquad\qquad \dfrac{\;x - 0 = 5\;}{}$

$\qquad\qquad \dfrac{\;x = 5\;}{}$

$\underline{(5,0)}$ is the *x*-intercept.

---

notes

▪ In general, the intercepts are easy to find because substituting 0 for **x** or **y** leads to an easy solution for the other variable. However, when
▪ the intercepts result in a point with fractional (or decimal) coordinates and estimation is involved, then a third point that satisfies the equation
▪ should be found to verify that the line is graphed correctly.

Practice Problems

1. Find the missing coordinate of each ordered pair so that it belongs to the solution set of the equation $2x + y = 4$:

$$(0,\ \ ),\ (\ \ ,0),\ (\ \ ,8),\ (-1,\ \ ).$$

2. Does the ordered pair $\left(1, \dfrac{3}{2}\right)$ satisfy the equation $3x + 2y = 6$?

3. Find the *x*-intercept and *y*-intercept of the equation $-3x + y = 9$.

4. Graph the linear equation $x - 2y = 3$.

Practice Problem Answers

1. $(0,4),(2,0),(-2,8),(-1,6)$ 2. yes

3. *x*-intercept $= (-3,0)$,   *y*-intercept $= (0,9)$ **4.**

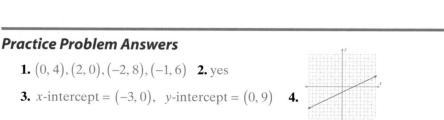

Exercises 9.2

Use your knowledge of *x*-intercepts and *y*-intercepts to match the equations in Exercises 1 through 6 with graphs a. through f.

1. $4x + 3y = 12$

2. $4x - 3y = 12$

3. $x + 2y = 8$

4. $-x + 2y = 8$

5. $x + 4y = 0$

6. $5x - y = 10$

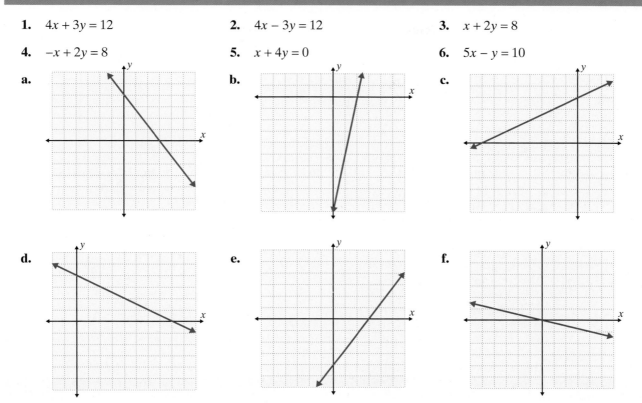

Locate at least two ordered pairs of real numbers that satisfy each of the linear equations and graph the corresponding line in the Cartesian coordinate system.

7. $x + y = 3$

8. $x + y = 4$

9. $y = x$

10. $2y = x$

11. $2x + y = 0$

12. $3x + 2y = 0$

13. $2x + 3y = 7$

14. $4x + 3y = 11$

15. $3x - 4y = 12$

16. $2x - 5y = 10$

17. $y = 4x + 4$

18. $y = x + 2$

19. $3y = 2x - 4$ **20.** $4x = 3y + 8$ **21.** $3x + 5y = 6$ **22.** $2x + 7y = -4$

23. $2x + 3y = 1$ **24.** $5x - 3y = -1$ **25.** $5x - 2y = 7$ **26.** $3x + 4y = 7$

27. $\dfrac{2}{3}x - y = 4$ **28.** $x + \dfrac{3}{4}y = 6$ **29.** $2x + \dfrac{1}{2}y = 3$ **30.** $\dfrac{2}{5}x - 3y = 5$

31. $5x = y + 2$ **32.** $4x = 3y - 5$

Graph the following linear equations by locating the *x*-intercept and the *y*-intercept.

33. $x + y = 6$ **34.** $x + y = 4$ **35.** $x - 2y = 8$ **36.** $x - 3y = 6$

37. $4x + y = 8$ **38.** $x + 3y = 9$ **39.** $x - 4y = -6$ **40.** $x - 6y = 3$

41. $y = 4x - 10$ **42.** $y = 2x - 9$ **43.** $3x - 2y = 6$ **44.** $5x + 2y = 10$

45. $2x + 3y = 12$ **46.** $3x + 7y = -21$ **47.** $3x - 7y = -21$ **48.** $3x + 2y = 15$

49. $5x + 3y = 7$ **50.** $2x + 3y = 5$ **51.** $y = \dfrac{1}{2}x - 4$ **52.** $y = -\dfrac{1}{3}x + 3$

53. $\dfrac{2}{3}x - 3y = 4$ **54.** $\dfrac{1}{2}x + 2y = 3$ **55.** $\dfrac{1}{2}x - \dfrac{3}{4}y = 6$ **56.** $\dfrac{2}{3}x + \dfrac{4}{3}y = 8$

Writing & Thinking

57. Explain, in your own words, why it is sufficient to find the *x*-intercept and *y*-intercept to graph a line (assuming that they are not the same point).

58. Explain, in your own words, how you can determine if an ordered pair is a solution to an equation.

9.3 The Slope-Intercept Form: $y = mx + b$

Objective A The Meaning of Slope

If you ride a bicycle up a mountain road, you certainly know when the **slope** (a measure of steepness called the **grade** for roads) increases because you have to pedal harder. The contractor who built the road was aware of the **slope** because trucks traveling the road must be able to control their downhill speed and be able to stop in a safe manner. A carpenter given a set of house plans calling for a roof with a **pitch** of 7 : 12 knows that for every 7 feet of rise (vertical distance) there are 12 feet of run (horizontal distance). That is, the ratio of rise to run is $\dfrac{rise}{run} = \dfrac{7}{12}$.

Figure 1

Note that this ratio can be in units other than feet, such as inches or meters. (See Figure 2.)

$$\frac{rise}{run} = \frac{7 \text{ inches}}{12 \text{ inches}} = \frac{3.5 \text{ feet}}{6 \text{ feet}} = \frac{14 \text{ feet}}{24 \text{ feet}}$$

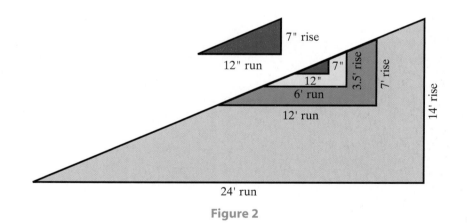

Figure 2

For a line, the **ratio of rise to run** is called the **slope of the line**. The graph of the linear equation $y = \dfrac{1}{3}x + 2$ is shown in Figure 3. What do you think is the slope of the line? Do you think that the slope is positive or negative? Do you think the slope might be $\dfrac{1}{3}$? $\dfrac{3}{1}$? 2?

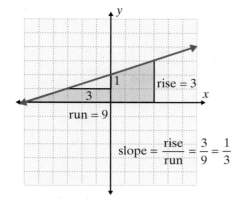

Figure 3

The concept of slope also relates to situations that involve **rate of change**. For example, the graphs in Figure 4 illustrate slope as miles per hour that a car travels and as pages per minute that a printer prints.

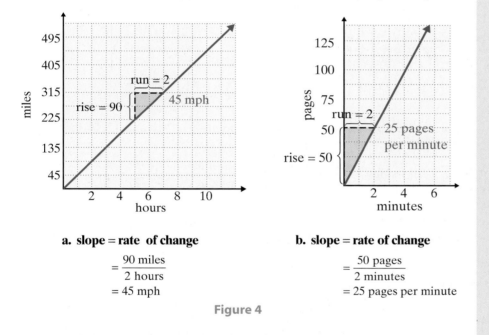

a. slope = rate of change

$$= \frac{90 \text{ miles}}{2 \text{ hours}}$$

$$= 45 \text{ mph}$$

b. slope = rate of change

$$= \frac{50 \text{ pages}}{2 \text{ minutes}}$$

$$= 25 \text{ pages per minute}$$

Figure 4

In general, the ratio of a change in one variable (say y) to a change in another variable (say x) is called the **rate of change of y with respect to x**. Figure 5 shows how the rate of change (the slope) can change over periods of time.

Source: Consumer Electronics Association   **Source:** U.S. Dept. of Commerce

Figure 5

Consider the line $y = 2x + 3$ and two points on the line $P_1(-2, -1)$ and $P_2(2, 7)$ as shown in Figure 6. (**Note:** In the notation P_1, 1 is called a **subscript** and P_1 is read "P sub 1". Similarly, P_2 is read "P sub 2." Subscripts are used in "labeling" and are not used in calculations.)

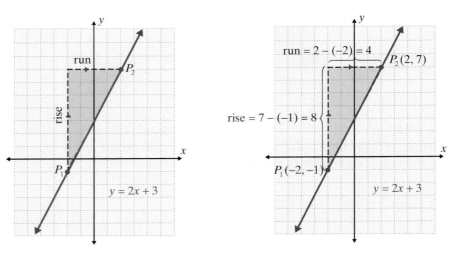

Figure 6

For the line $y = 2x + 3$ and using the points $(-2, -1)$ and $(2, 7)$ that are on the line,

$$\textbf{slope} = \frac{\text{rise}}{\text{run}} = \frac{\text{difference in } y\text{-values}}{\text{difference in } x\text{-values}} = \frac{7 - (-1)}{2 - (-2)} = \frac{8}{4} = 2.$$

From similar illustrations and the use of subscript notation, we can develop the following formula for the slope of any line.

Slope

Let $P_1(x_1, y_1)$ and $P_2(x_2, y_2)$ be two points on a line. The **slope** can be calculated as follows:

$$\textbf{slope} = m = \frac{\textbf{rise}}{\textbf{run}} = \frac{y_2 - y_1}{x_2 - x_1}.$$

Note: The letter m is standard notation for representing the slope of a line.

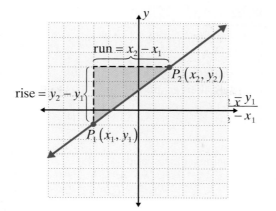

Figure 7

Example 1

Finding the Slope of a Line

Find the slope of the line that contains the points $(-1, 2)$ and $(3, 5)$, and then graph the line.

Solution

Using $(-1, 2)$ and $(3, 5)$, slope $= m = \dfrac{y_2 - y_1}{x_2 - x_1}$

$(x_1, y_1) \qquad (x_2, y_2)$

$\qquad\qquad = \dfrac{5 - 2}{3 - (-1)}$

$\qquad\qquad = \dfrac{3}{4}$

Or, using $(3, 5)$ and $(-1, 2)$,

$(x_1, y_1) \qquad (x_2, y_2)$

slope $= m = \dfrac{y_2 - y_1}{x_2 - x_1}$

$\qquad = \dfrac{2 - 5}{-1 - 3}$

$\qquad = \dfrac{-3}{-4}$

$\qquad = \dfrac{3}{4}$

1. Find the slope of the line that contains the points $(3, -1)$ and $(4, 2)$, and then graph the line.

Now work margin exercise **1.**

As we see in Example 1, **the slope is the same even if the order of the points is reversed**. The important part of the procedure is that **the coordinates must be subtracted in the same order in both the numerator and the denominator**.

In general,

$$\text{slope} = \frac{y_2 - y_1}{x_2 - x_1} = \frac{y_1 - y_2}{x_1 - x_2}.$$

2. Find the slope of the line that contains the points $(0, 5), (4, 2)$.

Example 2

Finding the Slope of a Line

Find the slope of the line that contains the points $(1, 3)$ and $(5, 1)$, and then graph the line.

Solution

Using $(1, 3)$ and $(5, 1)$,

$$(x_1, y_1) \quad (x_2, y_2)$$

$$\text{slope} = m = \frac{1 - 3}{5 - 1}$$

$$= \frac{-2}{4}$$

$$= -\frac{1}{2}$$

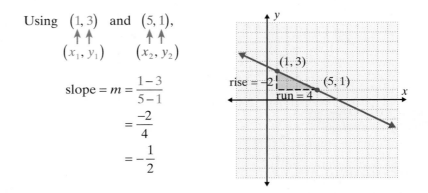

Now work margin exercise **2.**

---

notes

■ Lines with **positive slope go up** (increase) as we move along the line from left to right.

■

■ Lines with **negative slope go down** (decrease) as we move along the line from left to right.

Objective C   **Slopes of Horizontal and Vertical Lines**

Suppose that two points on a line have the same y-coordinate, such as $(-2, 3)$ and $(5, 3)$. Then the line through these two points will be **horizontal** as shown in Figure 8. In this case, the y-coordinates of the horizontal line are all 3, and the equation of the line is simply $y = 3$. The slope is

$$m = \frac{3 - 3}{5 - (-2)} = \frac{0}{7} = 0.$$

For any horizontal line, all of the y-values will be the same. Consequently, the formula for slope will always have 0 in the numerator. Therefore, **the slope of every horizontal line is 0**.

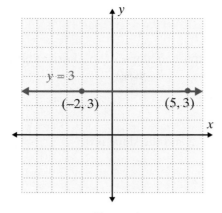

Figure 8

If two points have the same x-coordinates, such as $(1, 3)$ and $(1, -2)$, then the line through these two points will be **vertical** as in Figure 9. The x-coordinates for every point on the vertical line are all 1, and the equation of the line is simply $x = 1$. The slope is

$$m = \frac{-2-3}{1-1} = \frac{-5}{0}, \text{ which is } \textbf{undefined}.$$

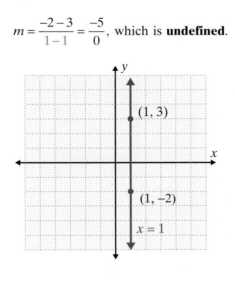

Figure 9

Horizontal and Vertical Lines

The following two general statements are true for horizontal and vertical lines:
1. For **horizontal lines** (of the form $y = b$), the **slope is 0**.
2. For **vertical lines** (of the form $x = a$), the **slope is undefined**.

3. a. Find the equation and slope of the horizontal line through the point $(3, -2)$.

b. Find the equation and slope of the vertical line through the point $(2, 4)$.

Example 3

Slopes of Horizontal and Vertical Lines

a. Find the equation and slope of the horizontal line through the point $(-2, 5)$.

Solution

The equation is $y = 5$ and the slope is 0.

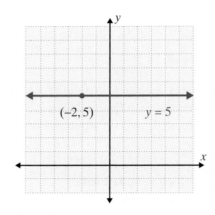

b. Find the equation and slope of the vertical line through the point $(3, 2)$.

Solution

The equation is $x = 3$ and the slope is undefined.

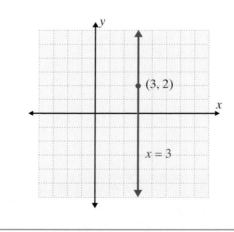

Now work margin exercise 3.

Objective D **Slope-Intercept Form:** $y = mx + b$

There are certain relationships between the coefficients in the equation of a line and the graph of that line. For example, consider the equation

$$y = 5x - 7.$$

First, find two points on the line and calculate the slope. $(0, -7)$ and $(2, 3)$ both satisfy the equation.

$$\text{slope} = m = \frac{3 - (-7)}{2 - 0} = \frac{10}{2} = 5$$

Observe that the slope, $m = 5$, is the same as the coefficient of x in the equation $y = 5x - 7$. This is not just a coincidence. In fact, if a linear equation is solved for y, then the coefficient of x will always be the slope of the line.

For $y = mx + b$, m is the Slope

For an equation in the form $y = mx + b$, the slope of the line is m.

For the line $y = mx + b$, the point where $x = 0$ is the point where the line crosses the y-axis. Recall that this point is called the **y-intercept**. By letting $x = 0$, we get

$$y = mx + b$$
$$y = m \cdot 0 + b$$
$$y = b.$$

Thus the point $(0, b)$ is the y-intercept. The concepts of slope and y-intercept lead to the following definition.

Slope-Intercept Form

$y = mx + b$ is called the **slope-intercept form** for the equation of a line, where m is the **slope** and $(0, b)$ is the **y-intercept**.

As illustrated in Example 4, an equation in the **standard form**

$$Ax + By = C \quad \text{with } B \neq 0$$

can be written in the slope-intercept form by solving for y.

Example 4

Using the Form $y = mx + b$

a. Find the slope and y-intercept of $-2x + 3y = 6$ and graph the line.

Solution

Solve for y.

$$-2x + 3y = 6$$
$$3y = 2x + 6$$
$$\frac{3y}{3} = \frac{2x}{3} + \frac{6}{3}$$
$$y = \frac{2}{3}x + 2$$

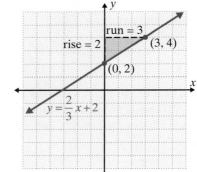

Thus $m = \dfrac{2}{3}$, which is the slope, and b is 2 making the y-intercept equal $(0, 2)$.

As shown in the graph, if we "rise" 2 units up and "run" 3 units to the right **from the y-intercept $(0, 2)$**, we locate another point $(3, 4)$. The line can be drawn through these two points. **Note:** As shown in the graph on the right, we could also first "run" 3 units right and "rise" 2 units up from the y-intercept to locate the point $(3, 4)$ on the graph.

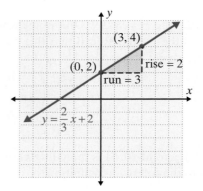

b. Find the slope and y-intercept of $x + 2y = -6$ and graph the line.

Solution

Solve for y.

$$x + 2y = -6$$
$$2y = -x - 6$$
$$\frac{2y}{2} = \frac{-x}{2} - \frac{6}{2}$$
$$y = -\frac{1}{2}x - 3$$

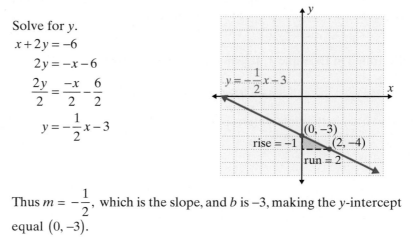

Thus $m = -\dfrac{1}{2}$, which is the slope, and b is -3, making the y-intercept equal $(0, -3)$.

We can treat $m = -\dfrac{1}{2}$ as $m = \dfrac{-1}{2}$ and the "rise" as –1 and the "run" as 2. Moving from $(0, -3)$ as shown in the graph on the previous page, we locate another point $(2, -4)$ on the graph and draw the line.

c. Find the equation of the line through the point $(0, -2)$ with slope $\dfrac{1}{2}$.

Solution

Because the x-coordinate is 0, we know that the point $(0, -2)$ is the y-intercept. So $b = -2$. The slope is $\dfrac{1}{2}$. So $m = \dfrac{1}{2}$. Substituting in slope-intercept form $y = mx + b$ gives the result: $y = \dfrac{1}{2}x - 2$.

Now work margin exercise 4.

4. a. Find the slope and y-intercept of $-4x + 2y = 12$ and graph the line.

b. Find the slope and y-intercept of $3x + 2y = -10$ and graph the line.

c. Find the equation of the line through the point $(0, -3)$ with a slope of $\dfrac{2}{3}$.

Practice Problems

1. Find the slope of the line through the two points $(1, 3)$ and $(4, 6)$. Graph the line.

2. Find the equation of the line through the point $(0, 5)$ with slope $-\dfrac{1}{3}$.

3. Find the slope and y-intercept for the line $2x + y = 7$.

4. Write the equation for the horizontal line through the point $(-1, 3)$. What is the slope of this line?

5. Write the equation for the vertical line through the point $(-1, 3)$. What is the slope of this line?

Practice Problem Answers

1. $m = 1$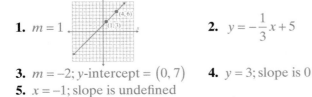

2. $y = -\dfrac{1}{3}x + 5$

3. $m = -2$; y-intercept $= (0, 7)$ **4.** $y = 3$; slope is 0

5. $x = -1$; slope is undefined

Exercises 9.3

Find the slope of the line determined by each set of points.

1. $(2, 4); (1, -1)$

2. $(1, -2); (1, 4)$

3. $(-6, 3); (1, 2)$

4. $(-3, 7); (4, -1)$

5. $(-5, 8); (3, 8)$

6. $(-2, 3); (-2, -1)$

7. $(5, 1); (3, 0)$

8. $(0, 0); (-2, -3)$

9. $\left(\dfrac{3}{4}, \dfrac{3}{2} \right); (1, 2)$

10. $\left(4, \dfrac{1}{2} \right); (-1, 2)$

11. $\left(\dfrac{3}{2}, \dfrac{4}{5} \right); \left(-2, \dfrac{1}{10} \right)$

12. $\left(\dfrac{7}{2}, \dfrac{3}{4} \right); \left(\dfrac{1}{2}, -3 \right)$

Determine whether each equation represents a horizontal line or vertical line and give its slope. Graph the line.

13. $y = 5$

14. $y = -2$

15. $x = -3$

16. $x = 1.7$

17. $3y = -18$

18. $4x = 2.4$

19. $-3x + 21 = 0$

20. $2y + 5 = 0$

21. $y = 2x - 1$

22. $y = 3x - 4$

23. $y = 5 - 4x$

24. $y = 4 - x$

25. $y = \dfrac{2}{3}x - 3$

26. $y = \dfrac{2}{5}x + 2$

27. $x + y = 5$

28. $x - 2y = 6$

29. $x + 5y = 10$

30. $4x + y = 0$

31. $4x + y + 3 = 0$

32. $2x + 7y + 7 = 0$

33. $2y - 8 = 0$

34. $3y - 9 = 0$

35. $2x = 3y$

36. $4x = y$

37. $3x + 9 = 0$

38. $4x + 7 = 0$

39. $5x - 6y = 18$

40. $3x + 6 = 6y$

41. $5 - 3x = 4y$

42. $5x = 11 - 2y$

43. $6x + 4y = -8$

44. $7x + 2y = 4$

45. $6y = -6 + 3x$

46. $4x = 3y - 7$

47. $5x - 2y + 5 = 0$

48. $6x + 5y = -15$

In reference to the equation $y = mx + b$, sketch the graph of three lines for each of the two characteristics listed below.

49. $m > 0$ and $b > 0$ 50. $m < 0$ and $b > 0$ 51. $m > 0$ and $b < 0$ 52. $m < 0$ and $b < 0$

Find an equation in slope-intercept form for the line passing through the given point with the given slope.

53. $(0, 3)$; $m = -\dfrac{1}{2}$ 54. $(0, 2)$; $m = \dfrac{1}{3}$ 55. $(0, -3)$; $m = \dfrac{2}{5}$ 56. $(0, -6)$; $m = \dfrac{4}{3}$

57. $(0, -5)$; $m = 4$ 58. $(0, 9)$; $m = -1$ 59. $(0, -4)$; $m = 1$ 60. $(0, 6)$; $m = -5$

61. $(0, -3)$; $m = -\dfrac{5}{6}$ 62. $(0, -1)$; $m = -\dfrac{3}{2}$

The graph of a line is shown with two points highlighted. Find a. the slope, b. the y-intercept (if there is one), and c. the equation of the line in slope-intercept form.

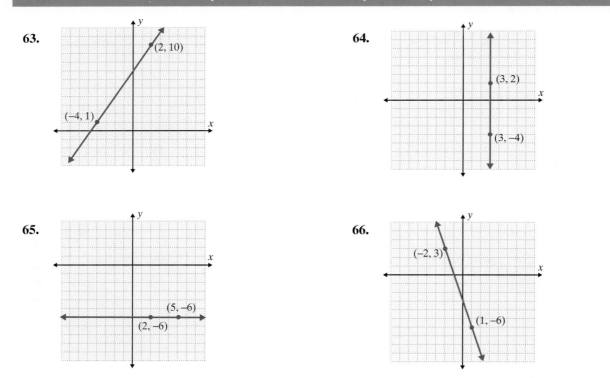

63.

64.

65.

66.

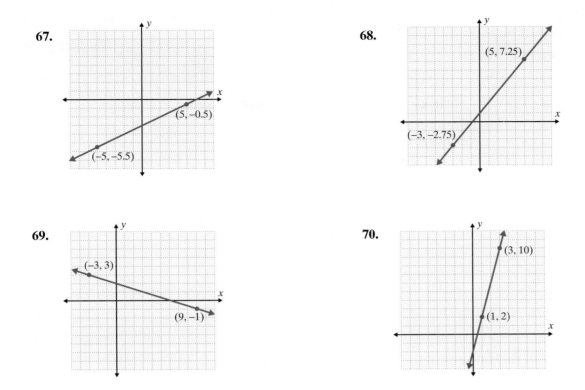

67. (5, −0.5) (−5, −5.5)

68. (5, 7.25) (−3, −2.75)

69. (−3, 3) (9, −1)

70. (3, 10) (1, 2)

Points are said to be *collinear* if they are on a straight line. If points are collinear, then the slope of the line through any two of them must be the same (because the line is the same line). Use this idea to determine whether or not the three point in each of the sets are collinear.

71. $\{(-1, 3), (0, 1), (5, -9)\}$

72. $\{(-2, -4), (0, 2), (3, 11)\}$

73. $\{(-2, 0), (0, 30), (1.5, 5.25)\}$

74. $\{(-1, -7), (1, 1), (2.5, 7)\}$

75. $\left\{\left(\frac{2}{3}, \frac{1}{2}\right), \left(0, \frac{5}{6}\right), \left(-\frac{3}{4}, \frac{29}{24}\right)\right\}$

76. $\left\{\left(\frac{3}{2}, -\frac{1}{3}\right), \left(0, \frac{1}{6}\right), \left(-\frac{1}{2}, \frac{3}{4}\right)\right\}$

Solve the following word problems.

77. Buying a New Car: John bought his new car for $35,000 in the year 2007. He knows that the value of his car has depreciated linearly. If the value of the car in 2010 was $23,000, what was the annual rate of depreciation of his car? Show this information on a graph. (When graphing, use years as the *x*-coordinates and the corresponding values of the car as the *y*-coordinates.)

78. Cell Phone Usage: The number of people in the United States with mobile cellular phones was about 180 million in 2004 and about 286 million in 2009. If the growth in mobile cellular phones was linear, what was the approximate rate of growth per year from 2004 to 2009. Show this information on a graph. (When graphing, use years as the *x*-coordinates and the corresponding number of users as the *y*-coordinates.)

79. **Internet Usage:** The given table shows the estimated number of internet users from 2004 to 2008. The number of users for each year is shown in millions.

   a. Plot these points on a graph.

   b. Connect the points with line segments.

   c. Find the slope of each line segment.

   d. Interpret each slope as a rate of change.
      Source: International Telecommunications Union Yearbook of Statistics

Year	Internet Users (in millions)
2004	185
2005	198
2006	210
2007	220
2008	231

80. **Urban Growth:** The following table shows the urban growth from 1850 to 2000 in New York, NY.

   a. Plot these points on a graph.

   b. Connect the points with line segments.

   c. Find the slope of each line segment.

   d. Interpret each slope as a rate of change.
      Source: U.S. Census Bureau

Year	Population
1850	515,547
1900	3,437,202
1950	7,891,957
2000	8,008,278

81. **Military:** The following graph shows the number of female active duty military personnel over a span from 1945 to 2009. The number of women listed includes both officers and enlisted personnel from the Army, the Navy, the Marine Corps, and the Air Force.

   a. Plot these points on a graph.

   b. Connect the points with line segments.

   c. Find the slope of each line segment.

   d. Interpret each slope as a rate of change.
      Source: U.S. Dept. of Defense

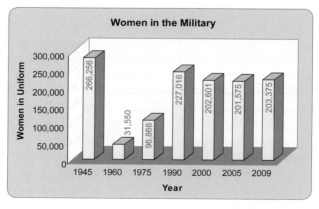

82. Marriage: The following graph shows the rates of marriage per 1000 people in the U.S., over a span from 1920 to 2008.

    a. Plot these points on a graph.

    b. Connect the points with line segments.

    c. Find the slope of each line segment.

    d. Interpret each slope as a rate of change.

        Source : U.S. National Center for Health Statistics

Collaborative Learning

83. The class should be divided into teams of 2 or 3 students. Each team will need access to a digital camera, a printer, and a ruler.

    a. Take pictures of 8 things with a defined slope. (**Suggestions:** A roof, a stair railing, a beach umbrella, a crooked tree, etc. Be creative!)

    b. Print each picture.

    c. Use a ruler to draw a coordinate system on top of each picture. You will probably want to use increments of in. or cm, depending on the size of your picture.

    d. Identify the line in each picture whose slope you are calculating and then use the coordinate systems you created to identify the coordinates of two points on each line .

    e. Use the points you just found to calculate the slope of the line in each picture.

    f. Share your findings with the class.

84. a. Explain in your own words why the slope of a horizontal line must be 0.

  b. Explain in your own words why the slope of a vertical line must be undefined.

85. a. Describe the graph of the line $y = 0$.

  b. Describe the graph of the line $x = 0$.

86. In the formula $y = mx + b$ explain the meaning of m and the meaning of b.

87. The slope of a road is called a **grade**. A steep grade is cause for truck drivers to have slow speed limits in mountains. What do you think that a "grade of 12%" means? Draw a picture of a right triangle that would indicate a grade of 12%.

9.4 The Point-Slope Form: $y - y_1 = m(x - x_1)$

Objective A — Graphing a Line Given a Point and the Slope

Lines represented by equations in the **standard form** $Ax + By = C$ and in the **slope-intercept form** $y = mx + b$ have been discussed in Sections 9.2 and 9.3. Previously we graphed lines using the y-intercept $(0, b)$ and the slope by moving vertically and then horizontally (or by moving horizontally and then vertically) from the y-intercept. This same technique can be used to graph lines if the given point is on the line but is not the y-intercept. Consider the following example.

Example 1

Graphing a Line Given a Point and the Slope

Graph the line with slope $m = -\dfrac{3}{4}$ and which passes through the point $(2, 5)$.

Solution

Start from the point $(2, 5)$ and locate another point on the line using the slope as $\dfrac{rise}{run} = \dfrac{-3}{4}$ or $\dfrac{3}{-4}$.

Two methods to locate another point on the line are shown here.

1. Move 4 units right and 3 units down, or
2. Move 3 units down and 4 units right.

Either way, you arrive at the same point $(6, 2)$.

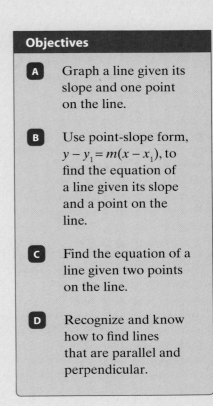

This means that we can move from the given point either with the rise first or the run first.

Note: Any numbers in the ratio of -3 to 4 can be used for the moves, such as -6 to 8 or 9 to -12.

Now work margin exercise 1.

Objective B — Point-Slope Form: $y - y_1 = m(x - x_1)$

Now consider finding the equation of the line given the point (x_1, y_1) on the line and the slope m. If (x, y) is **any other point** on the line, then the slope formula gives the equation

$$\frac{y - y_1}{x - x_1} = m.$$

Objectives

A Graph a line given its slope and one point on the line.

B Use point-slope form, $y - y_1 = m(x - x_1)$, to find the equation of a line given its slope and a point on the line.

C Find the equation of a line given two points on the line.

D Recognize and know how to find lines that are parallel and perpendicular.

1. Graph the line with the slope $m = -\dfrac{1}{3}$ and which passes through the point $(3, 4)$.

Multiplying both sides of this equation by the denominator (assuming the denominator is not 0) gives

$$y - y_1 = m(x - x_1)$$ which is called the **point-slope form**.

For example, suppose that a point $(x_1, y_1) = (8, 3)$ and the slope $m = -\dfrac{3}{4}$ are given.

If (x, y) represents any point on the line other than $(8, 3)$, then substituting into the formula for slope gives

$$\dfrac{y - y_1}{x - x_1} = m$$ formula for slope

$$\dfrac{y - 3}{x - 8} = -\dfrac{3}{4}$$ substitute the given information.

$$(x - 8)\left(\dfrac{y - 3}{x - 8}\right) = -\dfrac{3}{4}(x - 8)$$ multiply both sides by $(x - 8)$.

$$y - 3 = -\dfrac{3}{4}(x - 8).$$ point-slope form: $y - y_1 = m(x - x_1)$

From this point-slope form, we can manipulate the equation to get the other two forms:

$$y - 3 = -\dfrac{3}{4}(x - 8)$$

$$y - 3 = -\dfrac{3}{4}x + 6$$

or $$y = -\dfrac{3}{4}x + 9$$ slope-intercept form: $y = mx + b$

or $$4(y) = 4\left(-\dfrac{3}{4}x + 9\right)$$

$$4y = -3x + 36$$

$$3x + 4y = 36.$$ standard form: $Ax + By = C$

Point-Slope Form

An equation of the form

$$y - y_1 = m(x - x_1)$$

is called the **point-slope form** for the equation of a line that contains the point (x_1, y_1) and has slope m.

Example 2

Finding Equations of Lines Using the Slope and a Point

Find the equation of the line with a slope of $-\dfrac{1}{2}$ and which passes through the point $(2, 3)$. Graph the line using the point and slope.

Solution

Substitute the values into the point-slope form.

$$y - y_1 = m(x - x_1)$$

$$y - 3 = -\frac{1}{2}(x - 2) \qquad \text{point-slope form}$$

$$y - 3 = -\frac{1}{2}x + 1$$

or, $\qquad y = -\dfrac{1}{2}x + 4 \qquad$ slope-intercept form

or, $\qquad 2y = -x + 8$

$$x + 2y = 8 \qquad\qquad \text{standard form}$$

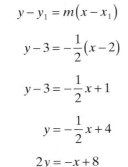

The point one unit down and two units right from $(2, 3)$ will be on the line because the slope is

$$m = \frac{\text{rise}}{\text{run}} = \frac{-1}{2} = -\frac{1}{2}.$$

With a negative slope, either the rise is negative and the run is positive, or the rise is positive and the run is negative. In either case, as the previous figure and the following figure illustrate, the line is the same.

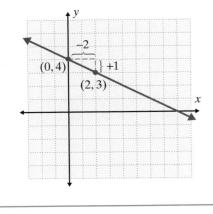

The point one unit up and two units to the left from $(2, 3)$ is on the line because the slope is

$$m = \frac{\text{rise}}{\text{run}} = \frac{1}{-2} = -\frac{1}{2}.$$

Now work margin exercise 2.

2. Find the equation of the line with a slope of $-\dfrac{2}{3}$ and which passes through the point $(3, 1)$. Graph the line using the point and slope.

In Example 2, the equation of the line is written in all three forms: point-slope form, slope-intercept form, and standard form. Generally any one of these forms is sufficient. However, there are situations in which one form is preferred over the others. Therefore, manipulation among the forms is an important skill. Also, if the answer in the text is in one form and your answer is in another form, you should be able to recognize that the answers are equivalent.

Objective C   **Finding the Equations of a Line Given Two Points**

Given two points that lie on a line, the equation of the line can be found using the following method.

Finding the Equation of a Line Given Two Points

To find the equation of a line given two points on the line:

1. Use the formula $m = \dfrac{y_2 - y_1}{x_2 - x_1}$ to find the slope.

2. Use this slope, m, and either point in the point-slope formula $y - y_1 = m(x - x_1)$ to find the equation.

Example 3

Using Two Points to Find the Equation of a Line

Find the equation of the line containing the two points $(-1, 2)$ and $(4, -2)$.

Solution

First, find the slope.

$$
\begin{aligned}
m &= \frac{y_2 - y_1}{x_2 - x_1} \\[2mm]
&= \frac{-2 - 2}{4 - (-1)} \\[2mm]
&= \frac{-4}{5} \\[2mm]
&= -\frac{4}{5}
\end{aligned}
$$

Now use one of the given points and the point-slope form for the equation of a line. (**Note:** $(-1, 2)$ and $(4, -2)$ are used on the next page to illustrate that either point may be used.)

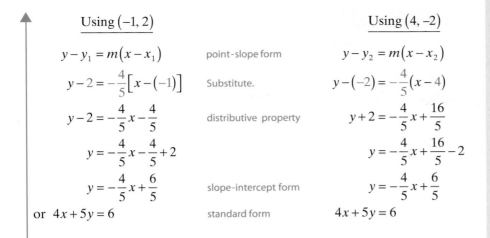

Using $(-1, 2)$		Using $(4, -2)$
$y - y_1 = m(x - x_1)$	point-slope form	$y - y_2 = m(x - x_2)$
$y - 2 = -\dfrac{4}{5}\left[x - (-1)\right]$	Substitute.	$y - (-2) = -\dfrac{4}{5}(x - 4)$
$y - 2 = -\dfrac{4}{5}x - \dfrac{4}{5}$	distributive property	$y + 2 = -\dfrac{4}{5}x + \dfrac{16}{5}$
$y = -\dfrac{4}{5}x - \dfrac{4}{5} + 2$		$y = -\dfrac{4}{5}x + \dfrac{16}{5} - 2$
$y = -\dfrac{4}{5}x + \dfrac{6}{5}$	slope-intercept form	$y = -\dfrac{4}{5}x + \dfrac{6}{5}$
or $4x + 5y = 6$	standard form	$4x + 5y = 6$

Now work margin exercise 3.

3. Find the equation of the line containing the two points $(-2, 4)$ and $(0, -1)$.

Objective D **Parallel Lines and Perpendicular Lines**

Parallel and Perpendicular Lines

Parallel lines are lines that never intersect (never cross each other) and these lines have the **same slope**. **Note:** All vertical lines (undefined slopes) are parallel to one another.

Perpendicular lines are lines that intersect at 90° (right) angles and whose slopes are **negative reciprocals** of each other. Horizontal lines are perpendicular to vertical lines.

As illustrated in Figure 1, the lines $y = 2x + 1$ and $y = 2x - 3$ are **parallel**. They have the same slope, 2. The lines $y = \dfrac{2}{3}x + 1$ and $y = -\dfrac{3}{2}x - 2$ are **perpendicular**. Their slopes are negative reciprocals of each other, $\dfrac{2}{3}$ and $-\dfrac{3}{2}$.

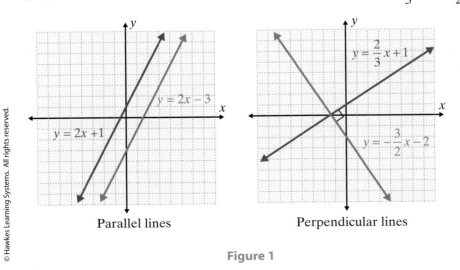

Parallel lines

Perpendicular lines

Figure 1

Example 4

Finding the Equations of Parallel Lines

Find the equation of the line through the point $(2, 3)$ and parallel to the line $5x + 3y = 1$. Graph both lines.

Solution

First, solve for y to find the slope of the given line.

$$5x + 3y = 1$$
$$3y = -5x + 1$$
$$y = -\frac{5}{3}x + \frac{1}{3}$$

Thus any line parallel to this line has slope $-\frac{5}{3}$.

Now use the point-slope form $y - y_1 = m(x - x_1)$ with $m = -\frac{5}{3}$ and $(x_1, y_1) = (2, 3)$.

$$y - 3 = -\frac{5}{3}(x - 2) \qquad \text{point-slope form}$$
$$3(y - 3) = -5(x - 2) \qquad \text{Multiply both sides by the LCD, 3.}$$
$$3y - 9 = -5x + 10 \qquad \text{Simplify.}$$
$$5x + 3y = 19 \qquad \text{standard form}$$
$$\text{or} \qquad y = -\frac{5}{3}x + \frac{19}{3} \qquad \text{slope-intercept form}$$

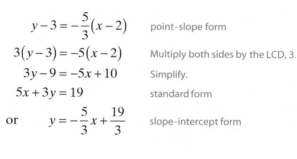
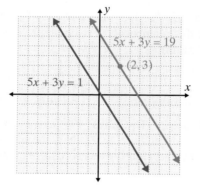

Example 5

Finding the Equations of Perpendicular Lines

Find the equation of the line through the point $(2, 3)$ and perpendicular to the line $5x + 3y = 1$. Graph both lines.

Solution

We know from Example 2 that the slope of the line $5x + 3y = 1$ is $-\dfrac{5}{3}$. Thus any line perpendicular to this line must have slope $m = \dfrac{3}{5}$ (the negative reciprocal of $-\dfrac{5}{3}$).

Now using the point-slope form $y - y_1 = m(x - x_1)$ with $m = \dfrac{3}{5}$, and $(x_1, y_1) = (2, 3)$, we have

$$y - 3 = \frac{3}{5}(x - 2)$$ point-slope form

$$5(y - 3) = 3(x - 2)$$ Multiply both sides by the LCD, 5.

$$5y - 15 = 3x - 6$$ Simplify.

$$3x - 5y = -9$$ standard form

or $$y = \frac{3}{5}x + \frac{9}{5}$$ slope-intercept form

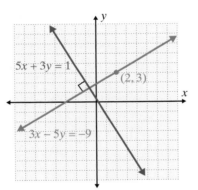

Now work margin exercises 4 and 5.

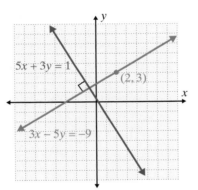 *(margin)*

4. Find the equation of the line through the point $(0, 2)$ and parallel to the line $3x + 2y = -2$. Graph both lines.

5. Find the equation of the line through the point $(0, 2)$ and perpendicular to the line $3x + 2y = -2$. Graph both lines.

Summary of Formulas and Properties of Straight Lines

1. $Ax + By = C$ Standard Form

2. $m = \dfrac{y_2 - y_1}{x_2 - x_1}$ Slope of a Line

3. $y = mx + b$ Slope-intercept Form

4. $y - y_1 = m(x - x_1)$ Point-slope Form

5. $y = b$ Horizontal Line, slope 0

6. $x = a$ Vertical Line, undefined slope

7. Parallel lines   Have the same slope.

8. Perpendicular lines   Have slopes that are negative reciprocals of each other.

Practice Problems

Find a linear equation in standard form whose graph satisfies the given conditions.

1. Passes through the point $(4, -1)$ with $m = 2$

2. Parallel to $y = -3x + 4$ and contains the point $(-1, 5)$

3. Perpendicular to $2x + y = 1$ and passes through the origin $(0, 0)$

4. Contains the two points $(6, -2)$ and $(2, 0)$

Practice Problem Answers

1. $2x - y = 9$ **2.** $3x + y = 2$ **3.** $x - 2y = 0$ **4.** $x + 2y = 2$

Exercises 9.4

Find a. the slope, b. a point on the line, and c. the graph of the line for the given equations in point-slope form.

1. $y - 1 = 2(x - 3)$

2. $y - 4 = \frac{1}{2}(x - 1)$

3. $y + 2 = -5(x)$

4. $y = -(x + 8)$

5. $y - 3 = -\frac{1}{4}(x + 2)$

6. $y + 6 = \frac{1}{3}(x - 7)$

Find an equation in standard form for the line passing through the given point with the given slope. Graph the line.

7. $(-2, 1)$; $m = -2$

8. $(3, 4)$; $m = 3$

9. $(5, -2)$; $m = 0$

10. $(0, 0)$; $m = -3$

11. $(-3, 6)$; $m = \frac{1}{2}$

12. $(-3, -1)$; m is undefined

13. $(2, 7)$; $m = \frac{3}{5}$

14. $(-1, -1)$; $m = -\frac{1}{4}$

15. $\left(-2, \frac{1}{3}\right)$; $m = \frac{2}{3}$

16. $\left(\frac{5}{2}, \frac{1}{2}\right)$; $m = -\frac{4}{3}$

17. $(-5, 2); (3, 6)$ **18.** $(-3, 4); (2, 1)$ **19.** $(-5, 1); (2, 0)$ **20.** $(-4, -4); (3, 1)$

21. $(0, 2); \left(1, \dfrac{3}{4}\right)$ **22.** $\left(\dfrac{5}{2}, 0\right); \left(2, -\dfrac{1}{3}\right)$ **23.** $(2, -5); (4, -5)$ **24.** $(0, 4); \left(1, \dfrac{1}{2}\right)$

25. $(-2, 6); (3, 1)$ **26.** $(8, 2); (0, 0)$

27. Find an equation for the horizontal line through the point $(-2, 6)$.

28. Find an equation for the vertical line through the point $(-1, -4)$.

29. Write an equation for the line parallel to the x-axis and containing the point $(2, 7)$.

30. Find an equation for the line parallel to the y-axis and containing the point $(2, -4)$.

31. Find an equation for the line perpendicular to $x = 4$ and that passes through $(-1, 7)$.

32. Find an equation for the line parallel to the line $-6y = 1$ and containing the point $(-3, 2)$.

33. Write an equation for the line parallel to the line $2x - y = 4$ and containing the origin. Graph both lines.

34. Find an equation for the line parallel to $7x - 3y = 1$ and containing the point $(1, 0)$. Graph both lines.

35. Write an equation for the line parallel to $5x = 7 + y$ and through the point $(-1, -3)$. Graph both lines.

36. Write an equation for the line that contains the point $(2, 2)$ and is perpendicular to the line $4x + 3y = 4$. Graph both lines.

37. Find an equation for the line that passes through the point $(4, -1)$ and is perpendicular to the line $5x - 3y + 4 = 0$. Graph both lines.

38. Write an equation for the line that is perpendicular to $8 - 3x - 2y = 0$ and passes through the point $(-4, -2)$.

39. Write an equation for the line through the origin that is perpendicular to $3x - y = 4$.

40. Find an equation for the line that is perpendicular to $2x + y = 5$ and that passes through $(6, -1)$.

41. Write an equation for the line that is perpendicular to $2x - y = 7$ and has the same y-intercept as $x - 3y = 6$.

42. Find an equation for the line with the same y-intercept as $5x + 4y = 12$ and that is perpendicular to $3x - 2y = 4$.

43. Show that the points $A(-2, 4)$, $B(0, 0)$, $C(6, 3)$, and $D(4, 7)$ are the vertices of a rectangle. (Plot the points and show that opposite sides are parallel and that adjacent sides are perpendicular.)

44. Show that the points $A(0, -1)$, $B(3, -4)$, $C(6, 3)$, and $D(9, 0)$ are the vertices of a parallelogram. (Plot the points and show that opposite sides are parallel.)

Determine whether each pair of lines is parallel, perpendicular, or neither. Graph both lines. (Hint: Write the equations in slope-intercept form and then compare slopes and y-intercepts.)

45. $\begin{cases} y = -2x + 3 \\ y = -2x - 1 \end{cases}$

46. $\begin{cases} y = 3x + 2 \\ y = -\dfrac{1}{3}x + 6 \end{cases}$

47. $\begin{cases} 4x + y = 4 \\ x - 4y = 8 \end{cases}$

48. $\begin{cases} 2x + 3y = 5 \\ 3x + 2y = 10 \end{cases}$

49. $\begin{cases} 2x + 2y = 9 \\ 2x - y = 6 \end{cases}$

50. $\begin{cases} 3x - 4y = 16 \\ 4x + 3y = 15 \end{cases}$

Writing & Thinking

51. Handicapped Access: Ramps for persons in wheelchairs or otherwise handicapped are now built into most buildings and walkways. (If ramps are not present in a building, then there must be elevators.) What do you think that the slope of a ramp should be for handicapped access? Look in your library or contact your local building permit office to find the recommended slope for such ramps.

52. Discuss the difference between the concepts of a line having slope of 0 and a line having undefined slope.

Objectives

A   Understand the concept of a function.

B   Find the domain and range of a relation or function.

C   Determine whether a relation is a function or not.

D   Use the vertical line test to determine whether a graph is or is not the graph of a function.

E   Understand the concept of a linear function.

F   Determine the domain of nonlinear functions.

G   Write a function using function notation.

H   Use a graphing calculator to graph functions.

9.5 Introduction to Functions and Function Notation

Objective A   Functions

Everyday use of the term **function** is not far from the technical use in mathematics. For example, distance traveled is a function of time; profit is a function of sales; heart rate is a function of exertion; and interest earned is a function of principal invested. In this sense, one variable "depends on" (or "is a function of") another.

Mathematicians distinguish between graphs of ordered pairs of real numbers as those that represent **functions** and those that do not. For example, every equation of the form $y = mx + b$ represents a function and we say that y "is a function of" x. Thus lines that are not vertical are the graphs of functions. As the following discussion indicates, vertical lines do not represent functions.

notes

The ordered pairs discussed in this text are ordered pairs of real numbers. However, more generally, ordered pairs might be other types of pairs such as (child, mother), (city, state), or (name, batting average).

Objective B   Domain and Range

Relation, Domain, and Range

A **relation** is a set of ordered pairs of real numbers.

The **domain**, D, of a relation is the set of all first coordinates in the relation.

The **range**, R, of a relation is the set of all second coordinates in the relation.

In the graph of a relation, the horizontal axis (the x-axis) is called the **domain axis**, and the vertical axis (the y-axis) is called the **range axis**.

Example 1

Finding the Domain and Range

Find the domain and range for each of the following relations.

a. $g = \left\{ (5, 7), (6, 2), (6, 3), (-1, 2) \right\}$

Solution

$D = \{5, 6, -1\}$ the set of all the first coordinates in g

$R = \{7, 2, 3\}$ the set of all the second coordinates in g

Note that 6 is written only once in the domain and 2 is written only once in the range, even though each appears more than once in the relation.

b. $f = \{(-1, 1), (1, 5), (0, 3)\}$

Solution

$D = \{-1, 1, 0\}$ the set of all the first coordinates in f

$R = \{1, 5, 3\}$ the set of all the second coordinates in f

Example 2

Reading the Domain and Range from the Graph of a Relation

Identify the domain and range from the graph of each relation.

Solution

a. The domain consists of the set of x-values for all points on the graph. In this case the domain is the interval $[-1, 3]$. The range consists of the set of y-values for all points on the graph. In this case, the range is the interval $[0, 6]$.

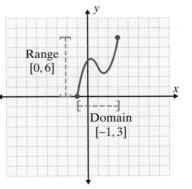

Solution

b. There is no restriction on the x-values which means that for every real number there is a point on the graph with that number as its x-value. Thus the domain is the interval $(-\infty, \infty)$. The y-values begin at -2 and then increase to infinity. The range is the interval $[-2, \infty)$.

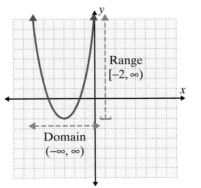

Now work margin exercises 1 and 2.

---

1. Find the domain and range of each of the following relations.

a. $g = \begin{Bmatrix} (4, 5), (7, 3), \\ (3, 6), (7, 5) \end{Bmatrix}$

b. $f = \begin{Bmatrix} (-2, 3), (-4, -3), \\ (0, 0) \end{Bmatrix}$

2. Identify the domain and range from the graph of each relation.

a.

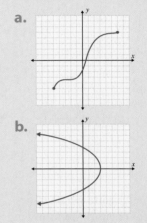

b.

Objective C Relations and Functions

The relation $f = \{(-1, 1), (1, 5), (0, 3)\}$, used in Example 1b, meets a particular condition in that each first coordinate has a unique corresponding second coordinate. Such a relation is called a **function**. Notice that g in Example 1a is **not** a function because the first coordinate 6 has two corresponding second coordinates, 2 and 3. Also, for ease in discussion and understanding, the relations illustrated in Examples 1 and 3 have only a finite number of ordered pairs. The graphs of these relations are isolated dots or points. As we will see, the graphs of most relations and functions have an infinite number of points, and their graphs are smooth curves. (**Note:** Lines are also deemed to be curves in mathematics.)

Functions

A **function** is a relation in which each domain element has exactly one corresponding range element.

The definition can also be stated in the following ways:

1. A function is a relation in which each first coordinate appears only once.
2. A function is a relation in which no two ordered pairs have the same first coordinate.

3. Determine whether or not each of the following relations is a function.

a. $s = \left\{ \begin{matrix} (4,2), (5,7), \\ (3,8), (4,\sqrt{6}) \end{matrix} \right\}$

b. $t = \left\{ \begin{matrix} (3,\sqrt{4}), (7,4), \\ (-3,\sqrt{4}), (\sqrt{3},4) \end{matrix} \right\}$

Example 3

Functions

Determine whether or not each of the following relations is a function.

a. $s = \left\{(2, 3), (1, 6), \left(2, \sqrt{5}\right), (0, -1)\right\}$

Solution

s is not a function. The number 2 appears as a first coordinate more than once.

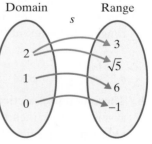

b. $t = \left\{(1, 5), (3, 5), \left(\sqrt{2}, 5\right), (-1, 5), (-4, 5)\right\}$

Solution

t is a function. Each first coordinate appears only once. The fact that the second coordinates are all the same has no effect on the concept of a function.

Now work margin exercise 3.

Objective D   Vertical Line Test

If one point on the graph of a relation is directly above or below another point on the graph, then these points have the same first coordinate (or *x*-coordinate). Such a relation is **not** a function. Therefore, the **vertical line test** can be used to tell whether or not a graph represents a function. (See Figures 1 and 2.)

Vertical Line Test

If **any** vertical line intersects the graph of a relation at more than one point, then the relation is **not** a function.

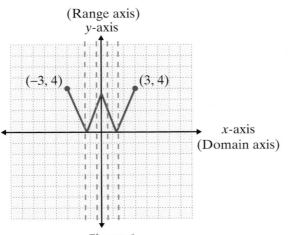

Figure 1

The vertical lines in Figure 1 indicate that this graph represents a function. From the graph, we see that the domain of the function is the interval of real numbers $[-3, 3]$ and the range of the function is the interval of real numbers $[0, 4]$.

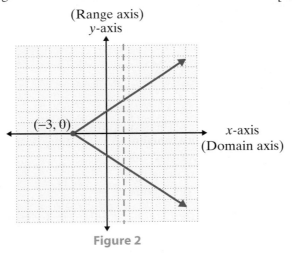

Figure 2

The relation in Figure 2 is **not** a function because the vertical line drawn intersects the graph at more than one point. Thus for that *x*-value, there is more than one corresponding *y*-value. Here $D = [-3, \infty)$ and $R = (-\infty, \infty)$.

Example 4

Vertical Line Test

Use the vertical line test to determine whether or not each of the following graphs represents a function. Then list the domain and range of each graph.

Solution

a. The relation is **not a function** since a vertical line can be drawn that intersects the graph at more than one point. Listing the ordered pairs shows that several x-coordinates appear more than once.

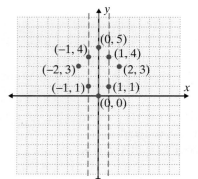

$$r = \begin{Bmatrix} (-2,3),(-1,1),(-1,4), \\ (0,0),(0,5),(1,1),(1,4),(2,3) \end{Bmatrix}$$

Here $D = \{-2,-1,0,1,2\}$ and $R = \{0,1,3,4,5\}$.

Solution

b. The relation **is a function.** No vertical line will intersect the graph at more than one point. Several vertical lines are drawn to illustrate this.

For this function, we see from the graph that $D = [-2,2]$ and $R = [0,2]$.

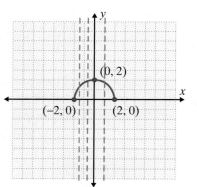

Solution

c. The relation is **not a function**. At least one vertical line (drawn) intersects the graph at more than one point.

Here $D = [-3, \infty)$

and $R = (-\infty, \infty)$.

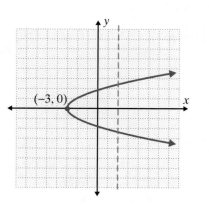

Solution

d. The relation is **not a function**. At least one vertical line intersects the graph at more than one point.

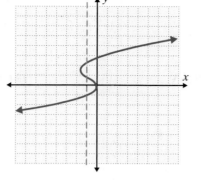

Here $D = (-\infty, \infty)$

and $R = (-\infty, \infty)$.

Now work margin exercise 4.

4. Use the vertical line test to determine whether or not each of the following graphs represents a function. Then list the domain and range of each graph.

a.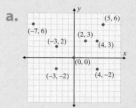

b.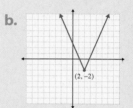

c.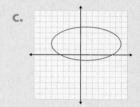

Objective E **Linear Functions**

All non-vertical lines represent functions. Thus we have the following definition for a linear function.

Linear Function

A **linear function** is a function represented by an equation of the form

$$y = mx + b.$$

The domain of a linear function is the set of all real numbers:

$$D = (-\infty, \infty).$$

If the graph of a linear function is not a horizontal line, then the range is also the set of all real numbers. If the line is horizontal, then the domain is still the set of all real numbers; however, the range is a set containing just a single number. For example, the graph of the linear equation $y = 5$ is a horizontal line. The domain of the function is the set of all real numbers and the range is the set containing only the number 5 (written $\{5\}$). Figures 3 and 4 show two linear functions and the domain and range of each function.

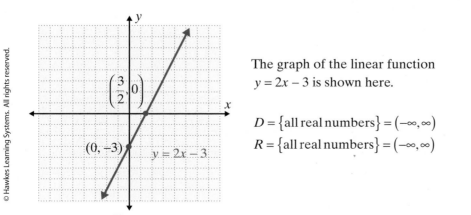

The graph of the linear function $y = 2x - 3$ is shown here.

$D = \{\text{all real numbers}\} = (-\infty, \infty)$

$R = \{\text{all real numbers}\} = (-\infty, \infty)$

Figure 3

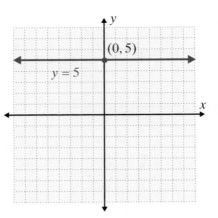

The graph of the linear function $y = 5$ is shown here.

$D = \{\text{all real numbers}\} = (-\infty, \infty)$

$R = \{5\}$

Figure 4

Objective F **Domains of Nonlinear Functions**

As we have seen, the domain for a linear function is the set of all real numbers. Now for the nonlinear function

$$y = \frac{2}{x-1}$$

we say that the domain (all possible values for x) is every real number for which the expression $\frac{2}{x-1}$ is **defined**. Because the denominator cannot be 0, the domain consists of all real numbers except 1. That is, $D = (-\infty, 1) \cup (1, \infty)$ or simply $x \neq 1$. We adopt the following rule concerning equations and domains:

Unless a finite domain is explicitly stated, the domain will be implied to be the set of all real x-values for which the given function is defined. That is, the domain consists of all values of x that give real values for y.

> ## notes
>
> ■ In determining the domain of a function, one fact to remember at this stage is that **no denominator can equal 0**. In future chapters we will
> ■ discuss other nonlinear functions with limited domains.

---

5. Find the domain for the function $y = \dfrac{3x-2}{x+3}$.

Example 5

Domain

Find the domain for the function $y = \dfrac{2x+1}{x-5}$.

Solution

The domain is all real numbers for which the expression $\dfrac{2x+1}{x-5}$ is defined. Thus $D = (-\infty, 5) \cup (5, \infty)$ or $x \neq 5$, because the denominator is 0 when $x = 5$.

Note: Here interval notation tells us that x can be any real number except 5.

Now work margin exercise 5.

Function Notation

We have used the ordered pair notation (x, y) to represent points in relations and functions. As the vertical line test will show, linear equations of the form

$$y = mx + b$$

where the equation is solved for y, represent **linear functions**. Another notation, called **function notation**, is more convenient for indicating calculations of values of a function and indicating operations performed with functions. In function notation, instead of writing y, write $f(x)$, read "*f* of *x*."

The letter f is the name of the function. The letters f, g, h, F, G, and H are commonly used in mathematics, but any letter other than x will do. We have used r, s, and t in previous examples.

The linear equation $y = -3x + 2$ represents a linear function and we can replace y with $f(x)$ as follows:

$$f(x) = -3x + 2.$$

Now in function notation, $f(4)$ means to replace x with 4 in the function.

$$f(4) = -3 \cdot (4) + 2 = -12 + 2 = -10$$

Thus the ordered pair $(4, -10)$ can be written as $\left(4, f(4)\right)$.

Example 6

Function Evaluation

For the function $g(x) = 4x + 5$, find:

a. $g(2)$

Solution

$$g(2) = 4(2) + 5 = 13$$

b. $g(-1)$

Solution

$$g(-1) = 4(-1) + 5 = 1$$

c. $g(0)$

Solution

$$g(0) = 4(0) + 5 = 5$$

6. For the function
$g(x) = 3x - 2$, find:

a. $g(3)$.

b. $g(-2)$.

c. $g(0)$.

7. For the function
$f(y) = y^3 - 2y + 5$, find:

a. $f(1)$.

b. $f(0)$.

c. $f(-3)$.

Function notation is valid for a wide variety of types of functions. Example 7 illustrates the use of function notation with a nonlinear function.

Example 7

Nonlinear Function Evaluation

For the function $h(x) = x^2 - 3x + 2$, find:

a. $h(4)$

Solution

$$h(4) = (4)^2 - 3(4) + 2 = 16 - 12 + 2 = 6$$

b. $h(0)$

Solution

$$h(0) = (0)^2 - 3(0) + 2 = 0 - 0 + 2 = 2$$

c. $h(-3)$

Solution

$$h(-3) = (-3)^2 - 3(-3) + 2 = 9 + 9 + 2 = 20$$

Now work margin exercises 6 and 7.

Objective H **Using a TI-84 Plus Graphing Calculator to Graph Functions**

There are many types and brands of graphing calculators available. For convenience and so that directions can be specific, only the TI-84 Plus graphing calculator is used in the related discussions in this text. Other graphing calculators may be used, but the steps required may be different from those indicated in the text. If you choose to use another calculator, be sure to read the manual for your calculator and follow the relevant directions.

In any case, remember that a calculator is just a tool to allow for fast calculations and to help in understanding some abstract concepts. A calculator does not replace your ability to think and reason or the need for algebraic knowledge and skills.

You should practice and experiment with your calculator until you feel comfortable with the results. **Do not be afraid of making mistakes. Note: that CLEAR or 2ND QUIT will get you out of most trouble and allow you to start over.**

Some Basics about the TI-84 Plus

1. **MODE** : Turn the calculator **ON** and press the **MODE** key. The screen should be highlighted as shown below. If it is not, use the arrow keys in the upper right corner of the keyboard to highlight the correct words and press **ENTER**. It is particularly important that Func is highlighted. This stands for function. See the manual for the meanings of the rest of the terms.

```
NORMAL  SCI  ENG
FLOAT  0123456789
RADIAN  DEGREE
FUNC  PAR  POL  SEQ
CONNECTED  DOT
SEQUENTIAL  SIMUL
REAL  a+bi  re^θi
FULL  HORIZ  G-T
SET CLOCK 05/05/10 10:26
```

2. **WINDOW** : Press the **WINDOW** key and the standard window will be displayed. By default, the standard window displays a graph with x-values and y-values ranging from −10 to 10 with tic marks on the axis every 1 unit.

```
WINDOW
 Xmin=-10
 Xmax=10
 Xscl=1
 Ymin=-10
 Ymax=10
 Yscl=1
 Xres=1
```

This window can be changed at any time by changing the individual numbers or pressing the **ZOOM** key and selecting an option from the menu displayed. Because of the shape of the display screen, the standard screen is not a square screen (one unit along the x-axis looks longer than one unit along the y-axis). Be aware that the slopes of lines are not truly depicted unless the screen is in a scale of about 3 : 2. A square screen can be attained by pressing zoom and 5 : ZSquare or by pressing the window key and setting Xmin $= -15$ and Xmax $= 15$ to give the x-axis a length of 30 and the y-axis a length of 20 (a ratio of 3:2).

3. **Y=** : The **Y=** key is in the upper left corner of the keyboard. This key will allow ten different functions to be entered. These functions are labeled as $Y_1, \ldots, Y_{10}$. The variable x may be entered using the **X,T,θ,n** key. The **^** key is used to indicate exponents. (Also note that the negative sign (−) is next to the **ENTER** key.) For example, the equation $y = x^2 + 3x$ would be entered as:

$$Y_1 = X{\wedge}2 + 3X$$

To change an entry, practice with the keys **DEL** (delete), **CLEAR**, and **2ND** INS (insert).

```
Plot1 Plot2 Plot3
\Y1◘X^2+3X
\Y2=
\Y3=
\Y4=
\Y5=
\Y6=
\Y7=
```

4. **GRAPH** : If this key is pressed, then the screen will display the graph of whatever functions are indicated in the **Y=** list with the = sign highlighted using the minimum and maximum values indicated in the current WINDOW. In many cases the WINDOW must be changed to accommodate the domain and range of the function or to show a point where two functions intersect.

5. **TRACE** : The **TRACE** key will display the current graph even if it is not already displayed and give the *x*- and *y*- coordinates of a point highlighted on the graph. The curve may be traced by pressing the left and right arrow keys. At each point on the graph, the corresponding *x*- and *y*- coordinates are indicated at the bottom of the screen. **(Remember that because of the limitations of the pixels (lighted dots) on the screen, these *x*- and *y*- coordinates are generally only approximations.)**

6. CALC: The CALC key (press **2ND** **TRACE**) gives a menu with seven items. Items 1 – 5 are used with graphs.

```
CALCULATE
1:value
2:zero
3:minimum
4:maximum
5:intersect
6:dy/dx
7:∫f(x)dx
```

After displaying a graph, select CALC. Then press **2**

and follow the steps outlined below to locate the point where the graph crosses the *x*-axis (the *x*-intercept). The graph must actually cross the axis. The *x*-value of this point is called a **zero** of the function because the corresponding *y*-value will be 0.

Step 1: With the left arrow, move the cursor to the left of the x-intercept on the graph. Press **ENTER** in response to the question "LeftBound?".

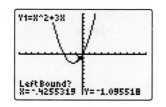

Step 2: With the right arrow, move the cursor to the right of the x-intercept on the graph. Press **ENTER** in response to the question "RightBound?".

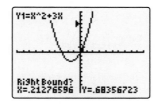

Step 3: With the left arrow, move the cursor near the x-intercept. Press **ENTER** in response to the question "Guess?".

The calculator's estimate of the zero will appear at the bottom of the display.

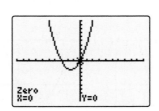

Example 8

Graphing Functions with a TI-84 Plus

Use a TI-84 Plus graphing calculator to find the graphs of each of the following functions. Use the CALC key to find the point where each graph intersects the x-axis. Changing the WINDOW may help you get a "better" or "more complete" picture of the function. This is a judgement call on your part.

8. Use a graphing calculator to find the graphs of each of the following functions. Use the **CALC** key to find the point where each graph intersects the x-axis.

a. $2x + 3y = -8$

b. $y = x^3 + 2x$

c. $y = 4x - 2;$
$y = 4x + 3;$
$y = 4x + 7$

a. $3x + y = -1$

Solution

To have the calculator graph a nonvertical straight line, you must first solve the equation for y. Solving for y gives,

$$y = -3x - 1.$$

(It is important that the 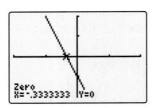 key be used to indicate the negative sign in front of $3x$. This is not the same as the subtraction key.)

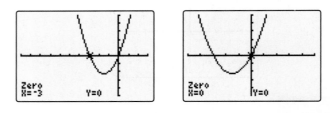

Note: Vertical lines are not functions and cannot be graphed by the calculator in function mode.

b. $y = x^2 + 3x$

Solution

Since the graph of this function has two x-intercepts, we have shown the graph twice. Each graph shows the coordinates of a distinct x-intercept.

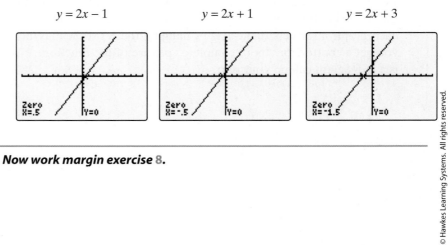

c. $y = 2x - 1; y = 2x + 1; y = 2x + 3$

Solution

| $y = 2x - 1$ | $y = 2x + 1$ | $y = 2x + 3$ |

Now work margin exercise 8.

notes

The standard window shows 96 pixels across the window and 64 pixels up and down the window. This gives a ratio of 3 to 2 and can give a slightly distorted view of the actual graph because the vertical pixels are squeezed into a smaller space. For Example 8c, the graphs of all three functions are in the standard window. Experiment by changing the window to a square window, say −9 to 9 for x and −6 to 6 for y. Then graph the functions and notice the slight differences (and better representation) in the appearances on the display.

Practice Problems

1. State the domain and range of the relation

   a. $\{(5, 6), (7, 8), (9, 0.5), (11, 0.3)\}$.

   b. Is the relation a function? Explain briefly.

2. Use the vertical line test to determine whether the graph on the right represents a function.

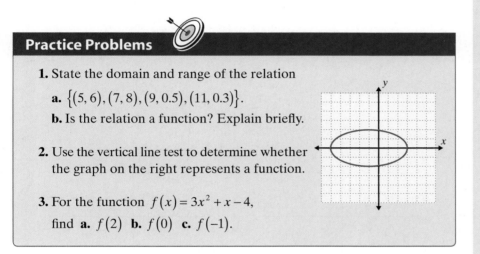

3. For the function $f(x) = 3x^2 + x - 4$,

   find **a.** $f(2)$ **b.** $f(0)$ **c.** $f(-1)$.

Practice Problem Answers

1. $D = \{5, 7, 9, 11\}; R = \{0.3, 0.5, 6, 8\}$. Yes, the relation is a function because each x-coordinate appears only once.

2. not a function   **3. a.** 10 **b.** −4 **c.** −2

List the sets of ordered pairs that correspond to the points. State the domain and range and indicate which of the relations are also functions.

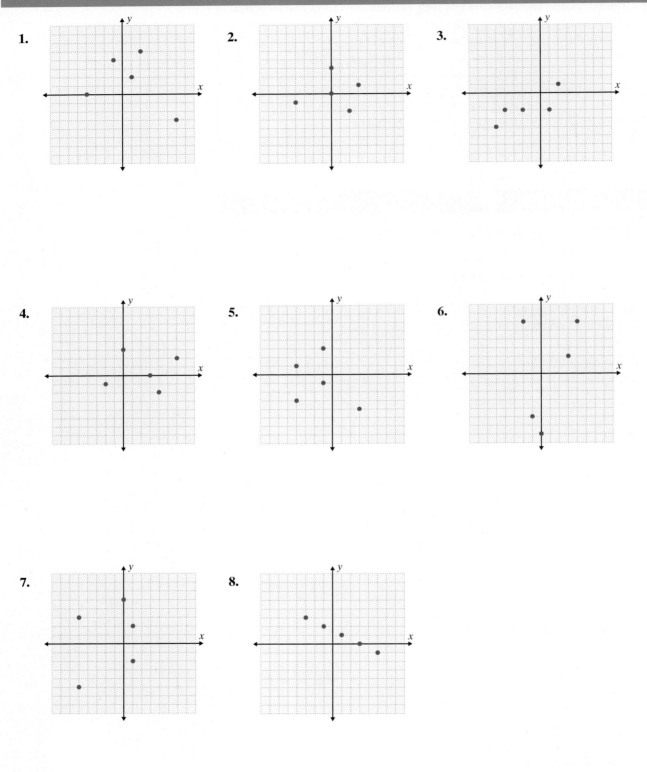

1.

2.

3.

4.

5.

6.

7.

8.

9. $f = \{(0,0),(1,6),(4,-2),(-3,5),(2,-1)\}$

10. $h = \{(1,-5),(2,-3),(-1,-3),(0,2),(4,3)\}$

11. $g = \{(-4,4),(-3,4),(1,4),(2,4),(3,4)\}$

12. $f = \{(-3,-3),(0,1),(-2,1),(3,1),(5,1)\}$

13. $s = \{(0,2),(-1,1),(2,4),(3,5),(-3,5)\}$

14. $t = \{(-1,-4),(0,-3),(2,-1),(4,1),(1,1)\}$

15. $f = \{(-1,4),(-1,2),(-1,0),(-1,6),(-1,-2)\}$

16. $g = \{(0,0),(-2,-5),(2,0),(4,-6),(5,2)\}$

Use the vertical line test to determine whether or not each graph represents a function. State the domain and range using interval notation.

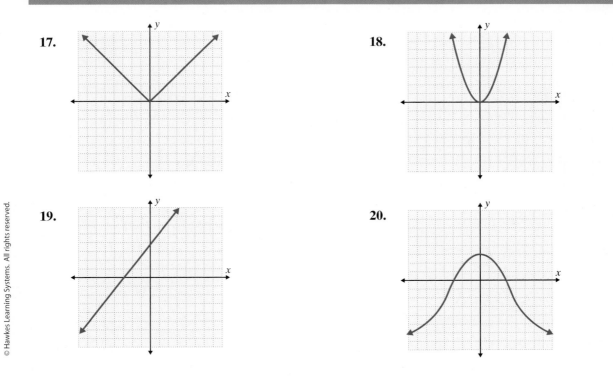

17.

18.

19.

20.

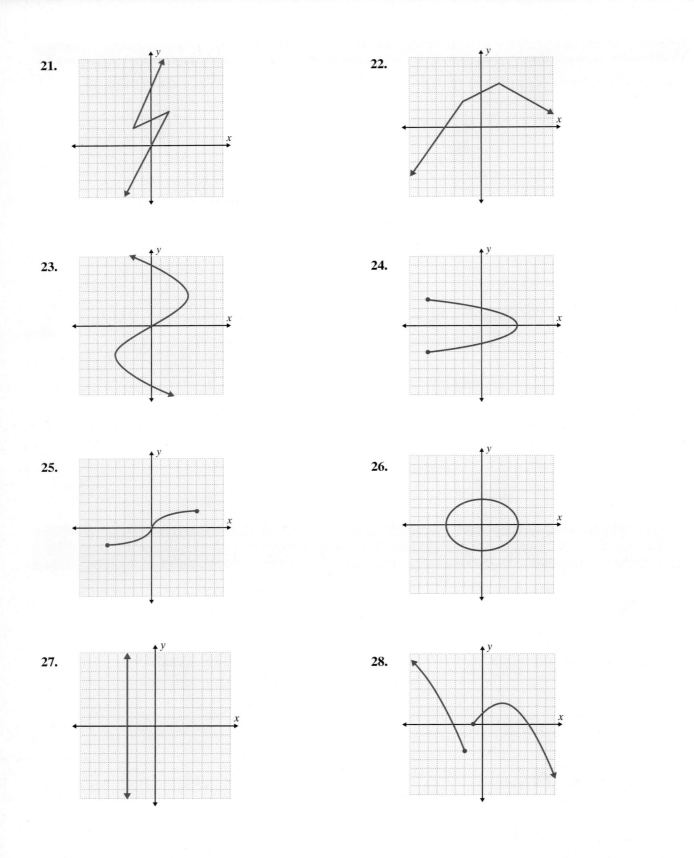

21.

22.

23.

24.

25.

26.

27.

28.

Express the function as a set of ordered pairs for the given equation and given domain. (Hint: Substitute each domain element for *x* and find the corresponding *y*-coordinate.)

29. $y = 3x + 1; D = \left\{-9, -\frac{1}{3}, 0, \frac{4}{3}, 2\right\}$

30. $y = -\frac{3}{4}x + 2; D = \left\{-4, -2, 0, 3, 4\right\}$

31. $y = 1 - 3x^2; D = \left\{-2, -1, 0, 1, 2\right\}$

32. $y = x^3 - 4x; D = \left\{-1, 0, \frac{1}{2}, 1, 2\right\}$

State the domains of the functions.

33. $y = -5x + 10$

34. $2x + y = 14$

35. $g(x) = \dfrac{8}{x}$

36. $h(x) = \dfrac{7}{3x}$

37. $y = \dfrac{13x^2 - 5x + 8}{x - 3}$

38. $f(x) = \dfrac{35}{x - 6}$

Find the values of the functions as indicated.

39. $f(x) = 3x - 10$
  a. $f(2)$ **b.** $f(-2)$ **c.** $f(0)$

40. $g(x) = -4x + 7$
  a. $g(-3)$ **b.** $g(6)$ **c** $g(0)$

41. $G(x) = x^2 + 5x + 6$
  a. $G(-2)$ **b.** $G(1)$ **c.** $G(5)$

42. $F(x) = 6x^2 - 10$
  a. $F(0)$ **b.** $F(-4)$ **c.** $F(4)$

43. $h(x) = x^3 - 8x$
  a. $h(-3)$ **b.** $h(0)$ **c.** $h(3)$

44. $P(x) = x^2 + 4x + 4$
  a. $P(-2)$ **b.** $P(10)$ **c.** $P(-5)$

Use a graphing calculator to graph the functions. Use the CALC features to find *x*-intercepts, if any. (The value of *y* will be 0 at those points.) For absolute value functions, select the MATH menu, then the NUM menu, and then 1: abs(. Remember to press) after entering the absolute value.

45. $y = 6$

46. $y = 4x$

47. $y = x + 5$

48. $y = -2x + 3$

49. $y = x^2 - 4x$

50. $y = 1 + 2x - x^2$

51. $y = -|3x|$

52. $y = |x + 2|$

53. $y = |x^2 - 3x|$

54. $y = 2x^3 - 5x^2 + 1$

55. $y = -x^3 + 3x - 1$

56. $y = x^4 - 10x^2 + 9$

Use the CALC features of the calculator to find the coordinates of any points of intersection of the graphs. (Hint: Item 5 on the CALC menu 5: intersect will help in finding that point (or points) of intersection of two functions, if there is one.) In the Y = menu use both Y₁ = and Y₂ = to be able to graph both functions at the same time.

57. $y = 3x + 2$
$y = 4 - x$

58. $y = 2 - x$
$y = x$

59. $y = 2x - 1$
$y = x^2$

60. $y = x + 3$
$y = -x^2 + x + 7$

The calculator display shows an incorrect graph for the corresponding equation. Explain how you know, by just looking at the graph, that a mistake has been made.

61. $y = 2x + 5$

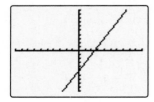

62. $y = -3x + 4$

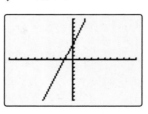

63. $y = \dfrac{2}{3}x - 2$

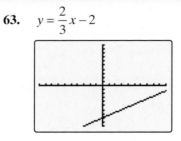

64. $y = -4x$

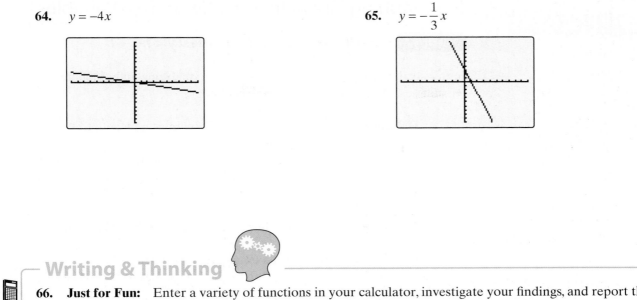

65. $y = -\dfrac{1}{3}x$

Writing & Thinking

66. **Just for Fun:** Enter a variety of functions in your calculator, investigate your findings, and report these to your class. Certainly, interesting discussions will follow!

9.6 Graphing Linear Inequalities in Two Variables

Objective A **Graphing Linear Inequalities: $y < mx + b$**

In this section, we will develop techniques for analyzing and graphing linear inequalities. We need the following terminology:

Half-plane	A straight line separates a plane into two **half-planes**. The points on one side of the line are in one of the half-planes, and the points on the other side of the line are in the other half-plane.
Boundary line	The line itself is called the **boundary line**.
Closed half-plane	If the boundary line is included in the solution set, then the half-plane is said to be **closed**.
Open half-plane	If the boundary line is not included in the solution set, then the half-plane is said to be **open**.

Figure 1 shows both **a.** an open half-plane and **b.** a closed half-plane with the line $2x - 3y = 10$ as the boundary line. The solution set is the shaded region.

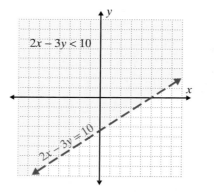

The points on the line $2x - 3y = 10$ are not included in the solution set so the line is dashed. The half-plane is open.

The points on the line $2x - 3y = 10$ are included in the solution set so the line is solid. The half-plane is closed.

Figure 1

Graphing Linear Inequalities

1. First, graph the boundary line (dashed if the inequality is < or >, solid if the inequality is ≤ or ≥).

2. Next, determine which side of the line to shade using one of the following methods.

   Method 1
   a. Test any one point obviously on one side of the line.
   b. If the test-point satisfies the inequality, shade the half-plane on that side of the line. Otherwise, shade the other half-plane.

   Note: The point $(0, 0)$, if it is not on the boundary line, is usually the easiest point to test.

   Method 2
   a. Solve the inequality for y (assuming that the line is not vertical).
   b. If the solution shows $y <$ or $y \leq$, then shade the half-plane below the line.
   c. If the solution shows $y >$ or $y \geq$, then shade the half-plane above the line.

   Note: If the boundary line is vertical, then solve for x. If the solution shows $x >$ or $x \geq$, then shade the half-plane to the right. If the solution shows $x <$ or $x \leq$, then shade the half-plane to the left.

3. The shaded half-plane (and the line if it is solid) is the solution to the inequality.

Example 1

Graphing Linear Inequalities

a. Graph the half-plane that satisfies the inequality $2x + y \leq 6$.

Solution

Method 1 is used in this example.

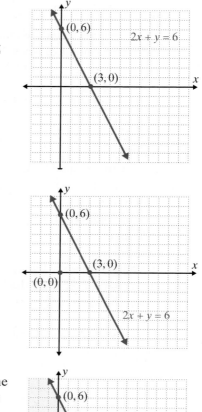

Step 1: Graph the boundary line $2x + y = 6$ as a solid line because the inequality is $\leq$ (less than or **equal to**).

Step 2: Test any point on one side of the line. In this example, we have chosen $(0, 0)$.

$$2(0) + (0) \leq 6$$
$$0 \leq 6$$

This is a true statement.

Step 3: Shade the half-plane on the same side of the line as the point $(0, 0)$. (The shaded half-plane and the boundary line is the solution set to the inequality.)

b. Graph the solution set to the inequality $y > 2x$.

Solution

Since the inequality is already solved for y, Method 2 is easy to apply.

Step 1: Graph the boundary line $y = 2x$ as a dashed line because the inequality is $>$.

Step 2: The solution shows >, so by Method 2, the graph consists of those points above the line. Shade the half-plane above the line.

Note: As a check, we see that the point $(-3, 0)$ gives $0 > -6$, a true statement. Thus we know we have shaded the correct half-plane.

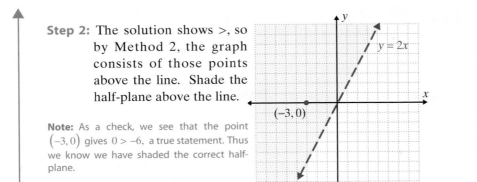

1. Graph the solution set to the following inequalities.

a. $9x - 3y \leq 21$

b. $x > -2$

c. $y < 4$

c. Graph the half-plane that satisfies the inequality $y > 1$.

Solution

Again, the inequality is already solved for y and Method 2 is used.

Step 1: Graph the boundary line $y = 1$ as a dashed line because the inequality is >. (The boundary line is a horizontal line.)

Step 2: By Method 2, shade the half-plane above the line.

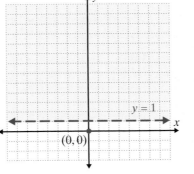

d. Graph the solution set to the inequality $x \leq 0$.

Solution

The boundary line is a vertical line and Method 1 is used.

Step 1: Graph the boundary line $x = 0$ as a solid line because the inequality is $\leq$ (less than or **equal to**). Note that this is the y-axis.

Step 2: Test the point $(-2, 1)$.

$-2 \leq 0$

This statement is true.

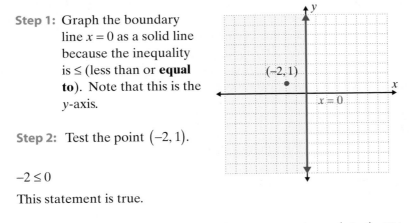

Step 3: Shade the half-plane on the same side of the line as $(-2, 1)$. This half-plane consists of the points with x-coordinates 0 or negative.

Now work margin exercise 1.

Objective B   Using a TI-84 Plus Graphing Calculator to Graph Linear Inequalities

The first step in using the TI-84 Plus (or any other graphing calculator) to graph a linear inequality is to solve the inequality for y. This is necessary because this is the way that the boundary line equation can be graphed as a function. Thus Method 2 for graphing the correct half-plane is appropriate.

Note that when you press the ⬬ Y= ⬬ key on the calculator, a slash (\ ) appears to the left of the Y expression as in \Y_1 =. This slash is actually a command to the calculator to graph the corresponding function as a solid line or curve. If you move the cursor to position it over the slash and hit **ENTER** repeatedly, the following options will appear.

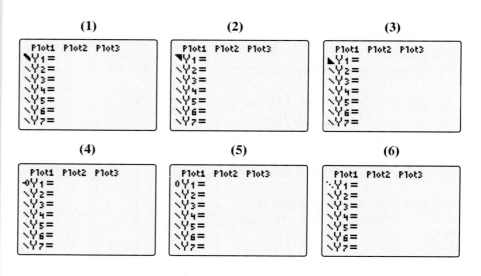

Figure 2

If the slash (which is actually four dots if you look closely) becomes a set of three dots **(6)**, then the corresponding graph of the function will be dotted. By setting the shading above the slash **(2)**, the corresponding graph on the display will show shading above the line or curve. By setting the shading below the slash **(3)**, the corresponding graph on the display will show shading below the line or curve. (The solid line occurs only when the slash is four dots, so the calculator is not good for determining whether the boundary curve is included or not.) Options 1, 4, and 5 are not used when graphing linear inequalities. The following examples illustrate two situations.

Example 2

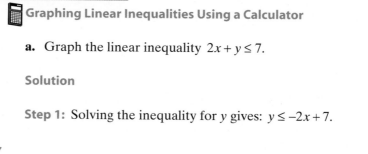 **Graphing Linear Inequalities Using a Calculator**

a. Graph the linear inequality $2x + y \le 7$.

Solution

Step 1: Solving the inequality for y gives: $y \le -2x + 7$.

Step 2: Press the key Y= and enter the function: $\backslash Y_1 = -2X + 7$.

Step 3: Go to the \ and hit ENTER three times so that the display appears as follows:

```
Plot1 Plot2 Plot3
\Y1◻-2X+7
\Y2=
\Y3=
\Y4=
\Y5=
\Y6=
\Y7=
```

Step 4: Press GRAPH and (using the standard WINDOW settings) the following graph should appear on the display.

b. Graph the linear inequality $-5x + 4y > -8$.

Solution

Step 1: Solving the inequality for y gives: $y > \dfrac{5}{4}x - 2$.

Step 2: Press the key Y= and enter the function:
$\backslash Y_1 = (5/4)X - 2$.

Step 3: Go to the \ and hit ENTER two times so that the display appears as follows:

```
Plot1 Plot2 Plot3
◥Y1◻(5/4)X-2
\Y2=
\Y3=
\Y4=
\Y5=
\Y6=
\Y7=
```

Step 4: Press GRAPH and (using the standard WINDOW settings) the following graph should appear on the display. (**Note:** The boundary line should actually be dotted.)

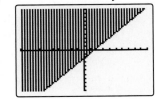

Now work margin exercise 2.

2. Use a graphing calculator to graph the following inequalities.

a. $5x - 8 > y - 6$

b. $6x - 3y \le 7$

Practice Problems

1. Which of the following points satisfy the inequality $x + y < 3$?

    a. $(2, 1)$ **b.** $\left(\dfrac{1}{2}, 3\right)$ **c.** $(0, 5)$ **d.** $(-5, 2)$

2. Which of the following points satisfy the inequality $x - 2y \geq 0$?

    a. $(2, 1)$ **b.** $(1, 3)$ **c.** $(4, 2)$ **d.** $(3, 1)$

3. Which of the following points satisfy the inequality $x < 3$?

    a. $(1, 0)$ **b.** $(0, 1)$ **c.** $(4, -1)$ **d.** $(2, 3)$

Practice Problem Answers

1. d   **2.** a, c, d   **3.** a, b, d

Graph the solution set of each of the linear inequalities.

1. $x + y \leq 7$

2. $x - y > -2$

3. $x - y > 4$

4. $x + y \leq 6$

5. $y < 4x$

6. $y < -2x$

7. $y \geq -3x$

8. $y > x$

9. $x - 2y > 8$

10. $x + 3y \leq 3$

11. $4x + y \geq 2$

12. $5x - y < 4$

13. $y \leq 5 - 3x$

14. $y \geq 8 - 2x$

15. $2y - x \leq 0$

16. $x + y > 0$

17. $x + 4 \geq 0$

18. $x - 5 \leq 0$

19. $y \geq -2$

20. $y + 3 < 0$

21. $4x < -3y + 9$

22. $3x < 2y - 4$

23. $3y > 4x + 6$

24. $5x < 2y - 6$

25. $x + 3y < 7$

26. $3x + 4y > 11$

27. $\frac{1}{2}x - y > 1$

28. $\frac{1}{3}x + y \geq 3$

29. $\dfrac{2}{3}x + y \geq 4$ **30.** $2x - \dfrac{4}{3}y > 8$

Use a graphing calculator to graph each of the linear inequalities.

31. $y > \dfrac{1}{2}x$ **32.** $2x \geq -6y$ **33.** $x - y \leq 5$ **34.** $x + 2y > 8$

35. $y \geq -3$ **36.** $y \leq -4$ **37.** $2x + y \leq 6$ **38.** $x - 3y \geq 9$

39. $3x + 2y \geq 12$ **40.** $3x - 4y > 15$

Writing & Thinking

41. Explain in your own words how to test to determine which side of the graph of an inequality should be shaded.

42. Describe the difference between a closed and an open half-plane.

Section 9.1: The Cartesian Coordinate System

Cartesian Coordinate System   page 688
Ordered Pairs (x, y)
    As solutions to linear equations
    First coordinate (the independent variable)
    Second coordinate (the dependent variable)
Quadrants
Origin $(0, 0)$
Horizontal Axis (x-axis)
Vertical Axis (y-axis)

One-To-One Correspondence   page 690
There is a **one-to-one correspondence** between points in a
plane and ordered pairs of real numbers.

Graphs of Points   page 690

Section 9.2: Graphing Linear Equations in Two Variables: $Ax + By = C$

Standard Form of a Linear Equation   page 704
Any equation of the form $Ax + By = C$, where A, B, and C are real
numbers and where A and B are not both 0, is called the **standard form**
of a **linear equation**.

To Graph a Linear Equation in Two Variables   page 704
1. Locate any two points that satisfy the equation.
2. Plot these two points on a Cartesian coordinate system.
3. Draw a line through these two points.
   Note: Every point on that line will satisfy the equation.
4. To check, locate a third point that satisfies the equation and check
   to see that it does indeed lie on the line.

y-intercept   page 706
The **y-intercept** is the point on the graph where the line crosses the
y-axis. The x-coordinate will be 0, which makes the y-intercept of the
form $(0, y)$.

x-intercept   page 706
The **x-intercept** is the point on the graph where the line crosses the
x-axis. The y-coordinate will be 0, which makes the x-intercept of the
form $(x, 0)$.

Section 9.3: The Slope-Intercept Form: $y = mx + b$

Slope as a Rate of Change   page 713

Slope of a Line   page 714

Let $P_1(x_1, y_1)$ and $P_2(x_2, y_2)$ be two points on a line. The **slope** can be calculated as follows:

$$\text{slope} = m = \frac{\text{rise}}{\text{run}} = \frac{y_2 - y_1}{x_2 - x_1}.$$

Positive and Negative Slopes   page 716

Lines with **positive slope** go up as we move along the line from left to right.

Lines with **negative slope** go down as we move along the line from left to right.

Horizontal and Vertical Lines   page 717

The following two general statements are true for horizontal and vertical lines:
1. For **horizontal lines** (of the form $y = b$), the **slope is 0**.
2. For **vertical lines** (of the form $x = a$), the **slope is undefined**.

Slope-Intercept Form   page 719

Any equation of the form $y = mx + b$ is called the **slope-intercept form** for the equation of a line. m is the slope and $(0, b)$ is the y-intercept.

Section 9.4: The Point-Slope Form: $y - y_1 = m(x - x_1)$

Graphing a Line Given a Point and the Slope

                                                                page 729

Point-Slope Form

An equation of the form $y - y_1 = m(x - x_1)$ is called the **point-slope form** for the equation of a line that contains the point (x_1, y_1) and has slope m.

Finding the Equation of a Line Given Two Points   page 730

Parallel Lines   page 732

Parallel lines are lines that never intersect (cross each other) and these lines have the same slope. (**Note:** All vertical lines (undefined slopes) are parallel to one another.)

Perpendicular Lines   page 733

Perpendicular lines are lines that intersect at $90°$ (right) angles and whose slopes are negative reciprocals of each other. Horizontal lines are perpendicular to vertical lines.

A **relation** is a set of ordered pairs of real numbers. The **domain**, **D**, of a relation is the set of all first coordinates in the relation. The **range**, **R**, of a relation is the set of all second coordinates in the relation.

A **function** is a relation in which each domain element has exactly one corresponding range element. The definition can also be stated in the following ways:
1. A function is a relation in which each first coordinate appears only once.
2. A function is a relation in which no two ordered pairs have the same first coordinate.

If **any** vertical line intersects the graph of a relation at more than one point, then the relation graphed is **not** a function.

A **linear function** is a function represented by an equation of the form $y = mx + b$. The domain of a linear function is the set of all real numbers, $D = (-\infty, \infty)$.

In **function notation**, instead of writing y, write $f(x)$, read "f of x." The letter f is the name of the function. The notation $f(4)$ means to replace x with 4 in the function.

Terminology for Graphing Linear Inequalities   page 760
    Half-plane
    Boundary line
    Closed Half-plane
    Open Half-plane

Graphing Linear Inequalities   page 761
    1. First, graph the boundary line (dashed if the inequality is < or >, solid if the inequality is ≤ or ≥).
    2. Next, determine which side of the line to shade using one of the following methods.
    Method 1
        a. Test any one point obviously on one side of the line.
        b. If the test-point satisfies the inequality, shade the half-plane on that side of the line. Otherwise, shade the other half-plane.
    Method 2
        a. Solve the inequality for y (assuming that the line is not vertical).
        b. If the solution shows $y <$ or $y \leq$, then shade the half-plane below the line.
        c. If the solution shows $y >$ or $y \geq$, then shade the half-plane above the line.
    Note: If the boundary line is vertical, then solve for x. If the solution shows $x >$ or $x \geq$, then shade the half-plane to the right. If the solution shows $x <$ or $x \leq$, then shade the half-plane to the left.
    3. The shaded half-plane (and the line if it is solid) is the solution to the inequality.

Using a Graphing Calculator to Graph Linear Inequalities   page 764

Chapter 9: Review

Section 9.1: The Cartesian Coordinate System

List the sets of ordered pairs corresponding to the points on the graphs. Assume that the grid lines are marked one unit apart.

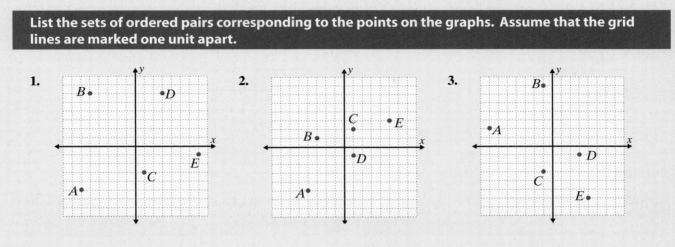

1.

2.

3.

Graph the sets of ordered pairs and label the points.

4. $\{E(-3, -4), F(0, 0), G(5, -1), H(1.5, -3)\}$

5. $\{M(6, 2), N(3, -1), O(0, -2), P(-4, -5)\}$

6. $\{Q(-4, -2), R(-4, 0), S(-4, 3), T(-4, 5)\}$

Determine which, if any, of the ordered pairs satisfy the given equations.

7. $3x - y = -2$
   a. $(1, 1)$
   b. $(-1, 1)$
   c. $(2, 2)$
   d. $(-2, 2)$

8. $x + 2y = 3$
   a. $(-3, 1)$
   b. $(2, 1)$
   c. $\left(2, \frac{1}{2}\right)$
   d. $(1, -2)$

9. $y = 4x$
   a. $(1, 4)$
   b. $(-1, 4)$
   c. $\left(\frac{1}{4}, 1\right)$
   d. $(-3, -12)$

10. $2x - 3y = 5$
    a. $(1, -1)$
    b. $\left(2, -\frac{1}{3}\right)$
    c. $(2.5, 0)$
    d. $(0, -2)$

Complete the tables so that each order pair will satisfy the given equation. Graph the resulting sets of ordered pairs.

11. $y = 5x$

x	y
0	
	−5
−2	
	5

12. $3x + y = −5$

x	y
0	
	−1
−3	
	−0.5

Section 9.2: Graphing Linear Equations in Two Variables: $Ax + By = C$

Use your knowledge of x-intercepts and y-intercepts to match each of the following equations with its graph.

13. $y = 2x + 4$

14. $y = 6 − x$

15. $y = 3 − 3x$

16. $−5x + y = 10$

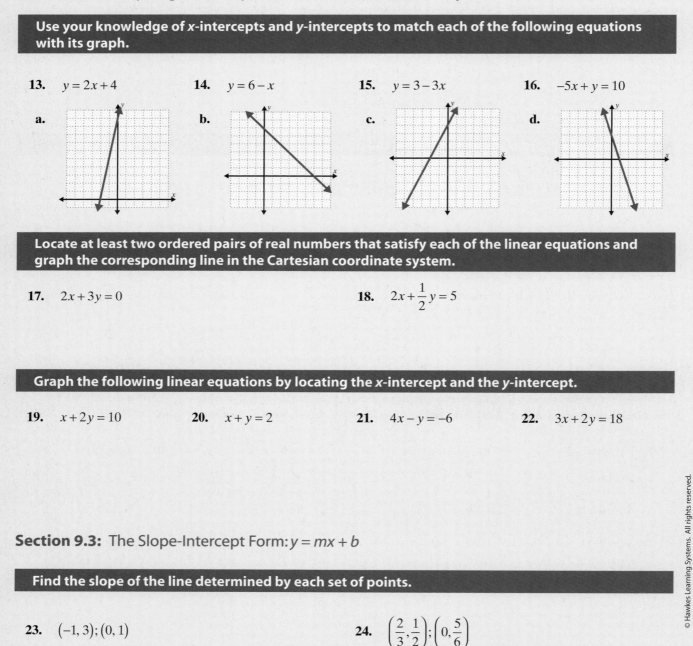

a.

b.

c.

d.

Locate at least two ordered pairs of real numbers that satisfy each of the linear equations and graph the corresponding line in the Cartesian coordinate system.

17. $2x + 3y = 0$

18. $2x + \dfrac{1}{2} y = 5$

Graph the following linear equations by locating the x-intercept and the y-intercept.

19. $x + 2y = 10$

20. $x + y = 2$

21. $4x − y = −6$

22. $3x + 2y = 18$

Section 9.3: The Slope-Intercept Form: $y = mx + b$

Find the slope of the line determined by each set of points.

23. $(−1, 3); (0, 1)$

24. $\left(\dfrac{2}{3}, \dfrac{1}{2} \right); \left(0, \dfrac{5}{6} \right)$

Determine whether each equation represents a horizontal line or vertical line and give its slope. Graph the line.

25. $3x = -5.4$

26. $3y - 4 = 0$

Write each equation in slope-intercept form. Find the slope and the *y*-intercept, and then use them to draw the graph.

27. $y = -\dfrac{2}{5}x + 1$

28. $y = \dfrac{4}{3}x - 2$

29. $x - 2y = 4$

30. $2x + y + 8 = 0$

31. $2y - 6 = 0$

32. $3x = -9$

Find an equation in slope-intercept form for the line passing through the given point with the given slope

33. $(0, 4)$; $m = -\dfrac{1}{3}$

34. $(0, -2)$; $m = 3$

The graph of a line is shown with two points highlighted. Find a. the slope, b. the *y*-intercept (if there is one), and c. the equation of the line in slope-intercept form.

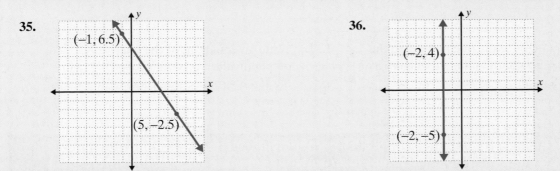

35. $(-1, 6.5)$ $(5, -2.5)$

36. $(-2, 4)$ $(-2, -5)$

Section 9.4: The Point-Slope Form: $y - y_1 = m\left(x - x_1\right)$

Find a. the slope, b. a point on the line, and c. the graph of the line for the given equations in point-slope form.

37. $y - 7 = 3(x + 1)$

38. $y + 3 = -(x - 1)$

39. $m = -3; (-4, 1)$

40. $m = 0; \left(-6, -\dfrac{3}{2}\right)$

41. $(-3, 5); (1, -3)$

42. $(4, 2); (6, 1)$

43. $\left(\dfrac{2}{3}, 0\right); \left(-2, \dfrac{1}{3}\right)$

44. $\left(0, -\dfrac{1}{2}\right); \left(\dfrac{3}{4}, \dfrac{1}{5}\right)$

45. Write an equation for the horizontal line through the point $(-3, 5)$.

46. Write an equation for the vertical line through the point $(5, 7)$.

47. Write an equation for the line through the point $(-6, -2)$ and parallel to the line $x - 4y = 4$. Graph both lines.

48. Write an equation for the line through the point $(1, 1)$ and perpendicular to the line $2x + y = 9$. Graph both lines.

49. Write an equation for the line parallel to the line $2x + 3y = 5$ with the same y-intercept as the line $x - 2y = 8$.

50. Write an equation for the line perpendicular to the line $3x - y = 2$ with the same y-intercept as the line $2x + y = 5$.

51. $\begin{cases} 5x + y = 5 \\ x - 5y = 10 \end{cases}$

52. $\begin{cases} 2x + y = 8 \\ y = -\dfrac{1}{2}x \end{cases}$

Section 9.5: Introduction to Functions and Function Notation

List the sets of ordered pairs that correspond to the points. State the domain and range and indicate which of the relations are also functions.

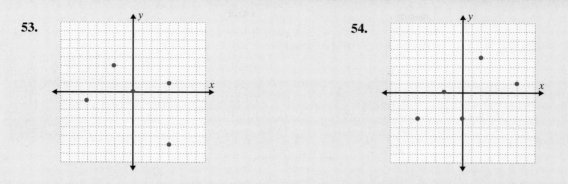

53.

54.

Graph the relations. State the domain and range and indicate which of the relations are functions.

55. $g = \{(-5, 1), (-3, -2), (-3, 2), (0, 0), (1, 1)\}$

56. $f = \{(-4, 3.2), (-3, 0.1), (0, 0.1), (0, -2), (2, 3.2)\}$

Use the vertical line test to determine whether or not each graph represents a function. State the domain and range using interval notation.

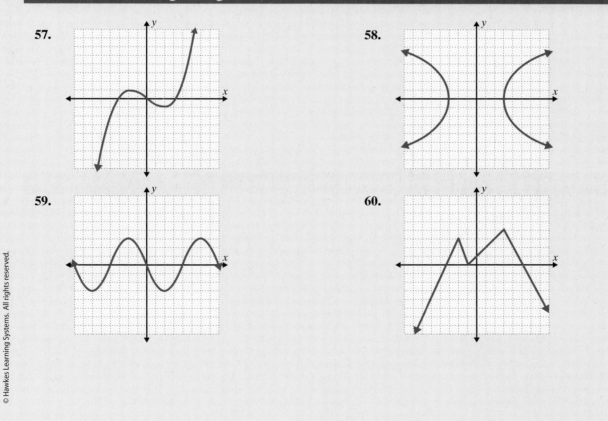

57.

58.

59.

60.

Find the values of the functions as indicated.

61. Given that $g(x) = 5x - 12$, find **a.** $g(6)$ **b.** $g(-1)$ and **c.** $g\left(\dfrac{3}{5}\right)$.

62. For $R(x) = x^2 - 5x + 6$, find **a.** $R(5)$ **b.** $R(-2)$ and **c.** $R(0)$.

Use a graphing calculator to graph the functions. Use the CALC features to find x-intercepts, if any. (The value of y will be 0 at those points.) For absolute value functions, select the MATH menu, then the NUM menu, and then 1 : abs (. Remember to press) after entering the absolute value.

63. $y = -3x$

64. $y = 3x + 1$

65. $y = \left|x^2 - 2x\right|$

66. $y = x^3 - 4x - 1$

Section 9.6: Graphing Linear Inequalities in Two Variables

Graph the solution set of each of the linear inequalities.

67. $y \le 2x - 1$

68. $y > -5$

69. $y < 3x$

70. $y - 3 > 6x$

71. $2x + 4 \ge 2y$

72. $3x + 2y > 12$

73. $\dfrac{3}{5}x + 6 \le y$

74. $y > \dfrac{1}{4}x - 3$

Use a graphing calculator to graph each of the linear inequalities.

75. $y > 2x$

76. $x + y \le 5$

Chapter 9: Test

List the set of ordered pairs corresponding to the points on the graph. Assume that the grid lines are marked one unit apart.

1.

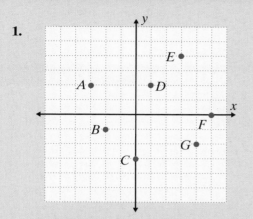

Graph the following set of ordered pairs and label the points.

2. $\{L(0, 2), M(4, -1), N(-3, 2), O(-1, -5), P(2, 1.5)\}$

Determine the missing coordinate in each of the ordered pairs so that the point will satisfy the equation given.

3. $3x + y = 2$
   a. $(0, \quad)$
   b. $(\quad, 0)$
   c. $(-2, \quad)$
   d. $(\quad, -7)$

4. $x - 5y = 6$
   a. $(0, \quad)$
   b. $(\quad, 0)$
   c. $(11, \quad)$
   d. $(\quad, -2)$

Locate at least two ordered pairs of real numbers that satisfy each of the linear equations and graph the corresponding line in the Cartesian coordinate system.

5. $x + 4y = 5$

6. $2x - 5y = 1$

Graph the following linear equation by locating the x-intercept and the y-intercept.

7. $5x - 3y = 9$

8. $\dfrac{4}{3}x + 2y = 8$

Find the slope of the line determined by each pair of points. Graph the line.

9. $(1, -2), (9, 7)$

10. $(-2, 5), (8, 3)$

Write each equation in slope-intercept form. Find the slope and y-intercept, and then use them to draw the graph.

11. $x - 3y = 4$

12. $4x + 3y = 3$

Solve the following word problems.

13. Bicycling: The following table shows the number of miles Sam rode his bicycle from one hour to another.

a. Plot these points on a graph.

b. Connect the points with line segments.

c. Find the slope of each line segment.

d. Interpret the slope as a rate of change.

Hour	Miles
First	20
Second	31
Third	17

14. Find an equation in standard form for the line passing through the point $(3, 7)$ and with the slope $m = -\dfrac{5}{3}$. Graph the line.

15. Find an equation in standard form for the line passing through the points $(-4, 6)$ and $(3, -2)$. Graph the line.

Find an equation in slope-intercept form for each of the equations described.

16. Horizontal and passing through $(-1, 6)$

17. Parallel to $3x + 2y = -1$ and passing through $(2, 4)$

18. Perpendicular to the y-axis and passing through the point $(3, -2)$

Answer the following question.

19. Write the definition of a function.

List the set of ordered pairs that corresponds to the points. State the domain and range and indicate whether or not the relation is a function.

20.

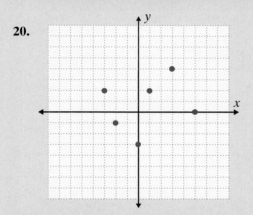

Use the vertical line test to determine whether or not the following graph represents a function. State the domain and range.

21.

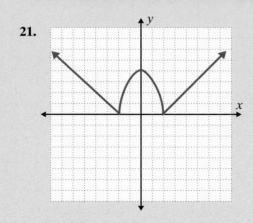

Solve the following problems.

22. Given that $f(x) = x^2 - 2x + 5$, find

    a. $f(-2)$ **b.** $f(0)$ **c.** $f(1)$.

23. **a.** Use a graphing calculator to graph the function $y = 5 + 3x - 2x^2$.

    b. Use the CALC features of the calculator to find x-intercepts, if any.

Graph the solution set of each of the linear inequalities.

24. $3x - 5y \leq 10$ **25.** $3x + 4y > 7$

Use a graphing calculator to graph the following linear inequality.

26. $y - 2x \leq -5$

Cumulative Review: Chapters 1 - 9

Find the LCM of each set of numbers or algebraic expressions.

1. $\{12, 15, 45\}$

2. $\{6, 20, 25, 35\}$

Solve the following problems.

3. Change the following values to percents:

    a. 0.275   **b.** $\dfrac{4}{5}$ **c.** 1.34

4. Find the mean of the set of numbers: $\{3.6, 8.9, 14.7, 25.3\}$

5. Fernando scored 80, 88, and 82 on three exams in statistics. What was his average score on these exams (to the nearest tenth)?

6. **Rectangles:** Find

    a. the perimeter and

    b. the area of the rectangle shown below with sides 10 ft and 20 ft.

7. **Triangles:** Find the length of the hypotenuse of the right triangle with legs of 4 in. and 6 in.

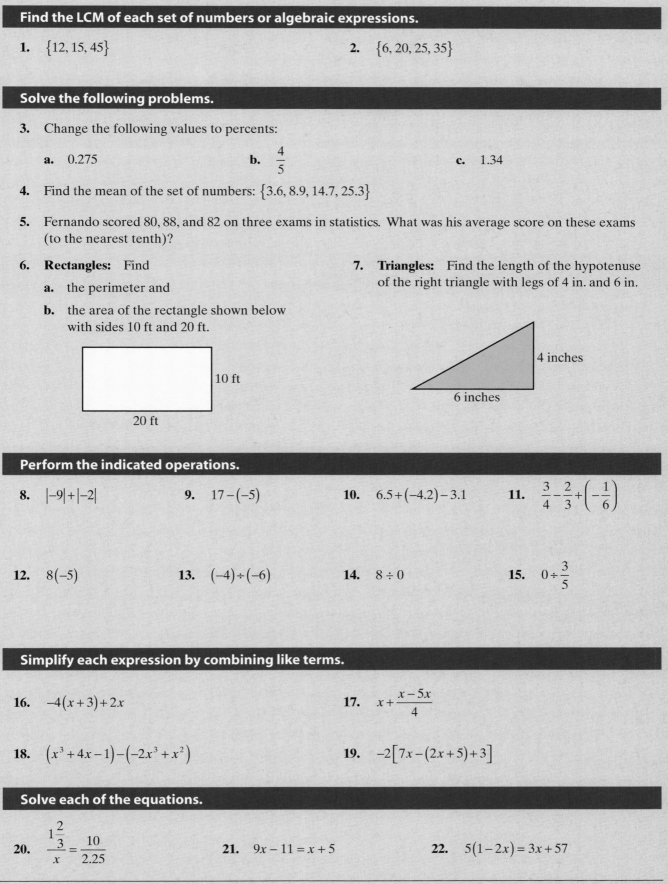

10 ft

20 ft

4 inches

6 inches

Perform the indicated operations.

8. $|-9| + |-2|$

9. $17 - (-5)$

10. $6.5 + (-4.2) - 3.1$

11. $\dfrac{3}{4} - \dfrac{2}{3} + \left(-\dfrac{1}{6}\right)$

12. $8(-5)$

13. $(-4) \div (-6)$

14. $8 \div 0$

15. $0 \div \dfrac{3}{5}$

Simplify each expression by combining like terms.

16. $-4(x+3) + 2x$

17. $x + \dfrac{x - 5x}{4}$

18. $(x^3 + 4x - 1) - (-2x^3 + x^2)$

19. $-2[7x - (2x+5) + 3]$

Solve each of the equations.

20. $\dfrac{1\frac{2}{3}}{x} = \dfrac{10}{2.25}$

21. $9x - 11 = x + 5$

22. $5(1 - 2x) = 3x + 57$

23. $5(2x+3) = 3(x-4)-1$

24. $\dfrac{7x}{8} + 5 = \dfrac{x}{4}$

25. Solve each equation for the indicated variable.

    a. Solve for n: $A = \dfrac{m+n}{2}$

    b. Solve for f: $\omega = 2\pi f$

Solve each inequality and graph the solution set on a real number line. Write each solution in interval notation. Assume that x is a real number.

26. $5x - 7 > x + 9$

27. $x + 8 - 5x \geq 2(x-2)$

28. $\dfrac{2x+1}{3} \leq \dfrac{3x}{5}$

Determine the missing coordinate in each of the ordered pairs so that the point will satisfy the equation given.

29. $2x - y = 4$

    a. $(0,\ \ )$

    b. $(\ \ ,0)$

    c. $(1,\ \ )$

    d. $(\ \ ,2)$

Graph the following linear equations by locating the x-intercept and the y-intercept.

30. $x + 2y = 6$

31. $3x - 4y = 6$

Write each equation in slope-intercept form. Find the slope and the y-intercept, and use them to draw the graph.

32. $x + 5y = 10$

33. $3x + y = 1$

Find an equation in standard form for the line determined by the given point and slope or two points.

34. $(0,0), m = 2$

35. $(5,2)$, m is undefined

36. $(0,3),(5,-1)$

Find an equation in slope-intercept form for each of the equations described.

37. Parallel to the y-axis and passing through $(1, -7)$

38. Perpendicular to $4x + 3y = 5$ and passing through $(4, 0)$

List the ordered pairs corresponding to the points in the given graph. State the domain and range and whether or not the relation is a function.

39.

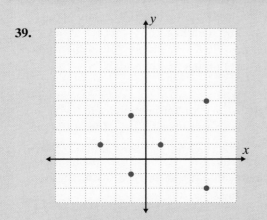

Use the vertical line test to determine whether or not each graph represents a function. State the domain and range.

40.

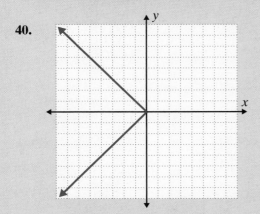

41. For $g(x) = x^2 - 4x + 7$, find $g(3)$.

42. For $G(x) = 5x^3 - 4x$, find $G(-2)$.

Graph the solution set of each of the linear inequalities.

43. $3x + y < 2$

44. Consecutive Even Integers: Find three consecutive even integers such that the sum of the second and third is equal to three times the first decreased by 14.

45. Rental Cost: The local supermarket charges a flat rate of $5, plus $3 per hour for rental of a carpet cleaner. If it cost Ron $26 to rent the machine, how many hours did he keep it?

46. Number Problem: The difference between twice a number and 3 is equal to the difference between five times the number and 2. Find the number.

47. Buying Online: Sarah found a purse she would like to buy at an online auction. If she submits a winning bid of $72 then with shipping and handling, the total cost of the purchase is $76.80. What percent of the total cost is the shipping and handling?

48. Rectangles: The length of a rectangle is 9 cm more than its width. The perimeter is 82 cm. Find the dimensions of the rectangle.

49. Interest: Christopher opened a sandwich shop and wants to invest some of his earnings. He decides to invest a total of $4500, part at 4.5% and the other at 7%. How much money does he need to invest at each rate so that the interest made is $230?

50. World Series of Poker: The following table shows the number of entrants in the main event of the World Series of Poker from 2003 to 2008.

Year	Number of Entrants
2003	839
2004	2576
2005	5619
2006	8773
2007	6358
2008	6844

  a. Plot these points on a graph.

  b. Connect the points with line segments.

  c. Find the slope of each line segment.

  d. Interpret the slope as a rate of change.

Source: Harrah's License Company

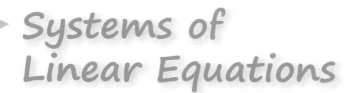

Systems of Linear Equations

Mathematics at Work!

Many times, one equation does not provide enough information for us to find the best solution. In this case, we must construct more than one equation, and use information from both to find an appropriate solution. Consider the following situation. (See Section 10.4.)

A carpenter is building a fenced-in pasture for a farmer. The carpenter has 900 feet of fencing to use. The farmer wants the length of the pasture to be twice the width. What should the dimensions of the pasture be? Recall that the perimeter of a rectangle can be found by using the equation $P = 2l + 2w$.

10.1 Systems of Linear Equations: Solutions by Graphing

Many applications involve two (or more) quantities, and by using two (or more) variables, we can form linear equations using the information given. Such a set of equations is called a **system**, and in this chapter we will develop techniques for solving systems of linear equations.

Graphing systems of two equations in two variables is helpful in visualizing the relationships between the equations. However, this approach is somewhat limited in finding solutions since numbers might be quite large, or solutions might involve fractions or decimals that must be estimated on the graph. Therefore, algebraic techniques are necessary to accurately solve systems of linear equations. Graphing in three dimensions will be left to later courses.

Objective A **Systems of Linear Equations**

Many applications, as we will see in Sections 10.4 and 10.5, involve solving pairs of linear equations. Such pairs are said to form **systems of equations** or **sets of simultaneous equations**. For example,

$$\begin{cases} x - 2y = 0 \\ 3x + y = 7 \end{cases} \quad \text{and} \quad \begin{cases} y = -x + 4 \\ y = 2x + 1 \end{cases}$$

are two systems of linear equations.

Each individual equation has an infinite number of solutions. That is, there are an infinite number of ordered pairs that satisfy each equation. But, we are interested in finding ordered pairs that satisfy **both** equations.

A **solution of a system** of linear equations is an ordered pair (or point) that satisfies **both** equations. To determine whether or not a particular ordered pair is a solution to a system, substitute the values for x and y in **both** equations. If the results for both equations are true statements, then the ordered pair is a solution to the system.

Example 1

Solution of a System

Show that $(2, 1)$ is a solution to the system $\begin{cases} x - 2y = 0 \\ 3x + y = 7 \end{cases}$

Solution

Substitute $x = 2$ and $y = 1$ into **both** equations. In the first equation:

$$2 - 2(1) \overset{?}{=} 0$$
$$2 - 2 = 0 \qquad \text{true statement}$$

In the second equation:

$$3(2) + 1 \overset{?}{=} 7$$
$$6 + 1 = 7 \qquad \text{true statement}$$

Because $(2, 1)$ satisfies both equations, $(\mathbf{2, 1})$ **is a solution to the system**.

Now work margin exercise 1.

1. Show that $(6, 4)$ is a solution to the system
$$\begin{cases} x - 3y = -6 \\ 3x + 2y = 26 \end{cases}$$

Example 2

Not a Solution of a System

Show that $(0, 4)$ is not a solution to the system $\begin{cases} y = -x + 4 \\ y = 2x + 1 \end{cases}$

Solution

Substitute $x = 0$ and $y = 4$ into **both** equations. In the first equation:

$$4 \overset{?}{=} -(0) + 4$$
$$4 = 0 + 4 \qquad \text{true statement}$$

In the second equation:

$$4 \overset{?}{=} 2(0) + 1$$
$$4 = 0 + 1 \qquad \text{false statement}$$

Because $(0, 4)$ does not satisfy **both** equations, $(\mathbf{0, 4})$ **is NOT a solution to the system**.

Now work margin exercise 2.

2. Show that $(3, 2)$ is not a solution to the system
$$\begin{cases} y = -2x + 8 \\ y = 3x - 8 \end{cases}$$

Objective B **Solving Systems of Linear Equations by Graphing**

The following questions concerning systems of equations need to be addressed:
How do we find the solution, if there is one?
Will there always be a solution to a system of linear equations?
Can there be more than one solution?

In this chapter, we will discuss three methods (graphing, substitution, and addition) for solving a system of linear equations. The first method is to **solve by graphing**. In this method, both equations are graphed and the point of intersection (if there is one) is the solution to the system. Table 1 illustrates the three possibilities for a system of two linear equations. Each system will be classified as one of the following:

1. **Consistent**   (has exactly one solution)
2. **Inconsistent**   (has no solution)
3. **Dependent**   (has an infinite number of solutions)

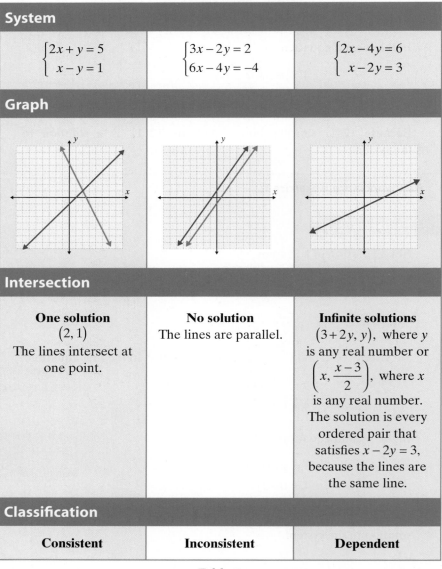

System		
$\begin{cases} 2x + y = 5 \\ x - y = 1 \end{cases}$	$\begin{cases} 3x - 2y = 2 \\ 6x - 4y = -4 \end{cases}$	$\begin{cases} 2x - 4y = 6 \\ x - 2y = 3 \end{cases}$
Graph		
Intersection		
One solution $(2, 1)$ The lines intersect at one point.	**No solution** The lines are parallel.	**Infinite solutions** $(3 + 2y, y)$, where y is any real number or $\left(x, \dfrac{x-3}{2} \right)$, where x is any real number. The solution is every ordered pair that satisfies $x - 2y = 3$, because the lines are the same line.
Classification		
Consistent	**Inconsistent**	**Dependent**

Table 1

To Solve a System of Linear Equations by Graphing

1. Graph both linear equations on the same set of axes.
2. Observe the point of intersection (if there is one).
    a. If the slopes of the two lines are different, then the lines intersect in one and only one point. The system has a single point as its solution.
    b. If the lines have the same slope and different y-intercepts, then the lines are parallel. The system has no solution.
    c. If the lines are the same line, then all the points on the line constitute the solution. There are an infinite number of solutions.
3. Check the solution (if there is one) in both of the original equations.

Solving by graphing can involve estimating the solutions whenever the intersection of the two lines is at a point not represented by a pair of integers. (There is nothing wrong with this technique. Just be aware that at times it can lack accuracy. See Example 6.)

You might want to review graphing lines in Chapter 9. Recall that lines can be graphed by plotting intercepts (as illustrated in Example 3). Other methods include finding a point on the line and using the slope to find another point (see Examples 4 and 5) or finding two points (see Example 6).

Example 3

A System with One Solution

Solve the following system of linear equations by graphing.

$$\begin{cases} x + y = 6 \\ \quad y = x + 4 \end{cases}$$

Solution

The two lines intersect at the point $(1, 5)$. Thus the solution to the system is $(1, 5)$ or $x = 1$ and $y = 5$.

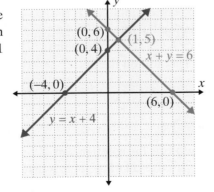

Check: Substitution shows that $(1, 5)$ satisfies **both** of the equations in the system.

$$
\begin{array}{cc}
x + y = 6 & y = x + 4 \\
\overset{?}{} & \overset{?}{} \\
(1) + (5) = 6 & (5) = (1) + 4 \\
6 = 6 & 5 = 5
\end{array}
$$

Example 4

A System with No Solution

Solve the following system of linear equations by graphing.

$$\begin{cases} y = 3x \\ y - 3x = -4 \end{cases}$$

Solution

The lines are parallel with the same slope, 3, and there are no points of intersection.

There is no solution to the system.

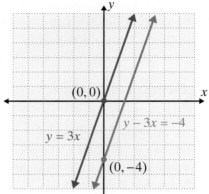

Example 5

A System with Infinitely Many Solutions

Solve the following system of linear equations by graphing.

$$\begin{cases} x + 2y = 6 \\ y = -\dfrac{1}{2}x + 3 \end{cases}$$

Solution

All points that lie on one line also lie on the other line. For example, $(4, 1)$ is a point on the line $x + 2y = 6$ since $4 + 2(1) = 6$. The point $(4, 1)$ is also on the line

$$y = -\frac{1}{2}x + 3$$

since $1 = -\dfrac{1}{2}(4) + 3$.

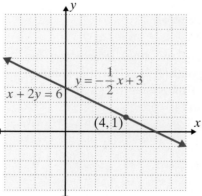

Therefore, the solution can be stated as all of the points that satisfy the equation $x + 2y = 6$.

The solution is the set of all points of the form $\left(x, -\dfrac{1}{2}x + 3\right)$, an infinite set of solutions. (Or, solving the first equation for x we have $(6 - 2y, y)$ for all values of y.)

Example 6

A System that Requires Estimation

Solve the following system of linear equations by graphing.

$$\begin{cases} x - 3y = 4 \\ 2x + y = 3 \end{cases}$$

Solution

The two lines intersect at one point. However, we can only estimate the point of intersection as $\left(2, -\frac{1}{2}\right)$. In this situation be aware that, although graphing gives a good "estimate," finding exact solutions to the system is not likely.

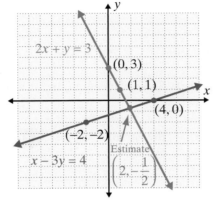

Check: Substitute $x = 2$ and $y = -\frac{1}{2}$.

$$2 - 3\left(-\frac{1}{2}\right) \overset{?}{=} 4 \qquad \text{and} \qquad 2(2) + \left(-\frac{1}{2}\right) \overset{?}{=} 3$$

$$\frac{7}{2} \neq 4 \qquad\qquad\qquad \frac{7}{2} \neq 3$$

Thus checking shows that the estimated solution $\left(2, -\frac{1}{2}\right)$ does not satisfy either equation. The estimated point of intersection is just that, an estimate. The following sections will provide algebraic techniques for solving systems of equations that would give the exact solution as $\left(\frac{13}{7}, -\frac{5}{7}\right)$.

Now work margin exercises 3 through 6.

Solve the following systems of linear equations by graphing.

3. $\begin{cases} y = 2x + 6 \\ x + y = -3 \end{cases}$

4. $\begin{cases} -\dfrac{1}{2}x + 1 = y \\ x + 2y = 8 \end{cases}$

5. $\begin{cases} 3x = y + 4 \\ 6x - 8 = 2y \end{cases}$

6. $\begin{cases} y = 4x - 3 \\ y - 2x = 0 \end{cases}$

---

notes

1. To use the graphing method, graph the lines as accurately as you can.

2. Be sure to check your solution by substituting it back into both of the original equations. (Of course, fractional estimates may not check exactly.)

Objective C Using a TI-84 Plus Graphing Calculator to Solve a System of Linear Equations

A graphing calculator can be used to locate (or estimate) the point of intersection of two lines (and therefore the solution to the system).

7. Use a graphing calculator to solve the following system of linear equations.

$$\begin{cases} x + y = -5 \\ x - 3y = -1 \end{cases}$$

Example 7

Using a Graphing Calculator

Use a graphing calculator to solve the system $\begin{cases} 2x + y = 8 \\ x - y = 1 \end{cases}$ as follows.

Solution

Step 1: Solve each equation for y. For this system $\begin{cases} y = -2x + 8 \\ y = x - 1 \end{cases}$

Step 2: Press ⬛ Y= and enter the two expressions for y. To get the variable X, press ⬛ X,T,θ,n . The display screen will appear as follows:

```
Plot1 Plot2 Plot3
\Y1◼-2X+8
\Y2◼X-1
\Y3=
\Y4=
\Y5=
\Y6=
\Y7=
```

Step 3: Press ⬛ GRAPH . (Both lines should appear. If not, you may need to adjust the ⬛ WINDOW .)

Step 4: Press ⬛ 2ND and CALC. Select 5: intersect. The cursor will appear on one of the lines. Use the right or left arrow to get near the point of intersection and press ⬛ ENTER . Then move the up or down arrow to get to the other line. Now use the right or left arrow to move closer to the point of intersection on this line and press ⬛ ENTER . Follow the directions for Guess? by moving the cursor to the point of intersection and pressing ⬛ ENTER .

Step 5: The answer $x = 3$ and $y = 2$ will appear at the bottom of the display screen.

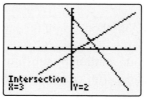

Note: Step 4 may seem somewhat complicated, but TRY IT. It is fun and accurate! (**Note:** If the lines are parallel (an inconsistent system) the calculator will give an error message.)

Now work margin exercise 7.

1. Determine which of the given points, if any, lie on both of the lines in the given system of equations by substituting each point into both equations.

$$\begin{cases} x + 2y = 8 \\ 2x - y = 1 \end{cases}$$

   a. $(8, 0)$ **b.** $(2, 3)$ **c.** $(0, -1)$ **d.** $(4, 2)$

2. Show that the following system of equations is inconsistent by determining the slope of each line and the y-intercept. Explain your reasoning.

$$\begin{cases} 3x + y = 1 \\ 6x + 2y = 7 \end{cases}$$

3. Solve the following system graphically.

$$\begin{cases} y = 2x + 5 \\ 4x - 2y = -10 \end{cases}$$

Practice Problem Answers

1. b.

2. $m_1 = -3, b_1 = 1, m_2 = -3, b_2 = \dfrac{7}{2}$; The lines do not intersect because they are parallel; they have the same slope, but different y-intercepts.

3. $(x, 2x + 5)$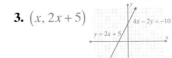

Exercises 10.1

Determine which of the given points, if any, lie on both of the lines in the systems of equations by substituting each point into both equations. See Examples 1 and 2.

1. $\begin{cases} x - y = 6 \\ 2x + y = 0 \end{cases}$

a. $(1, -2)$
b. $(4, -2)$
c. $(2, -4)$
d. $(-1, 2)$

2. $\begin{cases} x + 3y = 5 \\ 3y = 4 - x \end{cases}$

a. $(2, 1)$
b. $(2, -2)$
c. $(-1, 2)$
d. $(4, 0)$

3. $\begin{cases} 2x + 4y - 6 = 0 \\ 3x + 6y - 9 = 0 \end{cases}$

a. $(1, 1)$
b. $(2, 0)$
c. $\left(0, \dfrac{3}{2}\right)$
d. $(-1, 3)$

4. $\begin{cases} 5x - 2y - 5 = 0 \\ 5x = -3y \end{cases}$

a. $(1, 0)$
b. $\left(\dfrac{3}{5}, -1\right)$
c. $(0, 0)$
d. $(1, 4)$

The graphs of the lines represented by each system of equations are given. Determine the solution of the system by looking at the graph. Check your solution by substituting into both equations. See Example 3.

5. $\begin{cases} x + 2y = 4 \\ x - y = -2 \end{cases}$

6. $\begin{cases} x + 2y = 1 \\ 2x + y = -1 \end{cases}$

7. $\begin{cases} 2x - y = 6 \\ 3x + y = 14 \end{cases}$

8. $\begin{cases} x - y = 4 \\ 3x - y = 6 \end{cases}$

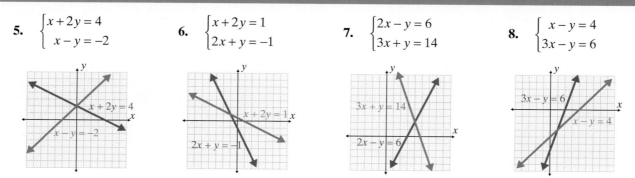

Show that each system of equations is inconsistent by determining the slope of each line and the *y*-intercept. (That is, show that the lines are parallel and do not intersect.) See Example 4.

9. $\begin{cases} 2x + y = 3 \\ 4x + 2y = 5 \end{cases}$

10. $\begin{cases} 3x - 5y = 1 \\ 6x - 10y = 4 \end{cases}$

11. $\begin{cases} y = \dfrac{1}{2}x + 3 \\ x - 2y = 1 \end{cases}$

12. $\begin{cases} 3x - y = 8 \\ x - \dfrac{1}{3}y = 2 \end{cases}$

Solve the following systems graphically. See Examples 3–6.

13. $\begin{cases} y = x + 1 \\ y + x = -5 \end{cases}$

14. $\begin{cases} y = 2x + 5 \\ 4x - 2y = 7 \end{cases}$

15. $\begin{cases} 2x - y = 4 \\ 3x + y = 6 \end{cases}$

16. $\begin{cases} 2x + 3y = 6 \\ 4x + 6y = 12 \end{cases}$ **17.** $\begin{cases} x + y - 5 = 0 \\ \quad\; x = 5 - y \end{cases}$ **18.** $\begin{cases} y = 4x - 3 \\ x = 2y - 8 \end{cases}$ **19.** $\begin{cases} 3x - y = 6 \\ \quad\; y = 3x \end{cases}$

20. $\begin{cases} \quad\; y = 2x \\ 2x + y = 4 \end{cases}$ **21.** $\begin{cases} x - y = 5 \\ \quad x = -3 \end{cases}$ **22.** $\begin{cases} x - 2y = 4 \\ \quad\;\; x = 4 \end{cases}$ **23.** $\begin{cases} \;x + 2y = 8 \\ 3x - 2y = 0 \end{cases}$

24. $\begin{cases} 2x + y = 0 \\ 4x + 2y = -8 \end{cases}$ **25.** $\begin{cases} 5x - 4y = 5 \\ \quad 8y = 10x - 10 \end{cases}$ **26.** $\begin{cases} x + y = 8 \\ \;5y = 2x + 5 \end{cases}$ **27.** $\begin{cases} 4x + 3y + 7 = 0 \\ 5x - 2y + 3 = 0 \end{cases}$

28. $\begin{cases} \;4x - 2y = 10 \\ -6x + 3y = -15 \end{cases}$ **29.** $\begin{cases} x = 5 \\ y = -1 \end{cases}$ **30.** $\begin{cases} y = 7 \\ x = 8 \end{cases}$ **31.** $\begin{cases} \dfrac{1}{2}x + 2y = 7 \\ \quad 2x = 4 - 8y \end{cases}$

32. $\begin{cases} 4x + y = 6 \\ 2x + \dfrac{1}{2}y = 3 \end{cases}$ **33.** $\begin{cases} 7x - 2y = 1 \\ \qquad y = 3 \end{cases}$ **34.** $\begin{cases} \quad\;\; x = 1.5 \\ x - 3y = 9 \end{cases}$ **35.** $\begin{cases} \quad\; y = \dfrac{1}{2}x + 2 \\ x - 2y + 4 = 0 \end{cases}$

36. $\begin{cases} 2x - 5y = 6 \\ y = \dfrac{2}{5}x + 1 \end{cases}$ 　　**37.** $\begin{cases} 2x + 3y = 5 \\ 3x - 2y = 1 \end{cases}$ 　　**38.** $\begin{cases} \dfrac{2}{3}x + y = 2 \\ x - 4y = 3 \end{cases}$ 　　**39.** $\begin{cases} x - y = 4 \\ 2y = 2x - 4 \end{cases}$

40. $\begin{cases} x + y = 4 \\ 2x - 3y = 3 \end{cases}$ 　　**41.** $\begin{cases} \dfrac{1}{2}x + \dfrac{1}{3}y = \dfrac{1}{6} \\ \dfrac{1}{4}x + \dfrac{1}{4}y = 0 \end{cases}$ 　　**42.** $\begin{cases} \dfrac{1}{4}x - y = \dfrac{13}{4} \\ \dfrac{1}{3}x + \dfrac{1}{6}y = -\dfrac{1}{6} \end{cases}$

Each of the following application problems has been modeled using a system of equations. Solve the system graphically.

43. The sum of two numbers is 25 and their difference is 15. What are the two numbers?

Let x = one number and y = the other number.

The corresponding modeling system is $\begin{cases} x + y = 25 \\ x - y = 15 \end{cases}$

44. Rectangles: The perimeter of a rectangle is 50 m and the length is 5 m longer than the width. Find the dimensions of the rectangle.

Let x = the length and y = the width.

The corresponding modeling system is $\begin{cases} 2x + 2y = 50 \\ x - y = 5 \end{cases}$

45. Swimming Pools: OSHA recommends that swimming pool owners clean their pool decks with a solvent composed of a 12% chlorine solution and a 3% chlorine solution. Fifteen gallons of the solvent consists of 6% chlorine. How much of each of the mixing solutions were used?

Let x = the number of gallons of the 12% solution

and y = the number of gallons of the 3% solution.

The corresponding modeling system is $\begin{cases} x + y = 15 \\ 0.12x + 0.03y = 0.06(15) \end{cases}$

46. School Supplies: A student bought a calculator and a textbook for a course in algebra. He told his friend that the total cost was $170 (without tax) and that the calculator cost $20 more than twice the cost of the textbook. What was the cost of each item?

Let x = the cost of the calculator

and y = the cost of the textbook.

The corresponding modeling system is $\begin{cases} x + y = 170 \\ x = 2y + 20 \end{cases}$

47. $\begin{cases} x + 2y = 9 \\ x - 2y = -7 \end{cases}$

48. $\begin{cases} x - 3y = 0 \\ 2x + y = 7 \end{cases}$

49. $\begin{cases} y = 2 \\ 2x - 3y = -3 \end{cases}$

50. $\begin{cases} 2x - 3y = 0 \\ 3x + 3y = \dfrac{5}{2} \end{cases}$

51. $\begin{cases} y = -3 \\ 2x + y = 0 \end{cases}$

52. $\begin{cases} 2x - 3y = 1.25 \\ x + 2y = 5 \end{cases}$

53. $\begin{cases} x + y = 3.5 \\ -2x + 5y = 7.7 \end{cases}$

54. $\begin{cases} 4x + y = -0.5 \\ x + 2y = -8 \end{cases}$

Writing & Thinking

55. Explain, in your own words, why the answer to a consistent system of linear equations can be written as an ordered pair.

10.2 Systems of Linear Equations: Solutions by Substitution

Objectives

A Solve systems of linear equations by using the method of substitution.

Objective A Solving Systems of Linear Equations by Substitution

As we discussed in Section 10.1, solving systems of linear equations by graphing is somewhat limited in accuracy. The graphs must be drawn very carefully and even then the points of intersection (if there are any) can be difficult to estimate accurately.

In this section, we will develop an algebraic method called the **method of substitution**. The objective in the substitution method is to eliminate one of the variables so that a new equation is formed with just one variable. If this new equation has one solution then the solution to the system is a single point. If this new equation is never true, then the system has no solution. If this new equation is always true, then the system has an infinite number of solutions.

To Solve a System of Linear Equations by Substitution

1. Solve one of the equations for one of the variables.
2. Substitute the resulting expression into the other equation.
3. Solve this new equation, if possible, and then substitute back into one of the original equations to find the value of the other variable. (This is known as **back substitution**.)
4. Check the solution in both of the original equations.

To illustrate the substitution method, consider the following system.

$$\begin{cases} y = -2x + 5 \\ x + 2y = 1 \end{cases}$$

How would you substitute? Since the first equation is already solved for y, a reasonable substitution would be to put $-2x + 5$ for y in the second equation. Try this and see what happens.

First equation already solved for y: $\qquad\qquad y = -2x + 5$

Substitute into second equation: $\qquad\qquad x + 2y = 1$

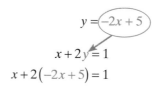

$$x + 2(-2x + 5) = 1$$

We now have one equation in only one variable, namely x. The problem has been reduced from one of solving two equations in two variables to solving one equation in one variable. Solve this equation for x. Then find the corresponding y-value by substituting this x-value into **either of the two original equations**.

$$x + 2(-2x + 5) = 1$$
$$x - 4x + 10 = 1$$
$$-3x = -9$$
$$x = 3$$

Back substitute $x = 3$ into $y = -2x + 5$.

$$y = -2(3) + 5$$
$$y = -6 + 5$$
$$y = -1$$

Thus the solution to the system is the point $(3, -1)$.

Substitution is not the only algebraic technique for solving a system of linear equations. It does work in all cases but is most often used when one of the equations is easily solved for one variable.

In the following examples, note how the results in Example 2 indicate that the system has no solution and the results in Example 3 indicate that the system has an infinite number of solutions.

Example 1

A System with One Solution

Solve the following system of linear equations by using substitution.

$$\begin{cases} y = \dfrac{5}{6}x + 2 \\ \dfrac{1}{6}x + y = 8 \end{cases}$$

Solution

The first equation is already solved for y. Substituting $\dfrac{5}{6}x + 2$ for y in the second equation gives the following.

$$\frac{1}{6}x + \left(\frac{5}{6}x + 2\right) = 8$$

$$6 \cdot \frac{1}{6}x + 6 \cdot \left(\frac{5}{6}x + 2\right) = 6 \cdot 8 \qquad \text{Multiply both sides by 6, the LCD.}$$

$$x + 5x + 12 = 48$$
$$6x + 12 = 48$$
$$6x = 36$$
$$x = 6$$

Substituting 6 for x in the first equation gives the corresponding value for y.

$$y = \frac{5}{6}(6) + 2 = 5 + 2 = 7$$

The solution to the system is $(6, 7)$.

Check: Substitution shows that $(6, 7)$ satisfies **both** of the equations in the system.

$$y = \frac{5}{6}x + 2 \qquad \frac{1}{6}x + y = 8$$

$$7 \overset{?}{=} \frac{5}{6}(6) + 2 \qquad \frac{1}{6}(6) + 7 \overset{?}{=} 8$$

$$7 = 7 \qquad\qquad 8 = 8$$

Example 2

A System with No Solution

Solve the following system of linear equations by using substitution.

$$\begin{cases} 3x + y = 1 \\ 6x + 2y = 3 \end{cases}$$

Solution

Solving the first equation for y gives $y = 1 - 3x$. Substituting $1 - 3x$ for y in the second equation gives the following.

$$6x + 2(1 - 3x) = 3$$
$$6x + 2 - 6x = 3$$
$$2 = 3$$

This last equation $(2 = 3)$ is **never true**. This tells us that the system has **no solution**. Graphically, the lines are parallel and there is no intersection.

Example 3

A System with Infinitely Many Solutions

Solve the following system of linear equations by using substitution.

$$\begin{cases} x - 2y = 1 \\ 3x - 6y = 3 \end{cases}$$

Solution

Solving the first equation for x gives $x = 1 + 2y$. Substituting $1 + 2y$ for x in the second equation gives the following.

$$3(1 + 2y) - 6y = 3$$
$$3 + 6y - 6y = 3$$
$$3 = 3$$

This last equation $(3 = 3)$ is **always true**. This tells us that the system has an **infinite number of solutions** and that the solutions are of the form $(1 + 2y, y)$ for all values of y. (Or, solving one of the equations for y we have $\left(x, \frac{1}{2}x - \frac{1}{2}\right)$ for all values of x.)

Now work margin exercises 1 through 3.

> ## notes
>
> ■ As illustrated in Example 3, the solution can take two forms for a system with an infinite number of solutions. In one form we can solve for x and then use this expression for x in the ordered pair format (x, y). For example, in Example 3, solving either equation for x gives $x = 1 + 2y$ and we can write $(1 + 2y, y)$ to represent all solutions.
>
> Alternatively, we can solve for y which gives $y = \frac{1}{2}x - \frac{1}{2}$ and we have the form $\left(x, \frac{1}{2}x - \frac{1}{2}\right)$ for all solutions. Try substituting various values for x and y in these expressions and you will see that all results satisfy both equations.

Example 4

A System with Decimals

Solve the following system of linear equations by using substitution.

$$\begin{cases} x + y = 5 \\ 0.2x + 0.3y = 0.9 \end{cases}$$

Solution

Solving the first equation for y gives $y = 5 - x$. Substituting $5 - x$ for y in the second equation gives the following.

$$0.2x + 0.3y = 0.9$$
$$0.2x + 0.3(5 - x) = 0.9$$
$$0.2x + 1.5 - 0.3x = 0.9$$
$$-0.1x = -0.6$$
$$\frac{-0.1x}{-0.1} = \frac{-0.6}{-0.1}$$
$$x = 6$$
$$y = 5 - x = 5 - 6 = -1$$

Note that you could multiply each term by 10 to eliminate the decimal.

The solution to the system is $(6, -1)$.

To check, substitute $x = 6$ and $y = -1$ in both of the original equations.

Now work margin exercise 4.

Use the method of addition to solve the following systems.

1. $\begin{cases} y = \dfrac{2}{5}x + 3 \\ \dfrac{3}{5}x + y = 10 \end{cases}$

2. $\begin{cases} 2x + y = 1 \\ 10x + 5y = 4 \end{cases}$

3. $\begin{cases} -2x + y = 1 \\ 10x - 5y = -5 \end{cases}$

4. Solve the following system of linear equations by using substitution.

$\begin{cases} x + y = 6 \\ 0.8x + 0.4y = 1.8 \end{cases}$

5. Solve the following system by using the method of substitution.

$$\begin{cases} x - y = -4 \\ 4x + 3y = 5 \end{cases}$$

Completion Example 5

Solve by Substitution

Solve the following system by using the method of substitution.

$$\begin{cases} x + y = 3 \\ 2x - y = 12 \end{cases}$$

Solutions

Solving the first equation for x gives $x =$ _____. Substituting _____ for x in the second equation gives the following.

$$2(\underline{\hspace{1cm}}) - y = 12$$
$$\underline{\hspace{1cm}} - y = 12$$
$$6 - \underline{\hspace{1cm}} = 12$$
$$-3y = \underline{\hspace{1cm}}$$
$$y = \underline{\hspace{1cm}}$$
$$x = 3 - (\underline{\hspace{1cm}}) = \underline{\hspace{1cm}}$$

The solution to the system is _____.

Now work margin exercise **5.**

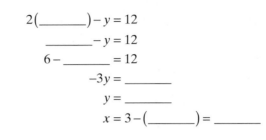

Practice Problems

Solve the following systems by using the method of substitution.

1. $\begin{cases} x + y = 3 \\ y = 2x \end{cases}$　　**2.** $\begin{cases} y = -3 + 2x \\ 4x - 2y = 6 \end{cases}$　　**3.** $\begin{cases} x + 2y = -1 \\ x - 4y = -4 \end{cases}$

4. $\begin{cases} y = 4 - 3x \\ y = -3x + 6 \end{cases}$　　**5.** $\begin{cases} y = 3x - 1 \\ 2x + y = 4 \end{cases}$

Completion Example Answers

5. Solving the first equation for x gives $x = 3 - y$. Substituting $3 - y$ for x in the second equation gives the following:

$$2(3 - y) - y = 12$$
$$6 - 2y - y = 12$$
$$6 - 3y = 12$$
$$-3y = 6$$
$$y = -2$$
$$x = 3 - (-2) = 5$$

The solution to the system is $(5, -2)$.

Practice Problem Answers

1. $(1, 2)$　**2.** $(x, -3 + 2x)$　**3.** $\left(-2, \dfrac{1}{2}\right)$　**4.** no solution　**5.** $(1, 2)$

Exercises 10.2

Solve each system by using the substitution method. See Examples 1–5.

1. $\begin{cases} x+y=6 \\ y=2x \end{cases}$

2. $\begin{cases} 5x+2y=21 \\ x=y \end{cases}$

3. $\begin{cases} 3x-7=y \\ 2y=6x-14 \end{cases}$

4. $\begin{cases} y=3x+4 \\ 2y=3x+5 \end{cases}$

5. $\begin{cases} x=3y \\ 3y-2x=6 \end{cases}$

6. $\begin{cases} 4x=y \\ 4x-y=7 \end{cases}$

7. $\begin{cases} x-5y+1=0 \\ x=7-3y \end{cases}$

8. $\begin{cases} 2x+5y=15 \\ x=y-3 \end{cases}$

9. $\begin{cases} 7x+y=9 \\ y=4-7x \end{cases}$

10. $\begin{cases} 3y+5x=5 \\ y=3-2x \end{cases}$

11. $\begin{cases} 3x-y=7 \\ x+y=5 \end{cases}$

12. $\begin{cases} 4x-2y=5 \\ y=2x+3 \end{cases}$

13. $\begin{cases} 3x+5y=-13 \\ y=3-2x \end{cases}$

14. $\begin{cases} 15x+5y=20 \\ y=-3x+4 \end{cases}$

15. $\begin{cases} x-y=5 \\ 2x+3y=0 \end{cases}$

16. $\begin{cases} 4x=8 \\ 3x+y=8 \end{cases}$

17. $\begin{cases} 2y=5 \\ 3x-4y=-4 \end{cases}$

18. $\begin{cases} x+y=8 \\ 3x+2y=8 \end{cases}$

19. $\begin{cases} y=2x-5 \\ 2x+y=-3 \end{cases}$

20. $\begin{cases} 2x+3y=5 \\ x-6y=0 \end{cases}$

21. $\begin{cases} x+5y=1 \\ x-3y=5 \end{cases}$

22. $\begin{cases} 3x+8y=-2 \\ x+2y=-1 \end{cases}$

23. $\begin{cases} 9x+3y=6 \\ 3x=2-y \end{cases}$

24. $\begin{cases} 5x+2y=-10 \\ 10x=-3-4y \end{cases}$

25. $\begin{cases} x-2y=-4 \\ 3x+y=-5 \end{cases}$

26. $\begin{cases} x+4y=3 \\ 3x-4y=7 \end{cases}$

27. $\begin{cases} 3x-y=-1 \\ 7x-4y=0 \end{cases}$

28. $\begin{cases} x+5y=-1 \\ 2x+7y=1 \end{cases}$

29. $\begin{cases} x+3y=5 \\ 3x+2y=7 \end{cases}$

30. $\begin{cases} 3x-4y-39=0 \\ 2x-y-13=0 \end{cases}$

31. $\begin{cases} \dfrac{1}{4}x-\dfrac{3}{2}y=-5 \\ -x+6y=20 \end{cases}$

32. $\begin{cases} \dfrac{-4}{3}x+2y=7 \\ \dfrac{8}{3}x-4y=-5 \end{cases}$

33. $\begin{cases} 6x-y=15 \\ 0.2x+0.5y=2.1 \end{cases}$

34. $\begin{cases} x+2y=3 \\ 0.4x+y=0.6 \end{cases}$

35. $\begin{cases} 0.2x-0.1y=0 \\ y=x+10 \end{cases}$

36. $\begin{cases} 0.1x-0.2y=1.4 \\ 3x+y=14 \end{cases}$

37. $\begin{cases} 3x-2y=5 \\ y=1.5x+2 \end{cases}$

38. $\begin{cases} x=2y-7.5 \\ 2x+4y=-15 \end{cases}$

39. $\begin{cases} \dfrac{1}{2}x+\dfrac{1}{3}y=4 \\ 3x+2y=24 \end{cases}$

40. $\begin{cases} \dfrac{1}{3}x+\dfrac{1}{7}y=2 \\ 7x+3y=42 \end{cases}$

41. $\begin{cases} \dfrac{x}{3}+\dfrac{y}{5}=1 \\ x+6y=12 \end{cases}$

42. $\begin{cases} \dfrac{x}{5}+\dfrac{y}{4}-3=0 \\ \dfrac{x}{10}-\dfrac{y}{2}+1=0 \end{cases}$

43. The sum of two numbers is 25 and their difference is 15. What are the two numbers? Let x = one number and y = the other number.

The corresponding modeling system is

$$\begin{cases} x + y = 25 \\ x - y = 15 \end{cases}$$

44. Rectangles: The perimeter of a rectangle is 50 meters and the length is 5 meters longer than the width. Find the dimensions of the rectangle.

Let x = the length and y = the width.

The corresponding modeling system is

$$\begin{cases} 2x + 2y = 50 \\ x - y = 5 \end{cases}$$

45. Swimming Pools: OSHA recommends that swimming pool owners clean their pool decks with a solvent composed of a 12% chlorine solution and a 3% chlorine solution. Fifteen gallons of the solvent consists of 6% chlorine. How much of each of the mixing solutions were used?

Let x = the number of gallons of the 12% solution

and y = the number of gallons of the 3% solution.

The corresponding modeling system is

$$\begin{cases} x + y = 15 \\ 0.12x + 0.03y = 0.06(15) \end{cases}$$

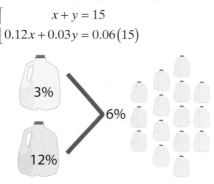

46. School Supplies: A student bought a calculator and a textbook for a course in algebra. He told his friend that the total cost was $170 (without tax) and that the calculator cost $20 more than twice the cost of the textbook. What was the cost of each item?

Let x = the cost of the calculator

and y = the cost of the textbook.

The corresponding modeling system is

$$\begin{cases} x + y = 170 \\ x = 2y + 20 \end{cases}$$

Writing & Thinking

47. Explain the advantages of solving a system of linear equations
   a. by graphing,
   b. by substitution.

10.3 Systems of Linear Equations: Solutions by Addition

Objective A   Solving Systems of Linear Equations by Addition

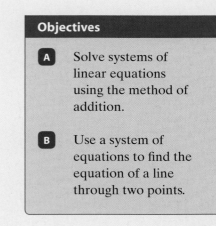

Objectives

A Solve systems of linear equations using the method of addition.

B Use a system of equations to find the equation of a line through two points.

We have discussed two methods for solving systems of linear equations:

1. **graphing**
2. **substitution**

We know that solutions found by graphing are not necessarily exact (estimation may be involved); and, in some cases, the method of substitution can lead to complicated algebraic steps. In this section, we consider a third method:

3. **addition** (or **method of elimination**).

In the **method of addition**, as with the method of substitution, the objective is to eliminate one of the variables so that a new equation is found with just one variable, if possible. If this new equation:

1. has one solution, the solution to the system is a **single point**.
2. is never true, the system has **no solution**.
3. is always true, the system has **infinite solutions**.

Consider solving the following system where both equations are written in standard form.

$$\begin{cases} x - 2y = -9 \\ x + 2y = 11 \end{cases}$$

In the **method of addition**, we write one equation under the other so that like terms are aligned vertically. (Note that, in this example, the coefficients of y are opposites, namely -2 and $+2$.)

Then add like terms as follows.

$$\begin{aligned} x - 2y &= -9 \\ x + 2y &= 11 \\ \hline 2x \phantom{{}-2y} &= 2 \\ x \phantom{{}-2y} &= 1 \end{aligned}$$

The y terms are eliminated because the coefficients are opposites.

Now substitute $x = 1$ into either of the original equations and solve for y.

$$\begin{aligned} 1 - 2y &= -9 \\ -2y &= -10 \\ y &= 5 \end{aligned} \quad \textbf{OR} \quad \begin{aligned} 1 + 2y &= 11 \\ 2y &= 10 \\ y &= 5 \end{aligned}$$

Therefore the **solution** to the system is $(1, 5)$.

This example was relatively easy because the coefficients for y were opposites. The general procedure can be outlined as follows.

To Solve a System of Linear Equations by Addition

1. Write the equations in **standard form**, one under the other, so that **like terms are aligned**.
2. Multiply all terms of one equation by a constant (and possibly all terms of the other equation by another constant) so that **two like terms have opposite coefficients**.
3. Add the two equations by **combining like terms** and solve the resulting equation, if possible.
4. **Back substitute into one of the original equations** to find the value of the other variable.
5. Check the solution (if there is one) in both of the original equations.

Example 1

A System with One Solution

Use the method of addition to solve the following system.

$$\begin{cases} 3x + 5y = -3 \\ -7x + 2y = 7 \end{cases}$$

Solution

Multiply each term in the first equation by 2 and each term in the second equation by -5. This will result in the y-coefficients being opposites. Add the two equations by combining like terms which will eliminate y. Solve for x.

$$\begin{cases} 3x + 5y = -3 \\ -7x + 2y = 7 \end{cases}$$

$$\begin{cases} [2](3x + 5y = -3) \longrightarrow \quad 6x + 10y = -6 \\ [-5](-7x + 2y = 7) \longrightarrow \quad 35x - 10y = -35 \end{cases}$$

$$\overline{41x = -41} \qquad \text{\small y is eliminated.}$$
$$x = -1$$

Substitute $x = -1$ into either one of the original equations.

$$
\begin{array}{ccc}
3x + 5y = -3 & & -7x + 2y = 7 \\
3(-1) + 5y = -3 & & -7(-1) + 2y = 7 \\
-3 + 5y = -3 & \text{OR} & 7 + 2y = 7 \\
5y = 0 & & 2y = 0 \\
y = 0 & & y = 0
\end{array}
$$

The solution is $(-1, 0)$.

Check: Substitution shows that $(-1, 0)$ satisfies **both** of the equations in the system.

$$3x + 5y = -3 \qquad\qquad -7x + 2y = 7$$

$$3(-1) + 5(0) \overset{?}{=} -3 \qquad\qquad -7(-1) + 2(0) \overset{?}{=} 7$$

$$-3 = -3 \qquad\qquad\qquad 7 = 7$$

notes

In Example 1, we could have eliminated x instead of y by multiplying the terms in the first equation by 7 and the terms in the second equation by 3. The solution would be the same. Try this yourself to confirm this method.

Example 2

A System with Infinitely Many Solutions

Solve the following system by using the method of addition.

$$\begin{cases} 3x - \dfrac{1}{2}y = 6 \\ 6x - y = 12 \end{cases}$$

Solution

Multiply the first equation by -2 so that the y-coefficients will be opposites.

$$\begin{cases} [-2]\left(3x - \dfrac{1}{2}y = 6\right) \longrightarrow -6x + y = -12 \\ \qquad (6x - y = 12) \longrightarrow \dfrac{6x - y = 12}{0 = 0} \end{cases}$$

Because this last equation, $0 = 0$, is **always true**, the system has infinitely many solutions. The solution set consists of all points that satisfy the equation $6x - y = 12$. Solving for y gives $y = 6x - 12$, and we can write the solution in the general form $(x, 6x - 12)$. Or, solving for x, $x = \dfrac{1}{6}y + 2$, and the solution can be written in the form $\left(\dfrac{1}{6}y + 2, y\right)$.

Example 3

A System with Decimals

Solve the following system by using the method of addition.

$$\begin{cases} x + 0.4y = 3.08 \\ 0.1x - y = 0.1 \end{cases}$$

Solution

Multiply the second equation by -10 so that the x-coefficients will be opposites.

$$\begin{cases} (x + 0.4y = 3.08) \longrightarrow & 1.0x + 0.4y = 3.08 \\ [-10](0.1x - y = 0.1) \longrightarrow & \underline{-1.0x + 10.0y = -1.0} \\ & 10.4y = 2.08 \quad \text{x is eliminated.} \\ & y = 0.2 \end{cases}$$

Substitute $y = 0.2$ into one of the original equations.

$$x + 0.4y = 3.08$$
$$x + 0.4(0.2) = 3.08$$
$$x + 0.08 = 3.08$$
$$x = 3$$

The solution is $(3, 0.2)$.

To check, substitute $x = 3$ and $y = 0.2$ in both of the original equations.

Completion Example 4

Solve by Addition

Solve the following system by using the method of addition.

$$\begin{cases} x - y = -1 \\ 2x + 3y = 33 \end{cases}$$

Solution

Multiply each term in the first equation by 3 and leave each term in the second equation as is. Now the y-coefficients will be opposites. Add the two equations by combining like terms. This will eliminate y. Solve for x.

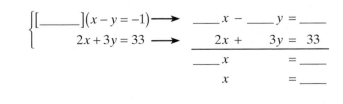

Substitute $x =$ _____ into either one of the original equations. Choosing the first equation gives the following

$$\underline{\quad\quad} - y = -1$$
$$-y = \underline{\quad\quad}$$
$$y = \underline{\quad\quad}$$

The solution is _____.

***Now work margin exercises* 1 through 4.**

Objective B   Using Systems of Equations to Find the Equation of a Line

Example 5

Using a System to Find the Equation of a Line

Using the formula $y = mx + b$, find the equation of the line determined by the two points $(3, 5)$ and $(-6, 2)$.

Solution

Write two equations in m and b by substituting the coordinates of the points for x and y. This gives the system,

$$\begin{cases} 5 = 3m + b \\ 2 = -6m + b \end{cases}$$

Multiply the second equation by -1 so that the b-coefficients will be opposites.

$$\begin{cases} (5 = 3m + b) & \longrightarrow & 5 = 3m + b \\ [-1]\,(2 = -6m + b) & \longrightarrow & \dfrac{-2 = 6m - b}{3 = 9m} \end{cases}$$

$$\frac{1}{3} = m$$

Use the method of addition to solve the following systems.

1. $\begin{cases} 3x + 2y = 0 \\ 5x - 2y = 16 \end{cases}$

2. $\begin{cases} 6x + 3y = 15 \\ 4x + 2y = 10 \end{cases}$

3. $\begin{cases} x + 0.5y = 0.1 \\ 0.5x - 2y = 10.4 \end{cases}$

4. $\begin{cases} 6x + y = -7 \\ -3x + 2y = 16 \end{cases}$

Completion Example Answers

4.

$$\begin{cases} [3](x - y = -1) & \longrightarrow & 3x - 3y = -3 \\ 2x + 3y = 33 & \longrightarrow & \dfrac{2x + 3y = 33}{5x \quad\;\; = 30} \end{cases}$$

$$x \quad\;\; = 6$$

Substitute $x = 6$ into either one of the original equations and solve for y. Choosing the first equation gives the following.

$$6 - y = -1$$
$$-y = -7$$
$$y = 7$$

The solution is $(6, 7)$.

5. Find the equation of the line determined by the two points $(4, -2)$ and $(6, 3)$.

Substitute $m = \dfrac{1}{3}$ into one of the original equations.

$$5 = 3m + b$$
$$5 = 3\left(\dfrac{1}{3}\right) + b$$
$$5 = 1 + b$$
$$4 = b$$

The equation of the line is $y = \dfrac{1}{3}x + 4$.

Now work margin exercise **5.**

Now that you know three methods for solving a system of linear equations (graphing, substitution, and addition), which method should you use? Consider the following guidelines when making your decision.

Guidelines for Deciding which Method to Use when Solving a System of Linear Equations

1. The graphing method is helpful in "seeing" the geometric relationship between the lines and finding approximate solutions. A calculator can be very helpful here.
2. Both the substitution method and the addition method give exact solutions.
3. The substitution method may be reasonable and efficient if one of the coefficients of one of the variables is 1.
4. The addition method is particularly efficient if the coefficients for one of the variables are opposites.

Practice Problems

Solve the following systems by using the method of addition.

1. $\begin{cases} 2x + 2y = 4 \\ x - y = -3 \end{cases}$
  2. $\begin{cases} 3x + 4y = 12 \\ \dfrac{1}{3}x - 8y = -5 \end{cases}$
  3. $\begin{cases} 2x - 5y = 6 \\ 4x - 10y = -2 \end{cases}$

4. $\begin{cases} 0.02x + 0.06y = 1.48 \\ 0.03x - 0.02y = 0.02 \end{cases}$
  5. $\begin{cases} y = 3x + 15 \\ 6x - 2y = -30 \end{cases}$

Practice Problem Answers

1. $\left(-\dfrac{1}{2}, \dfrac{5}{2}\right)$ **2.** $\left(3, \dfrac{3}{4}\right)$ **3.** no solution **4.** $(14, 20)$ **5.** $(x, 3x + 15)$

Exercises 10.3

Solve each system by using the addition method. See Examples 1 through 4.

1. $\begin{cases} 8x - y = 29 \\ 2x + y = 11 \end{cases}$

2. $\begin{cases} x + 3y = 9 \\ x - 7y = -1 \end{cases}$

3. $\begin{cases} 3x + 2y = 0 \\ 5x - 2y = 8 \end{cases}$

4. $\begin{cases} 12x - 3y = 21 \\ 4x - y = 7 \end{cases}$

5. $\begin{cases} 2x + 2y = 5 \\ x + y = 3 \end{cases}$

6. $\begin{cases} 2x - y = 7 \\ x + y = 2 \end{cases}$

7. $\begin{cases} 3x + 3y = 9 \\ x + y = 3 \end{cases}$

8. $\begin{cases} 9x + 2y = -42 \\ 5x - 6y = -2 \end{cases}$

9. $\begin{cases} \dfrac{1}{2}x + y = -4 \\ 3x - 4y = 6 \end{cases}$

10. $\begin{cases} x + y = 1 \\ x - \dfrac{1}{3}y = \dfrac{11}{3} \end{cases}$

11. $\begin{cases} x + y = 12 \\ 0.05x + 0.25y = 1.6 \end{cases}$

12. $\begin{cases} x + 0.1y = 8 \\ 0.1x + 0.01y = 0.64 \end{cases}$

Solve each system by using either the substitution method or the addition method. See Examples 1 through 4.

13. $\begin{cases} x = 11 + 2y \\ 2x - 3y = 17 \end{cases}$

14. $\begin{cases} 6x - 3y = 6 \\ y = 2x - 2 \end{cases}$

15. $\begin{cases} x - 2y = 4 \\ y = \dfrac{1}{2}x - 2 \end{cases}$

16. $\begin{cases} 2x + y = 3 \\ 4x + 2y = 7 \end{cases}$

17. $\begin{cases} x = 3y + 4 \\ y = 6 - 2x \end{cases}$

18. $\begin{cases} y = 2x + 14 \\ x = 14 - 3y \end{cases}$

19. $\begin{cases} 7x - y = 16 \\ 2y = 2 - 3x \end{cases}$

20. $\begin{cases} 3x + y = -10 \\ 2y - 1 = x \end{cases}$

21. $\begin{cases} 4x - 2y = 8 \\ 2x - y = 4 \end{cases}$

22. $\begin{cases} x + y = 6 \\ 2x + y = 16 \end{cases}$

23. $\begin{cases} 3x + 2y = 4 \\ x + 5y = -3 \end{cases}$

24. $\begin{cases} x + 2y = 0 \\ 2x = 4y \end{cases}$

25. $\begin{cases} 4x + 3y = 2 \\ 3x + 2y = 3 \end{cases}$

26. $\begin{cases} x - 3y = 4 \\ 3x - 9y = 10 \end{cases}$

27. $\begin{cases} 5x - 2y = 17 \\ 2x - 3y = 9 \end{cases}$

28. $\begin{cases} \dfrac{1}{2}x + 2y = 9 \\ 2x - 3y = 14 \end{cases}$

29. $\begin{cases} 3x + 2y = 14 \\ 7x + 3y = 26 \end{cases}$

30. $\begin{cases} 4x + 3y = 28 \\ 5x + 2y = 35 \end{cases}$

31. $\begin{cases} 2x + 7y = 2 \\ 5x + 3y = -24 \end{cases}$

32. $\begin{cases} 7x - 6y = -1 \\ 5x + 2y = 37 \end{cases}$

33. $\begin{cases} 10x + 4y = 7 \\ 5x + 2y = 15 \end{cases}$

34. $\begin{cases} 6x - 5y = -40 \\ 8x - 7y = -54 \end{cases}$

35. $\begin{cases} 0.5x - 0.3y = 7 \\ 0.3x - 0.4y = 2 \end{cases}$

36. $\begin{cases} 0.6x + 0.5y = 5.9 \\ 0.8x + 0.4y = 6 \end{cases}$

37. $\begin{cases} 2.5x + 1.8y = 7 \\ 3.5x - 2.7y = 4 \end{cases}$

38. $\begin{cases} 0.75x - 0.5y = 2 \\ 1.5x - 0.75y = 7.5 \end{cases}$

39. $\begin{cases} \dfrac{2}{3}x - \dfrac{1}{2}y = \dfrac{2}{3} \\ \dfrac{8}{3}x - 2y = \dfrac{17}{6} \end{cases}$

40. $\begin{cases} \dfrac{3}{4}x + \dfrac{1}{4}y = \dfrac{3}{8} \\ \dfrac{3}{2}x + \dfrac{1}{2}y = \dfrac{3}{4} \end{cases}$

41. $\begin{cases} \dfrac{1}{6}x - \dfrac{1}{12}y = -\dfrac{13}{6} \\ \dfrac{1}{5}x + \dfrac{1}{4}y = 2 \end{cases}$

42. $\begin{cases} \dfrac{5}{3}x - \dfrac{2}{3}y = -\dfrac{29}{30} \\ 2x + 5y = 0 \end{cases}$

Write an equation for the line determined by the two given points by using the formula $y = mx + b$ to set up a system of equations with m and b as the unknowns. See Example 5.

43. $(2, 3), (1, -2)$

44. $(0, 6), (-3, -3)$

45. $(1, -3), (5, -3)$

46. $(5, 3), (5, -4)$

47. $(1, 2), (-3, 0)$

48. $(-4, 2), (5, -1)$

A word problem is stated with equations given that represent a mathematical model for the problem. Solve the system by using either the method of substitution or the method of addition.

49. Investing: Georgia had \$10,000 to invest, and she put the money into two accounts. One of the accounts will pay 6% interest and the other will pay 10%. How much did she put in each account if the interest from the 10% account exceeded the interest from the 6% account by \$40?

Let x = amount in 10% account
and y = amount in 6% account.

The system that models the problem is

$$\begin{cases} x + y = 10,000 \\ 0.10x - 0.06y = 40 \end{cases}$$

50. Baseball: A minor league baseball team has a game attendance of 4500 people. Tickets cost \$5 for children and \$8 for adults. The total revenue made at this game was \$26,100. How many adults and how many children attended the game?

Let x = number of adults
and y = number of children.

The system that models the problem is

$$\begin{cases} x + y = 4500 \\ 8x + 5y = 26,100 \end{cases}$$

51. Acid Solutions: How many liters each of a 30% acid solution and a 40% acid solution must be used to produce 100 liters of a 36% acid solution?

Let x = amount of 30% solution
and y = amount of 40% solution.

The system that models the problem is

$$\begin{cases} x + y = 100 \\ 0.30x + 0.40y = 0.36(100) \end{cases}$$

52. Traveling by Car: Two cars leave Denver at the same time traveling in opposite directions. One travels at an average speed of 55 mph and the other at 65 mph. In how many hours will they be 420 miles apart?

Let x = time of travel for first car
and y = time of travel for second car.

The system that models the problem is

$$\begin{cases} x = y \\ 55x + 65y = 420 \end{cases}$$

65 mph

Denver

55 mph

Writing & Thinking

53. Explain, in your own words, why the answer to a system with infinite solutions is written as an ordered pair with variables.

10.4 Applications: Distance-Rate-Time, Number Problems, Amounts, and Costs

Objective A   **Distance-Rate-Time Problems**

Systems of equations occur in many practical situations such as:
> supply and demand in business,
> velocity and acceleration in engineering,
> money and interest in investments, and
> mixture in physics and chemistry.

Many of the problems in earlier sections illustrated these ideas. However, in those exercises, the system of equations was given. As you study the applications in Sections 10.4 and 10.5, you may want to refer to some of those exercises, as well as the examples, as guides in solving the given applications. **For these applications you will need to create your own systems of equations.**

Remember that the emphasis in all application problems is to develop your reasoning as well as your reading skills. You must learn how to transfer English phrases into algebraic expressions. This is the THINKING PART.

Example 1

Distance-Rate-Time (Rates Unknown)

A small plane flew 300 miles in 2 hours flying with the wind. Then on the return trip, flying against the wind, it traveled only 200 miles in 2 hours. What were the wind speed and the speed of the plane? (**Note:** The "speed of the plane" means how fast the plane would be flying with no wind.)

Solution

Let s = speed of plane
and w = wind speed.

When flying with the wind, the plane's actual rate will increase to $s + w$. When flying against the wind, the plane's actual rate will decrease to $s - w$. (**Note:** If the wind had been strong enough, the plane could actually have been flying backward, away from its destination.)

	Rate	×	Time	=	Distance
With the wind	$s + w$		2		$2(s + w)$
Against the wind	$s - w$		2		$2(s - w)$

1. Bob ran 5 miles in 30 minutes running with the wind and then turned around. For the next 30 minutes, running against the wind, he ran 4.2 miles. What were the wind speed and the speed Bob was running? Assume Bob maintained the same speed for his entire run. (**Hint:** Convert minutes to hours.)

2. One man takes the eastbound line at 9 A.M., and his wife takes the westbound line at 9:30 A.M. The husband's train averages 25 mph, while his wife's train averages 30 mph. At what time will the husband and wife be 122.5 miles apart?

The system of linear equations is as follows.

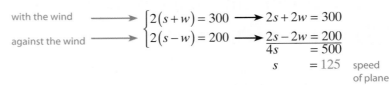

with the wind $\longrightarrow \begin{cases} 2(s+w)=300 \\ 2(s-w)=200 \end{cases}$

against the wind

$$\begin{aligned} 2s+2w &= 300 \\ \underline{2s-2w} &= \underline{200} \\ 4s &= 500 \\ s &= 125 \quad \text{speed of plane} \end{aligned}$$

Substitute $s=125$ into one of the original equations.

$$2(125+w)=300$$
$$125+w=150$$
$$w=25 \qquad \text{wind speed}$$

The speed of the plane was 125 mph, and the wind speed was 25 mph.

Now work margin exercise 1.

Example 2

Distance-Rate-Time (Times Unknown)

Two buses leave a bus station traveling in opposite directions. One leaves at noon and the second leaves at 1 P.M. The first one travels at an average speed of 55 mph and the second one at an average speed of 59 mph. At what time will the buses be 226 miles apart?

Solution

Let x = time of travel for first bus
and y = time of travel for second bus.

	Rate	×	Time	=	Distance
Bus 1	55		x		$55x$
Bus 2	59		y		$59y$

The system of linear equations is as follows.

$$\begin{cases} x = y+1 & \text{The first bus travels 1 hour longer than the second.} \\ 55x+59y=226 & \text{The sum of the distances is 226 miles.} \end{cases}$$

Substitution gives the following.

$$55(y+1)+59y=226$$
$$55y+55+59y=226$$
$$114y=171$$
$$y=1.5$$

This gives $x=1.5+1=2.5$.
Thus the first bus travels 2.5 hours and the second bus travels 1.5 hours. The buses will be 226 miles apart at 2:30 P.M.

Now work margin exercise 2.

Number Problems

Example 3

Number Problem

The sum of two numbers is 80 and their difference is 10. What are the two numbers?

Solution

Let x = one number
and y = the other number.
The system of linear equations is as follows.

$$\begin{cases} x + y = 80 & \text{The sum is 80.} \\ x - y = 10 & \text{The difference is 10.} \end{cases}$$

Solving by addition gives the following.

$$
\begin{aligned}
x + y &= 80 \\
\underline{x - y} &= \underline{10} \\
2x \quad &= 90 \\
x \quad &= 45
\end{aligned}
$$

Back substitute 45 for x in the first equation.
$$45 + y = 80$$
$$y = 80 - 45 = 35$$

The two numbers are 45 and 35.

Check:

$$45 + 35 = 80 \text{ and } 45 - 35 = 10$$

Now work margin exercise 3.

3. The sum of two numbers is 150 and their difference is 36. What are the two numbers?

Objective C **Amounts and Costs Problems**

Example 4

Counting Coins

Mike has $1.05 worth of change in nickels and quarters. If he has twice as many nickels as quarters, how many of each type of coin does he have?

Solution

We use two equations–one relating the number of coins and the other relating the value of the coins. (The value of each nickel is 5 cents and the value of each quarter is 25 cents.)

Applications: Distance-Rate-Time, Number Problems, Amounts, and Costs **Section 10.4** **817**

4. Seth has $3.85 worth of change in nickels and dimes. If he has 3 times as many dimes as nickels, how many of each type of coin does he have?

Let n = number of nickels
and q = number of quarters.

Coins	# of coins	Value	Total value
Nickels	n	0.05	$0.05n$
Quarters	q	0.25	$0.25q$

The system of linear equations is as follows.

$$\begin{cases} n = 2q \\ 0.05n + 0.25q = 1.05 \end{cases}$$

There are two times as many nickels as quarters.

The total amount of money is $1.05.

Note carefully that in the first equation q is multiplied by 2 because the number of nickels is twice the number of quarters. Therefore, n is bigger.

The first equation is already solved for n, so substituting $2q$ for n in the second equation gives the following.

$$0.05(2q) + 0.25q = 1.05$$
$$0.10q + 0.25q = 1.05$$
$$10q + 25q = 105 \qquad \text{Multiply the equation by 100.}$$
$$35q = 105$$
$$q = 3 \qquad \text{number of quarters}$$
$$n = 2q = 6 \qquad \text{number of nickels}$$

Mike has 3 quarters and 6 nickels.

Now work margin exercise 4.

Example 5

Calculating Age

Kathy is 6 years older than her sister, Sue. In 3 years, she will be twice as old as Sue. How old is each girl now?

Solution

We use two equations–one relating their ages now and the other relating their ages in 3 years.

Let K = Kathy's age now
and S = Sue's age now.

Then the system of linear equations is as follows.

$$\begin{cases} K - S = 6 \\ K + 3 = 2(S + 3) \end{cases}$$

The difference in their ages is 6 years.

In 3 years each age is increased by 3 and Kathy is twice as old as Sue.

Rewrite the second equation in standard form and solve by addition.

$$K + 3 = 2(S + 3)$$
$$K + 3 = 2S + 6$$
$$K - 2S = 3$$

Second equation in standard form

$$\begin{cases} (K - S = 6) \longrightarrow & K - S = 6 \\ [-1](K - 2S = 3) \longrightarrow & \underline{-K + 2S = -3} \\ & \qquad\quad S = 3 \quad \text{Sue's current age} \end{cases}$$

Back substitute $S = 3$ into one of the original equations.

$K - 3 = 6$

$\quad K = 6 + 3$

$\quad K = 9$ Kathy's current age

Kathy is 9 years old; Sue is 3 years old.

Now work margin exercise 5.

5. Enrique is 9 years older than his sister Maria. In 5 years, he will be twice as old as Maria. How old are Enrique and Maria now?

Example 6

Amounts and Costs

Three hot dogs and two orders of French fries cost $10.30. Four hot dogs and four orders of fries cost $15.60. What is the cost of a hot dog? What is the cost of an order of fries?

Solution

Let x = cost of one hot dog
and y = cost of one order of fries.
Then the system of linear equations is

$$\begin{cases} 3x + 2y = 10.30 & \text{Three hot dogs and two orders of French fries cost \$10.30.} \\ 4x + 4y = 15.60 & \text{Four hot dogs and four orders of fries cost \$15.60.} \end{cases}$$

Both equations are in standard form. Solve using the addition method.

$$\begin{cases} [-2]\,(3x + 2y = 10.30) \longrightarrow & -6x - 4y = -20.60 \\ \quad\;\; (4x + 4y = 15.60) \longrightarrow & \underline{\;4x + 4y = \;\;15.60\;} \\ & -2x \qquad\;\; = -5.00 \\ & \quad x \qquad\;\; = \;\;2.50 \quad \text{cost of one hot dog} \end{cases}$$

Back substitute $x = 2.50$ into one of the original equations.

$$3(2.50) + 2y = 10.30$$
$$7.50 + 2y = 10.30$$
$$2y = 2.80$$
$$y = 1.40 \quad \text{cost of one order of fries}$$

One hot dog costs $2.50 and one order of fries costs $1.40.

Now work margin exercise 6.

6. Seven sodas and six water bottles cost $13.55. Three sodas and four water bottles cost $6.95. What is the cost of a soda? What is the cost of a water bottle?

notes

■ You should consider making tables similar to those illustrated in Examples 1, 2, and 4 when working with applications. These tables
■ can help you organize the information in a more understandable form.

Exercises 10.4

1. The sum of two numbers is 56. Their difference is 10. Find the numbers.

2. The sum of two numbers is 40. The sum of twice the larger and 4 times the smaller is 108. Find the numbers.

3. The sum of two numbers is 36. Three times the smaller plus twice the larger is 87. Find the two numbers.

4. The sum of two integers is 102, and the larger number is 10 more than three times the smaller. Find the two integers.

5. The difference between two integers is 13, and their sum is 87. What are the two integers?

6. The difference between two numbers is 17. Four times the smaller is equal to 7 more than the larger. What are the numbers?

7. **Supplementary Angles:** Two angles are supplementary if the sum of their measures is 180°. Find two supplementary angles such that the smaller is 30° more than one half of the larger.

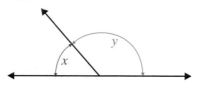

8. **Complementary Angles:** Two angles are complementary if the sum of their measures is 90°. Find two complementary angles such that one is 15° less than six times the other.

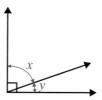

9. **Triangles:** The sum of the measures of the three angles of a triangle is 180°. In an isosceles triangle, two of the angles have the same measure. What are the measures of the angles of an isosceles triangle in which one angle measures 15° more than each of the other two equal angles?

10. **Triangles:** The sum of the measures of the three angles of a triangle is 180°. In an isosceles triangle, two of the angles have the same measure. What are the measures of the angles of an isosceles triangle in which each of the two equal angles measures 15° more than the third angle?

11. **Boating:** Liam makes a 4-mile motorboat trip downstream in 20 minutes $\left(\frac{1}{3} \text{ hr}\right)$. The return trip takes 30 minutes $\left(\frac{1}{2} \text{ hr}\right)$. Find the rate of the boat in still water and the rate of the current.

12. **Flying an Airplane:** Mr. McKelvey finds that flying with the wind he can travel 1188 miles in 6 hours. However, when flying against the wind, he travels only $\frac{2}{3}$ of the distance in the same amount of time. Find the speed of the plane in still air and the wind speed.

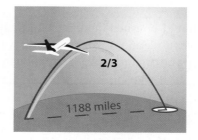

13. **Running:** Usain Bolt, the world-record holder in the 100 meter dash, ran 100 meters in 9.69 seconds with no wind. He later ran the same distance in 9.58 seconds with the wind. What was his speed and what was the wind speed?

14. **Boating:** Jessica drove her speedboat upriver this morning. It took her 1 hour going upriver and 54 minutes going down river. If she traveled 36 miles each way, what would have been the rate of the boat in still water and what was the rate of the current (in miles per hour)?

15. **Traveling by Car:** Randy made a business trip of 190 miles. He averaged 52 mph for the first part of the trip and 56 mph for the second part. If the total trip took $3\frac{1}{2}$ hours, how long did he travel at each rate?

16. **Traveling by Car:** Marian drove to a resort 335 miles from her home. She averaged 60 mph for the first part of her trip and 55 mph for the second part. If her total driving time was $5\frac{3}{4}$ hours, how long did she travel at each rate?

17. **Traveling by Car:** Marcos lives 364 miles away from his cousin Cana. They start driving at the same time and travel toward each other. Cana's speed is 11 mph faster than Marcos' speed. If they meet in 4 hrs, find their speeds.

18. **Traveling by Car:** Naomi and Linda live 324 miles apart. They start at the same time and travel toward each other. Naomi's speed is 8 mph greater than Linda's. If they meet in 3 hours, find their speeds.

19. **Traveling by Car:** Steve travels 4 times as fast as Tim. Starting at the same point, but traveling in opposite directions, they are 105 miles apart after 3 hours. Find their rates of travel.

20. **Traveling by Car:** Bella travels 5 mph less than twice as fast as June. Starting at the same point and traveling in the same direction, they are 80 miles apart after 4 hours. Find their speeds.

21. **Traveling by Train:** Two trains leave Dallas at the same time. One train travels east and the other travels west. The speed of the westbound train is 5 mph greater than the speed of the eastbound train. After 6 hours, they are 510 miles a part. Find the rate of each train. Assume the trains travel in a straight line in opposite directions.

22. **Boating:** A boat left Dana Point Marina at 11:00 am traveling at 10 knots (nautical miles per hour). Two hours later, a Coast Guard boat left the same marina traveling at 14 knots trying to catch the first boat. If both boats traveled the same course, at what time did the Coast Guard captain anticipate overtaking the first boat?

23. **Jogging:** A jogger runs into the countryside at a rate of 10 mph. He returns along the same route at 6 mph. If the total trip took 1 hour 36 minutes, how far did he jog?

24. **Biking:** A cyclist traveled to her destination at an average rate of 15 mph. By traveling 3 mph faster, she took 30 minutes less to return. What distance did she travel each way?

25. Coin Collecting: Sonja has some nickels and dimes. If she has 30 coins worth a total of $2.00, how many of each type of coin does she have?

26. Coin Collecting: Conner has a total of 27 coins consisting of quarters and dimes. The total value of the coins is $5.40. How many of each type of coin does he have?

27. Coin Collecting: A bag contains pennies and nickels only. If there are 182 coins in all and their value is $3.90, how many pennies and how many nickels are in the bag?

28. Coin Collecting: Your friend challenges you to figure out how many dimes and quarters are in a cash register. He tells you that there are 65 coins and that their value is $11.90. How many dimes and how many quarters are in the register?

29. Basketball Admission: Tickets for the local high school basketball game were priced at $3.50 for adults and $2.50 for students. If the income for one game was $9550 and the attendance was 3500, how many adults and how many students attended that game?

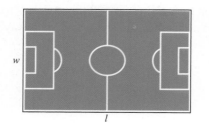

30. Rectangles: The width of a rectangle is $\frac{3}{4}$ of its length. If the perimeter of the rectangle is 140 feet, what are the dimensions of the rectangle?

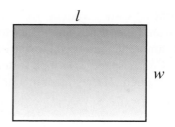

31. Rectangles: The length of a rectangle is 10 meters more than one half of the width. If the perimeter is 44 meters, what are the length and width?

32. Building a Fence: A farmer has 260 meters of fencing to build a rectangular corral. He wants the length to be 3 times as long as the width. What dimensions should he make his corral?

33. Soccer: At present, the length of a rectangular soccer field is 55 yards longer than the width. The city wants to rearrange the area containing the soccer field into two square playing fields. A math teacher on the council told them that if the width of the current field were to be increased by 5 yards and the length cut in half, the resulting field would be a square. What are the dimensions of the field currently?

34. Perimeter: Consider a square and a regular hexagon (a six-sided figure with sides of equal length). One side of the square is 5 feet longer than a side of the hexagon, and the two figures have the same perimeter. What are the lengths of the sides of each figure?

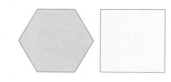

35. Rectangles: The length of a rectangle is 1 meter less than twice the width. If each side is increased by 4 meters, the perimeter will be 116 meters. Find the length and the width of the original rectangle.

36. Age: Ava is 8 years older than her brother Curt. Four years from now, Ava will be twice as old as Curt. How old is each at the present time?

37. Age: When they got married, Elvis Presley was 11 years older than his wife Priscilla. One year later, Priscilla was two-thirds of Elvis' age. How old was each of them when they got married?

38. Charity Admission: A Christmas charity party sold tickets for $45.00 for adults and $25.00 for children. The total number of tickets sold was 320 and the total for the ticket sales was $13,000. How many adult and how many children's tickets were sold?

39. Buying Books: Joan went to a book sale on campus and bought paperback books for $0.25 each and hardback books for $1.75 each. If she bought a total of 15 books for $11.25, how many of each type of book did she buy?

40. Recycling: Morton took some old newspapers and aluminum cans to the recycling center. Their total weight was 180 pounds. He received 1.5¢ per pound for the newspapers and 30¢ per pound for the cans. The total received was $14.10. How many pounds of each did Morton have?

41. Baseball Admission: Admission to the baseball game is $2.00 for general admission and $3.50 for reserved seats. The receipts were $36,250 for 12,500 paid admissions. How many of each ticket, general and reserved, were sold?

42. Going to the Theater: 70 children and 160 adults attended a play. The total receipts were $620. One adult ticket and 2 children's tickets cost $7. Find the price of each type of ticket.

43. Surfing: Last summer, Ernie sold surfboards. One style sold for $625 and the other sold for $550. He sold a total of 47 surfboards. How many of each style did he sell if the sales from each style were equal?

44. Selling Candy: The Candy Shack sells a particular candy in two different size packages. One size sells for $1.25 and the other sells for $1.75. If the store received $65.50 for 42 packages of candy, how many of each size were sold?

45. Golfing: The pro shop at the Divots Country Club ordered two brands of golf balls. Titleless balls cost $1.80 each and the Done Lob balls cost $1.50 each. The total cost of Titleless balls exceeded the total cost of the Done Lob balls by $108. If equal numbers of each brand were ordered, how many dozen of each brand were ordered?

46. Real Estate: Sellit Realty Company gets a 6% fee for selling improved properties and 10% for selling unimproved land. Last week, the total sales were $220,000 and their total fees were $16,400. What were the sales from each of the two types of properties?

47. Shopping: A men's clothing store sells two styles of sports jackets, one selling for $95 and one selling for $120. Last month, the store sold 40 jackets, with receipts totaling $4250. How many of each style did the store sell?

48. Shopping: Frank bought 2 shirts and 1 pair of dress pants for a total of $55. If he had bought 1 shirt and 2 pairs of dress pants, he would have paid $68. What was the price of each shirt and each pair of dress pants?

49. Fast Food: At McDonalds, 3 Big Macs and 5 orders of medium French fries cost $19.69. 6 Big Macs and 2 orders of medium French fries cost $25.06. What is the price of a Big Mac? What is the price of one order of medium French fries?

50. Stamp Collecting: The postal service charges 42¢ for letters that weigh 1 ounce or less and 17¢ more for letters that weigh between 1 and 2 ounces. Jeff, testing his father's math skills, gave his father $42.10 and asked him to purchase 80 stamps for his stamp collection, some 42¢ stamps and some 59¢ stamps. How many of each type of stamp did his dad buy if he used all the money?

51. Manufacturing: A small manufacturer produces two kinds of radios, model X and model Y. Model X takes 4 hours to produce and costs $8 each to make. Model Y takes 3 hours to produce and costs $7 each to make. If the manufacturer decides to allot a total of 58 hours and $126 each week, how many of each model will be produced?

52. Furniture Building: A furniture shop makes dining room chairs. Employees can build two styles of chairs. Style I takes 1 day and the materials cost $60. Style II takes $1\frac{1}{2}$ days but the materials only costs $30. If, during the last two months, they spent 36 days and $1200 building chairs, how many chairs of each style did they build?

Writing & Thinking

53. A two digit number can be written as ab, where a and b are the digits. We do **not** mean that the digits are multiplied, but the value of the number is $10a + b$. For example, the two digit number 34 has a value of $10 \cdot 3 + 4$. Set up and solve a system of equations for the following problem.

The sum of the digits of a two digit number is 13. If the digits are reversed, then the value of the number is increased by 45. What is the number?

10.5 Applications: Interest and Mixture

In this section we will study two more types of applications that can be solved using systems of linear equations: interest on money invested and mixture. These applications can be "wordy" and you will need to read carefully and analyze the information thoroughly to be able to translate it into a system involving two variables.

Objective A   Interest Problems

People in business and banking know several formulas for calculating interest. The formula used depends on the method of payment (monthly or yearly) and the type of interest (simple or compound). Also, penalties for late payments and even penalties for early payments might be involved. In any case, standard notation is the following:

$P \longrightarrow$ the principal (amount of money invested or borrowed)
$r \longrightarrow$ the rate of interest (an annual rate)
$t \longrightarrow$ the time (in one year or part of a year)
$I \longrightarrow$ the interest (paid or earned)

In this section, we will use only the basic formula for simple interest

$$I = Prt$$

with interest calculated on an annual basis. In this special case, we have $t = 1$ and the formula becomes

$$I = Pr.$$

Example 1

Interest

James has two investment accounts, one pays 6% interest and the other pays 10% interest. He has $1000 more in the 10% account than he has in the 6% account. In one year, the interest from the 10% account is $260 more than the interest from the 6% account. How much does he have in each account?

Solution

Careful reading indicates two types of information:
1. He has two accounts.
2. He earns two amounts of interest.

Let x = amount (principal) invested at 6%
and y = amount (principal) invested at 10%,
then,
$0.06x$ = interest earned on first account
$0.10y$ = interest earned on second account.

Now set up two equations.

$$\begin{cases} y = x + 1000 \\ 0.10y - 0.06x = 260 \end{cases}$$

y is larger than x by \$1000.

Interest from the 10% account is \$260 more than interest from the 6% account.

Because the first equation is already solved for y, we use the substitution method and substitute for y in the second equation.

$$0.10(x + 1000) - 0.06x = 260$$
$$10(x + 1000) - 6x = 26,000 \quad \text{Multiply by 100.}$$
$$10x + 10,000 - 6x = 26,000$$
$$4x = 16,000$$
$$x = 4000 \quad \text{amount at 6\%}$$

Back substitute $x = 4000$ into one of the original equations to find y.

$$y = x + 1000 = 4000 + 1000 = 5000 \quad \text{amount at 10\%}$$

James has \$4000 invested at 6% and \$5000 invested at 10%.

Example 2

Interest

Lila has \$7000 to invest. She decides to separate her funds into two accounts. One yields interest at the rate of 7% and the other at 12%. If she wants a total annual income from both accounts to be \$690, how should she split the money? (**Note:** The higher interest account is considered more risky. Otherwise, she would put the entire \$7000 into that account.)

Solution

Again, careful reading indicates two types of information:
1. She has two accounts.
2. She earns two amounts of interest.

Let x = amount (principal) invested at 7%
and y = amount (principal) invested at 12%,
then
$\quad 0.07x$ = interest earned on first account
$\quad 0.12y$ = interest earned on second account.

Now set up two equations.

$$\begin{cases} x + y = 7000 \\ 0.07x + 0.12y = 690 \end{cases}$$

The total amount invested is \$7000.

The total interest from both accounts is \$690.

Both equations are in standard form. Solve by addition. Multiply the first equation by -7 and the second by 100 to get opposite coefficients for x as follows.

$$\begin{cases} [-7] & (x + y = 7000) \\ [100] & (0.07x + 0.12y = 690) \end{cases} \longrightarrow \begin{array}{r} -7x - 7y = -49{,}000 \\ 7x + 12y = 69{,}000 \\ \hline 5y = 20{,}000 \\ y = 4000 \quad \text{amount at 12\%} \end{array}$$

Substitute $y = 4000$ into one of the original equations.

$$x + 4000 = 7000$$
$$x = 3000 \quad \text{amount at 7\%}$$

She should invest \$3000 at 7% and \$4000 at 12%.

Now work margin exercises 1 and 2.

Objective B **Mixture Problems**

Problems involving mixtures occur in physics and chemistry and in such places as candy stores or coffee shops. Two or more items of a different percentage of concentration of a chemical such as salt, chlorine, or antifreeze are to be mixed; or two or more types of food such as coffee, nuts, or candy are to be mixed to form a final mixture that satisfies certain conditions of percentage of concentration.

The basic plan is to write an equation that deals with only one part of the mixture (such as the salt in the mixture). The following examples explain how this can be accomplished.

Example 3

Mixture

How many ounces each of a 10% salt solution and a 15% salt solution must be used to produce 50 ounces of a 12% salt solution?

Solution

Let x = amount of 10% solution
and y = amount of 15% solution.

	Amount of solution	×	Percent of salt	=	Amount of salt
10% solution	x		0.10		$0.10x$
15% solution	y		0.15		$0.15y$
12% solution	50		0.12		$0.12(50)$

1. Fergus has two investment accounts for his toupee company. One pays 9% interest and the other pays 12% interest. He has $800 more in the 12% account than he has in the 9% account. In one year, the interest from the 12% account is $246 more than the interest from the 9% account. How much does Fergus have in each account?

2. Darnell has $9000 to invest. He decides to separate his money into two accounts. One yields interest at the rate of 5% and the other at 9%. If he wants a total annual income from both accounts to be $550, how should he split up the money?

Then the system of linear equations is as follows.

$$\begin{cases} x + y = 50 \\ 0.10x + 0.15y = 0.12(50) \end{cases}$$

The sum of the two amounts must be 50 ounces.

The sum of the amounts of salt from the two solutions equals the total amount of salt in the final solution.

Multiplying the first equation by -10 and the second by 100 gives

$$\begin{cases} [-10] \quad (x + \quad y = 50) \\ [100](0.10x + 0.15y = 0.12(50)) \end{cases} \longrightarrow \begin{array}{l} -10x - 10y = -500 \\ \underline{10x + 15y = 600} \\ \quad\quad 5y = 100 \\ \quad\quad\quad y = 20 \quad \text{amount of 15\%} \end{array}$$

Substitute $y = 20$ into one of the original equations.

$$x + 20 = 50$$
$$x = 30 \quad \text{amount of 10\%}$$

Use 30 ounces of the 10% solution and 20 ounces of the 15% solution.

Example 4

Mixture

How many gallons of a 20% acid solution should be mixed with a 30% acid solution to produce 100 gallons of a 23% solution?

Solution

Let x = amount of 20% solution
and y = amount of 30% solution.

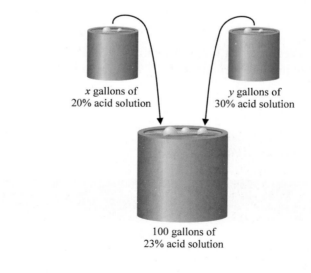

x gallons of
20% acid solution

y gallons of
30% acid solution

100 gallons of
23% acid solution

	Amount of solution	×	Percent of acid	=	Amount of acid
20% solution	x		0.20		$0.20x$
30% solution	y		0.30		$0.30y$
23% solution	100		0.23		$0.23(100)$

Then the system of linear equations is

$$\begin{cases} x + y = 100 \\ 0.20x + 0.30y = 0.23(100) \end{cases}$$

The sum of the two amounts must be 100 gallons.
The sum of the amounts of acid from the two solutions equals the total amount of acid in the final solution.

Multiplying the first equation by -20 and the second by 100 gives,

$$\begin{cases} [-20] & (x + y = 100) \\ [100](0.20x + 0.30y = 0.23(100)) \end{cases} \longrightarrow \begin{array}{r} -20x - 20y = -2000 \\ 20x + 30y = 2300 \\ \hline 10y = 300 \\ y = 30 \quad \text{amount of 30\%} \end{array}$$

Substitute $y = 30$ into one of the original equations.

$$x + 30 = 100$$
$$x = 70 \quad \text{amount of 20\%}$$

70 gallons of the 20% solution should be added to 30 gallons of the 30% solution. This will produce 100 gallons of a 23% solution.

***Now work margin exercises* 3 and 4.**

3. How many ounces each of a 12% chlorine solution and a 18% chlorine solution must be used to produce 150 ounces of a 14% chlorine solution?

4. How many gallons of a 17% soap solution should be mixed with a 22% soap solution to produce 120 gallons of a 21% solution?

1. **Investing:** Carmen invested $9000, part in a 6% passbook account and the rest in a 10% certificate account. If her annual interest was $680, how much did she invest at each rate?

2. **Investing:** Mr. Brown has $12,000 invested. Part is invested at 6% and the remainder at 8%. If the annual interest from the 6% investment is $230 more than the annual interest from the 8% investment, how much is invested at each rate?

3. **Investing:** Ten thousand dollars is invested, part at 5.5% and part at 6%. The annual interest from the 5.5% investment is $251 more than the annual interest from the 6% investment. How much is invested at each rate?

4. **Investing:** On two investments totaling $9500, Darius lost 3% on one and earned 6% on the other. If his net annual receipts were $282, how much was each investment?

5. **Investing:** Merideth has money in two savings accounts. One rate is 8% and the other is 10%. If she has $200 more in the 10% account, how much is invested at 8% if the total annual interest is $101?

6. **Investing:** Money is invested at two rates. One rate is 9% and the other is 13%. If there is $700 more invested at 9%, find the amount invested at each rate if the total annual interest is $239.

7. **Investing:** Ethan has half of his investments in stock paying an 11% dividend and the other half in a debentured stock paying 13% interest. If his total annual interest is $840, how much does he have invested?

8. **Investing:** Betty invested some of her money at 12% interest. She invested $300 more than twice that amount at 10%. How much is invested at each rate if her interest income is $318 annually?

9. **Investing:** GFA invested some money in a development yielding 24% and $9000 less in a development yielding 18%. If the first investment produces $2820 more per year than the second, how much is invested in each development?

10. **Investing:** Victoria invests a certain amount of money at 7% annual interest and three times that amount at 8%. If her annual interest income is $232.50, how much does she have invested at each rate?

11. **Investing:** Jamal has a certain amount of money invested at 5% annual interest and $500 more than twice that amount invested in bonds yielding 7%. His total annual income from interest is $187. How much does he have invested at each rate?

12. **Investing:** A total of $6000 is invested, part at 8% and the remainder at 12%. How much is invested at each rate if the annual interest is $620?

13. **Investing:** Ms. Merriman has $12,000 invested. Part is invested at 9% and the remainder at 11%. If the interest from the 9% investment is $380 more than the interest from the 11% investment, how much is invested at each rate?

14. **Investing:** Eight thousand dollars is invested, part at 15% and the remainder at 12%. If the annual interest income from the 15% investment is $66 more than the annual interest income from the 12% investment, how much is invested at each rate?

15. **Investing:** Morgan inherited $124,000 from her Uncle Edward. She invested a portion in bonds and the remainder in a long-term certificate account. The amount invested in bonds was $24,000 less than 3 times the amount invested in certificates. How much was invested in bonds and how much in certificates?

16. **Investing:** Sang has invested $48,000, part at 6% and the rest in a higher risk investment at 10%. How much did she invest at each rate to receive $4000 in interest after one year?

17. **Manufacturing:** A metallurgist has one alloy containing 20% copper and another containing 70% copper. How many pounds of each alloy must he use to make 50 pounds of a third alloy containing 50% copper?

18. **Manufacturing:** A manufacturer has received an order for 24 tons of a 60% copper alloy. His stock contains only alloys of 80% copper and 50% copper. How much of each will he need to fill the order?

19. **Tobacco:** A tobacco shop wants 50 ounces of tobacco that is 24% rare Turkish blend. How much each of a 30% Turkish blend and a 20% Turkish blend will be needed?

20. **Acid Solutions:** How many liters each of a 40% acid solution and a 55% acid solution must be used to produce 60 liters of a 45% acid solution?

21. **Dairy Production:** A dairy man wants to mix a 35% protein supplement and a standard 15% protein ration to make 1800 pounds of a high-grade 20% protein ration. How many pounds of each should he use?

22. **Manufacturing:** To meet the government's specifications, a certain alloy must be 65% aluminum. How many pounds each of a 70% aluminum alloy and a 54% aluminum alloy will be needed to produce 640 pounds of the 65% aluminum alloy?

23. **Food Science:** A meat market has ground beef that is 40% fat and extra lean ground beef that is only 15% fat. How many pounds of each will be needed to obtain 50 pounds of lean ground beef that is 25% fat?

24. **Gasoline:** George decides to mix grades of gasoline in his truck. He puts in 8 gallons of regular and 12 gallons of premium for a total cost of $55.80. If premium gasoline costs $0.15 more per gallon than regular, what was the price of each grade of gasoline?

25. **Acid Solutions:** How many grams of pure acid (100% acid) and how many grams of a 40% solution should be mixed together to get a total of 30 grams of a 60% solution?

26. **Coffee:** Pure 100% dark coffee beans are to be mixed with a mixture that is 60% dark beans. How much of each (pure dark and 60% dark beans) should be used to get a mixture of 50 pounds that contains 70% of the dark beans?

27. Salt Solutions: Pure salt is to be added to a 4% salt solution. How many ounces of salt and how many ounces of the 4% solution should be mixed together to get 60 ounces of a 20% salt solution?

28. Chemistry: How many liters each of a 12% iodine solution and a 30% iodine solution must be used to produce a total mixture of 90 liters of a 22% iodine solution?

29. Food Science: A candymaker is making truffles using a mixture of a melted dark chocolate that is 72% cocoa and milk chocolate that is 42% cocoa. If she wants 6 pounds of melted chocolate that is 52% cocoa, how much of each type of chocolate does she need?

30. Dairy Farming: A dairy needs 360 gallons of milk containing 4% butterfat. How many gallons each of milk containing 5% butterfat and milk containing 2% butterfat must be used to obtain the desired 360 gallons?

31. Body Piercings: It is recommended that one cleans new body piercings with a 1% salt solution for the first few weeks with the new piercing. You have a 0.5% solution and a 5% solution and you need to make 8 ounces of the 1% solution. How much of the 0.5% and 5% solutions will you need? (Round your answers to the nearest hundredth.)

32. Pharmacy: A druggist has two solutions of alcohol. One is 25% alcohol. The other is 45% alcohol. He wants to mix these two solutions to get 36 ounces that will be 30% alcohol. How many ounces of each of these two solutions should he mix together?

33. Your friend has $20,000 to invest and decided to invest part at 4% interest and the rest at 10% interest. Why might you advise him (or her) to invest all of it
   a. at 4%?
   b. at 10%?

Chapter 10: Index of Key Terms and Ideas

Section 10.1: Systems of Linear Equations: Solutions by Graphing

System of Linear Equations   page 788
A pair of linear equations considered together is called a **system of linear equations** (or a **set of simultaneous equations**).

Solution to a System of Equations   page 788
A **solution of a system** of linear equations is an ordered pair (or point) that satisfies **both** equations.

Solutions by Graphing   page 791
1. Graph both linear equations on the same set of axes.
2. Observe the point of intersection (if there is one).
   a. If the slopes of the two lines are different, then the lines intersect in one and only one point. The system has a single point as its solution.
   b. If the lines are distinct and have the same slope, then the lines are parallel. The system has no solution.
   c. If the lines are the same line, then all the points on the line constitute the solution. There are an infinite number of solutions.
3. Check the solution (if there is one) in both of the original equations.

Using a Graphing Calculator to Solve a System of Linear Equations   page 794

Section 10.2: Systems of Linear Equations: Solutions by Substitution

Solutions by Substitution   page 800
1. Solve one of the equations for one of the variables.
2. Substitute the resulting expression into the other equation.
3. Solve this new equation, if possible, and then substitute back into one of the original equations to find the value of the other variable.
4. Check the solution in both of the original equations.

Key Terms and Ideas   Chapter 10   833

Section 10.3: Systems of Linear Equations: Solutions by Addition

Solutions by Addition page 808
1. Write the equations in **standard form,** one under the other so that **like terms are aligned**.
2. Multiply all terms of one equation by a constant (and possibly all terms of the other equation by another constant) so that **two like terms have opposite coefficients**.
3. Add the two equations by **combining like terms** and solve the resulting equation, if possible.
4. **Back substitute into one of the original equations** to find the value of the other variable.
5. Check the solution (if there is one) in both of the original equations.

Guidelines for Deciding which Method to Use when Solving a System of Linear Equations page 812
1. The graphing method is helpful in "seeing" the geometric relationship between the lines and finding approximate solutions. A calculator can be very helpful here.
2. Both the substitution method and the addition method give exact solutions.
3. The substitution method may be reasonable and efficient if one of the coefficients of one of the variables is 1.
4. The addition method is particularly efficient if the coefficients for one of the variables are opposites.

Section: 10.4:
Applications: Distance-Rate-Time, Number Problems, Amounts, and Costs

Applications page 815
Distance-Rate-Time
Number Problems
Amounts and Costs

Section 10.5: Applications: Interest and Mixture

Applications page 825
Interest
Mixture

Chapter 10: Review

Section 10.1: Systems of Linear Equations: Solutions by Graphing

Determine which of the given points, if any, lie on both of the lines in the systems of equations by substituting each point into both equations.

1. $\begin{cases} y - x = 1 \\ 3y - x = -6 \end{cases}$ **a.** $\left(-\dfrac{9}{2}, -\dfrac{7}{2}\right)$ **b.** $(7, 8)$ **c.** $(6, 0)$ **d.** $\left(-5, \dfrac{1}{4}\right)$

2. $\begin{cases} 6x + 2y - 10 = 0 \\ 12x - 20 = -4y \end{cases}$ **a.** $(7, -5)$ **b.** $(4, -7)$ **c.** $(-1, 8)$ **d.** $(3, -4)$

Show that each system of equations is inconsistent by determining the slope of each line and the y-intercept. (That is, show that the lines are parallel and do not intersect.)

3. $\begin{cases} y = 2x \\ -2x + y = 5 \end{cases}$

4. $\begin{cases} 3x + y = 4 \\ 6x + 2y = -1 \end{cases}$

Solve the following systems graphically.

5. $\begin{cases} y = -2x + 5 \\ y = x + 2 \end{cases}$

6. $\begin{cases} y = -x + 10 \\ x + y = 4 \end{cases}$

7. $\begin{cases} y = -\dfrac{1}{2}x \\ 3x - 2y = 8 \end{cases}$

8. $\begin{cases} 5x + y = 4 \\ 10x + 2y = 8 \end{cases}$

9. $\begin{cases} -3x + y = 7 \\ -6x + 2y = 9 \end{cases}$

10. $\begin{cases} x + y = 3 \\ 2x + y = 2 \end{cases}$

11. $\begin{cases} x + 2y = 12 \\ 3x - y = -6 \end{cases}$

12. $\begin{cases} x + 4y = 6 \\ y + \dfrac{1}{4}x = \dfrac{3}{2} \end{cases}$

Use a graphing calculator to find the solutions to the given systems of linear equations. If necessary, round your answer to the nearest ten-thousandth.

13. $\begin{cases} 2x + 3y = 10.5 \\ x - y = -1 \end{cases}$

14. $\begin{cases} x - 3y = 2.5 \\ 2x + y = -5.5 \end{cases}$

15. $\begin{cases} y = 7 + x \\ x + 3y = 21 \end{cases}$

16. $\begin{cases} x - 5y = -1.5 \\ x - y = 4.3 \end{cases}$

Section 10.2: Systems of Linear Equations: Solutions by Substitution

Solve each system by using the method of substitution.

17. $\begin{cases} y = 2x \\ x - y = -1 \end{cases}$

18. $\begin{cases} y = 4x + 1 \\ 8x - 2y = -2 \end{cases}$

19. $\begin{cases} 3x - 2y = 4 \\ y - 6 = \dfrac{3}{2}x \end{cases}$

20. $\begin{cases} y = -2x + 5 \\ 4x = -2y - 1 \end{cases}$

21. $\begin{cases} 5x - y = 6 \\ 2y = 10x - 12 \end{cases}$

22. $\begin{cases} y = -\dfrac{1}{8}x \\ x - 4y = 4 \end{cases}$

23. $\begin{cases} 2x - y = 2 \\ 2x + y = -2 \end{cases}$

24. $\begin{cases} x + \dfrac{1}{2}y = 8 \\ y = x + 1 \end{cases}$

25. $\begin{cases} y = 7x - 5 \\ x - 3y = 0 \end{cases}$

26. $\begin{cases} 6x + y = -4 \\ 3x + \dfrac{1}{2}y = \dfrac{1}{2} \end{cases}$

Section 10.3: Systems of Linear Equations: Solutions by Addition

Solve each system by using the method of addition.

27. $\begin{cases} 5x + 3y = -11 \\ -5x + 2y = -4 \end{cases}$

28. $\begin{cases} 4x - 5y = 25 \\ 3x - 2y = 17 \end{cases}$

29. $\begin{cases} -4x + 3y = 9 \\ 8x - 2y = -14 \end{cases}$

30. $\begin{cases} 2x - y = 12 \\ 2x + \dfrac{1}{2}y = 15 \end{cases}$

31. $\begin{cases} 7x + 8y = 6 \\ \dfrac{7}{2}x + 16y = 12 \end{cases}$

32. $\begin{cases} x - 2y = 18 \\ 3x - 2y = 4 \end{cases}$

33. $\begin{cases} -9y - 3x = 2 \\ -3y - x = 6 \end{cases}$

34. $\begin{cases} 0.75 + 2y = 1.5x \\ 6 + 16y = 12x \end{cases}$

Solve each system by using either the substitution method or the addition method (whichever seems better to you).

35. $\begin{cases} x + y = 1 \\ x - y = 3 \end{cases}$

36. $\begin{cases} 3x - 2y = 5 \\ x + 3y = 17 \end{cases}$

37. $\begin{cases} x + 2y = 1 \\ 2x - 3y = 0 \end{cases}$

38. $\begin{cases} 5x + y = 1 \\ 3x + y = 3 \end{cases}$

39. $\begin{cases} -3x + y = 6 \\ 6x - 2y = -12 \end{cases}$

40. $\begin{cases} 3x + 4y = 8 \\ -\dfrac{3}{4}x + 4 = y \end{cases}$

41. $\begin{cases} y = -\dfrac{1}{2}x + 10 \\ 2y = 6 - x \end{cases}$

42. $\begin{cases} y = \dfrac{1}{3}x + 10 \\ x = -30 + 3y \end{cases}$

Write an equation for the line determined by the two given points by using the formula
$y = mx + b$ **to set up a system of equations with** m **and** b **as the unknowns.**

43. $(2, -9), (1, -5)$

44. $(8, -1), (0, 2)$

Section 10.4: Applications: Distance-Rate-Time, Number Problems, Amounts and Costs

Solve the following word problems.

45. Number Problem: The sum of two numbers is -12. One number is 3 more than twice the other.

46. Number Problem: The difference between two numbers is 2. Three times the smaller is equal to 8 more than the larger. What are the numbers?

47. Age: Alice is 8 years older than her brother John. In 5 years she will be twice as old as John. How old are Alice and John now?

48. Making Change: Jorge has $2.10 worth of change in quarters and dimes. If he has eight times as many dimes as quarters how many of each type of coin does he have?

49. Hiking: Brandon decided to hike in the mountains a total distance of 14.4 miles (on a trail he knew). He averaged 2 mph for the first part of the hike and 3 mph for the second part. If the total hike took 6 hours, how many hours did he hike at each rate?

50. Airport Taxi: Sam catches a taxi for a ride to the airport 36 miles away. The taxi maintained an average speed of 40 mph for the first part of the trip, however closer to the airport traffic became heavier and its speed was reduced to 20 mph. How long was spent at each speed if the total trip took 1 hour and 12 minutes?

51. Rectangles: The width of a rectangle is 5 meters less than the length. If the perimeter of the rectangle is 80 meters, what are the length and width?

52. Supplementary Angles: Two angles are supplementary if the sum of their measures is $180°$. Find two supplementary angles such that one angle is $20°$ less than three times the other.

53. Shopping: A golf pro shop sells two particular types of shirts, the first for $110 each and the other for $65 each. Last month the shop sold 50 of these shirts for a total of $4600. How many of each type of shirt did they sell?

54. Fair Food: While at the county fair, a family decides to grab a sweet snack. If they buy 5 boxes of candied popcorn and 3 funnel cakes the cost of the snack would be $26.05. If they buy 2 boxes of popcorn and 4 funnel cakes, the cost would be $21.90. Find the price of the box of popcorn and the price of the funnel cake.

Solve the following word problems.

55. **Investing:** Mr. Smith has $20,000 invested, part at 6% and the remainder at 8%. How much does he have invested at each rate if the annual interest is $1300?

56. **Investing:** Mary Jane invests a certain amount of money at 8% and twice as much at 5%. If her annual income from the two investments is $720, how much does she have invested at each rate?

57. **Investing:** If you invest half of your savings at 7% and the other half at 10% and you make $1275 annually, how much do you have invested at each rate?

58. **Investing:** A financial advisor advises Jennifer to invest her money equally in stocks and bonds. If the return on bonds is 4.5% and the return on the stocks is anticipated to be 7%, how much should she invest in each type of investment to make $2530 in one year?

59. **Salt Solutions:** How many gallons each of a 25% salt solution and a 40% salt solution must be used to produce 60 gallons of a 35% solution?

60. **Grocery Shopping:** A butcher decides to mix hamburger so it can be sold at a competitive price. If he mixes some that is 22% fat with some that is 10% fat, how many pounds of each type should he mix to get 80 pounds of hamburger that is 16% fat?

61. **Acid Solutions:** How many ounces of pure acid and how many ounces of a 10% acid solution should be mixed to make a total of 50 ounces of a 64% acid solution?

62. **Manufacturing:** A manufacturer decided to mix two aluminum alloys, one is 20% aluminum and the other is 60% aluminum. How many tons of each type should he use to get a total of 100 tons of 28% aluminum alloy?

Chapter 10: Test

Determine which of the given points, if any, lie on both of the lines in the systems of equations by substituting each point into both equations.

1. $\begin{cases} 3x - 7y = 5 \\ 5x - 2y = -11 \end{cases}$ **a.** $(1, 8)$ **b.** $(4, 1)$ **c.** $(-3, -2)$ **d.** $(13, 10)$

2. $\begin{cases} x - 2y = 7 \\ 2x - 3y = 5 \end{cases}$ **a.** $(0, 3)$ **b.** $(7, 0)$ **c.** $(1, -1)$ **d.** $(-11, -9)$

Solve the systems by graphing.

3. $\begin{cases} 2x = -3y + 9 \\ 4x + 6y = 18 \end{cases}$

4. $\begin{cases} x - y = 3 \\ x + 2y = 6 \end{cases}$

Solve the systems by using the method of substitution.

5. $\begin{cases} 5x - 2y = 0 \\ y = 3x + 4 \end{cases}$

6. $\begin{cases} 4x - y = -1 \\ -3x + 2y = 7 \end{cases}$

7. $\begin{cases} x = \dfrac{1}{3}y - 4 \\ 2x + \dfrac{3}{2}y = 5 \end{cases}$

Solve the systems by using the method of addition.

8. $\begin{cases} 3x + y = 5 \\ -12x - 4y = -7 \end{cases}$

9. $\begin{cases} 3x + 4y = 10 \\ x + 6y = 1 \end{cases}$

10. $\begin{cases} 5x + 2y = 1 \\ 4x + y = 5 \end{cases}$

Solve the systems by using any method.

11. $\begin{cases} x + y = 2 \\ y = -2x - 1 \end{cases}$

12. $\begin{cases} 6x + 2y - 8 = 0 \\ y = -3x \end{cases}$

13. $\begin{cases} 7x + 5y = -9 \\ 6x + 2y = 6 \end{cases}$

14. $\begin{cases} x + 3y = -2 \\ 2x = -6y - 4 \end{cases}$

Solve the following word problems.

15. **Coffee:** Every time Tish goes to a coffee shop she either buys a soy latte or an iced coffee. In one month she spent a total of $79.10. If the money she spent on soy lattes is $3.50 plus three times the money she spent on iced coffees, how much money did she spend on each drink?

16. **Boating:** Pete's boat can travel 48 miles upstream in 4 hours. The return trip takes 3 hours. Find the speed of the boat in still water and the speed of the current.

17. **Acid Solutions:** How many gallons of a 25% acid solution and a 50% acid solution must be used to produce 30 gallons of a 40% acid solution.

18. **Investing:** Gary has two investments yielding a total annual interest of $185.60. The amount invested at 8% is $320 less than twice the amount invested at 6%. How much is invested at each rate?

19. **Rectangles:** The perimeter of a rectangle is 60 inches and the length is 4 inches longer than the width. Find the dimensions of the rectangle.

20. **Manufacturing:** A metallurgist needs 2000 pounds of an alloy that is 80% copper. In stock, he has only alloys of 83% copper and 68% copper. How many pounds of each must be used?

21. **Making Change:** Sonia has a bag of coins with only nickels and quarters. She wants you to figure out how many of each type of coin she has and tells you that she has 105 coins and that the value of the coins is $17.25. Show her that you know algebra and determine how many nickels and how many quarters she has.

22. **School Supplies:** Eight pencils and two pens cost $2.22. Three pens and four pencils cost $2.69. What is the price of each pen and each pencil?

Cumulative Review: Chapters 1 - 10

Solve the following problems.

1. Write the number two hundred thousand, sixteen and four hundredths in decimal notation.

2. Round 17.986 to the nearest hundredth.

3. Find the decimal equivalent of $\dfrac{14}{35}$.

4. Write $\dfrac{9}{5}$ as a percent.

5. Write $1\dfrac{1}{2}\%$ in decimal form.

Find the value of each expression by performing the indicated operations.

6. $\dfrac{2}{15}+\dfrac{11}{15}+\dfrac{7}{15}$

7. $4-\dfrac{10}{11}$

8. $\left(\dfrac{2}{5}\right)\left(-\dfrac{5}{6}\right)\left(\dfrac{4}{7}\right)$

9. $2\dfrac{4}{15}+3\dfrac{1}{6}+4\dfrac{7}{10}$

10. $70\dfrac{1}{4}-23\dfrac{5}{6}$

11. $\left(-4\dfrac{5}{7}\right)\left(-2\dfrac{6}{11}\right)$

12. $6\div3\dfrac{1}{3}$

13. $(700)(-800)$

14. $40.3-67.2$

15. $71+\left|-0.35\right|+4.39$

16. $(0.27)(0.043)^2$

17. $27.404\div(-0.34)$

Use the rules for order of operations to evaluate each expression.

18. $\left(36\div3^2\cdot2\right)+12\div4-2^2$

19. $1.7^2-3\dfrac{1}{2}\div\dfrac{1}{4}$

Solve the following problems.

20. Evaluate $3x^2-5x-17$ for $x=-1.5$.

21. Find the prime factorization of 396.

22. Find (Use $\pi=3.14$.)
    a. the circumference, and
    b. the area of a circle with a radius of 10 feet.

23. Find the length of the hypotenuse of a right triangle if one of its legs is 5 inches long and the other leg is 10 inches long.

Solve each of the equations.

24. $5x-4=11$

25. $3x=7x-16$

26. $4(3x-1)=2(2x-5)-3$

27. $\dfrac{4x-1}{3}+\dfrac{x-5}{2}=2$

Solve the inequalities and graph the solution sets. Write each solution using interval notation.

28. $x + 4 - 3x \geq 2x + 5$

29. $\dfrac{x}{5} - 3.4 > \dfrac{x}{2} + 1.6$

Identify the slope and *y*-intercept for the following lines. If a line is horizontal or vertical, state it in your answer.

30. $4x = 8$

31. $3x - 2 = 5y$

32. $4x - y = 1$

33. $3 - 2y = 0$

Solve the following equations as indicated.

34. Find an equation in standard form for the line that passes through the point $(3, -4)$ and is parallel to the line $2x - y = 5$. Graph both lines.

35. Find an equation in slope-intercept form for the line passing through the points $(1, -4)$ and $(4, 2)$. Graph the line.

36. Three graphs are shown below. Use the vertical line test to determine whether or not each graph represents a function. State each graph's domain and range.

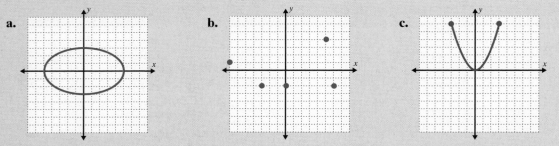

Solve the systems of linear equations by the stated method.

37. Determine which of the given points, if any, lie on both of the lines in the system of equations by substituting each point into both equations.

$$\begin{cases} x - 2y = 6 \\ \quad y = \dfrac{1}{2}x - 3 \end{cases}$$

a. $(0, 6)$ **b.** $(6, 0)$ **c.** $(2, -2)$ **d.** $\left(3, -\dfrac{3}{2}\right)$

38. Solve the system of linear equations by graphing: $\begin{cases} 2x + y = 6 \\ 3x - 2y = -5 \end{cases}$

39. Solve the system of linear equations by the addition method: $\begin{cases} x + 3y = 10 \\ 5x - y = 2 \end{cases}$

40. Solve the system of linear equations by the substitution method: $\begin{cases} x + y = -4 \\ 2x + 7y = 2 \end{cases}$

Solve the systems of linear equations using any method.

41. $\begin{cases} 5x + 2y = 3 \\ \quad\quad y = 4 \end{cases}$ **42.** $\begin{cases} 2x + 4y = 9 \\ 3x + 6y = 8 \end{cases}$ **43.** $\begin{cases} x + 5y = 10 \\ y - 2 = -\dfrac{1}{5}x \end{cases}$ **44.** $\begin{cases} 2x + y = 7 \\ 2x - y = 1 \end{cases}$

Solve the following word problems.

45. **Utilities:** Over the last 6 months, Tom has paid the following amounts for electricity:

48.60	59.51	74.85
65.44	67.27	55.13

Using these values, find the average amount Tom has paid per month for electricity. (Round your answer to the nearest hundredth.)

46. **Picking Clothes:** Tristan has acquired 10 shirts from playing 10 seasons of ultimate frisbee in a recreational league. Of these shirts, 4 are blue, 3 are red, 2 are green, and 1 is black. If he picks one of these shirts at random out of his drawer, find the probability that the shirt he selects is blue.

47. The sum of two numbers is 14. Twice the larger number added to three times the smaller is equal to 31. Find both numbers.

48. **Travel** Over spring break it took Leo 8 hours to drive to Florida. On the way back from Florida his average speed was 4 mph faster and it only took him 7.5 hours. What was Leo's average speed going to and coming from Florida?

49. **Investing:** Alicia has $7000 invested, some at 7% and the remainder at 8%. After one year, the interest from the 7% investment exceeds the interest from the 8% investment by $70. How much is invested at each rate?

50. **Baking:** Karl makes two kinds of cookies. Choc-O-Nut requires 4 oz of peanuts for each 10 oz of chocolate chips. Chocolate Krunch requires 12 oz of peanuts per 8 oz of chocolate chips. How many batches of each can he make if he has 36 oz of peanuts and 46 oz of chocolate chips?

Exponents and Polynomials

Mathematics at Work!

Consider the formula $F = ma$ (force equals mass times acceleration). Now, suppose a bowling ball has a mass of 5 kg and the ball's acceleration is represented by the polynomial $a = 4t^2 - 12t + 7$ m/s^2 where t represents time (in seconds). Using the formula $F = ma$, find a polynomial that represents the force of the ball at time t. (Note: The force is measured in Newtons, N.) (See Section 11.5.)

11.1 Exponents

Objective A **The Product Rule**

Previously, an **exponent** was defined as a number that tells how many times a number (called the **base**) is used in multiplication. This definition is limited because it is valid only if the exponents are positive integers. In this section, we will develop four properties of exponents that will help in simplifying algebraic expressions and expand your understanding of exponents to include variable bases and negative exponents. (In later sections you will study fractional exponents.)

From earlier, we know that

Exponent

$$6^2 = 6 \cdot 6 = 36$$

Base

and

$$6^3 = 6 \cdot 6 \cdot 6 = 216.$$

Also, the base may be a variable so that

$$x^3 = x \cdot x \cdot x \quad \text{and} \quad x^5 = x \cdot x \cdot x \cdot x \cdot x.$$

Now to find the products of expressions such as $6^2 \cdot 6^3$ or $x^3 \cdot x^5$ and to simplify these products, we can write down all the factors as follows.

$$6^2 \cdot 6^3 = (6 \cdot 6) \cdot (6 \cdot 6 \cdot 6) = 6^5$$
$$\text{and}$$
$$x^3 \cdot x^5 = (x \cdot x \cdot x) \cdot (x \cdot x \cdot x \cdot x \cdot x) = x^8.$$

With these examples in mind, what do you think would be a simplified form for the product $3^4 \cdot 3^3$? You were right if you thought 3^7. That is, $3^4 \cdot 3^3 = 3^7$. Notice that in each case, **the base stays the same**.

The preceding discussion, along with the basic concept of whole-number exponents, leads to the following **product rule for exponents**.

The Product Rule for Exponents

If a is a nonzero real number and m and n are integers, then

$$a^m \cdot a^n = a^{m+n}.$$

In words, to multiply two powers with the same base, keep the base and add the exponents.

notes

■ Reminder about the **exponent 1**: if a variable or constant has no exponent written, the exponent is understood to be 1.

■

For example, $y = y^1$ and $7 = 7^1$.

■

In general, for any real number a, $a = a^1$.

■

Example 1

Product Rule for Exponents

Use the product rule for exponents to simplify the following expressions.

a. $x^2 \cdot x^4$

Solution

$x^2 \cdot x^4 = x^{2+4} = x^6$

b. $y \cdot y^6$

Solution

$y \cdot y^6 = y^1 \cdot y^6 = y^{1+6} = y^7$

c. $4^2 \cdot 4$

Solution

$4^2 \cdot 4 = 4^{2+1} = 4^3 = 64$

Note that the base stays 4.
That is, the bases are not multiplied.

d. $2^3 \cdot 2^2$

Solution

$2^3 \cdot 2^2 = 2^{3+2} = 2^5 = 32$

Note that the base stays 2.
That is, the bases are not multiplied.

e. $(-2)^4 (-2)^3$

Solution

$(-2)^4 (-2)^3 = (-2)^{4+3} = (-2)^7 = -128$

1. a. $x^3 \cdot x^2$

   b. $y^5 \cdot y^4$

   c. $5^2 \cdot 5$

   d. $3^2 \cdot 3^3$

   e. $(-3)^5 (-3)^2$

2. a. $4y^3 \cdot 5y^6$

   b. $\left(-2x^4\right)\left(5x^3\right)$

   c. $\left(-5ab^3\right)\left(7ab^4\right)$

To multiply terms that have numerical coefficients and variables with exponents, **the coefficients are multiplied** as usual and **the exponents are added by using the product rule.** Generally variables and constants may need to be rearranged using the commutative and associative properties of multiplication. Example 2 illustrates these concepts.

Example 2

Product Rule for Exponents

Use the product rule for exponents when simplifying the following expressions.

a. $2y^2 \cdot 3y^9$

Solution

$$2y^2 \cdot 3y^9 = 2 \cdot 3 \cdot y^2 \cdot y^9$$
$$= 6y^{2+9}$$
$$= 6y^{11}$$

Coefficients 2 and 3 are multiplied and exponents 2 and 9 are added.

b. $\left(-3x^3\right)\left(-4x^3\right)$

Solution

$$\left(-3x^3\right)\left(-4x^3\right) = (-3)(-4) \cdot x^3 \cdot x^3$$
$$= 12x^{3+3}$$
$$= 12x^6$$

Coefficients −3 and −4 are multiplied and exponents 3 and 3 are added.

c. $\left(-6ab^2\right)\left(8ab^3\right)$

Solution

$$\left(-6ab^2\right)\left(8ab^3\right) = (-6) \cdot 8 \cdot a^1 \cdot a^1 \cdot b^2 \cdot b^3$$
$$= -48 \cdot a^{1+1} \cdot b^{2+3}$$
$$= -48a^2 b^5$$

Coefficients −6 and 8 are multiplied and exponents on each variable are added.

Now work margin exercises **1 and 2.**

Objective B **The Exponent 0**

The product rule is stated for m and n as **integer** exponents. This means that the rule is also valid for 0 and for negative exponents. As an aid for understanding 0 as an exponent, consider the following patterns of exponents for powers of 2, 3, and 10.

Powers of 2	Powers of 3	Powers of 10
$2^5 = 32$	$3^5 = 243$	$10^5 = 100,000$
$2^4 = 16$	$3^4 = 81$	$10^4 = 10,000$
$2^3 = 8$	$3^3 = 27$	$10^3 = 1000$
$2^2 = 4$	$3^2 = 9$	$10^2 = 100$
$2^1 = 2$	$3^1 = 3$	$10^1 = 10$
$2^0 = ?$	$3^0 = ?$	$10^0 = ?$

Do you notice that the patterns indicate that the exponent 0 gives the same value for the last number in each column? That is, $2^0 = 1$, $3^0 = 1$, and $10^0 = 1$.

Another approach to understanding 0 as an exponent is to consider the product rule. Remember that the product rule is stated for **integer exponents**. Applying this rule with the exponent 0 and the fact that 1 is the multiplicative identity gives results such as the following equations.

$$5^0 \cdot 5^2 = 5^{0+2} = 5^2 \qquad \text{This implies that } 5^0 = 1.$$
$$4^0 \cdot 4^3 = 4^{0+3} = 4^3 \qquad \text{This implies that } 4^0 = 1.$$

This discussion leads directly to the **rule for 0 as an exponent**.

The Exponent 0

If a is a nonzero real number, then

$$a^0 = 1.$$

The expression 0^0 is undefined.

notes

- Throughout this text, unless specifically stated otherwise, we will assume that the bases of exponents are nonzero.

Example 3

The Exponent 0

Simplify the following expressions using the rule for 0 as an exponent.

a. 10^0

Solution

$10^0 = 1$

3. Simplify the following expressions using the rule for 0 as an exponent.

a. 1587^0

b. $x^0 \cdot x^2$

c. $(-17)^0$

b. $x^0 \cdot x^3$

Solution

$x^0 \cdot x^3 = x^{0+3} = x^3$ or $x^0 \cdot x^3 = 1 \cdot x^3 = x^3$

c. $(-6)^0$

Solution

$(-6)^0 = 1$

Now work margin exercise 3.

Objective C **The Quotient Rule**

Now consider a fraction in which the numerator and denominator are powers with the same base, such as $\dfrac{5^4}{5^2}$ or $\dfrac{x^5}{x^2}$. We can write:

$$\frac{5^4}{5^2} = \frac{\cancel{5} \cdot \cancel{5} \cdot 5 \cdot 5}{\cancel{5} \cdot \cancel{5} \cdot 1} = \frac{5^2}{1} = 25 \quad \text{or} \quad \frac{5^4}{5^2} = 5^{4-2} = 5^2 = 25$$

and

$$\frac{x^5}{x^2} = \frac{\cancel{x} \cdot \cancel{x} \cdot x \cdot x \cdot x}{\cancel{x} \cdot \cancel{x} \cdot 1} = \frac{x^3}{1} = x^3 \quad \text{or} \quad \frac{x^5}{x^2} = x^{5-2} = x^3.$$

In fractions, as just illustrated, the exponents can be subtracted. Again, the base remains the same. We now have the following **quotient rule for exponents**.

Quotient Rule for Exponents

If a is a nonzero real number and m and n are integers, then

$$\frac{a^m}{a^n} = a^{m-n}.$$

In words, to divide two powers with the same base, keep the base and subtract the exponents. (Subtract the denominator exponent from the numerator exponent.)

Example 4

Quotient Rule for Exponents

Use the quotient rule for exponents to simplify the following expressions.

a. $\dfrac{x^6}{x}$

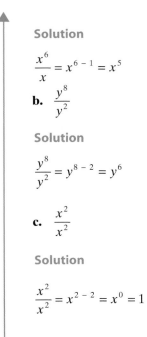

Solution

$$\frac{x^6}{x} = x^{6-1} = x^5$$

b. $\dfrac{y^8}{y^2}$

Solution

$$\frac{y^8}{y^2} = y^{8-2} = y^6$$

c. $\dfrac{x^2}{x^2}$

Solution

$$\frac{x^2}{x^2} = x^{2-2} = x^0 = 1$$

Note how this example shows another way to justify the idea that $a^0 = 1$. Since the numerator and denominator are the same and not 0, it makes sense that the fraction is equal to 1.

Now work margin exercise 4.

In division, with terms that have numerical coefficients, the **coefficients are divided** as usual and any **exponents are subtracted** by using the quotient rule. These ideas are illustrated in Example 5.

Example 5

Dividing Terms with Coefficients

Use the quotient rule for exponents when simplifying the following expressions.

a. $\dfrac{15x^{15}}{3x^3}$

Solution

$$\frac{15x^{15}}{3x^3} = \frac{15}{3} \cdot \frac{x^{15}}{x^3}$$

Coefficients 15 and 3 are divided and exponents 15 and 3 are subtracted.

$$= 5 \cdot x^{15-3}$$

$$= 5x^{12}$$

b. $\dfrac{20x^{10}y^6}{2x^2y^3}$

Solution

$$\frac{20x^{10}y^6}{2x^2y^3} = \frac{20}{2} \cdot \frac{x^{10}}{x^2} \cdot \frac{y^6}{y^3}$$

Coefficients 20 and 2 are divided and exponents on each variable are subtracted.

$$= 10 \cdot x^{10-2} \cdot y^{6-3}$$

$$= 10x^8y^3$$

Now work margin exercise 5.

4. Use the quotient rule for exponents to simplify the following expressions.

a. $\dfrac{x^5}{x^3}$

b. $\dfrac{y^7}{y^4}$

c. $\dfrac{x^5}{x^5}$

5. Use the quotient rule for exponents when simplifying the following expressions.

a. $\dfrac{12x^{12}}{4x^4}$

b. $\dfrac{21x^8y^7}{3x^2y^3}$

Negative Exponents

The quotient rule for exponents leads directly to the development of an understanding of negative exponents. In Examples 4 and 5, for each base, the exponent in the numerator was larger than or equal to the exponent in the denominator. Therefore, when the exponents were subtracted using the quotient rule, the result was either a positive exponent or the exponent 0. But, what if the larger exponent is in the denominator and we still apply the quotient rule?

For example, applying the quotient rule to $\dfrac{4^3}{4^5}$ gives

$$\frac{4^3}{4^5} = 4^{3-5} = 4^{-2}$$

which results in a negative exponent.

But, simply reducing $\dfrac{4^3}{4^5}$ gives

$$\frac{4^3}{4^5} = \frac{\cancel{4} \cdot \cancel{4} \cdot \cancel{4} \cdot 1}{\cancel{4} \cdot \cancel{4} \cdot \cancel{4} \cdot 4 \cdot 4} = \frac{1}{4 \cdot 4} = \frac{1}{4^2}.$$

This means that $4^{-2} = \dfrac{1}{4^2}$.

Similar discussions will show that $2^{-1} = \dfrac{1}{2}$, $5^{-2} = \dfrac{1}{5^2}$, and $x^{-3} = \dfrac{1}{x^3}$.

The **rule for negative exponents** follows.

Rule for Negative Exponents

If a is a nonzero real number and n is an integer, then

$$a^{-n} = \frac{1}{a^n}.$$

Example 6

Negative Exponents

Use the rule for negative exponents to simplify each expression so that it contains only positive exponents.

a. 5^{-1}

Solution

$5^{-1} = \dfrac{1}{5^1} = \dfrac{1}{5}$ using the rule for negative exponents

b. x^{-3}

Solution

$$x^{-3} = \frac{1}{x^3} \qquad \text{using the rule for negative exponents}$$

c. $x^{-9} \cdot x^7$

Solution

Here we use the product rule first and then the rule for negative exponents.

$$x^{-9} \cdot x^7 = x^{-9+7} = x^{-2} = \frac{1}{x^2}$$

Now work margin exercise 6.

Each of the expressions in Example 7 is simplified by using the appropriate rule for exponents. Study each example carefully. In each case, **the expression is considered simplified if each base appears only once and each base has only positive exponents**.

---

notes

- There is nothing wrong with negative exponents. In fact, negative exponents are preferred in some courses in mathematics and science.
- However, so that all answers are the same, in this course we will consider expressions to be simplified if:
    1. all exponents are positive and
    2. each base appears only once.

Example 7

Combining Rules for Exponents

Simplify each expression so that it contains only positive exponents.

a. $2^{-5} \cdot 2^8$

Solution

$$2^{-5} \cdot 2^8 = 2^{-5+8} \qquad \text{using the product rule with positive and negative exponents}$$
$$= 2^3$$
$$= 8$$

6. Use the rule for negative exponents to simplify each expression so that it contains only positive exponents.

a. 7^{-1}

b. x^{-7}

c. $x^{-11} \cdot x^6$

7. Simplify each expression so that it contains only positive exponents.

a. $3^{-7} \cdot 3^{10}$

b. $\dfrac{x^5}{x^{-1}}$

c. $\dfrac{8^{-4}}{8^{-2}}$

d. $\dfrac{x^7 y^3}{x^4 y^6}$

e. $\dfrac{8x^9 \cdot 3x^4}{6x^{16}}$

b. $\dfrac{x^6}{x^{-1}}$

Solution

$$\dfrac{x^6}{x^{-1}} = x^{6-(-1)}$$ using the quotient rule with positive and negative exponents

$$= x^{6+1}$$

$$= x^7$$

c. $\dfrac{10^{-5}}{10^{-2}}$

Solution

$$\dfrac{10^{-5}}{10^{-2}} = 10^{-5-(-2)}$$ using the quotient rule with negative exponents

$$= 10^{-5+2}$$

$$= 10^{-3}$$

$$= \dfrac{1}{10^3} \text{ or } \dfrac{1}{1000}$$ using the rule for negative exponents

d. $\dfrac{x^6 y^3}{x^2 y^5}$

Solution

$$\dfrac{x^6 y^3}{x^2 y^5} = x^{6-2} y^{3-5}$$ using the quotient rule with two variables

$$= x^4 y^{-2}$$

$$= \dfrac{x^4}{y^2}$$ using the rule for negative exponents

e. $\dfrac{15x^{10} \cdot 2x^2}{3x^{15}}$

Solution

$$\dfrac{15x^{10} \cdot 2x^2}{3x^{15}} = \dfrac{(15 \cdot 2)x^{10+2}}{3x^{15}}$$ using the product rule

$$= \dfrac{30x^{12}}{3x^{15}}$$

$$= 10x^{12-15}$$ using the quotient rule

$$= 10x^{-3}$$

$$= \dfrac{10}{x^3}$$ using the rule for negative exponents

Now work margin exercise 7.

Hawkes Learning Systems. All rights reserved.

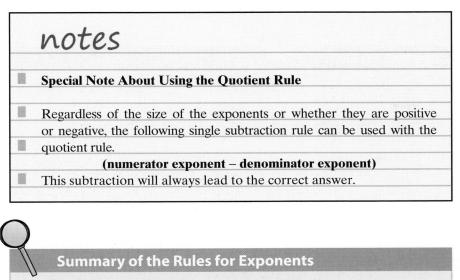

notes

Special Note About Using the Quotient Rule

Regardless of the size of the exponents or whether they are positive or negative, the following single subtraction rule can be used with the quotient rule.

(numerator exponent – denominator exponent)

This subtraction will always lead to the correct answer.

Summary of the Rules for Exponents

For any nonzero real number a and integers m and n:

1. The exponent 1: $a = a^1$
2. The exponent 0: $a^0 = 1$
3. The product rule: $a^m \cdot a^n = a^{m+n}$
4. The quotient rule: $\dfrac{a^m}{a^n} = a^{m-n}$
5. Negative exponents: $a^{-n} = \dfrac{1}{a^n}$

Objective E **Using a TI-84 Plus Graphing Calculator to Evaluate Expressions with Exponents**

On a TI-84 Plus graphing calculator, the caret key is used to indicate an exponent. Example 8 illustrates the use of a graphing calculator in evaluating expressions with exponents.

Example 8

Evaluating Expressions with Exponents

Use a graphing calculator to evaluate each expression.

a. 2^{-3} **b.** 23.18^0 **c.** $(-3.2)^3 (1.5)^2$

Solutions

The following solutions show how the caret key is used to indicate exponents. Be careful to use the negative sign key $(-)$ (and not the minus sign key) for negative numbers and negative exponents.

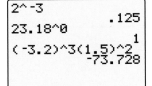

```
2^-3
              .125
23.18^0
              1
(-3.2)^3(1.5)^2
          -73.728
```

Now work margin exercise 8.

8. Use a graphing calculator to evaluate each expression.

a. 5^{-2}

b. 17.06^0

c. $(-4.7)^2 (1.2)^3$

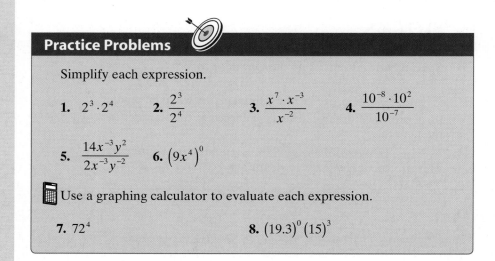

Practice Problems

Simplify each expression.

1. $2^3 \cdot 2^4$ **2.** $\dfrac{2^3}{2^4}$ **3.** $\dfrac{x^7 \cdot x^{-3}}{x^{-2}}$ **4.** $\dfrac{10^{-8} \cdot 10^2}{10^{-7}}$

5. $\dfrac{14x^{-3}y^2}{2x^{-3}y^{-2}}$ **6.** $\left(9x^4\right)^0$

Use a graphing calculator to evaluate each expression.

7. 72^4 **8.** $(19.3)^0 (15)^3$

Practice Problem Answers

1. $2^7 = 128$ **2.** $\dfrac{1}{2}$ **3.** x^6 **4.** 10

5. $7y^4$ **6.** 1 **7.** $26{,}873{,}856$ **8.** 3375

Exercises 11.1

Simplify each expression. The final form of the expressions with variables should contain only positive exponents. Assume that all variables represent nonzero numbers. See Examples 1 though 7.

1. $3^2 \cdot 3$

2. $7^2 \cdot 7^3$

3. $8^3 \cdot 8^0$

4. $5^0 \cdot 5^2$

5. 3^{-1}

6. 4^{-2}

7. 5^{-2}

8. 6^{-3}

9. $(-2)^4 (-2)^0$

10. $(-4)^3 (-4)^0$

11. $3(2^3)$

12. $6(3^2)$

13. $-4(5^3)$

14. $-2(3^3)$

15. $3(2^{-3})$

16. $4(3^{-2})$

17. $-3(5^{-2})$

18. $-5(2^{-2})$

19. $x^2 \cdot x^3$

20. $x^3 \cdot x$

21. $y^2 \cdot y^0$

22. $y^3 \cdot y^8$

23. x^{-3}

24. y^{-2}

25. $2x^{-1}$

26. $5y^{-4}$

27. $-8y^{-2}$

28. $-10x^{-3}$

29. $5x^6 y^{-4}$

30. $x^0 y^{-2}$

31. $3x^0 + y^0$

32. $5y^0 - 3x^0$

33. $\dfrac{7^3}{7}$

34. $\dfrac{9^5}{9^2}$

35. $\dfrac{10^3}{10^4}$

36. $\dfrac{10}{10^5}$

37. $\dfrac{2^3}{2^6}$

38. $\dfrac{5^7}{5^4}$

39. $\dfrac{x^4}{x^2}$

40. $\dfrac{x^6}{x^3}$

41. $\dfrac{x^3}{x}$

42. $\dfrac{y^7}{y^2}$

43. $\dfrac{x^7}{x^3}$

44. $\dfrac{x^8}{x^3}$

45. $\dfrac{x^{-2}}{x^2}$ **46.** $\dfrac{x^{-3}}{x}$ **47.** $\dfrac{x^4}{x^{-2}}$ **48.** $\dfrac{x^5}{x^{-1}}$

49. $\dfrac{x^{-3}}{x^{-5}}$ **50.** $\dfrac{x^{-4}}{x^{-1}}$ **51.** $\dfrac{y^{-2}}{y^{-4}}$ **52.** $\dfrac{y^3}{y^{-3}}$

53. $3x^3 \cdot x^0$ **54.** $3y \cdot y^4$ **55.** $x^3 \cdot x^2 \cdot x^{-1}$ **56.** $x^{-3} \cdot x^0 \cdot x^2$

57. $\left(4x^3\right)\left(9x^0\right)$ **58.** $\left(5x^2\right)\left(3x^4\right)$ **59.** $\left(-2x^2\right)\left(7x^3\right)$ **60.** $\left(3y^3\right)\left(-6y^2\right)$

61. $\left(-4x^5\right)\left(3x\right)$ **62.** $\left(6y^4\right)\left(5y^5\right)$ **63.** $\dfrac{8y^3}{2y^2}$ **64.** $\dfrac{12x^4}{3x}$

65. $\dfrac{9y^5}{3y^3}$ **66.** $\dfrac{-10x^5}{2x}$ **67.** $\dfrac{-8y^4}{4y^2}$ **68.** $\dfrac{12x^6}{-3x^3}$

69. $\dfrac{x^{-1}x^2}{x^3}$ **70.** $\dfrac{x \cdot x^3}{x^{-3}}$ **71.** $\dfrac{10^4 \cdot 10^{-3}}{10^{-2}}$ **72.** $\dfrac{10 \cdot 10^{-1}}{10^2}$

73. $\left(9x^2\right)^0$ **74.** $\left(-2x^{-3}y^5\right)^0$ **75.** $\left(9x^2y^3\right)\left(-2x^3y^4\right)$ **76.** $\left(-3xy\right)\left(-5x^2y^{-3}\right)$

77. $\dfrac{-8x^2y^4}{4x^3y^2}$ **78.** $\dfrac{-8x^{-2}y^4}{4x^2y^{-2}}$ **79.** $\left(3a^2b^4\right)\left(4ab^5c\right)$ **80.** $\left(-6a^3b^4\right)\left(4a^{-2}b^8\right)$

81. $\dfrac{36a^5b^0c}{-9a^{-5}b^{-3}}$ **82.** $\dfrac{7x^2y^{-2}}{28x^0yz^{-2}}$ **83.** $\dfrac{25y^6 \cdot 3y^{-2}}{15xy^4}$ **84.** $\dfrac{12a^{-2} \cdot 18a^4}{36a^2b^{-5}}$

Use a graphing calculator to evaluate each expression. Round quotients to the nearest ten-thousandth, if necessary. See Example 8.

85. $\left(2.16\right)^0$ **86.** $\left(-5.06\right)^2$ **87.** $\left(1.6\right)^{-2}$

88. $\left(2.1\right)^{-3}$ **89.** $\left(6.4\right)^4\left(2.3\right)^2$ **90.** $\left(-14.8\right)^2\left(21.3\right)^2$

11.2 Exponents and Scientific Notation

The summary of the rules for exponents given in Section 11.1 is repeated here for easy reference.

A Use the power rule for exponents to simplify expressions.

B Use the rule for a power of a product to simplify expressions.

C Use the rule for a power of a quotient to simplify expressions.

D Use combinations of rules for exponents to simplify expressions.

E Write decimal numbers in scientific notation.

F Operate with decimal numbers by using scientific notation.

Summary of the Rules for Exponents

For any nonzero real number a and integers m and n:
1. The exponent 1: $a = a^1$
2. The exponent 0: $a^0 = 1$
3. The product rule: $a^m \cdot a^n = a^{m+n}$
4. The quotient rule: $\dfrac{a^m}{a^n} = a^{m-n}$
5. Negative exponents: $a^{-n} = \dfrac{1}{a^n}$

Objective A **Power Rule**

Now consider what happens when a power is raised to a power. For example, to simplify the expressions $\left(x^2\right)^3$ and $\left(2^5\right)^2$, we can write

$$\left(x^2\right)^3 = x^2 \cdot x^2 \cdot x^2 = x^{2+2+2} = x^6 \text{ and } \left(2^5\right)^2 = 2^5 \cdot 2^5 = 2^{5+5} = 2^{10}.$$

However, this technique can be quite time-consuming when the exponent is large such as in $\left(3y^3\right)^{17}$. The **power rule for exponents** gives a convenient way to handle powers raised to powers.

Power Rule for Exponents

If a is a nonzero real number and m and n are integers, then

$$\left(a^m\right)^n = a^{mn}.$$

In other words, the value of a power raised to a power can be found by multiplying the exponents and keeping the base.

1. Simplify each expression by using the power rule for exponents.

a. $\left(x^3\right)^5$

b. $\left(x^4\right)^{-3}$

c. $\left(y^{-5}\right)^3$

d. $\left(3^2\right)^{-3}$

Example 1

Power Rule for Exponents

Simplify each expression by using the power rule for exponents.

a. $\left(x^2\right)^4$

Solution

$$\left(x^2\right)^4 = x^{2(4)} = x^8$$

b. $\left(x^5\right)^{-2}$

Solution

$$\left(x^5\right)^{-2} = x^{5(-2)} = x^{-10} = \frac{1}{x^{10}}$$

or $\left(x^5\right)^{-2} = \dfrac{1}{\left(x^5\right)^2} = \dfrac{1}{x^{5(2)}} = \dfrac{1}{x^{10}}$

c. $\left(y^{-7}\right)^2$

Solution

$$\left(y^{-7}\right)^2 = y^{(-7)(2)} = y^{-14} = \frac{1}{y^{14}}$$

d. $\left(2^3\right)^{-2}$

Solution

$$\left(2^3\right)^{-2} = 2^{3(-2)} = 2^{-6} = \frac{1}{2^6}$$ Evaluating gives $\dfrac{1}{2^6} = \dfrac{1}{64}$.

Another approach, because we have a numerical base, would be

$$\left(2^3\right)^{-2} = \left(8\right)^{-2} = \frac{1}{8^2}.$$ Evaluating gives $\dfrac{1}{8^2} = \dfrac{1}{64}$.

We see that while the base and exponent may be different, the value is the same.

Now work margin exercise 1.

Objective B **Rule for Power of a Product**

If the base of an exponent is a product, we will see that each factor in the product can be raised to the power indicated by the exponent. For example, $(10x)^3$ indicates that the product of 10 and x is to be raised to the 3rd power and $\left(-2x^2y\right)^5$ indicates that the product of -2, x^2, and y is to be raised to the 5th power. We can simplify these expressions as follows.

$$(10x)^3 = 10x \cdot 10x \cdot 10x$$
$$= 10 \cdot 10 \cdot 10 \cdot x \cdot x \cdot x$$
$$= 10^3 \cdot x^3$$
$$= 1000x^3$$

and $(-2x^2y)^5 = (-2x^2y)(-2x^2y)(-2x^2y)(-2x^2y)(-2x^2y)$
$$= (-2)(-2)(-2)(-2)(-2) \cdot x^2 \cdot x^2 \cdot x^2 \cdot x^2 \cdot x^2 \cdot y \cdot y \cdot y \cdot y \cdot y$$
$$= (-2)^5 (x^2)^5 \cdot y^5$$
$$= -32x^{10}y^5$$

We can simplify expressions such as these in a much easier fashion by using the following **power of a product rule for exponents**.

Rule for Power of a Product

If a and b are nonzero real numbers and n is an integer then

$$(ab)^n = a^n b^n.$$

In words, a power of a product is found by raising each factor to that power.

Example 2

Rule for Power of a Product

Simplify each expression by using the rule for power of a product.

a. $(5x)^2$

Solution

$$(5x)^2 = 5^2 \cdot x^2 = 25x^2$$

b. $(xy)^3$

Solution

$$(xy)^3 = x^3 \cdot y^3 = x^3 y^3$$

c. $(-7ab)^2$

Solution

$$(-7ab)^2 = (-7)^2 a^2 b^2 = 49a^2 b^2$$

2. Simplify each expression by using the rule for power of a product.

a. $(4x)^2$

---

b. $(xy)^7$

c. $(-9ab)^2$

d. $(ab)^{-3}$

e. $\left(x^3 y^{-4}\right)^3$

d. $(ab)^{-5}$

Solution

$$(ab)^{-5} = a^{-5} \cdot b^{-5} = \frac{1}{a^5} \cdot \frac{1}{b^5} = \frac{1}{a^5 b^5}$$

or, using the rule of negative exponents first and then the rule for the power of a product,

$$(ab)^{-5} = \frac{1}{(ab)^5} = \frac{1}{a^5 b^5}$$

e. $\left(x^2 y^{-3}\right)^4$

Solution

$$\left(x^2 y^{-3}\right)^4 = \left(x^2\right)^4 \left(y^{-3}\right)^4 = x^8 \cdot y^{-12} = \frac{x^8}{y^{12}}$$

Now work margin exercise 2.

notes

■ **Special Note about Negative Numbers and Exponents**

■ In an expression such as $-x^2$, we know that -1 is understood to be the coefficient of x^2. That is,

■ $$-x^2 = -1 \cdot x^2.$$

The same is true for expressions with numbers such as -7^2. That is,

$$-7^2 = -1 \cdot 7^2 = -1 \cdot 49 = -49.$$

We see that the exponent refers to 7 and **not** to -7. For the exponent to refer to -7 as the base, -7 **must be in parentheses** as follows:

$$(-7)^2 = (-7)(-7) = +49.$$

■ As another example,

$$-2^0 = -1 \cdot 2^0 = -1 \cdot 1 = -1 \quad \text{and} \quad (-2)^0 = 1.$$

Objective C **Rule for Power of a Quotient**

If the base of an exponent is a quotient (in fraction form), we will see that both the numerator and denominator are raised to the power indicated by the exponent. For example, in the expression $\left(\dfrac{2}{x}\right)^3$ the quotient (or fraction) $\dfrac{2}{x}$ is raised to the 3rd power.

We can simplify this expression as follows:

$$\left(\frac{2}{x}\right)^3 = \frac{2}{x} \cdot \frac{2}{x} \cdot \frac{2}{x} = \frac{2 \cdot 2 \cdot 2}{x \cdot x \cdot x} = \frac{2^3}{x^3} = \frac{8}{x^3}.$$

Or, we can simplify the expression in a much easier manner by applying the following **rule for the power of a quotient**.

Rule for Power of a Quotient

If a and b are nonzero real numbers and n is an integer, then

$$\left(\frac{a}{b}\right)^n = \frac{a^n}{b^n}.$$

In words, a power of a quotient (in fraction form) is found by raising both the numerator and the denominator to that power.

Example 3

Rule for Power of a Quotient

Simplify each expression by using the rule for the power of a quotient.

a. $\left(\dfrac{y}{x}\right)^5$

Solution: $\left(\dfrac{y}{x}\right)^5 = \dfrac{y^5}{x^5}$

c. $\left(\dfrac{2}{a}\right)^4$

Solution: $\left(\dfrac{2}{a}\right)^4 = \dfrac{2^4}{a^4} = \dfrac{16}{a^4}$

b. $\left(\dfrac{3}{4}\right)^3$

Solution: $\left(\dfrac{3}{4}\right)^3 = \dfrac{3^3}{4^3} = \dfrac{27}{64}$

d. $\left(\dfrac{x}{7}\right)^2$

Solution: $\left(\dfrac{x}{7}\right)^2 = \dfrac{x^2}{7^2} = \dfrac{x^2}{49}$

Now work margin exercise 3.

3. Simplify each expression by using the rule for the power of a quotient.

a. $\left(\dfrac{x}{y}\right)^7$

b. $\left(\dfrac{5}{6}\right)^2$

c. $\left(\dfrac{3}{a}\right)^3$

d. $\left(\dfrac{x}{6}\right)^3$

Objective D **Using Combinations of Rules for Exponents**

There may be more than one way to apply the various rules for exponents. As illustrated in the following examples, if you apply the rules correctly (even in a different sequence), the answer will be the same in every case.

4. Simplify each expression by using the appropriate rules for exponents.

a. $\left(\dfrac{-3x}{y^3}\right)^3$

b. $\left(\dfrac{4a^3b^2}{a^4b}\right)^2$

Example 4

Using Combinations of Rules for Exponents

Simplify each expression by using the appropriate rules for exponents.

a. $\left(\dfrac{-2x}{y^2}\right)^3$

Solution

$$\left(\frac{-2x}{y^2}\right)^3 = \frac{(-2x)^3}{\left(y^2\right)^3} = \frac{(-2)^3 x^3}{y^6} = \frac{-8x^3}{y^6}$$

b. $\left(\dfrac{3a^2b}{a^3b^2}\right)^2$

Solution

Method 1: Simplify inside the parentheses first.

$$\left(\frac{3a^2b}{a^3b^2}\right)^2 = \left(3a^{2-3}b^{1-2}\right)^2 = \left(3a^{-1}b^{-1}\right)^2 = 3^2 a^{-2}b^{-2} = \frac{9}{a^2b^2}$$

Method 2: Apply the power of a quotient rule first.

$$\left(\frac{3a^2b}{a^3b^2}\right)^2 = \frac{\left(3a^2b\right)^2}{\left(a^3b^2\right)^2} = \frac{3^2 a^{2(2)}b^2}{a^{3(2)}b^{2(2)}} = \frac{9a^4b^2}{a^6b^4} = 9a^{4-6}b^{2-4} = 9a^{-2}b^{-2} = \frac{9}{a^2b^2}$$

Note that the answer is the same even though the rules were applied in a different order.

Now work margin exercise 4.

Another general approach with fractions involving negative exponents is to note that

$$\left(\frac{a}{b}\right)^{-n} = \frac{a^{-n}}{b^{-n}} = \frac{b^n}{a^n} = \left(\frac{b}{a}\right)^n.$$

notes

▪ **In effect, there are two basic shortcuts with negative exponents and fractions:**

  ▪ **1.** Taking the reciprocal of a fraction changes the sign of any exponent in the fraction.

  ▪ **2.** Moving any term from numerator to denominator, or vice versa, changes the sign of the corresponding exponent.

▪

Example 5

Two Approaches with Fractional Expressions and Negative Exponents

Simplify: $\left(\dfrac{x^3}{y^5}\right)^{-4}$

Solution

Method 1: Use the ideas of reciprocals first.

$$\left(\frac{x^3}{y^5}\right)^{-4} = \left(\frac{y^5}{x^3}\right)^{4} = \frac{y^{5(4)}}{x^{3(4)}} = \frac{y^{20}}{x^{12}}$$

Method 2: Apply the power of a quotient rule first.

$$\left(\frac{x^3}{y^5}\right)^{-4} = \frac{\left(x^3\right)^{-4}}{\left(y^5\right)^{-4}} = \frac{x^{3(-4)}}{y^{5(-4)}} = \frac{x^{-12}}{y^{-20}} = \frac{y^{20}}{x^{12}}$$

Now work margin exercise 5.

Example 6

A More Complex Example

This example involves the application of a variety of steps. Study it carefully and see if you can get the same result by following a different sequence of steps.

Simplify: $\left(\dfrac{2x^2y^3}{3xy^{-2}}\right)^{-2}\left(\dfrac{4x^2y^{-1}}{3x^{-5}y^3}\right)^{-1}$

Solution

$$\left(\frac{2x^2y^3}{3xy^{-2}}\right)^{-2}\left(\frac{4x^2y^{-1}}{3x^{-5}y^3}\right)^{-1} = \frac{2^{-2}\cdot x^{2(-2)}\cdot y^{3(-2)}}{3^{-2}\cdot x^{-2}y^{-2(-2)}}\cdot\frac{4^{-1}\cdot x^{2(-1)}\cdot y^{-1(-1)}}{3^{-1}\cdot x^{-5(-1)}\cdot y^{3(-1)}}$$

$$= \frac{3^2}{2^2}\cdot\frac{x^{-4}}{x^{-2}}\cdot\frac{y^{-6}}{y^4}\cdot\frac{3}{4}\cdot\frac{x^{-2}}{x^5}\cdot\frac{y^1}{y^{-3}}$$

$$= \frac{9\cdot 3\cdot x^{-4-2}y^{-6+1}}{4\cdot 4\cdot x^{-2+5}y^{4-3}} = \frac{27x^{-6}y^{-5}}{16x^3y^1}$$

$$= \frac{27x^{-6-3}y^{-5-1}}{16} = \frac{27x^{-9}y^{-6}}{16} = \frac{27}{16x^9y^6}$$

Now work margin exercise 6.

5. Simplify.

$\left(\dfrac{x^6}{y^3}\right)^{-5}$

6. Simplify.

$\left(\dfrac{3x^3y^2}{4xy^{-3}}\right)^{-2}\left(\dfrac{5x^3y^{-2}}{2x^{-4}y^2}\right)^{-2}$

A complete summary of the rules for exponents includes the following eight rules.

Summary of the Rules for Exponents

For any nonzero real numbers a and b and integers m and n:

1. The exponent 1: $a = a^1$

2. The exponent 0: $a^0 = 1$

3. The product rule: $a^m \cdot a^n = a^{m+n}$

4. The quotient rule: $\dfrac{a^m}{a^n} = a^{m-n}$

5. Negative exponents: $a^{-n} = \dfrac{1}{a^n}$

6. Power rule: $\left(a^m\right)^n = a^{mn}$

7. Power of a product: $(ab)^n = a^n b^n$

8. Power of a quotient: $\left(\dfrac{a}{b}\right)^n = \dfrac{a^n}{b^n}$

Objective E **Scientific Notation**

A basic application of integer exponents occurs in scientific disciplines, such as astronomy and biology, when very large and very small numbers are involved. For example, the distance from the earth to the sun is approximately 93,000,000 miles, and the approximate radius of a carbon atom is 0.0000000077 centimeters.

In **scientific notation** (an option in all scientific and graphing calculators), **decimal numbers are written as the product of a number greater than or equal to 1 and less than 10, and an integer power of 10**. In scientific notation there is just one digit to the left of the decimal point. For example,

$$250{,}000 = 2.5 \times 10^5 \quad \text{and} \quad 0.000000345 = 3.45 \times 10^{-7}.$$

The exponent tells how many places the decimal point is to be moved and in what direction. If the exponent is positive, the decimal point is moved to the right.

$$5.6 \times 10^4 = 5.6{,}000. \quad \text{4 places right}$$

A negative exponent indicates that the decimal point should move to the left.

$$4.9 \times 10^{-3} = 0.004.9 \quad \text{3 places left}$$

Scientific Notation

If N is a decimal number, then in **scientific notation**

$$N = a \times 10^n \text{ where } 1 \le a < 10 \text{ and } n \text{ is an integer.}$$

Example 7

Decimals in Scientific Notation

Write the following decimal numbers in scientific notation.

a. 8,720,000

Solution

$$8,720,000 = 8.72 \times 10^6 \qquad \text{8.72 is between 1 and 10.}$$

To check, move the decimal point 6 places to the right and get the original number:

$$8.72 \times 10^6 = 8.720000. = 8,720,000$$
$$\qquad\qquad\qquad\;\; 1\;2\;3\;4\;5\;6$$

b. 0.000000376

Solution

$$0.000000376 = 3.76 \times 10^{-7} \qquad \text{3.76 is between 1 and 10.}$$

To check, move the decimal point 7 places to the left and get the original number:

$$3.76 \times 10^{-7} = 0.0000003.76 = 0.000000376$$
$$\qquad\qquad\qquad\quad 7\;6\;5\;4\;3\;2\;1$$

Now work margin exercise 7.

Write the following decimal numbers in scientific notation.

7. a. 63,900,000

   b. 0.00000245

Objective F **Operations Using Scientific Notation**

Example 8

Scientific Notation and Properties of Exponents

Simplify the following expressions by first writing the decimal numbers in scientific notation and then using the properties of exponents.

a. $\dfrac{(0.085)(41,000)}{0.00017}$

Simplify the following expressions by first writing the decimal numbers in scientific notation and then using the properties of exponents.

8. a. $\dfrac{(2600)(0.0036)}{520,000}$

b. $\dfrac{(108,000)(3400)}{(0.006)(510)}$

c. One mole, a value often used in physics and chemistry, equals 6.02×10^{23} particles for all substances. How many particles would be in 8 moles of carbon?

Solution

$$\frac{(0.085)(41,000)}{0.00017} = \frac{\left(8.5 \times 10^{-2}\right)\left(4.1 \times 10^{4}\right)}{1.7 \times 10^{-4}}$$

$$= \frac{\overset{5}{\cancel{(8.5)}}\,(4.1)}{\cancel{1.7}} \times \frac{\left(10^{-2}\right)\left(10^{4}\right)}{10^{-4}}$$

$$= 20.5 \times \frac{10^{2}}{10^{-4}}$$

$$= 2.05 \times 10^{1} \times \frac{10^{2}}{10^{-4}}$$

$$= 2.05 \times 10^{1 + 2 - (-4)}$$

$$= 2.05 \times 10^{7}$$

b. $\dfrac{(11,100)(0.064)}{(8,000,000)(370)}$

Solution

$$\frac{(11,100)(0.064)}{(8,000,000)(370)} = \frac{\left(1.11 \times 10^{4}\right)\left(6.4 \times 10^{-2}\right)}{\left(8.0 \times 10^{6}\right)\left(3.7 \times 10^{2}\right)}$$

$$= \frac{\overset{0.3}{\cancel{(1.11)}}\,\overset{0.8}{\cancel{(6.4)}}}{\cancel{(8.0)}\,\cancel{(3.7)}} \times \frac{10^{2}}{10^{8}}$$

$$= 0.24 \times 10^{2 - 8}$$

$$= 2.4 \times 10^{-1} \times 10^{-6}$$

$$= 2.4 \times 10^{-1 - 6}$$

$$= 2.4 \times 10^{-7}$$

c. Light travels approximately 3×10^{8} meters per second. How many meters per minute does light travel?

Solution

Since there are 60 seconds in one minute, multiply by 60.

$$3 \times 10^{8} \times 60 = 180 \times 10^{8}$$

$$= 1.8 \times 10^{2} \times 10^{8}$$

$$= 1.8 \times 10^{10}$$

Thus light travels 1.8×10^{10} meters per minute.

*Now work margin exercise **8**.*

Example 9

Scientific Notation and Calculators

a. Use a graphing calculator to evaluate the expression
$\dfrac{0.0042 \cdot 3{,}000{,}000}{0.21}$. Leave the answer in scientific notation.

Note: You can press the **MODE** key and select SCI on the first line to have all decimal calculations in scientific notation.

Solution

With a TI-84 Plus calculator (set in scientific notation mode) the display should appear as shown below. Note that the E in the display indicates an exponent with base 10.

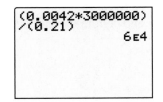

```
(0.0042*3000000)
/(0.21)
              6E4
```

b. A light-year is the distance light travels in one year. Use a graphing calculator to find the length of a light-year in scientific notation if light travels 186,000 miles per second.

Solution

60 seconds = 1 minute
60 minutes = 1 hour
24 hours = 1 day
365 days = 1 year

Multiplication gives the following display on your calculator.

```
186000*60*60*24*
365
      5.865696E12
```

Thus a light-year is $5.865696 \cdot 10^{12}$, or, 5,865,696,000,000 miles (5 trillion, 865 billion, 696 million miles).

9. a. Use a graphing calculator to evaluate the expression

$$\frac{(4{,}000{,}000)(0.00056)}{0.032}.$$

Leave the answer in scientific notation.

b. If light travels meters 3×10^{8} meters per second, how many meters does light travel in one day?

c. Use a calculator to evaluate the expression

$$\frac{(54{,}000)(5\times10^{3})}{4.5\times10^{-5}}.$$

Leave the answer in scientific notation.

c. Use a calculator to evaluate the expression $\dfrac{8600\left(3.0\times10^{5}\right)}{1.5\times10^{-6}}$.
Leave the answer in scientific notation.

Note: Remember, the caret key 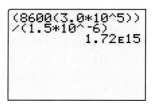 is used to indicate an exponent.

Solution

With a graphing calculator (set in scientific notation mode) the display should appear as shown below.

```
(8600(3.0*10^5))
/(1.5*10^-6)
          1.72E15
```

Note: The numerator and denominator must be set in parentheses.

Now work margin exercise 9.

Practice Problems

Simplify each expression.

1. $\dfrac{x^{-2}x^{5}}{x^{-7}}$ **2.** $\left(\dfrac{a^{2}b^{3}}{4}\right)^{0}$ **3.** $\dfrac{-3^{2}\cdot5}{2\cdot(-3)^{2}}$ **4.** $\left(\dfrac{5x}{3b}\right)^{-2}$

5. Write the number in scientific notation: 186,000 miles per second (speed of light in miles per second).

Practice Problem Answers

1. x^{10} **2.** 1   **3.** $-\dfrac{5}{2}$ **4.** $\dfrac{9b^{2}}{25x^{2}}$

5. 1.86×10^{5} miles per second

Exercises 11.2

Use the rules for exponents to simplify each of the expressions. Assume that all variables represent nonzero real numbers. See Examples 1 through 6.

1. -3^4

2. -5^2

3. -2^4

4. -20^2

5. $(-10)^6$

6. $(-4)^6$

7. $(6x^3)^2$

8. $(-3x^4)^2$

9. $4(-3x^2)^3$

10. $7(2y^{-2})^4$

11. $5(x^2 y^{-1})$

12. $-3(7xy^2)^0$

13. $-2(3x^5 y^{-2})^{-3}$

14. $-4(5x^{-3}y)^{-1}$

15. $\left(\dfrac{3x}{y}\right)^3$

16. $\left(\dfrac{-4x}{y^2}\right)^2$

17. $\left(\dfrac{6m^3}{n^5}\right)^0$

18. $\left(\dfrac{3x^2}{y^3}\right)^2$

19. $\left(\dfrac{-2x^2}{y^{-2}}\right)^2$

20. $\left(\dfrac{2x}{y^5}\right)^{-2}$

21. $\left(\dfrac{x}{y}\right)^{-2}$

22. $\left(\dfrac{2a}{b}\right)^{-1}$

23. $\left(\dfrac{3x}{y^{-2}}\right)^{-1}$

24. $\left(\dfrac{4a^2}{b^{-3}}\right)^{-3}$

25. $\left(\dfrac{-3}{xy^2}\right)^{-3}$

26. $\left(\dfrac{5xy^3}{y}\right)^2$

27. $\left(\dfrac{m^2 n^3}{mn}\right)^2$

28. $\left(\dfrac{2ab^3}{b^2}\right)^4$

29. $\left(\dfrac{-7^2 x^2 y}{y^3}\right)^{-1}$

30. $\left(\dfrac{2ab^4}{b^2}\right)^{-3}$

31. $\left(\dfrac{5x^3 y}{y^2}\right)^2$

32. $\left(\dfrac{2x^2 y}{y^3}\right)^{-4}$

33. $\left(\dfrac{x^3 y^{-1}}{y^2}\right)^2$

34. $\left(\dfrac{2a^2 b^{-1}}{b^2}\right)^3$

35. $\left(\dfrac{6y^5}{x^2 y^{-2}}\right)^2$

36. $\left(\dfrac{3x^4}{x^{-2} y^{-4}}\right)^3$

37. $\dfrac{\left(7x^{-2}y\right)^2}{\left(xy^{-1}\right)^2}$

38. $\dfrac{\left(-5x^3y^4\right)^2}{\left(3x^{-3}y\right)^2}$

39. $\dfrac{\left(3x^2y^{-1}\right)^{-2}}{\left(6x^{-1}y\right)^{-3}}$

40. $\dfrac{\left(2x^{-3}\right)^{-3}}{\left(5y^{-2}\right)^{-2}}$

41. $\dfrac{\left(4x^{-2}\right)\left(6x^5\right)}{(9y)\left(2y^{-1}\right)}$

42. $\dfrac{\left(5x^2\right)\left(3x^{-1}\right)^2}{\left(25y^3\right)\left(6y^{-2}\right)}$

43. $\left(\dfrac{3xy^3}{4x^2y^{-3}}\right)^{-1}\left(\dfrac{2x^3y^{-1}}{9x^{-3}y^{-1}}\right)^2$

44. $\left(\dfrac{5a^4b^{-2}}{6a^{-4}b^3}\right)^{-2}\left(\dfrac{5a^3b^4}{2^{-2}a^{-2}b^{-2}}\right)^3$

45. $\left(\dfrac{6x^{-4}yz^{-2}}{4^{-1}x^{-4}y^3z^{-2}}\right)^{-1}\left(\dfrac{2^{-2}xyz^{-3}}{12x^2y^2z^{-1}}\right)^{-2}$

46. $\left(\dfrac{3^{-5}a^5b^3c^{-1}}{3^{-2}abc}\right)^{-2}\left(\dfrac{7^{-1}a^{-4}bc^2}{7^{-2}a^{-3}bc^{-2}}\right)^{-2}$

Write the following numbers in scientific notation.

47. 86,000

48. 927,000

49. 0.0362

50. 0.0061

51. 18,300,000

52. 376,000,000

Write the following numbers in decimal form.

53. 4.2×10^{-2}

54. 8.35×10^{-3}

55. 7.56×10^6

56. 6.132×10^{-5}

57. 8.515×10^8

58. 9.374×10^7

First write each of the numbers in scientific notation. Then perform the indicated operations and leave your answer in scientific notation.

59. $300\cdot0.00015$

60. $0.000024\cdot40,000$

61. $0.0003\cdot0.0000025$

62. $0.00005\cdot0.00013$

63. $\dfrac{3900}{0.003}$

64. $\dfrac{4800}{12,000}$

65. $\dfrac{125}{50,000}$

66. $\dfrac{0.0046}{230}$

67. $\dfrac{0.02\cdot3900}{0.013}$

68. $\dfrac{0.0084\cdot0.003}{0.21\cdot60}$

69. $\dfrac{0.005\cdot650\cdot3.3}{0.0011\cdot2500}$

70. $\dfrac{5.4 \cdot 0.003 \cdot 50}{15 \cdot 0.0027 \cdot 200}$

71. $\dfrac{\left(1.4 \times 10^{-2}\right)(922)}{\left(3.5 \times 10^{3}\right)\left(2.0 \times 10^{6}\right)}$

72. $\dfrac{(4300)\left(3.0 \times 10^{2}\right)}{\left(1.5 \times 10^{-3}\right)\left(860 \times 10^{-2}\right)}$

Solve the following word problems.

73. **Atomic Mass:** The mass of a hydrogen atom is approximately 0.00000000000000000000000167 grams. Write this number in scientific notation.

74. **Circumference of the Earth:** The circumference of the earth is approximately 1,580,000,000 inches. Express this circumference in scientific notation.

75. **Anatomy:** There are approximately 6×10^{13} cells in an adult human body. Express this number in decimal form.

76. **World Population:** The world population is approximately 6.7×10^{9} people. Write this number in decimal form.

77. **Mass:** The mass of the earth is about 5,980,000,000,000,000,000,000,000,000 grams. Write this number in scientific notation.

78. **Time:** One year is approximately 31,500,000 seconds. Express this time in scientific notation.

79. **Speed of Light:** One light-year is approximately 9.46×10^{15} meters. The distance to a certain star is 4.3 light-years. How many meters is this?

80. **Speed of Light:** Light travels approximately 3×10^{10} centimeters per second. How many centimeters would this be per minute? Per hour? Express your answers in scientific notation.

81. **Atomic Weight:** An atom of gold weighs approximately 3.25×10^{-22} grams. What would be the weight of 2000 atoms of gold? Express your answer in scientific notation.

82. **Atomic Weight:** An ounce of gold contains 5×10^{22} atoms. All the gold ever taken out of the earth is estimated to be 3.0×10^{31} atoms. How many ounces of gold is this

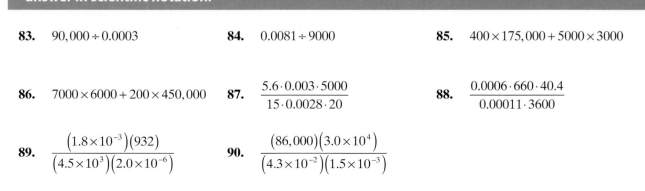

Use your calculator (set in scientific notation mode) to evaluate each expression. Leave the answer in scientific notation.

83. $90,000 \div 0.0003$

84. $0.0081 \div 9000$

85. $400 \times 175,000 + 5000 \times 3000$

86. $7000 \times 6000 + 200 \times 450,000$

87. $\dfrac{5.6 \cdot 0.003 \cdot 5000}{15 \cdot 0.0028 \cdot 20}$

88. $\dfrac{0.0006 \cdot 660 \cdot 40.4}{0.00011 \cdot 3600}$

89. $\dfrac{\left(1.8 \times 10^{-3}\right)(932)}{\left(4.5 \times 10^{3}\right)\left(2.0 \times 10^{-6}\right)}$

90. $\dfrac{(86,000)\left(3.0 \times 10^{4}\right)}{\left(4.3 \times 10^{-2}\right)\left(1.5 \times 10^{-3}\right)}$

11.3 Introduction to Polynomials

A Define a polynomial, and learn how to classify polynomials.

B Evaluate a polynomial for given values of the variable.

Objective A Definition of a Polynomial

A **term** is an expression that involves only multiplication and/or division with constants and/or variables. Remember that a number written next to a variable indicates multiplication, and the number is called the **numerical coefficient** (or **coefficient**) of the variable. For example,

$$3x, \quad -5y^2, \quad 17, \quad \text{and} \quad \frac{x}{y}$$

are all algebraic terms.

In the term

$3x$, 3 is the coefficient of x,

$-5y^2$, -5 is the coefficient of y^2, and

$\dfrac{x}{y}$, 1 is the coefficient.

A term that consists of only a number, such as 17, is also called a **constant** or a **constant term**.

Monomial

A **monomial in x** is a term of the form

$$kx^n$$

where k is a real number and n is a whole number.

n is called the **degree** of the term, and k is called the **coefficient**.

A monomial may have more than one variable, and the **degree of such a monomial is the sum of the degrees of its variables**. For example,

$$4x^2y^3 \text{ is a fifth-degree monomial in } x \text{ and } y.$$

However, in this chapter, only monomials of one variable (note that any variable may be used in place of x) will be discussed.

Since $x^0 = 1$, a nonzero constant can be multiplied by x^0 without changing its value. Thus we say that a **nonzero constant is a monomial of degree 0**. For example,

$$17 = 17x^0 \quad \text{and} \quad -6 = -6x^0$$

which means that the constants 17 and -6 are monomials of degree 0. However, for the special number 0, we can write

$$0 = 0x^2 = 0x^5 = 0x^{13}$$

and we say that **the constant 0 is a monomial of no degree.**

Monomials may have fractional or negative coefficients; however, **monomials may not have fractional or negative exponents**. These facts are part of the definition since in the expression kx^n, k (the coefficient) can be any real number, but n (the exponent) must be a whole number.

Expressions that **are not** monomials: $3\sqrt{x}$, $-15x^{\frac{2}{3}}$, $4a^{-2}$

Expressions that **are** monomials: 17, $3x$, $5y^2$, $\frac{2}{7}a^4$, πx^2, $\sqrt{2}x^3$

A **polynomial** is a monomial or the indicated sum or difference of monomials. Examples of polynomials are

$$3x, \quad y+5, \quad 4x^2 - 7x + 1, \quad \text{and} \quad a^{10} + 5a^3 - 2a^2 + 6.$$

Polynomial

A **polynomial** is a monomial or the indicated sum or difference of monomials.

The **degree of a polynomial** is the largest of the degrees of its terms.

The coefficient of the term of the largest degree is called the **leading coefficient**.

Special Terminology for Polynomials

Term	Definition	Examples
Monomial:	polynomial with one term	$-2x^3$ and $4a^5$
Binomial:	polynomial with two terms	$3x+5$ and a^2+3
Trinomial:	polynomial with three terms	x^2+6x-7 and a^3-8a^2+12a

Polynomials with four or more terms are simply referred to as **polynomials**.

Examples of polynomials are:

$-1.4x^5$ a fifth-degree monomial in x (leading coefficient is -1.4),

$4z^3 - 7.5z^2 - 5z$ a third-degree trinomial in z (leading coefficient is 4),

$\frac{3}{4}y^4 - 2y^3 + 4y - 6$ a fourth-degree polynomial in y (leading coefficient is $\frac{3}{4}$).

In each of these examples, the terms have been written so that the exponents on the variables decrease in order from left to right. We say that the terms are written in **descending order**. If the exponents on the terms increase in order from left to right, we say that the terms are written in **ascending order**. **As a general rule, for consistency and style in operating with polynomials, the polynomials in this chapter will be written in descending order.**

Example 1

Simplifying Polynomials

Simplify each of the following polynomials by combining **like terms**. (To review the definition of like terms, see Section 7.7.) Write the polynomial in descending order and state the degree and type of the polynomial.

a. $5x^3 + 7x^3$

Solution

$$5x^3 + 7x^3 = (5 + 7)x^3 = 12x^3 \qquad \text{third-degree monomial}$$

b. $5x^3 + 7x^3 - 2x$

Solution

$$5x^3 + 7x^3 - 2x = 12x^3 - 2x \qquad \text{third-degree binomial}$$

c. $\dfrac{1}{2}y + 3y - \dfrac{2}{3}y^2 - 7$

Solution

$$\dfrac{1}{2}y + 3y - \dfrac{2}{3}y^2 - 7 = -\dfrac{2}{3}y^2 + \dfrac{7}{2}y - 7 \quad \text{second-degree trinomial}$$

d. $x^2 + 8x - 15 - x^2$

Solution

$$x^2 + 8x - 15 - x^2 = 8x - 15 \qquad \text{first-degree binomial}$$

e. $-3y^4 + 2y^2 + y^{-1}$

Solution

This expression is not a polynomial since y has a negative exponent.

Completion Example 2

Simplifying Polynomials

Simplify the polynomial, then state the degree and type of the polynomial: $x^2 + 7x - 3x + 6 - 3x^2$.

Solution

$$x^2 + 7x - 3x + 6 - 3x^2$$
$$= \underline{\quad\quad} x^2 + \underline{\quad\quad} x + 6 \quad \underline{\quad\quad}\text{-degree trinomial}$$

Now work margin exercises 1 and 2.

Objective B **Evaluating Polynomials**

To evaluate a polynomial for a given value of the variable, substitute the value for the variable wherever it occurs in the polynomial and follow the rules for order of operations. A convenient notation for evaluating polynomials is the function notation, $p(x)$ (read "p of x") discussed in Section 9.5.

For example,

$$\text{if} \quad p(x) = x^2 - 4x + 13,$$

$$\text{then} \quad p(5) = (5)^2 - 4(5) + 13 = 25 - 20 + 13 = 18.$$

Example 3

Evaluating Polynomials

a. Given $p(x) = 4x^2 + 5x - 15$, find $p(3)$.

Solution

For $p(x) = 4x^2 + 5x - 15$,   Substitute 3 for *x*.
$$p(3) = 4(3)^2 + 5(3) - 15$$
$$= 4(9) + 15 - 15$$
$$= 36 + 15 - 15$$
$$= 36$$

Simplify each of the following polynomials by combining like terms. Also, write the polynomials in decending order and state the degree and type of the polynomial.

1. a. $4x^2 + 5x^2$

   b. $4x^2 + 5x^2 - 5x$

   c. $\dfrac{1}{3}y^2 + 4y^3 - 3y^2 + 8$

   d. $3x^3 + 4x^2 - 8 - 3x^3$

   e. $-4y^{-3} + 2y^{-1} + 3y^2$

2. $x^3 + 6x^2 - 5 + 12x^3 + 9 - 8x^2$

Completion Example Answers
   2. $-2x^2 + 4x + 6$ second-degree trinomial

b. Given $p(y) = 5y^3 + y^2 - 3y + 8$, find $p(-2)$.

Solution

For $p(y) = 5y^3 + y^2 - 3y + 8$,

$p(-2) = 5(-2)^3 + (-2)^2 - 3(-2) + 8$ Note the use of parentheses around –2.

$ = 5(-8) + 4 - 3(-2) + 8$

$ = -40 + 4 + 6 + 8$

$ = -22$

c. Rewrite the polynomial expression $f(x) = 6x + 13$ by substituting for x as indicated by the function notation $f(2a+1)$.

Solution

$f(2a+1) = 6(2a+1) + 13$

$ = 12a + 6 + 13$

$ = 12a + 19$

---

3. a. Given $p(x) = 3x^3 - 5x^2 - 13$, find $p(3)$.

b. Given $p(y) = 4y^3 + 2y^2 - 8y + 2$ find $p(-2)$.

c. Rewrite the polynomial expression $f(x) = 8x - 27$ by substituting for x as indicated by the function notation $f(3a-1)$.

4. Given the polynomial $R(y) = 8y^3 - 9y + 22$ find $R(2)$.

Completion Example 4

Evaluating Polynomials

Given the polynomial $p(x) = 5x^2 + 6x - 10$ find $p(3)$.

Solution

$$p(3) = 5 \cdot \underline{\quad}^2 + 6 \cdot \underline{\quad} - 10 = \underline{\quad} + \underline{\quad} - 10 = \underline{\quad}$$

***Now work margin exercises* 3 and 4.**

Practice Problems

Combine like terms and state the degree and type of the polynomial.

1. $8x^3 - 3x^2 - x^3 + 5 + 3x^2$ **2.** $5y^4 + y^4 - 3y^2 + 2y^2 + 4$

3. For the polynomial $p(x) = x^2 - 5x - 5$, find **a.** $p(3)$ and **b.** $p(-1)$.

Completion Example Answers

4. $5 \cdot 3^2 + 6 \cdot 3 - 10 = 45 + 18 - 10 = 53$

Practice Problem Answers

1. $7x^3 + 5$; third-degree binomial

2. $6y^4 - y^2 + 4$; fourth-degree trinomial

3. a. -11 **b.** 1

Exercises 11.3

Identify the expression as a monomial, binomial, trinomial, or not a polynomial. See Example 1.

1. $3x^4$

2. $5y^2 - 2y + 1$

3. $8x^3 - 7$

4. $-2x^{-2}$

5. $14a^7 - 2a - 6$

6. $17x^{\frac{2}{3}} + 5x^2$

7. $6a^3 + 5a^2 - a^{-3}$

8. $-3y^4 + 2y^2 - 9$

9. $\frac{1}{2}x^3 - \frac{2}{5}x$

10. $\frac{5}{8}x^5 + \frac{2}{3}x^4$

Simplify each polynomial. Write the polynomial in descending order and state the degree and type of the simplified polynomial. For each polynomial, also state the leading coefficient. See Examples 1 and 2.

11. $y + 3y$

12. $4x^2 - x + x^2$

13. $x^3 + 3x^2 - 2x$

14. $3x^2 - 8x + 8x$

15. $x^4 - 4x^2 + 2x^2 - x^4$

16. $2 - 6y + 5y - 2$

17. $-x^3 + 6x + x^3 - 6x$

18. $11x^2 - 3x + 2 - 7x^2$

19. $6a^5 + 2a^2 - 7a^3 - 3a^2$

20. $2x^2 - 3x^2 + 2 - 4x^2 - 2 + 5x^2$

21. $4y - 8y^2 + 2y^3 + 8y^2$

22. $2x + 9 - x + 1 - 2x$

23. $5y^2 + 3 - 2y^2 + 1 - 3y^2$

24. $13x^2 - 6x - 9x^2 - 4x$

25. $7x^3 + 3x^2 - 2x + x - 5x^3 + 1$

26. $-3y^5 + 7y - 2y^3 - 5 + 4y^2 + y^2$

27. $x^4 + 3x^4 - 2x + 5x - 10 - x^2 + x$

28. $a^3 + 2a^2 - 6a + 3a^3 + 2a^2 + 7a + 3$

29. $2x + 4x^2 + 6x + 9x^3$

30. $15y - y^3 + 2y^2 - 10y^2 + 2y - 16$

31. Given $p(x) = x^2 + 14x - 3$, find $p(-1)$.

32. Given $p(x) = -5x^2 - 8x + 7$, find $p(-3)$.

33. Given $p(x) = 3x^3 - 9x^2 - 10x - 11$, find $p(3)$.

34. Given $p(y) = y^3 - 5y^2 + 6y + 2$, find $p(2)$.

35. Given $p(y) = -4y^3 + 5y^2 + 12y - 1$, find $p(-10)$.

36. Given $p(a) = a^3 + 4a^2 + a + 2$, find $p(-5)$.

37. Given $p(a) = 2a^4 + 3a^2 - 8a$, find $p(-1)$.

38. Given $p(x) = 8x^4 + 2x^3 - 6x^2 - 7$, find $p(-2)$.

39. Given $p(x) = x^5 - x^3 + x - 2$, find $p(-2)$.

40. Given $p(x) = 3x^6 - 2x^5 + x^4 - x^3 - 3x^2 + 2x - 1$, find $p(1)$.

A polynomial is given. Rewrite the polynomial by substituting for the variable as indicated by the function notation. See Examples 3 and 4.

41. Given $p(x) = 3x^4 + 5x^3 - 8x^2 - 9x$, find $p(a)$.

42. Given $p(x) = 6x^5 + 5x^2 - 10x + 3$, find $p(c)$.

43. Given $f(x) = 3x + 5$, find $f(a + 2)$.

44. Given $f(x) = -4x + 6$, find $f(a - 2)$.

45. Given $g(x) = 5x - 10$, find $g(2a + 7)$.

46. Given $g(x) = -4x - 8$, find $g(3a + 1)$.

First-degree polynomials are also called **linear polynomials**, second-degree polynomials are called **quadratic polynomials**, and third-degree polynomials are called **cubic polynomials**. The related functions are called linear functions, quadratic functions, and cubic functions, respectively.

47. Use a graphing calculator to graph the following linear functions.

   a. $p(x) = 2x + 3$ **b.** $p(x) = -3x + 1$ **c.** $p(x) = \dfrac{1}{2}x$

48. Use a graphing calculator to graph the following quadratic functions.

   a. $p(x) = x^2$ **b.** $p(x) = x^2 + 6x + 9$ **c.** $p(x) = -x^2 + 2$

49. Use a graphing calculator to graph the following cubic functions.

   a. $p(x) = x^3$ **b.** $p(x) = x^3 - 4x$ **c.** $p(x) = x^3 + 2x^2 - 5$

50. Make up a few of your own linear, quadratic, and cubic functions and graph these functions with your calculator. Using the results from Exercises 47, 48, and 49, and your own functions, describe in your own words:

   a. the general shape of the graphs of linear functions.

   b. the general shape of the graphs of quadratic functions.

   c. the general shape of the graphs of cubic functions.

11.4 Addition and Subtraction with Polynomials

Objective A **Addition with Polynomials**

The **sum** of two or more polynomials is found by combining **like terms**. Remember that like terms (or similar terms) are constants or terms that contain the same variables raised to the same powers. For example,

$3x^2, -10x^2,$ and $1.4x^2$ are all **like terms**. Each has the same variable raised to the same power.

$5a^3$ and $-7a$ are **not** like terms. Each has the same variable, but the powers are different.

When adding polynomials, the polynomials may be written horizontally or vertically. For example,

$$\left(x^2 - 5x + 3\right) + \left(2x^2 - 8x - 4\right) + \left(3x^3 + x^2 - 5\right)$$
$$= 3x^3 + \left(x^2 + 2x^2 + x^2\right) + \left(-5x - 8x\right) + \left(3 - 4 - 5\right)$$
$$= 3x^3 + 4x^2 - 13x - 6.$$

If the polynomials are written in a vertical format, we align like terms, one beneath the other, in a column format and combine like terms in each column.

$$
\begin{array}{r}
x^2 - 5x + 3 \\
2x^2 - 8x - 4 \\
3x^3 + x^2 \quad\;\; - 5 \\
\hline
3x^3 + 4x^2 - 13x - 6
\end{array}
$$

Example 1

Adding Polynomials

a. Add as indicated:

$$\left(5x^3 - 8x^2 + 12x + 13\right) + \left(-2x^2 - 8\right) + \left(4x^3 - 5x + 14\right)$$

Solution

$$\left(5x^3 - 8x^2 + 12x + 13\right) + \left(-2x^2 - 8\right) + \left(4x^3 - 5x + 14\right)$$
$$= \left(5x^3 + 4x^3\right) + \left(-8x^2 - 2x^2\right) + \left(12x - 5x\right) + \left(13 - 8 + 14\right)$$
$$= 9x^3 - 10x^2 + 7x + 19$$

b. Find the sum: $\left(x^3 - x^2 + 5x\right) + \left(4x^3 + 5x^2 - 8x + 9\right)$

Solution

$$
\begin{array}{r}
x^3 - x^2 + 5x \\
4x^3 + 5x^2 - 8x + 9 \\
\hline
5x^3 + 4x^2 - 3x + 9
\end{array}
$$

Completion Example 2

Addition with Polynomials

Find the sum: $\left(4x^3 + 15x - 7\right) + \left(x^3 - 4x^2 + 6x + 2\right)$

Solution

$$\left(4x^3 + 15x - 7\right) + \left(x^3 - 4x^2 + 6x + 2\right)$$

$$= \left(4x^3 + \underline{\quad}\right) + \left(-4x^2\right) + \left(15x + \underline{\quad}\right) + \left(\underline{\quad} + \underline{\quad}\right)$$

$$= \underline{\quad} x^3 - 4x^2 + \underline{\quad} x - \underline{\quad}$$

***Now work margin exercises* 1 and 2.**

Find the following sums.

1. a. $\left(6x^3 + 3x^2 - 8x + 4\right)$
$+\left(2x^3 - x^2 + 2x - 8\right)$
$+\left(2x^2 - 7\right)$

b. $-3x^3 - 5x^2 + 7x - 7$
$+ \quad 7x^3 + 2x^2 + 9$

2. $\left(6x^2 - 8x + 3\right) + \left(x^2 + 4x + 9\right)$

Objective B **Subtraction with Polynomials**

A negative sign written in front of a polynomial in parentheses indicates the **opposite of the entire polynomial**. The opposite can be found by changing the sign of every term in the polynomial.

$$-\left(2x^2 + 3x - 7\right) = -2x^2 - 3x + 7$$

We can also think of the opposite of a polynomial as −1 times the polynomial, applying the distributive property as follows.

$$\begin{aligned}
-\left(2x^2 + 3x - 7\right) &= -1\left(2x^2 + 3x - 7\right) \\
&= -1\left(2x^2\right) - 1(3x) - 1(-7) \\
&= -2x^2 - 3x + 7
\end{aligned}$$

The result is the same with either approach. So the **difference** between two polynomials can be found by changing the sign of each term of the second polynomial and then combining like terms.

$$\begin{aligned}
\left(5x^2 - 3x - 7\right) - \left(2x^2 + 5x - 8\right) &= 5x^2 - 3x - 7 - 2x^2 - 5x + 8 \\
&= \left(5x^2 - 2x^2\right) + \left(-3x - 5x\right) + \left(-7 + 8\right) \\
&= 3x^2 - 8x + 1
\end{aligned}$$

If the polynomials are written in a vertical format, one beneath the other, we change the signs of the terms of the polynomial being subtracted and then combine like terms.

Subtract:

$$\begin{array}{ll}
5x^2 - 3x - 7 & 5x^2 - 3x - 7 \\
\underline{-\left(2x^2 + 5x - 8\right)} \longrightarrow & \underline{-2x^2 - 5x + 8} \\
& 3x^2 - 8x + 1
\end{array}$$

Completion Example Answers

2. $\left(4x^3 + x^3\right) + \left(-4x^2\right) + \left(15x + 6x\right) + \left(-7 + 2\right) = 5x^3 - 4x^2 + 21x - 5$

Example 3

Subtracting Polynomials

a. Subtract as indicated:

$$\left(9x^4 - 22x^3 + 3x^2 + 10\right) - \left(5x^4 - 2x^3 - 5x^2 + x\right)$$

Solution

$$\left(9x^4 - 22x^3 + 3x^2 + 10\right) - \left(5x^4 - 2x^3 - 5x^2 + x\right)$$
$$= 9x^4 - 22x^3 + 3x^2 + 10 - 5x^4 + 2x^3 + 5x^2 - x$$
$$= \left(9x^4 - 5x^4\right) + \left(-22x^3 + 2x^3\right) + \left(3x^2 + 5x^2\right) - x + 10$$
$$= 4x^4 - 20x^3 + 8x^2 - x + 10$$

b. Find the difference:

$$8x^3 + 5x^2 - 14$$
$$\underline{-\left(-2x^3 + x^2 + 6x\right)}$$

Solution

$$
\begin{array}{ll}
8x^3 + 5x^2 - 14 & \quad 8x^3 + 5x^2 + 0x - 14 \\
\underline{-\left(-2x^3 + x^2 + 6x\right)} \longrightarrow & \quad \underline{2x^3 - x^2 - 6x + 0} \\
& \quad 10x^3 + 4x^2 - 6x - 14
\end{array}
$$

Write in 0s for missing powers to help with alignment of like terms.

Find the following differences.

3. a. $\left(7x^4 - 18x^3 + 8x^2 - 11\right)$
$-\left(-8x^4 - 15x^3 + 3x^2 - 12\right)$

b. $9x^3 - 6x^2 + 1$
$\underline{-\left(3x^3 + 3x^2 + 8x\right)}$

4. $\left(6x^4 - 5x + 11\right)$
$-\left(2x^4 + 3x^3 - 4x + 8\right)$

Completion Example 4

Subtracting Polynomials

Find the difference: $\left(6x^3 - 4x^2 + 8x + 14\right) - \left(2x^3 - x^2 + 3x - 9\right)$

Solution

$$\left(6x^3 - 4x^2 + 8x + 14\right) - \left(2x^3 - x^2 + 3x - 9\right)$$
$$= 6x^3 - 4x^2 + 8x + 14 - 2x^3 \underline{} x^2 \underline{} 3x \underline{} 9$$
$$= \left(6x^3 \underline{}\right) + \left(-4x^2 \underline{}\right) + \left(8x \underline{}\right) + \left(14 \underline{}\right)$$
$$= \underline{} x^3 - \underline{} x^2 + \underline{} x + \underline{}$$

Now work margin exercises 3 and 4.

Completion Example Answers

4. $6x^3 - 4x^2 + 8x + 14 - 2x^3 + x^2 - 3x + 9$
$$= \left(6x^3 - 2x^3\right) + \left(-4x^2 + x^2\right) + \left(8x - 3x\right) + \left(14 + 9\right)$$
$$= 4x^3 - 3x^2 + 5x + 23$$

If an algebraic expression contains more than one pair of grouping symbols, such as parentheses ( ), brackets [ ], or braces { }, simplify by working to remove the innermost pair of symbols first. **Apply the rules for order of operations** just as if the variables were numbers and proceed to combine like terms.

Example 5

Simplifying Algebraic Expressions

Simplify each of the following expressions.

a. $5x - \left[2x + 3(4 - x) + 1\right] - 9$

Solution

$5x - \left[2x + 3(4 - x) + 1\right] - 9$ Work with the parentheses first since they are
$= 5x - \left[2x + 12 - 3x + 1\right] - 9$ included inside the brackets.
$= 5x - \left[-x + 13\right] - 9$
$= 5x + x - 13 - 9$
$= 6x - 22$

b. $-3(x - 4) + 2\left[x + 3(x - 3)\right]$

Solution

$-3(x - 4) + 2\left[x + 3(x - 3)\right]$ Work with the parentheses first since they are
$= -3(x - 4) + 2\left[x + 3x - 9\right]$ included inside the brackets.
$= -3(x - 4) + 2\left[4x - 9\right]$
$= -3x + 12 + 8x - 18$
$= 5x - 6$

Now work margin exercise 5.

5. Simplify each of the following expressions.

a. $3x - \left[1 + 4(5 - 2x) + 3x\right]$

b. $-5(x - 3) + 3\left[x + 2(x - 3)\right]$

Practice Problems

1. Add: $(15x + 4) + (3x^2 - 9x - 5)$

2. Subtract: $(-5x^3 - 3x + 4) - (3x^3 - x^2 + 4x - 7)$

3. Simplify: $2 - \left[3a - (4 - 7a) + 2a\right]$

Practice Problem Answers

1. $3x^2 + 6x - 1$ **2.** $-8x^3 + x^2 - 7x + 11$ **3.** $-12a + 6$

Exercises 11.4

1. $\left(2x^2 + 5x - 1\right) + \left(x^2 + 2x + 3\right)$

2. $\left(x^2 + 3x - 8\right) + \left(3x^2 - 2x + 4\right)$

3. $\left(x^2 + 7x - 7\right) + \left(x^2 + 4x\right)$

4. $\left(x^2 + 2x - 3\right) + \left(x^2 + 5\right)$

5. $\left(2x^2 - x - 1\right) + \left(x^2 + x + 1\right)$

6. $\left(3x^2 + 5x - 4\right) + \left(2x^2 + x - 6\right)$

7. $\left(-2x^2 - 3x + 9\right) + \left(3x^2 - 2x + 8\right)$

8. $\left(x^2 + 6x - 7\right) + \left(3x^2 + x - 1\right)$

9. $\left(-4x^2 + 2x - 1\right) + \left(3x^2 - x + 2\right) + \left(x - 8\right)$

10. $\left(8x^2 + 5x + 2\right) + \left(-3x^2 + 9x - 4\right) + \left(2x^2 + 6\right)$

11. $\left(x^2 + 2x - 1\right) + \left(3x^2 - x + 2\right) + \left(2x^3 - 4x - 8\right)$

12. $\left(x^3 + 2x - 9\right) + \left(x^2 - 5x + 2\right) + \left(x^3 - 4x^2 + 1\right)$

13.
$$
\begin{array}{r}
x^2 + 4x - 4 \\
\underline{-2x^2 + 3x + 1}
\end{array}
$$

14.
$$
\begin{array}{r}
2x^2 + 4x - 3 \\
\underline{3x^2 - 9x + 2}
\end{array}
$$

15.
$$
\begin{array}{r}
x^3 + 3x^2 \;+\; x \\
\underline{-2x^3 - \; x^2 \;+ 2x - 4}
\end{array}
$$

16.
$$
\begin{array}{r}
4x^3 + 5x^2 \qquad + 11 \\
\underline{2x^3 - 2x^2 - 3x - \; 6}
\end{array}
$$

17.
$$
\begin{array}{r}
7x^3 + 5x^2 \;+\; x \;-\; 6 \\
-3x^2 + 4x + 11 \\
\underline{-3x^3 - \; x^2 - 5x \;+\; 2}
\end{array}
$$

18.
$$
\begin{array}{r}
x^3 + 5x^2 + 7x - 3 \\
4x^2 + 3x - 9 \\
\underline{4x^3 + 2x^2 \qquad - 2}
\end{array}
$$

19.
$$
\begin{array}{r}
x^3 + 3x^2 \qquad - 4 \\
7x^2 + 2x + 1 \\
\underline{x^3 + \; x^2 - 6x}
\end{array}
$$

20.
$$
\begin{array}{r}
x^3 + 2x^2 \qquad - 5 \\
-2x^3 \qquad + x - 9 \\
\underline{x^3 - 2x^2 \qquad + 14}
\end{array}
$$

Find the indicated differences. See Examples 3 and 4.

21. $\left(2x^2 + 4x + 8\right) - \left(x^2 + 3x + 2\right)$

22. $\left(3x^2 + 7x - 6\right) - \left(x^2 + 2x + 5\right)$

23. $\left(x^2 - 9x + 2\right) - \left(4x^2 - 3x + 4\right)$

24. $\left(6x^2 + 11x + 2\right) - \left(4x^2 - 2x - 7\right)$

25. $\left(2x^2 - x - 10\right) - \left(-x^2 + 3x - 2\right)$

26. $\left(7x^2 + 4x - 9\right) - \left(-2x^2 + x - 9\right)$

27. $\left(x^4 + 8x^3 - 2x^2 - 5\right) - \left(2x^4 + 10x^3 - 2x^2 + 11\right)$

28. $\left(x^3 + 4x^2 - 3x - 7\right) - \left(3x^3 + x^2 + 2x + 1\right)$

29. $\left(-3x^4 + 2x^3 - 7x^2 + 6x + 12\right) - \left(x^4 + 9x^3 + 4x^2 + x - 1\right)$

30. $\left(2x^5 + 3x^3 - 2x^2 + x - 5\right) - \left(3x^5 - 2x^3 + 5x^2 + 6x - 1\right)$

31. $\left(9x^2 - 5\right) - \left(13x^2 - 6x - 6\right)$

32. $\left(8x^2 + 9\right) - \left(4x^2 - 3x - 2\right)$

33. $\left(3x^4 - 2x^3 - 8x - 1\right) - \left(5x^3 - 3x^2 - 3x - 10\right)$

34. $\left(x^5 + 6x^3 - 3x^2 - 5\right) - \left(2x^5 + 8x^3 + 5x + 17\right)$

35. $\begin{aligned} &14x^2 - 6x + 9 \\ -&\left(8x^2 + x - 9\right) \\ \hline \end{aligned}$

36. $\begin{aligned} &9x^2 - 3x + 2 \\ -&\left(4x^2 - 5x - 1\right) \\ \hline \end{aligned}$

37. $\begin{aligned} &5x^4 + 8x^2 + 11 \\ -&\left(-3x^4 + 2x^2 - 4\right) \\ \hline \end{aligned}$

38. $\begin{aligned} &11x^2 + 5x - 13 \\ -&\left(-3x^2 + 5x + 2\right) \\ \hline \end{aligned}$

39. $\begin{aligned} &x^3 + 6x^2 - 3 \\ -&\left(-x^3 + 2x^2 - 3x + 7\right) \\ \hline \end{aligned}$

40. $\begin{aligned} &3x^3 + 9x - 17 \\ -&\left(x^3 + 5x^2 - 2x - 6\right) \\ \hline \end{aligned}$

Simplify each of the algebraic expressions, and write the polynomials in descending order. See Example 5.

41. $5x + 2(x - 3) - (3x + 7)$

42. $-4(x - 6) - (8x + 2) - 3x$

43. $11 + \left[3x - 2(1 + 5x)\right]$

44. $2 + \left[9x - 4(3x + 2)\right]$

45. $8x - \left[2x + 4(x - 3) - 5\right]$

46. $17 - \left[-3x + 6(2x - 3) + 9\right]$

47. $3x^3 - \left[5 - 7\left(x^2 + 2\right) - 6x^2\right]$

48. $10x^3 - \left[8 - 5\left(3 - 2x^2\right) - 7x^2\right]$

49. $\left(2x^2 + 4\right) - \left[-8 + 2\left(7 - 3x^2\right) + x\right]$

50. $-\left[6x^2 - 3(4 + 2x) + 9\right] - \left(x^2 + 5\right)$

51. $2\left[3x + (x - 8) - (2x + 5)\right] - (x - 7)$

52. $-3\left[-x + (10 - 3x) - (8 - 3x)\right] + (2x - 1)$

53. $\left(x^2-1\right)+2\left[4+\left(3-x\right)\right]$

54. $\left(4-x^2\right)+3\left[\left(2x-3\right)-5\right]$

55. $-\left(x-5\right)+\left[6x-2\left(4-x\right)\right]$

56. $2\left(2x+1\right)-\left[5x-\left(2x+3\right)\right]$

57. Find the sum of $4x^2-3x$ and $6x+5$.

58. Subtract $2x^2-4x$ from $7x^3+5x$.

59. Subtract $3\left(x+1\right)$ from $5\left(2x-3\right)$.

60. Find the sum of $10x-2\left(3x+5\right)$ and $3\left(x-4\right)+16$.

61. Subtract $3x^2-4x+2$ from the sum of $4x^2+x-1$ and $6x-5$.

62. Subtract $-2x^2+6x+12$ from the sum of $2x^2+3x-1$ and $x^2-13x+2$.

63. Add $5x^3-8x+1$ to the difference between $2x^3+14x-3$ and x^2+6x+5.

64. Add $2x^3+4x^2+1$ to the difference between $-x^2+10x-3$ and x^3+2x^2+4x.

Writing & Thinking

65. Write the definition of a polynomial.

66. Explain, in your own words, how to subtract one polynomial from another.

67. Describe what is meant by the degree of a polynomial in x.

68. Give two examples that show how the sum of two binomials might not be a binomial.

11.5 Multiplication with Polynomials

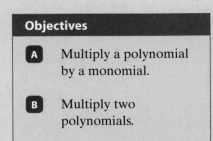
Up to this point, we have multiplied terms such as $5x^2 \cdot 3x^4 = 15x^6$ by using the product rule for exponents. Also, we have applied the distributive property to expressions such as $5(2x+3) = 10x+15$.

Now we will use both the product rule for exponents and the distributive property to multiply polynomials. We discuss three cases here:

    a. the product of a monomial with a polynomial of two or more terms,
    b. the product of two binomials, and
    c. the product of a binomial with a polynomial of more than two terms.

Objective A Multiplying a Polynomial by a Monomial

Using the distributive property, $a(b+c) = ab+ac$, with multiplication indicated on the left, we can find the product of a monomial with a polynomial of two or more terms as follows.

$$5x(2x+3) = 5x \cdot 2x + 5x \cdot 3 = 10x^2 + 15x$$

$$3x^2(4x-1) = 3x^2 \cdot 4x + 3x^2(-1) = 12x^3 - 3x^2$$

$$-4a^5(a^2 - 8a + 5) = -4a^5 \cdot a^2 - 4a^5(-8a) - 4a^5(5) = -4a^7 + 32a^6 - 20a^5$$

Objective B Multiplying Two Polynomials

Now suppose that we want to multiply two binomials, say, $(x+3)(x+7)$. We will apply the distributive property in the following way with multiplication indicated on the right of the parentheses.

Compare $(x+3)(x+7)$ to $(a+b)c = ac + bc$.

Think of $(x+7)$ as taking the place of c. Thus it takes the form,

$$
\begin{array}{ccccc}
(a+b)c & = & ac & + & bc \\
\downarrow\downarrow\downarrow & & \downarrow\downarrow & & \downarrow\downarrow \\
(x+3)(x+7) & = & x(x+7) & + & 3(x+7).
\end{array}
$$

Completing the products on the right, using the distributive property twice again, gives

$$
\begin{aligned}
(x+3)(x+7) &= x(x+7) + 3(x+7) \\
&= x \cdot x + x \cdot 7 + 3 \cdot x + 3 \cdot 7 \\
&= x^2 + 7x + 3x + 21 \\
&= x^2 + 10x + 21.
\end{aligned}
$$

In the same manner,

$$(x+2)(3x+4) = x(3x+4)+2(3x+4)$$
$$= x \cdot 3x + x \cdot 4 + 2 \cdot 3x + 2 \cdot 4$$
$$= 3x^2 + 4x + 6x + 8$$
$$= 3x^2 + 10x + 8.$$

Similarly,

$$(2x-1)(x^2+x-5) = 2x(x^2+x-5)-1(x^2+x-5)$$
$$= 2x \cdot x^2 + 2x \cdot x + 2x \cdot (-5) - 1 \cdot x^2 - 1 \cdot x - 1(-5)$$
$$= 2x^3 + 2x^2 - 10x - x^2 - x + 5$$
$$= 2x^3 + x^2 - 11x + 5.$$

The product of two polynomials can also be found by writing one polynomial under the other. **The distributive property is applied by multiplying each term of one polynomial by each term of the other.** Consider the product $(2x^2+3x-4)(3x+7)$. Now writing one polynomial under the other and applying the distributive property, we obtain the following.

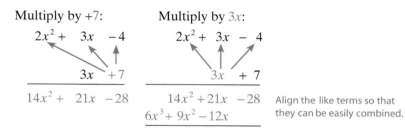

Multiply by +7:

$$2x^2 + \quad 3x \quad -4$$
$$\underline{\qquad\qquad 3x \quad +7}$$
$$14x^2 + \quad 21x \quad -28$$

Multiply by $3x$:

$$2x^2 + \quad 3x \quad -4$$
$$\underline{\qquad\qquad 3x \quad +7}$$
$$14x^2 + 21x \quad -28$$
$$\underline{6x^3 + 9x^2 - 12x}$$

Align the like terms so that they can be easily combined.

Finally, combine like terms:

$$2x^2 \quad + 3x \quad - \quad 4$$
$$\underline{\qquad\qquad\quad 3x \quad + \quad 7}$$
$$14x^2 \quad + 21x \quad - \quad 28$$
$$\underline{6x^3 + \quad 9x^2 \quad - 12x}$$
$$6x^3 + 23x^2 \quad + 9x \quad - \quad 28$$

Combine like terms.

Example 1

Multiplication with Polynomials

Find each product.

a. $-4x(x^2-3x+12)$

Solution

$$-4x(x^2-3x+12) = -4x \cdot x^2 - 4x(-3x) - 4x \cdot 12$$
$$= -4x^3 + 12x^2 - 48x$$

b. $(2x-4)(5x+3)$

Solution

$$
\begin{aligned}
(2x-4)(5x+3) &= 2x(5x+3)-4(5x+3)\\
&= 2x\cdot 5x + 2x\cdot 3 - 4\cdot 5x - 4\cdot 3\\
&= 10x^2 + 6x - 20x - 12\\
&= 10x^2 - 14x - 12
\end{aligned}
$$

c. $7y^2 - 3y + 2$
$\qquad\quad 2y + 3$

Solution

$$
\begin{array}{l}
7y^2 - 3y + 2\\
2y + 3\\
\hline
21y^2 - 9y + 6 \qquad \text{Multiply by 3.}\\
14y^3 - 6y^2 + 4y \qquad \text{Multiply by } 2y.\\
\hline
14y^3 + 15y^2 - 5y + 6 \qquad \text{Combine like terms.}
\end{array}
$$

d. $x^2 + 3x - 1$
$\qquad x^2 - 3x + 1$

Solution

$$
\begin{array}{l}
x^2 + 3x - 1\\
x^2 - 3x + 1\\
\hline
x^2 + 3x - 1 \qquad \text{Multiply by 1.}\\
-3x^3 - 9x^2 + 3x \qquad \text{Multiply by } -3x.\\
x^4 + 3x^3 - x^2 \qquad \text{Multiply by } x^2.\\
\hline
x^4 - 9x^2 + 6x - 1 \qquad \text{Combine like terms.}
\end{array}
$$

e. $(x-5)(x+2)(x-1)$

Solution

First multiply $(x-5)(x+2)$, then multiply this result by $(x-1)$.

$$
\begin{aligned}
(x-5)(x+2) &= x(x+2)-5(x+2)\\
&= x^2 + 2x - 5x - 10\\
&= x^2 - 3x - 10
\end{aligned}
$$

$$
\begin{aligned}
(x^2 - 3x - 10)(x-1) &= (x^2 - 3x - 10)x + (x^2 - 3x - 10)(-1)\\
&= x^3 - 3x^2 - 10x - x^2 + 3x + 10\\
&= x^3 - 4x^2 - 7x + 10
\end{aligned}
$$

Find each product.

1. a. $-6x\left(3x^2 - x - 3\right)$

   b. $\left(3x+3\right)\left(4x-3\right)$

   c. $4x^2 + 8x - 2$
$$\underline{\qquad\quad 3x+1}$$

   d. $x^2 + 4x - 2$
$$\underline{x^2 - 4x + 2}$$

   e. $\left(x-2\right)\left(x+3\right)\left(x-6\right)$

2. a. $5x^3\left(6x^4 - 3x^2 + 12\right)$

   b. $\left(x-4\right)\left(7x+5\right)$

Completion Example 2

Multiplication with Polynomials

Find each product.

a. $3x^2\left(x^2 + 12x - 5\right)$

Solution

$$3x^2\left(x^2 + 12x - 5\right) = \underline{\quad}\cdot x^2 + \underline{\quad}\cdot 12x + \underline{\quad}(-5)$$
$$= \underline{\quad}x^4 + \underline{\quad}x^3 - \underline{\quad}x^2$$

b. $\left(3x+8\right)\left(5x+4\right)$

Solution

$$\left(3x+8\right)\left(5x+4\right) = \underline{\quad}\cdot\left(5x+4\right) + \underline{\quad}\cdot\left(5x+4\right)$$
$$= \underline{\quad}\cdot 5x + \underline{\quad}\cdot 4 + \underline{\quad}\cdot 5x + \underline{\quad}\cdot 4$$
$$= \underline{\quad}x^2 + \underline{\quad}x + \underline{\quad}x + \underline{\quad}$$
$$= \underline{\quad}x^2 + \underline{\quad}x + \underline{\quad}$$

Now work margin exercises **1** and **2**.

Practice Problems

Find each product.

1. $2x\left(3x^2 + x - 1\right)$ **2.** $\left(x+3\right)\left(x-7\right)$

3. $\left(x-1\right)\left(x^2 + x - 4\right)$ **4.** $\left(x+1\right)\left(x-2\right)\left(x+2\right)$

Completion Example Answers

2. a. $3x^2 \cdot x^2 + 3x^2 \cdot 12x + 3x^2\left(-5\right) = 3x^4 + 36x^3 - 15x^2$

   b. $3x\cdot\left(5x+4\right) + 8\cdot\left(5x+4\right) = 3x\cdot 5x + 3x\cdot 4 + 8\cdot 5x + 8\cdot 4$
$$= 15x^2 + 12x + 40x + 32 = 15x^2 + 52x + 32$$

Practice Problem Answers

1. $6x^3 + 2x^2 - 2x$ **2.** $x^2 - 4x - 21$

3. $x^3 - 5x + 4$ **4.** $x^3 + x^2 - 4x - 4$

Exercises 11.5

1. $-3x^2\left(2x^3+5x\right)$

2. $5x^2\left(-4x^2+6\right)$

3. $4x^5\left(x^2-3x+1\right)$

4. $9x^3\left(2x^3-x^2+5x\right)$

5. $-1\left(y^5-8y+2\right)$

6. $-7\left(2y^4+3y^2+1\right)$

7. $-4x^3\left(x^5-2x^4+3x\right)$

8. $-2x^4\left(x^3-x^2+2x\right)$

9. $5x^3\left(5x^2-x+2\right)$

10. $-2x^2\left(x^3+5x-4\right)$

11. $a^2\left(a^5+2a^4-5a+1\right)$

12. $7t^3\left(-t^3+5t^2+2t+1\right)$

13. $3x\left(2x+1\right)-2\left(2x+1\right)$

14. $x\left(3x+4\right)+7\left(3x+4\right)$

15. $3a\left(3a-5\right)+5\left(3a-5\right)$

16. $6x\left(x-1\right)+5\left(x-1\right)$

17. $5x\left(-2x+7\right)-2\left(-2x+7\right)$

18. $y\left(y^2+1\right)-1\left(y^2+1\right)$

19. $x\left(x^2+3x+2\right)+2\left(x^2+3x+2\right)$ **20.** $4x\left(x^2-x+1\right)+3\left(x^2-x+1\right)$

21. $\left(x+4\right)\left(x-3\right)$

22. $\left(x+7\right)\left(x-5\right)$

23. $\left(a+6\right)\left(a-8\right)$

24. $\left(x+2\right)\left(x-4\right)$

25. $\left(x-2\right)\left(x-1\right)$

26. $\left(x-7\right)\left(x-8\right)$

27. $3\left(t+4\right)\left(t-5\right)$

28. $-4\left(x+6\right)\left(x-7\right)$

29. $x\left(x+3\right)\left(x+8\right)$

30. $t\left(t-4\right)\left(t-7\right)$

31. $\left(2x+1\right)\left(x-4\right)$

32. $\left(3x-1\right)\left(x+4\right)$

33. $\left(6x-1\right)\left(x+3\right)$

34. $\left(8x+15\right)\left(x+1\right)$

35. $\left(2x+3\right)\left(2x-3\right)$

36. $(3t+5)(3t-5)$

37. $(4x+1)(4x+1)$

38. $(5x-2)(5x-2)$

39. $(y+3)(y^2-y+4)$

40. $(2x+1)(x^2-7x+2)$

41. $\begin{array}{r} 3x+7 \\ \underline{x-5} \end{array}$

42. $\begin{array}{r} 2x+6 \\ \underline{x+3} \end{array}$

43. $\begin{array}{r} x^2+3x+1 \\ \underline{5x-9} \end{array}$

44. $\begin{array}{r} 8x^2+3x-2 \\ \underline{-2x+7} \end{array}$

45. $\begin{array}{r} 2x^2+3x+5 \\ \underline{x^2+2x-3} \end{array}$

46. $\begin{array}{r} 6x^2-x+8 \\ \underline{2x^2+5x+6} \end{array}$

Find the product and simplify if possible. See Examples 1 and 2.

47. $(3x-4)(x+2)$

48. $(t+6)(4t-7)$

49. $(2x+5)(x-1)$

50. $(5a-3)(a+4)$

51. $(7x+1)(x-2)$

52. $(x-2)(3x+8)$

53. $(2x+1)(3x-8)$

54. $(3x+7)(2x-5)$

55. $(2x+3)(2x+3)$

56. $(5y+2)(5y+2)$

57. $(x+3)(x^2-4)$

58. $(y^2+2)(y-4)$

59. $(2x+7)(2x-7)$

60. $(3x-4)(3x+4)$

61. $(x+1)(x^2-x+1)$

62. $(x-2)(x^2+2x+4)$

63. $(7a-2)(7a-2)$

64. $(5a-6)(5a-6)$

65. $(2x+3)(x^2-x-1)$

66. $(3x+1)(x^2-x+9)$

67. $(x+1)(x+2)(x+3)$

68. $(t-1)(t-2)(t-3)$

69. $(a^2+a-1)(a^2-a+1)$

70. $(y^2+y+2)(y^2+y-2)$

71. $(t^2+3t+2)^2$

72. $(a^2-4a+1)^2$

73. $(y+6)(y-6)+(y+5)(y-5)$

74. $(y-2)(y+2)+(y-1)(y+1)$

75. $(2a+1)(a-5)+(a-4)(a-4)$

76. $(x+4)(2x+1)+(x-3)(x-2)$

77. $(x-3)(x+5)-(x+3)(x+2)$

78. $(t+3)(t+3)-(t-2)(t-2)$

79. $(2a+3)(a+1)-(a-2)(a-2)$

80. $(4t-3)(t+4)-(t-2)(3t+1)$

Writing & Thinking

81. We have seen how the distributive property is used to multiply polynomials.

    a. Show how the distributive property can be used to find the product $\begin{array}{r} 75 \\ \times\,93 \\ \hline \end{array}$

    (**Hint:** $75 = 70 + 5$ and $93 = 90 + 3$)

    b. In the multiplication algorithm for multiplying whole numbers (as in the product above), we are told to "move to the left" when multiplying. For example

$$\begin{array}{r} 75 \\ \times\,93 \\ \hline 15 \\ 21 \\ 45 \\ 63 \\ \hline \end{array}$$

Why are the 21 and 45 moved one place to the left in the alignment?

When 9 and 7 are multiplied, we move the 63 two places left. Why?

11.6 Special Products of Binomials

Objectives

A Multiply polynomials using the FOIL method.

B Multiply binomials, finding products that are the difference of squares.

C Multiply binomials, finding products that are perfect square trinomials.

Objective A **The FOIL Method**

In the case of the **product of two binomials** such as $(2x+5)(3x-7)$, the **FOIL** method is useful. **F-O-I-L** is a mnemonic device (memory aid) to help in remembering which terms of the binomials to multiply together. First, by using the distributive property we can see how the terms are multiplied.

$$(2x+5)(3x-7) = 2x(3x-7)+5(3x-7)$$
$$= 2x \cdot 3x + 2x \cdot (-7) + 5 \cdot 3x + 5 \cdot (-7)$$

First terms	Outside terms	Inside terms	Last terms
F	O	I	L

Now we can use the FOIL method and then combine like terms to go directly to the answer.

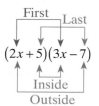

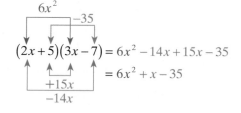

Example 1

FOIL Method

Use the FOIL method to find the products of the given binomials.

a. $(x+3)(2x+8)$

Solution

$$(x+3)(2x+8) = 2x^2 + 8x + 6x + 24$$
$$= 2x^2 + 14x + 24$$

b. $(2x-3)(3x-5)$

Solution

$$(2x-3)(3x-5) = 6x^2 - 10x - 9x + 15$$
$$= 6x^2 - 19x + 15$$

c. $(x+7)(x-7)$

Solution

$(x+7)(x-7) = x^2 - 7x + 7x - 49$ Apply the FOIL method mentally.

$\qquad\qquad\qquad = x^2 - 49$

Note that in this special case the two middle terms are opposites and their sum is 0.

Completion Example 2

FOIL Method

Use the FOIL method to find the product of the given binomial.
$(x+11)(3x-2)$

Solution

$(x+11)(3x-2) = x \cdot \underline{\quad} + x \cdot (\underline{\quad}) + 11 \cdot \underline{\quad} + 11 \cdot (\underline{\quad})$

$\qquad\qquad\qquad = \underline{\quad} x^2 + \underline{\quad} x - \underline{\quad}$

Now work margin exercises 1 and 2.

Objective B **The Difference of Two Squares:** $(x+a)(x-a) = x^2 - a^2$

In Example 1c, the middle terms, $-7x$ and $+7x$, are opposites of each other and their sum is 0. Therefore the resulting product has only two terms and each term is a square.

$$(x+7)(x-7) = x^2 - 49$$

In fact, when two binomials are in the form of the sum and difference of the same two terms, the product will always be the difference of the squares of the terms. The product is called the **difference of two squares**.

Difference of Two Squares

$$(x+a)(x-a) = x^2 - a^2$$

Use the FOIL method to find the products of the given binomials

1. a. $(x+7)(3x+4)$

    b. $(4x-5)(2x-2)$

    c. $(x+4)(x-4)$

2. $(2x-8)(5x-9)$

Completion Example Answers

2. $x \cdot 3x + x \cdot (-2) + 11 \cdot 3x + 11 \cdot (-2) = 3x^2 + 31x - 22$

In order to quickly recognize the difference of two squares, you should memorize the following squares of the positive integers from 1 to 20. The squares of integers are called **perfect squares.**

Perfect Squares from 1 to 400

1, 4, 9, 16, 25, 36, 49, 64, 81, 100, 121, 144, 169, 196, 225, 256, 289, 324, 361, 400

Example 3

Difference of Two Squares

Find the following products.

a. $(x+4)(x-4)$

Solution

The two binomials represent the sum and difference of x and 4. So, the product is the difference of their squares.

$$(x+4)(x-4) = x^2 - 4^2$$
$$= x^2 - 16 \qquad \text{difference of two squares}$$

b. $(3y+7)(3y-7)$

Solution

$$(3y+7)(3y-7) = (3y)^2 - 7^2 \qquad \textbf{Note: } (3y)^2 = 3y \cdot 3y = 9y^2$$
$$= 9y^2 - 49 \qquad \text{difference of two squares}$$

c. $(x^3-6)(x^3+6)$

Solution

$$(x^3-6)(x^3+6) = (x^3)^2 - 6^2 \qquad \textbf{Note: } (x^3)^2 = x^3 \cdot x^3 = x^6$$
$$= x^6 - 36 \qquad \text{difference of two squares}$$

Completion Example 4

Difference of Two Squares

Find the product: $(4x+9)(4x-9)$.

Solution

$$(4x+9)(4x-9) = (\underline{\hspace{1cm}})^2 - (\underline{\hspace{1cm}})^2$$
$$= \underline{\hspace{1cm}} - \underline{\hspace{1cm}}$$

Now work margin exercises 3 and 4.

Find the following products.

3. a. $(x+6)(x-6)$

b. $(4y+3)(4y-3)$

c. $(x^4-3)(x^4+3)$

4. $(2x^2+6)(2x^2-6)$

Objective C **Squares of Binomials:** $\begin{cases}(x+a)^2 = x^2 + 2ax + a^2 \\ (x-a)^2 = x^2 - 2ax + a^2\end{cases}$

Now we consider the case where the two binomials being multiplied are **the same**. That is, we want to consider the **square of a binomial**. The following examples, using the distributive property, illustrate two patterns that, after some practice, allow us to go directly to the products.

$(x+3)^2 = (x+3)(x+3) = x^2 + 3x + 3x + 9$

$\qquad\qquad\qquad = x^2 + 2\cdot 3x + 9$ The middle term is doubled.
$\qquad\qquad\qquad\qquad\qquad\qquad\qquad\qquad 3x + 3x = 2\cdot 3x$

$\qquad\qquad\qquad = x^2 + 6x + 9$ perfect square trinomial

$(x-11)^2 = (x-11)(x-11) = x^2 - 11x - 11x + 121$

$\qquad\qquad\qquad\qquad = x^2 - 2\cdot 11x + 121$ The middle term is doubled.
$\qquad\qquad\qquad\qquad\qquad\qquad\qquad\qquad\qquad -11x - 11x = 2(-11x) = -2\cdot 11x$

$\qquad\qquad\qquad\qquad = x^2 - 22x + 121$ perfect square trinomial

Note that in each case **the result of squaring the binomial is a trinomial**. These trinomials are called **perfect square trinomials**.

Squares of Binomials (Perfect Square Trinomials)

$(x+a)^2 = x^2 + 2ax + a^2$ **Square of a Binomial Sum**

$(x-a)^2 = x^2 - 2ax + a^2$ **Square of a Binomial Difference**

Completion Example Answers

4. $(4x)^2 - (9)^2 = 16x^2 - 81$

Example 5

Squares of Binomials

Find the following products.

a. $(2x + 3)^2$

Solution

The pattern for squaring a binomial gives the following.

$$(2x + 3)^2 = (2x)^2 + 2(3)(2x) + (3)^2 \quad \textbf{Note:} \, (2x)^2 = 2x \cdot 2x = 4x^2$$
$$= 4x^2 + 12x + 9$$

b. $(5x - 1)^2$

Solution

$$(5x - 1)^2 = (5x)^2 - 2(1)(5x) + (1)^2$$
$$= 25x^2 - 10x + 1$$

c. $(9 - x)^2$

Solution

$$(9 - x)^2 = (9)^2 - 2(x)(9) + x^2$$
$$= 81 - 18x + x^2$$
$$= x^2 - 18x + 81$$

d. $(y^3 + 1)^2$

Solution

$$(y^3 + 1)^2 = (y^3)^2 + 2(1)(y^3) + 1^2 \quad \textbf{Note:} \, (y^3)^2 = y^3 \cdot y^3 = y^{3+3} = y^6$$
$$= y^6 + 2y^3 + 1$$

Completion Example 6

Squares of Binomials

Find the product: $(3x + 10)^2$.

Solution

$$(3x + 10)^2 = (\underline{\quad})^2 + 2 \cdot \underline{\quad} \cdot \underline{\quad} + (\underline{\quad})^2$$
$$= \underline{\quad}x^2 + \underline{\quad}x + \underline{\quad}$$

Now work margin exercises 5 and 6.

Find the following products.

5. a. $(3x + 5)^2$

   b. $(7x - 2)^2$

   c. $(8 - 3x)^2$

   d. $(2y^3 - 1)^2$

6. $(5x - 2)^2$

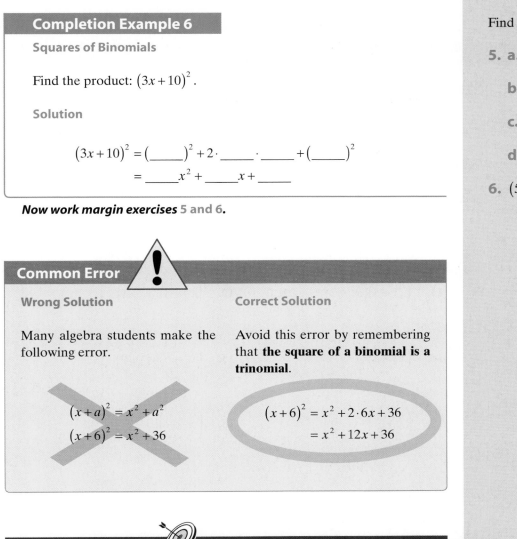

Common Error

Wrong Solution

Many algebra students make the following error.

$$(x + a)^2 = x^2 + a^2$$
$$(x + 6)^2 = x^2 + 36$$

Correct Solution

Avoid this error by remembering that **the square of a binomial is a trinomial**.

$$(x + 6)^2 = x^2 + 2 \cdot 6x + 36$$
$$= x^2 + 12x + 36$$

Practice Problems

Find the indicated products.

1. $(x + 10)(x - 10)$ **2.** $(x + 3)^2$ **3.** $(2x - 1)(x + 3)$

4. $(2x - 5)^2$ **5.** $(x^2 + 4)(x^2 - 3)$

Completion Example Answers

6. $(3x)^2 + 2 \cdot 3x \cdot 10 + (10)^2 = 9x^2 + 60x + 100$

Practice Problem Answers

1. $x^2 - 100$ **2.** $x^2 + 6x + 9$ **3.** $2x^2 + 5x - 3$

4. $4x^2 - 20x + 25$ **5.** $x^4 + x^2 - 12$

Exercises 11.6

1. $(x+2)(5x+1)$

2. $(7x-2)(x-3)$

3. $(4x-3)(x+4)$

4. $(x+11)(x-8)$

5. $(3x-7)(x-6)$

6. $(x+7)(2x+9)$

7. $(5+x)(5+x)$

8. $(3-x)(3-x)$

9. $(x^2+1)(x^2-1)$

10. $(x^2+5)(x^2-5)$

11. $(x^2+3)(x^2+3)$

12. $(x^3+8)(x^3+8)$

13. $(x^3-2)^2$

14. $(x^2-4)^2$

15. $(x^2-6)(x^2+9)$

16. $(x^2+3)(x^2-5)$

17. $\left(x+\dfrac{2}{3}\right)\left(x-\dfrac{2}{3}\right)$

18. $\left(x-\dfrac{1}{2}\right)\left(x+\dfrac{1}{2}\right)$

19. $\left(x+\dfrac{3}{4}\right)\left(x-\dfrac{3}{4}\right)$

20. $\left(x+\dfrac{3}{8}\right)\left(x-\dfrac{3}{8}\right)$

21. $\left(x+\dfrac{3}{5}\right)\left(x+\dfrac{3}{5}\right)$

22. $\left(x+\dfrac{4}{3}\right)\left(x+\dfrac{4}{3}\right)$

23. $\left(x-\dfrac{5}{6}\right)^2$

24. $\left(x-\dfrac{2}{7}\right)^2$

25. $\left(x+\dfrac{1}{4}\right)\left(x-\dfrac{1}{2}\right)$

26. $\left(x-\dfrac{1}{5}\right)\left(x+\dfrac{2}{3}\right)$

27. $\left(x+\dfrac{1}{3}\right)\left(x+\dfrac{1}{2}\right)$

28. $\left(x-\dfrac{4}{5}\right)\left(x-\dfrac{3}{10}\right)$

29. $(x-7)^2$

30. $(x-5)^2$

31. $(x+4)(x+4)$

32. $(x+8)(x+8)$

33. $(x+3)(x-3)$

34. $(x-6)(x+6)$

35. $(x+9)(x-9)$

36. $(x+12)(x-12)$

37. $(2x+3)(x-1)$

38. $(3x+1)(2x+5)$

39. $(3x-4)^2$

40. $(3x+1)^2$

41. $(5x+2)(5x-2)$

42. $(2x+1)(2x-1)$

43. $(3x-2)(3x-2)$

44. $(3+x)^2$

45. $(8-x)(8-x)$

46. $(5-x)(5-x)$

47. $(4x+5)(4x-5)$

48. $(11-x)(11+x)$

49. $(5x-9)(5x+9)$

50. $(9x+2)(9x-2)$

51. $(4-x)^2$

52. $(3x+2)^2$

53. $(2x+7)(2x-7)$

54. $(6x+5)(6x-5)$

55. $(5x^2+2)(2x^2-3)$

56. $(4x^2+7)(2x^2+1)$

57. $(1+7x)^2$

58. $(2-5x)^2$

59. $(x+1.4)(x-1.4)$

60. $(x-2.1)(x+2.1)$

61. $(x-2.5)^2$

62. $(x+1.7)^2$

63. $(x+2.15)(x-2.15)$

64. $(x+1.36)(x-1.36)$

65. $(x+1.24)^2$

66. $(x-1.45)^2$

67. $(1.42x+9.6)^2$

68. $(0.46x-0.71)^2$

69. $(11.4x+3.5)(11.4x-3.5)$

70. $(2.5x+11.4)(1.3x-16.9)$

71. $(12.6x-6.8)(7.4x+15.3)$

72. $(3.4x+6)(3.4x-6)$

Solve the following word problems.

73. Squares: A square is 20 inches on each side. A square x inches on each side is cut from each corner of the square.

a. Represent the area of the remaining portion of the square in the form of a polynomial function $A(x)$.

b. Represent the perimeter of the remaining portion of the square in the form of a polynomial function $P(x)$.

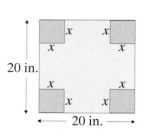

75. Rectangles: A rectangle has sides $(x+3)$ ft and $(x+5)$ ft. If a square x ft on a side is cut from the rectangle, represent the remaining area in the form of a polynomial function $A(x)$.

74. Probability: In the case of binomial probabilities, if x is the probability of success in one trial of an event, then the expression $f(x) = 15x^4(1-x)^2$ is the probability of 4 successes in 6 trials where $0 \le x \le 1$.

a. Represent the expression $f(x)$ as a single polynomial by multiplying the polynomials.

b. If a fair coin is tossed, the probability of heads occurring is $\dfrac{1}{2}$. That is, $x = \dfrac{1}{2}$. Find the probability of 4 heads occurring in 6 tosses.

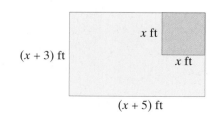

76. Architecture: The Americans with Disabilities Act requires sidewalks to be x feet wide in order for wheelchairs to fit on them. At the bottom, the Empire State Building is 425 feet long and 190 feet wide and a regulation sidewalk surrounds the building.

a. Represent the area covered by the building and the sidewalk in the form of a polynomial function.

b. Represent the area covered by the sidewalk only in the form of a polynomial function.

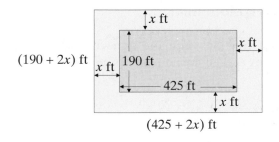

77. Boxes: A rectangular piece of cardboard that is 10 inches by 15 inches has squares of length x inches on a side cut from each corner. (Assume that $0 < x < 5$.)

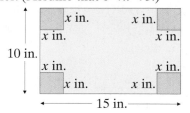

a. Represent the remaining area in the form of a polynomial function $A(x)$.

b. Represent the perimeter of the remaining figure in the form of a polynomial function $P(x)$.

c. If the flaps of the cardboard are folded up, an open box is formed. Represent the volume of this box in the form of a polynomial function $V(x)$.

78. Museums: The world's largest single aquarium habitat at the Georgia Aquarium is 284 feet long, 126 feet wide, and 30 feet deep. Another aquarium is attempting to make a tank that is x feet longer, wider, and deeper. Represent the volume of the new tank as a polynomial function $V(x)$.

Writing & Thinking

79. A square with sides of length $(x+5)$ can be broken up as shown in the diagram. The sums of the areas of the interior rectangles and squares is equal to the total area of the square: $(x+5)^2$. Show how this fits with the formula for the square of a sum.

	5	x
x	$5x$	x^2
5	5^2	$5x$

Objectives

A Divide a polynomial by a monomial.

B Divide polynomials by using the division algorithm.

11.7 Division with Polynomials

Fractions, such as $\dfrac{127}{2}$ and $\dfrac{1}{8}$, in which the numerator and denominator are integers are called rational numbers. Fractions in which the numerator and denominator are polynomials are called **rational expressions**. (No denominator can be 0.)

Rational Expressions: $\dfrac{x^3 - 6x^2 + 2x}{3x}, \dfrac{6x^2 - 7x - 2}{2x - 1}, \dfrac{5x^3 - 3x + 1}{x + 3}$, and $\dfrac{1}{10x}$

In this section, we will treat a rational expression as a division problem. With this basis, there are two situations to consider:

1. the denominator (divisor) is a monomial, or
2. the denominator (divisor) is not a monomial.

Objective A **Dividing by a Monomial**

We know that the sum of fractions with the same denominator can be written as a single fraction by adding the numerators and using the common denominator. For example,

$$\frac{5}{a} + \frac{3b}{a} + \frac{2c}{a} = \frac{5 + 3b + 2c}{a}.$$

If instead of adding the fractions, we start with the sum and want to divide the numerator by the denominator (with a monomial in the denominator), we divide each term in the numerator by the monomial denominator and simplify each fraction.

$$\frac{4x^3 + 8x^2 - 12x}{4x} = \frac{4x^3}{4x} + \frac{8x^2}{4x} - \frac{12x}{4x} = x^2 + 2x - 3$$

1. Divide each polynomial by the monomial denominator by writing each fraction as the sum or difference of fractions. Simplify if possible.

a. $\dfrac{18x^5 - 6x^4 + 9x}{3x}$

b. $\dfrac{5y^4 - 6y^3 - 8y}{2y}$

Example 1

Dividing by a Monomial

Divide each polynomial by the **monomial denominator** by writing each fraction as the sum (or difference) of fractions. Simplify each fraction, if possible.

a. $\dfrac{x^3 - 6x^2 + 2x}{3x}$

Solution

$$\frac{x^3 - 6x^2 + 2x}{3x} = \frac{x^3}{3x} - \frac{6x^2}{3x} + \frac{2x}{3x} = \frac{x^2}{3} - 2x + \frac{2}{3}$$

b. $\dfrac{15y^4 - 20y^3 + 5y^2}{5y^2}$

Solution

$$\frac{15y^4 - 20y^3 + 5y^2}{5y^2} = \frac{15y^4}{5y^2} - \frac{20y^3}{5y^2} + \frac{5y^2}{5y^2} = 3y^2 - 4y + 1$$

Now work margin exercise 1.

Objective B **The Division Algorithm**

In arithmetic, the **division algorithm** (called **long division**) is the process (or series of steps) that we follow when dividing two numbers. By this division algorithm, we can find $64 \div 5$ as follows.

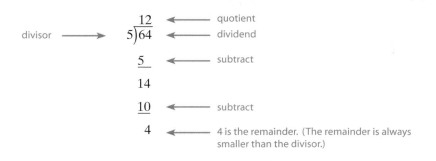

Check: $5 \cdot 12 + 4 = 60 + 4 = 64$ (Multiply the divisor times the quotient and add the remainder. The result should be the original dividend.)

We can also write the division in fraction form with the remainder over the divisor, giving a mixed number.

$$64 \div 5 = \frac{64}{5} = 12 + \frac{4}{5} = 12 \frac{4}{5}$$

In algebra, **the division algorithm with polynomials** is quite similar. In dividing one polynomial by another, with the degree of the divisor smaller than the degree of the dividend, the quotient will be another polynomial with a remainder.

$$\underline{\text{For } 64 \div 5} \qquad \underline{\text{For } P \div D}$$

$$64 = 12 \cdot 5 + 4 \qquad P = Q \cdot D + R \qquad \text{where,}$$

or

$$\frac{64}{5} = 12 + \frac{4}{5} \qquad \frac{P}{D} = Q + \frac{R}{D}$$

P is the dividend,
D is the divisor,
Q is the quotient, and
R is the remainder.

The remainder must be of smaller degree than the divisor. If the remainder is 0, then the divisor and quotient are factors of the dividend.

The Division Algorithm

For polynomials P and D, the division algorithm gives

$$\frac{P}{D} = Q + \frac{R}{D}, \qquad D \neq 0$$

where Q and R are polynomials and the **degree of $R <$ degree of D**.

The actual process of long division is not clear from this abstract definition. Although you are familiar with the process of long division with decimal numbers, this same procedure appears more complicated with polynomials. The **division algorithm** (**long division**) is illustrated in a step-by-step form in the following example. Study it carefully.

Example 2

The Division Algorithm

Simplify $\dfrac{6x^2 - 7x - 2}{2x - 1}$ by using long division.

Solution

	Calculation	Explanation

Step 1: $2x-1\overline{\smash{)}6x^2 - 7x - 2}$

Write both polynomials in order of descending powers. **If any powers are missing, fill in with 0s.**

Step 2:
$$\begin{array}{r} 3x \\ 2x-1\overline{\smash{)}6x^2 - 7x - 2} \end{array}$$

Mentally divide $6x^2$ by $2x$:
$\dfrac{6x^2}{2x} = 3x$. Write $3x$ above $6x^2$.

Step 3:
$$\begin{array}{r} 3x \\ 2x-1\overline{\smash{)}6x^2 - 7x - 2} \\ -\left(6x^2 - 3x\right) \end{array}$$

Multiply $3x$ times $(2x-1)$ and write the terms of the product, $6x^2 - 3x$, under the like terms in the dividend. Use a '−' sign to indicate that the product is to be subtracted.

Step 4:
$$\begin{array}{r} 3x \\ 2x-1\overline{\smash{)}6x^2 - 7x - 2} \\ -6x^2 + 3x \\ \hline -4x \end{array}$$

Subtract $6x^2 - 3x$ by changing signs and adding.

Step 5:
$$\begin{array}{r} 3x \\ 2x-1\overline{\smash{)}6x^2 - 7x - 2} \\ -6x^2 + 3x \\ \hline -4x - 2 \end{array}$$

Bring down the -2.

Step 6:
$$\begin{array}{r} 3x \; - \; 2 \\ 2x-1\overline{\smash{)}6x^2 - 7x - 2} \\ -6x^2 + 3x \\ \hline -4x - 2 \end{array}$$

Mentally divide $-4x$ by $2x$:
$\dfrac{-4x}{2x} = -2$.
Write -2 in the quotient.

Step 7:
$$\begin{array}{r} 3x \; - \; 2 \\ 2x-1\overline{\smash{)}6x^2 - 7x - 2} \\ -6x^2 + 3x \\ \hline -4x - 2 \\ -\left(-4x + 2\right) \end{array}$$

Multiply -2 times $(2x-1)$ and write the terms of the product, $-4x + 2$, under the like terms in the expression $-4x - 2$. Use a '−' sign to indicate that the product is to be subtracted.

Step 8:

$$2x-1\overline{\smash{\big)}\ 6x^2-7x-2}$$

$$3x\ -\ 2$$

$$\underline{-6x^2+\ 3x}$$

$$-\ 4x-2$$

$$\underline{+\ 4x-2}$$

$$-4$$

Subtract $-4x+2$ by changing signs and adding.

Thus the quotient is $3x-2$ and the remainder is -4.

In the form $Q+\dfrac{R}{D}$ we can write $\dfrac{6x^2-7x-2}{2x-1}=3x-2-\dfrac{4}{2x-1}$.

Check:

Show $Q\cdot D+R=P$.

$$(3x-2)(2x-1)-4=6x^2-3x-4x+2-4=6x^2-7x-2$$

Now work margin exercise 2.

Example 3

Long Division (Remander 0)

Divide $(25x^3-5x^2+3x+1)\div(5x+1)$ by using long division.

Solution

$$5x+1\overline{\smash{\big)}\ 25x^3-5x^2+3x+1}$$

$$5x^2-2x\ +\ 1$$

$$\underline{-(25x^3+5x^2)}$$

$$-10x^2+3x$$

$$\underline{-(-10x^2-2x)}$$

$$5x+1$$

$$\underline{-(5x+1)}$$

$$0$$

There is no remainder, thus the quotient is simply $5x^2-2x+1$.

Now work margin exercise 3.

In Example 3, because the remainder is 0, both $(5x+1)$ and $(5x^2-2x+1)$ are **factors** of $25x^3-5x^2+3x+1$. That is

$$(5x^2-2x+1)(5x+1)=25x^3-5x^2+3x+1 \text{ or } Q\cdot D=P.$$

Factoring polynomials will be discussed in detail in Sections 12.2, 12.3, and 12.4.

2. Simplify

$$\frac{18x^3+6x^2-10x-11}{6x+4}$$

by using long division.

3. Divide

$$\left(8x^3-2x^2+21x+45\right)\div\left(4x+5\right)$$

by using long division.

4. Simplify $\dfrac{7x^3+4x-9}{x-2}$ using long division.

Example 4

Long Division (Terms Missing)

Simplify $\dfrac{x^4+9x^2-3x+5}{x^2-x+2}$ by using long division.

Solution

Note that 0 is written as a placeholder for any missing powers of the variable. In this way, like terms are easily aligned vertically.

$$
\begin{array}{r}
x^2 + x + 8 \\
x^2-x+2\overline{\big)x^4+0x^3+9x^2-3x+5} \\
\underline{-\left(x^4-x^3+2x^2\right)} \\
x^3+7x^2-3x \\
\underline{-\left(x^3-x^2+2x\right)} \\
8x^2-5x+5 \\
\underline{-\left(8x^2-8x+16\right)} \\
3x-11
\end{array}
$$

Note that the remainder is of smaller degree than the divisor.

Thus the quotient is x^2+x+8 and the remainder is $3x-11$.

In the form $Q+\dfrac{R}{D}$ we can write $x^2+x+8+\dfrac{3x-11}{x^2-x+2}$.

Now work margin exercise 4.

Practice Problems

1. Express the quotient as a sum of fractions in simplified form:
$$\frac{8x^2+6x+1}{2x}$$

Use the division algorithm to divide. Write the answer in the form $Q+\dfrac{R}{D}$.

2. $\dfrac{3x^2-8x+5}{x+2}$

3. $\left(x^3+4x^2-10\right)\div\left(x^2+x-1\right)$

Answers to Practice Problems:

1. $4x+3+\dfrac{1}{2x}$ **2.** $3x-14+\dfrac{33}{x+2}$ **3.** $x+3+\dfrac{-2x-7}{x^2+x-1}$

11.7 Exercises

Express each quotient as a sum (or difference) of fractions and simplify if possible. See Example 1.

1. $\dfrac{8y^3 - 16y^2 + 24y}{8y}$

2. $\dfrac{18x^4 + 24x^3 + 36x^2}{6x^2}$

3. $\dfrac{34x^5 - 51x^4 + 17x^3}{17x^3}$

4. $\dfrac{14y^4 + 28y^3 + 12y^2}{2y^2}$

5. $\dfrac{110x^4 - 121x^3 + 11x^2}{11x}$

6. $\dfrac{15x^7 + 30x^6 - 45x^3}{15x^3}$

7. $\dfrac{-56x^4 + 98x^3 - 35x^2}{14x^2}$

8. $\dfrac{108x^6 - 72x^5 + 63x^4}{18x^4}$

9. $\dfrac{16y^6 - 56y^5 - 120y^4 + 64y^3}{16y^3}$

10. $\dfrac{20y^5 - 14y^4 + 21y^3 + 42y^2}{4y^2}$

Divide by using the division algorithm. Write the answers in the form $Q + \dfrac{R}{D}$, where the degree of R < the degree of D. See Examples 2 through 4.

11. $\dfrac{x^2 - 2x - 20}{x + 4}$

12. $\dfrac{x^2 + 9x - 5}{x - 1}$

13. $\dfrac{6x^2 - 11x - 3}{2x - 1}$

14. $\dfrac{10x^2 + 16x + 5}{5x + 3}$

15. $\dfrac{21x^2 + 25x - 3}{7x - 1}$

16. $\dfrac{15x^2 - 14x - 11}{3x - 4}$

17. $\dfrac{x^2 - 12x + 27}{x - 3}$

18. $\dfrac{x^2 - 12x + 35}{x - 5}$

19. $\dfrac{x^3 - 9x^2 + 8x - 3}{x - 8}$

20. $\dfrac{x^3 - 6x^2 + 8x - 5}{x - 2}$

21. $\dfrac{4x^3 + 2x^2 - 3x + 1}{x + 2}$

22. $\dfrac{3x^3 + 6x^2 + 8x - 5}{x + 1}$

23. $\dfrac{x^3 + 6x + 3}{x - 7}$

24. $\dfrac{2x^3 + 3x - 2}{x - 1}$

25. $\dfrac{2x^3 - 5x^2 + 6}{x + 2}$

26. $\dfrac{4x^3 - x^2 + 13}{x - 1}$

27. $\dfrac{21x^3 + 41x^2 + 13x + 5}{3x + 5}$

28. $\dfrac{6x^3 - 7x^2 + 14x - 8}{3x - 2}$

29. $\dfrac{2x^3 + 7x^2 + 10x - 6}{2x + 3}$

30. $\dfrac{6x^3 - 4x^2 + 5x - 7}{x - 2}$

31. $\dfrac{x^3 - x^2 - 10x - 10}{x - 4}$

32. $\dfrac{2x^3 - 3x^2 + 7x + 4}{2x - 1}$

33. $\dfrac{10x^3 + 11x^2 - 12x + 9}{5x + 3}$

34. $\dfrac{6x^3 + 19x^2 - 3x - 7}{6x + 1}$

35. $\dfrac{2x^3 - 7x + 2}{x + 4}$

36. $\dfrac{2x^3 + 4x^2 - 9}{x + 3}$

37. $\dfrac{9x^3 - 19x + 9}{3x - 2}$

38. $\dfrac{4x^3 - 8x^2 - 9x}{2x - 3}$

39. $\dfrac{6x^3 + 11x^2 + 25}{2x + 5}$

40. $\dfrac{16x^3 + 7x + 12}{4x + 3}$

41. $\dfrac{x^4 - 3x^3 + 2x^2 - x + 2}{x - 3}$

42. $\dfrac{x^4 + x^3 - 4x^2 + x - 3}{x + 6}$

43. $\dfrac{x^4 + 2x^2 - 3x + 5}{x - 2}$

44. $\dfrac{3x^4 + 2x^3 - 2x^2 - 1}{x + 1}$

45. $\dfrac{x^4 - x^2 + 3}{x - \dfrac{1}{2}}$

46. $\dfrac{x^3 + 2x^2 + 1}{x - \dfrac{2}{3}}$

47. $\dfrac{3x^3 + 5x^2 + 7x + 9}{x^2 + 2}$

48. $\dfrac{2x^4 + 2x^3 + 3x^2 + 6x - 1}{2x^2 + 3}$

49. $\dfrac{x^4 + x^3 - 4x + 1}{x^2 + 4}$

50. $\dfrac{2x^4 + x^3 - 8x^2 + 3x - 2}{x^2 - 5}$

51. $\dfrac{6x^3 + 5x^2 - 8x + 3}{3x^2 - 2x - 1}$

52. $\dfrac{x^3 - 9x^2 + 20x - 38}{x^2 - 3x + 5}$

53. $\dfrac{3x^4 - 7x^3 + 5x^2 + x - 2}{x^2 + x + 1}$

54. $\dfrac{2x^4 - x^3 - 10x^2 - 3x - 1}{x^2 - 3x + 1}$

55. $\dfrac{x^4 + 3x - 7}{x^2 + 2x - 3}$

56. $\dfrac{3x^4 - 4x^2 + 3}{x^2 + x - 1}$

57. $\dfrac{x^3 - 27}{x - 3}$

58. $\dfrac{x^3 + 125}{x + 5}$

59. $\dfrac{x^6 - 1}{x + 1}$

60. $\dfrac{x^6 - 1}{x - 1}$

61. $\dfrac{x^5 + 1}{x - 1}$

62. $\dfrac{x^6 + 1}{x + 1}$

63. $\dfrac{x^5 - x^3 + x}{x + \dfrac{1}{2}}$

64. $\dfrac{x^4 - 2x^3 + 4}{x + \dfrac{4}{5}}$

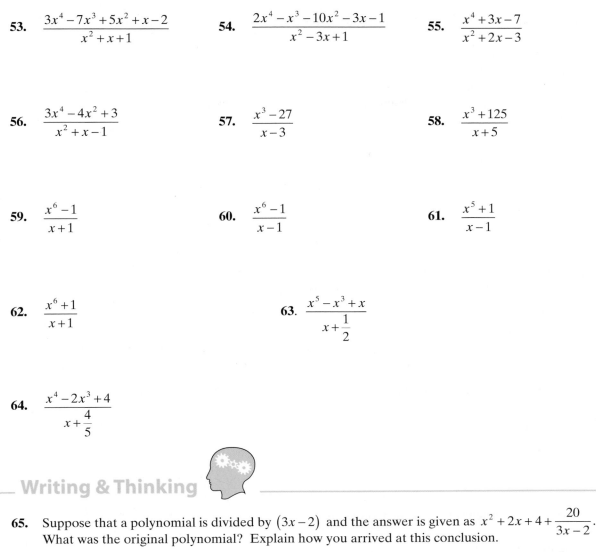

Writing & Thinking

65. Suppose that a polynomial is divided by $(3x - 2)$ and the answer is given as $x^2 + 2x + 4 + \dfrac{20}{3x - 2}$. What was the original polynomial? Explain how you arrived at this conclusion.

66. Suppose that a polynomial is divided by $(x + 5)$ and the answer is given as $x^2 - 3x + 2 - \dfrac{6}{x + 5}$. What was the original polynomial? Explain how you arrived at this conclusion.

67. Given that $P(x) = 2x^3 - 8x^2 + 10x + 15$.

    a. Find $P(2)$ then divide $P(x)$ by $x - 2$.

    b. Find $P(-1)$ then divide $P(x)$ by $x + 1$.

    c. Find $P(4)$ then divide $P(x)$ by $x - 4$.

    Do you see any pattern in the values of $P(a)$ for $x = a$ and the remainders you found in the division process? (**Hint:** Check the appendix section on Synthetic Division and the Remainder Theorem.)

Section 11.1: Exponents

Summary of the Rules for Exponents   page 855

For any nonzero real number a and integers m and n:

1. The exponent 1: $a = a^1$

2. The exponent 0: $a^0 = 1 \quad (a \neq 0)$

3. The product rule: $a^m \cdot a^n = a^{m+n}$

4. The quotient rule: $\dfrac{a^m}{a^n} = a^{m-n}$

5. Negative exponents: $a^{-n} = \dfrac{1}{a^n}$

Using a Calculator to Evaluate Expressions with Exponents   page 855

Section 11.2: Exponents and Scientific Notation

Summary of the Power Rules for Exponents   page 866

1. Power rule: $\left(a^m\right)^n = a^{mn}$

2. Power of a product: $\left(ab\right)^n = a^n b^n$

3. Power of a quotient: $\left(\dfrac{a}{b}\right)^n = \dfrac{a^n}{b^n}$

Using Combinations of Rules for Exponents   page 863

Scientific Notation   page 866

If N is a decimal number, then in **scientific notation** $N = a \times 10^n$ where $1 \leq a < 10$ and n is an integer.

Section 11.3: Introduction to Polynomials

Monomial   page 874

A **monomial in x** is a term of the form kx^n where k is a real number and n is a whole number. n is called the **degree** of the term, and k is called the **coefficient**.

Polynomial   page 875

A **polynomial** is a monomial or the indicated sum or difference of monomials. The **degree of a polynomial** is the largest of the degrees of its terms. The coefficient of the term of the largest degree is called the **leading coefficient**.

Special Terminology for Polynomials   page 875

Monomial: polynomial with one term
Binomial: polynomial with two terms
Trinomial: polynomial with three terms

Evaluation of Polynomials   page 877

A polynomial can be treated as a function and the notation $p(x)$ can be used.

Section 11.4: Addition and Subtraction with Polynomials

Addition with Polynomials   page 882

The **sum** of two or more polynomials is found by combining **like terms.**

Subtraction with Polynomials   page 883

The **difference** between two polynomials can be found by changing the sign of each term of the second polynomial and then combining like terms.

Simplifying Algebraic Expressions   page 885

If an algebraic expression contains more than one pair of grouping symbols, such as parentheses ( ), brackets [ ], or braces { }, simplify by working to remove the innermost pair of symbols first. **Apply the rules for order of operations** just as if the variables were numbers and proceed to combine like terms.

Section 11.5: Multiplication with Polynomials

Multiplying a Polynomial by a Monomial   page 889

Using the distributive property, $a(b+c)=ab+ac$, with multiplication indicated on the left, we can find the product of a monomial with a polynomial of two or more terms.

Multiplying Two Polynomials   page 889

The **distributive property** is applied by multiplying each term of one polynomial by each term of the other.

Chapter 11: Review

Section 11.1: Exponents

Simplify each expression. The final form of the expressions with variables should contain only positive exponents. Assume that all variables represent nonzero numbers.

1. $5^2 \cdot 5$

2. $4^3 \cdot 4^0$

3. $-3 \cdot 2^{-3}$

4. $5 \cdot 3^{-2}$

5. y^{-3}

6. $x^3 \cdot x^4$

7. $y^{-2} \cdot y^0$

8. $\left(6x^2\right)\left(-2x^2\right)$

9. $\dfrac{-12x^7}{2x^3}$

10. $\left(-3a^3\right)^0$

11. $\left(3a^3b^2c\right)\left(-2ab^2c\right)$

12. $\dfrac{35y^5 \cdot 2y^{-2}}{14xy^4}$

Section 11.2: Exponents and Scientific Notation

Use the rules for exponents to simplify each of the expressions. Assume that all variables represent nonzero real numbers.

13. $17\left(x^{-3}\right)^2$

14. $\left(\dfrac{y}{x}\right)^{-1}$

15. $\left(\dfrac{3x}{y^{-2}}\right)^2$

16. $\left(\dfrac{mn^{-1}}{m^3n^2}\right)^{-3}$

17. $\dfrac{\left(5x^{-2}\right)\left(-2x^6\right)}{\left(y^2\right)^{-3}}$

18. $\dfrac{\left(6x^{-2}\right)\left(5x^{-1}\right)}{\left(3y\right)\left(2y^{-1}\right)}$

19. $\left(\dfrac{3a^4bc^{-1}}{7ab^{-1}c^2}\right)^{-2}\left(\dfrac{9abc^{-1}}{7b^2c^2}\right)^2$

Write the following numbers in scientific notation.

20. 29,300,000

21. 0.0075

Write the following numbers in decimal form.

22. 7.24×10^{-4}

23. 9.485×10^7

First write each of the numbers in scientific notation. Then perform the indicated operations and leave your answer in scientific notation.

24. $\dfrac{0.0058}{290}$

25. $\dfrac{2.7 \cdot 0.002 \cdot 25}{54 \cdot 0.0005}$

Solve the following word problem.

26. **Altitude:** A commercial airplane has a cruising speed of 555 mph at an altitude of 35,000 ft. Express the altitude of the plane in scientific notation.

Section 11.3: Introduction to Polynomials

27. $2x^2 + 3y^{\frac{1}{2}} + 3$

28. $3 + 5y^3 + 20y$

29. $10y - \dfrac{4}{7}y^3$

30. $5x^2 - x + x^2$

31. $3a + 4a^2 + 6a - 10a^3$

32. $6x^2 + 4 - 2x^2 + 1 - 4x^2$

33. $a^3 - 2a^2 - 8a + 3a^3 - 2a^2 + 6a + 9$

34. Given $p(x) = x^2 - 13x + 5$, find $p(-1)$.

35. Given $p(x) = 3x^2 + 4x - 6$, find $p(2)$.

36. Given $p(y) = 3y^3 - 3y^2 + 10$, find $p(4)$.

37. Given $p(y) = y^2 - 8y + 13$, find $p(-5)$.

38. Given $p(x) = x^2 + 6x + 9$, find $p(a)$.

39. Given $p(x) = 2x - 7$, find $p(4a - 1)$.

Section 11.4: Addition and Subtraction with Polynomials

40. $\left(-5x^2 + 2x - 3\right) + \left(4x^2 - x + 6\right)$

41. $\left(x^2 + 7x + 12\right) + \left(2x - 8\right) + \left(x^2 + 10x\right)$

42.
$$x^3 + 4x^2 + x$$
$$\underline{-3x^3 - x^2 + 2x + 7}$$

43.
$$x^3 + 4x^2 - 6$$
$$8x^2 + 3x - 1$$
$$\underline{x^3 + x^2 - 4x}$$

44. $\left(x^4 + 7x^3 - x + 13\right) - \left(-x^4 + 6x^2 - x + 3\right)$

45. $\left(7x^2 - 14\right) - \left(4x^2 - 2x - 12\right)$

46.
$$13x^2 - 6x + 7$$
$$\underline{-\left(9x^2 + x + 5\right)}$$

47.
$$11x^3 + 10x - 20$$
$$\underline{-\left(x^3 + 5x^2 - 10x - 22\right)}$$

Simplify each of the algebraic expressions, and write the polynomials in descending order.

48. $6x + 2(x-7) - (3x+4)$

49. $3x + \left[8x - 4(3x+1) - 9\right]$

50. $\left(x^2 + 4\right) - \left[-6 + 2\left(3 - 2x^2\right) + x\right]$

51. $2\left[4x + 5(x-8) - (2x+3)\right] + (3x-5)$

52. Subtract $3x^2 - 2x + 4$ from $8x^2 - 10x + 3$.

53. Subtract $4x^2 - 8x + 1$ from the sum of $2x^2 - 5x + 2$ and $x^2 - 14x + 11$.

Section 11.5: Multiplication with Polynomials

Multiply as indicated and simplify if possible.

54. $7a^3\left(-5a^2 + 3a - 1\right)$

55. $3a(-2a+6) + 5(-2a+6)$

56. $(7t+3)(t-2)$

57. $(3a-3)(3a-3)$

58. $-4(x+3)(x-5)$

59. $(2y+1)\left(y^2 - 3y + 2\right)$

60. $y^2 + 4y + 1$
$\underline{\, 3y - 8}$

61. $x^2 + 3x + 5$
$\underline{x^2 + 3x - 2}$

Simplify.

62. $(x+4)(x-2) + (3x-1)(x-5)$

63. $(x+2)(x+5) - (x-6)(x+6)$

Section 11.6 Special Products of Binomials

Find each product and identify those that are the difference of two squares or perfect square trinomials.

64. $(2x+9)^2$

65. $(y+8)(y-8)$

66. $(3x+5)(2x+5)$

67. $(3x-1)^2$

68. $\left(y^2+6\right)\left(y^2-4\right)$

69. $(3x-7)(3x+7)$

70. $(7-3x)^2$

71. $\left(t^3 + \dfrac{1}{4}\right)\left(t^3 - \dfrac{1}{4}\right)$

72. $\left(y + \dfrac{2}{3}\right)\left(y - \dfrac{1}{2}\right)$

Section 11.7: Division with Polynomials

Express each quotient as a sum (or difference) of fractions and simplify if possible.

73. $\dfrac{4y^2 + 16y + 20}{2y^2}$

74. $\dfrac{12y^4 + 6y^3 - 3y^2 + 11y}{3y^2}$

Divide by using the division algorithm. Write the answers in the form $Q + \dfrac{R}{D}$, where the degree of $R <$ the degree of D.

75. $\dfrac{x^2 - 2x - 15}{x - 5}$

76. $\dfrac{8y^2 + 10y + 5}{y + 3}$

77. $\dfrac{4x^3 - 20x + 3}{x - 2}$

78. $\dfrac{64x^3 - 125}{4x - 5}$

79. $\dfrac{10x^3 + 17x^2 - 2x + 45}{2x + 5}$

80. $\dfrac{x^3 - 8x^2 + 20x - 50}{x^2 - 2x + 3}$

Chapter 11: Test

Use the rules for exponents to simplify each expression. Each answer should have only positive exponents. Assume that all variables represent nonzero numbers.

1. $\left(5a^2b^5\right)\left(-2a^3b^{-5}\right)$

2. $\left(-7x^4y^{-3}\right)^0$

3. $\dfrac{\left(-8x^2y^{-3}\right)^2}{16xy}$

4. $\dfrac{3x^{-2}}{9x^{-3}y^2}$

5. $\left(\dfrac{4xy^2}{x^3}\right)^{-1}$

6. $\left(\dfrac{2x^0y^3}{x^{-1}y}\right)^2$

Solve the following problems.

7. Write each of the following numbers in decimal notation.

    a. 1.35×10^5

    b. 2.7×10^{-6}

8. First write each of the numbers in scientific notation. Then perform the indicated operations and leave your answer in scientific notation.

    a. $250\cdot500{,}000$

    b. $\dfrac{65\cdot0.012}{1500}$

Simplify each polynomial. Write the polynomial in descending order and state the degree, type, and leading coefficient of the simplified polynomial.

9. $3x+4x^2-x^3+4x^2+x^3$

10. $2x^2+3x-x^3+x^2-1$

11. $17-14+4x^5-3x+2x^4+x^5-8x$

12. For the polynomial $P(x)=2x^3-5x^2+7x+10$, find

    a. $P(2)$

    b. $P(-3)$

Simplify the following expressions.

13. $7x+\left[2x-3(4x+1)+5\right]$

14. $12x-2\left[5-(7x+1)+3x\right]$

Perform the indicated operations and simplify each expression. Tell which, if any, products are the difference of two squares or perfect square trinomials.

15. $\left(5x^3 - 2x + 7\right) + \left(-x^2 + 8x - 2\right)$

16. $\left(x^4 + 3x^2 + 9\right) - \left(-6x^4 - 11x^2 + 5\right)$

17. $5x^2\left(3x^5 - 4x^4 + 3x^3 - 8x^2 - 2\right)$

18. $\left(7x + 3\right)\left(7x - 3\right)$

19. $\left(4x + 1\right)^2$

20. $\left(6x - 5\right)\left(6x - 5\right)$

21. $\left(2x + 5\right)\left(6x - 3\right)$

22. $3x\left(x - 7\right)\left(2x - 9\right)$

23. $\left(3x + 1\right)\left(3x - 1\right) - \left(2x + 3\right)\left(x - 5\right)$

24.
$$\begin{array}{r} 2x^3 - 3x - 7 \\ \times \quad 5x + 2 \\ \hline \end{array}$$

Express each quotient as a sum (or difference) of fractions and simplify if possible.

25. $\dfrac{4x^3 + 3x^2 - 6x}{2x^2}$

26. $\dfrac{5a^3 + 6a^4 + 3a}{3a^2}$

Divide by using the division algorithm. Write the answers in the form $Q + \dfrac{R}{D}$ where the degree of R < the degree of D.

27. $\left(2x^2 - 9x - 20\right) \div \left(2x + 3\right)$

28. $\dfrac{x^3 - 8x^2 + 3x + 15}{x^2 + x - 3}$

Solve the following word problems.

29. Subtract $5x^2 - 3x + 4$ from the sum of $4x^2 + 2x + 1$ and $3x^2 - 8x - 10$.

30. Rectangles: A sheet of metal is in the shape of a rectangle with width 12 inches and length 20 inches. A slot (see the figure) of width x inches and length $x + 3$ inches is cut from the top of the rectangle.

a. Write a polynomial function $A(x)$ that represents the area of the remaining figure.

b. Write a polynomial function $P(x)$ that represents the perimeter of the remaining figure.

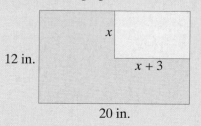

Cumulative Review: Chapters 1 - 11

Perform the indicated operations. Reduce all answers to lowest terms.

1. $4.12 - 7.9 + 13.2$

2. $\dfrac{2}{5} + \dfrac{4}{15} - \dfrac{2}{3} + \dfrac{7}{10}$

3. $\dfrac{21}{8} \cdot \dfrac{16}{49} \cdot \dfrac{35}{18}$

4. $61\overline{)1952}$

Use the rules for order of operations to evaluate the expressions.

5. $\dfrac{2}{3} \div \dfrac{1}{5} + \dfrac{4}{9}$

6. $\dfrac{8}{11} \div \left(-\dfrac{4}{5}\right) \cdot \dfrac{1}{2} + \dfrac{3}{22}$

Use the distributive property to complete the expressions.

7. $3x + 45 = 3($ $)$

8. $6x + 16 = 2($ $)$

Find the LCM for each set of numbers or terms.

9. $12, 15, 54$

10. $6a^2, 24ab^3, 30ab, 40a^2b^2$

The graphs of sets of real numbers are given. a. Use set-builder notation to indicate the set of numbers shown in each graph. b. Use interval notation to represent the graph. c. Tell what type of interval is illustrated.

11.
$-3 \quad\quad 0 \;\; 1.5 \;\; 3$

12.
$-4 \quad\quad -2 \quad\quad 0 \quad\quad 2$

Solve the following equations.

13. $13x - 5x + 7 = 2 - 19$

14. $7(4 - x) = 3(x + 4)$

15. $-2(5x + 1) + 2x = 4(x + 1)$

16. $\dfrac{2}{3}x + \dfrac{1}{2} = \dfrac{3}{4}$

17. $1.5x - 3.7 = 3.6x + 2.6$

Use your knowledge of linear equations to solve the following problems.

18. Graph the linear equation $8x - 3y = 24$ by locating the x-intercept and the y-intercept.

19. Write an equation in standard form for the line that has slope $m = \dfrac{2}{3}$ and contains the point $(-1, 2)$. Graph the line.

20. Write an equation in slope-intercept form for the line passing through the two points $(-5, 1)$ and $(3, -4)$. Graph the line.

21. Write an equation in standard form for the line that is parallel to the line $x + 6y = 6$ and passes through the point $(0, 7)$. Graph both lines.

Graph the following inequalities and tell what type of half-plane is graphed.

22. $3x + y < 10$

23. $4x + 3y \geq 9$

Simplify each expression so that it has no exponents or only positive exponents.

24. $\dfrac{4x^3}{2x^{-2}x^4}$

25. $\left(4x^2y\right)^3$

26. $\left(a^{-3}b^2\right)^{-2}$

27. $\left(\dfrac{3xy^4}{x^2}\right)^{-1}$

28. $\left(\dfrac{8x^{-3}y^2}{xy^{-1}}\right)^0$

29. $\left(\dfrac{3^{-1}x^3y^{-1}}{x^{-1}y^2}\right)^2$

Write each number in decimal notation.

30. **a.** 2.8×10^{-7}

b. 3.51×10^4

First write each of the numbers in scientific notation. Then perform the indicated operations and leave your answer in scientific notation.

31. $\dfrac{840}{0.00021}$

32. $\dfrac{0.005 \cdot 77}{0.011 \cdot 3500}$

For the following function, a. simplify the polynomial, b. find $P(3)$ and c. find $P(a)$.

33. $P(x) = 9x - x^3 + 3x^2 - x + x^3$

Simplify the polynomial. Write the polynomial in descending order and state the degree and type of the simplified polynomial. Also state the leading coefficient.

34. $3x^2 - 8(x - 5) + x^4 - 2x^3 + x^2 - 2x$

Perform the indicated operations and simplify each expression.

35. $\left(6x^2 + x - 10\right) - \left(x^3 - x^2 + x - 4\right)$

36. $\left(x^2 + 2x + 6\right) + \left(5x^2 - x - 2\right) - \left(8x + 3\right)$

37. $-3x\left(x^2-4x+1\right)$ **38.** $(x+4)(x-3)$ **39.** $(3x+7)(3x+7)$ **40.** $(2x-1)(2x+1)$

41. $(2x-9)(x+4)$ **42.** $(3x-4)(2x+3)$ **43.** $(4x+1)\left(x^2-x\right)$ **44.** $(y-5)\left(y^2+5y+25\right)$

Express the quotient as a sum (or difference) of fractions and simplify if possible.

45. $\dfrac{8x^2-14x+6}{2x}$

Divide by using the division algorithm and express each answer in the form $Q+\dfrac{R}{D}$, where the degree of R < the degree of D.

46. $\dfrac{x^2+7x-18}{x+9}$ **47.** $\dfrac{2x^3+5x^2+7}{x+3}$

Use the vertical line test to determine whether or not each graph represents a function. State the domain and range.

48.

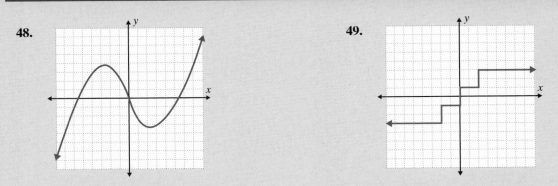

49.

Solve the following word problems.

50. **Consecutive Integers:** Find three consecutive even integers such that the sum of the second and third is equal to three times the first decreased by 14.

51. **Rectangle:** The length of a rectangle is 9 centimeters more than its width. The perimeter is 82 centimeters. Find the dimensions of the rectangle.

52. **Number Problem:** Twice the difference between a number and 16 is equal to 4 times the number increased by 6. What is the number?

53. **Triangles:** Find the length of the hypotenuse (to the nearest tenth of an inch) of a right triangle with legs of length 36 in. and 50 in.

54. Investing: Sylvester has two investments that total $100,000. One investment earns interest at 6% and the other investment earns interest at 8%. If the total amount of interest from the two investments is $6700 in one year, how much money does he have invested at each rate?

55. Biking: Rosa went bike riding. She biked from her home to the beach at a rate of 10 mph and returned at a rate 5 mph faster. (She had a tailwind coming back.) If her return trip took 20 minutes less than the trip to the beach, how far is her home from the beach?

56. Geometry: Find **a.** the perimeter and **b.** the area of the figure shown.

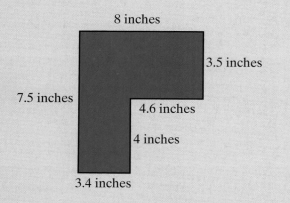

8 inches

3.5 inches

7.5 inches

4.6 inches

4 inches

3.4 inches

57. Square: A square is 30 inches on each side. A square x inches on each side is cut from each corner of the square.

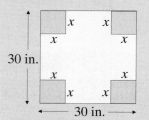

30 in.

30 in.

a. Represent the area of the remaining portion of the square in the form of a polynomial function, $A(x)$.

b. Represent the perimeter of the remaining portion of the square in the form of a polynomial function, $P(x)$.

Factoring Polynomials and Solving Quadratic Equations

Mathematics at Work!

Many concepts in physics can be represented by quadratic polynomials. For example, Joey is driving his car to the store. The velocity function for his car can be represented by $v = 24t^2 - 32t + 8$, where t represents the number of minutes he has been driving. Use factoring methods to find the times when Joey's car is not moving. (See Section 12.7.)

12.1 Greatest Common Factor and Factoring by Grouping

The result of multiplication is called the **product** and the numbers or expressions being multiplied are called **factors** of the product. The reverse of multiplication is called **factoring**. That is, given a product, we want to find the factors.

Multiplying Polynomials	**Factoring Polynomials**
$3x(x+5) = 3x^2 + 15x$	$3x^2 + 15x = 3x(x+5)$

factors   product   product   factors

Factoring polynomials relies heavily on the multiplication techniques developed in Section 11.5 and 11.6. You must remember how to multiply in order to be able to factor. Furthermore, you will find that the skills used in factoring polynomials are necessary when simplifying rational expressions (in a later chapter) and when solving equations. In other words, study this section and Sections 12.2, 12.3 and 12.4, with extra care.

Objective A **Greatest Common Factor of a Set of Terms**

The **greatest common factor** (**GCF**) of two or more integers is the largest integer that is a factor (or divisor) of all of the integers. For example, the GCF of 30 and 40 is 10. Note that 5 is also a common factor of 30 and 40, but 5 is not the **greatest** common factor. The number 10 is the largest number that will divide into both 30 and 40.

One way of finding the GCF is to use the prime factorization of each number. For example, to find the GCF for 36 and 60, we can write

$$36 = 4 \cdot 9 = 2 \cdot 2 \cdot 3 \cdot 3$$
$$60 = 4 \cdot 15 = 2 \cdot 2 \cdot 3 \cdot 5$$

The common factors are $2, 2,$ and 3 and their product is the GCF.

$$\text{GCF} = 2 \cdot 2 \cdot 3 = 12$$

Writing the prime factorizations using exponents gives

$$36 = 2^2 \cdot 3^2 \quad \text{and} \quad 60 = 2^2 \cdot 3 \cdot 5.$$

We can see that the GCF is the product of the greatest power of each prime factor that is common to both numbers. That is,

$$\text{GCF} = 2^2 \cdot 3 = 12.$$

This procedure can be used to find the GCF for any set of integers or algebraic terms with integer exponents.

Procedure for Finding the GCF of a Set of Terms

1. Find the prime factorization of all integers and integer coefficients.
2. List all the factors that are common to all terms, including variables.
3. Choose the greatest power of each factor that is common to all terms.
4. Multiply these powers to find the GCF.

Note: If there is no common prime factor or variable, then the GCF is 1.

Example 1

Finding the GCF

Find the GCF for each of the following sets of algebraic terms.

a. $\{30, 45, 75\}$

Solution

Find the prime factorization of each number:
$$30 = 2 \cdot 3 \cdot 5, \quad 45 = 3^2 \cdot 5, \quad \text{and} \quad 75 = 3 \cdot 5^2.$$

The common factors are 3 and 5 and the greatest power of each that is common to all numbers is 3^1 and 5^1.
Thus $\text{GCF} = 3^1 \cdot 5^1 = 15$.

b. $\left\{20x^4 y, 15x^3 y, 10x^5 y^2\right\}$

Solution

Writing each integer coefficient in prime factored form gives:

$$20x^4 y = 2^2 \cdot 5 \cdot x^4 \cdot y,$$
$$15x^3 y = 3 \cdot 5 \cdot x^3 \cdot y,$$
$$10x^5 y^2 = 2 \cdot 5 \cdot x^5 \cdot y^2.$$

The common factors are 5, x, and y and after finding the greatest power of each that is common to all three terms, we have 5^1, x^3, and y^1.
Thus $\text{GCF} = 5^1 \cdot x^3 \cdot y^1 = 5x^3 y$.

Now work margin exercise 1.

1. Find the GCF for each of the following sets of algebraic terms.

a. $\{5, 10, 20\}$

b. $\left\{150xy, 250y^2 x^3, 100x^2 y^2\right\}$

Now consider the polynomial $3n + 15$. We want to write this polynomial as a product of two factors. Since

$$3n + 15 = 3 \cdot n + 3 \cdot 5,$$

we see that 3 is a common factor of the two terms in the polynomial. By using the distributive property, we can write

$$3n + 15 = 3 \cdot n + 3 \cdot 5 = 3(n + 5).$$

In this way, the polynomial $3n + 15$ has been **factored** into the product of 3 and $(n + 5)$.

For a more general approach to finding factors of a polynomial, we can use the quotient rule for exponents.

$$\frac{a^m}{a^n} = a^{m-n}$$

This property is used when dividing terms. For example,

$$\frac{35x^8}{5x^2} = 7x^6 \quad \text{and} \quad \frac{16a^5}{-8a} = -2a^4.$$

To divide a polynomial by a monomial, each term in the polynomial is divided by the monomial. For example,

$$\frac{8x^3 - 14x^2 + 10x}{2x} = \frac{8x^3}{2x} - \frac{14x^2}{2x} + \frac{10x}{2x} = 4x^2 - 7x + 5.$$

With practice, this division can be done mentally.

This concept of dividing each term by a monomial is part of finding a monomial factor of a polynomial. Finding the greatest common monomial factor of a polynomial means to **find the GCF of the terms of the polynomial**. This monomial will be one factor, and the sum of the various quotients found by dividing each term by the GCF will be the other factor. Thus using the quotients just found and the fact that $2x$ is the GCF of the terms,

$$8x^3 - 14x^2 + 10x = 2x(4x^2 - 7x + 5).$$

We say that $2x$ is **factored out** and $2x$ and $(4x^2 - 7x + 5)$ are the factors of $8x^3 - 14x^2 + 10x$.

Factoring Out the GCF

To find a monomial that is the greatest common factor (GCF) of a polynomial:

1. Find the variable(s) of highest degree and the largest integer coefficient that are factors of each term of the polynomial. (The product of these is the greatest common monomial factor.)
2. Divide this monomial factor into each term of the polynomial, resulting in another polynomial factor.

Now factor $24x^6 - 12x^4 - 18x^3$. The GCF is $6x^3$ and is factored out as follows:

$$24x^6 - 12x^4 - 18x^3 = 6x^3 \cdot 4x^3 + 6x^3(-2x) + 6x^3(-3)$$
$$= 6x^3(4x^3 - 2x - 3).$$

The factoring can be checked by multiplying $6x^3$ and $4x^3 - 2x - 3$. The product should be the expression we started with:

$$6x^3(4x^3 - 2x - 3) = 24x^6 - 12x^4 - 18x^3.$$

By definition, the GCF of a polynomial will have a positive coefficient. **However, if the leading coefficient is negative, we may choose to factor out the negative of the GCF (or $-1 \cdot \text{GCF}$).** This technique will leave a positive coefficient for the first term of the other polynomial factor. For example,

the GCF for $-10a^4b + 15a^4$ is $5a^4$ and we can factor as follows.

$$-10a^4b + 15a^4 = 5a^4(-2b + 3)$$

or $\quad -10a^4b + 15a^4 = -5a^4(2b - 3)$

Both answers are correct.

Suppose we factor $8n + 16$ as follows.

$$8n + 16 = 4(2n + 4)$$

But note that $2n + 4$ has 2 as a common factor. Therefore, we have not factored **completely**. **An expression is factored completely if none of its factors can be factored.**

Factoring polynomials means to find factors that are integers (other than +1) or polynomials with integer coefficients. If this cannot be done, we say that the polynomial is **not factorable**. For example, the polynomials $x^2 + 36$ and $3x + 17$ are not factorable.

Example 2

Factoring Out the GCF of a Polynomial

Factor each polynomial by factoring out the greatest common monomial factor.

a. $6n + 30$

Solution

$$6n + 30 = 6 \cdot n + 6 \cdot 5 = 6(n+5)$$

Checking by multiplying gives $6(n+5) = 6n + 30$, the original expression.

b. $x^3 + x$

Solution

$$x^3 + x = x \cdot x^2 + x \cdot 1 = x(x^2 + 1) \qquad \text{+1 is the coefficient of } x.$$

c. $5x^3 - 15x^2$

Solution

$$5x^3 - 15x^2 = 5x^2 \cdot x + 5x^2(-3) = 5x^2(x - 3)$$

d. $2x^4 - 3x^2 + 2$

Solution

$2x^4 - 3x^2 + 2$
This polynomial has no common monomial factor other than 1. In fact, this polynomial is **not factorable**.

e. $-4a^5 + 2a^3 - 6a^2$

Solution

The GCF is $2a^2$ and we can factor as follows.
$$-4a^5 + 2a^3 - 6a^2 = 2a^2(-2a^3 + a - 3)$$
However, the leading coefficient is negative and we can also factor as follows.
$$-4a^5 + 2a^3 - 6a^2 = -2a^2(2a^3 - a + 3)$$
Both answers are correct. However, we will see later that having a positive leading coefficient for the polynomial in parentheses may make that polynomial easier to factor.

A polynomial may be in more than one variable. For example, $5x^2y + 10xy^2$ is in the two variables x and y. Thus the GCF may have more than one variable.

$$5x^2y + 10xy^2 = 5xy \cdot x + 5xy \cdot 2y$$
$$= 5xy(x + 2y)$$

Similarly,

$$4xy^3 - 2x^2y^2 + 8xy^2 = 2xy^2 \cdot 2y + 2xy^2(-x) + 2xy^2 \cdot 4$$
$$= 2xy^2(2y - x + 4).$$

Example 3

Factoring Out the GCF of a Polynomial

Factor each polynomial by finding the GCF (or $-1 \cdot$ GCF).

a. $4ax^3 + 4ax$

Solution

$$4ax^3 + 4ax = 4ax(x^2 + 1) \qquad \text{Note that } 4ax = 1 \cdot 4ax.$$

Checking by multiplying gives $4ax(x^2 + 1) = 4ax^3 + 4ax$, the original expression.

b. $3x^2y^2 - 6xy^2$

Solution

$$3x^2y^2 - 6xy^2 = 3xy^2(x - 2)$$

c. $-14by^3 - 7b^2y + 21by^2$

Solution

$$-14by^3 - 7b^2y + 21by^2 = -7by(2y^2 + b - 3y)$$

Note that by factoring out a negative term, in this case $-7by$, the leading coefficient in parentheses is positive.

d. $13a^4b^5 - 3a^2b^9 - 4a^6b^3$

Solution

$$13a^4b^5 - 3a^2b^9 - 4a^6b^3 = a^2b^3(13a^2b^2 - 3b^6 - 4a^4)$$

Now work margin exercises 2 and 3.

Factor out the GCF of the following polynomials (or $-1 \cdot$ GCF).

2. a. $7n + 21$

   b. $3y^3 + y^4$

   c. $9x + 54x^2$

   d. $6x^2 + 2x - 9$

   e. $-9a^3 - 18a + 9a^2$

3. a. $10xy + 30xy^2$

   b. $2a^2b^2 - 16ab^2$

   c. $5xz^2 + 15xz^3 - 20x^2z$

   d. $-3b^2d + 12b^2 - 15b^3d^2$

Objective C — Factoring by Grouping

Consider the expression

$$y(x+4)+2(x+4)$$

as the sum of two terms, $y(x+4)$ and $2(x+4)$. Each of these "terms" has the common **binomial** factor $(x+4)$. Factoring out this common binomial factor by using the distributive property gives

$$y(x+4)+2(x+4)=(x+4)(y+2).$$

Similarly,

$$3(x-2)-a(x-2)=(x-2)(3-a).$$

Now consider the product

$$(x+3)(y+5)=(x+3)y+(x+3)5$$
$$=xy+3y+5x+15$$

which has four terms and no like terms. Yet the product has two factors, namely $(x+3)$ and $(y+5)$. Factoring polynomials with four or more terms can sometimes be accomplished by grouping the terms and using the distributive property, as in the above discussion and the following examples. Keep in mind that the common factor can be a binomial or other polynomial.

4. Factor each polynomial.

a. $x^2(2x-y)+2(2x-y)$

b. $6y(x-u)+(x-u)$

Example 4

Factoring Out a Common Binomial Factor

Factor each polynomial.

a. $3x^2(5x+1)-2(5x+1)$

Solution

$$3x^2(5x+1)-2(5x+1)=(5x+1)(3x^2-2)$$

b. $7a(2x-3)+(2x-3)$

Solution

$$7a(2x-3)+(2x-3)=7a(2x-3)+1\cdot(2x-3) \qquad \text{1 is the understood coefficient of } (2x-3).$$
$$=(2x-3)(7a+1)$$

Now work margin exercise 4.

In the examples just discussed, the common binomial factor was in parentheses. However, many times the expression to be factored is in a form with four or more terms. For example, by multiplying in Example 4a, we get the following expression.

$$3x^2(5x+1) - 2(5x+1) = 15x^3 + 3x^2 - 10x - 2$$

The expression $15x^3 + 3x^2 - 10x - 2$ has four terms with no common monomial factor; yet, we know that it has the two binomial factors $(5x+1)$ and $(3x^2 - 2)$. We can find the binomial factors by **grouping**. This means looking for common factors in each group and then looking for common binomial factors. The process is illustrated in Example 5.

Example 5

Factoring by Grouping

Factor each polynomial by grouping.

a. $xy + 5x + 3y + 15$

Solution

$$xy + 5x + 3y + 15 = (xy + 5x) + (3y + 15) \qquad \text{Group terms that have a common monomial factor.}$$
$$= x(y+5) + 3(y+5) \qquad \text{using the distributive property}$$
$$= (y+5)(x+3) \qquad (y+5) \text{ is a common binomial factor.}$$

Checking gives $(y+5)(x+3) = xy + 5x + 3y + 15$.

b. $x^2 - xy - 5x + 5y$

Solution

$$x^2 - xy - 5x + 5y = (x^2 - xy) + (-5x + 5y)$$
$$= x(x-y) + 5(-x+y)$$

This does not work because $(x-y) \neq (-x+y)$. However, these two expressions are **opposites**. Thus we can find a common factor by factoring −5 instead of +5 from the last two terms.

$$x^2 - xy - 5x + 5y = (x^2 - xy) + (-5x + 5y)$$
$$= x(x-y) - 5(x-y)$$
$$= (x-y)(x-5) \qquad \text{Success!}$$

c. $x^2 + ax + 3x + 3y$

Solution

$$x^2 + ax + 3x + 3y = (x^2 + ax) + (3x + 3y)$$
$$= x(x+a) + 3(x+y)$$

But $x + a \neq x + y$ and there is no common factor.
So $x^2 + ax + 3x + 3y$ is **not factorable**.

5. Factor each polynomial by grouping.

    a. $xy + 2y + 6x + 12$

    b. $xy - 3y - 2x + 6$

    c. $2x + 6 + xy - x^2$

    d. $x^2 + xy + x + y$

    e. $8xy - 6wy - 12xz + 9wz$

---

d. $xy + 5x + y + 5$

Solution

$$\begin{aligned}
xy + 5x + y + 5 &= (xy + 5x) + (y + 5) \\
&= x(y + 5) + 1 \cdot (y + 5) \\
&= (y + 5)(x + 1)
\end{aligned}$$

1 is the understood coefficient of $(y + 5)$.

e. $5xy + 6uv - 3vy - 10ux$

Solution

In the expression $5xy + 6uv - 3vy - 10ux$ there is no common factor in the first two terms. However, the first and third terms have a common factor so we rearrange the terms as follows.

$$\begin{aligned}
5xy + 6uv - 3vy - 10ux &= (5xy - 3vy) + (6uv - 10ux) \\
&= y(5x - 3v) + 2u(3v - 5x)
\end{aligned}$$

Now we see that $5x - 3v$ and $3v - 5x$ are opposites and we factor out $-2u$ from the last two terms. The result is as follows.

$$\begin{aligned}
5xy + 6uv - 3vy - 10xu &= (5xy - 3vy) + (6uv - 10ux) \\
&= y(5x - 3v) - 2u(-3v + 5x) \\
&= y(5x - 3v) - 2u(5x - 3v) \quad \text{Note } 5x - 3v = -3v + 5x. \\
&= (5x - 3v)(y - 2u)
\end{aligned}$$

*Now work margin exercise **5**.*

Practice Problems

Factor each expression completely.

  1. $2x - 16$ **2.** $-5x^2 - 5x$

  3. $7ax^2 - 7ax$ **4.** $a^4b^5 + 2a^3b^2 - a^3b^3$

  5. $9x^2y^2 + 12x^2y - 6x^3$ **6.** $6a^3(x + 3) - (x + 3)$

  7. $5x + 35 - xy - 7y$

Practice Problem Answers

  1. $2(x - 8)$ **2.** $-5x(x + 1)$ **3.** $7ax(x - 1)$

  4. $a^3b^2(ab^3 + 2 - b)$ **5.** $3x^2(3y^2 + 4y - 2x)$

  6. $(x + 3)(6a^3 - 1)$ **7.** $(5 - y)(x + 7)$

Find the GCF for each set of terms. See Example 1.

1. $\{10, 15, 20\}$

2. $\{25, 30, 75\}$

3. $\{16, 40, 56\}$

4. $\{30, 42, 54\}$

5. $\{9, 14, 22\}$

6. $\{44, 66, 88\}$

7. $\{30x^3, 40x^5\}$

8. $\{15y^4, 25y\}$

9. $\{8a^3, 16a^4, 20a^2\}$

10. $\{36xy, 48xy, 60xy\}$

11. $\{26ab^2, 39a^2b, 52a^2b^2\}$

12. $\{28c^2d^3, 14c^3d^2, 42cd^2\}$

13. $\{45x^2y^2z^2, 75xy^2z^3\}$

14. $\{21a^5b^4c^3, 28a^3b^4c^3, 35a^3b^4c^2\}$

Simplify the expressions.

15. $\dfrac{x^7}{x^3}$

16. $\dfrac{x^8}{x^3}$

17. $\dfrac{-8y^3}{2y^2}$

18. $\dfrac{12x^2}{2x}$

19. $\dfrac{9x^5}{3x^2}$

20. $\dfrac{-10x^5}{2x}$

21. $\dfrac{4x^3y^2}{2xy}$

22. $\dfrac{21x^4y^3}{-3xy^2}$

Complete the factoring of the polynomial as indicated.

23. $3m + 27 = 3(\quad)$

24. $2x + 18 = 2(\quad)$

25. $5x^2 - 30x = 5x(\quad)$

26. $6y^3 - 24y^2 = 6y^2(\quad)$

27. $13ab^2 + 13ab = 13ab(\quad)$

28. $8x^2y - 4xy = 4xy(\quad)$

29. $-15xy^2 - 20x^2y - 5xy = -5xy(\quad)$

30. $-9m^3 - 3m^2 - 6m = -3m(\quad)$

31. $11x - 121$

32. $14x + 21$

33. $16y^3 + 12y$

34. $-3x^2 + 6x$

35. $-6ax + 9ay$

36. $4ax - 8ay$

37. $10x^2y - 25xy$

38. $16x^4y - 14x^2y$

39. $-18y^2z^2 + 2yz$

40. $-14x^2y^3 - 14x^2y$

41. $8y^2 - 32y + 8$

42. $5x^2 - 15x - 5$

43. $2xy^2 - 3xy - x$

44. $ad^2 + 10ad + 25a$

45. $8m^2x^3 - 12m^2y + 4m^2z$

46. $36t^2x^4 - 45t^2x^3 + 24t^2x^2$

47. $-56x^4z^3 - 98x^3z^4 - 35x^2z^5$

48. $34x^4y^6 - 51x^3y^5 + 17x^5y^4$

49. $15x^4y^2 + 24x^6y^6 - 32x^7y^3$

50. $-3x^2y^4 - 6x^3y^4 - 9x^2y^3$

51. $7y^2(y+3) + 2(y+3)$

52. $6a(a-7) - 5(a-7)$

53. $3x(x-4) + 2(x-4)$

54. $2x^2(x+5) + 7(x+5)$

55. $4x^3(x-2) - (x-2)$

56. $9a(x+1) - (x+1)$

57. $10y(2y+3) - 7(2y+3)$

58. $a(x+5) + b(x+5)$

59. $a(x-2) - b(x-2)$

60. $3a(x-10) + 5b(x-10)$

Factor each of the polynomials by grouping. If a polynomial cannot be factored, write "not factorable." See Example 5.

61. $bx + b + cx + c$

62. $3x + 3y + ax + ay$

63. $x^3 + 3x^2 + 6x + 18$

64. $2z^3 - 14z^2 + 3z - 21$

65. $10a^2 - 5az + 2a + z$

66. $x^2 - 4x + 6xy - 24y$

67. $3x + 3y - bx - by$

68. $ax + 5ay + 3x + 15y$

69. $5xy + yz - 20x - 4z$

70. $x - 3xy + 2z - 6zy$

71. $z^2 + 3 + az^2 + 3a$

72. $x^2 - 5 + x^2y + 5y$

73. $6ax + 12x + a + 2$

74. $4xy + 3x - 4y - 3$

75. $xy + x + y + 1$

76. $xy + x - y - 1$

77. $10xy - 2y^2 + 7yz - 35xz$

78. $7xy - 3y + 2x^2 - 3x$

79. $3xy - 4uy - 6vx + 8uv$

80. $xy + 5vy + 6ux + 30uv$

81. $3ab + 4ac + 2b + 6c$

82. $24y - 3yz + 2xz - 16x$

83. $6ac - 9ad + 2bc - 3bd$

84. $2ac - 3bc + 6ad - 9bd$

Writing & Thinking

85. Explain why the GCF of $-3x^2 + 3$ is 3 and not -3.

12.2 Factoring Trinomials: $x^2 + bx + c$

In Chapter 11 we learned to use the FOIL method to multiply two binomials. In many cases the simplified form of the product was a trinomial. In this section we will learn to factor trinomials by reversing the FOIL method. In particular, we will focus on factoring trinomials in one variable with leading coefficient 1.

Objective A   **Factoring Trinomials with Leading Coefficient 1**

Using the FOIL method to multiply $(x+5)$ and $(x+3)$, we find

$$\overset{\text{F}\quad\text{O}\quad\text{I}\quad\text{L}}{(x+5)(x+3) = x^2 + 3x + 5x + 3\cdot 5 = x^2 + 8x + 15}$$

where $8x$ comes from $3+5$ and 15 comes from $3\cdot 5$.

We see that the leading coefficient (coefficient of x^2) is 1, the coefficient 8 is the sum of 5 and 3, and the constant term 15 is the product of 5 and 3.

More generally,

$$(x+a)(x+b) = x^2 + bx + ax + ab$$
$$= x^2 + (b+a)x + ab$$

where $(b+a)$ is the sum of constants a and b, and ab is the product of constants a and b.

Now given a trinomial with leading coefficient 1, we want to find the binomial factors, if any. Reversing the relationship between a and b, as shown above, we can proceed as follows:

To factor a trinomial with leading coefficient 1, find two factors of the constant term whose sum is the coefficient of the middle term. (If these factors do not exist, the trinomial is **not factorable**.)

For example, to factor $x^2 + 11x + 30$, we need positive factors of $+30$ whose sum is $+11$.

Positive Factors of 30		Sums of These Factors
1 · 30	→	$1 + 30 = 31$
2 · 15	→	$2 + 15 = 17$
3 · 10	→	$3 + 10 = 13$
5 · 6	→	$5 + 6 = 11$

Now because $5\cdot 6 = 30$ and $5 + 6 = 11$, we have
$$x^2 + 11x + 30 = (x+5)(x+6).$$

Example 1

Factoring Trinomials with Leading Coefficient 1

Completely factor the following trinomials.

a. $x^2 + 8x + 12$

Solution

12 has three pairs of positive integer factors as illustrated in the following table. Of these 3 pairs, only $2 + 6$ is equal to 8.

Factors of 12

$$1 \cdot 12$$
$$2 \cdot 6 \longrightarrow 2 + 6 = 8 \quad \text{Thus } x^2 + 8x + 12 = (x+2)(x+6).$$
$$3 \cdot 4$$

Note: If the middle term had been $-8x$, then we would have wanted pairs of negative integer factors to find a sum of -8.

b. $y^2 - 8y - 20$

Solution

We want a pair of integer factors of -20 whose sum is -8. In this case, because the product is negative, one of the factors must be positive and the other negative.

Factors of -20

$$-1 \cdot 20$$
$$1 \cdot -20$$
$$-2 \cdot 10$$
$$2 \cdot -10 \longrightarrow 2 + (-10) = -8$$
$$-4 \cdot 5$$
$$4 \cdot -5$$

We have listed all the pairs of integer factors of -20. You can see that 2 and -10 are the only two whose sum is -8. Thus listing all the pairs is not necessary. This stage is called the **trial-and-error stage**. That is, you can **try** different pairs (mentally or by making a list) until you find the correct pair. If such a pair does not exist, the polynomial is **not factorable**.

In this case, we have

$$y^2 - 8y - 20 = (y+2)(y-10).$$

Note that by the commutative property of multiplication, the order of the factors does not matter. That is, we can also write

$$y^2 - 8y - 20 = (y-10)(y+2).$$

Completely factor the following trinomials.

1. a. $x^2 + 10x + 21$

   b. $x^2 - x - 20$

2. $x^2 - 5x + 6$

Completion Example 2

Factoring Trinomials

Completely factor the following trinomial.
$$y^2 + 10y + 16.$$

Solution

$$y^2 + 10y + 16 = (y + \underline{\hspace{1cm}})(y + \underline{\hspace{1cm}})$$

Now work margin exercises **1** and **2**.

To Factor Trinomials of the Form $x^2 + bx + c$

To factor $x^2 + bx + c$, if possible, find an integer pair of factors of c whose sum is b.

1. If c is positive, then both factors must have the same sign.
   a. Both will be positive if b is positive.
      Example: $x^2 + 5x + 4 = (x + 4)(x + 1)$
   b. Both will be negative if b is negative.
      Example: $x^2 - 5x + 4 = (x - 4)(x - 1)$
2. If c is negative, then one factor must be positive and the other negative.
   Examples: $x^2 + 6x - 7 = (x + 7)(x - 1)$ and $x^2 - 6x - 7 = (x - 7)(x + 1)$

Objective B **Finding a Common Monomial Factor First**

If a trinomial does not have a leading coefficient of 1, then we look for a common monomial factor. **If there is a common monomial factor, factor out this common monomial factor first** and then factor the remaining trinomial factor, if possible. (We will discuss factoring trinomials with leading coefficients other than 1 in the next section.) A polynomial is **completely factored** if none of its factors can be factored.

Completion Example Answers
   2. $(y + 8)(y + 2)$

Example 3

Finding a Common Monomial Factor

Completely factor the following trinomials by first factoring out the GCF in the form of a common monomial factor.

a. $5x^3 - 15x^2 + 10x$

Solution

First factor out the GCF, $5x$.

$5x^3 - 15x^2 + 10x = 5x(x^2 - 3x + 2)$ factored, but not completely factored

Now factor the trinomial $x^2 - 3x + 2$. Look for factors of $+2$ that add up to -3. Because $(-1)(-2) = +2$ and $(-1) + (-2) = -3$, we have

$$5x^3 - 15x^2 + 10x = 5x(x^2 - 3x + 2)$$
$$= 5x(x - 1)(x - 2).$$ completely factored

Check:

The factoring can be checked by multiplying the factors.

$$5x(x - 1)(x - 2) = 5x(x^2 - 2x - x + 2)$$
$$= 5x(x^2 - 3x + 2)$$
$$= 5x^3 - 15x^2 + 10x$$ the original expression

b. $10y^5 - 20y^4 - 80y^3$

Solution

First factor out the GCF, $10y^3$.

$$10y^5 - 20y^4 - 80y^3 = 10y^3(y^2 - 2y - 8)$$

Now factor the trinomial $y^2 - 2y - 8$. Look for factors of -8 that add up to -2. Because $(-4)(+2) = -8$ and $(-4) + (+2) = -2$, we have

$$10y^5 - 20y^4 - 80y^3 = 10y^3(y^2 - 2y - 8) = 10y^3(y - 4)(y + 2).$$

The factoring can be checked by multiplying the factors.

Completely factor the following trinomial by first factoring out the GCF in the form of a common monomial factor.

3. a. $7y^3 + 14y^2 - 49y$

b. $11x^3 + 22x^2 - 33x$

4. $6x^2 + 12x - 48$

Completion Example 4

Finding a Common Monomial Factor

Completely factor the following trinomial by first factoring out the GCF in the form of a common monomial factor.

$$5a^2 + 25a - 180$$

Solution

$$5a^2 + 25a - 180 = 5\left(a^2 + \underline{}a - \underline{}\right)$$
$$= 5\left(a + \underline{}\right)\left(a - \underline{}\right)$$

Now work margin exercises 3 and 4.

notes

■ When factoring polynomials, always look for a common monomial factor first. Then, if there is one, remember to include this common monomial factor as part of the answer. Not all polynomials are factorable. For example, no matter what combinations are tried, $x^2 + 3x + 4$ does not have two binomial factors with integer coefficients. (There are no factors of +4 that will add to +3.) We say that the polynomial is **not factorable** (or **prime**). **A polynomial is not factorable if it cannot be factored as the product of polynomials with integer coefficients.**

Practice Problems

Completely factor each polynomial. If the polynomial cannot be factored, write "not factorable."

1. $x^2 - 3x - 28$ **2.** $x^2 + 4x + 3$ **3.** $y^2 - 12y + 35$

4. $a^2 - a + 1$ **5.** $2y^3 + 24y^2 + 64y$ **6.** $4a^4 + 48a^3 + 108a^2$

Completion Example Answers
4. $5\left(a^2 + 5a - 36\right) = 5(a + 9)(a - 4)$

Practice Problem Answers

1. $(x - 7)(x + 4)$ **2.** $(x + 3)(x + 1)$ **3.** $(y - 5)(y - 7)$

4. not factorable   **5.** $2y(y + 8)(y + 4)$ **6.** $4a^2(a + 3)(a + 9)$

Exercises 12.2

List all pairs of integer factors for the given integer. Remember to include negative integers as well as positive integers.

1. 15

2. 12

3. 20

4. 30

5. −6

6. −7

7. 16

8. 18

9. −10

10. −25

Find the pair of integers whose product is the first integer and whose sum is the second integer.

11. 12, 7

12. 25, 26

13. −14, −5

14. −30, −1

15. −8, 7

16. −40, 6

17. 36, −12

18. 16, −10

19. 20, −9

20. 4, −5

Complete the factorization.

21. $x^2 + 6x + 5 = (x + 5)(\quad)$

22. $y^2 - 7y + 6 = (y - 1)(\quad)$

23. $p^2 - 9p - 10 = (p + 1)(\quad)$

24. $m^2 + 4m - 45 = (m - 5)(\quad)$

25. $a^2 + 12a + 36 = (a + 6)(\quad)$

26. $n^2 - 2n - 3 = (n - 3)(\quad)$

Completely factor each of the given trinomials. If a trinomial cannot be factored, write "not factorable." See Examples 1 and 2.

27. $x^2 - x - 12$

28. $x^2 - 6x - 27$

29. $y^2 + y - 30$

30. $x^2 + 6x - 36$

31. $m^2 + 3m - 1$

32. $x^2 + 3x - 18$

33. $x^2 - 8x + 16$

34. $a^2 + 10a + 25$

35. $x^2 + 7x + 12$

36. $a^2 + a + 2$

37. $y^2 - 3y + 2$

38. $y^2 - 14y + 24$

39. $x^2 + 3x + 5$ **40.** $y^2 + 12y + 35$ **41.** $x^2 - x - 72$

42. $y^2 + 8y + 7$ **43.** $z^2 - 15z + 54$ **44.** $a^2 + 4a - 21$

Completely factor each of the given polynomials. If a polynomial cannot be factored, write "not factorable." See Examples 3 and 4.

45. $x^3 + 10x^2 + 21x$ **46.** $x^3 + 8x^2 + 15x$ **47.** $5x^2 - 5x - 60$

48. $6x^2 + 24x + 18$ **49.** $10y^3 - 10y^2 - 60y$ **50.** $7y^3 - 70y^2 + 168y$

51. $4p^4 + 36p^3 + 32p^2$ **52.** $15m^5 - 30m^4 + 15m^3$ **53.** $2x^4 - 14x^3 - 36x^2$

54. $3y^6 + 33y^5 + 90y^4$ **55.** $2x^2 - 2x - 72$ **56.** $3x^2 - 18x + 30$

57. $2a^4 - 8a^3 - 120a^2$ **58.** $2a^4 + 24a^3 + 54a^2$ **59.** $3y^5 - 21y^4 - 24y^3$

60. $4y^5 + 28y^4 + 24y^3$ **61.** $x^3 - 10x^2 + 16x$ **62.** $x^3 - 2x^2 - 3x$

63. $5a^2 + 10a - 30$ **64.** $6a^2 + 24a + 12$ **65.** $20a^4 + 40a^3 + 20a^2$

66. $6x^4 - 12x^3 + 6x^2$

Using the information given, factor the polynomials functions to find the requested lengths.

67. Triangles: The area of a triangle is $\dfrac{1}{2}$ the product of its base and its height. If the area of the triangle shown is given by the function $A(x) = \dfrac{1}{2}x^2 + 24x,$ find representations for the lengths of its base and its height (where the base is longer than the height).

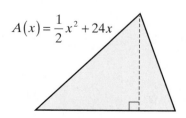

$$A(x) = \frac{1}{2}x^2 + 24x$$

68. Triangles: The area of a triangle is $\frac{1}{2}$ the product of its base and its height. If the area of the triangle shown is given by the function $A(x) = \frac{1}{2}x^2 + 14x$, find representations for the lengths of its base and its height (where the height is longer than the base).

$$A(x) = \frac{1}{2}x^2 + 14x$$

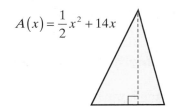

69. Rectangles: The area of the rectangle shown is given by the polynomial function $A(x) = 4x^2 + 20x$. If the width of the rectangle is $4x$, what is the length?

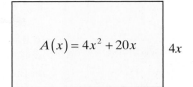

$$A(x) = 4x^2 + 20x$$

$4x$

70. Rectangles: The area of the rectangle shown is given by the polynomial function $A(x) = x^2 + 11x + 24$. If the length of the rectangle is $(x+8)$, what is the width?

$$A(x) = x^2 + 11x + 24$$

$x + 8$

Writing & Thinking

71. It is true that $2x^2 + 10x + 12 = (2x+6)(x+2) = (2x+4)(x+3)$. Explain how the trinomial can be factored in two ways. Is there some kind of error?

72. It is true that $5x^2 - 5x - 30 = (5x-15)(x+2)$. Explain why this is not the completely factored form of the trinomial.

12.3 Factoring Trinomials: $ax^2 + bx + c$

Second-degree trinomials in the variable x are of the general form

$ax^2 + bx + c$ where the coefficients $a, b,$ and c are real numbers and $a \neq 0$.

In this section we will discuss two methods of factoring trinomials of the form $ax^2 + bx + c$ in which the coefficients are restricted to integers. These two methods are the **trial-and-error method** (a reverse of the **FOIL** method of multiplication) and the ***ac*-method** (a form of **grouping**).

Objective A   **The Trial-and-Error Method of Factoring**

To help in understanding this method we review the FOIL method of multiplying two binomials as follows.

$$(2x+5)(3x+1) = 6x^2 + 17x + 5$$

$$(2x+5)(3x+1) = 2x \cdot 3x + 2x \cdot 1 + 5 \cdot 3x + 5 \cdot 1$$

$$= 6x^2 \quad + \quad 17x \quad + 5$$

F   The product of the **first** two terms is $6x^2$.

O
I   The sum of the **inner** and **outer** products is $17x$.

L   The product of the **last** two terms is 5.

Now consider the problem of factoring

$$6x^2 + 23x + 7$$

as the product of two binomials.

$$\mathbf{F} = 6x^2$$
$$\mathbf{L} = +7$$
$$6x^2 + 23x + 7 = (\quad)(\quad)$$

For $\mathbf{F} = 6x^2$, we know that $6x^2 = 6x \cdot x$ and $6x^2 = 3x \cdot 2x$.

For $\mathbf{L} = +7$, we know that $+7 = (+1)(+7)$ and $+7 = (-1)(-7)$.

Now we use various combinations for $\mathbf{F}$ and $\mathbf{L}$ in the **trial-and-error method** as follows.

1. List all the possible combinations of factors of $6x^2$ and $+7$ in their respective $\mathbf{F}$ and $\mathbf{L}$ positions. (See the following list.)
2. Check the sum of the products in the $\mathbf{O}$ and $\mathbf{I}$ positions until you find the sum to be $+23x$.
3. If none of these sums is $+23x$, the trinomial is not factorable.

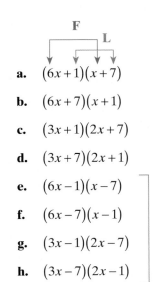

a. $(6x+1)(x+7)$

b. $(6x+7)(x+1)$

c. $(3x+1)(2x+7)$

d. $(3x+7)(2x+1)$

e. $(6x-1)(x-7)$

f. $(6x-7)(x-1)$

g. $(3x-1)(2x-7)$

h. $(3x-7)(2x-1)$

We really don't need to check these last four because the **O** and **I** would be negative, and we are looking for $+23x$. In this manner, the trial-and-error method is more efficient than it first appears to be.

Now investigating only the possibilities in the list with positive constants, we need to check the sums of the outer (**O**) and inner (**I**) products to find $+23x$.

a. $(6x+1)(x+7)$: $\mathbf{O}+\mathbf{I} = 42x+x = 43x$

x

$42x$

b. $(6x+7)(x+1)$: $\mathbf{O}+\mathbf{I} = 6x+7x = 13x$

$7x$

$6x$

c. $(3x+1)(2x+7)$: $\mathbf{O}+\mathbf{I} = 21x+2x = \boxed{23x}$ ← We found 23x! Yeah! With a little luck we could have found this first.

$2x$

$21x$

The correct factors, $(3x+1)$ and $(2x+7)$, have been found so we need not take the time to try the next product in the list, $(3x+7)(2x+1)$. Thus even though the list of possibilities of factors may be long, the actual time involved may be quite short if the correct factors are found early in the trial-and-error method. So we have

$$6x^2 + 23x + 7 = (3x+1)(2x+7).$$

Look at the constant term to determine what signs to use for the constants in the factors. The following guidelines will help limit the trial-and-error search.

Guidelines for the Trial-and-Error Method

1. If the sign of the constant term is positive (+), the signs in both factors will be the same, both positive if the middle term is positive or both negative if the middle term is negative.
2. If the sign of the constant term is negative (−), the signs in the factors will be different, one positive and one negative.

Example 1

Using the Trial-and-Error Method

Factor by using the trial-and-error method.

a. $x^2 + 6x + 5$

Solution

Since the middle term is $+6x$ and the constant is 5, we know that the two factors of 5 must both be positive, $+5$ and $+1$.

$$x^2 + 6x + 5 = (x+5)(x+1)$$

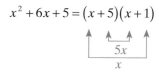
$$5x$$
$$x$$

b. $4x^2 - 4x - 15$

Solution

For **F**: $4x^2 = 4x \cdot x$ and $4x^2 = 2x \cdot 2x$.
For **L**: $-15 = -15 \cdot 1, -15 = -1 \cdot 15, -15 = -3 \cdot 5,$ and $-15 = -5 \cdot 3$.

Trials

$$(2x-15)(2x+1)$$ $2x - 30x = -28x$ is the wrong middle term.
$$-30x$$
$$2x$$

$$(2x-3)(2x+5)$$ $10x - 6x = +4x$ is the wrong middle term only because the sign is wrong. So just switch the signs and the factors will be right.
$$-6x$$
$$10x$$

$$(2x+3)(2x-5)$$ $-10x + 6x = -4x$ is the right middle term.
$$6x$$
$$-10x$$

Now that we have the answer, there is no need to try all the possibilities with $(4x\quad)(x\quad)$.

c. $6a^2 - 31a + 5$

Solution

Since the middle term is $-31a$ and the constant is $+5$, we know that the two factors of 5 must both be negative, -5 and -1. We try $\mathbf{F} = 6a^2 = 6a \cdot a$.

$$6a^2 - 31a + 5 = (6a - 1)(a - 5)$$

$-30a - a = -31a$

$-a$

$-30a$

So we found the correct factors on the first try.

Now work margin exercise 1.

notes

Reminder: To factor completely means to find factors of the polynomial, none of which are themselves factorable. Thus
$$2x^2 + 12x + 10 = (2x + 10)(x + 1)$$

is not factored completely because $2x + 10 = 2(x + 5)$.
We could write
$$2x^2 + 12x + 10 = (2x + 10)(x + 1)$$
$$= 2(x + 5)(x + 1).$$

This problem can be avoided by first factoring out the GCF (in this case, 2).

Example 2

Factoring Completely

Factor completely. Be sure to look first for the greatest common monomial factor.

a. $6x^3 - 8x^2 + 2x$

Solution

$$6x^3 - 8x^2 + 2x = 2x(3x^2 - 4x + 1) = 2x(3x - 1)(x - 1)$$

1. Use the trial-and-error method to factor.

a. $x^2 + 8x + 12$

b. $8y^2 - 2y - 21$

2. Factor completely.

　a. $8x^3 - 12x^2 + 4x$

　b. $21x^3 + 49x^2 - 7x$

Check:

Check the factorization by multiplying.

$$2x(3x-1)(x-1) = 2x(3x^2 - 3x - x + 1)$$
$$= 2x(3x^2 - 4x + 1)$$
$$= 6x^3 - 8x^2 + 2x \qquad \text{the original polynomial}$$

b. $-2x^2 - x + 6$

Solution

$$-2x^2 - x + 6 = -1(2x^2 + x - 6) = -1(2x - 3)(x + 2)$$

Note that factoring out −1 gives a positive leading coefficient for the trinomial.

c. $10x^3 + 5x^2 + 5x$

Solution

$$10x^3 + 5x^2 + 5x = 5x(2x^2 + x + 1)$$

Now consider the trinomial:

$$2x^2 + x + 1 = (2x + ?)(x + ?) \cdot$$

The factors of +1 need to be +1 and +1, but

$$(2x + 1)(x + 1) = 2x^2 + \underline{3x} + 1 \, .$$

So there is no way to factor and get a middle term of $+x$ for the product. This trinomial, $2x^2 + x + 1$, is **not factorable**.

We have
$$10x^3 + 5x^2 + 5x = 5x(2x^2 + x + 1) \qquad \text{factored completely}$$

Now work margin exercise 2.

Objective B　**The *ac*-Method of Factoring**

The ***ac*-method** of factoring is in reference to the coefficients a and c in the general form $ax^2 + bx + c$ and involves the method of factoring by grouping discussed in Section 12.1. The method is best explained by analyzing an example and explaining each step as follows.

We consider the problem of factoring the trinomial

$$2x^2 + 9x + 10 \quad \text{where } a = 2, b = 9, \text{ and } c = 10.$$

Analysis of Factoring by the *ac*-Method

General Method	Example
$ax^2 + bx + c$	$2x^2 + 9x + 10$

Step 1: Multiply $a \cdot c$.

Multiply $2 \cdot 10 = 20$.

Step 2: Find two integers whose product is ac and whose sum is b. If this is not possible, then the trinomial is **not factorable**.

Find two integers whose product is 20 and whose sum is 9. (In this case, $4 \cdot 5 = 20$ and $4 + 5 = 9$.)

Step 3: Rewrite the middle term (bx) using the two numbers found in Step 2 as coefficients.

Rewrite the middle term $(+9x)$ using $+4$ and $+5$ as coefficients.
$$2x^2 + 9x + 10 = 2x^2 + 4x + 5x + 10$$

Step 4: Factor by grouping the first two terms and the last two terms.

Factor by grouping the first two terms and the last two terms.
$$2x^2 + 4x + 5x + 10 = \left(2x^2 + 4x\right) + \left(5x + 10\right)$$
$$= 2x(x + 2) + 5(x + 2)$$

Step 5: Factor out the common binomial factor. This will give two binomial factors of the trinomial $ax^2 + bx + c$.

Factor out the common binomial factor $(x + 2)$. Thus
$$2x^2 + 9x + 10 = 2x^2 + 4x + 5x + 10$$
$$= \left(2x^2 + 4x\right) + \left(5x + 10\right)$$
$$= 2x(x + 2) + 5(x + 2)$$
$$= (x + 2)(2x + 5).$$

Example 3

Using the *ac*-Method

a. Factor $4x^2 + 33x + 35$ using the *ac*-method.

Solution

$a = 4, b = 33, c = 35$

Step 1: Find the product ac: $4 \cdot 35 = 140$.

Step 2: Find two integers whose product is 140 and whose sum is 33.

$(+5)(+28) = 140$ and $(+5) + (+28) = +33$

Note: This step may take some experimenting with factors. You might try prime factoring.

For example

$$140 = 10 \cdot 14 = 2 \cdot 5 \cdot 2 \cdot 7.$$

With combinations of these prime factors we can write the following.

Factors of 140		Sum
1	· 140	$1 + 140 = 141$
2	· 70	$2 + 70 = 72$
4	· 35	$4 + 35 = 39$
5	· 28	$5 + 28 = 33$ We can stop here!

5 and 28 are the desired coefficients.

Step 3: Rewrite $+33x$ as $+5x + 28x$, giving

$$4x^2 + 33x + 35 = 4x^2 + 5x + 28x + 35.$$

Step 4: Factor by grouping.

$$4x^2 + 33x + 35 = 4x^2 + 5x + 28x + 35$$
$$= x(4x + 5) + 7(4x + 5)$$

Step 5: Factor out the common binomial factor $(4x + 5)$.

$$4x^2 + 33x + 35 = 4x^2 + 5x + 28x + 35$$
$$= x(4x + 5) + 7(4x + 5)$$
$$= (4x + 5)(x + 7)$$

Note that in Step 3 we could have written $+33x$ as $+28x + 5x$. Try this to convince yourself that the result will be the same two factors.

b. Factor $12y^3 - 26y^2 + 12y$ using the *ac*-method.

Solution

First factor out the greatest common factor $2y$.

$$12y^3 - 26y^2 + 12y = 2y(6y^2 - 13y + 6)$$

Now factor the trinomial $6y^2 - 13y + 6$ with $a = 6, b = -13$, and $c = 6$.

Step 1: Find the product ac: $6(6) = 36$.

Step 2: Find two integers whose product is 36 and whose sum is -13.

Note: This may take some time and experimentation. We do know that both numbers must be negative because the product is positive and the sum is negative.

$$(-9)(-4) = +36 \quad \text{and} \quad -9 + (-4) = -13$$

Steps 3 and 4: Factor by grouping.

$$6y^2 - 13y + 6 = 6y^2 - 9y - 4y + 6$$
$$= 3y(2y - 3) - 2(2y - 3)$$

Note: −2 is factored from the last two terms so that there will be a common binomial factor $(2y - 3)$.

Step 5: Factor out the common binomial factor $(2y - 3)$.

$$6y^2 - 13y + 6 = 6y^2 - 9y - 4y + 6$$
$$= 3y(2y - 3) - 2(2y - 3)$$
$$= (2y - 3)(3y - 2)$$

Thus for the original expression,

$$12y^3 - 26y^2 + 12y = 2y(6y^2 - 13y + 6)$$
$$= 2y(2y - 3)(3y - 2)$$

Do not forget to write the common monomial factor, 2y, in the answer.

c. Factor $4x^2 - 5x - 6$ using the *ac*-method.

Solution

$a = 4, b = -5, c = -6$

Step 1: Find the product *ac*: $4(-6) = -24$.

Step 2: Find two integers whose product is −24 and whose sum is −5.

Note: We know that one number must be positive and the other negative because the product is negative.

$$(+3)(-8) = -24 \text{ and } (+3) + (-8) = -5.$$

Steps 3 and 4: Factor by grouping.

$$4x^2 - 5x - 6 = 4x^2 + 3x - 8x - 6$$
$$= x(4x + 3) - 2(4x + 3)$$

Step 5: Factor out the common binomial factor $(4x + 3)$.

$$4x^2 - 5x - 6 = 4x^2 + 3x - 8x - 6$$
$$= x(4x + 3) - 2(4x + 3)$$
$$= (4x + 3)(x - 2).$$

Now work margin exercise 3.

3. Use the *ac*-method to factor.

a. $5a^2 + 24a + 27$

b. $3b^2 + 6b - 24$

4. Completely factor each of the following trinomials. Be sure to look for the greatest common factor.

a. $5x^2 - 32x - 21$

b. $24x^2 + 87x - 36$

Completion Example 4

Factoring Trinomials

Completely factor each of the following trinomials. Be sure to look for the greatest common factor.

a. $15x^2 + 38x + 7$

Solution

$$15x^2 + 38x + 7 = (5x + \underline{\quad})(3x + \underline{\quad})$$

b. $4y^2 + 6y - 108$

Solution

$$4y^2 + 6y - 108 = 2(\underline{\quad} y^2 + \underline{\quad} y - \underline{\quad})$$
$$= 2(2y - \underline{\quad})(y + \underline{\quad})$$

Now work margin exercise 4.

notes

- No matter which method you use (the *ac*-method or the trial-and-error method), factoring trinomials takes time. With practice you will become more efficient with either method. Make sure to be patient and observant.

___

Completion Example Answers

4. a. $(5x + 1)(3x + 7)$

b. $2(2y^2 + 3y - 54) = 2(2y - 9)(y + 6)$

Practice Problems

Factor completely.

1. $3x^2 + 7x - 6$ **2.** $2x^2 + 6x - 8$ **3.** $3x^2 + 15x + 18$

4. $10x^2 - 41x - 18$ **5.** $x^2 + 11x + 28$ **6.** $-3x^3 + 6x^2 - 3x$

7. $3x^2 + 4x - 8$

Practice Problem Answers

1. $(3x - 2)(x + 3)$ **2.** $2(x - 1)(x + 4)$ **3.** $3(x + 2)(x + 3)$

4. $(5x + 2)(2x - 9)$ **5.** $(x + 7)(x + 4)$ **6.** $-3x(x - 1)(x - 1)$

7. not factorable

Exercises 12.3

Completely factor each of the given polynomials. If a polynomial cannot be factored, write "not factorable." See Examples 1 through 4.

1. $x^2 + 5x + 6$

2. $x^2 - 6x + 8$

3. $2x^2 - 3x - 5$

4. $3x^2 - 4x - 7$

5. $6x^2 + 11x + 5$

6. $4x^2 - 11x + 6$

7. $-x^2 + 3x - 2$

8. $-x^2 - 5x - 6$

9. $x^2 - 3x - 10$

10. $x^2 - 11x + 10$

11. $-x^2 + 13x + 14$

12. $-x^2 + 12x - 36$

13. $x^2 + 8x + 64$

14. $x^2 + 2x + 3$

15. $-2x^3 + x^2 + x$

16. $-2y^3 - 3y^2 - y$

17. $4t^2 - 3t - 1$

18. $2x^2 - 3x - 2$

19. $5a^2 - a - 6$

20. $3a^2 + 4a + 1$

21. $7x^2 + 5x - 2$

22. $4x^2 + 23x + 15$

23. $8x^2 - 10x - 3$

24. $6x^2 + 23x + 21$

25. $9x^2 - 3x - 20$

26. $4x^2 + 40x + 25$

27. $12x^2 - 38x + 20$

28. $12b^2 - 12b + 3$

29. $3x^2 - 7x + 2$

30. $7x^2 - 11x - 6$

31. $9x^2 - 6x + 1$

32. $4x^2 + 4x + 1$

33. $6y^2 + 7y + 2$

34. $12y^2 - 7y - 12$

35. $x^2 - 46x + 45$

36. $x^2 + 6x - 16$

37. $3x^2 + 9x + 5$

38. $5a^2 - 7a + 2$

39. $8a^2b - 22ab + 12b$

40. $12m^3n - 50m^2n + 8mn$

41. $x^2 + x + 1$

42. $x^2 + 2x + 2$

43. $16x^2 - 8x + 1$

44. $3x^2 - 11x - 4$

45. $64x^2 - 48x + 9$

46. $9x^2 - 12x + 4$

47. $6x^2 + 2x - 20$

48. $12y^2 - 15y + 3$

49. $10x^2 + 35x + 30$

50. $24y^2 + 4y - 4$

51. $-18x^2 + 72x - 8$

52. $7x^4 - 5x^3 + 3x^2$

53. $-45y^2 + 30y + 120$

54. $-12m^2 + 22m + 4$

55. $12x^2 - 60x + 75$

56. $32y^2 + 50$

57. $6x^3 + 9x^2 - 6x$

58. $-5y^2 + 40y - 60$

59. $12x^3 - 108x^2 + 243x$

60. $30a^3 + 51a^2 + 9a$

61. $9x^3y^3 + 9x^2y^3 + 9xy^3$

62. $48x^2y - 354xy + 126y$

63. $48xy^3 - 100xy^2 + 48xy$

64. $24a^2x^2 + 72a^2x + 243x$

65. $21y^4 - 98y^3 + 56y^2$

66. $72a^3 - 306a^2 + 189a$

Writing & Thinking

67. Discuss, in your own words, how the sign of the constant term determines what signs will be used in the factors when factoring trinomials.

68. **Volume of a Box:** The volume of an open box is found by cutting equal squares (x inches on a side) from a sheet of cardboard that is 5 inches by 25 inches. The function representing this volume is $V(x) = 4x^3 - 60x^2 + 125x$, where $0 < x < 2.5$. Factor this function and use the factors to explain, in your own words, how the function represents the volume.

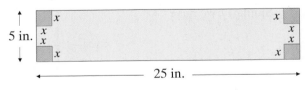

12.4 Special Factoring Techniques

In Section 11.6 we discussed the following three products.

I. $(x+a)(x-a) = x^2 - a^2$ difference of two squares

II. $(x+a)^2 = x^2 + 2ax + a^2$ square of a binomial sum

III. $(x-a)^2 = x^2 - 2ax + a^2$ square of a binomial difference

Two objectives in this section are to learn to factor products of these types (difference of two squares and squares of binomials) without using to the trial-and-error method. That is, with practice you will learn to recognize the "form" of these products and go directly to the factors. **Memorize the products and their names listed above.**

For easy reference to squares, Table 1 lists the squares of the integers from 1 to 20.

Perfect Squares from 1 to 400										
Integer n	1	2	3	4	5	6	7	8	9	10
Square n^2	1	4	9	16	25	36	49	64	81	100
Integer n	11	12	13	14	15	16	17	18	19	20
Square n^2	121	144	169	196	225	256	289	324	361	400

Table 1

Objective A   Difference of Two Squares: $x^2 - a^2 = (x+a)(x-a)$

Consider the polynomial $x^2 - 25$. By recognizing this expression as the **difference of two squares**, we can go directly to the factors:

$$x^2 - 25 = (x)^2 - (5)^2 = (x+5)(x-5).$$

Similarly, we have

$$9 - y^2 = (3)^2 - (y)^2 = (3+y)(3-y)$$

$$\text{and } 49x^2 - 36 = (7x)^2 - (6)^2 = (7x+6)(7x-6).$$

Remember to **look for a common monomial factor first**. For example,

$$6x^2y - 24y = 6y(x^2 - 4) = 6y(x+2)(x-2).$$

Example 1

Factoring the Difference of Two Squares

Factor completely.

a. $3a^2b - 3b$

Solution

$$3a^2b - 3b = 3b(a^2 - 1)$$ Factor out the GCF, 3b.
$$= 3b(a+1)(a-1)$$ difference of two squares
Don't forget that 3b is a factor.

b. $x^6 - 400$

Solution

Even powers, such as x^6, can always be treated as squares: $x^6 = (x^3)^2$.

$$x^6 - 400 = (x^3)^2 - 20^2$$
$$= (x^3 + 20)(x^3 - 20)$$ difference of two squares

Now work margin exercise 1.

1. Factor completely.

a. $7ax^2 - 343a$

b. $y^6 - 100$

Sum of Two Squares

The **sum of two squares** is an expression of the form $x^2 + a^2$ and is **not factorable**. For example, $x^2 + 36$ is the sum of two squares and is not factorable. There are no factors with integer coefficients whose product is $x^2 + 36$. To understand this situation, write

$$x^2 + 36 = x^2 + 0x + 36$$

and note that there are no factors of +36 that will add to 0.

Example 2

Using the Sum of Two Squares

Factor completely. Be sure to look first for the greatest common monomial factor.

a. $y^2 + 64$

Solution

$y^2 + 64$ is the sum of two squares and is not factorable.

2. Factor completely.

  a. $x^2 + 25$

  b. $45x^2 + 20$

b. $4x^2 + 100$

Solution

$$4x^2 + 100 = 4(x^2 + 25) \quad \text{factored completely}$$

We see that 4 is the greatest common monomial factor and $x^2 + 25$ is the sum of two squares and not factorable.

Now work margin exercise 2.

Objective B **Perfect Square Trinomials:** $\begin{cases} x^2 + 2ax + a^2 = (x+a)^2 \\ x^2 - 2ax + a^2 = (x-a)^2 \end{cases}$

We know that **squaring a binomial** leads to a **perfect square trinomial**. Therefore, factoring a perfect square trinomial gives the square of a binomial. We simply need to recognize whether or not a trinomial fits the "form."

In a perfect square trinomial, both the first and last terms of the trinomial must be perfect squares. If the first term is of the form x^2 and the last term is of the form a^2, then the middle term must be of the form $2ax$ or $-2ax$, as shown in the following examples.

$$x^2 + 8x + 16 = (x+4)^2 \quad \text{Here } 8x = 2 \cdot 4 \cdot x = 2ax \quad \text{and} \quad 16 = 4^2 = a^2.$$

$$x^2 - 6x + 9 = (x-3)^2 \quad \text{Here } -6x = -2 \cdot 3 \cdot x = -2ax \quad \text{and} \quad 9 = 3^2 = a^2.$$

Example 3

Factoring Perfect Square Trinomials

Factor completely.

a. $z^2 - 12z + 36$

Solution

In the form $x^2 - 2ax + a^2$ we have $x = z$ and $a = 6$.

$$z^2 - 12z + 36 = z^2 - 2 \cdot 6z + 6^2$$
$$= (z-6)^2$$

b. $4y^2 + 12y + 9$

Solution

In the form $x^2 + 2ax + a^2$ we have $x = 2y$ and $a = 3$.

$$4y^2 + 12y + 9 = (2y)^2 + 2 \cdot 3 \cdot 2y + 3^2$$
$$= (2y+3)^2$$

c. $2x^3 - 8x^2y + 8xy^2$

Solution

Factor out the GCF first. Then factor the **perfect square trinomial**.

$$2x^3 - 8x^2y + 8xy^2 = 2x\left(x^2 - 4xy + 4y^2\right)$$
$$= 2x\left[x^2 - 2\cdot 2y\cdot x + \left(2y\right)^2\right]$$
$$= 2x\left(x - 2y\right)^2$$

d. $\left(x^2 + 6x + 9\right) - y^2$

Solution

Treat $x^2 + 6x + 9$ as a perfect square trinomial, then factor the **difference of two squares**.

$$\left(x^2 + 6x + 9\right) - y^2 = \left(x + 3\right)^2 - y^2$$
$$= \left(x + 3 + y\right)\left(x + 3 - y\right)$$

Now work margin exercise 3.

3. Factor completely.

a. $z^2 + 40z + 400$

b. $3x^2z - 18xyz + 27y^2z$

c. $\left(y^2 + 8y + 16\right) - z^2$

Objective C **Sums and Differences of Two Cubes:**

$$\begin{cases} \boldsymbol{x^3 + a^3 = \left(x + a\right)\left(x^2 - ax + a^2\right)} \\ \boldsymbol{x^3 - a^3 = \left(x - a\right)\left(x^2 + ax + a^2\right)} \end{cases}$$

The formulas for the sums and differences of two cubes are new, and we can proceed to show that they are indeed true as follows.

$$\left(x + a\right)\left(x^2 - ax + a^2\right) = x\cdot x^2 - x\cdot ax + x\cdot a^2 + a\cdot x^2 - a\cdot ax + a\cdot a^2$$
$$= x^3 - ax^2 + a^2x + ax^2 - a^2x + a^3$$
$$= x^3 + a^3 \qquad \text{sum of two cubes}$$

$$\left(x - a\right)\left(x^2 + ax + a^2\right) = x\cdot x^2 + x\cdot ax + x\cdot a^2 - a\cdot x^2 - a\cdot ax - a\cdot a^2$$
$$= x^3 + ax^2 + a^2x - ax^2 - a^2x - a^3$$
$$= x^3 - a^3 \qquad \text{difference of two cubes}$$

notes

Important Notes About the Sum and Difference of Two Cubes

1. In each case, after multiplying the factors the middle terms drop out and only two terms are left.
2. The trinomials in parentheses are not perfect square trinomials. These trinomials are not factorable.
3. The sign in the binomial agrees with the sign in the result.

Because we know that factoring is the reverse of multiplication, we use these formulas to factor the sums and differences of two cubes. For example,

$$x^3 + 27 = (x)^3 + (3)^3$$
$$= (x+3)\left((x)^2 - (3)(x) + (3)^2\right)$$
$$= (x+3)(x^2 - 3x + 9)$$

and

$$x^6 - 125 = (x^2)^3 - (5)^3$$
$$= (x^2 - 5)\left[(x^2)^2 + (5)(x^2) + (5)^2\right]$$
$$= (x^2 - 5)(x^4 + 5x^2 + 25).$$

As an aid in factoring sums and differences of two cubes, Table 2 contains the cubes of the integers from 1 to 10. These cubes are called perfect cubes.

Perfect Cubes from 1 to 1000									
Integer n	1	2	3	4	5	6	7	8	9	10
Cube n^3	1	8	27	64	125	216	343	512	729	1000

Table 2

Example 4

Factoring Sums and Differences of Two Cubes

Factor completely.

a. $x^3 - 8$

Solution

$$x^3 - 8 = x^3 - 2^3$$
$$= (x-2)(x^2 + 2 \cdot x + 2^2)$$
$$= (x-2)(x^2 + 2x + 4)$$

Note: Remember that the second polynomial is not a perfect square trinomial and cannot be factored.

b. $x^6 + 64y^3$

Solution

$$x^6 + 64y^3 = (x^2)^3 + (4y)^3$$
$$= (x^2 + 4y)\left[(x^2)^2 - 4y \cdot x^2 + (4y)^2\right]$$
$$= (x^2 + 4y)(x^4 - 4x^2y + 16y^2)$$

c. $16y^{12} - 250$

Solution

Factor out the GCF first. Then factor the **difference of two cubes**.

$$16y^{12} - 250 = 2\left(8y^{12} - 125\right)$$
$$= 2\left[\left(2y^4\right)^3 - 5^3\right]$$
$$= 2\left(2y^4 - 5\right)\left[\left(2y^4\right)^2 + (5)\left(2y^4\right) + 5^2\right]$$
$$= 2\left(2y^4 - 5\right)\left(4y^8 + 10y^4 + 25\right)$$

Now work margin exercise 4.

4. Factor completely.

a. $y^3 - 27$

b. $48x^{12} - 750$

Practice Problems

Completely factor each of the following polynomials. If a polynomial cannot be factored, write "not factorable."

1. $5x^2 - 80$ **2.** $9x^2 - 12x + 4$ **3.** $4x^2 + 20x + 25$

4. $2x^3 - 250$ **5.** $\left(y^2 - 4y + 4\right) - x^4$ **6.** $x^3 + 8y^3$

Practice Problem Answers

1. $5(x-4)(x+4)$ **2.** $(3x-2)^2$

3. $(2x+5)^2$ **4.** $2(x-5)\left(x^2 + 5x + 25\right)$

5. $\left(y-2-x^2\right)\left(y-2+x^2\right)$ **6.** $(x+2y)\left(x^2 - 2xy + 4y^2\right)$

Exercises 12.4

Completely factor each of the given polynomials. If a polynomial cannot be factored, write "not factorable." See Examples 1 through 4.

1. $x^2 - 25$ **2.** $y^2 - 121$ **3.** $81 - y^2$

4. $25 - z^2$ **5.** $2x^2 - 128$ **6.** $3x^2 - 108$

7. $4x^4 - 64$ **8.** $5x^4 - 125$ **9.** $y^2 + 100$

10. $4x^2 + 49$ **11.** $y^2 - 16y + 64$ **12.** $z^2 + 18z + 81$

13. $-4x^2 + 100$ **14.** $-12x^4 + 3$ **15.** $9x^2 - 25$

16. $4x^2 - 49$ **17.** $y^2 - 10y + 25$ **18.** $x^2 + 12x + 36$

19. $4x^2 - 4x + 1$ **20.** $49x^2 - 14x + 1$ **21.** $25x^2 + 30x + 9$

22. $9y^2 + 12y + 4$ **23.** $16x^2 - 40x + 25$ **24.** $9x^2 - 12x + 4$

25. $4x^3 - 64x$ **26.** $50x^3 - 8x$ **27.** $2x^3y + 32x^2y + 128xy$

28. $3x^2y - 30xy + 75y$ **29.** $y^2 + 6y + 9$ **30.** $y^2 + 4y + 4$

31. $x^2 - 20x + 100$ **32.** $25x^2 - 10x + 1$ **33.** $x^4 + 10x^2y + 25y^2$

34. $16x^4 + 8x^2y + y^2$ **35.** $x^3 - 125$ **36.** $x^3 - 64$

37. $y^3 + 216$ **38.** $y^3 + 1$ **39.** $x^3 + 27y^3$

40. $8x^3 + 1$ **41.** $x^2 + 64y^2$ **42.** $3x^3 + 81$

43. $4x^3 - 32$

44. $64x^3 + 27y^3$

45. $54x^3 - 2y^3$

46. $3x^4 + 375xy^3$

47. $x^3y + y^4$

48. $x^4y^3 - x$

49. $x^2y^2 - x^2y^5$

50. $2x^2 - 16x^2y^3$

51. $24x^4y + 81xy^4$

52. $x^6 - 64y^3$

53. $x^6 - y^9$

54. $64x^2 + 1$

55. $27x^3 + y^6$

56. $x^3 + 64z^3$

57. $8x^3 + y^3$

58. $x^3 + 125y^3$

59. $8y^3 - 8$

60. $36x^3 + 36$

61. $9x^2 - y^2$

62. $x^2 - 4y^2$

63. $x^4 - 16y^4$

64. $81x^4 - 1$

65. $(x - y)^2 - 81$

66. $(x + 2y)^2 - 25$

67. $(x^2 - 2xy + y^2) - 36$

68. $(x^2 + 4xy + 4y^2) - 25$

69. $(16x^2 + 8x + 1) - y^2$

70. $x^2 - (y^2 + 6y + 9)$

Use the information given to answer the following problems.

71. **a.** Represent the area of the shaded region of the square shown below as the difference of two squares.

    b. Use the factors of the expression in part **a.** to draw (and label the sides of) a rectangle that has the same area as the shaded region.

72. **a.** Use a polynomial function to represent the area of the shaded region of the square.

    b. Use a polynomial function to represent the perimeter of the shaded figure.

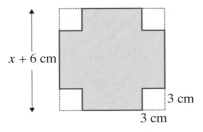

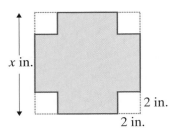

73. **a.** Show that the sum of the areas of the rectangles and squares in the figure is a perfect square trinomial.

b. Rearrange the rectangles and squares in the form of a square and represent its area as the square of a binomial.

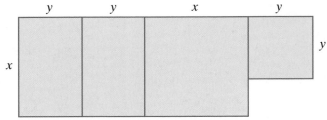

74. **Compound Interest:** Compound interest is interest earned on interest. If a principal P is invested and compounded annually (once a year) at a rate of r, then the amount, A_1 accumulated in one year is $A_1 = P + Pr$.

In factored form, we have: $A_1 = P + Pr = P(1+r)$.

At the end of the second year the amount accumulated is $A_2 = (P + Pr) + (P + Pr)r$.

a. Write the expression for A_2 in factored form similar to that for A_1.

b. Write an expression for the amount accumulated in three years, A_3, in factored form.

c. Write an expression for A_n the amount accumulated in n years.

d. Use the formula you developed in part **c.** and your calculator to find the amount accumulated if $10,000 is invested at 6% and compounded annually for 20 years.

75. You may have heard of (or studied) the following rules for division of an integer by 3 and 9:

1. An integer is divisible by 3 if the sum of its digits is divisible by 3.

2. An integer is divisible by 9 if the sum of its digits is divisible by 9.

The proofs of both **1.** and **2.** can be started as follows.

Let abc represent a three-digit integer.

$$\text{Then } abc = 100a + 10b + c$$
$$= (99 + 1)a + (9 + 1)b + c$$
$$= (\text{now you finish the proofs})$$

Use the pattern just shown and prove both **1.** and **2.** for a four-digit integer.

12.5 Additional Factoring Practice

Objective A **General Guidelines for Factoring Polynomials**

As we have seen so far in Chapter 12, factoring is a basic algebraic skill. Factoring is useful in solving equations and, as we will see in Chapter 13, necessary for simplifying algebraic fractions. This section is provided as extra practice in factoring. The following guidelines are provided for easy reference.

Objectives

A Know the guidelines for deciding what method to use when factoring a polynomial.

General Guidelines for Factoring Polynomials

1. **Always look for a common monomial factor first.** If the leading coefficient is negative, factor out a negative monomial even if it is just -1.
2. **Check the number of terms.**
   a. Two terms:
   1. difference of two squares? – factorable
   2. sum of two squares? – not factorable
   3. difference of two cubes? – factorable
   4. sum of two cubes? – factorable
   b. Three terms:
   1. perfect square trinomial?
   2. use trial-and-error method?
      Guidelines for the trial-and-error method
      a. If the sign of the constant term is positive, the signs in both factors will be the same, either both positive or both negative.
      b. If the sign of the constant term is negative, the signs in the factors will be different, one positive and one negative.
   3. use *ac*-method?
      Guidelines for the *ac*-method
      a. Multiply $a \cdot c$.
      b. Find two integers whose product is ac and whose sum is b. If this is not possible, the trinomial is not factorable.
      c. Rewrite the middle term (bx) using the two numbers found in step **b.** as coefficients.
      d. Factor by grouping.
   c. Four terms:
   Group terms with a common factor and factor out any common binomial factor.
3. **Check the possibility of factoring any of the factors**.

Checking: Factoring can be checked by multiplying the factors. The product should be the original expression.

Exercises 12.5

Completely factor each of the given polynomials. If a polynomial cannot be factored, write "not factorable."

1. $m^2 + 7m + 6$

2. $a^2 - 4a + 3$

3. $x^2 + 11x + 18$

4. $y^2 + 8y + 15$

5. $x^2 - 100$

6. $n^2 - 8n + 12$

7. $m^2 - m - 6$

8. $y^2 - 49$

9. $a^2 + 2a + 24$

10. $-x^2 - 12x - 35$

11. $64a^2 - 1$

12. $49x^2 + 4$

13. $x^2 + 10x + 25$

14. $x^2 + 3x - 10$

15. $x^2 + 9x - 36$

16. $x^2 + 16x + 64$

17. $3a^2 + 12a - 36$

18. $-2y^2 + 24y - 70$

19. $-5x^2 + 70x - 240$

20. $7t^2 + 14t - 168$

21. $64 + 49t^2$

22. $3x^2 - 147$

23. $x^3 - 4x^2 - 12x$

24. $3n^3 + 15n^2 + 18n$

25. $112a - 2a^2 - 2a^3$

26. $200x + 20x^2 - 4x^3$

27. $16x^3 - 100x$

28. $48x^3 - 27x$

29. $-3x^2 + 17x - 10$

30. $2x^2 + 7x + 3$

31. $6x^2 - 11x + 4$

32. $12x^2 - 32x + 5$

33. $12m^2 + m - 6$

34. $6t^2 + t - 35$

35. $4x^2 - 14x + 6$

36. $-4x^2 + 18x - 20$

37. $8x^2 + 6x - 35$

38. $12x^2 + 5x - 3$

39. $20x^2 - 21x - 54$

40. $21x^2 - x - 10$

41. $14 + 11x - 15x^2$

42. $24 + x - 3x^2$

43. $-8a^2 + 22a - 15$

44. $63x^2 - 40x - 12$

45. $20y^2 + 9y - 20$

46. $35x^2 - x - 6$

47. $18x^2 - 15x + 2$

48. $12x^2 - 47x + 11$

49. $-150x^2 + 96$

50. $252x - 175x^3$

51. $12n^2 - 60n - 75$

52. $-12x^3 - 2x^2 - 70x$

53. $21a^3 - 13a^2 - 2a$

54. $13x^3 + 120x^2 + 100x$

55. $36x^3 + 21x^2 - 30x$

56. $63x - 3x^2 - 30x^3$

57. $16x^3 - 52x^2 + 22x$

58. $24y^3 - 4y^2 - 160y$

59. $75 + 10m + 120m^2$

60. $144x^3 - 10x^2 - 50x$

61. $xy + 3y - 4x - 12$

62. $2xz + 10x + z + 5$

63. $x^2 + 2xy - 6x - 12y$

64. $2y^2 + 6yz + 5y + 15z$

65. $-x^3 + 8x^2 + 5x - 40$

66. $2x^3 - 14x^2 - 3x + 21$

67. $x^3 + 125$

68. $y^3 - 1000$

69. $x^4y^3 - x^4$

70. $x^6y^3 - x^3$

71. $8a^6 + 27b^6$

72. $a^9 + 64b^3$

73. $x^6y^3 - 125$

74. $x^3y^3 + 216$

75. $x^3 + 7x^2 - 9x - 63$

76. $x^5 + 5x^4 - 4x - 20$

77. $9x^2 - (y+6)^2$

78. $(x+2)^2 - 25a^2$

79. $(y^2 + 20y + 100) - 49x^2$

80. $(t^2 + 22t + 121) - 16s^2$

12.6 Solving Quadratic Equations by Factoring

In this chapter the emphasis has been on second-degree polynomials and functions. Such polynomials are particularly evident in physics with the path of thrown objects (or projectiles) affected by gravity, in mathematics involving area (area of a circle, rectangle, square, and triangle), and in many situations in higher-level mathematics. The methods of factoring we have studied (*ac*-method, trial-and-error method, difference of two squares, and perfect square trinomials) have been related, in general, to second-degree polynomials. Second-degree polynomials are called **quadratic polynomials** (or **quadratics**) and, as just discussed, they play a major role in many applications of mathematics. In fact, quadratic polynomials and techniques for solving quadratic equations are central topics in the first two courses in algebra.

Objective A Solving Quadratic Equations by Factoring

Quadratic Equations

Quadratic equations are equations that can be written in the form $ax^2 + bx + c = 0$ where a, b, and c are real numbers and $a \neq 0$.

The form $ax^2 + bx + c = 0$ is called the **standard form** (or **general form**) of a **quadratic equation**. In the standard form, a quadratic polynomial is on one side of the equation and 0 is on the other side. For example,

$x^2 - 8x + 12 = 0$ is a quadratic equation in standard form,

while

$3x^2 - 2x = 27$ and $x^2 = 25x$ are quadratic equations, just not in standard form.

These last two equations can be manipulated algebraically so that 0 is on one side. When solving quadratic equations by factoring, having 0 on one side before we factor is necessary because of the following **zero-factor property**.

Zero-Factor Property

If the product of two (or more) factors is 0, then at least one of the factors must be 0. That is, for real numbers a and b,

if $a \cdot b = 0$, then $a = 0$ or $b = 0$ or both.

Example 1

Solving Factored Quadratic Equations

Solve the following quadratic equation.

$(x-5)(2x-7)=0$

Solution

Since the quadratic is already factored and the other side of the equation is 0, we use the zero-factor property and set each factor equal to 0. This process yields two linear equations, which can then be solved.

$$x - 5 = 0 \qquad \text{or} \qquad 2x - 7 = 0$$
$$x = 5 \qquad\qquad\qquad 2x = 7$$
$$x = \frac{7}{2}$$

Thus the **two solutions** (or **roots**) to the original equation are $x = 5$ and $x = \frac{7}{2}$. Or, we can say that the solution set is $\left\{\frac{7}{2}, 5\right\}$.

The solutions can be **checked** by substituting them one at a time for x in the equation. That is, there will be two "checks."

Substituting $x = 5$ gives

$$(x-5)(2x-7) = (5-5)(2\cdot5-7)$$
$$= (0)(3)$$
$$= 0.$$

Substituting $x = \frac{7}{2}$ gives

$$(x-5)(2x-7) = \left(\frac{7}{2}-5\right)\left(2\cdot\frac{7}{2}-7\right)$$
$$= \left(-\frac{3}{2}\right)(7-7)$$
$$= \left(-\frac{3}{2}\right)(0)$$
$$= 0.$$

Therefore, both 5 and $\frac{7}{2}$ are solutions to the original equation.

Now work margin exercise 1.

1. Find all possible solutions to the following equation.

$$(y-7)(3y-5)=0$$

In Example 1 the polynomial was factored and the other side of the equation was 0. The equation was solved by setting each factor, in turn, equal to 0 and solving the resulting linear equations. These solutions tell us that the original equation has two solutions. **In general, a quadratic equation has two solutions.** In the special cases where the two factors are the same, there is only one solution and it is called a **double solution** (or **double root**). The following examples show how to solve quadratic equations by factoring. Remember that the equation must be in standard form with one side of the equation equal to 0. Study these examples carefully.

Example 2

Solving Quadratic Equations by Factoring

Solve the following equations by writing the equation in standard form with one side 0 and factoring the polynomial. Then set each factor equal to 0 and solve. Checking is left as an exercise for the student.

a. $3x^2 = 6x$

Solution

$$3x^2 = 6x$$

$$3x^2 - 6x = 0$$ Write the equation in standard form with 0 on one side by subtracting $6x$ from both sides.

$$3x(x-2) = 0$$ Factor out the common monomial, $3x$.

$3x = 0$ or $x - 2 = 0$ Set each factor equal to 0.

    $x = 0$ $x = 2$ Solve each linear equation.

The solutions are 0 and 2. Or, we can say that the solution set is $\{0, 2\}$.

b. $x^2 - 8x + 16 = 0$

Solution

$$x^2 - 8x + 16 = 0$$

$$(x - 4)^2 = 0$$ The trinomial is a perfect square.

$x - 4 = 0$ or $x - 4 = 0$ Both factors are the same, so there is only

    $x = 4$ $x = 4$ one distinct solution.

The only solution is 4, and it is called a **double solution** (or **double root**). The solution set is $\{4\}$.

c. $4x^2 - 4x = 24$

Solution

$$4x^2 - 4x = 24$$

$$4x^2 - 4x - 24 = 0$$ Add -24 to both sides so that one side is 0.

$$4(x^2 - x - 6) = 0$$ Factor out the common monomial, 4.

$$4(x - 3)(x + 2) = 0$$ Factor

$x - 3 = 0$ or $x + 2 = 0$ The constant factor 4 can never be 0 and does not

    $x = 3$ $x = -2$ affect the solution.

The solutions are 3 and -2. Or, we can say the solution set is $\{-2, 3\}$.

d. $(x + 5)^2 = 36$

Solution

$$(x + 5)^2 = 36$$
$$x^2 + 10x + 25 = 36 \qquad \text{Expand } (x+5)^2.$$
$$x^2 + 10x - 11 = 0 \qquad \text{Add } -36 \text{ to both sides so that one side is } 0.$$
$$(x + 11)(x - 1) = 0 \qquad \text{Factor.}$$
$$x + 11 = 0 \qquad \text{or} \quad x - 1 = 0$$
$$x = -11 \qquad\qquad x = 1$$

The solutions are −11 and 1.

e. $3x(x-1) = 2(5-x)$

Solution

$$3x(x - 1) = 2(5 - x)$$
$$3x^2 - 3x = 10 - 2x \qquad \text{Use the distributive property.}$$
$$3x^2 - 3x + 2x - 10 = 0 \qquad \text{Arrange terms so that } 0 \text{ is on one side.}$$
$$3x^2 - x - 10 = 0 \qquad \text{Simplify.}$$
$$(3x + 5)(x - 2) = 0 \qquad \text{Factor.}$$
$$3x + 5 = 0 \qquad \text{or} \qquad x - 2 = 0$$
$$3x = -5 \qquad\qquad x = 2$$
$$x = -\frac{5}{3}$$

The solutions are $-\dfrac{5}{3}$ and 2.

f. $\dfrac{2x^2}{15} - \dfrac{x}{3} = -\dfrac{1}{5}$

Solution

$$\frac{2x^2}{15} - \frac{x}{3} = -\frac{1}{5}$$
$$15 \cdot \frac{2x^2}{15} - \frac{x}{3} \cdot 15 = -\frac{1}{5} \cdot 15 \qquad \text{Multiply each term by 15, the LCM of the denominators, to get integer coefficients.}$$
$$2x^2 - 5x = -3 \qquad \text{Simplify.}$$
$$2x^2 - 5x + 3 = 0 \qquad \text{Add 3 to both sides so that one side is } 0.$$
$$(2x - 3)(x - 1) = 0 \qquad \text{Factor.}$$

$$2x - 3 = 0 \quad \text{or} \quad x - 1 = 0$$
$$2x = 3 \qquad\qquad x = 1$$
$$x = \frac{3}{2}$$

The solutions are $\frac{3}{2}$ and 1.

Solve the following quadratic equations by factoring.

2. a. $4x^2 = 8x$

  b. $x^2 - 6x + 9 = 0$

  c. $9x^2 - 27x = 36$

  d. $x(6x + 21) = 2(4x + 14)$

  e. $\dfrac{x^2}{6} - \dfrac{x}{3} = \dfrac{1}{2}$

3. $(x + 4)^2 = 9$

Completion Example 3

Solving Quadratic Equations by Factoring

Solve the following quadratic equation by factoring.
$$(x - 2)^2 = 64$$

Solution

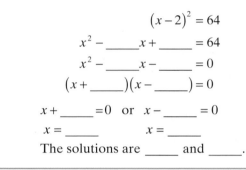

$$(x - 2)^2 = 64$$
$$x^2 - \underline{\quad} x + \underline{\quad} = 64$$
$$x^2 - \underline{\quad} x - \underline{\quad} = 0$$
$$(x + \underline{\quad})(x - \underline{\quad}) = 0$$
$$x + \underline{\quad} = 0 \quad \text{or} \quad x - \underline{\quad} = 0$$
$$x = \underline{\quad} \qquad\qquad x = \underline{\quad}$$

The solutions are $\underline{\quad}$ and $\underline{\quad}$.

Now work margin exercises **2** and **3.**

In the next example, we show that factoring can be used to solve equations of degrees higher than second degree. There will be more than two factors, but just as with quadratics, if the product is 0, then the solutions are found by setting each factor equal to 0.

Completion Example Answers

   3. $x^2 - 4x + 4 = 64$; $x^2 - 4x - 60 = 0$

     $(x + 6)(x - 10) = 0$; $x + 6 = 0$ or $x - 10 = 0$

     $x = -6 \qquad\qquad x = 10$

     The solutions are -6 and 10.

Example 4

Solving Higher Degree Equations

Solve the following equation: $2x^3 - 4x^2 - 6x = 0$.

Solution

$$2x^3 - 4x^2 - 6x = 0$$

$$2x\left(x^2 - 2x - 3\right) = 0 \qquad \text{Factor out the common monomial, } 2x.$$

$$2x(x - 3)(x + 1) = 0 \qquad \text{Factor the trinomial.}$$

$$2x = 0 \quad \text{or} \quad x - 3 = 0 \quad \text{or} \quad x + 1 = 0 \qquad \text{Set each factor equal to 0 and solve.}$$

$$x = 0 \qquad\qquad x = 3 \qquad\qquad x = -1$$

The solutions are 0, 3, and −1.
Or, we can say that the solution set is $\{-1, 0, 3\}$.

Now work margin exercise 4.

The procedure to solve a quadratic equation can be summarized as follows.

To Solve a Quadratic Equation by Factoring

1. Add or subtract terms as necessary so that 0 is on one side of the equation and the equation is in the standard form $ax^2 + bx + c = 0$ where $a, b,$ and c are real numbers and $a \neq 0$.
2. Factor completely. (If there are any fractional coefficients, multiply each term by the least common denominator first so that all coefficients will be integers.)
3. Set each nonconstant factor equal to 0 and solve each linear equation for the unknown.
4. Check each solution, one at a time, in the original equation.

Special Comment: All of the quadratic equations in this section can be solved by factoring. That is, all of the quadratic polynomials are factorable. However, as we have seen in some of the previous sections, not all polynomials are factorable. In Chapter 15 we will develop techniques (other than factoring) for solving quadratic equations whether the quadratic polynomial is factorable or not.

4. Solve the following equation.

$$3x^3 - 9x^2 - 30x = 0$$

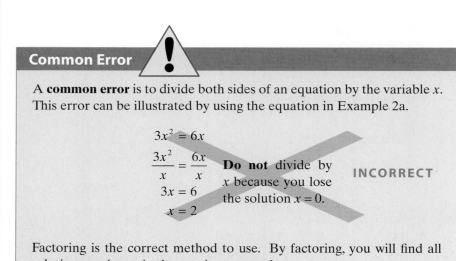

Common Error

A **common error** is to divide both sides of an equation by the variable x. This error can be illustrated by using the equation in Example 2a.

$$3x^2 = 6x$$
$$\frac{3x^2}{x} = \frac{6x}{x}$$
$$3x = 6$$
$$x = 2$$

Do not divide by x because you lose the solution $x = 0$.

INCORRECT

Factoring is the correct method to use. By factoring, you will find all solutions as shown in the previous examples.

Objective B **Finding an Equation Given the Roots**

To help develop a complete understanding of the relationships between factors, factoring, and solving equations, we reverse the process of finding solutions. That is, we want to **find an equation that has certain given solutions (or roots)**. For example, to find an equation that has the roots

$$x = 4 \quad \text{and} \quad x = -7$$

we proceed as follows.

1. Rewrite the linear equations with 0 on one side:
$$x - 4 = 0 \quad \text{and} \quad x + 7 = 0.$$

2. Form the product of the factors and set this product equal to 0:
$$(x - 4)(x + 7) = 0.$$

3. Multiply the factors. The resulting quadratic equation must have the two given roots (because we know the factors):
$$x^2 + 3x - 28 = 0.$$

notes

■ This equation can be multiplied by any nonzero constant and a new equation will be formed, but it will still have the same solutions, namely 4 and −7. Thus technically, there are many equations with these two roots.

■

The formal reasoning is based on the following theorem called the **factor theorem**.

Factor Theorem

If $x = c$ is a root of a polynomial equation in the form $P(x) = 0$, then $x - c$ is a factor of the polynomial $P(x)$.

Example 5

Using the Factor Theorem to Find Equations with Given Roots

Find a polynomial equation with integer coefficients that has the given roots:

$$x = 3 \text{ and } x = -\frac{2}{3}.$$

Solution

Form the linear equations and then find the product of the factors.

$x = 3$	$x = -\dfrac{2}{3}$
$x - 3 = 0$	$x + \dfrac{2}{3} = 0$ Rewrite the equations with 0 on one side.
	$3x + 2 = 0$ Multiply by 3 to get integer coefficients.

Form the equation by setting the product of the factors equal to 0 and multiplying.

$(x - 3)(3x + 2) = 0$ The equation has integer coefficients.

$3x^2 - 7x - 6 = 0$ This quadratic equation has the two given roots.

Now work margin exercise 5.

5. Use the factor theorem to find an equation with integer coefficients that has the given roots:

$$x = 4 \text{ and } x = \frac{3}{2}$$

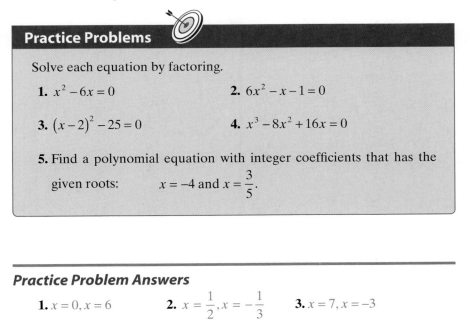

Practice Problems

Solve each equation by factoring.

1. $x^2 - 6x = 0$

2. $6x^2 - x - 1 = 0$

3. $(x - 2)^2 - 25 = 0$

4. $x^3 - 8x^2 + 16x = 0$

5. Find a polynomial equation with integer coefficients that has the given roots: $\quad x = -4 \text{ and } x = \dfrac{3}{5}.$

Practice Problem Answers

1. $x = 0, x = 6$

2. $x = \dfrac{1}{2}, x = -\dfrac{1}{3}$

3. $x = 7, x = -3$

4. $x = 0, x = 4$

5. $5x^2 + 17x - 12 = 0$

1. $(x-3)(x-2)=0$ **2.** $(x+5)(x-2)=0$ **3.** $(2x-9)(x+2)=0$ **4.** $(x+7)(3x-4)=0$

5. $0=(x+3)(x+3)$ **6.** $0=(x+10)(x-10)$ **7.** $(x+5)(x+5)=0$ **8.** $(x+5)(x-5)=0$

9. $2x(x-2)=0$ **10.** $3x(x+3)=0$ **11.** $(x+6)^2=0$ **12.** $5(x-9)^2=0$

13. $x^2-3x-4=0$ **14.** $x^2+7x+12=0$ **15.** $x^2-x-12=0$

16. $x^2-11x+18=0$ **17.** $0=x^2+3x$ **18.** $0=x^2-3x$

19. $x^2+8=6x$ **20.** $x^2=x+30$ **21.** $2x^2+2x-24=0$

22. $9x^2+63x+90=0$ **23.** $0=2x^2-5x-3$ **24.** $0=2x^2-x-3$

25. $3x^2-4x-4=0$ **26.** $3x^2-8x+5=0$ **27.** $2x^2-7x=4$

28. $4x^2+8x=-3$ **29.** $-2x=3x^2-8$ **30.** $6x^2+2=-7x$

31. $4x^2-12x+9=0$ **32.** $25x^2-60x+36=0$ **33.** $8x=5x^2$

34. $15x=3x^2$ **35.** $9x^2-36=0$ **36.** $4x^2-16=0$

37. $5x^2=10x-5$ **38.** $2x^2=4x+6$ **39.** $8x^2+32=32x$

40. $6x^2=18x+24$ **41.** $\dfrac{x^2}{9}=1$ **42.** $\dfrac{x^2}{2}=8$

43. $\dfrac{x^2}{5}-x-10=0$ **44.** $\dfrac{2}{3}x^2+2x-\dfrac{20}{3}=0$ **45.** $\dfrac{x^2}{8}+x+\dfrac{3}{2}=0$

46. $\dfrac{x^2}{6} - \dfrac{1}{2}x - 3 = 0$

47. $x^2 - x + \dfrac{1}{4} = 0$

48. $\dfrac{x^2}{3} - 2x + 3 = 0$

49. $x^3 + 8x = 6x^2$

50. $x^3 = x^2 + 30x$

51. $6x^3 + 7x^2 = -2x$

52. $3x^3 = 8x - 2x^2$

53. $0 = x^2 - 100$

54. $0 = x^2 - 121$

55. $3x^2 - 75 = 0$

56. $5x^2 - 45 = 0$

57. $x^2 + 8x + 16 = 0$

58. $x^2 + 14x + 49 = 0$

59. $3x^2 = 18x - 27$

60. $5x^2 = 10x - 5$

61. $(x - 1)^2 = 4$

62. $(x - 3)^2 = 1$

63. $(x + 5)^2 = 9$

64. $(x + 4)^2 = 16$

65. $(x + 4)(x - 1) = 6$

66. $(x - 5)(x + 3) = 9$

67. $27 = (x + 2)(x - 4)$

68. $-1 = (x + 2)(x + 4)$

69. $x(x + 7) = 3(x + 4)$

70. $x(x + 9) = 6(x + 3)$

71. $3x(x + 1) = 2(x + 1)$

72. $2x(x - 1) = 3(x - 1)$

73. $x(2x + 1) = 6(x + 2)$

74. $3x(x + 3) = 2(2x - 1)$

Write a polynomial equation with integer coefficients that has the given roots. See Example 5.

75. $y = 3, y = -2$

76. $x = 5, x = 7$

77. $x = -5, x = -\dfrac{1}{2}$

78. $x = \dfrac{1}{4}, x = -1$

79. $x = \dfrac{1}{2}, x = \dfrac{3}{4}$

80. $y = \dfrac{2}{3}, y = \dfrac{1}{6}$

81. $x = 0, x = 3, x = -2$

82. $y = 0, y = -4, y = 1$

83. $y = -2, y = 3, y = 3$ (3 is a double root.)

84. $x = -1, x = -1, x = -1$ (−1 is a triple root.)

85. When solving equations by factoring, one side of the equation must be 0. Explain why this is so.

86. In solving the equation $(x+5)(x-4)=6,$ why can't we just put one factor equal to 3 and the other equal to 2? Certainly $3 \cdot 2 = 6$.

87. A ball is dropped from the top of a building that is 784 feet high. The height of the ball above ground level is given by the polynomial function $h(t) = -16t^2 + 784$ where t is measured in seconds.

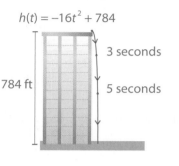

$h(t) = -16t^2 + 784$

784 ft

3 seconds

5 seconds

   a. How high is the ball after 3 seconds? 5 seconds?

   b. How far has the ball traveled in 3 seconds? 5 seconds?

   c. When will the ball hit the ground? Explain your reasoning in terms of factors.

12.7 Applications of Quadratic Equations

Whether or not application problems cause you difficulty depends a great deal on your personal experiences and general reasoning abilities. These abilities are developed over a long period of time. A problem that is easy for you, possibly because you have had experience in a particular situation, might be quite difficult for a friend and vice versa.

Most problems do not say specifically to add, subtract, multiply, or divide. You are to know from the nature of the problem what to do. You are to ask yourself, "What information is given? What am I trying to find? What tools, skills, and abilities do I need to use?"

Word problems should be approached in an orderly manner. You should have an "attack plan."

Attack Plan for Application Problems

1. Read the problem carefully at least twice.
2. Decide what is asked for and assign a variable or variable expression to the unknown quantities.
3. Organize a chart, table, or diagram relating all the information provided.
4. Form an equation. (A formula of some type may be necessary.)
5. Solve the equation.
6. Check your solution with the wording of the problem to be sure it makes sense.

Several types of problems lead to quadratic equations. The problems in this section are set up so that the equations can be solved by factoring. More general problems and approaches to solving quadratic equations are discussed in Chapter 15.

Objectives

A Use quadratic equations to solve problems related to numbers and geometry.

B Use quadratic equations to solve problems involving consecutive integers.

C Use quadratic equations to solve problems related to the Pythagorean Theorem.

Example 1

Applications of Quadratic Equations

a. One number is four more than another and the sum of their squares is 296. What are the numbers?

Solution

Let x = smaller number. Then $x + 4$ = larger number.

$$x^2 + (x+4)^2 = 296 \quad \text{Add the squares.}$$
$$x^2 + x^2 + 8x + 16 = 296 \quad \text{Expand } (x+4)^2.$$
$$2x^2 + 8x - 280 = 0 \quad \text{Write the equation in standard form.}$$
$$2(x^2 + 4x - 140) = 0 \quad \text{Factor out the GCF.}$$
$$2(x+14)(x-10) = 0 \quad \text{Factor the trinomial.}$$

$$x + 14 = 0 \qquad \text{or} \qquad x - 10 = 0$$
$$x = -14 \qquad\qquad x = 10$$
$$x + 4 = -10 \qquad\qquad x + 4 = 14$$

There are two sets of answers to the problem: 10 and 14 or −14 and −10.

Check:
$$10^2 + 14^2 = 100 + 196 = 296$$
$$\text{and } (-14)^2 + (-10)^2 = 196 + 100 = 296$$

b. In an orange grove, there are 10 more trees in each row than there are rows. How many rows are there if there are 96 trees in the grove?

Solution

Let r = number of rows.
Then $r + 10$ = number of trees per row.
Set up the equation and solve.

$$r(r + 10) = 96$$
$$r^2 + 10r = 96$$
$$r^2 + 10r - 96 = 0$$
$$(r - 6)(r + 16) = 0$$

$$r - 6 = 0 \qquad \text{or} \qquad r + 16 = 0$$
$$r = 6 \qquad\qquad r = -16$$

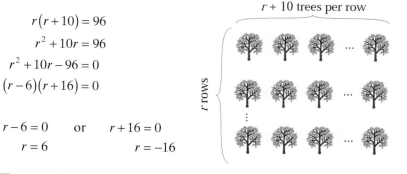

There are 6 rows in the grove ($6 \cdot 16 = 96$ trees).

Note: While −16 is a solution to the equation, −16 does not fit the conditions of the problem and is discarded. You cannot have −16 rows.

c. A rectangle has an area of 135 square meters and a perimeter of 48 meters. What are the dimensions of the rectangle?

Solution

The area of a rectangle is the product of its length and width ($A = lw$).
The perimeter of a rectangle is given by $P = 2l + 2w$.
Since the perimeter is 48 meters, then the length plus the width must be 24 meters (one half of the perimeter).

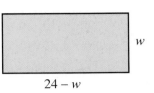

$24 - w$

Let w = width. Then $24 - w$ = length.
Set up an equation for the area and solve.

$$(24 - w) \cdot w = 135$$
$$24w - w^2 = 135$$
$$0 = w^2 - 24w + 135$$
$$0 = (w - 9)(w - 15)$$

$w - 9 = 0$ or $w - 15 = 0$
$w = 9$ $w = 15$
$l = 24 - 9 = 15$ $l = 24 - 15 = 9$

The dimensions are 9 meters by 15 meters ($9 \cdot 15 = 135$).

d. A man wants to build a fence on three sides of a rectangular-shaped lot he owns. If 180 feet of fencing is needed and the area of the lot is 4000 square feet, what are the dimensions of the lot?

Solution

Let x = one of two equal sides. Then $180 - 2x$ = third side.
Set up an equation for the area and solve.

$$x(180 - 2x) = 4000$$
$$180x - 2x^2 = 4000$$
$$0 = 2x^2 - 180x + 4000$$
$$0 = 2(x^2 - 90x + 2000)$$
$$0 = 2(x - 50)(x - 40)$$

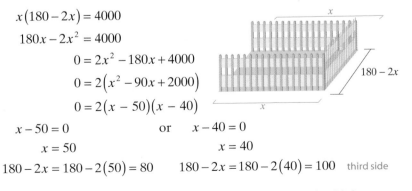

$x - 50 = 0$ or $x - 40 = 0$
$x = 50$ $x = 40$
$180 - 2x = 180 - 2(50) = 80$ $180 - 2x = 180 - 2(40) = 100$ third side

Thus there are two possible answers: the lot is 50 feet by 80 feet or the lot is 40 feet by 100 feet.

Now work margin exercise 1.

1. a. One number is five less than another and the sum of their squares is 325. What are the numbers?

b. In a theater, there are four less seats in a row than there are rows. How many rows are there if there are 357 seats in the theater?

c. A rectangle has an area of 288 square feet and a perimeter of 68 feet. What are the dimensions of the rectangle?

d. A husband and wife want to screen in their back patio. If the perimeter of the three sides is 38 feet and the area of the patio is 168 square feet, what are the dimenisions of the patio?

Objective B **Consecutive Integers**

In Section 8.4, we discussed applications with **consecutive integers**, **consecutive even integers**, and **consecutive odd integers**. Because the applications involved only addition and subtraction, the related equations were first degree. In this section the applications involve squaring expressions and solving quadratic equations. For convenience, the definitions and representations of the various types of integers are repeated here.

Consecutive Integers

Integers are **consecutive** if each is 1 more than the previous integer. Three consecutive integers can be represented as n, $n + 1$, and $n + 2$.

For example: 5, 6, 7

Consecutive Even Integers

Even integers are consecutive if each is 2 more than the previous even integer. Three consecutive even integers can be represented as n, $n + 2$, and $n + 4$, where n is an even integer.

For example: 24, 26, 28

Consecutive Odd Integers

Odd integers are consecutive if each is 2 more than the previous odd integer. Three consecutive odd integers can be represented as n, $n + 2$, and $n + 4$, where n is an odd integer.

For example: 41, 43, 45.

Note that consecutive even and consecutive odd integers are represented in the same way. The value of the first integer n determines whether the remaining integers are even or odd.

2. a. Find two consecutive negative even integers such that the sum of their squares is 340.

b. Find three consecutive even integers such that the product of the second and third is 112 more than the first,

Example 2

Consecutive Integers

a. Find two consecutive positive integers such that the sum of their squares is 265.

Solution

Let n = first integer.
Then $n + 1$ = next consecutive integer.
Set up and solve the related equation.

$$n^2 + (n+1)^2 = 265$$
$$n^2 + n^2 + 2n + 1 = 265$$
$$2n^2 + 2n - 264 = 0$$
$$2(n^2 + n - 132) = 0$$
$$2(n+12)(n-11) = 0$$

$n + 12 = 0$	or	$n - 11 = 0$
$n = -12$		$n = 11$
$n + 1 = -11$		$n + 1 = 12$

Consider the solution $n = -12$. The next consecutive integer, $n + 1$, is -11. While it is true that the sum of their squares is 265, we must remember that the problem calls for **positive** consecutive integers. Therefore, we can only consider positive solutions. Hence, the two integers are 11 and 12.

b. Find three consecutive odd integers such that the product of the first and second is 68 more than the third.

Solution

Let n = first odd integer
and $n + 2$ = second consecutive odd integer
and $n + 4$ = third consecutive odd integer.
Set up and solve the related equation.

$$n(n+2) = (n+4) + 68$$
$$n^2 + 2n = n + 72$$
$$n^2 + n - 72 = 0$$
$$(n+9)(n-8) = 0$$

$n + 9 = 0$	or	$n - 8 = 0$
$n = -9$		$n = 8$
$n + 2 = -7$		$n + 2 = 10$
$n + 4 = -5$		$n + 4 = 12$

The three consecutive odd integers are -9, -7, and -5. Note that 8, 10, and 12 are even and therefore cannot be considered a solution to the problem.

Now work margin exercise 2.

Objective C **The Pythagorean Theorem**

A geometric topic that often generates quadratic equations is right triangles. Right triangles and the Pythagorean Theorem were discussed in section 5.7. Remember, in a **right triangle**, one of the angles is a right angle (measures 90°), and the side opposite this angle (the longest side) is called the **hypotenuse**. The other two sides are called **legs**.

The Pythagorean Theorem

In a right triangle, the square of the length of the hypotenuse is equal to the sum of the squares of the lengths of the two legs:

$$c^2 = a^2 + b^2.$$

---

3. A guyline for a portable beach volleyball set is 10 feet long from the top of the net to where it is staked to the ground. The distance from the base of the pole to the point of attachment on the beach is 2 feet less than the height of the volleyball net. What is the height of the net?

Example 3

The Pythagorean Theorem

A support wire is 25 feet long and stretches from a tree to a point on the ground. The point of attachment on the tree is 5 feet higher than the distance from the base of the tree to the point of attachment on the ground. How far up the tree is the point of attachment?

Solution

Let $x =$ distance from base of tree to point of attachment on ground; then $x + 5 =$ height of point of attachment on tree. By the Pythagorean theorem, we have the following.

$$(x+5)^2 + x^2 = 25^2$$
$$x^2 + 10x + 25 + x^2 = 625$$
$$2x^2 + 10x - 600 = 0$$
$$2(x^2 + 5x - 300) = 0$$
$$2(x-15)(x+20) = 0$$
$$x - 15 = 0 \text{ or } x + 20 = 0$$
$$x = 15 \qquad x = -20$$

Because distance must be positive, -20 is not a possible solution. The solution is $x = 15$ and $x + 5 = 20$. Thus the point of attachment is 20 feet up the tree.

Now work margin exercise 3.

Exercises 12.7

1. One number is eight more than another. Their product is −16. What are the numbers?

2. One number is 10 more than another. If their product is −25, find the numbers.

3. The square of an integer is equal to seven times the integer. Find the integer.

4. The square of an integer is equal to twice the integer. Find the integer.

5. If the square of a positive integer is added to three times the integer, the result is 28. Find the integer.

6. If the square of a positive integer is added to three times the integer, the result is 54. Find the integer.

7. One number is seven more than another. Their product is 78. Find the numbers.

8. One positive number is three more than twice another. If the product is 27, find the numbers.

9. One positive number is six more than another. The sum of their squares is 260. What are the numbers?

10. One number is five less than another. The sum of their squares is 97. Find the numbers.

11. The difference between two positive integers is 8. If the smaller is added to the square of the larger, the sum is 124. Find the integers.

12. One positive number is 3 more than twice another. If the square of the smaller is added to the larger, the sum is 51. Find the numbers.

13. The product of a negative integer and 5 less than twice the integer equals the integer plus 56. Find the integer.

14. Find a positive integer such that the product of the integer with a number three less than the integer is equal to the integer increased by 32.

15. The product of two consecutive positive integers is 72. Find the integers.

16. Find two consecutive integers whose product is 110.

17. Find two consecutive positive integers such that the sum of their squares is 85.

18. Find two consecutive positive integers such that the sum of their squares is 145.

19. The product of two consecutive odd integers is 63. Find the integers.

20. The product of two consecutive even integers is 168. Find the integers.

21. Find two consecutive positive integers such that the square of the second integer added to four times the first is equal to 41.

22. Find two consecutive negative integers such that 6 times the first plus the square of the second equals 34.

23. Find three consecutive positive integers such that twice the product of the two smaller integers is 88 more than the product of the two larger integers.

24. Find three consecutive odd integers such that the product of the first and third is 71 more than 10 times the second.

25. Four consecutive integers are such that if the product of the first and third is multiplied by 6, the result is equal to the sum of the second and the square of the fourth. What are the integers?

26. Find four consecutive even integers such that the square of the sum of the first and second is equal to 516 more than twice the product of the third and fourth.

27. **Rectangles:** The length of a rectangle is twice the width. The area is 72 square inches. Find the length and width of the rectangle.

28. **Rectangles:** The length of a rectangle is three times the width. If the area is 147 square centimeters, find the length and width of the rectangle.

29. **Rectangles:** The length of a rectangle is four times the width. If the area is 64 square feet, find the length and width of the rectangle.

30. **Rectangles:** The length of a rectangle is five times the width. If the area is 180 square inches, find the length and width of the rectangle.

31. **Rectangles:** The width of a rectangle is 4 feet less than the length. The area is 117 square feet. Find the length and width of the rectangle.

32. **Rectangles:** The length of a rectangular yard is 12 meters greater than the width. If the area of the yard is 85 square meters, find the length and width of the yard.

33. **Triangles:** The height of a triangle is 4 feet less than the base. The area of the triangle is 16 square feet. Find the length of the base and the height of the triangle.

34. **Triangles:** The base of a triangle exceeds the height by 5 meters. If the area is 42 square meters, find the length of the base and the height of the triangle.

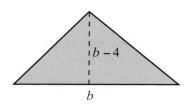

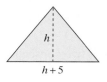

35. Triangles: The base of a triangle is 15 inches greater than the height. If the area is 63 square inches, find the length of the base.

36. Triangles: The base of a triangle is 6 feet less than the height. The area is 56 square feet. Find the height.

37. Rectangles: The perimeter of a rectangle is 32 inches. The area of the rectangle is 48 square inches. Find the dimensions of the rectangle.

38. Rectangles: The area of a rectangle is 24 square centimeters. If the perimeter is 20 centimeters, find the length and width of the rectangle.

39. Orchards: An orchard has 140 apple trees. The number of rows exceeds the number of trees per row by 13. How many trees are there in each row?

40. Military: One formation for a drill team is rectangular. The number of members in each row exceeds the number of rows by 3. If there is a total of 108 members in the formation, how many rows are there?

41. Theater: A theater can seat 144 people. The number of rows is 7 less than the number of seats in each row. How many rows of seats are there?

42. College Sports: An empty field on a college campus is being used for overflow parking for a football game. It currently has 187 cars in it. If the number of rows of cars is six less than the number of cars in each row, how many rows are there?

43. Parking: The parking garage at Baltimore-Washington International Airport contains 8400 parking spaces. The number of cars that can be parked on each floor exceeds the number of floors by 1675. How many floors are there in the parking garage?

44. Library Books: One bookshelf in the public library can hold 175 books. The number of books on each shelf exceeds the number of shelves by 18. How many books are on each shelf?

45. Rectangles: The length of a rectangle is 7 centimeters greater than the width. If 4 centimeters are added to both the length and width, the new area would be 98 square centimeters. Find the dimensions of the original rectangle.

46. Rectangles: The width of a rectangle is 5 meters less than the length. If 6 meters are added to both the length and width, the new area will be 300 square meters. Find the dimensions of the original rectangle.

47. Gardening: Susan is going to fence a rectangular flower garden in her back yard. She has 50 feet of fencing and she plans to use the house as the fence on one side of the garden. If the area is 300 square feet, what are the dimensions of the flower garden?

48. Ranching: A rancher is going to build a corral with 52 yards of fencing. He is planning to use the barn as one side of the corral. If the area is 320 square yards, what are the dimensions?

49. Communication: A telephone pole is to have a guy wire attached to its top and anchored to the ground at a point that is at a distance 34 feet less than the height of the pole from the base. If the wire is to be 2 feet longer than the height of the pole, what is the height of the pole?

50. Trees: Lucy is standing next to the General Sherman tree in Sequoia National Park, home of some of the largest trees in the world. The distance from Lucy to the base of the tree is 71 m less than the height of the tree. If the distance from Lucy to the top of the tree is 1 m more than the height of the tree, how tall is the General Sherman?

51. Holiday decorating: A Christmas tree is supported by a wire that is 1 foot longer than the height of the tree. The wire is anchored at a point whose distance from the base of the tree is 49 feet shorter than the height of the tree. What is the height of the tree?

52. Architecture: An architect wants to draw a rectangle with a diagonal of 13 inches. The length of the rectangle is to be 2 inches more than twice the width. What dimensions should she make the rectangle?

53. Gymnastics: Incline mats, or triangle mats, are offered with different levels of incline to help gymnasts learn basic moves. As the name may suggest, two sides of the mat are right triangles. If the height of the mat is 28 inches shorter than the length of the mat and the hypotenuse is 8 inches longer than the length of the mat, what is the length of the mat?

54. Laser show: Bill uses mirrors to augment the "laser experience" at a laser show. At one show he places three mirrors, A, B, C, in a right triangular form. If the distance between A and B is 15 m more than the distance between A and C, and the distance between B and C is 15 m less than the distance between A and C, what is the distance between mirror A and mirror C?

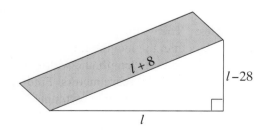

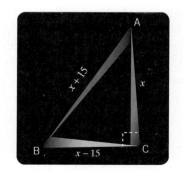

The demand for a product is the number of units of the product, x, that consumers are willing to buy when the market price is p dollars. The consumers' total expenditure for the product, S, is found by multiplying the price times the demand. $(S = px)$

55. During the summer at a local market, a farmer will sell $8p + 588$ pounds of peaches at p dollars per pound. If he sold \$900 worth of peaches this summer, what was the price per pound of the peaches?

56. On a hot afternoon, fans at a stadium will buy $2000 - 100p$ drinks for p dollars each. If the total sales after a game were \$4375, what was the price per drink?

57. When fishing reels are priced at p dollars, local consumers will buy $36 - p$ fishing reels. What is the price if total sales were \$320?

58. A manufacturer can sell $100 - 2p$ lamps at p dollars each. If the receipts from the lamps total \$1200, what is the price of the lamps?

Writing & Thinking

59. If three positive integers satisfy the Pythagorean Theorem, they are called a **Pythagorean triple**. For example, 3, 4, and 5 are a **Pythagorean triple** because $3^2 + 4^2 = 5^2$. There are an infinite number of such triples. To see how some triples can be found, fill out the following table and verify that the numbers in the rightmost three columns are indeed Pythagorean triples.

u	v	2uv	u² − v²	u² + v²
2	1	4	3	5
3	2	12	5	13
5	2			
4	3			
7	1			
6	5			

60. The pattern in Kara's linoleum flooring is in the shape of a square 8 inches on a side with right triangles of sides x inches placed on each side of the original square so that a new larger square is formed. What is the area of the new square? Explain why you do not need to find the value of x.

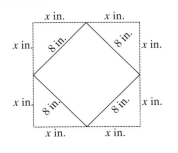

Section 12.1: Greatest Common Factor and Factoring by Grouping

Greatest Common Factor (GCF)   page 928
  The **greatest common factor (GCF)** of two or more integers is the
  largest integer that is a factor (or divisor) of all of the integers.

Procedure for Finding the GCF   page 929
  1. Find the prime factorization of all integers and integer coefficients.
  2. List all the factors that are common to all terms, including variables.
  3. Choose the greatest power of each factor common to all terms.
  4. Multiply these powers to find the GCF.
  Note: If there is no common prime factor or variable, then the GCF is 1.

Factoring Out the GCF   page 931
  To find a monomial that is the GCF of a polynomial:
  1. Find the variable(s) of highest degree and the largest integer
     coefficient that is a factor of each term of the polynomial. (This is
     one factor.)
  2. Divide this monomial factor into each term of the polynomial,
     resulting in another polynomial factor.

Factoring by Grouping   page 934
  To **factor by grouping**, look for common factors in each group and then
  look for common binomial factors.

Section 12.2: Factoring Trinomials: $x^2 + bx + c$

To Factor Trinomials of the Form $x^2 + bx + c$ page 942
  To factor $x^2 + bx + c$, if possible, find an integer pair of factors of c
  whose sum is b.
  1. If c is positive, then both factors must have the same sign.
     a. Both will be positive if b is positive.
     b. Both will be negative if b is negative.
  2. If c is negative, then one factor must be positive and the other
     negative.

Not Factorable (or Prime)   page 944
  A polynomial is **not factorable** if it cannot be factored as the product of
  polynomials with integer coefficients.

Section 12.3: Factoring Trinomials: $ax^2 + bx + c$

Trial-and-Error Method　　　　　　　　　　　　　　　　　　　　page 950
1. If the sign of the constant term is positive (+), the signs in both factors will be the same, either both positive or both negative.
2. If the sign of the constant term is negative (−), the signs in the factors will be different, one positive and one negative.

The *ac*-Method (Grouping)　　　　　　　　　　　　　　　　　page 953
1. Multiply $a \cdot c$.
2. Find two integers whose product is ac and whose sum is b. If this is not possible, then the trinomial is **not factorable**.
3. Rewrite the middle term (bx) using the two numbers found in Step 2 as coefficients.
4. Factor by grouping the first two terms and the last two terms.
5. Factor out the common binomial factor. This will give two binomial factors of the trinomial $ax^2 + bx + c$.

Section 12.4: Special Factoring Techniques

Special Factoring Techniques　　　　　　　　　　　　　　　　page 960
1. $x^2 - a^2 = (x+a)(x-a)$:　　Difference of two squares
2. $x^2 + 2ax + a^2 = (x+a)^2$:　　Perfect square trinomial
3. $x^2 - 2ax + a^2 = (x-a)^2$:　　Perfect square trinomial
4. $x^3 + a^3 = (x+a)(x^2 - ax + a^2)$:　　Sum of two cubes
5. $x^3 - a^3 = (x-a)(x^2 + ax + a^2)$:　　Difference of two cubes

Sum of Two Squares　　　　　　　　　　　　　　　　　　　　page 961
The **sum of two squares** is an expression of the form $x^2 + a^2$ and is **not factorable**.

General Guidelines for Factoring Polynomials   page 969

1. **Always look for a common monomial factor first.** If the leading coefficient is negative, factor out a negative monomial even if it is just −1.

2. Check the number of terms.

   a. **Two terms:**
      1. difference of two squares? – factorable
      2. sum of two squares? – not factorable
      3. difference of two cubes? – factorable
      4. sum of two cubes? – factorable

   b. **Three terms:**
      1. perfect square trinomial?
      2. use trial-and-error method?
         Guidelines for the trial-and-error method
         a. If the sign of the constant term is positive, the signs in both factors will be the same, either both positive or both negative.
         b. If the sign of the constant term is negative, the signs in the factors will be different, one positive and one negative.
      3. use *ac*-method?
         Guidelines for the ac-method
         a. Multiply $a \cdot c$.
         b. Find two integers whose product is ac and whose sum is b. If this is not possible, the trinomial is not factorable.
         c. Rewrite the middle term (bx) using the two numbers found in step b. as coefficients.
         d. Factor by grouping.

   c. **Four terms:**
      Group terms with a common factor and factor out any common binomial factor.

3. **Check the possibility of factoring any of the factors.**

Checking: Factoring can be checked by multiplying the factors. The product should be the original expression.

Quadratic Equations   page 972

    Quadratic equations are equations that can be written in the form $ax^2 + bx + c = 0$ where a, b, and c are real numbers and $a \neq 0$.

Zero-Factor Property   page 972

    If the product of two (or more) factors is 0, then at least one of the factors must be 0. That is, for real numbers a and b,

$$\text{if } a \cdot b = 0, \text{ then } a = 0 \text{ or } b = 0 \text{ or both.}$$

Solving Quadratic Equations by Factoring   page 977

    1. Add or subtract terms as necessary so that 0 is on one side of the equation and the equation is in the standard form $ax^2 + bx + c = 0$, where a, b, and c are real numbers and $a \neq 0$.

    2. Factor completely.

    3. Set each nonconstant factor equal to 0 and solve each linear equation for the unknown.

    4. Check each solution, one at a time, in the original equation.

Finding an Equation Given the Roots   page 978

Factor Theorem   page 979

    If $x = c$ is a root of a polynomial equation in the form $P(x) = 0$, then $x - c$ is a factor of the polynomial $P(x)$.

Attack Plan for Word Problems   page 983

    1. Read the problem carefully at least twice.

    2. Decide what is asked for and assign a variable or variable expression to the unknown quantities.

    3. Organize a chart, table, or diagram relating all the information provided.

    4. Form an equation. (A formula of some type may be necessary.)

    5. Solve the equation.

    6. Check your solution with the wording of the problem to be sure it makes sense.

Consecutive Integers   page 986

The Pythagorean Theorem   page 988

    In a right triangle, if c is the length of the hypotenuse and a and b are the lengths of the legs, then $c^2 = a^2 + b^2$.

Chapter 12: Review

Section 12.1: Greatest Common Factor and Factoring by Grouping

Complete the factoring of the polynomial as indicated.

1. $4x^2y^2 - 12xy^3 + 8xy^2 = 4xy^2(\underline{\hspace{1cm}})$

2. $14x^2y + 21x^2 = 7x^2(\underline{\hspace{1cm}})$

Factor each of the polynomials by finding the GCF (or $-1 \cdot$ GCF).

3. $11x - 22$

4. $-4y^2 + 28y$

5. $-7t^3x^4 + 98t^4x^3 + 35t^5x$ **6.** $16x^3y - 12x^2y$

Factor each expression by factoring out the common binomial factor.

7. $3a(a+7) - 2(a+7)$

8. $2a(x-10) + 3b(x-10)$

Factor each of the polynomials by grouping. If a polynomial cannot be factored, write "not factorable."

9. $ax - a + cx - c$

10. $x - 4xy + 2z - 8zy$

11. $z^2 + 5 + cz^2 + 5c$

12. $x^2 - x^2y + 6 + 6y$

Section 12.2: Factoring Trinomials: $x^2 + bx + c$

Find the pair of integers whose product is the first integer and whose sum is the second integer.

13. $-42, -1$

14. $27, 12$

Completely factor each of the given polynomials. If a polynomial cannot be factored, write "not factorable."

15. $m^2 + 7m + 6$

16. $y^2 + 2y + 24$

17. $n^2 - 8n + 12$

18. $a^2 - 5a - 50$

19. $c^2 + 3c + 10$

20. $3m^2 + 12m + 12$

21. $2y^3 + 14y^2 + 20y$

22. $7x^4 + 21x^3 - 28x^2$

23. $9x^5 + 81x^4 + 180x^3$

24. $11a^3 - 110a^2 - 121a$

25. $a^2 + 11ab + 30b^2$

26. $4y^2 - 28xy - 4x^2$

Section 12.3: Factoring Trinomials: $ax^2 + bx + c$

Completely factor each of the given polynomials. If a polynomial cannot be factored, write "not factorable."

27. $x^2 + 12x + 35$

28. $-12x^2 + 32x - 5$

29. $6y^2 - 11y + 4$

30. $12y^2 - 11y + 4$

31. $63x^2 - 3x - 30$

32. $16x^2 + 12x - 70$

33. $24 + x - 3x^2$

34. $2y^2 - 13y + 5$

35. $3y^2 + 10y + 6$

36. $200 + 20x - 4x^2$

37. $7xy^2 + 14xy - 168x$

38. $18x^3 - 15x^2 + 2x$

Section 12.4: Special Factoring Techniques

Completely factor each of the given polynomials. If a polynomial cannot be factored, write "not factorable."

39. $x^6 - 100$

40. $16x^2 - 25$

41. $x^2 + 144$

42. $4y^2 + 32y + 64$

43. $4y^2 + 169$

44. $8x^2 - 242$

45. $3x^3 - 48x$

46. $-3x^2 - 12x - 12$

47. $8x^9 - y^6$

48. $4x^4 + 500x$

49. $216x^3 - 27$

Section 12.5: Additional Factoring Practice

Completely factor each of the given polynomials. If a polynomial cannot be factored, write "not factorable."

50. $z^2 - 5z - 36$

51. $6y^2 + 23$

52. $-3x^2 - 10x + 8$

53. $10x^3 + 35x^2 + 30x$

54. $49x^2 - 4y^2$

55. $2x^3 - 54y^3$

56. $3x^3 - 9x^2 + x - 3$

Section 12.6: Solving Quadratic Equations by Factoring

Solve the equations by factoring.

57. $x^2 + 5x = 0$ **58.** $3x^2 - 17x + 10 = 0$ **59.** $y^2 + 2y - 35 = 0$ **60.** $16x^3 - 100x = 0$

61. $x^3 - 4x^2 - 12x = 0$ **62.** $5x^2 = 125$ **63.** $x^2 = 2x + 35$ **64.** $2a^2 + 28a + 98 = 0$

65. $(x + 3)^2 = 9$ **66.** $(x - 5)^2 = 64$ **67.** $(x - 2)(x - 6) = 5$ **68.** $(x + 3)(2x + 1) = -3$

Write a polynomial equation with integer coefficients that has the given roots.

69. $x = 4,\ x = -5$ **70.** $x = -\dfrac{3}{4},\ x = \dfrac{5}{8}$ **71.** $y = 1,\ y = -4,\ y = -4$
(−4 is a double root.)

Section 12.7: Applications of Quadratic Equations

Solve the following word problems.

72. Number Problem: One positive number is three more than another and the sum of their squares is 269. What are the numbers?

73. Number Problem: One number is 20 more than another. Their product is −84. What are the numbers?

74. Construction: A tract of new homes is arranged in rectangular fashion with 5 more houses on each street than there are streets. How many streets are there if there are 126 houses in the tract?

75. Rectangles: A rectangle has an area of 450 square yards and a perimeter of 90 yards. What are the dimensions of the rectangle?

76. Consecutive Integers: The sum of the squares of two consecutive odd positive integers is 290. Find the integers.

77. Consecutive Integers: The product of two consecutive even integers is 72 more than 4 times the larger integer. Find the integers.

78. Triangles: The base of a triangle is 12 inches greater than the height. If the area is 80 square inches, find the length of the base.

79. Theater: A small theater can seat 600 people. The number of rows is 10 less than the number of seats in each row. How many rows of seats are there? How many seats are in each row?

80. Rectangles: The diagonal of a rectangle is 29 meters long. The length of the rectangle is 1 meter longer than the width. Find the length and width of the rectangle.

Chapter 12: Test

Factor each of the polynomials by finding the GCF (or $-1 \cdot$ GCF).

1. $28ab^2x - 21ab^2y$

2. $18yz^3 - 6y^2z^3 + 12yz^2$

Completely factor each of the given polynomials. If the polynomial cannot be factored, write "not factorable."

3. $x^2 - 9x + 20$

4. $-x^2 - 14x - 49$

5. $xy - 7x + 35 - 5y$

6. $6x^2 - 6$

7. $12x^2 + 2x - 10$

8. $3x^2 + x - 24$

9. $16x^2 - 25y^2$

10. $2x^3 - x^2 - 3x$

11. $6x^2 - 13x + 6$

12. $2xy - 3y + 14x - 21$

13. $4x^2 + 25$

14. $-3x^3 + 6x^2 - 6x$

Solve the equations.

15. $x^2 - 7x - 8 = 0$

16. $-3x^2 = 18x$

17. $0 = 4x^2 - 17x - 15$

18. $(2x - 7)(x + 1) = 6x - 19$

Solve the following word problems.

19. Polynomial Equation: Find a polynomial equation with integer coefficients that has $x = 3$ and $x = -8$ as solutions.

20. Number Problem: One number is 10 less than five times another number. Their product is 120. Find the numbers.

21. Rectangles: The length of a rectangle is 7 centimeters less than twice the width. If the area of the rectangle is 165 square centimeters, find the length and width.

22. Consecutive Integers: The product of two consecutive positive integers is 342. Find the two integers.

23. Number Problem: The difference between two positive numbers is 9. If the smaller is added to the square of the larger, the result is 147. Find the numbers.

24. Architecture: The average staircase has steps with a width that is 6 cm more than the height and a diagonal that is 12 cm more than the height. How tall is the average step?

25. Squares: The function $A(x) = 9x^2 + 30x + 25$. can be used to represent the area of a square. Write a polynomial function $P(x)$ that represents the perimeter of the square. (**Hint:** Factor the expression to find the length and the width.)

Cumulative Review: Chapters 1 - 12

Find the LCM for each set of terms.

1. $\{20, 12, 24\}$

2. $\{8x^2, 14x^2y, 21xy\}$

Find the value of each expression by using the rules for order of operations.

3. $2 \cdot 3^2 \div 6 \cdot 3 - 3$

4. $6 + 3\left[4 - 2\left(3^3 - 1\right)\right]$

Perform the indicated operations. Reduce all answers to lowest terms.

5. $\dfrac{7}{12} + \dfrac{9}{16}$

6. $\dfrac{11}{15a} - \dfrac{5}{12a}$

7. $\dfrac{6x}{25} \cdot \dfrac{5}{4x}$

8. $\dfrac{40}{92} \div \dfrac{2}{15x}$

Perform the indicated operations. Reduce all answers to lowest terms.

9. 42% of _____ is 105.

10. 30% of 410 is _____.

Solve each of the equations.

11. $4(2x - 3) + 2 = 5 - (2x + 6)$

12. $\dfrac{4x}{7} - 3 = 9$

13. $\dfrac{2x + 3}{6} - \dfrac{x + 1}{4} = 2$

Determine whether each of the following equations is a conditional equations, an identity, or a contradiction.

14. $4(x + 2) + 5x - 5 = 2 - (x + 4)$

15. $5(3x - 2.3) + 1.1 = 4.5(2x - 0.1) + 6x$

Solve each formula for the indicated variable.

16. $y = mx + b$; solve for x.

17. $3x + 5y = 10$; solve for y.

Solve the following inequalities and graph the solution sets. Write each solution in interval notation. Assume that x is a real number.

18. $5x + 3 \geq 2x - 15$

19. $-16 < 3x + 5 < 17$

Solve the following problems.

20. Write the equation $3x + 7y = -14$ in slope-intercept form. Find the slope and the y-intercept, and then use them to draw the graph.

21. Find the equation for the line passing through the point $(-4, 7)$ with slope 0. Graph the line.

22. Find the equation in standard form of the line determined by the two points $(-5, 3)$ and $(2, -4)$. Graph the line.

23. Find the equation in slope-intercept form of the line parallel to the line $y = 3x - 7$ and passing through the point $(-1, 1)$. Graph both lines.

24. Given the relation $r = \{(2, -3), (3, -2), (5, 0), (7.1, 3.2)\}$.

   a. Graph the relation.

   b. State the domain of the relation.

   c. State the range of the relation.

   d. Is the relation a function? Explain.

Evaluate the function at the given values.

25. For the function $f(x) = x^2 - 3x + 4$, find

   a. $f(-6)$ **b.** $f(0)$ **c.** $f\left(\dfrac{1}{2}\right)$

Graph the inequality.

26. $3x + 2y \geq 10$

Use the properties of exponents to simplify each of the expressions. Answers should contain only positive exponents. Assume that all variables represent nonzero real numbers.

27. $\left(-2x^3 y^2\right)^{-3}$ **28.** $\left(\dfrac{6x^2}{y^5}\right)^2$ **29.** $\dfrac{\left(xy^0\right)^3}{\left(x^3 y^{-1}\right)^2}$ **30.** $\left(\dfrac{x^2 y^{-2}}{3y^5}\right)^{-2}$

Write each number in the following expression in scientific notation and simplify. Show the steps you use. Do not use a calculator. Leave the answer in scientific notation.

31. 0.00000056×0.0003

Perform the indicated operations and simplify by combining like terms.

32. $\left(2x^2 + 6x - 7\right) + \left(2x^2 - x - 1\right)$

33. $-(x+6)(3x-1)$

34. $(2x-7)(2x-7)$

35. $2x(x+5) - (x+1)(x-3)$

36. Express the following quotient as a sum of fractions and simplify, if possible.

$$\frac{8x^2y^2 - 5xy^2 + 4xy}{4xy^2}$$

Factor each expression as completely as possible.

37. $8x - 20$

38. $xy + 3x + 2y + 6$

39. $x^2 - 9x + 18$

40. $6x^2 - x - 12$

41. $16x^2y - 24xy$

42. $3x^2 - 48y^2$

43. $x^2 + x + 3$

44. $3x^2 + 5x + 2$

45. $2x^3 - 20x^2 + 50x$

46. $5x^3 - 135y^3$

Solve the equations.

47. $2x(x+5)(x-2) = 0$

48. $21x - 3x^2 = 0$

49. $x^2 = 3x + 28$

50. $8x = 12x + 2x^2$

Solve the following problems.

51. Find an equation that has $x = -10$ and $x = -5$ as roots.

52. What is the percent of profit if:

a. $520 is made on an investment of $4000?

b. $385 is made on an investment of $3500?

c. Which was the better investment based on return rate?

53. **Traveling by Car:** For winter break, Lindsey and Sloan decided to make the 510 mile trip from Charlotte, NC to Philadelphia, PA together. Sloan drove first at an average speed of 63 mph for 4.5 hours. Lindsey drove for 3.5 hours. What was Lindsey's average speed? Please round to the nearest mph.

54. **Investing:** Karl invested a total of $10,000 in two separate accounts. One account paid 6% interest and the other paid 8% interest. If the annual income from both accounts was $650, how much did he have invested in each account?

55. Rectangles: The perimeter of a rectangle is 60 inches. The area of the rectangle is 221 square inches. Find the dimensions of the rectangle. (**Hint:** After substituting and simplifying, you will have a quadratic equation.)

56. Number Problem: The difference between two positive numbers is 7. If the square of the smaller is added to the square of the larger, the result is 137. Find the numbers.

57. Consecutive Integers: Find two consecutive integers such that the sum of their squares is equal to 145.

58. Water Slides: A pool's straight water slide is 8 m longer than the height of the slide's ladder. If the distance from the base of the ladder to the end of the slide is 7 m longer than the height of the ladder, what is the height of the ladder?

Rational Expressions

Mathematics at Work!

Collaboration is a very important concept, and can help us meet deadlines. If we know how much time individuals can do a job alone, rational functions allow us to calculate how much time can be saved by collaboration. (See Section 13.5.)

Sarah and Jeff are raking leaves. Sarah can rake the yard in 3 hours by herself, and Jeff can rake the yard in $2\frac{5}{8}$ hours working alone. How long will it take the two of them to rake the yard if they work together?

13.1 Multiplication and Division with Rational Expressions

Objective A **Introduction to Rational Expressions**

The term **rational number** is the technical name for a fraction in which both the numerator and denominator are integers. Similarly, the term **rational expression** is the technical name for a fraction in which both the numerator and denominator are polynomials.

Rational Expressions

A **rational expression** is an algebraic expression that can be written in the form

$$\frac{P}{Q}$$ where P and Q are polynomials and $Q \neq 0$.

Examples of rational expressions are

$$\frac{4x^2}{9}, \quad \frac{y^2-25}{y^2+25}, \quad \text{and} \quad \frac{x^2+7x-6}{x^2-5x-14}.$$

As with rational numbers, the denominators of rational expressions cannot be 0. If a numerical value is substituted for a variable in a rational expression and the denominator assumes a value of 0, we say that the expression is **undefined** for that value of the variable. (These values are called **restrictions** on the variable.)

notes

- **Remember, the denominator of a rational expression can never be 0.** Division by 0 is undefined.
-

Example 1

Finding Restrictions on the Variable

Determine what values of the variable, if any, will make the rational expression undefined.

a. $\dfrac{5}{3x-1}$

Solution

$3x - 1 = 0$ Set the denominator equal to 0.

$3x = 1$ Solve the equation.

$x = \dfrac{1}{3}$

Thus the expression $\dfrac{5}{3x-1}$ is undefined for $x = \dfrac{1}{3}$. Any other real number may be substituted for x in the expression. We write $x \neq \dfrac{1}{3}$ to indicate the restriction on the variable.

b. $\dfrac{x^2 - 4}{x^2 - 5x - 6}$

Solution

$x^2 - 5x - 6 = 0$ Set the denominator equal to 0.

$(x - 6)(x + 1) = 0$ Solve the equation by factoring.

$x - 6 = 0$ or $x + 1 = 0$

$x = 6$ $x = -1$

Thus there are two restrictions on the variable: 6 and −1. We write $x \neq -1, 6$.

c. $\dfrac{x + 3}{x^2 + 36}$

Solution

$x^2 + 36 = 0$ Set the denominator equal to 0.

$x^2 = -36$ Solve the equation.

However there is no real number whose square is −36. Thus there are **no restrictions** on the variable.

Now work margin exercise 1.

1. Determine which values of the variable make the expression undefined.

a. $\dfrac{9}{5x-1}$

b. $\dfrac{x^2 - 25}{x^2 - 7x + 12}$

c. $\dfrac{x + 8}{x^2 + 16}$

notes

- **Comments about the Numerator being 0**

- **If the numerator of a rational expression has a value of 0 and the denominator is not 0 for that value of the variable, then the expression is defined and has a value of 0.** If both numerator and denominator are 0, then the expression is **undefined** just as in the case where only the denominator is 0.

The rules for operating with rational expressions are essentially the same as those for operating with fractions in arithmetic. That is, simplifying, multiplying, and dividing rational expressions involve factoring and reducing. Addition and subtraction of rational expressions require common denominators. The basic rules for fractions were discussed in earlier sections and are summarized here for easy reference.

Summary of Arithmetic Rules for Rational Numbers (or Fractions)

A **fraction** (or **rational number**) is a number that can be written in the form $\frac{a}{b}$ where a and b are integers and $b \neq 0$. (Remember, no denominator can be 0.)

The Fundamental Principle: $\dfrac{a}{b} = \dfrac{a \cdot k}{b \cdot k}$ where $b, k \neq 0$

The **reciprocal** of $\dfrac{a}{b}$ is $\dfrac{b}{a}$ and $\dfrac{a}{b} \cdot \dfrac{b}{a} = 1$ where $a, b \neq 0$.

Multiplication: $\dfrac{a}{b} \cdot \dfrac{c}{d} = \dfrac{a \cdot c}{b \cdot d}$ where $b, d \neq 0$

Division: $\dfrac{a}{b} \div \dfrac{c}{d} = \dfrac{a}{b} \cdot \dfrac{d}{c}$ where $b, c, d \neq 0$

Addition: $\dfrac{a}{b} + \dfrac{c}{b} = \dfrac{a+c}{b}$ where $b \neq 0$

Subtraction: $\dfrac{a}{b} - \dfrac{c}{b} = \dfrac{a-c}{b}$ where $b \neq 0$

For rational expressions, each rule can be restated by replacing a and b with P and Q where P and Q represent polynomials. In particular, the fundamental principle can be restated as follows:

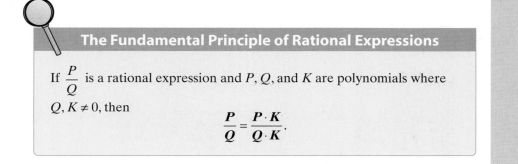

The Fundamental Principle of Rational Expressions

If $\dfrac{P}{Q}$ is a rational expression and $P, Q,$ and K are polynomials where $Q, K \neq 0,$ then

$$\frac{P}{Q} = \frac{P \cdot K}{Q \cdot K}.$$

Objective B **Reducing (or Simplifying) Rational Expressions**

The fundamental principle can be used to **reduce** (or **simplify**) a rational expression to **lower terms,** for multiplication or division and to **build** a rational expression to **higher terms,** for addition or subtraction. Just as with rational numbers, a rational expression is said to be **reduced to lowest terms** if the numerator and denominator have no common factors other than 1 and −1.

Example 2

Reducing Rational Expressions

Use the fundamental principle to reduce each expression to lowest terms. State any restrictions on the variable by using the fact that no denominator can be 0. This restriction applies to denominators **before and after** a rational expression is reduced.

a. $\dfrac{2x - 10}{3x - 15}$

Solution

$$\frac{2x - 10}{3x - 15} = \frac{2(x - 5)}{3(x - 5)} = \frac{2}{3} \qquad (x \neq 5)$$

Note that $x - 5$ is a common **factor.** The key word here is **factor.** We reduce using **factors** only.

b. $\dfrac{x^3 - 64}{x^2 - 16}$

Solution

$$\frac{x^3 - 64}{x^2 - 16} = \frac{(x - 4)(x^2 + 4x + 16)}{(x + 4)(x - 4)}$$

Reduce. The common **factor** is $x - 4$. Note that $x^3 - 64$ is the difference of two cubes. Also, note that $x^2 + 4x + 16$ is not factorable.

$$= \frac{x^2 + 4x + 16}{x + 4} \qquad (x \neq -4, 4)$$

2. Reduce the expression to lowest terms. State any restrictions on the variable.

a. $\dfrac{2x-6}{5x-15}$

b. $\dfrac{x^3-125}{x^2-25}$

c. $\dfrac{x-5}{5-x}$

c. $\dfrac{y-10}{10-y}$

Solution

$$\dfrac{y-10}{10-y} = \dfrac{y-10}{-y+10}$$

Note that the expression $10-y$ is the opposite of $y-10$. When **nonzero opposites** are divided, the quotient is always -1.

$$= \dfrac{1\cancel{(y-10)}}{-1\cancel{(y-10)}}$$

$$= \dfrac{1}{-1} = -1 \qquad (y \neq 10)$$

Now work margin exercise **2.**

In Example 2c, the result was -1. The expression $10-y$ is the opposite of $y-10$ for any value of y. That is,

$$10-y = -y+10 = -1(y-10) = -(y-10).$$

When nonzero opposites are divided, the quotient is always -1. For example,

$$\dfrac{-9}{9} = -1, \qquad \dfrac{23}{-23} = -1, \qquad \dfrac{x-5}{5-x} = \dfrac{\cancel{(x-5)}}{-1\cancel{(x-5)}} = \dfrac{1}{-1} = -1 \ (x \neq 5).$$

Opposites in Rational Expressions

For a polynomial P, $\dfrac{-P}{P} = -1$ where $P \neq 0$.

In particular, $\dfrac{a-x}{x-a} = \dfrac{-(x-a)}{x-a} = -1$ where $x \neq a$.

Remember that the key word when reducing is **factor.** Many students make mistakes similar to the following when working with rational expressions.

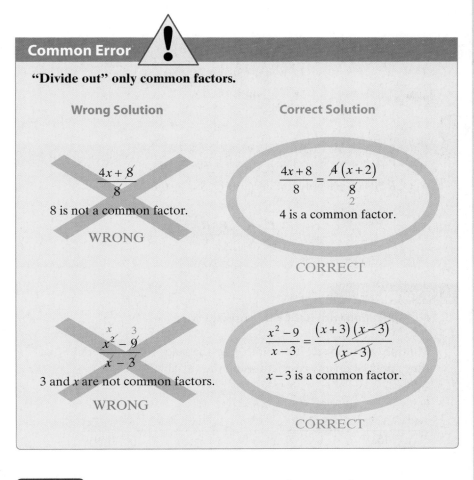

Common Error

"Divide out" only common factors.

Wrong Solution	Correct Solution
$\dfrac{4x + \cancel{8}}{\cancel{8}}$	$\dfrac{4x+8}{8} = \dfrac{\cancel{4}\,(x+2)}{\cancel{8}_{\;2}}$
8 is not a common factor.	4 is a common factor.
WRONG	CORRECT
$\dfrac{\cancel{x^2} - \cancel{9}}{x-3}$	$\dfrac{x^2-9}{x-3} = \dfrac{(x+3)\,\cancel{(x-3)}}{\cancel{(x-3)}}$
3 and x are not common factors.	$x-3$ is a common factor.
WRONG	CORRECT

Objective C **Multiplication with Rational Expressions**

To Multiply Rational Expressions

To multiply any two (or more) rational expressions,

1. Completely factor each numerator and denominator.
2. Multiply the numerators and multiply the denominators, keeping the expressions in factored form.
3. "Divide out" any common factors from the numerators and denominators. Remember that no denominator can have a value of 0.

No factoring is necessary in this example,

$$\frac{2x}{x-6} \cdot \frac{x+5}{x-4} = \frac{2x(x+5)}{(x-6)(x-4)} = \frac{2x^2+10x}{(x-6)(x-4)}. \qquad (x \neq 4, 6)$$

However, in this case the numerators and denominators must be factored.

$$\frac{y^2-4}{y^3} \cdot \frac{y^2-3y}{y^2-y-6} = \frac{\cancel{(y+2)}\,(y-2)\,\cancel{(y)}\,\cancel{(y-3)}}{\underset{y^2}{\cancel{y^3}}\,\cancel{(y-3)}\,\cancel{(y+2)}} = \frac{y-2}{y^2} \quad (y \neq -2, 0, 3)$$

Multiplication with Rational Expressions

If $P, Q, R,$ and S are polynomials and $Q, S \neq 0,$ then

$$\frac{P}{Q} \cdot \frac{R}{S} = \frac{P \cdot R}{Q \cdot S}.$$

Example 3

Multiplication with Rational Expressions

Multiply and reduce, if possible. Use the rules for exponents when they apply. State any restrictions on the variable(s).

a.

$$\frac{5x^2 y}{9xy^3} \cdot \frac{6x^3 y^2}{15xy^4} = \frac{\cancel{5} \cdot 2 \cdot \cancel{3} \cdot x^5 \cdot y^3}{3 \cdot 3 \cdot \cancel{3} \cdot \cancel{5} \cdot x^2 \cdot y^7} = \frac{2x^{5-2} y^{3-7}}{9} = \frac{2x^3 y^{-4}}{9} = \frac{2x^3}{9y^4}$$

$$(x \neq 0, y \neq 0)$$

b. $\quad \dfrac{x}{x-2} \cdot \dfrac{x^2-4}{x^2} = \dfrac{\cancel{x}\,(x+2)\,\cancel{(x-2)}}{\cancel{(x-2)}\,\underset{x}{\cancel{x^2}}} = \dfrac{x+2}{x}$

$$(x \neq 0, 2)$$

c.

$$\frac{3x-3}{x^2+x} \cdot \frac{x^2+2x+1}{3x^2-6x+3} = \frac{3\,\cancel{(x-1)}\,\overset{(x+1)}{\cancel{(x+1)^2}}}{x\,\cancel{(x+1)} \cdot 3\,\underset{(x-1)}{\cancel{(x-1)^2}}} = \frac{x+1}{x(x-1)} \text{ or } \frac{x+1}{x^2-x}$$

$$(x \neq -1, 0, 1)$$

d.

$$\frac{x^2-7x+12}{2x+6} \cdot \frac{x^2-4}{x^2-2x-8} = \frac{\cancel{(x-4)}\,(x-3)\,(x+2)\,(x-2)}{2(x+3)\,\cancel{(x-4)}\,\cancel{(x+2)}}$$

$$= \frac{(x-3)(x-2)}{2(x+3)} \text{ or } \frac{x^2-5x+6}{2(x+3)} \text{ or } \frac{x^2-5x+6}{2x+6}$$

$$(x \neq -3, -2, 4)$$

notes

As shown in Examples 3c and 3d there may be more than one correct form for an answer. After a rational expression has been reduced, the numerator and denominator may be multiplied out or left in factored form. **Generally, the denominator will be left in factored form and the numerator multiplied out.** As we will see in the next section, this form makes the results easier to add and subtract. However, be aware that this form is just an option, and multiplying out the denominator is not an error.

Completion Example 4

Multiplication with Rational Expressions

Multiply and reduce, if possible. State any restrictions on the variable.

$$\frac{x^2 - 36}{x^4} \cdot \frac{x}{x-6}$$

Solution

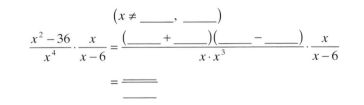

$$\left(x \neq \underline{\quad}, \underline{\quad}\right)$$

$$\frac{x^2 - 36}{x^4} \cdot \frac{x}{x-6} = \frac{\left(\underline{\quad} + \underline{\quad}\right)\left(\underline{\quad} - \underline{\quad}\right)}{x \cdot x^3} \cdot \frac{x}{x-6}$$

$$= \underline{\overline{\quad}}$$

Now work margin exercises 3 and 4.

Objective D ### Division with Rational Expressions

To divide any two rational expressions, multiply the first fraction by the **reciprocal** of the second fraction.

Division with Rational Expressions

If P, Q, R, and S are polynomials with $Q, R, S \neq 0$, then

$$\frac{P}{Q} \div \frac{R}{S} = \frac{P}{Q} \cdot \frac{S}{R}.$$

Note: that $\dfrac{S}{R}$ is the reciprocal of $\dfrac{R}{S}$.

Completion Example Answers

4. $(x \neq 0, 6)$; $\dfrac{(x+6)(x-6)}{x \cdot x^3} \cdot \dfrac{x}{x-6} = \dfrac{x+6}{x^3}$

Multiply and reduce, if possible. State any restrictions on the variable(s).

3. a. $\dfrac{7x^4 y}{9x^2 y^6} \cdot \dfrac{3x^3 y^3}{14xy^4}$

b. $\dfrac{x}{x-3} \cdot \dfrac{x^2 - 9}{x^3}$

c. $\dfrac{2x+2}{x^2 - x} \cdot \dfrac{x^2 - 2x + 1}{2x^2 + 4x + 2}$

d. $\dfrac{x^2 - 8x + 12}{3x + 9} \cdot \dfrac{x^2 - 25}{x^2 - 7x + 10}$

4. $\dfrac{x^2 + 2x - 3}{x^2} \cdot \dfrac{x}{x+3}$

Example 5

Division with Rational Expressions

Divide and reduce, if possible. Assume that no denominator has a value of 0.

a. $\dfrac{12x^2 y}{10xy^2} \div \dfrac{3x^4 y}{xy^3}$

Solution

$$\dfrac{12x^2 y}{10xy^2} \div \dfrac{3x^4 y}{xy^3} = \dfrac{12x^2 y}{10xy^2} \cdot \dfrac{xy^3}{3x^4 y}$$

Note that in this example we have used the quotient rule for exponents.

$$= \dfrac{\cancel{2} \cdot 2 \cdot \cancel{3} \cdot x^3 \cdot y^4}{\cancel{2} \cdot 5 \cdot \cancel{3} \cdot x^5 \cdot y^3}$$

$$= \dfrac{2x^{3-5} y^{4-3}}{5}$$

$$= \dfrac{2x^{-2} y}{5} = \dfrac{2y}{5x^2}$$

b. $\dfrac{x^3 - y^3}{x^3} \div \dfrac{y-x}{xy}$

Solution

$$\dfrac{x^3 - y^3}{x^3} \div \dfrac{y-x}{xy} = \dfrac{x^3 - y^3}{x^3} \cdot \dfrac{xy}{y-x}$$

Note that $\dfrac{x-y}{y-x} = -1$.

$$= \dfrac{\overset{-1}{\cancel{(x-y)}}\left(x^2 + xy + y^2\right)\cancel{x}y}{\underset{x^2}{\cancel{x^3}}\,\cancel{(y-x)}}$$

$$= \dfrac{-y\left(x^2 + xy + y^2\right)}{x^2} = \dfrac{-x^2 y - xy^2 - y^3}{x^2}$$

c. $\dfrac{x^2 - 8x + 15}{2x^2 + 11x + 5} \div \dfrac{2x^2 - 5x - 3}{4x^2 - 1}$

Solution

$$\dfrac{x^2 - 8x + 15}{2x^2 + 11x + 5} \div \dfrac{2x^2 - 5x - 3}{4x^2 - 1} = \dfrac{x^2 - 8x + 15}{2x^2 + 11x + 5} \cdot \dfrac{4x^2 - 1}{2x^2 - 5x - 3}$$

$$= \dfrac{\cancel{(x-3)}(x-5)(2x-1)\cancel{(2x+1)}}{(2x+1)(x+5)\cancel{(x-3)}\cancel{(2x+1)}}$$

$$= \dfrac{(x-5)(2x-1)}{(2x+1)(x+5)} = \dfrac{2x^2 - 11x + 5}{(2x+1)(x+5)}$$

Remember that you have the option of leaving the numerator and/or denominator in factored form.

Completion Example 6

Division with Rational Expressions

Divide and reduce, if possible. Assume that no denominator has a value of 0.

$$\frac{x^2-4}{2x-1} \div \frac{x^2-2x}{2x^2+x-1}$$

Solution

$$\frac{x^2-4}{2x-1} \div \frac{x^2-2x}{2x^2+x-1} = \frac{x^2-4}{2x-1} \cdot \frac{2x^2+x-1}{x^2-2x}$$

$$= \frac{(\underline{\quad})(\underline{\quad})}{2x-1} \cdot \frac{(\underline{\quad})(\underline{\quad})}{x(\underline{\quad})}$$

$$= \frac{(\underline{\quad})(\underline{\quad})}{\underline{\quad}}$$

$$= \frac{\overline{\quad\quad}}{\underline{\quad}}$$

Now work margin exercises 5 and 6.

Divide and reduce, if possible. Assume that no denominator has a value of 0.

5. a. $\dfrac{15x^3y}{12xy^3} \div \dfrac{5x^5y}{xy^2}$

b. $\dfrac{x^3+y^3}{y^3} \div \dfrac{-x-y}{xy}$

c. $\dfrac{x^2-9x+18}{3x^2+19x+6} \div \dfrac{3x^2-17x-6}{x^2+x-30}$

6. $\dfrac{3x+1}{x-4} \div \dfrac{3x^2-11x-4}{x^2-16}$

Practice Problems

Reduce to lowest terms. State any restrictions on the variables.

1. $\dfrac{5x+20}{7x+28}$ **2.** $\dfrac{6-3x}{3x-6}$ **3.** $\dfrac{x^2+x-2}{x^2+3x+2}$

Perform the following operations and simplify the results. Assume that no denominator has a value of 0.

4. $\dfrac{x-7}{x^3} \cdot \dfrac{x^2}{49-x^2}$ **5.** $\dfrac{y^2-y-6}{y^2-5y+6} \cdot \dfrac{y^2-4}{y^2+4y+4}$

6. $\dfrac{x^3+3x}{2x+1} \div \dfrac{x^2+3}{x+1}$ **7.** $\dfrac{x^2+2x-3}{x^2-3x-10} \cdot \dfrac{2x^2-9x-5}{x^2-2x+1} \div \dfrac{4x+2}{x^2-x}$

Completion Example Answers

6. $\dfrac{(x+2)(x-2)}{2x-1} \cdot \dfrac{(2x-1)(x+1)}{x(x-2)} = \dfrac{(x+2)(x+1)}{x}$

$$= \dfrac{x^2+3x+2}{x}$$

Practice Problem Answers

1. $\dfrac{5}{7}$, $x \neq -4$ **2.** -1, $x \neq 2$ **3.** $\dfrac{x-1}{x+1}$, $x \neq -2, -1$ **4.** $\dfrac{-1}{x(x+7)}$

5. 1 **6.** $\dfrac{x^2+x}{2x+1}$ **7.** $\dfrac{x^2+3x}{2(x+2)}$

Exercises 13.1

Reduce to lowest terms. State any restrictions on the variable(s). See examples 1 and 2.

1. $\dfrac{9x^2y^3}{12xy^4}$

2. $\dfrac{18xy^4}{27x^2y}$

3. $\dfrac{20x^5}{30x^2y^3}$

4. $\dfrac{15y^4}{20x^3y^2}$

5. $\dfrac{x}{x^2-3x}$

6. $\dfrac{3x}{x^2+5x}$

7. $\dfrac{7x-14}{x-2}$

8. $\dfrac{4-2x}{2x-4}$

9. $\dfrac{9-3x}{4x-12}$

10. $\dfrac{2x-8}{16-4x}$

11. $\dfrac{6x^2+4x}{3xy+2y}$

12. $\dfrac{1+3y}{4x+12xy}$

13. $\dfrac{x^2+6x}{x^2+5x-6}$

14. $\dfrac{x^2-y^2}{3x^2+3xy}$

15. $\dfrac{x^3+27}{x^2-9}$

16. $\dfrac{x^3-8}{x^2-4}$

17. $\dfrac{xy-3y+2x-6}{y^2-4}$

18. $\dfrac{3x^2+14x-24}{18-9x-2x^2}$

19. $\dfrac{x^3-8}{5x-2y+xy-10}$

20. $\dfrac{x^3+64}{2x^2+x-28}$

Perform the indicated operations and reduce to lowest terms. Assume that no denominator has a value of 0. See examples 3 through 6.

21. $\dfrac{3ax^2}{4b}\cdot\dfrac{6b^2}{27x^2y}$

22. $\dfrac{18x^3}{5y^2}\cdot\dfrac{30y^3}{9x^4}$

23. $\dfrac{24x^3}{25y^2}\cdot\dfrac{10y^5}{18x}$

24. $\dfrac{16x^8}{3y^{11}}\cdot\dfrac{-21y^9}{10x^7}$

25. $\dfrac{x^2-9}{x^2+2x}\cdot\dfrac{x+2}{x-3}$

26. $\dfrac{16x^2-9}{3x^2-15x}\cdot\dfrac{6}{4x+3}$

27. $\dfrac{x^2+2x-3}{x^2+3x} \cdot \dfrac{x}{x+1}$

28. $\dfrac{4x+16}{x^2-16} \cdot \dfrac{x-4}{x}$

29. $\dfrac{x^2+6x-16}{x^2-64} \cdot \dfrac{1}{2-x}$

30. $\dfrac{4-x^2}{x^2-4x+4} \cdot \dfrac{3}{x+2}$

31. $\dfrac{x^2-5x+6}{x^2-4x} \cdot \dfrac{x-4}{x-3}$

32. $\dfrac{2x^2+x-3}{x^2+4x} \cdot \dfrac{2x+8}{x-1}$

33. $\dfrac{2x^2+10x}{3x^2+5x+2} \cdot \dfrac{6x+4}{x^2}$

34. $\dfrac{x+3}{x^2-16} \cdot \dfrac{x^2-3x-4}{x^2-1}$

35. $\dfrac{x}{x^2+7x+12} \cdot \dfrac{x^2-2x-24}{x^2-7x+6}$

36. $\dfrac{x^2-2x-3}{x+5} \cdot \dfrac{x^2-5x-14}{x^2-x-6}$

37. $\dfrac{8-2x-x^2}{x^2-2x} \cdot \dfrac{x-4}{x^2-3x-4}$

38. $\dfrac{3x^2+21x}{x^2-49} \cdot \dfrac{x^2-5x+4}{x^2+3x-4}$

39. $\dfrac{(x-2y)^2}{x^2-5xy+6y^2} \cdot \dfrac{x+2y}{x^2-4xy+4y^2}$

40. $\dfrac{4x^2+6x}{x^2+3x-10} \cdot \dfrac{x^2+4x-12}{x^2+5x-6}$

41. $\dfrac{2x^2+5x+2}{3x^2+8x+4} \cdot \dfrac{3x^2-x-2}{4x^3-x}$

42. $\dfrac{x^2+5x}{4x^2+12x+9} \cdot \dfrac{6x^2+7x-3}{x^2+10x+25}$

43. $\dfrac{x^2+x+1}{x^2-1} \cdot \dfrac{x^2-2x+1}{x^3-1}$

44. $\dfrac{x^2-9}{2x^2+4x+8} \cdot \dfrac{x^3-8}{x^2-5x+6}$

45. $\dfrac{x-2}{x^2-2x+4} \cdot \dfrac{x^3+8}{x^2-4x+4}$

46. $\dfrac{2x^2-7x+3}{x^2-9} \cdot \dfrac{3x^2+8x-3}{6x^2+x-1}$

47. $\dfrac{12x^2y}{9xy^9} \div \dfrac{4x^4y}{x^2y^3}$

48. $\dfrac{35xy^3}{24x^3y} \div \dfrac{15x^4y^3}{84xy^4}$

49. $\dfrac{45xy^4}{21x^2y^2} \div \dfrac{40x^4}{112xy^5}$

50. $\dfrac{x-3}{15x} \div \dfrac{4x-12}{5}$

51. $\dfrac{x-1}{6x+6} \div \dfrac{2x-2}{x^2+x}$

52. $\dfrac{7x-14}{x^2} \div \dfrac{x^2-4}{x^3}$

53. $\dfrac{6x^2-54}{x^4} \div \dfrac{x-3}{x^2}$

54. $\dfrac{x^2-25}{6x+30} \div \dfrac{x-5}{x}$

55. $\dfrac{2x-1}{x^2+2x} \div \dfrac{10x^2-5x}{6x^2+12x}$

56. $\dfrac{x+3}{x^2+3x-4} \div \dfrac{x+2}{x^2+x-2}$

57. $\dfrac{6x^2-7x-3}{x^2-1} \div \dfrac{2x-3}{x-1}$

58. $\dfrac{x^2-9}{2x^2+7x+3} \div \dfrac{x^2-3x}{2x^2+11x+5}$

59. $\dfrac{x^2-6x+9}{x^2-4x+3} \div \dfrac{2x^2-7x+3}{x^2-3x+2}$

60. $\dfrac{x^3+2x^2}{x^3+64} \div \dfrac{4x^2}{x^2-4x+16}$

61. $\dfrac{2x+1}{4x-x^2} \div \dfrac{4x^2-1}{x^2-16}$

62. $\dfrac{x^2-4x+4}{x^2+5x+6} \div \dfrac{x^2+2x-8}{x^2+7x+12}$

63. $\dfrac{x^2-x-6}{x^2+6x+8} \div \dfrac{x^2-4x+3}{x^2+5x+4}$

64. $\dfrac{x^2-x-12}{6x^2+x-9} \div \dfrac{x^2-6x+8}{3x^2-x-6}$

65. $\dfrac{6x^2+5x+1}{4x^3-3x^2} \div \dfrac{3x^2-2x-1}{3x^2-2x+1}$

66. $\dfrac{8x^2 + 2x - 15}{3x^2 + 13x + 4} \div \dfrac{2x^2 + 5x + 3}{6x^2 - x - 1}$

67. $\dfrac{3x^2 + 13x + 14}{4x^3 - 3x^2} \div \dfrac{6x^2 - x - 35}{4x^2 + 5x - 6}$

68. $\dfrac{3x^2 + 2x}{9x^2 - 4} \div \dfrac{27x^3 - 8}{9x^2 - 6x + 4}$

69. $\dfrac{x^2 - 8x + 15}{x^2 - 9x + 14} \div \dfrac{x^2 + 4x - 21}{x - 1}$

70. $\dfrac{6 - 11x - 10x^2}{2x^2 + x - 3} \div \dfrac{5x^3 - 2x^2}{3x^2 - 5x + 2}$

71. $\dfrac{x - 6}{x^2 - 7x + 6} \cdot \dfrac{x^2 - 3x}{x + 3} \cdot \dfrac{x^2 - 9}{x^2 - 4x + 3}$

72. $\dfrac{3x^2 + 11x + 10}{2x^2 + x - 6} \cdot \dfrac{x^2 + 2x - 3}{2x - 1} \cdot \dfrac{2x - 3}{3x^2 + 2x - 5}$

73. $\dfrac{x^3 + 3x^2}{x^2 + 7x + 12} \cdot \dfrac{2x^2 + 7x - 4}{2x^2 - x} \cdot \dfrac{x^2 + 4x - 5}{2x^2 - x - 1}$

74. $\dfrac{x^2 + 2x - 3}{x^2 + 10x + 21} \cdot \dfrac{x^2 + 6x + 5}{x^2 - 7x - 8} \cdot \dfrac{x^2 - x - 56}{x^2 - 3x - 40}$

75. $\dfrac{2x^2 - 5x + 2}{4xy - 2y + 6x - 3} \div \dfrac{xy - 2y + 3x - 6}{2y^2 + 9y + 9}$

76. $\dfrac{2xy - 12x + y - 6}{y^2 - 2y - 24} \div \dfrac{2x^2 + 11x + 5}{xy + 5y + 4x + 20}$

Answer the following word problems.

77. Rectangles: The area of a rectangle (in square feet) is represented by the polynomial function $A(x) = 4x^2 - 4x - 15$. If the length of the rectangle is $(2x + 3)$ feet, find a representation for the width.

$$A(x) = 4x^2 - 4x - 15.$$

$$(2x + 3)$$

78. Rectangles: The area of a rectangle (in square feet) is represented by the polynomial function $A(x) = 3x^2 - x - 10$. If the length of the rectangle is $(3x + 5)$ feet, find a representation for the width.

$$A(x) = 3x^2 - x - 10.$$

$$(3x + 5)$$

79. **a.** Define rational expression.

b. Give an example of a rational expression that is undefined for $x = -2$ and $x = 3$ and has a value of 0 for $x = 1$. Explain how you determined this expression.

c. Give an example of a rational expression that is undefined for $x = -5$ and never has a value of 0. Explain how you determined this expression.

80. Write the opposite of each of the following expressions.

    a. $3 - x$ **b.** $2x - 7$ **c.** $x + 5$ **d.** $-3x - 2$

81. Given the rational function $f(x) = \dfrac{x - 4}{x^2 - 100}$:

    a. For what values, if any, will $f(x) = 0$?

    b. For what values, if any, is $f(x)$ undefined?

13.2 Addition and Subtraction with Rational Expressions

Objectives

A Add rational expressions.

B Subtract rational expressions.

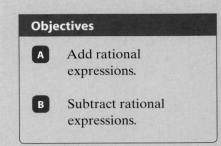

Objective A — Addition with Rational Expressions

To add rational expressions with a common denominator, proceed just as with fractions: add the numerators and keep the common denominator. For example,

$$\frac{5}{x+2}+\frac{6}{x+2}=\frac{5+6}{x+2}=\frac{11}{x+2}. \qquad (x \neq -2)$$

In some cases the sum can be reduced:

$$\frac{x^2+6}{x+2}+\frac{5x}{x+2}=\frac{x^2+5x+6}{x+2}=\frac{\cancel{(x+2)}(x+3)}{\cancel{x+2}}=x+3. \qquad (x \neq -2)$$

Addition with Rational Expressions

For polynomials P, Q, and R, with $Q \neq 0$,

$$\frac{P}{Q}+\frac{R}{Q}=\frac{P+R}{Q}.$$

Example 1

Adding Rational Expressions with a Common Denominator

Find each sum and reduce if possible. (Note the importance of the factoring techniques we studied in Chapter 12.)

a. $\dfrac{x}{x^2-1}+\dfrac{1}{x^2-1}$

Solution

$$\frac{x}{x^2-1}+\frac{1}{x^2-1}=\frac{x+1}{x^2-1}$$

$$=\frac{\overset{1}{\cancel{x+1}}}{\cancel{(x+1)}(x-1)}$$

$$=\frac{1}{x-1} \qquad (x \neq -1, 1)$$

Remember, if we use the entire expression in the numerator (or denominator) to reduce, we are left with a factor of 1.

1. Find each sum and reduce if possible. State any restrictions on the variable.

a. $\dfrac{x}{x^2-25}+\dfrac{5}{x^2-25}$

b. $\dfrac{2}{x^2+8x+15}+\dfrac{3x+7}{x^2+8x+15}$

---

b. $\dfrac{1}{x^2+7x+10}+\dfrac{2x+3}{x^2+7x+10}$

Solution

$$\dfrac{1}{x^2+7x+10}+\dfrac{2x+3}{x^2+7x+10}=\dfrac{2x+4}{x^2+7x+10}$$

$$=\dfrac{2\,\cancel{(x+2)}}{(x+5)\,\cancel{(x+2)}}$$

$$=\dfrac{2}{x+5}\qquad (x\neq-5,-2)$$

Now work margin exercise 1.

The rational expressions in Example 1 had common denominators. To add expressions with different denominators, we need to find the least common multiple (LCM) of the denominators. The LCM was discussed in Section 2.3. The procedure is stated here for polynomials.

To Find the LCM for a Set of Polynomials

1. Completely factor each polynomial (including prime factors for numerical factors).
2. Form the product of all factors that appear, using each factor the most number of times it appears in any one polynomial.

The LCM of a set of denominators is called the **least common denominator (LCD)**. To add fractions with different denominators, begin by changing each fraction to an equivalent fraction with the LCD as the denominator. This is called **building the fraction to higher terms**.

Use the following procedure when adding rational expressions with different denominators.

Procedure for Adding Rational Expressions with Different Denominators

1. Find the LCD (the LCM of the denominators).
2. Rewrite each fraction in an equivalent form with the LCD as the denominator.
3. Add the numerators and keep the common denominator.
4. Reduce if possible.

Example 2

Adding Rational Expressions with Different Denominators

Find each sum and reduce if possible. Assume that no denominator has a value of 0.

a. $\dfrac{y}{y-3} + \dfrac{6}{y+4}$

Solution

In this case, neither denominator can be factored so the LCD is the product of these factors. That is, $\text{LCD} = (y-3)(y+4)$.
Now, using the fundamental principle, we have

$$\dfrac{y}{y-3} + \dfrac{6}{y+4} = \dfrac{y(y+4)}{(y-3)(y+4)} + \dfrac{6(y-3)}{(y+4)(y-3)}$$

$$= \dfrac{(y^2+4y)+(6y-18)}{(y-3)(y+4)}$$

$$= \dfrac{y^2+10y-18}{(y-3)(y+4)}. \qquad \text{The numerator is not factorable and the expression is reduced.}$$

b. $\dfrac{1}{x^2+6x+9} + \dfrac{1}{x^2-9} + \dfrac{1}{2x+6}$

Solution

First, find the LCD.

Step 1: Factor each expression completely.

$$x^2+6x+9 = (x+3)^2$$
$$x^2-9 = (x+3)(x-3)$$
$$2x+6 = 2(x+3)$$

Step 2: Form the product of 2, $(x+3)^2$, and $(x-3)$. That is, use each factor the most number of times it appears in any one factorization.

$$\text{LCD} = 2(x+3)^2(x-3)$$

Now use the LCD and add as follows.

$$\dfrac{1}{x^2+6x+9} + \dfrac{1}{x^2-9} + \dfrac{1}{2x+6}$$

$$= \dfrac{1}{(x+3)^2} + \dfrac{1}{(x+3)(x-3)} + \dfrac{1}{2(x+3)}$$

$$= \dfrac{1\cdot 2(x-3)}{(x+3)^2 \cdot 2(x-3)} + \dfrac{1\cdot 2(x+3)}{(x+3)(x-3)\cdot 2(x+3)} + \dfrac{1\cdot(x+3)(x-3)}{2(x+3)\cdot(x+3)(x-3)}$$

$$= \dfrac{(2x-6)+(2x+6)+(x^2-9)}{2(x+3)^2(x-3)} = \dfrac{x^2+4x-9}{2(x+3)^2(x-3)}$$

Find each sum and reduce if possible. Assume that no denominator has a value of 0.

2. a. $\dfrac{x}{x+2}+\dfrac{3}{x+3}$

 b. $\dfrac{1}{x^2+10x+25}+\dfrac{1}{x^2-25}+\dfrac{1}{2x+10}$

3. $\dfrac{s}{s+3}+\dfrac{4}{s+1}$

Completion Example 3

Addition with Rational Expressions

Perform the indicated addition and reduce if possible. Assume that no denominator is 0.

$$\frac{y}{y-5}+\frac{3}{y+6}$$

Solution

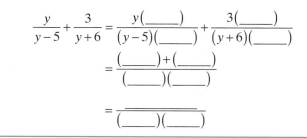

Now work margin exercises 2 and 3.

notes

Important Note about the Form of Answers

In Examples 2a, 2b, and 3, each denominator is left in factored form as a convenience for possibly reducing or adding to some other expression later. You may choose to multiply out these factors. Either form is correct. For consistency, denominators are left in factored form in the answers in the back of the text.

Objective B ## Subtraction with Rational Expressions

When subtracting fractions, the placement of negative signs can be critical. For example, note how –2 can be indicated in three different forms:

$$-\frac{6}{3}=-2,\quad \frac{-6}{3}=-2,\ \text{ and }\ \frac{6}{-3}=-2.$$

Thus we have $-\dfrac{6}{3}=\dfrac{-6}{3}=\dfrac{6}{-3}.$

With polynomials we have the following statement about the placement of negative signs which can be **very useful in subtraction**. We seldom leave the negative sign in the denominator.

Completion Example Answers

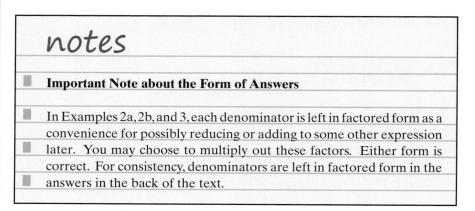

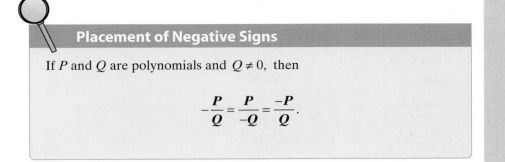

Placement of Negative Signs

If P and Q are polynomials and $Q \neq 0,$ then

$$-\frac{P}{Q} = \frac{P}{-Q} = \frac{-P}{Q}.$$

To subtract rational expressions with a common denominator, proceed just as with fractions: subtract the numerators and keep the common denominator. For example,

$$\frac{17}{x+7} - \frac{23}{x+7} = \frac{17-23}{x+7} = \frac{-6}{x+7} \quad \text{or} \quad -\frac{6}{x+7}.$$

Subtraction with Rational Expressions

For polynomials $P, Q,$ and $R,$ with $Q \neq 0,$

$$\frac{P}{Q} - \frac{R}{Q} = \frac{P-R}{Q}.$$

Example 4

Subtracting Rational Expressions with a Common Denominator

Find each difference and reduce if possible. Assume that no denominator has a value of 0.

a. $\dfrac{2x-5y}{x+y} - \dfrac{3x-7y}{x+y}$

Solution

$$\frac{2x-5y}{x+y} - \frac{3x-7y}{x+y} = \frac{2x-5y-(3x-7y)}{x+y} \qquad \text{Subtract the entire numerator.}$$

$$= \frac{2x-5y-3x+7y}{x+y}$$

$$= \frac{-x+2y}{x+y}$$

4. Find the difference and reduce if possible. Assume that no denominator has a value of 0.

a. $\dfrac{4x-7y}{3x-y}-\dfrac{3x-9y}{3x-y}$

b. $\dfrac{x^2}{x^2+6x+9}-\dfrac{2x+15}{x^2+6x+9}$

c. $\dfrac{x}{x-4}-\dfrac{5}{4-x}$

b. $\dfrac{x^2}{x^2+4x+4}-\dfrac{2x+8}{x^2+4x+4}$

Solution

$$\dfrac{x^2}{x^2+4x+4}-\dfrac{2x+8}{x^2+4x+4}=\dfrac{x^2-(2x+8)}{x^2+4x+4} \quad \text{Subtract the entire numerator.}$$

$$=\dfrac{x^2-2x-8}{x^2+4x+4}$$

$$=\dfrac{(x-4)\,\cancel{(x+2)}}{(x+2)\,\cancel{(x+2)}} \quad \text{Factor and reduce.}$$

$$=\dfrac{x-4}{x+2}$$

c. $\dfrac{x}{x-5}-\dfrac{3}{5-x}$

Solution

Each denominator is the **opposite** of the other. Multiply both the numerator and denominator of the second fraction by -1 so that both denominators will be the same, in this case $x-5$.

$$\dfrac{x}{x-5}-\dfrac{3}{5-x}=\dfrac{x}{x-5}-\dfrac{3}{(5-x)}\cdot\dfrac{(-1)}{(-1)}$$

$$=\dfrac{x}{x-5}-\dfrac{-3}{x-5}$$

$$=\dfrac{x-(-3)}{x-5}$$

$$=\dfrac{x+3}{x-5}$$

Now work margin exercise 4.

Many beginning students make a mistake when subtracting rational expressions by not subtracting the entire numerator. They make a mistake similar to the following.

Wrong Solution

$$\frac{10}{x+5} - \frac{3-x}{x+5} = \frac{10-3-x}{x+5} = \frac{7-x}{x+5}$$

WRONG

By using parentheses, you can avoid such mistakes.

Correct Solution

$$\frac{10}{x+5} - \frac{3-x}{x+5} = \frac{10-(3-x)}{x+5} = \frac{10-3+x}{x+5} = \frac{7+x}{x+5} = \frac{x+7}{x+5}$$

CORRECT

As with addition, if the rational expressions do not have the same denominator, find the LCM of the denominators (the LCD) and use the fundamental principle to **build each fraction to higher terms**, if necessary, so that each has the LCD as the denominator.

Example 5

Subtracting Rational Expressions with Different Denominators

Find each difference and reduce if possible. Assume that no denominator has a value of 0.

a. $\dfrac{x+5}{x-5} - \dfrac{100}{x^2-25}$

Solution

$$\left.\begin{array}{l} x - 5 = x - 5 \\ x^2 - 25 = (x+5)(x-5) \end{array}\right\} \quad \text{LCD} = (x+5)(x-5)$$

$$\frac{x+5}{x-5} - \frac{100}{x^2-25} = \frac{(x+5)(x+5)}{(x-5)(x+5)} - \frac{100}{(x+5)(x-5)}$$

$$= \frac{(x^2+10x+25)-100}{(x+5)(x-5)}$$

$$= \frac{x^2+10x+25-100}{(x+5)(x-5)}$$

$$= \frac{x^2+10x-75}{(x+5)(x-5)}$$

$$= \frac{(x+15)\,(\cancel{x-5})}{(x+5)\,(\cancel{x-5})}$$

$$= \frac{x+15}{x+5}$$

b. $\dfrac{x+y}{(x-y)^2} - \dfrac{x}{2x^2-2y^2}$

Solution

$$\left.\begin{array}{l} (x-y)^2 = (x-y)^2 \\ 2x^2-2y^2 = 2(x-y)(x+y) \end{array}\right\} \quad \text{LCD} = 2(x-y)^2(x+y)$$

$$\frac{x+y}{(x-y)^2} - \frac{x}{2x^2-2y^2}$$

$$= \frac{(x+y)\cdot 2(x+y)}{(x-y)^2 \cdot 2(x+y)} - \frac{x(x-y)}{2(x-y)(x+y)(x-y)}$$

$$= \frac{2(x^2+2xy+y^2)-(x^2-xy)}{2(x-y)^2(x+y)}$$

$$= \frac{2x^2+4xy+2y^2-x^2+xy}{2(x-y)^2(x+y)}$$

$$= \frac{x^2+5xy+2y^2}{2(x-y)^2(x+y)}$$

c. $\dfrac{3x-12}{x^2+x-20} - \dfrac{x^2+5x}{x^2+9x+20}$

Hint: In this problem, both expressions can be reduced before looking for the LCD.

Solution

$$\frac{3x-12}{x^2+x-20} - \frac{x^2+5x}{x^2+9x+20}$$

$$= \frac{3(\cancel{x-4})}{(x+5)(\cancel{x-4})} - \frac{x(\cancel{x+5})}{(\cancel{x+5})(x+4)}$$

$$= \frac{3}{x+5} - \frac{x}{x+4}$$

Now subtract these two expressions with $\text{LCD} = (x+5)(x+4)$.

$$\frac{3}{x+5} - \frac{x}{x+4} = \frac{3(x+4)}{(x+5)(x+4)} - \frac{x(x+5)}{(x+4)(x+5)}$$

$$= \frac{(3x+12) - (x^2+5x)}{(x+5)(x+4)}$$

$$= \frac{3x+12-x^2-5x}{(x+5)(x+4)}$$

$$= \frac{-x^2-2x+12}{(x+5)(x+4)}$$

d. $\dfrac{x+1}{xy-3y+4x-12} - \dfrac{x-3}{xy+6y+4x+24}$

Solution

$$\begin{aligned} xy-3y+4x-12 &= y(x-3)+4(x-3) \\ &= (x-3)(y+4) \\ xy+6y+4x+24 &= y(x+6)+4(x+6) \\ &= (x+6)(y+4) \end{aligned} \left. \begin{aligned} & \\ & \\ & \end{aligned} \right\} \ \text{LCD} = (x-3)(y+4)(x+6)$$

$$\frac{x+1}{xy-3y+4x-12} - \frac{x-3}{xy+6y+4x+24}$$

$$= \frac{(x+1)(x+6)}{(y+4)(x-3)(x+6)} - \frac{(x-3)(x-3)}{(x+6)(y+4)(x-3)}$$

$$= \frac{x^2+7x+6 - (x^2-6x+9)}{(y+4)(x-3)(x+6)}$$

$$= \frac{x^2+7x+6-x^2+6x-9}{(y+4)(x-3)(x+6)}$$

$$= \frac{13x-3}{(y+4)(x-3)(x+6)}$$

5. Find the difference and reduce if possible. Assume that no denominator has a value of 0.

a. $\dfrac{x+6}{x-6} - \dfrac{144}{x^2-36}$

b. $\dfrac{5x+y}{(x+y)^2} - \dfrac{y}{3x^2-3y^2}$

c. $\dfrac{4x-8}{x^2+4x-12} - \dfrac{x^2+6x}{x^2+9x+18}$

d.

$\dfrac{y+4}{xy-2y+x-2} - \dfrac{y-5}{xy-2y+3x-6}$

6. Perform the indicated subtraction and reduce if possible. Assume that no denominator is 0.

$\dfrac{x-4}{x^2-2x+1} - \dfrac{5}{x-1}$

Completion Example 6

Additional Practice with Rational Expressions

Perform the indicated subtraction and reduce if possible. Assume that no denominator is 0.

$$\frac{x}{x^2-x-2} - \frac{1}{x-2}$$

Solution

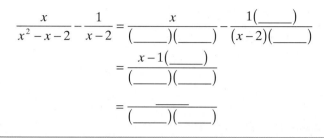

$$\frac{x}{x^2-x-2} - \frac{1}{x-2} = \frac{x}{(\underline{\quad})(\underline{\quad})} - \frac{1(\underline{\quad})}{(x-2)(\underline{\quad})}$$

$$= \frac{x-1(\underline{\quad})}{(\underline{\quad})(\underline{\quad})}$$

$$= \frac{\overline{\underline{\quad}}}{(\underline{\quad})(\underline{\quad})}$$

Now work margin exercises 5 and 6.

Practice Problems

Perform the indicated operations and reduce if possible. Assume that no denominator is 0.

1. $\dfrac{1}{1-y} + \dfrac{2}{y^2-1}$

2. $\dfrac{x+3}{x^2+x-6} + \dfrac{x-2}{x^2+4x-12}$

3. $\dfrac{1}{y+2} - \dfrac{1}{y^3+8}$

4. $\dfrac{x}{x^2-1} - \dfrac{1}{x-1}$

Completion Example Answers

6. $\dfrac{x}{(x-2)(x+1)} - \dfrac{1(x+1)}{(x-2)(x+1)} = \dfrac{x-1(x+1)}{(x-2)(x+1)} = \dfrac{-1}{(x-2)(x+1)}$

Practice Problem Answers

1. $\dfrac{-1}{y+1}$ **2.** $\dfrac{2x+4}{(x-2)(x+6)}$ **3.** $\dfrac{y^2-2y+3}{(y+2)(y^2-2y+4)}$ **4.** $\dfrac{-1}{(x+1)(x-1)}$

Exercises 13.2

Perform the indicated operations and reduce if possible. Assume that no denominator has a value of 0. See examples 1 through 6.

1. $\dfrac{3x}{x+4} + \dfrac{12}{x+4}$

2. $\dfrac{7x}{x+5} + \dfrac{35}{x+5}$

3. $\dfrac{x-1}{x+6} + \dfrac{x+13}{x+6}$

4. $\dfrac{3x-1}{2x-6} + \dfrac{x-11}{2x-6}$

5. $\dfrac{3x+1}{5x+2} + \dfrac{2x+1}{5x+2}$

6. $\dfrac{x^2+3}{x+1} + \dfrac{4x}{x+1}$

7. $\dfrac{x-5}{x^2-2x+1} + \dfrac{x+3}{x^2-2x+1}$

8. $\dfrac{2x^2+5}{x^2-4} + \dfrac{3x-1}{x^2-4}$

9. $\dfrac{13}{7-x} - \dfrac{1}{x-7}$

10. $\dfrac{6x}{x-6} + \dfrac{36}{6-x}$

11. $\dfrac{3x}{x-4} + \dfrac{16-x}{4-x}$

12. $\dfrac{20}{x-10} - \dfrac{3}{10-x}$

13. $\dfrac{x^2+2}{x^2+x-12} + \dfrac{x+1}{12-x-x^2}$

14. $\dfrac{10}{x^2-x-6} - \dfrac{5x}{6+x-x^2}$

15. $\dfrac{x^2+2}{x^2-4} - \dfrac{4x-2}{x^2-4}$

16. $\dfrac{2x+5}{2x^2-x-1} - \dfrac{4x+2}{2x^2-x-1}$

17. $\dfrac{x+3}{7x-2} + \dfrac{2x-1}{14x-4}$

18. $\dfrac{3x+1}{4x+10} + \dfrac{4-x}{2x+5}$

19. $\dfrac{5}{x-3} + \dfrac{x}{x^2-9}$

20. $\dfrac{x+1}{x^2-3x-10} + \dfrac{x}{x-5}$

21. $\dfrac{x}{x-1} - \dfrac{4}{x+2}$

22. $\dfrac{x-1}{3x-1} - \dfrac{8+4x}{x+2}$

23. $\dfrac{x+2}{x+3} - \dfrac{4}{3-x}$

24. $\dfrac{x-1}{4-x} + \dfrac{3x}{x+5}$

25. $\dfrac{x+2}{3x+9} + \dfrac{2x-1}{2x-6}$

26. $\dfrac{x}{4x-8} - \dfrac{3x+2}{3x+6}$

27. $\dfrac{3x}{6+x} - \dfrac{2x}{x^2-36}$

28. $\dfrac{4}{x+5} - \dfrac{2x+3}{x^2+4x-5}$

29. $\dfrac{4x+1}{7-x} + \dfrac{x-1}{x^2-8x+7}$

30. $\dfrac{3x-4}{x^2-x-20} - \dfrac{2}{5-x}$

31. $\dfrac{4x}{x^2+3x-28} + \dfrac{3}{x^2+6x-7}$

32. $\dfrac{3x}{x^2+2x+1} - \dfrac{x}{x^2+4x+4}$

33. $\dfrac{3x+4}{2x^2-23x+30} - \dfrac{x+5}{2x^2-19x+24}$

34. $\dfrac{x+1}{x^2-3x+2} + \dfrac{6}{x^2-6x+8}$

35. $\dfrac{4x-1}{x^2-5x+4} + \dfrac{2x+7}{x^2-11x+28}$

36. $\dfrac{7x+3}{5x^2+27x+36} + \dfrac{3x-2}{5x^2+22x+24}$

37. $\dfrac{x-6}{7x^2-3x-4} + \dfrac{7-x}{7x^2+18x+8}$

38. $\dfrac{x+10}{x^2+5x+4} - \dfrac{4}{x^2+6x+8}$

39. $\dfrac{x-3}{4x^2-5x-6} - \dfrac{4x+10}{2x^2+x-10}$

40. $\dfrac{2x+1}{8x^2-37x-15} + \dfrac{2-x}{8x^2+11x+3}$

41. $\dfrac{3x}{4-x} + \dfrac{7x}{x+4} - \dfrac{x-3}{x^2-16}$

42. $\dfrac{x}{x+3} + \dfrac{x+1}{3-x} + \dfrac{x^2+4}{x^2-9}$

43. $-\dfrac{1}{2} + \dfrac{x-5}{x-3} + \dfrac{x-1}{x^2-5x+6}$

44. $-4 + \dfrac{1-2x}{x+6} + \dfrac{x^2+1}{x^2+4x-12}$

45. $\dfrac{2}{x^2-4}-\dfrac{3}{x^2-3x+2}+\dfrac{x-1}{x^2+x-2}$

46. $\dfrac{5}{x^2+3x+2}+\dfrac{4}{x^2+6x+8}-\dfrac{6}{x^2+5x+4}$

47. $\dfrac{x}{x^2+4x-21}+\dfrac{1-x}{x^2+8x+7}+\dfrac{3x}{x^2-2x-3}$

48. $\dfrac{x+2}{9x^2-6x+4}+\dfrac{10x-5x^2}{27x^3+8}-\dfrac{2}{3x+2}$

49. $\dfrac{3x+9}{x^2-5x+4}+\dfrac{49}{12+x-x^2}+\dfrac{3x+21}{x^2+2x-3}$

50. $\dfrac{5x+22}{x^2+8x+15}+\dfrac{4}{x^2+4x+3}+\dfrac{6}{x^2+6x+5}$

51. $\dfrac{x}{xy+x-2y-2}+\dfrac{x+2}{xy+x+y+1}$

52. $\dfrac{4x}{xy-3x+y-3}+\dfrac{x+2}{xy+2y-3x-6}$

53. $\dfrac{3y}{xy+2x+3y+6}+\dfrac{x}{x^2-2x-15}$

54. $\dfrac{2}{xy-4x-2y+8}+\dfrac{5y}{y^2-3y-4}$

55. $\dfrac{x+6}{x^2+x+1}-\dfrac{3x^2+x-4}{x^3-1}$

56. $\dfrac{2x-5}{8x^2-4x+2}+\dfrac{x^2-2x+5}{8x^3+1}$

57. $\dfrac{x+1}{x^3-3x^2+x-3}+\dfrac{x^2-5x-8}{x^4-8x^2-9}$

58. $\dfrac{x+4}{x^3-5x^2+6x-30}-\dfrac{x-7}{x^3-2x^2+6x-12}$

59. $\dfrac{x+1}{2x^2-x-1}+\dfrac{2x}{2x^2+5x+2}-\dfrac{2x}{3x^2+4x-4}$

60. $\dfrac{x-6}{3x^2+10x+3}-\dfrac{2x}{5x^2-3x-2}+\dfrac{2x}{3x^2-2x-1}$

Writing & Thinking

61. Discuss the steps in the process you go through when adding two rational expressions with different denominators. That is, discuss how you find the least common denominator when adding rational expressions and how you use this LCD to find equivalent rational expressions that you can add.

13.3 Complex Fractions

Objective A **Simplifying Complex Fractions (First Method)**

A **complex fraction** is a fraction in which the numerator and/or denominator are themselves fractions or the sum or difference of fractions. Examples of complex fractions are:

$$\frac{\dfrac{1}{x+3}-\dfrac{1}{x}}{1+\dfrac{3}{x}} \qquad \text{and} \qquad \frac{x+y}{x^{-1}+y^{-1}} = \frac{x+y}{\dfrac{1}{x}+\dfrac{1}{y}}.$$

The objective here is to develop techniques for simplifying complex fractions so that they are written in the form of a single reduced rational expression.

In a complex fraction such as $\dfrac{\dfrac{1}{x+3}-\dfrac{1}{x}}{1+\dfrac{3}{x}}$, the large fraction bar is a symbol of inclusion.

The expression could also be written as follows.

$$\frac{\dfrac{1}{x+3}-\dfrac{1}{x}}{1+\dfrac{3}{x}} = \left(\frac{1}{x+3}-\frac{1}{x}\right) \div \left(1+\frac{3}{x}\right)$$

Thus a complex fraction indicates that the numerator is to be divided by the denominator.

To Simplify Complex Fractions (First Method)

1. Simplify the numerator so that it is a single rational expression.
2. Simplify the denominator so that it is a single rational expression.
3. Divide the numerator by the denominator and reduce to lowest terms.

This method is used to simplify the complex fractions in Examples 1 and 2. Study the examples closely so that you understand what happens at each step.

Example 1

First Method for Simplifying Complex Fractions

Simplify the complex fraction $\dfrac{\dfrac{3x}{y^2}}{\dfrac{12x}{7y}}$.

Solution

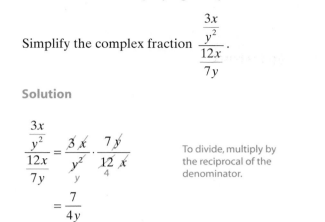

$$\frac{\dfrac{3x}{y^2}}{\dfrac{12x}{7y}} = \frac{\cancel{3}\,\cancel{x}}{y^{\cancel{2}}} \cdot \frac{7\cancel{y}}{\cancel{12}\,\cancel{x}}$$

To divide, multiply by the reciprocal of the denominator.

$$= \frac{7}{4y}$$

Example 2

First Method for Simplifying Complex Fractions

Simplify the following complex fractions.

a. $\dfrac{\dfrac{1}{x+3} - \dfrac{1}{x}}{1 + \dfrac{3}{x}}$

Solution

$$\frac{\dfrac{1}{x+3} - \dfrac{1}{x}}{1 + \dfrac{3}{x}} = \frac{\dfrac{1 \cdot x}{(x+3) \cdot x} - \dfrac{1(x+3)}{x(x+3)}}{\dfrac{x}{x} + \dfrac{3}{x}}$$

Combine the fractions in the numerator and in the denominator separately.

Note that $1 = \dfrac{x}{x}$

$$= \frac{\dfrac{x-(x+3)}{x(x+3)}}{\dfrac{x+3}{x}}$$

$$= \frac{\dfrac{x-x-3}{x(x+3)}}{\dfrac{x+3}{x}}$$

$$= \frac{\dfrac{-3}{x(x+3)}}{\dfrac{x+3}{x}}$$

$$= \frac{-3}{\cancel{x}(x+3)} \cdot \frac{\cancel{x}}{x+3}$$

To divide, multiply by the reciprocal of the denominator.

$$= \frac{-3}{(x+3)^2}$$

Simplify the following complex fractions:

1. $\dfrac{\dfrac{5x}{y^3}}{\dfrac{15x}{y^2}}$

2. a. $\dfrac{\dfrac{1}{x+6}-\dfrac{1}{x}}{1+\dfrac{6}{x}}$

b. $\dfrac{3x-3y}{\left(3x\right)^{-1}-\left(3y\right)^{-1}}$

b. $\dfrac{x+y}{x^{-1}+y^{-1}}$

Solution

$$\dfrac{x+y}{x^{-1}+y^{-1}}=\dfrac{x+y}{\dfrac{1}{x}+\dfrac{1}{y}}$$

Recall that $x^{-1}=\dfrac{1}{x}$ and $y^{-1}=\dfrac{1}{y}$.

$$=\dfrac{\dfrac{x+y}{1}}{\dfrac{1}{x}\cdot\dfrac{y}{y}+\dfrac{1}{y}\cdot\dfrac{x}{x}}$$

Add the two fractions in the denominator.

$$=\dfrac{\dfrac{x+y}{1}}{\dfrac{y}{xy}+\dfrac{x}{xy}}$$

$$=\dfrac{\dfrac{x+y}{1}}{\dfrac{y+x}{xy}}$$

$$=\dfrac{x+y}{1}\cdot\dfrac{xy}{y+x}$$

Multiply by the reciprocal of the denominator.

$$=\dfrac{xy}{1}$$

$$=xy$$

Now work margin exercise **1 and 2.**

Objective B **Simplifying Complex Fractions (Second Method)**

A second method is to find the LCM of the denominators in the fractions in both the original numerator and the original denominator and then multiply **both** the numerator and denominator by this LCM.

To Simplify Complex Fractions (Second Method)

1. Find the LCM of all the denominators in the numerator and denominator of the complex fraction.
2. Multiply both the numerator and denominator of the complex fraction by this LCM.
3. Simplify **both** the numerator and denominator and reduce to lowest terms.

Example 3

Second Method for Simplifying Complex Fractions

Simplify the following complex fractions.

a. $\dfrac{\dfrac{1}{x+3} - \dfrac{1}{x}}{1 + \dfrac{3}{x}}$

Solution

$\dfrac{\dfrac{1}{x+3} - \dfrac{1}{x}}{1 + \dfrac{3}{x}} = \dfrac{\left(\dfrac{1}{x+3} - \dfrac{1}{x}\right) \cdot x(x+3)}{\left(1 + \dfrac{3}{x}\right) \cdot x(x+3)}$

Multiply by $x(x+3)$, the LCM of $\{x, x+3\}$. This multiplication can be done because the net effect is that the fraction is multiplied by 1.

$= \dfrac{\dfrac{1}{x+3} \cdot x(x+3) - \dfrac{1}{x} \cdot x(x+3)}{1 \cdot x(x+3) + \dfrac{3}{x} \cdot x(x+3)}$

$= \dfrac{x - (x+3)}{x(x+3) + 3(x+3)}$

$= \dfrac{x - x - 3}{(x+3)(x+3)}$

$= \dfrac{-3}{(x+3)^2}$

Note that this matches the result found in Example 2a using Method 1.

b. $\dfrac{x+y}{x^{-1} + y^{-1}}$

Solution

$\dfrac{x+y}{x^{-1} + y^{-1}} = \dfrac{\dfrac{x+y}{1}}{\dfrac{1}{x} + \dfrac{1}{y}}$

$= \dfrac{\left(\dfrac{x+y}{1}\right)xy}{\left(\dfrac{1}{x} + \dfrac{1}{y}\right)xy}$

Multiply by xy, the LCM of $\{1, x, y\}$.

$= \dfrac{(x+y)xy}{\dfrac{1}{x} \cdot xy + \dfrac{1}{y} \cdot xy}$

$= \dfrac{(x+y)xy}{y+x} = xy$

Now work margin exercise 3.

---

3. Simplify the following complex fractions, using the second method for simplifying complex fractions:

a. $\dfrac{\dfrac{1}{x+6} - \dfrac{1}{x}}{1 + \dfrac{6}{x}}$

b. $\dfrac{3x - 3y}{(3x)^{-1} - (3y)^{-1}}$

Simplifying Complex Algebraic Expressions

A **complex algebraic expression** is an expression that involves rational expressions and more than one operation. In simplifying such expressions, the rules for order of operations apply. As with complex fractions, the objective is to simplify the expression so that it is written in the form of a single reduced rational expression.

4. Simplify the following expression:

$$\frac{6}{x+4}+\frac{x}{x+4}\div\frac{x}{x-4}$$

Example 4

Simplifying Complex Algebraic Expressions

Simplify the following expression.

$$\frac{4-x}{x+3}+\frac{x}{x+3}\div\frac{x}{x-3}$$

Solution

In a complex algebraic expression such as

$$\frac{4-x}{x+3}+\frac{x}{x+3}\div\frac{x}{x-3}$$

the rules for order of operations indicate that the division is to be done first.

$$\frac{4-x}{x+3}+\frac{x}{x+3}\div\frac{x}{x-3}=\frac{4-x}{x+3}+\frac{\cancel{x}}{x+3}\cdot\frac{x-3}{\cancel{x}}$$

$$=\frac{4-x}{x+3}+\frac{x-3}{x+3}$$

$$=\frac{4-x+x-3}{x+3}$$

$$=\frac{1}{x+3}$$

Now work margin exercise 4.

Practice Problems

Simplify each of the following expressions.

1. $\dfrac{\dfrac{1}{x}}{1+\dfrac{1}{x}}$

2. $\dfrac{\dfrac{1}{x+2}-\dfrac{1}{x}}{1+\dfrac{2}{x}}$

3. $\dfrac{1+\dfrac{3}{x-3}}{x-\dfrac{x^2}{x-3}}$

4. $\dfrac{\dfrac{1}{x+y}-\dfrac{1}{x-y}}{\dfrac{2y}{x^2-y^2}}$

5. $\dfrac{5}{x}-\dfrac{3}{x-2}\div\dfrac{x}{x-2}$

Practice Problem Answers

1. $\dfrac{1}{x+1}$

2. $\dfrac{-2}{(x+2)^2}$

3. $-\dfrac{1}{3}$

4. -1

5. $\dfrac{2}{x}$

Exercises 13.3

1. $\dfrac{\dfrac{2x}{3y^2}}{\dfrac{5x^2}{6y}}$

2. $\dfrac{\dfrac{6x^2}{5y}}{\dfrac{x}{10y^2}}$

3. $\dfrac{\dfrac{12x^3}{7y^2}}{\dfrac{3x^5}{2y}}$

4. $\dfrac{\dfrac{9x^2}{7y^3}}{\dfrac{3xy}{14}}$

5. $\dfrac{\dfrac{x+3}{2x}}{\dfrac{2x-1}{4x^2}}$

6. $\dfrac{\dfrac{x-2}{6x}}{\dfrac{x+3}{3x^2}}$

7. $\dfrac{\dfrac{2x-1}{x}}{\dfrac{2}{x}+3}$

8. $\dfrac{2-\dfrac{3}{x}}{\dfrac{x^2-4}{x}}$

9. $\dfrac{\dfrac{3}{x}+\dfrac{1}{2x}}{1+\dfrac{2}{x}}$

10. $\dfrac{\dfrac{3}{x}+\dfrac{5}{2x}}{\dfrac{1}{x}+4}$

11. $\dfrac{1+\dfrac{1}{x}}{1-\dfrac{1}{x^2}}$

12. $\dfrac{\dfrac{2}{y}+1}{\dfrac{4}{y^2}-1}$

13. $\dfrac{\dfrac{1}{x}+\dfrac{1}{3x}}{\dfrac{x+6}{x^2}}$

14. $\dfrac{\dfrac{3}{x}-\dfrac{6}{x^2}}{\dfrac{x-2}{x^2}}$

15. $\dfrac{\dfrac{7}{x}-\dfrac{14}{x^2}}{\dfrac{1}{x}-\dfrac{4}{x^3}}$

16. $\dfrac{\dfrac{3}{x}-\dfrac{6}{x^2}}{\dfrac{1}{x}-\dfrac{2}{x^2}}$

17. $\dfrac{\dfrac{1}{3}+\dfrac{1}{x}}{\dfrac{1}{2}-\dfrac{1}{x}}$

18. $\dfrac{\dfrac{x}{y}-\dfrac{1}{3}}{\dfrac{6}{y}-\dfrac{2}{x}}$

19. $\dfrac{\dfrac{2}{x}+\dfrac{3}{4y}}{\dfrac{3}{2x}-\dfrac{5}{3y}}$

20. $\dfrac{\dfrac{4}{3x}-\dfrac{5}{y}}{\dfrac{1}{3}+\dfrac{3}{y}}$

21. $\dfrac{1+x^{-1}}{1-x^{-2}}$

22. $\dfrac{x^{-3}+1}{1-x^{-1}}$

23. $\dfrac{1}{x^{-1}+y^{-1}}$

24. $\dfrac{x-y}{x^{-2}-y^{-2}}$

25. $\dfrac{x^{-1}+y^{-1}}{x+y}$

26. $\dfrac{y^{-2}-x^{-2}}{x+y}$

27. $\dfrac{x^{-1}+y^{-1}}{x^{-1}-y^{-1}}$

28. $\dfrac{x^{-1}+y^{-1}}{x^{-2}-y^{-2}}$

29. $\dfrac{\dfrac{4}{x}-1}{1-\dfrac{1}{x-3}}$

30. $\dfrac{x+\dfrac{3}{x-4}}{1-\dfrac{1}{x}}$

31. $\dfrac{1-\dfrac{4}{x+3}}{1-\dfrac{2}{x+1}}$

32. $\dfrac{1+\dfrac{4}{2x-3}}{1+\dfrac{x}{x+1}}$

33. $\dfrac{\dfrac{1}{x+h}-\dfrac{1}{x}}{h}$

34. $\dfrac{\dfrac{1}{(x+h)^2}-\dfrac{1}{x^2}}{h}$

35. $\dfrac{\left(2+\dfrac{1}{x+h}\right)-\left(2+\dfrac{1}{x}\right)}{h}$

36. $\dfrac{\left(\dfrac{1}{(x+h)^2}-3\right)-\left(\dfrac{1}{x^2}-3\right)}{h}$

37. $\dfrac{x^2-4y^2}{1-\dfrac{2x+y}{x-y}}$

38. $\dfrac{\dfrac{8x^2-2y^2}{4x-1}}{\dfrac{4x-1}{x-y}-2}$

39. $\dfrac{\dfrac{x+1}{x-1}-\dfrac{x-1}{x+1}}{\dfrac{x+1}{x-1}+\dfrac{x-1}{x+1}}$

40. $\dfrac{\dfrac{1}{x^2-1}-\dfrac{1}{x+1}}{\dfrac{1}{x-1}+\dfrac{1}{x^2-1}}$

41. $\dfrac{\dfrac{x}{x-4}-\dfrac{1}{x-1}}{\dfrac{x}{x-1}+\dfrac{2}{x-3}}$

42. $\dfrac{\dfrac{1}{x+1}-\dfrac{x}{x+2}}{\dfrac{x}{x+2}-\dfrac{2}{x-1}}$

Write each of the expressions as a single fraction reduced to lowest terms. See example 4.

43. $\dfrac{1}{x+1}-\dfrac{3}{2x}\cdot\dfrac{4x}{x+1}$

44. $\dfrac{4}{x}-\dfrac{2}{x^2-2x}\cdot\dfrac{x-2}{5}$

45. $\left(\dfrac{8}{x}-\dfrac{3}{4x}\right)\div\dfrac{4x+5}{x}$

46. $\left(\dfrac{2}{x}+\dfrac{5}{x-3}\right)\div\dfrac{x}{2x-6}$

47. $\dfrac{x}{x-1}-\dfrac{3}{x-1}\cdot\dfrac{x+2}{x}$

48. $\dfrac{x+3}{x+2}+\dfrac{x}{x+2}\cdot\dfrac{x-3}{x^2}$

49. $\dfrac{x-1}{x+4}+\dfrac{x-6}{x^2+3x-4}\div\dfrac{x-4}{x-1}$

50. $\dfrac{x}{x+3}-\dfrac{3}{x+5}\div\dfrac{x-2}{x^2+3x-10}$

51. Some complex fractions involve the sum (or difference) of complex fractions. Beginning with the outermost denominator, simplify each of the following expressions.

a. $1 + \cfrac{1}{1 + \cfrac{1}{1 + \cfrac{1}{1 + 1}}}$

b. $2 - \cfrac{1}{2 - \cfrac{1}{2 - \cfrac{1}{2 - 1}}}$

c. $x + \cfrac{1}{x + \cfrac{1}{x + \cfrac{1}{x + 1}}}$

A Solve equations that involve rational expressions.

B Use the process of solving equations to manipulate formulas.

C Use proportions to relate similar triangles.

13.4 Solving Equations with Rational Expressions

Objective A **Equations with Rational Expressions**

Remember, from Chapter 4, that a **proportion** is an equation stating that two ratios are equal. For example, $\frac{4}{8} = \frac{3}{6}$ is a proportion. The equation $\frac{3}{x-6} = \frac{5}{x}$ is a proportion with rational expressions. An equation such as

$$\frac{3}{x} + \frac{1}{8} = \frac{13}{4x}$$

involves the sum of rational expressions and **is not a proportion.** To solve this equation we "clear" the fractions by multiplying both sides of the equation by the LCD of the fractions. In this case, the LCD is $8x$ and we can proceed as follows:

$$\frac{3}{x} + \frac{1}{8} = \frac{13}{4x}$$ **Note:** $x \neq 0$.

$$8x\left(\frac{3}{x} + \frac{1}{8}\right) = 8x \cdot \frac{13}{4x}$$ Multiply both sides by $8x$, the LCD.

$$8x \cdot \frac{3}{x} + 8x \cdot \frac{1}{8} = 8x \cdot \frac{13}{4x}$$ Use the distributive property and reduce.

$$24 + x = 26$$ Simplify.

$$x = 2.$$

As we have seen, rational expressions may contain variables in either the numerator or denominator or both. In any case, a general approach to solving equations that contain rational expressions is as follows.

To Solve an Equation with Rational Expressions

1. Find the LCD of the fractions.
2. Multiply both sides of the equation by this LCD and simplify.
3. Solve the resulting equation. (This equation will have only polynomials on both sides.)
4. Check each solution in the **original equation**. (Remember that no denominator can be 0 and any solution that gives a 0 denominator is to be discarded.)

Checking is particularly important when equations have rational expressions. Multiplying by the LCD may introduce solutions that are not solutions to the original equation. Such solutions are called **extraneous solutions** or **extraneous roots** and occur because multiplication by a variable expression may, in effect, be multiplying the original equation by 0.

Example 1

Solving Equations with Rational Expressions

State any restrictions on the variable, and then solve the equation.

a. $\dfrac{x-5}{2x} = \dfrac{6}{3x}$

Solution

First find the LCD of the fractions, and then multiply both sides of the equation by the LCD.

$$\left.\begin{array}{c} 2x \\ 3x \end{array}\right\} \quad \text{LCD} = 2 \cdot 3 \cdot x = 6x$$

$$\dfrac{x-5}{2x} = \dfrac{6}{3x} \qquad (x \neq 0)$$

$$6x \cdot \dfrac{x-5}{2x} = 6x \cdot \dfrac{6}{3x} \qquad \text{Multiply both sides by the LCD, } 6x.$$

$$3(x-5) = 2(6)$$

$$3x - 15 = 12$$

$$3x = 27$$

$$x = 9$$

Since 9 is not a restriction, there is one solution, $x = 9$.

b. $\dfrac{1}{x-4} = \dfrac{3}{x^2 - 5x}$

Solution

First find the LCD of the fractions, and then multiply both sides of the equation by the LCD.

$$\left.\begin{array}{c} x-4 = x-4 \\ x^2 - 5x = x(x-5) \end{array}\right\} \quad \text{LCD} = x(x-5)(x-4)$$

$$x(x-5)\,\cancel{(x-4)} \cdot \dfrac{1}{\cancel{x-4}} = \cancel{x}\,\cancel{(x-5)}(x-4) \cdot \dfrac{3}{\cancel{x}\,\cancel{(x-5)}} \qquad (x \neq 0, 4, 5)$$

$$x(x-5) = 3(x-4)$$

$$x^2 - 5x = 3x - 12$$

$$x^2 - 8x + 12 = 0$$

$$(x-6)(x-2) = 0$$

$$x - 6 = 0 \quad \text{or} \quad x - 2 = 0$$

$$x = 6 \qquad\qquad x = 2$$

Since 6 and 2 are not restrictions, there are two solutions, $x = 6$ and $x = 2$.

c. $\dfrac{x}{x-2} + \dfrac{x-6}{x(x-2)} = \dfrac{5x}{x-2} - \dfrac{10}{x-2}$

Solution

First find the LCD of the fractions and then multiply each term on both sides of the equation by the LCD.

$$\left.\begin{array}{c} x-2 \\ x(x-2) \end{array}\right\} \quad \mathrm{LCD} = x(x-2)$$

$$x(x-2) \cdot \dfrac{x}{x-2} + x(x-2) \cdot \dfrac{x-6}{x(x-2)} = x(x-2) \cdot \dfrac{5x}{x-2} - x(x-2) \cdot \dfrac{10}{x-2}$$

$$x^2 + x - 6 = 5x^2 - 10x \qquad (x \neq 0, 2)$$
$$0 = 4x^2 - 11x + 6$$
$$0 = (4x-3)(x-2)$$
$$4x - 3 = 0 \quad \text{or} \quad x - 2 = 0$$
$$4x = 3 \qquad\qquad \cancel{x = 2}$$
$$x = \dfrac{3}{4}$$

The only solution is $x = \dfrac{3}{4}$, since 2 is a restricted value $(x \neq 0, 2)$ and thus **not** a solution. No denominator can be 0.

d. $\dfrac{2}{x^2-9} = \dfrac{1}{x^2} + \dfrac{1}{x^2-3x}$

Solution

First find the LCD of the fractions and then multiply each term on both sides of the equation by the LCD.

$$\left.\begin{array}{c} x^2 - 9 = (x+3)(x-3) \\ x^2 = x^2 \\ x^2 - 3x = x(x-3) \end{array}\right\} \quad \mathrm{LCD} = x^2(x+3)(x-3)$$

$$x^2(x+3)(x-3) \cdot \dfrac{2}{(x+3)(x-3)}$$
$$= x^2(x+3)(x-3) \cdot \dfrac{1}{x^2} + x^2(x+3)(x-3) \cdot \dfrac{1}{x(x-3)} \qquad (x \neq 0, -3, 3)$$

$$2x^2 = (x+3)(x-3) + x(x+3)$$
$$2x^2 = x^2 - 9 + x^2 + 3x$$
$$2x^2 = 2x^2 + 3x - 9$$
$$9 = 3x$$
$$\cancel{3 = x} \qquad\qquad \text{Note that 3 is one of the restrictions.}$$

There is no solution. The solution set is the empty set, $\varnothing$. The original equation is a contradiction.

Completion Example 2

Solving Equations with Rational Expressions

State any restrictions on the variable then solve the equation:

$$\frac{x}{x-1} - \frac{3x+1}{x^2+4x-5} = \frac{x+2}{x+5}.$$

Solution

$$x \neq \underline{\quad} , \underline{\quad}$$

$$\frac{x}{x-1} - \frac{3x+1}{x^2+4x-5} = \frac{x+2}{x+5}$$

$$(x+5)(\underline{\quad}) \cdot \frac{x}{x-1} - (x+5)(\underline{\quad}) \cdot \frac{3x+1}{(x+5)(x-1)} = (x+5)(\underline{\quad}) \cdot \frac{x+2}{x+5}$$

$$(x+5) \cdot x - (\underline{\quad}) = (\underline{\quad})(x+2)$$

$$x^2 + 5x - \underline{\quad} - \underline{\quad} = \underline{\quad\quad}$$

$$\underline{\quad} = \underline{\quad}$$

$$x = \underline{\quad}$$

Now work margin exercises 1 *and* 2.

Objective B **Formulas**

As discussed previously, many formulas are equations relating more than one variable. The equation is solved for one variable in terms of another variable (maybe more than one). The process of solving equations can be used to manipulate the formula so that it is solved for one of the other variables. As illustrated in Example 3, the resulting equation may involve a rational expression.

State any restrictions on the variable, and then solve the equation:

1. a. $\dfrac{x-7}{2x} = \dfrac{8}{4x}$

b. $\dfrac{1}{x+2} = \dfrac{6}{x^2+7x}$

c.
$$\frac{x}{x-2} + \frac{4x-4}{x(x-2)} = \frac{6x}{x-2} - \frac{8}{x-2}$$

d. $\dfrac{4}{x^2-25} = \dfrac{2}{x^2} + \dfrac{2}{x^2-5x}$

2. $\dfrac{4y^2}{y^2+3y-10} = \dfrac{y+1}{y+5} + \dfrac{3y}{y-2}$

Completion Example Answers

2.
$$x \neq -5, 1$$

$$(x+5)(x-1) \cdot \frac{x}{x-1} - (x+5)(x-1) \cdot \frac{3x+1}{(x+5)(x-1)} = (x+5)(x-1) \cdot \frac{x+2}{x+5}$$

$$(x+5) \cdot x - (3x+1) = (x-1)(x+2)$$

$$x^2 + 5x - 3x - 1 = x^2 + x - 2$$

$$2x - 1 = x - 2$$

$$x = -1$$

3. The surface area of a rectangular box is given by the formula, $SA = 2(lh) + 2(wh) + 2(lw)$. (Here SA is treated as a single variable.) Solve the formula for l.

Example 3

Solving a Formula for a Specified Variable

The formula $S = 2\pi r^2 + 2\pi rh$ is used to find the surface area (S) of a right circular cylinder, where r is the radius of the cylinder and h is the height of the cylinder. Solve the formula for h.

Solution

$$S = 2\pi r^2 + 2\pi rh \qquad \text{Write the formula.}$$

$$S - 2\pi r^2 = 2\pi rh \qquad \text{Add } -2\pi r^2 \text{ to both sides of the equation.}$$

$$\frac{S - 2\pi r^2}{2\pi r} = h \qquad \text{Divide both sides by } 2\pi r.$$

Thus the formula solved for h is: $h = \dfrac{S - 2\pi r^2}{2\pi r} \quad \left(\text{or } h = \dfrac{S}{2\pi r} - r \right).$

Now work margin exercise 3.

Objective C **Similar Triangles**

Similar triangles, previously discussed in Chapter 5, are triangles that meet the following two conditions:

1. The measures of the corresponding angles are equal.
2. The lengths of the corresponding sides are proportional.

In similar triangles, corresponding sides are those sides opposite the equal angles. (See Figure 1.)

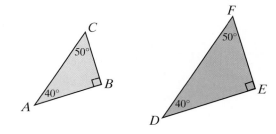

Figure 1

We write $\triangle ABC \sim \triangle DEF$. ($\sim$ is read "**is similar to**".) The lengths of the corresponding sides are proportional. Thus,

$$\frac{AB}{DE} = \frac{BC}{EF} \quad \text{and} \quad \frac{AB}{DE} = \frac{AC}{DF} \quad \text{and} \quad \frac{BC}{EF} = \frac{AC}{DF}.$$

In a pair of similar triangles, we can often find the length of an unknown side by setting up a proportion and solving. Example 4 illustrates such a situation.

Example 4

Similar Triangles

In the figure shown, $\triangle ABC \sim \triangle PQR$. Find the lengths of the sides AB and QR.

4. Looking at the picture in Example 3, assume in triangle ABC that side AB has a value of $x - 4$ and side BC has a value of 5. In triangle PQR side PQ has a value of 9 and side RQ has a value of x. Solve for x.

Solution

Set up a proportion involving corresponding sides and solve for x.

$$\frac{x-4}{10} = \frac{6}{x}$$

$$10x \cdot \frac{x-4}{10} = 10x \cdot \frac{6}{x}$$

$$x(x-4) = 10 \cdot 6$$

$$x^2 - 4x = 60$$

$$x^2 - 4x - 60 = 0$$

$$(x-10)(x+6) = 0$$

$$x - 10 = 0 \quad \text{or} \quad x + 6 = 0$$

$$x = 10 \qquad \qquad x = -6$$

Because the length of a side cannot be negative, the only acceptable solution is $x = 10$. Thus $QR = 10$. Substituting 10 for x gives $AB = 10 - 4 = 6$.

Now work margin exercise 4.

Practice Problems

State any restrictions on the variable and then solve the equations.

1. $\dfrac{16}{x} = \dfrac{4}{x-5}$

2. $\dfrac{5}{3x+2} = \dfrac{4}{3x+1}$

3. $\dfrac{4x}{x+3} = \dfrac{1}{x}$

4. $\dfrac{x}{x-3} - \dfrac{2x+3}{x^2+x-12} = \dfrac{x-1}{x+4}$

5. Solve the formula $A = P + Pr$ for r.

Practice Problem Answers

1. $x \neq 0, 5; x = \dfrac{20}{3}$

2. $x \neq -\dfrac{1}{3}, -\dfrac{2}{3}; x = \dfrac{13}{3}$

3. $x \neq -3, 0; x = -\dfrac{3}{4}$ or 1

4. $x \neq -4, 3; \ x = 1$

5. $r = \dfrac{A-P}{P}$

Exercises 13.4

1. $\dfrac{4x}{7} = \dfrac{x+5}{3}$

2. $\dfrac{3x+1}{4} = \dfrac{2x+1}{3}$

3. $\dfrac{10}{x} = \dfrac{5}{x-2}$

4. $\dfrac{8}{x-3} = \dfrac{12}{2x-3}$

5. $\dfrac{4}{x-4} = \dfrac{2}{x+3}$

6. $\dfrac{3}{x+5} = \dfrac{6}{x-2}$

7. $\dfrac{x+2}{5x} = \dfrac{x-6}{3x}$

8. $\dfrac{x-4}{3x} = \dfrac{x-2}{5x}$

9. $\dfrac{5x+2}{x-6} = \dfrac{11}{4}$

10. $\dfrac{x+9}{3x+2} = \dfrac{5}{8}$

11. $\dfrac{5x}{4} - \dfrac{1}{2} = -\dfrac{3}{16}$

12. $\dfrac{x}{6} - \dfrac{1}{42} = \dfrac{1}{7}$

13. $\dfrac{3x-1}{6} - \dfrac{x+3}{4} = \dfrac{7}{12}$

14. $\dfrac{x-2}{3} - \dfrac{x-3}{5} = \dfrac{13}{15}$

15. $\dfrac{2+x}{4} - \dfrac{5x-2}{12} = \dfrac{8-2x}{5}$

16. $\dfrac{4x+1}{5} = \dfrac{2x+3}{2} - \dfrac{x+2}{4}$

17. $\dfrac{2}{3x} = \dfrac{1}{4} - \dfrac{1}{6x}$

18. $\dfrac{1}{x} - \dfrac{8}{21} = \dfrac{3}{7x}$

19. $\dfrac{3}{5x} - \dfrac{1}{5} = \dfrac{3}{4x}$

20. $\dfrac{3}{8x} - \dfrac{7}{10} = \dfrac{1}{5x}$

21. $\dfrac{3}{4x} - \dfrac{1}{2} = \dfrac{7}{8x} + \dfrac{1}{6}$

22. $\dfrac{5}{3x} + \dfrac{1}{2} = \dfrac{7}{9x} - \dfrac{5}{6}$

23. $\dfrac{2}{4x+1} = \dfrac{4}{x^2+9x}$

24. $\dfrac{3}{4x-1} = \dfrac{4}{x^2+x}$

25. $\dfrac{9}{x^2-6x} = \dfrac{5}{2x-3}$

26. $\dfrac{-9}{x^2+5x} = \dfrac{8}{4-9x}$

27. $\dfrac{x}{x-4} - \dfrac{4}{2x-1} = 1$

28. $\dfrac{x}{x+3} + \dfrac{1}{x+2} = 1$

29. $\dfrac{x+2}{x+1} + \dfrac{x+2}{x+4} = 2$

30. $\dfrac{3x-2}{x+4} + \dfrac{2x+5}{x-1} = 5$

31. $\dfrac{2}{4x-1} + \dfrac{1}{x+1} = \dfrac{3}{x+1}$

32. $\dfrac{x-2}{x+4} - \dfrac{3}{2x+1} = \dfrac{x-7}{x+4}$

33. $\dfrac{x-2}{x-3} + \dfrac{x-3}{x-2} = \dfrac{2x^2}{x^2-5x+6}$

34. $\dfrac{x}{x-4} - \dfrac{12x}{x^2+x-20} = \dfrac{x-1}{x+5}$

35. $\dfrac{3x+5}{3x+2} + \dfrac{8x+16}{3x^2-4x-4} = \dfrac{x+2}{x-2}$

36. $\dfrac{3x+5}{3x+2} - \dfrac{4-2x}{3x^2+8x+4} = \dfrac{x+4}{x+2}$

37. $\dfrac{3}{3x-1} + \dfrac{1}{x+1} = \dfrac{4}{2x-1}$

38. $\dfrac{2}{x+1} + \dfrac{4}{2x-3} = \dfrac{4}{x-5}$

Solve each of the formulas for the specified variables. Assume no denominator has a value of 0. See Example 3.

39. $S = \dfrac{a}{1-r}$; solve for r (formula for the sum of an infinite geometric sequence)

40. $z = \dfrac{x-\bar{x}}{s}$; solve for x (formula used in statistics)

41. $z = \dfrac{x-\bar{x}}{s}$; solve for s (formula used in statistics)

42. $a_n = a_1 + (n-1)d$; solve for d (formula for the nth term in an arithmetic sequence)

43. $m = \dfrac{y-y_1}{x-x_1}$; solve for y (formula for the slope of a line)

44. $v_{avg} = \dfrac{d_2 - d_1}{t_2 - t_1}$; solve for d_2 (formula for average velocity)

45. $\dfrac{1}{R_{total}} = \dfrac{1}{R_1} + \dfrac{1}{R_2}$; solve for R_{total} (formula used in electronics)

46. $\dfrac{1}{x} = \dfrac{1}{t_1} + \dfrac{1}{t_2}$; solve for x (formula used in mathematics)

47. $A = P + Pr$; solve for P (formula used for compound interest)

48. $y = \dfrac{ax+b}{cx+d}$; solve for x (formula used in mathematics)

The following exercises show pairs of similar triangles. Find the lengths of the sides labeled with variables. See example 4.

49. △JKL ~ △JTB

50. △QRP ~ △TUS

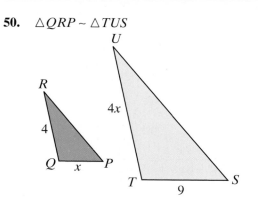

51. △ABC ~ △RST

52. △FED ~ △FGH

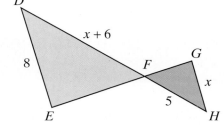

53. △SUT ~ △PRQ

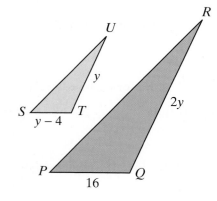

54. △ABC ~ △DEC

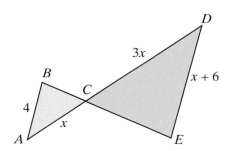

55. Parallelograms: In the parallelogram $ABCD$, $AB = CD = 10$ in. Diagonal $AC = 12$ in. The point M on AB is 6 in. from A. Point P is the intersection of DM with AC. The triangles APM and CPD are similar. (Symbolically, $\triangle APM \sim \triangle CPD$.) What are the lengths of AP and PC?

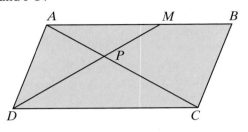

56. Parallelograms: If, in the same figure discussed in Exercise 55, the point P is the point of intersection of the two diagonals, AC and DB, what are the lengths of AP and PC?

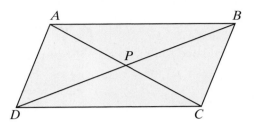

Writing & Thinking

In simplifying rational expressions, the result is a rational or polynomial expression. However, in solving equations with rational expressions, the goal is to find a value (or values) for the variable that will make the equation a true statement. Many students confuse these two ideas. To avoid confusing the techniques for adding and subtracting rational expressions with the techniques for solving equations, simplify the expression in part **a.** and solve the equation in part **b.** Explain, in your own words, the differences in your procedures. Assume no denominator has a value of 0.

57. a. $\dfrac{10}{x} + \dfrac{31}{x-1} + \dfrac{4x}{x-1}$ **b.** $\dfrac{10}{x} + \dfrac{31}{x-1} = \dfrac{4x}{x-1}$

58. a. $\dfrac{-4}{x^2-16} + \dfrac{x}{2x+8} - \dfrac{1}{4}$ **b.** $\dfrac{-4}{x^2-16} + \dfrac{x}{2x+8} = \dfrac{1}{4}$

59. a. $\dfrac{3x}{x^2-4} + \dfrac{5}{x+2} + \dfrac{2}{x-2}$ **b.** $\dfrac{3x}{x^2-4} + \dfrac{5}{x+2} = \dfrac{2}{x-2}$

60. a. $\dfrac{7}{5x} + \dfrac{2}{x-4} - \dfrac{3}{5x}$ **b.** $\dfrac{7}{5x} + \dfrac{2}{x-4} = \dfrac{3}{5x}$

61. a. $\dfrac{2}{x+9} - \dfrac{2}{x-9} + \dfrac{1}{2}$ **b.** $\dfrac{2}{x+9} - \dfrac{2}{x-9} = \dfrac{1}{2}$

A Solve applied problems related to fractions.

B Solve applied problems related to work.

C Solve applications involving distance, rate, and time.

13.5 Applications

The following strategy for solving word problems is valid for all word problems that involve algebraic equations.

To Solve a Word Problem Containing Rational Expressions

1. Read the problem carefully. Read it several times if necessary.
2. Decide what is asked for and assign a variable to the unknown quantity.
3. Draw a diagram or set up a chart whenever possible as a visual aid.
4. Form an equation that relates the information provided.
5. Solve the equation.
6. Check your solution with the wording of the problem to be sure it makes sense.

Objective A **Problems Related to Fractions**

We now introduce word problems involving rational expressions with problems relating the numerator and denominator of a fraction.

Example 1

Fractions

The denominator of a fraction is 8 more than the numerator. If both the numerator and denominator are increased by 3, the new fraction is equal to $\frac{1}{2}$. Find the original fraction.

Solution

Reread the problem to be sure that you understand all terminology used. Assign variables to the unknown quantities.

Let, n = original numerator

$n + 8$ = original denominator

$\dfrac{n}{n+8}$ = original fraction

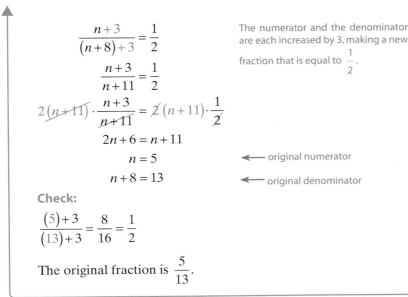

$$\frac{n+3}{(n+8)+3} = \frac{1}{2}$$

The numerator and the denominator are each increased by 3, making a new fraction that is equal to $\frac{1}{2}$.

$$\frac{n+3}{n+11} = \frac{1}{2}$$

$$2(n+11) \cdot \frac{n+3}{n+11} = 2(n+11) \cdot \frac{1}{2}$$

$$2n+6 = n+11$$

$$n = 5 \quad \longleftarrow \text{original numerator}$$

$$n+8 = 13 \quad \longleftarrow \text{original denominator}$$

Check:

$$\frac{(5)+3}{(13)+3} = \frac{8}{16} = \frac{1}{2}$$

The original fraction is $\frac{5}{13}$.

Now work margin exercise 1.

1. The denominator of a fraction is 3 more than the numerator. If both the numerator and denominator are increased by 4, the new fraction is equal to $\frac{3}{4}$. Find the original fraction.

Objective B **Problems Related to Work**

Problems involving work usually translate into equations involving rational expressions. The basic idea is to **represent what part of the work is done in one unit of time**. For example, if a man can dig a ditch in 3 hours, what part (of the ditch-digging job) can he do in one hour? The answer is $\frac{1}{3}$ of the work in one hour. If a fence was painted in 2 days, then $\frac{1}{2}$ of the work of painting the fence was done in 1 day. (These ideas assume a steady working pace.) In general, if the total work took x hours, then $\frac{1}{x}$ of the total work would be done in one hour.

Example 2

Work Problems

a. A carpenter can build a certain type of patio cover in 6 hours. His partner takes 8 hours to build the same cover. How long would it take them working together to build this type of patio cover?

Solution

Let x = number of hours to build the cover working together.

Person(s)	Time of Work (in Hours)	Part of Work Done in 1 Hour
Carpenter	6	$\frac{1}{6}$
Partner	8	$\frac{1}{8}$
Together	x	$\frac{1}{x}$

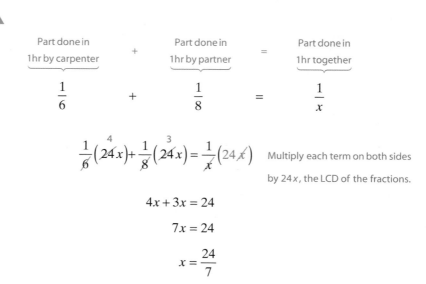

$$\frac{1}{\cancel{6}}\left(\overset{4}{\cancel{24}}x\right) + \frac{1}{\cancel{8}}\left(\overset{3}{\cancel{24}}x\right) = \frac{1}{\cancel{x}}\left(24\,\cancel{x}\right)$$

Multiply each term on both sides by 24x, the LCD of the fractions.

$$4x + 3x = 24$$

$$7x = 24$$

$$x = \frac{24}{7}$$

Together, they can build the patio cover in $\frac{24}{7}$ hours, or $3\frac{3}{7}$ hours.

(Note that this answer is reasonable because the time is less than either person would take working alone.)

b. A man can wax his car three times faster than his daughter can. Together they can do the job in 4 hours. How long would it take each of them working alone?

Solution

Let t = number of hours for the man alone to wax the car and $3t$ = number of hours for the daughter alone to wax the car.

Person(s)	Time of Work (in Hours)	Part of Work Done in 1 Hour
Man	t	$\frac{1}{t}$
Daughter	$3t$	$\frac{1}{3t}$
Together	4	$\frac{1}{4}$

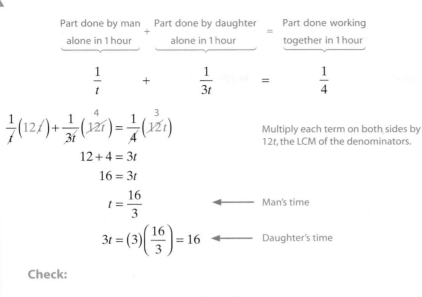

$$\underbrace{\frac{1}{t}}_{\substack{\text{Part done by man} \\ \text{alone in 1 hour}}} + \underbrace{\frac{1}{3t}}_{\substack{\text{Part done by daughter} \\ \text{alone in 1 hour}}} = \underbrace{\frac{1}{4}}_{\substack{\text{Part done working} \\ \text{together in 1 hour}}}$$

$$\frac{1}{t}(12t) + \frac{1}{3t}\overset{4}{(12t)} = \frac{1}{4}\overset{3}{(12t)}$$ Multiply each term on both sides by $12t$, the LCM of the denominators.

$$12 + 4 = 3t$$

$$16 = 3t$$

$$t = \frac{16}{3} \qquad \longleftarrow \text{Man's time}$$

$$3t = (3)\left(\frac{16}{3}\right) = 16 \qquad \longleftarrow \text{Daughter's time}$$

Check:

$$\text{Man's part in 1 hr} = \frac{1}{t} = \frac{1}{\frac{16}{3}} = \frac{3}{16}$$

$$\text{Daughter's part in 1 hr} = \frac{1}{3t} = \frac{1}{3 \cdot \frac{16}{3}} = \frac{1}{16}$$

$$\text{Man's part in 4 hr} = \frac{3}{16} \cdot 4 = \frac{3}{4}$$

$$\text{Daughter's part in 4 hr} = \frac{1}{16} \cdot 4 = \frac{1}{4}$$

$$\frac{3}{4} + \frac{1}{4} = 1 \text{ car waxed in 4 hours.}$$

Working alone, the man takes $\frac{16}{3}$ or $5\frac{1}{3}$ hours, and his daughter takes 16 hours.

c. A man was told that his new pool would fill through an inlet valve in 3 hours. He knew something was wrong when the pool took 8 hours to fill. He found he had left the drain valve open. How long would it take to drain the pool once it is completely filled and only the drain valve is open?

Solution

Let t = time to drain pool with only the drain valve open.

Note: We use the information gained when the pool was filled with both valves open. In that situation, the inlet and outlet valves worked against each other.

2. Solve the following work per unit of time problems:

a. An electrician can wire a house in 12 hours. Her assistant takes 18 hours to wire the same size house. How long would it take them to wire the house together?

b. A mother can carve a pumpkin twice as fast as her eight year old son. Together they can carve the pumpkin in 3 hours. How long would it take each of them working alone?

c. Janet's pool normally takes 4 hours to fill. Last time it took 10 hours to fill and she found that she had left the drain open. How long would it take to drain the pool once it is completely filled and only the drain valve is open?

Valves	Hours to Fill or Drain	Part Filled or Drained in 1 Hour
Inlet	3	$\dfrac{1}{3}$
Outlet	t	$\dfrac{1}{t}$
Together	8	$\dfrac{1}{8}$

Part filled by inlet in 1 hour	−	Part emptied by outlet in 1 hour	=	Part filled together in 1 hour
$\dfrac{1}{3}$	−	$\dfrac{1}{t}$	=	$\dfrac{1}{8}$

$$\frac{1}{3}\left(\overset{8}{24}t\right) - \frac{1}{t}\left(24t\right) = \frac{1}{8}\left(\overset{3}{24}t\right)$$

$$8t - 24 = 3t$$

$$5t = 24$$

$$t = \frac{24}{5}$$

The pool would drain in $\dfrac{24}{5}$ or $4\dfrac{4}{5}$ hours. (Note that this is more time than the inlet valve would take to fill the pool. If the outlet valve worked faster than the inlet valve, then the pool would never have filled in the first place.)

Now work margin exercise 2.

Objective C **Problems Related to Distance-Rate-Time: $d = rt$**

You may recall that the basic formula involving distance, rate, and time is $d = rt$. This relationship can also be stated in the forms $t = \dfrac{d}{r}$ and $r = \dfrac{d}{t}$.

If distance and rate are known or can be represented, then $t = \dfrac{d}{r}$ is the way to represent time. Similarly, if the distance and time are known or can be represented, then $r = \dfrac{d}{t}$ is the way to represent rate.

Example 3

Distance-Rate-Time

a. On Lake Itasca a man can row his boat 5 miles per hour. On the nearby Mississippi River it takes him the same time to row 5 miles downstream as it does to row 3 miles upstream. What is the speed of the river current in miles per hour?

Solution

Let c = the speed of the current.
Distance and rate are represented first in the table below. Then the time going downstream and coming back upstream is represented in terms of distance and rate. Since the rate is in miles per hour, the distance is in miles and the time is in hours.

	Distance d	Rate r	Time $t = \dfrac{d}{r}$
Downstream	5	$5 + c$	$\dfrac{5}{5+c}$
Upstream	3	$5 - c$	$\dfrac{3}{5-c}$

$$\frac{5}{5+c} = \frac{3}{5-c} \qquad \text{The times are equal.}$$

$$(5+c)(5-c) \cdot \frac{5}{5+c} = (5+c)(5-c) \cdot \frac{3}{5-c}$$

$$25 - 5c = 15 + 3c$$

$$-8c = -10$$

$$c = \frac{5}{4}$$

Check:

$$\text{Time downstream} = \frac{5}{5 + \dfrac{5}{4}} = \frac{5}{\dfrac{20}{4} + \dfrac{5}{4}} = \frac{5}{\dfrac{25}{4}} = 5 \cdot \frac{4}{25} = \frac{4}{5} \text{ hours}$$

$$\text{Time upstream} = \frac{3}{5 - \dfrac{5}{4}} = \frac{3}{\dfrac{20}{4} - \dfrac{5}{4}} = \frac{3}{\dfrac{15}{4}} = 3 \cdot \frac{4}{15} = \frac{4}{5} \text{ hours}$$

The times are equal. The rate of the river current is $\dfrac{5}{4}$ mph, or $1\dfrac{1}{4}$ mph.

3. Solve the following distance-rate-time problems:

a. On Lake Wespanee a man can paddle his kayak at a rate of 6 miles per hour. On the nearby Millitonka River it takes him the same time to paddle 6 miles downstream as it does to paddle 3 miles upstream. What is the speed of the current of the river?

b. If a commercial airplane travels twice as fast as a private aircraft, and the private aircraft takes $2\frac{1}{2}$ hours longer to travel 900 miles, what is the speed of each airplane?

b. If a passenger train travels three times as fast as a freight train, and the freight train takes 4 hours longer to travel 210 miles, what is the speed of each train?

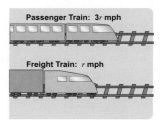

Passenger Train: $3r$ mph

Freight Train: r mph

Solution

Let r = rate of freight train in miles per hour
    $3r$ = rate of passenger train in miles per hour

	Distance d	Rate r	Time $t = \dfrac{d}{r}$
Freight	210	r	$\dfrac{210}{r}$
Passenger	210	$3r$	$\dfrac{210}{3r}$

Note: If the rate is faster, then the time is shorter. Thus the fraction $\dfrac{210}{3r}$ is smaller than the fraction $\dfrac{210}{r}$.

$$\frac{210}{r} - \frac{210}{3r} = 4 \qquad \text{The difference between their times is 4 hours.}$$

$$\frac{210}{r} - \frac{70}{r} = 4$$

$$\frac{210}{\cancel{r}} \cdot \cancel{r} - \frac{70}{\cancel{r}} \cdot \cancel{r} = 4 \cdot r$$

$$210 - 70 = 4r$$

$$140 = 4r$$

$$35 = r$$

$$105 = 3r$$

Check:

Time for freight train = $\dfrac{210}{35} = 6$ hours

Time for passenger train = $\dfrac{210}{105} = 2$ hours

$6 - 2 = 4$ hours difference in time

The freight train travels 35 mph, and the passenger train travels 105 mph.

Now work margin exercise 3.

Exercises 13.5

1. The sum of two numbers is 117 and they are in the ratio of 8 to 5. Find the two numbers.

2. If 4 is subtracted from a certain number and the difference is divided by 2, the result is 1 more than $\frac{1}{5}$ of the original number. Find the original number.

3. What number must be added to both the numerator and denominator of $\frac{16}{21}$ to make the resulting fraction equal to $\frac{5}{6}$?

4. Find the number that can be subtracted from both the numerator and denominator of the fraction $\frac{69}{102}$ so that the result is $\frac{5}{8}$.

5. The denominator of a fraction exceeds the numerator by 7. If the numerator is increased by 3 and the denominator is increased by 5, the resulting fraction is equal to $\frac{1}{2}$. Find the original fraction.

6. The numerator of a fraction exceeds the denominator by 5. If the numerator is decreased by 4 and the denominator is increased by 3, the resulting fraction is equal to $\frac{4}{5}$. Find the original fraction.

7. One number is $\frac{3}{4}$ of another number. Their sum is 63. Find the numbers.

8. The sum of two numbers is 24. If $\frac{2}{5}$ of the larger number is equal to $\frac{2}{3}$ of the smaller number, find the numbers.

9. One number exceeds another by 5. The sum of their reciprocals is equal to 19 divided by the product of the two numbers. Find the numbers.

10. One number is 3 less than another. The sum of their reciprocals is equal to 7 divided by the product of the two numbers. Find the numbers.

11. **Shirt Sales:** A manufacturer sold a group of shirts for $1026. One-fifth of the shirts were priced at $18 each and the remainder at $24 each. How many shirts were sold?

12. **Paying Bills:** Luis spent $\frac{1}{5}$ of his monthly salary for rent and $\frac{1}{6}$ of his monthly salary for his car payment. If $950 was left, what was his monthly salary?

13. **Painting:** Suppose that an artist expects that for every 9 special brushes she orders, 7 will be good and 2 will be defective. If she orders 54 brushes, how many will she expect to be defective?

14. **Travel by Car:** It takes Rosa, traveling at 30 mph, 30 minutes longer to go a certain distance than it takes Melody traveling at 50 mph. Find the distance traveled.

15. **Travel by Plane:** It takes a plane flying at 450 mph 25 minutes longer to travel a certain distance than it takes a second plane to fly the same distance at 500 mph. Find the distance.

16. **Landscaping:** Toni needs 4 hours to complete the yard work. Her husband, Sonny, needs 6 hours to do the work. How long will the job take if they work together?

17. **Manufacturing:** In 1921, automated wrapping machines were used to aid in the wrapping of Hershey Kisses in the Hershey chocolate factory. The machine could wrap the candies 100 times faster than a person could. Together they can wrap a crate full of Kisses in 5 minutes. How long would it take each of them working alone?

18. **Mass Mailings:** Ben's secretary can address the weekly newsletters in $4\frac{1}{2}$ hours. Charlie's secretary needs only 3 hours. How long will it take if they both work on the job?

19. **Shoveling Snow:** Working together, Greg and Cindy can clean the snow from the driveway in 20 minutes. It would have taken Cindy, working alone, 36 minutes. How long would it have taken Greg alone?

20. **Carpentry:** A carpenter and his partner can put up a patio cover in $3\frac{3}{7}$ hours. If the partner needs 8 hours to complete the patio alone, how long would it take the carpenter working alone?

21. **Travel:** Beth can travel 208 miles in the same length of time it takes Anna to travel 192 miles. If Beth's speed is 4 mph greater than Anna's, find both rates.

22. **Biking:** Kirk can bike 32 miles in the same amount of time that his twin brother Karl can bike 24 miles. If Kirk bikes 2 mph faster than Karl, how fast does each man bike?

23. **Plane Speeds:** A commercial airliner can travel 750 miles in the same amount of time that it takes a private plane to travel 300 miles. The speed of the airliner is 60 mph more than twice the speed of the private plane. Find the speed of each aircraft.

24. **Car Speeds:** Gabriela drives her car 350 miles and averages a certain speed. If the average speed had been 9 mph less, she could have traveled only 300 miles in the same length of time. What was her average speed?

25. **Boating:** A family travels 18 miles down river and returns. It takes 8 hours to make the round trip. Their rate in still water is twice the rate of the river's current. How long will the return trip take?

26. Boat Speed: Cruise ships travel 5 times faster than sailboats (in optimal wind conditions). If it takes 16 hours longer for the sailboat to travel 100 miles from Charleston, SC to Savannah, GA, what is the speed of each boat?

27. Wind Speed: An airplane can fly 650 mph in still air. If it can travel 2800 miles with the wind in the same time it can travel 2400 miles against the wind, find the wind speed. (**Note**: A tailwind increases the speed of the plane and a headwind decreases the speed of the plane.)

28. Wind Speed: A one-engine plane can fly 120 mph in still air. If it can fly 490 miles with a tailwind in the same time that it can fly 350 miles against a headwind, what is the speed of the wind?

29. Filling a Pool: Using a small inlet pipe it takes 9 hours to fill a pool. Using a large inlet pipe it only takes 3 hours. If both are used simultaneously, how long will it take to fill the pool?

30. Filling a Pool: An inlet pipe on a swimming pool can be used to fill the pool in 36 hours. The drain pipe can be used to empty the pool in 40 hours. If the pool is $\frac{2}{3}$ filled using the inlet pipe and then the drain pipe is accidentally opened, how long from that time will it take to fill the pool?

31. Clearing Land: A contractor hires two bulldozers to clear the trees from a 20-acre tract of land. One works twice as fast as the other. It takes them 3 days to clear the tract working together. How long would it take each of them alone?

32. Store Maintenance: John, Ralph, and Denny, working together, can clean their bait and tackle store in 6 hours. Working alone, Ralph takes twice as long to clean the store as does John. Denny needs three times as long as does John. How long would it take each man working alone?

33. Boating: Francois rode his jet ski 36 miles downstream and then 36 miles back. The round trip took $5\frac{1}{4}$ hours. Find the speed of the jet ski in still water and the speed of the current if the speed of the current is $\frac{1}{7}$ the speed of the jet ski.

34. Boating: Momence, IL is 12 miles upstream on the same side of the river from Kankakee, IL on the Kankakee River. A motorboat that can travel 8 mph in still water leaves Momence and travels downstream toward Kankakee. At the same time, another boat that can travel 10 mph leaves Kankakee and travels upstream toward Momence. Each boat completes the trip in the same amount of time. Find the rate of the current.

35. Skiing: Samantha rides the ski lift to the top of Blue Mountain, a distance of $1\frac{3}{4}$ kilometers (a little more than 1 mile). She then skis directly down the slope. If she skis five times as fast as the lift travels and the total trip takes 45 minutes, find the rate at which she skis.

Writing & Thinking

36. If n is any integer, then $2n$ is an even integer and $2n + 1$ is an odd integer. Use these ideas to solve the following problems.

  a. Find two consecutive odd integers such that the sum of their reciprocals is $\frac{12}{35}$.

  b. Find two consecutive even integers such that the sum of the first and the reciprocal of the second is $\frac{9}{4}$.

13.6 Variation

Objective A **Direct Variation**

Suppose that you ride your bicycle at a steady rate of 15 miles per hour (not quite as fast as Lance Armstrong, but you are enjoying yourself). If you ride for 1 hour, the distance you travel would be 15 miles. If you ride for two hours, the distance you travel would be 30 miles. This relationship can be written in the form of the

formula $d = 15t$ (or $\dfrac{d}{t} = 15$) where d is the distance traveled and t is the time in hours. We say that distance and time **vary directly** (or are in **direct variation** or are **directly proportional**). The term proportional implies that the ratio is constant. In this example, 15 is the constant and is called the **constant of variation**. When two variables vary directly, an **increase in the value of one variable indicates an increase in the other**, and the ratio of the two quantities is constant.

> ### Direct Variation
>
> A variable quantity y **varies directly as** (or is **directly proportional to**) a variable x if there is a constant k such that
>
> $$\frac{y}{x} = k \text{ or } y = kx.$$
>
> The constant k is called the **constant of variation**.

Example 1

Direct Variation

If y varies directly as x, and $y = 6$ when $x = 2$, find y if $x = 6$.

Solution

$y = kx$ — general formula for direct variation
$6 = 2k$ — Substitute the known values and solve for k.
$3 = k$ — Use this value for k in the general formula.

So $y = 3x$. Thus, if $x = 6$ then, $y = 3 \cdot 6 = 18$.

Example 2

Direct Variation

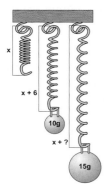

A spring will stretch a greater distance as more weight is placed on the end of the spring. The distance (*d*) the spring stretches varies directly as the weight (*w*) placed at the end of the spring. This is a property of springs studied in physics and is known as Hooke's Law. If a weight of 10 g stretches a certain spring 6 cm, how far will the spring stretch with a weight of 15 g? (**Note:** We assume that the weight is not so great as to break the spring.)

1. If *y* varies directly as *x*, and *y* = 10 when *x* = 5, find *y* if *x* = 2.

2. If a 25 g weight is placed on the same spring, mentioned in Example 2, calculate how far the spring will stretch.

Solution

Because the two variables are directly proportional, the relationship can be indicated with the formula

$$d = k \cdot w$$ where *d* = distance spring stretches in cm,

w = weight in g, and

k = constant of variation.

First, substitute the given information to find the value for *k*. (The value of *k* will depend on the particular spring. Springs made of different material or which are wound more tightly will have different values for *k*.)

$$d = k \cdot w$$ Substitute the known values into the formula.

$$6 = k \cdot 10$$ Use this value for *k* in the general formula.

$$\frac{3}{5} = k$$ The constant of variation is $\frac{3}{5}$ (or 0.6).

So, $$d = \frac{3}{5}w.$$

If *w* = 15, we have $d = \frac{3}{5} \cdot 15 = 9$.

The spring will stretch 9 cm if a weight of 15 g is placed at its end.

Now work margin exercises 1 and 2.

Listed here are several formulas involving direct variation.

$$d = \frac{3}{5}w$$ Hooke's Law for a spring where $k = \frac{3}{5}$.

$$C = 2\pi r$$ The circumference of a circle is directly proportional to the radius.

$$A = \pi r^2$$ The area of a circle varies directly as the radius squared.

$$P = 625d$$ Water pressure is proportional to the depth of the water.

Inverse Variation

When two variables vary in such a way that their product is constant, we say that the two variables **vary inversely** or are **inversely proportional**. For example, if a gas is placed in a container as in an automobile engine and pressure is increased on the gas, then the product of the pressure and the volume of gas will remain constant. That is, pressure and volume are related by the formula $V \cdot P = k$ or, $V = \dfrac{k}{P}$.

Note: that if a product of two variables is to remain constant, then **an increase in the value of one variable must be accompanied by a decrease in the other**. Or, in the case of a fraction with a constant numerator, if the denominator increases in value, then the fraction decreases in value. For the gas in an engine, an increase in pressure indicates a decrease in the volume of gas.

Inverse Variation

A variable quantity y **varies inversely as** (or is **inversely proportional to**) a variable x if there is a constant k such that

$$x \cdot y = k \text{ or } y = \frac{k}{x}.$$

The constant k is called the **constant of variation**.

Example 3

Inverse Variation

If y varies inversely as the cube of x, and $y = -1$ when $x = 3$, find y if $x = -3$.

Solution

$$y = \frac{k}{x^3} \qquad \text{general formula for inverse variation}$$

$$-1 = \frac{k}{3^3} \qquad \text{Substitute the known values and solve for } k.$$

$$-1 = \frac{k}{27}$$

$$-27 = k \qquad \text{Use this value for } k \text{ in the general formula.}$$

So $y = \dfrac{-27}{x^3}$. Thus, if $x = -3$, then $y = \dfrac{-27}{(-3)^3} = \dfrac{-27}{-27} = 1$.

Example 4

Inverse Variation

The gravitational force (F) between an object and the Earth is inversely proportional to the square of the distance (d) from the object to the center of the Earth. Hence we have the formula

$$F \cdot d^2 = k \text{ or } F = \frac{k}{d^2},$$

where F = force, d = distance, and k = constant variation

(As the distance of an object from the Earth becomes larger, the gravitational force exerted by the Earth on the object becomes smaller.)

If an astronaut weighs 200 pounds on the surface of the Earth, what will he weigh 100 miles above the Earth? Assume that the radius of the Earth is 4000 miles.

Solution

We know when

$F = 200 = 2 \times 10^2$ pounds
$d = 4000 = 4 \times 10^3$ miles.

Use scientific notation to make values simpler to work with in the calculations.

$$2 \times 10^2 = \frac{k}{\left(4 \times 10^3\right)^2}$$

Substitute and solve for k.

$$k = 2 \times 10^2 \times 16 \times 10^6 = 32 \times 10^8 = 3.2 \times 10^9$$

So,

$$F = \frac{3.2 \times 10^9}{d^2}.$$

When the astronaut is 100 miles above the Earth, $d = 4100 = 4.1 \times 10^3$ miles. Then,

$$F = \frac{3.2 \times 10^9}{16.81 \times 10^6} \approx 0.190 \times 10^3 = 190 \text{ pounds}.$$

That is, 100 miles above the Earth the astronaut will weigh about 190 pounds.

Now work margin exercises 3 and 4.

3. If y varies inversely as x raised to the fifth power, and $y = -2$ when $x = 2$, find y when $x = -2$.

4. Assuming the same astronaut is on his way to the moon and is 300 miles from the surface of the Earth and calculate his weight.

Combined Variation

If a variable varies either directly or inversely as more than one other variable, the variation is said to be **combined variation**. If the combined variation is all direct variation (the variables are multiplied), then it is called **joint variation**. For example, the volume of a cylinder varies jointly as its height and the square of its radius.

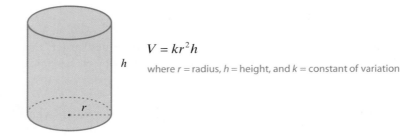

$$V = kr^2h$$

where r = radius, h = height, and k = constant of variation

Using this information, what is the value of k, the constant of variation, if a cylinder has the approximate measurements $V = 198$ cubic feet, $r = 3$ feet, and $h = 7$ feet?

$$V = k \cdot r^2 \cdot h$$ V **varies jointly** as r^2 and h.

$$198 = k \cdot 3^2 \cdot 7$$ Substitute the known values.

$$\frac{198}{9 \cdot 7} = k$$ We know from experience that $k = \pi$. Since the measurements are only approximate,

$$k = \frac{22}{7} \approx 3.14$$ the estimate for k is only approximate.

Substituting the constant of variation, the formula is $V = \pi r^2 h$.

5. If z varies jointly as x^3 and y^2, and $z = 108$ when $x = 2$ and $y = 3$, what is z when $x = 3$ and $y = 4$?

Example 5

Joint Variation

If z varies jointly as x^2 and y, and $z = 18$ when $x = 2$ and $y = 4$, what is z when $x = 4$ and $y = 3$?

Solution

$$z = k \cdot x^2 \cdot y$$

$$18 = k \cdot 2^2 \cdot 4$$ Substitute the known values and solve for k.

$$18 = k \cdot 16$$

$$\frac{9}{8} = k$$

So, $z = \frac{9}{8} x^2 y.$ Substitute $\frac{9}{8}$ for k in the general formula.

If, $x = 4$ and $y = 3$ then,

$$z = \frac{9}{8} \cdot 4^2 \cdot 3$$

$$z = 54.$$

Now work margin exercise 5.

Example 6

More Variation

a. The distance an object falls varies directly as the square of the time it falls (until it hits the ground and assuming little or no air resistance). If an object fell 64 feet in two seconds, how far would it have fallen by the end of three seconds?

Solution

$d = k \cdot t^2$ where d = distance, t = time (in seconds), and k = constant of variation

$64 = k \cdot 2^2$ Substitute the known values and solve for k.
$16 = k$

So, $d = 16t^2$.   Substitute 16 for k in the general formula.

If, $t = 3$ then,

$$d = 16 \cdot 3^2$$
$$d = 144.$$

The object would have fallen 144 feet in 3 seconds.

b. The volume of a gas in a container varies inversely as the pressure on the gas. If a gas has a volume of 200 cubic inches under pressure of 5 pounds per square inch, what will be its volume if the pressure is increased to 8 pounds per square inch?

Solution

$V = \dfrac{k}{P}$ where V = volume, P = pressure, and k = constant variation

$200 = \dfrac{k}{5}$ Substitute the known values and solve for k.

$k = 1000$

So, $V = \dfrac{1000}{P}$ Substitute 1000 for k in the general formula.

$V = \dfrac{1000}{8} = 125.$

The volume will be 125 cubic inches.

6. Solve the following problems on variation.

a. The distance an object falls varies directly as the square of the time it falls. If an object fell 64 feet in two seconds, how many feet will it fall in 5 seconds?

b. A volume of a gas varies inversely as the pressure on a gas. If a gas has a volume of 300 cubic inches under pressure of 8 pounds per square inch, what will be its volume if the pressure increases to 12 pounds per square inch?

c. The safe load a bridge support beam can carry varies jointly as the width w and the square of the depth d and inversely as the length l. A 5 inch wide steel beam that is 12 inches deep and 15 feet long can support a load of 9 tons (18000 pounds). What is the safe load of a beam of the same material that is 4 inches wide, 8 inches deep, and 12 feet long?

c. The safe load L of a wooden beam supported at both ends varies jointly as the width w and the square of the depth d and inversely as the length l. A 3 in. wide by 10 in. deep beam that is 8 ft long supports a load of 9600 lb safely. What is the safe load of a beam of the same material that is 4 in. wide, 9 in. deep, and 12 ft long?

Solution

$$L = \frac{k \cdot w \cdot d^2}{l}$$

where L = safe load, w = width, d = depth, and l = length

$$9600 = \frac{k \cdot 3 \cdot 10^2}{8}$$

$$9600 = \frac{k \cdot 300}{8}$$

$$k = \frac{9600 \cdot 8}{300}$$

Substitute the known values and solve for k.

$$k = 256$$

So, $$L = \frac{256 \cdot w \cdot d^2}{l}$$

Substitute 256 for k in the general formula.

$$L = \frac{256 \cdot 4 \cdot 9^2}{12} = \frac{256 \cdot 4 \cdot 81}{12} = 6912.$$

The safe load will be 6912 lb.

Now work margin exercise 6.

Practice Problems

1. The length that a hanging spring stretches varies directly as the weight placed on the end of the spring. If a weight of 5 mg stretches a certain spring 3 cm, how far will the spring stretch with a weight of 6 mg?

2. The volume of propane in a container varies inversely as the pressure on the gas. If the propane has a volume of 200 in.^{3} under a pressure of 4 lb per in.^{2}, what will be its volume if the pressure is increased to 5 lb per in.^{2}?

Practice Problem Answers

1. $\frac{18}{5}$ cm

2. 160 in.^{3}

Exercises 13.6

1. If y varies directly as x, and $y = 3$ when $x = 9$, find y if $x = 7$.

2. If y is directly proportional to x^2, and $y = 3$ when $x = 2$, what is y when $x = 8$?

3. If y varies inversely as x, and $y = 5$ when $x = 8$, find y if $x = 20$.

4. If y is inversely proportional to x, and $y = 5$ when $x = 4$, what is y when $x = 2$?

5. If y varies inversely as x^2, and $y = -8$ when $x = 2$, find y if $x = 3$.

6. If y is inversely proportional to x^3, and $y = 40$ when $x = \dfrac{1}{2}$, what is y when $x = \dfrac{1}{3}$?

7. If y is directly proportional to the square root of x, and $y = 6$ when $x = \dfrac{1}{4}$, what is y when $x = 9$?

8. If y is directly proportional to the square of x, and $y = 80$ when $x = 4$, what is y when $x = 6$?

9. z varies jointly as x and y, and $z = 60$ when $x = 2$ and $y = 3$. Find z if $x = 3$ and $y = 4$.

10. z varies jointly as x and y, and $z = -6$ when $x = 5$ and $y = 8$. Find z if $x = 12$ and $y = 15$.

11. z varies jointly as x and y^2, and $z = 63$ when $x = 5$ and $y = 3$. Find z if $x = \dfrac{10}{3}$ and $y = 2$.

12. z varies jointly as x^2 and y, and $z = 20$ when $x = 2$ and $y = 3$. Find z if $x = 4$ and $y = \dfrac{7}{10}$.

13. z varies directly as x and inversely as y^2. If $z = 5$ when $x = 1$ and $y = 2$, find z if $x = 2$ and $y = 1$.

14. z varies directly as x^3 and inversely as y^2. If $z = 24$ when $x = 2$ and $y = 2$, find z if $x = 3$ and $y = 2$.

15. z varies directly as $\sqrt{x}$ and inversely as y. If $z = 24$ when $x = 4$ and $y = 3$, find z if $x = 9$ and $y = 2$.

16. z varies directly as x^2 and inversely as $\sqrt{y}$. If $z = 108$ when $x = 6$ and $y = 4$, find z if $x = 4$ and $y = 9$.

17. s varies directly as the sum of r and t and inversely as w. If $s = 24$ when $r = 7$ and $t = 8$ and $w = 9$, find s if $r = 9$ and $t = 3$ and $w = 18$.

18. s varies directly as r and inversely as the difference of t and u. If $s = 36$ when $r = 12$ and $t = 9$ and $u = 6$, find s if $r = 18$ and $t = 11$ and $u = 8$.

19. L varies jointly as m and n and inversely as p. If $L = 6$ when $m = 7$ and $n = 8$ and $p = 12$, find L if $m = 15$ and $n = 14$ and $p = 10$.

20. W varies jointly as x and y and inversely as z. If $W = 10$ when $x = 6$ and $y = 5$ and $z = 2$, find W if $x = 12$ and $y = 6$ and $z = 3$.

21. **Free Falling Object:** The distance a free falling object falls is directly proportional to the square of the time it falls (before it hits the ground). If an object fell 256 feet in 4 seconds, how far will it have fallen by the end of 5 seconds?

22. **Stretching a Spring:** The length a hanging spring stretches varies directly with the weight placed on the end. If a spring stretches 5 in. with a weight of 10 lb, how far will the spring stretch if the weight is increased to 12 lb?

23. Gas Prices: The total price (P) of gasoline purchased varies directly with the number of gallons purchased. If 10 gallons are purchased for $39.80, what will be the price of 15 gallons?

24. Economics: Research shows that the value of gold and the value of the dollar are inversely proportional. In 2008, gold cost $900 per ounce and the dollar had a rating of 75 on the US dollar index. In 2010, the cost of gold was $1100 per ounce. What is the current rating of the dollar? (Round your answer to the nearest hundredth.)

25. Pizza: The circumference of a circle varies directly as the diameter. A circular pizza pie with a diameter of 1 foot has a circumference of 3.14 feet. What will be the circumference of a pizza pie with a diameter of 1.5 feet?

26. Pizza: The area of a circle varies directly as the square of its radius. A circular pizza pie with a radius of 6 in. has an area of 113.04 in.^{2} What will be the area of a pizza pie with a radius of 9 in.?

27. Triangles: Several triangles have the same area. In this set of triangles the height and base are inversely proportional. In one such triangle the height is 5 m and the base is 12 m. Find the height of the triangle in this set with a base of 10 m.

28. Weight in Space: If an astronaut weighs 250 pounds on the surface of the earth, what will the astronaut weigh 150 miles above the earth? Assume that the radius of the earth is 4000 miles, and round to the nearest tenth. (See Example 4.)

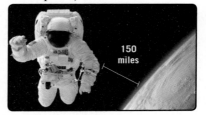

29. Elongation of a Wire: The elongation (E) in a wire, when a mass (m) is hung at its free end, varies jointly as the mass and the length (l) of the wire and inversely as the cross-sectional area (A) of the wire. The elongation is 0.0055 cm when a mass of 120 g is attached to a wire 330 cm long, with a cross-sectional area of 0.4 cm^2. Find the elongation if a mass of 160 g is attached to the same wire. Round your answer to the nearest ten-thousandth.

30. Elongation of a Wire: When a mass of 240 oz is suspended by a wire 49 in. long whose cross-sectional area is 0.035 in.^{2}, the elongation of the wire is 0.016 in. Find the elongation if the same mass is suspended by a 28 in. wire of the same material with a cross-sectional area of 0.04 in.^{2} (See Exercise 29.)

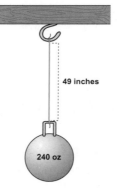

49 inches

240 oz

31. Safe Load of a Wooden Beam: The safe load (*L*) of a wooden beam supported at both ends varies jointly as the width (*w*) and the square of the depth (*d*) and inversely as the length (*l*). A 4 in. wide 6 in. deep beam 12 ft long supports a load of 4800 lb safely. What is the safe load of a beam of the same material that is 6 in. wide 10 in. deep and 15 ft long? (See Example 6c.)

32. Safe Load of a Wooden Beam: A wooden beam 2 in. wide, 8 in. deep, and 14 ft long holds up to 2400 lb. What load would a beam 3 in. wide 6 in. deep and 15 ft long, of the same material, support?

33. Gravitational Force: The gravitational force of attraction, *F*, between two bodies varies directly as the product of their masses, m_1 and m_2, and inversely as the square of the distance, *d*, between them. The gravitational force between a 5-kg mass and a 2-kg mass 1 m apart is 1.5×10^{-10} N. Find the force between a 24-kg mass and a 9-kg mass that are 6 m apart. (N represents a unit of force called a newton.)

34. Gravitational Force: In Exercise 33, what is the force if the distance between the 24-kg mass and the 9-kg mass is cut in half?

Solve the following problems.

Lifting Force

The lifting force (or lift), *L*, in pounds exerted by the atmosphere on the wings of an airplane is related to the area, *A*, of the wings in square feet and the speed (or velocity), *v*, of the plane in miles per hour by the formula

$$L = kAv^2, \text{ where } k \text{ is the constant of variation.}$$

35. If the lift is 9600 lb for a wing area of 120 ft^2 and a speed of 80 mph, find the lift of the same airplane at a speed of 100 mph.

36. The lift for a wing of area 280 ft^2 is 34,300 lb when the plane is traveling at 210 mph. What is the lift if the speed is decreased to 180 mph?

37. The lift for a wing with an area of 144 ft^2 is 10,000 lb when the plane is traveling at 150 mph. What is the lift if the speed is decreased to 120 mph?

38. A plane traveling 140 mph with wing area 195 ft^2 has 12,500 lb of lift exerted on the wings. Find the lift for the same plane traveling at 168 mph.

Solve the following problems.

Pressure

Boyle's Law states that if the temperature of a gas sample remains the same, the pressure, *P*, of the gas is related to the volume *V* by the formula

$$P = \frac{k}{V}, \text{ where } k \text{ is the constant of proportionality.}$$

39. A pressure of 1600 lb per ft^2 is exerted by 2 ft^3 of air in a cylinder. If a piston is pushed into the cylinder until the pressure is 1800 lb per ft^2, what will be the volume of the air? Round to the nearest tenth.

40. The volume of gas in a container is 300 cm^3 when the pressure on the gas is 20 g per cm^2. What will be the volume if the pressure is increased to 30 g per cm^2?

41. The pressure in a canister of gas is 1360 g per in.^{2} when the volume of gas is 5 in.^{3} If the volume is reduced to 4 in.^{3}, what is the pressure?

42. A scuba diver is using a diving tank that can hold 6 liters of air. If the tank has a pressure rating of 220 bar when full, what is the pressure rating when the volume of gas is 4 liters?

Solve the following problems.

Electricity

The resistance, R (in ohms), in a wire is given by the formula

$$R = \frac{kL}{d^2},$$

where k is the constant of variation, L is the length of the wire and d is the diameter.

43. The resistance of a wire 500 ft long with a diameter of 0.01 in. is 20 ohms. What is the resistance of a wire 1500 ft long with a diameter of 0.02 in.?

44. The resistance is 2.6 ohms when the diameter of a wire is 0.02 in. and the wire is 10 ft long. Find the resistance of the same type of wire with a diameter of 0.01 in. and a length of 5 ft.

45. Tristan's car stereo uses a 5 ft audio wire with diameter 0.025 in. and resistance of 1.6 ohms. What is the resistance of 8 ft of the same type of audio wire?

46. Nicole purchased a spool of wire with diameter 0.01 in. for the speakers in her home audio system. If the resistance of 15 ft of this wire is 6 ohms, what is the resistance of 25 ft of the wire?

Solve the following problems.

Levers

If a lever is balanced with weight on opposite sides of its balance point, then the following proportion exists:

$$\frac{W_1}{W_2} = \frac{L_2}{L_1} \quad \text{or} \quad W_1 L_1 = W_2 L_2$$

where $L_1 + L_2 = L$, the total length of the lever.

47. How much weight can be raised at one end of a bar 8 ft long by the downward force of 60 lb when the balance point is $\frac{1}{2}$ ft from the unknown weight?

48. Where should the balance point of a bar 12 ft long be located if a 120 lb force is to raise a load weighing 960 lb?

49. Find the location of the balance point of a 25 ft board that can raise a 300 lb package with a downward force of 75 lb.

50. How much weight can be raised on one end of a 17 meter board by 90 kilograms, if the balance point is 5 meters from the unknown weight?

51. Explain, in your own words, the meaning of the terms

    a. direct variation,

    b. inverse variation,

    c. joint variation, and

    d. combined variation.

Discuss an example of each type of variation that you have observed in your daily life.

Section 13.1: Multiplication and Division with Rational Expressions

Rational Expression   page 1008

A **rational expression** is an algebraic expression of the form $\frac{P}{Q}$ where P and Q are polynomials and $Q \neq 0$.

Restrictions on a Variable   page 1008

Values of the variable that make a rational expression undefined are called **restrictions** on the variable.

Arithmetic Rules for Rational Numbers (or Fractions)   page 1010

The Fundamental Principle of Rational Expressions   page 1011

If $\frac{P}{Q}$ is a rational expression and P, Q and K are polynomials where $Q, K \neq 0$, then $\frac{P}{Q} = \frac{P \cdot K}{Q \cdot K}$.

Opposites in Rational Expressions   page 1012

For a polynomial P, $\frac{-P}{P} = -1$ where $P \neq 0$.

In particular, $\frac{a-x}{x-a} = \frac{-(x-a)}{x-a} = -1$ where $x \neq a$.

Multiplication with Rational Expressions   page 1013

$\frac{P}{Q} \cdot \frac{R}{S} = \frac{P \cdot R}{Q \cdot S}$ where $P, Q, R,$ and S are polynomials with $Q, S \neq 0$.

Division with Rational Expressions   page 1015

$\frac{P}{Q} \div \frac{R}{S} = \frac{P}{Q} \cdot \frac{S}{R}$ where $P, Q, R,$ and S are polynomials with $Q, R, S \neq 0$.

Section 13.2: Addition and Subtraction with Rational Expressions

Addition and Subtraction with Rational Expressions   page 1023

$\frac{P}{Q} + \frac{R}{Q} = \frac{P+R}{Q}$ and $\frac{P}{Q} - \frac{R}{Q} = \frac{P-R}{Q}$ where $Q \neq 0$.

To Find the LCM for a Set of Polynomials   page 1024

1. Completely factor each polynomial.
2. Form the product of all factors that appear, using each factor the most number of times it appears in any one polynomial.

Procedure for Adding (or Subtracting) Rational Expressions   page 1024
with Different Denominators
    1. Find the LCD (the LCM of the denominators).
    2. Rewrite each fraction in an equivalent form with the LCD as the denominator.
    3. Add (or subtract) the numerators and keep the common denominator.
    4. Reduce if possible.

Placement of Negative Signs   page 1027
$$-\frac{P}{Q} = \frac{P}{-Q} = \frac{-P}{Q} \quad \text{where } Q \neq 0$$

Section 13.3: Complex Fractions

Complex Fractions   page 1036
    A **complex fraction** is a fraction in which the numerator and/or denominator are themselves fractions or the sum or difference of fractions.

To Simplify Complex Fractions   page 1036
    First Method
    1. Simplify the numerator so that it is a single rational expression.
    2. Simplify the denominator so that it is a single rational expression.
    3. Divide the numerator by the denominator and reduce to lowest terms.

    Second Method   page 1038
    1. Find the LCM of all the denominators in the numerator and denominator of the complex fraction.
    2. Multiply both the numerator and denominator of the complex fraction by this LCM.
    3. Simplify both the numerator and denominator and reduce to lowest terms.

Simplifying Complex Algebraic Expressions   page 1040
    A **complex algebraic expression** is an expression that involves rational expressions and more than one operation.

To Solve an Equation Containing Rational Expressions   page 1044
1. Find the LCD of the fractions.
2. Multiply both sides of the equation by this LCD and simplify.
3. Solve the resulting equation.
4. Check each solution in the original equation. (Remember that no denominator can equal 0.)

Checking for Extraneous Solutions   page 1044

Similar Triangles   page 1048
  Similar triangles are triangles that meet the following two conditions:
1. The measures of the corresponding angles are equal.
2. The lengths of the corresponding sides are proportional.

Strategy for Solving Word Problems   page 1054
1. Read the problem carefully. Read it several times if necessary.
2. Decide what is asked for and assign a variable to the unknown quantity.
3. Draw a diagram or set up a chart whenever possible as a visual aid.
4. Form an equation that relates the information provided.
5. Solve the equation.
6. Check your solution with the wording of the problem to be sure it makes sense.

Applications   page 1054
  Number problems related to fractions
  Work problems
  Distance-Rate-Time problems

Direct Variation   page 1064
  A variable quantity y **varies directly** as (or is **directly proportional to**) a variable x if there is a constant k such that

$$\frac{y}{x} = k \text{ or } y = kx.$$

  The constant k is called the **constant of variation**.

Inverse Variation   page 1066

A variable quantity y **varies inversely as** (or is **inversely proportional to**) a variable x if there is a constant k such that

$$x \cdot y = k \text{ or } y = \frac{k}{x}.$$

The constant k is called the **constant of variation**.

Combined Variation   page 1068

If a variable varies either directly or inversely with more than one other variable, the variation is said to be **combined variation**.

Joint Variation   page 1068

If the combined variation is all direct variation (the variables are multiplied), then it is called **joint variation**.

Chapter 13: Review

Section 13.1: Multiplication and Division with Rational Expressions

Reduce to lowest terms. State any restrictions on the variable(s).

1. $\dfrac{18xy^3}{96x^2y^3}$

2. $\dfrac{4x-12}{9-3x}$

3. $\dfrac{3x^2-75}{x^2-10x+25}$

4. $\dfrac{2x^2+x-3}{2x^2+13x+15}$

Perform the indicated operations and reduce to lowest terms. Assume that no denominator has a value of 0.

5. $\dfrac{2x-6}{3x-9}\cdot\dfrac{2x}{2x-4}$

6. $\dfrac{25x^2-9}{15x^2-9x}\cdot\dfrac{6x^2}{5x-3}$

7. $\dfrac{24x^2y}{9xy^6}\div\dfrac{4x^4y}{3x^2y^3}$

8. $\dfrac{x-1}{7x+7}\div\dfrac{3x-3}{x^2+x}$

9. $\dfrac{x^2-8x+15}{x^2+5x-14}\div\dfrac{x^2-9}{x^2-49}$

10. $\dfrac{x-3}{x^2-3x-4}\div\dfrac{x^2-x-2}{3x+3}$

11. $\dfrac{x^2+2x-8}{4x^3}\cdot\dfrac{5x^2}{2x^2-5x+2}$

12. $\dfrac{x-1}{x^2+7x+6}\cdot\dfrac{x^2-4x}{x+4}\cdot\dfrac{x^2-16}{x^2-5x+4}$

Section 13.2: Addition and Subtraction with Rational Expressions

Perform the indicated operations and reduce if possible. Assume that no denominator has a value of 0.

13. $\dfrac{6x}{x+2}+\dfrac{12}{x+2}$

14. $\dfrac{8x}{x+6}+\dfrac{48}{x+6}$

15. $\dfrac{5}{x-2}+\dfrac{x}{x^2-4}$

16. $\dfrac{x+1}{x^2-10x-11}+\dfrac{x}{x-11}$

17. $\dfrac{x^2}{x^2+2x+1}-\dfrac{2x-1}{x^2-2x+1}$

18. $\dfrac{x}{3x^2+4x+1}+\dfrac{2x+1}{2x^2+5x+3}$

19. $\dfrac{6}{2x-3}-\dfrac{5}{3-2x}$

20. $\dfrac{6x+24}{x^2+8x+16}+\dfrac{3x-12}{x^2-16}$

21. $\dfrac{x+2}{x^2+x+1}-\dfrac{x-2}{x^3-1}$

22. $\dfrac{3x+2}{x^2+2x-3}-\dfrac{x}{3x^2-2x-1}$

Section 13.3: Complex Fractions

Simplify the following complex fractions.

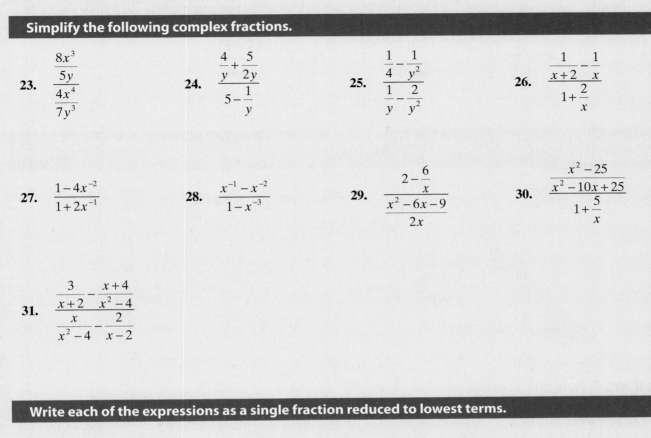

23. $\dfrac{\dfrac{8x^3}{5y}}{\dfrac{4x^4}{7y^3}}$

24. $\dfrac{\dfrac{4}{y}+\dfrac{5}{2y}}{5-\dfrac{1}{y}}$

25. $\dfrac{\dfrac{1}{4}-\dfrac{1}{y^2}}{\dfrac{1}{y}-\dfrac{2}{y^2}}$

26. $\dfrac{\dfrac{1}{x+2}-\dfrac{1}{x}}{1+\dfrac{2}{x}}$

27. $\dfrac{1-4x^{-2}}{1+2x^{-1}}$

28. $\dfrac{x^{-1}-x^{-2}}{1-x^{-3}}$

29. $\dfrac{2-\dfrac{6}{x}}{\dfrac{x^2-6x-9}{2x}}$

30. $\dfrac{\dfrac{x^2-25}{x^2-10x+25}}{1+\dfrac{5}{x}}$

31. $\dfrac{\dfrac{3}{x+2}-\dfrac{x+4}{x^2-4}}{\dfrac{x}{x^2-4}-\dfrac{2}{x-2}}$

Write each of the expressions as a single fraction reduced to lowest terms.

32. $\dfrac{1}{x+2}-\dfrac{5}{2x}\cdot\dfrac{4x}{x+2}$

33. $\left(\dfrac{4}{y}-\dfrac{2}{3y}\right)\div\dfrac{3y+2}{6y}$

34. $\left(\dfrac{4}{y}+\dfrac{2}{y+3}\right)\div\dfrac{y^2-9}{2y}$

Section 13.4: Solving Equations with Rational Expressions

State any restrictions on x, and then solve the equations.

35. $\dfrac{4}{x+3}=\dfrac{1}{x-3}$

36. $\dfrac{x-4}{3x}=\dfrac{2}{x-3}$

37. $\dfrac{2}{5x}+\dfrac{3}{8}=\dfrac{3}{10x}-\dfrac{1}{4}$

38. $\dfrac{6}{7x-4}=\dfrac{-5}{3x+2}$

39. $\dfrac{5}{x}+\dfrac{2}{x-1}=\dfrac{2x}{x-1}$

40. $\dfrac{9}{x^2-9}+\dfrac{x}{3x+9}=\dfrac{1}{3}$

41. $\dfrac{x}{x-4}-\dfrac{6x}{x^2-x-12}=\dfrac{7x}{x+3}$

42. $\dfrac{2x+7}{15}-\dfrac{x+2}{x+6}=\dfrac{2x-6}{5}$

43. $\dfrac{y - y_1}{x - x_1} = m$; solve for x_1

44. $a_n = \dfrac{a_1\left(1 - r^n\right)}{1 - r}$; solve for a_1

The following exercises show pairs of similar triangles. Find the lengths of the sides labeled with variables.

45. $\triangle ABC \sim \triangle DEF$

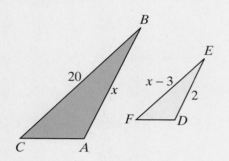

46. $\triangle STU \sim \triangle VWU$

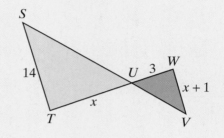

Section 13.5: Applications

Solve the following word problems.

47. Ratio: The ratio of two numbers is 2 to 3. If the sum of the numbers is 45, find the numbers.

48. Education: In a statistics class the ratio of men to women is 6 to 5. If there are 33 students in the class, how many women are in the class?

49. Fractions: The numerator of a fraction is two less than the denominator. If both the numerator and denominator are decreased by 2, the new fraction will reduce to $\dfrac{2}{3}$. What was the original fraction? (Do not reduce answer.)

50. Swimming Pools: One pipe can fill a swimming pool twice as fast as another. When both pipes are used, the pool can be filled in 2 hours. How long would it take each pipe alone to fill the pool?

51. Construction: Working together, a bricklayer and his partner can lay a brick sidewalk in 4.8 hours. If he could have completed the job in 8 hours working alone, how long would his partner have needed to complete the job working alone?

52. Accounting: Working together, Alice and Judy can prepare an account report in 2.4 hours. Working alone, Alice would need two hours longer to prepare the same report than Judy would. How long would it take each of them working alone?

53. Painting a room: It takes Tyler four hours less to paint a room than his son would take to paint the same sized room. If they can paint this size room in $1\frac{1}{2}$ hours working together, how long would each of them take to paint such a room alone?

54. Biking: Garrett rides his bike to the top of Mount High, a distance of 12 miles. He then rides back down the mountain four times as fast as he rode up the mountain. If the total ride took 2.5 hours, find the rate at which he rode up the mountain.

55. Boating: A boat can travel 10 miles downstream in the same time it travels 6 miles upstream. If the rate of the current is 6 mph, find the speed of the boat in still water.

56. Car speed: Mr. Lukin had a golf game scheduled with a friend 80 miles from his home. After traveling 30 miles in heavy traffic, he needed to increase his speed by 25 mph to make the scheduled tee off time. If he traveled the same length of time at each rate, find the rates.

57. Plane speed: After flying 760 miles, the pilot increased the speed by 30 mph and continued on for another 817 miles. If the same amount of time was spent at each rate, find the rates.

58. Research projects: It takes Matilde twice as long to complete a certain type of research project as it takes Raphael to do the same type of project. They found that if they worked together they could finish similar projects in 1 hour. How long would it take each of them working alone to finish this type of project?

Section 13.6: Variation

Use the information given to find the unknown values.

59. If y varies directly as x and $y = 4$ when $x = 20$, find y if $x = 30$.

60. If y is inversely proportional to x and $y = 8$ when $x = 10$, find y when $x = 16$.

61. If y is directly proportional to the square of x, and $y = 10$ when x is 3, what is y when x is 7?

62. If y is inversely proportional to x^4, and $y = 50$ when $x = \frac{1}{2}$, what is y when $x = \frac{1}{10}$?

Solve the following word problems.

63. Springs: The length a hanging spring stretches varies directly with the weight placed on the end. If a spring stretches 10 in. when a weight of 14 lb is placed at its end, how far will the spring stretch if the weight is increased to 20 lb?

64. Pizza: The circumference of a circle varies directly as the diameter of the circle. A circular pizza with a diameter of 4 in. has a circumference of 12.56 in. What will be the circumference of a pizza with a diameter of 6 in.?

65. Pizza: The area of a circle varies directly as the square of its radius. A circular pizza with a radius of 4 in. has an area of 50.24 in.^{2} What is the area of a pizza with a radius of 6 in.?

66. Gears: A gear that has 48 teeth is meshed with a gear that has 16 teeth. Find the speed (in revolutions per minute) of the small gear if the large one is rotating at 54 rpm. (**Hint:** $T_1 R_1 = T_2 R_2$ where T_1 and T_2 are the number of teeth and R_1 and R_2 are the number of rotations per minute.)

67. Free Falling Object: The distance a free falling object falls is directly proportional to the square of the time it falls (before it hits the ground). If an object fell 256 feet in 4 seconds, how far will it have fallen by the end of 6 seconds?

68. Resistance in a Wire: The resistance, R (in ohms), in a wire is given by the formula $R = \dfrac{kL}{d^2}$ where k is the constant of variation, L is the length of the wire and d is the diameter of the wire. The resistance of a wire 600 feet long with a diameter of 0.01 in. is 22 ohms. What is the resistance of the same type of wire 1800 feet long with a diameter of 0.02 in.?

69. Volume of a Circular Cylinder: The volume of a circular cylinder varies jointly as its height and the square of its radius. If a cylinder with $r = 2$ feet and $h = 4$ feet holds 50.24 cubic feet of water, what is the volume of water in the cylinder when $h = 3$ feet?

70. Selling Lemonade: Billy realizes that the sale of his homemade lemonade varies directly with the amount of sugar squared and inversely with the amount of lemon juice in the recipe. If he sells 54 glasses when he uses 9 cups of sugar and 6 cups of lemon juice how much lemon juice does he use when he uses 10 cups of sugar and sells 80 glasses of lemonade?

Chapter 13: Test

Reduce to lowest terms. State any restrictions on the variable(s).

1. $\dfrac{x^2+3x}{x^2+7x+12}$

2. $\dfrac{2x+5}{4x^2+20x+25}$

Perform the indicated operations and reduce to lowest terms. Assume that no denominator has a value of 0.

3. $\dfrac{x+3}{x^2+3x-4}\cdot\dfrac{x^2+x-2}{x+2}$

4. $\dfrac{6x^2-x-2}{12x^2+5x-2}\div\dfrac{4x^2-1}{8x^2-6x+1}$

5. $\dfrac{x}{x^2+3x-10}+\dfrac{3x}{4-x^2}$

6. $\dfrac{x-4}{3x^2+5x+2}-\dfrac{x-1}{x^2-3x-4}$

7. $\dfrac{x^2-16}{x^2-4x}\cdot\dfrac{x^2}{x+4}\div\dfrac{x-1}{2x^2-2x}$

8. $\dfrac{x}{x+3}-\dfrac{x+1}{x-3}+\dfrac{x^2+4}{x^2-9}$

Simplify the complex fractions.

9. $\dfrac{\dfrac{5y}{3x^2}}{\dfrac{10y^4}{9x}}$

10. $\dfrac{\dfrac{4}{3x}+\dfrac{1}{6x}}{\dfrac{1}{x^2}-\dfrac{1}{2x}}$

11. $\dfrac{x^{-1}-y^{-1}}{x-y}$

12. $\dfrac{\dfrac{4}{x}+\dfrac{2}{x+3}}{\dfrac{x^2-9}{2x}}$

13. Simplify the expression in part **a.** and solve the equation in part **b.**

a. $\dfrac{3}{x}-\dfrac{2}{x+1}+\dfrac{5}{2x}$

b. $\dfrac{3}{x}-\dfrac{2}{x+1}=\dfrac{5}{2x}$

State any restrictions on x and then solve the equations.

14. $\dfrac{x-1}{x+4}=\dfrac{4}{5}$

15. $\dfrac{4}{7}-\dfrac{1}{2x}=1+\dfrac{1}{x}$

16. $\dfrac{4}{x+4}+\dfrac{3}{x-1}=\dfrac{1}{x^2+3x-4}$

17. $\dfrac{x}{x+2}=\dfrac{1}{x+1}-\dfrac{x}{x^2+3x+2}$

Solve each of the formulas for the specified variables.

18. $S = \dfrac{n(a_1 + a_n)}{2}$; solve for n

19. $y = mx + b$; solve for x

Solve the following problems as indicated.

20. Triangles ABC and DEC are similar. (Symbolically, $\triangle ABC \sim \triangle DEC$.) Find the lengths of sides AC and DC.

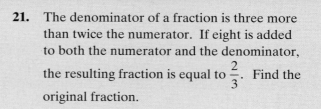

21. The denominator of a fraction is three more than twice the numerator. If eight is added to both the numerator and the denominator, the resulting fraction is equal to $\dfrac{2}{3}$. Find the original fraction.

22. z varies directly as x^3. If $z = 3$ when $x = 2$, find z if $x = 4$.

23. z varies directly as x^2 and inversely as $\sqrt{y}$. If $z = 24$ when $x = 3$ and $y = 4$, find z if $x = 5$ and $y = 9$.

Solve the following word problems.

24. House cleaning: Sonya can clean the apartment in 6 hours. It takes Lucy 12 hours to clean it. If they work together, how long will it take them?

25. Travel: Mario can travel 228 miles in the same time that Carlos travels 168 miles. If Mario's speed is 15 mph faster than Carlos', find their rates.

26. Boating: Bob travels 4 miles upstream. In the same time, he could have traveled 7 miles downstream. If the speed of the current is 3 mph, find the speed of the boat in still water.

27. Stretching a spring: Hooke's Law states that the distance a spring will stretch vertically is directly proportional to the weight placed at its end. If a particular spring will stretch 5 cm when a weight of 4 g is placed at its end, how far will the spring stretch if a weight of 6 g is placed at its end?

28. Candy: The volume of a gumball is directly proportional to the radius cubed. A gumball with a diameter of 1.75 inches has an approximate volume of 2.805 inches cubed. What is the volume of a gumball that has a diameter of 2.14 inches? (Round your answer to the nearest hundredth.)

Cumulative Review: Chapters 1 - 13

Solve the following problems as indicated.

1. Find the mean, median, mode, and range for the following set of data.

   1.5   2.5   1.0   1.3   1.5   1.9   0.9   1.8

2. A triangle with at least two equal sides is called an _____ triangle, while a triangle with all sides equal is called an _____ triangle.

3. Graph the set of integers less than −5 on a number line.

4. Perform the indicated operations and reduce.

   a. $\dfrac{5}{6} + \dfrac{3}{4}$

   b. $\dfrac{1}{2} - \dfrac{5}{7}$

   c. $\dfrac{1}{2} \cdot \dfrac{1}{2}$

   d. $\dfrac{7}{32} \div \dfrac{21}{8}$

5. Write an algebraic expression for the following English phrase:

   "5 less than twice the sum of a number and 6."

Perform the indicated operations and simplify the expressions.

6. $(3x + 7)(x - 4)$

7. $(2x - 5)^2$

8. $-4(x + 2) + 5(2x - 3)$

9. $x(x + 7) - (x + 1)(x - 4)$

Solve the following problems as indicated.

10. Solve the inequality $\dfrac{x+5}{2} < \dfrac{3x}{4} + 1$ and graph the solution set. Write the solution set using interval notation. Assume that x is a real number.

11. Given the relation $r = \{(-1, 5), (0, 2), (-1, 0), (5, 0), (6, 0)\}$.

    a. Graph the relation.

    b. What is the domain of the relation?

    c. What is the range of the relation?

    d. Is the relation a function? Explain.

12. For the function $f(x) = x^3 - 2x^2 - 1$, find: **a.** $f(3)$ **b.** $f(0)$ **c.** $f\left(-\dfrac{2}{3}\right)$

13. Graph the equation $6x + 4y = 18$ by locating the x-intercept and the y-intercept.

14. Find the equation in slope-intercept form of the line that has slope $\dfrac{3}{4}$ and passes through the point $(-2, 4)$. Graph the line.

15. Find the equation in slope-intercept form of the line passing through the two points $(-5, -1)$ and $(4, 2)$. Graph the line.

16. Find the equation of the line passing through the point $(-3, 7)$ with slope undefined. Graph the line.

Graph the following inequality.

17. $3x + y > 4$.

Simplify each expression. The final form of the expressions with variables should contain only positive exponents. Assume that all variables represent nonzero numbers.

18. $\dfrac{32x^3 y^2}{4xy^2}$

19. $\dfrac{15xy^{-1}}{3x^{-2}y^{-3}}$

Perform the indicated operations and simplify each expression.

20. $2(4x + 3) + 5(x - 1)$

21. $(x + 1) - (4x^2 + 3x - 2)$

22. $(2x - 7)(x + 4)$

23. $(x + 6)^2$

Completely factor each polynomial.

24. $3xy + y^2 + 3x + y$

25. $4x^2 - 4x - 15$

26. $6x^2 - 7x + 2$

27. $6x^3 - 22x^2 - 8x$

28. $9x^6 - 4y^2$

29. $8x^3 + 125$

30. Find an equation that has $x = -1$, $x = 0$ and $x = -3$ as roots.

Express each quotient as a sum (or difference) of fractions and simplify if possible.

31. $\dfrac{7x^2y^2 - 21x^2y^3 + 8x^2y^4}{7x^2y^2}$

Divide by using the long division algorithm.

32. $\dfrac{2x^3 + 5x^2 + 7}{x + 3}$

Reduce the rational expressions to lowest terms and indicate any restrictions on the variable.

33. $\dfrac{x}{x^2 + x}$

34. $\dfrac{x^2 + 2x - 15}{2x^2 - 12x + 18}$

Perform the indicated operations and simplify. Assume that no denominator is 0.

35. $\dfrac{x^2}{x+y} - \dfrac{y^2}{x+y}$

36. $\dfrac{4x}{3x+3} - \dfrac{x}{x+1}$

37. $\dfrac{4x}{x-4} \div \dfrac{12x^2}{x^2-16}$

38. $\dfrac{x^2 + 3x + 2}{x + 3} \cdot \dfrac{3x^2 + 6x}{x + 1}$

39. $\dfrac{2x+1}{x^2+5x-6} \cdot \dfrac{x^2+6x}{x}$

40. $\dfrac{8}{x^2+x-6} + \dfrac{2x}{x^2-3x+2}$

41. $\dfrac{x}{x^2+3x-4} + \dfrac{x+1}{x^2-1}$

42. $\dfrac{x+1}{x^2+4x+4} \div \dfrac{x^2-x-2}{x^2-2x-8}$

Simplify the complex algebraic fractions.

43. $\dfrac{\dfrac{3}{x} + \dfrac{1}{6x}}{\dfrac{7}{3x}}$

44. $\dfrac{\dfrac{1}{4x} + \dfrac{1}{x^2}}{\dfrac{1}{2x} + \dfrac{1}{x}}$

Solve each of the equations.

45. $9y + 3y - 50 = 22$

46. $4.78 - 0.3x + 0.5x = 2.31$

47. $4(x+2) - 7 = -2(3x+1) - 3$

48. $4x^2 + 20x + 25 = 0$

49. $x^3 - x^2 = 20x$

50. $0 = x^2 - 7x + 10$

51. $\dfrac{7}{2x-1} = \dfrac{3}{x+6}$

52. **a.** Simplify the following expression: $\dfrac{3}{x} - \dfrac{5}{x+3}$.

b. Solve the following equation: $\dfrac{3}{x} = \dfrac{5}{x+3}$.

53. **a.** Simplify the following expression: $\dfrac{3x}{x+2} + \dfrac{2}{x-4} + 3$.

b. Solve the following equation: $\dfrac{3x}{x+2} + \dfrac{2}{x-4} = 3$.

54. In the formula $A = P + Prt$, solve for t.

Solve the following word problems.

55. **Investing:** How long will it take for an investment of $600 at a rate of 5% to be worth $615? (**Hint:** $I = Prt$)

56. **Circular Cylinders:** Find the volume of a right circular cylinder with a radius of 6 in. and a height of 11 in. (Use $\pi = 3.14$.)

57. **Airplanes:** An airplane can travel 1035 miles in the same time that a train travels 270 miles. The speed of the plane is 50 mph more than three times the speed of the train. Find the speed of each.

58. **Waxing a car:** A man can wax his car three times as fast as his daughter can. Together they can complete the job in 2 hours. How long would it take each of them working alone?

59. **Baseball:** Louis is averaging 13 hits for every 50 times at bat. If he maintains this average, how many at bats will he need in order to get 156 hits?

60. **Springs:** The weight on a spring varies directly as the length the spring stretches. If a hanging spring stretches 5 cm when a weight of 13 g is placed at its end, how far will the spring stretch if a weight of 20 g is placed at its end?

Radicals

Mathematics at Work!

When you watch a sunset, how far are you from the horizon? Because of the curvature of the earth, this distance can be deceiving. To calculate the distance in miles between you and the horizon, use the formula $d = 1.32\sqrt{h}$ where h is your eye's height above the ground in feet. If you are watching the sunset from the roof of a two-story building, and your eye is 28 feet above the ground, how far away is the horizon? (See Section 14.2.)

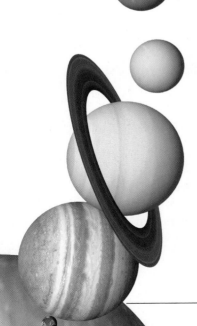

14.1 Roots and Radicals

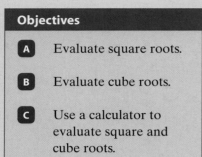
You are probably familiar with the concept of **square roots** and the square root symbol $\left(\text{or } \textbf{radical sign } \sqrt{}\right)$ from the discussions of the Pythagorean theorem in Chapter 5 and real numbers in Chapter 7. For example, $\sqrt{3}$ represents the square root of 3 and is the number whose square is 3.

Objective A   Perfect Squares and Square Roots

A number is **squared** when it is multiplied by itself. For example,

$$6^2 = 6 \cdot 6 = 36 \qquad \text{and} \qquad (-1.5)^2 = (-1.5)(-1.5) = 2.25.$$

If an integer is squared, the result is called a **perfect square**. The squares for the integers from 1 to 20 are shown in Table 1 for easy reference.

Squares of Integers from 1 to 20 (Perfect Squares)										
Integers (n)	1	2	3	4	5	6	7	8	9	10
Perfect Squares (n^2)	1	4	9	16	25	36	49	64	81	100
Integers (n)	11	12	13	14	15	16	17	18	19	20
Perfect Squares (n^2)	121	144	169	196	225	256	289	324	361	400

Table 1

Now we want to reverse the process of squaring. That is, given a number, we want to find a number that when squared will result in the given number. This is called **finding a square root** of the given number. In general,

if $b^2 = a$, then b is a square root of a.

For example,

- because $5^2 = 25$, 5 is a **square root** of 25 and we write $\sqrt{25} = 5$.

- because $9^2 = 81$, 9 is a **square root** of 81 and we write $\sqrt{81} = 9$.

Radical Terminology

The symbol $\sqrt{}$ is called a **radical sign**.

The number under the radical sign is called the **radicand**.

The complete expression, such as $\sqrt{64}$, is called a **radical** or **radical expression**.

Every positive real number has two square roots, one positive and one negative. The positive square root is called the **principal square root**. For example,

- because $(8)^2 = 64$, $\sqrt{64} = 8$. ⟵ the **principal square root**

- because $(-8)^2 = 64$, $-\sqrt{64} = -8$. ⟵ the **negative square root**

The number 0 has only one square root, namely 0.

Square Root

If a is a nonnegative real number, then

$$\sqrt{a} \text{ is the \textbf{principal square root} of } a,$$

and

$$-\sqrt{a} \text{ is the \textbf{negative square root} of } a.$$

notes

■ Square roots of negative numbers are not real numbers. For example, $\sqrt{-4}$ is not a real number. There is no real number whose square is ■ −4. Numbers of this type are discussed in Appendices A.8 and A. 9.

Example 1

Evaluating Square Roots

a. 64 has two square roots, one positive and one negative. The $\sqrt{}$ sign is understood to represent the **positive square root** (or the **principal square root**) and $-\sqrt{}$ represents the **negative square root**. Therefore, we have

$$\sqrt{64} = 8 \qquad \text{and} \qquad -\sqrt{64} = -8.$$

b. Because $11^2 = 121$, we have $\sqrt{121} = 11$ and $-\sqrt{121} = -11$.

c. Because $0^2 = 0$, $\sqrt{0^2} = 0$.

d. $\sqrt{-25}$ is not a real number.

Now work margin exercise **1.**

1. If real square roots exist, state them.

a. 81

b. 196

c. 49

d. −36

2. Evaluate the square roots.

a. $\sqrt{\dfrac{9}{64}}$

b. $\sqrt{0.0016}$

3. The square root of 67 is approximately 8.1854. State why or why not this is a reasonable estimate.

Example 2

Evaluating Square Roots

a. Because $\left(\dfrac{4}{5}\right)^2 = \dfrac{16}{25}$, we know that $\sqrt{\dfrac{16}{25}} = \dfrac{4}{5}$.

b. $-\sqrt{0.0009} = -0.03$ because $(0.03)^2 = 0.0009$.

Now work margin exercise **2.**

Recall that **rational numbers** are numbers that can be expressed as the quotient of integers and are of the form $\dfrac{a}{b}$ where a and b are integers and $b \neq 0$. In decimal form, rational numbers are of the form of terminating decimals or repeating infinite decimals. Square roots of perfect square radicands simplify to rational numbers.

No rational number will square to give 2. So, $\sqrt{2}$ is an **irrational number** and a decimal representation is an infinite nonrepeating decimal. Your calculator will show

$$\sqrt{2} = 1.414213562\ldots \qquad \text{accurate to 9 decimal places}$$

$$\sqrt{2} \approx 1.4142 \qquad \text{rounded to 4 decimal places}$$

To get a better idea of $\sqrt{2}$ we can compare as follows:

$$1 < 2 < 4 \qquad \text{Note that 1 and 4 are perfect squares.}$$
$$\text{and } \sqrt{1} < \sqrt{2} < \sqrt{4}$$

which gives $1 < \sqrt{2} < 2$

and we see that indeed $\sqrt{2}$ is between 1 and 2 and the value 1.4142 is reasonable.

Example 3

Estimating Square Roots

A calculator will give $\sqrt{30} \approx 5.4772$ rounded to the nearest ten-thousandth. Check that this is a reasonable estimate.

Solution

Because $25 < 30 < 36$, we have $\sqrt{25} < \sqrt{30} < \sqrt{36}$ and $5 < \sqrt{30} < 6$. The approximation 5.4772 is between 5 and 6 and is reasonable.

Another approach is to square as follows:
$$(5.4772)^2 = 29.99971984 \text{ which is close to 30.}$$

Now work margin exercise **3.**

Cube Roots

A number is **cubed** when it is used as a factor 3 times. For example,

$$5^3 = 5 \cdot 5 \cdot 5 = 125 \qquad \text{and} \qquad (-30)^3 = (-30) \cdot (-30) \cdot (-30) = -27,000.$$

If an integer is cubed, the result is called a **perfect cube**. The cubes for the integers from 1 to 10 are shown here for easy reference.

Cubes of Integers from 1 to 10 (Perfect Cubes)				
$1^3 = 1$	$2^3 = 8$	$3^3 = 27$	$4^3 = 64$	$5^3 = 125$
$6^3 = 216$	$7^3 = 343$	$8^3 = 512$	$9^3 = 729$	$10^3 = 1000$

Table 2

The reverse of cubing is finding the **cube root**, symbolized $\sqrt[3]{}$. For example, the cube root of 125 is 5 and we write $\sqrt[3]{125} = 5$.

Cube Root

If a is a real number, then $\sqrt[3]{a}$ is the **cube root** of a.

notes

In the cube root expression $\sqrt[3]{a}$ the number 3 is called the **index**. In a square root expression such as $\sqrt{a}$ the index is understood to be 2 and **is not written.** Expressions with square roots and cube roots (as well as other roots) are called **radical expressions.**

Example 4

Evaluating Cube Roots

a. Because $2^3 = 8$, $\sqrt[3]{8} = 2$.

b. Because $(-6)^3 = -216$, $\sqrt[3]{-216} = -6$.

Note that the cube root of a negative number is a real number and is negative.

c. Because $\left(\dfrac{1}{3}\right)^3 = \dfrac{1}{27}$, $\sqrt[3]{\dfrac{1}{27}} = \dfrac{1}{3}$.

Now work margin exercise 4.

4. Evaluate the following radical expressions.

a. $\sqrt[3]{64}$

b. $\sqrt[3]{-125}$

c. $\sqrt[3]{\dfrac{1}{1000}}$

Using a TI-84 Plus Graphing Calculator to Evaluate Expressions with Radicals

A TI-84 Plus graphing calculator can be used to find decimal approximations for radicals and expressions containing radicals. The displays in Example 5 illustrate the advantage of being able to see the entire expression being evaluated.

Example 5

Evaluating Radical Expressions with a Calculator

The following radical expressions are evaluated by using a TI-84 Plus graphing calculator. In each example the steps (or keys to press) are shown. The TI-84 Plus gives answers rounded to nine decimal places. You may choose (through the **MODE** key) to have answers rounded to fewer than nine places.

a. $\sqrt{17}$

Solution

Step 1: Press **2ND** x^2 to get the square root symbol $\sqrt{}$. (**Note:** When the $\sqrt{}$ symbol appears, it will appear with a left-hand parenthesis. You should press the right-hand parenthesis to close the square root operation.)

Step 2: Enter **1** **7** and the right-hand parenthesis **)** .

Step 3: Press **ENTER** .

The display will appear as follows:

```
√(17)
        4.123105626
```

b. $3\sqrt{20}$ **Note:** This expression represents 3 times $\sqrt{20}$.

Solution

To find $3\sqrt{20}$ proceed as follows:

Step 1: Enter **3** .

Step 2: Press **2ND** x^2 . (This gives the $\sqrt{}$ symbol.)

Step 3: Enter **2** **0** and the right-hand parenthesis **)** .

Step 4: Press **ENTER** .

The display will appear as follows:

Note: The calculator is programmed to follow the rules for order of operations.

Now work margin exercise 5.

5. Evaluate the following radical expressions with a calculator.

a. $\sqrt{45}$

b. $5\sqrt{30}$

Practice Problems

Simplify the following square roots and cube roots.

1. $\sqrt{49}$ **2.** $\sqrt[3]{125}$ **3.** $-\sqrt{25}$ **4.** $\sqrt{196}$

Use your knowledge of square roots and cube roots to determine whether each number is rational, irrational, or nonreal.

5. $\sqrt{\dfrac{9}{16}}$ **6.** $\sqrt{324}$ **7.** $\sqrt{-25}$ **8.** $\sqrt{3}$

Use your calculator to find the value (rounded to the nearest ten-thousandth) of each of the following radical expressions.

9. $5\sqrt{2}$ **10.** $6-4\sqrt{7}$

Practice Problem Answers

1. 7 **2.** 5 **3.** −5 **4.** 14

5. rational **6.** rational **7.** nonreal **8.** irrational

9. 7.0711 **10.** −4.5830

Exercises 14.1

Simplify the following square roots and cube roots. See Examples 1, 2, and 4.

1. $\sqrt{9}$ **2.** $\sqrt{49}$ **3.** $\sqrt{81}$ **4.** $\sqrt{36}$ **5.** $\sqrt{289}$

6. $\sqrt{121}$ **7.** $\sqrt{169}$ **8.** $\sqrt{361}$ **9.** $\sqrt[3]{1}$ **10.** $\sqrt[3]{1000}$

11. $\sqrt[3]{125}$ **12.** $\sqrt[3]{343}$ **13.** $\sqrt[3]{216}$ **14.** $\sqrt[3]{512}$ **15.** $\sqrt{\dfrac{1}{4}}$

16. $\sqrt{\dfrac{9}{16}}$ **17.** $\sqrt[3]{\dfrac{27}{64}}$ **18.** $\sqrt[3]{\dfrac{1}{8}}$ **19.** $\sqrt{0.04}$ **20.** $\sqrt{0.0081}$

21. $-\sqrt{100}$ **22.** $-\sqrt{144}$ **23.** $-\sqrt{0.0016}$ **24.** $-\sqrt{0.000004}$ **25.** $\sqrt[3]{-27}$

26. $\sqrt[3]{-64}$ **27.** $\sqrt[3]{-125}$ **28.** $\sqrt[3]{729}$ **29.** $\sqrt{\dfrac{9}{25}}$ **30.** $\sqrt{\dfrac{25}{81}}$

Estimates (rounded to the nearest ten-thousandth) of radicals are given. Show that these are reasonable estimates. See Example 3.

31. $\sqrt{74} \approx 8.6023$ **32.** $\sqrt{18} \approx 4.2426$

33. $\sqrt{32} \approx 5.6569$ **34.** $\sqrt{110} \approx 10.4881$

Use your knowledge of square roots and cube roots to determine whether each number is rational, irrational, or nonreal.

35. $\sqrt{4}$ **36.** $\sqrt{17}$ **37.** $\sqrt{169}$ **38.** $\sqrt[3]{8}$

39. $\sqrt{\dfrac{2}{9}}$ **40.** $-\sqrt{\dfrac{1}{4}}$ **41.** $\sqrt{-36}$ **42.** $\sqrt[3]{-27}$

43. $-\sqrt[3]{125}$　　　**44.** $\sqrt{-10}$　　　📟**45.** $\sqrt{1.68}$　　　📟**46.** $\sqrt{5.29}$

📟 **Use a calculator to find the value of each radical expression rounded to the nearest ten-thousandth. See Example 5.**

47. $\sqrt{39}$　　　**48.** $\sqrt{150}$　　　**49.** $\sqrt{6.23}$　　　**50.** $\sqrt{9.6}$

51. $\sqrt{\dfrac{1}{5}}$　　　**52.** $\sqrt{\dfrac{3}{8}}$　　　**53.** $4\sqrt{5}$　　　**54.** $6\sqrt{3}$

55. $-2\sqrt{17}$　　　**56.** $-3\sqrt{6}$

Writing & Thinking

57. Discuss, in your own words, why the square root of a negative number is not a real number.

58. Discuss, in your own words, why the cube root of a negative number is a negative number.

14.2 Simplifying Radicals

Objectives

A Simplify algebraic expressions that contain square roots.

B Simplify radical expressions that contain variables.

C Simplify algebraic expressions that contain cube roots.

Objective A   Algebraic Expressions with Square Roots

Various roots can be related to solutions of equations, and we want such numbers to be in a **simplified form** for easier calculations and algebraic manipulations. We need the two properties of radicals stated here for square roots.

Properties of Square Roots

If a and b are **positive** real numbers, then

1. $\sqrt{ab} = \sqrt{a}\sqrt{b}$

2. $\sqrt{\dfrac{a}{b}} = \dfrac{\sqrt{a}}{\sqrt{b}}$

As an example, we know that $\sqrt{144} = 12$. However, in a situation where you may have forgotten this, you can proceed as follows using property 1 of square roots:

$$\sqrt{144} = \sqrt{36} \cdot \sqrt{4} = 6 \cdot 2 = 12.$$

Similarly, using property 2, we can write

$$\sqrt{\frac{49}{36}} = \frac{\sqrt{49}}{\sqrt{36}} = \frac{7}{6}.$$

Simplest Form for Square Roots

A square root is considered to be in **simplest form** when the radicand has no perfect square as a factor.

The number 200 is not a perfect square and using a calculator we will find $\sqrt{200} \approx 14.1421$. Now to simplify $\sqrt{200}$, we can use property 1 of square roots and any of the following three approaches.

Approach 1:   Factor 200 as $4 \cdot 50$ because 4 is a perfect square. This gives,

$$\sqrt{200} = \sqrt{4 \cdot 50} = \sqrt{4} \cdot \sqrt{50} = 2\sqrt{50}.$$

However, $2\sqrt{50}$ is **not in simplest form** because 50 has a perfect square factor, 25. Thus to complete the process, we have

$$\sqrt{200} = 2\sqrt{50} = 2\sqrt{25 \cdot 2} = 2\sqrt{25} \cdot \sqrt{2} = 2 \cdot 5 \cdot \sqrt{2} = 10\sqrt{2}.$$

Approach 2: Note that 100 is a perfect square factor of 200 and $200 = 100 \cdot 2$.

$$\sqrt{200} = \sqrt{100 \cdot 2} = \sqrt{100} \cdot \sqrt{2} = 10\sqrt{2}$$

Approach 3: Use prime factors.

$$\begin{aligned}
\sqrt{200} &= \sqrt{2 \cdot 2 \cdot 2 \cdot 5 \cdot 5} \\
&= \sqrt{2 \cdot 2 \cdot 5 \cdot 5} \cdot \sqrt{2} \\
&= \sqrt{2 \cdot 2} \cdot \sqrt{5 \cdot 5} \cdot \sqrt{2} \\
&= 2 \cdot 5 \cdot \sqrt{2} \\
&= 10\sqrt{2}
\end{aligned}$$

notes

Of these three approaches, the second appears to be the easiest because it has the fewest steps. However, "seeing" the largest perfect square factor may be difficult. If you do not immediately see a perfect square factor, proceed by finding other factors or prime factors as illustrated.

Example 1

Simplifying Numerical Expressions with Square Roots

Simplify each numerical expression so that there are no perfect square factors in the radicand.

a. $\sqrt{48}$

Solution

$\sqrt{48} = \sqrt{16 \cdot 3} = \sqrt{16} \cdot \sqrt{3} = 4\sqrt{3}$ 16 is the largest perfect square factor.

b. $\sqrt{63}$

Solution

$\sqrt{63} = \sqrt{9 \cdot 7} = \sqrt{9} \cdot \sqrt{7} = 3\sqrt{7}$ 9 is the largest perfect square factor.

c. $\sqrt{\dfrac{75}{16}}$

Solution

$$\sqrt{\frac{75}{16}} = \frac{\sqrt{75}}{\sqrt{16}} = \frac{\sqrt{25 \cdot 3}}{\sqrt{16}} = \frac{\sqrt{25} \cdot \sqrt{3}}{\sqrt{16}} = \frac{5\sqrt{3}}{4}$$

Now work margin exercise 1.

1. Simplify each numerical expression so that there are no perfect square factors in the radicand.

   a. $\sqrt{98}$

   b. $\sqrt{45}$

   c. $\sqrt{\dfrac{12}{25}}$

To simplify square root expressions that contain variables, such as $\sqrt{x^2}$, we must be aware of whether the variable represents a positive real number $(x > 0)$, zero $(x = 0)$, or a negative number $(x < 0)$.

For example,

$$\text{if } x = 0, \text{ then } \sqrt{x^2} = \sqrt{0^2} = \sqrt{0} = 0 = x.$$
$$\text{If } x = 5, \text{ then } \sqrt{x^2} = \sqrt{5^2} = \sqrt{25} = 5 = x.$$
$$\text{But, if } x = -5, \text{ then } \sqrt{x^2} = \sqrt{(-5)^2} = \sqrt{25} = 5 \neq x.$$
$$\text{In fact, if } x = -5, \text{ then } \sqrt{x^2} = \sqrt{(-5)^2} = \sqrt{25} = 5 = |-5| = |x|.$$

Thus simplifying radical expressions with variables involves more detailed analysis than simplifying radical expressions with only constants. The following definition indicates the correct way to simplify $\sqrt{x^2}$.

Square Root of x^2

If x is a real number, then $\sqrt{x^2} = |x|$.

Note: If $x \geq 0$ is given, then we can write $\sqrt{x^2} = x$.

Although using the absolute value when simplifying square roots is correct mathematically, we can avoid some confusion by assuming that the variable under the radical sign represents only positive real numbers or 0. This eliminates the need for absolute value signs.

Therefore, for the remainder of this text, we will assume that $x > 0$ and write $\sqrt{x^2} = x$, unless specifically stated otherwise, in all square root expressions.

Example 2

Simplifying Square Roots with Variables

Simplify each of the following radical expressions. Assume that all variables represent positive real numbers. (Note that, by making this assumption, we need not be concerned about the absolute value sign.)

a. $\sqrt{16y^2}$

Solution

$$\sqrt{16y^2} = 4y$$

b. $\sqrt{72a^2}$

Solution

$$\sqrt{72a^2} = \sqrt{36a^2} \cdot \sqrt{2} = 6a\sqrt{2}$$

c. $\sqrt{12x^2y^2}$

Solution

$$\sqrt{12x^2y^2} = \sqrt{4x^2y^2} \cdot \sqrt{3} = 2xy\sqrt{3}$$

Now work margin exercise 2.

To find the square root of an expression with even exponents, divide the exponents by 2. For example,

$$x^2 \cdot x^2 = x^4 \qquad a^3 \cdot a^3 = a^6 \qquad y^5 \cdot y^5 = y^{10}$$

and $\quad \sqrt{x^4} = x^2 \qquad \sqrt{a^6} = a^3 \qquad \sqrt{y^{10}} = y^5.$

To find the square root of an expression with odd exponents, factor the expression into two terms, one with exponent 1 and the other with an even exponent. For example,

$$x^3 = x^2 \cdot x \qquad \text{and} \qquad y^9 = y^8 \cdot y$$

which means that

$$\sqrt{x^3} = \sqrt{x^2 \cdot x} = \sqrt{x^2} \cdot \sqrt{x} = x \cdot \sqrt{x} \qquad \text{and} \qquad \sqrt{y^9} = \sqrt{y^8 \cdot y} = \sqrt{y^8} \cdot \sqrt{y} = y^4 \sqrt{y}.$$

Example 3

Simplifying Square Roots

Simplify each of the following radical expressions. Look for perfect square factors and even powers of the variables. Assume that all variables represent positive real numbers.

a. $\sqrt{81x^4}$

Solution

$$\sqrt{81x^4} = 9x^2 \qquad\qquad \text{The exponent 4 is divided by 2.}$$

b. $\sqrt{64x^5y}$

Solution

$$\sqrt{64x^5y} = \sqrt{64x^4} \cdot \sqrt{xy} = 8x^2\sqrt{xy}$$

2. Simplify each of the following radical expressions. Assume that all variables represent positive real numbers.

a. $\sqrt{36z^2}$

b. $\sqrt{75b^2}$

c. $\sqrt{45c^2d^2}$

3. Simplify each of the following radical expressions. Assume that all variables represent positive real numbers.

a. $\sqrt{16x^8}$

b. $\sqrt{100x^3y^3}$

c. $\sqrt{12x^8y^{12}}$

d. $\sqrt{\dfrac{25z^{18}}{y^8}}$

---

c. $\sqrt{18a^4b^6}$

Solution

$$\sqrt{18a^4b^6} = \sqrt{9a^4b^6} \cdot \sqrt{2} = 3a^2b^3\sqrt{2}$$

Each exponent is divided by 2.

d. $\sqrt{\dfrac{9a^{13}}{b^4}}$

Solution

$$\sqrt{\dfrac{9a^{13}}{b^4}} = \dfrac{\sqrt{9a^{13}}}{\sqrt{b^4}} = \dfrac{\sqrt{9a^{12}} \cdot \sqrt{a}}{\sqrt{b^4}} = \dfrac{3a^6\sqrt{a}}{b^2}$$

Recall $a, b > 0$.

Now work margin exercise 3.

Objective C **Algebraic Expressions with Cube Roots**

When simplifying expressions with cube roots, we need to be aware of perfect cube numbers and variables with exponents that are multiples of 3. (Multiples of 3 are 3, 6, 9, 12, 15, and so on.) **Thus exponents are divided by 3 in simplifying cube root expressions.** For example,

$$x^2 \cdot x^2 \cdot x^2 = x^6 \qquad a^3 \cdot a^3 \cdot a^3 = a^9 \qquad y^5 \cdot y^5 \cdot y^5 = y^{15}$$

and $\qquad \sqrt[3]{x^6} = x^2 \qquad \sqrt[3]{a^9} = a^3 \qquad \sqrt[3]{y^{15}} = y^5$

Simplest Form for Cube Roots

A cube root is considered to be in **simplest form** when the radicand has no perfect cube as a factor.

When finding cube roots, we need not be concerned about positive and negative values for variables because cube roots of negative numbers are defined to be negative. For example,

we have $\qquad (-2)^3 = -8$ and therefore $\sqrt[3]{-8} = -2$.

Similarly, $\qquad (-5)^3 = -125$ and therefore, $\sqrt[3]{-125} = -5$.

Thus $\sqrt[3]{x^3} = x$ whether $x \geq 0$ or $x < 0$.

Example 4

Simplifying Expressions with Cube Roots

Simplify each of the following radical expressions. Look for perfect cube factors and powers of the variables that are multiples of 3.

a. $\sqrt[3]{54x^6}$

Solution

$$\sqrt[3]{54x^6} = \sqrt[3]{27x^6} \cdot \sqrt[3]{2} = 3x^2\sqrt[3]{2}$$

Note: 27 is a perfect cube and the exponent 6 is divisible by 3.

b. $\sqrt[3]{-40x^4y^{13}}$

Solution

$$\sqrt[3]{-40x^4y^{13}} = \sqrt[3]{-8x^3y^{12}} \cdot \sqrt[3]{5xy} = -2xy^4\sqrt[3]{5xy}$$

Note: −8 is a perfect cube and the exponents on the variables are separated so that one exponent on each variable is divisible by 3.

c. $\sqrt[3]{250a^8b^{11}}$

Solution

$$\sqrt[3]{250a^8b^{11}} = \sqrt[3]{125a^6b^9} \cdot \sqrt[3]{2a^2b^2} = 5a^2b^3\sqrt[3]{2a^2b^2}$$

Note: 125 is a perfect cube and the exponents on the variables are separated so that one exponent on each variable is divisible by 3.

Now work margin exercise 4.

4. Simplify each of the following radical expressions.

   a. $\sqrt[3]{48z^3}$

   b. $\sqrt[3]{-81a^8b^{12}}$

   c. $\sqrt[3]{686x^6y^{11}}$

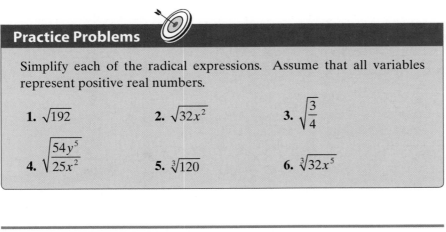

Practice Problems

Simplify each of the radical expressions. Assume that all variables represent positive real numbers.

1. $\sqrt{192}$ **2.** $\sqrt{32x^2}$ **3.** $\sqrt{\dfrac{3}{4}}$

4. $\sqrt{\dfrac{54y^5}{25x^2}}$ **5.** $\sqrt[3]{120}$ **6.** $\sqrt[3]{32x^5}$

Practice Problem Answers

1. $8\sqrt{3}$ **2.** $4x\sqrt{2}$ **3.** $\dfrac{\sqrt{3}}{2}$

4. $\dfrac{3y^2\sqrt{6y}}{5x}$ **5.** $2\sqrt[3]{15}$ **6.** $2x\sqrt[3]{4x^2}$

Exercises 14.2

Simplify each of the radical expressions. Assume all variables represent positive real numbers. See Examples 1 through 4.

1. $\sqrt{12}$

2. $-\sqrt{45}$

3. $\sqrt{288}$

4. $-\sqrt{63}$

5. $-\sqrt{72}$

6. $\sqrt{98}$

7. $-\sqrt{56}$

8. $\sqrt{162}$

9. $-\sqrt{125}$

10. $-\sqrt{121}$

11. $\sqrt{\dfrac{1}{4}}$

12. $\sqrt{\dfrac{32}{49}}$

13. $-\sqrt{\dfrac{11}{64}}$

14. $-\sqrt{\dfrac{125}{100}}$

15. $\sqrt{\dfrac{28}{25}}$

16. $\sqrt{\dfrac{147}{100}}$

17. $\sqrt{36x^2}$

18. $\sqrt{49y^2}$

19. $\sqrt{8x^3}$

20. $\sqrt{18a^5}$

21. $\sqrt{24x^{11}y^2}$

22. $\sqrt{20x^{15}y^3}$

23. $\sqrt{125x^3y^6}$

24. $\sqrt{8x^5y^4}$

25. $-\sqrt{18x^2y^2}$

26. $-\sqrt{32x^4y^8}$

27. $\sqrt{12ab^2c^3}$

28. $\sqrt{45a^2b^3c^4}$

29. $\sqrt{75x^4y^6z^8}$

30. $\sqrt{200x^2y^2z^2}$

31. $\sqrt{\dfrac{5x^4}{9}}$

32. $-\sqrt{\dfrac{7y^6}{16x^4}}$

33. $\sqrt{\dfrac{32a^5}{81b^{16}}}$

34. $\sqrt{\dfrac{75x^8}{121y^{12}}}$

35. $\sqrt{\dfrac{200x^8}{289}}$

36. $\sqrt{\dfrac{32x^{15}y^{10}}{169}}$

37. $\sqrt[3]{216}$

38. $\sqrt[3]{1}$

39. $\sqrt[3]{56}$

40. $\sqrt[3]{72}$

41. $\sqrt[3]{-1}$

42. $\sqrt[3]{-125}$

43. $\sqrt[3]{-128}$

44. $\sqrt[3]{-250}$

45. $\sqrt[3]{125x^4}$

46. $\sqrt[3]{64a^{12}}$

47. $\sqrt[3]{-8x^8}$

48. $\sqrt[3]{-512a^5}$

49. $\sqrt[3]{72a^6b^4}$

50. $\sqrt[3]{108ab^9}$

51. $\sqrt[3]{216x^6y^5}$

52. $\sqrt[3]{64x^9y^2}$

53. $\sqrt[3]{24x^5y^7z^9}$

54. $\sqrt[3]{250x^6y^9z^{15}}$

55. $\dfrac{\sqrt[3]{81}}{6}$

56. $\dfrac{\sqrt[3]{192}}{10}$

57. $\sqrt[3]{\dfrac{375}{8}}$

58. $\sqrt[3]{\dfrac{-48}{125}}$

59. $\sqrt[3]{\dfrac{125y^{12}}{27x^6}}$

60. $\sqrt[3]{\dfrac{x^6z^3}{64y^9}}$

Use the formula $s = \sqrt[3]{V}$, **which relates the length of the sides of a cube and the volume** V, **to answer the questions below.**

61. **Puzzle Cube:** The volume of a puzzle cube is 343 cubic inches. What is the length of one side?

62. **Building Blocks:** Three cubic blocks of different volumes were stacked on top of each other. The top block was 216 cubic centimeters. The middle block was 343 cubic centimeters, and the bottom block was 512 cubic centimeters. How tall was the stack of blocks?

Use the following two formulas used in electricity to answer the questions below.

$$I = \sqrt{\dfrac{P}{R}}$$
$$E = \sqrt{PR}$$

P = power (in watts)
I = current (in amperes)
E = voltage (in volts)
R = resistance (in ohms, Ω)

63. **Electricity:**   What is the current in amperes of a light bulb that produces 150 watts of power and has a 25 Ω resistance?

64. **Electricity:**   If a light bulb has a resistance of 30 Ω and produces 90 watts of power, what is its current in amperes?

65. **Electricity:**   How many volts of electricity would Meghan need to produce 48 Ω of resistance from a 300 watt lamp?

66. **Electricity:**   A 5000 Ω resistor is rated at 2.5 watts. What is the maximum voltage of electricity that should be connected across it?

Writing & Thinking

67.   Under what conditions is the expression $\sqrt{a}$ not a real number?

68.   Explain why the expression $\sqrt[3]{y}$ is a real number regardless of whether $y > 0$, $y < 0$, or $y = 0$.

14.3 Addition, Subtraction, and Multiplication with Radicals

Objective A   Addition and Subtraction with Radicals

Recall that to find the sum $2x^2 + 3x^2 - 8x^2$, you can use the distributive property and write

$$2x^2 + 3x^2 - 8x^2 = (2 + 3 - 8)x^2$$
$$= -3x^2.$$

Recall that the terms $2x^2, 3x^2,$ and $-8x^2$ are called **like terms** because each term contains the same variable expression, x^2. Similarly,

$$2\sqrt{5} + 3\sqrt{5} - 8\sqrt{5} = (2 + 3 - 8)\sqrt{5}$$
$$= -3\sqrt{5}$$

and $2\sqrt{5}, 3\sqrt{5},$ and $-8\sqrt{5}$ are called **like radicals** because each term contains the same radical expression, $\sqrt{5}$. **Like radicals** have the same index and radicand or they can be simplified so that they have the same index and radicand.

The terms $2\sqrt{3}$ and $2\sqrt{7}$ are **not** like radicals because the radicands are not the same, and neither expression can be simplified. Therefore, a sum such as

$$2\sqrt{3} + 2\sqrt{7}$$

cannot be simplified. That is, the terms cannot be combined.

In some cases, radicals that are not like radicals can be simplified, and the results may lead to like radicals. For example, $4\sqrt{12}, \sqrt{75},$ and $-\sqrt{108}$ are not like radicals. However, simplification of each radical allows the sum of these radicals to be found as follows:

$$4\sqrt{12} + \sqrt{75} - \sqrt{108} = 4\sqrt{4 \cdot 3} + \sqrt{25 \cdot 3} - \sqrt{36 \cdot 3}$$
$$= 4 \cdot 2\sqrt{3} + 5\sqrt{3} - 6\sqrt{3}$$
$$= 8\sqrt{3} + 5\sqrt{3} - 6\sqrt{3}$$
$$= (8 + 5 - 6)\sqrt{3}$$
$$= 7\sqrt{3}$$

Example 1

Addition and Subtraction with Radicals

Perform the indicated operation and simplify, if possible. Assume that all variables represent positive real numbers.

a. $\sqrt{32x} + \sqrt{18x}$

Solution

$$\sqrt{32x} + \sqrt{18x} = \sqrt{16 \cdot 2x} + \sqrt{9 \cdot 2x}$$
$$= 4\sqrt{2x} + 3\sqrt{2x}$$
$$= (4+3)\sqrt{2x}$$
$$= 7\sqrt{2x}$$

b. $\sqrt{12} + \sqrt{18} + \sqrt{27}$

Solution

$$\sqrt{12} + \sqrt{18} + \sqrt{27}$$
$$= \sqrt{4 \cdot 3} + \sqrt{9 \cdot 2} + \sqrt{9 \cdot 3}$$
$$= 2\sqrt{3} + 3\sqrt{2} + 3\sqrt{3}$$
$$= (2+3)\sqrt{3} + 3\sqrt{2}$$
$$= 5\sqrt{3} + 3\sqrt{2}$$

Note that $\sqrt{3}$ and $\sqrt{2}$ are **not** like radicals. Therefore, the last expression is already fully simplified.

c. $\sqrt[3]{5x} - \sqrt[3]{40x}$

Solution

$$\sqrt[3]{5x} - \sqrt[3]{40x} = \sqrt[3]{5x} - \sqrt[3]{8 \cdot 5x}$$
$$= \sqrt[3]{5x} - 2\sqrt[3]{5x}$$
$$= (1-2)\sqrt[3]{5x}$$
$$= -\sqrt[3]{5x}$$

d. $x\sqrt{4y^3} - 5\sqrt{x^2 y^3}$

Solution

$$x\sqrt{4y^3} - 5\sqrt{x^2 y^3} = x\sqrt{4y^2}\sqrt{y} - 5\sqrt{x^2 y^2}\sqrt{y}$$
$$= 2xy\sqrt{y} - 5xy\sqrt{y}$$
$$= -3xy\sqrt{y}$$

Completion Example 2

Addition and Subtraction with Radicals

Perform the indicated operations and simplify, if possible. Assume that all variables represent positive real numbers.

a. $3\sqrt{6} + 5\sqrt{6} - \sqrt{6}$

Solution

$$3\sqrt{6} + 5\sqrt{6} - \sqrt{6} = (\underline{\qquad})\sqrt{6} = \underline{\qquad}\sqrt{6}$$

b. $\sqrt{a} + 4\sqrt{b} + 8\sqrt{a} + 3\sqrt{b}$

Solution

$$\sqrt{a} + 4\sqrt{b} + 8\sqrt{a} + 3\sqrt{b} = (\underline{\qquad})\sqrt{a} + (\underline{\qquad})\sqrt{b}$$
$$= \underline{\qquad}\sqrt{a} + \underline{\qquad}\sqrt{b}$$

Now work margin exercises 1 and 2.

Perform the indicated operation and simplify, if possible. Assume that all variables are positive.

1. a. $\sqrt{75a} + \sqrt{48a}$

   b. $\sqrt{45} + \sqrt{20} + \sqrt{27}$

   c. $\sqrt[3]{9x} - \sqrt[3]{72x}$

2. a. $4\sqrt{7} - 6\sqrt{7} + 8\sqrt{7}$

   b. $8\sqrt{x} - 5\sqrt{3} - 2\sqrt{3} - 6\sqrt{x}$

Objective B **Multiplication with Radical Expressions**

To find a product such as $\left(\sqrt{3} + 5\right)\left(\sqrt{3} - 7\right)$ treat the two expressions as two binomials and multiply just as with polynomials. For example, using the FOIL method, we get

$$\left(\sqrt{3} + 5\right)\left(\sqrt{3} - 7\right) = \left(\sqrt{3}\right)^2 - 7\sqrt{3} + 5\sqrt{3} + 5(-7)$$
$$= 3 - 7\sqrt{3} + 5\sqrt{3} - 35$$
$$= 3 - 35 + (-7 + 5)\sqrt{3}$$
$$= -32 - 2\sqrt{3}.$$

Example 3

Multiplication with Radicals

Multiply and simplify the following expressions.

a. $\sqrt{5} \cdot \sqrt{15}$

Solution

$$\sqrt{5} \cdot \sqrt{15} = \sqrt{5} \cdot \sqrt{5} \cdot \sqrt{3} = \left(\sqrt{5}\right)^2 \cdot \sqrt{3} = 5\sqrt{3}$$

Completion Example Answers
2. a. $(3 + 5 - 1)\sqrt{6} = 7\sqrt{6}$
b. $(1 + 8)\sqrt{a} + (4 + 3)\sqrt{b} = 9\sqrt{a} + 7\sqrt{b}$

b. $\sqrt{7}\left(\sqrt{7} - \sqrt{14}\right)$

Solution

$$\sqrt{7}\left(\sqrt{7} - \sqrt{14}\right) = \sqrt{7} \cdot \sqrt{7} - \sqrt{7} \cdot \sqrt{14}$$
$$= \left(\sqrt{7}\right)^2 - \sqrt{7}\left(\sqrt{7} \cdot \sqrt{2}\right)$$
$$= \left(\sqrt{7}\right)^2 - \left(\sqrt{7}\right)^2 \sqrt{2}$$
$$= 7 - 7\sqrt{2}$$

c. $\left(3\sqrt{7} - 2\right)\left(\sqrt{7} + 3\right)$

Solution

$$\left(3\sqrt{7} - 2\right)\left(\sqrt{7} + 3\right) = 3\left(\sqrt{7}\right)^2 + 3 \cdot 3\sqrt{7} - 2\sqrt{7} - 2 \cdot 3$$
$$= 3 \cdot 7 + 9\sqrt{7} - 2\sqrt{7} - 6$$
$$= 21 - 6 + (9 - 2)\sqrt{7}$$
$$= 15 + 7\sqrt{7}$$

d. $\left(\sqrt{6} + \sqrt{2}\right)^2$

Solution

$$\left(\sqrt{6} + \sqrt{2}\right)^2 = \left(\sqrt{6}\right)^2 + 2\sqrt{6}\sqrt{2} + \left(\sqrt{2}\right)^2 \qquad (a + b)^2 = a^2 + 2ab + b^2$$
$$= 6 + 2\sqrt{12} + 2$$
$$= 8 + 2\sqrt{4} \cdot \sqrt{3}$$
$$= 8 + 2 \cdot 2\sqrt{3}$$
$$= 8 + 4\sqrt{3}$$

e. $\left(\sqrt{2x} + 5\right)\left(\sqrt{2x} - 5\right)$

Solution

$$\left(\sqrt{2x} + 5\right)\left(\sqrt{2x} - 5\right) = \left(\sqrt{2x}\right)^2 - (5)^2 \qquad (a + b)(a - b) = a^2 - b^2$$
$$= 2x - 25$$

Completion Example 4

Multiplication with Radicals

Multiply and simplify the following expressions. Assume that all variables represent positive numbers.

a. $6\sqrt{3}\left(\sqrt{3}+2\sqrt{7}\right)$

Solution

$$6\sqrt{3}\left(\sqrt{3}+2\sqrt{7}\right)=\underline{\hspace{1cm}}\cdot\sqrt{3}+\underline{\hspace{1cm}}\cdot 2\sqrt{7}=\underline{\hspace{1cm}}+\underline{\hspace{1cm}}$$

b. $\left(\sqrt{x}+8\right)\left(\sqrt{x}-4\right)$

Solution

$$\left(\sqrt{x}+8\right)\left(\sqrt{x}-4\right)=\sqrt{x}\cdot\underline{\hspace{1cm}}+\sqrt{x}\cdot\left(\underline{\hspace{1cm}}\right)+8\cdot\underline{\hspace{1cm}}+8\cdot\left(\underline{\hspace{1cm}}\right)$$

$$=\underline{\hspace{1cm}}+\underline{\hspace{1cm}}-\underline{\hspace{1cm}}$$

Now work margin exercises 3 and 4.

Objective C **Evaluating Radical Expressions with a TI-84 Plus Graphing Calculator**

Techniques for using a TI-84 Plus graphing calculator to evaluate radical expressions were illustrated in Section 14.1. These same basic techniques are used to evaluate numerical expressions that contain sums, differences, products, and quotients of radicals. Be careful to use parentheses to ensure that the rules for order of operations are maintained. In particular, sums and differences in numerators and denominators of fractions must be enclosed in parentheses. Study the following examples carefully.

Multiply and simplify the following expressions. Assume that all variables represent positive real numbers.

3. a. $\sqrt{3}\cdot\sqrt{75}$

b. $\sqrt{13}\left(\sqrt{26}-\sqrt{13}\right)$

c. $\left(2\sqrt{5}-1\right)\left(\sqrt{5}+2\right)$

d. $\left(\sqrt{7}-\sqrt{5}\right)^2$

e. $\left(\sqrt{3z}-\sqrt{5}\right)\left(\sqrt{3z}+\sqrt{5}\right)$

4. a. $4\sqrt{5}\left(\sqrt{6}-\sqrt{5}\right)$

b. $\left(\sqrt{s}-7\right)\left(9+\sqrt{s}\right)$

Completion Example Answers

4 a. $6\sqrt{3}\cdot\sqrt{3}+6\sqrt{3}\cdot 2\sqrt{7}=18+12\sqrt{21}$

b. $\sqrt{x}\cdot\sqrt{x}+\sqrt{x}\cdot(-4)+8\cdot\sqrt{x}+8\cdot(-4)=x+4\sqrt{x}-32$

5. Use a TI-84 Plus graphing calculator to evaluate each expression. Round answers to the nearest ten-thousandth.

a. $4 + 5\sqrt{2}$

b. $\left(\sqrt{8} - 1\right)\left(\sqrt{8} + 1\right)$

Example 5

Using a TI-84 Graphing Calculator to Evaluate Radical Expressions

Use a TI-84 Plus graphing calculator to evaluate each expression. Round answers to the nearest ten-thousandth.

a. $3 + 2\sqrt{5}$

Solution

The display should appear as follows.

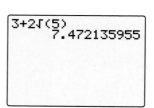

Thus $3 + 2\sqrt{5} = 7.4721359... \approx 7.4721.$ rounded to the nearest ten-thousandth

b. $\left(\sqrt{2} + 5\right)\left(\sqrt{2} - 5\right)$

Solution

The display should appear as follows.

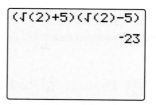

Note: The right parenthesis on 2 must be included. Otherwise, the calculator will interpret the expression as $\sqrt{(2+5)}$ (or $\sqrt{7}$) which is not intended.

Thus, $\left(\sqrt{2} + 5\right)\left(\sqrt{2} - 5\right) = -23.$

Now work margin exercise 5.

Practice Problems

Simplify each expression. Assume that all variables represent positive real numbers.

1. $2\sqrt{10} - 6\sqrt{10}$ **2.** $\sqrt{5} + \sqrt{45} - \sqrt{15}$ **3.** $\sqrt{8x} - 3\sqrt{2x} + \sqrt{18x}$

4. $\sqrt[3]{x^5} + x\sqrt[3]{27x^2}$ **5.** $\left(\sqrt{3} + \sqrt{2}\right)^2$ **6.** $\left(3 + \sqrt{2}\right)^2$

7. $\left(\sqrt{3} + \sqrt{8}\right)^2$

Practice Problem Answers

1. $-4\sqrt{10}$ **2.** $4\sqrt{5} - \sqrt{15}$ **3.** $2\sqrt{2x}$

4. $4x\sqrt[3]{x^2}$ **5.** $5 + 2\sqrt{6}$ **6.** $11 + 6\sqrt{2}$ **7.** $11 + 4\sqrt{6}$

Exercises 14.3

Simplify the following radical expressions. Assume that all variables represent positive real numbers. See Examples 1 and 2.

1. $3\sqrt{2} + 5\sqrt{2}$

2. $7\sqrt{3} - 2\sqrt{3}$

3. $4\sqrt{11} - 3\sqrt{11}$

4. $6\sqrt{5} + \sqrt{5}$

5. $8\sqrt{10} - 11\sqrt{10}$

6. $6\sqrt{17} - 9\sqrt{17}$

7. $4\sqrt[3]{3} + 9\sqrt[3]{3}$

8. $11\sqrt[3]{14} - 6\sqrt[3]{14}$

9. $6\sqrt{11} - 5\sqrt{11} - 2\sqrt{11}$

10. $\sqrt{7} + 6\sqrt{7} - 2\sqrt{7}$

11. $\sqrt{a} + 4\sqrt{a} - 2\sqrt{a}$

12. $2\sqrt{x} - 3\sqrt{x} + 7\sqrt{x}$

13. $5\sqrt{x} + 3\sqrt{x} - \sqrt{x}$

14. $6\sqrt{xy} - 10\sqrt{xy} + \sqrt{xy}$

15. $3\sqrt{2} + 5\sqrt{3} - 2\sqrt{3} + \sqrt{2}$

16. $\sqrt{5} + \sqrt{4} - 2\sqrt{5} + 6$

17. $2\sqrt{a} + 7\sqrt{b} - 6\sqrt{a} + \sqrt{b}$

18. $4\sqrt{x} - 3\sqrt{x} + 2\sqrt{y} + 2\sqrt{x}$

19. $6\sqrt[3]{x} - 4\sqrt[3]{y} + 7\sqrt[3]{x} + 2\sqrt[3]{y}$

20. $5\sqrt[3]{x} + 9\sqrt[3]{y} - 10\sqrt[3]{y} + 4\sqrt[3]{x}$

21. $\sqrt{12} + \sqrt{27}$

22. $\sqrt{32} - \sqrt{18}$

23. $3\sqrt{5} - \sqrt{45}$

24. $2\sqrt{7} + 5\sqrt{28}$

25. $3\sqrt[3]{54} + 8\sqrt[3]{2}$

26. $2\sqrt[3]{128} + 5\sqrt[3]{-54}$

27. $\sqrt{50} - \sqrt{18} - 3\sqrt{12}$

28. $2\sqrt{48} - \sqrt{54} + \sqrt{27}$

29. $2\sqrt{20} - \sqrt{45} + \sqrt{36}$

30. $\sqrt{18} - 2\sqrt{12} + 5\sqrt{2}$

31. $\sqrt{8} - 2\sqrt{3} + \sqrt{27} - \sqrt{72}$

32. $\sqrt{80} + \sqrt{8} - \sqrt{45} + \sqrt{50}$

33. $5\sqrt[3]{16} - 4\sqrt[3]{24} + \sqrt[3]{-250}$

34. $\sqrt[3]{192} - 2\sqrt[3]{128} + \sqrt[3]{-81}$

35. $6\sqrt{2x} - \sqrt{8x}$

36. $5\sqrt{3x} + 2\sqrt{12x}$

37. $5y\sqrt{2y} - y\sqrt{18y}$

38. $9x\sqrt{xy} - x\sqrt{16xy}$

39. $4x\sqrt{3xy} - x\sqrt{12xy} - 2x\sqrt{27xy}$

40. $x\sqrt{32x} - x\sqrt{50x} + 2x\sqrt{18x}$

41. $\sqrt{36x^3} + \sqrt{81x^3}$

42. $\sqrt{4a^2b} + \sqrt{9a^2b}$

43. $\sqrt{16x^3y^4} - \sqrt{25x^3y^4}$

44. $\sqrt{72x^{12}y^{15}} + \sqrt{18x^{12}y^{15}} + \sqrt{2x^{12}y^{15}}$

45. $\sqrt{12x^{10}y^{20}} + \sqrt{27x^{10}y^{20}} - \sqrt{3x^{10}y^{20}}$

46. $\sqrt[3]{8a^{12}} + \sqrt[3]{1000a^{12}}$

47. $\sqrt[3]{-27x^{24}y^6} + \sqrt[3]{-125x^{24}y^6}$

48. $\sqrt[3]{27a^{15}b} + \sqrt[3]{8a^{15}b} + \sqrt[3]{64a^{15}b}$

49. $\sqrt[3]{-16x^9y^{12}} - \sqrt[3]{16x^{12}y^9} + \sqrt[3]{54x^3y^6}$

50. $\sqrt[3]{54x^{13}y^3} + \sqrt[3]{8x^{23}y^6} + \sqrt[3]{3x^{13}y^3}$

Multiply the following radical expressions and then simplify the result. Assume that all variables represent positive real numbers. See Examples 3 and 4.

51. $\sqrt{2}\left(3 - 4\sqrt{2}\right)$

52. $2\sqrt{7}\left(\sqrt{7} + 3\sqrt{2}\right)$

53. $3\sqrt{18} \cdot \sqrt{2}$

54. $2\sqrt{10} \cdot \sqrt{5}$

55. $-2\sqrt{6} \cdot \sqrt{8}$

56. $2\sqrt{15} \cdot 5\sqrt{6}$

57. $\sqrt{3}\left(\sqrt{2} + 2\sqrt{12}\right)$

58. $\sqrt{2}\left(\sqrt{3} - \sqrt{6}\right)$

59. $\sqrt{y}\left(\sqrt{x} + 2\sqrt{y}\right)$

60. $\sqrt{x}\left(\sqrt{x} - 3\sqrt{y}\right)$

61. $\left(3 + \sqrt{2}\right)\left(5 - \sqrt{2}\right)$

62. $\left(\sqrt{6} + 2\right)\left(\sqrt{6} - 2\right)$

63. $\left(\sqrt{3x} - 8\right)\left(\sqrt{3x} - 1\right)$

64. $\left(6 + \sqrt{2x}\right)\left(4 + \sqrt{2x}\right)$

65. $\left(2\sqrt{7} + 4\right)\left(\sqrt{7} - 3\right)$

66. $\left(5\sqrt{3} - 2\right)\left(2\sqrt{3} - 7\right)$

67. $\left(\sqrt{5} + 2\sqrt{2}\right)^2$

68. $\left(2\sqrt{5} + 3\sqrt{2}\right)^2$

69. $\left(\sqrt{2} + \sqrt{3}\right)\left(\sqrt{5} - \sqrt{3}\right)$

70. $\left(\sqrt{6} + \sqrt{5}\right)\left(\sqrt{6} - \sqrt{2}\right)$

71. $\left(\sqrt{x} + \sqrt{6}\right)\left(\sqrt{x} - 3\sqrt{6}\right)$

72. $\left(\sqrt{11} + \sqrt{3}\right)\left(\sqrt{11} - 2\sqrt{3}\right)$

73. $\left(3\sqrt{7} + \sqrt{5}\right)\left(3\sqrt{7} - \sqrt{5}\right)$

74. $\left(7\sqrt{x} + \sqrt{2}\right)\left(7\sqrt{x} - \sqrt{2}\right)$

75. $\left(\sqrt{x} + 5\sqrt{y}\right)^2$

76. $\left(3\sqrt{x} + \sqrt{y}\right)^2$

77. $13 - \sqrt{75}$

78. $5 - \sqrt{67}$

79. $\sqrt{900} + \sqrt{2.56}$

80. $\sqrt{1600} - \sqrt{1.69}$

81. $\left(\sqrt{7} + 8\right)\left(\sqrt{7} - 8\right)$

82. $\left(\sqrt{8} - \sqrt{5}\right)\left(\sqrt{8} + \sqrt{5}\right)$

83. $\left(2\sqrt{3} + 5\sqrt{2}\right)\left(\sqrt{10} - 3\sqrt{5}\right)$

84. $\left(6\sqrt{5} + 5\sqrt{7}\right)\left(3\sqrt{2} - \sqrt{6}\right)$

Solve the following word problems.

85. **Radio Circuits:** For a complete radio circuit, $d = \sqrt{2g} + \sqrt{2h}$, where d equals the visual horizon distance and g and h are the heights of the radio antennas at the respective stations. What is d when $g = 75$ ft and $h = 85$ ft?

86. **Tile Patterns:** Mary is making a tile decoration for her wall. Using square tiles of different sizes, Mary created one decoration that is five tiles across, with sides touching. The first tile is 10 in.^{2}, the second is 20 in.^{2}, the third is 30 in.^{2}, the fourth is 20 in.^{2}, and the fifth is 10 in.^{2} What is the length of the decoration?

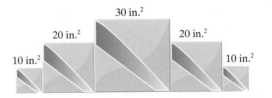

14.4 Rationalizing Denominators

Objective A **Rationalizing Denominators with One Term in the Denominator**

Each of the expressions

$$\frac{5}{\sqrt{3}}, \quad \frac{\sqrt{7}}{\sqrt{8}}, \quad \text{and} \quad \frac{2}{3-\sqrt{2}}$$

contain a radical in the denominator that is an irrational number. Such expressions are not considered in simplest form because they are difficult to operate with algebraically. Calculations of sums and differences are much easier if the denominators are rational expressions. So, in simplifying, **the objective is to find an equivalent fraction that has a rational number or an expression with no radicals for a denominator**.

That is, we want to simplify the expression by **rationalizing the denominator**. The following examples illustrate the method. Each fraction is multiplied by 1 in the form $\dfrac{k}{k}$, where multiplication by k rationalizes the denominator.

To Rationalize a Denominator Containing a Square Root or a Cube Root

1. If the denominator contains a square root, multiply both the numerator and denominator by an expression that will give a denominator with no square roots.

2. If the denominator contains a cube root, multiply both the numerator and denominator by an expression that will give a denominator with no cube roots.

Example 1

Rationalizing Denominators

a. $\dfrac{5}{\sqrt{3}}$

Solution

$$\frac{5}{\sqrt{3}} = \frac{5 \cdot \sqrt{3}}{\sqrt{3} \cdot \sqrt{3}} = \frac{5\sqrt{3}}{3}$$

Multiply the numerator and denominator by $\sqrt{3}$ because $\sqrt{3} \cdot \sqrt{3} = 3$, a rational number.

b. $\dfrac{4}{\sqrt{x}}$ $\left(\text{Assume } x > 0\right)$

Solution

$$\frac{4}{\sqrt{x}} = \frac{4 \cdot \sqrt{x}}{\sqrt{x} \cdot \sqrt{x}} = \frac{4\sqrt{x}}{x}$$

Multiply the numerator and denominator by $\sqrt{x}$ because $\sqrt{x} \cdot \sqrt{x} = x$. There is no guarantee that x is rational, but the radical sign does not appear in the denominator of the result.

c. $\dfrac{3}{7\sqrt{2}}$

Solution

$$\dfrac{3}{7\sqrt{2}} = \dfrac{3 \cdot \sqrt{2}}{7\sqrt{2} \cdot \sqrt{2}} = \dfrac{3\sqrt{2}}{7 \cdot 2} = \dfrac{3\sqrt{2}}{14}$$

Multiply the numerator and denominator by $\sqrt{2}$ because $\sqrt{2} \cdot \sqrt{2} = 2$ is a rational number.

d. $\dfrac{\sqrt{7}}{\sqrt{8}}$

Solution

$$\dfrac{\sqrt{7}}{\sqrt{8}} = \dfrac{\sqrt{7} \cdot \sqrt{2}}{\sqrt{8} \cdot \sqrt{2}} = \dfrac{\sqrt{14}}{\sqrt{16}} = \dfrac{\sqrt{14}}{4}$$

Multiply the numerator and denominator by $\sqrt{2}$ because $\sqrt{8} \cdot \sqrt{2} = \sqrt{16} = 4$ is a rational number. (**Note:** $8 \cdot 2 = 16$ and 16 is a perfect square.)

Now work margin exercise 1.

In Example 1d, if the numerator and denominator are multiplied by $\sqrt{8}$, the results will be the same, but one more step will be added because the fraction can be reduced.

$$\dfrac{\sqrt{7}}{\sqrt{8}} = \dfrac{\sqrt{7} \cdot \sqrt{8}}{\sqrt{8} \cdot \sqrt{8}} = \dfrac{\sqrt{56}}{8} = \dfrac{\sqrt{4} \cdot \sqrt{14}}{8} = \dfrac{2\sqrt{14}}{8} = \dfrac{\sqrt{14}}{4}$$

Objective B **Rationalizing Denominators with a Sum or Difference in the Denominator**

If the denominator has a radical expression with **a sum or difference involving square roots** such as

$$\dfrac{2}{4 - \sqrt{2}} \quad \text{or} \quad \dfrac{12}{3 + \sqrt{5}},$$

then a different method is used for rationalizing the denominator. In this method, we think of the denominator in the form of $a - b$ or $a + b$. Thus

if $a - b = 4 - \sqrt{2}$, then $a + b = 4 + \sqrt{2}$

and

if $a + b = 3 + \sqrt{5}$, then $a - b = 3 - \sqrt{5}$.

The two expressions $(a - b)$ and $(a + b)$ are called **conjugates** of each other, and, as we know, their product $(a - b)(a + b)$ results in the **difference of two squares**:

$$(a - b)(a + b) = a^2 - b^2.$$

1. Rationalize each denominator.

a. $\dfrac{8}{\sqrt{5}}$

b. $\dfrac{17}{\sqrt{y}}$ (Assume $y > 0$)

c. $\dfrac{4}{9\sqrt{3}}$

d. $\dfrac{\sqrt{11}}{\sqrt{18}}$

Example 2

Rationalizing a Denominator with a Sum or Difference Involving a Square Root

Simplify each expression by rationalizing the denominator.

a. $\dfrac{2}{4 - \sqrt{2}}$

Solution

Multiply the numerator and denominator by $4 + \sqrt{2}$.

$$\frac{2}{4 - \sqrt{2}} = \frac{2\left(4 + \sqrt{2}\right)}{\left(4 - \sqrt{2}\right)\left(4 + \sqrt{2}\right)}$$ If $a - b = 4 - \sqrt{2}$, then $a + b = 4 + \sqrt{2}$.

$$= \frac{2\left(4 + \sqrt{2}\right)}{4^2 - \left(\sqrt{2}\right)^2}$$ The denominator is the difference of two squares.

$$= \frac{2\left(4 + \sqrt{2}\right)}{16 - 2}$$

$$= \frac{\overset{1}{\cancel{2}}\left(4 + \sqrt{2}\right)}{\underset{7}{\cancel{14}}}$$

$$= \frac{4 + \sqrt{2}}{7}$$ The denominator is a rational number. Note that the numerator is now irrational. However, this is generally preferred over having an irrational denominator.

b. $\dfrac{31}{6 + \sqrt{5}}$

Solution

Multiply the numerator and denominator by $6 - \sqrt{5}$.

$$\frac{31}{6+\sqrt{5}} = \frac{31\left(6-\sqrt{5}\right)}{\left(6+\sqrt{5}\right)\left(6-\sqrt{5}\right)}$$

$$= \frac{31\left(6-\sqrt{5}\right)}{36-5}$$

$$= \frac{\cancel{31}\left(6-\sqrt{5}\right)}{\cancel{31}}$$

$$= 6-\sqrt{5}$$

c. $\dfrac{1}{\sqrt{7}-\sqrt{2}}$

Solution

Multiply the numerator and denominator by $\sqrt{7}+\sqrt{2}$.

$$\frac{1}{\sqrt{7}-\sqrt{2}} = \frac{1\left(\sqrt{7}+\sqrt{2}\right)}{\left(\sqrt{7}-\sqrt{2}\right)\left(\sqrt{7}+\sqrt{2}\right)}$$

$$= \frac{\sqrt{7}+\sqrt{2}}{7-2} = \frac{\sqrt{7}+\sqrt{2}}{5}$$

d. $\dfrac{6}{1+\sqrt{x}}$

Solution

Multiply the numerator and denominator by $1-\sqrt{x}$.

$$\frac{6}{1+\sqrt{x}} = \frac{6\left(1-\sqrt{x}\right)}{\left(1+\sqrt{x}\right)\left(1-\sqrt{x}\right)}$$

$$= \frac{6\left(1-\sqrt{x}\right)}{1-x}$$

$$= \frac{6-6\sqrt{x}}{1-x}$$

e. $\dfrac{x-y}{\sqrt{x}-\sqrt{y}}$

Solution

Multiply the numerator and denominator by $\sqrt{x}+\sqrt{y}$.

$$\frac{x-y}{\sqrt{x}-\sqrt{y}} = \frac{\left(x-y\right)\left(\sqrt{x}+\sqrt{y}\right)}{\left(\sqrt{x}-\sqrt{y}\right)\left(\sqrt{x}+\sqrt{y}\right)}$$

$$= \frac{\cancel{\left(x-y\right)}\left(\sqrt{x}+\sqrt{y}\right)}{\cancel{x-y}}$$

$$= \sqrt{x}+\sqrt{y}$$

Simplify each expression by rationalizing the denominator.

2. a. $\dfrac{5}{\sqrt{3}-5}$

b. $\dfrac{15}{7+\sqrt{6}}$

c. $\dfrac{1}{\sqrt{8}+\sqrt{3}}$

d. $\dfrac{8}{1-\sqrt{z}}$

e. $\dfrac{x+y}{\sqrt{x}+\sqrt{y}}$

3. $\dfrac{11}{8-\sqrt{3}}$

Completion Example 3

Rationalizing Denominators

Rationalize the denominator and simplify: $\dfrac{74}{9+\sqrt{7}}$.

Solution

$$\frac{74}{9+\sqrt{7}} = \frac{74(\underline{\quad})}{(9+\sqrt{7})(\underline{\quad})} = \frac{74(\underline{\quad})}{\underline{\quad}^2 - (\underline{\quad})^2}$$

$$= \frac{74(\underline{\quad})}{\underline{\quad} - (\underline{\quad})} = \frac{74(\underline{\quad})}{\underline{\quad}} = \underline{\quad}$$

Now work margin exercises 2 and 3.

Practice Problems

Simplify each expression. Assume that all variables represent positive real numbers.

1. $\sqrt{\dfrac{3}{8a^2}}$

2. $\dfrac{\sqrt{3ab}}{\sqrt{5a^3 b^6}}$

3. $\dfrac{4}{\sqrt{2}+\sqrt{6}}$

4. $\dfrac{x-5}{\sqrt{x}-\sqrt{5}}$

5. $\dfrac{\sqrt{5}-3\sqrt{2}}{\sqrt{6}+\sqrt{10}}$

Completion Example Answers

3. $\dfrac{74\left(9-\sqrt{7}\right)}{\left(9+\sqrt{7}\right)\left(9-\sqrt{7}\right)} = \dfrac{74\left(9-\sqrt{7}\right)}{9^2 - \left(\sqrt{7}\right)^2} = \dfrac{74\left(9-\sqrt{7}\right)}{81-7} = \dfrac{\cancel{74}\left(9-\sqrt{7}\right)}{\cancel{74}} = 9-\sqrt{7}$

Practice Problem Answers

1. $\dfrac{\sqrt{6}}{4a}$

2. $\dfrac{\sqrt{15b}}{5ab^3}$

3. $\sqrt{6}-\sqrt{2}$

4. $\sqrt{x}+\sqrt{5}$

5. $\dfrac{5\sqrt{2}+6\sqrt{3}-6\sqrt{5}-\sqrt{30}}{4}$

Exercises 14.4

1. $\dfrac{5}{\sqrt{2}}$

2. $\dfrac{7}{\sqrt{5}}$

3. $\dfrac{-3}{\sqrt{7}}$

4. $\dfrac{-10}{\sqrt{2}}$

5. $\dfrac{6}{\sqrt{3}}$

6. $\dfrac{8}{\sqrt{2}}$

7. $\dfrac{\sqrt{18}}{\sqrt{2}}$

8. $\dfrac{\sqrt{25}}{\sqrt{3}}$

9. $\dfrac{\sqrt{27x}}{\sqrt{3x}}$

10. $\dfrac{\sqrt{45y}}{\sqrt{5y}}$

11. $\dfrac{\sqrt{ab}}{\sqrt{9ab}}$

12. $\dfrac{\sqrt{5}}{\sqrt{12}}$

13. $\dfrac{\sqrt{4}}{\sqrt{3}}$

14. $\sqrt{\dfrac{3}{8}}$

15. $\sqrt{\dfrac{9}{2}}$

16. $\sqrt{\dfrac{3}{5}}$

17. $\sqrt{\dfrac{1}{x}}$

18. $\sqrt{\dfrac{x}{y}}$

19. $\sqrt{\dfrac{2x}{y}}$

20. $\sqrt{\dfrac{x}{4y}}$

21. $\dfrac{2}{\sqrt{2y}}$

22. $\dfrac{-10}{3\sqrt{5}}$

23. $\dfrac{21}{5\sqrt{7}}$

24. $\dfrac{x}{5\sqrt{x}}$

25. $\dfrac{-2y}{5\sqrt{2y}}$

26. $\dfrac{3}{1+\sqrt{2}}$

27. $\dfrac{2}{\sqrt{6}-2}$

28. $\dfrac{-11}{\sqrt{3}-4}$

29. $\dfrac{1}{\sqrt{5}-3}$

30. $\dfrac{7}{3-2\sqrt{2}}$

31. $\dfrac{-6}{5-3\sqrt{2}}$

32. $\dfrac{11}{2\sqrt{3}+1}$

33. $\dfrac{-\sqrt{3}}{\sqrt{2}+5}$

34. $\dfrac{\sqrt{2}}{\sqrt{7}+4}$

35. $\dfrac{7}{1-3\sqrt{5}}$

36. $\dfrac{-3\sqrt{3}}{6+\sqrt{3}}$ **37.** $\dfrac{1}{\sqrt{3}-\sqrt{5}}$ **38.** $\dfrac{-4}{\sqrt{7}-\sqrt{3}}$ **39.** $\dfrac{-5}{\sqrt{2}+\sqrt{3}}$ **40.** $\dfrac{7}{\sqrt{2}+\sqrt{5}}$

41. $\dfrac{4}{\sqrt{x}+1}$ **42.** $\dfrac{-7}{\sqrt{x}-3}$ **43.** $\dfrac{5}{6+\sqrt{y}}$ **44.** $\dfrac{x}{\sqrt{x}+2}$ **45.** $\dfrac{8}{2\sqrt{x}+3}$

46. $\dfrac{3\sqrt{x}}{\sqrt{2x}-5}$ **47.** $\dfrac{\sqrt{4y}}{\sqrt{5y}-\sqrt{3}}$ **48.** $\dfrac{\sqrt{3x}}{\sqrt{2}+\sqrt{3x}}$ **49.** $\dfrac{3}{\sqrt{x}-\sqrt{y}}$ **50.** $\dfrac{4}{2\sqrt{x}+\sqrt{y}}$

51. $\dfrac{x}{\sqrt{x}+2\sqrt{y}}$ **52.** $\dfrac{y}{\sqrt{x}-\sqrt{3y}}$ **53.** $\dfrac{\sqrt{3}+1}{\sqrt{3}-2}$ **54.** $\dfrac{\sqrt{2}+4}{5-\sqrt{2}}$ **55.** $\dfrac{\sqrt{5}-2}{\sqrt{5}+3}$

56. $\dfrac{1+\sqrt{3}}{3-\sqrt{3}}$ **57.** $\dfrac{\sqrt{x}+1}{\sqrt{x}-1}$ **58.** $\dfrac{\sqrt{x}-4}{\sqrt{x}+3}$ **59.** $\dfrac{\sqrt{x}+2}{\sqrt{3x}+y}$ **60.** $\dfrac{3-\sqrt{x}}{2\sqrt{x}+y}$

Writing & Thinking

61. In your own words, explain how to rationalize the denominator of a fraction containing the sum or difference of square roots in the denominator. Why does this work?

14.5 Equations with Radicals

Objectives

A Solve equations that contain one or more radical expressions.

Objective A **Solving Equations that Contain One or More Radical Expressions**

Each of the following equations involves at least one radical expression.

$$x + 3 = \sqrt{x + 5} \qquad \sqrt{x} - \sqrt{2x - 14} = 1 \qquad \sqrt[3]{x + 1} = 5$$

If the radicals are square roots, we solve by squaring both sides of the equations. If the radical is some other root and this root can be isolated on one side of the equation, we solve by raising both sides of the equation to the integer power corresponding to the index of the radical. For example, with a cube root both sides are raised to the third power.

Squaring both sides of an equation may introduce new solutions. For example, the first-degree equation $x = -3$ has only one solution, namely, -3. However, squaring both sides gives the quadratic equation

$$x^2 = (-3)^2 \qquad \text{or} \qquad x^2 = 9.$$

The quadratic equation $x^2 = 9$ has two solutions, 3 and -3. Thus a new solution that is not a solution to the original equation has been introduced. Such a solution is called an **extraneous solution**.

When both sides of an equation are raised to a power, an extraneous solution may be introduced. Be sure to check all solutions in the original equation.

The following examples illustrate a variety of situations involving radicals. The steps used are related to the following general method.

Method for Solving Equations with Radicals

1. Isolate one of the radicals on one side of the equation. (An equation may have more than one radical.)
2. Raise both sides of the equation to the power corresponding to the index of the radical.
3. If the equation still contains a radical, repeat steps 1 and 2.
4. Solve the equation after all the radicals have been eliminated.
5. Be sure to check all possible solutions in the original equation and eliminate any extraneous solutions.

Example 1

Equations with One Radical

Solve the following equations.

a. $\sqrt{x^2 + 13} = 7$

Solution

The radical is by itself on one side of the equation, so square both sides.

$$\sqrt{x^2 + 13} = 7$$
$$\left(\sqrt{x^2 + 13}\right)^2 = 7^2 \qquad \text{Square both sides.}$$
$$x^2 + 13 = 49 \qquad \text{This new equation contains no radical.}$$
$$x^2 - 36 = 0$$
$$(x + 6)(x - 6) = 0 \qquad \text{Solve by factoring.}$$
$$x = -6 \quad \text{or} \quad x = 6$$

Check: **Check both answers** in the original equation.

$$\sqrt{(-6)^2 + 13} \overset{?}{=} 7 \qquad \sqrt{(6)^2 + 13} \overset{?}{=} 7$$
$$\sqrt{36 + 13} \overset{?}{=} 7 \qquad \sqrt{36 + 13} \overset{?}{=} 7$$
$$\sqrt{49} \overset{?}{=} 7 \qquad \sqrt{49} \overset{?}{=} 7$$
$$7 = 7 \qquad 7 = 7$$

Both −6 and 6 are solutions.

b. $\sqrt{y^2 - 10y - 11} = 1 + y$

Solution

Since there is only one radical and it is by itself on one side of the equation, square both sides.

$$\sqrt{y^2 - 10y - 11} = 1 + y$$
$$\left(\sqrt{y^2 - 10y - 11}\right)^2 = (1 + y)^2 \qquad \text{Square both sides.}$$
$$y^2 - 10y - 11 = 1 + 2y + y^2$$
$$-12y - 12 = 0 \qquad \text{Simplifying gives a first-degree equation.}$$
$$-12y = 12$$
$$y = -1$$

Check:

$$\sqrt{(-1)^2 - 10(-1) - 11} \stackrel{?}{=} 1 + (-1)$$

$$\sqrt{1 + 10 - 11} \stackrel{?}{=} 0$$

$$\sqrt{0} \stackrel{?}{=} 0$$

$$0 = 0$$

There is one solution, -1.

c. $\sqrt{3x + 13} + 3 = 2x$

Solution

$$\sqrt{3x + 13} + 3 = 2x$$

$$\sqrt{3x + 13} = 2x - 3 \qquad \text{Isolate the radical.}$$

$$\left(\sqrt{3x + 13}\right)^2 = (2x - 3)^2 \qquad \text{Square both sides.}$$

$$3x + 13 = 4x^2 - 12x + 9$$

$$0 = 4x^2 - 15x - 4$$

$$0 = (4x + 1)(x - 4) \qquad \text{Solve by factoring.}$$

$$x = -\frac{1}{4} \quad \text{or} \quad x = 4$$

Check: **Check both answers** in the original equation.

$$\sqrt{3\left(-\frac{1}{4}\right) + 13} + 3 \stackrel{?}{=} 2\left(-\frac{1}{4}\right) \qquad\qquad \sqrt{3(4) + 13} + 3 \stackrel{?}{=} 2(4)$$

$$\sqrt{\frac{49}{4}} + 3 \stackrel{?}{=} -\frac{1}{2} \qquad\qquad\qquad \sqrt{25} + 3 \stackrel{?}{=} 8$$

$$\frac{7}{2} + 3 \stackrel{?}{=} -\frac{1}{2} \qquad\qquad\qquad 5 + 3 \stackrel{?}{=} 8$$

$$\frac{13}{2} \neq -\frac{1}{2} \qquad\qquad\qquad 8 = 8$$

$-\dfrac{1}{4}$ is **not** a solution. The only solution is 4.

d. $\sqrt{x + 1} = -3$

Solution

We could stop right here. There is **no real solution** to this equation because the radical on the left is nonnegative and cannot possibly equal -3, a negative number. Suppose we did not notice this relationship. Then proceeding as usual, we will find an answer that is not a solution to the original equation.

1. Solve the following equations

    a. $\sqrt{x^2 - 48} = 4$

    b. $\sqrt{y^2 + y - 1} = y + 3$

    c. $\sqrt{-x^2 + x + 12} - 3 = x$

    d. $\sqrt{x + 5} = -9$

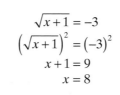

$$\sqrt{x + 1} = -3$$
$$\left(\sqrt{x + 1}\right)^2 = \left(-3\right)^2 \qquad \text{Square both sides.}$$
$$x + 1 = 9$$
$$x = 8$$

Check:

$$\sqrt{(8) + 1} \overset{?}{=} -3$$
$$\sqrt{9} \overset{?}{=} -3$$
$$3 \neq -3$$

So, 8 does **not** check, therefore there is **no solution**.

Now work margin exercise 1.

notes

It is possible that after checking the answers you may find that none of the answers are solutions. In this case the answer is **no solution**.

Completion Example 2

Solving Equations with Radicals

Solve the following equation: $\sqrt{5y + 6} = 3y - 2$.

Solution

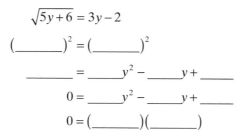

$$\sqrt{5y + 6} = 3y - 2$$
$$(\underline{\hspace{1cm}})^2 = (\underline{\hspace{1cm}})^2$$
$$\underline{\hspace{1cm}} = \underline{\hspace{0.5cm}}y^2 - \underline{\hspace{0.5cm}}y + \underline{\hspace{0.5cm}}$$
$$0 = \underline{\hspace{0.5cm}}y^2 - \underline{\hspace{0.5cm}}y + \underline{\hspace{0.5cm}}$$
$$0 = (\underline{\hspace{1cm}})(\underline{\hspace{1cm}})$$

Completion Example Answers

2. $\left(\sqrt{5y + 6}\right)^2 = \left(3y - 2\right)^2$

$$5y + 6 = 9y^2 - 12y + 4$$
$$0 = 9y^2 - 17y - 2$$
$$0 = \left(9y + 1\right)\left(y - 2\right)$$

$$\underline{\hspace{1.5cm}} = 0 \text{ or } \underline{\hspace{1.5cm}} = 0$$

$$y = \underline{\hspace{0.8cm}} \qquad\qquad y = \underline{\hspace{0.8cm}}$$

The number _____ does not check. The number _____ does check.

So the solution is _____.

Now work margin exercise 2.

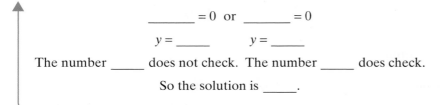

2. Solve the following equation:
$$\sqrt{3x+4} = 2x+1.$$

Example 3

Equations with Two Radicals

Solve the following equations with two radicals. You may need to rearrange terms to square twice.

a. $\sqrt{x+4} = \sqrt{3x-2}$

Solution

There are two radicals on opposite sides of the equation. Squaring both sides will give a new equation with no radicals.

$$\sqrt{x+4} = \sqrt{3x-2}$$
$$\left(\sqrt{x+4}\right)^2 = \left(\sqrt{3x-2}\right)^2$$
$$x+4 = 3x-2$$
$$6 = 2x \qquad \text{Simplifying gives a first-degree equation.}$$
$$3 = x$$

Check: $\sqrt{(3)+4} \overset{?}{=} \sqrt{3(3)-2}$

$$\sqrt{7} = \sqrt{7}$$

There is one solution, 3.

Completion Example Answers

2. $9y+1 = 0$ or $y-2 = 0$; $y = -\dfrac{1}{9}$; $y = 2$

The number $-\dfrac{1}{9}$ does not check. The number 2 does check.

So the solution is 2.

3. Solve the following equation with two radicals.

a. $\sqrt{2x+5} = \sqrt{x+7}$

b. $2\sqrt{x} - \sqrt{3x-11} = 2$

b. $\sqrt{x} - \sqrt{2x-14} = 1$

Solution

Where there is a sum or difference of radicals, squaring is easier if the radicals are on different sides of the equation. Also, squaring both sides of the equation is easier if one of the radicals is by itself on one side of the equation.

$$\sqrt{x} - \sqrt{2x-14} = 1$$
$$\sqrt{x} = 1 + \sqrt{2x-14} \qquad \text{Isolate one of the radicals.}$$
$$\left(\sqrt{x}\right)^2 = \left(1+\sqrt{2x-14}\right)^2 \qquad \text{Square both sides.}$$
$$x = 1 + 2\sqrt{2x-14} + (2x-14) \qquad \begin{array}{l}\text{Treat the right-hand side}\\\text{as the square of a binomial.}\end{array}$$
$$x = 2\sqrt{2x-14} + 2x - 13$$
$$-x + 13 = 2\sqrt{2x-14} \qquad \begin{array}{l}\text{Simplify so that the radical}\\\text{is on one side by itself.}\end{array}$$
$$\left(-x+13\right)^2 = \left(2\sqrt{2x-14}\right)^2 \qquad \text{Square both sides again.}$$
$$x^2 - 26x + 169 = 4(2x-14)$$
$$x^2 - 26x + 169 = 8x - 56$$
$$x^2 - 34x + 225 = 0$$
$$(x-9)(x-25) = 0 \qquad \text{Solve by factoring.}$$
$$x = 9 \quad \text{or} \quad x = 25$$

Check:

Check both answers in the original equation.

$$\sqrt{(9)} - \sqrt{2(9)-14} \overset{?}{=} 1 \qquad\qquad \sqrt{(25)} - \sqrt{2(25)-14} \overset{?}{=} 1$$
$$3 - \sqrt{4} \overset{?}{=} 1 \qquad\qquad 5 - \sqrt{36} \overset{?}{=} 1$$
$$3 - 2 \overset{?}{=} 1 \qquad\qquad 5 - 6 \overset{?}{=} 1$$
$$1 = 1 \qquad\qquad -1 \neq 1$$

25 is **not** a solution. The only solution is 9.

Now work margin exercise 3.

Example 4

Equations containing a Cube Root

Solve the following equation containing a cube root: $\sqrt[3]{2x+1}+1=3$

Solution

First, get the radical by itself on one side of the equation. Then since this radical is a cube root, cube both sides of the equation.

$$\sqrt[3]{2x+1}+1=3$$
$$\sqrt[3]{2x+1}=2 \qquad \text{Add} -1 \text{ to both sides.}$$
$$\left(\sqrt[3]{2x+1}\right)^3=2^3 \qquad \text{Cube both sides.}$$
$$2x+1=8 \qquad \text{Solve the equation.}$$
$$x=\frac{7}{2}$$

Check:

Check in the original equation.

$$\sqrt[3]{2\left(\frac{7}{2}\right)+1}+1\overset{?}{=}3$$
$$\sqrt[3]{7+1}+1\overset{?}{=}3$$
$$\sqrt[3]{8}+1\overset{?}{=}3$$
$$2+1\overset{?}{=}3$$
$$3=3$$

There is one solution, $\frac{7}{2}$.

Now work margin exercise 4.

4. Solve the following equation containing a cube root.

$$\sqrt[3]{3x+3}+2=5$$

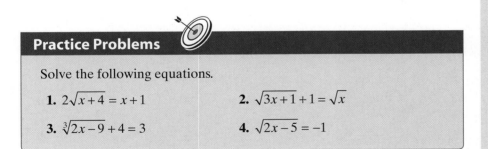

Practice Problems

Solve the following equations.

1. $2\sqrt{x+4}=x+1$

2. $\sqrt{3x+1}+1=\sqrt{x}$

3. $\sqrt[3]{2x-9}+4=3$

4. $\sqrt{2x-5}=-1$

Practice Problem Answers

1. $x=5$ **2.** no solution **3.** $x=4$ **4.** no solution

Exercises 14.5

Solve the following equations. Be sure to check your answers in the original equation. See Examples 1 through 4.

1. $\sqrt{8x+1} = 5$

2. $\sqrt{7x+1} = 6$

3. $\sqrt{3x+4} = -5$

4. $\sqrt{4x-3} = 7$

5. $\sqrt{6-x} = 3$

6. $\sqrt{11-x} = 5$

7. $\sqrt{5x-6} = 8$

8. $\sqrt{2x-5} = -1$

9. $\sqrt{5x+4} = 7$

10. $\sqrt{3x-2} = 4$

11. $\sqrt{x-4}+6 = 2$

12. $\sqrt{6x+4}+2 = 10$

13. $\sqrt{4x+1}+4 = 9$

14. $\sqrt{2x-7}+5 = 3$

15. $\sqrt{x(x+3)} = 2$

16. $\sqrt{x(x-5)} = 6$

17. $\sqrt{x(2x+5)} = 5$

18. $\sqrt{x(3x-14)} = 7$

19. $\sqrt{x+6} = x+4$

20. $\sqrt{x+7} = 2x-1$

21. $\sqrt{x-2} = x-2$

22. $\sqrt{x+3} = x+3$

23. $x-2 = \sqrt{3x-6}$

24. $x+6 = \sqrt{2x+12}$

25. $\sqrt{x^2-16} = 3$

26. $\sqrt{x^2-25} = 12$

27. $5+\sqrt{x+5}-2x = 0$

28. $x-2-\sqrt{x+4} = 0$

29. $2x = \sqrt{7x-3}+3$

30. $x-\sqrt{3x-8} = 4$

31. $\sqrt{2x+5} = \sqrt{4x-1}$

32. $\sqrt{5x-1} = \sqrt{x+7}$

33. $\sqrt{3x+2} = \sqrt{9x-10}$

34. $\sqrt{2+x} = \sqrt{2x-7}$

35. $\sqrt{2x-1} = \sqrt{x+1}$

36. $\sqrt{3x+2} = \sqrt{x+4}$

37. $\sqrt{x+2} = \sqrt{2x-5}$

38. $\sqrt{2x-5} = \sqrt{3x-9}$

39. $\sqrt{4x-3} = \sqrt{2x+5}$

40. $\sqrt{4x-6} = \sqrt{3x-1}$

41. $\sqrt{3x+1} = 1-\sqrt{x}$

42. $\sqrt{x} = \sqrt{x+16} - 2$

43. $\sqrt{x+4} = \sqrt{x+11} - 1$

44. $\sqrt{1-x} + 2 = \sqrt{13-x}$

45. $\sqrt{x+1} = \sqrt{x+6} + 1$

46. $\sqrt{x+4} = \sqrt{x+20} - 2$

47. $\sqrt{x+5} + \sqrt{x} = 5$

48. $\sqrt{x} + \sqrt{x-3} = 3$

49. $\sqrt{2x+3} = 1+\sqrt{x+1}$

50. $\sqrt{5x-18} - 4 = \sqrt{5x+6}$

51. $\sqrt{3x+1} - \sqrt{x+4} = 1$

52. $\sqrt{3x+4} - \sqrt{x+5} = 1$

53. $\sqrt{5x-1} = 4-\sqrt{x-1}$

54. $\sqrt{2x-5} - 2 = \sqrt{x-2}$

55. $\sqrt{2x-1} + \sqrt{x+3} = 3$

56. $\sqrt{2x+3} - \sqrt{x+5} = 1$

57. $\sqrt[3]{4+3x} = -2$

58. $\sqrt[3]{2+9x} = 9$

59. $\sqrt[3]{5x+4} = 4$

60. $\sqrt[3]{7x+1} = -5$

14.6 Rational Exponents

Objective A nth **Roots**

In Section 14.1 we restricted our discussions to radicals involving square roots and cube roots. In this section we will expand on those ideas by discussing radicals indicating n^{th} roots in general and how to relate radical expressions to expressions with rational (fractional) exponents. For example, the fifth root of x can be written in radical form as $\sqrt[5]{x}$ or with a fractional exponent as $x^{\frac{1}{5}}$.

To understand roots in general, consider the following notation. (**Note:** In this discussion, we assume that n is a positive integer.)

Type of Root	Radical Notation and Exponential Notation
For square roots,	if $b = \sqrt{a}$, then $b = a^{\frac{1}{2}}$.
For cube roots,	if $b = \sqrt[3]{a}$, then $b = a^{\frac{1}{3}}$.
For fourth roots,	if $b = \sqrt[4]{a}$, then $b = a^{\frac{1}{4}}$.
For n^{th} roots,	if $b = \sqrt[n]{a}$, then $b = a^{\frac{1}{n}}$.

Table 1

For example,

- because $\sqrt[4]{16} = 2$, we can say that $2 = 16^{\frac{1}{4}}$.

- because $\sqrt[5]{243} = 3$, we can say that $3 = 243^{\frac{1}{5}}$.

The following notation is used for all radical expressions.

Radical Notation

If n is a positive integer, then $\sqrt[n]{a} = a^{\frac{1}{n}}$ (assuming $\sqrt[n]{a}$ is a real number).

The expression $\sqrt[n]{a}$ is called a **radical**.

The symbol $\sqrt[n]{}$ is called a **radical sign**.

n is called the **index**.

a is called the **radicand**.

Note: If no index is given, it is understood to be 2. For example, $\sqrt{3} = \sqrt[2]{3} = 3^{\frac{1}{2}}$.

Objectives

A Understand the meaning of nth root.

B Translate expressions using radicals into expressions using rational exponents and translate expressions using rational exponents into expressions using radicals.

C Simplify expressions using the properties of rational exponents.

D Evaluate expressions of the form $a^{\frac{m}{n}}$ with a calculator.

notes

■ **Special Notes about the Index *n*:**

■ For the expression $\sqrt[n]{a}$ (or $a^{\frac{1}{n}}$) to be a real number:
    1. when *a* is nonnegative, *n* can be any index, and
■   **2.** when *a* is negative, *n* must be odd.
        (If *a* is negative and *n* is even, then $\sqrt[n]{a}$ is nonreal.)
■

Example 1

Evaluating Principal *n*th Roots

a. $49^{\frac{1}{2}} = \sqrt{49} = 7,$ because $7^2 = 49.$

b. $81^{\frac{1}{4}} = \sqrt[4]{81} = 3,$ because $3^4 = 81.$

c. $(-8)^{\frac{1}{3}} = \sqrt[3]{-8} = -2,$ because $(-2)^3 = -8.$

d. $(0.00001)^{\frac{1}{5}} = \sqrt[5]{0.00001} = 0.1,$ because $(0.1)^5 = 0.00001.$

e. $(-16)^{\frac{1}{2}} = \sqrt{-16}$ is not a real number. Any even root of a negative number is nonreal.

Now work margin exercise 1.

Objective B **Rational Exponents of the Form** $a^{\frac{n}{m}}$

In Chapter 11 we discussed the rules of exponents using only integer exponents. These same rules of exponents apply to rational exponents (fractional exponents) as well and are repeated here for easy reference.

---

1. Evaluate the *n*th root.

a. $36^{\frac{1}{2}}$

b. $16^{\frac{1}{4}}$

c. $(-27)^{\frac{1}{3}}$

d. $(0.0016)^{\frac{1}{4}}$

e. $(-25)^{\frac{1}{2}}$

Now consider the problem of evaluating the expression $8^{\frac{2}{3}}$ where the exponent, $\dfrac{2}{3}$, is of the form $\dfrac{m}{n}$. Using the power rule for exponents, we can write

$$8^{\frac{2}{3}} = \left(8^{\frac{1}{3}}\right)^2 = (2)^2 = 4$$

or,

$$8^{\frac{2}{3}} = \left(8^2\right)^{\frac{1}{3}} = (64)^{\frac{1}{3}} = 4.$$

The result is the same with either approach. That is, we can take the cube root first and then square the answer. Or, we can square first and then take the cube root. In general, for an exponent of the form $\dfrac{m}{n}$, taking the nth root first and then raising this root to the exponent m is easier because the numbers are smaller.

For example,

$$81^{\frac{3}{4}} = \left(81^{\frac{1}{4}}\right)^3 = (3)^3 = 27$$

is easier to calculate and work with than

$$81^{\frac{3}{4}} = \left(81^3\right)^{\frac{1}{4}} = (531,441)^{\frac{1}{4}} = 27.$$

The fourth root of 81 is more commonly known than the fourth root of 531,441.

The General Form $a^{\frac{m}{n}}$

If n is a positive integer, m is any integer, and $a^{\frac{1}{n}}$ is a real number, then

$$a^{\frac{m}{n}} = \left(a^{\frac{1}{n}}\right)^m = \left(a^m\right)^{\frac{1}{n}}.$$

In radical notation: $\quad a^{\frac{m}{n}} = \left(\sqrt[n]{a}\right)^m = \sqrt[n]{a^m}$

Example 2

Conversion Between Exponential Notation and Radical Notation

Assume that each variable represents a positive real number. Each expression is changed to an equivalent expression in either radical or exponential notation.

a. $x^{\frac{2}{3}} = \sqrt[3]{x^2}$ The index, 3, is the denominator in the rational exponent.

b. $3x^{\frac{4}{5}} = 3\sqrt[5]{x^4}$ The coefficient, 3, is not affected by the exponent.

c. $-a^{\frac{3}{2}} = -\sqrt{a^3}$ −1 is the understood coefficient.

d. $\sqrt[6]{a^5} = a^{\frac{5}{6}}$ The index, 6, is the denominator of the rational exponent.

e. $5\sqrt{x} = 5x^{\frac{1}{2}}$ In a square root the index is understood to be 2.

f. $-\sqrt[3]{4} = -4^{\frac{1}{3}}$ The coefficient, −1, is not affected by the exponent.
                 Also, we could write $-4^{\frac{1}{3}} = -1 \cdot 4^{\frac{1}{3}}$.

Now work margin exercise 2.

2. Convert each expression to an equivalent radical or exponential form. Assume all variables represent positive, real numbers.

a. $x^{\frac{2}{5}}$

b. $8z^{\frac{6}{7}}$

c. $-b^{\frac{5}{4}}$

d. $\sqrt[7]{x^5}$

e. $3\sqrt{s}$

f. $-\sqrt[4]{5}$

Objective C   **Simplifying Expressions with Rational Exponents**

Expressions with rational exponents such as

$$x^{\frac{2}{3}} \cdot x^{\frac{1}{6}}, \quad \frac{x^{\frac{3}{4}}}{x^{\frac{1}{3}}}, \quad \text{and} \quad \left(2a^{\frac{1}{4}}\right)^3$$

can be simplified using the rules for exponents.

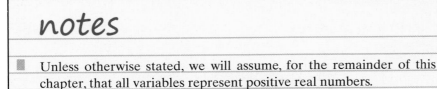

> ## notes
>
> ▪ Unless otherwise stated, we will assume, for the remainder of this chapter, that all variables represent positive real numbers.

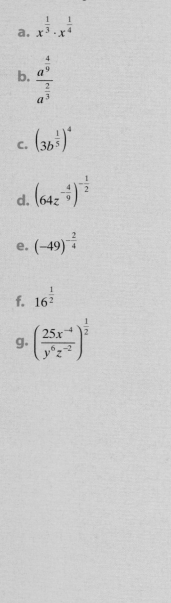

3. Simplify each expression using one or more of the rules for exponents.

a. $x^{\frac{1}{3}} \cdot x^{\frac{1}{4}}$

b. $\dfrac{a^{\frac{4}{9}}}{a^{\frac{2}{3}}}$

c. $\left(3b^{\frac{1}{5}}\right)^{4}$

d. $\left(64z^{-\frac{4}{9}}\right)^{-\frac{1}{2}}$

e. $(-49)^{-\frac{2}{4}}$

f. $16^{\frac{1}{2}}$

g. $\left(\dfrac{25x^{-4}}{y^{6}z^{-2}}\right)^{\frac{1}{2}}$

Example 3

Simplifying Expressions with Rational Exponents

Each expression is simplified using one or more of the rules for exponents.

a. $x^{\frac{2}{3}} \cdot x^{\frac{1}{6}} = x^{\frac{2}{3}+\frac{1}{6}}$

  Find a common denominator and add the exponents.

$\qquad = x^{\frac{4}{6}+\frac{1}{6}} = x^{\frac{5}{6}}$

b. $\dfrac{x^{\frac{3}{4}}}{x^{\frac{1}{3}}} = x^{\frac{3}{4}-\frac{1}{3}}$

  Find a common denominator and subtract the exponents.

$\qquad = x^{\frac{9}{12}-\frac{4}{12}} = x^{\frac{5}{12}}$

c. $\left(2a^{\frac{1}{4}}\right)^{3} = 2^{3} \cdot a^{\frac{1}{4}(3)} = 8a^{\frac{3}{4}}$

d. $\left(27y^{-\frac{9}{10}}\right)^{-\frac{1}{3}} = 27^{-\frac{1}{3}} \cdot y^{-\frac{9}{10}\left(-\frac{1}{3}\right)}$

  Multiply the exponents of y and reduce the fraction to $\dfrac{3}{10}$.

$\qquad = \dfrac{y^{\frac{3}{10}}}{27^{\frac{1}{3}}} = \dfrac{y^{\frac{3}{10}}}{3}$

e. $(-36)^{-\frac{1}{2}} = \dfrac{1}{(-36)^{\frac{1}{2}}}$

  This is not a real number because $(-36)^{\frac{1}{2}} = \sqrt{-36}$ is not real.

f. $9^{\frac{2}{4}} = 9^{\frac{1}{2}} = 3$

  The exponent can be reduced as long as the expression is real.

g. $\left(\dfrac{49x^{6}y^{-2}}{z^{-4}}\right)^{\frac{1}{2}} = \dfrac{\left(49x^{6}y^{-2}\right)^{\frac{1}{2}}}{\left(z^{-4}\right)^{\frac{1}{2}}}$

  Study this example carefully.

$\qquad = \dfrac{49^{\frac{1}{2}}x^{6\left(\frac{1}{2}\right)}y^{-2\left(\frac{1}{2}\right)}}{z^{-4\left(\frac{1}{2}\right)}}$

  Use the power rule four times.

$\qquad = \dfrac{7x^{3}y^{-1}}{z^{-2}}$

  Simplify exponents.

$\qquad = \dfrac{7x^{3}z^{2}}{y}$

  Use the properties of negative exponents.

Now work margin exercise 3.

Example 4 shows how to use fractional exponents to simplify rather complicated looking radical expressions. The results may seem surprising at first.

Example 4

Simplifying Radical Notation by Changing to Exponential Notation

Simplify each expression by first changing it into an equivalent expression with rational exponents. Then rewrite the answer in simplified radical form.

a. $\sqrt[4]{\sqrt[3]{x}} = \left(\sqrt[3]{x}\right)^{\frac{1}{4}} = \left(x^{\frac{1}{3}}\right)^{\frac{1}{4}}$ Note that $\dfrac{1}{3} \cdot \dfrac{1}{4} = \dfrac{1}{12}$.

$\qquad\quad = x^{\frac{1}{12}} = \sqrt[12]{x}$

b. $\sqrt[3]{a}\sqrt{a} = a^{\frac{1}{3}} \cdot a^{\frac{1}{2}}$

$\qquad\quad = a^{\frac{1}{3}+\frac{1}{2}} = a^{\frac{2}{6}+\frac{3}{6}}$

$\qquad\quad = a^{\frac{5}{6}} = \sqrt[6]{a^5}$

c. $\dfrac{\sqrt{x^3}\sqrt[3]{x^2}}{\sqrt[5]{x^2}} = \dfrac{x^{\frac{3}{2}} \cdot x^{\frac{2}{3}}}{x^{\frac{2}{5}}} = \dfrac{x^{\frac{3}{2}+\frac{2}{3}}}{x^{\frac{2}{5}}} = \dfrac{x^{\frac{9}{6}+\frac{4}{6}}}{x^{\frac{2}{5}}} = \dfrac{x^{\frac{13}{6}}}{x^{\frac{2}{5}}}$

$\qquad\quad = x^{\frac{13}{6}-\frac{2}{5}} = x^{\frac{65}{30}-\frac{12}{30}} = x^{\frac{53}{30}}$ Note that

$\qquad\quad = x^{\frac{30}{30}} \cdot x^{\frac{23}{30}} = x \cdot x^{\frac{23}{30}} = x\sqrt[30]{x^{23}}$ $\dfrac{53}{30} = \dfrac{30}{30} + \dfrac{23}{30} = 1 + \dfrac{23}{30}$.

Now work margin exercise 4.

Objective D **Evaluating Roots with a TI-84 Plus Calculator (The ⌃ Key)**

The caret key ⌃ on the TI-84 Plus calculator (and most graphing calculators) is used to indicate exponents. Using this key, roots of real numbers can be calculated with up to nine digit accuracy. To set the number of decimal places you wish in any calculations, press the **MODE** key and highlight the digit opposite the word FLOAT that indicates the desired accuracy. If no digit is highlighted, then the accuracy will be to nine decimal places (in some cases ten decimal places).

4. Rewrite the expression in simplified radical form.

a. $\sqrt[5]{\sqrt[2]{x}}$

b. $\sqrt[4]{x} \cdot \sqrt{x}$

c. $\dfrac{\sqrt[3]{x^2} \cdot \sqrt{x^3}}{\sqrt[6]{x^5}}$

5. Evaluate the following expressions using a TI-84 Plus graphing calculator.

a. $64^{\frac{4}{3}}$

b. $42^{\frac{5}{6}}$

To Find the Value of $a^{\frac{m}{n}}$ with a TI-84 Plus Graphing Calculator

1. Enter the value of the base, a.
2. Press the caret key ⌃.
3. Enter the fractional exponent enclosed in parentheses. (This exponent may be positive or negative.)
4. Press ENTER.

Example 5

Evaluating Rational Exponents with a Calculator

Evaluate the following expressions using a graphing calculator.

a. $125^{\frac{4}{3}}$

Solution

To find $125^{\frac{4}{3}}$ proceed as follows.

Step 1: Enter the base, 125.
Step 2: Press the caret key ⌃.
Step 3: Enter the exponent in parentheses, $\frac{4}{3}$.
Step 4: Press ENTER.

The display should read as follows.

```
125^(4/3)
              625
```

b. $36^{\frac{3}{5}}$

Solution

To find $36^{\frac{3}{5}}$ proceed as follows.

Step 1: Enter the base, 36.
Step 2: Press the caret key ⌃.
Step 3: Enter the exponent in parentheses, $\frac{3}{5}$.
Step 4: Press ENTER.

The display should read as follows.

```
36^(3/5)
         8.585814487
```

Now work margin exercise 5.

Practice Problems

Simplify each of the following expressions. Leave the answers with rational exponents.

1. $64^{\frac{2}{3}}$

2. $x^{\frac{3}{4}} \cdot x^{\frac{1}{5}} \cdot x^{\frac{1}{2}}$

3. $\dfrac{x^{\frac{1}{6}} \cdot y^{\frac{1}{2}}}{x^{\frac{1}{3}} \cdot y^{\frac{1}{4}}}$

4. $\left(16^{\frac{3}{4}}\right)^{-2}$

5. $-81^{\frac{1}{4}}$

Simplify each expression by first changing to an equivalent expression with rational exponents. Rewrite the answer in simplified radical form.

6. $\sqrt[4]{x} \cdot \sqrt{x}$

7. $\sqrt[5]{\sqrt[3]{a^2}}$

8. $\dfrac{\sqrt{36x^5}}{\sqrt[3]{8x^3}}$

Use a graphing calculator to find the following values rounded to the nearest ten-thousandth.

9. $128^{\frac{1}{5}}$

10. $100^{-\frac{1}{4}}$

Practice Problem Answers

1. 16

2. $x^{\frac{29}{20}}$

3. $\dfrac{y^{\frac{1}{4}}}{x^{\frac{1}{6}}}$

4. $\dfrac{1}{64}$

5. -3

6. $\sqrt[4]{x^3}$

7. $\sqrt[15]{a^2}$

8. $3x\sqrt{x}$

9. 2.6390

10. 0.3162

Exercises 14.6

1. $9^{\frac{1}{2}}$

2. $121^{\frac{1}{2}}$

3. $100^{-\frac{1}{2}}$

4. $25^{-\frac{1}{2}}$

5. $-64^{\frac{3}{2}}$

6. $(-64)^{\frac{3}{2}}$

7. $(-64)^{\frac{1}{3}}$

8. $-(64)^{\frac{1}{3}}$

9. $\left(-\frac{4}{25}\right)^{\frac{1}{2}}$

10. $-\left(\frac{4}{25}\right)^{\frac{1}{2}}$

11. $\left(\frac{9}{49}\right)^{\frac{1}{2}}$

12. $\left(\frac{225}{144}\right)^{\frac{1}{2}}$

13. $64^{\frac{2}{3}}$

14. $8^{-\frac{2}{3}}$

15. $(-216)^{-\frac{1}{3}}$

16. $(-125)^{\frac{1}{3}}$

17. $\left(\frac{8}{125}\right)^{-\frac{1}{3}}$

18. $-\left(\frac{16}{81}\right)^{-\frac{3}{4}}$

19. $\left(-\frac{1}{32}\right)^{\frac{2}{5}}$

20. $\left(\frac{27}{64}\right)^{\frac{2}{3}}$

21. $3\cdot16^{-\frac{3}{4}}$

22. $2\cdot25^{-\frac{1}{2}}$

23. $-100^{-\frac{3}{2}}$

24. $-49^{-\frac{5}{2}}$

25. $\left[\left(\frac{1}{32}\right)^{\frac{2}{5}}\right]^{-3}$

26. $\left[(-27)^{\frac{2}{3}}\right]^{-2}$

27. $25^{\frac{2}{3}}$

28. $81^{\frac{7}{4}}$

29. $100^{\frac{7}{2}}$

30. $100^{\frac{1}{3}}$

31. $250^{\frac{5}{6}}$

32. $2000^{\frac{2}{3}}$

33. $24^{-\frac{3}{4}}$

34. $18^{-\frac{3}{2}}$

35. $\sqrt[9]{72}$

36. $\sqrt[8]{63}$

37. $\sqrt[4]{0.0025}$

38. $\sqrt[5]{0.00032}$

39. $\sqrt[4]{3600}$

40. $\sqrt[6]{4500}$

41. $\sqrt[5]{35.4}$

42. $\sqrt[10]{1.8}$

Simplify each algebraic expression. Assume that all variables represent positive real numbers. Leave the answers in rational exponent form. See Example 3.

43. $\left(2x^{\frac{1}{3}}\right)^3$

44. $\left(3x^{\frac{1}{2}}\right)^4$

45. $\left(9a^4\right)^{-\frac{1}{2}}$

46. $\left(16a^3\right)^{-\frac{1}{4}}$

47. $8x^2 \cdot x^{\frac{1}{2}}$

48. $3x^3 \cdot x^{\frac{2}{3}}$

49. $5a^2 \cdot a^{-\frac{1}{3}} \cdot a^{\frac{1}{2}}$

50. $a^{\frac{2}{3}} \cdot a^{-\frac{3}{5}} \cdot a^0$

51. $\dfrac{x^{\frac{3}{4}}}{x^{\frac{1}{6}}}$

52. $\dfrac{a^{\frac{2}{3}}}{a^{\frac{1}{9}}}$

53. $\dfrac{x^{\frac{2}{5}}}{x^{-\frac{1}{10}}}$

54. $\dfrac{a^{\frac{1}{2}}}{a^{-\frac{2}{3}}}$

55. $\dfrac{a^{\frac{3}{4}} \cdot a^{\frac{1}{8}}}{a^2}$

56. $\dfrac{x^{\frac{2}{3}} \cdot x^{\frac{4}{3}}}{x^2}$

57. $\dfrac{a^{\frac{1}{2}} \cdot a^{-\frac{3}{4}}}{a^{-\frac{1}{2}}}$

58. $\dfrac{x^{\frac{2}{3}}x^{-1}}{x^{-\frac{3}{2}}}$

59. $\dfrac{a^{\frac{3}{2}}b^{\frac{4}{5}}}{a^{-\frac{1}{2}}b^2}$

60. $\dfrac{a^{\frac{3}{4}}b^{-\frac{1}{3}}}{a^{\frac{3}{2}}b^{\frac{1}{6}}}$

61. $\left(2x^{\frac{1}{2}}y^{\frac{1}{3}}\right)^3$

62. $\left(a^{\frac{1}{2}}a^{\frac{1}{3}}\right)^6$

63. $\left(4x^{-\frac{3}{4}}y^{\frac{1}{5}}\right)^{-2}$

64. $\left(81a^{-8}b^2\right)^{-\frac{1}{4}}$

65. $\left(-x^3y^6z^{-6}\right)^{\frac{2}{3}}$

66. $\left(9x^2y^{-4}z^{-3}\right)^{\frac{3}{2}}$

67. $\left(\dfrac{x^2y^{-3}}{z^4}\right)^{-\frac{1}{2}}$

68. $\left(\dfrac{27a^3b^6}{c^9}\right)^{-\frac{1}{3}}$

69. $\left(\dfrac{16a^{-4}b^3}{c^4}\right)^{\frac{3}{4}}$

70. $\left(\dfrac{-27a^2b^3}{c^{-3}}\right)^{\frac{1}{3}}$

71. $\dfrac{\left(x^{\frac{1}{4}}y^{\frac{1}{2}}\right)^3}{x^{\frac{1}{2}}y^{\frac{1}{4}}}$

72. $\dfrac{\left(x^{\frac{1}{2}}y\right)^{-\frac{1}{3}}}{x^{\frac{2}{3}}y^{-1}}$

73. $\dfrac{\left(8x^2y\right)^{-\frac{1}{3}}}{\left(5x^{\frac{1}{3}}y^{-\frac{1}{2}}\right)^2}$

74. $\dfrac{\left(25a^4b^{-1}\right)^{\frac{1}{2}}}{\left(2a^{\frac{1}{5}}b^{\frac{3}{5}}\right)^3}$

75. $\left(\dfrac{a^{-3}b^{\frac{1}{3}}}{a^{\frac{1}{2}}b}\right)^{\frac{1}{2}} \cdot \left(\dfrac{ab^{\frac{1}{2}}}{a^{-\frac{2}{3}}b^{-1}}\right)^{\frac{1}{2}}$

76. $\left(\dfrac{x^2 y^{\frac{1}{3}}}{x^{\frac{1}{2}} y^{\frac{3}{2}}}\right)^{\frac{1}{2}} \cdot \left(\dfrac{x^{-\frac{1}{2}} y^{\frac{2}{3}}}{x^{-1} y^{\frac{3}{4}}}\right)^{2}$

77. $\dfrac{\left(27xy^{\frac{1}{2}}\right)^{\frac{1}{3}} \cdot \left(x^{\frac{1}{2}}y\right)^{\frac{1}{6}}}{\left(25x^{-\frac{1}{2}}y\right)^{\frac{1}{2}} \left(16x^{\frac{1}{3}}y\right)^{\frac{1}{2}}}$

78. $\dfrac{\left(4a^{-6}b\right)^{\frac{1}{2}}}{\left(7a^2b^3\right)^{-1}} \cdot \dfrac{\left(49a^4b^3\right)^{-\frac{1}{2}}}{\left(64a^{-3}b^6\right)^{\frac{2}{3}}}$

> **Simplify each expression by first changing it into an equivalent expression with rational exponents. Rewrite the answer in simplified radical form. See Examples 2 and 4.**

79. $\sqrt[3]{a} \cdot \sqrt{a}$

80. $\sqrt[3]{x^2} \cdot \sqrt[5]{x^3}$

81. $\dfrac{\sqrt[4]{y^3}}{\sqrt[6]{y}}$

82. $\dfrac{\sqrt[3]{x^4}}{\sqrt[4]{x}}$

83. $\dfrac{\sqrt[3]{x^2}\,\sqrt[5]{x^6}}{\sqrt{x^3}}$

84. $\dfrac{a\sqrt[4]{a}}{\sqrt[3]{a}\,\sqrt{a}}$

85. $\sqrt{\sqrt[3]{y}}$

86. $\sqrt[5]{\sqrt{x}}$

87. $\sqrt[3]{\sqrt[3]{x}}$

88. $\sqrt{\sqrt{a}}$

89. $\sqrt[15]{(7a)^5}$

90. $\sqrt[21]{(3x)^7}$

91. $\sqrt[4]{\sqrt[3]{\sqrt{x}}}$

92. $\sqrt[5]{\sqrt[4]{\sqrt[3]{x}}}$

93. $\left(\sqrt[3]{a^4bc^2}\right)^{15}$

94. $\left(\sqrt[4]{a^3b^6c}\right)^{12}$

Writing & Thinking

95. Is $\sqrt[5]{a} \cdot \sqrt{a}$ the same as $\sqrt[5]{a^2}$? Explain why or why not.

96. Assume that x represents a positive real number. Describe what kind of number the exponent n must be for x^n to mean

   a. a product.

   b. a quotient.

   c. 1.

   d. a radical.

14.7 Functions with Radicals

Objective A **Review of Functions and Function Notation**

The concept of functions is among the most important and useful ideas in all of mathematics. Functions were introduced earlier along with function notation, such as $f(x)$ (read "f of x"). We also discussed **linear functions** and the use of function notation in evaluating functions. Function notation was used again in where $p(x)$ was used to represent polynomials. In this section the function concept is expanded to include **radical functions** (functions with radicals). The definitions of relations and functions and the vertical line test are restated here for review and easy reference.

Relation, Domain, and Range

A **relation** is a set of ordered pairs of real numbers.

The **domain D** of a relation is the set of all first coordinates in the relation.

The **range R** of a relation is the set of all second coordinates in the relation.

Function

A **function** is a relation in which each domain element has exactly one corresponding range element.

The definition can also be stated in the following ways:

1. A function is a relation in which each first coordinate appears only once.
2. A function is a relation in which no two ordered pairs have the same first coordinate.

Vertical Line Test

If **any** vertical line intersects the graph of a relation at more than one point, then the relation is **not** a function.

Objectives

A Briefly review the concepts of functions and function notation.

B Find the domain and range of radical functions.

C Evaluate radical functions.

D Graph radical functions.

E Use a graphing calculator to graph radical functions.

We have used the ordered pair notation (x, y) to represent points on the graphs of relations and functions. For example, $y = 2x - 5$ represents a linear function and its graph is a line (as shown in the figure below).

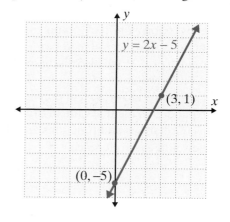

$y = 2x - 5$

$(3, 1)$

$(0, -5)$

Figure 1

Objective B **Finding the Domain and Range of Radical Functions**

We define **radical functions** (functions with radical expressions) as follows.

Radical Function

A **radical function** is a function of the form $y = \sqrt[n]{g(x)}$ in which the radicand contains a variable expression.

The **domain** of such a function depends on the index, n:

1. If n is an even number, the domain is the set of all x such that $g(x) \geq 0$.
2. If n is an odd number, the domain is the set of all real numbers, $(-\infty, \infty)$.

Examples of radical functions are

$$y = 3\sqrt{x}, \quad f(x) = \sqrt{2x + 3}, \quad \text{and} \quad y = \sqrt[3]{x - 7}.$$

Example 1

Finding the Domain of a Radical Function

Determine the domain of each radical function.

a. $f(x) = \sqrt{2x + 3}$

Solution

Because the index is 2, the radicand must be nonnegative. (That is, the expression under the radical sign cannot be negative.)

Thus, we must have

$$2x + 3 \geq 0$$
$$2x \geq -3$$
$$x \geq -\frac{3}{2}$$

and the domain is the interval of real numbers $\left[-\frac{3}{2}, \infty\right)$.

b. $y = \sqrt[3]{x-7}$

Solution

Because the index is 3, an odd number, the radicand may be any real number.
Thus the domain is the set of all real numbers $(-\infty, \infty)$.

Now work margin exercise 1.

1. Determine the domain of each radical function.

a. $f(x) = \sqrt{5x+10}$

b. $y = \sqrt[3]{-x-9}$

Objective C **Evaluating Radical Functions**

As illustrated earlier, function notation is particularly useful when evaluating functions for specific values of the variable. For example,

$$\text{if, } f(x) = \sqrt{2x+3} \text{ then, } f\left(\frac{1}{2}\right) = \sqrt{2\left(\frac{1}{2}\right)+3} = \sqrt{1+3} = \sqrt{4} = 2$$
$$\text{and, } f(0) = \sqrt{2(0)+3} = \sqrt{0+3} = \sqrt{3}.$$

A calculator can be used to find decimal approximations. Such approximations are helpful when estimating the locations of points on a graph. For example,

$$\text{if, } \quad f(x) = \sqrt{x-5},$$
$$\text{then, } f(8) = \sqrt{(8)-5} = \sqrt{3} \approx 1.7321 \qquad \text{to the nearest ten-thousandth}$$
$$\text{and } \quad f(25) = \sqrt{(25)-5} = \sqrt{20} \approx 4.4721. \quad \text{to the nearest ten-thousandth}$$

2. Find the corresponding
   $f(x)$ values for the given
   values of x.

   a. Given $f(x) = 5\sqrt{2x}$, find
   $f(x)$ values for $x = 0, 2, 8$.

   b. Given $f(x) = \sqrt[3]{2x-3}$, find
   $f(x)$ values for $x = 1, 2, 15$.

Example 2

Evaluating Radical Functions

Complete each table by finding the corresponding $f(x)$ values for the given values of x.

a. $f(x) = \sqrt[3]{x-7}$

x	f(x)
7	?
6	?
−1	?

Solution

x	f(x)
7	$\sqrt[3]{7-7} = \sqrt[3]{0} = 0$
6	$\sqrt[3]{6-7} = \sqrt[3]{-1} = -1$
−1	$\sqrt[3]{-1-7} = \sqrt[3]{-8} = -2$

b. $f(x) = 3\sqrt{x}$

x	f(x)
0	?
4	?
6	?

Solution

x	f(x)
0	$3\sqrt{0} = 0$
4	$3\sqrt{4} = 3 \cdot 2 = 6$
6	$3\sqrt{6} \approx 7.3485$

Now work margin exercise 2.

Objective D **Graphing Radical Functions**

To graph a radical function, we need to be aware of its domain and to plot at least a few points to see the nature of the resulting curve. Example 3 shows how to begin to graph the radical function $y = \sqrt{x+5}$.

Example 3

Graphing a Radical Function

Graph the function $y = \sqrt{x+5}$.

Solution

For the domain we have

$$x + 5 \geq 0$$
$$x \geq -5.$$

To see the nature of the graph we select a few values for x in the domain and find the corresponding values of y. Then we plot the points on a graph.

x	y
–5	0
–4	1
–3	$\sqrt{2} \approx 1.41$
0	$\sqrt{5} \approx 2.24$
4	3

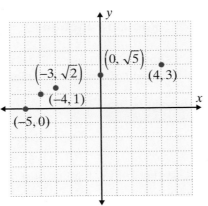

Note that $\sqrt{x+5}$ is the principal square root. This means that $y \geq 0$. Thus the point $(-5, 0)$ is on the x-axis and the remaining points on the graph are above the x-axis. So, we can complete the graph by drawing a smooth curve that passes through the selected points. The graph of the function is shown here.

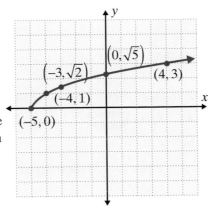

We see that the domain is $[-5, \infty)$ and the range is $[0, \infty)$.

3. Graph the function
$y = -\sqrt{2x+1}$.

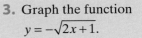

To use a TI-84 Plus graphing calculator to graph this function,

Step 1: Press  and enter the function as shown:

Step 2: Press ⬛GRAPH. (You may need to adjust the window.) The result will be the graph as shown here:

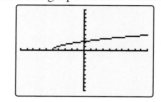

Now work margin exercise 3.

4. a. Use the TABLE feature of a TI-84 Plus graphing calculator to locate five points on the graph of the function $y = \sqrt[3]{2 - 4x}$.

x	y_1

b. Use a TI-84 Plus graphing calculator to graph the function.

Objective E **Using a TI-84 Plus Graphing Calculator to Graph Radical Functions**

Example 4 shows how to use a TI-84 Plus graphing calculator to find many points on the graph of a radical function and then how to graph the function.

Example 4

Using a TI-84 Plus to Graph a Radical Function

a. Use the TABLE feature of a TI-84 Plus graphing calculator to locate many points on the graph of the function $y = \sqrt[3]{2x - 3}$.

Solution

Using the TABLE feature of a TI-84 Plus:

Step 1: Press ⬛Y= and enter the function as follows:

    a. Press **MATH**.

    b. Choose 4: $\sqrt[3]{\ }$ (.

    c. Enter $2x - 3$) and press **ENTER**

Step 2: Press TBLSET (which is [2ND] [WINDOW]) and set the display as shown here:

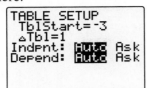

Step 3: Press TABLE (which is [2ND] [GRAPH]) and the display will appear as follows:

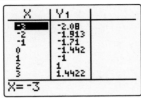

b. Plot several points (approximately) on a graph and then connect them with a smooth curve.

Once your calculator is displaying the table, you may scroll up and down the display to find as many points as you like. A few are shown here to see the nature of the graph.

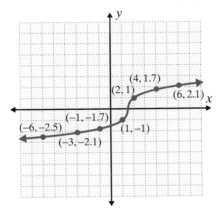

c. Use a TI-84 Plus graphing calculator to graph the function.

Solution

Press [GRAPH] and the display will appear with the curve as follows:

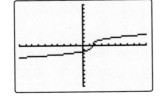

Now work margin exercise 4.

Find the indicated value of the function.

1. For $f(x) = \sqrt{x+7}$, find $f(5)$.

2. For $g(x) = \sqrt{3x-1}$, find $g\left(\dfrac{2}{3}\right)$.

3. For $h(x) = \sqrt[3]{x+5}$, find $h(-13)$.

Use a calculator to estimate the value of the function to the nearest ten-thousandth.

4. Estimate $f(2)$ for $f(x) = \sqrt{4x-1}$.

5. Estimate $g(-3)$ for $g(x) = \sqrt[3]{1-3x}$.

Practice Problem Answers

1. $2\sqrt{3}$ **2.** 1 **3.** -2

4. 2.6458 **5.** 2.1544

Exercises 14.7

Find each function value as indicated and write the answers in both radical notation and decimal notation. If necessary, round decimal values to the nearest ten-thousandth. See Example 2.

1. Given $f(x) = \sqrt{2x+1}$, find

   a. $f(2)$ b. $f(4)$ c. $f(24.5)$ d. $f(1.5)$

2. Given $f(x) = \sqrt{5-3x}$, find

   a. $f(0)$ b. $f(-2)$ c. $f\left(-\dfrac{20}{3}\right)$ d. $f(-2.4)$

3. Given $g(x) = \sqrt[3]{x+6}$, find

   a. $g(21)$ b. $g(-7)$ c. $g(-14)$ d. $g(18)$

4. Given $h(x) = \sqrt[3]{4-x}$, find

   a. $h(4)$ b. $h(-4)$ c. $h(3.999)$ d. $h(-2.5)$

Use interval notation to indicate the domain of each radical function. See Example 1.

5. $y = \sqrt{x+8}$ 6. $y = \sqrt{2x-1}$ 7. $y = \sqrt{2.5-5x}$ 8. $y = \sqrt{1-3x}$

9. $f(x) = \sqrt[3]{x+4}$ 10. $f(x) = \sqrt[3]{6x}$ 11. $g(x) = \sqrt[4]{x}$ 12. $g(x) = \sqrt[4]{7-x}$

13. $y = \sqrt[5]{4x-1}$ 14. $y = \sqrt[5]{8+x}$

15. $y = \sqrt{x-2}$

16. $y = \sqrt{2-x}$

17. $y = -\sqrt{x-3}$

18. $y = -\sqrt{3-x}$

19. $y = \sqrt{x+4}$

20. $y = \sqrt{x-4}$

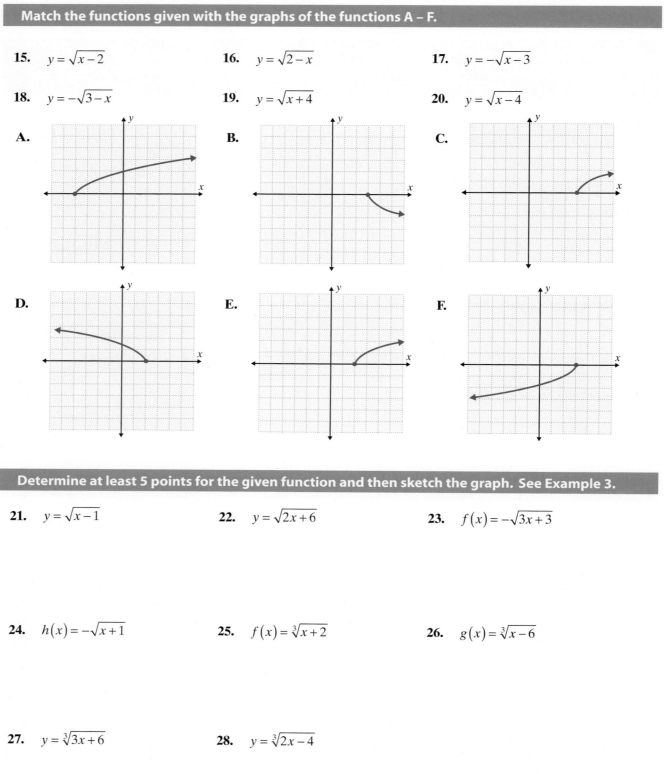

A.

B.

C.

D.

E.

F.

Determine at least 5 points for the given function and then sketch the graph. See Example 3.

21. $y = \sqrt{x-1}$

22. $y = \sqrt{2x+6}$

23. $f(x) = -\sqrt{3x+3}$

24. $h(x) = -\sqrt{x+1}$

25. $f(x) = \sqrt[3]{x+2}$

26. $g(x) = \sqrt[3]{x-6}$

27. $y = \sqrt[3]{3x+6}$

28. $y = \sqrt[3]{2x-4}$

29. $y = 3\sqrt{x+2}$

30. $y = 2\sqrt{3-x}$

31. $g(x) = -\sqrt{2x}$

32. $f(x) = \sqrt{3x}$

33. $f(x) = -\sqrt{x+4}$

34. $f(x) = -\sqrt{5-x}$

35. $y = -\sqrt[3]{x+2}$

36. $y = -\sqrt[3]{3x+4}$

37. $g(x) = -\sqrt[4]{x+5}$

38. $y = \sqrt[4]{2x+6}$

39. $y = \sqrt[5]{2x+1}$

40. $y = \sqrt[5]{x+7}$

Writing & Thinking

41. The graph of the radical function $f(x) = \sqrt{x}$ is shown with two values of x on the x-axis, 3 and $3 + h$.

a. Rationalize the numerator of the expression $\dfrac{f(3+h) - f(3)}{h} = \dfrac{\sqrt{3+h} - \sqrt{3}}{h}$ by multiplying both the numerator and denominator by the conjugate of the numerator. Then simplify the resulting expression.

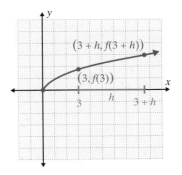

b. What do you think this expression represents graphically?
(**Hint:** Two points determine a line.)

c. Using your results from parts **a.** and **b.**, what do you see happening on the graph if the value of h shrinks slowly to 0?

d. Using your analysis from part **c.**, what happens to the value of your simplified expression in part **a.** and what do you think this value represents?

Section 14.1: Roots and Radicals

Perfect Squares   page 1092

   The square of an integer is called a **perfect square**.

Radical Terminology   page 1092

   The symbol $\sqrt{}$ is called a **radical sign**.
   The number under the radical sign is called the **radicand**.
   The complete expression, such as $\sqrt{64}$, is called a **radical** or **radical expression**.

Square Root   page 1092

   If a is a nonnegative real number, then $\sqrt{a}$ is the **principal square root** of a and $-\sqrt{a}$ is the **negative square root** of a.

Cube Root   page 1095

   If a is a real number, then $\sqrt[3]{a}$ is the **cube root** of a.

Evaluating Radical Expressions Using a Calculator   page 1096

Section 14.2: Simplifying Radicals

Properties of Square Roots   page 1100

   If a and b are positive real numbers, then

   1. $\sqrt{ab} = \sqrt{a}\sqrt{b}$

   2. $\sqrt{\dfrac{a}{b}} = \dfrac{\sqrt{a}}{\sqrt{b}}$

Simplest Form of a Square Root   page 1100

   A square root is considered to be in simplest form when the radicand has no perfect square as a factor.

Square Root of x^2   page 1102

   If x is a real number, then $\sqrt{x^2} = |x|$.

   Note: If $x \geq 0$ is given, then we can write $\sqrt{x^2} = x$.

Simplest Form of a Cube Root   page 1104

   A cube root is considered to be in simplest form when the radicand has no perfect cube as a factor.

Section 14.5: Equations with Radicals

Extraneous Solution   page 1125

An **extraneous solution** is a number that is found when solving an equation but that does not satisfy the original equation. They can be unintentionally introduced by raising both sides of an equation to a power.

Method for Solving Equations with Radicals   page 1125

1. Isolate one of the radicals on one side of the equation. (An equation may have more than one radical.)
2. Raise both sides of the equation to the power corresponding to the index of the radical.
3. If the equation still contains a radical, repeat steps 1 and 2.
4. Solve the equation after all the radicals have been eliminated.
5. Be sure to check all possible solutions in the original equation and eliminate any extraneous solutions.

Section 14.6: Rational Exponents

Radical Notation and Fractional Exponents   page 1134

If n is a positive integer, then $\sqrt[n]{a} = a^{\frac{1}{n}}$ (assuming $\sqrt[n]{a}$ is a real number).

Special Notes about the Index n   page 1135

For the expression $\sqrt[n]{a}$ or $a^{\frac{1}{n}}$ to be a real number:

1. when a is nonnegative, n can be any index, and
2. when a is negative, n must be odd.
   (If a is negative and n is even, then $\sqrt[n]{a}$ is nonreal.)

Review of the Rules for Exponents   page 1136

The General Form $a^{\frac{m}{n}}$ page 1137

If n is a positive integer, m is any integer, and $a^{\frac{1}{n}}$ is a real number, then

$$a^{\frac{m}{n}} = \left(a^{\frac{1}{n}}\right)^{m} = \left(a^{m}\right)^{\frac{1}{n}}.$$

In radical notation: $a^{\frac{m}{n}} = \left(\sqrt[n]{a}\right)^{m} = \sqrt[n]{a^{m}}$

Simplifying Expressions with Rational Exponents   page 1137

Evaluating Roots with a Graphing Calculator   page 1139

Radical Function   page 1146

A **radical function** is a function of the form $y = \sqrt[n]{g(x)}$ in which the radicand contains a variable expression.

The **domain** of such a function depends on the index, n:

1. If n is an even number, the domain is the set of all x such that $g(x) \geq 0$.

2. If n is an odd number, the domain is the set of all real numbers $(-\infty, \infty)$.

Evaluating Radical Functions   page 1147

Graphing Radical Functions   page 1149

Using a Graphing Calculator to Graph Radical Functions   page 1150

Chapter 14: Review

Section 14.1: Roots and Radicals

Simplify the following square roots and cube roots.

1. $\sqrt{36}$

2. $\sqrt{196}$

3. $\sqrt[3]{729}$

4. $\sqrt{0.0009}$

5. $\sqrt[3]{-343}$

6. $\sqrt[3]{\dfrac{125}{27}}$

Use your knowledge of square roots and cube roots to determine whether each number is rational, irrational, or nonreal.

7. $\sqrt{-20}$

8. $\sqrt{3}$

9. $-\sqrt{196}$

10. $\sqrt{\dfrac{1}{9}}$

Use a calculator to find the value of each radical expression rounded to the nearest ten-thousandth.

11. $\sqrt{21}$

12. $-4\sqrt{18}$

Section 14.2: Simplifying Radicals

Simplify each of the radical expressions. Assume that all variables represent positive real numbers.

13. $-\sqrt{225}$

14. $\sqrt{9x^3}$

15. $\sqrt{8a^4}$

16. $\sqrt{50x^3y^2}$

17. $-\sqrt{81x^2y}$

18. $\sqrt[3]{40}$

19. $\sqrt[3]{-81x^5y^7}$

20. $\sqrt[3]{54a^4b^2}$

21. $\sqrt{\dfrac{75a^3}{9}}$

22. $\sqrt[3]{\dfrac{8y^{12}}{27x^{15}}}$

Section 14.3: Addition, Subtraction, and Multiplication with Radicals

Simplify the following radical expressions. Assume that all variables represent positive real numbers.

23. $\sqrt{11} - 5\sqrt{11}$

24. $3\sqrt{7} + 4\sqrt{28}$

25. $6\sqrt{x} + \sqrt{x} - 2\sqrt{x}$

26. $2\sqrt{12} - 6\sqrt{75} + \sqrt{50}$

27. $2x\sqrt{y} - \sqrt{4x^2y}$

28. $\sqrt{3}\left(\sqrt{3}+4\sqrt{2}\right)$ **29.** $\left(2\sqrt{6}+5\right)\left(2\sqrt{6}-5\right)$ **30.** $\left(\sqrt{5}+\sqrt{2}\right)^2$

31. $\left(\sqrt{x}-\sqrt{y}\right)\left(\sqrt{y}+\sqrt{x}\right)$ **32.** $\left(3\sqrt{x}+\sqrt{2}\right)\left(5\sqrt{x}+\sqrt{2}\right)$

🖩 **Use a graphing calculator to evaluate each expression. Round answers to the nearest ten-thousandth, if necessary.**

33. $23+\sqrt{6}$ **34.** $\left(1+4\sqrt{3}\right)\left(2-\sqrt{6}\right)$

Section 14.4: Rationalizing Denominators

Rationalize the denominator and simplify. Assume that all variables represent positive real numbers.

35. $\dfrac{-8}{\sqrt{5}}$ **36.** $\sqrt{\dfrac{9}{24}}$ **37.** $\dfrac{28}{5\sqrt{7}}$ **38.** $\dfrac{a}{3\sqrt{a}}$

39. $\dfrac{x}{\sqrt{x}+3}$ **40.** $\dfrac{\sqrt{20}}{3+\sqrt{5}}$ **41.** $\dfrac{5}{\sqrt{7}+\sqrt{3}}$ **42.** $\dfrac{\sqrt{5}+\sqrt{3}}{2\sqrt{5}-\sqrt{3}}$

Section 14.5: Equations with Radicals

Solve the following equations. Be sure to check your answers in the original equation.

43. $\sqrt{3x+4}=5$ **44.** $\sqrt{x-4}-6=2$ **45.** $\sqrt{x^2-17}=8$

46. $x+3=\sqrt{x+9}$ **47.** $x-3=\sqrt{2x-6}$ **48.** $\sqrt{2x-7}=3-x$

49. $\sqrt{4x+1}=\sqrt{2x+5}$ **50.** $\sqrt{x-3}+\sqrt{x}=3$ **51.** $\sqrt{3x+1}=-4$

52. $\sqrt[3]{x\left(x-6\right)}=-2$

Section 14.6: Rational Exponents

Simplify each numerical expression.

53. $81^{\frac{1}{2}}$ **54.** $-36^{-\frac{1}{2}}$ **55.** $\left(\dfrac{81}{16}\right)^{-\frac{1}{4}}$ **56.** $\left(\dfrac{27}{125}\right)^{\frac{2}{3}}$

Use a graphing calculator to find the value of each numerical expression accurate to the nearest ten-thousandth, if necessary.

57. $36^{\frac{2}{3}}$

58. $\sqrt[5]{0.00081}$

Simplify each algebraic expression. Assume that all variables represent positive real numbers. Leave the answers in rational exponent form.

59. $\left(3x^{\frac{2}{3}}\right)^3$

60. $5y^3 \cdot y^{\frac{1}{3}}$

61. $\dfrac{x^{\frac{5}{6}} \cdot x^{-\frac{1}{3}}}{x^3}$

62. $\left(\dfrac{27a^6b^3}{c^3}\right)^{\frac{2}{3}}$

Simplify each expression by first changing it into an equivalent expression with rational exponents. Rewrite the answer in simplified radical form.

63. $\dfrac{\sqrt[3]{x}\,\sqrt[6]{x^5}}{\sqrt[4]{x^3}}$

64. $\sqrt{\sqrt[3]{64x^2}}$

Section 14.7: Functions with Radicals

Find each function value as indicated and write the answers in both radical notation and decimal notation. (If necessary, round decimal values to the nearest ten-thousandth.)

65. Given $g(x) = \sqrt{1-4x}$, find **a.** $g\left(\dfrac{1}{4}\right)$ and **b.** $g(-6)$.

66. Given $h(x) = \sqrt[3]{6-x}$, find **a.** $h(0)$ and **b.** $h(-10)$.

Use interval notation to indicate the domain of each radical function.

67. $y = \sqrt{2x+1}$

68. $f(x) = \sqrt[3]{4x}$

Find and label at least 5 points on the graph of the function and then sketch the graph of the function.

69. $y = \sqrt{x-3}$

70. $y = \sqrt[3]{x+1}$

Use a graphing calculator to graph each of the functions.

71. $f(x) = -2\sqrt[3]{x+3}$

72. $y = \sqrt[4]{x+1}$

Chapter 14: Test

Simplify the expressions. Assume that all variables represent positive real numbers.

1. $\sqrt{112}$

2. $\sqrt[3]{8}$

3. $\sqrt{\dfrac{144}{49}}$

4. $\sqrt{120xy^4}$

5. $\sqrt[3]{48x^2y^5}$

6. $\sqrt[3]{\dfrac{343x^{18}y^4}{8z^6}}$

Perform the indicated operations and simplify. Assume that all variables represent positive real numbers.

7. $2\sqrt{75} + 3\sqrt{27} - \sqrt{12}$

8. $5x\sqrt{y^3} - 2\sqrt{x^2y^3} - 4y\sqrt{x^2y}$

9. $\left(\sqrt{3} - \sqrt{2}\right)^2$

10. $\left(6 + \sqrt{3x}\right)\left(5 - 2\sqrt{3x}\right)$

Rationalize each denominator and simplify, if possible.

11. $\sqrt{\dfrac{5y^2}{8x^3}}$

12. $\dfrac{1-x}{1-\sqrt{x}}$

Solve the following equations. Be sure to check your answers in the original equation.

13. $\sqrt{9x-5} - 3 = 8$

14. $\sqrt[3]{3x+4} = -2$

15. $\sqrt{x+8} - 2 = x$

16. $\sqrt{5x+1} = 1 + \sqrt{3x+4}$

Find the domain of the following radical functions. Write the answer in interval notation.

17. $y = \sqrt{3x+4}$

18. $f(x) = \sqrt[3]{2x+5}$

Solve the following problems.

19. Use a graphing calculator to graph the radical function $f(x) = -\sqrt{x+3}$. Sketch the graph and label 3 points on the graph.

20. Write $(2x)^{\frac{2}{3}}$ in radical notation.

21. Write $\sqrt[6]{8x^2 y^4}$ as an equivalent, simplified expression with rational exponents.

Simplify each numerical expression. Assume that all variables represent positive real numbers. Leave the answers in rational exponent form.

22. $(-8)^{\frac{2}{3}}$

23. $4x^{\frac{1}{2}} \cdot x^{\frac{2}{3}}$

24. $\left(\dfrac{16x^{-4}y}{y^{-1}} \right)^{\frac{3}{4}}$

Use a calculator to find the value of each expression rounded to the nearest ten-thousandth.

25. $\sqrt[5]{119}$

26. $32^{-\frac{3}{5}}$

27. $\left(\sqrt{2} + 6 \right)\left(\sqrt{2} - 1 \right)$

Cumulative Review: Chapters 1 - 14

Solve the following problems.

1. Write the number 65,080,005.39 in its English word equivalent.

2. Convert the following values to percents.

    a. 6.43   **b.** 0.333   **c.** $\dfrac{19}{25}$

Perform the following operations. Reduce all fractions to lowest terms, and express all improper fractions as mixed numbers.

3. $\dfrac{2}{3} + \dfrac{5}{6} + \dfrac{2}{9}$ **4.** $17\dfrac{5}{8} - 12\dfrac{7}{10}$

5. $6.09 + 10.6 + 7$ **6.** $(-2.03) + (16.7)(-5.6)$

Solve each of the following equations.

7. $15x + 6(x+1) = 7$ **8.** $4(5x - 1) = 8(2x + 3)$

Solve the inequality and graph the solution set. Write the solution using interval notation. Assume that x is a real number.

9. $6x - 2 < 4x + 10$

Solve the following problems.

10. Graph the linear equation $4x - 2y = -8$ by locating the y-intercept and x-intercept.

11. Write the equation $x + 6y = 18$ in slope-intercept form. Then find the slope and the y-intercept, and use them to draw the graph.

12. Find an equation for the vertical line through the point $(5, -7)$.

13. Find an equation in standard form for the line passing through the point $\left(\dfrac{1}{4}, -3\right)$ with slope $m = -4$.

14. Given $P(x) = x^3 - 8x^2 + 19x - 12$, find

    a. $P(0)$ **b.** $P(4)$ **c.** $P(-3)$

15. Write the expression $\dfrac{0.008 \times 40{,}000}{320 \times 0.001}$ in scientific notation and simplify.

Perform the indicated operations.

16. $\left(x^2 + 7x - 5\right) - \left(-2x^3 + 5x^2 - x - 1\right)$

17. $(2x - 7)(3x - 1)$

18. $(5x + 2)(4 - x)$

19. $(7x - 2)^2$

Factor completely.

20. $28 + x - 2x^2$

21. $64y^4 + 32y^3 + 4y^2$

22. $5x^3 - 320$

23. $x^3 + 4x^2 - x - 4$

Solve each quadratic equation by factoring.

24. $x^2 - 13x - 48 = 0$

25. $x = 2x^2 - 6$

26. $0 = 15x^2 - 11x + 2$

Find the equation described below.

27. Find an equation that has $x = 4$ and $x = 7$ as roots.

Solve the following system of linear equations by graphing both equations and locating the point of intersection.

28. $\begin{cases} 3x - 2y = 7 \\ x + 3y = -5 \end{cases}$

Solve each system by using either the substitution method or the addition method.

29. $\begin{cases} -2y = x - 26 \\ y = 2x - 22 \end{cases}$

30. $\begin{cases} x + y = -5 \\ \dfrac{1}{2}x + \dfrac{1}{2}y = \dfrac{7}{2} \end{cases}$

31. $\begin{cases} x + y \le 4 \\ 3x - y \le 2 \end{cases}$

Simplify the expressions. Assume that all variables represent positive real numbers.

32. $\sqrt{80}$

33. $\sqrt{288x^5}$

34. $\sqrt[3]{125}$

35. $\sqrt[3]{16x^6y^{10}}$

Perform the indicated operations and simplify. Assume that all variables represent positive real numbers.

36. $\sqrt{27} + \sqrt{75} - \sqrt{108}$

37. $\left(\sqrt{2} - y\right)\left(x + \sqrt{3}\right)$

Rationalize the denominator and simplify if possible.

38. $\dfrac{\sqrt{5}}{\sqrt{2y}}$

39. $\dfrac{\sqrt{5} - \sqrt{6}}{\sqrt{5} + \sqrt{6}}$

Solve the radical equations given. Be sure to check your answers in the original equation.

40. $\sqrt{x + 10} + 1 = x + 5$

41. $\sqrt[3]{x - 7} + 3 = 1$

Use interval notation to indicate the domain of the radical function.

42. $\sqrt[4]{6 - 2x}$

Solve the following word problems.

43. **Circles:** Find
    a. the circumference and
    b. the area of a circle with a radius of 8 feet.

44. **Triangles:** Find the length of the hypotenuse of a right triangle with legs of length 10 meters and 20 meters. Write the answer in both simplified radical form and decimal form (accurate to three decimal places).

45. **Interest:** If $1550 is deposited in an account paying 8% compounded monthly:
    a. What will be the total amount in the account at the end of 3 years (to the nearest dollar)?
    b. How much interest will be earned?

46. Find two consecutive integers such that 3 more than twice the smaller is equal to 13 less than the larger.

47. **Antifreeze:** In a mixture, there are 3 parts antifreeze for each 8 parts water. If you want a total of 220 gallons of mixture, how many gallons of antifreeze are needed?

48. **Investing:** Harold has $50,000 that he wants to invest in two accounts. One pays 6% interest, and the other (at a higher risk) pays 10% interest. If he wants a $3600 annual return on these two investments, how much should he put into each account?

49. Eating Out: For lunch, Jason had one burrito and 2 tacos. Matt had 2 burritos and 3 tacos. If Jason spent $5.30 and Matt spent $9.25, find the price of one burrito and the price of one taco.

50. Selling Candy: A grocer plans to make up a special mix of two popular kinds of candy for Halloween. He wants to mix a total of 100 pounds to sell for $1.75 per pound. Individually, the two types sell for $1.25 and $2.50 per pound. How many pounds of each of the two kinds should he put in the mix?

51. Pizza: The Local Pizza Hotspot Restaurant offers a cheese pizza in three sizes: 7 in. diameter for $3.25, 12 in. diameter for $8.95, 14 in. diameter for $10.95. (Use $\pi = 3.14$.)

   a. What is the price per sq in. for each size pizza? Round your answers to the nearest tenth of a cent.

   b. Which is the best buy? Why?

52. Docking a Boat: A boat is being pulled to the shore from a dock. When the rope to the boat is 40 meters long, the boat is 30 meters from the dock. What is the height of the dock (to the nearest tenth of a meter) above the deck of the boat?

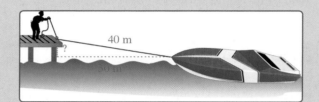

Quadratic Equations

Mathematics at Work!

The Cooper family is installing a pool in their backyard. The pool they have chosen measures 10 ft by 15 ft. Mr. Cooper is going to put a concrete deck around the pool. If the area the pool and deck together can occupy is 414 square feet, how wide will the concrete deck be? (See Section 15.4.)

15.1 Quadratic Equations: The Square Root Method

Objectives

A Review solving quadratic equations by factoring.

B Solve quadratic equations using the definition of square root.

C Solve problems related to right triangles and the Pythagorean Theorem.

Objective A **Review of Solving Quadratic Equations by Factoring**

Factoring polynomials and solving quadratic equations by factoring were discussed in Chapter 12. While this chapter deals with three new techniques for solving quadratic equations, the method of solving by factoring is so basic and important that an example is given here for emphasis and review.
Note: Now would be a good time to review all of Chapter 12.

Example 1

Factoring Method

Solve the following quadratic equations by factoring.

a. $x^2 + 7x = 18$

Solution

$$x^2 + 7x = 18 \qquad \text{original equation}$$
$$x^2 + 7x - 18 = 0 \qquad \text{Standard form with 0 on one side.}$$
$$(x+9)(x-2) = 0 \qquad \text{Factor.}$$
$$x + 9 = 0 \quad \text{or} \quad x - 2 = 0 \qquad \text{Set each factor equal to 0.}$$
$$x = -9 \qquad\qquad x = 2 \qquad \text{Solve each linear equation.}$$

The solutions are -9 and 2.

b. $2x^2 - x = 3$

Solution

$$2x^2 - x = 3 \qquad \text{original equation}$$
$$2x^2 - x - 3 = 0 \qquad \text{standard form with 0 on one side}$$
$$(2x-3)(x+1) = 0 \qquad \text{Factor.}$$
$$2x - 3 = 0 \quad \text{or} \quad x + 1 = 0 \qquad \text{Set each factor equal to 0.}$$
$$x = \frac{3}{2} \qquad\qquad x = -1 \qquad \text{Solve each linear equation.}$$

The solutions are $\frac{3}{2}$ and -1.

1. Solve the following quadratic equations by factoring.

a. $x^2 = 20x$

b. $2x^2 + 10x - 28 = 0$

Now work margin exercise **1.**

Solving Quadratic Equations Using the Square Root Method

Solving quadratic equations by factoring is a very good technique, and we want to use it much of the time. However, there are quadratic expressions that are not easily factored and some that are not factorable at all using real numbers. For example, consider solving the equation $x^2 = 5$ by factoring. As the following steps show, the factors involve square roots that are not integers.

$$x^2 = 5$$

$$x^2 - 5 = 0 \qquad \text{Get 0 on one side.}$$

$$x^2 - \left(\sqrt{5}\right)^2 = 0 \qquad \text{Difference of two squares with } 5 = \left(\sqrt{5}\right)^2.$$

$$\left(x + \sqrt{5}\right)\left(x - \sqrt{5}\right) = 0 \qquad \text{Factor.}$$

$$x + \sqrt{5} = 0 \quad \text{or} \quad x - \sqrt{5} = 0 \qquad \text{Set each factor equal to 0.}$$

$$x = -\sqrt{5} \qquad\qquad x = \sqrt{5} \qquad \text{There are two irrational solutions.}$$

A simpler and more direct way to solve this equation is by **taking square roots of both sides**, as shown in the following statement.

Square Root Method

For a quadratic equation in the form $x^2 = c$ where c is nonnegative,

$$x = \sqrt{c} \quad \text{or} \quad x = -\sqrt{c}.$$

This can be written as $x = \pm\sqrt{c}$.

---

notes

In this method, we do not set a polynomial expression equal to 0. Instead we set a squared expression equal to a nonnegative real number.

Example 2

The Square Root Method

Solve the following quadratic equations by using the square root method. Write each radical in simplest form.

a. $3x^2 = 51$

Solution

$$3x^2 = 51 \qquad \text{Divide both sides by 3 so that the coefficient of } x^2 \text{ is 1.}$$

$$x^2 = 17$$

$$x = \pm\sqrt{17} \qquad \text{Keep in mind that the expression } x = \pm\sqrt{17} \text{ represents}$$
$$\text{the two equations } x = \sqrt{17} \text{ and } x = -\sqrt{17}.$$

b. $(x+4)^2 = 21$

Solution

$(x+4)^2 = 21$

$x + 4 = \pm\sqrt{21}$

$x = -4 \pm \sqrt{21}$ There are two solutions: $-4 + \sqrt{21}$ and $-4 - \sqrt{21}$.

c. $(x-2)^2 = 50$

Solution

$(x-2)^2 = 50$

$x - 2 = \pm\sqrt{50}$ Simplify the radical $\left(\sqrt{50} = \sqrt{25}\sqrt{2} = 5\sqrt{2}\right)$.

$x = 2 \pm 5\sqrt{2}$ There are two solutions: $2 + 5\sqrt{2}$ and $2 - 5\sqrt{2}$.

d. $(2x+4)^2 = 72$

Solution

$(2x+4)^2 = 72$

$2x + 4 = \pm\sqrt{72}$

$2x = -4 \pm 6\sqrt{2}$ Simplify the radical $\left(\sqrt{72} = \sqrt{36}\sqrt{2} = 6\sqrt{2}\right)$.

$x = \dfrac{-4 \pm 6\sqrt{2}}{2}$

$x = \dfrac{\cancel{2}\left(-2 \pm 3\sqrt{2}\right)}{\cancel{2}}$ Factor and simplify.

$x = -2 \pm 3\sqrt{2}$ There are two solutions: $-2 + 3\sqrt{2}$ and $-2 - 3\sqrt{2}$.

e. $(x+7)^2 + 10 = 8$

Solution

$(x+7)^2 + 10 = 8$

$(x+7)^2 = -2$

There is no real solution. The square of a real number cannot be negative.

Completion Example 3

The Square Root Method

Solve the following quadratic equation by using the square root method.

$$4(x+2)^2 = 80$$

Solution

$$4(x+2)^2 = 80$$
$$(x+2)^2 = \underline{\quad\quad}$$
$$x+2 = \pm\underline{\quad\quad}$$
$$x = \underline{\quad\quad} \pm \underline{\quad\quad}$$

*Now work margin exercises **2** and **3**.*

Objective C   **The Pythagorean Theorem**

The Pythagorean Theorem was first discussed in Section 5.7 and again in 12.7. In this section, we show how the Pythagorean Theorem can be used to solve a variety of applications which involve solving quadratic equations. The theorem is stated again for easy reference and to emphasize its importance.

The Pythagorean Theorem

In a right triangle, the square of the length of the hypotenuse is equal to the sum of the squares of the lengths of the two legs.

$$c^2 = a^2 + b^2$$

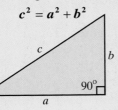

Solve the following quadratic equations by using the square root method. Write the radicals in simplest form.

2. a. $5x^2 = 60$

   b. $(x+5)^2 = 40$

   c. $(x+1)^2 = 75$

   d. $(3x-6)^2 = 18$

   e. $(x+5)^2 + 6 = 4$

3. $5(x-4)^2 = 60$

Completion Example Answers

3. $20;\ \pm 2\sqrt{5};\ -2 \pm 2\sqrt{5}$

 a. If the hypotenuse of a right triangle is 26 feet long and one leg is 24 feet long, what is the length of the other leg?

b. The diagonal of a rectangle is three times the width. The length of the rectangle is 10 inches. Find the width of the rectangle.

Example 4

The Pythagorean Theorem

a. If the hypotenuse of a right triangle is 15 cm long and one leg is 10 cm long, what is the length of the other leg?

Solution

Using the Pythagorean Theorem,

$$a^2 + 10^2 = 15^2$$

$$a^2 + 100 = 225$$

$$a^2 = 125$$

$$a = \sqrt{125}$$

$$= \sqrt{25} \cdot \sqrt{5} = 5\sqrt{5} \ (\approx 11.18\,\text{cm})$$

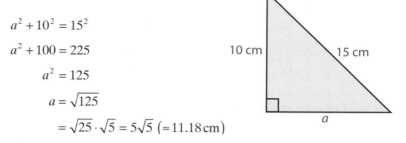

The other leg is $5\sqrt{5}$ cm long (or approximately 11.18 cm long).

(The negative solution to the quadratic equation is not considered because length is not negative.)

b. The diagonal of a rectangle is twice the width. The length of the rectangle is 6 feet. Find the width of the rectangle.

Solution

Let x = width of the rectangle
and $2x$ = length of the diagonal.

Then, by using the Pythagorean Theorem, we have

$$(2x)^2 = x^2 + 6^2$$

$$4x^2 = x^2 + 36$$

$$3x^2 = 36$$

$$x^2 = 12$$

$$x = \sqrt{12} = \sqrt{4} \cdot \sqrt{3} = 2\sqrt{3} \ (\approx 3.46).$$

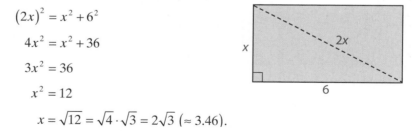

The width of the rectangle is $2\sqrt{3}$ feet or approximately 3.46 feet. (The negative solution to the quadratic equation is not considered because length is not negative.)

Now work margin exercise 4.

Practice Problems

Solve the following quadratic equations by using the square root method. Write the radicals in simplest form.

1. $x^2 = 14$ **2.** $5x^2 = 320$ **3.** $(x-4)^2 = 12$ **4.** $(3x+1)^2 = 54$

Practice Problem Answers
1. $x = \pm\sqrt{14}$ **2.** $x = \pm 8$ **3.** $x = 4 \pm 2\sqrt{3}$ **4.** $x = \dfrac{-1 \pm 3\sqrt{6}}{3}$

Exercises 15.1

1. $x^2 = 11x$

2. $x^2 - 10x + 16 = 0$

3. $x^2 = -15x - 36$

4. $2x^2 + 36x + 34 = 0$

5. $9x^2 + 6x - 15 = 0$

6. $5x^2 + 17x = -6$

7. $(x+3)(x-1) = 4x$

8. $(x-7)(x-2) = 6$

9. $(2x-3)(2x+1) = 3x - 6$

10. $(x-2)(5x+4) = 3x^2 - 15x - 12$

Solve the following quadratic equations by using the square root method. Write each radical in simplest form. See Examples 2 and 3.

11. $x^2 = 121$

12. $x^2 = 81$

13. $3x^2 = 108$

14. $5x^2 = 245$

15. $x^2 = 35$

16. $x^2 = 42$

17. $x^2 - 62 = 0$

18. $x^2 - 75 = 0$

19. $x^2 - 45 = 0$

20. $x^2 - 98 = 0$

21. $3x^2 = 54$

22. $5x^2 = 60$

23. $9x^2 = 4$

24. $4x^2 = 25$

25. $(x-1)^2 = 4$

26. $(x+3)^2 = 9$

27. $(x+2)^2 = -25$

28. $(x-5)^2 = 36$

29. $(x+1)^2 = \dfrac{1}{4}$

30. $(x-9)^2 = -\dfrac{9}{25}$

31. $(x-3)^2 = \dfrac{4}{9}$

32. $(x-2)^2 = \dfrac{1}{16}$

33. $(x-6)^2 = 18$

34. $(x+8)^2 = 75$

35. $2(x-7)^2 = 24$

36. $3(x+11)^2 = 60$

37. $(3x+4)^2 = 27$

38. $(2x+1)^2 = 48$

39. $(5x - 2)^2 = 63$ **40.** $(4x - 3)^2 = 125$

The lengths of two sides are given for each of the following right triangles. Determine the length of the missing side. See Example 4.

41. $a = 9, b = 12, c = ?$

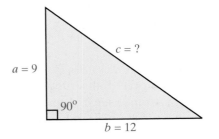

42. $a = 10, b = 24, c = ?$

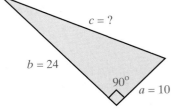

43. $a = 6, c = 12, b = ?$

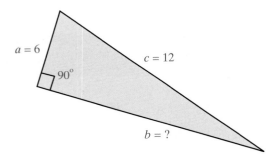

44. $b = 10, c = 30, a = ?$

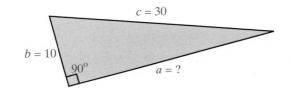

45. $a = 1, b = ?, c = \sqrt{2}$

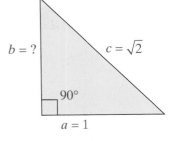

46. $b = 4, c = 4\sqrt{2}, a = ?$

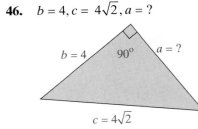

Solve the following word problems. See Example 4.

47. **Right Triangles:** The hypotenuse of a right triangle is twice the length of one of the legs. The length of the other leg is $4\sqrt{3}$ feet. Find the length of the leg and the hypotenuse.

48. **Right Triangles:** One leg of a right triangle is three times the length of the other. The length of the hypotenuse is 20 cm. Find the lengths of the legs.

49. Right Triangles: The two legs of a right triangle are the same length. The hypotenuse is 6 cm long. Find the length of the legs.

50. Right Triangles: The two legs of a right triangle are the same length. The hypotenuse is $4\sqrt{2}$ m long. Find the length of the legs.

51. Distance: The top of a telephone pole is 40 feet above the ground and a guy wire is to be stretched from the top of the pole to a point on the ground 10 feet from the base of the pole. How long (to the nearest tenth of a foot) should the wire be if, for connecting purposes, 6 feet of wire is to be added after the distance is calculated?

52. Distance: The library is located "around the corner" from the bank. If the library is 100 yards from the street corner and the bank is 75 yards around the corner on another street, what is the distance between the library and the bank "as the crow flies?"

53. Time: A ball is dropped from the top of a building that is known to be 144 feet high. The formula for finding the height of the ball at any time is $h = 144 - 16t^2$ where t is measured in seconds. How many seconds will it take for the ball to hit the ground?

54. Distance: An oil spill from a ruptured pipeline is circular in shape and covers an area of about 8 square miles. About how many miles long (to the nearest tenth of a mile) is the diameter of the oil spill? (Use $\pi = 3.14$.)

Use a calculator to solve the following quadratic equations. Round your answers to the nearest hundredth.

55. $x^2 = 647$

56. $x^2 = 378$

57. $19x^2 = 523$

58. $14x^2 = 795$

59. $6x^2 = 17.32$

60. $15x^2 = 229.63$

61. $2.1x^2 = 35.82$

62. $4.7x^2 = 118.34$

Use a calculator to find the two values of each expression accurate to the nearest ten-thousandth.

63. $2 \pm \sqrt{5}$

64. $-4 \pm \sqrt{89}$

65. $\dfrac{2 \pm \sqrt{7}}{2}$

66. $\dfrac{5 \pm \sqrt{13}}{10}$

15.2 Quadratic Equations: Completing the Square

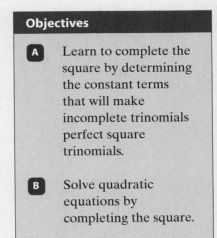

Objective A **Completing the Square**

In Chapters 11 and 12, we discussed perfect square trinomials:

$$(x+a)^2 = x^2 + 2ax + a^2 \quad \text{and} \quad (x-a)^2 = x^2 - 2ax + a^2$$

In particular, in Section 12.4, we discussed factoring perfect square trinomials and how to add terms to binomials so that the resulting trinomial will be a perfect square. This procedure is called **completing the square**. The discussion here is how to solve quadratic equations by completing the square.

As review, we complete the square for $x^2 + 12x$ and $a^2 - 20a$. Both binomials have a leading coefficient of 1.

$$x^2 + 12x + \underline{\hspace{1cm}} = (\hspace{1cm})^2$$

$$\frac{1}{2}(12) = 6 \text{ and } 6^2 = 36$$

Therefore, the number 36 will complete the square.

$$x^2 + 12x + \underline{36} = (x + 6)^2$$

Similarly, for

$$a^2 - 20a + \underline{\hspace{1cm}} = (\hspace{1cm})^2$$

we have

$$\frac{1}{2}(-20) = -10 \text{ and } (-10)^2 = 100.$$

So,

$$a^2 - 20a + \underline{100} = (a - 10)^2.$$

If the leading coefficient is **not 1**, we proceed to factor out the leading coefficient as follows:

$$2x^2 + 20x + \underline{\hspace{1cm}} = 2(\hspace{1cm})^2$$

$$2(x^2 + 10x + \underline{\hspace{1cm}}) = 2(\hspace{1cm})^2 \quad \text{Factor out 2.}$$

$$2(x^2 + 10x + \underline{25}) = 2(x + 5)^2 \quad \text{Complete the square of the expression inside the parentheses, } \frac{1}{2}(10) = 5 \text{ and } 5^2 = 25.$$

$$2x^2 + 20x + 50 = 2(x + 5)^2 \quad \text{The number to be added to the original expression is } 2 \cdot 25 = 50.$$

Objective B **Solving Quadratic Equations by Completing the Square**

The following procedure is used in the examples to solve quadratic equations by completing the square.

Example 1

Completing The Square

Solve the following quadratic equations by completing the square.

a. $x^2 - 6x + 4 = 0$

Solution

$x^2 - 6x + 4 = 0$

$\quad x^2 - 6x = -4$ Add −4 to both sides of the equation.

$x^2 - 6x + 9 = -4 + 9$ Add 9 to both sides of the equation. The left side is now a **perfect square trinomial**. $\frac{1}{2}(-6) = -3$ and $(-3)^2 = 9$.

$\quad\quad (x-3)^2 = 5$ Simplify.

$\quad\quad\quad x - 3 = \pm\sqrt{5}$ Find square roots of both sides.

$\quad\quad\quad\quad x = 3 \pm \sqrt{5}$ Solve for x. Remember, there are two solutions: $3 + \sqrt{5}$ and $3 - \sqrt{5}$.

b. $x^2 + 5x = 7$

Solution

$\quad\quad x^2 + 5x = 7$

$x^2 + 5x + \dfrac{25}{4} = 7 + \dfrac{25}{4}$ Complete the square on the left: $\dfrac{1}{2} \cdot 5 = \dfrac{5}{2}$ and $\left(\dfrac{5}{2}\right)^2 = \dfrac{25}{4}$.

$\quad\quad \left(x + \dfrac{5}{2}\right)^2 = \dfrac{53}{4}$ Simplify: $\left(7 + \dfrac{25}{4} = \dfrac{28}{4} + \dfrac{25}{4} = \dfrac{53}{4}\right)$.

$\quad\quad\quad\quad x + \dfrac{5}{2} = \pm\sqrt{\dfrac{53}{4}}$ Find square roots.

$$x = -\frac{5}{2} \pm \sqrt{\frac{53}{4}}$$ Solve for x.

$$x = -\frac{5}{2} \pm \frac{\sqrt{53}}{2}$$ Special property of square roots: $\sqrt{\frac{a}{b}} = \frac{\sqrt{a}}{\sqrt{b}}$ for $a > 0$ and $b > 0$.

or $x = \dfrac{-5 \pm \sqrt{53}}{2}$ Combine the fractions.

c. $6x^2 + 12x - 9 = 0$

Solution

$6x^2 + 12x - 9 = 0$

$6x^2 + 12x = 9$ Add 9 to both sides of the equation.

$\dfrac{6x^2}{6} + \dfrac{12x}{6} = \dfrac{9}{6}$ Divide each term by 6 **so that the leading coefficient will be 1**.

$x^2 + 2x = \dfrac{3}{2}$ The leading coefficient is 1.

$x^2 + 2x + 1 = \dfrac{3}{2} + 1$ Complete the square: $\frac{1}{2} \cdot 2 = 1$ and $1^2 = 1$.

$(x + 1)^2 = \dfrac{5}{2}$ Simplify.

$x + 1 = \pm \sqrt{\dfrac{5}{2}}$ Find square roots.

$x = -1 \pm \sqrt{\dfrac{5}{2}}$ Solve for x.

$x = -1 \pm \dfrac{\sqrt{5}}{\sqrt{2}} \cdot \dfrac{\sqrt{2}}{\sqrt{2}}$ Rationalize the denominator.

$x = -1 \pm \dfrac{\sqrt{10}}{2}$ Simplify.

or $x = \dfrac{-2 \pm \sqrt{10}}{2}$ Combine the fractions with a common denominator.

d. $2x^2 + 5x - 8 = 0$

Solution

$2x^2 + 5x - 8 = 0$

$2x^2 + 5x = 8$ Add 8 to both sides of the equation.

$\dfrac{2x^2}{2} + \dfrac{5x}{2} = \dfrac{8}{2}$ Divide each term by 2 **so that the leading coefficient will be 1**.

$x^2 + \dfrac{5}{2}x = 4$ Simplify.

$$x^2 + \frac{5}{2}x + \frac{25}{16} = 4 + \frac{25}{16}$$ Complete the square: $\frac{1}{2} \cdot \frac{5}{2} = \frac{5}{4}$ and $\left(\frac{5}{4}\right)^2 = \frac{25}{16}$.

$$\left(x + \frac{5}{4}\right)^2 = \frac{89}{16}$$ Simplify $\left(4 + \frac{25}{16} = \frac{64}{16} + \frac{25}{16} = \frac{89}{16}\right)$.

$$x + \frac{5}{4} = \pm\sqrt{\frac{89}{16}}$$ Find the square roots.

$$x = -\frac{5}{4} \pm \frac{\sqrt{89}}{4}$$ Solve for x. Also, $\sqrt{\frac{89}{16}} = \frac{\sqrt{89}}{\sqrt{16}} = \frac{\sqrt{89}}{4}$.

$$\text{or} \quad x = \frac{-5 \pm \sqrt{89}}{4}$$ Combine the fractions.

e. $x^2 + 8x = -4$

Solution

$$x^2 + 8x = -4$$

$$x^2 + 8x + 16 = -4 + 16$$ Complete the square.

$$(x + 4)^2 = 12$$ Simplify.

$$x + 4 = \pm\sqrt{12}$$ Find square roots.

$$x = -4 \pm \sqrt{12}$$ Solve for x.

$$x = -4 \pm 2\sqrt{3}$$ $\sqrt{12} = \sqrt{4} \cdot \sqrt{3} = 2\sqrt{3}$ by special property of radicals $\sqrt{ab} = \sqrt{a} \cdot \sqrt{b}$ for $a > 0$ and $b > 0$.

Completion Example 2

Completing the Square

Solve the following quadratic equation by completing the square:

$$2x^2 - 12x + 2 = 0$$

Solutions

$$2x^2 - 12x + 2 = 0$$
$$2x^2 - 12x = \underline{}$$
$$x^2 - 6x = \underline{}$$
$$x^2 - 6x + \underline{} = \underline{} + \underline{}$$
$$\left(x - \underline{}\right)^2 = \underline{}$$
$$x - \underline{} = \pm\underline{}$$
$$x = \underline{} \pm \underline{}$$

***Now work margin exercises* 1 and 2.**

Solving quadratic equations by completing the square is a good technique and it provides practice with several algebraic operations. However, do not forget that, in general, factoring is the preferred technique whenever the factors are relatively easy to find. Some of the following exercises allow the options of solving by factoring or by completing the square.

Solve the following quadratic equations by completing the square.

1. **a.** $x^2 - 8x - 4 = 0$

 b. $x^2 + 9x = 3$

 c. $3x^2 + 6x - 4 = 0$

 d. $2x^2 - 6x = 1$

 e. $x^2 - 16x = -14$

2. $0 = 3x^2 - 6x - 15$

Completion Example Answers

2. $-2; \ -1;$
$$x^2 - 6x + \underline{9} = \underline{-1} + \underline{9}$$
$$\left(x - \underline{3}\right)^2 = \underline{8}$$
$$x - \underline{3} = \pm\underline{2\sqrt{2}}$$
$$x = \underline{3} \pm \underline{2\sqrt{2}}$$

Exercises 15.2

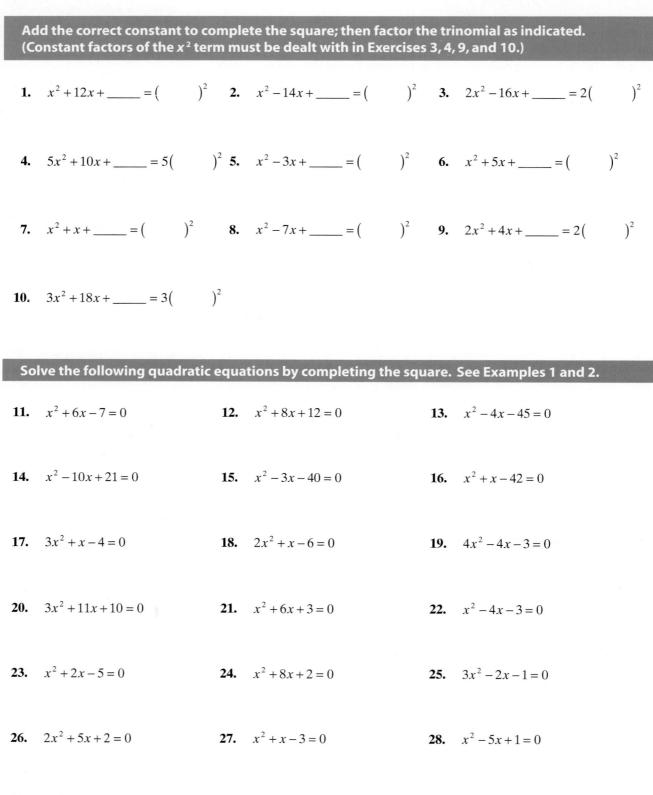

1. $x^2 + 12x + \underline{\hspace{1cm}} = (\quad)^2$ **2.** $x^2 - 14x + \underline{\hspace{1cm}} = (\quad)^2$ **3.** $2x^2 - 16x + \underline{\hspace{1cm}} = 2(\quad)^2$

4. $5x^2 + 10x + \underline{\hspace{1cm}} = 5(\quad)^2$ **5.** $x^2 - 3x + \underline{\hspace{1cm}} = (\quad)^2$ **6.** $x^2 + 5x + \underline{\hspace{1cm}} = (\quad)^2$

7. $x^2 + x + \underline{\hspace{1cm}} = (\quad)^2$ **8.** $x^2 - 7x + \underline{\hspace{1cm}} = (\quad)^2$ **9.** $2x^2 + 4x + \underline{\hspace{1cm}} = 2(\quad)^2$

10. $3x^2 + 18x + \underline{\hspace{1cm}} = 3(\quad)^2$

Solve the following quadratic equations by completing the square. See Examples 1 and 2.

11. $x^2 + 6x - 7 = 0$ **12.** $x^2 + 8x + 12 = 0$ **13.** $x^2 - 4x - 45 = 0$

14. $x^2 - 10x + 21 = 0$ **15.** $x^2 - 3x - 40 = 0$ **16.** $x^2 + x - 42 = 0$

17. $3x^2 + x - 4 = 0$ **18.** $2x^2 + x - 6 = 0$ **19.** $4x^2 - 4x - 3 = 0$

20. $3x^2 + 11x + 10 = 0$ **21.** $x^2 + 6x + 3 = 0$ **22.** $x^2 - 4x - 3 = 0$

23. $x^2 + 2x - 5 = 0$ **24.** $x^2 + 8x + 2 = 0$ **25.** $3x^2 - 2x - 1 = 0$

26. $2x^2 + 5x + 2 = 0$ **27.** $x^2 + x - 3 = 0$ **28.** $x^2 - 5x + 1 = 0$

29. $2x^2 + 3x - 1 = 0$ **30.** $3x^2 - 4x - 2 = 0$

31. $x^2 - 9x + 2 = 0$

32. $x^2 - 8x - 20 = 0$

33. $x^2 + 7x - 14 = 0$

34. $x^2 + 5x + 3 = 0$

35. $x^2 - 11x - 26 = 0$

36. $x^2 + 6x - 4 = 0$

37. $2x^2 + x - 2 = 0$

38. $6x^2 - 2x - 2 = 0$

39. $3x^2 - 6x + 3 = 0$

40. $5x^2 + 15x - 5 = 0$

41. $4x^2 + 20x - 8 = 0$

42. $2x^2 - 7x + 4 = 0$

43. $6x^2 - 8x + 1 = 0$

44. $5x^2 - 10x + 3 = 0$

45. $2x^2 + 7x + 4 = 0$

46. $3x^2 - 5x - 3 = 0$

47. $4x^2 - 2x - 3 = 0$

48. $2x^2 - 9x + 7 = 0$

49. $3x^2 + 8x + 5 = 0$

50. $5x^2 + 11x - 1 = 0$

51. $2x^2 + 5x + 3 = 0$

52. $2x^2 - 12x + 10 = 0$

53. $3x^2 + 24x + 21 = 0$

54. $4x^2 + 6x - 10 = 0$

55. $6x^2 - x - 1 = 0$

Objectives

A   Write quadratic equations in standard form $ax^2 + bx + c = 0$.

B   Solve quadratic equations by using the quadratic formula.

Objective A   **The Quadratic Equation:** $ax^2 + bx + c = 0$

Now we are interested in developing a formula that will be useful in solving quadratic equations of any form. **This formula will always work**, but do not forget the factoring and completing the square techniques because in some cases they are easier to apply than the formula.

General Form of a Quadratic Equation

The general **quadratic equation** is

$$ax^2 + bx + c = 0$$

where a, b, and c are real constants and $a \neq 0$.

We want to solve the general quadratic equation for x in terms of the constant coefficients a, b, and c. The technique is to **complete the square** (Section 15.2).

Development of the Quadratic Formula

$ax^2 + bx + c = 0$	Begin with the general quadratic equation.
$ax^2 + bx = -c$	Add $-c$ to both sides.
$\dfrac{ax^2}{a} + \dfrac{bx}{a} = \dfrac{-c}{a}$	Divide each term by a so that the leading coefficient will be 1.
$x^2 + \dfrac{b}{a}x = \dfrac{-c}{a}$	Rewrite: $\dfrac{bx}{a} = \dfrac{b}{a}x$.
$x^2 + \dfrac{b}{a}x + \left(\dfrac{b}{2a}\right)^2 = \left(\dfrac{b}{2a}\right)^2 + \dfrac{-c}{a}$	Complete the square: $\dfrac{1}{2}\left(\dfrac{b}{a}\right) = \dfrac{b}{2a}$.
$\left(x + \dfrac{b}{2a}\right)^2 = \dfrac{b^2}{4a^2} + \dfrac{-c}{a}$	Simplify.
$\left(x + \dfrac{b}{2a}\right)^2 = \dfrac{b^2}{4a^2} + \dfrac{-c \cdot 4a}{a \cdot 4a}$	Common denominator is $4a^2$.
$\left(x + \dfrac{b}{2a}\right)^2 = \dfrac{b^2 - 4ac}{4a^2}$	Simplify.
$x + \dfrac{b}{2a} = \pm\sqrt{\dfrac{b^2 - 4ac}{4a^2}}$	Find the square roots of both sides.
$x + \dfrac{b}{2a} = \pm\dfrac{\sqrt{b^2 - 4ac}}{\sqrt{4a^2}}$	Use the relationship $\sqrt{\dfrac{a}{b}} = \dfrac{\sqrt{a}}{\sqrt{b}}$ if $a, b > 0$.

$$x + \frac{b}{2a} = \pm \frac{\sqrt{b^2 - 4ac}}{2a}$$ Simplify.

$$x = \frac{-b}{2a} \pm \frac{\sqrt{b^2 - 4ac}}{2a}$$ Solve for x.

$$x = \frac{-b \pm \sqrt{b^2 - 4ac}}{2a}$$ The Quadratic Formula

Or, we can write the two solutions separately as:

$$x = \frac{-b + \sqrt{b^2 - 4ac}}{2a} \quad \text{and} \quad x = \frac{-b - \sqrt{b^2 - 4ac}}{2a}$$

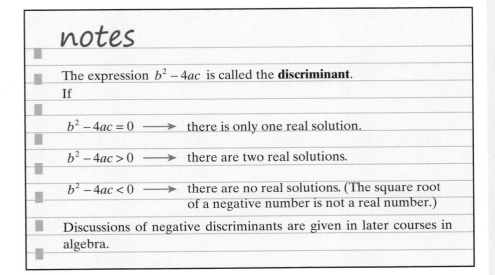

Quadratic Formula

The solutions of the general quadratic equation $ax^2 + bx + c = 0$, where $a \neq 0$, are

$$x = \frac{-b \pm \sqrt{b^2 - 4ac}}{2a}.$$

notes

The expression $b^2 - 4ac$ is called the **discriminant**. If

$b^2 - 4ac = 0 \longrightarrow$ there is only one real solution.

$b^2 - 4ac > 0 \longrightarrow$ there are two real solutions.

$b^2 - 4ac < 0 \longrightarrow$ there are no real solutions. (The square root of a negative number is not a real number.)

Discussions of negative discriminants are given in later courses in algebra.

Objective B **Solving Quadratic Equations by Using the Quadratic Formula**

Now the solutions to quadratic equations can be found by going directly to the formula.

Example 1

The Quadratic Formula

Solve the following quadratic equations by using the quadratic formula:

$$x = \frac{-b \pm \sqrt{b^2 - 4ac}}{2a}.$$

a. $2x^2 + x - 2 = 0$

Solution

$a = 2, b = 1, c = -2$

$$x = \frac{-1 \pm \sqrt{1^2 - 4(2)(-2)}}{2 \cdot 2} = \frac{-1 \pm \sqrt{1 + 16}}{4} = \frac{-1 \pm \sqrt{17}}{4}$$

In many practical applications of quadratic equations, we want to know a decimal approximation of the solutions. Using a calculator, we find the following approximate values to the solutions of the equation $2x^2 + x - 2 = 0$:

$$x = \frac{-1 + \sqrt{17}}{4} \approx 0.78078 \quad \text{or} \quad x = \frac{-1 - \sqrt{17}}{4} \approx -1.2808$$

b. $-3x^2 = -5x + 1$

Solution

$-3x^2 = -5x + 1$

$-3x^2 + 5x - 1 = 0$ First, rewrite the equation so that one side is 0.

Now we see that $a = -3, b = 5,$ and $c = -1$.

Substituting in the quadratic formula gives

$$x = \frac{-5 \pm \sqrt{5^2 - 4(-3)(-1)}}{2(-3)} = \frac{-5 \pm \sqrt{25 - 12}}{-6} = \frac{-5 \pm \sqrt{13}}{-6} = \frac{5 \pm \sqrt{13}}{6}.$$

This shows that the quadratic formula works correctly even though the leading coefficient a is negative. We could also multiply all the terms on both sides of the equation by -1 and solve the new equation. **The solutions will be the same.**

$$-3x^2 + 5x - 1 = 0$$

$$3x^2 - 5x + 1 = 0 \quad \text{Multiply every term by } -1.$$

Now $a = 3$, $b = -5$, and $c = 1$.

$$x = \frac{-(-5) \pm \sqrt{(-5)^2 - 4(3)(1)}}{2 \cdot 3} = \frac{5 \pm \sqrt{25 - 12}}{6} = \frac{5 \pm \sqrt{13}}{6}$$

Again, if needed, a calculator will give approximate values for the solutions:

$$x = \frac{5 + \sqrt{13}}{6} \approx 1.4343 \quad \text{and} \quad x = \frac{5 - \sqrt{13}}{6} \approx 0.2324.$$

c. $\dfrac{1}{6}x^2 - x + \dfrac{1}{2} = 0$

Solution

$$6 \cdot \frac{1}{6}x^2 - 6 \cdot x + 6 \cdot \frac{1}{2} = 6 \cdot 0 \qquad \text{Multiply each term by 6, the}$$
$$\text{least common denominator.}$$

$$x^2 - 6x + 3 = 0 \qquad \text{Integer coefficients are much}$$
$$\text{easier to use in the formula.}$$

So $a = 1$, $b = -6$, and $c = 3$.

$$x = \frac{-(-6) \pm \sqrt{(-6)^2 - 4(1)(3)}}{2 \cdot 1} = \frac{6 \pm \sqrt{36 - 12}}{2}$$

$$= \frac{6 \pm \sqrt{24}}{2} = \frac{6 \pm 2\sqrt{6}}{2} = \frac{2(3 \pm \sqrt{6})}{2}$$

$$= 3 \pm \sqrt{6}$$

d. $2x^2 - 25 = 0$

Solution

The square root method could be applied by adding 25 to both sides, dividing by 2, and then taking square roots. The result is the same by using the quadratic formula with $b = 0$.

$$2x^2 - 25 = 0 \ \left(\text{or } 2x^2 + 0x - 25 = 0\right)$$

So, $a = 2$, $b = 0$, and $c = -25$.

$$x = \frac{-(0) \pm \sqrt{0^2 - 4(2)(-25)}}{2 \cdot 2} = \frac{\pm\sqrt{200}}{4} = \frac{\pm 10\sqrt{2}}{4} = \frac{\pm 5\sqrt{2}}{2}$$

Use the quadratic formula to solve the following quadratic equations.

1. a. $x^2 + 4x - 1 = 0$

   b. $-4x^2 = 2x - 3$

   c. $\frac{1}{3}x^2 + x + \frac{1}{2} = 0$

   d. $3x^2 - 40 = 0$

2. $4x^2 + 6x + 1 = 0$

Completion Example 2

The Quadratic Formula

Use the quadratic formula to solve the following quadratic equation:

$2x^2 - 5x - 3 = 0$

Solution

In this case, $a =$ _____, $b =$ _____, and $c =$ _____

Substituting in the quadratic formula gives

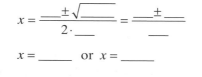

$x =$ _____ or $x =$ _____

Now work margin exercises 1 and 2.

Example 3

The Quadratic Formula

Use the quadratic formula to solve the following quadratic equation:

a. $(3x - 1)(x + 2) = 4x$

Solution

$(3x - 1)(x + 2) = 4x$

$3x^2 + 5x - 2 = 4x$ Multiply on the left-hand side.

$3x^2 + x - 2 = 0$ Subtract $4x$ from both sides so that one side is 0.

Completion Example Answers

2. In this case, $a = \underline{2}$, $b = \underline{-5}$, and $c = \underline{-3}$

Substituting in the quadratic formula gives

$$x = \frac{-(-5) \pm \sqrt{(-5)^2 - 4(2)(-3)}}{2 \cdot 2} = \frac{5 \pm 7}{4}$$

$x = \underline{3}$ or $x = -\frac{1}{\underline{2}}$

Now we see that $a = 3$, $b = 1$, and $c = -2$.

Substituting in the quadratic formula gives

$$x = \frac{-1 \pm \sqrt{1^2 - 4(3)(-2)}}{2 \cdot 3} = \frac{-1 \pm \sqrt{25}}{6} = \frac{-1 \pm 5}{6}$$

$$x = \frac{-1 + 5}{6} = \frac{4}{6} = \frac{2}{3} \quad \text{or} \quad x = \frac{-1 - 5}{6} = \frac{-6}{6} = -1$$

Note: Whenever the solutions are rational numbers, the equation can be solved by factoring. In this example, we could have solved as follows:

$$3x^2 + x - 2 = 0$$

$$(3x - 2)(x + 1) = 0$$

$$3x - 2 = 0 \quad \text{or} \quad x + 1 = 0$$

$$x = \frac{2}{3} \qquad x = -1$$

b. $4x^2 + 12x + 9 = 0$

Solution

$a = 4$, $b = 12$, $c = 9$

$$x = \frac{-12 \pm \sqrt{12^2 - 4(4)(9)}}{2 \cdot 4}$$

$$= \frac{-12 \pm \sqrt{144 - 144}}{8} = \frac{-12 \pm \sqrt{0}}{8} = -\frac{12}{8} = -\frac{3}{2}$$

Note that when the discriminant is 0, there is only one solution. This equation could also be solved by factoring in the following manner:

$4x^2 + 12x + 9 = 0$ The quadratic expression is a **perfect square trinomial**.

$(2x + 3)^2 = 0$

$2x + 3 = 0$ The factor $2x + 3$ is repeated.

$2x = -3$

$x = -\dfrac{3}{2}$ $-\dfrac{3}{2}$ is called a **double solution** or a **double root**.

Now work margin exercise 3.

3. Use the quadratic formula to solve the following quadratic equations.

a. $(x + 1)(x - 1) = 3x$

b. $4x^2 + 20x + 25 = 0$

Quadratic equations should be solved by factoring whenever possible. Factoring is generally the easiest method. This fact was illustrated in Examples 3a and 3b. However, the quadratic formula will always give the solutions whether they are rational, irrational, or nonreal (as shown in Example 4). Thus the quadratic formula is a very useful tool. **In future courses in mathematics every text and instructor will assume that you know or can easily remember the quadratic formula.**

Example 4

Nonreal Solutions

Solve the following equation: $x^2 + x + 1 = 0$.

Solution

$a = 1, \; b = 1, \; c = 1$

$$x = \frac{-1 \pm \sqrt{1^2 - 4(1)(1)}}{2 \cdot 1} = \frac{-1 \pm \sqrt{1-4}}{2} = \frac{-1 \pm \sqrt{-3}}{2}$$

There is no real solution. This example illustrates the fact that not every equation has real solutions. None of the exercises in this text have this kind of answer. Such solutions are called **nonreal complex numbers** and will be discussed in detail in the next course in algebra.

Now work margin exercise 4.

4. Solve the following quadratic equation.

$x^2 + 3x + 5 = 0$

Practice Problems

Solve the equations by using the quadratic formula.

1. $x^2 + 5x + 1 = 0$ **2.** $4x^2 - x = 1$

3. $(x+2)(x-2) = 4x$ **4.** $-2x^2 + 3 = 0$ (**Note:** Here $b = 0$.)

5. $5x^2 - x = 0$ (**Note:** Here $c = 0$.)

Solve the following quadratic equations and write the answers in decimal form accurate to four decimal places.

6. $x^2 + 10x + 5 = 0$ **7.** $5x^2 - x - 1 = 0$

Practice Problem Answers

1. $x = \dfrac{-5 \pm \sqrt{21}}{2}$ **2.** $x = \dfrac{1 \pm \sqrt{17}}{8}$

3. $x = 2 \pm 2\sqrt{2}$ **4.** $x = \dfrac{0 \pm 2\sqrt{6}}{-4}$ or $x = \dfrac{\pm\sqrt{6}}{2}$

5. $x = 0, \dfrac{1}{5}$ (This problem could be solved by factoring.)

6. $x = -0.5279, -9.4721$ **7.** $x = -0.3583, 0.5583$

Exercises 15.3

Rewrite each of the quadratic equations in the form $ax^2 + bx + c = 0$ with $a > 0$; then identify the constants a, b, and c.

1. $x^2 - 3x = 2$ **2.** $x^2 + 2 = 5x$ **3.** $x = 2x^2 + 6$ **4.** $5x^2 = 3x - 1$

5. $4x + 3 = 7x^2$ **6.** $x = 4 - 3x^2$ **7.** $4 = 3x^2 - 9x$ **8.** $6x + 4 = 3x^2$

9. $x^2 + 5x = 3 - x^2$ **10.** $x^2 + 4x - 1 = 2x + 3x^2$

Solve each of the quadratic equations by using the quadratic formula. See Examples 1 and 2.

11. $x^2 - 4x - 1 = 0$ **12.** $x^2 - 3x + 1 = 0$ **13.** $x^2 - 3x = 4$ **14.** $x^2 + 5x = 2$

15. $-2x^2 + x = -1$ **16.** $-3x^2 + x = -1$ **17.** $5x^2 + 3x - 2 = 0$ **18.** $-2x^2 + 5x - 1 = 0$

19. $9x^2 = 3x$ **20.** $4x^2 - 81 = 0$ **21.** $x^2 - 7 = 0$ **22.** $2x^2 + 5x - 3 = 0$

23. $x^2 + 4x = x - 2x^2$ **24.** $x^2 - 2x + 1 = 2 - 3x^2$ **25.** $3x^2 + 4x = 0$ **26.** $4x^2 - 10 = 0$

27. $\dfrac{2}{5}x^2 + x - 1 = 0$ **28.** $2x^2 + 3x - \dfrac{3}{4} = 0$ **29.** $-\dfrac{x^2}{2} - 2x + \dfrac{1}{3} = 0$ **30.** $\dfrac{x^2}{3} - x - \dfrac{1}{5} = 0$

31. $(2x + 1)(x + 3) = 2x + 6$ **32.** $(x + 5)(x - 1) = -3$ **33.** $\dfrac{5x + 2}{3x} = x - 1$ **34.** $-6x^2 = -3x - 1$

35. $4x^2 = 7x - 3$ **36.** $9x^2 - 6x - 1 = 0$

37. $2x^2 + 7x + 3 = 0$ **38.** $5x^2 - x - 4 = 0$ **39.** $3x^2 - 7x + 1 = 0$ **40.** $2x^2 - 2x - 1 = 0$

41. $10x^2 = x + 24$ **42.** $9x^2 + 12x = -2$ **43.** $3x^2 - 11x = 4$ **44.** $5x^2 = 7x + 5$

45. $-4x^2 + 11x - 5 = 0$ **46.** $-4x^2 + 12x - 9 = 0$ **47.** $10x^2 + 35x + 30 = 0$ **48.** $6x^2 + 2x = 20$

49. $25x^2 + 4 = 20$ **50.** $3x^2 - 4x + \dfrac{1}{3} = 0$ **51.** $\dfrac{3}{4}x^2 - 2x + \dfrac{1}{8} = 0$ **52.** $\dfrac{11}{2}x + 1 = 3x^2$

53. $\dfrac{3}{7}x^2 = \dfrac{1}{2}x + 1$ **54.** $\dfrac{35}{4x} = x - 1$

55. $x^2 - x - 1 = 0$ **56.** $3x^2 - 8x - 13 = 0$ **57.** $(2x+1)(2x-1) = 3x$

58. $\dfrac{-x-1}{3x} = x + 1$ **59.** $5x^2 - 100 = 0$ **60.** $(3x+1)(3x-2) = 10x$

15.4 Applications

Applications Involving Quadratic Equations

Objectives

A Solve applied problems using quadratic equations.

Application problems are designed to teach you to read carefully, to think clearly, and to translate from English to algebraic expressions and equations. The problems do not tell you directly to add, subtract, multiply, divide, or square. You must decide on a method of attack based on the wording of the problem and your previous experience and knowledge. Study the following examples and read the explanations carefully. They are similar to some, but not all, of the problems in the exercises. Example 1 makes use of the Pythagorean Theorem that was discussed in Sections 5.7, 12.7, and 15.1.

Example 1

The Pythagorean Theorem

The length of a rectangle is 7 feet longer than the width. If one diagonal measures 13 feet, what are the dimensions of the rectangle?

Solution

Draw a diagram for problems involving geometric figures whenever possible.

Let x = width of the rectangle
and $x + 7$ = length of the rectangle.

Then, by the Pythagorean Theorem:

$$(x+7)^2 + x^2 = 13^2$$
$$x^2 + 14x + 49 + x^2 = 169$$
$$2x^2 + 14x + 49 - 169 = 0$$
$$2x^2 + 14x - 120 = 0$$
$$2(x^2 + 7x - 60) = 0$$
$$2(x-5)(x+12) = 0$$

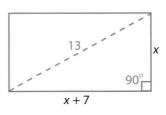

$x - 5 = 0$ or $x + 12 = 0$

$x = 5$ $\quad$ ~~$x = -12$~~ A negative number does not fit the conditions of the problem.

Thus $x = 5$ and $x + 7 = 12$.

Check: $5^2 + 12^2 = 13^2$
$25 + 144 = 169$
$169 = 169$ $\quad$ true statement

The width is 5 feet and the length is 12 feet.

1. In a right triangle, the hypotenuse is 10 ft long. One of the legs is 2 ft longer than the other leg. What are the lengths of the legs?

Now work margin exercise **1.**

2. Working together, Henry and his son, Bill, can paint one room in 2 hours. When Bill paints a room of the same size by himself, he takes 3 hours longer than his father. How long does it take his father to paint a room of this size?

Example 2

Work

Working for a janitorial service, a woman and her daughter can clean a building in 5 hours. If the daughter were to do the job by herself, she would take 24 hours longer than her mother would take. How long would it take her mother to clean the building without the daughter's help?

(**Note:** For reference, this problem is similar to those discussed in Section 13.5 on work. The difference is that in this section the resulting equations are quadratic.)

Solution

Let $\quad x \quad = \quad$ time for mother alone
and $\quad x + 24 \quad = \quad$ time for daughter alone.

	Hours to Complete	Part Completed in 1 Hour
Mother	x	$\dfrac{1}{x}$
Daughter	$x + 24$	$\dfrac{1}{x + 24}$
Together	5	$\dfrac{1}{5}$

part done by mother in 1 hour	+	part done by daughter in 1 hour	=	part done working together in 1 hour
$\dfrac{1}{x}$	+	$\dfrac{1}{x + 24}$	=	$\dfrac{1}{5}$

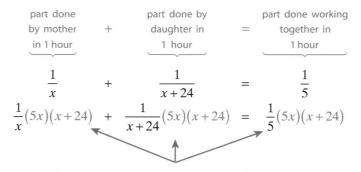

$$\frac{1}{x}(5x)(x+24) + \frac{1}{x+24}(5x)(x+24) = \frac{1}{5}(5x)(x+24)$$

Multiply each term by the LCM of the denominators.

$$5(x+24)+5x = x(x+24)$$

$$5x+120+5x = x^2 + 24x$$

$$0 = x^2 + 24x - 10x - 120$$

$$0 = x^2 + 14x - 120$$

$$0 = (x-6)(x+20)$$

$$x-6=0 \ \text{ or } \ x+20=0$$

$$x = 6 \qquad\qquad \cancel{x = -20}$$

The mother could do the job alone in 6 hours.

Now work margin exercise **2.**

Example 3

Distance-Rate-Time

A small plane travels at a speed of 200 mph in still air. Flying with a tailwind, the plane is clocked over a distance of 960 miles. Flying against a headwind, it takes 2 hours more time to complete the return trip. What was the wind velocity?

Solution

The basic formula is $d = rt$ (distance = rate · time).

Also, $t = \dfrac{d}{r}$ and $r = \dfrac{d}{t}$.

Let $\qquad x$ = wind velocity.

Then $\quad 200 + x$ = speed of airplane going with the wind (tailwind),

$\qquad\quad 200 - x$ = speed of airplane returning against the wind (headwind),

and $\qquad 960$ = distance each way.

We know distance and can represent rate (or speed), so the formula $t = \dfrac{d}{r}$ is used for representing time.

	Rate	Time	Distance
Going	$200 + x$	$\dfrac{960}{200 + x}$	960
Returning	$200 - x$	$\dfrac{960}{200 - x}$	960

$$\underbrace{\text{time}}_{\text{returning}} \quad - \quad \underbrace{\text{time}}_{\text{going}} \quad = \quad \underbrace{\text{difference}}_{\text{in time}}$$

$$\frac{960}{200 - x} \quad - \quad \frac{960}{200 + x} \quad = \quad 2$$

$$(200 - x)(200 + x)\frac{960}{200 - x} - (200 - x)(200 + x)\frac{960}{200 + x} = (200 - x)(200 + x)2$$

$$960(200 + x) - 960(200 - x) = 2(40{,}000 - x^2)$$

$$192{,}000 + 960x - 192{,}000 + 960x = 80{,}000 - 2x^2$$

$$2x^2 + 1920x - 80{,}000 = 0$$

$$x^2 + 960x - 40{,}000 = 0$$

$$(x - 40)(x + 1000) = 0$$

$$x - 40 = 0 \quad \text{or} \quad x + 1000 = 0$$

$$x = 40 \qquad\qquad \cancel{x = -1000}$$

The wind velocity was 40 mph.

3. A motorboat travels 10 mph in still water. The boat takes 4 hours longer to travel 48 miles going upstream than it does to travel 24 miles going downstream. Find the rate of the current.

Check:

$$\frac{960}{200-40} - \frac{960}{200+40} \stackrel{?}{=} 2$$

$$\frac{960}{160} - \frac{960}{240} \stackrel{?}{=} 2$$

$$6-4 \stackrel{?}{=} 2$$

$$2 = 2 \quad \text{true statement}$$

Now work margin exercise 3.

Example 4

Geometry

a. A square piece of cardboard has a small square, 2 in. by 2 in., cut from each corner. The edges are then folded up to form a box with a volume of 5000 in.^{3} What are the dimensions of the box? (**Hint:** The volume is the product of the length, width, and height: $V = lwh$.)

Solution

Draw a diagram illustrating the information.

Let x = one side of the square.

$$2(x-4)(x-4) = 5000$$
$$2(x^2 - 8x + 16) = 5000$$
$$x^2 - 8x + 16 = 2500$$
$$x^2 - 8x - 2484 = 0$$
$$(x-54)(x+46) = 0$$

$$x - 54 = 0 \quad \text{or} \quad x + 46 = 0$$
$$x = 54 \qquad \quad \cancel{x = -46}$$
$$x - 4 = 50$$

The dimensions of the box are 50 in. by 50 in. by 2 in.

b. A little league baseball field is in the shape of a square with sides of 60 feet. What is the distance (to the nearest tenth of a foot) the catcher must throw from home plate to second base?

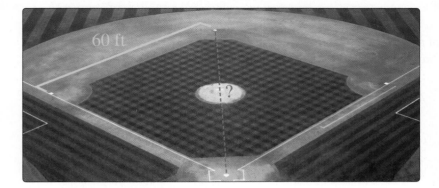

Solution

Since the distance from home plate to second base is the hypotenuse of a right triangle with sides of length 60 feet, we can use the Pythagorean Theorem as follows:

$$x^2 = 60^2 + 60^2$$

$$x^2 = 3600 + 3600$$

$$x^2 = 7200$$

$x = \sqrt{7200}$ We want only the positive square root.

$x = 84.9$ ft   rounded to the nearest tenth of a foot

The catcher must throw about 84.9 feet.

Now work margin exercise 4.

4. A square has sides of length 24 ft. Find the length of the diagonal of the square to the nearest tenth of a foot. (The **diagonal** of a square is a line connecting opposite corners of the square.)

Exercises 15.4

1. **Number Problem:** The sum of a positive number and its square is 132. Find the number.

2. **Rectangles:** The dimensions of a rectangle can be represented by two consecutive even integers. The area of the rectangle is 528 cm². Find the width and the length of the rectangle.

3. **Rectangles:** A rectangle has a length 5 m less than twice its width. If the area is 63 m², find the dimensions of the rectangle.

4. **Number Problem:** The sum of a positive number and its square is 992. Find the number.

5. **Farming:** The area of a rectangular field is 198 m². If it takes 58 m of fencing to enclose the field, what are the dimensions of the field? (**Hint:** The length plus the width is 29 meters.)

6. **Rectangles:** The length of a rectangle is 2 cm less than 3 times its width. If the area of the rectangle is 225 cm², find the dimensions of the original rectangle.

7. **Swimming Pools:** The Wilsons have a rectangular swimming pool that is 10 ft longer than it is wide. The pool is completely surrounded by a concrete deck that is 6 ft wide. The total area of the pool and the deck is 1344 ft². Find the dimensions of the pool.

8. **Consecutive Integers:** The product of two positive consecutive odd integers exceeds their sum by 287. Find the positive integers.

9. **Consecutive Integers:** The sum of the squares of two positive consecutive integers is 221. Find the positive integers.

10. **Number Problems** The difference between two positive numbers is 9. If the smaller number is added to the square of the larger number, the result is 147. Find the numbers.

11. **Number Problem:** The difference between a positive number and 3 is four times the reciprocal of the number. Find the number. (**Hint:** The reciprocal of x is $\frac{1}{x}$.)

12. **Number Problem:** The sum of a positive number and 5 is fourteen times the reciprocal of the number. Find the number. (**Hint:** The reciprocal of x is $\frac{1}{x}$.)

13. **Squares:** Each side of a square is increased by 10 cm. The area of the resulting square is 9 times the area of the original square. Find the length of the sides of the original square.

14. **Squares:** If 5 m are added to each side of a square, the area of the resulting square is four times the area of the original square. Find the length of the sides of the original square.

15. **Rectangles:** The diagonal of a rectangle is 13 m. The length is 2 m more than twice the width. Find the dimensions of the rectangle.

16. **Rectangles:** The length of a rectangle is 4 m more than its width. If the diagonal is 20 m, what are the dimensions of the rectangle?

17. **Triangles:** A right triangle has two equal sides. If the hypotenuse is 12 cm, what is the length of the equal sides?

18. **Rental Units:** Mr. Prince owns a 15-unit apartment complex. If all units are rented, the rent for each apartment is $200 per month. Each time the rent is increased by $20, he will lose 1 tenant. What is the rental rate if he receives $3120 monthly in rent? (Hint: Let x = number of empty units.)

19. **Group Travel:** The Ski Club is planning to charter a bus to a ski resort. The cost will be $900 and each member will share the cost equally. If the club had 15 more members, the cost per person would be $10 less. How many are in the club now? (**Hint:** If x = number in club now, $\frac{900}{x}$ = cost per person.)

20. **Sales:** A sporting goods store owner estimates that if he sells a certain model of basketball shoes for x dollars a pair, he will be able to sell $125 - x$ pairs. Find the price if his sales are $3750. Is there more than one possible answer?

21. Travel: Mr. Green traveled to a city 200 miles from his home to attend a meeting. Due to car trouble, his average speed returning was 10 mph less than his speed going. If the total driving time for the round trip was 9 hours, at what rate of speed did he travel to the city?

	Rate	Time	Distance
Going	x	?	200
Returning	$x-10$	?	200

22. Boating: A motorboat takes a total of 2 hours to travel 8 miles downstream and 4 miles back on a river that is flowing at a rate of 2 mph. Find the rate of the boat in still water.

23. Boating: A small motorboat travels 12 mph in still water. It takes 2 hours longer to travel 45 miles going upstream than it does going downstream. Find the rate of the current. (**Hint:** $12 + c$ = rate going downstream and $12 - c$ = rate going upstream.)

24. Travel: Recently Mr. and Mrs. Roberts spent their vacation in San Francisco, which is 540 miles from their home. Being a little reluctant to return home, the Roberts took 2 hours longer on their return trip and their average speed was 9 mph slower than when they were going. What was their average rate of speed as they traveled from home to San Francisco?

25. Club Membership: The Blumin Garden Club planned to give their president a gift of appreciation costing $120 and to divide the cost evenly. In the meantime, 5 members dropped out of the club. If it now costs each of the remaining members $2 more than originally planned, how many members initially participated in the gift buying? (**Hint:** If x = number in club initially, $\dfrac{120}{x}$ = cost per member.)

26. Boxes: A rectangular sheet of metal is 6 in. longer than it is wide. A box is to be made by cutting out 3 in. squares at each corner and folding up the sides. If the box has a volume of 336 in.^{3}, what were the original dimensions of the sheet metal? (See Example 4.)

27. Boxes: A box is to be made out of a square piece of cardboard by cutting out 2 in. squares at each corner and folding up the sides. If the box has a volume of 162 in.^{3}, how big was the piece of cardboard? (See Example 4.)

28. Painting: A woman and her daughter can paint their cabin in 3 hours. Working alone it would take the daughter 8 hours longer than it would the mother. How long would it take the mother to paint the cabin alone?

29. **Filling a Tank:** Two pipes can fill a tank in 8 min if both are turned on. If only one is used it would take 30 minutes longer for the smaller pipe to fill the tank than the larger pipe. How long will it take the smaller pipe to fill the tank?

30. **Plowing:** A farmer and his son can plow a field with two tractors in 4 hours. If it would take the son 6 hours longer than the father to plow the field alone, how long would it take each if they worked alone?

31. **Throwing a Ball:** A ball is thrown upward with an initial velocity of 32 ft/sec from the edge of a cliff near the beach. The cliff is 50 ft above the beach and the height of the ball can be found by using the equation

$$h = -16t^2 + 32t + 50$$

where t is measured in seconds.

a. When will the ball be 66 feet above the beach?

b. When will the ball be 30 feet above the beach?

c. In about how many seconds will the ball hit the beach? Round your answer to the nearest tenth.

32. **Shooting a Projectile:** The height of a projectile fired upward from the ground with a velocity of 128 ft/sec is given by the formula $h = -16t^2 + 128t$.

a. When will the projectile be 256 feet above the ground?

b. Will the projectile ever be 300 feet above the ground? Explain.

c. When will the projectile be 240 feet above the ground?

d. In how many seconds will the projectile hit the ground?

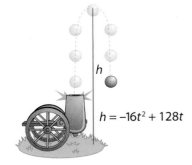

$$h = -16t^2 + 128t$$

33. Ladder Height: A ladder is 30 ft long and you want to place the base of the ladder 10 ft from the base of a building. About how far up the building (to the nearest tenth of a foot) will the ladder reach?

34. Flag Poles: A flag pole is on top of a building and is held in place by steel cables attached to the top of the pole. If one such cable is 40 ft long and is attached at a point on the roof of the building 20 ft from the base of the flag pole, what is the length of the flag pole (to the nearest tenth of a foot)?

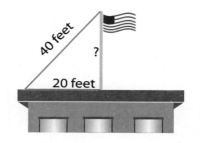

Writing & Thinking

35. Suppose that you are to solve an applied problem and the solution leads to a quadratic equation. You decide to use the quadratic formula to solve the equation. Explain what restrictions you must be aware of when you use the formula.

15.5 Quadratic Functions: $y = ax^2 + bx + c$

In Chapter 9, we discussed linear equations and linear functions in great detail. For example,

$$y = 2x + 5 \text{ is a linear function}$$

and its graph is a straight line with slope 2 and y-intercept $(0, 5)$. The equation can also be written in the function form

$$f(x) = 2x + 5$$

In general, we can write linear functions in the form

$$y = mx + b \quad \text{or} \quad f(x) = mx + b$$

where the slope is m and the y-intercept is b.

Objective A   Quadratic Functions

Now consider the function

$$y = x^2 - 2x - 3$$

This function is **not** linear, so its graph is **not** a straight line. The nature of the graph can be seen by plotting several points, as shown in Figure 1.

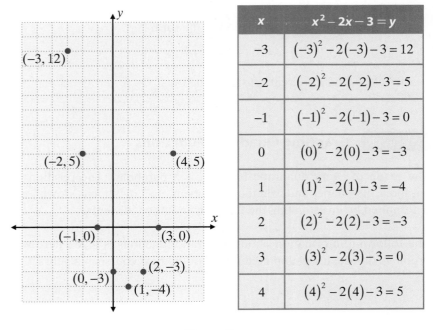

x	$x^2 - 2x - 3 = y$
–3	$(-3)^2 - 2(-3) - 3 = 12$
–2	$(-2)^2 - 2(-2) - 3 = 5$
–1	$(-1)^2 - 2(-1) - 3 = 0$
0	$(0)^2 - 2(0) - 3 = -3$
1	$(1)^2 - 2(1) - 3 = -4$
2	$(2)^2 - 2(2) - 3 = -3$
3	$(3)^2 - 2(3) - 3 = 0$
4	$(4)^2 - 2(4) - 3 = 5$

Figure 1

The points do not lie on a straight line. The graph can be seen by drawing a smooth curve called a **parabola** through the points, as shown in Figure 2. The point $(1, -4)$ is the "turning point" of the curve and is called the **vertex** of the parabola. The line $x = 1$ is the **line of symmetry** or the **axis of symmetry** for the parabola. The curve is a "mirror image" of itself on either side of the line $x = 1$.

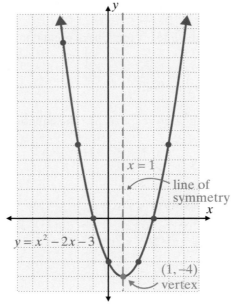

The graph of $y = x^2 - 2x - 3$ is a parabola.

$x = 1$ is the line of symmetry.

The point $(1, -4)$ is the vertex. This is the lowest point on the curve and -4 is the **minimum value** for y.

$x = 1$

line of symmetry

$y = x^2 - 2x - 3$

$(1, -4)$ vertex

Figure 2

Parabolas (or parabolic arcs) occur frequently in describing information in real life. For example, the paths of projectiles (thrown balls, artillery shells, arrows) and a revenue function in business can be illustrated using parabolas.

Quadratic Function

A **quadratic function** is a function of the form

$$y = ax^2 + bx + c$$

where a, b, and c are real constants and $a \neq 0$.

Objective B **Graphing Quadratic Functions**

The graph of every quadratic function is a parabola. The position of the parabola, its shape, and whether it "opens up" or "opens down" can be determined by investigating the function and its coefficients.

The following information about quadratic functions is given here without proof. A thorough development is part of the next course in algebra.

General Information on Quadratic Functions

For the **quadratic function** $y = ax^2 + bx + c$:

1. If $a > 0$, the parabola "opens upward."

2. If $a < 0$, the parabola "opens downward."

3. $x = -\dfrac{b}{2a}$ is the **line of symmetry**.

4. The **vertex** (turning point) occurs where $x = -\dfrac{b}{2a}$. Substitute this value for x in the function and find the y-value of the vertex. The vertex is the lowest point on the curve if the parabola opens upward or it is the highest point on the curve if the parabola opens downward.

Several graphs related to the basic form

$$y = ax^2 \quad \text{where } b = 0 \text{ and } c = 0$$

are shown in Figure 3.

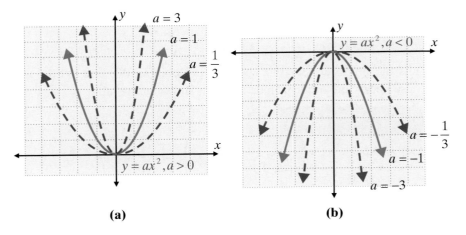

(a)   **(b)**

Figure 3

In the cases illustrated in Figure 3, where $y = ax^2$, we have $b = 0$, and

$$x = -\frac{b}{2a} = -\frac{0}{2a} = 0.$$

So the vertex occurs at the origin, $(0,0)$. In addition, the axis of symmetry is the y-axis (the line $x = 0$).

1. For the following quadratic function, find **a.** its vertex, **b.** its line of symmetry, and **c.** its x-intercepts. Plot a few specific points and graph the parabola.

$$y = x^2 - 4$$

Example 1

Graphing a Quadratic Function ($b = 0$)

For the quadratic function $y = x^2 - 1$, find **a.** its vertex, **b.** its line of symmetry, and **c.** its x-intercepts. Plot a few specific points and graph the parabola.

Solution

For $y = x^2 - 1$, we have $a = 1, b = 0,$ and $c = -1$.

a. The vertex is at $x = -\dfrac{0}{2 \cdot 1} = 0$. Substituting 0 for x gives

$$y = 0^2 - 1 = -1.$$

The vertex is the point $(0, -1)$.

b. The line of symmetry is $x = -\dfrac{0}{2(1)} = 0$.

c. The x-intercepts can be found by setting $y = 0$ and solving the resulting quadratic equation. Thus

$$x^2 - 1 = 0$$
$$x^2 = 1$$
$$x = \pm\sqrt{1}$$
$$x = \pm 1.$$

Thus the graph crosses the x-axis at the points $(1, 0)$ and $(-1, 0)$. These points and others can be found as shown in the table.

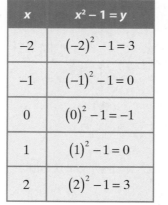

x	$x^2 - 1 = y$
-2	$(-2)^2 - 1 = 3$
-1	$(-1)^2 - 1 = 0$
0	$(0)^2 - 1 = -1$
1	$(1)^2 - 1 = 0$
2	$(2)^2 - 1 = 3$

-1 is called the minimum value of the quadratic function.

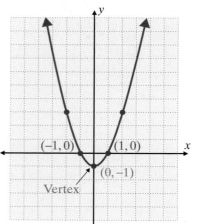

Now work margin exercise 1.

Example 2

Graphing a Quadratic Function

For the quadratic function $f(x) = x^2 - 6x + 1$, find **a.** its vertex, **b.** its line of symmetry, and **c.** its x-intercepts. Plot a few specific points and graph the parabola.

Solution

For $f(x) = x^2 - 6x + 1$, $a = 1$, $b = -6$, and $c = 1$.

a. The vertex is at $x = -\dfrac{-6}{2 \cdot 1} = 3$. Substituting 3 for x gives

$y = 3^2 - 6 \cdot 3 + 1 = -8$. The vertex is the point $(3, -8)$.

b. The line of symmetry is $x = -\dfrac{-6}{2(1)} = 3$.

c. The x-intercepts can be found by setting $y = 0$ and solving the resulting quadratic equation. Thus, using the quadratic formula, we have

$x^2 - 6x + 1 = 0$

$x = \dfrac{-(-6) \pm \sqrt{(-6)^2 - 4(1)(1)}}{2 \cdot 1} = \dfrac{6 \pm \sqrt{32}}{2} = \dfrac{6 \pm 4\sqrt{2}}{2} = 3 \pm 2\sqrt{2}.$

The x-intercepts are at $\left(3 + 2\sqrt{2}, 0\right)$ and $\left(3 - 2\sqrt{2}, 0\right)$.

Estimating these values with a calculator gives the x-intercepts at about $(5.83, 0)$ and $(0.17, 0)$.

Thus the graph crosses the x-axis at about $(5.83, 0)$ and $(0.17, 0)$.

Other points can be found as shown in the table.

x	$x^2 - 6x + 1 = y$
1	$(1)^2 - 6(1) + 1 = -4$
3	$(3)^2 - 6(3) + 1 = -8$
5	$(5)^2 - 6(5) + 1 = -4$
$3 + 2\sqrt{2}$	$\left(3 + 2\sqrt{2}\right)^2 - 6\left(3 + 2\sqrt{2}\right) + 1 = 0$
$3 - 2\sqrt{2}$	$\left(3 - 2\sqrt{2}\right)^2 - 6\left(3 - 2\sqrt{2}\right) + 1 = 0$

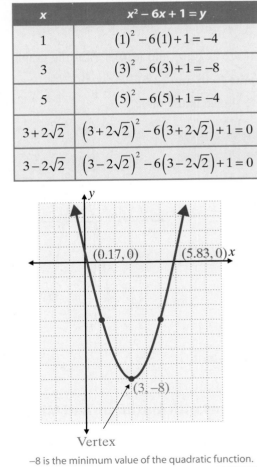

-8 is the minimum value of the quadratic function.

Now work margin exercise 2.

2. For the following quadratic function, find **a.** its vertex, **b.** its line of symmetry, and **c.** its x-intercepts. Plot a few specific points and graph the parabola.

$f(x) = x^2 - 4x + 2$

3. For the following quadratic function, find **a.** its vertex, **b.** its line of symmetry, and **c.** its x-intercepts. Plot a few specific points and graph the parabola.

$$y = -3x^2 + 6x$$

Example 3

Graphing a Quadratic Function ($c = 0$)

For the quadratic function $y = -2x^2 + 4x$, find **a.** its vertex, **b.** its line of symmetry, and **c.** its x-intercepts. Plot a few specific points and graph the parabola.

Solution

For $y = -2x^2 + 4x$, $a = -2$, $b = 4$, and $c = 0$.

a. The vertex is at $x = -\dfrac{4}{2(-2)} = 1$. Substituting 1 for x gives

$$y = -2 \cdot 1^2 + 4 \cdot 1 = 2.$$

The vertex is the point $(1, 2)$.

b. The line of symmetry is $x = -\dfrac{4}{2(-2)} = 1$.

c. The x-intercepts can be found by setting $y = 0$ and solving the resulting quadratic equation. Thus, by factoring, we have

$$-2x^2 + 4x = 0$$
$$-2x(x - 2) = 0$$
$$x = 0 \text{ and } x = 2$$

The x-intercepts are at $(0, 0)$ and $(2, 0)$. Thus the graph crosses the x-axis at the points $(0, 0)$ and $(2, 0)$.

These points are shown in the table.

x	$-2x^2 + 4x = y$
0	$-2(0)^2 + 4(0) = 0$
1	$-2(1)^2 + 4(1) = 2$
2	$-2(2)^2 + 4(2) = 0$

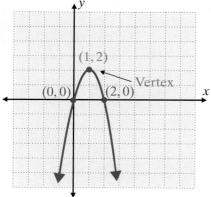

2 is called the maximum value of the quadratic function.

Now work margin exercise **3**.

Objective C **Applications with Maximum and Minimum Values**

The vertex of a vertical parabola is either the lowest point or the highest point on the parabola.

Minimum and Maximum Values

For a parabola with its equation in the form $y = a(x-h)^2 + k$,

1. If $a > 0$, then the parabola opens upward, (h, k) is the lowest point, and the y-value k is called the **minimum value** of the function.
2. If $a < 0$, then the parabola opens downward, (h, k) is the highest point, and y-value k is called the **maximum value** of the function.

If the function is in the general quadratic form $y = ax^2 + bx + c$, then the maximum or minimum value can be found by letting $x = -\dfrac{b}{2a}$ and solving for y.

The concepts of maximum and minimum values of a function help not only in graphing but also in solving many types of applications. Applications involving quadratic functions are discussed here. Other types of applications are discussed in more advanced courses in mathematics.

Example 4

Minimum and Maximum Values

a. Find the dimensions of the rectangle with maximum area if the perimeter is 40 m.

Solution

Let A = area
l = length
w = width.

$A = lw$ and $2l + 2w = 40$

$$2w = 40 - 2l$$

$$w = 20 - l$$

So, substituting for w,

$$A = l(20 - l)$$

$$A = 20l - l^2$$

$$A = -l^2 + 20l.$$

This is a quadratic function with $a = -1, b = 20,$ and $c = 0$. The maximum area occurs at the vertex of the corresponding parabola where

$$l = -\frac{b}{2a} = -\frac{20}{2(-1)} = 10$$

and $w = 20 - l = 20 - 10 = 10.$

Thus the maximum area occurs when $l = 10$ m and $w = 10$ m. (The rectangle is, in fact, a square and the area = 100 m².)

4. Solve the following minimum and maximum value equations.

a. Find the dimensions of the rectangle with maximum area if the perimeter is 72 in.

b. A minor league hockey team sells tickets for games at a price of $8.00 per ticket. On average they sell 4000 tickets per home game. The team estimates that for each $1.00 increase in ticket price they will sell 200 fewer tickets. What price should they charge to maximize their revenue (profit) per game? What will be their maximum revenue?

c. A towing company is going to build a rectangular fence against the side of a building to store impounded cars. The manager has 360 yards of fencing and wants to enclose the maximum area possible inside the lot. What are the dimensions of the lot with the maximum area and what is this area?

b. A sandwich company sells hot dogs at the local baseball stadium for $3.00 each. On average they sell 2000 hot dogs per game. The company estimates that each time the price is raised by 25¢, they will sell 100 fewer hot dogs. What price should they charge to maximize their revenue (income) per game? What will be the maximum revenue?

Solution

Let x = number of 25¢ increases in price.
Then $3.00 + 0.25x$ = price per hot dog,
and $2000 - 100x$ = number of hot dogs sold.

$$\text{Revenue}(R) = (\text{price per unit}) \cdot (\text{number of units sold})$$

So, $R = (3.00 + 0.25x)(2000 - 100x)$
$$= 6000 - 300x + 500x - 25x^2$$
$$= -25x^2 + 200x + 6000.$$

The revenue is represented by a quadratic function and the maximum revenue occurs at the point where

$$x = -\frac{b}{2a} = -\frac{200}{-50} = 4.$$

For $x = 4$,

$$\text{price per hot dog} = 3.00 + 0.25(4) = \$4.00,$$

and

$$\text{revenue} = (4)(2000 - 100 \cdot 4) = \$6400.$$

Thus the company will make its maximum revenue of $6400 by charging $4 per hot dog.

c. A rancher is going to build three sides of a rectangular corral next to a river. He has 240 feet of fencing and wants to enclose the maximum area possible inside the corral. What are the dimensions of the corral with the maximum area and what is this area?

Solution

Let x = length of one of the two equal sides of the rectangular corral.
Then $240 - 2x$ = length of third side of the rectangular corral.
Since area equals length times width, the area, A, of the corral is represented by the quadratic function $A = x(240 - 2x) = 240x - 2x^2$,

the maximum area occurs at the point where $x = -\dfrac{b}{2a} = -\dfrac{240}{-4} = 60.$

Two sides of the rectangle are 60 feet and the third side is $240 - 2(60) = 120$ feet.

The maximum area possible is $60(120) = 7200$ square feet.

Now work margin exercise 4.

Objective D   **Using a Calculator to Graph Quadratic Functions**

We illustrate how to graph a quadratic function on a TI-84 Plus graphing calculator by graphing the function $y = x^2 + 2x - 3$ in the following step-by-step manner.

Step 1: Press the key marked ⬤ Y= .

Step 2: Enter the function to be graphed: $\backslash Y_1 = X^2 + 2X - 3$.

   (**Note:** The variable X is found by pressing ⬤ X,T,θ,n .

   X^2 is then found by pressing ⬤ x^2 .)

   (Or, you can enter $\backslash Y_1 = X \wedge 2 + 2X - 3$.)

Step 3: Press ⬤ GRAPH in the upper right hand corner of the keyboard. The graph of the parabola should appear on the display as shown here.

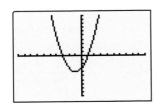

If the graph does not appear as shown, you may need to adjust your window as follows:

a.   Press ⬤ WINDOW and set the numbers as needed, or

b.   Press ⬤ ZOOM and then press ⬤ **6** to get the standard window to get the window from -10 to 10 on the x-axis and from -10 to 10 on the y-axis.

To estimate the location of various points on the graph of the parabola, you can press ⬤ TRACE and then use the left and right arrows to move along the parabola. The function $Y_1 = X^2 + 2X - 3$ will appear at the top of the display and the coordinates of the point indicated by the cursor will appear at the bottom of the display. An example is shown here.

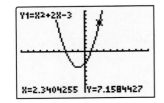

5. Use your graphing calculator to graph the quadratic function $y = -2x^2 + 3x + 6$. Then use the trace key to estimate the location of three points on the graph.

Example 5

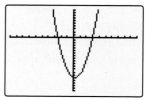 **Graphing with a Calculator**

Use your graphing calculator to graph the quadratic function $y = 2x^2 - x - 15$. Then use the trace key to estimate the location of three points on the graph. (As shown in the following diagrams, set the **WINDOW** so that **xmin** = −10, **xmax** = 10, **ymin** = −20 and **ymax** = 10.)

Solution

Step 1: Press the key marked ⬭ Y= .

Step 2: Enter the function to be graphed: $\backslash Y_1 = 2X^2 - X - 15$.

Step 3: Press ⬭ GRAPH in the upper right hand corner of the keyboard. The graph of the parabola should appear on the display as shown here.

Now, press the ⬭ TRACE key and move the cursor around to estimate the location of points. Three such locations are shown here.

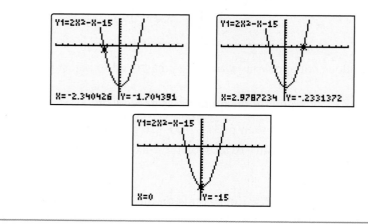

Now work margin exercise 5.

Exercises 15.5

For each quadratic function, **a.** find its vertex, **b.** find its line of symmetry, **c.** locate (or estimate) its *x*-intercepts, and **d.** graph the function. (Note: If solving the quadratic equation results in nonreal solutions, then the graph does not cross the *x*-axis.) See Examples 1 through 3.

1. $y = x^2 + 4$

2. $y = x^2 - 6$

3. $y = 8 - x^2$

4. $y = -2 - x^2$

5. $y = x^2 - 2x - 3$

6. $y = x^2 - 4x + 5$

7. $y = x^2 + 6x$

8. $y = x^2 - 4x$

9. $y = -x^2 - 4x + 2$

10. $y = -x^2 - 5x + 4$

11. $y = 2x^2 - 10x + 3$

12. $y = 2x^2 - 12x - 5$

13. $y = x^2 + 7x - 4$ **14.** $y = x^2 + 5x + 4$ **15.** $y = -x^2 + x - 3$

16. $y = -x^2 - 7x + 3$ **17.** $y = 2x^2 + 7x - 4$ **18.** $y = 2x^2 - x - 3$

19. $y = 3x^2 + 5x + 2$ **20.** $y = 3x^2 - 9x + 5$

Solve the following word problems. See Example 4.

21. The perimeter of a rectangle is 60 yd. What are the dimensions of the rectangle with maximum area?

22. The perimeter of a rectangle is 56 ft. What are the dimensions of the rectangle with maximum area?

23. A ball is thrown vertically upward from the ground with an initial velocity of 112 ft per sec.

    a. When will the ball reach its maximum height?

    b. What will be the maximum height?

24. A ball is thrown vertically upward from the ground with an initial velocity of 104 ft per sec.

    a. When will the ball reach its maximum height?

    b. What will be the maximum height?

25. A stone is projected vertically upward from a platform that is 32 ft high at a rate of 128 ft per sec.

    a. When will the stone reach its maximum height?

    b. What will be the maximum height?

26. A stone is projected vertically upward from a platform that is 20 ft high at a rate of 160 ft per sec.

    a. When will the stone reach its maximum height?

    b. What will be the maximum height?

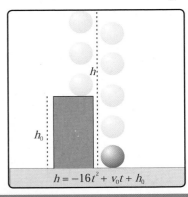

$$h = -16t^2 + v_0 t + h_0$$

In business, the term revenue represents income. The revenue (income) is found by multiplying the number of units sold times the price per unit. [Revenue = (price) · (units sold)]

27. A store owner estimates that by charging x dollars each for a certain lamp, he can sell $40 - x$ lamps each week. What price will yield maximum revenue?

    a. What is the revenue function $R(x)$?

    b. What price will yield a maximum revenue?

28. When fishing reels are priced at x dollars each, local consumers will buy $36 - x$ fishing reels.

    a. What is the revenue function $R(x)$?

    b. What price will yield a maximum revenue?

29. A manufacturer produces calculators. She estimates that by selling them for x dollars each, she will be able to sell $80 - 2x$ calculators each week.

    a. What is the revenue function $R(x)$?

    b. What price will yield a maximum revenue?

    c. What will be the maximum revenue?

30. A manufacturer produces radios. He estimates that by selling them for x dollars each, he will be able to sell $100 - x$ radios each month.

    a. What is the revenue function $R(x)$?

    b. What price will yield a maximum revenue?

    c. What will be the maximum revenue?

31. Use your graphing calculator to graph each of the following quadratic functions.

    a. Graph each equation one at a time.

    b. Graph all three parabolas at the same time (all three curves should appear on the display).

      i. $y = x^2 + 1$ **ii.** $y = x^2 + 3$ **iii.** $y = x^2 - 4$

32. In your own words, state the effect on the graph of $y = x^2$ by changing the constant term k in the function $y = x^2 + k$.

33. Use your graphing calculator to graph each of the following quadratic functions

    a. Graph each equation one at a time.

    b. Graph all three parabolas at the same time (all three curves should appear on the display).

      i. $y = (x - 3)^2$ **ii.** $y = (x - 5)^2$ **iii.** $y = (x + 2)^2$

34. In your own words, state the effect on the graph of $y = x^2$ by changing the constant term h in the function $y = (x - h)^2$.

35. Discuss the general relationship of the graph of a function of the form $y = (x - h)^2 + k$ to the graph of the function $y = x^2$.

Chapter 15: Index of Key Terms and Ideas

Quadratic Functions   page 1206

A quadratic function is a function of the form
$$y = ax^2 + bx + c$$
where a, b, and c are real constants and $a \neq 0$.

General Information on Quadratic Functions   page 1207

For the quadratic function $y = ax^2 + bx + c$:
1. If $a > 0$, the parabola "opens upward."
2. If $a < 0$, the parabola "opens downward."
3. $x = -\dfrac{b}{2a}$ is the **line of symmetry**.
4. The **vertex** (turning point) occurs where $x = -\dfrac{b}{2a}$.

   Substitute this value for x in the function and find the y-value of the vertex. The vertex is the lowest point on the curve if the parabola opens upward or it is the highest point on the curve if the parabola opens downward.

Minimum and Maximum Values   page 1211

For a parabola with its equation in the form $y = a(x - h)^2 + k$,

1. If $a > 0$, then the parabola opens upward, (h, k) is the lowest point, and the y-value k is called the **minimum value** of the function.
2. If $a < 0$, then the parabola opens downward, (h, k) is the highest point, and y-value k is called the **maximum value** of the function.

Using a Calculator to Graph Quadratic Functions   page 1213

Chapter 15: Review

Section 15.1: Quadratic Equations: The Square Root Method

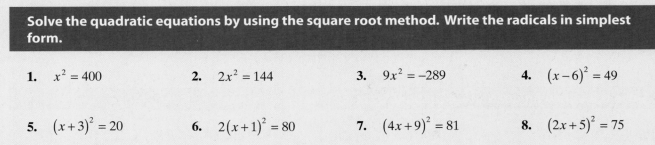

Solve the quadratic equations by using the square root method. Write the radicals in simplest form.

1. $x^2 = 400$

2. $2x^2 = 144$

3. $9x^2 = -289$

4. $(x-6)^2 = 49$

5. $(x+3)^2 = 20$

6. $2(x+1)^2 = 80$

7. $(4x+9)^2 = 81$

8. $(2x+5)^2 = 75$

The lengths of two sides are given for each of the following right triangles. Determine the length of the missing side.

9. $a = 8$ $b = 8$ $c = ?$

10. $a = ?$ $b = 20$ $c = 29$

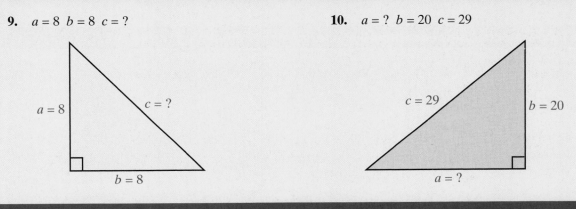

Solve the following word problems.

11. Perimeter: The hypotenuse of a right triangle is 2 in. longer than one of the legs. The length of the other leg is 10 in. Find the perimeter of the triangle.

12. Time: A ball is dropped from the top of a building known to be 100 meters high. The formula for finding the height (in meters) of the ball at time t (in seconds) is $h = -4.9t^2 + 100$. In how many seconds (to the nearest tenth of a second) will the ball hit the ground?

Use a graphing calculator to solve the following equations. Round your answers to the nearest hundredth.

13. $x^2 = 250$

14. $20x^2 = 52.5$

Section 15.2: Quadratic Equations: Completing the Square

Add the correct constant to complete the square, then factor the trinomial as indicated.

15. $x^2 + 10x + \underline{\qquad} = (\underline{\qquad})^2$

16. $x^2 - x + \underline{\qquad} = (\underline{\qquad})^2$

Solve the following quadratic equations by completing the square.

17. $x^2 - 8x - 9 = 0$

18. $x^2 + 3x + 2 = 0$

19. $9x^2 - 4 = 0$

20. $x^2 - 4x + 1 = 0$

21. $x^2 + x - 4 = 0$

22. $2x^2 + 3x - 2 = 0$

Solve the following quadratic equations by factoring or by completing the square.

23. $16x^2 - 25 = 0$

24. $x^2 + 8x - 20 = 0$

25. $6x^2 - 12x + 5 = 0$

26. $3x^2 - 4x - 3 = 0$

27. $5x^2 - 10x + 1 = 0$

28. $3x^2 + 2x - 5 = 0$

Section 15.3: Quadratic Equations: The Quadratic Formula

Solve each of the quadratic equations by using the quadratic formula.

29. $2x^2 - x - 2 = 0$

30. $3x^2 - 6x + 2 = 0$

31. $2x^2 - 49 = 0$

32. $\dfrac{1}{3}x^2 - x + \dfrac{1}{2} = 0$

33. $(2x + 1)(x - 1) = 8$

34. $(3x - 1)(x + 2) = 6x$

Solve each of the quadratic equations by using any method (factoring, completing the square, or the quadratic formula).

35. $x^2 + x - 2 = 0$

36. $x^2 + 6x = 10$

37. $7x^2 - 3x = 0$

38. $\dfrac{11}{2}x - 1 = 2x^2$

39. $\dfrac{3}{4}x^2 + 2x = 3$

40. $-4x^2 + 3x + 1 = 0$

Solve the following quadratic equations and use a calculator to write the answers in decimal form to the nearest ten-thousandth.

41. $x^2 - 4x - 1 = 0$

42. $4x^2 - 8x - 11 = 0$

Section 15.4: Applications

Solve the following word problems.

43. The sum of the squares of two consecutive positive integers is 85. Find these integers.

44. The length of a rectangle is 6 ft longer than the width. The area of the rectangle is 720 ft^2. Find the dimensions of the rectangle.

45. The length of a rectangle is 5 m less than twice the width. If both dimensions are decreased by 3 m, the area is 84 m^2. Find the length and width of the original rectangle.

46. The product of two consecutive even integers is 10 less than 5 times their sum. Find the integers.

47. Right Triangles: A right triangle has two equal sides. If the hypotenuse is 20 meters, what is the length of each of the equal sides?

48. Average Speed: Christine drove her car to a mathematics teachers' conference in Sacramento, 288 miles from her home. Because she had a late start, she drove a little faster than usual. On the trip home her average speed dropped by 8 miles per hour. If her total driving time for the round trip was 8.5 hours, what was her average speed going to the conference?

49. Boating: Jeremy and Thomas rowed their boat down a river for 12 miles then rowed back to their dock. The round trip took 8 hours. If the rate of the current was 2 mph, at what rate can Jeremy and Thomas row in still water?

50. Swimming Pools: Two pipes can be used to fill a swimming pool. If both are turned on, the pool will fill in 2.4 hours. If only one of the pipes is used, the larger pipe would fill the pool 2 hours faster than the smaller pipe. How many hours would it take for the pool to be filled by each of the pipes working alone?

51. Dimensions of a Box: A square piece of tin has a small square, 5 cm by 5 cm, cut from each corner. The edges are then folded up to form a box with a volume of 40,500 cm^3. What are the dimensions of the box?

52. Flying in a Plane: A small private plane can fly at 200 mph when the air is calm. The plane would take 24 minutes longer to fly 396 miles against the wind than flying that distance with the wind. What is the rate of the wind?

Section 15.5: Quadratic Functions: $y = ax^2 + bx + c$

For each quadratic function, a. find its vertex, b. find its line of symmetry, c. locate (or estimate) its x-intercepts, and d. graph the function.

53. $y = x^2 - 4$

54. $y = 16 - x^2$

55. $y = x^2 - 6x + 5$

56. $y = x^2 - 10x$

57. $y = -x^2 + 2x - 3$

58. $y = 3x^2 - 10x + 2$

Solve the following word problems.

59. **Rectangles:** The perimeter of a rectangle is 80 yards. What are the dimensions of the rectangle with maximum area?

60. **Jewelry Sales:** Michael estimates that by charging x dollars for each of his pieces of jewelry he can sell $80 - x$ pieces each week.

   a. Write his revenue function for his jewelry sales.

   b. What price should he charge to maximize his revenue from these pieces of jewelry?

   c. What will be his maximum revenue from these items?

For Exercises 61 and 62, use the formula $h = -16t^2 + v_0 t + h_0$ where h is the height (in feet) of a projectile in t seconds. Its initial velocity is v_0 and its initial height is h_0.

61. **Toy Rocket:** A toy rocket is projected from the ground with an initial velocity of 48 ft/sec.

   a. When will the rocket reach its maximum height?

   b. What will be the maximum height?

   c. In how many seconds will the rocket hit the ground?

62. **Throwing a Ball:** A ball is thrown upward at 64 ft/sec from the edge of a cliff known to be 80 ft above the beach.

   a. When will the ball reach its maximum height?

   b. What will be the maximum height?

   c. In how many seconds will the ball hit the ground?

Chapter 15: Test

Solve the following quadratic equations by factoring.

1. $x^2 - 3x + 2 = 0$

2. $5x^2 + 12x = 0$

3. $3x^2 - x - 10 = 0$

Solve the following quadratic equations.

4. $8x^2 = 96$

5. $(x + 5)^2 = 49$

6. $3(x - 1)^2 = 240$

The lengths of two sides are given for the following right triangle. Determine the length of the missing side.

7. $a = 5 \quad b = ? \quad c = 13$

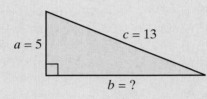

Add the correct constant to complete the square, then factor the trinomial as indicated.

8. $x^2 - 24x + \underline{\qquad} = (\qquad)^2$

9. $3x^2 + 9x + \underline{\qquad} = 3(\qquad)^2$

Solve the following quadratic equations by completing the square.

10. $x^2 + 6x + 8 = 0$

11. $x^2 - 5x = 6$

12. $2x^2 - 8x - 4 = 0$

Solve the following quadratic equations by using the quadratic formula.

13. $x^2 + 8x + 4 = 0$

14. $3x^2 + 8x + 2 = 0$

15. $2x^2 = 4x + 3$

16. $\frac{1}{2}x^2 - x = 2$

Solve the following quadratic equations by using any method.

17. $x^2 + 6x = 2$

18. $3x^2 - 7x + 2 = 0$

19. $2x^2 - 3x = 1$

For each quadratic function, a. find its vertex, b. find its line of symmetry, c. locate (or estimate) its x-intercepts, and d. graph the function.

20. $y = x^2 - 5$ **21.** $y = -x^2 + 4x$ **22.** $y = 2x^2 + 3x + 1$

Solve the following word problems.

23. Squares: The diagonal of a square is 36 inches long. Find the length of each side.

24. Boat: A small boat travels 8 mph in still water. It takes 10 hours longer to travel 60 miles upstream than it takes to travel 60 miles downstream. Find the rate of the current.

25. Toy Rocket: A toy rocket is projected upward from the ground with an initial velocity of 144 feet per second. The height of the rocket can be found by using the equation $h = -16t^2 + 144t$ where t is measured in seconds.

  a. In approximately how many seconds will the rocket be 96 ft above the ground?

  b. What is the maximum height of the rocket?

  c. In approximately how many seconds will the rocket hit the ground?

26. Rectangular Sheet: A rectangular sheet of metal is 3 in. longer than it is wide. A box is to be made by cutting out 4 in. squares at each corner and folding up the sides. If the box has a volume of 720 in.^{3}, what are the dimensions of the sheet of metal?

(**Hint:** Sketch a diagram of the rectangle and a diagram of the box.)

Cumulative Review: Chapters 1 - 15

Simplify each of the following expressions.

1. $13.1 + 109.7 - 35.8$

2. $\dfrac{4}{3} + 2\dfrac{5}{6} - 5\dfrac{1}{9}$

3. $1.2(16) - 22 \div 17.6 + (1.5)^2$

4. $\dfrac{8}{9} \div \dfrac{14}{33} \cdot \dfrac{5}{22}$

5. $\left(-3x^2y\right)^2$

6. $\dfrac{\left(2x^2\right)\left(-6x^3\right)}{3x^{-1}}$

Simplify the following expressions by combining like terms.

7. $4x + 3 - 2(x + 4) + (2 - x)$

8. $4x - \left[3x - 2(x + 5) + 7 - 5x\right]$

Perform each of the indicated operations and simplify.

9. $\left(5x^2 + 3x - 7\right) + \left(2x^2 - 3x - 7\right)$

10. $\left(2x^2 - 7x + 3\right) + \left(5x^2 - 3x + 4\right)$

11. $\dfrac{x^2}{x^2 - 4} \cdot \dfrac{x^2 - 3x + 2}{x^2 - x}$

12. $\dfrac{3x}{x^2 - 6x - 7} \div \dfrac{2x^2}{x^2 - 8x + 7}$

13. $\dfrac{2x}{x^2 - 25} - \dfrac{x + 1}{x + 5}$

14. $\dfrac{4x}{x^2 - 3x - 4} + \dfrac{x - 2}{x^2 + 3x + 2}$

Solve each of the following equations.

15. $13 = x + 17$

16. $\dfrac{3}{4}x - \dfrac{7}{12}x - \dfrac{2}{3} = 4$

17. $6y - 4.5 = y + 4.5$

18. $5(x - 8) - 3(x - 6) = 0$

19. $x^2 - 8x + 15 = 0$

20. $x^2 + 5x - 36 = 0$

21. $\dfrac{1}{x} + \dfrac{2}{3x} = 1$

22. $\dfrac{3}{x - 1} + \dfrac{5}{x + 1} = 2$

23. $x + 3 = 2\sqrt{2x + 6}$

24. $(x - 4)^2 = 9$

25. $x^2 + 4x - 2 = 0$

26. $2x^2 - 3x - 4 = 0$

Solve each inequality and graph the solution set on a real number line.

27. $15 - 3x + 1 \geq x + 4$

28. $2(x - 5) - 4 < x - 3(x - 1)$

Solve each system of linear equations.

29. $\begin{cases} 2x - 5y = 11 \\ 5x - 3y = -1 \end{cases}$

30. $\begin{cases} x + 2y = 6 \\ y = -\dfrac{1}{2}x + 3 \end{cases}$

31. $\begin{cases} y = -3x + 4 \\ 3x + y = 7 \end{cases}$

Solve each formula for the indicated letter.

32 $P = a + 2b$, solve for b

33. $A = \dfrac{h}{2}(a + b)$, solve for a

Write the following number in scientific notation.

34. 32,000,000

Factor the following expressions.

35. $x^2 - 12x + 36$

36. $25x^2 - 49$

37. $3x^2 + 10x + 3$

38. $3x^2 + 6x - 72$

39. $x^3 + 4x^2 - 8x - 32$

Simplify the following radical expressions and then use a calculator to find the value of each expression to the nearest ten-thousandth.

40. $\sqrt{63}$

41. $\sqrt{243}$

42. $\sqrt[3]{250}$

43. $\sqrt{\dfrac{25}{12}}$

44. $\sqrt{147} + \sqrt{48}$

45. $\left(2\sqrt{3} + 1\right)\left(\sqrt{3} - 4\right)$

Determine the value of the function at the given values.

46. Given the function $f(x) = 5x - 3$, find **a.** $f(-1)$, **b.** $f(6)$, and **c.** $f(0)$.

Graph each linear function and state its slope and its *y*-intercept.

47. $y = \dfrac{2}{3}x + 4$

48. $4x - 3y = 6$

49. $2x + 4y = -5$

Solve the following problems.

50. Given the two points $(-5, 2)$ and $(3, 5)$, find:

   a. the slope of the line through the two points,

   b. the distance between the two points,

   c. the midpoint of the line segment joining the points, and

   d. an equation of the line through the two points.

51. Find the equation in slope-intercept form of the line perpendicular to the line $y = 3x - 7$ and passing through the point $(-1, 1)$. Graph both lines.

For each quadratic function, a. find its vertex, b. find its line of symmetry, c. locate (or estimate) its *x*-intercepts, and d. graph the function.

52. $y = x^2 - 8x + 7$

53. $y = 2(x + 3)^2 - 6$

54. $y = x^2 + x + 1$

Solve the following word problems.

55. **Games:** Andrea got the following scores playing eight rounds of of the game Boggle.

12	8	9	17
16	6	11	9

Find her mean score.

56. **College Enrollment:** At a small college, the ratio of women to men is $9:7$. If there is a total enrollment of 6400 students, how many women and how many men are enrolled?

57. **Investing:** Anna has money in two separate savings accounts; one pays 5% and one pays 8%. The amount invested at 5% exceeds the amount invested at 8% by $800. The total annual interest is $209. How much is invested at each rate?

58. **Time:** Tim can clean the weeds from a vacant lot in 10 hours. Beth can do the same job in 8 hours. How long would it take if they worked together?

59. **Spheres:** What is the surface area area of a sphere with a radius of 20 ft? (Use $\pi = 3.14$.)

60. **Rectangles:** The length of a rectangle is five more than twice the width. The area is 52 cm². Find the width and the length.

61. **Shopping:** George bought 7 pairs of socks. Some cost $2.50 per pair and some cost $3.00 per pair. The total cost was $19.50. How many pairs of each type did he buy?

62. **Boating:** Anand traveled 6 miles downstream and returned. His speed downstream was 5 mph faster than his speed upstream. If the total trip took 1 hour, find his speed in each direction.

63. **Throwing a Ball:** A ball is thrown upward from the edge of a building with an initial velocity of 96 feet per second. The building is 120 feet tall and the height of the ball above the ground can be found by using the equation $h = -16t^2 + 96t + 120$ where t is measured in seconds.

   a. When will the ball be 200 feet above the ground?

   b. When will the ball be 60 feet above the ground? Round your answer to the nearest hundreth.

   c. In about how many seconds will the ball hit the ground? Round your answer to the nearest hundreth.

64. **Ladders:** A ladder is 40 ft long and the base of the ladder is to be set 15 ft from the base of a building under construction. About how far up the building (to the nearest tenth of a foot) will the ladder reach?

Appendices:
Further Topics in Algebra

Mathematics at Work!

Measurements surround us as we go about our daily lives. Every day when you drive to a friend's house, fill a car with gasoline, or cook, measurements play a role in how you interact with the world around you. Knowing how different units of measurement work together can help you be more successful.

Jennifer is making flan. The recipe calls for 1 liter of milk. However, she only has measuring cups that measure fluid ounces. How many fluid ounces of milk will she need to add to the flan? (See Sections A.1, A.2, and A.3.)

A.1 U.S. Measurements

Objective A **Basic Units of Measure in the U.S. Customary System**

You measure things every day: the distance you drive to work, the time it takes you to get to school, how much gas you use in a week, the area you mow in your yard, the weight of the apples you just bought. Possible measurements you use are miles, gallons, square feet, and pounds. These are units of measure in the **U. S. customary system**. Most of the rest of the world uses the metric system with meters, liters, and grams. (See the next section.)

In the **U. S. customary system**, the units are not systematically related. Historically, some of the units were associated with parts of the body, which would vary from person to person. For example, a foot was the length of a person's foot and a yard was the distance from the tip of one's nose to the tip of one's fingers with the arm outstretched.

There is considerably more consistency now because the official weights and measures are monitored by the government. Table 1 below shows relationships between some of the basic units used for length, weight, capacity, and time in the U. S. customary system.

Relationships Between Measurements in the U.S. Customary System	
U.S. Units of Length	
12 inches (in.) = 1 foot (ft)	3 feet = 1 yard (yd)
36 inches = 1 yard	5280 feet = 1 mile (mi)
U.S. Units of Weight	
16 ounces (oz) = 1 pound (lb)	2000 pounds = 1 ton (T)
U.S. Units of Capacity	
8 fluid ounces (fl oz) = 1 cup (c)	2 pints = 1 quart (qt)
2 cups = 1 pint (pt) = 16 fluid ounces	4 quarts = 1 gallon (gal)
Units of Time	
60 seconds (sec) = 1 minute (min)	24 hours = 1 day
60 minutes = 1 hour (hr)	7 days = 1 week

Table 1

There is no simple way to convert from one unit of measurement to another. You must simply memorize the information in Table 1. You are probably familiar with many of the conversions.

Example 1

Basic Conversions

Some basic conversions from Table 1.

a. 1 gal = _____ qt

Solution

1 gal = 4 qt

b. 3 ft = _____ yd

Solution

3 ft = 1 yd

c. 60 min = _____ hr

Solution

60 min = 1 hr

d. 1 T = _____ lb

Solution

1 T = 2000 lb

Now work margin exercise 1.

Objective B **Conversion by Using Multiplication or Division**

To convert from one unit of measure to another, not in Table 1, you must decide to either multiply or divide. In any case, you still must have memorized the conversions in the table. The following two statements will help.

Converting from one size unit to another

1. **Multiply** to convert to smaller units.
   (There will be more smaller units.)

2. **Divide** to convert to larger units.
   (There will be fewer larger units.)

1. Use table 1 to perform the following conversions.

a. 16 oz = _____ lb
b. 1 qt = _____ pt
c. 1 hr = _____ min
d. 1 yd = _____ in.

2. Use multiplication or division to perform the following conversions.

a. 3 yd = _____ in

b. 4 c = _____ fl oz

c. 28 days = _____ weeks

d. 60 in. = _____ ft

Example 2

Using Multiplication/Division to Convert Measures

a. 3 c = _____ fl oz

Solution

You are converting from a *larger* unit to a *smaller* unit (a *cup* is larger than a *fluid ounce*), so multiply. Because there are 8 fluid ounces in 1 cup, multiply by 8.

$3 \text{ c} = 8 \cdot 3 \text{ fl oz} = 24 \text{ fl oz}$

b. 5 gal = _____ qt

Solution

You are converting from a *larger* unit to a *smaller* unit (a *gallon* is larger than a *quart*), so multiply. Because there are 4 quarts in 1 gallon, multiply by 4.

$5 \text{ gal} = 4 \cdot 5 \text{ qt} = 20 \text{ qt}$

c. 150 min = _____ hr

Solution

You are converting from a *smaller* unit to a *larger* unit (a *minute* is smaller than an *hour*), so divide. Because there are 60 minutes in 1 hour, divide by 60.

$150 \text{ min} = \dfrac{150}{60} \text{ hr} = 2.5 \text{ hr} \left(\text{ or } 2\dfrac{1}{2} \text{ hr} \right)$

d. 39 in. = _____ ft

Solution

You are converting from a *smaller* unit to a *larger* unit (an *inch* is smaller than a *foot*), so divide. Because there are 12 inches in 1 foot, divide by 12.

$39 \text{ in.} = \dfrac{39}{12} \text{ ft} = 3.25 \text{ ft} \ \left(\text{or } 3\dfrac{1}{4} \text{ ft} \right)$

Now work margin exercise **2.**

Objective C Conversion by Using Unit Fractions

A second method of conversion involves the use of unit fractions. A **unit fraction** is a fraction equivalent to 1. For example,

$$\frac{3 \text{ ft}}{3 \text{ ft}} = \frac{\overset{1}{\cancel{3}} \, \cancel{\text{ft}}}{\underset{1}{\cancel{3}} \, \cancel{\text{ft}}} = 1$$

But from Table 1, we know that 3 ft = 1 yd. With this fact we can write two unit fractions with numerator or denominator 1 as follows:

$$\frac{1 \text{ yd}}{3 \text{ ft}} = 1 \quad \text{and} \quad \frac{3 \text{ ft}}{1 \text{ yd}} = 1$$

Similarly,

$$\frac{1 \text{ min}}{60 \text{ sec}} = 1 \quad \text{and} \quad \frac{60 \text{ sec}}{1 \text{ min}} = 1$$

This type of fraction is useful in making a conversion. Because the value of the fraction is 1, **multiplication by a unit fraction** does not change the value of the expression being converted. Use the following guidelines in determining which type of fraction to use.

Using Unit Fractions for Converting Measures

1. The numerator should be in the measurement units of the result.

2. The denominator should be in the measurement units to be converted.

Example 3

Using Unit Fractions to Convert Measures

a. 21 ft = _____ yd

Solution

Choose the unit fraction with yards in the numerator and feet in the denominator.

$$\frac{1 \text{ yd}}{3 \text{ ft}} = 1$$

Now multiply by this fraction as follows:

$$21 \text{ ft} = 21 \text{ ft} \cdot \frac{1 \text{ yd}}{3 \text{ ft}} = \overset{7}{\cancel{21}} \, \cancel{\text{ft}} \cdot \frac{1 \text{ yd}}{\underset{1}{\cancel{3}} \, \cancel{\text{ft}}} = 7 \cdot 1 \text{ yd} = 7 \text{ yd}$$

Note that the measure label of feet (ft) divides out and the result is in yards (yd).

3. Use unit fractions to perform the following conversions.

   a. 10,000 lb = _____ T

   b. 5 days = _____ hr

   c. 128 fl oz = _____ c

   d. 3 mi = _____ ft

b. 15 hr = _____ min

Solution

Choose the unit fraction with minutes in the numerator and hours in the denominator:

$$\frac{60 \text{ min}}{1 \text{ hr}} = 1$$

Now multiply by this fraction as follows:

$$15 \text{ hr} = 15 \cancel{\text{hr}} \cdot \frac{60 \text{ min}}{1 \cancel{\text{hr}}} = 15 \cdot 60 \text{ min} = 900 \text{ min}$$

Note that, as with the multiplication/division method, the number became larger because minutes are a smaller unit than hours.

c. $6\frac{1}{2}$ qt = _____ pt

Solution

Choose the unit fraction with pints in the numerator and quarts in the denominator:

$$\frac{2 \text{ pt}}{1 \text{ qt}} = 1$$

Now multiply by this fraction as follows:

$$6\frac{1}{2} \text{ qt} = 6\frac{1}{2} \cancel{\text{qt}} \cdot \frac{2 \text{ pt}}{1 \cancel{\text{qt}}} = \frac{13}{\cancel{2}} \cdot \cancel{2}\,\text{pt} = 13 \text{ pt}$$

d. 40 oz = _____ lb

Solution

Choose the unit fraction with pounds in the numerator and ounces in the denominator:

$$\frac{1 \text{lb}}{16 \text{oz}} = 1$$

$$40 \text{ oz} = 40 \text{ oz} \cdot \frac{1 \text{ lb}}{16 \text{ oz}} = 40 \cancel{\text{oz}} \cdot \frac{1 \text{ lb}}{16 \cancel{\text{oz}}} = \frac{\cancel{8} \cdot 5 \cdot 1}{\cancel{8} \cdot 2} \text{ lb} = \frac{5}{2} \text{ lb} = 2\frac{1}{2} \text{ lb}$$

Now work margin exercise **3.**

Exercises A.1

1. 1 ft = _____ in.

2. 1 week = _____ days

3. 24 hr = _____ day

4. 2 c = _____ pt

5. 4 qt = _____ gal

6. 36 in. = _____ yd

7. 1 mi = _____ ft

8. 1 lb = _____ oz

9. 2000 lb = _____ T

10. 12 in. = _____ ft

11. 3 ft = _____ in.

12. 10,560 ft = _____ mi

13. 3 weeks = _____ days

14. 5 min = _____ sec

15. 90 min = _____ hr

16. 9 qt = _____ gal

17. 6000 lb = _____ T

18. 5 yd = _____ ft

19. 72 in. = _____ yd

20. 48 in. = _____ ft

21. 7 yd = _____ ft

22. 3 mi = _____ ft

23. 3 pt = _____ fl oz

24. 6 pt = _____ qt

25. 18 in. = _____ ft

26. 7920 ft = _____ mi

27. 7 weeks = _____ days

28. 32 fl oz = _____ c

29. 5 qt = _____ pt

30. 24 oz = _____ lb

31. 3 yd = _____ ft

32. 4 min = _____ sec

33. 96 hr = _____ days

34. 5.5 lb = _____ oz

35. 3.5 ft = _____ in.

36. 6 qt = _____ gal

37. 16 T = _____ lb

38. 2.5 min = _____ sec

39. 150 min = _____ hr

40. 13 yd = _____ ft

A.2 The Metric System

Objectives

A Know the metric units of measurement for length.

B Know the metric units of measurement for mass.

C Know the metric units of measurement for volume.

D Know the metric units of measurement for liquid volume.

Objective A **Metric Units of Length**

The metric system of measurement is used by about 90% of the people in the world. The United States is the only major industrialized country still committed to the U.S. customary system (formerly called the English system). Even in the United States, the metric system is used in many fields of study and business, such as medicine, science, the computer industry, and the military. Industries involved in international trade must be familiar with the metric system.

The **meter** is the basic unit of length in the metric system. Smaller and larger units are named by putting a prefix in front of the basic unit; for example, **centi**meter and **kilo**meter. The prefixes we will use are shown in boldface print in Table 1. Other prefixes that indicate extremely small units are micro-, nano-, pico-, femto- and atto-. Prefixes that indicate extremely large units are mega-, giga-, and tera-. You are probably familiar with computer terms such as megabytes and gigabytes.

Metric Measures of Length		
1 **milli**meter (mm)	= 0.001 meter	1 m = 1000 mm
1 **centi**meter (cm)	= 0.01 meter	1 m = 100 cm
1 **deci**meter (dm)	= 0.1 meter	1 m = 10 dm
1 meter (m)	= 1.0 meter (the basic unit)	
1 **deka**meter (dam)	= 10 meters	
1 **hecto**meter (hm)	= 100 meters	
1 **kilo**meter (km)	= 1000 meters	

Table 1

As indicated in Table 1, the metric units of length are related to each other by powers of 10. That is, simply multiply by 10 to get the equivalent measure expressed in the next lower (or smaller) unit. Thus you will have **more** of a **smaller** unit. For example,

$$1 \text{ m} = 10 \text{ dm} = 100 \text{ cm} = 1000 \text{ mm}.$$

Conversely, divide by 10 to get the equivalent measure expressed in the next higher (or larger) unit. Thus you will have **less** of a **larger** unit. For example,

$$1 \text{ mm} = 0.1 \text{ cm} = 0.01 \text{ dm} = 0.001 \text{ m}.$$

Changing Metric Measures of Length

Procedure	Example
To change to a measure of length that is: one unit smaller, multiply by 10,	**Larger to Smaller** 6 cm = 60 mm
two units smaller, multiply by 100,	3 m = 300 cm
three units smaller, multiply by 1000, and so on.	12 m = 12 000 mm
To change to a measure of length that is: one unit larger, divide by 10,	**Smaller to Larger** 35 mm = 3.5 cm
two units larger, divide by 100,	250 cm = 2.5 m
three units larger, divide by 1000, and so on.	32 m = 0.032 km

Example 1

Converting Metric Units of Length

The following examples illustrate how to change from larger to smaller units of length in the metric system.

a. 5.6 m = 100 (5.6) cm = 560 cm

b. 5.6 m = 1000 (5.6) mm = 5600 mm

c. 23.5 cm = 10 (23.5) mm = 235 mm

d. 1.42 km = 1000 (1.42) m = 1420 m

Example 2

Converting Metric Units of Length

The following examples illustrate how to change from smaller to larger units of length in the metric system.

a. 375 cm = $\left(\dfrac{375}{100}\right)$ m = 3.75 m

b. 375 mm = $\left(\dfrac{375}{1000}\right)$ m = 0.375 m

c. 1055 m = $\left(\dfrac{1055}{1000}\right)$ km = 1.055 km

*Now work margin exercises **1** and **2**.*

Change the following units as indicated.

1. a. 35 m = _____ cm

   b. 6.4 cm = _____ mm

   c. 1.23 km = _____ m

   d. 6 m = _____ mm

2. a. 5.9 m = _____ km

   b. 320 mm = _____ m

   c. 6500 cm= _____ m

   d. 70 m= _____ dam

Another technique for changing units in the metric system can be illustrated with the concept of a number line with the metric prefixes below in order.

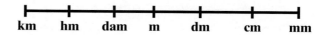

Simply move the decimal point in the direction of change.

Example 3

Converting Metric Units of Length

Change 56 cm to the equivalent measure in meters.

Solution

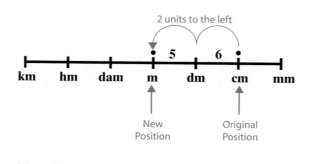

Thus, 56 cm = 0.56 m

Convert the following length measurements as indicated.

3. Change 65 decimeters to the equivalent measure in dekameters.

4. Change 43 meters to an equivalent measure in centimeters.

Example 4

Converting Metric Units of Length

Change 13.5 m to an equivalent measure in millimeters.

Solution

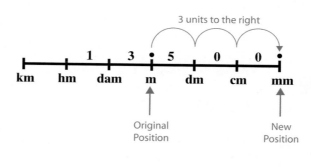

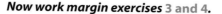

Thus, 13.5 m = 13 500 mm

Now work margin exercises 3 and 4.

> ## notes
>
> ▪ In the metric system,
>
> ▪ **1.** A 0 is written to the left of the decimal point if there is no whole number part (0.287 m).
>
> ▪ **2.** No commas are used in writing numbers. If a number has more than four digits (left or right of the decimal point), the digits are grouped in threes from the decimal point with a space between the groups (25 000 m or 0.000 34 m).

Objective B **Mass (Weight)**

Mass is the amount of material in an object. Regardless of where the object is in space, its mass remains the same. (See Figure 1.) **Weight** is the force of the Earth's gravitational pull on an object. The farther an object is from Earth, the less the gravitational pull of the Earth. Thus astronauts experience weightlessness in space, but their mass is unchanged.

The two objects have the same mass and balance on an equal arm balance, regardless of their location in space.

Figure 1

Because most of us do not stray far from the Earth's surface, weight and mass will be used interchangeably in this text. Thus a **mass** of 20 kilograms will be said to **weigh** 20 kilograms.

The basic unit of mass in the metric system is the **kilogram***, about 2.2 pounds. In some fields, such as medicine, the **gram** (about the mass of a paper clip) is more convenient as a basic unit than the kilogram.

Large masses, such as loaded trucks and railroad cars, are measured by the **metric ton** (1000 kilograms, or about 2200 pounds). (See Tables 2 and 3.)

---

* Technically, a kilogram is the mass of a certain cylinder of platinum-iridium alloy kept by the International Bureau of Weights and Measures in Paris.
Originally, the basic unit was a gram, defined to be the mass of 1 cm³ of distilled water at 4° Celsius. This mass is still considered accurate for many purposes, so that
    1 cm³ of water has a mass of 1 g.
    1 dm³ of water has a mass of 1 kg.
    1 m³ of water has a mass of 1000 kg, or 1 metric ton.

Measures of Mass	
1 **milli**gram (mg)	= 0.001 gram
1 **centi**gram (cg)	= 0.01 gram
1 **deci**gram (dg)	= 0.1 gram
1 gram (g)	= 1.0 gram
1 **deka**gram (dag)	= 10 grams
1 **hecto**gram (hg)	= 100 grams
1 **kilo**gram (kg)	= 1000 grams
1 metric ton (t)	= 1 000 000 grams

Table 2

Equivalent Measures of Mass	
1000 mg = 1 g	0.001 g = 1 mg
1000 g = 1 kg	0.001 kg = 1 g
1000 kg = 1 t	0.001 t = 1 kg
1t = 1000 kg = 1 000 000 g = 1 000 000 000 mg	

Table 3

Units of Mass

A chart can be used to change from one unit of mass to another.

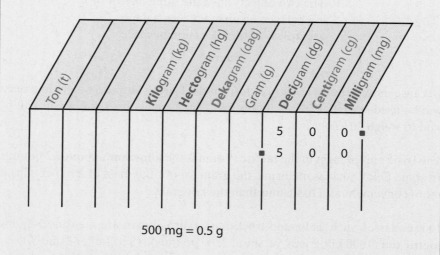

500 mg = 0.5 g

1. List each unit across the top.
2. Enter the given number so that there is one digit in each column with the decimal point on the given unit line.
3. Move the decimal point to the desired unit line.
4. Fill in the spaces with 0s using one digit per column.

The use of the chart below shows how the following equivalent measures can be found.

Example 5

Equivalent Measures of Mass

a. 23 mg = 0.023 g

b. 6 g = 6000 mg

c. 49 kg = 49 000 g

d. 5 t = 5000 kg

e. 70 kg = 0.07 t

Ton			Kilogram	Hectogram	Dekagram	Gram	Decigram	Centigram	Milligram	
								2	3 ∎	**a.** Move the decimal point to g and fill in 0s. 23 mg = 0.023 g
						∎ 0	2	3		
						6 ∎				**b.** Move the decimal point to mg and fill in 0s. 6 g = 6000 mg
						6	0	0	0 ∎	
		4	9 ∎							**c.** Move the decimal point to g and fill in 0s. 49 kg = 49 000 g
		4	9	0	0	0 ∎				
5 ∎										**d.** Move the decimal point to kg and fill in 0s. 5 t = 5000 kg
5	0	0	0 ∎							
		7	0 ∎							**e.** Move the decimal point to t and fill in 0s. 70 kg = 0.070 t
	∎ 0	7	0							

Make your own chart on a piece of paper and see if you agree with the results in Example 6.

Example 6

Equivalent Measures of Mass

a. 60 mg = 0.06 g

b. 135 mg = 0.135 g

c. 5700 kg = 5.7 t

d. 100 g = 0.1 kg

e. 78 kg = 78 000 g

Now work margin exercises 5 and 6.

Convert the following mass measurements as indicated.

5. a. Change 121 kilograms to grams.

b. Change 3500 kilograms to tons.

c. Change 4 576 000 grams to tons.

d. Change 6700 milligrams to grams.

6. a. Change 43 kilograms to grams.

b. Change 250 kilograms to tons.

c. Change 23 grams to milligrams.

Volume is a measure of the space enclosed by a three-dimensional figure and is measured in **cubic units**. The volume or space contained within a cube that is 1 centimeter on each edge is **one cubic centimeter**, or 1 cm³, as shown in Figure 2. A cubic centimeter is about the size of a sugar cube.

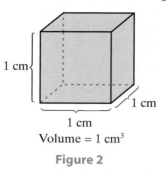

1 cm

1 cm

1 cm

Volume = 1 cm³

Figure 2

A rectangular solid that has edges of 3 cm, 2 cm and 5 cm has a volume of 3 cm × 2 cm × 5 cm = 30 cm³. We can think of the rectangular solid as being three layers of ten cubic centimeters, as shown in Figure 3.

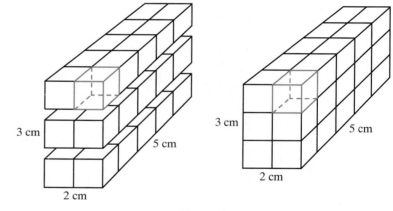

3 cm

5 cm

2 cm

3 cm

5 cm

2 cm

Figure 3

If a cube is 1 decimeter along each edge, then the volume of the cube is 1 cubic decimeter (or 1 dm³). In terms of centimeters, this same cube has volume,

$$10 \text{ cm} \times 10 \text{ cm} \times 10 \text{ cm} = 1000 \text{ cm}^3.$$

That is, as shown in Figure 4,

$$1 \text{ dm}^3 = 1000 \text{ cm}^3.$$

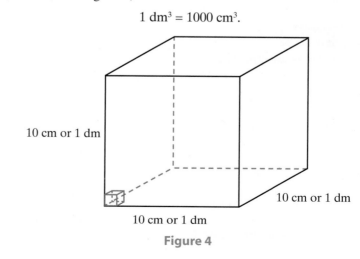

10 cm or 1 dm

10 cm or 1 dm

10 cm or 1 dm

Figure 4

This relationship is true of cubic units in the metric system: equivalent cubic units can be found by multiplying the larger units by 1000 to get the smaller units. Again, we can use a chart; however, this time **there must be three digits in each column.**

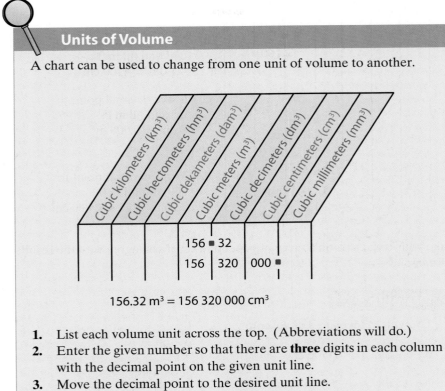

Units of Volume

A chart can be used to change from one unit of volume to another.

Cubic kilometers (km³) Cubic hectometers (hm³) Cubic dekameters (dam³) Cubic meters (m³) Cubic decimeters (dm³) Cubic centimeters (cm³) Cubic millimeters (mm³)

156 ∎ 32

156 | 320 | 000 ∎

156.32 m³ = 156 320 000 cm³

1. List each volume unit across the top. (Abbreviations will do.)
2. Enter the given number so that there are **three** digits in each column with the decimal point on the given unit line.
3. Move the decimal point to the desired unit line.
4. Fill in the spaces with 0s using three digits per column.

Example 7

Equivalent Measures of Volume

a. 15 cm³   =   15 000 mm³

b. 4.1 dm³   =   4100 cm³

c. 8 dm³   =   0.008 m³

d. 22.6 m³   =   22 600 000 cm³

The chart on the following page shows how these equivalent measures were found.

Convert the following volume measurements as indicated.

7. a. Change 750 cubic millimeters to cubic centimeters.

b. Change 19 cubic decimeters to cubic centimeters.

c. Change 1.6 cubic meters to cubic centimeters.

d. Change 63.7 cubic meters to cubic decimeters.

8. a. Change 6.3 cubic centimeters to cubic meters.

b. Change 192 cubic decimeters to cubic centimeters.

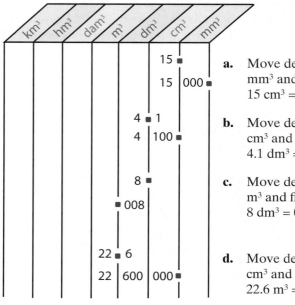

a. Move decimal point to mm^3 and fill in 0s.
$15\ cm^3 = 15\ 000\ mm^3$

b. Move decimal point to cm^3 and fill in 0s.
$4.1\ dm^3 = 4100\ cm^3$

c. Move decimal point to m^3 and fill in 0s.
$8\ dm^3 = 0.008\ m^3$

d. Move decimal point to cm^3 and fill in 0s.
$22.6\ m^3 = 22\ 600\ 000\ cm^3$

Make your own chart on a piece of paper and see if you agree with the results in Example 8.

Example 8

Equivalent Measures of Volume

a. $3.7\ dm^3$ = $3700\ cm^3$

b. $0.8\ m^3$ = $0.0008\ dam^3$

c. $4\ m^3$ = $4000\ dm^3$ = $4\ 000\ 000\ cm^3$ = $4\ 000\ 000\ 000\ mm^3$

Now work margin exercises 7 and 8.

Volumes measured in cubic kilometers are so large that they are not used in everyday situations. Many scientists work with these large volumes. More practically, we are only interested in m^3, dm^3, cm^3, and mm^3.

Objective D Liquid Volume

Liquid volume is measured in **liters** (abbreviated L). You are probably familiar with 1 L and 2 L bottles of soda on your grocer's shelf. A **liter** is the volume enclosed in a cube that is 10 cm on each edge. So 1 liter is equal to

$$10\ cm \times 10\ cm \times 10\ cm = 1000\ cm^3 \quad or \quad 1\ liter = 1000\ cm^3$$

That is, the cubic box shown in Figure 4 would hold 1 liter of liquid.

The prefixes kilo-, hecto-, deka-, deci-, centi-, and milli- all indicate the same part of a liter as they do of the meter. **One digit per column** will be helpful for changing units. The centiliter (cL), deciliter(dL), and dekaliter (daL) are not commonly used and are not included in the tables or exercises.

Measures of Liquid Volume		
1 **milli**liter (mL)	=	0.001 liter
1 liter (L)	=	1.0 liter
1 **hecto**liter (hL)	=	100 liters
1 **kilo**liter (kL)	=	1000 liters

Table 4

Equivalent Measures of Volume			
1000 mL = 1 L		1 mL = 1 cm^3	
1000 L = 1 kL		1 L = 1 dm^3	
10 hL = 1 kL		1 kL = 1 m^3	

Table 5

The use of the chart below shows how the following equivalent measures can be found.

Example 9

Equivalent Measures of Liquid Volume

a. 6 L = 6000 mL = 0.06 hL

b. 500 mL = 0.5 L

c. 3 kL = 3000 L

d. 72 hL = 7.2 kL

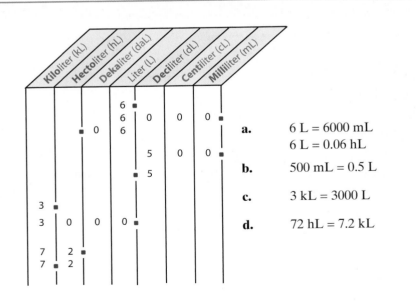

a. 6 L = 6000 mL
6 L = 0.06 hL

b. 500 mL = 0.5 L

c. 3 kL = 3000 L

d. 72 hL = 7.2 kL

There is an interesting "crossover" relationship between liquid volume measures and cubic volume measures. Since,

$$1\ L = 1000\ mL \text{ and } 1\ L = 1000\ cm^3$$

we have

$$1\ mL = 1\ cm^3.$$

Also,

$$1\ kL = 1000\ L = 1\ 000\ 000\ cm^3 \text{ and } 1\ 000\ 000\ cm^3 = 1\ m^3.$$

This gives

$$1\ kL = 1000\ L = 1\ m^3.$$

Use Table 4 to confirm the following conversions.

Example 10

Equivalent Measures of Liquid Volume

a. 6000 mL = 6 L

b. 3.2 L = 3200 mL

c. 60 hL = 6 kL

d. 637 mL = 0.637 L

e. 70 mL = 70 cm^3

f. 3.8 kL = 3.8 m^3

***Now work margin exercises* 9 and 10.**

Convert the following liquid volume measurements as indicated.

9. **a.** Change 9.37 liters to hectoliters.

   b. Change 353 milliliters to liters.

   c. Change 12 kiloliters to liters.

10. **a.** Change 1952 milliliters to liters.

   b. Change 124 milliliters to cubic centimeters.

   c. Change 19.75 kiloliters to cubic meters.

Practice Problems

Change the following units as indicated.

1. 500 mg = _____ g

2. 43 g = _____ mg

3. 62 g = _____ kg

4. 18 cm^3 = _____ mm^3

5. 7.9 dm^3 = _____ m^3

6. 2 mL = _____ L

7. 500 mL = _____ L

8. 16 mL = _____ cm^3

Practice Problem Answers

1. 0.5 g   **2.** 43 000 mg   **3.** 0.062 kg   **4.** 18 000 mm^3

5. 0.0079 m^3 **6.** 0.002 L   **7.** 0.5 L   **8.** 16 cm^3

Exercises A.2

1. 3 m = _____ cm

2. 0.8 m = _____ cm

3. 1.5 m = _____ mm

4. 1.9 cm = _____ mm

5. 45 cm = _____ mm

6. 27 dm = _____ cm

7. 13.6 km = _____ m

8. 4.38 km = _____ m

9. 18.25 m = _____ dm

10. 60 m = _____ dm

11. 4.8 mm = _____ cm

12. 36 mm = _____ cm

13. 6.5 cm = _____ m

14. 82 cm = _____ m

15. 1300 mm = _____ m

16. 750 mm = _____ m

17. 5.25 cm = _____ m

18. 185 m = _____ km

19. 5500 m = _____ km

20. 140 cm = _____ dm

21. Change 245 mm to meters.

22. Convert 23 cm to meters.

23. How many kilometers are in 10 000 m?

24. How many meters are in 1.5 km?

25. What number of meters is equivalent to 20 000 cm?

26. Express 4.73 m in centimeters.

27. How many centimeters are in 3.2 mm?

28. Change 87 mm to meters.

29. Express 17.35 m in millimeters.

30. What number of kilometers is equivalent to 140 000 m?

Change the following units of mass (weight) as indicated. See Examples 5 and 6.

31. 2 g = _____ mg

32. 7 kg = _____ g

33. 3700 kg = _____ t

34. 34.5 mg = _____ g

35. 5600 g = _____ kg

36. 4000 kg = _____ t

37. 91 kg = _____ t

38. 73 kg = _____ mg

39. 0.7 g = _____ mg

40. 0.54 g = _____ mg

41. How many kilograms are there in 5 tons?

42. How many kilograms are there in 17 tons?

43. Change 2 tons to kilograms.

44. Change 896 mg to grams.

45. Express 896 g in milligrams.

46. Express 342 kg in grams.

47. Convert 75 000 g to kilograms.

48. Convert 3000 mg to grams.

49. Convert 7 tons to grams.

50. Convert 0.4 t to grams.

51. Change 0.34 g to kilograms.

52. Change 0.78 g to milligrams.

53. How many grams are in 16 mg?

54. How many milligrams are in 2.5 g?

55. 92.3 g = _____ kg

56. 3.94 g = _____ mg

57. 7.58 t = _____ kg

58. 5.6 t = _____ kg

59. 2963 kg = _____ t

60. 3547 kg = _____ t

Fill in the blanks below. See Examples 7 and 8.

61. 1 cm^3 = _____ mm^3

1 dm^3 = _____ cm^3

1 m^3 = _____ dm^3

1 km^3 = _____ m^3

Change the following units of volume as indicated. See Examples 7 and 8.

62. 73 m^3 = _____ dm^3

63. 0.9 m^3 = _____ dm^3

64. 525 cm^3 = _____ m^3

65. 400 m^3 = _____ cm^3

66. 8.7 m^3 = _____ cm^3

67. 63 dm^3 = _____ m^3

68. How many cm^3 are in 45 mm^3?

69. How many mm^3 are in 3.1 cm^3

70. Change 19 mm^3 to dm^3.

71. Change 5 cm^3 to mm^3.

72. Convert 2 dm^3 to cm^3.

73. Convert 76.4 cm^3 to m^3.

Change the following units of liquid volume as indicated. See Examples 9 and 10.

74. Change 5.3 L to milliliters.

75. Change 30 cm^3 to milliliters.

76. Change 30 cm^3 to liters.

77. Change 5.3 mL to liters.

78. 48 kL = _____ L

79. 72 000 L = _____ kL

80. 290 L = _____ kL

81. 569 mL = _____ L

82. 80 L = _____ mL = _____ cm^3

83. 7.3 L = _____ mL = _____ cm^3

A.3 U.S. to Metric Conversions

Objective A **Temperature**

We begin the following discussion of equivalent measures between the U.S. customary and metric systems with measures of temperature.

Temperature

Temperature: U.S. customary measure is in **degrees Fahrenheit** (°F).

Metric measure is in **degrees Celsius** (°C).

The two scales are shown here on thermometers. Approximate conversions can be found by reading along a ruler or the edge of a piece of paper held horizontally across the page.

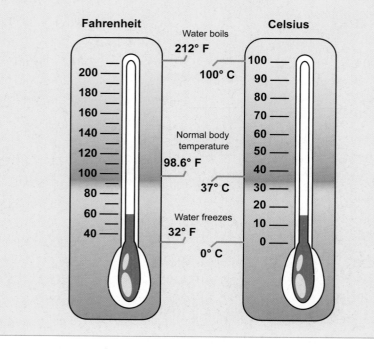

Example 1

Equivalent Measures of Temperature

Hold a straight edge horizontally across the two thermometers and you will read:

$100°C = 212°F$ Water boils at sea level.
$40°C = 104°F$ Hot day in the desert.
$20°C = 68°F$ Comfortable room temperature.

Objectives

A Learn the U.S. customary and metric equivalents for measures of temperature.

B Learn the U.S. customary and metric equivalents for measures of length.

C Learn the U.S. customary and metric equivalents for measures of area.

D Learn the U.S. customary and metric equivalents for measures of volume.

E Learn the U.S. customary and metric equivalents for measures of mass.

Two formulas that give exact conversions are given here.

F = Fahrenheit temperature and C = Celsius temperature.

$$C = \frac{5(F-32)}{9} \qquad F = \frac{9 \cdot C}{5} + 32$$

A calculator will give answers accurate to nine digits. Answers that are not exact may be rounded to whatever place of accuracy you choose.

Now work margin exercise 1.

Use a straight edge across the two thermometers to convert each temperature.

1. a. 0°C

   b. 50°F

Example 2

Converting Measures of Temperature

Let $F = 86°$ and find the equivalent measure in Celsius.

Solution

$$C = \frac{5(86-32)}{9} = \frac{5(54)}{9} = 30 \qquad \text{Thus } 86°F = 30°C.$$

Example 3

Converting Measures of Temperature

Let $C = 40°$ and convert this to degrees Fahrenheit.

Solution

$$F = \frac{9 \cdot 40}{5} + 32 = 72 + 32 = 104 \qquad \text{Thus } 40°C = 104°F.$$

Now work margin exercises 2 and 3.

Use the conversion formulas given to convert each temperature.

2. 77°F

3. 45°C

Objective B **Length**

In the tables of Length Equivalents, Area Equivalents, Volume Equivalents, and Mass Equivalents (Tables 1 – 4) the equivalent measures are rounded. Any calculations with these measures (with or without a calculator) cannot be any more accurate than the measure in the table. Table 1 shows some length equivalents.

Length Equivalents	
U.S. to Metric	**Metric to U.S.**
1 in. = 2.54 cm (exact)	1 cm = 0.394 in.
1 ft = 0.305 m	1 m = 3.28 ft
1 yd = 0.914 m	1 m = 1.09 yd
1 mi = 1.61 km	1 km = 0.62 mi

Table 1

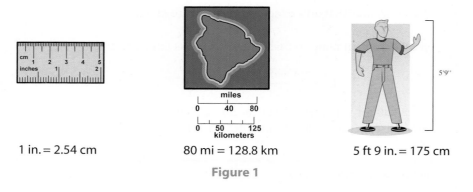

1 in. = 2.54 cm   80 mi = 128.8 km   5 ft 9 in. = 175 cm

Figure 1

In Examples 4 – 7, use Table 1 to convert measurements as indicated.

Example 4

Converting Measures of Length

6 ft = _____ cm

Solution

6 ft = 72 in. = 72(2.54 cm) = 183 cm (rounded)

or

6 ft = 6(0.305 m) = 1.83 m = 183 cm

Example 5

Converting Measures of Length

25 mi = _____ km

Solution

25 mi = 25(1.61 km) = 40.25 km

Example 6

Converting Measures of Length

30 m = _____ ft

Solution

30 m = 30(3.28 ft) = 98.4 ft

Example 7

Converting Measures of Length

10 km = _____ mi

Solution

10 km = 10(0.62 mi) = 6.2 mi

Now work margin exercises 4 through 7.

Use Table 1 to convert each of the following measurements as indicated.

4. 84 in. = _____ m

5. 100 yd = _____ m

6. 75 cm = _____ in.

7. 500 m = _____ ft

Objective C **Area**

Area Equivalents	
U.S. to Metric	**Metric to U.S.**
$1 \text{ in.}^2 = 6.45 \text{ cm}^2$	$1 \text{ cm}^2 = 0.155 \text{ in.}^2$
$1 \text{ ft}^2 = 0.093 \text{ m}^2$	$1 \text{ m}^2 = 10.764 \text{ ft}^2$
$1 \text{ yd}^2 = 0.836 \text{ m}^2$	$1 \text{ m}^2 = 1.196 \text{ yd}^2$
$1 \text{ acre} = 0.405 \text{ ha}$	$1 \text{ ha} = 2.47 \text{ acres}$

Table 2

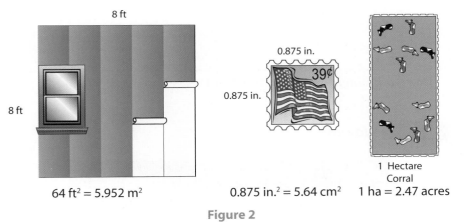

64 ft² = 5.952 m²   0.875 in.² = 5.64 cm²   1 ha = 2.47 acres

1 Hectare Corral

Figure 2

In Examples 8 – 11, use Table 2 to convert the measures as indicated.

Example 8

Converting Measures of Area

$40 \text{ yd}^2 =$ _____ m^2

Solution

$40 \text{ yd}^2 = 40\left(0.836 \text{ m}^2\right) = 33.44 \text{ m}^2$

Example 9

Converting Measures of Area

5 acres = _____ ha

Solution

5 acres = 5(0.405 ha) = 2.025 ha

Example 10

Converting Measures of Area

5 ha = _____ acres

Solution

5 ha = 5(2.47 acres) = 12.35 acres

Use Table 2 to convert each of the following measurements as indicated.

8. 53 in.^{2} = _____ cm^2

9. 16 acres = _____ ha

10. 3 ha = _____ acres

11. 38 m^2 = _____ ft^2

Example 11

Converting Measures of Area

$100 \text{ cm}^2 =$ _____ in.^{2}

Solution

$100 \text{ cm}^2 = 100\left(0.155 \text{ in.}^2\right) = 15.5 \text{ in.}^2$

***Now work margin exercises* 8 through 11.**

Objective D **Volume**

Volume Equivalents	
U.S. to Metric	**Metric to U.S.**
$1 \text{ in.}^3 = 16.387 \text{ cm}^3$	$1 \text{ cm}^3 = 0.06 \text{ in.}^3$
$1 \text{ ft}^3 = 0.028 \text{ m}^3$	$1 \text{ m}^3 = 35.315 \text{ ft}^3$
$1 \text{ qt} = 0.946 \text{ L}$	$1 \text{ L} = 1.06 \text{ qt}$
$1 \text{ gal} = 3.785 \text{ L}$	$1 \text{ L} = 0.264 \text{ gal}$

Table 3

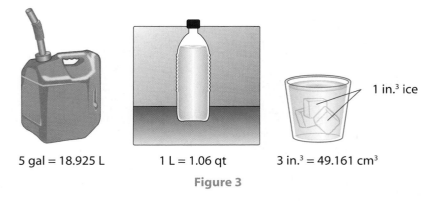

5 gal = 18.925 L   1 L = 1.06 qt   1 in.³ ice   3 in.³ = 49.161 cm³

Figure 3

In Examples 12 – 15, use Table 3 to convert the measures as indicated.

Example 12

Converting Measures of Liquid Volume

20 gal = _____ L

Solution

20 gal = 20(3.785 L) = 75.7 L

Example 13

Converting Measures of Liquid Volume

42 L = _____ gal

Solution

42 L = 42(0.264 gal) = 11.088 gal
or
42 L = 11.1 gal (rounded)

Example 14

Converting Measures of Liquid Volume

6 qt = _____ L

Solution

6 qt = 6(0.946 L) = 5.676 L
or
6 qt = 5.7 L (rounded)

Use Table 3 to convert each of the following measurements as indicated.

12. 4 gal = _____ L

13. 16 L = _____ qt

14. 28 qt = _____ L

15. 12 m³ = _____ ft³

Example 15

Converting Measures of Volume

$$10 \text{ cm}^3 = \underline{\hspace{1cm}} \text{ in.}^3$$

Solution

$$10 \text{ cm}^3 = 10\left(0.06 \text{ in.}^3\right) = 0.6 \text{ in.}^3$$

Now work margin exercises **12 through 15.**

Objective E **Mass**

Mass Equivalents	
U.S. to Metric	**Metric to U.S.**
1 oz = 28.35 g	1 g = 0.035 oz
1 lb = 0.454 kg	1 kg = 2.205 lb

Table 4

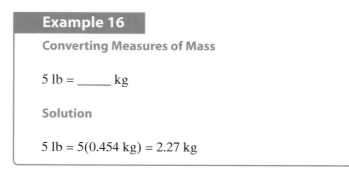
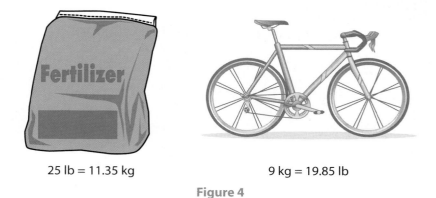

25 lb = 11.35 kg   9 kg = 19.85 lb

Figure 4

In Examples 16 and 17, use Table 4 to convert the measures as indicated.

Example 16

Converting Measures of Mass

$$5 \text{ lb} = \underline{\hspace{1cm}} \text{ kg}$$

Solution

$$5 \text{ lb} = 5(0.454 \text{ kg}) = 2.27 \text{ kg}$$

Example 17

Converting Measures of Mass

$15 \text{ kg} = \underline{\hspace{1cm}} \text{ lb}$

Solution

$15 \text{ kg} = 15(2.205 \text{ lb}) = 33.075 \text{ lb}$

or

$15 \text{ kg} = 33.1 \text{ lb (rounded)}$

***Now work margin exercises* 16 and 17.**

Use Table 4 to convert each of the following measurements as indicated.

16. $3 \text{ oz} = \underline{\hspace{1cm}} \text{ g}$

17. $42 \text{ g} = \underline{\hspace{1cm}} \text{ oz}$

Practice Problems

Convert the following measures as indicated. You may need to refer to the tables throughout this lesson.

1. $23°\text{F} = \underline{\hspace{1cm}} °\text{C}$

2. $14 \text{ in.} = \underline{\hspace{1cm}} \text{ cm}$

3. $36 \text{ ft}^2 = \underline{\hspace{1cm}} \text{ m}^2$

4. $24 \text{ cm}^3 = \underline{\hspace{1cm}} \text{ in.}^3$

5. $33 \text{ kg} = \underline{\hspace{1cm}} \text{ lb}$

Practice Problem Answers

1. $-5°\text{C}$ **2.** 35.56 cm **3.** 3.348 m^2

4. 1.44 in.^3 **5.** 72.765 lb

Exercises A.3

1. $25°C =$ _____ $°F$

2. $80°C =$ _____ $°F$

3. $50°C =$ _____ $°F$

4. $35°C =$ _____ $°F$

5. $50°F =$ _____ $°C$

6. $100°F =$ _____ $°C$

7. Change $32°F$ to degrees Celsius.

8. Change $41°F$ to degrees Celsius.

9. How many meters are in 3 yds?

10. How many meters are in 5 yds?

11. Change 60 miles to kilometers.

12. Change 100 miles to kilometers.

13. Convert 200 kilometers to miles.

14. Convert 65 kilometers to miles.

15. How many inches are in 50 cm?

16. How many inches are in 100 cm?

17. $3 \text{ in.}^2 =$ _____ cm^2

18. $16 \text{ in.}^2 =$ _____ cm^2

19. $600 \text{ ft}^2 =$ _____ m^2

20. $300 \text{ ft}^2 =$ _____ m^2

21. $100 \text{ yd}^2 =$ _____ m^2

22. $250 \text{ yd}^2 =$ _____ m^2

23. $1000 \text{ acres} =$ _____ ha

24. $250 \text{ acres} =$ _____ ha

25. How many acres are in 300 ha?

26. How many acres are in 400 ha?

27. Change 5 m^2 to square feet.

28. Change 10 m^2 to square feet.

29. Change 30 cm^2 to square inches.

30. Change 50 cm^2 to square inches.

31. $10 \text{ qt} =$ _____ L

32. $20 \text{ qt} =$ _____ L

33. $10 \text{ L} =$ _____ qt

34. $25 \text{ L} =$ _____ qt

35. $42 \text{ L} =$ _____ gal

36. $50 \text{ L} =$ _____ gal

37. 10 lb = _____ kg

38. 500 kg = _____ lb

39. 16 oz = _____ g

40. 100 g = _____ oz

Solve the following word problems.

41. **Area:** Suppose that the home you are considering buying sits on a rectangular shaped lot that is 270 feet by 121 feet. Convert this area to square meters.

42. **Area:** A new manufacturing building covers an area of 3 acres. How many hectares of ground does the new building cover?

43. A painting of a landscape is on a rectangular canvas that measures 3 feet by 4 feet.

    a. How many square centimeters of wall space will the painting cover when it is hanging?

    b. How many square meters?

A.4 Absolute Value Equations and Inequalities

Objectives

A Solve absolute value equations.

B Solve equations with two absolute value expressions.

C Solve absolute value inequalities.

Objective A **Absolute Value Equations**

The definition of **absolute value** was given in Section 7.1 and is stated again here for easy reference. Remember, the absolute value of a number is its distance from 0 on a number line.

Absolute Value

For any real number x,
$$|x| = \begin{cases} x & \text{if } x \geq 0 \\ -x & \text{if } x < 0 \end{cases}$$

Equations involving absolute value may have more than one solution (all of which must be included when giving an answer). For example, suppose that $|x| = 3$. Since $|3| = 3$ and $|-3| = -(-3) = 3$, we have either $x = 3$ or $x = -3$. We can say that the solution set is $\{3, -3\}$. In general, **any number and its opposite have the same absolute value**.

Solving Absolute Value Equations

For $c > 0$:

a. If $|x| = c$, then $x = c$ or $x = -c$.

b. If $|ax + b| = c$, then $ax + b = c$ or $ax + b = -c$.

Note: If the absolute value expression is isolated on one side of the equation, we say that the equation is in **standard form**. You may need to manipulate the absolute value equation to get it into standard form before you can solve it. (See Example 1d.)

Example 1

Solving Absolute Value Equations

Solve the following equations involving absolute value.

a. $|x| = 5$

Solution

$x = 5$ or $x = -5$

b. $|3x - 4| = 5$

Solution

$$3x - 4 = 5 \quad \text{or} \quad 3x - 4 = -5$$
$$3x = 9 \qquad\qquad 3x = -1$$
$$x = 3 \qquad\qquad x = -\frac{1}{3}$$

c. $|4x - 1| = -8$

Solution

There is no number that has a negative absolute value. Therefore, this equation has no solution. (The solution is $\varnothing$ and the equation is a contradiction.)

d. $5|3x + 17| - 4 = 51$

Solution

$$5|3x + 17| - 4 = 51 \qquad \text{Write the equation.}$$
$$5|3x + 17| = 55 \qquad \text{Add 4 to both sides.}$$
$$|3x + 17| = 11 \qquad \begin{array}{l}\text{Divide both sides by 5. } \textbf{We must put}\\ \textbf{the equation in standard form.}\end{array}$$
$$3x + 17 = 11 \quad \text{or} \quad 3x + 17 = -11$$
$$3x = -6 \qquad\qquad 3x = -28$$
$$x = -2 \qquad\qquad x = -\frac{28}{3}$$

Now work margin exercise 1.

1. Solve the following equations involving absolute value.

a. $|x| = 8$

b. $|5x - 2| = 8$

c. $|2x + 5| = -6$

d. $7|4x + 1| + 5 = 54$

Objective B **Equations with Two Absolute Value Expressions**

If two numbers have the same absolute value, then either they are equal or they are opposites of each other. This fact can be used to solve equations that involve two absolute value expressions.

Solving Equations with Two Absolute Value Expressions

If $|a| = |b|$, then either $a = b$ or $a = -b$.

More generally,

if $|ax + b| = |cx + d|$, then either $ax + b = cx + d$ or $ax + b = -(cx + d)$.

2. Solve $|4x-5|=|3x-16|$.

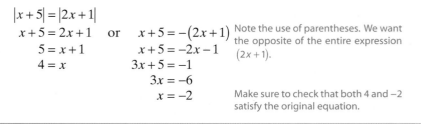

Example 2

Solving Equations with Two Absolute Value Expressions

Solve $|x+5|=|2x+1|$.

Solution

In this case, the two expressions $(x+5)$ and $(2x+1)$ are equal to each other or are opposites of each other.

$$|x+5|=|2x+1|$$

$x+5=2x+1$ or	$x+5=-(2x+1)$
$5=x+1$	$x+5=-2x-1$
$4=x$	$3x+5=-1$
	$3x=-6$
	$x=-2$

Note the use of parentheses. We want the opposite of the entire expression $(2x+1)$.

Make sure to check that both 4 and −2 satisfy the original equation.

Now work margin exercise 2.

Objective C Absolute Value Inequalities

Now consider an inequality with absolute value such as $|x|<3$. For a number to have an absolute value less than 3, it must be within 3 units of 0. That is, the numbers between −3 and 3 have their absolute values less than 3 because they are within 3 units of 0. Thus for $|x|<3$.

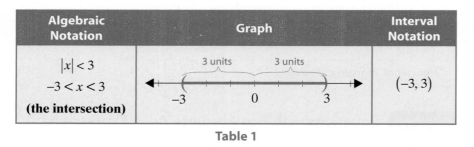

Algebraic Notation	Graph	Interval Notation
$|x|<3$ $-3<x<3$ **(the intersection)**		$(-3,3)$

Table 1

The inequality $|x-5|<3$ means that the distance between x and 5 is less than 3. That is, we want all the values of x that are within 3 units of 5. The inequality is solved algebraically as follows.

$$|x-5|<3$$
$$-3<x-5<3$$
$$-3+5<x-5+5<3+5$$

$x-5$ is between -3 and 3.

Add 5 to each part of the expression, just as in solving linear inequalities.

$$2<x<8$$
$$x \text{ is in } (2,8)$$

Simplify each expression.

Use interval notation.

The values for x are between 2 and 8 and are within 3 units of 5.

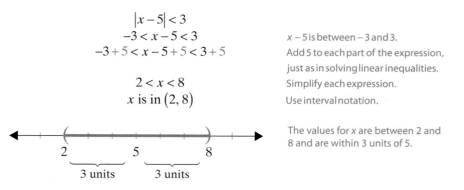

Solving Absolute Value Inequalities with < (or ≤)

For $c > 0$:

   a. If $|x| < c$, then $-c < x < c$.

   b. If $|ax + b| < c$, then $-c < ax + b < c$.

The inequalities in **a.** and **b.** are also true if < is replaced by ≤.

Note: If the absolute value expression is isolated on one side of the inequality, we say that the inequality is in **standard form**. You may need to manipulate the absolute value inequality to get it into standard form before you can solve it. (See Examples 3c and 3e.)

Example 3

Solving Absolute Value Inequalities

Solve the following absolute value inequalities and graph the solution sets.

a. $|x| \le 6$

Solution

$$|x| \le 6$$
$$-6 \le x \le 6$$
or x is in $[-6, 6]$

b. $|x + 3| < 2$

Solution

$$|x + 3| < 2$$
$$-2 < x + 3 < 2$$
$$-2 - 3 < x + 3 - 3 < 2 - 3$$
$$-5 < x < -1$$
or x is in $(-5, -1)$

3. Solve the following absolute value inequalities and graph the solution sets. Write each solution using interval notation.

a. $|x| < 3$

b. $|x + 1| \leq 4$

c. $4|2x + 3| < 20$

d. $|4x + 3| \leq -3$

e. $|3x - 4| + 1 \leq 9$

c. $3|2x - 7| < 15$

Solution

$$3|2x - 7| < 15$$
$$|2x - 7| < 5$$
$$-5 < 2x - 7 < 5$$
$$-5 + 7 < 2x - 7 + 7 < 5 + 7$$
$$2 < 2x < 12$$
$$1 < x < 6$$
or x is in $(1, 6)$

Divide both sides by 3 in order to get the inequality in standard form. Remember, we must get the expression in **standard form** before solving any absolute value inequality.

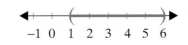

d. $|x + 9| < -\dfrac{1}{2}$

Solution

Since absolute value is always nonnegative (greater than or equal to 0), no number has an absolute value less than $-\dfrac{1}{2}$. Thus there is **no solution**, ∅.

e. $|2x + 4| + 4 < 7$

Solution

$$|2x + 4| + 4 < 7$$
$$|2x + 4| < 3$$
$$-3 < 2x + 4 < 3$$
$$-7 < 2x < -1$$
$$-\dfrac{7}{2} < x < -\dfrac{1}{2}$$
So x is in $\left(-\dfrac{7}{2}, -\dfrac{1}{2}\right)$.

Add −4 to both sides in order to get the expression in standard form.

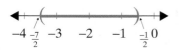

Now work margin exercise 3.

We have been discussing inequalities in which the absolute value is less than some positive number. Now consider an inequality where the absolute value is greater than some positive number, such as $|x| > 3$. For a number to have an absolute value greater than 3, its distance from 0 must be greater than 3. That is, numbers that are greater than 3 **or** less than −3 will have absolute values greater than 3. Thus for $|x| > 3$:

Algebraic Notation	Graph	Interval Notation
$\lvert x \rvert > 3$ $x > 3$ **or** $x < -3$ **(the union)**	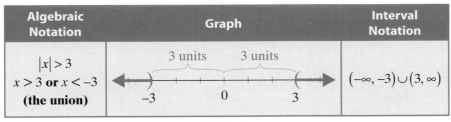	$(-\infty, -3) \cup (3, \infty)$

Table 2

The inequality $\lvert x - 5 \rvert > 6$ means that the distance between x and 5 is more
than 6. That is, we want all values of x that are more than 6 units from 5. The
inequality is solved algebraically as follows.

$\lvert x - 5 \rvert > 6$ indicates that

$x - 5 < -6$ **or** $x - 5 > 6$. $x - 5$ is less than -6 or greater than 6

Solving both inequalities gives

$x - 5 + 5 < -6 + 5$ **or** $x - 5 + 5 > 6 + 5$,   Add 5 to each side, just as

in solving linear inequalities.

$x < -1$ **or** $x > 11$.   Simplify.

So x is in $(-\infty, -1) \cup (11, \infty)$.

Note : The values for x less than -1
or greater than 11 are more than 6
units from 5. Thus we can interpret
the inequality $\lvert x-5 \rvert > 6$ to mean that
the distance from x to 5 is greater than 6.

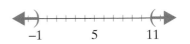

Solving Absolute Value Inequalities with > (or ≥)

For $c > 0$:

a. If $\lvert x \rvert > c$, then $x < -c$ **or** $x > c$.

b. If $\lvert ax + b \rvert > c$, then $ax + b < -c$ **or** $ax + b > c$.

The inequalities in **a.** and **b.** are true if > is replaced by ≥.

Example 4

Solving Absolute Value Inequalities

Solve the following absolute value inequalities and graph the solution set.

a. $|x| \geq 5$

Solution

$|x| \geq 5$

$x \leq -5 \quad \text{or} \quad x \geq 5$

So x is in $(-\infty, -5] \cup [5, \infty)$.

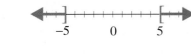

b. $|4x - 3| > 2$

Solution

$|4x - 3| > 2$

$4x - 3 < -2 \quad \text{or} \quad 4x - 3 > 2$

$4x < 1 \quad \text{or} \quad 4x > 5$

$x < \dfrac{1}{4} \quad \text{or} \quad x > \dfrac{5}{4}$

So x is in $\left(-\infty, \dfrac{1}{4}\right) \cup \left(\dfrac{5}{4}, \infty\right)$.

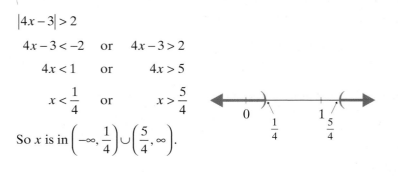

c. $|3x - 8| > -6$

Solution

There is nothing to do here except observe that no matter what is substituted for x, the absolute value will be greater than -6. Absolute value is always nonnegative (greater than or equal to 0). The solution to the inequality is **all real numbers**, so shade the entire number line. In interval notation, x is in $(-\infty, \infty)$.

d. $|2x-5|-5 \geq 4$

Solution

$$|2x-5|-5 \geq 4$$

$$|2x-5| \geq 9$$

$$2x-5 \leq -9 \quad \text{or} \quad 2x-5 \geq 9$$

$$2x \leq -4 \quad \text{or} \quad 2x \geq 14 \qquad \text{Add 5 to both sides in order to get the}$$
$$\qquad\qquad\qquad\qquad\qquad\qquad\qquad \text{inequality in } \textbf{standard form.}$$

$$x \leq -2 \quad \text{or} \quad x \geq 7$$

So x is in $(-\infty, -2] \cup [7, \infty)$.

$-2 \quad 0 \qquad\qquad 7$

Now work margin exercise 4.

Practice Problems

Solve the following equations.

1. $|x| = 10$

2. $|2x-1| = 8.2$

3. $|4x+1| = -5$

4. $|x+4| = |3x-2|$

Solve the following absolute value inequalities and graph the solution sets.

5. $|x-6| \leq 7$

6. $|4x+1| < 13$

7. $3|2x+6| \geq 20-2$

4. Solve the following absolute value inequalities and graph the solution sets. Write each solution using interval notation.

a. $|x| > 2$

b. $|2x+7| \geq 3$

c. $|5x-1| \geq -1$

d. $|3x+2|+1 \geq 8$

Practice Problem Answers

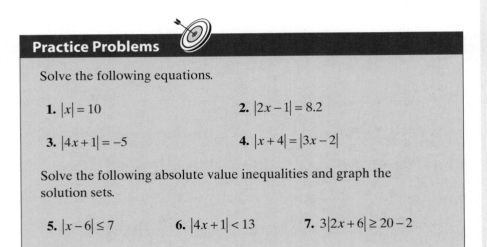

1. $x = -10$ or $x = 10$

2. $x = -3.6$ or $x = 4.6$

3. no solution

4. $x = -\dfrac{1}{2}$ or $x = 3$

5. $[-1, 13]$

6. $\left(-\dfrac{7}{2}, 3\right)$

7. $(-\infty, -6] \cup [0, \infty)$

Exercises A.4

1. $|x| = 8$

2. $|x| = 6$

3. $|z| = -\dfrac{1}{5}$

4. $|z| = \dfrac{1}{5}$

5. $|x + 3| = 2$

6. $|y + 5| = -7$

7. $|6x - 1| = 9$

8. $|3x + 1| = 8$

9. $|6n + 4| = 8$

10. $|3x - 5| = 10$

11. $|3x + 4| = -9$

12. $|-2x + 1| = -3$

13. $|-5x + 10| = 0$

14. $|6y + 4| = 0$

15. $|-4x + 1| = 7$

16. $|-3x + 4| = 7$

17. $|5x - 2| + 4 = 7$

18. $|2x - 7| - 1 = 0$

19. $|-3x + 4| - 2 = 3$

20. $|-x + 5| + 1 = 9$

21. $\left|\dfrac{1}{4}x - \dfrac{1}{2}\right| = 6$

22. $\left|\dfrac{1}{5}y - \dfrac{2}{3}\right| = \dfrac{2}{3}$

23. $5\left|\dfrac{x}{2} + 1\right| - 7 = 8$

24. $6\left|\dfrac{x}{5} - 2\right| + 5 = 11$

25. $3\left|\dfrac{x}{3} + 1\right| - 5 = -2$

26. $2\left|\dfrac{x}{4} - 3\right| + 6 = 10$

27. $|2x - 1| = |x + 2|$

28. $|2x - 5| = |x - 3|$

29. $|x + 3| = |x - 5|$

30. $|x - 8| = |x + 4|$

31. $|3x + 1| = |4 - x|$

32. $|5x + 4| = |1 - 3x|$

33. $\left|\dfrac{3x}{2} + 2\right| = \left|\dfrac{x}{4} + 3\right|$

34. $\left|\dfrac{x}{3} - 4\right| = \left|\dfrac{5x}{6} + 1\right|$

35. $\left|\dfrac{2x}{5} - 3\right| = \left|\dfrac{x}{2} - 1\right|$

36. $\left|\dfrac{4x}{3} + 7\right| = \left|\dfrac{x}{4} + 2\right|$

37. $|x| \geq -2$

38. $|x| \geq 3$

39. $|x| \leq \dfrac{4}{5}$

40. $|x| \geq \dfrac{7}{2}$

41. $|x - 3| > 2$

42. $|y - 4| \leq 5$

43. $|x + 6| \leq 4$

44. $|x + 2| \leq -4$

45. $|x + 5| \geq 3$

46. $|x - 1| < 6$

47. $|2x - 1| \geq 2$

48. $|3x + 4| > -8$

49. $|3 - 2x| < -2$

50. $|4 + 3x| > 5$

51. $|5 + 4x| \leq 3$

52. $|5x - 2| < 8$

53. $|3x + 4| - 1 < 0$

54. $|2x - 3| - 3 \leq 0$

55. $\left|\dfrac{3x}{2} - 4\right| \geq 5$

56. $\left|\dfrac{3}{7}y + \dfrac{1}{2}\right| > 2$

57. $|2x - 9| - 7 \leq 4$

58. $|3x - 7| + 4 \leq 4$

59. $-4 < |6x - 1| + 4$

60. $4 \leq |3x + 1| - 6$

61. $5 > |4 - 2x| + 2$

62. $7 > |8 - 5x| + 3$

63. $3|4x + 5| - 5 > 10$

64. $6|4x - 7| + 7 > 19$

65. $4|7x + 9| - 3 < 17$

66. $2|7x - 3| + 4 \geq 12$

A set of real numbers is described. **a.** Sketch a graph of the set on a real number line. **b.** Represent each set using absolute value notation. **c.** Represent each set using interval notation. If the set is one interval, state what type of interval it is.

67. The set of real numbers between −10 and 10, inclusive

68. The set of real numbers within 7 units of 4

69. The set of real numbers more than 6 units from 8

70. The set of real numbers greater than or equal to 3 units from −1

71. The set of real numbers within 2 units of −5

A.5 Synthetic Division and the Remainder Theorem

Objective A **Synthetic Division**

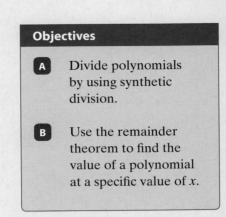

Objectives

A Divide polynomials by using synthetic division.

B Use the remainder theorem to find the value of a polynomial at a specific value of x.

In Section 11.7, polynomials are divided by using the division algorithm (long division). In the special case **when the divisor of a rational expression is a first-degree binomial with leading coefficient 1**, long division can be simplified by omitting the variables entirely and writing only certain coefficients. The procedure is called **synthetic division**. The following analysis describes how the procedure works for $\dfrac{5x^3 + 11x^2 - 3x + 1}{x + 3}$. (Note that $x + 3$ is first-degree with leading coefficient 1.)

a. With Variables

$$
\begin{array}{r}
5x^2 - 4x + 9 \\
x+3\overline{)5x^3 + 11x^2 - 3x + 1} \\
5x^3 + 15x^2 \\
-4x^2 - 3x \\
-4x^2 - 12x \\
9x + 1 \\
9x + 27 \\
-26
\end{array}
$$

b. Without Variables

$$
\begin{array}{r}
5 \quad -4 \quad +9 \\
1+3\overline{)5 \quad +11 \quad -3 \quad +1} \\
\boxed{5} + 15 \\
-4 \quad \boxed{-3} \\
\boxed{-4} \quad -12 \\
9 \quad \boxed{+1} \\
\boxed{9} \quad +27 \\
-26
\end{array}
$$

The boxed numbers in step **b.** can be omitted since they are repetitions of the numbers directly above them.

c. Boxed numbers omitted

$$
\begin{array}{r}
5 \quad -4 \quad +9 \\
1+3\overline{)5 \quad +11 \quad -3 \quad +1} \\
+15 \\
-4 \\
-12 \\
9 \\
+27 \\
-26
\end{array}
$$

d. Numbers moved up to fill in spaces

$$
\begin{array}{r}
5 \quad -4 \quad +9 \\
1+3\overline{)5 \quad +11 \quad -3 \quad +1} \\
+15 \quad -12 \quad +27 \\
-4 \quad +9 \quad -26
\end{array}
$$

Next, omit the 1 in the divisor, change +3 to −3, and write the opposites of the boxed numbers (because the quotient coefficient will now be multiplied by −3 instead of +3), as shown in steps **e.** and **f.** This allows the numbers to be added instead of subtracted. The number 5 is written on the bottom line, and the top line is omitted. The quotient and remainder can now be read from the bottom line.

e.
$$\begin{array}{r} 5 \quad -4 \quad +9 \\ 1+3{\overline{\smash{\big)}\,5 \quad +11 \quad -3 \quad +1}} \\ \underline{+15 \quad -12 \quad +27} \\ -4 \quad +9 \quad -26 \end{array}$$

f.
$$\begin{array}{r} -3{\overline{\smash{\big)}\,5 \quad +11 \quad -3 \quad +1}} \\ \downarrow -15+12-27 \\ \underline{5 \; - \; 4+ \; 9-26} \end{array}$$

This represents

$$5x^2 - 4x + 9 + \frac{-26}{x+3}.$$

The numbers on the bottom now represent the coefficients of a polynomial of **one degree less than the dividend**, along with the remainder. The last number to the right is the remainder.

In summary, synthetic division can be accomplished as follows:

1. Write only the coefficients of the dividend and the opposite of the constant in the divisor.

$$\begin{array}{c|cccc} -3 & 5 & 11 & -3 & 1 \\ \hline & & & & \end{array}$$

2. Rewrite the first coefficient (5) as the first coefficient in the quotient.

$$\begin{array}{c|cccc} -3 & 5 & 11 & -3 & 1 \\ & \downarrow & & & \\ \hline & 5 & & & \end{array}$$

3. Multiply the coefficient (5) by the constant divisor (−3) and **add** this product (−15) to the second coefficient.

$$\begin{array}{c|cccc} -3 & 5 & 11 & -3 & 1 \\ & \downarrow & -15 & & \\ \hline & 5 & -4 & & \end{array}$$

4. Continue to multiply each new coefficient by the constant divisor and add this product to the next coefficient in the dividend.

$$\begin{array}{c|cccc} -3 & 5 & 11 & -3 & 1 \\ & \downarrow & -15 & 12 & -27 \\ \hline & 5 & -4 & 9 & -26 \end{array}$$

5. The constants on the bottom line are the coefficients of the quotient and the remainder.

$$\frac{5x^3 + 11x^2 - 3x + 1}{x+3} = 5x^2 - 4x + 9 + \frac{-26}{x+3}$$

$$= 5x^2 - 4x + 9 - \frac{26}{x+3}$$

1. Use synthetic division to write each expression in the form $Q + \dfrac{R}{D}$.

a. $\dfrac{3x^3 - 5x^2 + 2x + 1}{x - 1}$

b. $\dfrac{x^4 + 2x^3 + 5x - 9}{x + 3}$

Example 1

Synthetic Division

Use synthetic division to write each expression in the form $Q + \dfrac{R}{D}$.

a. $\dfrac{4x^3 + 10x^2 + 11}{x + 5}$

Solution

$$
\begin{array}{r|rrrr}
-5 & 4 & 10 & 0 & 11 \\
  &   & -20 & 50 & -250 \\
\hline
  & 4 & -10 & 50 & -239
\end{array}
$$

Since there is no x-term, 0 is the coefficient. The coefficient is 0 for any missing term.

$$\frac{4x^3 + 10x^2 + 11}{x + 5} = 4x^2 - 10x + 50 + \frac{-239}{x + 5}$$

$$= 4x^2 - 10x + 50 - \frac{239}{x + 5}$$

b. $\dfrac{2x^4 - x^3 - 5x^2 - 2x + 7}{x - 2}$

Solution

$$
\begin{array}{r|rrrrr}
2 & 2 & -1 & -5 & -2 & 7 \\
  &   & 4 & 6 & 2 & 0 \\
\hline
  & 2 & 3 & 1 & 0 & 7
\end{array}
$$

$$\frac{2x^4 - x^3 - 5x^2 - 2x + 7}{x - 2} = 2x^3 + 3x^2 + x + \frac{7}{x - 2}$$

Now work margin exercise 1.

notes

Remember that synthetic division is used only when the divisor is a first-degree polynomial of the form $(x + c)$ or $(x - c)$.

Objective B   The Remainder Theorem

Synthetic division can be used for several purposes, one of which is to find the value of a polynomial for a particular value of x. For example, we know (from Section 11.3) that if

$$P(x) = x^3 - 5x^2 + 7x - 10,$$

then

$$P(2) = 2^3 - 5 \cdot 2^2 + 7 \cdot 2 - 10 = -8.$$

Using synthetic division to divide $x^3 - 5x^2 + 7x - 10$ by $x - 2$ we have

$$\begin{array}{r|rrrr} 2 & 1 & -5 & 7 & -10 \\ & & 2 & -6 & 2 \\ \hline & 1 & -3 & 1 & -8 \end{array}$$ ← remainder

The fact that the remainder is the same as $P(2)$ is not an accident. In fact, as the following theorem states, the remainder when a polynomial is divided by a first-degree factor of the form $(x-c)$ will always be $P(c)$.

The Remainder Theorem

If a polynomial $P(x)$ is divided by $(x-c)$, then the remainder will be $P(c)$.

Proof:

By the division algorithm we know that $\dfrac{P(x)}{x-c} = Q(x) + \dfrac{R}{x-c}$ where R is a constant.

(Remember that the degree of the remainder must be less than the degree of the divisor.)

Multiplying through by $(x-c)$, we have

$$P(x) = (x-c) \cdot Q(x) + R$$

and substituting $x = c$ gives

$$\begin{aligned} P(c) &= (c-c) \cdot Q(c) + R \\ &= 0 \cdot Q(c) + R \\ &= 0 + R \\ &= R. \end{aligned}$$

The proof is complete.

Example 2

The Remainder Theorem and Synthetic Division

a. Use synthetic division to find $P(5)$ given $P(x) = -2x^2 + 15x - 50$.

Solution

$$\begin{array}{r|rrr} 5 & -2 & 15 & -50 \\ & & -10 & 25 \\ \hline & -2 & 5 & -25 \end{array}$$ ← remainder $= P(5)$

Thus $P(5) = -25$.

(Checking shows $P(5) = -2 \cdot 5^2 + 15 \cdot 5 - 50 = -50 + 75 - 50 = -25$.)

b. Use synthetic division to find $P(-3)$, given

$$P(x) = 3x^4 + 10x^3 - 5x^2 + 125.$$

Note: To evaluate $P(-3)$, think of the divisor in the form $(x+3) = (x-(-3))$. That is, in the form $(x-c)$, $c = -3$.

Solution

```
-3│  3   10   -5    0   125
        -9   -3   24   -72
     3    1   -8   24    53  ← remainder = P(-3)
```

Thus $P(-3) = 53$.

c. Use synthetic division to show that $(x-6)$ is a factor of $P(x) = x^3 - 14x^2 + 53x - 30$.

Solution

```
6│  1  -14    53   -30
         6   -48    30
    1   -8     5     0  ← remainder = P(6)
```

Thus the remainder is $P(6) = 0$ and $(x-6)$ **is a factor of** $P(x)$.

Note: The coefficients in the quotient tell us that $x^2 - 8x + 5$ is also a factor of $P(x)$.

Now work margin exercise 2.

2. a. Use synthetic division to find $P(-2)$ given $4x^2 + 3x - 7$.

b. Use synthetic division to find $P(3)$ given $-x^3 + 4x^2 - 2x + 5$.

Exercises A.5

Divide the following expressions using synthetic division. **a.** Write the answer in the form $Q + \dfrac{R}{D}$ where R is a constant. **b.** In each exercise, $D = (x - c)$. State the value of c and the value of $P(c)$. (Assume $P(x)$ is the numerator of the fraction.) See Examples 1 and 2.

1. $\dfrac{x^2 - 12x + 27}{x - 3}$

2. $\dfrac{x^2 - 12x + 35}{x - 5}$

3. $\dfrac{x^3 + 4x^2 + x - 1}{x + 8}$

4. $\dfrac{x^3 - 6x^2 + 8x - 5}{x - 2}$

5. $\dfrac{4x^3 + 2x^2 - 3x + 1}{x + 2}$

6. $\dfrac{3x^3 + 6x^2 + 8x - 5}{x + 1}$

7. $\dfrac{x^3 + 6x + 3}{x - 7}$

8. $\dfrac{2x^3 - 7x + 2}{x + 4}$

9. $\dfrac{2x^3 + 4x^2 - 9}{x + 3}$

10. $\dfrac{4x^3 - x^2 + 13}{x - 1}$

11. $\dfrac{x^4 - 3x^3 + 2x^2 - x + 2}{x - 3}$

12. $\dfrac{x^4 + x^3 - 4x^2 + x - 3}{x + 6}$

13. $\dfrac{x^4 + 2x^2 - 3x + 5}{x - 2}$

14. $\dfrac{3x^4 + 2x^3 + 2x^2 + x - 1}{x + 1}$

15. $\dfrac{x^4 - x^2 + 3}{x - \dfrac{1}{2}}$

16. $\dfrac{x^3 + 2x^2 + 1}{x - \dfrac{2}{3}}$

17. $\dfrac{x^5 - 1}{x - 1}$

18. $\dfrac{x^5 - x^3 + x}{x + \dfrac{1}{2}}$

19. $\dfrac{x^4 - 2x^3 + 4}{x + \dfrac{4}{5}}$

20. $\dfrac{x^6 + 1}{x + 1}$

Collaborative Learning

21. With the class divided into teams of 3 or 4 students, each team should develop answers to the following questions and be prepared to discuss the answers in class.

a. First use long division to divide the polynomial $P(x) = 2x^3 - 8x^2 + 10x + 15$ by $2x - 1$.

Then use synthetic division to divide the same polynomial by $x - \dfrac{1}{2}$.

Do the same process with two or three other polynomials and divisors. Next compare the corresponding long and synthetic division answers and explain how the answers are related.

b. Use the results from part **a.** and explain algebraically the relationship of the answers when a polynomial is divided (using long division) by $ax - b$ and (using synthetic division) by $x - \dfrac{b}{a}$.

c. Show how the remainder theorem should be restated if $x - c$ is replaced by $ax - b$.

A.6 Graphing Systems of Linear Inequalities

Objective A Solving Systems of Linear Inequalities Graphically

In some branches of mathematics, in particular a topic called game theory, the solution to a very sophisticated problem can involve the set of points that satisfies a system of several **linear inequalities**. In business these ideas relate to problems such as minimizing the cost of shipping goods from several warehouses to distribution outlets. In this section we will consider graphing the solution sets to only two inequalities. We will leave the problem solving techniques to another course.

In an earlier section we graphed linear inequalities of the form $y < mx + b$ (or $y \le mx + b$ or $y > mx + b$ or $y \ge mx + b$). These graphs are called **half-planes**. The line $y = mx + b$ is called the **boundary line** and the half-planes are **open** (the boundary line is not included) or **closed** (the boundary line is included).

In this section, we will develop techniques for graphing (and therefore solving) **systems of two linear inequalities**. The **solution set** (if there are any solutions) to such a system of linear inequalities consists of the points in the **intersection** of two half-planes and portions of boundary lines indicated by the inequalities. The following procedure may be used to solve a system of linear inequalities.

To Solve a System of Two Linear Inequalities

1. For each inequality, graph the boundary line and shade the appropriate half-plane.
2. Determine the region of the graph that is common to both half-planes (the region where the shading overlaps).
   (This region is called the **intersection** of the two half-planes and is the **solution set** of the system.)
3. To check, pick one test-point in the intersection and verify that it satisfies both inequalities.

Note: If there is no intersection, then the system has no solution.

Example 1

Graphing Systems of Linear Inequalities

a. Graph the points that satisfy the system of inequalities:
$$\begin{cases} x \le 2 \\ y \ge -x + 1 \end{cases}$$

Solution

Step 1: For $x \le 2$, the points are to the left of and on the line $x = 2$.

Step 2: For $y \geq -x + 1$, the points are above and on the line $y = -x + 1$.

Step 3: Determine the region that is common to both half-planes. In this case, we test the point $(0, 3)$. On the graph below, the solution is the purple-shaded region and its boundary lines.

$0 \leq 2$ a true statement

$3 \geq -0 + 1$ a true statement

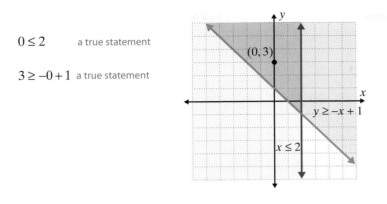

b. Solve the system of linear inequalities graphically: $\begin{cases} 2x + y \leq 6 \\ x + y < 4 \end{cases}$

Solution

Step 1: Solve each inequality for y: $\begin{cases} y \leq -2x + 6 \\ y < -x + 4 \end{cases}$

Step 2: For $y \leq -2x + 6$, the points are below and on the line $y = -2x + 6$.

Step 3: For $y < -x + 4$, the points are below but not on the line $y = -x + 4$.

Step 4: Determine the region that is common to both half-planes. Note that the line $y = -x + 4$ is dashed to indicate that the points on the line are not included. In this case, we test the point $(0, 0)$. On the graph below, the solution is the purple-shaded region and its boundary lines.

$2 \cdot 0 + 0 \leq 6$ a true statement

$0 + 0 < 4$ a true statement

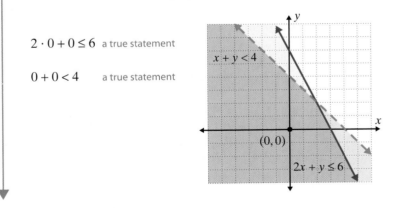

1. Graph the points that satisfy the following systems of inequalities:

a. $\begin{cases} y \geq 2 \\ x - y < 4 \end{cases}$

b. $\begin{cases} x + 2y < 3 \\ 3x + y \geq -6 \end{cases}$

c. $\begin{cases} y < -3x \\ y > -3x - 4 \end{cases}$

c. Solve the system of linear inequalities graphically: $\begin{cases} y \geq x \\ y \leq x + 2 \end{cases}$

Solution

Step 1: For $y \geq x$, the points are above and on the line $y = x$.

Step 2: For $y \leq x + 2$, the points are below and on the line $y = x + 2$.

Step 3: The solution set consists of the boundary lines and the region between them.

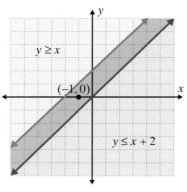

Now work margin exercise 1.

notes

- When the boundary lines are parallel there are three possibilities:
    1. The common region will be in the form of a strip between two lines (as in Example 1c).
    2. The common region will be a half-plane, as the solution to one inequality will be entirely contained within the solution of the other inequality.
    3. There will be no common region which means there is no solution.

Objective B **Using a TI-84 Plus Graphing Calculator to Graph Systems of Linear Inequalities**

To graph (and therefore solve) a system of linear inequalities (with no vertical line) with a TI-84 Plus graphing calculator, proceed as follows:

1. Solve each inequality for y.
2. Press the [Y=] key and enter the two expressions for Y_1 and Y_2.
3. Move to the left of Y_1 and Y_2 and press [ENTER] until the desired graphing symbol appears.
4. Press [GRAPH] and the desired region will be graphed as a cross-hatched area.
   (You may need to reset the [WINDOW] so that both regions appear.)

Example 2 illustrates how this can be done.

Example 2

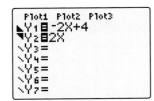 **Using a Graphing Calculator to Solve Systems of Linear Inequalities**

Use a TI-84 Plus graphing calculator to graph the following system of linear inequalities:

$$\begin{cases} 2x + y < 4 \\ 2x - y \leq 0 \end{cases}$$

Solution

Step 1: Solve each inequality for y: $\begin{cases} y < -2x + 4 \\ y \geq 2x \end{cases}$

Note: Solving $2x - y \leq 0$ for y can be written as $2x \leq y$ and then as $y \geq 2x$.

Steps 2 and 3: Press the [Y=] key and enter both functions and the corresponding symbols as they appear here:

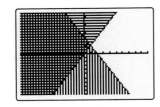

```
Plot1  Plot2  Plot3
\Y1◘-2X+4
▼Y2◘2X
\Y3=
\Y4=
\Y5=
\Y6=
\Y7=
```

(**Remember:** To shade your graphs, position the cursor over the slash next to Y_1 (or Y_2) and hit [ENTER] repeatedly until the appropriate shading is displayed.)

Step 4: Press [GRAPH]. The display should appear as follows. The solution is the cross-hatched region and the points on the line $2x - y = 0$.

Now work margin exercise 2.

Practice Problems

Solve the systems of two linear inequalities graphically.

1. $\begin{cases} y \leq -3 \\ y \geq x - 3 \end{cases}$ **2.** $\begin{cases} y > 2 \\ x < 2 \end{cases}$ **3.** $\begin{cases} x + y \geq 0 \\ x - 2y \geq 4 \end{cases}$

Practice Problem Answers

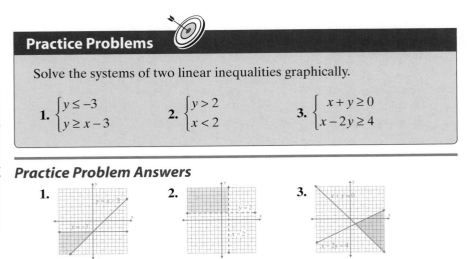

1.   **2.**   **3.**

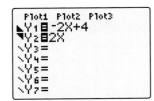 **2.** Use a TI-84 Plus graphing calculator to graph the following system of linear inequalities:

$$\begin{cases} y - 2x \geq -5 \\ y + 4x < 7 \end{cases}$$

Exercises A.6

1. $\begin{cases} y > 2 \\ x \geq -3 \end{cases}$

2. $\begin{cases} 2x + 5 < 0 \\ y \geq 2 \end{cases}$

3. $\begin{cases} x < 3 \\ y > -x + 2 \end{cases}$

4. $\begin{cases} y \leq -5 \\ y \geq x - 5 \end{cases}$

5. $\begin{cases} x \leq 3 \\ 2x + y > 7 \end{cases}$

6. $\begin{cases} 2x - y > 4 \\ y < -1 \end{cases}$

7. $\begin{cases} x - 3y \leq 3 \\ x < 5 \end{cases}$

8. $\begin{cases} 3x - 2y \geq 8 \\ y \geq 0 \end{cases}$

9. $\begin{cases} x - y \geq 0 \\ 3x - 2y \geq 4 \end{cases}$

10. $\begin{cases} y \geq x - 2 \\ x + y \geq -2 \end{cases}$

11. $\begin{cases} 3x + y \leq 10 \\ 5x - y \geq 6 \end{cases}$

12. $\begin{cases} y > 3x + 1 \\ -3x + y < -1 \end{cases}$

13. $\begin{cases} 3x + 4y \geq -7 \\ y < 2x + 1 \end{cases}$

14. $\begin{cases} 2x - 3y \geq 0 \\ 8x - 3y < 36 \end{cases}$

15. $\begin{cases} x + y < 4 \\ 2x - 3y < 3 \end{cases}$

16. $\begin{cases} 2x + 3y < 12 \\ 3x + 2y > 13 \end{cases}$

17. $\begin{cases} x + y \geq 0 \\ x - 2y \geq 6 \end{cases}$

18. $\begin{cases} y \geq 2x + 3 \\ y \leq x - 2 \end{cases}$

19. $\begin{cases} x + 3y \leq 9 \\ x - y \geq 5 \end{cases}$

20. $\begin{cases} x - y \geq -2 \\ x + 2y < -1 \end{cases}$

21. $\begin{cases} y \leq x + 3 \\ x - y \leq -5 \end{cases}$

22. $\begin{cases} y \geq 2x - 5 \\ 3x + 2y > -3 \end{cases}$

23. $\begin{cases} y \leq -2x \\ y > -2x - 6 \end{cases}$

24. $\begin{cases} y > x - 4 \\ y < x + 2 \end{cases}$

25. $\begin{cases} y \geq 0 \\ 3x - 5y \leq 10 \end{cases}$　　**26.** $\begin{cases} y \leq 0 \\ 3x + y \leq 11 \end{cases}$　　**27.** $\begin{cases} 4x - 3y \geq 6 \\ 3x - y \leq 3 \end{cases}$　　**28.** $\begin{cases} 3x + 2y \leq 15 \\ 2x + 5y \geq 10 \end{cases}$

29. $\begin{cases} 3x - 4y \geq -6 \\ 3x + 2y \leq 12 \end{cases}$　　**30.** $\begin{cases} 3y \leq 2x + 2 \\ x + 2y \leq 11 \end{cases}$　　**31.** $\begin{cases} x + y \leq 8 \\ 3x - 2y \geq -6 \end{cases}$　　**32.** $\begin{cases} x + y \leq 7 \\ 2x - y \leq 8 \end{cases}$

33. $\begin{cases} y \leq x \\ y < 2x + 1 \end{cases}$　　**34.** $\begin{cases} x - y \geq -2 \\ 4x - y < 16 \end{cases}$

Writing & Thinking

35. Graph the inequalities and explain how you can tell that there is no solution.

$$\begin{cases} y \leq 2x - 5 \\ y \geq 2x + 3 \end{cases}$$

Objective A **Solving Systems of Linear Equations in Three Variables**

The equation $2x + 3y - z = 16$ is called a **linear equation in three variables**. The general form is

$$Ax + By + Cz = D \quad \text{where } A, B, \text{ and } C \text{ are not all equal to } 0.$$

The solutions to such equations are called **ordered triples** and are of the form (x_0, y_0, z_0). One ordered triple that satisfies the equation $2x + 3y - z = 16$ is $(1, 4, -2)$. To check this, substitute $x = 1$, $y = 4$, and $z = -2$ into the equation to see if the result is 16:

$$2(1) + 3(4) - (-2) = 2 + 12 + 2$$
$$= 16.$$

There are an infinite number of ordered triples that satisfy any linear equation in three variables in which at least two of the coefficients are nonzero. Any two values may be substituted for two of the variables, and then the value for the third variable can be calculated. For example, by letting $x = -1$ and $y = 5$ in the equation $2x + 3y - z = 16$, we find:

$$2(-1) + 3(5) - z = 16$$
$$-2 + 15 - z = 16$$
$$-z = 3$$
$$z = -3.$$

Hence, the ordered triple $(-1, 5, -3)$ also satisfies the equation $2x + 3y - z = 16$. Graphs can be drawn in three dimensions by using a coordinate system involving three mutually perpendicular number lines labeled as the x-axis, y-axis, and z-axis. Three planes are formed: the xy-plane, the xz-plane, and the yz-plane. The three axes separate space into eight regions called **octants**. You can "picture" the first octant as the region bounded by the floor of a room and two walls with the axes meeting in a corner. The floor is the xy-plane. The axes can be ordered in a "right-hand" or "left-hand" format. Figure 1 shows the point represented by the ordered triple $(2, 3, 1)$ in a right-hand system.

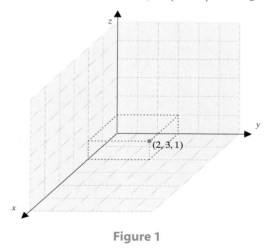

Figure 1

The graphs of linear equations in three variables are planes in three dimensions. A portion of the graph of $2x + 3y - z = 16$ appears in Figure 2.

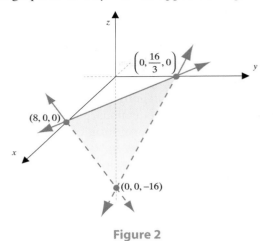

Figure 2

Two distinct planes will either be parallel or they will intersect. If they intersect, their intersection will be a straight line. If three distinct planes intersect, they will intersect in a straight line or in a single point represented by an ordered triple.

The graphs of systems of three linear equations in three variables can be both interesting and informative, but they can be difficult to sketch and points of intersection difficult to estimate. Also, most graphing calculators are limited to graphs in two dimensions, so they are not useful in graphically analyzing systems of linear equations in three variables.

Therefore, in this text, only algebraic techniques for solving these systems will be discussed. Figure 3 illustrates four different possibilities for the relative positions of three planes.

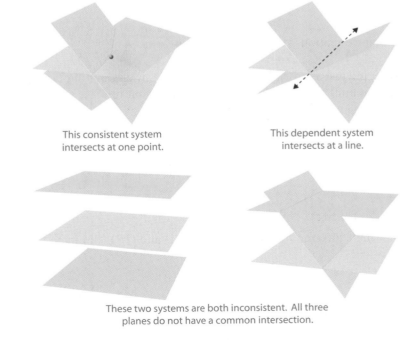

This consistent system intersects at one point.

This dependent system intersects at a line.

These two systems are both inconsistent. All three planes do not have a common intersection.

Figure 3

> ## To Solve a System of Three Linear Equations in Three Variables
>
> 1. Select two equations and eliminate one variable by using the addition method.
> 2. Select a different pair of equations and eliminate the **same** variable.
> 3. Steps 1 and 2 give **two** linear equations in **two** variables. Solve these equations by either addition or substitution as discussed in Sections 7.2 and 7.3.
> 4. Back substitute the values found in Step 3 into any one of the original equations to find the value of the third variable.
> 5. Check the solution (if one exists) in all three of the original equations.

The solution possibilities for a system of three equations in three variables are as follows:

1. There will be exactly one ordered triple solution. (Graphically, the three planes intersect at one point.)
2. There will be an infinite number of solutions. (Graphically, the three planes intersect in a line or are the same plane.)
3. There will be no solutions. (Graphically, there are no points common to all 3 planes.)

The technique is illustrated with the following system.

$$\begin{cases} 2x + 3y - z = 16 & \text{(I)} \\ x - y + 3z = -9 & \text{(II)} \\ 5x + 2y - z = 15 & \text{(III)} \end{cases}$$

Step 1: Using equations (I) and (II), eliminate y.

Note: We could just as easily have chosen to begin by eliminating x or z. To be sure that you understand the process you might want to solve the system by first eliminating x and then again by first eliminating z. In all cases, the answer will be the same.

$$\begin{array}{ll} \text{(I)} & \left\{ \; (2x + 3y - z = 16) \longrightarrow 2x + 3y - z = 16 \right. \\ \text{(II)} & \left. [3] \; (x - y + 3z = -9) \longrightarrow 3x - 3y + 9z = -27 \right. \\ & \hline 5x + 8z = -11 \; \text{(IV)} \end{array}$$

Step 2: Using a different pair of equations, (II) and (III), eliminate the **same** variable, y.

$$\begin{array}{ll} \text{(II)} & \left\{ [2] \; (x - y + 3z = -9) \longrightarrow 2x - 2y + 6z = -18 \right. \\ \text{(III)} & \left. \; (5x + 2y - z = 15) \longrightarrow 5x + 2y - z = 15 \right. \\ & \hline 7x + 5z = -3 \; \text{(V)} \end{array}$$

Step 3: Using the results of Steps 1 and 2, solve the two equations for x and z.

$$
\begin{array}{lll}
(\text{IV}) & [-7] & (5x + 8z = -11) \longrightarrow \\
(\text{V}) & [5] & (7x + 5z = -3) \longrightarrow
\end{array}
$$

$$
\begin{array}{rcr}
-35x - 56z &=& 77 \\
35x + 25z &=& -15 \\
\hline
-31z &=& 62 \\
z &=& -2
\end{array}
$$

Back substitute $z = -2$ into equation (IV) to find x.

$$
\begin{aligned}
5x + 8(-2) &= -11 \qquad &&\text{Using equation (IV).} \\
5x - 16 &= -11 \\
5x &= 5 \\
x &= 1
\end{aligned}
$$

Step 4: Using $x = 1$ and $z = -2$, back substitute to find y.

$$
\begin{aligned}
1 - y + 3(-2) &= -9 \qquad &&\text{Using equation (II).} \\
-y - 5 &= -9 \\
-y &= -4 \\
y &= 4
\end{aligned}
$$

The solution is $(1, 4, -2)$. The solution can be checked by substituting the results into **all three** of the original equations.

$$
\begin{cases}
2(1) + 3(4) - (-2) = 16 & (\text{I}) \\
(1) - (4) + 3(-2) = -9 & (\text{II}) \\
5(1) + 2(4) - (-2) = 15 & (\text{III})
\end{cases}
$$

Example 1

Three Variables (a System with One Solution)

Solve the following system of linear equations.

$$
\begin{cases}
x - y + 2z = -4 & (\text{I}) \\
2x + 3y + z = \dfrac{1}{2} & (\text{II}) \\
x + 4y - 2z = 4 & (\text{III})
\end{cases}
$$

Solution

Using equations (I) and (III), eliminate z.

$$
\begin{array}{llrcrcrcr}
(\text{I}) & x & - & y & + & 2z & = & -4 \\
(\text{III}) & x & + & 4y & - & 2z & = & 4 \\
\hline
& 2x & + & 3y & & & = & 0 & (\text{IV})
\end{array}
$$

Using equations (I) and (II), eliminate z.

$$
\begin{array}{ll}
(\text{I}) & \\
(\text{II}) & [-2]
\end{array}
\begin{cases}
(x - y + 2z = -4) \longrightarrow \\
\left(2x + 3y + z = \dfrac{1}{2}\right) \longrightarrow
\end{cases}
$$

$$
\begin{array}{rcrcrcr}
x & - & y & + & 2z & = & -4 \\
-4x & - & 6y & - & 2z & = & -1 \\
\hline
-3x & - & 7y & & & = & -5
\end{array}
$$

Eliminate the variable x using the two equations in x and y.

$$\begin{array}{ll} \text{(IV)} & [3] \quad (2x + 3y = 0) \longrightarrow \quad 6x + 9y = 0 \\ \text{(V)} & [2] \quad (-3x - 7y = -5) \longrightarrow \underline{-6x - 14y = -10} \\ & \qquad\qquad\qquad\qquad\qquad\qquad\qquad -5y = -10 \\ & \qquad\qquad\qquad\qquad\qquad\qquad\qquad\quad\ y = 2 \end{array}$$

Back substituting into equation (IV) to find x yields:

$$2x + 3(2) = 0$$
$$2x + 6 = 0$$
$$2x = -6$$
$$x = -3.$$

Finally, using $x = -3$ and $y = 2$, back substitute into (I).

$$(-3) - (2) + 2z = -4$$
$$-5 + 2z = -4$$
$$2z = 1$$
$$z = \frac{1}{2}$$

The solution is $\left(-3, 2, \dfrac{1}{2}\right)$.

Check:

The solution can be checked by substituting $\left(-3, 2, \dfrac{1}{2}\right)$ into all three of the original equations.

$$\begin{cases} (-3) - (2) + 2\left(\dfrac{1}{2}\right) = -4 & \text{(I)} \\[2mm] 2(-3) + 3(2) + \left(\dfrac{1}{2}\right) = \dfrac{1}{2} & \text{(II)} \\[2mm] (-3) + 4(2) - 2\left(\dfrac{1}{2}\right) = 4 & \text{(III)} \end{cases}$$

Example 2

Three Variables (a System with No Solution)

Solve the following system of linear equations.
$$\begin{cases} 3x - 5y + z = 6 & \text{(I)} \\ x - y + 3z = -1 & \text{(II)} \\ 2x - 2y + 6z = 5 & \text{(III)} \end{cases}$$

Solution

Using equations (I) and (II), eliminate z.

$$\begin{array}{l} \text{(I)} \\ \text{(II)} \end{array} \left\{ \begin{array}{l} [-3] \ (3x \ - \ 5y \ + \ z \ = \ 6) \longrightarrow \ -9x \ + \ 15y \ - \ 3z \ = \ -18 \\ \ \ (x \ - \ y \ + \ 3z \ = \ -1) \longrightarrow \ x \ - \ y \ + \ 3z \ = \ -1 \end{array} \right. $$
$$\underline{}$$
$$-8x \ + \ 14y \ \ = \ -19 \ \text{(IV)}$$

Using equations (II) and (III), eliminate z.

$$\begin{array}{l} \text{(II)} \\ \text{(III)} \end{array} \left\{ \begin{array}{l} [-2] \ \ (x \ - \ y \ + \ 3z \ = \ -1) \longrightarrow -2x \ + \ 2y \ - \ 6z \ = \ 2 \\ \ \ (2x \ - \ 2y \ + \ 6z \ = \ 5) \longrightarrow \ \ 2x \ - \ 2y \ + \ 6z \ = \ 5 \end{array} \right.$$
$$\underline{}$$
$$0 \ = \ 7 \ \text{(V)}$$

This last equation is **false**. Thus the system has **no solution**.

Example 3

Three Variables (a System with Infinitely Many Solutions)

Solve the following system of linear equations.

$$\begin{cases} 3x - 2y + 4z = 1 & \text{(I)} \\ 3x - y + 7z = -1 & \text{(II)} \\ 6x - 5y + 5z = 4 & \text{(III)} \end{cases}$$

Solution

Using equations (I) and (II), eliminate x.

$$\begin{array}{l} \text{(I)} \\ \text{(II)} \end{array} \left\{ \begin{array}{l} \ (3x \ - \ 2y \ + \ 4z \ = \ 1) \longrightarrow \ 3x \ - \ 2y \ + \ 4z \ = 1 \\ [-1](3x \ - \ y \ + \ 7z \ = \ -1) \longrightarrow -3x \ + \ y \ - \ 7z \ = 1 \end{array} \right.$$
$$\underline{}$$
$$-y \ - \ 3z \ = \ 2 \quad \text{(IV)}$$

Using equations (I) and (III), eliminate x.

$$\begin{array}{l} \text{(I)} \\ \text{(III)} \end{array} \left\{ \begin{array}{l} [-2] \ (3x \ - \ 2y \ + \ 4z \ = \ 1) \longrightarrow -6x \ + \ 4y \ - \ 8z \ = -2 \\ \ (6x \ - \ 5y \ + \ 5z \ = \ 4) \longrightarrow \ \ 6x \ - \ 5y \ + \ 5z \ = \ 4 \end{array} \right.$$
$$\underline{}$$
$$-y \ - \ 3z \ = \ 2 \quad \text{(V)}$$

Using equations (IV) and (V), eliminate y.

$$\begin{array}{l} \text{(IV)} \\ \text{(V)} \end{array} \left\{ \begin{array}{l} \ (-y \ - \ 3z \ = \ 2) \longrightarrow -y \ - \ 3z \ = \ 2 \\ [-1](-y \ - \ 3z \ = \ 2) \longrightarrow \ \ y \ + \ 3z \ = -2 \end{array} \right.$$
$$\underline{}$$
$$0 \ = \ 0$$

Because this last equation, $0 = 0$, is **always true**, the system has an **infinite number of solutions**.

Now work margin exercises 1 through 3.

Solve the following systems of linear equations.

1. $\begin{cases} 2x + y + z = 4 \\ x + 2y + z = 1 \\ 3x + y - z = -3 \end{cases}$

2. $\begin{cases} 7x + 2y - z = 7 \\ 3x - y - 2z = 1 \\ 9x - 3y - 6z = -3 \end{cases}$

3. $\begin{cases} y + 3z = 1 \\ x + y = 3 \\ x + 2y + 3z = 4 \end{cases}$

4. A piggy bank contains $21.70 in $1 coins, quarters and dimes. There are 37 coins in all and 5 more quarters than dimes. How many coins of each kind are there?

Example 4

Three Variables (Application)

A cash register contains $341 in $20, $5, and $2 bills. There are twenty-eight bills in all and three more $2 bills than $5 bills. How many bills of each kind are there?

Solution

Let x = number of $20 bills

y = number of $5 bills

z = number of $2 bills

$$\begin{cases} x + y + z = 28 & (\text{I}) \\ 20x + 5y + 2z = 341 & (\text{II}) \\ z = y + 3 & (\text{III}) \end{cases}$$

There are twenty-eight bills.

The total value is $341.

There are three more $2 bills than $5 bills.

Using equations (I) and (II), eliminate x.

$$\begin{array}{ll} (\text{I}) & [-20] \quad (x + y + z = 28) \rightarrow -20x - 20y - 20z = -560 \\ (\text{II}) & \quad\quad (20x + 5y + 2z = 341) \rightarrow \underline{20x + 5y + 2z = 341} \\ & \quad\quad\quad\quad\quad\quad\quad\quad\quad\quad\quad\quad\quad\quad -15y - 18z = -219 \;(\text{IV}) \end{array}$$

As equation (III) already has no x term, we rewrite it in the form $y - z = -3$ and use this equation along with the equation just found.

$$\begin{array}{ll} (\text{III}) & [15] \quad (y - z = -3) \rightarrow 15y - 15z = -45 \\ (\text{IV}) & \quad\quad (-15y - 18z = -219) \rightarrow \underline{-15y - 18z = -219} \\ & \quad\quad\quad\quad\quad\quad\quad\quad\quad\quad\quad\quad\quad\quad -33z = -264 \\ & \quad\quad\quad\quad\quad\quad\quad\quad\quad\quad\quad\quad\quad\quad\quad\quad z = 8 \end{array}$$

Back substituting into equation (III) to solve for y gives:

$$8 = y + 3$$
$$5 = y.$$

Now we can substitute the values $z = 8$ and $y = 5$ into equation (I).

$$x + 5 + 8 = 28$$
$$x + 13 = 28$$
$$x = 15$$

There are fifteen $20 bills, five $5 bills, and eight $2 bills.

Now work margin exercise 4.

Practice Problems

Solve the following system of linear equations: $\begin{cases} 2x + y + z = 4 \\ x + 2y + z = 1 \\ 3x + y - z = -3 \end{cases}$

Practice Problem Answers

1. $(1, -2, 4)$

Exercises A.7

1. $\begin{cases} x + y - z = 0 \\ 3x + 2y + z = 4 \\ x - 3y + 4z = 5 \end{cases}$

2. $\begin{cases} x - y + 2z = 3 \\ -6x + y + 3z = 7 \\ x + 2y - 5z = -4 \end{cases}$

3. $\begin{cases} 2x - y - z = 1 \\ 2x - 3y - 4z = 0 \\ x + y - z = 4 \end{cases}$

4. $\begin{cases} y + z = 6 \\ x + 5y - 4z = 4 \\ x - 3y + 5z = 7 \end{cases}$

5. $\begin{cases} x + y - 2z = 4 \\ 2x + y = 1 \\ 5x + 3y - 2z = 6 \end{cases}$

6. $\begin{cases} x - y + 5z = -6 \\ x + 2z = 0 \\ 6x + y + 3z = 0 \end{cases}$

7. $\begin{cases} y + z = 2 \\ x + z = 5 \\ x + y = 5 \end{cases}$

8. $\begin{cases} 2y + z = -4 \\ 3x + 4z = 11 \\ x + y = -2 \end{cases}$

9. $\begin{cases} x - y + 2z = -3 \\ 2x + y - z = 5 \\ 3x - 2y + 2z = -3 \end{cases}$

10. $\begin{cases} x - y - 2z = 3 \\ x + 2y + z = 1 \\ 3y + 3z = -2 \end{cases}$

11. $\begin{cases} 2x - y + 5z = -2 \\ x + 3y - z = 6 \\ 4x + y + 3z = -2 \end{cases}$

12. $\begin{cases} 2x - y + 5z = 5 \\ x - 2y + 3z = 0 \\ x + y + 4z = 7 \end{cases}$

13. $\begin{cases} 3x + y + 4z = -6 \\ 2x + 3y - z = 2 \\ 5x + 4y + 3z = 2 \end{cases}$

14. $\begin{cases} 2x + y - z = -3 \\ -x + 2y + z = 5 \\ 2x + 3y - 2z = -3 \end{cases}$

15. $\begin{cases} x - 2y + z = 7 \\ x - y - 4z = -4 \\ x + 4y - 2z = -5 \end{cases}$

16. $\begin{cases} 2x - 2y + 3z = 4 \\ x - 3y + 2z = 2 \\ x + y + z = 1 \end{cases}$

17. $\begin{cases} 2x + 3y + z = 4 \\ 3x - 5y + 2z = -5 \\ 4x - 6y + 3z = -7 \end{cases}$

18. $\begin{cases} x + y + z = 3 \\ 2x - y - 2z = -3 \\ 3x + 2y + z = 4 \end{cases}$

19. $\begin{cases} 2x - 3y + z = -1 \\ 6x - 9y - 4z = 4 \\ 4x + 6y - z = 5 \end{cases}$

20. $\begin{cases} x + 6y + z = 6 \\ 2x + 3y - 2z = 8 \\ 2x + 4z = 3 \end{cases}$

21. The sum of three integers is 67. The sum of the first and second integers is 13 more than the third integer. The third integer is 7 less than the first. Find the three integers.

22. The sum of three integers is 189. The first integer is 28 less than the second. The second integer is 21 less than the sum of the first and third integers. Find the three integers.

23. **Money:** A wallet contains $218 in $10, $5, and $1 bills. There are forty-six bills in all and four more fives than tens. How many bills of each kind are there?

24. **Money:** Sally is trying to get her brother Robert to learn to think algebraically. She tells him that she has 23 coins in her purse, including nickels, dimes, and quarters. She has two more dimes than quarters, and the total value of the coins is $2.50. How many of each kind of coin does she have?

25. **Perimeter of a Triangle:** The perimeter of a triangle is 73 cm. The longest side is 13 cm less than the sum of the other two sides. The shortest side is 11 cm less than the longest side. Find the lengths of the three sides.

26. **Triangles:** The sum of the measures of the three angles of a triangle is 180°. In one particular triangle, the largest angle is 10° more than three times the smallest angle, and the remaining angle is one-half the largest angle. What are the measures of the three angles?

27. **Theater:** A theater has three types of seats for Broadway plays: main floor, balcony, and mezzanine. Main floor tickets are $60, balcony tickets are $45, and mezzanine tickets are $30. Opening night the sales totaled $27,600. Main floor sales were 20 more than the total of balcony and mezzanine sales. Mezzanine sales were 40 more than two times balcony sales. How many of each type of ticket was sold?

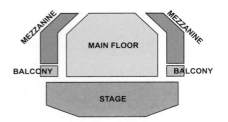

28. **Fruit Stand:** At Steve's Fruit Stand, 4 pounds of bananas, 2 pounds of apples, and 3 pounds of grapes cost $16.40. Five pounds of bananas, 4 pounds of apples, and 2 pounds of grapes cost $16.60. Two pounds of bananas, 3 pounds of apples, and 1 pound of grapes cost $9.60. Find the price per pound of each kind of fruit.

29. **Flower Arranging:** A florist is creating bridesmaids' bouquets for a wedding. Each 16 flower bouquet will have a mixture of lilies, roses, and daisies and cost $92. Lilies cost $10, roses cost $6, and daisies cost $4 a stem. If each bouquet will have as many daisies as it will roses and lilies combined, how many of each type of flower will be in the bouquet?

30. **Construction:** The Tates are having a house built. The cost is split up into three parts: the house, the lot, and the improvements. The cost of building the house is $16,000 more than three times the cost of the lot. The cost of the improvements (the landscaping, sidewalks, and upgrades) is one-third the cost of the lot. If the total cost is $159,000, what is the cost of each part of the construction?

31. Investing: Kirk inherited $100,000 from his aunt and decided to invest in three different accounts: savings, bonds, and stocks. The amount in his bond account was $10,000 more than three times the amount in his stock account. At the end of the first year, the savings account returned 5%, the bond 8%, and the stocks 10% for total interest of $7400. How much did he invest in each account?

32. Stock Market: Melissa has saved a total of $30,000 and wants to invest in three different stocks: PepsiCo, IBM, and Microsoft. She wants the PepsiCo amount to be $1000 less than twice the IBM amount and the Microsoft amount to be $2000 more than the total in the other two stocks. How much should she invest in each stock?

33. Chemistry: A chemist wants to mix 9 liters of a 25% acid solution. Because of limited amounts on hand, the mixture is to come from three different solutions, one with 10% acid, another with 30% acid, and a third with 40% acid. The amount of the 10% solution must be twice the amount of the 40% solution, and the amount of the 30% solution must equal the total amount of the other two solutions. How much of each solution must be used?

34. Manufacturing: An appliance company makes three versions of their popular stand mixer. The standard model has production costs of $100, the deluxe model of $175, and the premium model of $250. On any given day the factory line makes 140 mixers and spends $21,500 on production costs. If the number of standard mixers made equals the sum of the premium and deluxe models made, how many of each kind of mixer are made each day?

Writing & Thinking

35. Is it possible for three linear equations in three unknowns to have exactly two solutions? Explain your reasoning in detail.

36. In geometry, three non-collinear points determine a plane. (That is, if three points are not on a line, then there is a unique plane that contains all three points.) Find the values of A, B, and C (and therefore the equation of the plane) given $Ax + By + Cz = 3$ and the three points on the plane $(0, 3, 2), (0, 0, 1)$ and $(-3, 0, 3)$. Sketch the plane in three dimensions as best you can by locating the three given points.

37. As stated in Exercise 36, three non-collinear points determine a plane. Find the values of A, B, and C (and therefore the equation of the plane) given $Ax + By + Cz = 10$ and the three points on the plane $(2, 0, -2), (3, -1, 0)$ and $(-1, 5, -4)$. Sketch the plane in three dimensions as best you can by locating the three given points.

A.8 Introduction to Complex Numbers

Square Roots of Negative Numbers

One of the properties of real numbers is that the square of any real number is nonnegative. That is, for any real number x, $x^2 \geq 0$. The square roots of negative numbers, such as $\sqrt{-4}$ and $\sqrt{-5}$, are not real numbers. However, they can be defined as part of the system of **complex numbers**.

Complex numbers include all the real numbers and the even roots of negative numbers. Earlier we saw how these numbers occur as solutions to quadratic equations. At first such numbers seem to be somewhat impractical because they are difficult to picture in any type of geometric setting and they are not solutions to the types of word problems that are familiar. However, complex numbers do occur quite naturally in trigonometry and higher level mathematics and have practical applications in such fields as electrical engineering.

The first step in the development of complex numbers is to define $\sqrt{-1}$.

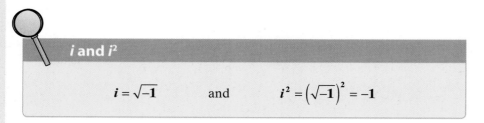

i and i^2

$$i = \sqrt{-1} \qquad \text{and} \qquad i^2 = \left(\sqrt{-1}\right)^2 = -1$$

Using the definition of $\sqrt{-1}$, the following definition for the square root of a negative number can be made.

$\sqrt{-a} = \sqrt{a}\, i$

If a is a positive real number, then

$$\sqrt{-a} = \sqrt{a} \cdot \sqrt{-1} = \sqrt{a}\, i.$$

Note: The number i is not under the radical sign. To avoid confusion, we sometimes write $i\sqrt{a}$.

Objectives

A Simplify square roots of negative numbers.

B Identify the real parts and the imaginary parts of complex numbers.

C Solve equations with complex numbers by setting the real parts and the imaginary parts equal to each other.

D Add and subtract with complex numbers.

Example 1

Square Roots of Negative Numbers

Simplify the following radicals.

a. $\sqrt{-25} = \sqrt{-1}\sqrt{25} = i \cdot 5 = 5i$ $(5i)^2 = 5^2 i^2 = 25(-1) = -25$

b. $\sqrt{-36} = \sqrt{-1}\sqrt{36} = i \cdot 6 = 6i$

c. $\sqrt{-24} = \sqrt{-1}\sqrt{4 \cdot 6} = i \cdot 2 \cdot \sqrt{6} = 2\sqrt{6}\,i$ (or $2i\sqrt{6}$)

d. $\sqrt{-45} = \sqrt{-1}\sqrt{9 \cdot 5} = i \cdot 3 \cdot \sqrt{5} = 3\sqrt{5}\,i$ (or $3i\sqrt{5}$)

We can write $2\sqrt{6}\,i$ and $3\sqrt{5}\,i$ as long as we take care not to include the i under the radical sign.

Now work margin exercise 1.

Objective B   **Real and Imaginary Parts of Complex Numbers**

Complex Numbers

The **standard form** of a **complex number** is $a + bi$, where a and b are real numbers. a is called the **real part** and b is called the **imaginary part.**

If $b = 0$, then $a + bi = a + 0i = a$ is a **real number.**

If $a = 0$, then $a + bi = 0 + bi = bi$ is called a **pure imaginary number** (or an **imaginary number**).

Complex Number: $a + bi$

real part   imaginary part

notes

The term "imaginary" is somewhat misleading. Complex numbers and imaginary numbers are no more "imaginary" than any other type of number. In fact, all the types of numbers that we have studied (whole numbers, integers, rational numbers, irrational numbers, and real numbers) are products of human imagination.

1. Simplify the following radicals.

a. $\sqrt{-100}$

b. $\sqrt{-49}$

c. $\sqrt{-18}$

d. $\sqrt{-72}$

2. Identify the real and imaginary parts of each complex number.

a. $5i$

b. $14 + \sqrt{7}i$

c. $\dfrac{6 - 11i}{5}$

d. -13

Example 2

Real and Imaginary Parts

Identify the real and imaginary parts of each complex number.

a. $4 - 2i$ 4 is the real part; -2 is the imaginary part.

b. $\dfrac{5 + 2i}{3}$ $\dfrac{5 + 2i}{3} = \dfrac{5}{3} + \dfrac{2}{3}i$ in standard form

Thus $\dfrac{5}{3}$ is the real part; $\dfrac{2}{3}$ is the imaginary part.

c. 7 $7 = 7 + 0i$ in standard form

Thus 7 is the real part; 0 is the imaginary part. (Remember, if $b = 0$, the complex number is a real number.)

d. $-\sqrt{3}\,i$ $-\sqrt{3}\,i = 0 - \sqrt{3}\,i$ in standard form

Thus 0 is the real part; $-\sqrt{3}$ is the imaginary part. (If $a = 0$ and $b \neq 0$, then the complex number is a pure imaginary number.)

Now work margin exercise 2.

In general, if a is a real number, then we can write $a = a + 0i$. This means that a is a complex number. **Thus every real number is a complex number.** Figure 1 illustrates the relationships among the various types of numbers we have studied.

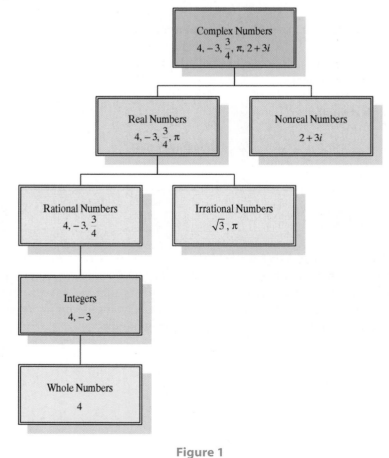

Figure 1

Solving Equations with Complex Numbers

If two complex numbers are equal, then the real parts are equal and the imaginary parts are equal. For example, if

$$x + yi = 7 + 2i,$$

then

$$x = 7 \quad \text{and} \quad y = 2.$$

This relationship can be used to solve equations involving complex numbers.

Equality of Complex Numbers

For complex numbers $a + bi$ and $c + di$,
$$\text{if } \boldsymbol{a + bi = c + di}, \text{ then } \boldsymbol{a = c} \text{ and } \boldsymbol{b = d}.$$

Example 3

Solving Equations

Solve each equation for x and y.

a. $(x + 3) + 2yi = 7 - 6i$

Solution

Equate the real parts and the imaginary parts, and solve the resulting equations.
$$x + 3 = 7 \quad \text{and} \quad 2y = -6$$
$$x = 4 \qquad\qquad y = -3$$

b. $2y + 3 - 8i = 9 + 4xi$

Solution

Equate the real parts and the imaginary parts, and solve the resulting equations.
$$2y + 3 = 9 \quad \text{and} \quad -8 = 4x$$
$$2y = 6 \qquad\qquad -2 = x$$
$$y = 3$$

Now work margin exercise 3.

3. Solve each equation for x and y.

a. $(x - 7) + 4yi = 3 - 8i$

b. $4y + 7 - 5i = 11 + xi$

Objective D **Addition and Subtraction with Complex Numbers**

Adding and subtracting complex numbers is similar to adding and subtracting polynomials. Simply combine like terms. For example,
$$(2 + 3i) + (9 - 8i) = 2 + 9 + 3i - 8i$$
$$= (2 + 9) + (3 - 8)i$$
$$= 11 - 5i.$$

Similarly,

$$(5-2i)-(6+7i) = 5-2i-6-7i$$
$$= (5-6)+(-2-7)i$$
$$= -1-9i.$$

Addition and Subtraction with Complex Numbers

For complex numbers $a + bi$ and $c + di$,

$$(a+bi)+(c+di) = (a+c)+(b+d)i$$
and
$$(a+bi)-(c+di) = (a-c)+(b-d)i.$$

4. Find each sum or difference as indicated.

a. $(5-3i)+(2+7i)$

b. $\left(-7-\sqrt{2}i\right)-\left(-7+3\sqrt{2}i\right)$

c. $\left(4-\sqrt{5}i\right)+\left(\sqrt{2}+8i\right)$

Example 4

Addition and Subtraction with Complex Numbers

Find each sum or difference as indicated.

a. $(6-2i)+(1-2i)$

Solution

$$(6-2i)+(1-2i) = (6+1)+(-2-2)i$$
$$= 7-4i$$

b. $\left(-8-\sqrt{2}\,i\right)-\left(-8+\sqrt{2}\,i\right)$

Solution

$$\left(-8-\sqrt{2}\,i\right)-\left(-8+\sqrt{2}\,i\right) = \left(-8-(-8)\right)+\left(-\sqrt{2}-\sqrt{2}\right)i$$
$$= (-8+8)+\left(-2\sqrt{2}\right)i$$
$$= 0-2\sqrt{2}i$$
$$= -2\sqrt{2}\,i \text{ (or } -2i\sqrt{2})$$

c. $\left(\sqrt{3}-2i\right)+\left(1+\sqrt{5}\,i\right)$

Solution

$$\left(\sqrt{3}-2i\right)+\left(1+\sqrt{5}\,i\right) = \left(\sqrt{3}+1\right)+\left(-2+\sqrt{5}\right)i$$
$$= \left(\sqrt{3}+1\right)+\left(\sqrt{5}-2\right)i$$

Note: Here, the coefficients do not simplify. This means that the real part is $\sqrt{3}+1$ and the imaginary part is $\sqrt{5}-2$.

Now work margin exercise 4.

Practice Problems

1. Find the real part and the imaginary part of $2 - \sqrt{39}i$.

Add or subtract as indicated. Simplify your answers.

2. $\left(-7 + \sqrt{3}i\right) + \left(5 - 2i\right)$

3. $\left(4 + i\right) - \left(5 + 2i\right)$

Solve for x and y.

4. $x + yi = \sqrt{2} - 7i$

5. $3y + \left(x - 7\right)i = -9 + 2i$

Practice Problem Answers

1. real part is 2, imaginary part is $-\sqrt{39}$

2. $-2 + \left(\sqrt{3} - 2\right)i$

3. $-1 - i$

4. $x = \sqrt{2}$ and $y = -7$

5. $x = 9$ and $y = -3$

Exercises A.8

1. $4 - 3i$

2. $\dfrac{3}{4} + i$

3. $-11 + \sqrt{2}\, i$

4. $6 + \sqrt{3}\, i$

5. $\dfrac{3}{8}$

6. $\dfrac{4}{7} i$

7. $\dfrac{4 + 7i}{5}$

8. $\dfrac{2 - i}{4}$

9. $\dfrac{2}{3} + \sqrt{17}\, i$

10. $-\sqrt{5} + \dfrac{\sqrt{2}}{2} i$

Simplify the following radicals. See Example 1.

11. $\sqrt{-49}$

12. $\sqrt{-121}$

13. $-\sqrt{-64}$

14. $-\sqrt{-169}$

15. $\sqrt{147}$

16. $\sqrt{128}$

17. $2\sqrt{-150}$

18. $4\sqrt{-99}$

19. $-2\sqrt{-108}$

20. $2\sqrt{175}$

21. $\sqrt{242}$

22. $\sqrt{-192}$

23. $\sqrt{-1000}$

24. $\sqrt{-243}$

Solve the equations for *x* and *y*. See Example 3.

25. $x + 3i = 6 - yi$

26. $2x - 8i = -2 + 4yi$

27. $\sqrt{5} - 2i = y + xi$

28. $\sqrt{2} - 2yi = 3x + 6i$

29. $\sqrt{2} + i - 3 = x + yi$

30. $\sqrt{5}\, i - 3 + 4i = x + yi$

31. $3x + 2 - 7i = i - 2yi + 5$

32. $x + yi + 8 = 2i + 4 - 3yi$

33. $x + 2i = 5 - yi - 3 - 4i$

34. $x + 2i = 5 - yi - 3 - 4i$

35. $2 + 3i + x = 5 - 7i + yi$

36. $11i - 2x + 4 = 10 - 3i + 2yi$

37. $2x - 2yi + 6 = 6i - x + 2$

38. $x + 4 - 3x + i = 8 + yi$

Find each sum or difference as indicated. See Example 4.

39. $(2 + 3i) + (4 - i)$

40. $(7 - i) + (3 + 6i)$

41. $(4 + 5i) - (3 - 2i)$

42. $(-3+2i)-(6+2i)$

43. $(4-3i)+(2-3i)$

44. $(7+5i)+(6-2i)$

45. $(8+9i)-(8-5i)$

46. $(-6+i)-(2+3i)$

47. $(\sqrt{5}-2i)+(3-4i)$

48. $(4+3i)-(\sqrt{2}+3i)$

49. $(7+\sqrt{6}\,i)+(-2+i)$

50. $(\sqrt{11}+2i)+(5-7i)$

51. $(\sqrt{3}+\sqrt{2}\,i)-(5+\sqrt{2}\,i)$

52. $(\sqrt{5}+\sqrt{3}\,i)+(1-i)$

53. $(5+\sqrt{-25})-(7+\sqrt{-100})$

54. $(1+\sqrt{-36})-(-4-\sqrt{-49})$

55. $(13-3\sqrt{-16})+(-2-4\sqrt{-1})$

56. $(7+\sqrt{-9})-(3-2\sqrt{-25})$

57. $(4+i)+(-3-2i)-(-1-i)$

58. $(-2-3i)+(6+i)-(2+5i)$

59. $(7+3i)+(2-4i)-(6-5i)$

60. $(-5+7i)+(4-2i)-(3-5i)$

Writing & Thinking

61. Answer the following questions and give a brief explanation of your answer.

    a. Is every real number a complex number?

    b. Is every complex number a real number?

62. List 5 numbers that do and 5 numbers that do not fit each of the following categories (if possible).

    a. rational number   **b.** integer

    c. real number   **d.** pure imaginary number

    e. complex number   **f.** irrational number

A.9 Multiplication and Division with Complex Numbers

Objective A   **Multiplication with Complex Numbers**

The product of two complex numbers can be found by using the FOIL method for multiplying two binomials. (See previous section.) **Remember that $i^2 = -1$.** For example,

$$(3 + 5i)(2 + i) = 6 + 3i + 10i + 5i^2$$
$$= 6 + 13i - 5 \qquad \qquad 5i^2 = 5(-1) = -5$$
$$= 1 + 13i.$$

1. Find the following products.

   a. $(4i)(1 - 7i)$

   b. $(4 + 2i)(3 + 5i)$

   c. $\left(\sqrt{2} - i\right)\left(\sqrt{2} + i\right)$

Example 1

Multiplication with Complex Numbers

Find the following products.

a. $(3i)(2 - 7i)$

Solution

$$(3i)(2 - 7i) = 6i - 21i^2$$
$$= 6i - 21(-1)$$
$$= 21 + 6i$$

b. $(5 + i)(2 + 6i)$

Solution

$$(5 + i)(2 + 6i) = 5(2 + 6i) + i(2 + 6i)$$
$$= 10 + 30i + 2i + 6i^2$$
$$= 10 + 32i - 6$$
$$= 4 + 32i$$

c. $\left(\sqrt{2} - i\right)\left(\sqrt{2} - i\right)$

Solution

$$\left(\sqrt{2} - i\right)\left(\sqrt{2} - i\right) = \left(\sqrt{2}\right)^2 - \sqrt{2} \cdot i - \sqrt{2} \cdot i + i^2$$
$$= 2 - 2\sqrt{2}\,i - 1 \qquad \qquad \text{Remember } i^2 = -1.$$
$$= 1 - 2i\sqrt{2}$$

d. $(-1 + i)(2 - i)$

Solution

$$(-1 + i)(2 - i) = -2 + i + 2i - i^2$$
$$= -2 + 3i + 1$$
$$= -1 + 3i$$

Now work margin exercise 1.

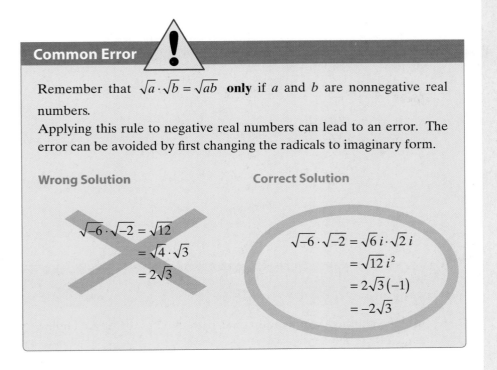

Common Error

Remember that $\sqrt{a} \cdot \sqrt{b} = \sqrt{ab}$ **only** if a and b are nonnegative real numbers.

Applying this rule to negative real numbers can lead to an error. The error can be avoided by first changing the radicals to imaginary form.

Wrong Solution

$$\sqrt{-6} \cdot \sqrt{-2} = \sqrt{12}$$
$$= \sqrt{4} \cdot \sqrt{3}$$
$$= 2\sqrt{3}$$

Correct Solution

$$\sqrt{-6} \cdot \sqrt{-2} = \sqrt{6}\,i \cdot \sqrt{2}\,i$$
$$= \sqrt{12}\,i^2$$
$$= 2\sqrt{3}\,(-1)$$
$$= -2\sqrt{3}$$

Objective B   **Division with Complex Numbers**

The two complex numbers $a + bi$ and $a - bi$ are called **complex conjugates** or simply **conjugates** of each other. As the following steps show, **the product of two complex conjugates will always be a nonnegative real number**.

$$(a+bi)(a-bi) = a^2 - abi + abi - b^2 i^2$$
$$= a^2 - b^2 i^2$$
$$= a^2 + b^2$$

The resulting product, $a^2 + b^2$, is a real number, and it is nonnegative since it is the sum of the squares of real numbers.

Remember that the form $a + bi$ is called the **standard form** of a complex number. The standard form allows for easy identification of the real and imaginary parts. Thus

$$\frac{1+3i}{5} = \frac{1}{5} + \frac{3}{5}i \text{ in standard form.}$$

The real part is $\dfrac{1}{5}$ and the imaginary part is $\dfrac{3}{5}$.

To write the fraction $\dfrac{1+i}{2-3i}$ in standard form, multiply both the numerator and denominator by $2 + 3i$, the complex conjugate of the denominator, and simplify. This will give a positive real number in the denominator.

$$\frac{1+i}{2-3i} = \frac{(1+i)(2+3i)}{(2-3i)(2+3i)}$$

2 + 3*i* is the conjugate of the denominator.

$$= \frac{2+3i+2i+3i^2}{2^2+6i-6i-3^2i^2}$$

$$= \frac{2+5i-3}{4-(9)(-1)}$$

Reminder: $i^2 = -1$.

$$= \frac{-1+5i}{13}$$

$$= -\frac{1}{13} + \frac{5}{13}i$$

Writing Fractions with Complex Numbers in Standard Form

1. Multiply both the numerator and denominator by the complex conjugate of the denominator.
2. Simplify the resulting products in both the numerator and denominator.
3. Write the simplified result in standard form.

Remember the following special product. We restate it here to emphasize its importance.

$$(a+bi)(a-bi) = a^2 + b^2$$

Example 2

Division with Complex Numbers

Write the following fractions in standard form.

a. $\dfrac{4}{-1-5i}$

Solution

$$\frac{4}{-1-5i} = \frac{4(-1+5i)}{(-1-5i)(-1+5i)}$$

$$= \frac{-4+20i}{(-1)^2-(5i)^2} = \frac{-4+20i}{1+25}$$

$$= \frac{-4+20i}{26} = -\frac{4}{26} + \frac{20}{26}i$$

$$= -\frac{2}{13} + \frac{10}{13}i$$

b. $\dfrac{\sqrt{3}+i}{\sqrt{3}-i}$

Solution

$$\dfrac{\sqrt{3}+i}{\sqrt{3}-i}=\dfrac{\left(\sqrt{3}+i\right)\left(\sqrt{3}+i\right)}{\left(\sqrt{3}-i\right)\left(\sqrt{3}+i\right)}$$

$$=\dfrac{3+2\sqrt{3}\,i+i^2}{\left(\sqrt{3}\right)^2-i^2}=\dfrac{2+2\sqrt{3}\,i}{3+1}$$

$$=\dfrac{2+2\sqrt{3}\,i}{4}=\dfrac{2}{4}+\dfrac{2\sqrt{3}}{4}\,i$$

$$=\dfrac{1}{2}+\dfrac{\sqrt{3}}{2}\,i$$

c. $\dfrac{6+i}{i}$

Solution

$$\dfrac{6+i}{i}=\dfrac{\left(6+i\right)\left(-i\right)}{i\left(-i\right)}$$ Since $i=0+i$ and $-i=0-i$, the number $-i$ is the conjugate of i.

$$=\dfrac{-6i-i^2}{-i^2}=\dfrac{-6i+1}{1}$$

$$=1-6i$$

d. $\dfrac{\sqrt{2}+i}{-\sqrt{2}+i}$

Solution

$$\dfrac{\sqrt{2}+i}{-\sqrt{2}+i}=\dfrac{\left(\sqrt{2}+i\right)\left(-\sqrt{2}-i\right)}{\left(-\sqrt{2}+i\right)\left(-\sqrt{2}-i\right)}$$

$$=\dfrac{-\left(\sqrt{2}\right)^2-\sqrt{2}\,i-\sqrt{2}\,i-i^2}{\left(-\sqrt{2}\right)^2-i^2}$$

$$=\dfrac{-2-2\sqrt{2}\,i+1}{2+1}=\dfrac{-1-2\sqrt{2}\,i}{3}$$

$$=-\dfrac{1}{3}-\dfrac{2\sqrt{2}}{3}\,i$$

Now work margin exercise 2.

2. Write the following fractions in standard form.

a. $\dfrac{8}{1-2i}$

b. $\dfrac{\sqrt{5}-2i}{\sqrt{5}+2i}$

c. $\dfrac{4-i}{4i}$

d. $\dfrac{\sqrt{3}+4i}{-\sqrt{3}+4i}$

Powers of _i_

The powers of _i_ form an interesting pattern. Regardless of the particular integer exponent, there are only four possible values for any power of _i_:

$$i, \quad -1, \quad -i, \quad \text{and} \quad 1.$$

The fact that these are the only four possibilities for powers of _i_ becomes apparent from studying the following powers.

$$i^1 = i$$
$$i^2 = -1$$
$$i^3 = i^2 \cdot i = -1 \cdot i = -i$$
$$i^4 = i^2 \cdot i^2 = (-1)(-1) = 1$$
$$i^5 = i^4 \cdot i = 1 \cdot i = i$$
$$i^6 = i^4 \cdot i^2 = (1)(-1) = -1$$
$$i^7 = i^4 \cdot i^3 = (1)(-i) = -i$$
$$i^8 = i^4 \cdot i^4 = 1 \cdot 1 = 1$$

Higher powers of _i_ can be simplified using the fact that when _i_ is raised to a power that is a multiple of 4, the result is 1. Thus if _n_ is a positive integer, then

$$i^{4n} = \left(i^4\right)^n = 1^n = 1$$
$$i^{4n+1} = i^{4n} \cdot i = 1 \cdot i = i$$
$$i^{4n+2} = i^{4n} \cdot i^2 = (1) \cdot (-1) = -1$$
$$i^{4n+3} = i^{4n} \cdot i^3 = (1) \cdot (-i) = -i.$$

3. Simplify each power of _i_.

a. i^{37}

b. i^{14}

c. i^{-8}

Example 3

Powers of _i_

Simplify each power of _i_.

a. $i^{45} = i^{44} \cdot i = \left(i^4\right)^{11} \cdot i = 1^{11} \cdot i = i$ $i = 0 + i$ in standard form.

b. $i^{58} = i^{56} \cdot i^2 = \left(i^4\right)^{14} \cdot i^2 = 1^{14} \cdot (-1) = -1$ $1 = -1 + 0\,i$ in standard form.

c. $i^{-7} = \dfrac{1}{i^7} = \dfrac{1}{i^7} \cdot \dfrac{i}{i} = \dfrac{i}{i^8} = \dfrac{i}{1} = i$ $i = 0 + i$ in standard form.

Now work margin exercise 3.

Practice Problems

Write each of the following numbers in standard form.

1. $-2i(3-i)$ **2.** $(2+4i)(1+i)$ **3.** i^{17}

4. i^{-2} **5.** $\dfrac{2}{1+5i}$ **6.** $\dfrac{7+i}{2-i}$

Practice Problem Answers

1. $-2-6i$ **2.** $-2+6i$ **3.** $0+i$ **4.** $-1+0i$

5. $\dfrac{1}{13}-\dfrac{5}{13}i$ **6.** $\dfrac{13}{5}+\dfrac{9}{5}i$

Exercises A.9

1. $8(2+3i)$

2. $-3(7-4i)$

3. $-7(\sqrt{2}-i)$

4. $\sqrt{3}(\sqrt{3}+2i)$

5. $3i(4-i)$

6. $-4i(6-7i)$

7. $-i(\sqrt{3}+i)$

8. $2i(\sqrt{5}+2i)$

9. $\sqrt{3}i(2-\sqrt{3}i)$

10. $5i(2-\sqrt{2}i)$

11. $(5+3i)(1+i)$

12. $(2+7i)(6+i)$

13. $(-3+5i)(-1+2i)$

14. $(6+2i)(3-i)$

15. $(2-3i)(2+3i)$

16. $(4+5i)(4-5i)$

17. $(4+3i)(7-2i)$

18. $(-2+5i)(i-1)$

19. $(5+7i)^2$

20. $(3+2i)^2$

21. $(\sqrt{3}+i)(\sqrt{3}-2i)$

22. $(2\sqrt{5}+3i)(\sqrt{5}-i)$

23. $(5-\sqrt{2}i)(5-\sqrt{2}i)$

24. $(\sqrt{7}+3i)(\sqrt{7}+i)$

25. $(4+\sqrt{5}i)(4-\sqrt{5}i)$

26. $(7+2\sqrt{3}i)(7-2\sqrt{3}i)$

27. $(\sqrt{5}+2i)(\sqrt{2}-i)$

28. $(2\sqrt{3}+i)(4+3i)$

29. $(3+\sqrt{5}i)(3+\sqrt{6}i)$

30. $(2-\sqrt{3}i)(3-\sqrt{2}i)$

31. $\dfrac{-3}{i}$

32. $\dfrac{7}{i}$

33. $\dfrac{5}{4i}$

34. $\dfrac{-3}{2i}$

35. $\dfrac{2+i}{-4i}$

36. $\dfrac{3-4i}{3i}$

37. $\dfrac{-4}{1+2i}$

38. $\dfrac{7}{5-2i}$

39. $\dfrac{6}{4-3i}$

40. $\dfrac{-8}{6+i}$

41. $\dfrac{2i}{5-i}$

42. $\dfrac{-4i}{1+3i}$

43. $\dfrac{2-i}{2+5i}$

44. $\dfrac{6+i}{3-4i}$

45. $\dfrac{2-3i}{-1+5i}$

46. $\dfrac{-3+i}{7-2i}$

47. $\dfrac{1+4i}{\sqrt{3}+i}$

48. $\dfrac{9-2i}{\sqrt{5}+i}$

49. $\dfrac{\sqrt{3}+2i}{\sqrt{3}-2i}$ **50.** $\dfrac{\sqrt{6}-3i}{\sqrt{6}+3i}$

Simplify the following powers of i and write each result in standard form. Assume k is a positive integer. See Example 3.

51. i^{13} **52.** i^{20} **53.** i^{30} **54.** i^{15} **55.** i^{-3}

56. i^{-5} **57.** i^{4k} **58.** i^{4k+2} **59.** i^{4k+3} **60.** i^{4k+1}

Find the indicated products and simplify.

61. $(x+3i)(x-3i)$ **62.** $(y+5i)(y-5i)$ **63.** $\left(x+\sqrt{2}\,i\right)\left(x-\sqrt{2}\,i\right)$

64. $\left(2x+\sqrt{7}\,i\right)\left(2x-\sqrt{7}\,i\right)$ **65.** $\left(\sqrt{5}y+2i\right)\left(\sqrt{5}y-2i\right)$ **66.** $\left(y-\sqrt{3}\,i\right)\left(y+\sqrt{3}\,i\right)$

67. $\left[(x+2)+6i\right]\left[(x+2)-6i\right]$ **68.** $\left[(x+1)-\sqrt{8}\,i\right]\left[(x+1)+\sqrt{8}\,i\right]$ **69.** $\left[(y-3)+2i\right]\left[(y-3)-2i\right]$

70. $\left[(x-1)+5i\right]\left[(x-1)-5i\right]$

Writing & Thinking

71. Explain why the product of every complex number and its conjugate is a nonnegative real number.

72. Explain why $\sqrt{-4}\cdot\sqrt{-4}\neq 4$. What is the correct value of $\sqrt{-4}\cdot\sqrt{-4}$?

73. What condition is necessary for the conjugate of a complex number, $a+bi$, to be equal to the reciprocal of this number?

Answers

Chapter 1: Whole Numbers

Margin Exercises **1.** $400 + 60 + 3$ **2.** $7000 + 300 + 0 + 2$ **3.** $20,000 + 9000 + 500 + 20 + 4$ **4. a.** 9 **b.** 6 **c.** 1
5. a. thirty-two million, four hundred fifty thousand, ninety **b.** five thousand seven hundred eighty-four **6. a.** 6041 **b.** 1,483,007
Exercises **1.** left; decimal point **3.** 4 **5.** 0, 1, 2, 3, 4, 5, 6, 7, 8, 9 **7. a.** 6 **b.** 3 **c.** 1 **9.** $30 + 7$ **11.** $50 + 6$ **13.** $1000 + 800 + 90 + 2$
15. $20,000 + 5000 + 600 + 50 + 8$ **17.** eighty-three **19.** ten thousand, five hundred
21. five hundred ninety-two thousand, three hundred **23.** seventy-one million, five hundred thousand **25.** 580 **27.** 2005
29. 10,011 **31.** 400,736 **33.** 82,700,000 **35.** 281,300,501 **37.** ninety-three million; one hundred forty-nine million, seven
hundred thirty thousand **39.** three hundred fifty-two thousand, one hundred forty-three; nine hundred twelve thousand, fifty
41. sixty-three million, eight hundred thousand; fourteen thousand, forty

Margin Exercises **1. a.** 197 **b.** 157 **2.** 501 **3.** 1614 **4.** $1377 **5. a.** 17; associative property of addition **b.** 29; commutative
property of addition **c.** 39; additive identity property **6.** 38 cm **7.** 58 cm **8.** 233 **9.** 155 **10.** 6577 **11.** 386 **12.** $36
13. $3640 **14.** $15,230
Exercises **1.** 78 **3.** 7495 **5.** 14 **7.** 6682 **9.** 361 **11.** 9577 **13.** 10,500 **15.** 16,072 **17.** 329, 134 **19.** 1,238,914 **21.** 177,209
23. 42,121,646 **25.** commutative property of addition **27.** associative property of addition **29.** associative property of
addition **31.** additive identity **33.** commutative property of addition **35.** 2762 miles **37.** $1053 **39.** 17,740 appliances
41. 12 in. **43.** 15 ft **45.** 22 in. **47.** 31 m **49.** 156 in. **51.** 20 **53.** 94 **55.** 531 **57.** 5209 **59.** 1531 **61.** 1569 **63.** 694
65. 2,806,644 **67.** 3,800,559 **69.** 7,352,439 **71.** 140 **73.** 2152 miles **75.** $250,404 **77.** 16 years

Margin Exercises **1.** 90 **2.** 884 **3.** 480 **4. a.** 96; commutative property of multiplication **b.** 0; multiplication property of 0
c. 35; multiplicative identity property **d.** 42; associative property of multiplication **5.** $3 \cdot 10 + 3 \cdot 6 = 30 + 18 = 48$
6. 3796 square yards
Exercises **1.** 162 **3.** 252 **5.** 0 **7.** 56 **9.** 336 **11.** 4146 **13.** 324 **15.** 2976 **17.** 15,035 **19.** 45,279 **21.** 14,070 **23.** 166,986
25. 2352 **27.** 416 **29.** 4233 **31.** 2100 **33.** 17,632 **35.** 7632 **37.** 21,238 **39.** 24,300 **41.** 81,180 **43.** 240,600 **45.** 43,680
47. 89,664 **49.** 63,210 **51.** 3,305,565 **53.** 378,400 **55.** 1,500,000 **57.** 6,036,009 **59.** 8,262,880 **61.** 108,000,000 **63.** 0 **65.** 0
67. 8300 **69.** 0 **71.** associative property of multiplication **73.** multiplication property of 0 **75.** $3 \cdot 9 + 3 \cdot 7 = 27 + 21 = 48$
77. $9 \cdot 2 + 9 \cdot 9 = 18 + 81 = 99$ **79.** 196 in.^{2} **81.** 1200 ft^2 **83.** 648 sq ft **85.** 384 sq in. **87.** 9250 calories **89.** $3569
91. $15,200,000

Margin Exercises **1.** 5 R3 **2. a.** 4 **b.** 7 **c.** 1 **d.** 86 **3. a.** undefined **b.** 0 **4.** 2 R5 **5.** 69 R1 **6.** 21 R4 **7.** 308 R23

8. 155 R9 **9.** 214 R24 **10.**

$$14\overline{)294} \qquad 21\overline{)294}$$

21	14
28	21
14	84
14	84
0	0

11. $107 **12.** $238 **13.** 234 calories

Exercises **1.** 3 **3.** 9 **5.** undefined **7.** 1 **9.** 6 **11.** 5 **13.** 32 **15.** 7 **17.** 14 **19.** 30 **21.** 6 R4 **23.** 24 **25.** 10 R11 **27.** 15 R5
29. 401 **31.** 23 **33.** 300 R13 **35.** 301 R4 **37.** 268 **39.** 654 **41.** 1234 **43.** 1208 R45 **45.** 124 R61 **47.** 441 R75
49. 196 R370 **51.** 407 **53.** 14 **55.** 3 cookies **57.** 22 teams **59.** 17 grams **61.** $7020 per year
63. 3 pieces per student with 10 pieces left **65.** 377 days **67.** $2574 **69.** 720 pounds

Margin Exercises   1. a. 580;   **b.** 300;   **c.** 10,000;
d. 9700;   **2. a.** 7400 **b.** 30,000 **c.** 100,000 **d.** 28,000,000   **3.** 290 feet   **4.** estimate: 370; sum: 316
5. estimate: 9000; sum: 8861   **6.** estimate: 2200; difference: 1838   **7.** estimate: 1400 bags; difference: 1001 bags
8. estimate: 1400; product: 1332   **9.** estimate: $100,000; product: $87,320   **10.** estimate: 20; quotient: 21
11. estimate: 11; quotient: 11 R46   **12.** estimate: 25; quotient: 37 R8; difference: 12   **13.** estimate: 25 buses; quotient: 21 buses
Exercises   1. 760 **3.** 90 **5.** 5350 **7.** 3500 **9.** 4500 **11.** 12,600 **13.** 7000 **15.** 800 **17.** 5000 **19.** 7000 **21.** 10,000
23. 100,000 **25.** 80,000 **27.** 300,000 **29.** 306,000,000 **31.** $16,000,000,000 **33.** $25,000 **35.** $7100 **37.** 100,000 miles
39. estimate: 800; sum: 789   **41.** estimate: 70,000; sum: 71,671   **43.** estimate: 6000; difference: 5467
45. estimate: 5000; difference: 5931   **47.** estimate: 160,000; product: 174,045   **49.** estimate: 400,000; product: 341,240
51. estimate: 350; quotient: 357   **53.** estimate: 71; quotient: 74   **55.** estimate: $40,000; actual: $35,316
57. estimate: $7500; actual: $7304   **59.** estimate: $2410; actual: $2174   **61.** estimate: $18,000; actual: $21,575
63. estimate: $6000; actual: $7634   **65.** estimate: 40 vans; actual: 36 vans   **67.** estimate: $3000; actual: $3160

Margin Exercises   1. a. base: 8; exponent: 3 **b.** base: 14; exponent: 6   **2. a.** "seven squared is equal to forty-nine" **b.** "four cubed
is equal to sixty-four" **c.** "three to the fifth power is equal to two hundred forty-three"   **3. a.** $8^3 = 512$ **b.** $9^2 = 81$ **c.** $4^4 = 256$
d. $2^7 = 128$ **4. a.** 1 **b.** 1 **c.** 813 **d.** 1   **5.** 8   **6.** 11   **7.** 5   **8.** 68   **9.** 11   **10.** 19
Exercises   1. 12, 7 **3.** 10, 4 **5.** 19, 0 **7.** 4, 72 **9.** 2^2 **11.** 9^3 **13.** 7^4 **15.** 8^6 **17.** $2^2 \cdot 7^2$ **19.** $5^3 \cdot 11^2$ **21.** 8 **23.** 81 **25.** 20
27. undefined **29.** 1 **31.** 8 **33.** 1 **35.** 216 **37.** 900 **39.** 45 **41.** 21 **43.** 28 **45.** 3 **47.** 18 **49.** 5 **51.** 5 **53.** 10 **55.** 35
57. 3 **59.** 36 **61.** 2 **63.** 32 **65.** 23 **67.** 0 **69.** 14 **71.** 15 **73.** 61 **75.** 22 **77.** 10 **79.** 24 **81.** 34 **83.** 18 **85.** 9 **87.** 50
89. 21 **91.** 524 **93.** 1, 8, 27, 64, 125, 216, 343, 512, 729, 1000

Margin Exercises   1. 47   **2.** 258   **3.** 43   **4.** $25,125   **5.** $868   **6.** 21   **7.** 36 home runs   **8.** 200,000 people   **9.** 12 points
Exercisess   1. 1570 **3.** 1800 **5.** 140 **7.** $160 **9.** $786 **11.** $874 **13. a.** $18,000 **b.** $3700 **c.** $47,900
15. a. Virginia **b.** Maine **c.** 4300 people **17.** $485 **19.** $316 **21.** 103 **23.** 6 **25.** 485 **27.** $61 **29.** 21 points **31.** 85
33. $26/share; $500 profit **35. a.** $665 **b.** $2495 **37.** 71,904 people **39. a.** 27,593 students **b.** $16,555 **c.** $27,119 **d.** $9314

Margin Exercises   1. Yes, the last digit is 8, an even digit.   **2.** No, $7 + 9 + 1 + 2 = 19$, and 19 is not divisible by 3.
3. Yes, 76 is divisible by 4.   **4.** No, the last digit is not 0 or 5.   **5.** No, $1 + 5 + 7 + 6 = 19$, and 19 is not divisible by 3.
6. Yes, $4 + 6 + 5 + 3 = 18$, and 18 is divisible by 9.   **7.** Yes, the last digit is 0.   **8.** Yes, $6 + 1 + 2 = 9$, and 9 is divisible by 9.
9. 45 divides the product 36 times.   **10.** 55 divides the product 96 times.   **11.** 30 divides the product 21 times.
12. 21 does not divide the product.
Exercises   1. 552; 3998; 710; 116; 25,326; 7800; 36; 411,076   **3.** 552; 116; 7800; 36; 411,076   **5.** 552; 25,326; 7800; 36   **7.** 710; 7800
9. 3, 9 **11.** 3, 9 **13.** 3 **15.** 2 **17.** 3, 5 **19.** none **21.** 2, 3, 4, 5, 6, 10 **23.** 2, 5, 10 **25.** 2 **27.** 5 **29.** 3 **31.** 3, 5 **33.** 2 **35.** 5
37. 2, 3, 6 **39.** 2, 3, 4, 6 **41.** 3, 5 **43.** 2, 3, 4, 5, 6, 9, 10 **45.** 2, 3, 4, 6, 9 **47.** 2, 3, 4, 6, 9 **49.** none **51.** 2, 3, 5, 6, 9, 10 **53.** 2, 3, 5, 6, 10
55. 3, 5 **57.** 2 **59.** 2, 3, 4, 6 **61. a.** Yes, 14 divides the product. **b.** 15 times   **63. a.** No, 21 does not divide the product.
65. a. Yes, 12 divides the product. **b.** 42 times   **67. a.** Yes, 32 divides the product. **b.** 33 times
69. a. No, 81 does not divide the product.   **71. a.** Because 3 is a factor of 9, a number that is divisible by 9 will also be divisible
by 3; Answers will vary. **b.** 15, 93; Answers will vary.

Margin Exercises   1. a. prime; 13 has exactly two factors, 1 and 13. **b.** composite; 1, 5, and 25 are all factors of 25.
c. composite; 1, 2, 4, 8, 16, and 32 are all factors of 32.   **2.** composite; the factors of 404 are 1, 2, 4, 101, 202, and 404.
3. prime; test the prime numbers less than 11 to come to this conclusion.   **4.** composite; 1, 13, 19, and 247 are factors of 247.
5. 21 and 4   **6.** $2 \cdot 2 \cdot 3 \cdot 7$ or $2^2 \cdot 3 \cdot 7$ **7.** $2^4 \cdot 5$ **8.** $3 \cdot 5 \cdot 11$ **9.** 1, 2, 3, 6, 7, 14, 21, and 42.
10. 1, 2, 4, 5, 8, 10, 16, 20, 32, 40, 80, and 160.
Exercises   1. A number whose only factors are 1 and itself. Answers will vary.   **3.** 5, 10, 15, 20, 25   **5.** 13, 26, 39, 52, 65
7. 25, 50, 75, 100, 125   **9.** 31, 62, 93, 124, 155   **11.** 2, 3, 5, 7, 11, 13, 17, 19, 23, 29, 31, 37, 41, 43, 47, 53, 59, 61, 67, 71, 73, 79, 83, 89, 97

13. composite; $1, 2, 4, 8, 16, 32$ **15.** prime   **17.** composite; $1, 3, 17, 51$ **19.** prime   **21.** prime   **23.** composite; $1, 3, 19, 57$
25. composite; $1, 5, 41, 205$ **27.** composite; $1, 3, 9, 13, 39, 117$ **29.** $3, 4$ **31.** $1, 12$ **33.** $2, 25$ **35.** $3, 8$ **37.** $3, 12$
39. $3, 21$ **41.** $2^2 \cdot 7$ **43.** 2^4 **45.** $2^2 \cdot 3 \cdot 5$ **47.** $2 \cdot 3 \cdot 13$ **49.** $3 \cdot 5 \cdot 7$ **51.** $2^5 \cdot 5$ **53.** $2 \cdot 3 \cdot 5^2$ **55.** $2^3 \cdot 5$ **57.** $3 \cdot 31$ **59.** $2^3 \cdot 3^2 \cdot 5$
61. $1, 2, 3, 6, 9, 18$ **63.** $1, 2, 7, 14, 49, 98$ **65.** $1, 3, 5, 9, 15, 45$ **67.** $1, 2, 3, 6, 9, 18, 27, 54$ **69.** $1, 5, 11, 25, 55, 275$
71. a. 1 has exactly 1 factor, so it cannot be prime nor composite by the definitions. Answers will vary.   **b.** 0 is not a counting number, so it is inconsistent with both definitions. Answers will vary.   **73.** 28 and 496

Chapter 1: Review

1. $400 + 90 + 5$; four hundred ninety-five   **2.** $1000 + 900 + 70 + 5$; one thousand nine hundred seventy-five
3. $60,000 + 0 + 300 + 0 + 8$; sixty thousand, three hundred eight   **4.** $2,000,000 + 400,000 + 60,000 + 0 + 0 + 0 + 0$; two million, four hundred sixty thousand   **5.** 807   **6.** 4656   **7.** 17,002   **8.** 72,340,083   **9.** commutative property of addition
10. associative property of addition   **11.** additive identity property   **12.** 72   **13.** 10,541   **14.** 1674   **15.** 508   **16.** 2384
17. 2102   **18.** 126 inches   **19.** 0   **20.** 5600   **21.** 360,000   **22.** 17,394   **23.** 1558   **24.** 5096   **25.** 3,913,100   **26.** 24,185,000
27. 0   **28.** undefined   **29.** 15   **30.** 8   **31.** 295   **32.** 606   **33.** 563 R7   **34.** 135 R81   **35.** 630   **36.** 700   **37.** 2600   **38.** 15,000
39. estimate: 1600; sum: 1625   **40.** estimate: 5000; difference: 5931   **41.** estimate: 6000; product: 4872
42. estimate: 80; quotient: 67   **43.** base: 3; exponent: 5   **44.** $2 \cdot 3^3 \cdot 5^2$ **45.** 16   **46.** 1   **47.** 15   **48.** 46   **49.** 35   **50.** 13   **51.** 2
52. 2   **53.** 1059   **54.** 35   **55.** 9   **56.** $1485; $99   **57.** 83   **58.** 70   **59.** $7700   **60.** $2350   **61.** $3, 5, 9$ **62.** $2, 3, 4, 6, 9$
63. none   **64.** $2, 3, 4, 5, 6, 9, 10$ **65.** $2, 4$ **66.** 3   **67. a.** yes, 6 divides the product   **b.** 21 times   **68. a.** yes, 28 divides the product
b. 15 times   **69.** $3, 6, 9, 12, 15, 18, 21, 24, 27, 30$; Only 3 is prime.   **70.** yes, because the only factors of 223 are 1 and itself.
71. $2, 3, 5, 7, 11, 13, 17, 19, 23, 29, 31, 37, 41, 43, 47, 53, 59$ **72.** $4, 6$ **73.** $2 \cdot 3 \cdot 5^2$ **74.** $5 \cdot 13$ **75.** $2^7 \cdot 3$ **76.** $2 \cdot 3^2 \cdot 5 \cdot 11$

Chapter 1: Test

1. $8000 + 900 + 50 + 2$; eight thousand nine hundred fifty-two   **2.** identity   **3.** $4 \cdot 3 = 3 \cdot 4$; Answers will vary.   **4.** 1000
5. 140,000   **6.** 12,009   **7.** 1735   **8.** 4,057,750   **9.** 488   **10.** 1229   **11.** 5707   **12.** 2584   **13.** 220,405   **14.** 210,938   **15.** 403
16. 172 R388   **17.** 2005   **18.** 74   **19.** 54   **20. a.** $306   **b.** $51   **c.** $224   **21.** $2025   **22.** 80 miles   **23. a.** 152 in.   **b.** 1428 square in.
24. base; exponent; 343   **25. a.** $7, 11, 13, 17, 19$ **b.** $49, 121, 169, 289, 361$ **26.** 13   **27.** 48   **28.** 0   **29.** $2, 3, 5, 6, 9, 10$ **30.** $2, 3, 4, 6, 9$
31. $2, 4, 5, 10$ **32.** Yes, 42 divides the product; 30 times.   **33.** $2^2 \cdot 31$ **34.** $3 \cdot 5 \cdot 11$ **35.** $2^2 \cdot 37$

Chapter 2: Fractions and Mixed Numbers

Section 2.1: Introduction to Fractions and Mixed Numbers

Margin Exercises   1. a. $\dfrac{1}{6}$ **b.** $\dfrac{2}{5}$ **2.** $\dfrac{3}{8}$ **3.** $\dfrac{3}{2}$ (3 of 2 equal parts)   **4. a.** 0   **b.** undefined   **5.** $\dfrac{15}{32}$ **6. a.** $\dfrac{28}{5}$ **b.** 0   **c.** $\dfrac{9}{35}$
7. 8   **8.** 15   **9.** 36   **10.** $\dfrac{1}{160}$ **11. a.** $\dfrac{2}{9}$ **b.** $\dfrac{2}{3}$ **c.** $\dfrac{7}{4}$ **12.** $\dfrac{9}{5}$ **13.** $\dfrac{8}{11}$ **14.** $\dfrac{94}{9}$ **15.** $\dfrac{101}{32}$ **16.** $1\dfrac{9}{11}$

Exercises   1. $\dfrac{1}{4}$ **3.** $\dfrac{1}{4}$ **5.** 0   **7.** undefined   **9.** $\dfrac{3}{32}$ **11.** $\dfrac{9}{49}$ **13.** $\dfrac{15}{32}$ **15.** 0   **17.** $\dfrac{35}{12}$ **19.** $\dfrac{10}{1}$ or 10   **21.** $\dfrac{45}{2}$
23. $\dfrac{32}{15}$ **25.** $\dfrac{99}{20}$ **27.** $\dfrac{48}{455}$ **29.** $\dfrac{1}{10}$ **31.** $\dfrac{1}{8}$ **33.** $\dfrac{12}{35}$ **35.** 15   **37.** 10   **39.** 54   **41.** 40   **43.** 44   **45.** 32   **47.** 90   **49.** 49
51. $\dfrac{1}{3}$ **53.** $\dfrac{3}{4}$ **55.** $\dfrac{2}{5}$ **57.** $\dfrac{7}{18}$ **59.** 0   **61.** $\dfrac{2}{5}$ **63.** $\dfrac{5}{6}$ **65.** $\dfrac{2}{3}$ **67.** $\dfrac{2}{3}$ **69.** $\dfrac{6}{25}$ **71.** $\dfrac{2}{3}$ **73.** $\dfrac{12}{35}$ **75.** 4   **77.** $\dfrac{5}{4}$ **79.** $\dfrac{9}{8}$
81. $\dfrac{15}{11}$ **83.** $\dfrac{29}{8}$ **85.** $\dfrac{28}{5}$ **87.** $\dfrac{46}{3}$ **89.** $\dfrac{25}{2}$ **91.** $2\dfrac{1}{2}$ **93.** $2\dfrac{1}{8}$ **95.** $2\dfrac{1}{15}$ **97.** $3\dfrac{4}{9}$ **99.** $\dfrac{1}{12}$ **101.** $\dfrac{29}{35}$ **103.** $\dfrac{1}{8}$ **105.** $\dfrac{9}{20}$

Section 2.2: Multiplication and Division with Fractions and Mixed Numbers

Margin Exercises   1. $\dfrac{49}{6}$ or $8\dfrac{1}{6}$ **2.** $\dfrac{17}{10}$ or $1\dfrac{7}{10}$ **3.** $\dfrac{7}{18}$ **4.** $\dfrac{4}{3}$ or $1\dfrac{1}{3}$ **5.** 2   **6.** $\dfrac{77}{4}$ or $19\dfrac{1}{4}$ **7.** $\dfrac{1}{3}$ **8.** $\dfrac{16}{9}$ **9.** .375   **10.** $\dfrac{8}{7}$
11. $\dfrac{1}{16}$ **12.** $\dfrac{8}{9}$ **13.** $\dfrac{21}{40}$ **14.** $\dfrac{25}{16}$ or $1\dfrac{9}{16}$ **15.** $\dfrac{20}{63}$ **16. a.** yes   **b.** less than 3000   **c.** 4500 pounds

Exercises   1. $2\dfrac{34}{35}$ **3.** $12\dfrac{4}{9}$ **5.** 35   **7.** $\dfrac{8}{9}$ **9.** $\dfrac{5}{7}$ **11.** $\dfrac{1}{3}$ **13.** 1   **15.** $\dfrac{1}{6}$ **17.** $\dfrac{1}{12}$ **19.** $\dfrac{2}{25}$ **21.** $\dfrac{10}{3}$ or $3\dfrac{1}{3}$ **23.** $\dfrac{4}{15}$

25. $\dfrac{15}{23}$ **27.** $\dfrac{20}{9}$ or $2\dfrac{2}{9}$ **29.** $\dfrac{105}{4}$ or $26\dfrac{1}{4}$ **31.** 126 **33.** $\dfrac{18}{5}$ or $3\dfrac{3}{5}$ **35.** $\dfrac{7}{2}$ or $3\dfrac{1}{2}$ **37.** $\dfrac{8}{9}$ **39.** $\dfrac{5}{7}$ **41.** $\dfrac{7}{5}$ or $1\dfrac{2}{5}$ **43.** $\dfrac{1}{3}$

45. $\dfrac{3}{4}$ **47.** $\dfrac{4}{9}$ **49.** $\dfrac{5}{8}$ **51.** $\dfrac{39}{32}$ or $1\dfrac{7}{32}$ **53.** $\dfrac{10}{9}$ or $1\dfrac{1}{9}$ **55.** $\dfrac{32}{21}$ or $1\dfrac{11}{21}$ **57.** $\dfrac{10}{3}$ or $3\dfrac{1}{3}$ **59.** $\dfrac{16}{21}$ **61.** $\dfrac{16}{105}$ **63.** $\dfrac{63}{50}$ or $1\dfrac{13}{50}$

65. $\dfrac{7}{5}$ or $1\dfrac{2}{5}$ **67.** $\dfrac{5}{7}$ **69.** $\dfrac{8}{3}$ or $2\dfrac{2}{3}$ **71.** $\dfrac{3}{8}$ **73.** 448 **75.** $\dfrac{3}{8}$ **77. a.** $\dfrac{3}{8}$ **b.** $\dfrac{1}{4}$ **79. a.** $\dfrac{5}{16}$ **b.** $\dfrac{5}{4}$ **81.** $\dfrac{50}{27}$ or $1\dfrac{23}{27}$

83. a. $\dfrac{2}{5}$ **b.** $\dfrac{12}{25}$ **85.** $\dfrac{108}{5}$ or $21\dfrac{3}{5}$ **87.** 90 games **89.** $\dfrac{1}{5}; \dfrac{4}{5}$ **91.** 5 feet **93.** 1250 freshmen **95. a.** more **b.** less

c. 200 passengers **97.** 63 miles **99. a.** for the court **b.** pass by 5 votes **101. a.** more **b.** less **c.** 8000 steel rods per week

103. 975 students **105.** $15 \div 3 = 5; 15 \div \dfrac{1}{3} = 45; 5 \neq 45$

Section 2.3: Least Common Multiple (LCM)

Margin Exercises **1.** 300 **2.** 126 **3.** 120 **4.** 180 **5.** 441 **6.** 210; 14, 6, 5 **7.** 450; 45, 25, 6 **8. a.** 30 minutes
b. 5, 10, and 6, respectively

Exercises **1.** 5, 10, 15, 20, 25, 30 **3.** 10, 20, 30, 40, 50, 60 **5.** 25, 50, 75, 100, 125, 150 **7.** 24 **9.** 36 **11.** 110 **13.** 60 **15.** 200
17. 196 **19.** 240 **21.** 100 **23.** 700 **25.** 252 **27.** 8 **29.** 60 **31.** 240 **33.** 2250 **35.** 726 **37.** 2610 **39.** 675 **41.** 1560
43. 120 **45.** 120; 15, 12, 8 **47.** 120; 12, 8, 5 **49.** 270; 45, 15, 10, 6 **51.** 4410; 98, 70, 45 **53.** 14,157; 143, 99, 39
55. a. 60 minutes **b.** 4, 3, and 2 trips, respectively **57. a.** every 24 days **b.** every 24 days **59.** every 180 days
61. a. 840 seconds **b.** 24 laps, 21 laps, and 20 laps, respectively **63. a.** 6 hours **b.** 7 PM **65. a.** 40 min **b.** 20 min **c.** 40 min
67. 60 years **69.** number of waffles: 24 waffles; number of boxes: 4 boxes **71.** Answers will vary. Since the LCM is constructed using the prime factors of each number in the set, by definition, each number will divide the LCM. **73.** Answers will vary.

Section 2.4: Addition and Subtraction with Fractions

Margin Exercises **1.** $\dfrac{4}{7}$ **2.** $\dfrac{8}{9}$ **3.** $\dfrac{7}{15}$ **4.** $\dfrac{5}{12}$ **5.** $\dfrac{59}{40}$ lbs **6.** $13,000 **7.** $\dfrac{1}{5}$ **8.** $\dfrac{1}{6}$ **9.** $\dfrac{2}{55}$ **10.** $\dfrac{9}{20}$ **11.** $\dfrac{11}{75}$

12. The Cavaliers lost $\dfrac{1}{15}$ of their games by exactly 5 runs.

Exercises **1.** $\dfrac{5}{14}$ **3.** 2 **5.** $\dfrac{13}{18}$ **7.** $\dfrac{11}{8}$ **9.** $\dfrac{23}{56}$ **11.** $\dfrac{3}{2}$ **13.** 1 **15.** $\dfrac{17}{21}$ **17.** $\dfrac{23}{54}$ **19.** $\dfrac{173}{84}$ **21.** $\dfrac{5}{8}$ **23.** $\dfrac{22}{27}$

25. $\dfrac{317}{1000}$ **27.** $\dfrac{271}{10,000}$ **29.** $\dfrac{753}{1000}$ **31.** $\dfrac{613}{1000}$ **33.** $\dfrac{5}{3}$ **35.** $\dfrac{19}{100}$ **37.** $\dfrac{73}{96}$ **39.** $\dfrac{3}{5}$ **41.** $\dfrac{1}{3}$ **43.** $\dfrac{1}{2}$ **45.** $\dfrac{1}{12}$ **47.** $\dfrac{13}{20}$

49. $\dfrac{1}{40}$ **51.** $\dfrac{27}{8}$ **53.** $\dfrac{3}{50}$ **55.** $\dfrac{3}{20}$ **57.** $\dfrac{1}{50}$ **59.** $\dfrac{23}{20}$ inches **61.** $\dfrac{49}{80}$ inch **63. a.** $\dfrac{5}{18}$ **b.** $6000 **65.** $\dfrac{7}{24}$ inch **67.** $\dfrac{7}{30}$

69. $\dfrac{7}{8}$ teaspoon **71. a.** $\dfrac{477}{1000}$ **b.** $\dfrac{523}{1000}$ **73.** $\dfrac{29}{40}$ inch **75. a.** $\dfrac{19}{60}$ **b.** $950 **c.** $11,400

Section 2.5: Addition and Subtraction with Mixed Numbers

Margin Exercises **1.** $11\dfrac{4}{5}$ **2.** $8\dfrac{7}{10}$ **3.** $19\dfrac{3}{35}$ **4.** 17 meters **5.** $4\dfrac{1}{5}$ **6.** $7\dfrac{3}{14}$ **7.** $6\dfrac{4}{8} = 6\dfrac{1}{2}$ **8.** $3\dfrac{7}{8}$ **9.** $7\dfrac{3}{5}$

Exercises **1.** $14\dfrac{5}{7}$ **3.** 8 **5.** $12\dfrac{3}{4}$ **7.** $8\dfrac{2}{3}$ **9.** $34\dfrac{7}{18}$ **11.** $8\dfrac{13}{15}$ **13.** $18\dfrac{7}{8}$ **15.** $9\dfrac{7}{20}$ **17.** $1\dfrac{5}{8}$ **19.** 4 **21.** $4\dfrac{3}{5}$ **23.** $5\dfrac{1}{8}$

25. $7\dfrac{1}{16}$ **27.** $6\dfrac{1}{6}$ **29.** $3\dfrac{7}{8}$ **31.** $4\dfrac{1}{2}$ **33.** $10\dfrac{3}{8}$ **35.** $8\dfrac{7}{12}$ hours **37.** $89\dfrac{1}{8}$ ft **39.** $36\dfrac{3}{4}$ in. **41.** $1\dfrac{3}{20}$ hours

43. $6\dfrac{3}{5}$ minutes **45.** $3\dfrac{1}{4}$ pounds **47.** $4\dfrac{13}{20}$ parts

Section 2.6: Order of Operations with Fractions and Mixed Numbers

Margin Exercises **1.** $\dfrac{9}{22}$ is larger by $\dfrac{7}{66}$. **2.** $\dfrac{19}{24}$ is larger by $\dfrac{1}{72}$. **3.** $\dfrac{5}{9}, \dfrac{7}{12}, \dfrac{2}{3}; \dfrac{1}{9}$ **4.** $\dfrac{17}{45}$ **5.** $\dfrac{31}{108}$ **6.** $\dfrac{7}{12}$

7. $\dfrac{35}{12}$ or $2\dfrac{11}{12}$ **8.** $\dfrac{15}{8}$ or $1\dfrac{7}{8}$ **9.** 2 **10.** $\dfrac{35}{16}$ or $2\dfrac{3}{16}$ **11.** $\dfrac{83}{10}$ or $8\dfrac{3}{10}$ **12.** $\dfrac{61}{18}$ or $3\dfrac{7}{18}$

Exercises **1.** $\dfrac{3}{4}$ by $\dfrac{1}{12}$ **3.** $\dfrac{17}{20}$ by $\dfrac{1}{20}$ **5.** $\dfrac{13}{20}$ by $\dfrac{1}{40}$ **7.** equal **9.** $\dfrac{11}{48}$ by $\dfrac{1}{60}$ **11.** $\dfrac{2}{7}, \dfrac{1}{3}, \dfrac{3}{8}; \dfrac{5}{56}$ **13.** $\dfrac{11}{12}, \dfrac{19}{20}, \dfrac{7}{6}; \dfrac{1}{4}$

15. $\dfrac{1}{4}, \dfrac{1}{3}, \dfrac{1}{2}; \dfrac{1}{4}$ **17.** $\dfrac{13}{18}, \dfrac{7}{9}, \dfrac{31}{36}; \dfrac{5}{36}$ **19.** $\dfrac{20}{10,000}, \dfrac{3}{1000}, \dfrac{1}{100}; \dfrac{1}{125}$ **21.** $\dfrac{2}{3}$ **23.** $\dfrac{187}{32}$ or $5\dfrac{27}{32}$ **25.** $\dfrac{5}{54}$ **27.** $\dfrac{15}{64}$

29. $\dfrac{29}{36}$ **31.** $\dfrac{11}{70}$ **33.** $\dfrac{33}{4}$ or $8\dfrac{1}{4}$ **35.** $\dfrac{33}{2}$ or $16\dfrac{1}{2}$ **37.** $\dfrac{3}{28}$ **39.** $\dfrac{3}{2}$ or $1\dfrac{1}{2}$ **41.** $\dfrac{11}{8}$ or $1\dfrac{3}{8}$ **43.** $\dfrac{15}{4}$ or $3\dfrac{3}{4}$ **45.** $\dfrac{21}{22}$

47. 20 **49.** 4 **51.** 1 **53.** $\dfrac{1}{5}$ **55.** $\dfrac{23}{30}$ **57.** $\dfrac{172}{9}$ or $19\dfrac{1}{9}$ **59.** $\dfrac{3}{5}$ **61.** $\dfrac{543}{40}$ or $13\dfrac{23}{40}$ **63.** $\dfrac{47}{40}$ or $1\dfrac{7}{40}$ **65.** $\dfrac{223}{32}$ or $6\dfrac{31}{32}$

67. 43 inches **69.** $2\dfrac{1}{48}$ feet **71.** 375 **73.** 17 lb **75. a.** $\dfrac{1}{3}$ kg **b.** $6 **77. a.** Yes; if both fractions are greater than one half, the sum will be greater than one. Answers will vary. **b.** No; multiplying a number by a fraction between 0 and 1 results in a product that is less than the original number. Answers will vary. **79. a.** Answers will vary. **b.** Answers will vary.

Chapter 2: Review

1. 0 **2.** undefined **3.** 2 **4.** 54 **5.** 75 **6.** $\dfrac{1}{2}$ **7.** $\dfrac{9}{8}$ **8.** $\dfrac{5}{4}$ **9.** $\dfrac{51}{10}$ **10.** $\dfrac{35}{12}$ **11.** $\dfrac{67}{5}$ **12.** $6\dfrac{5}{8}$ **13.** 7 **14.** $3\dfrac{21}{50}$ **15.** $\dfrac{3}{49}$

16. $\dfrac{91}{9}$ or $10\dfrac{1}{9}$ **17.** $\dfrac{36}{25}$ or $1\dfrac{11}{25}$ **18.** $\dfrac{51}{20}$ or $2\dfrac{11}{20}$ **19.** $\dfrac{1}{30}$ **20.** $\dfrac{1}{12}$ **21.** $\dfrac{56}{3}$ or $18\dfrac{2}{3}$ **22.** 104 **23.** 1 **24.** $\dfrac{3}{5}$ **25.** $\dfrac{9}{4}$ or $2\dfrac{1}{4}$

26. $\dfrac{5}{4}$ or $1\dfrac{1}{4}$ **27.** 11, 22, 33, 44, 55, 66 **28.** 84; 21, 6, 4 **29.** 60; 10, 4, 1 **30.** 840; 105, 84, 56, 30 **31.** 1155; 385, 231, 165, 105

32. 81; 27, 9, 3, 1 **33. a.** every 36 days **b.** every 180 days **34.** $\dfrac{5}{7}$ **35.** $\dfrac{2}{3}$ **36.** $\dfrac{1}{4}$ **37.** $\dfrac{49}{72}$ **38.** $\dfrac{7}{22}$ **39.** $\dfrac{25}{54}$ **40.** $\dfrac{7}{20}$ **41.** $\dfrac{1}{3}$

42. $\dfrac{7}{8}$ **43.** $\dfrac{1}{2}$ pound **44.** 14 students **45.** $27\dfrac{2}{3}$ **46.** 9 **47.** $2\dfrac{1}{2}$ **48.** $12\dfrac{7}{8}$ **49.** $3\dfrac{3}{4}$ **50.** $6\dfrac{7}{12}$ **51.** $19\dfrac{1}{12}$ **52.** $120\dfrac{17}{18}$

53. $7\dfrac{5}{6}$ **54.** $\dfrac{4}{5}$ by $\dfrac{2}{15}$ **55.** $\dfrac{11}{20}, \dfrac{5}{9}, \dfrac{7}{12}, \dfrac{1}{30}$ **56.** $\dfrac{25}{112}$ **57.** $\dfrac{7}{5}$ or $1\dfrac{2}{5}$ **58.** $\dfrac{19}{30}$ **59.** $\dfrac{7}{18}$ **60.** $\dfrac{35}{16}$ or $2\dfrac{3}{16}$ **61.** $\dfrac{25}{24}$ or $1\dfrac{1}{24}$

62. $\dfrac{1}{10}$ **63.** $\dfrac{90}{7}$ or $12\dfrac{6}{7}$ **64.** $\dfrac{49}{15}$ or $3\dfrac{4}{15}$ **65.** $\dfrac{33}{4}$ or $8\dfrac{1}{4}$ **66.** $\dfrac{11}{70}$ **67.** $\dfrac{7}{4}$ or $1\dfrac{3}{4}$

Chapter 2: Test

1. $\dfrac{5}{6}$ **2.** $\dfrac{3}{4}$ **3.** $\dfrac{31}{10}$ **4.** $11\dfrac{2}{3}$ **5.** $\dfrac{2}{9}$ **6.** 25 **7.** $\dfrac{9}{80}$ **8.** 20 **9.** $\dfrac{15}{4}$ or $3\dfrac{3}{4}$ **10.** $\dfrac{1}{3}$ **11.** $\dfrac{3}{2}$ or $1\dfrac{1}{2}$ **12.** 700; 35, 28, 20 **13.** $\dfrac{1}{2}$

14. 2 **15.** 10 **16.** $1\dfrac{2}{5}$ **17.** $\dfrac{141}{20}$ or $7\dfrac{1}{20}$ **18.** $\dfrac{80}{27}$ or $2\dfrac{26}{27}$ **19.** $\dfrac{1}{16}$ **20.** $\dfrac{1}{2}$ **21.** $\dfrac{495}{32}$ or $15\dfrac{15}{32}$ **22.** $\dfrac{3}{8}$ **23.** $\dfrac{2}{3}, \dfrac{3}{4}, \dfrac{7}{8}, \dfrac{5}{24}$

24. a. $\dfrac{115}{4}$ or $28\dfrac{3}{4}$ miles **b.** $\dfrac{115}{12}$ or $9\dfrac{7}{12}$ miles **25.** 39 in.

Cumulative Review: Chapters 1 – 2

1. 50,000 + 3000 + 400 + 60 + 0; fifty-three thousand, four hundred sixty **2.** 270,000 **3.** commutative property of addition
4. 5,600,000 **5.** 158 **6.** 2, 3, 5, 7, 11, 13, 17, 19, 23, 29 **7. a.** $170 = 2 \cdot 5 \cdot 17$ **b.** $305 = 5 \cdot 61$ **8.** 0, undefined **9.** 43 **10.** 13,250

11. 2527 **12.** 49,776 **13.** 203 **14.** 182 **15.** $\dfrac{7}{48}$ **16.** $\dfrac{1}{20}$ **17.** $\dfrac{7}{24}$ **18.** $\dfrac{28}{15}$ or $1\dfrac{13}{15}$ **19.** $\dfrac{183}{35}$ or $5\dfrac{8}{35}$ **20.** $\dfrac{7}{4}$ or $1\dfrac{3}{4}$

21. $\dfrac{315}{2}$ or $157\dfrac{1}{2}$ **22.** 3 **23.** $\dfrac{61}{48}$ or $1\dfrac{13}{48}$ **24.** $\dfrac{123}{10}$ or $12\dfrac{3}{10}$ **25.** 8190 is divisible by 2, 3, 5, 6, 9, and 10. Reasons will vary. Use divisibility rules. **26.** 431 is prime **27. a.** 630; 35, 15, 7 **b.** 3360; 112, 56, 40, 35 **28.** $3680 **29.** 775 **30.** 17 in. **31.** 310 yards

32. $\dfrac{17}{36}$ meter **33.** 11 servings **34.** $\dfrac{7}{2}$ or $3\dfrac{1}{2}$ pounds **35. a.** $\dfrac{37}{12}$ or $3\dfrac{1}{2}$ gallons **b.** $\dfrac{29}{12}$ or $2\dfrac{5}{12}$ gallons

Chapter 3: Decimal Numbers

Section 3.1: Introduction to Decimal Numbers

Margin Exercises   **1.** 19.3; nineteen and three tenths   **2.** 39.0184; thirty-nine and one hundred eighty-four ten-thousandths
3. 1200.0035   **4.** 0.1235   **5.** 6.44   **6.** 9.251   **7.** 2.01, 2.1, 2.11   **8.** 8.6   **9.** 5.04   **10.** 0.018   **11.** 240
Exercises   **1.** 6.5   **3.** 19.075   **5.** 62.547   **7.** 13.02   **9.** 200.6   **11.** nine tenths   **13.** six and five hundredths
15. fifty and seven thousandths   **17.** eight hundred and nine thousandths   **19.** five thousand and five thousandths   **21.** 0.4
23. 0.23   **25.** 5.028   **27.** 600.66   **29.** 3495.342   **31.** 0.27   **33.** 0.01   **35.** 0.0499   **37.** 24.645   **39.** 0.003, 0.03, 0.33   **41.** 1.762,
2.3644, 2.51   **43.** 9.52, 9.523, 10.113, 11.4   **45. a.** 7 **b.** 8 **c.** 8; 7; 8; 8 **d.** 34.8   **47.** 89.0   **49.** 18.1   **51.** 20.0   **53.** 0.39   **55.** 8.00
57. 0.08   **59.** 0.057   **61.** 0.002   **63.** 32.458   **65.** 479   **67.** 164   **69.** 301   **71.** 16,000   **73.** 76,500   **75.** 500   **77.** 62,000
79. 104,000   **81.** 7,305,000   **83.** one hundred one and seventy-five hundredths   **85.** three and fourteen thousand, one hundred
fifty-nine hundred-thousandths   **87.** two and eight hundred twenty-five ten-thousandths   **89.** thirty-five and eight tenths;
twenty-six and nine tenths; eighteen and nine tenths; twelve and three tenths; seven and two tenths   **91.** nine hundred fourteen
thousandths; one and nine hundredths; thirty-nine and thirty-seven hundredths; three and thirty-seven hundredths
93. Answers will vary.

Section 3.2: Addition and Subtraction with Decimal Numbers

Margin Exercises   **1.** 28.96   **2.** 42.0667   **3.** 62.673   **4.** $156.58   **5.** 6.192   **6.** 141.467   **7.** 127 (estimate); 132.12 (sum)
Exercises   **1.** 2.3   **3.** 7.55   **5.** 72.31   **7.** 276.096   **9.** 44.6516   **11.** 118.333   **13.** 7.148   **15.** 93.877   **17.** 103.429   **19.** 137.150
21. 1.44   **23.** 15.89   **25.** 64.947   **27.** 4.7974   **29.** 2.9434   **31.** 13.51   **33.** 45.148   **35.** 37.147 **37.** 120 (estimate); 119.22 (sum)
39. 96 (estimate); 92.46 (sum)   **41.** 188 (estimate); 207.45 (sum)   **43.** 20 (estimate); 26.08 (difference)
45. 10 (estimate); 9.09 (difference)   **47.** 2.4 (estimate); 2.621 (difference)   **49.** 61.2 cm   **51.** 10 cm   **53.** 18.84 m   **55.** $4.50
57. 12.28 in.   **59. a.** 6.97 feet   **b.** 2.73 feet   **61. a.** 17 hours   **b.** 17.667 hours   **63. a.** 15 feet   **b.** 15.03 feet   **65.** 384.873 degrees
67. $6089.90   **69.** 12.1778 tons   **71. a.** $142.76   **b.** $185.10   **73. a.** 758.99909 sec **b.** 761.00073 sec   **75.** $9041.12
77. a. Answers will vary.   **b.** Answers will vary.

Section 3.3: Multiplication with Decimal Numbers

Margin Exercises   **1.** 0.16   **2.** 0.224   **3.** 11.38081   **4.** 3.306   **5. a.** 1.3 **b.** 8780   **6.** 4 (estimate); 5.022715 (product)   **7.** $600
Exercises   **1. i.** B **ii.** C **iii.** E **iv.** A **v.** D   **3.** 0.42   **5.** 0.04   **7.** 21.6   **9.** 0.42   **11.** 0.004   **13.** 0.108   **15.** 0.0276   **17.** 0.0486
19. 0.0006   **21.** 0.375   **23.** 1.4000   **25.** 1.725   **27.** 5.063   **29.** 0.080   **31.** 346   **33.** 782   **35.** 1610   **37.** 4.35   **39.** 18.6   **41.** 380
43. 50   **45.** 74,000   **47.** 0.5 (estimate); 0.535 (product)   **49.** 0.0009 (estimate); 0.0009222 (product)
51. 8 (estimate); 9.224655 (product)   **53.** 5.4 (estimate); 4.9077 (product)   **55.** 60 (estimate); 54.5034 (product)
57. 2.4 (estimate); 2.32986 (product)   **59. a.** $5000   **b.** $5959.80   **61.** $418.75   **63.** 26.6 cm (perimeter); 29.40 sq cm (area)
65. 18.8 mm (perimeter); 22.09 sq mm (area)   **67.** $61,625   **69.** $621.40   **71. a.** $250 **b.** $294.30
73. a. $900 **b.** $891.50 **c.** $10,698.00   **75.** $932.60   **77.** 2.8926 inches

Section 3.4: Division with Decimal Numbers

Margin Exercises   **1.** 1872   **2.** 5.2   **3.** 14.9   **4.** 1.55   **5. a.** 1.6 **b.** 0.08346 **c.** 0.732   **6.** 7.28   **7.** 8.65   **8. a.** 5 **b.** 5.5   **9.** 10 meters
per second   **10.** 2222 songs   **11.** $2.25
Exercises   **1. i.** B **ii.** E **iii.** D **iv.** C **v.** A   **3.** 2.34   **5.** 0.99   **7.** 0.08   **9.** 2056   **11.** 20   **13.** 56.9   **15.** 0.7   **17.** 0.1   **19.** 21.0
21. 0.01   **23.** 5.70   **25.** 2.74   **27.** 5.04   **29.** 0.784   **31.** 0.5036   **33.** 0.07385   **35.** 16.7   **37.** 0.785   **39.** 0.01699   **41.** 97.07
43. 15.99   **45.** 26.88   **47.** 16.11   **49.** 3.5 (estimate); 3.087 (quotient)   **51.** 0.22 (estimate); 0.285 (quotient)
53. 0.005 (estimate); 0.007 (quotient)   **55. a.** 400 miles   **b.** 442.8 miles   **57. a.** about $1 per pound   **b.** $1.25 per pound
59. $239.56   **61.** $295   **63.** 5466 at bats   **65.** 225 free throws   **67.** 42.2 kilometers   **69.** $39,811.74   **71.** 67.92
65. Answers will vary.   **67. a.** 25.20 mph **b.** 25.17 mph

Section 3.5: Decimal Numbers and Fractions

Margin Exercises **1.** $\frac{12}{25}$ **2.** $\frac{19}{20}$ **3.** $\frac{381}{500}$ **4.** $\frac{3}{100}$ **5.** $\frac{79}{5} = 15\frac{4}{5}$ **6.** 0.4 **7.** 0.65 **8.** 0.13333... **9.** 0.230769230769...
10. 17.03 **11.** $\frac{9}{16}$ is larger than 0.52. The difference is 0.0425.

Exercises **1.** $\frac{9}{10}$ **3.** $\frac{5}{10}$ **5.** $\frac{62}{100}$ **7.** $\frac{57}{100}$ **9.** $\frac{526}{1000}$ **11.** $\frac{16}{1000}$ **13.** $\frac{51}{10}$ **15.** $\frac{815}{100}$ **17.** $\frac{1}{8}$ **19.** $\frac{9}{50}$ **21.** $\frac{9}{40}$ **23.** $\frac{17}{100}$

25. $\frac{16}{5}$ or $3\frac{1}{5}$ **27.** $\frac{25}{4}$ or $6\frac{1}{4}$ **29.** $0.\overline{6}$ **31.** $0.\overline{63}$ **33.** 0.6875 **35.** $0.\overline{428571}$ **37.** $0.1\overline{6}$ **39.** $0.\overline{5}$ **41.** 0.292 **43.** 0.417

45. 0.031 **47.** 1.231 **49.** 1.429 **51.** 0.7 **53.** 1.635 **55.** 14.98 **57.** 1.125 **59.** 13.51 **61.** 0.089 **63.** 11.083 **65.** 2.638 **67.** 27.3

69. 2.3 is larger; 0.05 **71.** 0.28 is larger; $0.00\overline{72}$ **73.** 3.3 is larger; $0.15\overline{71428}$ **75.** $3\frac{2}{3}$ is larger; $0.1\overline{6}$ **77.** $\frac{7}{10}, \frac{3}{4}, 0.76$

79. $\frac{5}{16}, 0.3126, 0.314$ **81.** $83\frac{4}{5}$ **83.** $26\frac{3}{10}$; $24\frac{1}{10}$ **85.** $21\frac{1}{2}$ **87.** 0.14 **89.** 6.86 **91.** 3.3 **93.** 63.869 **95.** 2.25 **97.** 0.6667

99. 0.88 **101.** 44.4063

Chapter 3: Review

1. 62.09 **2.** 15.357 **3.** $81\frac{47}{100}$ **4.** $200\frac{1}{2}$ **5.** seven and eight hundredths

6. twelve and one hundred thirty-seven thousandths **7.** 84.075 **8.** 3000.003 **9.** 0.081 **10.** 2.358 **11.** 0.03, 0.0303, 0.033
12. 2.41, 2.4112, 2.412, 2.42 **13.** 5900 **14.** 7.6 **15.** 0.039 **16.** 2.06990 **17.** 26.82 **18.** 104.272 **19.** 124.43 **20.** 212.78
21. 93.418 **22.** 22.708 **23.** 9.02 **24.** 3.9623 **25.** 27.84 **26.** 139.35 **27.** 55.3 in. **28.** 0.72 **29.** 0.0064 **30.** 1.728
31. 171.55 **32.** 0.53219 **33.** 1.2772 **34.** 235 **35.** 1.7632 **36.** 5964.1 **37.** 1195.56 ft^2 **38.** 7.34 **39.** 0.91 **40.** 32.1 **41.** 10

42. 7.08 **43.** 33.85 **44.** 880.53 **45.** 19.07 **46.** 1.96 **47.** 0.389 **48.** 0.00567 **49.** $10.02 **50.** $\frac{7}{100}$ **51.** $2\frac{1}{40}$ **52.** $\frac{3}{200}$

53. $0.\overline{3}$ **54.** 0.625 **55.** $2.\overline{4}$ **56.** 0.882 **57.** 0.980 **58.** $568,650 **59.** $900 **60.** 39.89 **61.** 1.125 **62.** 15.75

Chapter 3: Test

1. a. thirty and six hundred fifty-seven thousandths **b.** $30\frac{657}{1000}$ **2.** 2.032 **3.** 0.3125 **4.** 0.619, 0.626, 0.72 **5.** $\frac{1}{5}, \frac{1}{4}, 0.275$

6. a. 216.7 **b.** 216.70 **c.** 216.705 **d.** 220 **7. a.** 73.0 **b.** 73.01 **c.** 73.015 **d.** 70 **8.** 103.758 **9.** 0.294 **10.** 1.888 **11.** 18.145
12. 17.83 **13.** 122.11 **14.** 12.80335 **15.** 1920 **16.** 0.03614 **17.** 37.313 **18.** 0.90395 **19.** 1.425 **20. a.** 90 **b.** 66.28

21. 392 miles **22.** $5\frac{17}{25}$ grams **23. a.** 11.79 hours **b.** 19.75 hours **24.** 184 candy bars

Cumulative Review: Chapters 1 – 3

1. a. five million, six hundred twelve thousand, nine **b.** five thousand, six hundred twelve and nine thousandths
2. a. commutative property of addition **b.** associative property of multiplication **c.** additive identity property **3. a.** 4591.06

b. 4591.1 **c.** 4591 **d.** 4600 **4.** $\frac{64}{5}$ **5.** $4\frac{3}{8}$ **6. a.** 0.875 **b.** $0.\overline{27}$ **7.** $2^2 \cdot 3 \cdot 5 \cdot 7$ **8. a.** 140 **b.** 64 **9.** 34 **10.** 1444 **11.** 25.5

12. 1215 **13.** $\frac{1}{30}$ **14.** $8\frac{21}{40}$ **15.** $55\frac{29}{40}$ **16.** $32\frac{7}{20}$ **17.** $\frac{7}{22}$ **18.** $\frac{32}{9}$ or $3\frac{5}{9}$ **19.** 30.385 **20.** 37.0001 **21.** 2.8952 **22.** 37.77

23. 1.5686 **24.** 2899 **25.** 35.8 **26.** 0.2134 **27.** 13 **28.** 28 **29.** 45 **30.** $\frac{11}{30}$ **31.** 23.4 **32.** 0.51 or $\frac{51}{100}$ **33.** 3841 **34.** 72

35. 18.9 **36.** 2.185 **37. a.** $200,000 **b.** $207,700 **38.** Pay cash because he can save $5760. **39.** $11.78 **40.** $300.64

41. $2\frac{9}{20}$ hr **42.** 200 meters

Chapter 4: Ratios and Proportions, Percent and Applications

Section 4.1: Ratios and Proportions

Margin Exercises **1.** 4 apples : 5 oranges or 4 apples to 5 oranges or $\dfrac{4 \text{ apples}}{5 \text{ oranges}}$ **2.** $\dfrac{500 \text{ washers}}{4000 \text{ bolts}} = \dfrac{1 \text{ washer}}{8 \text{ bolts}}$

3. a. $\dfrac{3 \text{ quarters}}{1 \text{ dollar}}$ **b.** $\dfrac{3}{4}$ **4.** true **5.** false

Exercises **1.** $\dfrac{1}{2}$ **3.** $\dfrac{4}{1}$ or 4 **5.** $\dfrac{50 \text{ miles}}{1 \text{ hour}}$ **7.** $\dfrac{25 \text{ miles}}{1 \text{ gallon}}$ **9.** $\dfrac{6 \text{ chairs}}{5 \text{ people}}$ **11.** $\dfrac{3}{4}$ **13.** $\dfrac{8}{7}$ **15.** $\dfrac{\$2 \text{ profit}}{\$5 \text{ invested}}$ **17.** $\dfrac{1}{1}$ or 1

19. $\dfrac{1 \text{ hit}}{4 \text{ times at bat}}$ **21.** $\dfrac{7}{25}$ **23.** $\dfrac{9}{41}$ **25.** $\dfrac{31 \text{ men}}{43 \text{ women}}$ **27.** $\dfrac{4}{5}$ **29.** $\dfrac{91 \text{ applicants}}{2 \text{ openings}}$ **31.** $\dfrac{25 \text{ miles}}{1 \text{ gallon}}$ **33.** $\dfrac{3}{2}$ **35.** true

37. true **39.** false **41.** true **43.** true **45.** true **47.** true **49.** true **51.** false **53.** false **55.** true **57.** true **59.** true **61.** false

63. true **65.** false **67.** true **69.** false **71.** equivalent **73.** yes

75. Multiplying both sides by 35 yields: $\dfrac{16}{\cancel{5}} \cdot \cancel{35}^{7} = \dfrac{22.4}{\cancel{7}} \cdot \cancel{35}^{5}$ Cross multiplying yields: $\dfrac{16}{5} = \dfrac{22.4}{7}$

$16 \cdot 7 = 22.4 \cdot 5$ $\qquad\qquad$ $16 \cdot 7 = 5 \cdot 22.4$

In general, this same technique will work for all proportions because if $\dfrac{a}{b} = \dfrac{c}{d}$ and we multiply both sides of the equation by bd then: $\dfrac{a}{\cancel{b}} \cdot \cancel{b}d = \dfrac{c}{\cancel{d}} \cdot b\cancel{d} \rightarrow ad = cb$ or $ad = bc$, which is the same as the equation found by cross multiplication.

Section 4.2: Solving Proportions

Margin Exercises **1.** $x = 20$ **2.** $R = 30$ **3.** $x = 60$ **4.** $z = \dfrac{3}{8}$ **5.** 6 pounds **6.** 126 voted in Precinct 2 **7.** 7 hours

Exercises **1.** $x = 12$ **3.** $B = 40$ **5.** $A = \dfrac{21}{2}$ **7.** $x = 1$ **9.** $w = 24$ **11.** $x = \dfrac{3}{2}$ **13.** $A = 2$ **15.** $R = \dfrac{100}{3}$ or $33\dfrac{1}{3}$ **17.** $x = 1$

19. $B = 7.8$ **21.** $R = 50$ **23.** $A = 27.3$ **25.** $B = 5800$ **27.** 7 gallons **29.** 20 yards **31.** \$12 **33.** \$34 **35.** 7.5 mph; 52.5 mph

37. 9 hours **39.** \$648 **41.** 90 lb **43.** 4700 grams **45.** 96 lb **47.** c

Section 4.3: Decimals and Percents

Margin Exercises **1. a.** 9% **b.** 1.25% **c.** 125% **d.** $6\dfrac{1}{4}$% **2. a.** 34% **b.** 0.82% **c.** 100% **d.** 579.9%

3. a. 0.40 **b.** 2.11 **c.** 0.006 **d.** 0.2937

Exercises **1.** 60% **3.** 65% **5.** 100% **7.** 20% **9.** 15% **11.** 53% **13.** 125% **15.** 336% **17.** 0.48% **19.** 2.14% **21.** 2%

23. 10% **25.** 36% **27.** 40% **29.** 2.5% **31.** 5.5% **33.** 110% **35.** 200% **37.** 0.02 **39.** 0.18 **41.** 0.3 **43.** 0.0026

45. 1.25 **47.** 2.32 **49.** 0.173 **51.** 0.132 **53.** 12% **55.** 28% **57.** 0.064 **59.** 0.3 **61.** 6.5% **63.** 4.76

Section 4.4: Fractions and Percents

Margin Exercises **1.** 81.25% **2.** 150% **3.** 27.3% **4.** $93\dfrac{1}{3}$% or 93.3% **5.** $\dfrac{4}{5}$ **6.** $2\dfrac{7}{20}$

Exercises **1.** 3% **3.** 7% **5.** 50% **7.** 25% **9.** 55% **11.** 30% **13.** 20% **15.** 80% **17.** 26% **19.** 48% **21.** $12\dfrac{1}{2}$%

23. $87\dfrac{1}{2}$% **25.** $8\dfrac{1}{3}$% **27.** 85% **29.** $141\dfrac{2}{3}$% **31.** $116\dfrac{2}{3}$% **33.** 105% **35.** 175% **37.** $137\dfrac{1}{2}$% **39.** 210% **41.** $\dfrac{1}{10}$

43. $\dfrac{3}{20}$ **45.** $\dfrac{1}{4}$ **47.** $\dfrac{1}{2}$ **49.** $\dfrac{3}{8}$ **51.** $\dfrac{1}{3}$ **53.** $\dfrac{33}{100}$ **55.** $\dfrac{1}{400}$ **57.** 1 **59.** $1\dfrac{1}{5}$ **61.** $\dfrac{3}{1000}$ **63.** $\dfrac{5}{8}$ **65.** $\dfrac{3}{400}$ **67.** 0.55; 55%

69. $\dfrac{7}{4}$ or $1\dfrac{3}{4}$; 175% **71.** $\dfrac{21}{200}$; 0.105 **73.** $3\dfrac{17}{20}$ **75.** $\dfrac{7}{20,000}$ **77. a.** $8\dfrac{1}{3}$% **b.** 25% **c.** $6\dfrac{1}{4}$% or 6.25%

79. $6\dfrac{1}{2}$% or 6.5% **81.** 85% **83.** 0.000075% **85.** 1300%

Section 4.5: Solving Percent Problems by Using Proportions: $\frac{P}{100} = \frac{A}{B}$

Margin Exercises   **1.** 12   **2.** 500   **3.** 22   **4.** 40%

Exercises   **1.** 47   **3.** 15   **5.** 900   **7.** 62   **9.** 150   **11.** 20%   **13.** 3.6   **15.** 70   **17.** 40   **19.** 46   **21.** 24   **23.** $33\frac{1}{3}\%$ **25.** 17.5

27. 160   **29.** 230   **31.** 120   **33.** 620   **35.** 200   **37.** 33.6   **39.** 76.5   **41.** 62.1   **43.** 66.5%   **45.** $16\frac{2}{3}\%$ **47.** 160   **49.** 105

51. 52,330 fans   **53.** 47.31%   **55.** $97,600   **57. a.** 93.75%   **b.** 6.25%; keep this player
59. Non-Resident Alien: 1453; Native American: 135; Hispanic: 2197; Asian or Pacific Islander: 3633; Black: 1081; White: 7975

Section 4.6: Solving Percent Problems by Using the Equation: $R \cdot B = A$

Margin Exercises   **1.** 13.7   **2.** 25   **3.** 120%   **4.** 90   **5.** 16

Exercises   **1.** 7   **3.** 9   **5.** 42   **7.** 700   **9.** 150   **11.** 20   **13.** 150   **15.** 20   **17.** 70   **19.** 36   **21.** 18   **23.** $33\frac{1}{3}\%$ **25.** 12.5   **27.** 110

29. 180%   **31.** 72   **33.** 520   **35.** 200   **37.** 16.32   **39.** 58.5   **41.** 11.4   **43.** 25%   **45.** 36   **47.** 80   **49.** 16   **51.** 10   **53.** 25
55. 28   **57.** 163.2   **59.** 6800   **61.** 4.8%   **63.** 34.29 g   **65.** 40
67. Texas: 55.556; Oakland: 50; Minnesota: 58.025; Chicago: 54.321; Tampa Bay: 59.259; New York: 58.642; San Francisco: 56.790; San Diego: 55.556; Cincinnati: 56.173; St. Louis: 53.086; Philadelphia: 59.877; Atlanta: 56.173

Section 4.7: Applications: Discount, Sales Tax, and Commission

Margin Exercises   **1. a.** $13   **b.** $39   **2.** $150   **3.** $41.73   **4.** $2682   **5.** 25%   **6.** 3%
Exercises   **1. a.** $465   **b.** $15,035   **3. a.** $1.81   **b.** $32.01   **5. a.** $750   **b.** $600   **c.** $636   **7. a.** $10.53   **b.** 60%   **9.** $4.30   **11.** $656.40
13. a. $161   **b.** $5474   **15.** 1.5%   **17.** $11,700   **19.** $7500   **21.** $28,000   **23.** $2240   **25.** $108,000   **27.** $112,700   **29.** 5.9%
31. 2.484%   **33. a.** $170.02   **b.** 28.34%   **35.** 12,349,458 people   **37.** 22%   **39. a.** $675 and $775, respectively
b. The $100 difference is 12.9% based on the second salesman's pay, or 14.8% based on the first salesman's pay.
41. a. $8800   **b.** Jerry   **c.** Wilma; explanations will vary. For sales over $8800, 8% of the sales is more than $500 plus 3% of the sales over $2000.

Section 4.8: Applications: Profit, Simple Interest, and Compound Interest

Margin Exercises   **1. a.** $9   **b.** 150%   **c.** 60%   **2.** $150   **3.** $35 interest   **4.** the principal is $25,000   **5.** 300 days or $\frac{5}{6}$ year
6. $22.84   **7.** $3216.22
Exercises   **1. a.** $500   **b.** 25%   **c.** 20%   **3. a.** $180   **b.** 40%   **c.** $28\frac{4}{7}\%$ or 28.6%   **5.** $30   **7.** $200   **9.** $833.33   **11.** 1 year

13. $108   **15.** $5.63   **17.** $1030   **19.** 10%   **21.** Principal: $100, Rate: 8.5%, Time: 30 days or $\frac{1}{12}$ year, Interest: $16   **23.** $315.25
25. $9181.50   **27. a.** $162.41   **b.** $4162.41   **29.** $813.45

Chapter 4: Review

1. $\frac{17 \text{ students}}{20 \text{ desks}}$ **2.** $\frac{67 \text{ miles}}{1 \text{ hour}}$ **3.** $\frac{1}{10}$ **4.** $\frac{1 \text{ car}}{4 \text{ tires}}$ **5.** $\frac{1}{2}$ **6.** $\frac{2}{3}$ **7.** false   **8.** false   **9.** false   **10.** true   **11.** $y = 2$ **12.** $B = 21$

13. $x = 48$ **14.** $y = 28$ **15.** $x = \frac{1}{2}$ **16.** $y = 21.6$ **17.** $B = 90$ **18.** $A = 35.7$ **19.** $120   **20.** 200 ft   **21.** 7.5   **22.** 30.48 cm

23. 27%   **24.** 7%   **25.** 0.6%   **26.** 425%   **27.** 3%   **28.** 36%   **29.** 0.5%   **30.** 163%   **31.** 0.08   **32.** 0.0042   **33.** 1.83   **34.** 0.158

35. 16%   **36.** 65%   **37.** 230%   **38.** 162.5%   **39.** $\frac{19}{100}$ **40.** $\frac{3}{5}$ **41.** $1\frac{3}{10}$ **42.** $\frac{7}{1000}$ **43.** $\frac{1}{3}$ **44.** $\frac{1}{1000}$ **45.** 12   **46.** 50

47. 40   **48.** 50   **49.** 70   **50.** 20.34   **51.** 75%   **52.** 400   **53.** 2250   **54.** 8   **55.** 40   **56.** 27   **57.** 20   **58.** 56   **59.** 12.96   **60.** 200%

61. 103.36%   **62.** 168,161   **63. a.** $97.50   **b.** $227.50   **64. a.** $100   **b.** 25%   **65.** $46,250   **66. a.** $2600   **b.** $2756   **67.** 1.2%

68. $3750   **69. a.** 25%   **b.** 20%   **70. a.** $375   **b.** $50   **c.** 20%   **d.** 16.67%   **71.** 1 year   **72.** $10,000   **73. a.** 2%   **b.** $5100   **74.** $352.91
75. a. $269.35   **b.** $10,269.35

Chapter 4: Test

1. $\frac{3}{5}$ **2.** $\frac{2}{5}$ **3.** $\frac{55 \text{ miles}}{1 \text{ hour}}$ **4.** $x = 27$ **5.** $x = \frac{5}{6}$ **6.** 24 miles   **7.** $295   **8.** 0.35   **9.** 0.071   **10.** 1.32   **11.** 85%   **12.** 16%

13. 160% **14.** $\dfrac{7}{50}$ **15.** 4 **16.** $\dfrac{1}{8}$ **17.** 15.6 **18.** 42.86 **19.** 90 **20.** 84.5 **21.** 192 **22.** $37\dfrac{1}{2}$% or 37.5% **23.** 200 **24.** 15

25. $11.93 **26. a.** 50% **b.** $33\dfrac{1}{3}$% **27.** $1950 **28.** $67.50 **29.** $800 **30.** 7%

Cumulative Review: Chapters 1 – 4

1. 200,016 **2.** 300.004 **3.** 17.00 **4.** $3\dfrac{7}{8}$ **5.** 0.4 **6.** 0.525 **7.** 180% **8.** 0.015 **9.** 10,159 **10.** $\dfrac{4}{3} = 1\dfrac{1}{3}$ **11.** $\dfrac{41}{11} = 3\dfrac{8}{11}$

12. $46\dfrac{5}{12}$ **13.** $\dfrac{152}{15} = 10\dfrac{2}{15}$ **14.** 398.988 **15.** 75.744 **16.** $\dfrac{8}{105}$ **17.** 0.01161 **18.** 54 **19.** 12 **20.** $\dfrac{9}{5} = 1\dfrac{4}{5}$ **21.** 5,600,000

22. 80.6 **23.** 7 **24.** 8 **25.** $3^4 \cdot 5^2 \cdot 7$ **26.** 2, 3, 4 and 6 **27.** $2^2 \cdot 3^2 \cdot 11$ **28.** 210 **29.** 0 **30.** true **31.** 50 **32.** 18.5

33. 250 **34.** $x = \dfrac{3}{8}$ **35.** 196 **36.** 325 **37. a.** $\dfrac{1}{8}$ **b.** 340 calories **38. a.** 173.2 **b.** 1716.13 **39.** 80 **40.** 2400 milligrams

41. $61.41 **42.** 10.6% **43.** $225 **44.** $4500

Chapter 5: Geometry

Section 5.1: Angles

Margin Exercises **1.** $m\angle 1 = 120°$; $m\angle 2 = 60°$ **2. a.** right **b.** obtuse **c.** straight **3.** neither; because $m\angle 2 + m\angle 3 = 110°$
4. a. 110° **b.** no **5.** $\angle ROS \cong \angle TOU$ and $\angle ROU \cong \angle SOT$ **6. a.** 40° **b.** 90° **c.** 50° **d.** 50°
7. a. $\angle VQZ$ or $\angle WQX$ **b.** $m\angle WQV = 75°$ **8.** $m\angle 4 = 80°$; $m\angle 6 = 80°$; $m\angle 5 = 100°$
Exercises **1.** an angle measuring 90° **3.** two lines that intersect to form right angles **5.** 35° **7.** 120° **9.** acute **11.** obtuse
13. right **15. a.** obtuse **b.** acute **c.** right **17. a.** 180° **b.** 90° **c.** 30° **d.** 150° **19.** 30° **21.** 30° **23.** 120° **25. a.** 135° **b.** 90°
c. 70° **d.** 45° **27. a.** $\angle AOF$ and $\angle FOE$; $\angle AOF$ and $\angle AOB$; $\angle FOC$ and $\angle COB$; $\angle DOF$ and $\angle DOB$; $\angle EOF$ and $\angle EOB$; $\angle EOB$
and $\angle BOA$; $\angle EOC$ and $\angle COA$; $\angle EOD$ and $\angle DOA$ **b.** $\angle AOB$ and $\angle BOC$; $\angle COD$ and $\angle DOE$; $\angle BOC$ and $\angle EOF$
29. a. $\angle CPB$ and $\angle BPD$; $\angle CPA$ and $\angle APD$; $\angle CPA$ and $\angle CPB$; $\angle BPD$ and $\angle DPA$ **b.** There are none.
31. a. $m\angle 2 = 70°$; $m\angle 3 = 90°$; $m\angle 4 = 20°$; $m\angle 5 = 70°$ **b.** $\angle 3$ **c.** $\angle 2$ and $\angle 5$ **33.** 125°; $\angle 3$ and $\angle 1$ are vertical angles
35. $m\angle 7 = 125°$; $\angle 6$ and $\angle 7$ are supplementary angles **37. a.** $\angle QPR$ and $\angle RPB$ **b.** $\angle APR$ **c.** $\angle APQ$ and $\angle QPB$ **d.** None
e. $\angle QPR$ and $\angle RPB$ **f.** $\angle APQ$ and $\angle QPB$; $\angle APR$ and $\angle RPB$ **g.** $\angle APQ$ and $\angle QPB$; $\angle APQ$ and $\angle QPR$; $\angle APR$ and $\angle RPB$;
$\angle QPR$ and $\angle RPB$

Section 5.2: Perimeter

Margin Exercises **1.** 32 ft **2.** 85 ft **3.** 130 in. **4.** 58 cm **5. a.** 29 meters **b.** $217.50 **6.** 64 m
Exercises **1.** A polygon is a closed plane figure, with three or more sides, in which each side is a line segment. **3. a.** true
b. false, not all rectangles have four equal sides. **c.** false, not all parallelograms have four right angles. **d.** true **5.** 44 m

7. 152 mm **9.** 166 ft **11.** 38.1 in. **13.** $12\dfrac{1}{6}$ ft **15.** 12 in. **17.** 15 ft **19.** 70 in. **21.** 28 in. **23.** 41 km **25.** 60 cm **27.** 60 cm

29. 76 yd **31.** 96 inches; stop signs **33. a.** 4605 ft **b.** 15.7 minutes **35.** 386 ft **37. a.** 4180 m **b.** 4180 m **c.** they are identical

Section 5.3: Area

Margin Exercises **1.** 28 mm² **2.** 36 cm² **3.** 30 ft² **4.** 80 cm² **5.** 48 m² **6. a.** 120 yd **b.** 900 yd²

Exercises **1.** 165 cm² **3.** 525 km² **5.** 81 ft² **7.** $\dfrac{5}{27}$ in.² **9.** 27.37 ft² **11.** 48 in.² **13.** 99 ft² **15.** 162 yd² **17.** 1925 cm²

19. 196 in.² **21.** 48 cm² **23.** 160 in.² **25.** 90 cm² **27.** 107 m² **29.** 99 in.² **31.** 220 mm² **33.** 36 cm² **35.** 192 m² **37.** 99 in.²
39. a. 12 ft; 6 ft² **b.** 30 cm; 30 cm² **c.** 48 in.; 96 in.² **41. a.** 70 cm **b.** 220 cm² **43.** 11.7 square miles **45. a.** 30 ft² **b.** 30 ft
47. a. 148.5 ft² **b.** $1410.75 **49.** The area of the triangle is one-half the area of the rectangle.

Section 5.4: Circles

Margin Exercises **1.** 69.08 m; 379.94 m² **2.** 47.1 ft; 176.625 ft² **3.** 66.82 in. **4.** 140.67 yd² **5. a.** 27.42 cm **b.** 21.87 cm²
Exercises **1.** $C = 31.4$ ft; $A = 78.5$ ft² **3.** $C = 439.6$ cm; $A = 15,386$ cm² **5.** $C = 19.468$ yd; $A = 30.1754$ yd²

7. $C = 43.96$ m; $A = 153.86$ m^2 **9.** $C = 4.71$ ft; $A = 1.76625$ ft^2 **11.** $P = 21.98$ m; $A = 38.465$ m^2 **13.** $P = 35.98$ in.; $A = 76.93$ in.^{2}
15. $P = 10.71$ cm; $A = 7.065$ cm^2 **17.** $P = 13.71$ in.; $A = 12.5325$ in.^{2} **19.** $P = 36.56$ m; $A = 38.88$ m^2 **21.** $P = 19.42$ cm; $A = 26.13$ cm^2
23. $P = 14.28$ m; $A = 11.14$ m^2 **25.** $P = 21.42$ m; $A = 32.13$ m^2 **27.** $P = 20.13$ yd; $A = 21.195$ yd^2 **29.** 7536 m^2

31. a. 61.68 in.   **b.** 226.08 in.^{2} **33.** $15{,}307.5$ square feet   **35.** 314.96 square inches   **37.** $\dfrac{C}{d} = \pi$

Section 5.5: Volume and Surface Area

Margin Exercises   **1.** 810 in.^{3} **2.** 113.04 ft^3 **3.** 84.78 m^3 **4.** 678.24 cm^3 **5.** 8 ft^3 **6.** 576 ft^2 **7.** 452.16 in.^{2}
Exercises   **1. a.** C **b.** E **c.** D **d.** B **e.** A   **3.** 11.775 in.^{3} **5.** 904.32 ft^3 **7.** 80 cm^3 **9.** 14.13 ft^3 **11.** 96 in.^{3} **13.** 2289.06 cm^3
15. 224 cm^3 **17.** 282.6 cm^3 **19.** 56.52 ft^3 **21.** 122 in.^{2} **23.** 653.12 yd^2 **25.** 50.24 cm^2 **27.** 3322.12 in.^{2} **29. a.** 3.375 ft^3 **b.** 13.5 ft^2
31. 10.39 in.^{3} **33.** 800 ft^3

Section 5.6: Triangles

Margin Exercises   **1.** scalene   **2.** yes   **3. a.** $90°$ **b.** right **c.** $\overline{BO}$ **d.** $\overline{RB}$ and $\overline{RO}$ **e.** yes; because $m\angle R = 90°$
4. Statement **b.** is correct.   **5.** 2.5 cm   **6.** These triangles are congruent by ASA.
Exercises   **1.** scalene   **3.** right   **5.** isosceles   **7.** isosceles and right   **9.** acute   **11.** acute   **13.** The triangles are not similar. The
corresponding sides are not proportional.   **15.** $\triangle ABC \sim \triangle EDC$. Corresponding angles have the same measure
17. $x = 50°$; $y = 60°$ **19.** $x = 7.5$; $y = 9$ **21.** $x = 95°$; $y = 20$ **23.** congruent by SAS   **25.** not congruent   **27.** 7.5 ft   **29.** 160 ft.
31. a. Each of the other four angles measures $75°$.   **b.** The triangles are similar since the three pairs of corresponding angles are
congruent.   **33. a.** Yes; because the sum of the lengths of two sides is greater than the length of the third side.   **b.** right triangle
and scalene   **c.** Answers will vary

Section 5.7: Square Roots and the Pythagorean Theorem

Margin Exercises   **1. a.** 324 **b.** 13 **2.** 2.2361 **3.** 2.7386 **4.** no, $8^2 \neq 7^2 + 4^2$ **5.** 17 cm   **6.** $\sqrt{208}$ or 14.42 yd
7. $\sqrt{2125}$ or 46.10 ft
Exercises   **1.** 144 **3.** 15 **5.** 6 **7.** 400 **9.** 13 **11.** $(1.732)^2 = 2.999824$ and $(1.733)^2 = 3.003289$ **13.** 3.4641 **15.** 4.8990
17. 6.9282 **19.** 19.0526 **21.** 0.5 **23.** 0.9 **25.** 1.2 **27.** 1.5 **29.** 90 **31.** 30 **33.** 0.05 **35.** 2.8284 **37.** 3.1623 **39.** 6.7082
41. 1.23 **43.** 3.01 **45.** 28.2843 **47.** 0.0548 **49.** 0.03 **51.** yes; $6^2 + 8^2 = 10^2$ **53.** no; $6^2 \neq 4^2 + 3^2$ **55.** $c = \sqrt{5}$ or 2.2361
57. $c = \sqrt{50}$ or 7.0711 **59.** $c = \sqrt{116}$ or 10.7703 **61.** 63.2 feet   **63.** 522.0 feet   **65. a.** 127.3 feet **b.** closer to home plate
67. 44.9 inches   **69.** 7.1 cm   **71. a.** 22.76 feet **b.** 91.04 feet   **73. a.** 17 feet **b.** 51 feet   **75.** 13.23 meters

Chapter 5: Review

1. $m\angle A = 75°$; $m\angle B = 45°$; $m\angle C = 60°$ **2.** $m\angle S = 75°$; $m\angle P = 110°$; $m\angle Q = 50°$; $m\angle R = 125°$ **3.** obtuse   **4.** acute   **5.** $45°$
6. $30°$ **7.** $120°$ **8.** $105°$ **9. a.** $70°$ **b.** $77°$ **10. a.** $90°$ **b.** $30°$ **11. a.** $m\angle 2 = 123°$; $m\angle 3 = 57°$; $m\angle 4 = 123°$ **b.** $\angle 1$ and $\angle 2$; $\angle 3$ and $\angle 4$
c. $\angle 1$ and $\angle 3$; $\angle 2$ and $\angle 4$ **12.** Answers will vary. For a triangle $P = a + b + c$; For a square $P = 4s$; For a rectangle $P = 2a + 2b$
13. $14\frac{5}{6}$ m   **14.** 110 cm   **15.** 24 cm   **16.** 16.4 ft   **17.** 21.2 yd   **18.** 86 cm   **19.** $27\frac{1}{2}$ ft   **20.** 37.24 cm   **21.** 84 cm   **22.** 30 ft
23. 50.8 in.   **24.** 256 ft^2 **25.** 95 yd^2 **26.** 96 in.^{2} **27.** 168.3 ft^2 **28.** 252 yd^2 **29.** 144 yd^2 **30.** 96 cm^2 **31.** 121.55 ft^2 **32.** 740 in.^{2}
33. 97.24 yd^2 **34.** 226 in.^{2} **35.** 62 yd^2 **36.** $C = 50.868$ yd; $A = 206.02$ yd^2 **37.** $C = 50.24$ ft; $A = 200.96$ ft^2
38. $C = 37.68$ ft; $A = 113.04$ ft^2 **39.** $C = 18.84$ ft; $A = 28.26$ ft^2 **40.** $P = 17.85$ cm; $A = 19.625$ cm^2 **41.** $P = 35.98$ in.; $A = 76.93$ in.^{2}
42. $P = 36.56$ in.; $A = 89.12$ in.^{2} **43.** $P = 54.84$ m; $A = 87.48$ m^2 **44.** $P = 20.85$ cm; $A = 24.8125$ cm^2
45. $P = 106.24$ km; $A = 648.96$ km^2 **46.** $P = 34.7$ in.; $A = 84.25$ in.^{2} **47.** $P = 127.49$ yd; $A = 850.155$ yd^2 **48.** 144 in.^{3}
49. 7234.56 ft^3 **50.** 150.72 dm^3 **51.** 200 cm^3 **52.** 56.028 in.^{3} **53.** 381.51 ft^3 **54.** 628 m^3 **55.** 264 in.^{3} **56.** 678.24 cm^3
57. 981.25 km^3 **58.** 715.92 m^2 **59.** 314 mm^2 **60.** 184 in.^{2} **61.** scalene   **62.** equilateral   **63.** right   **64.** obtuse   **65.** isosceles
66. $x = 35°$; $y = 105°$ **67.** $x = 15$ m; $y = 8$ m   **68.** congruent by SAS   **69.** congruent by SSS   **70.** not congruent   **71.** 121
72. 9 **73.** 7.7460 **74.** 3.2 **75.** 9.5917 **76.** 0.63 **77.** yes; $26^2 = 24^2 + 10^2$ **78.** 5.6569

Chapter 5: Test

1. a. right angle **b.** obtuse angle **c.** $\angle AOB$ and $\angle DOE$ (or $\angle BOD$ and $\angle AOE$) **d.** $\angle AOB$ and $\angle AOE$; $\angle AOB$ and $\angle BOD$
(or $\angle AOE$ and $\angle DOE$; $\angle DOE$ and $\angle DOB$)   **2. a.** $55°$ **b.** $165°$ **3. a.** 18.84 in. **b.** 28.26 in.^{2} **4. a.** 16.8 m **b.** 12m^2 **5. a.** $40\frac{1}{3}$ in.
b. $99\frac{1}{6}$ in.^{2} **6.** 196 in.^{2} **7.** 10 cm^2 **8.** 785 cm^3 **9.** 16 dm^3 **10. a.** isosceles **b.** obtuse **c.** right **d.** scalene

11. a. 30° **b.** obtuse **c.** $\overline{RT}$ **12.** $x = 2.5$ **13.** $m\angle BOA = m\angle DOC$ because they are vertical angles, and we are given that $\angle B \cong \angle D$. Since the sum of the angles must be 180°, $m\angle A = m\angle C$. Therefore the triangles are similar. **14.** $x = 3$; $y = 4$
15. a. 65.7 ft **b.** 239.25 ft^2 **16. a.** 60 m **b.** 177.75 m^2 **17.** yes; $74^2 = 70^2 + 24^2$ **a.** 168 m **b.** 840 m^2 **18.** 1256 ft^3 **19.** 251.2 cm^3
20. a. $16\frac{3}{5}$ cm **b.** $9\frac{9}{25}$ cm^2 **21. a.** 39 in. **b.** 92 sq in. **22. a.** 16 in. **b.** 12 sq in. **23. a.**

b. 60 ft **c.** 120 ft^2
24. a. the Pythagorean Theorem **b.** yes; $15^2 + 8^2 = 17^2$
25. a. 280 yards **b.** 4000 square yards **c.** 107.7 yards

Cumulative Review: Chapters 1 – 5

1. 120 **2.** $2^2 \cdot 19$ **3.** $4\frac{1}{6}$ **4.** 220% **5.** $1\frac{1}{2}$ **6.** $233\frac{1}{3}\%$ **7.** 75 **8.** Four hundred twenty-three and eighty-five hundredths

9. 200 **10.** undefined **11.** 0 **12.** 3,500,000 **13.** 14.64 **14.** $\frac{33}{20}$ **15.** estimate: 260; sum: 304.06

16. estimate: 2000; difference: 1523.1 **17.** estimate: 120; product: 118.26 **18.** 22 **19.** 219.8 **20.** $\frac{37}{30}$ **21. a.** 3.3166 **b.** 21

c. 44.7214 **d.** 1.8868 **22. a.** obtuse angle **b.** acute angle **c.** $\angle AOC$ and $\angle DOF$ **d.** Yes. They are vertical angles.

23. a. scalene **b.** isosceles and right **c.** equilateral and acute (and isosceles) **24.** 67.3° **25. a.** 240 minutes **b.** $\frac{3}{10}$
26. $1\frac{1}{2}$ wontons and $1\frac{1}{4}$ spring rolls **27.** $34.38 **28.** $57\frac{1}{7}$ miles (or 57.14 miles) **29.** $650.00 **30.** $940 **31.** 12.5%
32. 140 cm^2 **33.** 87.92 in. (or 7.37 ft) **34.** 180 mm^3 **35.** 276.32 in.^{2} **36. a.** 28 mm **b.** 36 mm^2 **37. a.** 76 mi **b.** 184 mi^2
38. $15\sqrt{2}$ ft and 21.21 ft **39.** $20^2 = 12^2 + 16^2$

Chapter 6: Statistics, Graphs, and Probability

Section 6.1: Statistics: Mean, Median, Mode, and Range

Margin Exercises **1.** 2.66 **2.** 35 **3.** mode: 35; range: 37
Exercises **1. a.** 79 **b.** 84 **c.** 85 **d.** 38 **3. a.** 6.2 hours **b.** 6 hours **c.** 6 hours **d.** 6 hours **5. a.** $400 **b.** $375 **c.** $325 **d.** $225
7. a. 81° **b.** 79° **c.** 88° **d.** 26° **9. a.** 22 **b.** 19 **c.** 18 **d.** 21 **11. a.** 18.5 in. **b.** 14.9 in. **c.** none **d.** 22.0 in. **13. a.** 80 **b.** 82 **c.** 85 **d.** 36
15. a. $48,625 **b.** $46,500 **c.** $63,000 **d.** $43,000 **17. a.** $16,480.60 **b.** $15,618.50 **c.** none **d.** $11,502
19. a. 1135 miles **b.** 980 miles **c.** none **d.** 2020 miles **21. a.** 3.3 **b.** 2.4 **c.** Answers will vary.

Section 6.2: Reading Graphs

Margin Exercises **1. a.** $125,000 **b.** March **c.** $75,000 **d.** 100% **2. a.** $13,750 **b.** $5500 **c.** $3850 **3. a.** 80° **b.** 60° **c.** 12°

4. a. first class **b.** $\frac{18}{50} = 36\%$ **c.** third and fourth classes

Exercises **1. a.** Social Science **b.** Chemistry & Phys, Humanities **c.** about 3300 **d.** about 21.2% **3. a.** Sue **b.** Bob and Sue
c. 85.7% **d.** Bob and Sue; Bob and Sue; Yes in most cases. **e.** No, the vertical scales represents two different types of quantities.
5. a. News: 300 min; Movies: 120 min; Sitcoms: 156 min; Soaps: 180 min; Drama: 144 min; Children's shows: 120 min;
Commercials: 180 min **b.** News **c.** 480 min **7. a.** February and May **b.** 6 inches **c.** March **d.** 3.58 inches
9. a. August **b.** 7 home runs **c.** April and June **d.** 5 home runs **e.** about 22% **f.** about 17%
11. a. West-5%; Northeast-28%; Midwest-35%; South-32% **b.** West-22%; Northeast-19%; Midwest-23%; South-36% **c.** South
d. 5% **e.** West **f.** 5%, 1900 **g.** 22%, 2000 **h.** Midwest **13. a.** 8 **b.** 3 **c.** eighth class **d.** 2 **e.** 27, 29 **f.** 50 **g.** 10 **h.** 16%
15. a. $530 **b.** Repairs: 15.1%; Gas 15.1%; Insurance 13.2%; Loan Payment 56.6% **c.** 7 : 8

Section 6.3: Constructing Graphs from Databases

Margin Exercises

1.

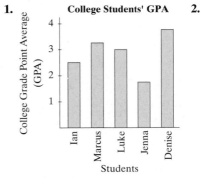

College Students' GPA

College Grade Point Average (GPA)

Ian, Marcus, Luke, Jenna, Denise

Students

2.

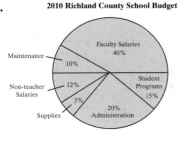

2010 Richland County School Budget

Faculty Salaries 40%
Maintenance 10%
Non-teacher Salaries 12%
Supplies 3%
Administration 20%
Student Programs 15%

Exercises

1.

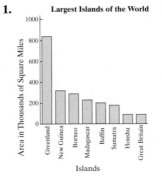

Largest Islands of the World

Area in Thousands of Square Miles

Greenland, New Guinea, Borneo, Madagascar, Baffin, Sumatra, Honshu, Great Britain

Islands

3.

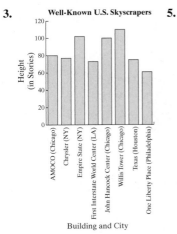

Well-Known U.S. Skyscrapers

Height (in Stories)

AMOCO (Chicago), Chrysler (NY), Empire State (NY), First Interstate World Center (LA), John Hancock Center (Chicago), Willis Tower (Chicago), Texas (Houston), One Liberty Place (Philadelphia)

Building and City

5.

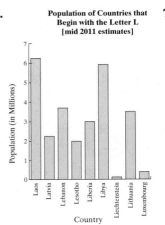

Population of Countries that Begin with the Letter L [mid 2011 estimates]

Population (in Millions)

Laos, Latvia, Lebanon, Lesotho, Liberia, Libya, Liechtenstein, Lithuania, Luxembourg

Country

7.

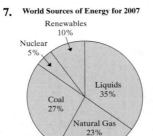

World Sources of Energy for 2007

Renewables 10%
Nuclear 5%
Coal 27%
Liquids 35%
Natural Gas 23%

9.

Cholesterol Levels of 100 Students

High Risk 10%
Recommended 35%
Moderate Risk 40%
Borderline 15%

11.

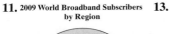

2009 World Broadband Subscribers by Region

North America 21%
Other 16%
South & East Asia 25%
Asia & Pacific 14%
Western Europe 24%

13.

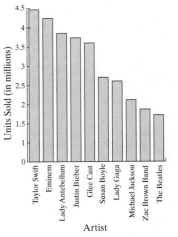

Top 10 Selling Musical Artists in 2010

Units Sold (in millions)

Taylor Swift, Eminem, Lady Antebellum, Justin Bieber, Glee Cast, Susan Boyle, Lady Gaga, Michael Jackson, Zac Brown Band, The Beatles

Artist

Section 6.4: Probability

Margin Exercises

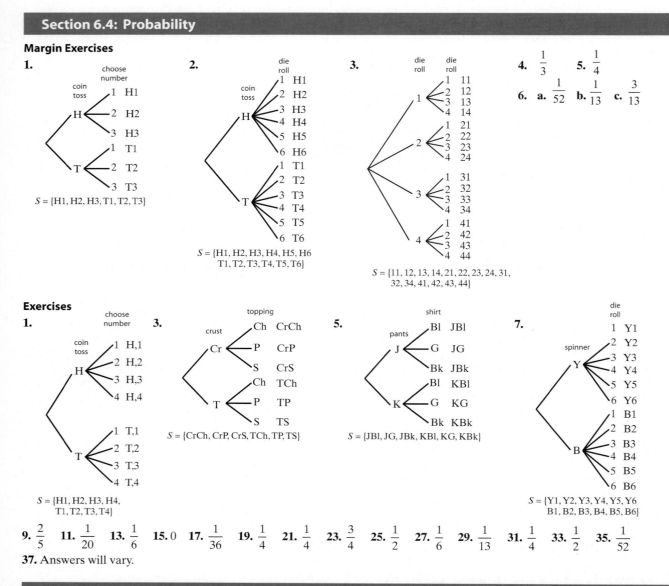

1.

```
       choose
        number
coin      1  H1
toss
     H    2  H2

          3  H3

          1  T1

     T    2  T2

          3  T3
```
$S = \{H1, H2, H3, T1, T2, T3\}$

2.

```
              die
               roll
coin      1  H1
toss      2  H2
     H    3  H3
          4  H4
          5  H5
          6  H6
          1  T1
          2  T2
     T    3  T3
          4  T4
          5  T5
          6  T6
```
$S = \{H1, H2, H3, H4, H5, H6$
$T1, T2, T3, T4, T5, T6\}$

3.

```
   die     die
   roll    roll
        1  11
   1    2  12
        3  13
        4  14
        1  21
   2    2  22
        3  23
        4  24
        1  31
   3    2  32
        3  33
        4  34
        1  41
   4    2  42
        3  43
        4  44
```
$S = \{11, 12, 13, 14, 21, 22, 23, 24, 31,$
$32, 34, 41, 42, 43, 44\}$

4. $\dfrac{1}{3}$ **5.** $\dfrac{1}{4}$

6. a. $\dfrac{1}{52}$ **b.** $\dfrac{1}{13}$ **c.** $\dfrac{3}{13}$

Exercises

1.

```
       choose
        number
coin      1  H,1
toss
     H    2  H,2

          3  H,3

          4  H,4

          1  T,1

     T    2  T,2

          3  T,3

          4  T,4
```
$S = \{H1, H2, H3, H4,$
$T1, T2, T3, T4\}$

3.

```
              topping
crust     Ch  CrCh
     Cr   P   CrP
          S   CrS
          Ch  TCh
     T    P   TP
          S   TS
```
$S = \{CrCh, CrP, CrS, TCh, TP, TS\}$

5.

```
              shirt
pants     Bl  JBl
     J    G   JG
          Bk  JBk
          Bl  KBl
     K    G   KG
          Bk  KBk
```
$S = \{JBl, JG, JBk, KBl, KG, KBk\}$

7.

```
                die
                 roll
spinner     1  Y1
            2  Y2
       Y    3  Y3
            4  Y4
            5  Y5
            6  Y6
            1  B1
            2  B2
       B    3  B3
            4  B4
            5  B5
            6  B6
```
$S = \{Y1, Y2, Y3, Y4, Y5, Y6$
$B1, B2, B3, B4, B5, B6\}$

9. $\dfrac{2}{5}$ **11.** $\dfrac{1}{20}$ **13.** $\dfrac{1}{6}$ **15.** 0 **17.** $\dfrac{1}{36}$ **19.** $\dfrac{1}{4}$ **21.** $\dfrac{1}{4}$ **23.** $\dfrac{3}{4}$ **25.** $\dfrac{1}{2}$ **27.** $\dfrac{1}{6}$ **29.** $\dfrac{1}{13}$ **31.** $\dfrac{1}{4}$ **33.** $\dfrac{1}{2}$ **35.** $\dfrac{1}{52}$

37. Answers will vary.

Chapter 6: Review

1. a. 45 **b.** 36 **c.** 36 **d.** 72   **2. a.** $301,124\,\text{m}^3$ **b.** $238,200\,\text{m}^3$ **c.** none **d.** $418,280\,\text{m}^3$ **3. a.** 8,698,125 **b.** 8,061,500 **c.** none
d. 6,711,000   **4. a.** 15.4 **b.** 14.5 **c.** 17 and 3 **d.** 44   **5. a.** 199.1 **b.** 91 **c.** 2 **d.** 1518   **6. a.** 9% **b.** Palm OS **c.** 16% more
7. a. Big Bang Theory **b.** Saturday Night Football **c.** 12,666,000 more people **d.** 14,809,000
8. a. 7 **b.** 30 minutes **c.** 7th **d.** 15 **e.** 519.5 and 549.5 **f.** 30.6% **g.** 57.1%   **9. a.** August **b.** February **c.** 10.9% **d.** 9.82%

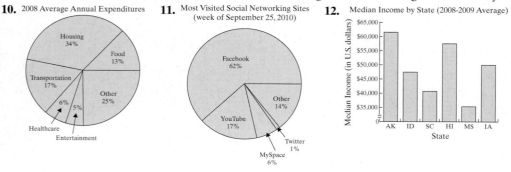

10. 2008 Average Annual Expenditures

11. Most Visited Social Networking Sites
(week of September 25, 2010)

12. Median Income by State (2008-2009 Average)

13.

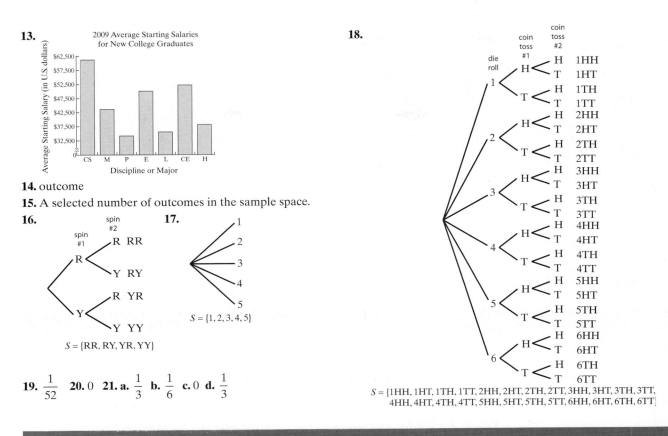

2009 Average Starting Salaries for New College Graduates

Average Starting Salary (in U.S. dollars) — x-axis: Discipline or Major (CS, M, P, E, L, CE, H)

14. outcome

15. A selected number of outcomes in the sample space.

16.

spin #1 / spin #2

R — R RR
R — Y RY
Y — R YR
Y — Y YY

$S = \{RR, RY, YR, YY\}$

17.

1
2
3
4
5

$S = \{1, 2, 3, 4, 5\}$

18.

die roll / coin toss #1 / coin toss #2

1 — H — H 1HH
1 — H — T 1HT
1 — T — H 1TH
1 — T — T 1TT
2 — H — H 2HH
2 — H — T 2HT
2 — T — H 2TH
2 — T — T 2TT
3 — H — H 3HH
3 — H — T 3HT
3 — T — H 3TH
3 — T — T 3TT
4 — H — H 4HH
4 — H — T 4HT
4 — T — H 4TH
4 — T — T 4TT
5 — H — H 5HH
5 — H — T 5HT
5 — T — H 5TH
5 — T — T 5TT
6 — H — H 6HH
6 — H — T 6HT
6 — T — H 6TH
6 — T — T 6TT

$S = \{$1HH, 1HT, 1TH, 1TT, 2HH, 2HT, 2TH, 2TT, 3HH, 3HT, 3TH, 3TT, 4HH, 4HT, 4TH, 4TT, 5HH, 5HT, 5TH, 5TT, 6HH, 6HT, 6TH, 6TT$\}$

19. $\dfrac{1}{52}$ **20.** 0 **21. a.** $\dfrac{1}{3}$ **b.** $\dfrac{1}{6}$ **c.** 0 **d.** $\dfrac{1}{3}$

Chapter 6: Test

1. a. 25.4 **b.** 25 **c.** 25 **d.** 14   **2. a.** 34.7 **b.** 4 **c.** 0, 3, 4 **d.** 157   **3. a.** New Orleans Saints **b.** Cleveland Browns **c.** 4163 yards
d. 938 more yards **e.** 18.0%   **4. a.** Nintendo DS **b.** 9% **c.** 43%   **5.**

7. $\dfrac{1}{4}$ **8.** $\dfrac{1}{36}$ **9.** $\dfrac{1}{2}$ **10. a.** $\dfrac{2}{3}$ **b.** $\dfrac{1}{3}$

5. Year Owner-Occupied Housing Units were Built

1939 or earlier 13%
2000 or later 12%
1940 to 1959 18%
1980 to 1999 30%
1960 to 1979 27%

6.

coin toss / choose number

H — 1 H1
H — 2 H2
T — 1 T1
T — 2 T2

$S = \{H1, H2, T1, T2\}$

Cumulative Review: Chapters 1 – 6

1. 6 **2.** 7 **3.** 18 **4.** 3 **5.** $\dfrac{36}{65}$ **6.** $\dfrac{17}{35}$ **7.** $\dfrac{79}{42}$ or $1\dfrac{37}{42}$ **8.** $\dfrac{6}{7}$ **9.** 67.872 **10.** 37.313 **11.** 1920 **12.** 39.5404 **13.** 2, 3, 4, 6

14. 3, 5 **15.** prime **16.** composite; 1, 3, 7, 9, 21, 63 **17.** 216 **18.** 600 **19.** $\dfrac{7}{8}$ **20.** $\dfrac{17}{4}$ **21.** 0.4375 **22.** 0.6 **23.** 25%

24. 138.4% **25.** 0.382 **26.** 2.59 **27.** $x = 5$ **28.** $y = 300$ **29.** 48 mm; 84 mm^2 **30.** $16\dfrac{1}{6}$ yd; $15\dfrac{17}{24}$ yd^2 **31.** 53.16 in.; 136.536 in.^{2}

32. a. $x = 20$ m; $y = 25$ m **b.** $x = 12$ in., $y = 7$ in. **33. a.** $m\angle x = 30°$ **b.** obtuse **c.** $\overline{RT}$ **34.** 81.2% **35.** $22.50
36. 440 yards **37.** $x = 72°$, $y = 48°$ **38. a.** 806,814.3 **b.** 410,635.5 **39. a.** $169,043.30 **b.** $170,945.50 **c.** $150,000 **d.** $63,958
40. Housing: $10,500; Food: $7000; Taxes: $3500; Transportation: $3500; Clothing: $3500; Savings: $1750; Entertainment: $2800;
Education: $2450 **41.** $4000 **42.** $\dfrac{3}{5}$ **43.** $18,000

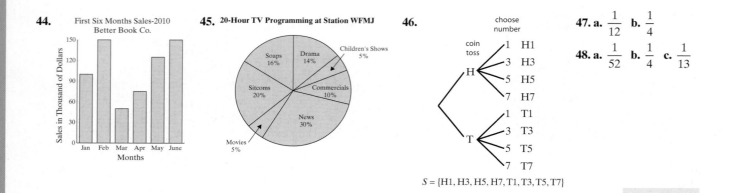

44. First Six Months Sales-2010
Better Book Co.

45. 20-Hour TV Programming at Station WFMJ

46.

$S = \{H1, H3, H5, H7, T1, T3, T5, T7\}$

47. a. $\dfrac{1}{12}$ **b.** $\dfrac{1}{4}$

48. a. $\dfrac{1}{52}$ **b.** $\dfrac{1}{4}$ **c.** $\dfrac{1}{13}$

Chapter 7: Introduction to Algebra

Section 7.1: The Real Number Line and Absolute Value

Margin Exercises **1. a.** +10 **b.** +8 **c.** −17 **2. a.** (number line) **b.** (number line) **3. a.** 1, 20 **b.** −6, 1, 20

c. $-6, -\dfrac{1}{7}, 1, 20$ **d.** All numbers in S are real numbers. **4. a.** true **b.** false **c.** true **d.** true **e.** false **5. a.** (number line)

b. (number line) **c.** (number line) **6. a.** 4 **b.** −7.4 **7.** true **8. a.** $z = -3, 3$ **b.** no solution

9. $\{\ldots, -3, -2, -1, 1, 2, 3, \ldots\}$ (number line) **10.** $\{-2, -1, 0, 1, 2\}$ (number line)

Exercises **1.** 4, 8 **3.** −7, −2, 0, 4, 8 **5.** $-7, -2, -\dfrac{5}{3}, -1.4, 0, \dfrac{3}{5}, 4, 5.9, 8$ **7.** (number line) **9.** (number line)

11. (number line) **13.** (number line) **15.** (number line) **17.** (number line)

19. (number line) **21.** (number line) **23.** < **25.** > **27.** < **29.** = **31.** > **33.** < **35.** = **37.** > **39.** true

41. false; $-9 < -8.5$ **43.** true **45.** true **47.** false; $-|-3| > -|4|$ **49.** $x = -5, 5$ **51.** $x = -2, 2$ **53.** no solution **55.** $x = -23, 23$

57. $\{4, 5, 6, \ldots\}$ (number line) **59.** $\{\ldots, -8, -7, -6, 6, 7, 8, \ldots\}$ (number line)

61. $\{-2, -1, 0, 1, 2\}$ (number line) **63.** Sometimes. Examples will vary. **65.** Sometimes. Examples will vary.

67. If y is a negative number then $-y$ represents a positive number. For example, if $y = -2$, then $-y = -(-2) = 2$.

Section 7.2: Addition with Integers

Margin Exercises **1. a.** −20 **b.** 15 **c.** 5.9 **d.** $-\dfrac{11}{15}$ **2. a.** 8 **b.** −7 **c.** −2.3 **d.** $\dfrac{4}{9}$ **3. a.** −5 **b.** −3.6 **4. a.** −7 **b.** −4.3 **5. a.** −34

b. 39 **6. a.** −5 is a solution. **b.** −5 is not a solution. **c.** −5 is not a solution. **d.** −5 is not a solution.

Exercises **1.** 13 **3.** −4 **5.** 0 **7.** 5 **9.** −13 **11.** −8 **13.** −10 **15.** 0 **17.** 17 **19.** −4.6 **21.** −29 **23.** −9 **25.** −15.4 **27.** −3

29. −54 **31.** −2.5 **33.** $\dfrac{3}{7}$ **35.** $\dfrac{5}{8}$ **37.** $-\dfrac{7}{4}$ **39.** −7 **41.** −24 **43.** 3.1 **45.** 76 **47.** −310 **49.** 27 **51.** 18 **53.** 14 **55.** 35

57. −2 is a solution. **59.** −4 is a solution. **61.** −0.6 is a solution. **63.** 18 is a solution. **65.** −2 is not a solution.

67. $\dfrac{5}{3}$ is not a solution. **69.** −72 is not a solution. **71.** Sometimes. Examples will vary. **73.** Sometimes. Examples will vary.

75. Never. Examples will vary. **77.** Never. Examples will vary. **79.** 84 **81.** 3807 **83.** −97,714 **85.** $|0| + |0| = 0$

Section 7.3: Subtraction with Integers

Margin Exercises **1. a.** −14 **b.** +9 **2. a.** 0 **b.** −11 **c.** 3.8 **d.** $-\dfrac{3}{4}$ **e.** $\dfrac{13}{10}$ **3. a.** −46 **b.** −107 **c.** 0 **d.** 11.9 **4. a.** −19°F **b.** −11,000 ft

5. a. 19 sales **b.** 221 lb   **6. a.** 3 is a solution. **b.** 15 is a solution. **c.** $-\dfrac{1}{12}$ is a solution.

Exercises **1.** -11 **3.** 6 **5.** -4.7 **7.** 0 **9.** $\dfrac{5}{7}$ **11.** 5 **13.** -10 **15.** 7 **17.** 3 **19.** 12 **21.** 24 **23.** 5.9 **25.** -4 **27.** 10 **29.** $-\dfrac{7}{20}$

31. $-\dfrac{1}{4}$ **33.** $\dfrac{3}{40}$ **35.** -15 **37.** -16 **39.** -57 **41.** -6.0 **43.** 1 **45.** -7 **47.** -26 **49.** 11 **51.** -15 **53.** -1 **55.** -3 **57.** 1 **59.** 0

61. $-8 < -5$ **63** $-3.1 < 3.1$ **65.** $\dfrac{19}{4} = \dfrac{19}{4}$ **67.** $0 > -27$ **69.** $-237 > -248$ **71.** $-18°F$ (a decrease of 18 degrees Fahrenheit)

73. $8 (an increase of 8 dollars) **75.** $-14,777$ ft (a decrease of 14,777 feet) **77.** 85 years old **79.** -8 is a solution.

81. -2 is a solution. **83.** $-\dfrac{2}{5}$ is a solution. **85.** 0.7 is not a solution. **87.** 1 is not a solution. **89.** -4 is a solution.

91. -5 is a solution. **93.** -28 is a solution. **95.** $1044 > -39$ **97.** $-15,254 > -35,090$ **99.** gain of 35 yards

101. lost 7 pounds; 203 pounds **103.** $-\$677$ **105.** $-\$115$ (a loss of 15 dollars)

107. Add the opposite of the second number to the first number.

Section 7.4: Multiplication and Division with Real Numbers

Margin Exercises **1. a.** -28 **b.** -32 **c.** -3 **d.** -6.4 **e.** $-\dfrac{7}{2}$ **2. a.** $+24$ **b.** $+\dfrac{22}{15}$ **c.** $+17.5$ **d.** $+48$ **3.** 0 **4. a.** 6 **b.** -6 **c.** -6 **d.** 6

e. undefined **f.** 0 **5. a.** 12 **b.** $-\dfrac{7}{4}$ **c.** -2.1 **d.** 2.9 **6. a.** $4°F$ **b.** 79 **c.** 47.7 mph

Exercises **1.** -12 **3.** 48 **5.** 56 **7.** -21 **9.** 56 **11.** 26 **13.** -70 **15.** -24 **17.** -288 **19.** 0 **21.** -9 **23.** -7.31 **25.** $-\dfrac{1}{6}$

27. $\dfrac{18}{5}$ **29.** $-\dfrac{12}{5}$ **31.** $-\dfrac{9}{2}$ **33.** 4 **35.** -6 **37.** 2 **39.** 0 **41.** -3 **43.** -17 **45.** undefined **47.** 5 **49.** -0.6 **51.** 20 **53.** -2.3

55. $-\dfrac{1}{12}$ **57.** $\dfrac{9}{22}$ **59.** $-\dfrac{5}{12}$ **61.** false; $(-4)(6) < 3 \cdot 8$ **63.** true **65.** false; $6(-3) = (-14)+(-4)$ **67.** true **69.** true

71. -12 **73.** $163 **75.** 26.6 calls **77.** 80 **79.** 78.6 inches **81.** 31.5 hours **83.** 18,902 square miles **85.** $-352,665$

87. -2671 **89.** -45.33 **91.** -10.29 **93.** 2.83 **95.** -90.365 **97.** Answers will vary.

Section 7.5: Order of Operations with Real Numbers

Margin Exercises **1. a.** -37 **b.** -22 **c.** -12 **2.** 17 **3.** $-\dfrac{11}{150}$

Exercises **1. a.** 36 **b.** 16 **3.** -25 **5.** -10 **7.** -45 **9.** -137 **11.** 152 **13.** -189 **15.** 143 **17.** -36 **19.** -10 **21.** $\dfrac{11}{30}$

23. $\dfrac{1}{24}$ **25.** $\dfrac{31}{24}$ **27.** $\dfrac{107}{24}$ **29.** $\dfrac{41}{32}$ **31.** $\dfrac{3}{5}$ **33.** $-\dfrac{341}{30}$ **35.** $-\dfrac{15}{17}$ **37.** 24 **39.** $\dfrac{7}{10}$ **41.** $5\dfrac{1}{12}$ or $\dfrac{61}{12}$ **43.** $-\dfrac{13}{64}$

45. 42.45 **47.** 67.77 **49.** 15.41 **51.** $(3^2 - 9) = 0$ and division by 0 is undefined. **53.** It will always be larger. If you square a negative number you get a positive number. All positive numbers are larger than all negative numbers.

Section 7.6: Properties of Real Numbers

Margin Exercises **1. a.** commutative property of addition **b.** associative property of addition **c.** zero-factor law
d. distributive property **e.** associative property of multiplication **2. a.** commutative property of addition
$(-7) + 21 = 14$ and $21 + (-7) = 14$ **b.** associative property of multiplication $(5 \cdot 4) \cdot 2 = 40$ and $5(4 \cdot 2) = 40$
c. distributive property $11(-4 + 3) = -11$ and $11(-4) + 33 = -11$

Exercises **1.** $3 + 7$ **3.** $4 \cdot 19$ **5.** $30 + 48$ **7.** $(2 \cdot 3) \cdot x$ **9.** $(3 + x) + 7$ **11.** 0 **13.** $x + 7$ **15.** $2x - 24$ **17.** 0

19. commutative property of addition **21.** multiplicative identity **23.** associative property of addition

25. commutative property of multiplication **27.** commutative property of multiplication **29.** multiplicative inverse

31. additive inverse **33.** multiplicative identity **35.** zero-factor law **37.** associative property of addition

39. $6(11) = 66$ and $6 \cdot 3 + 6 \cdot 8 = 66$ **41.** $10(-7) = -70$ and $10 \cdot 2 - 10 \cdot 9 = -70$ **43.** commutative property of multiplication;
$6 \cdot 4 = 4 \cdot 6 = 24$ **45.** associative property of addition; $8 + (5 + (-2)) = (8 + 5) + (-2) = 11$ **47.** distributive property;
$5(4 + 18) = 5(4) + 90 = 110$ **49.** associative property of multiplication; $(6 \cdot (-2)) \cdot 9 = 6 \cdot (-2 \cdot 9) = -108$

51. commutative property of addition; $3 + (-34) = (-34) + 3 = -31$ **53.** commutative property of addition;
$2(3 + 4) = 2(4 + 3) = 14$ **55.** commutative property of addition; $5 + (4 - 15) = (4 - 15) + 5 = -6$

57. associative property of multiplication; $(3 \cdot 4) \cdot 5 = 3 \cdot (4 \cdot 5) = 60$

59. a. Multiply each term that is part of the sum (in the parentheses) by a. **b.** An expression that "distributes addition over multiplication" would look like $a + (bc) = (a+b)(a+c)$, however this is not a true statement. Answers will vary.

Section 7.7: Simplifying and Evaluating Algebraic Expressions

Margin Exercises **1.** $-5.3, 9$, and 0 are like terms. $4y, -2y$, and $7y$ are like terms. $10xy^2$ and $-xy^2$ are like terms. **2. a.** $9x$ **b.** $1.1z$ **c.** $7x^2 + 5y$ **d.** $10a + 8$ **e.** $8x$ **3. a.** For $x = 4, 2x^2 = 32$. For $x = 5, 2x^2 = 50$. **b.** For $x = 4, -2x^2 = -32$. For $x = 5, -2x^2 = -50$ **4. a.** $7x + 6 = -22$ **b.** $-3ab - a = 12$ **c.** $5y - 16 = 4$

Exercises **1.** $-5, \frac{1}{6}$, and 8 are like terms; $7x$ and $9x$ are like terms. **3.** $-x^2$ and $2x^2$ are like terms; $5xy$ and $-6xy$ are like terms; $3x^2y$ and $5x^2y$ are like terms. **5.** $24, 8.3$, and -6 are like terms; $1.5xyz, -1.4xyz$, and xyz are like terms. **7.** 64 **9.** -121 **11.** $15x$ **13.** $3x$ **15.** $-2n$ **17.** $5y^2$ **19.** $34x^2$ **21.** $7x + 2$ **23.** $x - 4y$ **25.** $13x^2 - 2y$ **27.** $2n + 3$ **29.** $7a - 8b$ **31.** $8x + y$ **33.** $2x^2 - x$ **35.** $-2n^2 + 2n - 2$ **37.** $3x^2 - xy + y^2$ **39.** $-2y$ **41.** $\frac{8x}{3}$ **43. a.** $3x + 4$ **b.** 16 **45. a.** $-2x - 8$ **b.** -16 **47. a.** $5y + 1$ **b.** 16 **49. a.** $-3x - 7y$ **b.** -33 **51. a.** $3.6x^2 + 2$ **b.** 59.6 **53. a.** $3ab + b^2 + b^3$ **b.** 6 **55. a.** $3.7x + 1.1$ **b.** 15.9 **57. a.** $8a$ **b.** -16 **59. a.** $3b$ **b.** -3 **61. a.** $-x - 54$ **b.** -58 **63.** -5^2 is the square of 5 multiplied by -1 by the order of operations; while $(-5)^2$ is the square of -5. $(-5^2 = -25$ and $(-5)^2 = 25)$

Section 7.8: Translating English Phrases and Algebraic Expressions

Margin Exercises **1. a.** $7x$ **b.** $5 + n$ **c.** $4(y + 2)$ **d.** $2x + 3$ **e.** $9x - 4$ **2. a.** the product of 10 and a number **b.** four times a number increased by 7 **c.** seven times the difference between a number and 5 **3. a.** $\frac{x}{-9}$ **b.** $4n - 3$ **c.** $2(8 + a)$ **d.** $12f$ **e.** $25 + 0.33x$

Exercises **1.** 4 times a number **3.** 5 more than a number **5.** 7 times the sum of a number and 1.1 **7.** -2 times the difference between a number and 8 **9.** 6 divided by the difference between a number and 1 **11.** 5 times the sum of twice a number and 3 **13.** 3 times a number plus 7; 3 times the sum of a number and 7 **15.** the product of 7 and a number minus 3; 7 times the difference between a number and 3 **17.** $x + 6$ **19.** $x - 4$ **21.** $\frac{2x}{10}$ **23.** $3x - 5$ **25.** $8 - 2x$ **27.** $20 - 4.8x$ **29.** $9(x + 2)$ **31.** $4(x + 1) - 13$ **33.** $3(x + 6) + 8$ **35.** $3(7 - x) - 4$ **37. a.** $x - 6$ **b.** $6 - x$ **39. a.** $x - 20$ **b.** $20 - x$ **41.** $24d$ **43.** $3.15x$ **45.** $365y$ **47.** $7t + 3$ **49.** $7t + 3$ **51.** $20 + 0.15m$ **53.** $2w + 2(2w - 3) = 6w - 6$ **55.** A phrase whose meaning is not clear or for which there may be two or more interpretations.

Chapter 7: Review

1. a. $\frac{2}{1}$ **b.** $-2, 0, \frac{2}{1}$ **c.** $-9.3, -2, -1.343434..., -\frac{3}{4}, 0, \frac{2}{1}, \frac{10}{3}$ **d.** $\sqrt{2}$ **2.** [number line with points at -3, 0, 3, 4] **3.** [number line with points at -4, $-1\frac{1}{4}$, 2.5]
4. [number line with points $0\ 1\ 2\ 3\ 4\ 5$] **5.** $<$ **6.** $>$ **7.** true **8.** false; $|-2.3| > 0$ **9.** $x = -5, 5$ **10.** no solution
11. [number line from -5 to 3] **12.** [number line $-5, -4$ and $4, 5$] **13.** 15 **14.** 5 **15.** 0 **16.** -12.9 **17.** -18 **18.** $-\frac{11}{12}$ **19.** 8 **20.** -45
21. 2.1 is not a solution **22.** 3 is a solution **23.** -7 **24.** $-\frac{11}{15}$ **25.** -4 **26.** -0.9 **27.** -17 **28.** 37 **29.** -6000 feet **30.** $-33°$F
31. -7 is not a solution **32.** -7.1 is not a solution **33.** -25.6 **34.** 90 **35.** $-\frac{3}{4}$ **36.** 24 **37.** 5 **38.** undefined **39.** 0 **40.** -0.4
41. $-\frac{6}{5}$ **42.** 62 **43.** 65.8 inches **44.** 7.7 hurricanes **45.** 88 **46.** 18 **47.** -4 **48.** 26 **49.** 32 **50.** $-\frac{1}{360}$ **51.** $\frac{31}{28}$ **52.** $-\frac{9}{8}$
53. commutative property of addition **54.** associative property of addition **55.** additive identity **56.** additive inverse
57. commutative property of multiplication **58.** multiplicative identity **59.** multiplicative inverse **60.** distributive property
61. $6y$ **62.** $-3a^2$ **63.** $-3x + 2$ **64.** $3n + 4$ **65.** $y^2 + 7y$ **66.** $7a^2 + 4a$ **67. a.** $-a^2 + 4a + 7$ **b.** -5 **68. a.** $17x - 17$ **b.** -34
69. a. $-2y - 26$ **b.** -34 **70. a.** $7y$ **b.** 28 **71.** 3 more than 5 times a number **72.** -3 multiplied by the sum of a number and 2
73. 5 times the difference between a number and 3 **74.** 7 times a number divided by 33
75. 50 divided by the product of 6 and a number **76.** $28 - 6x$ **77.** $72 + 8(x + 2)$ **78.** $10x - 32$ **79.** $24x + 5$ **80.** $0.49y$

Chapter 7: Test

1. a. $-5, -1, 0$ **b.** $-5, -1, \frac{-1}{3}, 0, \frac{1}{2}, 3\frac{1}{4}, 7.121212...$ **c.** $-\pi$ **d.** all are real numbers **2.** $3, -23$; [number line with points at -23, -10, 3] **3. a.** $<$ **b.** $>$ **c.** $=$ **4. a.** $\{-2, -1, 0, 1, 2\}$ **b.** $\{..., -11, -10, -9, 9, 10, 11, ...\}$ **5.** 23 **6.** -1.3 **7.** 112 **8.** 0 **9.** -3 **10.** -0.3 **11.** $-\frac{4}{9}$ **12.** $\frac{7}{18}$

13. -44 **14.** 19 **15.** $\dfrac{29}{40}$ **16.** $\dfrac{17}{90}$ **17.** additive identity **18.** commutative property of multiplication

19. associative property of addition **20.** zero-factor law **21.** $5x^2 + 7x, 6$ **22.** $6y - 6, 12$ **23.** the product of 5 and a number

increased by 18 **24.** 3 multiplied by the sum of a number and 6 **25.** $6x - 3$ **26.** $2(x + 15) - 4$ **27.** $-\dfrac{33}{16}$ **28.** -22.5

29. a. 15 gallons **b.** No, you need \$46.50 to fill the tank so you are short \$6.50. **30.** 152 points **31.** $15°F$

Cumulative Review: Chapters 1 – 7

1. 19 **2.** 25 **3.** -72 **4.** 16 **5.** -1 **6.** 32 **7.** 16 **8.** $2, 3, 4, 6, 9$ **9.** none **10.** $2^3 \cdot 3 \cdot 11$ **11.** associative property of addition

12. a. $\dfrac{3}{4}$ **b.** $\dfrac{2}{3}$ **13.** $\dfrac{53}{6}$ **14.** 120 **15.** $\dfrac{3}{4}, \dfrac{13}{16}, \dfrac{6}{7}$ **16.** $\dfrac{7}{12}$ **17.** $-\dfrac{4}{27}$ **18.** $\dfrac{33}{14}$ or $2\dfrac{5}{14}$ **19.** $\dfrac{5}{2}$ or $2\dfrac{1}{2}$ **20.** $-\dfrac{5}{12}$

21. 8719.43 **22.** 6.4 **23.** 4.51 **24.** 5604.69 **25.** $0.\overline{6}$ **26.** $\dfrac{2}{9}$ **27.** 0.13 **28.** 15.68 **29.** $x = 30$ **30. a.** Toyota Prius

b. Chevrolet Camaro **c.** 11 MPG **31.** $m = 8$ cm; $n = 105°$ **32.** mean = 117.4 minutes; median = 112.5 minutes; mode = 108

minutes; range = 50 minutes **33.** **34.** **35. a.** $-2x - 10$ **b.** -4

36. a. $3y^2 + 6y$ **b.** 0 **37.** $5 + 2(x - 8)$ **38.** $12(x + 7)$ **39.** -24 **40.** -73 **41.** $\sqrt{208}$ in. ≈ 14.4 in. **42.** \$200 **43.** \$38.52

44. 40 feet **45. a.** 325.6 yards **b.** 6261.15 square yards **46.** 1944 cm^3 **47. a.** $\dfrac{1}{13}$ **b.** $\dfrac{1}{2}$ **48.** $-24°F$ **49.** never **50.** sometimes

Chapter 8: Solving Linear Equations and Inequalities

Section 8.1: Solving Linear Equations: $x + b = c$ and $ax = c$

Margin Exercises **1.** $x = 17$ **2.** $x = -12$ **3.** $x = 3.7$ **4.** $x = \dfrac{9}{8}$ **5.** $z = -5$ **6.** $z = 4.2$ **7.** $x = 11$ **8.** $x = 4$ **9.** $x = 15$

10. $x = \dfrac{6}{5}$ **11.** \$104.99

Exercises **1.** $x = 7$ **3.** $y = -4$ **5.** $x = -19$ **7.** $n = 37$ **9.** $z = -6$ **11.** $x = 5$ **13.** $y = -5.9$ **15.** $x = -1.2$ **17.** $x = \dfrac{11}{20}$ **19.** $x = 9$

21. $y = 8$ **23.** $x = 20$ **25.** $y = 10$ **27.** $x = -2$ **29.** $x = 12$ **31.** $n = 8$ **33.** $y = 2.1$ **35.** $x = \dfrac{20}{9}$ **37.** $x = -13.3$ **39.** $y = -12$

41. $x = -\dfrac{2}{5}$ **43.** $x = -4$ **45.** $x = \dfrac{5}{8}$ **47.** $n = 9.7$ **49.** $x = -4$ **51.** 1945 kanji characters **53.** 6.25 tons **55.** $y = -50.753$

57. $x = -17.214$ **59.** $x = 246$ **61.** $x = -153.17$ **63. a.** Yes; It is stating that $6 + 3$ is equal to 9. **b.** No; If we substitute 4 for x, we get the statement $9 = 10$, which is not true.

Section 8.2: Solving Linear Equations: $ax + b = c$

Margin Exercises **1.** $x = -13$ **2.** $y = 2$ **3.** $z = 0.2$ **4.** $x = 5$ **5.** $x = 2$ **6.** $x = 108$ **7.** $x = \dfrac{11}{12}$

Exercises **1.** $x = -3$ **3.** $x = 2$ **5.** $x = 2$ **7.** $x = 2$ **9.** $y = -1$ **11.** $t = -1$ **13.** $x = -0.12$ **15.** $x = 4$ **17.** $x = 0$ **19.** $y = 0$ **21.** $x = -2$

23. $y = -6$ **25.** $n = 6$ **27.** $n = 8$ **29.** $x = 0$ **31.** $x = -7$ **33.** $x = -\dfrac{1}{8}$ **35.** $x = -\dfrac{13}{2}$ **37.** $x = -\dfrac{21}{5}$ **39.** $x = -\dfrac{8}{15}$ **41.** $y = \dfrac{28}{5}$

43. $x = 2$ **45.** $y = \dfrac{7}{5}$ **47.** $x = -4.5$ **49.** $x = -44$ **51.** $x = 2$ **53.** $x = -4$ **55.** $y = 0.5$ **57.** $x = 1.5$ **59.** $x = 0.2$

61. 14,000 tickets per hour **63.** $36.5°C$ **65.** $x = 6.1$ **67.** $x = 1.12$

Section 8.3: Solving Linear Equations: $ax + b = cx + d$

Margin Exercises **1.** $x = -6$ **2.** $x = -7$ **3.** $y = 2.1$ **4.** $x = 4$ **5.** $y = -4$ **6.** $x = -6$ **7.** $x = 2$ **8.** contradiction **9.** identity

10. conditional

Exercises **1.** $x = -5$ **3.** $n = 3$ **5.** $y = 6$ **7.** $x = 3$ **9.** $n = 0$ **11.** $y = 0$ **13.** $z = -1$ **15.** $y = \dfrac{1}{5}$ **17.** $x = -3$ **19.** $x = -4$

21. $x = -21$ **23.** $y = 0$ **25.** $y = 1$ **27.** $x = -\dfrac{3}{2}$ **29.** $x = \dfrac{1}{4}$ **31.** $x = \dfrac{3}{17}$ **33.** $x = -\dfrac{1}{4}$ **35.** $x = \dfrac{8}{5}$ **37.** $x = \dfrac{2}{3}$ **39.** $x = 6$

41. $x = -11$ **43.** $x = \dfrac{1}{2}$ **45.** $x = -5$ **47.** $n = -1.5$ **49.** $x = 0$ **51.** conditional   **53.** conditional   **55.** contradiction   **57.** identity

59. conditional   **61.** 1800 square feet   **63.** 240 sundaes   **65.** $x = -50.21$ **67.** $x = 1.067$

Section 8.4: Applications: Number Problems and Consecutive Integers

Margin Exercises   **1.** 64   **2.** -5 **3.** 5, 23   **4.** $-5, -3,$ and -1 **5.** 12, 14, 16   **6.** $2800   **7.** bagpipes: $288.85; kilt: $218.45

Exercises   **1.** $x - 5 = 13 - x$; 9   **3.** $36 = 2x + 4$; 16   **5.** $7x = 2x + 35$; 7   **7.** $3x + 14 = 6 - x$; -2 **9.** $\dfrac{2x}{5} = x + 6$; -10

11. $4(x - 5) = x + 4$; 8   **13.** $\dfrac{2x + 5}{11} = 4 - x$; 3   **15.** $2x + 3x = 4(x + 3)$; 12   **17.** $n + (n + 2) = 60$; 29, 31

19. $n + (n + 1) + (n + 2) = 69$; 22, 23, 24   **21.** $n + (n + 1) + (n + 2) + (n + 3) = 74$; 17, 18, 19, 20

23. $171 - n = (n + 1) + (n + 2)$; 56, 57, 58   **25.** $208 - 3n = (n + 1) + (n + 2) + (n + 3) - 50$; 42, 43, 44, 45

27. $n + 2(n + 2) = 4(n + 4) - 54$; 42, 44, 46   **29.** $(n + 2) + (n + 4) - n = 66$; 60, 62, 64   **31.** $2n + 3(n + 2) = 2(n + 4) + 7$; 3, 5, 7

33. $c + (c + 49.50) = 125.74$; calculator: $38.12, textbook: $87.62   **35.** $2x + 90{,}000 = 310{,}000$; $110,000

37. $3x + 1500 = 12{,}000$; $3500   **39.** $(x + 56) + x = 542$; 243 boys   **41.** $(x + 68) + x = 158$; $45 million on electric guitars

43. $2x + 0.28(250) = 140$; $35   **45.** $x + x + 2 + x + 6 = 29$; 7 feet, 9 feet, 13 feet   **47.** $x + x + 25{,}000 = 275{,}000$; Lot: $125,000,

House: $150,000   **49.** The difference between five times a number and the number is equal to 8; $x = 2$

51. Find two consecutive integers whose sum is 33; $n = 16, 17$ **53.** Find two consecutive integers such that 3 times the second is 53 more than the first; $n = 25, 26$ **55. a.** $n, n + 2, n + 4, n + 6$ **b.** $n, n + 2, n + 4, n + 6$ **c.** Yes; Answers will vary.

Section 8.5: Working with Formulas

Margin Exercises   **1.** $2020   **2.** $F = 122°$ **3.** 1,012,500 lb   **4.** $30°$ **5.** $\dfrac{P}{V} = I$ **6.** $t = \dfrac{I}{Pr}$ **7.** $x = \dfrac{5}{2}y - 3$

8. a. $y = \dfrac{400 - 25z}{16}$ or $y = -\dfrac{25}{16}z + 25$ **b.** $z = \dfrac{400 - 16y}{25}$ or $z = -\dfrac{16}{25}y + 16$ **9.** $x = 4 - 2y - 3z$

Exercises   **1.** $120   **3.** $10,000   **5. a.** $183.75   **b.** $3683.75   **7.** 2 seconds   **9.** 4 milliliters   **11.** $1030   **13.** 14 rafters

15. 336 in. or 28 ft   **17.** $1030   **19.** 230 calculators   **21.** $5 million   **23.** $2400   **25.** 7 hours   **27.** 1.625 miles per hour

29. $b = P - a - c$ **31.** $m = \dfrac{F}{a}$ **33.** $w = \dfrac{A}{l}$ **35.** $n = \dfrac{R}{p}$ **37.** $P = A - I$ **39.** $m = 2A - n$ **41.** $t = \dfrac{I}{Pr}$ **43.** $b = \dfrac{P - a}{2}$

45. $\beta = 180° - \alpha - \gamma$ **47.** $h = \dfrac{V}{lw}$ **49.** $b = \dfrac{2A}{h}$ **51.** $\pi = \dfrac{A}{r^2}$ **53.** $g = \dfrac{mv^2}{2K}$ **55.** $y = \dfrac{6 - 2x}{3}$ **57.** $x = \dfrac{11 - 2y}{5}$

59. $b = \dfrac{2A - hc}{h}$ or $b = \dfrac{2A}{h} - c$ **61.** $x = \dfrac{8R + 36}{3}$ or $x = \dfrac{8R}{3} + 12$ **63.** $y = -x - 12$ **65.** $C = nt + 9$ **67.** $C = 325n + 5400$

69. a. 0; No, because the numerator will be zero and thus the whole fraction will be equal to zero for all values of s.   **b.** $x < 70$ **c.** Answers will vary.

Section 8.6: Applications: Distance-Rate-Time, Interest, Average

Margin Exercises   **1.** First part took $\dfrac{4}{3}$ or $1\dfrac{1}{3}$ hours; Second part took $\dfrac{8}{3}$ or $2\dfrac{2}{3}$ hours

2. $2300 in the low-risk stock; $12,700 in the high-risk stock   **3.** 53   **4.** $2300

Exercises   **1.** 1.68 mph   **3.** 8.75 hours   **5.** 3 hours   **7.** 60 mph; 300 miles   **9.** 7 hours   **11.** 36 mph; 60 mph

13. Day = 56 mph; Night = 69 mph   **15.** 4.5 miles   **17.** $14,000 at 5%; $11,000 at 6%   **19.** 4.5% on $10,000; 5.5% on $6000

21. 6.5% on $4000; 6% on $3000   **23.** $24,000 at 4.5%; $18,000 at 6%   **25.** $6720 at 5.5%; $5280 at 7%

27. $600 at 2.5%; $800 at 4%   **29.** $11,000 at 4%; $9500 at 5%   **31.** $40.50   **33.** $113.75   **35.** 71   **37.** 62 min   **39.** 6 hours   **41.** 67

Section 8.7: Linear Inequalities

Margin Exercises   **1.** 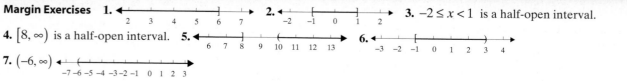   **2.**   **3.** $-2 \le x < 1$ is a half-open interval.

4. $[8, \infty)$ is a half-open interval.   **5.**   **6.**

7. $(-6, \infty)$

8. $\left[\dfrac{4}{5}, \infty\right)$ **9.** $\left[-\dfrac{1}{9}, \infty\right)$ **10.** $(-\infty, 2]$

11. $(-3, 1]$ **12.** $(-3, 2)$ **13.** $\left[3, \dfrac{27}{5}\right]$

14. The maximum final dosage that can be administered safely is 400 milligrams   **15.** Ashley can buy at most 8 rose centerpieces.

Exercises   **1.**   **3.**   **5.**   **7.** no solution   **9.** $\{x \mid 3 \le x < 5\}$

11. $\{x \mid x \ge -2.5\}$ **13. a.** $\{x \mid -3 \le x \le 1\}$; **b.** $[-3, 1]$; **c.** closed interval   **15. a.** $\{x \mid -8 < x \le -2\}$; **b.** $(-8, -2]$; **c.** half-open interval

17. a. $\{x \mid x \le 1\}$; **b.** $(-\infty, 1]$; **c.** half-open interval   **19. a.** $\{x \mid -4 < x < 4\}$; **b.** $(-4, 4)$; **c.** open interval

21. half-open interval   **23.** open interval

25. half-open interval   **27.** closed interval

29. half-open interval   **31.** $(-\infty, 1)$ **33.** $(2, \infty)$

35. $\left(\dfrac{8}{3}, \infty\right)$ **37.** $[-0.4, \infty)$ **39.** $(-\infty, 2)$

41. $[3, \infty)$ **43.** $(-\infty, 2]$ **45.** $\left(-\dfrac{1}{3}, \infty\right)$

47. $\left(-\infty, -\dfrac{11}{2}\right)$ **49.** $[1, \infty)$ **51.** $\left(\dfrac{9}{2}, \infty\right)$

53. $(-\infty, -0.6)$ **55.** $(6.5, \infty)$ **57.** $(-\infty, -13)$

59. $\left(-\infty, -\dfrac{10}{9}\right]$ **61.** $[-4, \infty)$ **63.** $(-\infty, 1]$

65. $(-17, \infty)$ **67.** $(-\infty, -30)$ **69.** $\left(-\infty, -\dfrac{29}{2}\right]$

71. $(-9, 1)$ **73.** $\left[\dfrac{1}{2}, \dfrac{3}{2}\right]$ **75.** $[3, 15]$

77. $(-10, -5)$ **79.** $(-2.8, -0.3)$

81. a. The student would need a score higher than 102 points, which is not possible. Thus he cannot earn an A in the course.
b. The student must score at least 192 points to earn an A in the course.   **83.** He must sell at least 10 cars.   **85.** Answers will vary.

Chapter 8: Review

1. $x = 7$ **2.** $x = 14$ **3.** $y = -11$ **4.** $n = -8$ **5.** $n = 6$ **6.** $y = -18$ **7.** $y = -30$ **8.** $y = 0.4$ **9.** $x = 7$ **10.** $t = -9$ **11.** $x = -5$

12. $x = -8$ **13.** $y = 9$ **14.** $y = -7$ **15.** $x = -25$ **16.** $y = -3$ **17.** $n = \dfrac{2}{5}$ **18.** $x = 7$ **19.** $x = -8.9$ **20.** $x = 12$ **21.** $y = 1.02$

22. $n = 0$ **23.** $x = 27$ **24.** $n = -12$ **25.** $x = 4$ **26.** $x = \dfrac{27}{5}$ **27.** $y = -2$ **28.** $x = -\dfrac{1}{9}$ **29.** $x = \dfrac{2}{15}$ **30.** $y = -35.6$

31. contradiction   **32.** identity   **33.** $42 = 4 + 2x$; 19   **34.** $3x - 7 = 17 - x$; 6   **35.** $\dfrac{2x}{9} = x - 7$; 9

36. $x + (x + 2) + (x + 4) = 279$; 91, 93, 95   **37.** $(x + 2) + 2x = (x + 4) + 18$; 10, 12, 14

38. $x + (x + 1) + (x + 2) + (x + 3) = 190$; 46, 47, 48, 49   **39.** $2x + 150{,}000 = 640{,}000$; \$245,000   **40.** $15{,}325 = 7621 + 642x$; 12 months

41. $5x + 36 = 61$; 25 houses   **42.** $16 + 82 + 6x = 200$; 17 packs   **43.** $C = 0°$ **44.** 47.5 mph   **45.** 40 meters   **46.** $P = 0.16$

47. $\pi = \dfrac{L}{2rh}$ **48.** $R = P + C$ **49.** $g = -\dfrac{v - v_0}{t}$ **50.** $m = \dfrac{2gK}{v^2}$ **51.** $y = 6 - 3x$ **52.** $x = \dfrac{10 + 2y}{5}$ **53.** $\dfrac{11}{3}$ hr; $\dfrac{7}{3}$ hr

54. 55 mph, 247.5 miles   **55.** George: 55 mph; Joshua: 50 mph   **56.** \$30,000 at 4%, \$10,000 at 6.5%

57. \$4000 at 6%, \$16,000 at 8%   **58.** \$80   **59.** \$220   **60.** 77 points   **61.** \$220   **62.**

63.   **64.** $\{x \mid -1 \le x < 6\}$ **65. a.** $\{x \mid -4 < x < 2\}$ **b.** $(-4, 2)$ **c.** open interval   **66. a.** $\{x \mid x \ge 3\}$

b. $[3, \infty)$ **c.** half-open interval   **67.** $[1, \infty)$ **68.** $(-\infty, 7]$

69. $(-\infty,-5)$ **70.** $\left(-\infty,\dfrac{5}{4}\right)$ **71.** $(-1,3)$

72. $[2,6)$

Chapter 8: Test

1. $x=-4$ **2.** $y=-29$ **3.** $x=-3$ **4.** $x=-6$ **5.** $x=5$ **6.** $y=0$ **7.** $a=-\dfrac{9}{8}$ **8.** $x=\dfrac{3}{2}$ **9.** $x=20$ **10.** $x=-2$ **11.** identity

12. conditional **13.** $(2y+5)+y=-22;\ -9,-13$ **14.** $3(n+2)=n+(n+4)+27;\ n=25,27,29$ **15.** $2n+3(n+1)=83;\ 16,17$

16. $2(1)+1(3)+2x=21;\ 8$ 2-pt shots **17.** $m=\dfrac{N-p}{rt}$ **18.** $y=\dfrac{7-5x}{3}$ **19.** \$29.00 **20. a.** 18 months **b.** \$490

21. \$12,000 at 5%, \$13,000 at 3.5% **22.** 4 hours **23.** \$332.50 **24.** 88 **25. a.** $\{x|x<5\}$ **b.** $(-\infty,5)$ **c.** open interval

26. a. $\{x|-1\le x\le 4\}$ **b.** $[-1,4]$ **c.** closed interval **27.** $(-5,\infty)$

28. $\left(-\infty,-\dfrac{7}{2}\right)$ **29.** $(-1,5)$

Cumulative Review: Chapters 1 – 8

1. 17.00 **2.** 0.4 **3.** 0.015 **4.** 2, 3, and 4 **5.** $2^2\cdot 3^2\cdot 11$ **6.** 210 **7.** 0 **8.** 7 and 160 **9.** 50 **10.** $x=\dfrac{3}{8}$ **11.** 196 **12.** 79

13. \$200 **14.** 120 ft **15.** 2565 cm³; 1486 cm² **16. a.** 25.12 ft **b.** 62.80 ft **c.** 263.76 ft^2 **17.** $x=50°,y=60°$ **18. a.** $\dfrac{7}{18}$ **b.** $\dfrac{13}{18}$

19. a. $\dfrac{1}{36}$ **b.** 0 **c.** $\dfrac{1}{12}$ **20. a.** 320 **b.** 350 **c.** 350 **d.** 200 **21.** Food: \$9000; Housing: \$11,250; Transportation: \$6750;

Miscellaneous: \$6750; Savings: \$2250; Education: \$4500; Taxes: \$4500 **22.** $\dfrac{2}{7}$ **23.** 113.2 **24.** -432 **25.** $\dfrac{9}{20}$ **26.** 10

27. 13 **28.** $y=6$ **29.** $x=18$ **30.** $x=12$ **31.** $x=-1$ **32.** $x=-11$ **33.** $y=\dfrac{24}{5}$ **34.** conditional **35.** contradiction

36. 18, 20, 22 **37.** Hard drive: \$84.60; Battery: \$46.15 **38.** $\{x|x\ge -8.3\}$ **39.** **40. a.** $[-3,3]$ **b.** $(6,8)$

c. $[0.3,2.7)$ **41.** $\left[\dfrac{13}{3},\infty\right)$ **42.** $(-6,2)$

Chapter 9: Linear Equations and Inequalities in Two Variables

Section 9.1: The Cartesian Coordinate System

Margin Exercises 1.a. **b.** **2. a.** $(0,-3),(2,5)$ satisfy the equation $y=4x-3$

b. $(0,3),(5,2)$ $(15,0),(30,-3)$

c.

x	y	(x,y)
0	2	$(0,2)$
$\dfrac{1}{3}$	1	$\left(\dfrac{1}{3},1\right)$
-2	8	$(-2,8)$
$\dfrac{2}{3}$	0	$\left(\dfrac{2}{3},0\right)$

3. a. For example: $(-5,-2)(0,1)(5,4)$ **b.** For example: $(-5,5)(0,1)(5,-3)$

Exercises 1. $\{A(-5,1),B(-3,3),C(-1,1),D(1,2),E(2,-2)\}$

3. $\{A(-3,-2),B(-1,-3),C(-1,3),D(0,0),E(2,1)\}$

5. $\{A(-4,4),B(-3,-4),C(0,-4),D(0,3),E(4,1)\}$ **7.** $\{A(-3,-5),B(-1,4),C(0,-1),D(3,1),E(6,0)\}$

9. $\{A(-5,0),B(-2,2),C(-1,-4),D(0,6),E(2,0)\}$

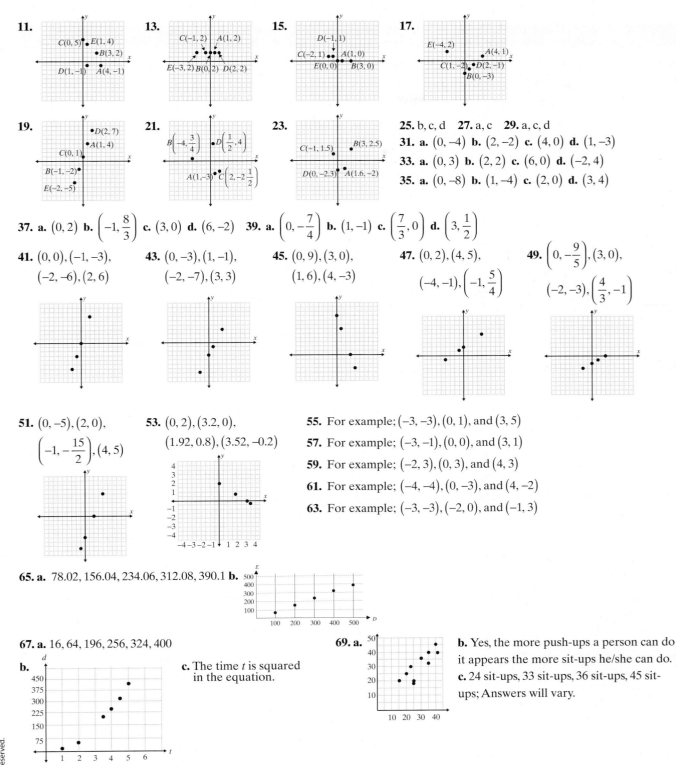

11.

$C(0, 5)$ $E(1, 4)$ $B(3, 2)$ $D(1, -1)$ $A(4, -1)$

13.

$C(-1, 2)$ $A(1, 2)$ $E(-3, 2)$ $B(0, 2)$ $D(2, 2)$

15.

$D(-1, 1)$ $C(-2, 1)$ $A(1, 0)$ $E(0, 0)$ $B(3, 0)$

17.

$E(-4, 2)$ $A(4, 1)$ $C(1, -2)$ $D(2, -1)$ $B(0, -3)$

19.

$D(2, 7)$ $A(1, 4)$ $C(0, 1)$ $B(-1, -2)$ $E(-2, -5)$

21.

$B\left(-4, \frac{3}{4}\right)$ $D\left(\frac{1}{2}, 4\right)$ $A(1, -3)$ $C\left(2, -2\frac{1}{2}\right)$

23.

$C(-1, 1.5)$ $B(3, 2.5)$ $D(0, -2.3)$ $A(1.6, -2)$

25. b, c, d   **27.** a, c   **29.** a, c, d

31. a. $(0, -4)$ **b.** $(2, -2)$ **c.** $(4, 0)$ **d.** $(1, -3)$

33. a. $(0, 3)$ **b.** $(2, 2)$ **c.** $(6, 0)$ **d.** $(-2, 4)$

35. a. $(0, -8)$ **b.** $(1, -4)$ **c.** $(2, 0)$ **d.** $(3, 4)$

37. a. $(0, 2)$ **b.** $\left(-1, \frac{8}{3}\right)$ **c.** $(3, 0)$ **d.** $(6, -2)$ **39. a.** $\left(0, -\frac{7}{4}\right)$ **b.** $(1, -1)$ **c.** $\left(\frac{7}{3}, 0\right)$ **d.** $\left(3, \frac{1}{2}\right)$

41. $(0, 0), (-1, -3),$ $(-2, -6), (2, 6)$

43. $(0, -3), (1, -1),$ $(-2, -7), (3, 3)$

45. $(0, 9), (3, 0),$ $(1, 6), (4, -3)$

47. $(0, 2), (4, 5),$ $(-4, -1), \left(-1, \frac{5}{4}\right)$

49. $\left(0, -\frac{9}{5}\right), (3, 0),$ $(-2, -3), \left(\frac{4}{3}, -1\right)$

51. $(0, -5), (2, 0),$ $\left(-1, -\frac{15}{2}\right), (4, 5)$

53. $(0, 2), (3.2, 0),$ $(1.92, 0.8), (3.52, -0.2)$

55. For example; $(-3, -3), (0, 1),$ and $(3, 5)$

57. For example; $(-3, -1), (0, 0),$ and $(3, 1)$

59. For example; $(-2, 3), (0, 3),$ and $(4, 3)$

61. For example; $(-4, -4), (0, -3),$ and $(4, -2)$

63. For example; $(-3, -3), (-2, 0),$ and $(-1, 3)$

65. a. 78.02, 156.04, 234.06, 312.08, 390.1 **b.**

67. a. 16, 64, 196, 256, 324, 400 **b.** **c.** The time t is squared in the equation.

69. a. **b.** Yes, the more push-ups a person can do it appears the more sit-ups he/she can do. **c.** 24 sit-ups, 33 sit-ups, 36 sit-ups, 45 sit-ups; Answers will vary.

71. Answers will vary. Not all scatterplots can be used to predict information related to the two variables graphed because not all variables are related.

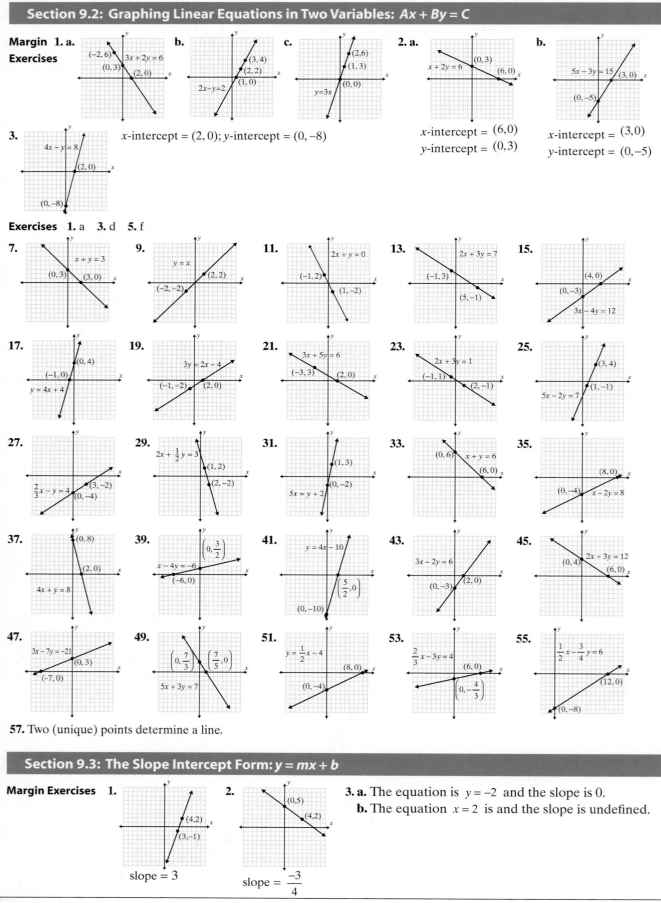

Margin Exercises 1. a. b. c. 2. a. b.

3. x-intercept = $(2, 0)$; y-intercept = $(0, -8)$

2. a. x-intercept = $(6, 0)$
y-intercept = $(0, 3)$

b. x-intercept = $(3, 0)$
y-intercept = $(0, -5)$

Exercises 1. a 3. d 5. f

7. 9. 11. 13. 15.

17. 19. 21. 23. 25.

27. 29. 31. 33. 35.

37. 39. 41. 43. 45.

47. 49. 51. 53. 55.

57. Two (unique) points determine a line.

Section 9.3: The Slope Intercept Form: $y = mx + b$

Margin Exercises 1. 2.

3. a. The equation is $y = -2$ and the slope is 0.
   b. The equation $x = 2$ is and the slope is undefined.

slope = 3

slope = $\dfrac{-3}{4}$

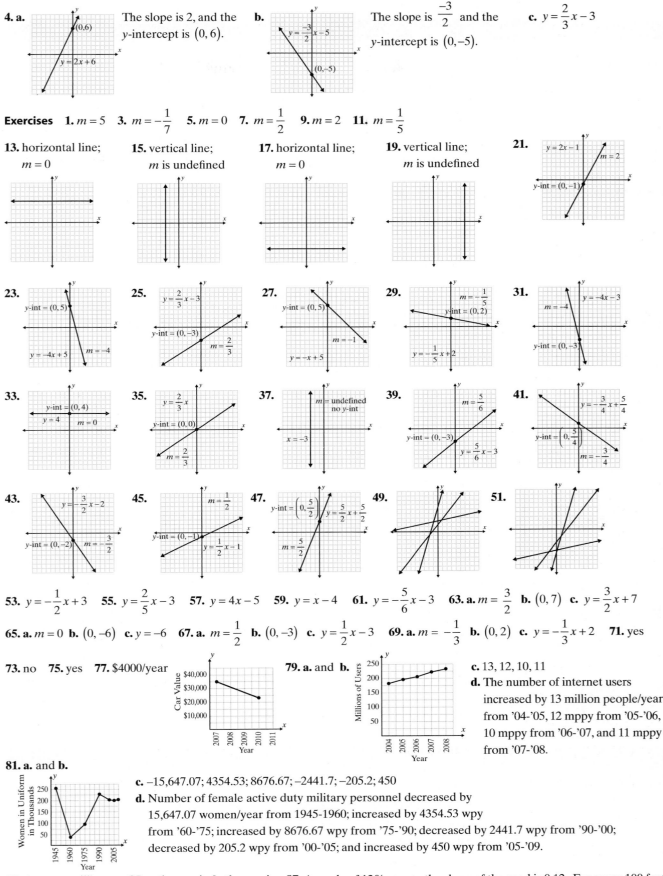

4. a. The slope is 2, and the y-intercept is $(0, 6)$.
$y = 2x + 6$

b. The slope is $\dfrac{-3}{2}$ and the y-intercept is $(0, -5)$.
$y = \dfrac{-3}{2}x - 5$

c. $y = \dfrac{2}{3}x - 3$

Exercises **1.** $m = 5$ **3.** $m = -\dfrac{1}{7}$ **5.** $m = 0$ **7.** $m = \dfrac{1}{2}$ **9.** $m = 2$ **11.** $m = \dfrac{1}{5}$

13. horizontal line; $m = 0$

15. vertical line; m is undefined

17. horizontal line; $m = 0$

19. vertical line; m is undefined

21. $y = 2x - 1$ $m = 2$ y-int $= (0, -1)$

23. y-int $= (0, 5)$ $y = -4x + 5$ $m = -4$

25. $y = \dfrac{2}{3}x - 3$ y-int $= (0, -3)$ $m = \dfrac{2}{3}$

27. y-int $= (0, 5)$ $m = -1$ $y = -x + 5$

29. $m = -\dfrac{1}{5}$ y-int $= (0, 2)$ $y = -\dfrac{1}{5}x + 2$

31. $m = -4$ $y = -4x - 3$ y-int $= (0, -3)$

33. y-int $= (0, 4)$ $y = 4$ $m = 0$

35. $y = \dfrac{2}{3}x$ y-int $= (0, 0)$ $m = \dfrac{2}{3}$

37. $m =$ undefined no y-int $x = -3$

39. $m = \dfrac{5}{6}$ y-int $= (0, -3)$ $y = \dfrac{5}{6}x - 3$

41. $y = -\dfrac{3}{4}x + \dfrac{5}{4}$ y-int $= \left(0, \dfrac{5}{4}\right)$ $m = -\dfrac{3}{4}$

43. $y = -\dfrac{3}{2}x - 2$ y-int $= (0, -2)$ $m = -\dfrac{3}{2}$

45. $m = \dfrac{1}{2}$ y-int $= (0, -1)$ $y = \dfrac{1}{2}x - 1$

47. y-int $= \left(0, \dfrac{5}{2}\right)$ $y = \dfrac{5}{2}x + \dfrac{5}{2}$ $m = \dfrac{5}{2}$

49.

51.

53. $y = -\dfrac{1}{2}x + 3$ **55.** $y = \dfrac{2}{5}x - 3$ **57.** $y = 4x - 5$ **59.** $y = x - 4$ **61.** $y = -\dfrac{5}{6}x - 3$ **63. a.** $m = \dfrac{3}{2}$ **b.** $(0, 7)$ **c.** $y = \dfrac{3}{2}x + 7$

65. a. $m = 0$ **b.** $(0, -6)$ **c.** $y = -6$ **67. a.** $m = \dfrac{1}{2}$ **b.** $(0, -3)$ **c.** $y = \dfrac{1}{2}x - 3$ **69. a.** $m = -\dfrac{1}{3}$ **b.** $(0, 2)$ **c.** $y = -\dfrac{1}{3}x + 2$ **71.** yes

73. no **75.** yes **77.** \$4000/year

79. a. and **b.**

c. 13, 12, 10, 11
d. The number of internet users increased by 13 million people/year from '04-'05, 12 mppy from '05-'06, 10 mppy from '06-'07, and 11 mppy from '07-'08.

81. a. and **b.**

c. −15,647.07; 4354.53; 8676.67; −2441.7; −205.2; 450
d. Number of female active duty military personnel decreased by 15,647.07 women/year from 1945-1960; increased by 4354.53 wpy from '60-'75; increased by 8676.67 wpy from '75-'90; decreased by 2441.7 wpy from '90-'00; decreased by 205.2 wpy from '00-'05; and increased by 450 wpy from '05-'09.

83. Answers will vary. **85. a.** the x-axis **b.** the y-axis **87.** A grade of 12% means the slope of the road is 0.12. For every 100 feet of horizontal distance (run) there is 12 feet of vertical distance (rise).

Section 9.4: The Point-Slope Form: $y - y_1 = m(x - x_1)$

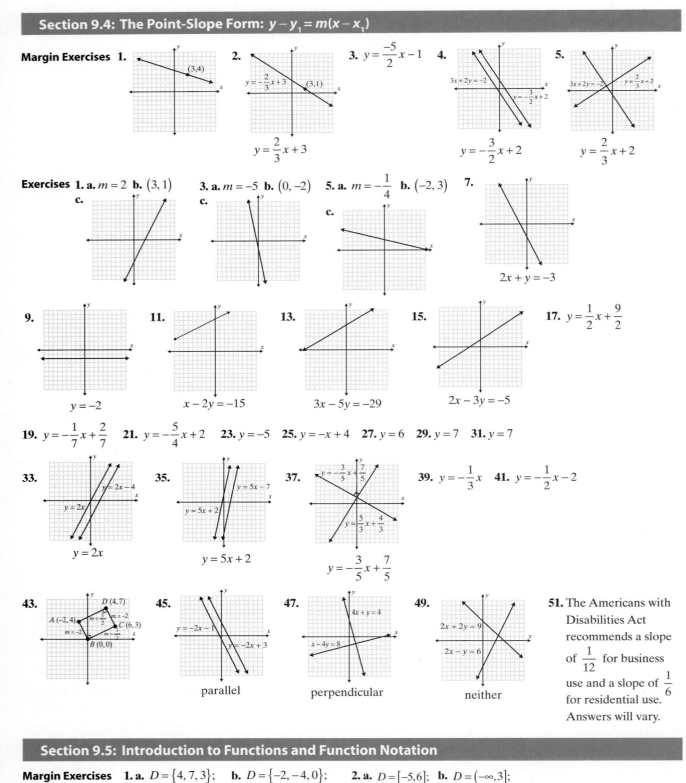

Margin Exercises 1. (3,4)

2. $y = -\dfrac{2}{3}x + 3$ (3,1) $y = \dfrac{2}{3}x + 3$

3. $y = \dfrac{-5}{2}x - 1$

4. $3x + 2y = -2$ $y = -\dfrac{3}{2}x + 2$ $y = -\dfrac{3}{2}x + 2$

5. $3x + 2y = -2$ $y = \dfrac{2}{3}x + 2$ $y = \dfrac{2}{3}x + 2$

Exercises 1. a. $m = 2$ **b.** $(3, 1)$ **c.**

3. a. $m = -5$ **b.** $(0, -2)$ **c.**

5. a. $m = -\dfrac{1}{4}$ **b.** $(-2, 3)$ **c.**

7. $2x + y = -3$

9. $y = -2$

11. $x - 2y = -15$

13. $3x - 5y = -29$

15. $2x - 3y = -5$

17. $y = \dfrac{1}{2}x + \dfrac{9}{2}$

19. $y = -\dfrac{1}{7}x + \dfrac{2}{7}$ **21.** $y = -\dfrac{5}{4}x + 2$ **23.** $y = -5$ **25.** $y = -x + 4$ **27.** $y = 6$ **29.** $y = 7$ **31.** $y = 7$

33. $y = 2x - 4$ $y = 2x$ $y = 2x$

35. $y = 5x - 7$ $y = 5x + 2$ $y = 5x + 2$

37. $y = -\dfrac{3}{5}x + \dfrac{7}{5}$ $y = \dfrac{5}{3}x + \dfrac{4}{3}$ $y = -\dfrac{3}{5}x + \dfrac{7}{5}$

39. $y = -\dfrac{1}{3}x$ **41.** $y = -\dfrac{1}{2}x - 2$

43. $D(4,7)$ $A(-2,4)$ $m = -\dfrac{1}{2}$ $m = -2$ $C(6,3)$ $m = -2$ $m = -\dfrac{1}{2}$ $B(0,0)$

45. $y = -2x - 1$ $y = -2x + 3$ parallel

47. $4x + y = 4$ $x - 4y = 8$ perpendicular

49. $2x + 2y = 9$ $2x - y = 6$ neither

51. The Americans with Disabilities Act recommends a slope of $\dfrac{1}{12}$ for business use and a slope of $\dfrac{1}{6}$ for residential use. Answers will vary.

Section 9.5: Introduction to Functions and Function Notation

Margin Exercises   1. a. $D = \{4, 7, 3\}$;   **b.** $D = \{-2, -4, 0\}$;   **2. a.** $D = [-5, 6]$;   **b.** $D = (-\infty, 3]$;
$R = \{5, 3, 6\}$ $R = \{3, -3, 0\}$ $R = [-5, 5]$ $R = (-\infty, \infty)$

3. a. not a function **b.** function   **4. a.** not a function; $D = \{-7, -3, 0, 2, 4, 5\}$; $R = \{-2, 0, 2, 3, 6\}$
   b. function $D = (-\infty, \infty)$; $R = [-2, \infty)$
   c. not a function $D = [-5, 7]$; $R = [-1, 5]$

5. $D = (-\infty, -3) \cup (-3, \infty)$ or $x \neq -3$ **6. a.** 7 **b.** -8 **c.** -2 **7. a.** 4 **b.** 5 **c.** -16

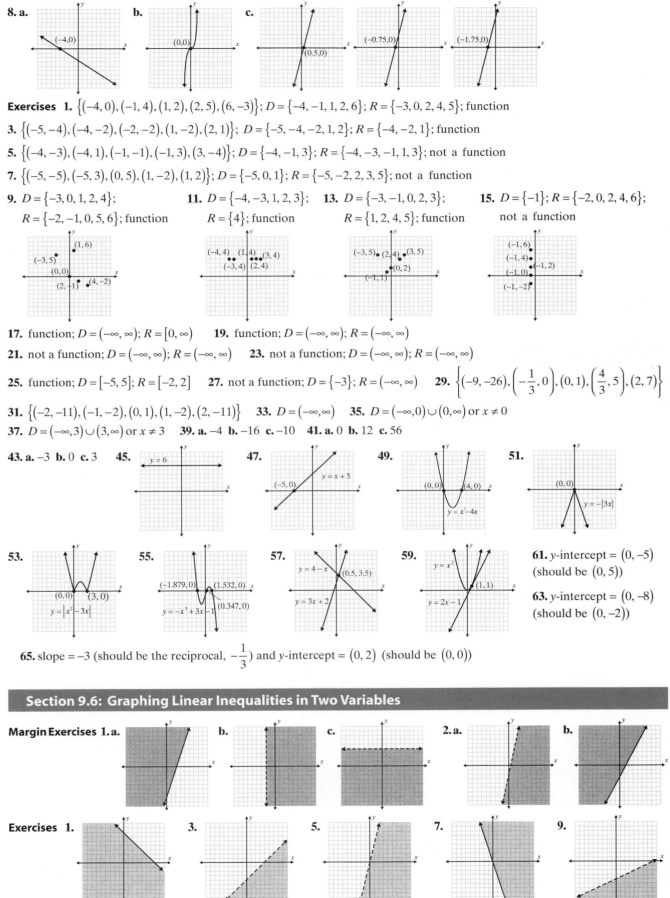

8. a. (−4,0)

b. (0,0)

c. (0.5,0) (−0.75,0) (−1.75,0)

Exercises **1.** $\{(-4,0),(-1,4),(1,2),(2,5),(6,-3)\}$; $D=\{-4,-1,1,2,6\}$; $R=\{-3,0,2,4,5\}$; function

3. $\{(-5,-4),(-4,-2),(-2,-2),(1,-2),(2,1)\}$; $D=\{-5,-4,-2,1,2\}$; $R=\{-4,-2,1\}$; function

5. $\{(-4,-3),(-4,1),(-1,-1),(-1,3),(3,-4)\}$; $D=\{-4,-1,3\}$; $R=\{-4,-3,-1,1,3\}$; not a function

7. $\{(-5,-5),(-5,3),(0,5),(1,-2),(1,2)\}$; $D=\{-5,0,1\}$; $R=\{-5,-2,2,3,5\}$; not a function

9. $D=\{-3,0,1,2,4\}$; **11.** $D=\{-4,-3,1,2,3\}$; **13.** $D=\{-3,-1,0,2,3\}$; **15.** $D=\{-1\}$; $R=\{-2,0,2,4,6\}$;
$R=\{-2,-1,0,5,6\}$; function $R=\{4\}$; function $R=\{1,2,4,5\}$; function   not a function

17. function; $D=(-\infty,\infty)$; $R=[0,\infty)$ **19.** function; $D=(-\infty,\infty)$; $R=(-\infty,\infty)$

21. not a function; $D=(-\infty,\infty)$; $R=(-\infty,\infty)$ **23.** not a function; $D=(-\infty,\infty)$; $R=(-\infty,\infty)$

25. function; $D=[-5,5]$; $R=[-2,2]$ **27.** not a function; $D=\{-3\}$; $R=(-\infty,\infty)$ **29.** $\left\{(-9,-26),\left(-\frac{1}{3},0\right),(0,1),\left(\frac{4}{3},5\right),(2,7)\right\}$

31. $\{(-2,-11),(-1,-2),(0,1),(1,-2),(2,-11)\}$ **33.** $D=(-\infty,\infty)$ **35.** $D=(-\infty,0)\cup(0,\infty)$ or $x\neq 0$

37. $D=(-\infty,3)\cup(3,\infty)$ or $x\neq 3$ **39. a.** −4 **b.** −16 **c.** −10   **41. a.** 0 **b.** 12 **c.** 56

43. a. −3 **b.** 0 **c.** 3   **45.** $y=6$ **47.** $(-5,0)$ $y=x+5$ **49.** $(0,0)$ $(4,0)$ $y=x^2-4x$ **51.** $(0,0)$ $y=-|3x|$

53. $(0,0)$ $(3,0)$ $y=|x^2-3x|$ **55.** $(-1.879,0)$ $(1.532,0)$ $(0.347,0)$ $y=-x^3+3x-1$ **57.** $y=4-x$ $(0.5,3.5)$ $y=3x+2$ **59.** $y=x^2$ $(1,1)$ $y=2x-1$ **61.** y-intercept $=(0,-5)$ (should be $(0,5)$)

63. y-intercept $=(0,-8)$ (should be $(0,-2)$)

65. slope $=-3$ (should be the reciprocal, $-\frac{1}{3}$) and y-intercept $=(0,2)$ (should be $(0,0)$)

Section 9.6: Graphing Linear Inequalities in Two Variables

Margin Exercises **1. a.** **b.** **c.** **2. a.** **b.**

Exercises **1.** **3.** **5.** **7.** **9.**

11. 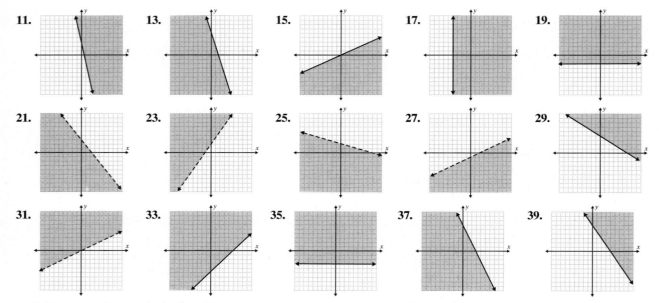 **13.** **15.** **17.** **19.**

21. **23.** **25.** **27.** **29.**

31. **33.** **35.** **37.** **39.**

41. Test any point not on the line. If the test point satisfies the inequality, shade the half-plane on that side of the line. Otherwise, shade the other half-plane.

Chapter 9: Review

1. $\{A(-6, -5), B(-5, 6), C(1, -3), D(3, 6), E(7, -1)\}$ **2.** $\{A(-4, -5), B(-3, 1), C(1, 2), D(1, -1), E(5, 3)\}$

3. $\{A(-7, 2), B(-1, 7), C(-1, -3), D(3, -1), E(4, -6)\}$ **4.** **5.** **6.**

7. none **8.** c **9.** a, c, d **10.** a, b, c

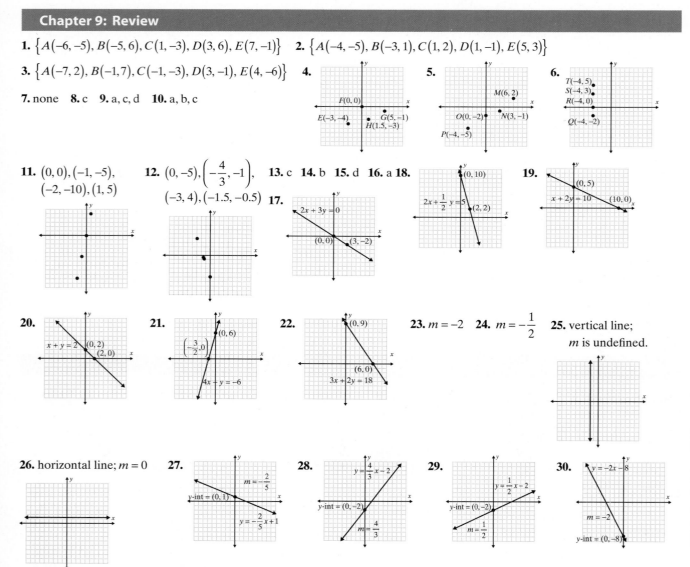

11. $(0, 0), (-1, -5),$
$(-2, -10), (1, 5)$

12. $(0, -5), \left(-\dfrac{4}{3}, -1\right),$
$(-3, 4), (-1.5, -0.5)$

13. c **14.** b **15.** d **16.** a

17. **18.** **19.**

20. **21.** **22.** **23.** $m = -2$ **24.** $m = -\dfrac{1}{2}$ **25.** vertical line; m is undefined.

26. horizontal line; $m = 0$ **27.** **28.** **29.** **30.**

31.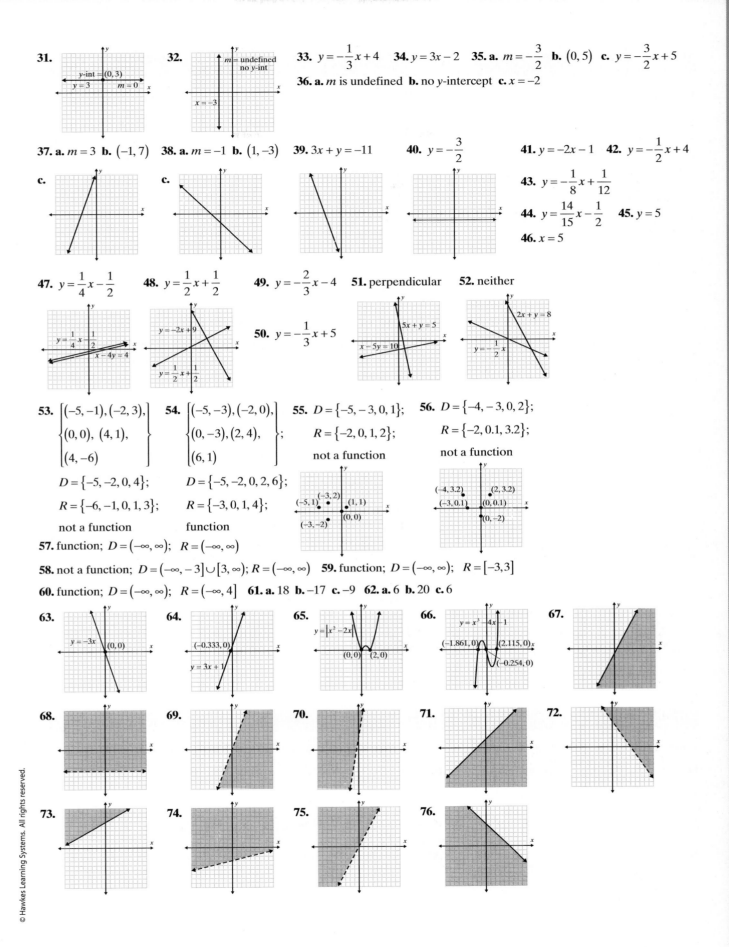
y-int = (0, 3)
y = 3   m = 0

32.
m = undefined
no y-int
x = -3

33. $y = -\dfrac{1}{3}x + 4$ **34.** $y = 3x - 2$ **35. a.** $m = -\dfrac{3}{2}$ **b.** $(0, 5)$ **c.** $y = -\dfrac{3}{2}x + 5$

36. a. m is undefined   **b.** no y-intercept   **c.** $x = -2$

37. a. $m = 3$ **b.** $(-1, 7)$ **38. a.** $m = -1$ **b.** $(1, -3)$ **39.** $3x + y = -11$ **40.** $y = -\dfrac{3}{2}$ **41.** $y = -2x - 1$ **42.** $y = -\dfrac{1}{2}x + 4$

c.   **c.**

43. $y = -\dfrac{1}{8}x + \dfrac{1}{12}$

44. $y = \dfrac{14}{15}x - \dfrac{1}{2}$ **45.** $y = 5$

46. $x = 5$

47. $y = \dfrac{1}{4}x - \dfrac{1}{2}$ **48.** $y = \dfrac{1}{2}x + \dfrac{1}{2}$ **49.** $y = -\dfrac{2}{3}x - 4$ **51.** perpendicular   **52.** neither

$y = \dfrac{1}{4}x - \dfrac{1}{2}$ $y = -2x + 9$ **50.** $y = -\dfrac{1}{3}x + 5$ $5x + y = 5$ $2x + y = 8$
$x - 4y = 4$ $y = \dfrac{1}{2}x + \dfrac{1}{2}$ $x - 5y = 10$ $y = -\dfrac{1}{2}x$

53. $\begin{Bmatrix} (-5, -1), (-2, 3), \\ (0, 0), (4, 1), \\ (4, -6) \end{Bmatrix}$ **54.** $\begin{Bmatrix} (-5, -3), (-2, 0), \\ (0, -3), (2, 4), \\ (6, 1) \end{Bmatrix}$;   **55.** $D = \{-5, -3, 0, 1\}$;   **56.** $D = \{-4, -3, 0, 2\}$;

$R = \{-2, 0, 1, 2\}$; $R = \{-2, 0.1, 3.2\}$;

$D = \{-5, -2, 0, 4\}$; $D = \{-5, -2, 0, 2, 6\}$;   not a function   not a function

$R = \{-6, -1, 0, 1, 3\}$; $R = \{-3, 0, 1, 4\}$;

not a function   function
$(-3, 2)$ $(-4, 3.2)$ $(2, 3.2)$
$(-5, 1)$ $(1, 1)$ $(-3, 0.1)$ $(0, 0.1)$
$(-3, -2)$ $(0, 0)$ $(0, -2)$

57. function; $D = (-\infty, \infty)$; $R = (-\infty, \infty)$

58. not a function; $D = (-\infty, -3] \cup [3, \infty)$; $R = (-\infty, \infty)$ **59.** function; $D = (-\infty, \infty)$; $R = [-3, 3]$

60. function; $D = (-\infty, \infty)$; $R = (-\infty, 4]$ **61. a.** 18   **b.** -17 **c.** -9 **62. a.** 6   **b.** 20   **c.** 6

63.   **64.**   **65.**   **66.**   **67.**
$y = -3x$ $(0, 0)$ $(-0.333, 0)$ $y = |x^2 - 2x|$ $y = x^3 - 4x - 1$
$y = 3x + 1$ $(0, 0)$ $(2, 0)$ $(-1.861, 0)$ $(2.115, 0)$
$(-0.254, 0)$

68.   **69.**   **70.**   **71.**   **72.**

73.   **74.**   **75.**   **76.**

Chapter 9: Test

1. $\{A(-3,2), B(-2,-1), C(0,-3), D(1,2), E(3,4), F(5,0), G(4,-2)\}$

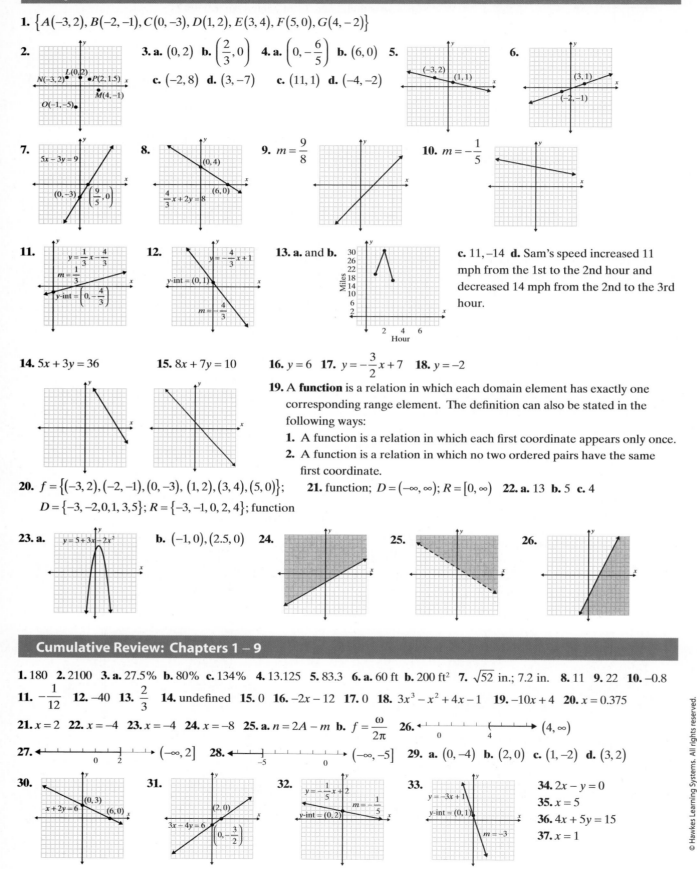

2.

3. a. $(0,2)$ **b.** $\left(\dfrac{2}{3},0\right)$ **4. a.** $\left(0,-\dfrac{6}{5}\right)$ **b.** $(6,0)$ **5.**

c. $(-2,8)$ **d.** $(3,-7)$ **c.** $(11,1)$ **d.** $(-4,-2)$

6.

7. $5x-3y=9$

8. $\dfrac{4}{3}x+2y=8$

9. $m=\dfrac{9}{8}$

10. $m=-\dfrac{1}{5}$

11. $y=\dfrac{1}{3}x-\dfrac{4}{3}$; $m=\dfrac{1}{3}$; y-int $=\left(0,-\dfrac{4}{3}\right)$

12. $y=-\dfrac{4}{3}x+1$; y-int $=(0,1)$; $m=-\dfrac{4}{3}$

13. a. and b.

c. $11,-14$ **d.** Sam's speed increased 11 mph from the 1st to the 2nd hour and decreased 14 mph from the 2nd to the 3rd hour.

14. $5x+3y=36$ **15.** $8x+7y=10$ **16.** $y=6$ **17.** $y=-\dfrac{3}{2}x+7$ **18.** $y=-2$

19. A **function** is a relation in which each domain element has exactly one corresponding range element. The definition can also be stated in the following ways:
1. A function is a relation in which each first coordinate appears only once.
2. A function is a relation in which no two ordered pairs have the same first coordinate.

20. $f=\{(-3,2),(-2,-1),(0,-3),(1,2),(3,4),(5,0)\}$; $D=\{-3,-2,0,1,3,5\}$; $R=\{-3,-1,0,2,4\}$; function

21. function; $D=(-\infty,\infty)$; $R=[0,\infty)$ **22. a.** 13 **b.** 5 **c.** 4

23. a. $y=5+3x-2x^2$ **b.** $(-1,0),(2.5,0)$ **24.**

25.

26.

Cumulative Review: Chapters 1 – 9

1. 180 **2.** 2100 **3. a.** 27.5% **b.** 80% **c.** 134% **4.** 13.125 **5.** 83.3 **6. a.** 60 ft **b.** 200 ft² **7.** $\sqrt{52}$ in.; 7.2 in. **8.** 11 **9.** 22 **10.** −0.8

11. $-\dfrac{1}{12}$ **12.** −40 **13.** $\dfrac{2}{3}$ **14.** undefined **15.** 0 **16.** $-2x-12$ **17.** 0 **18.** $3x^3-x^2+4x-1$ **19.** $-10x+4$ **20.** $x=0.375$

21. $x=2$ **22.** $x=-4$ **23.** $x=-4$ **24.** $x=-8$ **25. a.** $n=2A-m$ **b.** $f=\dfrac{\omega}{2\pi}$ **26.** $(4,\infty)$

27. $(-\infty,2]$ **28.** $(-\infty,-5]$ **29. a.** $(0,-4)$ **b.** $(2,0)$ **c.** $(1,-2)$ **d.** $(3,2)$

30. $x+2y=6$; $(0,3)$; $(6,0)$

31. $3x-4y=6$; $(2,0)$; $\left(0,-\dfrac{3}{2}\right)$

32. $y=-\dfrac{1}{5}x+2$; $m=-\dfrac{1}{5}$; y-int $=(0,2)$

33. $y=-3x+1$; y-int $=(0,1)$; $m=-3$

34. $2x-y=0$
35. $x=5$
36. $4x+5y=15$
37. $x=1$

38. $y = \dfrac{3}{4}x - 3$ **39.** $\{(-3,1),(-1,-1),(-1,3),(1,1),(4,-2),(4,4)\}$; $D = \{-3,-1,1,4\}$; $R = \{-2,-1,1,3,4\}$; not a function

40. not a function; $D = (-\infty, 0]$; $R = (-\infty, \infty)$ **41.** 4 **42.** -32 **43.**

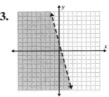

44. 20, 22, 24 **45.** 7 hours **46.** $-\dfrac{1}{3}$

47. 6.25% **48.** length $= 25$ cm; width $= 16$ cm

49. $3400 at 4.5%; $1100 at 7%

50. a. and **b.**

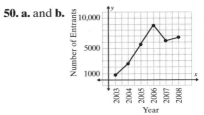

c. 1737; 3043; 3154; -2415; 486

d. Number of entrants increased by 1737 people/year from 2003-2004; increased by 3043 people/year from 2004-2005; increased by 3154 people/year from 2005-2006; decreased by 2415 people/year from 2006-2007; increased by 486 people/year from 2007-2008.

Chapter 10: Systems of Linear Equations

Section 10.1: Systems of Linear Equations: Solutions by Graphing

Margin Exercises 1. $6 - 12 = -6$ true statement; $18 + 8 = 26$ true statement **2.** $2 = -6 + 8$ true statement; $2 = 9 - 8$ false statement

3. $(-3, 0)$ **4.** no solution **5.** $(x, 3x - 4)$ **6.** $\left(\dfrac{3}{2}, 3\right)$ **7.** $(-4, -1)$

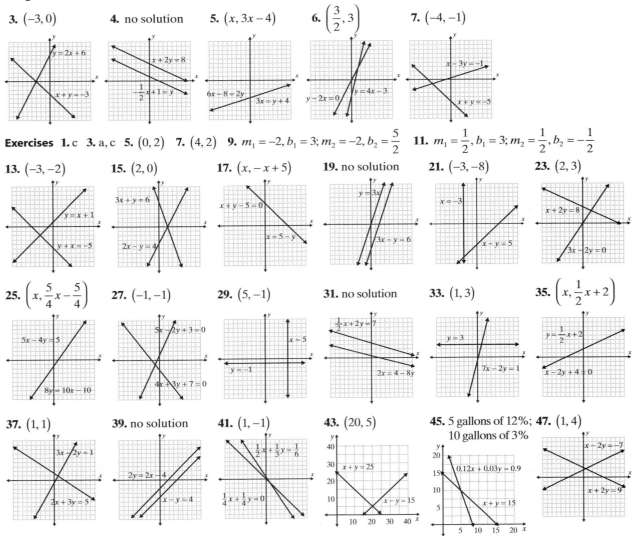

Exercises 1. c **3.** a, c **5.** $(0, 2)$ **7.** $(4, 2)$ **9.** $m_1 = -2, b_1 = 3; m_2 = -2, b_2 = \dfrac{5}{2}$ **11.** $m_1 = \dfrac{1}{2}, b_1 = 3; m_2 = \dfrac{1}{2}, b_2 = -\dfrac{1}{2}$

13. $(-3, -2)$ **15.** $(2, 0)$ **17.** $(x, -x + 5)$ **19.** no solution **21.** $(-3, -8)$ **23.** $(2, 3)$

25. $\left(x, \dfrac{5}{4}x - \dfrac{5}{4}\right)$ **27.** $(-1, -1)$ **29.** $(5, -1)$ **31.** no solution **33.** $(1, 3)$ **35.** $\left(x, \dfrac{1}{2}x + 2\right)$

37. $(1, 1)$ **39.** no solution **41.** $(1, -1)$ **43.** $(20, 5)$ **45.** 5 gallons of 12% 10 gallons of 3% **47.** $(1, 4)$

49. $(1.5, 2)$ **51.** $(1.5, -3)$ **53.** $(1.4, 2.1)$ **55.** The solution to a consistent system of linear equations is a single point, which is easily written as an ordered pair.

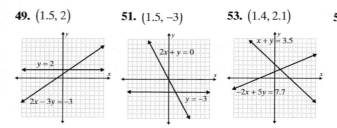

Section 10.2: Systems of Linear Equations: Solutions by Substitution

Margin Exercises 1. $\left(7, \dfrac{29}{5}\right)$ **2.** no solution.   **3.** $(x, 1+2x)$ or $\left(\dfrac{y-1}{2}, y\right)$ **4.** $(-1.5, 7.5)$ **5.** $(-1, 3)$

Exercises 1. $(2, 4)$ **3.** $(x, 3x-7)$ **5.** $(-6, -2)$ **7.** $(4, 1)$ **9.** no solution   **11.** $(3, 2)$ **13.** $(4, -5)$ **15.** $(3, -2)$ **17.** $\left(2, \dfrac{5}{2}\right)$

19. $\left(\dfrac{1}{2}, -4\right)$ **21.** $\left(\dfrac{7}{2}, -\dfrac{1}{2}\right)$ **23.** $(x, 2-3x)$ **25.** $(-2, 1)$ **27.** $\left(-\dfrac{4}{5}, -\dfrac{7}{5}\right)$ **29.** $\left(\dfrac{11}{7}, \dfrac{8}{7}\right)$ **31.** $\left(x, \dfrac{1}{6}x + \dfrac{10}{3}\right)$ **33.** $(3, 3)$ **35.** $(10, 20)$

37. no solution   **39.** $\left(x, -\dfrac{3}{2}x + 12\right)$ **41.** $\left(2, \dfrac{5}{3}\right)$ **43.** $(20, 5)$ **45.** 5 gallons of 12%; 10 gallons of 3%   **47.** Answers will vary.

Section 10.3: Systems of Linear Equations: Solutions by Addition

Margin Exercises 1. $(2, -3)$ **2.** $(x, -2x+5)$ **3.** $(2.4, -4.6)$ **4.** $(-2, 5)$ **5.** $y = \dfrac{5}{2}x - 12$

Exercises 1. $(4, 3)$ **3.** $\left(1, -\dfrac{3}{2}\right)$ **5.** no solution   **7.** $(x, 3-x)$ **9.** $(-2, -3)$ **11.** $(7, 5)$ **13.** $(1, -5)$ **15.** $\left(x, \dfrac{1}{2}x - 2\right)$

17. $\left(\dfrac{22}{7}, -\dfrac{2}{7}\right)$ **19.** $(2, -2)$ **21.** $(x, 2x-4)$ **23.** $(2, -1)$ **25.** $(5, -6)$ **27.** $(3, -1)$ **29.** $(2, 4)$ **31.** $(-6, 2)$ **33.** no solution

35. $(20, 10)$ **37.** $\left(2, \dfrac{10}{9}\right)$ **39.** no solution   **41.** $\left(-\dfrac{45}{7}, \dfrac{92}{7}\right)$ **43.** $y = 5x - 7; m = 5, b = -7$ **45.** $y = -3; m = 0, b = -3$

47. $y = \dfrac{1}{2}x + \dfrac{3}{2}; m = \dfrac{1}{2}, b = \dfrac{3}{2}$ **49.** \$4000 at 10%; \$6000 at 6%   **51.** 40 liters of 30% solution (x); 60 liters of 40% solution (y)

53. Answers will vary.

Section 10.4: Applications: Distance-Rate-Time, Number Problems, Amounts, and Costs

Margin Exercises 1. The wind speed was 0.8 miles per hour and Bob was running 9.2 miles per hour.   **2.** 11:30 AM   **3.** 93 and 57
4. 33 dimes and 11 nickels   **5.** Enrique is 13 and Maria is 4.   **6.** A soda costs \$1.25 and a water bottle costs \$0.80.
Exercises 1. 23, 33   **3.** 15, 21   **5.** 37, 50   **7.** 80°, 100°   **9.** 55°, 55°, 70°   **11.** Rate of boat = 10 mph; Rate of current = 2 mph
13. Bolt's speed was 10.32 meters per second and the wind speed was 0.12 meters per second.

15. He traveled $1\dfrac{1}{2}$ hours at 52 mph and 2 hours at 56 mph.   **17.** Marcos traveled at 40 mph and Cana traveled at 51 mph.

19. Steve traveled at 28 mph and Tim traveled at 7 mph.   **21.** The westbound train was traveling 45 mph and the eastbound
train was traveling 40 mph.   **23.** He jogged 12 miles.   **25.** 20 nickels and 10 dimes   **27.** 52 nickels and 130 pennies
29. 800 adults and 2700 students attended   **31.** $l = 14$ meters; $w = 8$ meters   **33.** 100 yards × 45 yards
35. $l = 33$ meters; $w = 17$ meters   **37.** Priscilla was 21 years old and Elvis was 32 years old.
39. She bought 10 paperbacks and 5 hardbacks.   **41.** 5000 general admission and 7500 reserved tickets were sold.
43. 22 at \$625 and 25 and \$550   **45.** 30 dozen (360 balls)   **47.** The store sold 22 of the \$95 jackets and 18 of the \$120 jackets.
49. One Big Mac costs \$3.58 and one order of medium French fries costs \$1.79.
51. They will produce 7 of Model X and 10 of Model Y.   **53.** The number is 49.

Section 10.5: Applications: Interest and Mixture

Margin Exercises 1. Fergus has invested \$5000 at 9% and \$5800 at 12%.   **2.** He should invest \$2500 at 9% and \$6500 at 5%.
3. 50 ounces of the 18% solution and 100 ounces of the 12% solution
4. 96 gallons of the 22% solution and 24 gallons of the 17% solution
Exercises 1. \$5500 at 6%; \$3500 at 10%   **3.** \$7400 at 5.5%; \$2600 at 6%   **5.** \$450   **7.** \$3500 in each or \$7000 total

9. $20,000 at 24%; $11,000 at 18% **11.** $800 at 5%; $2100 at 7% **13.** $8500 at 9%; $3500 at 11%
15. $87,000 in bonds; $37,000 in certificates **17.** 20 pounds of 20%; 30 pounds of 70% **19.** 20 ounces of 30%; 30 ounces of 20%
21. 450 pounds of 35%; 1350 pounds of 15% **23.** 20 pounds of 40%; 30 pounds of 15% **25.** 10 g of acid; 20 g of the 40% solution
27. 10 oz of salt; 50 oz of the 4% solution **29.** 2 lb of 72%; 4 lb of 42% **31.** 7.11 oz of 0.5% solution; 0.89 oz of 5% solution
33. a. Answers will vary. **b.** Answers will vary.

Chapter 10: Review

1. a **2.** b, c, d **3.** $m_1 = 2, b_1 = 0; m_2 = 2, b_2 = 5$ **4.** $m_1 = -3, b_1 = 4; m_2 = -3, b_2 = -\dfrac{1}{2}$

5. $(1, 3)$ **6.** no solution **7.** $(2, -1)$ **8.** $(x, 4 - 5x)$ **9.** no solution **10.** $(-1, 4)$

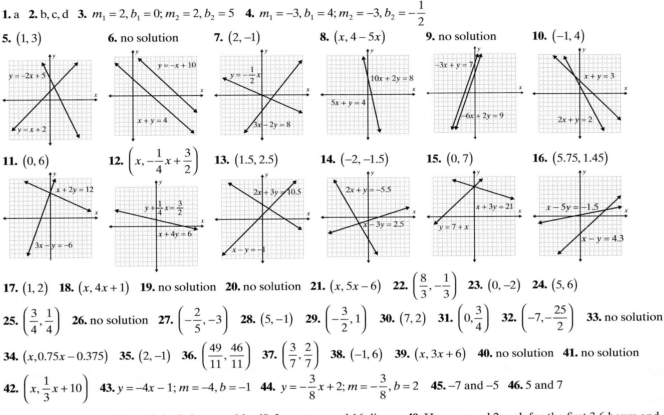

11. $(0, 6)$ **12.** $\left(x, -\dfrac{1}{4}x + \dfrac{3}{2}\right)$ **13.** $(1.5, 2.5)$ **14.** $(-2, -1.5)$ **15.** $(0, 7)$ **16.** $(5.75, 1.45)$

17. $(1, 2)$ **18.** $(x, 4x + 1)$ **19.** no solution **20.** no solution **21.** $(x, 5x - 6)$ **22.** $\left(\dfrac{8}{3}, -\dfrac{1}{3}\right)$ **23.** $(0, -2)$ **24.** $(5, 6)$

25. $\left(\dfrac{3}{4}, \dfrac{1}{4}\right)$ **26.** no solution **27.** $\left(-\dfrac{2}{5}, -3\right)$ **28.** $(5, -1)$ **29.** $\left(-\dfrac{3}{2}, 1\right)$ **30.** $(7, 2)$ **31.** $\left(0, \dfrac{3}{4}\right)$ **32.** $\left(-7, -\dfrac{25}{2}\right)$ **33.** no solution

34. $(x, 0.75x - 0.375)$ **35.** $(2, -1)$ **36.** $\left(\dfrac{49}{11}, \dfrac{46}{11}\right)$ **37.** $\left(\dfrac{3}{7}, \dfrac{2}{7}\right)$ **38.** $(-1, 6)$ **39.** $(x, 3x + 6)$ **40.** no solution **41.** no solution

42. $\left(x, \dfrac{1}{3}x + 10\right)$ **43.** $y = -4x - 1; m = -4, b = -1$ **44.** $y = -\dfrac{3}{8}x + 2; m = -\dfrac{3}{8}, b = 2$ **45.** -7 and -5 **46.** 5 and 7

47. Alice is 11 years old and John is 3 years old. **48.** 2 quarters and 16 dimes **49.** He averaged 2 mph for the first 3.6 hours and
3 mph for the last 2.4 hours. **50.** 36 minutes at 40 mph and 36 minutes at 20 mph **51.** length = 22.5 meters; width = 17.5 meters
52. 130°, 50° **53.** They sold 30 of the $110 shirt and 20 of the $65 dollar shirt. **54.** popcorn: $2.75; funnel cake: $4.10
55. $15,000 at 6%; $5000 at 8% **56.** $4000 at 8%, $8000 at 5% **57.** $7500 at each rate **58.** $22,000 in each type of investment
59. 20 gallons of 25% salt; 40 gallons of 40% salt **60.** 40 pounds of 22% fat; 40 pounds of 10% fat
61. 30 ounces of pure acid, 20 ounces of 10% acid **62.** 80 tons of 20% aluminum, 20 tons of 60% aluminum

Chapter 10: Test

1. c **2.** d **3.** $\left(x, -\dfrac{2}{3}x + 3\right)$ **4.** $(4, 1)$ **5.** $(-8, -20)$ **6.** $(1, 5)$ **7.** $(-2, 6)$ **8.** no solution

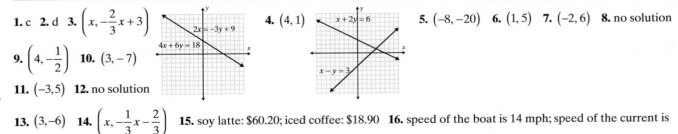

9. $\left(4, -\dfrac{1}{2}\right)$ **10.** $(3, -7)$

11. $(-3, 5)$ **12.** no solution

13. $(3, -6)$ **14.** $\left(x, -\dfrac{1}{3}x - \dfrac{2}{3}\right)$ **15.** soy latte: $60.20; iced coffee: $18.90 **16.** speed of the boat is 14 mph; speed of the current is
2 mph **17.** 12 gallons of 25% and 18 gallons of 50% **18.** $1600 at 8% and $960 at 6% **19.** 17 inches by 13 inches
20. 1600 lb of 83% and 400 lb of 68% **21.** nickels = 45; quarters = 60 **22.** pen price: $0.79; pencil price: $0.08

1. 200,016.04 **2.** 17.99 **3.** 0.4 **4.** 180% **5.** 0.015 **6.** $\frac{4}{3}$ **7.** $3\frac{1}{11}$ or $\frac{34}{11}$ **8.** $-\frac{4}{21}$ **9.** $10\frac{2}{15}$ **10.** $46\frac{5}{12}$ **11.** 12 **12.** $1\frac{4}{5}$

13. −560,000 **14.** −26.9 **15.** 75.74 **16.** 0.00049923 **17.** −80.6 **18.** 7 **19.** −11.11 **20.** −2.75 **21.** $2^2 \cdot 3^2 \cdot 11$ **22. a.** $C = 62.8$ ft

b. $A = 314$ ft^2 **23.** $5\sqrt{5}$ in. or 11.18 in. **24.** $x = 3$ **25.** $x = 4$ **26.** $x = -\frac{9}{8}$ **27.** $x = \frac{29}{11}$ **28.** $\left(-\infty, -\frac{1}{4}\right]$

29. $\left(-\infty, -\frac{50}{3}\right)$ **30.** m is undefined; no y-intercept; vertical line

31. $m = \frac{3}{5}$; y-intercept $= \left(0, -\frac{2}{5}\right)$ **32.** $m = 4$; y-intercept $= (0, -1)$ **33.** $m = 0$; y-intercept $= \left(0, \frac{3}{2}\right)$; horizontal line

34. $2x - y = 10$ **35.** $y = 2x - 6$ **36. a.** not a function; $D = [-5, 5]$; $R = [-3, 3]$ **b.** function;

$D = \{-7, -3, 0, 5, 6\}$; $R = \{-2, 1, 4\}$ **c.** function; $D = [-3, 3]$; $R = [0, 6]$

   37. b, c, d **38.** $(-1, 4)$ **39.** $(1, 3)$ **40.** $(-6, 2)$ **41.** $(-1, 4)$

42. no solution **43.** $\left(x, 2 - \frac{1}{5}x\right)$ **44.** $(2, 3)$ **45.** $61.80 **46.** $\frac{2}{5}$ **47.** 3 and 11 **48.** going to: 60 mph; coming from: 64 mph

49. $4200 at 7%; $2800 at 8% **50.** 3 batches of Choc-O-Nut; 2 batches of Chocolate Krunch

Chapter 11: Exponents and Polynomials

Section 11.1: Exponents

Margin Exercises 1. a. x^5 **b.** y^9 **c.** 125 **d.** 243 **e.** −2187 **2. a.** $20y^9$ **b.** $-10x^7$ **c.** $-35a^2b^7$ **3. a.** 1 **b.** x^2 **c.** 1 **4. a.** x^2 **b.** y^3

c. 1 **5. a.** $3x^8$ **b.** $7x^6y^4$ **6. a.** $\frac{1}{7}$ **b.** $\frac{1}{x^7}$ **c.** $\frac{1}{x^5}$ **7. a.** 3^3 or 27 **b.** x^6 **c.** $\frac{1}{8^2}$ or $\frac{1}{64}$ **d.** $\frac{x^3}{y^3}$ **e.** $\frac{4}{x^3}$ **8. a.** 0.04 **b.** 1 **c.** 38.17152

Exercises 1. 27 **3.** 512 **5.** $\frac{1}{3}$ **7.** $\frac{1}{25}$ **9.** 16 **11.** 24 **13.** −500 **15.** $\frac{3}{8}$ **17.** $-\frac{3}{25}$ **19.** x^5 **21.** y^2 **23.** $\frac{1}{x^3}$ **25.** $\frac{2}{x}$

27. $-\frac{8}{y^2}$ **29.** $\frac{5x^6}{y^4}$ **31.** 4 **33.** 49 **35.** $\frac{1}{10}$ **37.** $\frac{1}{8}$ **39.** x^2 **41.** x^2 **43.** x^4 **45.** $\frac{1}{x^4}$ **47.** x^6 **49.** x^2 **51.** y^2 **53.** $3x^3$

55. x^4 **57.** $36x^3$ **59.** $-14x^5$ **61.** $-12x^6$ **63.** $4y$ **65.** $3y^2$ **67.** $-2y^2$ **69.** $\frac{1}{x^2}$ **71.** 1000 **73.** 1 **75.** $-18x^5y^7$ **77.** $-\frac{2y^2}{x}$

79. $12a^3b^9c$ **81.** $-4a^{10}b^3c$ **83.** $\frac{5}{x}$ **85.** 1 **87.** 0.3906 **89.** 8875.147264

Section 11.2: Exponents and Scientific Notation

Margin Exercises 1. a. x^{15} **b.** $\frac{1}{x^{12}}$ **c.** $\frac{1}{y^{15}}$ **d.** $\frac{1}{3^6}$ or $\frac{1}{9^3}$ or $\frac{1}{729}$ **2. a.** $16x^2$ **b.** x^7y^7 **c.** $81a^2b^2$ **d.** $\frac{1}{a^3b^3}$ **e.** $\frac{x^9}{y^{12}}$ **3. a.** $\frac{x^7}{y^7}$

b. $\frac{25}{36}$ **c.** $\frac{27}{a^3}$ **d.** $\frac{x^3}{216}$ **4. a.** $\frac{-27x^3}{y^9}$ **b.** $\frac{16b^2}{a^2}$ **5.** $\frac{y^{15}}{x^{30}}$ **6.** $\frac{64}{225x^{18}y^2}$ **7. a.** 6.39×10^7 **b.** 2.45×10^{-6} **8. a.** 1.8×10^{-5}

b. 1.2×10^8 **c.** 4.816×10^{24} particles **9. a.** 7E4 **b.** 2.592E13 **c.** 6E12

Exercises 1. −81 **3.** −16 **5.** 1,000,000 **7.** $36x^6$ **9.** $-108x^6$ **11.** $\frac{5x^2}{y}$ **13.** $-\frac{2y^6}{27x^{15}}$ **15.** $\frac{27x^3}{y^3}$ **17.** 1 **19.** $4x^4y^4$ **21.** $\frac{y^2}{x^2}$

23. $\frac{1}{3xy^2}$ **25.** $-\frac{x^3y^6}{27}$ **27.** m^2n^4 **29.** $-\frac{y^2}{49x^2}$ **31.** $\frac{25x^6}{y^2}$ **33.** $\frac{x^6}{y^6}$ **35.** $\frac{36y^{14}}{x^4}$ **37.** $\frac{49y^4}{x^6}$ **39.** $\frac{24y^5}{x^7}$ **41.** $\frac{4x^3}{3}$ **43.** $\frac{16x^{13}}{243y^6}$

45. $96x^2y^4z^4$ **47.** 8.6×10^4 **49.** 3.62×10^{-2} **51.** 1.83×10^7 **53.** 0.042 **55.** $7{,}560{,}000$ **57.** $851{,}500{,}000$

59. $\left(3\times10^2\right)\left(1.5\times10^{-4}\right)$; 4.5×10^{-2} **61.** $\left(3\times10^{-4}\right)\left(2.5\times10^{-6}\right)$; 7.5×10^{-10} **63.** $\dfrac{3.9\times10^3}{3\times10^{-3}}$; 1.3×10^6 **65.** $\dfrac{1.25\times10^2}{5\times10^4}$; 2.5×10^{-3}

67. $\dfrac{\left(2\times10^{-2}\right)\left(3.9\times10^3\right)}{1.3\times10^{-2}}$; 6×10^3 **69.** $\dfrac{\left(5\times10^{-3}\right)\left(6.5\times10^2\right)\left(3.3\times10^0\right)}{\left(1.1\times10^{-3}\right)\left(2.5\times10^3\right)}$; 3.9×10^0 **71.** $\dfrac{\left(1.4\times10^{-2}\right)\left(9.22\times10^2\right)}{\left(3.5\times10^3\right)\left(2.0\times10^6\right)}$; 1.844×10^{-9}

73. 1.67×10^{-24} grams **75.** $60{,}000{,}000{,}000{,}000$ cells **77.** 5.98×10^{27} grams **79.** 4.0678×10^{16} m **81.** 6.5×10^{-19} grams **83.** 3E8

85. 8.5E7 **87.** 1E2 **89.** 1.864E2

Section 11.3: Introduction to Polynomials

Margin Exercises 1. a. $9x^2$; second-degree monomial **b.** $9x^2-5x$; second-degree binomial **c.** $4y^3-\dfrac{8}{3}y^2+8$; third-degree trinomial **d.** $4x^2-8$; second-degree binomial **e.** This expression is not a polynomial. **2.** $13x^3-2x^2+4$; third degree-trinomial **3. a.** 23 **b.** –6 **c.** $24a-35$ **4.** 68

Exercises 1. monomial **3.** binomial **5.** trinomial **7.** not a polynomial **9.** binomial

11. $4y$; first-degree monomial; leading coefficient 4 **13.** x^3+3x^2-2x; third-degree trinomial; leading coefficient 1

15. $-2x^2$; second-degree monomial; leading coefficient –2 **17.** 0; monomial of no degree; leading coefficient 0

19. $6a^5-7a^3-a^2$; fifth-degree trinomial; leading coefficient 6 **21.** $2y^3+4y$; third-degree binomial; leading coefficient 2

23. 4; monomial of degree 0; leading coefficient 4 **25.** $2x^3+3x^2-x+1$; third-degree polynomial; leading coefficient 2

27. $4x^4-x^2+4x-10$; fourth-degree polynomial; leading coefficient 4 **29.** $9x^3+4x^2+8x$; third-degree trinomial; leading coefficient 9 **31.** –16 **33.** –41 **35.** 4379 **37.** 13 **39.** –28 **41.** $3a^4+5a^3-8a^2-9a$ **43.** $3a+11$ **45.** $10a+25$

47. a.   **b.**   **c.**   **49. a.**   **b.**   **c.**

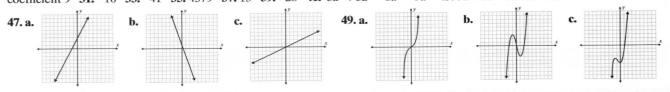

Section 11.4: Addition and Subtraction with Polynomials

Margin Exercises 1. a. $8x^3+4x^2-6x-11$ **b.** $4x^3-3x^2+7x+2$ **2.** $7x^2-4x+12$ **3. a.** $15x^4-3x^3+5x^2+1$

b. $6x^3-9x^2-8x+1$ **4.** $4x^4-3x^3-x+3$ **5. a.** $8x-21$ **b.** $4x-3$

Exercises 1. $3x^2+7x+2$ **3.** $2x^2+11x-7$ **5.** $3x^2$ **7.** $x^2-5x+17$ **9.** $-x^2+2x-7$ **11.** $2x^3+4x^2-3x-7$

13. $-x^2+7x-3$ **15.** $-x^3+2x^2+3x-4$ **17.** $4x^3+x^2+7$ **19.** $2x^3+11x^2-4x-3$ **21.** x^2+x+6 **23.** $-3x^2-6x-2$

25. $3x^2-4x-8$ **27.** $-x^4-2x^3-16$ **29.** $-4x^4-7x^3-11x^2+5x+13$ **31.** $-4x^2+6x+1$ **33.** $3x^4-7x^3+3x^2-5x+9$

35. $6x^2-7x+18$ **37.** $8x^4+6x^2+15$ **39.** $2x^3+4x^2+3x-10$ **41.** $4x-13$ **43.** $-7x+9$ **45.** $2x+17$ **47.** $3x^3+13x^2+9$

49. $8x^2-x-2$ **51.** $3x-19$ **53.** $x^2-2x+13$ **55.** $7x-3$ **57.** $4x^2+3x+5$ **59.** $7x-18$ **61.** $x^2+11x-8$ **63.** $7x^3-x^2-7$

65. any monomial or algebraic sum of monomials **67.** The largest of the degrees of its terms after like terms have been combined.

Section 11.5: Mutiplication with Polynomials

Margin Exercises 1. a. $-18x^3+6x^2+18x$ **b.** $12x^2+3x-9$ **c.** $12x^3+28x^2+2x-2$ **d.** $x^4-16x^2+16x-4$ **e.** $x^3-5x^2-12x+36$

2. a. $30x^7-15x^5+60x^3$ **b.** $7x^2-23x-20$

Exercises 1. $-6x^5-15x^3$ **3.** $4x^7-12x^6+4x^5$ **5.** $-y^5+8y-2$ **7.** $-4x^8+8x^7-12x^4$ **9.** $25x^5-5x^4+10x^3$

11. $a^7+2a^6-5a^3+a^2$ **13.** $6x^2-x-2$ **15.** $9a^2-25$ **17.** $-10x^2+39x-14$ **19.** x^3+5x^2+8x+4 **21.** x^2+x-12

23. $a^2-2a-48$ **25.** x^2-3x+2 **27.** $3t^2-3t-60$ **29.** x^3+11x^2+24x **31.** $2x^2-7x-4$ **33.** $6x^2+17x-3$ **35.** $4x^2-9$

37. $16x^2+8x+1$ **39.** y^3+2y^2+y+12 **41.** $3x^2-8x-35$ **43.** $5x^3+6x^2-22x-9$ **45.** $2x^4+7x^3+5x^2+x-15$ **47.** $3x^2+2x-8$

49. $2x^2+3x-5$ **51.** $7x^2-13x-2$ **53.** $6x^2-13x-8$ **55.** $4x^2+12x+9$ **57.** $x^3+3x^2-4x-12$ **59.** $4x^2-49$ **61.** x^3+1

63. $49a^2-28a+4$ **65.** $2x^3+x^2-5x-3$ **67.** $x^3+6x^2+11x+6$ **69.** a^4-a^2+2a-1 **71.** $t^4+6t^3+13t^2+12t+4$

73. $2y^2-61$ **75.** $3a^2-17a+11$ **77.** $-3x-21$ **79.** a^2+9a-1

Section 11.6: Special Products of Binomials

Margin Exercises 1. a. $3x^2 + 25x + 28$ **b.** $8x^2 - 18x + 10$ **c.** $x^2 - 16$ **2.** $10x^2 - 58x + 72$ **3. a.** $x^2 - 36$ **b.** $16y^2 - 9$ **c.** $x^8 - 9$
4. $4x^4 - 36$ **5. a.** $9x^2 + 30x + 25$ **b.** $49x^2 - 28x + 4$ **c.** $9x^2 - 48x + 64$ **d.** $4y^6 - 4y^3 + 1$ **6.** $25x^2 - 20x + 4$

Exercises 1. $5x^2 + 11x + 2$ **3.** $4x^2 + 13x - 12$ **5.** $3x^2 - 25x + 42$ **7.** $x^2 + 10x + 25$ **9.** $x^4 - 1$ **11.** $x^4 + 6x^2 + 9$

13. $x^6 - 4x^3 + 4$ **15.** $x^4 + 3x^2 - 54$ **17.** $x^2 - \dfrac{4}{9}$ **19.** $x^2 - \dfrac{9}{16}$ **21.** $x^2 + \dfrac{6}{5}x + \dfrac{9}{25}$ **23.** $x^2 - \dfrac{5}{3}x + \dfrac{25}{36}$ **25.** $x^2 - \dfrac{1}{4}x - \dfrac{1}{8}$

27. $x^2 + \dfrac{5}{6}x + \dfrac{1}{6}$ **29.** $x^2 - 14x + 49$; perfect square trinomial **31.** $x^2 + 8x + 16$; perfect square trinomial

33. $x^2 - 9$; difference of two squares **35.** $x^2 - 81$; difference of two squares **37.** $2x^2 + x - 3$

39. $9x^2 - 24x + 16$; perfect square trinomial **41.** $25x^2 - 4$; difference of two squares

43. $9x^2 - 12x + 4$; perfect square trinomial **45.** $x^2 - 16x + 64$; perfect square trinomial **47.** $16x^2 - 25$; difference of two squares

49. $25x^2 - 81$; difference of two squares **51.** $x^2 - 8x + 16$; perfect square trinomial **53.** $4x^2 - 49$; difference of two squares

55. $10x^4 - 11x^2 - 6$ **57.** $49x^2 + 14x + 1$; perfect square trinomial **59.** $x^2 - 1.96$ **61.** $x^2 - 5x + 6.25$ **63.** $x^2 - 4.6225$

65. $x^2 + 2.48x + 1.5376$ **67.** $2.0164x^2 + 27.264x + 92.16$ **69.** $129.96x^2 - 12.25$ **71.** $93.24x^2 + 142.46x - 104.04$

73. a. $A(x) = 400 - 4x^2$ **b.** $P(x) = 4(20 - 2x) + 8x = 80$ **75.** $A(x) = 8x + 15$ **77. a.** $A(x) = 150 - 4x^2$

b. $P(x) = 2(10 - 2x) + 2(15 - 2x) + 8x = 50$ **c.** $V(x) = 4x^3 - 50x^2 + 150x$

79. As indicated in the diagram, $(x + 5)^2 = x^2 + 2(5x) + 5^2$. Answers will vary.

Section 11.7: Division with Polynomials

Margin Exercises 1. a. $6x^4 - 2x^3 + 3$ **b.** $\dfrac{5y^3}{2} - 3y^2 - 4$ **2.** $3x^2 - x - 1 - \dfrac{7}{6x + 4}$ **3.** $2x^2 - 3x + 9$ **4.** $7x^2 + 14x + 32 + \dfrac{55}{x - 2}$

Exercises 1. $y^2 - 2y + 3$ **3.** $2x^2 - 3x + 1$ **5.** $10x^3 - 11x^2 + x$ **7.** $-4x^2 + 7x - \dfrac{5}{2}$ **9.** $y^3 - \dfrac{7}{2}y^2 - \dfrac{15}{2}y + 4$

11. $x - 6 + \dfrac{4}{x + 4}$ **13.** $3x - 4 - \dfrac{7}{2x - 1}$ **15.** $3x + 4 + \dfrac{1}{7x - 1}$ **17.** $x - 9$ **19.** $x^2 - x - \dfrac{3}{x - 8}$ **21.** $4x^2 - 6x + 9 - \dfrac{17}{x + 2}$

23. $x^2 + 7x + 55 + \dfrac{388}{x - 7}$ **25.** $2x^2 - 9x + 18 - \dfrac{30}{x + 2}$ **27.** $7x^2 + 2x + 1$ **29.** $x^2 + 2x + 2 - \dfrac{12}{2x + 3}$ **31.** $x^2 + 3x + 2 - \dfrac{2}{x - 4}$

33. $2x^2 + x - 3 + \dfrac{18}{5x + 3}$ **35.** $2x^2 - 8x + 25 - \dfrac{98}{x + 4}$ **37.** $3x^2 + 2x - 5 - \dfrac{1}{3x - 2}$ **39.** $3x^2 - 2x + 5$ **41.** $x^3 + 2x + 5 + \dfrac{17}{x - 3}$

43. $x^3 + 2x^2 + 6x + 9 + \dfrac{23}{x - 2}$ **45.** $x^3 + \dfrac{1}{2}x^2 - \dfrac{3}{4}x - \dfrac{3}{8} + \dfrac{45}{16\left(x - \dfrac{1}{2}\right)}$ **47.** $3x + 5 + \dfrac{x - 1}{x^2 + 2}$ **49.** $x^2 + x - 4 + \dfrac{-8x + 17}{x^2 + 4}$

51. $2x + 3 + \dfrac{6}{3x^2 - 2x - 1}$ **53.** $3x^2 - 10x + 12 + \dfrac{-x - 14}{x^2 + x + 1}$ **55.** $x^2 - 2x + 7 + \dfrac{-17x + 14}{x^2 + 2x - 3}$ **57.** $x^2 + 3x + 9$

59. $x^5 - x^4 + x^3 - x^2 + x - 1$ **61.** $x^4 + x^3 + x^2 + x + 1 + \dfrac{2}{x - 1}$ **63.** $x^4 - \dfrac{1}{2}x^3 - \dfrac{3}{4}x^2 + \dfrac{3}{8}x + \dfrac{13}{16} - \dfrac{13}{32\left(x + \dfrac{1}{2}\right)}$

65. $3x^3 + 4x^2 + 8x + 12$; Multiply the divisor $(3x - 2)$ by the quotient, $x^2 + 2x + 4 + \dfrac{20}{3x - 2}$, to get the original polynomial.

67. a. $19; 2x^2 - 4x + 2 + \dfrac{19}{x - 2}$ **b.** $-5; 2x^2 - 10x + 20 - \dfrac{5}{x + 1}$ **c.** $55; 2x^2 + 10 + \dfrac{55}{x - 4}$ Yes; $R = P(a)$ when $P(x)$ is divided by $x - a$.

Chapter 11: Review

1. 125 **2.** 64 **3.** $-\dfrac{3}{8}$ **4.** $\dfrac{5}{9}$ **5.** $\dfrac{1}{y^3}$ **6.** x^7 **7.** $\dfrac{1}{y^2}$ **8.** $-12x^4$ **9.** $-6x^4$ **10.** 1 **11.** $-6a^4b^4c^2$ **12.** $\dfrac{5}{xy}$ **13.** $\dfrac{17}{x^6}$ **14.** $\dfrac{x}{y}$

15. $9x^2y^4$ **16.** m^6n^9 **17.** $-10x^4y^6$ **18.** $\dfrac{5}{x^3}$ **19.** $\dfrac{9}{a^4b^6}$ **20.** 2.93×10^7 **21.** 7.5×10^{-3} **22.** 0.000724 **23.** $94,850,000$

24. $\dfrac{5.8 \times 10^{-3}}{2.9 \times 10^2}$; 2×10^{-5} **25.** $\dfrac{2.7 \times 10^0 \cdot 2 \times 10^{-3} \cdot 2.5 \times 10^1}{5.4 \times 10^1 \cdot 5 \times 10^{-4}}$; 5.0×10^0 **26.** 3.5×10^4 ft   **27.** not a polynomial   **28.** trinomial

29. binomial   **30.** $6x^2 - x$; second-degree binomial; leading coefficient 6

31. $-10a^3 + 4a^2 + 9a$; third-degree trinomial; leading coefficient -10 **32.** 5; monomial of 0 degree; leading coefficient 5

33. $4a^3 - 4a^2 - 2a + 9$; third-degree polynomial; leading coefficient 4   **34.** 19   **35.** 14   **36.** 154   **37.** 78   **38.** $a^2 + 6a + 9$ **39.** $8a - 9$

40. $-x^2 + x + 3$ **41.** $2x^2 + 19x + 4$ **42.** $-2x^3 + 3x^2 + 3x + 7$ **43.** $2x^3 + 13x^2 - x - 7$ **44.** $2x^4 + 7x^3 - 6x^2 + 10$ **45.** $3x^2 + 2x - 2$

46. $4x^2 - 7x + 2$ **47.** $10x^3 - 5x^2 + 20x + 2$ **48.** $5x - 18$ **49.** $-x - 13$ **50.** $5x^2 - x + 4$ **51.** $17x - 91$ **52.** $5x^2 - 8x - 1$

53. $-x^2 - 11x + 12$ **54.** $-35a^5 + 21a^4 - 7a^3$ **55.** $-6a^2 + 8a + 30$ **56.** $7t^2 - 11t - 6$ **57.** $9a^2 - 18a + 9$ **58.** $-4x^2 + 8x + 60$

59. $2y^3 - 5y^2 + y + 2$ **60.** $3y^3 + 4y^2 - 29y - 8$ **61.** $x^4 + 6x^3 + 12x^2 + 9x - 10$ **62.** $4x^2 - 14x - 3$ **63.** $7x + 46$

64. $4x^2 + 36x + 81$; perfect square trinomial   **65.** $y^2 - 64$; difference of two squares   **66.** $6x^2 + 25x + 25$

67. $9x^2 - 6x + 1$; perfect square trinomial   **68.** $y^4 + 2y^2 - 24$ **69.** $9x^2 - 49$; difference of two squares

70. $9x^2 - 42x + 49$; perfect square trinomial   **71.** $t^6 - \dfrac{1}{16}$; difference of two squares   **72.** $y^2 + \dfrac{1}{6}y - \dfrac{1}{3}$ **73.** $2 + \dfrac{8}{y} + \dfrac{10}{y^2}$

74. $4y^2 + 2y - 1 + \dfrac{11}{3y}$ **75.** $x + 3$ **76.** $8y - 14 + \dfrac{47}{y+3}$ **77.** $4x^2 + 8x - 4 - \dfrac{5}{x-2}$ **78.** $16x^2 + 20x + 25$ **79.** $5x^2 - 4x + 9$

80. $x - 6 + \dfrac{5x - 32}{x^2 - 2x + 3}$

Chapter 11: Test

1. $-10a^5$ **2.** 1   **3.** $\dfrac{4x^3}{y^7}$ **4.** $\dfrac{x}{3y^2}$ **5.** $\dfrac{x^2}{4y^2}$ **6.** $4x^2y^4$ **7. a.** 135,000   **b.** 0.0000027   **8. a.** $2.5 \times 10^2 \cdot 5 \times 10^5$; 1.25×10^8

b. $\dfrac{6.5 \times 10^1 \cdot 1.2 \times 10^{-2}}{1.5 \times 10^3}$; 5.2×10^{-4} **9.** $8x^2 + 3x$; second-degree binomial; leading coefficient 8

10. $-x^3 + 3x^2 + 3x - 1$; third-degree polynomial; leading coefficient -1

11. $5x^5 + 2x^4 - 11x + 3$; fifth-degree polynomial; leading coefficient 5   **12. a.** 20   **b.** -110 **13.** $-3x + 2$ **14.** $20x - 8$

15. $5x^3 - x^2 + 6x + 5$ **16.** $7x^4 + 14x^2 + 4$ **17.** $15x^7 - 20x^6 + 15x^5 - 40x^4 - 10x^2$ **18.** $49x^2 - 9$; difference of two squares

19. $16x^2 + 8x + 1$; perfect square trinomial   **20.** $36x^2 - 60x + 25$; perfect square trinomial   **21.** $12x^2 + 24x - 15$

22. $6x^3 - 69x^2 + 189x$ **23.** $7x^2 + 7x + 14$ **24.** $10x^4 + 4x^3 - 15x^2 - 41x - 14$ **25.** $2x + \dfrac{3}{2} - \dfrac{3}{x}$ **26.** $\dfrac{5a}{3} + 2a^2 + \dfrac{1}{a}$

27. $x - 6 - \dfrac{2}{2x+3}$ **28.** $x - 9 + \dfrac{15x - 12}{x^2 + x - 3}$ **29.** $2x^2 - 3x - 13$ **30. a.** $A(x) = 240 - (x^2 + 3x) = -(x^2 + 3x - 240)$ **b.** $P(x) = 64$

Cumulative Review: Chapters 1 – 11

1. 9.42   **2.** $\dfrac{7}{10}$ **3.** $\dfrac{5}{3}$ **4.** 32   **5.** $\dfrac{34}{9}$ **6.** $-\dfrac{7}{22}$ **7.** $x + 15$ **8.** $3x + 8$ **9.** 540   **10.** $120a^2b^3$ **11. a.** $\{x \mid x < 1.5\}$ **b.** $(-\infty, 1.5)$

c. open interval   **12. a.** $\{x \mid -2 \le x \le 0\}$ **b.** $[-2, 0]$ **c.** closed interval   **13.** $x = -3$ **14.** $x = \dfrac{8}{5}$ **15.** $x = -\dfrac{1}{2}$ **16.** $x = \dfrac{3}{8}$ **17.** $x = -3$

18.   **19.** $-2x + 3y = 8$ **20.** $y = -\dfrac{5}{8}x - \dfrac{17}{8}$ **21.** $x + 6y = 42$ **22.** open;

23. closed;

24. $2x$ **25.** $64x^6y^3$ **26.** $\dfrac{a^6}{b^4}$ **27.** $\dfrac{x}{3y^4}$ **28.** 1 **29.** $\dfrac{x^8}{9y^6}$ **30. a.** 0.00000028 **b.** $35{,}100$ **31.** $\dfrac{8.4\times10^2}{2.1\times10^{-4}}; 4\times10^6$

32. $\dfrac{5\times10^{-3}\cdot7.7\times10}{1.1\times10^{-2}\cdot3.5\times10^3}; 1\times10^{-2}$ **33. a.** $3x^2+8x$ **b.** 51 **c.** $3a^2+8a$ **34.** $x^4-2x^3+4x^2-10x+40$; fourth-degree polynomial; leading coefficient 1 **35.** $-x^3+7x^2-6$ **36.** $6x^2-7x+1$ **37.** $-3x^3+12x^2-3x$ **38.** x^2+x-12 **39.** $9x^2+42x+49$ **40.** $4x^2-1$ **41.** $2x^2-x-36$

42. $6x^2+x-12$ **43.** $4x^3-3x^2-x$ **44.** y^3-125 **45.** $4x-7+\dfrac{3}{x}$ **46.** $x-2$ **47.** $2x^2-x+3-\dfrac{2}{x+3}$ **48.** function; $D=(-\infty,\infty)$; $R=(-\infty,\infty)$

49. not a function; $D=(-\infty,\infty)$; $R=[-3,3]$ **50.** $20, 22,$ and 24 **51.** length $= 25$ cm; width $= 16$ cm **52.** -19 **53.** 61.6 in.

54. \$35,000 at 8%; \$65,000 at 6% **55.** 10 miles **56. a.** 31 in. **b.** 41.6 in.^{2} **57. a.** $A(x)=900-4x^2$ **b.** $P(x)=4(30-2x)+8x=120$

Chapter 12: Factoring Polynomials and Solving Quadratic Equations

Section 12.1: Greatest Common Factor and Factoring by Grouping

Margin Exercises 1. a. 5 **b.** $50xy$ **2. a.** $7(n+3)$ **b.** $y^3(3+y)$ **c.** $9x(1+6x)$ **d.** not factorable

e. $-9a(a^2+2-a)$ **3. a.** $10xy(1+3y)$ **b.** $2ab^2(a-8)$ **c.** $5xz(z+3z^2-4x)$ **d.** $-3b^2(d-4+5bd^2)$ **4. a.** $(2x-y)(x^2+2)$

b. $(x-u)(6y+1)$ **5. a.** $(x+2)(y+6)$ **b.** $(x-3)(y-2)$ **c.** not factorable **d.** $(x+y)(x+1)$ **e.** $(4x-3w)(2y-3z)$

Exercises 1. 5 **3.** 8 **5.** 1 **7.** $10x^3$ **9.** $4a^2$ **11.** $13ab$ **13.** $15xy^2z^2$ **15.** x^4 **17.** $-4y$ **19.** $3x^3$ **21.** $2x^2y$ **23.** $m+9$ **25.** $x-6$

27. $b+1$ **29.** $3y+4x+1$ **31.** $11(x-11)$ **33.** $4y(4y^2+3)$ **35.** $-3a(2x-3y)$ **37.** $5xy(2x-5)$ **39.** $-2yz(9yz-1)$ **41.** $8(y^2-4y+1)$

43. $x(2y^2-3y-1)$ **45.** $4m^2(2x^3-3y+z)$ **47.** $-7x^2z^3(8x^2+14xz+5z^2)$ **49.** $x^4y^2(15+24x^2y^4-32x^3y)$ **51.** $(y+3)(7y^2+2)$

53. $(x-4)(3x+2)$ **55.** $(x-2)(4x^3-1)$ **57.** $(2y+3)(10y-7)$ **59.** $(x-2)(a-b)$ **61.** $(b+c)(x+1)$ **63.** $(x^2+6)(x+3)$

65. not factorable **67.** $(3-b)(x+y)$ **69.** $(y-4)(5x+z)$ **71.** $(z^2+3)(a+1)$ **73.** $(6x+1)(a+2)$ **75.** $(x+1)(y+1)$

77. $(2y-7z)(5x-y)$ **79.** $(3x-4u)(y-2v)$ **81.** not factorable **83.** $(2c-3d)(3a+b)$

85. Although both can be factored out of $-3x^2+3$, 3 is greater than -3 making it the greatest common factor.

Section 12.2: Factoring Trinomials: x^2+bx+c

Margin Exercises 1. a. $(x+3)(x+7)$ **b.** $(x-5)(x+4)$ **2.** $(x-3)(x-2)$ **3. a.** $7y(y^2-7+2y)$ **b.** $11xy(x+3)(x-1)$

4. $6(x+4)(x-2)$

Exercises 1. $\{1,15\},\{-1,-15\},\{3,5\},\{-3,-5\}$ **3.** $\{1,20\},\{-1,-20\},\{4,5\},\{-4,-5\},\{2,10\},\{-2,-10\}$

5. $\{1,-6\},\{6,-1\},\{2,-3\},\{3,-2\}$ **7.** $\{1,16\},\{-1,-16\},\{4,4\},\{-4,-4\},\{8,2\},\{-8,-2\}$ **9.** $\{1,-10\},\{10,-1\},\{5,-2\},\{2,-5\}$

11. $4,3$ **13.** $-7,2$ **15.** $8,-1$ **17.** $-6,-6$ **19.** $-5,-4$ **21.** $x+1$ **23.** $p-10$ **25.** $a+6$ **27.** $(x-4)(x+3)$ **29.** $(y+6)(y-5)$

31. not factorable **33.** $(x-4)(x-4)$ **35.** $(x+4)(x+3)$ **37.** $(y-1)(y-2)$ **39.** not factorable **41.** $(x+8)(x-9)$

43. $(z-6)(z-9)$ **45.** $x(x+7)(x+3)$ **47.** $5(x-4)(x+3)$ **49.** $10y(y-3)(y+2)$ **51.** $4p^2(p+1)(p+8)$ **53.** $2x^2(x-9)(x+2)$

55. $2(x^2-x-36)$ **57.** $2a^2(a-10)(a+6)$ **59.** $3y^3(y-8)(y+1)$ **61.** $x(x-2)(x-8)$ **63.** $5(a^2+2a-6)$ **65.** $20a^2(a+1)(a+1)$

67. base $= x+48$; height $= x$ **69.** $x+5$ **71.** This is not an error, but the trinomial is not completely factored. The completely factored form of this trinomial is $2(x+2)(x+3)$.

Margin Exercises 1. a. $(x+6)(x+2)$ **b.** $(4y-7)(2y+3)$ **2. a.** $4x(2x-1)(x-1)$ **b.** $7x(3x^2+7x-1)$ **3. a.** $(5a+9)(a+3)$ **b.** $3(b-2)(b+4)$ **4. a.** $(5x+3)(x-7)$ **b.** $3(x+4)(8x-3)$

Exercises 1. $(x+2)(x+3)$ **3.** $(2x-5)(x+1)$ **5.** $(6x+5)(x+1)$ **7.** $-(x-2)(x-1)$ **9.** $(x-5)(x+2)$ **11.** $-(x-14)(x+1)$ **13.** not factorable **15.** $-x(2x+1)(x-1)$ **17.** $(t-1)(4t+1)$ **19.** $(5a-6)(a+1)$ **21.** $(7x-2)(x+1)$ **23.** $(2x-3)(4x+1)$ **25.** $(3x+4)(3x-5)$ **27.** $2(2x-5)(3x-2)$ **29.** $(3x-1)(x-2)$ **31.** $(3x-1)(3x-1)$ **33.** $(3y+2)(2y+1)$ **35.** $(x-1)(x-45)$ **37.** not factorable **39.** $2b(4a-3)(a-2)$ **41.** not factorable **43.** $(4x-1)(4x-1)$ **45.** $(8x-3)(8x-3)$ **47.** $2(3x-5)(x+2)$ **49.** $5(2x+3)(x+2)$ **51.** $-2(9x^2-36x+4)$ **53.** $-15(3y+4)(y-2)$ **55.** $3(2x-5)(2x-5)$ **57.** $3x(2x-1)(x+2)$ **59.** $3x(2x-9)(2x-9)$ **61.** $9xy^3(x^2+x+1)$ **63.** $4xy(3y-4)(4y-3)$ **65.** $7y^2(y-4)(3y-2)$

67. If the sign of the constant term is positive, the signs in the factors will both be positive or both be negative. If the sign of the constant term is negative, the sign in one factor will be positive and the sign in the other factor will be negative.

Margin Exercises 1. a. $7a(x-7)(x+7)$ **b.** $(y^3-10)(y^3+10)$ **2. a.** not factorable **b.** $5(9x^2+4)$ **3. a.** $(z+20)^2$ **b.** $3z(x-3y)^2$ **c.** $(y+4-z)(y+4+z)$ **4. a.** $(y-3)(y^2+3y+9)$ **b.** $6(2x^4-5)(4x^8+10x^4+25)$

Exercises 1. $(x-5)(x+5)$ **3.** $(9-y)(9+y)$ **5.** $2(x-8)(x+8)$ **7.** $4(x-2)(x+2)(x^2+4)$ **9.** not factorable **11.** $(y-8)^2$ **13.** $-4(x-5)(x+5)$ **15.** $(3x-5)(3x+5)$ **17.** $(y-5)^2$ **19.** $(2x-1)^2$ **21.** $(5x+3)^2$ **23.** $(4x-5)^2$ **25.** $4x(x-4)(x+4)$ **27.** $2xy(x+8)^2$ **29.** $(y+3)^2$ **31.** $(x-10)^2$ **33.** $(x^2+5y)^2$ **35.** $(x-5)(x^2+5x+25)$ **37.** $(y+6)(y^2-6y+36)$ **39.** $(x+3y)(x^2-3xy+9y^2)$ **41.** not factorable **43.** $4(x-2)(x^2+2x+4)$ **45.** $2(3x-y)(9x^2+3xy+y^2)$ **47.** $y(x+y)(x^2-xy+y^2)$ **49.** $x^2y^2(1-y)(1+y+y^2)$ **51.** $3xy(2x+3y)(4x^2-6xy+9y^2)$ **53.** $(x^2-y^3)(x^4+x^2y^3+y^6)$ **55.** $(3x+y^2)(9x^2-3xy^2+y^4)$ **57.** $(2x+y)(4x^2-2xy+y^2)$ **59.** $8(y-1)(y^2+y+1)$ **61.** $(3x-y)(3x+y)$ **63.** $(x-2y)(x+2y)(x^2+4y^2)$ **65.** $(x-y-9)(x-y+9)$ **67.** $(x-y-6)(x-y+6)$ **69.** $(4x+1-y)(4x+1+y)$ **71. a.** x^2-16 **b.**

$x-4$

$x+4$

73. a. $xy+xy+x^2+y^2 = x^2+2xy+y^2 = (x+y)^2$ **b.**

$(x+y)(x+y) = (x+y)^2$

75. For a 3-digit integer: $abc = 100a + 10b + c = (99+1)a + (9+1)b + c = 9(11a+b) + a + b + c$ So, if the sum $(a+b+c)$ is divisible by 3 (or 9), then the number abc will be divisible by 3 (or 9).

For a 4-digit integer: $abcd = 1000a + 100b + 10c + d = (999+1)a + (99+1)b + (9+1)c + d = 9(111a+11b+c) + a + b + c + d$ So, if the sum $(a+b+c+d)$ is divisible by 3 (or 9), then the number $abcd$ will be divisible by 3 (or 9).

Exercises 1. $(m+6)(m+1)$ **3.** $(x+9)(x+2)$ **5.** $(x-10)(x+10)$ **7.** $(m-3)(m+2)$ **9.** not factorable **11.** $(8a-1)(8a+1)$ **13.** $(x+5)^2$ **15.** $(x+12)(x-3)$ **17.** $3(a+6)(a-2)$ **19.** $-5(x-6)(x-8)$ **21.** not factorable **23.** $x(x-6)(x+2)$ **25.** $-2a(a+8)(a-7)$ **27.** $4x(2x-5)(2x+5)$ **29.** $-(x-5)(3x-2)$ **31.** $(2x-1)(3x-4)$ **33.** $(4m+3)(3m-2)$ **35.** $2(2x-1)(x-3)$ **37.** $(4x-7)(2x+5)$ **39.** $(5x+6)(4x-9)$ **41.** $-(5x-7)(3x+2)$ **43.** $-(2a-3)(4a-5)$ **45.** $(4y+5)(5y-4)$ **47.** $(6x-1)(3x-2)$ **49.** $-6(5x-4)(5x+4)$ **51.** $3(4n^2-20n-25)$ **53.** $a(21a^2-13a-2)$ **55.** $3x(3x-2)(4x+5)$ **57.** $2x(2x-1)(4x-11)$ **59.** $5(24m^2+2m+15)$ **61.** $(y-4)(x+3)$ **63.** $(x+2y)(x-6)$ **65.** $-(x^2-5)(x-8)$ **67.** $(x+5)(x^2-5x+25)$ **69.** $x^4(y-1)(y^2+y+1)$ **71.** $(2a^2+3b^2)(4a^4-6a^2b^2+9b^4)$ **73.** $(x^2y-5)(x^4y^2+5x^2y+25)$ **75.** $(x-3)(x+3)(x+7)$ **77.** $(3x+y+6)(3x-y-6)$ **79.** $(y+10+7x)(y+10-7x)$

Section 12.6: Solving Quadratic Equations by Factoring

Margin Exercises 1. $y = 7, \dfrac{5}{3}$ **2. a.** $x = 0, 2$ **b.** $x = 3$ (double root) **c.** $x = -1, 4$ **d.** $x = -\dfrac{7}{2}, \dfrac{4}{3}$ **e.** $x = -1, 3$ **3.** $x = -7, -1$
4. $x = -2, 0, 5$ **5.** $2x^2 - 11x + 12 = 0$

Exercises 1. $x = 2, 3$ **3.** $x = -2, \dfrac{9}{2}$ **5.** $x = -3$ **7.** $x = -5$ **9.** $x = 0, 2$ **11.** $x = -6$ **13.** $x = -1, 4$ **15.** $x = -3, 4$ **17.** $x = -3, 0$

19. $x = 2, 4$ **21.** $x = -4, 3$ **23.** $x = -\dfrac{1}{2}, 3$ **25.** $x = -\dfrac{2}{3}, 2$ **27.** $x = -\dfrac{1}{2}, 4$ **29.** $x = -2, \dfrac{4}{3}$ **31.** $x = \dfrac{3}{2}$ **33.** $x = 0, \dfrac{8}{5}$

35. $x = -2, 2$ **37.** $x = 1$ **39.** $x = 2$ **41.** $x = -3, 3$ **43.** $x = -5, 10$ **45.** $x = -6, -2$ **47.** $x = \dfrac{1}{2}$ **49.** $x = 0, 2, 4$ **51.** $x = -\dfrac{2}{3}, -\dfrac{1}{2}, 0$

53. $x = -10, 10$ **55.** $x = -5, 5$ **57.** $x = -4$ **59.** $x = 3$ **61.** $x = -1, 3$ **63.** $x = -8, -2$ **65.** $x = -5, 2$ **67.** $x = -5, 7$ **69.** $x = -6, 2$

71. $x = -1, \dfrac{2}{3}$ **73.** $x = -\dfrac{3}{2}, 4$ **75.** $y^2 - y - 6 = 0$ **77.** $2x^2 + 11x + 5 = 0$ **79.** $8x^2 - 10x + 3 = 0$ **81.** $x^3 - x^2 - 6x = 0$

83. $y^3 - 4y^2 - 3y + 18 = 0$ **85.** This allows for use of the zero-factor property which says that for the product to equal zero one of the factors must equal zero. Answers will vary. **87. a.** 640 ft; 384 ft **b.** 144 ft; 400 ft **c.** 7 seconds; $0 = -16(t + 7)(t - 7)$

Section 12.7: Applications of Quadratic Equations

Margin Exercises 1. a. $15, 10$ or $-15, -10$ **b.** 21 rows **c.** 16 feet by 18 feet **d.** 12 feet by 14 feet or 7 feet by 24 feet
2. a. $-14, -12$ **b.** $8, 10, 12$ **3.** 8 feet

Exercises 1. $x(x + 8) = -16$; $x = -4$, so the numbers are -4 and 4 **3.** $x^2 = 7x; x = 0, 7$ **5.** $x^2 + 3x = 28; x = 4$
7. $x(x + 7) = 78$; $x = -13, 6$; so the numbers are -13 and -6 or 13 and 6 **9.** $(x + 6)^2 + x^2 = 260$; $x = 8$; so the numbers are 8 and 14
11. $x + (x + 8)^2 = 124$; $x = 3$; so the integers are 3 and 11 **13.** $x(2x - 5) = x + 56$; $x = -4$
15. $x(x + 1) = 72$; $x = 8$; so the integers are 8 and 9 **17.** $x^2 + (x + 1)^2 = 85$; $x = 6$; so the integers are 6 and 7
19. $x(x + 2) = 63$; $x = -9, 7$; so the integers are -9 and -7 or 7 and 9 **21.** $4x + (x + 1)^2 = 41$; $x = 4$; so the integers are 4 and 5
23. $2x(x + 1) = (x + 1)(x + 2) + 88$; $x = 10$; so the integers are $10, 11$, and 12 **25.** $6x(x + 2) = (x + 1) + (x + 3)^2$; $x = -2, 1$; so the integers are $-2, -1, 0$, and 1 or $1, 2, 3$, and 4 **27.** $w(2w) = 72$; $w = 6$; so width is 6 in. and length is 12 in.
29. $w(4w) = 64$; $w = 4$; so width is 4 ft and length is 16 ft **31.** $l(l - 4) = 117$; $l = 13$; so width is 9 ft and length is 13 ft
33. $\dfrac{1}{2}b(b - 4) = 16$; $b = 8$; so base is 8 ft and height is 4 ft **35.** $\dfrac{1}{2}(h + 15)h = 63$; $h = 6$; so base is 21 in.
37. $w(16 - w) = 48$; $w = 4, 12$; so the rectangle is 4 in. by 12 in. **39.** $r(r + 13) = 140$; $r = 7$; so there are 7 trees in each row
41. $r(r + 7) = 144$; $r = 9$; so there are 9 rows **43.** $n(n + 1675) = 8400$; $n = 5$, so there are 5 floors
45. $(w + 11)(w + 4) = 98$; $w = 3$; so the rectangle is 3 cm by 10 cm **47.** $w(50 - 2w) = 300$; $w = 10, 15$; so width is 10 ft and length is 30 ft or width is 15 ft and length is 20 ft **49.** $h^2 + (h - 34)^2 = (h + 2)^2$; $h = 48$; so the height of the pole is 48 ft
51. $x^2 + (x - 49)^2 = (x + 1)^2$; $x = 60$; so height is 60 ft **53.** $l^2 + (l - 28)^2 = (l + 8)^2$; $l = 60$, so the length of the mat is 60 inches
55. $1.50 per pound **57.** $16 or $20 per reel **59.** $20^2 + 21^2 = 29^2; 24^2 + 7^2 = 25^2; 14^2 + 48^2 = 50^2; 60^2 + 11^2 = 61^2$

Chapter 12: Review

1. $x - 3y + 2$ **2.** $2y + 3$ **3.** $11(x - 2)$ **4.** $-4y(y - 7)$ **5.** $-7t^3x(x^3 - 14tx^2 - 5t^2)$ **6.** $4x^2y(4x - 3)$ **7.** $(a + 7)(3a - 2)$
8. $(x - 10)(2a + 3b)$ **9.** $(a + c)(x - 1)$ **10.** $(1 - 4y)(x + 2z)$ **11.** $(z^2 + 5)(1 + c)$ **12.** not factorable **13.** $-7, 6$ **14.** $3, 9$
15. $(m + 6)(m + 1)$ **16.** not factorable **17.** $(n - 6)(n - 2)$ **18.** $(a - 10)(a + 5)$ **19.** not factorable **20.** $3(m + 2)(m + 2)$
21. $2y(y + 5)(y + 2)$ **22.** $7x^2(x - 1)(x + 4)$ **23.** $9x^3(x + 5)(x + 4)$ **24.** $11a(a + 1)(a - 11)$ **25.** $(a + 5b)(a + 6b)$
26. $4(y^2 - 7xy - x^2)$ **27.** $(x + 7)(x + 5)$ **28.** $-(6x - 1)(2x - 5)$ **29.** $(2y - 1)(3y - 4)$ **30.** not factorable
31. $3(3x + 2)(7x - 5)$ **32.** $2(2x + 5)(4x - 7)$ **33.** $-(3x + 8)(x - 3)$ **34.** not factorable **35.** not factorable
36. $-4(x - 10)(x + 5)$ **37.** $7x(y - 4)(y + 6)$ **38.** $x(3x - 2)(6x - 1)$ **39.** $(x^3 - 10)(x^3 + 10)$ **40.** $(4x - 5)(4x + 5)$
41. not factorable **42.** $4(y + 4)^2$ **43.** not factorable **44.** $2(2x - 11)(2x + 11)$ **45.** $3x(x - 4)(x + 4)$ **46.** $-3(x + 2)^2$

47. $\left(2x^3 - y^2\right)\left(4x^6 + 2x^3y^2 + y^4\right)$ **48.** $4x(x+5)\left(x^2 - 5x + 25\right)$ **49.** $27(2x-1)\left(4x^2 + 2x + 1\right)$ **50.** $(z+4)(z-9)$

51. not factorable **52.** $-(3x-2)(x+4)$ **53.** $5x(x+2)(2x+3)$ **54.** $(7x+2y)(7x-2y)$ **55.** $2(x-3y)\left(x^2 + 3xy + 9y^2\right)$

56. $\left(3x^2 + 1\right)(x-3)$ **57.** $x = -5, 0$ **58.** $x = \dfrac{2}{3}, 5$ **59.** $y = 5, -7$ **60.** $x = -\dfrac{5}{2}, 0, \dfrac{5}{2}$ **61.** $x = -2, 0, 6$ **62.** $x = -5, 5$ **63.** $x = -5, 7$

64. $a = -7$ **65.** $x = -6, 0$ **66.** $x = -3, 13$ **67.** $x = 1, 7$ **68.** $x = -2, -\dfrac{3}{2}$ **69.** $x^2 + x - 20 = 0$

70. $32x^2 + 4x - 15 = 0$ **71.** $y^3 + 7y^2 + 8y - 16 = 0$ **72.** 13, 10 **73.** 6, −14 or 14, −6 **74.** 9 streets **75.** 15×30 yards **76.** 11, 13

77. 10, 12 or −8, −6 **78.** 20 inches **79.** 20 rows, 30 seats **80.** length = 21 meters; width = 20 meters

Chapter 12: Test

1. $7ab^2(4x - 3y)$ **2.** $6yz^2(3z - yz + 2)$ **3.** $(x-5)(x-4)$ **4.** $-(x+7)^2$ **5.** $(x-5)(y-7)$ **6.** $6(x+1)(x-1)$ **7.** $2(6x-5)(x+1)$

8. $(x+3)(3x-8)$ **9.** $(4x-5y)(4x+5y)$ **10.** $x(x+1)(2x-3)$ **11.** $(2x-3)(3x-2)$ **12.** $(y+7)(2x-3)$ **13.** not factorable

14. $-3x\left(x^2 - 2x + 2\right)$ **15.** $x = -1, 8$ **16.** $x = -6, 0$ **17.** $x = -\dfrac{3}{4}, 5$ **18.** $x = \dfrac{3}{2}, 4$ **19.** $x^2 + 5x - 24 = 0$ **20.** 6, 20 or −30, −4

21. length = 15 centimeters; width = 11 centimeters **22.** 18, 19 **23.** 3, 12 **24.** 18 cm **25.** $P(x) = 4(3x+5)$

Cumulative Review: Chapters 1 – 12

1. 120 **2.** $168x^2y$ **3.** 6 **4.** −138 **5.** $\dfrac{55}{48}$ **6.** $\dfrac{19}{60a}$ **7.** $\dfrac{3}{10}$ **8.** $\dfrac{75x}{23}$ **9.** 250 **10.** 123 **11.** $x = \dfrac{9}{10}$ **12.** $x = 21$ **13.** $x = 21$

14. conditional **15.** contradiction **16.** $x = \dfrac{y-b}{m}$ **17.** $y = \dfrac{10-3x}{5}$ **18.** $[-6, \infty)$

19. $(-7, 4)$

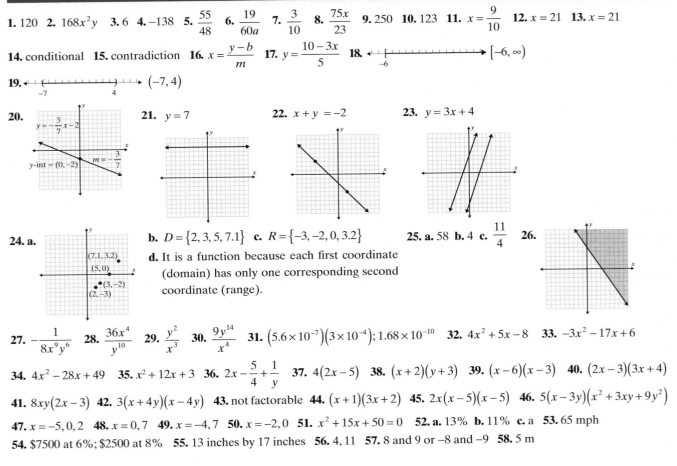

20. $y = -\dfrac{3}{7}x - 2$ y-int = $(0, -2)$ $m = \dfrac{3}{7}$

21. $y = 7$

22. $x + y = -2$

23. $y = 3x + 4$

24. a. (7.1, 3.2), (5, 0), (3, −2), (2, −3) **b.** $D = \{2, 3, 5, 7.1\}$ **c.** $R = \{-3, -2, 0, 3.2\}$ **d.** It is a function because each first coordinate (domain) has only one corresponding second coordinate (range).

25. a. 58 **b.** 4 **c.** $\dfrac{11}{4}$ **26.**

27. $-\dfrac{1}{8x^9y^6}$ **28.** $\dfrac{36x^4}{y^{10}}$ **29.** $\dfrac{y^2}{x^3}$ **30.** $\dfrac{9y^{14}}{x^4}$ **31.** $\left(5.6 \times 10^{-7}\right)\left(3 \times 10^{-4}\right); 1.68 \times 10^{-10}$ **32.** $4x^2 + 5x - 8$ **33.** $-3x^2 - 17x + 6$

34. $4x^2 - 28x + 49$ **35.** $x^2 + 12x + 3$ **36.** $2x - \dfrac{5}{4} + \dfrac{1}{y}$ **37.** $4(2x-5)$ **38.** $(x+2)(y+3)$ **39.** $(x-6)(x-3)$ **40.** $(2x-3)(3x+4)$

41. $8xy(2x-3)$ **42.** $3(x+4y)(x-4y)$ **43.** not factorable **44.** $(x+1)(3x+2)$ **45.** $2x(x-5)(x-5)$ **46.** $5(x-3y)\left(x^2 + 3xy + 9y^2\right)$

47. $x = -5, 0, 2$ **48.** $x = 0, 7$ **49.** $x = -4, 7$ **50.** $x = -2, 0$ **51.** $x^2 + 15x + 50 = 0$ **52. a.** 13% **b.** 11% **c.** a **53.** 65 mph

54. \$7500 at 6%; \$2500 at 8% **55.** 13 inches by 17 inches **56.** 4, 11 **57.** 8 and 9 or −8 and −9 **58.** 5 m

Chapter 13: Rational Expressions

Section 13.1: Multiplication and Division with Rational Expressions

Margin Exercises **1. a.** $x \neq \dfrac{1}{5}$ **b.** $x \neq 3, 4$ **c.** no restrictions **2. a.** $\dfrac{2}{5}, (x \neq 3)$ **b.** $\dfrac{x^2 + 5x + 25}{x + 5}, (x \neq -5, 5)$ **c.** $-1, (x \neq 5)$

3. a. $\dfrac{x^4}{6y^6}, (x \neq 0, y \neq 0)$ **b.** $\dfrac{x+3}{x^2}, (x \neq 0, 3)$ **c.** $\dfrac{x-1}{x(x+1)}$ or $\dfrac{x-1}{x^2+x}, (x \neq -1, 0, 1)$ **d.** $\dfrac{x^2 - x - 30}{3x + 9}, (x \neq -3, 2, 5)$

4. $\dfrac{x-1}{x}, (x \neq 0, -3)$ **5. a.** $\dfrac{1}{4x^2 y}$ **b.** $\dfrac{-x(x^2 - xy + y^2)}{y^2}$ or $\dfrac{-x^3 + x^2 y - xy^2}{y^2}$ **c.** $\dfrac{(x-3)(x-5)}{(3x+1)(3x+1)}$ or $\dfrac{x^2 - 8x + 15}{9x^2 + 6x + 1}$ **6.** $\dfrac{x+4}{x-4}$

Exercises **1.** $\dfrac{3x}{4y}; x \neq 0, y \neq 0$ **3.** $\dfrac{2x^3}{3y^3}; x \neq 0, y \neq 0$ **5.** $\dfrac{1}{x-3}; x \neq 0, 3$ **7.** $7; x \neq 2$ **9.** $-\dfrac{3}{4}; x \neq 3$ **11.** $\dfrac{2x}{y}; x \neq -\dfrac{2}{3}, y \neq 0$

13. $\dfrac{x}{x-1}; x \neq -6, 1$ **15.** $\dfrac{x^2 - 3x + 9}{x - 3}; x \neq -3, 3$ **17.** $\dfrac{x-3}{y-2}; y \neq -2, 2$ **19.** $\dfrac{x^2 + 2x + 4}{y + 5}; x \neq 2, y \neq -5$ **21.** $\dfrac{ab}{6y}$ **23.** $\dfrac{8x^2 y^3}{15}$ **25.** $\dfrac{x+3}{x}$

27. $\dfrac{x-1}{x+1}$ **29.** $-\dfrac{1}{x-8}$ **31.** $\dfrac{x-2}{x}$ **33.** $\dfrac{4x+20}{x(x+1)}$ **35.** $\dfrac{x}{(x+3)(x-1)}$ **37.** $-\dfrac{x+4}{x(x+1)}$ **39.** $\dfrac{x+2y}{(x-3y)(x-2y)}$ **41.** $\dfrac{x-1}{x(2x-1)}$

43. $\dfrac{1}{x+1}$ **45.** $\dfrac{x+2}{x-2}$ **47.** $\dfrac{1}{3xy^6}$ **49.** $\dfrac{6y^7}{x^4}$ **51.** $\dfrac{x}{12}$ **53.** $\dfrac{6x+18}{x^2}$ **55.** $\dfrac{6}{5x}$ **57.** $\dfrac{3x+1}{x+1}$ **59.** $\dfrac{x-2}{2x-1}$ **61.** $-\dfrac{x+4}{x(2x-1)}$ **63.** $\dfrac{x+1}{x-1}$

65. $\dfrac{6x^3 - x^2 + 1}{x^2(4x-3)(x-1)}$ **67.** $\dfrac{x^2 + 4x + 4}{x^2(2x-5)}$ **69.** $\dfrac{x^2 - 6x + 5}{(x-7)(x-2)(x+7)}$ **71.** $\dfrac{x^2 - 3x}{(x-1)^2}$ **73.** $\dfrac{x^2 + 5x}{2x+1}$ **75.** 1 **77.** $2x - 5$ feet

79. a. A rational expression is an algebraic expression that can be written in the form $\dfrac{P}{Q}$ where P and Q are polynomials and $Q \neq 0$. **b.** $\dfrac{x-1}{(x+2)(x-3)}$ Answers will vary. **c.** $\dfrac{1}{x+5}$ Answers will vary. **81. a.** $x = 4$ **b.** $x = -10, 10$

Section 13.2: Addition and Subtraction with Rational Expressions

Margin Exercises **1. a.** $\dfrac{1}{x-5}, (x \neq -5, 5)$ **b.** $\dfrac{3}{x+5}, (x \neq -5, -3)$ **2. a.** $\dfrac{x^2 + 6x + 6}{(x+3)(x+2)}$ **b.** $\dfrac{x^2 + 4x - 25}{2(x+5)^2(x-5)}$ **3.** $\dfrac{s^2 + 5s + 12}{(s+3)(s+1)}$

4. a. $\dfrac{x+2y}{3x-y}$ **b.** $\dfrac{x-5}{x+3}$ **c.** $\dfrac{x+5}{x-4}$ **5. a.** $\dfrac{x+18}{x+6}$ **b.** $\dfrac{15x^2 - 13xy - 4y^2}{3(x+y)^2(x-y)}$ **c.** $\dfrac{-x^2 - 2x + 12}{(x+6)(x+3)}$ **d.** $\dfrac{11y+17}{(x-2)(y+1)(y+3)}$ **6.** $\dfrac{-4x+1}{(x-1)^2}$

Exercises **1.** 3 **3.** 2 **5.** 1 **7.** $\dfrac{2}{x-1}$ **9.** $\dfrac{14}{7-x}$ **11.** 4 **13.** $\dfrac{x^2 - x + 1}{(x+4)(x-3)}$ **15.** $\dfrac{x-2}{x+2}$ **17.** $\dfrac{4x+5}{2(7x-2)}$ **19.** $\dfrac{6x+15}{(x+3)(x-3)}$

21. $\dfrac{x^2 - 2x + 4}{(x+2)(x-1)}$ **23.** $\dfrac{-x^2 - 3x - 6}{(x+3)(3-x)}$ **25.** $\dfrac{8x^2 + 13x - 21}{6(x+3)(x-3)}$ **27.** $\dfrac{3x^2 - 20x}{(x+6)(x-6)}$ **29.** $\dfrac{-4x}{x-7}$ **31.** $\dfrac{4x^2 - x - 12}{(x+7)(x-4)(x-1)}$

33. $\dfrac{x-6}{(x-10)(x-8)}$ **35.** $\dfrac{6x}{(x-1)(x-7)}$ **37.** $\dfrac{4x-19}{(7x+4)(x-1)(x+2)}$ **39.** $\dfrac{-7x-9}{(4x+3)(x-2)}$ **41.** $\dfrac{4x^2 - 41x + 3}{(x+4)(x-4)}$ **43.** $\dfrac{x-4}{2(x-2)}$

45. $\dfrac{x^2 - 4x - 6}{(x+2)(x-2)(x-1)}$ **47.** $\dfrac{3x^2 + 26x - 3}{(x+7)(x-3)(x+1)}$ **49.** $\dfrac{6x+2}{(x-1)(x+3)}$ **51.** $\dfrac{2x^2 + x - 4}{(x-2)(y+1)(x+1)}$ **53.** $\dfrac{2x + 4xy - 15y}{(x+3)(y+2)(x-5)}$

55. $\dfrac{-2x+2}{x^2 + x + 1}$ **57.** $\dfrac{2x^2 - x - 5}{(x-3)(x+3)(x^2+1)}$ **59.** $\dfrac{5x^3 - x^2 + 6x - 4}{(2x+1)(x-1)(x+2)(3x-2)}$ **61.** Answers will vary.

Section 13.3: Complex Fractions

Margin Exercises **1.** $\dfrac{1}{3y}$ **2. a.** $\dfrac{-6}{(x+6)^2}$ **b.** $-9xy$ **3. a.** $\dfrac{-6}{(x+6)^2}$ **b.** $-9xy$ **4.** $\dfrac{x+2}{x+4}$

Exercises 1. $\dfrac{4}{5xy}$ **3.** $\dfrac{8}{7x^2 y}$ **5.** $\dfrac{2x^2+6x}{2x-1}$ **7.** $\dfrac{2x-1}{2+3x}$ **9.** $\dfrac{7}{2(x+2)}$ **11.** $\dfrac{x}{x-1}$ **13.** $\dfrac{4x}{3(x+6)}$ **15.** $\dfrac{7x}{x+2}$ **17.** $\dfrac{2x+6}{3(x-2)}$

19. $\dfrac{24y+9x}{2(9y-10x)}$ **21.** $\dfrac{x}{x-1}$ **23.** $\dfrac{xy}{x+y}$ **25.** $\dfrac{1}{xy}$ **27.** $\dfrac{y+x}{y-x}$ **29.** $\dfrac{3-x}{x}$ **31.** $\dfrac{x+1}{x+3}$ **33.** $\dfrac{-1}{x(x+h)}$ **35.** $\dfrac{-1}{x(x+h)}$

37. $-(x-2y)(x-y)$ **39.** $\dfrac{2x}{x^2+1}$ **41.** $\dfrac{(x-3)(x^2-2x+4)}{(x-4)(x-2)(x+1)}$ **43.** $\dfrac{-5}{x+1}$ **45.** $\dfrac{29}{4(4x+5)}$ **47.** $\dfrac{x^2-3x-6}{x(x-1)}$ **49.** $\dfrac{x^2-4x-2}{(x-4)(x+4)}$

51. a. $\dfrac{8}{5}$ **b.** 1 **c.** $\dfrac{x^4+x^3+3x^2+2x+1}{x^3+x^2+2x+1}$

Section 13.4: Solving Equations with Rational Expressions

Margin Exercises 1. a. $x \neq 0$; $x=11$ **b.** $x \neq -7,-2,0$; $x=-4,3$ **c.** $x \neq 0,2$; $x=\dfrac{2}{5}$ **d.** $x \neq -5,0,5$; no solution **2.** $x \neq -5,2$; $x=\dfrac{1}{7}$

3. $l=\dfrac{SA-2wh}{2h+2w}$ **4.** $x=9$

Exercises 1. $x=7$ **3.** $x \neq 0,2$; $x=4$ **5.** $x \neq -3,4$; $x=-10$ **7.** $x \neq 0$; $x=18$ **9.** $x \neq 6$; $x=-\dfrac{74}{9}$ **11.** $x=\dfrac{1}{4}$ **13.** $x=6$ **15.** $x=4$

17. $x \neq 0$; $x=\dfrac{10}{3}$ **19.** $x \neq 0$; $x=-\dfrac{3}{4}$ **21.** $x \neq 0$; $x=-\dfrac{3}{16}$ **23.** $x \neq -9,-\dfrac{1}{4},0$; $x=-2,1$ **25.** $x \neq \dfrac{3}{2},0,6$; $x=\dfrac{3}{5},9$

27. $x \neq \dfrac{1}{2},4$; $x=-3$ **29.** $x \neq -4,-1$; $x=2$ **31.** $x \neq -1,\dfrac{1}{4}$; $x=\dfrac{2}{3}$ **33.** $x \neq 2,3$; $x=\dfrac{13}{10}$ **35.** $x \neq -\dfrac{2}{3},2$; no solution

37. $x \neq -1,\dfrac{1}{3},\dfrac{1}{2}$; $x=\dfrac{1}{5}$ **39.** $r=\dfrac{S-a}{S}$ **41.** $s=\dfrac{x-\bar{x}}{z}$ **43.** $y=m(x-x_1)+y_1$ **45.** $R_{\text{total}}=\dfrac{R_1 R_2}{R_1+R_2}$ **47.** $P=\dfrac{A}{1+r}$

49. $LK=12,JB=4$ **51.** $AC=2,ST=12$ **53.** $ST=8,TU=12,QR=24$ **55.** $AP=\dfrac{9}{2}$ in., $PC=\dfrac{15}{2}$ in. **57. a.** $\dfrac{4x^2+41x-10}{x(x-1)}$

b. $x=\dfrac{1}{4},10$ **59. a.** $\dfrac{10x-6}{(x-2)(x+2)}$ **b.** $x=\dfrac{7}{3}$ **61. a.** $\dfrac{x^2-153}{2(x-9)(x+9)}$ **b.** $x=3,-3$

Section 13.5: Applications

Margin Exercises 1. $\dfrac{5}{8}$ **2. a.** $\dfrac{36}{5}$ or $7\dfrac{1}{5}$ hours **b.** It takes the mom $\dfrac{9}{2}$ or $4\dfrac{1}{2}$ hours and her son takes 9 hours. **c.** The pool will

drain in $\dfrac{20}{3}$ or $6\dfrac{2}{3}$ hours. **3. a.** $c=2$ mph **b.** commercial airplane 360 mph; private airplane 180 mph

Exercises 1. 72, 45 **3.** 9 **5.** $\dfrac{6}{13}$ **7.** 36, 27 **9.** 7, 12 **11.** 45 shirts **13.** 12 brushes **15.** 1875 miles

17. person: 505 minutes; machine: 5.05 minutes **19.** 45 minutes **21.** Beth: 52 mph; Anna: 48 mph

23. commercial airliner: 300 mph; private plane: 120 mph **25.** 6 hours **27.** 50 mph **29.** $\dfrac{9}{4}$ or $2\dfrac{1}{4}$ hours

31. $\dfrac{9}{2}$ or $4\dfrac{1}{2}$ days; 9 days **33.** jet ski: 14 mph; current: 2 mph **35.** 14 kilometers per hour

Section 13.6: Variation

Margin Exercises 1. $y=4$ **2.** The spring will stretch 15 cm. **3.** $y=2$ **4.** 173 pounds **5.** $z=648$ **6. a.** 400 feet **b.** 200 cubic inches
c. 8000 pounds

Exercises 1. $\dfrac{7}{3}$ **3.** 2 **5.** $-\dfrac{32}{9}$ **7.** 36 **9.** 120 **11.** $\dfrac{56}{3}$ **13.** 40 **15.** 54 **17.** $\dfrac{48}{5}$ **19.** 27 **21.** 400 feet **23.** \$59.70
25. 4.71 feet **27.** 6 m **29.** 0.0073 cm **31.** 16,000 lb **33.** 9×10^{-11} N **35.** 15,000 lb **37.** 6400 lb **39.** 1.8 ft³ **41.** 1700 g per in.²
43. 15 ohms **45.** 2.56 ohms **47.** 900 lb **49.** 5 ft from the 300 lb weight (or 20 feet from the 75 lb weight)
51. a. When two variables vary directly, an increase in the value of one variable indicates an increase in the other, and the ratio of the two quantities is constant. **b.** When two variables vary inversely, an increase in the value of one variable indicates a decrease in the other, and the product of the two quantities is constant. **c.** Joint variation is when a variable varies directly with more than one other variable. **d.** Combined variation is when a variable varies directly or inversely with more than one variable.

1. $\dfrac{3}{16x}; x \neq 0, y \neq 0$ 2. $-\dfrac{4}{3}; x \neq 3$ 3. $\dfrac{3x+15}{x-5}; x \neq 5$ 4. $\dfrac{x-1}{x+5}; x \neq -5, -\dfrac{3}{2}$ 5. $\dfrac{2x}{3(x-2)}$ 6. $\dfrac{10x^2+6x}{5x-3}$ 7. $\dfrac{2}{xy^3}$ 8. $\dfrac{x}{21}$

9. $\dfrac{x^2-12x+35}{(x-2)(x+3)}$ 10. $\dfrac{3x-9}{(x-4)(x+1)(x-2)}$ 11. $\dfrac{5x+20}{4x(2x-1)}$ 12. $\dfrac{x^2-4x}{(x+1)(x+6)}$ 13. 6   14. 8   15. $\dfrac{6x+10}{(x+2)(x-2)}$ 16. $\dfrac{x+1}{x-11}$

17. $\dfrac{x^4-4x^3-2x^2+1}{(x-1)^2(x+1)^2}$ 18. $\dfrac{8x^2+8x+1}{(x+1)(3x+1)(2x+3)}$ 19. $\dfrac{11}{2x-3}$ 20. $\dfrac{9}{x+4}$ 21. $\dfrac{x^2}{(x-1)(x^2+x+1)}$ 22. $\dfrac{8x^2+6x+2}{(x-1)(x+3)(3x+1)}$

23. $\dfrac{14y^2}{5x}$ 24. $\dfrac{13}{2(5y-1)}$ 25. $\dfrac{y+2}{4}$ 26. $\dfrac{-2}{(x+2)^2}$ 27. $\dfrac{x-2}{x}$ 28. $\dfrac{x}{x^2+x+1}$ 29. $\dfrac{4x-12}{x^2-6x-9}$ 30. $\dfrac{x}{x-5}$ 31. $\dfrac{-2x+10}{x+4}$

32. $-\dfrac{9}{x+2}$ 33. $\dfrac{20}{3y+2}$ 34. $\dfrac{12y+24}{(y+3)^2(y-3)}$ 35. $x \neq -3, 3; x = 5$ 36. $x \neq 0, 3; x = 1, 12$ 37. $x \neq 0; x = -\dfrac{4}{25}$

38. $x \neq -\dfrac{2}{3}, \dfrac{4}{7}; x = \dfrac{8}{53}$ 39. $x \neq 0, 1; x = \dfrac{5}{2}$ 40. $x \neq -3, 3; x = 12$ 41. $x \neq -3, 4; x = 0, \dfrac{25}{6}$ 42. $x \neq -6; x = -\dfrac{15}{2}, 4$

43. $x_1 = x - \dfrac{y-y_1}{m}$ 44. $a_1 = \dfrac{a_n(1-r)}{1-r^n}$ 45. $AB = 8, FE = 5$ 46. $TU = 6, VW = 7$ 47. 18, 27   48. 15 women   49. $\dfrac{6}{8}$

50. 3 hours; 6 hours   51. 12 hours   52. Alice takes 6 hrs; Judy takes 4 hrs   53. Tyler takes 2 hrs; his son takes 6 hrs   54. 6 mph

55. 24 mph   56. Initial rate was 37.5 mph and final rate was 62.5 mph   57. Initial rate was 400 mph and final rate was 430 mph

58. Raphael takes 1.5 hrs; Matilde takes 3 hrs   59. 6   60. 5   61. $\dfrac{490}{9}$ 62. 31,250   63. $\dfrac{100}{7}$ in.   64. 18.84 in.   65. 113.04 in.^{2}

66. 162 rpm   67. 576 feet   68. 16.5 ohms   69. 37.68 cubic feet   70. 5 cups of lemon juice

1. $\dfrac{x}{x+4}; x \neq -4, -3$ 2. $\dfrac{1}{2x+5}; x \neq -\dfrac{5}{2}$ 3. $\dfrac{x+3}{x+4}$ 4. $\dfrac{3x-2}{3x+2}$ 5. $\dfrac{-2x^2-13x}{(x+5)(x+2)(x-2)}$ 6. $\dfrac{-2x^2-7x+18}{(3x+2)(x-4)(x+1)}$ 7. $2x^2$

8. $\dfrac{x^2-7x+1}{(x+3)(x-3)}$ 9. $\dfrac{3}{2xy^3}$ 10. $\dfrac{-3x}{x-2}$ 11. $-\dfrac{1}{xy}$ 12. $\dfrac{12x+24}{(x-3)(x+3)^2}$ 13. a. $\dfrac{7x+11}{2x(x+1)}$ b. $x = \dfrac{1}{3}$ 14. $x \neq -4; x = 21$

15. $x \neq 0; x = -\dfrac{7}{2}$ 16. $x \neq -4, 1; x = -1$ 17. $x \neq -2, -1; x = 1$ 18. $n = \dfrac{2S}{a_1+a_n}$ 19. $x = \dfrac{y-b}{m}$ 20. $\overline{AC} = 36, \overline{DC} = 24$ 21. $\dfrac{2}{7}$

22. $z = 24$ 23. $z = \dfrac{400}{9}$ 24. 4 hours   25. Carlos: 42 mph; Mario: 57 mph   26. 11 mph   27. $\dfrac{15}{2}$ cm   28. 5.13 in.^{3}

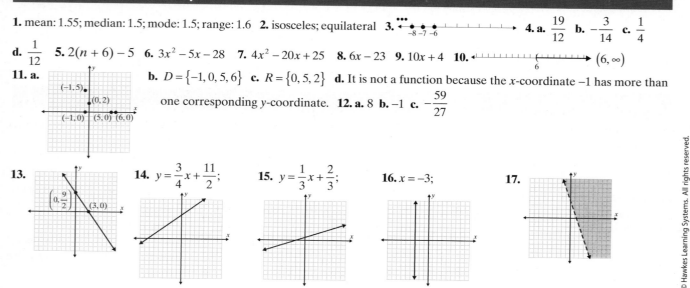

1. mean: 1.55; median: 1.5; mode: 1.5; range: 1.6   2. isosceles; equilateral   3. [number line]   4. a. $\dfrac{19}{12}$ b. $-\dfrac{3}{14}$ c. $\dfrac{1}{4}$

d. $\dfrac{1}{12}$ 5. $2(n+6)-5$ 6. $3x^2-5x-28$ 7. $4x^2-20x+25$ 8. $6x-23$ 9. $10x+4$ 10. [number line] $(6, \infty)$

11. a. [graph: points (-1.5), $(0,2)$, $(-1,0)$, $(5,0)$, $(6,0)$]   b. $D = \{-1, 0, 5, 6\}$ c. $R = \{0, 5, 2\}$ d. It is not a function because the x-coordinate -1 has more than

one corresponding y-coordinate.   12. a. 8   b. -1 c. $-\dfrac{59}{27}$

13. [graph: points $(0, \frac{9}{2})$, $(3,0)$]   14. $y = \dfrac{3}{4}x + \dfrac{11}{2}$; [graph]   15. $y = \dfrac{1}{3}x + \dfrac{2}{3}$; [graph]   16. $x = -3$; [graph]   17. [graph]

18. $8x^2$ **19.** $5x^3y^2$ **20.** $13x+1$ **21.** $-4x^2-2x+3$ **22.** $2x^2+x-28$ **23.** $x^2+12x+36$ **24.** $(3x+y)(y+1)$ **25.** $(2x-5)(2x+3)$

26. $(3x-2)(2x-1)$ **27.** $2x(3x+1)(x-4)$ **28.** $(3x^3+2y)(3x^3-2y)$ **29.** $(2x+5)(4x^2-10x+25)$ **30.** $x^3+4x^2+3x=0$

31. $1-3y+\dfrac{8}{7}y^2$ **32.** $2x^2-x+3-\dfrac{2}{x+3}$ **33.** $\dfrac{1}{x+1}; x \neq -1, 0$ **34.** $\dfrac{x+5}{2(x-3)}; x \neq 3$ **35.** $x-y$ **36.** $\dfrac{x}{3(x+1)}$ **37.** $\dfrac{x+4}{3x}$

38. $\dfrac{3x^3+12x^2+12x}{x+3}$ **39.** $\dfrac{2x+1}{x-1}$ **40.** $\dfrac{2x^2+14x-8}{(x+3)(x-1)(x-2)}$ **41.** $\dfrac{2x+4}{(x-1)(x+4)}$ **42.** $\dfrac{x-4}{(x+2)(x-2)}$ **43.** $\dfrac{19}{14}$ **44.** $\dfrac{x+4}{6x}$

45. $y=6$ **46.** $x=-12.35$ **47.** $x=-\dfrac{3}{5}$ **48.** $x=-\dfrac{5}{2}$ **49.** $x=-4,0,5$ **50.** $x=2,5$ **51.** $x=-45$ **52. a.** $\dfrac{9-2x}{x(x+3)}$ **b.** $x=\dfrac{9}{2}$

53. a. $\dfrac{6x^2-16x-20}{(x-4)(x+2)}$ **b.** $x=7$ **54.** $t=\dfrac{A-P}{Pr}$ **55.** $\dfrac{1}{2}$ year **56.** 1243.44 in³ **57.** train's speed is 60 mph; speed of airplane is

230 mph **58.** The father takes $\dfrac{8}{3}$ or $2\dfrac{2}{3}$ hours and the daughter takes 8 hours **59.** 600 times **60.** $\dfrac{100}{13}$ or $7\dfrac{9}{13}$ cm

Chapter 14: Radicals

Section 14.1: Roots and Radicals

Margin Exercises 1. a. $-9, 9$ **b.** $-14, 14$ **c.** $-7, 7$ **d.** Not a real number **2. a.** $\dfrac{3}{8}$ **b.** 0.04 **3.** Because $64 < 67 < 81$, we have

$\sqrt{64} < \sqrt{67} < \sqrt{81}$ and $8 < \sqrt{67} < 9$. The approximation 8.1854 is between 8 and 9 and is reasonable. **4. a.** 4 **b.** -5 **c.** $\dfrac{1}{10}$
5. a. 6.708203932 **b.** 27.386127875

Exercises 1. 3 **3.** 9 **5.** 17 **7.** 13 **9.** 1 **11.** 5 **13.** 6 **15.** $\dfrac{1}{2}$ **17.** $\dfrac{3}{4}$ **19.** 0.2 **21.** -10 **23.** -0.04 **25.** -3 **27.** -5 **29.** $\dfrac{3}{5}$

31. $\sqrt{64} < \sqrt{74} < \sqrt{81}$ and $8 < \sqrt{74} < 9$ because $64 < 74 < 81$ or $(8.6023)^2 = 73.99956529$

33. $\sqrt{25} < \sqrt{32} < \sqrt{36}$ and $5 < \sqrt{32} < 6$ because $25 < 32 < 36$ or $(5.6569)^2 = 32.00051761$ **35.** rational **37.** rational

39. irrational **41.** nonreal **43.** rational **45.** irrational **47.** 6.2450 **49.** 2.4960 **51.** 0.4472 **53.** 8.9443 **55.** -8.2462

57. There is no real number that results in a negative number when squared.

Section 14.2: Simplifying Radicals

Margin Exercises 1. a. $7\sqrt{2}$ **b.** $3\sqrt{5}$ **c.** $\dfrac{2\sqrt{3}}{5}$ **2. a.** $6z$ **b.** $5b\sqrt{3}$ **c.** $3cd\sqrt{5}$ **3. a.** $4x^4$ **b.** $10xy\sqrt{xy}$ **c.** $2x^4y^6\sqrt{3}$ **d.** $\dfrac{5z^9}{y^4}$

4. a. $2z\sqrt[3]{6}$ **b.** $-3a^2b^4\sqrt[3]{3a^2}$ **c.** $7x^2y^3\sqrt[3]{2y^2}$

Exercises 1. $2\sqrt{3}$ **3.** $12\sqrt{2}$ **5.** $-6\sqrt{2}$ **7.** $-2\sqrt{14}$ **9.** $-5\sqrt{5}$ **11.** $\dfrac{1}{2}$ **13.** $-\dfrac{\sqrt{11}}{8}$ **15.** $\dfrac{2\sqrt{7}}{5}$ **17.** $6x$ **19.** $2x\sqrt{2x}$

21. $2x^5y\sqrt{6x}$ **23.** $5xy^3\sqrt{5x}$ **25.** $-3xy\sqrt{2}$ **27.** $2bc\sqrt{3ac}$ **29.** $5x^2y^3z^4\sqrt{3}$ **31.** $\dfrac{x^2\sqrt{5}}{3}$ **33.** $\dfrac{4a^2\sqrt{2a}}{9b^8}$ **35.** $\dfrac{10x^4\sqrt{2}}{17}$

37. 6 **39.** $2\sqrt[3]{7}$ **41.** -1 **43.** $-4\sqrt[3]{2}$ **45.** $5x\sqrt[3]{x}$ **47.** $-2x^2\sqrt[3]{x^2}$ **49.** $2a^2b\sqrt[3]{9b}$ **51.** $6x^2y\sqrt[3]{y^2}$ **53.** $2xy^2z^3\sqrt[3]{3x^2y}$ **55.** $\dfrac{\sqrt[3]{3}}{2}$

57. $\dfrac{5\sqrt[3]{3}}{2}$ **59.** $\dfrac{5y^4}{3x^2}$ **61.** 7 inches **63.** $\sqrt{6} \approx 2.45$ amperes **65.** 120 volts **67.** When $a < 0$.

Section 14.3: Addition, Subtraction, and Multiplication with Radicals

Margin Exercises 1. a. $9\sqrt{3a}$ **b.** $5\sqrt{5}+3\sqrt{3}$ **c.** $-\sqrt[3]{9x}$ **2. a.** $6\sqrt{7}$ **b.** $2\sqrt{x}-7\sqrt{3}$ **3. a.** 15 **b.** $13\sqrt{2}-13$ **c.** $8+3\sqrt{5}$

d. $12-2\sqrt{35}$ **e.** $3z-5$ **4. a.** $4\sqrt{30}-20$ **b.** $s+2\sqrt{s}-63$ **5. a.** 11.0711 **b.** 7

Exercises 1. $8\sqrt{2}$ **3.** $\sqrt{11}$ **5.** $-3\sqrt{10}$ **7.** $13\sqrt[3]{3}$ **9.** $-\sqrt{11}$ **11.** $3\sqrt{a}$ **13.** $7\sqrt{x}$ **15.** $4\sqrt{2}+3\sqrt{3}$ **17.** $8\sqrt{b}-4\sqrt{a}$ **19.** $13\sqrt[3]{x}-2\sqrt[3]{y}$

21. $5\sqrt{3}$ **23.** 0 **25.** $17\sqrt[3]{2}$ **27.** $2\sqrt{2}-6\sqrt{3}$ **29.** $6+\sqrt{5}$ **31.** $\sqrt{3}-4\sqrt{2}$ **33.** $5\sqrt[3]{2}-8\sqrt[3]{3}$ **35.** $4\sqrt{2x}$ **37.** $2y\sqrt{2y}$ **39.** $-4x\sqrt{3xy}$

41. $15x\sqrt{x}$ **43.** $-xy^2\sqrt{x}$ **45.** $4x^5y^{10}\sqrt{3}$ **47.** $-8x^8y^2$ **49.** $xy^2\sqrt[3]{2}\left(-2x^2y^2-2x^3y+3\right)$ **51.** $3\sqrt{2}-8$ **53.** 18 **55.** $-8\sqrt{3}$

57. $12+\sqrt{6}$ **59.** $2y+\sqrt{xy}$ **61.** $13+2\sqrt{2}$ **63.** $3x-9\sqrt{3x}+8$ **65.** $2-2\sqrt{7}$ **67.** $13+4\sqrt{10}$ **69.** $\sqrt{10}+\sqrt{15}-\sqrt{6}-3$

71. $x - 2\sqrt{6x} - 18$ **73.** 58   **75.** $x + 10\sqrt{xy} + 25y$ **77.** 4.3397   **79.** 31.6   **81.** -57 **83.** -37.3569 **85.** $5\sqrt{6} + \sqrt{170} \approx 25.29$ ft

Section 14.4: Rationalizing Denominators

Margin Exercises   1. a. $\dfrac{8\sqrt{5}}{5}$ **b.** $\dfrac{17\sqrt{y}}{y}$ **c.** $\dfrac{4\sqrt{3}}{27}$ **d.** $\dfrac{\sqrt{22}}{6}$ **2. a.** $-\dfrac{5\sqrt{3}+25}{22}$ **b.** $\dfrac{105-15\sqrt{6}}{43}$ **c.** $\dfrac{2\sqrt{2}-\sqrt{3}}{5}$ **d.** $\dfrac{8+8\sqrt{z}}{1-z}$

e. $\dfrac{x\sqrt{x}-x\sqrt{y}+y\sqrt{x}-y\sqrt{y}}{x-y}$ **3.** $\dfrac{11\left(8+\sqrt{3}\right)}{61}$

Exercises   1. $\dfrac{5\sqrt{2}}{2}$ **3.** $\dfrac{-3\sqrt{7}}{7}$ **5.** $2\sqrt{3}$ **7.** 3   **9.** 3   **11.** $\dfrac{1}{3}$ **13.** $\dfrac{2\sqrt{3}}{3}$ **15.** $\dfrac{3\sqrt{2}}{2}$ **17.** $\dfrac{\sqrt{x}}{x}$ **19.** $\dfrac{\sqrt{2xy}}{y}$ **21.** $\dfrac{\sqrt{2y}}{y}$ **23.** $\dfrac{3\sqrt{7}}{5}$

25. $\dfrac{-\sqrt{2y}}{5}$ **27.** $\sqrt{6}+2$ **29.** $\dfrac{-\left(\sqrt{5}+3\right)}{4}$ **31.** $\dfrac{-6\left(5+3\sqrt{2}\right)}{7}$ **33.** $\dfrac{\sqrt{3}\left(\sqrt{2}-5\right)}{23}$ **35.** $\dfrac{-7\left(1+3\sqrt{5}\right)}{44}$ **37.** $\dfrac{-\left(\sqrt{3}+\sqrt{5}\right)}{2}$

39. $5\left(\sqrt{2}-\sqrt{3}\right)$ **41.** $\dfrac{4\left(\sqrt{x}-1\right)}{\left(x-1\right)}$ **43.** $\dfrac{5\left(6-\sqrt{y}\right)}{36-y}$ **45.** $\dfrac{8\left(2\sqrt{x}-3\right)}{4x-9}$ **47.** $\dfrac{2\sqrt{y}\left(\sqrt{5y}+\sqrt{3}\right)}{\left(5y-3\right)}$ **49.** $\dfrac{3\left(\sqrt{x}+\sqrt{y}\right)}{x-y}$

51. $\dfrac{x\left(\sqrt{x}-2\sqrt{y}\right)}{\left(x-4y\right)}$ **53.** $-\left(\sqrt{3}+1\right)\left(\sqrt{3}+2\right)$ **55.** $\dfrac{-\left(\sqrt{5}-2\right)\left(\sqrt{5}-3\right)}{4}$ **57.** $\dfrac{\left(\sqrt{x}+1\right)^2}{\left(x-1\right)}$ **59.** $\dfrac{\left(\sqrt{x}+2\right)\left(\sqrt{3x}-y\right)}{\left(3x-y^2\right)}$

61. Multiply both the numerator and the denominator by the conjugate of the denominator. This works because multiplying the denominator by its conjugate results in an expression with no square roots. Answers will vary.

Section 14.5: Equations with Radicals

Margin Exercises   1. a. $x = -8$ or $x = 8$ **b.** $y = -2$ **c.** $x = -\dfrac{1}{2}$ or $x = -3$ **d.** no solution   **2.** $x = \dfrac{3}{4}$ **3. a.** $x = 2$ **b.** $x = 9, 25$ **4.** $x = 8$

Exercises   1. $x = 3$ **3.** no solution   **5.** $x = -3$ **7.** $x = 14$ **9.** $x = 9$ **11.** no solution   **13.** $x = 6$ **15.** $x = -4, 1$ **17.** $x = -5, \dfrac{5}{2}$

19. $x = -2$ **21.** $x = 2, 3$ **23.** $x = 2, 5$ **25.** $x = -5, 5$ **27.** $x = 4$ **29.** $x = 4$ **31.** $x = 3$ **33.** $x = 2$ **35.** $x = 2$ **37.** $x = 7$ **39.** $x = 4$ **41.** $x = 0$
43. $x = 5$ **45.** no solution   **47.** $x = 4$ **49.** $x = -1, 3$ **51.** $x = 5$ **53.** $x = 2$ **55.** $x = 1$ **57.** $x = -4$ **59.** $x = 12$

Section 14.6: Rational Exponents

Margin Exercises   1. a. 6   **b.** 2   **c.** -3 **d.** 0.2   **e.** not a real number   **2. a.** $\sqrt[5]{x^2}$ **b.** $8\sqrt[7]{z^6}$ **c.** $-\sqrt[4]{b^5}$ **d.** $x^{\frac{5}{7}}$ **e.** $3s^{\frac{1}{2}}$ **f.** $-5^{\frac{1}{4}}$

3. a. $x^{\frac{7}{12}}$ **b.** $\dfrac{1}{a^{\frac{2}{9}}}$ **c.** $81b^{\frac{4}{5}}$ **d.** $\dfrac{z^{\frac{2}{9}}}{8}$ **e.** not a real number   **f.** 4   **g.** $\dfrac{5z}{x^2y^3}$ **4. a.** $\sqrt[10]{x}$ **b.** $\sqrt[4]{x^3}$ **c.** $\sqrt[3]{x^4}$ **5. a.** 256   **b.** 22.52722735

Exercises   1. 3   **3.** $\dfrac{1}{10}$ **5.** -512 **7.** -4 **9.** nonreal   **11.** $\dfrac{3}{7}$ **13.** 16   **15.** $-\dfrac{1}{6}$ **17.** $\dfrac{5}{2}$ **19.** $\dfrac{1}{4}$ **21.** $\dfrac{3}{8}$ **23.** $-\dfrac{1}{1000}$ **25.** 64   **27.** 8.5499

29. 10,000,000   **31.** 99.6055   **33.** 0.0922   **35.** 1.6083   **37.** 0.2236   **39.** 7.7460   **41.** 2.0408   **43.** $8x$ **45.** $\dfrac{1}{3a^2}$ **47.** $8x^{\frac{5}{2}}$ **49.** $5a^{\frac{13}{6}}$

51. $x^{\frac{7}{12}}$ **53.** $x^{\frac{1}{2}}$ **55.** $\dfrac{1}{a^{\frac{9}{8}}}$ **57.** $a^{\frac{1}{4}}$ **59.** $\dfrac{a^2}{b^{\frac{6}{5}}}$ **61.** $8x^{\frac{3}{2}}y$ **63.** $\dfrac{x^{\frac{3}{2}}}{16y^{\frac{2}{5}}}$ **65.** $\dfrac{x^2y^4}{z^4}$ **67.** $\dfrac{y^{\frac{3}{2}}z^2}{x}$ **69.** $\dfrac{8b^{\frac{9}{4}}}{a^3c^3}$ **71.** $x^{\frac{1}{4}}y^{\frac{5}{4}}$ **73.** $\dfrac{y^{\frac{2}{3}}}{50x^{\frac{4}{3}}}$ **75.** $\dfrac{b^{\frac{5}{12}}}{a^{\frac{11}{12}}}$

77. $\dfrac{3x^{\frac{1}{2}}}{20y^{\frac{2}{3}}}$ **79.** $\sqrt[6]{a^5}$ **81.** $\sqrt[12]{y^7}$ **83.** $\sqrt[30]{x^{11}}$ **85.** $\sqrt[6]{y}$ **87.** $\sqrt[9]{x}$ **89.** $\sqrt[3]{7a}$ **91.** $\sqrt[24]{x}$ **93.** $a^{20}b^5c^{10}$ **95.** No: $\sqrt[5]{a}\cdot\sqrt{a} = a^{\frac{7}{10}}$; $\sqrt[5]{a^2} = a^{\frac{2}{5}}$; $a^{\frac{7}{10}} \neq a^{\frac{2}{5}}$

Section 14.7: Functions with Radicals

Margin Exercises   1. a. $[-2, \infty)$ **b.** $(-\infty, \infty)$ **2. a.** $f(0) = 0, f(2) = 10, f(8) = 20$ **b.** $f(1) = -1, f(2) = 1, f(15) = 3$

3.

$y = \sqrt{2x+1}$

4. a.

x	$y1$
-7	3.1072
-1	1.8171
0	1.2599
2	-1.817
7	-2.962

b.

$(-7, 3.11)$ $(-1, 1.82)$ $(0, 1.26)$ $(2, -1.82)$ $(7, -2.96)$

Exercises 1. a. $\sqrt{5} \approx 2.2361$ **b.** 3 **c.** $5\sqrt{2} \approx 7.0711$ **d.** 2 **3. a.** 3 **b.** -1 **c.** -2 **d.** $2\sqrt[3]{3} \approx 2.8845$ **5.** $[-8, \infty)$ **7.** $\left(-\infty, \dfrac{1}{2}\right]$ **9.** $(-\infty, \infty)$

11. $[0, \infty)$ **13.** $(-\infty, \infty)$ **15.** E **17.** B **19.** A

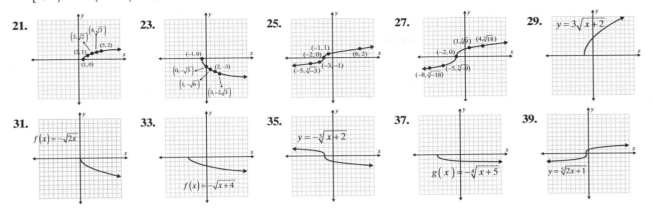

21. **23.** **25.** **27.** **29.**

31. **33.** **35.** **37.** **39.**

41. a. $\dfrac{1}{\sqrt{3+h}+\sqrt{3}}$ **b.** Slope of the line connecting $\left(3+h, f(3+h)\right)$ and $\left(3, f(3)\right)$. **c.** A line just touching the curve at one point.

d. $\dfrac{1}{2\sqrt{3}}$; represents the slope of the line tangent to $f(x)$ at $x = 3$.

Chapter 14: Review

1. 6 **2.** 14 **3.** 9 **4.** 0.03 **5.** -7 **6.** $\dfrac{5}{3}$ **7.** nonreal **8.** irrational **9.** rational **10.** rational **11.** 4.5826 **12.** -16.9706 **13.** -15

14. $3x\sqrt{x}$ **15.** $2a^2\sqrt{2}$ **16.** $5xy\sqrt{2x}$ **17.** $-9x\sqrt{y}$ **18.** $2\sqrt[3]{5}$ **19.** $-3xy^2\sqrt[3]{3x^2y}$ **20.** $3a\sqrt[3]{2ab^2}$ **21.** $\dfrac{5a\sqrt{3a}}{3}$ **22.** $\dfrac{2y^4}{3x^5}$

23. $-4\sqrt{11}$ **24.** $11\sqrt{7}$ **25.** $5\sqrt{x}$ **26.** $5\sqrt{2} - 26\sqrt{3}$ **27.** 0 **28.** $4\sqrt{6}+3$ **29.** -1 **30.** $7+2\sqrt{10}$ **31.** $x - y$

32. $8\sqrt{2x} + 15x + 2$ **33.** 25.4495 **34.** -3.5636 **35.** $\dfrac{-8\sqrt{5}}{5}$ **36.** $\dfrac{\sqrt{6}}{4}$ **37.** $\dfrac{4\sqrt{7}}{5}$ **38.** $\dfrac{\sqrt{a}}{3}$ **39.** $\dfrac{x\sqrt{x} - 3x}{x - 9}$ **40.** $\dfrac{3\sqrt{5} - 5}{2}$

41. $\dfrac{5\sqrt{7} - 5\sqrt{3}}{4}$ **42.** $\dfrac{13 + 3\sqrt{15}}{17}$ **43.** $x = 7$ **44.** $x = 68$ **45.** $x = -9, 9$ **46.** $x = 0$ **47.** $x = 3, 5$ **48.** no solution **49.** $x = 2$

50. $x = 4$ **51.** no solution **52.** $x = 2, 4$ **53.** 9 **54.** $-\dfrac{1}{6}$ **55.** $\dfrac{2}{3}$ **56.** $\dfrac{9}{25}$ **57.** 10.9027 **58.** 0.2408 **59.** $27x^2$ **60.** $5y^{\frac{10}{3}}$ **61.** $\dfrac{1}{x^{\frac{5}{2}}}$

62. $\dfrac{9a^4b^2}{c^2}$ **63.** $\sqrt[12]{x^5}$ **64.** $2\sqrt[3]{x}$ **65. a.** 0 **b.** 5 **66. a.** $\sqrt[3]{6} \approx 1.8171$ **b.** $2\sqrt[3]{2} \approx 2.5198$ **67.** $\left[-\dfrac{1}{2}, \infty\right)$ **68.** $(-\infty, \infty)$

69. **70.** **71.** **72.**

Chapter 14: Test

1. $4\sqrt{7}$ **2.** 2 **3.** $\dfrac{12}{7}$ **4.** $2y^2\sqrt{30x}$ **5.** $2y\sqrt[3]{6x^2y^2}$ **6.** $\dfrac{7x^6y\sqrt[3]{y}}{2z^2}$ **7.** $17\sqrt{3}$ **8.** $-xy\sqrt{y}$ **9.** $5 - 2\sqrt{6}$ **10.** $30 - 7\sqrt{3x} - 6x$

11. $\dfrac{y\sqrt{10x}}{4x^2}$ **12.** $1 + \sqrt{x}$ **13.** $x = 14$ **14.** $x = -4$ **15.** $x = 1$ **16.** $x = 7$ **17.** $\left[-\dfrac{4}{3}, \infty\right)$ **18.** $(-\infty, \infty)$ **19.**

20. $\sqrt[3]{4x^2}$ **21.** $2^{\frac{1}{2}}x^{\frac{1}{3}}y^{\frac{2}{3}}$ **22.** 4 **23.** $4x^{\frac{7}{6}}$ **24.** $\dfrac{8y^{\frac{3}{2}}}{x^3}$ **25.** 2.6008 **26.** 0.125 **27.** 3.0711

Cumulative Review: Chapters 1 – 14

1. sixty-five million, eighty thousand, five and thirty-nine hundredths **2. a.** 643% **b.** 33.3% **c.** 76% **3.** $1\dfrac{13}{18}$ **4.** $4\dfrac{37}{40}$ **5.** 23.69

6. -95.55 **7.** $x = \dfrac{1}{21}$ **8.** $x = 7$ **9.** $(-\infty, 6)$ **10.**

11.

12. $x = 5$ **13.** $4x + y = -2$ **14. a.** -12 **b.** 0 **c.** -168

15. $\dfrac{\left(8 \times 10^{-3}\right)\left(4 \times 10^{4}\right)}{\left(3.2 \times 10^{2}\right)\left(1 \times 10^{-3}\right)}; 1 \times 10^{3}$ **16.** $2x^{3} - 4x^{2} + 8x - 4$ **17.** $6x^{2} - 23x + 7$ **18.** $-5x^{2} + 18x + 8$ **19.** $49x^{2} - 28x + 4$

20. $(7 + 2x)(4 - x)$ **21.** $4y^{2}(4y + 1)^{2}$ **22.** $5(x - 4)\left(x^{2} + 4x + 16\right)$ **23.** $(x + 4)(x + 1)(x - 1)$ **24.** $(x - 16)(x + 3) = 0; \ x = -3, 16$

25. $(2x + 3)(x - 2) = 0; \ x = -\dfrac{3}{2}, 2$ **26.** $(5x - 2)(3x - 1) = 0; \ x = \dfrac{1}{3}, \dfrac{2}{5}$ **27.** $x^{2} - 11x + 28 = 0$ **28.**

29. $(14, 6)$ **30.** no solution **31.** $\left(\dfrac{3}{2}, \dfrac{5}{2}\right)$ **32.** $4\sqrt{5}$ **33.** $12x^{2}\sqrt{2x}$ **34.** 5 **35.** $2x^{2}y^{3}\sqrt[3]{2y}$

36. $2\sqrt{3}$ **37.** $x\sqrt{2} + \sqrt{6} - xy - y\sqrt{3}$ **38.** $\dfrac{\sqrt{10y}}{2y}$ **39.** $2\sqrt{30} - 11$ **40.** $x = -1$ **41.** $x = -1$

42. $(-\infty, 3]$ **43. a.** 16π ft (or 50.24 ft) **b.** 64π ft^{2} (or 200.96 ft^{2}) **44.** $10\sqrt{5}$ m; 22.361 m **45. a.** \$1969 **b.** \$419 **46.** $-15, -14$

47. 60 gallons **48.** \$35,000 in 6% and \$15,000 in 10% **49.** Burrito price is \$2.60 and taco price is \$1.35
50. 60 pounds of the first type (\$1.25 candy) and 40 pounds of the second type (\$2.50 candy) **51. a.** 8.4¢, 7.9¢, 7.1¢
b. 14 in. diameter for \$10.95 costs less per square inch. **52.** 26.5 meters

Chapter 15: Quadratic Equations

Section 15.1: Quadratic Equations: The Square Root Method

Margin Exercises 1. a. $x = 0, 20$ **b.** $x = -7, 2$ **2. a.** $x = \pm 2\sqrt{3}$ **b.** $x = -5 \pm 2\sqrt{10}$ **c.** $x = -1 \pm 5\sqrt{3}$ **d.** $x = 2 \pm \sqrt{2}$ **e.** no solution
3. $x = 4 \pm 2\sqrt{3}$ **4. a.** 10 ft **b.** $\dfrac{5\sqrt{2}}{2}$ in. $(\approx 3.54$ in.)

Exercises 1. $x = 0, 11$ **3.** $x = -12, -3$ **5.** $x = 1, -\dfrac{5}{3}$ **7.** $x = -1, 3$ **9.** $x = \dfrac{3}{4}, 1$ **11.** $x = \pm 11$ **13.** $x = \pm 6$ **15.** $x = \pm\sqrt{35}$

17. $x = \pm\sqrt{62}$ **19.** $x = \pm 3\sqrt{5}$ **21.** $x = \pm 3\sqrt{2}$ **23.** $x = \pm\dfrac{2}{3}$ **25.** $x = -1, 3$ **27.** no solution **29.** $x = -\dfrac{3}{2}, -\dfrac{1}{2}$ **31.** $x = \dfrac{7}{3}, \dfrac{11}{3}$

33. $x = 6 \pm 3\sqrt{2}$ **35.** $x = 7 \pm 2\sqrt{3}$ **37.** $x = \dfrac{-4 \pm 3\sqrt{3}}{3}$ **39.** $x = \dfrac{2 \pm 3\sqrt{7}}{5}$ **41.** $c = 15$ **43.** $b = 6\sqrt{3}$ **45.** $b = 1$

47. The length of the leg is 4 feet and the hypotenuse is 8 feet. **49.** $3\sqrt{2}$ cm **51.** 47.2 feet **53.** 3 seconds **55.** $x = \pm 25.44$
57. $x = \pm 5.25$ **59.** $x = \pm 1.70$ **61.** $x = \pm 4.13$ **63.** $4.2361, -0.2361$ **65.** $2.3229, -0.3229$

Section 15.2: Quadratic Equations: Completing the Square

Margin Exercises 1. a. $x = 4 \pm 2\sqrt{5}$ **b.** $x = \dfrac{-9 \pm \sqrt{93}}{2}$ **c.** $x = \dfrac{-3 \pm \sqrt{21}}{3}$ **d.** $x = \dfrac{3 \pm \sqrt{11}}{2}$ **e.** $x = 8 \pm 5\sqrt{2}$ **2.** $x = 1 \pm \sqrt{6}$

Exercises 1. $x^{2} + 12x + 36 = (x + 6)^{2}$ **3.** $2x^{2} - 16x + 32 = 2(x - 4)^{2}$ **5.** $x^{2} - 3x + \dfrac{9}{4} = \left(x - \dfrac{3}{2}\right)^{2}$ **7.** $x^{2} + x + \dfrac{1}{4} = \left(x + \dfrac{1}{2}\right)^{2}$

9. $2x^{2} + 4x + 2 = 2(x + 1)^{2}$ **11.** $x = -7, 1$ **13.** $x = -5, 9$ **15.** $x = -5, 8$ **17.** $x = -\dfrac{4}{3}, 1$ **19.** $x = -\dfrac{1}{2}, \dfrac{3}{2}$ **21.** $x = -3 \pm \sqrt{6}$

23. $x = -1 \pm \sqrt{6}$ **25.** $x = -\dfrac{1}{3}, 1$ **27.** $x = \dfrac{-1 \pm \sqrt{13}}{2}$ **29.** $x = \dfrac{-3 \pm \sqrt{17}}{4}$ **31.** $x = \dfrac{9 \pm \sqrt{73}}{2}$ **33.** $x = \dfrac{-7 \pm \sqrt{105}}{2}$ **35.** $x = -2, 13$

37. $x = \dfrac{-1 \pm \sqrt{17}}{4}$ **39.** $x = 1$ **41.** $x = \dfrac{-5 \pm \sqrt{33}}{2}$ **43.** $x = \dfrac{4 \pm \sqrt{10}}{6}$ **45.** $x = \dfrac{-7 \pm \sqrt{17}}{4}$ **47.** $x = \dfrac{1 \pm \sqrt{13}}{4}$ **49.** $x = -\dfrac{5}{3}, -1$

51. $x = -1, -\dfrac{3}{2}$ **53.** $x = -1, -7$ **55.** $x = -\dfrac{1}{3}, \dfrac{1}{2}$

Section 15.3: Quadratic Equations: The Quadratic Formula

Margin Exercises **1. a.** $x = -2 \pm \sqrt{5}$ **b.** $x = \dfrac{-1 \pm \sqrt{13}}{4}$ **c.** $x = \dfrac{-3 \pm \sqrt{3}}{2}$ **d.** $x = \pm\dfrac{2\sqrt{30}}{3}$ **2.** $x = \dfrac{-3 \pm \sqrt{5}}{4}$ **3. a.** $x = \dfrac{3 \pm \sqrt{13}}{2}$

b. $x = -\dfrac{5}{2}$ **4.** With $a = 1, b = 3$, and $c = 5$, $x = \dfrac{-3 \pm \sqrt{-11}}{2}$. So, there is no real solution.

Exercises **1.** $x^2 - 3x - 2 = 0; a = 1, b = -3, c = -2$ **3.** $2x^2 - x + 6 = 0; a = 2, b = -1, c = 6$ **5.** $7x^2 - 4x - 3 = 0; a = 7, b = -4, c = -3$

7. $3x^2 - 9x - 4 = 0; a = 3, b = -9, c = -4$ **9.** $2x^2 + 5x - 3 = 0; a = 2, b = 5, c = -3$ **11.** $x = 2 \pm \sqrt{5}$ **13.** $x = -1, 4$ **15.** $x = -\dfrac{1}{2}, 1$

17. $x = -1, \dfrac{2}{5}$ **19.** $x = 0, \dfrac{1}{3}$ **21.** $x = \pm\sqrt{7}$ **23.** $x = -1, 0$ **25.** $x = -\dfrac{4}{3}, 0$ **27.** $x = \dfrac{-5 \pm \sqrt{65}}{4}$ **29.** $x = \dfrac{-6 \pm \sqrt{42}}{3}$ **31.** $x = \dfrac{1}{2}, -3$

33. $x = \dfrac{4 \pm \sqrt{22}}{3}$ **35.** $x = 1, \dfrac{3}{4}$ **37.** $x = -3, -\dfrac{1}{2}$ **39.** $x = \dfrac{7 \pm \sqrt{37}}{6}$ **41.** $x = -\dfrac{3}{2}, \dfrac{8}{5}$ **43.** $x = 4, -\dfrac{1}{3}$ **45.** $x = \dfrac{11 \pm \sqrt{41}}{8}$

47. $x = -\dfrac{3}{2}, -2$ **49.** $x = \pm\dfrac{4}{5}$ **51.** $x = \dfrac{8 \pm \sqrt{58}}{6}$ **53.** $x = \dfrac{7 \pm \sqrt{385}}{12}$ **55.** $x = 1.6180, -0.6180$ **57.** $x = 1, -0.25$ **59.** $x = \pm 4.4721$

Section 15.4: Applications

Margin Exercises **1.** 6 ft and 8 ft **2.** 3 hours **3.** 2 mph **4.** 33.9 ft

Exercises **1.** $x^2 + x = 132; x = 11$ **3.** $w(2w - 5) = 63; w = 7$, the rectangle is 7 m by 9 m

5. $x(29 - x) = 198; x = 11$ or 18, the field is 11 m by 18 m **7.** $(w + 12)(w + 22) = 1344; w = 20$, the pool is 20 ft by 30 ft

9. $x^2 + (x + 1)^2 = 221; x = 10$, the numbers are 10 and 11 **11.** $x - 3 = \dfrac{4}{x}; x = 4$

13. $(x + 10)^2 = 9x^2; x = 5$, the side of the original square is 5 cm **15.** $w^2 + (2w + 2)^2 = 169; w = 5$, the rectangle is 5 m by 12 m

17. $2x^2 = 144$; the legs of the triangle are $6\sqrt{2}$ cm each **19.** $\dfrac{900}{x} - 10 = \dfrac{900}{x + 15}; x = 30$, there are 30 people in the club

21. $\dfrac{200}{x} + \dfrac{200}{x - 10} = 9; x = 50$, he traveled to the city at 50 mph **23.** $\dfrac{45}{12 - c} - \dfrac{45}{12 + c} = 2; c = 3$, the rate of the current is 3 mph

25. $\dfrac{120}{x} + 2 = \dfrac{120}{x - 5}; x = 20$, there were initially 20 members **27.** $2(x - 4)^2 = 162; x = 13$, the cardboard was 13 in. by 13 in.

29. $\dfrac{1}{x} + \dfrac{1}{x + 30} = \dfrac{1}{8}; x = 10$; the smaller pipe would take 40 min **31. a.** 1 sec **b.** 2.5 sec **c.** 3.0 sec **33.** 28.3 ft

35. a cannot be equal to zero and b^2 must be greater than or equal to $4ac$ to produce a real solution. Also, all the solutions found may not apply to the problem at hand. You must check that each answer makes sense in the context of the problem.

Section 15.5: Quadratic Functions: $y = ax^2 + bx + c$

Margin Exercises **1. a.** $(0, -4)$ **2. a.** $(2, -2)$ **3. a.** $(1, 3)$

b. $x = 0$ **b.** $x = 2$ **b.** $x = 1$

c. $(-2, 0)$ and $(2, 0)$ **c.** $(2 - \sqrt{2}, 0)$ and $(2 + \sqrt{2}, 0)$ **c.** $(0, 0)$ and $(2, 0)$

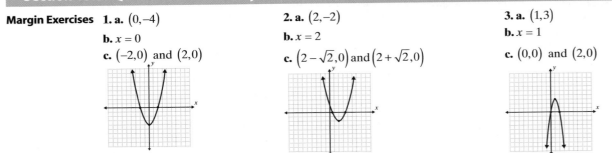

4. a. length = 18 in. width = 18 in. **b.** The hockey team will make its maximum revenue of $39,200 by charging a ticket price of $14 per ticket. **c.** Two sides of the lot are 90 yards and the third side is 180 yards for a maximum area of 16,200 yards. **5.** Answers will vary.

Exercises

1. a. $(0,4)$ **b.** $x = 0$
c. none
d.

3. a. $(0,8)$ **b.** $x = 0$
c. $\left(2\sqrt{2}, 0\right), \left(-2\sqrt{2}, 0\right)$
d.

5. a. $(1, -4)$ **b.** $x = 1$
c. $(-1, 0), (3, 0)$
d.

7. a. $(-3, -9)$ **b.** $x = -3$
c. $(-6, 0), (0, 0)$
d.

9. a. $(-2, 6)$ **b.** $x = -2$
c. $\left(-2 + \sqrt{6}, 0\right), \left(-2 - \sqrt{6}, 0\right)$
d.

11. a. $\left(\dfrac{5}{2}, -\dfrac{19}{2}\right)$ **b.** $x = \dfrac{5}{2}$
c. $\left(\dfrac{5 + \sqrt{19}}{2}, 0\right), \left(\dfrac{5 - \sqrt{19}}{2}, 0\right)$
d.

13. a. $\left(-\dfrac{7}{2}, -\dfrac{65}{4}\right)$ **b.** $x = -\dfrac{7}{2}$
c. $\left(\dfrac{-7 + \sqrt{65}}{2}, 0\right), \left(\dfrac{-7 - \sqrt{65}}{2}, 0\right)$
d.

15. a. $\left(\dfrac{1}{2}, -\dfrac{11}{4}\right)$
b. $x = \dfrac{1}{2}$ **c.** none
d.

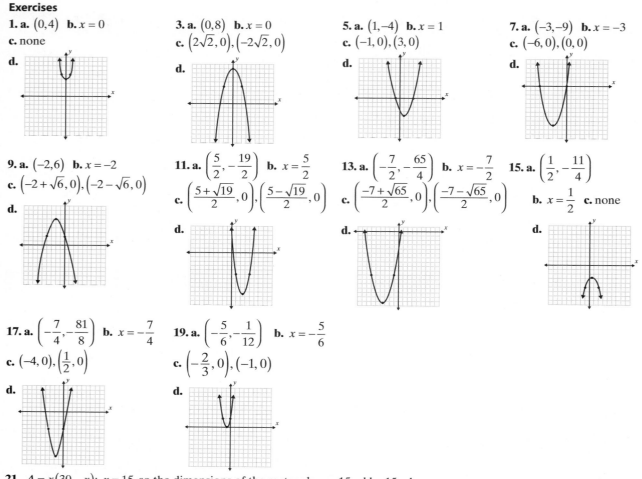

17. a. $\left(-\dfrac{7}{4}, -\dfrac{81}{8}\right)$ **b.** $x = -\dfrac{7}{4}$
c. $(-4, 0), \left(\dfrac{1}{2}, 0\right)$
d.

19. a. $\left(-\dfrac{5}{6}, -\dfrac{1}{12}\right)$ **b.** $x = -\dfrac{5}{6}$
c. $\left(-\dfrac{2}{3}, 0\right), (-1, 0)$
d.

21. $A = x(30 - x)$; $x = 15$, so the dimensions of the rectangle are 15 yd by 15 yd.
23. a. 3.5 sec **b.** 196 ft   **25. a.** 4 sec **b.** 288 ft   **27. a.** $R(x) = x(40 - x)$ **b.** $20   **29. a.** $R(x) = x(80 - 2x)$ **b.** $20 **c.** $800

31. a.

b.

33. a.

b.

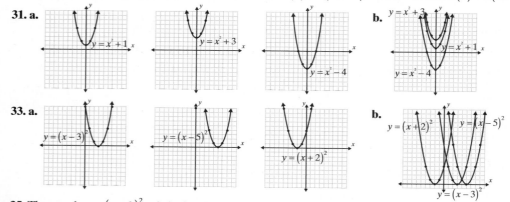

35. The graph $y = (x - h)^2 + k$ is the graph $y = x^2$ shifted k units vertically and h units horizontally.

Chapter 15: Review

1. $x = \pm 20$ **2.** $x = \pm 6\sqrt{2}$ **3.** no solution   **4.** $x = -1, 13$ **5.** $x = -3 \pm 2\sqrt{5}$ **6.** $x = -1 \pm 2\sqrt{10}$ **7.** $x = -\dfrac{9}{2}, 0$

8. $x = \dfrac{-5 \pm 5\sqrt{3}}{2}$ **9.** $c = 8\sqrt{2}$ **10.** $a = 21$ **11.** 60 in.   **12.** 4.5 seconds   **13.** ± 15.81 **14.** ± 1.62 **15.** $x^2 + 10x + \underline{25} = (x + 5)^2$

16. $x^2 - x + \dfrac{1}{4} = \left(x - \dfrac{1}{2}\right)^2$ **17.** $x = -1, 9$ **18.** $x = -2, -1$ **19.** $x = \pm\dfrac{2}{3}$ **20.** $x = 2 \pm \sqrt{3}$ **21.** $x = \dfrac{-1 \pm \sqrt{17}}{2}$ **22.** $x = -2, \dfrac{1}{2}$

23. $x = \pm\dfrac{5}{4}$ **24.** $x = -10, 2$ **25.** $x = \dfrac{6 \pm \sqrt{6}}{6}$ **26.** $x = \dfrac{2 \pm \sqrt{13}}{3}$ **27.** $x = \dfrac{5 \pm 2\sqrt{5}}{5}$ **28.** $x = \dfrac{-5}{3}, 1$ **29.** $x = \dfrac{1 \pm \sqrt{17}}{4}$

30. $x = \dfrac{3 \pm \sqrt{3}}{3}$ **31.** $x = \dfrac{\pm 7\sqrt{2}}{2}$ **32.** $x = \dfrac{3 \pm \sqrt{3}}{2}$ **33.** $x = \dfrac{1 \pm \sqrt{73}}{4}$ **34.** $x = -\dfrac{2}{3}, 1$ **35.** $x = -2, 1$ **36.** $x = -3 \pm \sqrt{19}$

37. $x = 0, \dfrac{3}{7}$ **38.** $x = \dfrac{11 \pm \sqrt{89}}{8}$ **39.** $x = \dfrac{-4 \pm 2\sqrt{13}}{3}$ **40.** $x = -\dfrac{1}{4}, 1$ **41.** $x = 4.2361, -0.2361$ **42.** $x = 2.9365, -0.9365$ **43.** $6, 7$

44. length is 30 ft, width is 24 ft **45.** length is 15 m, width is 10 m **46.** The numbers are 0 and 2 or 8 and 10 **47.** The length of the two identical sides is $10\sqrt{2}$ meters. **48.** 72 mph **49.** 4 mph **50.** 6 hours, 4 hours **51.** 90 cm × 90 cm × 5 cm **52.** 20 mph

53. a. $(0, -4)$ **54. a.** $(0, 16)$ **55. a.** $(3, -4)$ **56. a.** $(5, -25)$ **57. a.** $(1, -2)$
b. $x = 0$ **b.** $x = 0$ **b.** $x = 3$ **b.** $x = 5$ **b.** $x = 1$
c. $(2, 0), (-2, 0)$ **c.** $(4, 0), (-4, 0)$ **c.** $(1, 0), (5, 0)$ **c.** $(0, 0), (10, 0)$ **c.** none
d.

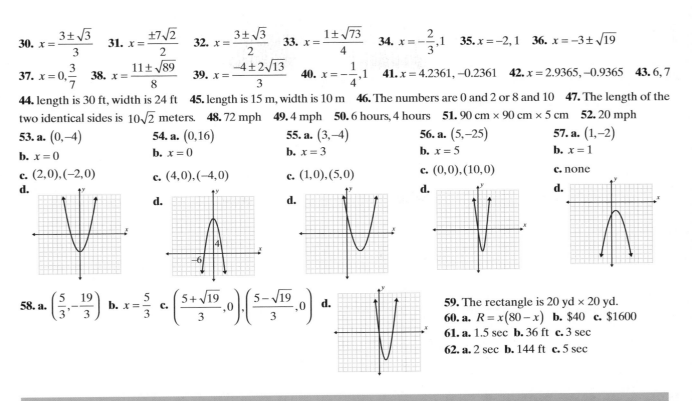

58. a. $\left(\dfrac{5}{3}, -\dfrac{19}{3}\right)$ **b.** $x = \dfrac{5}{3}$ **c.** $\left(\dfrac{5 + \sqrt{19}}{3}, 0\right), \left(\dfrac{5 - \sqrt{19}}{3}, 0\right)$ **d.**

59. The rectangle is 20 yd × 20 yd.
60. a. $R = x(80 - x)$ **b.** \$40 **c.** \$1600
61. a. 1.5 sec **b.** 36 ft **c.** 3 sec
62. a. 2 sec **b.** 144 ft **c.** 5 sec

Chapter 15: Test

1. $x = 1, 2$ **2.** $x = -\dfrac{12}{5}, 0$ **3.** $x = -\dfrac{5}{3}, 2$ **4.** $x = \pm 2\sqrt{3}$ **5.** $x = -12, 2$ **6.** $x = 1 \pm 4\sqrt{5}$ **7.** $b = 12$

8. $x^2 - 24x + 144 = (x - 12)^2$ **9.** $3x^2 + 9x + \dfrac{27}{4} = 3\left(x + \dfrac{3}{2}\right)^2$ **10.** $x = -4, -2$ **11.** $x = -1, 6$ **12.** $x = 2 \pm \sqrt{6}$ **13.** $x = -4 \pm 2\sqrt{3}$

14. $x = \dfrac{-4 \pm \sqrt{10}}{3}$ **15.** $x = \dfrac{2 \pm \sqrt{10}}{2}$ **16.** $x = 1 \pm \sqrt{5}$ **17.** $x = -3 \pm \sqrt{11}$ **18.** $x = \dfrac{1}{3}, 2$ **19.** $x = \dfrac{3 \pm \sqrt{17}}{4}$

20. a. $(0, -5)$ **b.** $x = 0$ **21. a.** $(2, 4)$ **b.** $x = 2$ **22. a.** $\left(-\dfrac{3}{4}, -\dfrac{1}{8}\right)$ **b.** $x = \dfrac{-3}{4}$
c. $\left(\sqrt{5}, 0\right), \left(-\sqrt{5}, 0\right)$ **d.**   **c.** $(0, 0), (4, 0)$ **d.**   **c.** $(-1, 0), \left(-\dfrac{1}{2}, 0\right)$ **d.**

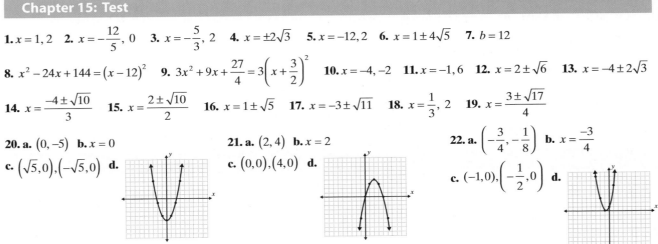

23. $18\sqrt{2}$ inches **24.** The rate of current is 4 mph **25. a.** 0.725 sec and 8.275 sec **b.** 324 ft **c.** 9 sec
26. The metal is 20 in. by 23 in.

Cumulative Review: Chapters 1 – 15

1. 87 **2.** $-\dfrac{17}{18}$ **3.** 20.2 **4.** $\dfrac{10}{21}$ **5.** $9x^4 y^2$ **6.** $-4x^6$ **7.** $x - 3$ **8.** $8x + 3$ **9.** $7x^2 - 14$ **10.** $7x^2 - 10x + 7$ **11.** $\dfrac{x}{x + 2}$

12. $\dfrac{3x - 3}{2x(x + 1)}$ **13.** $\dfrac{-x^2 + 6x + 5}{(x - 5)(x + 5)}$ **14.** $\dfrac{5x^2 + 2x + 8}{(x - 4)(x + 1)(x + 2)}$ **15.** $x = -4$ **16.** $x = 28$ **17.** $y = 1.8$ **18.** $x = 11$ **19.** $x = 3, 5$

20. $x = -9, 4$ **21.** $x = \dfrac{5}{3}$ **22.** $x = 0, 4$ **23.** $x = -3, 5$ **24.** $x = 1, 7$ **25.** $x = -2 \pm \sqrt{6}$ **26.** $x = \dfrac{3 \pm \sqrt{41}}{4}$

27. $x \le 3$ 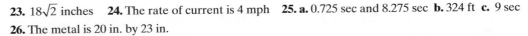 **28.** $x < \dfrac{17}{4}$ **29.** $x = -2, y = -3$ **30.** $\left(x, -\dfrac{1}{2}x + 3\right)$ **31.** no solution

32. $b = \dfrac{P - a}{2}$ **33.** $a = \dfrac{2A}{h} - b$ **34.** 3.2×10^7 **35.** $(x - 6)^2$ **36.** $(5x - 7)(5x + 7)$ **37.** $(x + 3)(3x + 1)$

38. $3(x - 4)(x + 6)$ **39.** $(x^2 - 8)(x + 4)$ **40.** $3\sqrt{7} = 7.9373$ **41.** $9\sqrt{3} = 15.5885$ **42.** $5\sqrt[3]{2} = 6.2996$

43. $\dfrac{5\sqrt{3}}{6} = 1.4434$ **44.** $11\sqrt{3} = 19.0526$ **45.** $2 - 7\sqrt{3} = -10.1244$ **46. a.** -8 **b.** 27 **c.** -3 **47.**

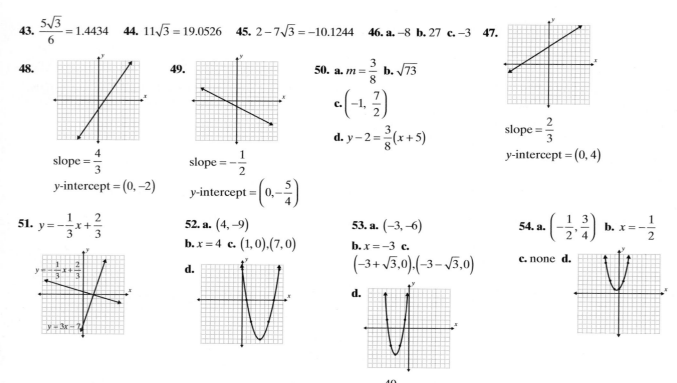

48.   slope $= \dfrac{4}{3}$ y-intercept $= (0, -2)$

49.   slope $= -\dfrac{1}{2}$ y-intercept $= \left(0, -\dfrac{5}{4}\right)$

50. a. $m = \dfrac{3}{8}$ **b.** $\sqrt{73}$
c. $\left(-1, \dfrac{7}{2}\right)$
d. $y - 2 = \dfrac{3}{8}(x + 5)$

slope $= \dfrac{2}{3}$
y-intercept $= (0, 4)$

51. $y = -\dfrac{1}{3}x + \dfrac{2}{3}$

52. a. $(4, -9)$
b. $x = 4$ **c.** $(1, 0), (7, 0)$
d.

53. a. $(-3, -6)$
b. $x = -3$ **c.** $\left(-3 + \sqrt{3}, 0\right), \left(-3 - \sqrt{3}, 0\right)$
d.

54. a. $\left(-\dfrac{1}{2}, \dfrac{3}{4}\right)$ **b.** $x = -\dfrac{1}{2}$
c. none **d.**

55. 11   **56.** 3600 women, 2800 men   **57.** \$2100 at 5%, \$1300 at 8%   **58.** $\dfrac{40}{9}$ hours   **59.** 5024 ft^2 **60.** width is 4 cm, length is 13 cm
61. 3 at \$2.50 and 4 at \$3.00   **62.** 10 mph upstream, 15 mph downstream   **63. a.** 1 second and 5 seconds **b.** 6.57 seconds
c. 7.06 seconds   **64.** 37.1 ft

Appendices: Further Topics in Algebra

A.1: U.S. Measurements

Margin Exercises **1. a.** 1 lb **b.** 2 pt **c.** 60 min **d.** 36 in.   **2. a.** 108 in. **b.** 32 fl oz **c.** 4 weeks **d.** 5 ft   **3. a.** 5 T **b.** 120 hr **c.** 16 c
d. 15,840 ft

Exercises **1.** 12 in.   **3.** 1 day   **5.** 1 gal   **7.** 5280 ft   **9.** 1 T   **11.** 36 in.   **13.** 21 days   **15.** 1.5 hr   **17.** 3 T   **19.** 2 yd   **21.** 21 ft
23. 48 fl oz   **25.** 1.5 ft   **27.** 49 days   **29.** 10 pt   **31.** 9 ft   **33.** 4 days   **35.** 42 in.   **37.** 32,000 lb   **39.** 2.5 hr

A.2: The Metric System

Margin Examples **1. a.** 3500 cm **b.** 64 mm **c.** 1230 m **d.** 6000 mm   **2. a.** 0.0059 **b.** 0.320 **c.** 65 **d.** 7   **3.** 0.65 dam   **4.** 4300 cm
5. a. 121 000 g **b.** 3.5 tons **c.** 4.576 tons **d.** 6.7 g   **6. a.** 43 000 grams **b.** 0.250 tons **c.** 23 000 milligrams   **7. a.** 0.75 cubic centimeters
b. 19 000 cubic centimeters **c.** 1 600 000 cubic centimeters **d.** 63 700 cubic decimeters   **8. a.** 0.000 006 3 cubic meters
b. 192 000 cubic centimeters   **9. a.** 0.0937 hectoliters **b.** 0.353 liters **c.** 12 000 liters   **10. a.** 1.952 liters **b.** 124 cubic centimeters
c. 19.75 cubic meters

Exercises **1.** 300   **3.** 1500   **5.** 450   **7.** 13 600   **9.** 182.5   **11.** 0.48   **13.** 0.065   **15.** 1.3   **17.** 0.0525   **19.** 5.5   **21.** 0.245 m   **23.** 10 km
25. 200 m   **27.** 0.32 cm   **29.** 17 350 mm   **31.** 2000   **33.** 3.7   **35.** 5.6   **37.** 0.091   **39.** 700   **41.** 5000 kg   **43.** 2000 kg   **45.** 896 000 mg
47. 75 kg   **49.** 7 000 000 g   **51.** 0.000 34 kg   **53.** 0.016 g   **55.** 0.0923   **57.** 7580   **59.** 2.963   **61.** 1000; 1000; 1000; 1 000 000 000
63. 900   **65.** 400 000 000   **67.** 0.063   **69.** 3100 mm^3 **71.** 5000 mm^3 **73.** 0.0000764 L   **75.** 30 mL   **77.** 0.0053 L   **79.** 72   **81.** 0.569
83. 7300; 7300

A.3: U.S. to Metric Conversions

Margin Examples **1. a.** 32°F **b.** 10°C   **2.** 25°C   **3.** 113°F   **4.** 2.1336 m   **5.** 91.4 m   **6.** 29.55 in.   **7.** 1640 ft   **8.** 341.85 cm^2 **9.** 6.48 ha

10. 7.41 acres   11. 409.032 ft^2 12. 15.14 L   13. 16.96 qt   14. 26.488 L   15. 423.78 ft^3 16. 85.05 g   17. 1.47 oz
Exercises   1. 77°F   3. 122°F   5. 10°C   7. 0°C   9. 2.74 m   11. 96.6 km   13. 124 mi   15. 19.7 in.   17. 19.35 cm^2 19. 55.8 m^2
21. 83.6 m^2 23. 405 ha   25. 741 acres   27. 53.82 ft^2 29. 4.65 in^2 31. 9.46 L   33. 10.6 qt   35. 11.088 gal   37. 4.54 kg   39. 453.6 g
41. 3038.31 m^2 43. a. 11 145.6 cm^2 b. 1.116 m^2

A.4: Absolute Value Equations and Inequalities

Margin Exercises   1. a. $x = -8, 8$ **b.** $x = -\dfrac{6}{5}, 2$ **c.** no solution   **d.** $x = -2, \dfrac{3}{2}$ **2.** $x = -11, 3$ **3. a.** $(-3, 3)$

b. $[-5, 3]$ **c.** $(-4, 1)$ **d.** no solution

e. $\left[-\dfrac{4}{3}, 4\right]$ **4. a.** $(-\infty, -2) \cup (2, \infty)$ **b.** $(-\infty, -5] \cup [-2, \infty)$

c. $(-\infty, \infty)$ **d.** $(-\infty, -3] \cup \left[\dfrac{5}{3}, \infty\right)$

Exercises   1. $x = -8, 8$ **3.** no solution   **5.** $x = -5, -1$ **7.** $x = -\dfrac{4}{3}, \dfrac{5}{3}$ **9.** $n = -2, \dfrac{2}{3}$ **11.** no solution   **13.** $x = 2$

15. $x = -\dfrac{3}{2}, 2$ **17.** $x = -\dfrac{1}{5}, 1$ **19.** $x = -\dfrac{1}{3}, 3$ **21.** $x = -22, 26$ **23.** $x = -8, 4$ **25.** $x = -6, 0$ **27.** $x = -\dfrac{1}{3}, 3$ **29.** $x = 1$

31. $x = -\dfrac{5}{2}, \dfrac{3}{4}$ **33.** $x = -\dfrac{20}{7}, \dfrac{4}{5}$ **35.** $x = -20, \dfrac{40}{9}$ **37.** $(-\infty, \infty)$

39. $\left[-\dfrac{4}{5}, \dfrac{4}{5}\right]$ **41.** $(-\infty, 1) \cup (5, \infty)$ **43.** $[-10, -2]$

45. $(-\infty, -8] \cup [-2, \infty)$ **47.** $\left(-\infty, -\dfrac{1}{2}\right] \cup \left[\dfrac{3}{2}, \infty\right)$

49. no solution   **51.** $\left[-2, -\dfrac{1}{2}\right]$ **53.** $\left(-\dfrac{5}{3}, -1\right)$

55. $\left(-\infty, -\dfrac{2}{3}\right] \cup [6, \infty)$ **57.** $[-1, 10]$ **59.** $(-\infty, \infty)$

61. $\left(\dfrac{1}{2}, \dfrac{7}{2}\right)$ **63.** $\left(-\infty, -\dfrac{5}{2}\right) \cup (0, \infty)$ **65.** $\left(-2, -\dfrac{4}{7}\right)$

67. a.   **b.** $|x| \le 10$ **c.** $[-10, 10]$, closed interval   **69. a.**   **b.** $|x - 8| > 6$

c. $(-\infty, 2) \cup (14, \infty)$ **71. a.**   **b.** $|x + 5| < 2$ **c.** $(-7, -3)$, open interval

A. 5: Synthetic Division and the Remainder Theorem

Margin Exercises   1. a. $3x^2 - 2x + \dfrac{1}{x - 1}$ **b.** $x^3 - x^2 + 3x - 4 + \dfrac{3}{x + 3}$ **2. a.** $P(-2) = 3$ **b.** $P(3) = 8$

Exercises   1. a. $x - 9$ **b.** $c = 3; P(3) = 0$ **3. a.** $x^2 - 4x + 33 - \dfrac{265}{x + 8}$ **b.** $c = -8; P(-8) = -265$ **5. a.** $4x^2 - 6x + 9 - \dfrac{17}{x + 2}$

b. $c = -2; P(-2) = -17$ **7. a.** $x^2 + 7x + 55 + \dfrac{388}{x - 7}$ **b.** $c = 7; P(7) = 388$ **9. a.** $2x^2 - 10x + 30 - \dfrac{99}{x + 3}$ **b.** $c = -3; P(-3) = -27$

11. a. $x^3 + 2x + 5 + \dfrac{17}{x - 3}$ **b.** $c = 3; P(3) = 17$ **13. a.** $x^3 + 2x^2 + 6x + 9 + \dfrac{23}{x - 2}$ **b.** $c = 2; P(2) = 23$

15. a. $x^3 + \dfrac{1}{2}x^2 - \dfrac{3}{4}x - \dfrac{3}{8} + \dfrac{45}{16\left(x - \dfrac{1}{2}\right)}$ **b.** $c = \dfrac{1}{2}; P\left(\dfrac{1}{2}\right) = \dfrac{45}{16}$ **17. a.** $x^4 + x^3 + x^2 + x + 1$ **b.** $c = 1; P(1) = 0$

19. a. $x^3 - \dfrac{14}{5}x^2 + \dfrac{56}{25}x - \dfrac{224}{125} + \dfrac{3396}{625\left(x + \dfrac{4}{5}\right)}$ **b.** $c = -\dfrac{4}{5}; P\left(-\dfrac{4}{5}\right) = \dfrac{3396}{625}$ **21. a.** $x^2 - \dfrac{7}{2}x + \dfrac{13}{4} + \dfrac{73}{4(2x - 1)}; 2x^2 - 7x + \dfrac{13}{2} + \dfrac{73}{4\left(x - \dfrac{1}{2}\right)}$

b. $\dfrac{P(x)}{x - \dfrac{b}{a}} = \dfrac{aP(x)}{ax - b}$ **c.** If a polynomial $P(x)$ is divided by $(ax - b)$, then the remainder will be $P\left(\dfrac{b}{a}\right)$.   Answers will vary.

A.6: Graphing Systems of Linear Inequalities

Margin Exercises 1. a. **b.** **c.** **2.**

Exercises 1. **3.** **5.** **7.** **9.**

11. **13.** **15.** **17.** **19.**

21. no solution **23.** **25.** **27.** **29.**

31. **33.** **35.** The solutions of the two inequalities do not overlap.

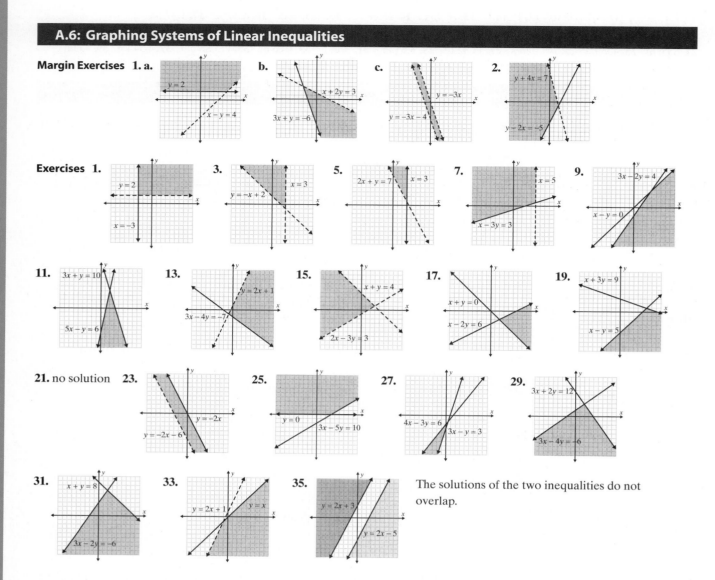

A.7: Systems of Linear Equations: Three Variables

Margin Exercises 1. $(1, -2, 4)$ **2.** no solution **3.** infinite number of solutions. **4.** 18 dollar coins, 12 quarters, and 7 dimes

Exercises 1. $(1, 0, 1)$ **3.** $(1, 2, -1)$ **5.** infinite number of solutions **7.** $(4, 1, 1)$ **9.** $(1, 2, -1)$ **11.** $(-2, 3, 1)$ **13.** no solution

15. $(3, -1, 2)$ **17.** $(2, 1, -3)$ **19.** $\left(\frac{1}{2}, \frac{1}{3}, -1 \right)$ **21.** 34, 6, 27 **23.** 18 ones, 16 fives, 12 tens **25.** 19 cm, 24 cm, 30 cm

27. 300 main floor, 200 mezzanine, and 80 balcony **29.** 3 lilies, 5 roses, and 8 daisies **31.** savings: \$30,000; bonds: \$55,000; stocks: \$15,000 **33.** 3 liters of 10%, 4.5 liters of 30%, 1.5 liters of 40% **35.** No. Graphically, the three planes intersect in one point (one solution) or in a line (infinitely many solutions) or they do not have a common intersection (no solution).

37. $A = 4, B = 2, C = -1$

A.8: Introduction to Complex Numbers

Margin Exercises 1. a. $10i$ **b.** $7i$ **c.** $3i\sqrt{2}$ **d.** $6i\sqrt{2}$ **2. a.** real: 0, imaginary: 5 **b.** real: 14, imaginary: $\sqrt{7}$ **c.** real: $\dfrac{6}{5}$, imaginary: $-\dfrac{11}{5}$ **d.** real: -13, imaginary: 0 **3. a.** $x = 10$ and $y = -2$ **b.** $y = 1$ and $x = -5$ **4. a.** $7 + 4i$ **b.** $-4\sqrt{2}\,i$ **c.** $(4 + \sqrt{2}) + (8 - \sqrt{5})i$

Exercises 1. real part is 4, imaginary part is -3 **3.** real part is -11, imaginary part is $\sqrt{2}$ **5.** real part is $\dfrac{3}{8}$, imaginary part is 0 **7.** real part is $\dfrac{4}{5}$, imaginary part is $\dfrac{7}{5}$ **9.** real part is $\dfrac{2}{3}$, imaginary part is $\sqrt{17}$ **11.** $7i$ **13.** $-8i$ **15.** $7\sqrt{3}$ **17.** $10i\sqrt{6}$ **19.** $-12i\sqrt{3}$ **21.** $11\sqrt{2}$ **23.** $10i\sqrt{10}$ **25.** $x = 6, y = -3$ **27.** $x = -2, y = \sqrt{5}$ **29.** $x = \sqrt{2} - 3, y = 1$ **31.** $x = 1, y = 4$ **33.** $x = 2, y = -6$ **35.** $x = 3, y = 10$ **37.** $x = -\dfrac{4}{3}, y = -3$ **39.** $6 + 2i$ **41.** $1 + 7i$ **43.** $6 - 6i$ **45.** $14i$ **47.** $(3 + \sqrt{5}) - 6i$ **49.** $5 + (\sqrt{6} + 1)i$ **51.** $\sqrt{3} - 5$ **53.** $-2 - 5i$ **55.** $11 - 16i$ **57.** 2 **59.** $3 + 4i$ **61. a.** yes **b.** no

A.9: Multiplication and Division with Complex Numbers

Margin Exercises 1. a. $28 + 4i$ **b.** $2 + 26i$ **c.** 3 **2. a.** $\dfrac{8}{5} + \dfrac{16}{5}i$ **b.** $\dfrac{1}{9} - \dfrac{4\sqrt{5}}{9}i$ **c.** $-\dfrac{1}{4} - i$ **d.** $\dfrac{13}{19} - \dfrac{8\sqrt{3}}{19}i$ **3. a.** i **b.** -1 **c.** 1

Exercises 1. $16 + 24i$ **3.** $-7\sqrt{2} + 7i$ **5.** $3 + 12i$ **7.** $1 - i\sqrt{3}$ **9.** $3 + 2i\sqrt{3}$ **11.** $2 + 8i$ **13.** $-7 - 11i$ **15.** $13 + 0i$ **17.** $34 + 13i$ **19.** $-24 + 70i$ **21.** $5 - i\sqrt{3}$ **23.** $23 - 10i\sqrt{2}$ **25.** $21 + 0i$ **27.** $(2 + \sqrt{10}) + (2\sqrt{2} - \sqrt{5})i$ **29.** $(9 - \sqrt{30}) + (3\sqrt{5} + 3\sqrt{6})i$ **31.** $0 + 3i$ **33.** $0 - \dfrac{5}{4}i$ **35.** $-\dfrac{1}{4} + \dfrac{1}{2}i$ **37.** $-\dfrac{4}{5} + \dfrac{8}{5}i$ **39.** $\dfrac{24}{25} + \dfrac{18}{25}i$ **41.** $-\dfrac{1}{13} + \dfrac{5}{13}i$ **43.** $-\dfrac{1}{29} - \dfrac{12}{29}i$ **45.** $-\dfrac{17}{26} - \dfrac{7}{26}i$ **47.** $\dfrac{4 + \sqrt{3}}{4} + \left(\dfrac{4\sqrt{3} - 1}{4}\right)i$ **49.** $-\dfrac{1}{7} + \dfrac{4\sqrt{3}}{7}i$ **51.** $0 + i$ **53.** $-1 + 0i$ **55.** $0 + i$ **57.** $1 + 0i$ **59.** $0 - i$ **61.** $x^2 + 9$ **63.** $x^2 + 2$ **65.** $5y^2 + 4$ **67.** $x^2 + 4x + 40$ **69.** $y^2 - 6y + 13$ **71.** Given complex number $(a + bi)$: $(a + bi)(a - bi) = a^2 - abi + abi - b^2 i^2 = a^2 + b^2$ which is the sum of squares of real numbers. Thus the product must be a positive real number. **73.** $a^2 + b^2 = 1$

Glossary

A

Absolute value   The distance a number is from 0 on a number line

Acute angle   An angle whose measure is between $0°$ and $90°$

Acute triangle   Triangle in which all three angles are acute

Addends   The numbers being added in an addition problem

Addition principle of equality   If the same algebraic expression is added to both sides of an equation, the new equation has the same solutions as the original equation.

Additive identity   The number 0 is called the additive identity.

Additive identity property   The sum of any number and 0 is equal to the number itself.

Additive inverse   The opposite of an integer; two integers are additive inverses (or opposites) if their sum is equal to 0

Adjacent angles   Two angles are adjacent if they have a common side.

Algebraic expression   A combination of variables and numbers using any of the operations of addition, subtraction, multiplication, or division, as well as exponents

Altitude of a triangle   The height of a triangle

Angle   Consists of two rays with a common endpoint (called a vertex)

Area   The measure of the interior, or enclosed region, of a plane surface

Ascending order   The exponents on the terms of a polynomial increase in order from left to right

Associative property of addition   The grouping of the numbers in addition can be changed.

Associative property of multiplication   The grouping of the numbers in multiplication can be changed.

Average   The sum of all the data divided by the number of data items. Also referred to as the mean or the arithmetic average.

B

Bar graph   Used to emphasize comparative amounts

Base   In the expression a^n, the number a is called the base.

Binomial   A polynomial with two terms

C

Change in value   To calculate the change in value, take the end value and subtract the beginning value.

Circle   The set of all points in a plane that are some fixed distance from a fixed point called the center of the circle.

Class   In a histogram, an interval (or range) of numbers that contain data items

Circle graph   Used to help in understanding percents or parts of a whole

Circumference   The perimeter of a circle

Class boundaries   In a histogram, numbers that are halfway between the upper limit of one class and the lower limit of the next class

Class width   In a histogram, the difference between the class boundaries of a class (the width of each bar)

Closed figure   A closed figure begins and ends at the same point.

Closed half-plane   If the boundary line is included in the solution set, then the half-plane is said to be closed.

Closed interval   Both endpoints of the interval are included

Coefficient   The number written next to a variable

Combined variation   If a variable varies either directly or inversely with more than one other variable, the variation is said to be combined variation.

Commission   A fee paid to an agent or salesperson for a service

Commutative property of addition   The order of the numbers in addition can be reversed.

Commutative property of multiplication   The order of the numbers in multiplication can be reversed.

Complementary angles   Two angles are complementary if the sum of their measures is $90°$.

Complex algebraic expression   Expression that involves rational expressions and more than one operation

Complex conjugates   The two complex numbers $a + bi$ and $a - bi$ are called complex conjugates or simply conjugates of each other.

Complex fraction   Fraction in which the numerator or denominator or both contain one or more fractions or mixed numbers

Complex numbers   Set of numbers that includes all the real numbers and the even roots of negative numbers

Composite number   A counting number with more than two different factors (or divisors)

Compound interest   Interest paid on interest earned

Compound inequality   A mathematical expression that uses inequality symbols to compare the order of three expressions or values

Conditional equation   An equation that has a finite number (a countable number) of solutions

Congruent angles   If two angles have the same measure, they are said to be congruent angles.

Congruent triangles   Two triangles are congruent if the corresponding angles have the same measure and the lengths of the corresponding sides are equal.

Conjugates   The two expressions $(a - b)$ and $(a + b)$ are called conjugates; the product of conjugates results in the difference of two squares.

Consecutive even integers   Even integers are consecutive if each is 2 more than the previous even integer.

Consecutive integers   Integers are consecutive if each is 1 more than the previous integer.

Consecutive odd integers   Odd integers are consecutive if each is 2 more than the previous odd integer.

Consistent system   A system of equations that has exactly one solution

Constant (or constant term)   A term that consists of only a number

Constant of variation   The constant multiplier in a relationship of direct or inverse variation

Contradiction   An equation that simplifies to a statement that is never true (such as $0 = 2$) and has no solution

Coordinate   Either of the numbers in an ordered pair. May also refer to the number that corresponds to a point on a number line

Cube   A rectangular solid in which the length, width, and height are all equal

Cube root   The cube root of a number equals another number that when cubed results in the original number

Cube of a number   In expressions with exponent 3, the base is said to be cubed.

D

Data   Value(s) measuring some characteristic of interest such as income, height, weight, grade point averages, scores on tests, and so on

Decimal notation   Decimal notation uses a decimal point, with whole numbers written to the left of the decimal point and fractions written to the right of the decimal point.

Decimal numbers   Numbers written in decimal notation are said to be decimal numbers (or simply decimals).

Decimal point   A period inserted between the whole number and fractional parts of a decimal number

Degree of a polynomial   The largest of the degrees of its terms

Degree of a term   The sum of the exponents on the variables

Denominator   The bottom number in a fraction

Dependent system   A system of equations that has an infinite number of solutions

Dependent variable   the second coordinate y in an ordered pair

Descending order   The exponents on the terms of a polynomial decrease in order from left to right

Diameter   The distance from one point on a circle through the center to the point directly opposite it

Difference   The missing addend found as result of subtraction

Difference of cubes   A binomial that can be written in the form $x^3 - a^3$

Difference of squares   A binomial that can be written in the form $x^2 - a^2$

Digit   A symbol used in our number system; namely 0, 1, 2, 3, 4, 5, 6, 7, 8, and 9

Direct variation   A variable quantity y varies directly as a variable x if there is a constant k such that $\frac{y}{x} = k$ or $y = kx$. When two variables vary directly, an increase in one indicates an increase in the other.

Discount   A reduction in the original selling price of an item; the difference between the original price and the sale price

Discriminant   In the quadratic formula, the expression $b^2 - 4ac$ is called the discriminant.

Distributive property   The product of a number and a sum is equal to the sum of the products of the number and each of the addends.

Dividend   The number being divided

Divisible   If a number can be divided by another number so that the remainder is 0, then the dividend is divisible by the divisor.

Division algorithm (or long division)   The process (or series of steps) that we follow when dividing two numbers or two polynomials

Divisor   The number doing the dividing

Domain of a function   The set of all first coordinates in a relation

Domain axis   In the graph of a relation, the horizontal axis (the x-axis)

Double solution The special cases where the two factors of a quadratic equation are the same, and there is only one solution.

E

Elements The items in the set

Empty set (null set) A set with absolutely no elements

Equation A statement that two algebraic expressions are equal

Equilateral triangle A triangle in which all three sides have equal lengths

Equivalent equations Equations with the same solution

Even number If an integer is divided by 2 and the remainder is 0, then the integer is even.

Event Some (or all) of the outcomes from the sample space

Experiment An activity in which the result is random in nature

Exponent A number placed above the base to show the number of times the base is multiplied by itself

Exponential notation Notation of the form a^n, where a is the base, and x is the exponent

Extraneous roots In rational expressions, solutions that are not solutions to the original equation; may be introduced by multiplying by the LCD

Extraneous solution A number that is found when solving an equation but that does not satisfy the original equation

F

Factor A number that is being multiplied; may also refer to a number that divides a given number

Factor theorem If $x = c$ is a root of a polynomial equation in the form $P(x) = 0$, then $x - c$ is a factor of the polynomial $P(x)$.

Factoring Given a product, find the factors

FOIL method Procedure for multiplying two binomials; multiply the first terms, the outside terms, the inside terms, and the last terms

Formula General statement (usually an equation) that relates two or more variables

Fraction A number that can represent parts of a whole, the ratio of two numbers, or division. Also called a rational number

Frequency In a histogram, the number of data items in a class

Function A relation in which each domain element has exactly one corresponding range element

Function notation Notation of the form $f(x)$, where f is the name of the function, and x is the input variable

G

Graph Visual representation of numerical information

Greatest common factor (GCF) The largest integer or algebraic term that is a factor (or divisor) of all of the numbers or terms

H

Half-open interval Only one endpoint of the interval is included

Half-plane A straight line separates a plane into two half-planes.

Hemisphere Half of a sphere

Histograms Used to indicate data in classes (a range or interval of numbers)

Horizontal line A line with a slope of 0

Hypotenuse The longest side of a right triangle; the side opposite the right angle.

I

Identity An equation that leads to a statement that is always true (such as $0 = 0$) and has an infinite number of solutions

Imaginary part Real number b in a complex number $a + bi$

Improper fraction A fraction in which the numerator is greater than or equal to the denominator

Inconsistent system A system of equations that has no solution

Independent variable The first coordinate x in an ordered pair

Index of a radical The index of the radical $\sqrt[n]{a}$ is the number n.

Inequality A mathematical expression that includes the symbols $<, >, \leq, \geq$, or $\neq$

Integers The set of numbers consisting of the whole numbers and their opposites

Interest Money paid for the use of money

Intersect Two lines intersect if there is one point on both lines

Intersection The intersection of two (or more) sets is the set of all elements that belong to both sets.

Interval   The set of all real numbers between two endpoints is called an interval of real numbers.

Interval notation   Notation to represent intervals of real numbers where brackets indicate that an endpoint is included and parentheses indicate that an endpoint is not included

Inverse Variation   A variable quantity varies inversely as a variable x if there is a constant k such that $x \cdot y = k$ or $y = \dfrac{k}{x}$. When two variables vary inversely, an increase in one indicates a decrease in the other.

Irrational numbers   Numbers that can be written as infinite nonrepeating decimals

Isosceles triangle   Triangle in which two or more sides have equal lengths.

J

Joint variation   If the combined variation is all direct variation (the variables are multiplied), then it is called joint variation.

L

Leading coefficient   The coefficient of the term with the largest degree of a polynomial

Least common denominator (LCD)   The least common multiple of the denominators of two or more fractions

Least common multiple (LCM)   The smallest number that is a multiple of each of the given numbers

Leg   Either of the two sides of a right triangle that are not the hypotenuse

Like radicals   Radicals that have the same index and radicand or can be simplified so that they have the same index and radicand

Like terms (similar terms)   Terms that are constants or terms that contain the same variables raised to the same powers

Line   A line has no beginning or end. Lines are labeled with small letters or by two points on the line.

Line graph   Used to indicate tendencies or trends over a period of time

Line of symmetry   The line through the vertex of a parabola that divides the graph into two symmetrical parts

Line segment   Consists of two points on a line and all the points between them

Linear equation in x   Equation that can be written in the form $ax + b = c$, where $a, b,$ and c are constants and $a \neq 0$.

Linear equation in three variables   Equation of the form $ax + by + cz = d$ where $a, b,$ and c are not all 0

Linear function   A function represented by an equation of the form $y = mx + b$

Linear inequality   Inequality that contain only constant or linear terms

Lower class limit   In a histogram, the smallest whole number that belongs to a class.

M

Mass   The amount of material in an object

Mean   The sum of all the data divided by the number of data items. Also referred to as the average or arithmetic average.

Measure of an angle   The size of the angle; measured in degrees

Median   The middle data item

Metric system   System of measurement used by about 90% of the world, but is not often used in the United States

Minuend   The number or quantity from which another (the subtrahend) is to be subtracted

Mixed number   The sum of a whole number and a proper fraction

Mode   The single data item that appears the most number of times

Monomial   A polynomial with one term

Multiples   To find the multiples of a number, multiply each of the counting numbers by that number.

Multiplication (or division) principle of equality   If both sides of an equation are multiplied by (or divided by) the same nonzero constant, the new equation has the same solutions as the original equation.

Multiplicative identity   The number 1 is called the multiplicative identity.

Multiplicative identity property   The product of any number and 1 is the number itself.

Multiplicative inverse   Reciprocal; two numbers are multiplicative inverses if their product is equal to 1

N

Natural (counting) numbers   The numbers $1, 2, 3, 4, \ldots$

Negative integers   The opposites of the natural numbers; they lie to the left of 0 on a number line

Net change   The net change in a measure is the algebraic sum of several numbers.

Nonreal complex numbers   Complex numbers of the form $a + bi$ where $b \neq 0$

Nonterminating decimal number   If the remainder of division is never 0, the decimal quotient is nonterminating.

Note   A loan for a period of 1 year or less

Numerator   The top number in a fraction

O

Obtuse angle   An angle whose measure is between 90° and 180°

Obtuse triangle   Triangle in which one angle is obtuse

Octant   One of the eight regions that are formed by the intersection of the xy-plane, the xz-plane, and the yz-plane.

Odd numbers   If an integer is divided by 2 and the remainder is 1, then the integer is odd.

Open half-plane   If the boundary line is not included in the solution set, then the half-plane is said to be open.

Open interval   Neither endpoint of the interval is included

Opposite   Two integers are opposites (or additive inverses) if their sum is equal to 0.

Ordered pair   Pair of numbers in the form (x, y) where the order of the numbers is critical

Ordered triple   Three numbers in the form (x, y, z) where the order of the numbers is critical

Origin   The point of intersection of the x-axis and the y-axis

Outcome   An individual result of an experiment

P

Parabola   The graph of a quadratic function is called a parabola.

Parallel lines   Lines that never intersect (cross each other) and whose slopes are equal

Parallelogram   A four-sided polygon with both pairs of opposite sides parallel

Pentagon   A 5-sided polygon

Percent   The ratio of a number to 100

Perfect cube   The cube of an integer

Perfect square   The square of an integer

Perfect square trinomial   The result of squaring a binomial

Perimeter   The distance around a figure; found by adding the lengths of the sides of the figure

Period   Group of three digits separated with commas

Perpendicular lines   Lines that intersect at 90° (right) angles and whose slopes are negative reciprocals of each other

Pi (π) The ratio of a circle's circumference to its diameter; approximated by 3.14

Plane   Flat surfaces, such as a table top or wall, represent planes.

Plane geometry   The study of the properties of figures in a plane

Point   A dot represents a point. Points are labeled with capital letters.

Point-slope form   The point-slope form for the equation of a line is $y - y_1 = m(x - x_1)$, where m is the slope of the line and (x_1, y_1) is any point on the line.

Polygon   A closed plane figure, with three or more sides, in which each side is a line segment

Polynomial   A polynomial is a monomial or the indicated sum or difference of monomials.

Positive integers   The natural numbers; they lie to the right of 0 on a number line

Prime factorization   The unique factorization of a composite number that contains only prime factors.

Prime number   A counting number greater than 1 that has exactly two different factors (or divisors) — itself and 1

Principal   The initial amount of money that is invested or borrowed

Principal square root   Every positive real number has two square roots, one positive and one negative. The positive square root is called the principal square root.

Probability   The likelihood of a particular result

Product   The result of multiplication

Profit   The difference between selling price and cost

Proper fraction   A fraction in which the numerator is less than the denominator

Proportion   A statement that two ratios are equal

Pythagorean theorem   In a right triangle, the legs, a and b, and the hypotenuse, c, have the following relationship: $a^2 + b^2 = c^2$.

Q

Quadrant   The x-axis and y-axis separate the Cartesian plane into four quadrants.

Quadratic equation   Equations that can be written in the form $ax^2 + bx + c = 0$ where a, b, and c are real numbers and $a \neq 0$

Quadratic formula   A formula that is used to find the solutions of the general quadratic equation $ax^2 + bx + c = 0$; the quadratic formula is $\dfrac{-b \pm \sqrt{b^2 - 4ac}}{2a}$

Quadratic function   A function of the form $y = ax^2 + bx + c$ where a, b, and c are real numbers and $a \neq 0$

Quotient   The result of division

R

Radical   The complete expression involving both the radical sign and the radicand

Radical sign   The symbol $\sqrt{}$

Radical function   Function of the form $y = \sqrt[n]{g(x)}$ in which the radicand contains a variable expression

Radicand   The number under the radical sign.

Radius   The distance from the center of a circle to any point on the circle

Range   The difference between the largest and smallest data items

Range of a function   The set of all second coordinates in the relation

Range axis   In the graph of a relation, the vertical axis (the y-axis)

Ratio   A comparison of two quantities by division

Rational expressions   Fractions in which the numerator and denominator are polynomials

Rational number   A number that can be written in the form $\frac{a}{b}$ where a and b are integers and $b \neq 0$

Rationalizing a denominator   The process used to remove radicals from the denominator of a rational expression

Ray   Consists of a point (called the endpoint) and all the points on a line on one side of that point

Real numbers   Consist of all rational and irrational numbers

Real part   Real number a in a complex number $a + bi$

Reciprocals   If the product of two nonzero fractions is 1, then the fractions are called reciprocals of each other.

Rectangle   A polygon with four sides in which adjacent sides are perpendicular (meet at a 90° angle)

Regular hexagon   A six-sided polygon where all sides have equal length

Regular octagon   An eight-sided polygon where all sides have equal length

Relation   A set of ordered pairs of real numbers

Remainder   The number left after division

Repeating decimal number   Decimal number that does not terminate, but has a repeating pattern to its digits

Restrictions on a variable   Values of the variable that make a rational expression undefined

Right angle   An angle whose measure is equal to 90°

Right triangle   A triangle containing one right angle

Roster form   The elements of a set are listed within braces

Rounding   To find another number close to the given number

S

Sale price   The new, reduced price of an item after a discount has been applied

Sales tax   A tax charged on the actual selling price of goods sold by retailers

Sample Space   The set of all possible outcomes of an experiment

Scalene triangle   Triangle in which no two sides are equal in length

Scientific notation   Decimal numbers are written as the product of a number greater than or equal to one and less than 10, and an integer power of 10

Semicircle   Half of a circle

Set   A collection of objects or numbers

Set-builder notation   The elements of a set are described by giving a condition (or restriction) for the variable

Similar triangles   Two triangles are similar if the measures of the corresponding angles are equal and the lengths of the corresponding sides are proportional.

Simple interest   Interest that involves only one payment at the end of the term of a loan.

Simplest form for cube roots   A cube root is considered to be in simplest form when the radicand has no perfect cube as a factor.

Simplest form for square roots   A square root is considered to be in simplest form when the radicand has no perfect square as a factor.

Slope   The ratio of rise to run of a line

Slope-intercept form   The slope-intercept form for the equation of a line is $y = mx + b$, where m is the slope of the line and $(0, b)$ is the y-intercept.

Solution   A solution to an equation is a number that gives a true statement when substituted for the variable in the equation

Solution set   The solutions to an equation form a solution set.

Sphere   All points in three dimensions that are the same distance from a fixed point; a ball is an example of a sphere

Square   A rectangle in which all four sides are the same length

Square root   The square root of a number equals another number that when squared results in the original number

Square of a number   In expressions with exponent 2, the base is said to be squared

Standard form of a linear equation Equation of the form $Ax + By = C$, where A, B, and C are real numbers and where A and B are not both 0.

Standard form of a quadratic equation Equation of the form $ax^2 + bx + c = 0$ where a, b, and c are real numbers and $a \neq 0$

Standard form of a complex number The standard form of a complex number is $a + bi$ where a and b are real numbers

Statistics The study of how to gather, organize, analyze, and interpret numerical information. In statistics, a particular measure or characteristic of a part, or sample, of a larger collection of items is called a *statistic*.

Straight angle An angle whose measure is equal to 180°

Subtrahend The number or quantity to be subtracted

Sum The result of addition

Sum of cubes A product that can be written in the form $x^3 + a^3$

Sum of squares Expression of the form $x^2 + a^2$ that is not factorable

Supplementary angles Two angles are supplementary if the sum of their measures is 180°.

Surface area A geometric solid is a measure of the outside surface in square units.

Synthetic division A simplified version of long division of polynomials in which the variables are omitted entirely and only coefficients are written

System of linear equations A set of two or more equations

System of linear inequalities A set of two or more inequalities

T

Term Any constant or variable, or the indicated product and/or quotient of constants and variables

Terminating decimal number If the remainder of division is eventually 0, the decimal quotient is said to be terminating.

Transversal A line in a plane that intersects two or more lines in that plane in different points

Trapezoid A four-sided polygon with one pair of opposite sides that are parallel

Tree diagram Used to "picture" the possible outcomes of an experiment

Triangle A polygon with three sides

Trinomial A polynomial with three terms

U

Union The union of two (or more) sets is the set of all elements that belong to either one set or the other set or to both sets.

Unit fraction A fraction equivalent to 1

Upper class limit In a histogram, the largest whole number that belongs to a class

V

Variable A symbol (generally a letter of the alphabet) that is used to represent an unknown number

Vertex of a parabola The "turning point" of the curve that represents a quadratic function

Vertex of a polygon Each point where two sides of a polygon meet is called a vertex.

Vertex of an angle Common endpoint of the rays that form the angle

Vertical angles The angles opposite each other created by two intersecting lines; vertical angles are congruent

Vertical lines A line whose slope is undefined

Vertical line test If any vertical line intersects the graph of a relation at more than one point, then the relation is not a function.

Volume The measure of the space enclosed by a three-dimensional figure

W

Weight Force of the Earth's gravitational pull on an object

Whole numbers The number 0 and the natural numbers

X

x-axis The horizontal number line

x-intercept The point on the graph where the line crosses the x-axis

Y

y-axis The vertical number line

y-intercept The point on the graph where the line crosses the y-axis

Z

Zero-factor law The product of any number and 0 is equal to 0.

Zero-factor property If the product of two (or more) factors is 0, then at least one of the factors must be 0.

B

Back substitution 800, 808, 1288
Bar graphs 76, 453
  constructing 467
Bar over repeating digits 251
Base 64, 846
  of percent 307
  of a proportion 300
Base ten 3
Binomials 875
  dividing polynomials by 907
  expansion of 889, 896
  multiplication of 889, 896
  multiplying using FOIL 896
  multiplying $(x + a)(x - a)$ 897
  squares of 899
Bisector, angle 362
Borrowing, in subtraction 15, 172
Boundary line 760, 1280
Boyle's Law 1073
Braces 548, 656, 885, 915
Brackets 548, 885, 915
Building fractions to higher terms 157, 161
Building rational expressions to higher terms 1011, 1024, 1029

C

Calculator commands
  scientific notation 870
Calculators. *See* Graphing calculators
Caret key 749, 870, 1139
Carrying a digit, in addition 10
Cartesian coordinate system 688
Celsius 1252
  conversion to Fahrenheit 1252
Celsius to Fahrenheit equation 634, 675
Center of a circle 385
Centi-
  gram 1242
  liter 1246
  meter 1238, 1244
Change in value 529
Changing decimal numbers to fractions 247
Changing decimal numbers to percents 287
Changing fractions to decimal numbers 249, 251
Changing fractions to higher terms 157, 160, 161
Changing fractions to percents 292
Changing mixed numbers to percents 292
Changing percents to decimal numbers 288

Changing percents to fractions 294, 309
Changing percents to mixed numbers 294
Checking account applications 74
Checking solutions 662, 676
Circle graphs 453, 455
  constructing 468
Circle(s)
  area of 386, 1065
  center of 385
  circumference of 385, 429, 1065
  defined 385, 386, 429
  diameter of 385, 429
  radius of 385
Circular cylinder
  surface area 1048
  volume 1068
Circumference, of a circle 385, 429, 1065
Class 457
Class boundaries, in histograms 457, 484
Class width, in histograms 457, 484
Closed figures 365
Closed half-plane 760
Closed interval 658
Coefficient 557, 874
  leading 875
  monomial 874
  of one 562
Collinear points 725
Combined variation 1068
Combining like radicals 1109
Combining like terms 563, 882, 915
Commission 318
Common denominator 155, 1023
Common errors 978, 1013
  adding fractions 158
  dividing with whole numbers 41
  exponents 65
  percents 295
Common factors 178, 1011
Commutative property
  of addition 13, 169, 556
  of multiplication 27, 556
Comparing decimal numbers 206
Comparing fractions 178, 272
Complementary angles 354, 820
Completing the square 1179, 1186
Complex conjugates 1305
Complex fractions 182, 1036
Complex numbers 1296
  addition 1299
  conjugates of 1305
  division 1305
  equations 1299
  fractions 1306
  imaginary part of 1297
  multiplication 1304
  powers of i 1308

  real part of 1297
  solving 1299
  standard form of 1297, 1305
  subtraction 1299
Components of an ordered pair 688
Composite numbers 93
  all factors of 99
  factoring 97
Compound inequalities 661, 665
  and 657
  graphing of 665, 667
  intersection of two sets 1264
  or 657
Compound interest 331, 968
Conditional equation 618
Condition on x 656
Congruent angles 356
Congruent triangles 409
  by angle-side-angle (ASA) 410
  by side-angle-side (SAS) 410
  by side-side-side (SSS) 409
  symbol for 409
Conjugates 1119
Consecutive integers 625, 986
  even 626, 986
  odd 626, 986
Consistent system of equations 790, 1287
Constant of variation 1064
Constant (or constant term) 562, 594, 874
  degree of zero 874
Consumer demand 993
Consumer items applications 74
Contradiction 619
Conversion
  area 1255
  by division 1233
  by multiplication 1233
  length 1253
  mass 1258
  temperature 1252
  volume 1256
  with unit fractions 1235
Coordinate 500
  first 688
  of a point on a number line 500
  second 688
Coordinate system, Cartesian 688
Corresponding angles 359, 406, 409
Corresponding sides 407, 409
Cost problems 817
Counting numbers 2
Cross multiplication 279, 307
Cross products 272, 279
Cubed 1095
Cubed values 64
Cube roots
  evaluating 1095

Order of operations, rules for 67, 180, 548
Origin 689
Original selling price 316
Outcome 477

P

Parabola 1205
  axis (line) of symmetry 1205
  maximum/minimum values 1206
  vertex of 1205
Parallel lines 358, 733, 790
Parallelogram 365
  area of 376
Parentheses 548, 565
PEMDAS 549
Pentagon 14, 368
Percent(s) 286, 300, 310
  amount 300, 307
  base 300, 307
  commission 318
  common errors 295
  decimal numbers changed from 288
  decimal numbers changed to 287
  of discount 316
  equation $R \cdot B = A$ 307
  fractions changed from 294, 309
  fractions changed to 292
  interest 328
  of profit 327
  sales tax 317
  sign 286
Percentage. *See also* Percent(s)
Percent of decrease 319
Percent of increase 319
Perfect cubes 1095
Perfect squares 416, 898, 1092
Perfect square trinomials 962, 899
Perimeter 13, 366
  circle (circumference) 386, 1065
  pentagon 14
  square 366
  triangle 171
  units of measurement 13, 366
Period 2
Perpendicular lines 358, 733
Pi 386, 634, 675
Pitch of a roof 712
Placement of negative signs 1027
Place of accuracy 48
Place value system
  decimal numbers 3
  whole numbers 204
Plane(s) 350, 1286
Plane geometry 350
Point 350, 500
Point on a graph 703
  collinear 725
Point-slope form 730
Pólya, George 315
  Pólya's four-step process for solving

problems 315
Polygons 365
  closed figures 365
  parallelograms 365
  pentagons 14, 368
  rectangles 365
  squares 365
  trapezoids 365
  triangles 365
Polynomials 875
  addition 882
  ascending order 875
  binomial 875
  cubic 881
  degree of 875
  of degree zero 874
  descending order 875
  distributive property 890, 915
  division
    division algorithm 907
    synthetic division 1273
  evaluating 877
  factoring
    by the *ac*-method 952
    by grouping 934
    by trial and error 941
    finding the least common multiple 1024
  irreducible 952
  leading coefficient 875
  linear 881
  monomial 874
  multiplied by monomials 889
  multiplying 889
    FOIL method 896
  of no degree 874
  perfect square trinomials 899
  prime 944
  quadratic 881
  simplifying 876
  special products 896
  subtracting 883
  zero-factor property 972
Population 446
Positive integers 501
Positive numbers
  absolute value of 507
Positive square root 1093
Power of an exponent 849
  combining like terms 563
Power of a product rule 861
Power of a quotient 863
Power of a term 64
Power rule for exponents 859
Powers of 10
  division by 237
  multiplying by 226
Powers of i 1308
Prime factorizations 97, 122, 145
  Fundamental theorem of arithmetic 97
Prime number(s)
  defined 93

determining whether a number is a 95
Prime polynomial 944
Principal 328, 635
Principal square root 1093
Principle(s)
  addition principle of equality 594
  division principle of equality 598
  multiplication principle of equality 598
Probability 477
Problem solving 72, 281
Product 24, 64, 571, 928
  of decimal numbers 224
  divisibility of a 89
  estimating a 54, 226
  of fractions 119
  of integers 538
  of mixed numbers 131
  power of a 859
  of whole numbers 24
Product rule for exponent(s) 846
Profit 327
Projectiles 1217
Proper fractions 117, 123
Properties
  of addition 12, 556
  additive identity 13, 556
  additive inverse 524, 556
  associative 556
    of addition 13, 169
    of multiplication 27
  commutative 556
    of addition 13, 169
    of multiplication 27
  distributive 28, 556
  of exponents 1136
  identity
    additive 13
    multiplicative 27
  of multiplication 27, 556
  multiplication with 0 27
  multiplicative identity 27, 556
  of square roots 1100
  of straight lines 735
  zero-factor 556
Proportions 272
  applications 281
  cross products used to solve 272
  problem solving with 281
  similar triangles 407
  solving 279
Protractor 351
Pure imaginary number 1297
$P(x)$ 877
Pythagoras 418, 988
Pythagorean Theorem 418, 988, 1195
Pythagorean triple 426, 993

Q

Quadrant 689

Relationships Between Measurements in the U.S. Customary System

Length

12 inches (in.) = 1 foot (ft) 36 inches = 1 yard	3 feet = 1 yard (yd) 5280 feet = 1 mile (mi)

Weight

16 ounces (oz) = 1 pound (lb)	2000 pounds = 1 ton (T)

Capacity

8 fluid ounces (fl oz) = 1 cup (c) 2 cups = 1 pint (pt) = 16 fluid ounces	2 pints = 1 quart (qt) 4 quarts = 1 gallon (gal)

Time

60 seconds (sec) = 1 minute (min) 60 minutes = 1 hour (hr)	24 hours = 1 day 7 days = 1 week

Relationships Between Measurements in the Metric System

Length

1 millimeter (mm)	= 0.001 meter	1 m = 1000 mm
1 centimeter (cm)	= 0.01 meter	1 m = 100 cm
1 decimeter (dm)	= 0.1 meter	1 m = 10 dm
1 meter (m)	= 1.0 meter	
1 dekameter (dam)	= 10 meters	
1 hectometer (hm)	= 100 meters	
1 kilometer (km)	= 1000 meters	

Liquid

1 milliliter (mL)	= 0.001 liter	1 L = 1000 mL
1 liter (L)	= 1.0 liter	
1 hectoliter (hL)	= 100 liters	
1 kiloliter (kL)	= 1000 liters	1 kL = 10 hL

Mass

1 milligram (mg)	= 0.001 gram	1 g = 1000 mg
1 centigram (cg)	= 0.01 gram	
1 decigram (dg)	= 0.1 gram	
1 gram (g)	= 1.0 gram	
1 dekagram (dag)	= 10 grams	
1 hectogram (hg)	= 100 grams	
1 kilogram (kg)	= 1000 grams	1 g = 0.001 kg
1 metric ton (t)	= 1000 kilograms	1 kg = 0.001 t

1t = 1000 kg = 1,000,000 g = 1,000,000,000 mg

U.S. Customary and Metric Equivalents

U.S. to Metric	Metric to U.S.
Length	
1 in. = 2.54 cm (exact) 1 ft = 0.305 m 1 yd = 0.914 m 1 mi = 1.61 km	1 cm = 0.394 in. 1 m = 3.28 ft 1 m = 1.09 yd 1 km = 0.62 mi
Area	
$1\ in.^2 = 6.45\ cm^2$ $1\ ft^2 = 0.093\ m^2$ $1\ yd^2 = 0.836\ m^2$ 1 acre = 0.405 ha	$1\ cm^2 = 0.155\ in.^2$ $1\ m^2 = 10.764\ ft^2$ $1\ m^2 = 1.196\ yd^2$ 1 ha = 2.47 acres
Volume	
$1\ in.^3 = 16.387\ cm^3$ $1\ ft^3 = 0.028\ m^3$ 1 qt = 0.946 L 1 gal = 3.785 L	$1\ cm^3 = 0.06\ in.^3$ $1\ m^3 = 35.315\ ft^3$ 1 L = 1.06 qt 1 L = 0.264 gal
Mass	
1 oz = 28.35 g 1 lb = 0.454 kg	1 g = 0.035 oz 1 kg = 2.205 lb

Temperature Equivalents

Celsius		Fahrenheit
100°	Water boils	212°
90°		194°
80°		176°
70°		158°
60°		140°
50°		122°
40°		104°
30°	Comfort	86°
20°	range	68°
10°		50°
0°	Water freezes	32°

Formulas and Definitions

For the Proportion $\dfrac{P}{100} = \dfrac{A}{B}$

$P\%$ = percent (written as the ratio $\dfrac{P}{100}$)

B = base (number that we are finding the percent of)

A = amount (a part of the base)

Terms Related to the Basic Equation $R \cdot B = A$

R = **rate** or percent (as a decimal or fraction)

B = **base** (number we are finding the percent of)

A = **amount** (a part of the base)

Terms Related to Profit

Profit: The difference between selling price and cost.

profit = selling price − cost

Percent of Profit:

1. Percent of profit **based on cost**: $\dfrac{\text{profit}}{\text{cost}}$

2. Percent of profit **based on selling price**: $\dfrac{\text{profit}}{\text{selling price}}$

Formula for Calculating Simple Interest

$I = P \cdot r \cdot t,$

where

I = interest (earned or paid)

P = principal (the amount invested or borrowed)

r = rate of interest (stated as an annual rate) in decimal or fraction form

t = time (one year or fraction of a year)

List of Common Formulas

Formula	Meaning
$C = \dfrac{5}{9}(F - 32)$	**Temperature** in degrees Celsius C equals $\dfrac{5}{9}$ times the difference between the Fahrenheit temperature F and 32.
$d = rt$	The **distance traveled** d equals the product of the rate of speed r and the time t.
$L = 2\pi rh$	The **lateral surface area** L (top and bottom not included) of a cylinder is equal to 2π times the radius r of the base times the height h.
$F = ma$	In physics, the **force** F acting on an object is equal to its mass m times acceleration a.

The Pythagorean Theorem

In a right triangle, the square of the length of the hypotenuse is equal to the sum of the squares of the lengths of the two legs: $c^2 = a^2 + b^2$

Probability of an Event

$$\text{probability of an event} = \dfrac{\text{number of outcomes in event}}{\text{number of outcomes in sample space}}$$